HANDBOOK OF ECONOMETRICS
VOLUME I

EHEHEHEHEHEHEHEHEHEHEHEHEHEHEHEHEHEHEHEHE

HANDBOOKS IN ECONOMICS

2

Series Editors

KENNETH J. ARROW
MICHAEL D. INTRILIGATOR

NORTH-HOLLAND
AMSTERDAM • LONDON • NEW YORK • TOKYO

EHEHEHEHEHEHEHEHEHEHEHEHEHEHEHEHEHEHEHEHE

HANDBOOK OF ECONOMETRICS

VOLUME I

Edited by

ZVI GRILICHES
Harvard University

and

MICHAEL D. INTRILIGATOR
University of California, Los Angeles

NORTH-HOLLAND
AMSTERDAM • LONDON • NEW YORK • TOKYO

ELSEVIER SCIENCE PUBLISHERS B.V.
Sara Burgerhartstraat 25
P.O. Box 1991, 1000 BZ Amsterdam
The Netherlands

ISBN for this volume 0 444 86185 8
ISBN for this set 0 444 86188 2

First edition: 1983
2nd printing: 1988
3rd printing: 1990
4th printing: 1992

Library of Congress Cataloging in Publication Data
Main entry under title:
Handbook of econometrics.
(Handbooks in economics; bk. 2)
Includes bibliographies.
I. Econometrics–Addresses, essays, lectures.
I. Griliches, Zvi. 1930- II. Intriligator,
Michael D. III. Series.
HB139.H36 1983 33O'.028 83-2396
ISBN 0-444-86185-8

This book is printed on acid-free paper.

PRINTED IN THE NETHERLANDS

INTRODUCTION TO THE SERIES

The aim of the *Handbooks in Economics* series is to produce Handbooks for various branches of economics, each of which is a definitive source, reference, and teaching supplement for use by professional researchers and advanced graduate students. Each Handbook provides self-contained surveys of the current state of a branch of economics in the form of chapters prepared by leading specialists on various aspects of this branch of economics. These surveys summarize not only received results but also newer developments, from recent journal articles and discussion papers. Some original material is also included, but the main goal is to provide comprehensive and accessible surveys. The Handbooks are intended to provide not only useful reference volumes for professional collections but also possible supplementary readings for advanced courses for graduate students in economics.

CONTENTS OF THE HANDBOOK

VOLUME I

PREFACE TO THE HANDBOOK

Purpose

The *Handbook of Econometrics* aims to serve as a source, reference, and teaching supplement for the field of econometrics, the branch of economics concerned with the empirical estimation of economic relationships. Econometrics is conceived broadly to include not only econometric models and estimation theory but also econometric data analysis, econometric applications in various substantive fields, and the uses of estimated econometric models. Our purpose has been to provide reasonably comprehensive and up-to-date surveys of recent developments and the state of various aspects of econometrics as of the early 1980s, written at a level intended for professional use by economists, econometricians, and statisticians and for use in advanced graduate econometrics courses.

Econometrics is the application of mathematics and statistical methods to the analysis of economic data. Mathematical models help us to structure our perceptions about the forces generating the data we want to analyze, while statistical methods help us to summarize the data, estimate the parameters of our models, and interpret the strength of the evidence for the various hypotheses that we wish to examine. The evidence provided by the data affects our ideas about the appropriateness of the original model and may result in significant revisions of such models. There is, thus, a continuous interplay in econometrics between mathematical-theoretical modeling of economic behavior, data collection, data summarizing, model fitting, and model evaluation. Theory suggests data to be sought and examined; data availability suggests new theoretical questions and stimulates the development of new statistical methods. The examination of theories in light of data leads to their revision. The examination of data in the light of theory leads often to new interpretations and sometimes to questions about its quality or relevance and to attempts to collect new and different data.

In this volume we review only a subset of what might be called "econometrics". The mathematical-theoretical tools required for model building are discussed primarily in the *Handbook of Mathematical Economics*. Issues of sampling theory, survey design, data collection and editing, and computer programming, all important aspects of the daily life of a practicing econometrician, had, by and large, to be left out of the scope of this *Handbook*. We concentrate, instead, on statistical problems and economic interpretation issues associated with the modeling and estimation of economic behavioral relationships from already assembled and often badly collected data. If economists had access to good experimental data, or were able to design and to perform the relevant economic experiments, the topics to be covered in such a *Handbook* would be quite different. The fact that the generation and collection of economic data is mostly outside the hands of the econometrician is the cause of many of the inferential problems which are discussed in this *Handbook*.

Organization

The organization of the *Handbook* follows in relatively systematic fashion the way an econometric study would proceed, starting from basic mathematical and statistical methods and econometric models, proceeding to estimation and computation, through testing, and ultimately to applications and uses. The *Handbook* also includes a fairly detailed development of time series topics and many other special topics. In particular:

Part 1 summarizes some basic tools used repeatedly in econometrics, including linear algebra, matrix methods, and statistical theory.

Part 2 deals with econometric models, their relationship to economic models, their identification, and the question of model choice and specification analysis.

Part 3 takes up more advanced topics in estimation and computation theory such as non-linear regression methods, biased estimation, and computational algorithms in econometrics. This part also includes a series of chapters on simultaneous equations models, their specification and estimation, distribution theory for such models, and their Bayesian analysis.

Part 4 considers testing of econometric estimators, including Wald, likelihood ratio, and Lagrange multiplier tests; multiple hypothesis testing; distribution theory for econometric estimators and associated test statistics; and Monte Carlo experimentation in econometrics.

Part 5 treats various topics in time series analysis, including time series and spectral methods in econometrics, dynamic specification, inference and casuality in economic time series models, continuous time stochastic models, random and changing coefficient models, and the analysis of panel data.

Parts 6 and 7 present discussions of various special topics in econometrics, including latent variable, limited dependent variable, and discrete choice models; functional forms in econometric model building; economic data issues including longitudinal data issues; and disequilibrium, self selection, and switching models.

Finally, *Part 8* covers selected applications and uses of econometrics. Because of the extremely wide range of applications of econometrics, we could select only a few of the more prominent applications. (Other applications will be treated in later volumes in the "Handbooks in Economics" Series.) Applications discussed here include demand analysis, production and cost analysis, and labor economics. This part includes also chapters on evaluating the predictive accuracy of models, econometric approaches to stabilization policy, the formulation and estimation of models with actors having rational expectations, and the use of econometric models for economic policy formation.

A brief history of econometrics

A brief review of the history of econometrics will put this *Handbook* in perspective. The historical evolution of econometrics was driven both by the increased availability of data and by the desire of generations of scientists to analyze such data in a rigorous and coherent fashion. There are many historical precursors to that which became "econometrics" in this century. Attempts to interpret economic data "scientifically" go back at least as far as Sir William Petty's "political arithmetic" in the seventeenth century and Engel's studies of household expenditures in the nineteenth. The results of the latter became known as Engel's Law, stating that the proportion of total expenditures devoted to food falls as income rises. This "Law" has been tested extensively for many countries over various time periods, as discussed in Houthakker's (1957) centenary article.

The development of statistical theory has played a critical role in the history of econometrics since econometric techniques are, to a large extent, based on multivariable statistics. Modern statistical theory starts with the work of Legendre and Gauss on least squares, motivated by the attempt to remove errors of observation in astronomy and geodesy. The next great impulse came from biology, in particular from evolutionary theory with, among others, Galton's work on regression (a term he later invented). Later developments in mathematical statistics included Yule's work on multiple regression, Karl Pearson's formulation of the notions of probable error and of testing hypotheses, the more rigourous small-sample theory of Student and R. A. Fisher, R. A. Fisher's work on the foundations of statistical inference, and the Neyman–Pearson theory of hypothesis testing. All of these developments in mathematical statistics had a significant influence on the development of econometrics.

In the first half of the twentieth century the increased availability of price and quantity data and the interest in price indexes aided by the development of family expenditure surveys generated interest both in theoretical modeling of demand structures and their empirical estimation. Particularly noteworthy were the demand studies of Moore (1914, 1917), Marschak (1931), and Schultz (1928, 1938) and studies of family expenditure by Allen and Bowley (1935). This period also witnessed the initial formulation of the identification problem in econometrics in E. Working (1927); studies of production functions by Cobb and Douglas (1928) [see also Douglas (1948)], and Marschak and Andrews (1944); studies of price determination in agricultural markets by H. Working (1922), Wright (1925), Hanau (1928), Bean (1928), and Waugh (1929), among others; and the statistical modeling of business cycles by Slutsky (1927) and Frisch (1933). Macroeconometric modeling also began in the 1930s by Tinbergen (1935, 1939) and was given additional impetus by the development of National Income Accounts in the United States and other countries and by Keynes' theoretical work.

The growth of data availability and the development of economic and statistical theory generated a demand for more extensive, more rigorous and higher quality data analysis efforts, stimulating significant research into the methodology of economic data analysis. Of great importance in this respect was the founding of the Econometric Society in 1930 and the publication, starting in 1933, of its journal *Econometrica*. Ragnar Frisch played a key role as the first editor of this Journal.

There was a great flourishing of econometric theory and applications in the period after World War II, particularly due to the work of the Cowles Commission at the University of Chicago. The development of the simultaneous equations model in Haavelmo (1943, 1944, 1947), Koopmans (1950), Hood and Koopmans (1953), Theil (1954) and Basmann (1957) provided econometricians with tools designed specifically for them, rather than for biologists and psychologists. The estimation of simultaneous equations and macroeconometric models in Klein (1950) and Klein and Goldberger (1955) started economic forecasting on a new path. This period also witnessed the important demand studies by Stone (1954a, 1954b) for the United Kingdom, and Wold and Jureen (1953) for Sweden, and the influential studies by Friedman (1957) of the consumption function, and by Theil (1958) of economic forecasts and policy. [For collections of historically important papers in econometrics see Zellner (1968), Hooper and Nerlove (1970), and Dowling and Glahe (1970).]

The more recent period of the 1960s and 1970s has witnessed many important developments in econometric theory and applications. Econometric theory has been refined and extended in many ways. Of particular note is the Bayesian approach to econometrics and the study of special features of econometric models, such as limited dependent variables, latent variables, and non-linear models. Great progress was also made in the statistical analysis of time series. In addition, the development of electronic computers, the great increase in computing power, and the development of sophisticated econometric software packages made it possible to pursue much more ambitious data analysis strategies. These developments expanded the range of applications of econometric methods greatly beyond the earlier applications to household expenditure, demand functions, production and cost functions, and macroeconometric models. Econometrics is now used in virtually every field of economics, including public finance, monetary economics, labor economics, international economics, economic history, health economics, studies of fertility, and studies of criminal behavior, just to mention a few. In all of these fields the greater use of econometric techniques, based in part on increased data availability and more powerful estimation techniques, has led to greater precision in the specification, estimation, and testing of economic data-based models.

Most of the important developments in econometric methods during the 1960s and 1970s are discussed in this *Handbook*. The significant topics under development in this period and the chapters treating them include:

(1) *Bayesian econometrics*, using Bayesian methods in the specification and estimation of econometric models. These topics are discussed in Chapter 2 by Zellner and Chapter 9 by Drèze and Richard.

(2) *Time series methods*, including specialized techniques and problems arising in the analysis of economic time series, such as spectral methods, dynamic specification, and causality. These techniques and problems are discussed in Part 5 on "Time Series Topics," including Chapter 17 by Granger and Watson; Chapter 18 by Hendry, Pagan and Sargan; Chapter 19 by Geweke; Chapter 20 by Bergstrom; and Chapter 21 by Chow. Related issues are discussed in Chapter 33 by Fair.

(3) *Discrete choice models*, in which there is a discrete choice of alternatives available, e.g. buy/don't buy decisions, yes/no responses, or alternative possibilities for urban transportation. Such models are discussed specifically in Chapter 27 by Dhrymes and Chapter 24 by McFadden and are also treated in Chapter 22 by Chamberlain, Chapter 28 by Maddala, and Chapter 29 by Heckman and Singer.

(4) *Latent variables models*, in which certain unmeasurable variables systematically influence measured phenomena, such as ability influencing earnings. This topic is treated in Chapter 23 by Aigner, Hsiao, Kapteyn, and Wansbeek and reappears in various guises in Chapter 22 by Chamberlain and Chapter 32 by Heckman and MaCurdy, among others.

(5) *Specification analysis*, involving problems of model choice and their specification and identification. These issues are treated in Chapter 3 by Intriligator, Chapter 4 by Hsiao, Chapter 5 by Leamer, Chapter 26 by Lau, and Chapter 28 by Maddala. This topic, of course, pervades many other chapters in this *Handbook* and overlaps with chapters which deal with testing and distribution theory.

(6) *Non-linear models and methods*, in which models that are intrinsically nonlinear are specified and estimated. Such models are discussed in Chapter 6 by Amemiya and Chapter 12 by Quandt and surface also in many of the other chapters of this *Handbook*.

(7) *Data analysis issues*, involving various problems with data and how they can be treated. These issues are treated in Chapter 10 by Judge and Bock, in Chapter 11 by Krasker, Kuh, and Welsch, Chapter 22 by Chamberlain, Chapter 25 by Griliches, and Chapter 29 by Heckman and Singer, among others.

(8) *Testing and small sample theory*, including various test procedures and Monte Carlo experimentation. These topics are treated in Part 4 on "Testing," including Chapter 13 by Engle, Chapter 14 by Savin, Chapter 15 by Rothenberg, and Chapter 16 by Hendry. Related issues are discussed in Chapter 8 by Phillips.

(9) *Rational expectations* models which treat economic agents as forming expectations in an optimal fashion, given the information available to them,

impose cross-equation constraints on parameters and lead to new problems of identification and estimation. This topic is discussed in Chapter 34 by Taylor.

ZVI GRILICHES
Harvard University

MICHAEL D. INTRILIGATOR
University of California, Los Angeles

References

Allen, R. G. D. and A. L. Bowley (1935) *Family Expenditure*. London: P. S. King
Basmann, R. L. (1957) "A Generalized Classical Method of Linear Estimation of Coefficients in a Structural Equation", *Econometrica*, 25, 77–83.
Bean, L. H. (1928) "Some Interrelationships between the Supply, Price, and Consumption of Cotton", *USDA*, mimeographed.
Cobb, C. W. and P. H. Douglas (1928) "A Theory of Production", *American Economic Review*, 18 (supplement), 139–165.
Douglas, P. H. (1948) "Are There Laws of Production?", *American Economic Review*, 38, 1–41.
Dowling, J. M. and F. R. Glahe (eds.) (1970) *Readings in Econometric Theory*. Boulder: Colorado Associated University Press.
Friedman, M. (1957) *A Theory of the Consumption Function*, National Bureau of Economic Research. Princeton: Princeton University Press.
Frisch, R. (1933) "Propagation Problems and Impulse Problems in Dynamic Economics", in: *Economic Essays in Honor of Gustav Cassel*. London: George Allen & Unwin, pp. 171–205.
Haavelmo, T. (1943) "The Statistical Implications of a System of Simultaneous Equations", *Econometrica*, 11, 1–12.
Haavelmo, T. (1944) "The Probability Approach in Econometrics", *Econometrica*, 12 (supplement), 1–115.
Haavelmo, T. (1947) "Methods of Measuring the Marginal Propensity to Consume", *Journal of the American Statistical Association*, 42, 105–122 (reprinted in Hood and Koopmans (eds) (1953)).
Hanau, A. (1928) "Die Prognose der Schweinepreise", *Vierteljahrshefte zur Konjukturforschung*, Sonderheft 7, Berlin.
Hood, W. C. and T. C. Koopmans (eds.) (1953) *Studies in Econometric Method*, Cowles Commission Monograph No. 14, New York: John Wiley & Sons.
Hooper, J. W. and M. Nerlove (eds.) (1970) *Selected Readings in Econometrics from Econometrica*. Cambridge: MIT Press.
Houthakker, H. S. (1957) "An International Comparison of Household Expenditure Patterns, Commemorating the Centenary of Engel's Law", *Econometrica*, 25, 532–551.
Klein, L. R. (1950) *Economic Fluctuations in the United States, 1921–1941*, Cowles Commission Monograph No. 11. New York: John Wiley & Sons.
Klein, L. R. and A. S. Goldberger (1955) *An Econometric Model of the United States, 1929–1952*. Amsterdam: North-Holland Publishing Co.
Koopmans, T. C. (ed.) (1950) *Statistical Inference in Dynamic Economic Models*, Cowles Commission Monograph No. 10. New York: John Wiley & Sons.
Marschak, J. (1931) *Elastizität der Nachfrage*. Tübingen: J. C. B. Mohr.
Marschak, J. and W. H. Andrews (1944) "Random Simultaneous Equations and the Theory of Production", *Econometrica* 12, 143–205.
Moore, H. L. (1914) *Economic Cycles: Their Law and Cause*. New York: The Macmillan Company.
Moore, H. L. (1917) *Forecasting the Yield and Price of Cotton*. New York: The Macmillan Company.
Schultz, H. (1928) *Statistical Laws of Demand and Supply*. Chicago: University of Chicago Press.
Schultz, H. (1938) *The Theory and Measurement of Demand*, Chicago: University of Chicago Press.

Slutsky, E. (1927) "The Summation of Random Causes as the Source of Cyclic Processes" (Russian with English summary), in: *Problems of Economic Conditions*, vol. 3. Moscow: Rev. English edn., 1937; *Econometrica*, 5, 105–146.
Stone, R. (1954a) "Linear Expenditure Systems and Demand Analysis: An Application to the Pattern of British Demand", *Economic Journal*, 64, 511–527.
Stone, R. (1954b) *The Measurement of Consumers' Expenditure and Behavior in the United Kingdom, 1920–1938*. New York: Cambridge University Press.
Theil, H. (1954) "Estimation of Parameters of Econometric Models", *Bulletin of the International Statistics Institute*, 34, 122–128.
Theil, H. (1958) *Economic Forecasts and Policy*. Amsterdam: North-Holland Publishing Co. (Second Edition, 1961).
Tinbergen, J. (1935) "Quantitative Fragen der Konjunkturpolitik", *Weltwirtschaftliches Archiv*, 42, 316–399.
Tinbergen, J. (1939) *Statistical Testing of Business Cycle Theories*. Vol. 1: *A Method and its Application to Investment Activity*; Vol. 2: *Business Cycles in the United States of America, 1919–1932*. Geneva: League of Nations.
Waugh, F. V. (1929) *Quality as a Determinant of Vegetable Prices*. New York: Columbia University Press.
Wold, H. and L. Jureen (1953) *Demand Analysis*. New York: John Wiley & Sons.
Working, E. J. (1927) "What do Statistical 'Demand Curves' Show?", *Quarterly Journal of Economics*, 41, 212–235.
Working, H. (1922) "Factors Determining the Price of Potatoes in St. Paul and Minneapolis", University of Minnesota Agricultural Experiment Station Technical Bulletin 10.
Wright, S. (1925) *Corn and Hog Correlations*, Washington, USDA, Bul. 1300.
Zellner, A., Ed. (1968) *Readings in Economic Statistics and Econometrics*. Boston: Little, Brown.

CONTENTS OF VOLUME I

Part 3 – ESTIMATION AND COMPUTATION

Chapter 12

Computational Problems and Methods

PART 1

MATHEMATICAL AND STATISTICAL METHODS IN ECONOMETRICS

Chapter 1

LINEAR ALGEBRA AND MATRIX METHODS IN ECONOMETRICS

HENRI THEIL*

University of Florida

Contents

*Research supported in part by NSF Grant SOC76-82718. The author is indebted to Kenneth Clements (Reserve Bank of Australia, Sydney) and Michael Intriligator (University of California, Los Angeles) for comments on an earlier draft of this chapter.

Handbook of Econometrics, Volume I, Edited by Z. Griliches and M.D. Intriligator

1. Introduction

Vectors and matrices played a minor role in the econometric literature published before World War II, but they have become an indispensable tool in the last several decades. Part of this development results from the importance of matrix tools for the statistical component of econometrics; another reason is the increased use of matrix algebra in the economic theory underlying econometric relations. The objective of this chapter is to provide a selective survey of both areas. Elementary properties of matrices and determinants are assumed to be known, including summation, multiplication, inversion, and transposition, but the concepts of linear dependence and orthogonality of vectors and the rank of a matrix are briefly reviewed in Appendix A. Reference is made to Dhrymes (1978), Graybill (1969), or Hadley (1961) for elementary properties not covered in this chapter.

Matrices are indicated by boldface italic upper case letters (such as $\boldsymbol{A}$), column vectors by boldface italic lower case letters ($\boldsymbol{a}$), and row vectors by boldface italic lower case letters with a prime added ($\boldsymbol{a}'$) to indicate that they are obtained from the corresponding column vector by transposition. The following abbreviations are used:

$$\begin{aligned} \text{LS} &= \text{least squares}, \\ \text{GLS} &= \text{generalized least squares}, \\ \text{ML} &= \text{maximum likelihood}, \\ \delta_{ij} &= \text{Kronecker delta } (=1 \text{ if } i=j,\ 0 \text{ if } i \neq j). \end{aligned}$$

2. Why are matrix methods useful in econometrics?

2.1. *Linear systems and quadratic forms*

A major reason why matrix methods are useful is that many topics in econometrics have a multivariate character. For example, consider a system of L simultaneous linear equations in L endogenous and K exogenous variables. We write $y_{\alpha l}$ and $x_{\alpha k}$ for the αth observation on the lth endogenous and the kth exogenous variable. Then the jth equation for observation α takes the form

$$\sum_{l=1}^{L} \gamma_{lj} y_{\alpha l} + \sum_{k=1}^{K} \beta_{kj} x_{\alpha k} = \varepsilon_{\alpha j}, \tag{2.1}$$

where $\varepsilon_{\alpha j}$ is a random disturbance and the γ's and β's are coefficients. We can write (2.1) for $j=1,\ldots,L$ in the form

$$y_\alpha'\Gamma + x_\alpha' B = \varepsilon_\alpha', \tag{2.2}$$

where $y_\alpha' = [y_{\alpha 1}\ldots y_{\alpha L}]$ and $x_\alpha' = [x_{\alpha 1}\ldots x_{\alpha K}]$ are observation vectors on the endogenous and the exogenous variables, respectively, $\varepsilon_\alpha' = [\varepsilon_{\alpha 1}\ldots\varepsilon_{\alpha L}]$ is a disturbance vector, and Γ and B are coefficient matrices of order $L\times L$ and $K\times L$, respectively:

$$\Gamma = \begin{bmatrix} \gamma_{11} & \gamma_{12}\cdots\gamma_{1L} \\ \gamma_{21} & \gamma_{22}\cdots\gamma_{2L} \\ \vdots & \vdots \quad \vdots \\ \gamma_{L1} & \gamma_{L2}\cdots\gamma_{LL} \end{bmatrix}, \qquad B = \begin{bmatrix} \beta_{11} & \beta_{12}\cdots\beta_{1L} \\ \beta_{21} & \beta_{22}\cdots\beta_{2L} \\ \vdots & \vdots \quad \vdots \\ \beta_{K1} & \beta_{K2}\cdots\beta_{KL} \end{bmatrix}.$$

When there are n observations ($\alpha = 1,\ldots,n$), there are Ln equations of the form (2.1) and n equations of the form (2.2). We can combine these equations compactly into

$$Y\Gamma + XB = E, \tag{2.3}$$

where Y and X are observation matrices of the two sets of variables of order $n\times L$ and $n\times K$, respectively:

$$Y = \begin{bmatrix} y_{11} & y_{12}\cdots y_{1L} \\ y_{21} & y_{22}\cdots y_{2L} \\ \vdots & \vdots \quad \vdots \\ y_{n1} & y_{n2}\cdots y_{nL} \end{bmatrix}, \qquad X = \begin{bmatrix} x_{11} & x_{12}\cdots x_{1K} \\ x_{21} & x_{22}\cdots x_{2K} \\ \vdots & \vdots \quad \vdots \\ x_{n1} & x_{n2}\cdots x_{nK} \end{bmatrix},$$

and E is an $n\times L$ disturbance matrix:

$$E = \begin{bmatrix} \varepsilon_{11} & \varepsilon_{12}\cdots\varepsilon_{1L} \\ \varepsilon_{21} & \varepsilon_{22}\cdots\varepsilon_{2L} \\ \vdots & \vdots \quad \vdots \\ \varepsilon_{n1} & \varepsilon_{n2}\cdots\varepsilon_{nL} \end{bmatrix}.$$

Note that Γ is square ($L\times L$). If Γ is also non-singular, we can postmultipy (2.3) by Γ^{-1}:

$$Y = -XB\Gamma^{-1} + E\Gamma^{-1}. \tag{2.4}$$

This is the *reduced form* for all n observations on all L endogenous variables, each of which is described linearly in terms of exogenous values and disturbances. By contrast, the equations (2.1) or (2.2) or (2.3) from which (2.4) is derived constitute the *structural form* of the equation system.

The previous paragraphs illustrate the convenience of matrices for linear systems. However, the expression "linear algebra" should not be interpreted in the sense that matrices are useful for linear systems only. The treatment of quadratic functions can also be simplified by means of matrices. Let $g(z_1,\dots,z_k)$ be a three times differentiable function. A Taylor expansion yields

$$\begin{aligned} g(z_1,\dots,z_k) &= g(\bar{z}_1,\dots,\bar{z}_k) + \sum_{i=1}^{k}(z_i-\bar{z}_i)\frac{\partial g}{\partial z_i} \\ &\quad + \frac{1}{2}\sum_{i=1}^{k}\sum_{j=1}^{k}(z_i-\bar{z}_i)\frac{\partial^2 g}{\partial z_i\,\partial z_j}(z_j-\bar{z}_j)+O_3, \end{aligned} \tag{2.5}$$

where O_3 is a third-order remainder term, while the derivatives $\partial g/\partial z_i$ and $\partial^2 g/\partial z_i\,\partial z_j$ are all evaluated at $z_1=\bar{z}_1,\dots,z_k=\bar{z}_k$. We introduce z and $\bar{z}$ as vectors with ith elements z_i and $\bar{z}_i$, respectively. Then (2.5) can be written in the more compact form

$$g(z)=g(\bar{z})+(z-\bar{z})'\frac{\partial g}{\partial z}+\frac{1}{2}(z-\bar{z})'\frac{\partial^2 g}{\partial z\,\partial z'}(z-\bar{z})+O_3, \tag{2.6}$$

where the column vector $\partial g/\partial z=[\partial g/\partial z_i]$ is the *gradient* of $g(\cdot)$ at $\bar{z}$ (the vector of first-order derivatives) and the matrix $\partial^2 g/\partial z\,\partial z'=[\partial^2 g/\partial z_i\,\partial z_j]$ is the *Hessian matrix* of $g(\cdot)$ at $\bar{z}$ (the matrix of second-order derivatives). A Hessian matrix is always symmetric when the function is three times differentiable.

2.2. *Vectors and matrices in statistical theory*

Vectors and matrices are also important in the statistical component of econometrics. Let $\boldsymbol{r}$ be a column vector consisting of the random variables $r_1,\dots,r_n$. The expectation $\mathcal{E}\boldsymbol{r}$ is defined as the column vector of expectations $\mathcal{E}r_1,\dots,\mathcal{E}r_n$. Next consider

$$(\boldsymbol{r}-\mathcal{E}\boldsymbol{r})(\boldsymbol{r}-\mathcal{E}\boldsymbol{r})'=\begin{bmatrix} r_1-\mathcal{E}r_1 \\ r_2-\mathcal{E}r_2 \\ \vdots \\ r_n-\mathcal{E}r_n \end{bmatrix}\begin{bmatrix} r_1-\mathcal{E}r_1 & r_2-\mathcal{E}r_2 \dots r_n-\mathcal{E}r_n \end{bmatrix}$$

and take the expectation of each element of this product matrix. When defining the expectation of a random matrix as the matrix of the expectations of the constituent elements, we obtain:

$$\mathcal{E}[(\boldsymbol{r}-\mathcal{E}\boldsymbol{r})(\boldsymbol{r}-\mathcal{E}\boldsymbol{r})'] = \begin{bmatrix} \operatorname{var} r_1 & \operatorname{cov}(r_1, r_2) & \cdots & \operatorname{cov}(r_1, r_n) \\ \operatorname{cov}(r_2, r_1) & \operatorname{var} r_2 & \cdots & \operatorname{cov}(r_2, r_n) \\ \vdots & \vdots & & \vdots \\ \operatorname{cov}(r_n, r_1) & \operatorname{cov}(r_n, r_2) & \cdots & \operatorname{var} r_n \end{bmatrix}.$$

This is the variance–covariance matrix (*covariance matrix*, for short) of the vector $\boldsymbol{r}$, to be written $\mathcal{V}(\boldsymbol{r})$. The covariance matrix is always symmetric and contains the variances along the diagonal. If the elements of $\boldsymbol{r}$ are pairwise uncorrelated, $\mathcal{V}(\boldsymbol{r})$ is a diagonal matrix. If these elements also have equal variances (equal to σ^2, say), $\mathcal{V}(\boldsymbol{r})$ is a *scalar matrix*, $\sigma^2\boldsymbol{I}$; that is, a scalar multiple σ^2 of the unit or identity matrix.

The multivariate nature of econometrics was emphasized at the beginning of this section. This will usually imply that there are several unknown parameters; we arrange these in a vector $\boldsymbol{\theta}$. The problem is then to obtain a "good" estimator $\hat{\boldsymbol{\theta}}$ of $\boldsymbol{\theta}$ as well as a satisfactory measure of how good the estimator is; the most popular measure is the covariance matrix $\mathcal{V}(\hat{\boldsymbol{\theta}})$. Sometimes this problem is simple, but that is not always the case, in particular when the model is non-linear in the parameters. A general method of estimation is maximum likelihood (ML) which can be shown to have certain optimal properties for large samples under relatively weak conditions. The derivation of the ML estimates and their large-sample covariance matrix involves the *information matrix*, which is (apart from sign) the expectation of the matrix of second-order derivatives of the log-likelihood function with respect to the parameters. The prominence of ML estimation in recent years has greatly contributed to the increased use of matrix methods in econometrics.

2.3. *Least squares in the standard linear model*

We consider the model

$$\boldsymbol{y} = \boldsymbol{X}\boldsymbol{\beta} + \boldsymbol{\varepsilon}, \tag{2.7}$$

where $\boldsymbol{y}$ is an n-element column vector of observations on the dependent (or endogenous) variable, $\boldsymbol{X}$ is an $n \times K$ observation matrix of rank K on the K independent (or exogenous) variables, $\boldsymbol{\beta}$ is a parameter vector, and $\boldsymbol{\varepsilon}$ is a disturbance vector. The *standard linear model* postulates that $\boldsymbol{\varepsilon}$ has zero expectation and covariance matrix $\sigma^2\boldsymbol{I}$, where σ^2 is an unknown positive parameter, and that the elements of $\boldsymbol{X}$ are all non-stochastic. Note that this model can be viewed as a special case of (2.3) for $\boldsymbol{\Gamma}=\boldsymbol{I}$ and $L=1$.

The problem is to estimate $\boldsymbol{\beta}$ and σ^2. The *least-squares* (LS) estimator of $\boldsymbol{\beta}$ is

$$\boldsymbol{b}=(\boldsymbol{X}'\boldsymbol{X})^{-1}\boldsymbol{X}'\boldsymbol{y} \tag{2.8}$$

which owes its name to the fact that it minimizes the residual sum of squares. To verify this proposition we write $\boldsymbol{e}=\boldsymbol{y}-\boldsymbol{X}\boldsymbol{b}$ for the residual vector; then the residual sum of squares equals

$$\boldsymbol{e}'\boldsymbol{e}=\boldsymbol{y}'\boldsymbol{y}-2\boldsymbol{y}'\boldsymbol{X}\boldsymbol{b}+\boldsymbol{b}'\boldsymbol{X}'\boldsymbol{X}\boldsymbol{b}, \tag{2.9}$$

which is to be minimized by varying $\boldsymbol{b}$. This is achieved by equating the gradient of (2.9) to zero. A comparison of (2.9) with (2.5) and (2.6), with $\boldsymbol{z}$ interpreted as $\boldsymbol{b}$, shows that the gradient of (2.9) equals $-2\boldsymbol{X}'\boldsymbol{y}+2\boldsymbol{X}'\boldsymbol{X}\boldsymbol{b}$, from which the solution (2.8) follows directly.

Substitution of (2.7) into (2.8) yields $\boldsymbol{b}-\boldsymbol{\beta}=(\boldsymbol{X}'\boldsymbol{X})^{-1}\boldsymbol{X}'\boldsymbol{\varepsilon}$. Hence, given $\mathcal{E}\boldsymbol{\varepsilon}=\boldsymbol{0}$ and the non-randomness of $\boldsymbol{X}$, $\boldsymbol{b}$ is an unbiased estimator of $\boldsymbol{\beta}$. Its covariance matrix is

$$\mathcal{V}(\boldsymbol{b})=(\boldsymbol{X}'\boldsymbol{X})^{-1}\boldsymbol{X}'\mathcal{V}(\boldsymbol{\varepsilon})\boldsymbol{X}(\boldsymbol{X}'\boldsymbol{X})^{-1}=\sigma^2(\boldsymbol{X}'\boldsymbol{X})^{-1} \tag{2.10}$$

because $\boldsymbol{X}'\mathcal{V}(\boldsymbol{\varepsilon})\boldsymbol{X}=\sigma^2\boldsymbol{X}'\boldsymbol{X}$ follows from $\mathcal{V}(\boldsymbol{\varepsilon})=\sigma^2\boldsymbol{I}$. The Gauss–Markov theorem states that $\boldsymbol{b}$ is a best linear unbiased estimator of $\boldsymbol{\beta}$, which amounts to an optimum LS property within the class of $\boldsymbol{\beta}$ estimators that are linear in $\boldsymbol{y}$ and unbiased. This property implies that each element of $\boldsymbol{b}$ has the smallest possible variance; that is, there exists no other linear unbiased estimator of $\boldsymbol{\beta}$ whose elements have smaller variances than those of the corresponding elements of $\boldsymbol{b}$. A more general formulation of the Gauss–Markov theorem will be given and proved in Section 6.

Substitution of (2.8) into $\boldsymbol{e}=\boldsymbol{y}-\boldsymbol{X}\boldsymbol{b}$ yields $\boldsymbol{e}=\boldsymbol{M}\boldsymbol{y}$, where $\boldsymbol{M}$ is the symmetric matrix

$$\boldsymbol{M}=\boldsymbol{I}-\boldsymbol{X}(\boldsymbol{X}'\boldsymbol{X})^{-1}\boldsymbol{X}' \tag{2.11}$$

which satisfies $\boldsymbol{M}\boldsymbol{X}=\boldsymbol{0}$; therefore, $\boldsymbol{e}=\boldsymbol{M}\boldsymbol{y}=\boldsymbol{M}(\boldsymbol{X}\boldsymbol{\beta}+\boldsymbol{\varepsilon})=\boldsymbol{M}\boldsymbol{\varepsilon}$. Also, $\boldsymbol{M}$ is *idempotent*, i.e. $\boldsymbol{M}^2=\boldsymbol{M}$. The LS residual sum of squares equals $\boldsymbol{e}'\boldsymbol{e}=\boldsymbol{\varepsilon}'\boldsymbol{M}'\boldsymbol{M}\boldsymbol{\varepsilon}=\boldsymbol{\varepsilon}'\boldsymbol{M}^2\boldsymbol{\varepsilon}$ and hence

$$\boldsymbol{e}'\boldsymbol{e}=\boldsymbol{\varepsilon}'\boldsymbol{M}\boldsymbol{\varepsilon}. \tag{2.12}$$

It is shown in the next paragraph that $\mathcal{E}(\boldsymbol{\varepsilon}'\boldsymbol{M}\boldsymbol{\varepsilon}) = \sigma^2(n-K)$ so that (2.12) implies that σ^2 is estimated unbiasedly by $\boldsymbol{e}'\boldsymbol{e}/(n-K)$: the LS residual sum of squares divided by the excess of the number of observations (n) over the number of coefficients adjusted (K).

To prove $\mathcal{E}(\boldsymbol{\varepsilon}'\boldsymbol{M}\boldsymbol{\varepsilon}) = \sigma^2(n-K)$ we define the *trace* of a square matrix as the sum of its diagonal elements: $\operatorname{tr}\boldsymbol{A} = a_{11} + \cdots + a_{nn}$. We use $\operatorname{tr}\boldsymbol{AB} = \operatorname{tr}\boldsymbol{BA}$ (if $\boldsymbol{AB}$ and $\boldsymbol{BA}$ exist) to write $\boldsymbol{\varepsilon}'\boldsymbol{M}\boldsymbol{\varepsilon}$ as $\operatorname{tr}\boldsymbol{M}\boldsymbol{\varepsilon}\boldsymbol{\varepsilon}'$. Next we use $\operatorname{tr}(\boldsymbol{A}+\boldsymbol{B}) = \operatorname{tr}\boldsymbol{A} + \operatorname{tr}\boldsymbol{B}$ (if $\boldsymbol{A}$ and $\boldsymbol{B}$ are square of the same order) to write $\operatorname{tr}\boldsymbol{M}\boldsymbol{\varepsilon}\boldsymbol{\varepsilon}'$ as $\operatorname{tr}\boldsymbol{\varepsilon}\boldsymbol{\varepsilon}' - \operatorname{tr}\boldsymbol{X}(\boldsymbol{X}'\boldsymbol{X})^{-1}\boldsymbol{X}'\boldsymbol{\varepsilon}\boldsymbol{\varepsilon}'$ [see (2.11)]. Thus, since $\boldsymbol{X}$ is non-stochastic and the trace is a linear operator,

$$\begin{aligned}\mathcal{E}(\boldsymbol{\varepsilon}'\boldsymbol{M}\boldsymbol{\varepsilon}) &= \operatorname{tr}\mathcal{E}(\boldsymbol{\varepsilon}\boldsymbol{\varepsilon}') - \operatorname{tr}\boldsymbol{X}(\boldsymbol{X}'\boldsymbol{X})^{-1}\boldsymbol{X}'\mathcal{E}(\boldsymbol{\varepsilon}\boldsymbol{\varepsilon}')\\ &= \sigma^2\operatorname{tr}\boldsymbol{I} - \sigma^2\operatorname{tr}\boldsymbol{X}(\boldsymbol{X}'\boldsymbol{X})^{-1}\boldsymbol{X}'\\ &= \sigma^2 n - \sigma^2\operatorname{tr}(\boldsymbol{X}'\boldsymbol{X})^{-1}\boldsymbol{X}'\boldsymbol{X},\end{aligned}$$

which confirms $\mathcal{E}(\boldsymbol{\varepsilon}'\boldsymbol{M}\boldsymbol{\varepsilon}) = \sigma^2(n-K)$ because $(\boldsymbol{X}'\boldsymbol{X})^{-1}\boldsymbol{X}'\boldsymbol{X} = \boldsymbol{I}$ of order $K \times K$.

If, in addition to the conditions listed in the discussion following eq. (2.7), the elements of $\boldsymbol{\varepsilon}$ are normally distributed, the LS estimator $\boldsymbol{b}$ of $\boldsymbol{\beta}$ is identical to the ML estimator; also, $(n-K)s^2/\sigma^2$ is then distributed as χ^2 with $n-K$ degrees of freedom and $\boldsymbol{b}$ and s^2 are independently distributed. For a proof of this result see, for example, Theil (1971, sec. 3.5).

If the covariance matrix of $\boldsymbol{\varepsilon}$ is $\sigma^2\boldsymbol{V}$ rather than $\sigma^2\boldsymbol{I}$, where $\boldsymbol{V}$ is a non-singular matrix, we can extend the Gauss–Markov theorem to Aitken's (1935) theorem. The best linear unbiased estimator of $\boldsymbol{\beta}$ is now

$$\hat{\boldsymbol{\beta}} = (\boldsymbol{X}'\boldsymbol{V}^{-1}\boldsymbol{X})^{-1}\boldsymbol{X}'\boldsymbol{V}^{-1}\boldsymbol{y}, \tag{2.13}$$

and its covariance matrix is

$$\mathcal{V}(\hat{\boldsymbol{\beta}}) = \sigma^2(\boldsymbol{X}'\boldsymbol{V}^{-1}\boldsymbol{X})^{-1}. \tag{2.14}$$

The estimator $\hat{\boldsymbol{\beta}}$ is the *generalized least-squares* (GLS) estimator of $\boldsymbol{\beta}$; we shall see in Section 7 how it can be derived from the LS estimator $\boldsymbol{b}$.

2.4. *Vectors and matrices in consumption theory*

It would be inappropriate to leave the impression that vectors and matrices are important in econometrics primarily because of problems of statistical inference. They are also important for the problem of how to specify economic relations. We shall illustrate this here for the analysis of consumer demand, which is one of the oldest topics in applied econometrics. References for the account which follows include Barten (1977), Brown and Deaton (1972), Phlips (1974), Theil (1975–76), and Deaton's chapter on demand analysis in this Handbook (Chapter 30).

Let there be N goods in the marketplace. We write $\boldsymbol{p} = [p_i]$ and $\boldsymbol{q} = [q_i]$ for the price and quantity vectors. The consumer's preferences are measured by a utility function $u(\boldsymbol{q})$ which is assumed to be three times differentiable. His problem is to maximize $u(\boldsymbol{q})$ by varying $\boldsymbol{q}$ subject to the budget constraints $\boldsymbol{p}'\boldsymbol{q} = M$, where M is the given positive amount of total expenditure (to be called income for brevity's sake). Prices are also assumed to be positive and given from the consumer's point of view. Once he has solved this problem, the demand for each good becomes a function of income and prices. What can be said about the derivatives of demand, $\partial q_i/\partial M$ and $\partial q_i/\partial p_j$?

Neoclassical consumption theory answers this question by constructing the Lagrangian function $u(\boldsymbol{q}) - \lambda(\boldsymbol{p}'\boldsymbol{q} - M)$ and differentiating this function with respect to the q_i's. When these derivatives are equated to zero, we obtain the familiar proportionality of marginal utilities and prices:

$$\frac{\partial u}{\partial q_i} = \lambda p_i, \qquad i = 1,\dots,N, \tag{2.15}$$

or, in vector notation, $\partial u/\partial \boldsymbol{q} = \lambda \boldsymbol{p}$: the gradient of the utility function at the optimal point is proportional to the price vector. The proportionality coefficient λ has the interpretation as the marginal utility of income.[1]

The proportionality (2.15) and the budget constraint $\boldsymbol{p}'\boldsymbol{q} = M$ provide $N+1$ equations in $N+1$ unknowns: $\boldsymbol{q}$ and λ. Since these equations hold identically in M and $\boldsymbol{p}$, we can differentiate them with respect to these variables. Differentiation of $\boldsymbol{p}'\boldsymbol{q} = M$ with respect to M yields $\sum_i p_i(\partial q_i/\partial M) = 1$ or

$$\boldsymbol{p}'\frac{\partial \boldsymbol{q}}{\partial M} = 1, \tag{2.16}$$

where $\partial \boldsymbol{q}/\partial M = [\partial q_i/\partial M]$ is the vector of income derivatives of demand. Differentiation of $\boldsymbol{p}'\boldsymbol{q} = M$ with respect to p_j yields $\sum_i p_i(\partial q_i/\partial p_j) + q_j = 0$ ($j = 1,\dots,N$) or

$$\boldsymbol{p}'\frac{\partial \boldsymbol{q}}{\partial \boldsymbol{p}'} = -\boldsymbol{q}', \tag{2.17}$$

where $\partial \boldsymbol{q}/\partial \boldsymbol{p}' = [\partial q_i/\partial p_j]$ is the $N \times N$ matrix of price derivatives of demand. Differentiation of (2.15) with respect to M and application of the chain rule

[1]Dividing both sides of (2.15) by p_i yields $\partial u/\partial(p_i q_i) = \lambda$, which shows that an extra dollar of income spent on any of the N goods raises utility by λ. This provides an intuitive justification for the interpretation. A more rigorous justification would require the introduction of the indirect utility function, which is beyond the scope of this chapter.

yields:

$$\sum_{k=1}^{N} \frac{\partial^2 u}{\partial q_i \partial q_k} \frac{\partial q_k}{\partial M} = p_i \frac{\partial \lambda}{\partial M}, \qquad i = 1, \ldots, N.$$

Similarly, differentiation of (2.15) with respect to p_j yields:

$$\sum_{k=1}^{N} \frac{\partial^2 u}{\partial q_i \partial q_k} \frac{\partial q_k}{\partial p_j} = p_i \frac{\partial \lambda}{\partial p_j} + \lambda \delta_{ij}, \qquad i, j = 1, \ldots, N,$$

where δ_{ij} is the Kronecker delta ($=1$ if $i = j$, 0 if $i \neq j$). We can write the last two equations in matrix form as

$$\boldsymbol{U} \frac{\partial \boldsymbol{q}}{\partial M} = \frac{\partial \lambda}{\partial M} \boldsymbol{p}, \qquad \boldsymbol{U} \frac{\partial \boldsymbol{q}}{\partial \boldsymbol{p}'} = \lambda \boldsymbol{I} + \boldsymbol{p} \frac{\partial \lambda}{\partial \boldsymbol{p}'}, \tag{2.18}$$

where $\boldsymbol{U} = \partial^2 u / \partial \boldsymbol{q} \partial \boldsymbol{q}'$ is the Hessian matrix of the consumer's utility function. We show at the end of Section 3 how the four equations displayed in (2.16)–(2.18) can be combined in partitioned matrix form and how they can be used to provide solutions for the income and price derivatives of demand under appropriate conditions.

3. Partitioned matrices

Partitioning a matrix into submatrices is one device for the exploitation of the mathematical structure of this matrix. This can be of considerable importance in multivariate situations.

3.1. *The algebra of partitioned matrices*

We write the left-most matrix in (2.3) as $\boldsymbol{Y} = [\boldsymbol{Y}_1 \quad \boldsymbol{Y}_2]$, where

$$\boldsymbol{Y}_1 = \begin{bmatrix} y_{11} & y_{12} \\ y_{21} & y_{22} \\ \vdots & \vdots \\ y_{n1} & y_{n2} \end{bmatrix}, \qquad \boldsymbol{Y}_2 = \begin{bmatrix} y_{13} & y_{14} \cdots y_{1L} \\ y_{23} & y_{24} \cdots y_{2L} \\ \vdots & \vdots \quad \vdots \\ y_{n3} & y_{n4} \cdots y_{nL} \end{bmatrix}.$$

The partitioning $\boldsymbol{Y} = [\boldsymbol{Y}_1 \quad \boldsymbol{Y}_2]$ is by sets of columns, the observations on the first two endogenous variables being separated from those on the others. Partitioning may take place by row sets and column sets. The addition rule for matrices can be applied in partitioned form,

$$\begin{bmatrix} A_{11} & A_{12} \\ A_{21} & A_{22} \\ A_{31} & A_{32} \end{bmatrix} + \begin{bmatrix} B_{11} & B_{12} \\ B_{21} & B_{22} \\ B_{31} & B_{32} \end{bmatrix} = \begin{bmatrix} A_{11}+B_{11} & A_{12}+B_{12} \\ A_{21}+B_{21} & A_{22}+B_{22} \\ A_{31}+B_{31} & A_{32}+B_{32} \end{bmatrix},$$

provided A_{ij} and B_{ij} have the same order for each (i, j). A similar result holds for multiplication,

$$\begin{bmatrix} P_{11} & P_{12} \\ P_{21} & P_{22} \end{bmatrix} \begin{bmatrix} Q_{11} & Q_{12} \\ Q_{21} & Q_{22} \end{bmatrix} = \begin{bmatrix} P_{11}Q_{11}+P_{12}Q_{21} & P_{11}Q_{12}+P_{12}Q_{22} \\ P_{21}Q_{11}+P_{22}Q_{21} & P_{21}Q_{12}+P_{22}Q_{22} \end{bmatrix},$$

provided that the number of columns of P_{11} and P_{21} is equal to the number of rows of Q_{11} and Q_{12} (similarly for $P_{12}, P_{22}, Q_{21}, Q_{22}$).

The inverse of a symmetric partitioned matrix is frequently needed. Two alternative expressions are available:

$$\begin{bmatrix} A & B \\ B' & C \end{bmatrix}^{-1} = \begin{bmatrix} D & -DBC^{-1} \\ -C^{-1}B'D & C^{-1}+C^{-1}B'DBC^{-1} \end{bmatrix}, \tag{3.1}$$

$$\begin{bmatrix} A & B \\ B' & C \end{bmatrix}^{-1} = \begin{bmatrix} A^{-1}+A^{-1}BEB'A^{-1} & -A^{-1}BE \\ -EB'A^{-1} & E \end{bmatrix}, \tag{3.2}$$

where $D = (A - BC^{-1}B')^{-1}$ and $E = (C - B'A^{-1}B)^{-1}$. The use of (3.1) requires that C be non-singular; for (3.2) we must assume that A is non-singular. The verification of these results is a matter of straightforward partitioned multiplication; for a constructive proof see Theil (1971, sec. 1.2).

The density function of the L-variate normal distribution with mean vector μ and non-singular covariance matrix Σ is

$$f(x) = \frac{1}{(2\pi)^{L/2}|\Sigma|^{1/2}} \exp\{-\tfrac{1}{2}(x-\mu)'\Sigma^{-1}(x-\mu)\}, \tag{3.3}$$

where $|\Sigma|$ is the determinant value of Σ. Suppose that each of the first L' variates is uncorrelated with all $L - L'$ other variates. Then μ and Σ may be partitioned,

$$\mu = \begin{bmatrix} \mu_1 \\ \mu_2 \end{bmatrix}, \qquad \Sigma = \begin{bmatrix} \Sigma_1 & 0 \\ 0 & \Sigma_2 \end{bmatrix}, \tag{3.4}$$

where (μ_1, Σ_1) contains the first- and second-order moments of the first L'

variates and $(\boldsymbol{\mu}_2, \boldsymbol{\Sigma}_2)$ those of the last $L - L'$. The density function (3.3) can now be written as the product of

$$f_1(\boldsymbol{x}_1) = \frac{1}{(2\pi)^{L'/2}|\boldsymbol{\Sigma}_1|^{1/2}} \exp\{-\tfrac{1}{2}(\boldsymbol{x}_1 - \boldsymbol{\mu}_1)'\boldsymbol{\Sigma}_1^{-1}(\boldsymbol{x}_1 - \boldsymbol{\mu}_1)\}$$

and analogous function $f_2(\boldsymbol{x}_2)$. Clearly, the L-element normal vector consists of two subvectors which are independently distributed.

3.2. *Block-recursive systems*

We return to the equation system (2.3) and assume that the rows of $\boldsymbol{E}$ are independent L-variate normal vectors with zero mean and covariance matrix $\boldsymbol{\Sigma}$, as shown in (2.4), $\boldsymbol{\Sigma}_1$ being of order $L' \times L'$. We also assume that $\boldsymbol{\Gamma}$ can be partitioned as

$$\boldsymbol{\Gamma} = \begin{bmatrix} \boldsymbol{\Gamma}_1 & \boldsymbol{\Gamma}_3 \\ \boldsymbol{0} & \boldsymbol{\Gamma}_2 \end{bmatrix}, \tag{3.5}$$

with $\boldsymbol{\Gamma}_1$ of order $L' \times L'$. Then we can write (2.3) as

$$[\boldsymbol{Y}_1 \quad \boldsymbol{Y}_2] \begin{bmatrix} \boldsymbol{\Gamma}_1 & \boldsymbol{\Gamma}_3 \\ \boldsymbol{0} & \boldsymbol{\Gamma}_2 \end{bmatrix} + \boldsymbol{X}[\boldsymbol{B}_1 \quad \boldsymbol{B}_2] = [\boldsymbol{E}_1 \quad \boldsymbol{E}_2]$$

or

$$\boldsymbol{Y}_1\boldsymbol{\Gamma}_1 + \boldsymbol{X}\boldsymbol{B}_1 = \boldsymbol{E}_1, \tag{3.6}$$

$$\boldsymbol{Y}_2\boldsymbol{\Gamma}_2 + [\boldsymbol{X} \quad \boldsymbol{Y}_1] \begin{bmatrix} \boldsymbol{B}_2 \\ \boldsymbol{\Gamma}_3 \end{bmatrix} = \boldsymbol{E}_2, \tag{3.7}$$

where $\boldsymbol{Y} = [\boldsymbol{Y}_1 \quad \boldsymbol{Y}_2]$, $\boldsymbol{B} = [\boldsymbol{B}_1 \quad \boldsymbol{B}_2]$, and $\boldsymbol{E} = [\boldsymbol{E}_1 \quad \boldsymbol{E}_2]$ with $\boldsymbol{Y}_1$ and $\boldsymbol{E}_1$ of order $n \times L'$ and $\boldsymbol{B}_1$ of order $K \times L'$.

There is nothing special about (3.6), which is an equation system comparable to (2.3) but of smaller size. However, (3.7) is an equation system in which the L' variables whose observations are arranged in $\boldsymbol{Y}_1$ can be viewed as exogenous rather than endogenous. This is indicated by combining $\boldsymbol{Y}_1$ with $\boldsymbol{X}$ in partitioned matrix form. There are two reasons why $\boldsymbol{Y}_1$ can be viewed as exogenous in (3.7). First, $\boldsymbol{Y}_1$ is obtained from the system (3.6) which does not involve $\boldsymbol{Y}_2$. Secondly, the random component $\boldsymbol{E}_1$ in (3.6) is independent of $\boldsymbol{E}_2$ in (3.7) because of the assumed normality with a block-diagonal $\boldsymbol{\Sigma}$. The case discussed here is that of a *block-recursive system*, with a block-triangular $\boldsymbol{\Gamma}$ [see (3.5)] and a block-diagonal $\boldsymbol{\Sigma}$ [see (3.4)]. Under appropriate identification conditions, ML estimation of the unknown elements of $\boldsymbol{\Gamma}$ and $\boldsymbol{B}$ can be applied to the two subsystems (3.6) and (3.7) separately.

3.3. Income and price derivatives revisited

It is readily verified that eqs. (2.16)–(2.18) can be written in partitioned matrix form as

$$\begin{bmatrix} \boldsymbol{U} & \boldsymbol{p} \\ \boldsymbol{p}' & 0 \end{bmatrix} \begin{bmatrix} \partial \boldsymbol{q}/\partial M & \partial \boldsymbol{q}/\partial \boldsymbol{p}' \\ -\partial\lambda/\partial M & -\partial\lambda/\partial \boldsymbol{p}' \end{bmatrix} = \begin{bmatrix} \boldsymbol{0} & \lambda \boldsymbol{I} \\ 1 & -\boldsymbol{q}' \end{bmatrix}, \tag{3.8}$$

which is Barten's (1964) fundamental matrix equation in consumption theory. All three partitioned matrices in (3.8) are of order $(N+1)\times(N+1)$, and the left-most matrix is the Hessian matrix of utility function bordered by prices. If $\boldsymbol{U}$ is non-singular, we can use (3.2) for the inverse of this bordered matrix:

$$\begin{bmatrix} \boldsymbol{U} & \boldsymbol{p} \\ \boldsymbol{p}' & 0 \end{bmatrix}^{-1} = \frac{1}{\boldsymbol{p}'\boldsymbol{U}^{-1}\boldsymbol{p}} \begin{bmatrix} (\boldsymbol{p}'\boldsymbol{U}^{-1}\boldsymbol{p})\boldsymbol{U}^{-1} - \boldsymbol{U}^{-1}\boldsymbol{p}(\boldsymbol{U}^{-1}\boldsymbol{p})' & \boldsymbol{U}^{-1}\boldsymbol{p} \\ (\boldsymbol{U}^{-1}\boldsymbol{p})' & -1 \end{bmatrix}.$$

Premultiplication of (3.8) by this inverse yields solutions for the income and price derivatives:

$$\frac{\partial \boldsymbol{q}}{\partial M} = \frac{1}{\boldsymbol{p}'\boldsymbol{U}^{-1}\boldsymbol{p}}\boldsymbol{U}^{-1}\boldsymbol{p}, \qquad \frac{\partial \lambda}{\partial M} = \frac{1}{\boldsymbol{p}'\boldsymbol{U}^{-1}\boldsymbol{p}} \tag{3.9}$$

$$\frac{\partial \boldsymbol{q}}{\partial \boldsymbol{p}'} = \lambda \boldsymbol{U}^{-1} - \frac{\lambda}{\boldsymbol{p}'\boldsymbol{U}^{-1}\boldsymbol{p}}\boldsymbol{U}^{-1}\boldsymbol{p}(\boldsymbol{U}^{-1}\boldsymbol{p})' - \frac{1}{\boldsymbol{p}'\boldsymbol{U}^{-1}\boldsymbol{p}}\boldsymbol{U}^{-1}\boldsymbol{p}\boldsymbol{q}'. \tag{3.10}$$

It follows from (3.9) that we can write the income derivatives of demand as

$$\frac{\partial \boldsymbol{q}}{\partial M} = \frac{\partial \lambda}{\partial M}\boldsymbol{U}^{-1}\boldsymbol{p}, \tag{3.11}$$

and from (3.9) and (3.10) that we can simplify the price derivatives to

$$\frac{\partial \boldsymbol{q}}{\partial \boldsymbol{p}'} = \lambda \boldsymbol{U}^{-1} - \frac{\lambda}{\partial\lambda/\partial M}\frac{\partial \boldsymbol{q}}{\partial M}\frac{\partial \boldsymbol{q}'}{\partial M} - \frac{\partial \boldsymbol{q}}{\partial M}\boldsymbol{q}'. \tag{3.12}$$

The last matrix, $-(\partial \boldsymbol{q}/\partial M)\boldsymbol{q}'$, represents the *income effect* of the price changes on demand. Note that this matrix has unit rank and is not symmetric. The two other matrices on the right in (3.12) are symmetric and jointly represent the substitution effect of the price changes. The first matrix, $\lambda \boldsymbol{U}^{-1}$, gives the *specific substitution effect* and the second (which has unit rank) gives the *general substitution effect*. The latter effect describes the general competition of all goods for an extra dollar of income. The distinction between the two components of the substitution effect is from Houthakker (1960). We can combine these components by writing (3.12) in the form

$$\frac{\partial \boldsymbol{q}}{\partial \boldsymbol{p}'} = \lambda \boldsymbol{U}^{-1}\left(\boldsymbol{I} - \boldsymbol{p}\frac{\partial \boldsymbol{q}'}{\partial M}\right) - \frac{\partial \boldsymbol{q}}{\partial M}\boldsymbol{q}', \tag{3.13}$$

which is obtained by using (3.11) for the first $\partial \boldsymbol{q}/\partial M$ that occurs in (3.12).

4. Kronecker products and the vectorization of matrices

A special form of partitioning is that in which all submatrices are scalar multiples of the same matrix $\boldsymbol{B}$ of order $p \times q$. We write this as

$$\boldsymbol{A}\otimes\boldsymbol{B} = \begin{bmatrix} a_{11}\boldsymbol{B} & a_{12}\boldsymbol{B}\dots a_{1n}\boldsymbol{B} \\ a_{21}\boldsymbol{B} & a_{22}\boldsymbol{B}\dots a_{2n}\boldsymbol{B} \\ \vdots & \vdots \quad \vdots \\ a_{m1}\boldsymbol{B} & a_{m2}\boldsymbol{B}\dots a_{mn}\boldsymbol{B} \end{bmatrix},$$

and refer to $\boldsymbol{A}\otimes\boldsymbol{B}$ as the Kronecker product of $\boldsymbol{A} = [a_{ij}]$ and $\boldsymbol{B}$. The order of this product is $mp \times nq$. Kronecker products are particularly convenient when several equations are analyzed simultaneously.

4.1. *The algebra of Kronecker products*

It is a matter of straightforward partitioned multiplication to verify that

$$(\boldsymbol{A}\otimes\boldsymbol{B})(\boldsymbol{C}\otimes\boldsymbol{D}) = \boldsymbol{AC}\otimes\boldsymbol{BD}, \tag{4.1}$$

provided $\boldsymbol{AC}$ and $\boldsymbol{BD}$ exist. Also, if $\boldsymbol{A}$ and $\boldsymbol{B}$ are square and non-singular, then

$$(\boldsymbol{A}\otimes\boldsymbol{B})^{-1} = \boldsymbol{A}^{-1}\otimes\boldsymbol{B}^{-1} \tag{4.2}$$

because (4.1) implies $(\boldsymbol{A}\otimes\boldsymbol{B})(\boldsymbol{A}^{-1}\otimes\boldsymbol{B}^{-1}) = \boldsymbol{AA}^{-1}\otimes\boldsymbol{BB}^{-1} = \boldsymbol{I}\otimes\boldsymbol{I} = \boldsymbol{I}$, where the three unit matrices will in general be of different order. We can obviously extend

(4.1) to

$$(A_1 \otimes B_1)(A_2 \otimes B_2)(A_3 \otimes B_3) = A_1 A_2 A_3 \otimes B_1 B_2 B_3$$

provided $A_1 A_2 A_3$ and $B_1 B_2 B_3$ exist.

Other useful properties of Kronecker products are:

$$(A \otimes B)' = A' \otimes B', \tag{4.3}$$

$$A \otimes (B + C) = A \otimes B + A \otimes C, \tag{4.4}$$

$$(B + C) \otimes A = B \otimes A + C \otimes A, \tag{4.5}$$

$$A \otimes (B \otimes C) = (A \otimes B) \otimes C. \tag{4.6}$$

Note the implication of (4.3) that $A \otimes B$ is symmetric when A and B are symmetric. Other properties of Kronecker products are considered in Section 7.

4.2. *Joint generalized least-squares estimation of several equations*

In (2.1) and (2.3) we considered a system of L linear equations in L endogenous variables. Here we consider the special case in which each equation describes one endogenous variable in terms of exogenous variables only. If the observations on all variables are $\alpha = 1, \dots, n$, we can write the L equations in a form similar to (2.7):

$$y_j = X_j \beta_j + \varepsilon_j, \qquad j = 1, \dots, L, \tag{4.7}$$

where $y_j = [y_{\alpha j}]$ is the observation vector on the jth endogenous variable, $\varepsilon_j = [\varepsilon_{\alpha j}]$ is the associated disturbance vector with zero expectation, X_j is the observation matrix on the K_j exogenous variables in the jth equation, and β_j is the K_j-element parameter vector.

We can write (4.7) for all j in partitioned matrix form:

$$\begin{bmatrix} y_1 \\ y_2 \\ \vdots \\ y_L \end{bmatrix} = \begin{bmatrix} X_1 & 0 \dots 0 \\ 0 & X_2 \dots 0 \\ \vdots & \vdots \quad \vdots \\ 0 & 0 \dots X_L \end{bmatrix} \begin{bmatrix} \beta_1 \\ \beta_2 \\ \vdots \\ \beta_L \end{bmatrix} + \begin{bmatrix} \varepsilon_1 \\ \varepsilon_2 \\ \vdots \\ \varepsilon_L \end{bmatrix} \tag{4.8}$$

or, more briefly, as

$$y = Z\beta + \varepsilon, \tag{4.9}$$

where $\boldsymbol{y}$ and $\boldsymbol{\varepsilon}$ are Ln-element vectors and $\boldsymbol{Z}$ contains Ln rows, while the number of columns of $\boldsymbol{Z}$ and that of the elements of $\boldsymbol{\beta}$ are both $K_1+\cdots+K_L$. The covariance matrix of $\boldsymbol{\varepsilon}$ is thus of order $Ln\times Ln$ and can be partitioned into L^2 submatrices of the form $\mathcal{E}(\boldsymbol{\varepsilon}_j\boldsymbol{\varepsilon}_l')$. For $j=l$ this submatrix equals the covariance matrix $\mathcal{V}(\boldsymbol{\varepsilon}_j)$. We assume that the n disturbances of each of the L equations have equal variance and are uncorrelated so that $\mathcal{V}(\boldsymbol{\varepsilon}_j)=\sigma_{jj}\boldsymbol{I}$, where $\sigma_{jj}=\operatorname{var}\varepsilon_{\alpha j}$ (each α). For $j\neq l$ the submatrix $\mathcal{E}(\boldsymbol{\varepsilon}_j\boldsymbol{\varepsilon}_l')$ contains the "contemporaneous" covariances $\mathcal{E}(\varepsilon_{\alpha j}\varepsilon_{\alpha l})$ for $\alpha=1,\dots,n$ in the diagonal. We assume that these covariances are all equal to σ_{jl} and that all non-contemporaneous covariances vanish: $\mathcal{E}(\varepsilon_{\alpha j}\varepsilon_{\eta l})=0$ for $\alpha\neq\eta$. Therefore, $\mathcal{E}(\boldsymbol{\varepsilon}_j\boldsymbol{\varepsilon}_l')=\sigma_{jl}\boldsymbol{I}$, which contains $\mathcal{V}(\boldsymbol{\varepsilon}_j)=\sigma_{jj}\boldsymbol{I}$ as a special case. The full covariance matrix of the Ln-element vector $\boldsymbol{\varepsilon}$ is thus:

$$\mathcal{V}(\boldsymbol{\varepsilon})=\begin{bmatrix}\sigma_{11}\boldsymbol{I} & \sigma_{12}\boldsymbol{I}\dots\sigma_{1L}\boldsymbol{I}\\ \sigma_{21}\boldsymbol{I} & \sigma_{22}\boldsymbol{I}\dots\sigma_{2L}\boldsymbol{I}\\ \vdots & \vdots \quad \vdots\\ \sigma_{L1}\boldsymbol{I} & \sigma_{L2}\boldsymbol{I}\dots\sigma_{LL}\boldsymbol{I}\end{bmatrix}=\boldsymbol{\Sigma}\otimes\boldsymbol{I}, \tag{4.10}$$

where $\boldsymbol{\Sigma}=[\sigma_{jl}]$ is the contemporaneous covariance matrix, i.e. the covariance matrix of $[\varepsilon_{\alpha 1}\dots\varepsilon_{\alpha L}]$ for $\alpha=1,\dots,n$.

Suppose that $\boldsymbol{\Sigma}$ is non-singular so that $\boldsymbol{\Sigma}^{-1}\otimes\boldsymbol{I}$ is the inverse of the matrix (4.10) in view of (4.2). Also, suppose that $\boldsymbol{X}_1,\dots,\boldsymbol{X}_L$ and hence $\boldsymbol{Z}$ have full column rank. Application of the GLS results (2.13) and (2.14) to (4.9) and (4.10) then yields

$$\hat{\boldsymbol{\beta}}=\left[\boldsymbol{Z}'(\boldsymbol{\Sigma}^{-1}\otimes\boldsymbol{I})\boldsymbol{Z}\right]^{-1}\boldsymbol{Z}'(\boldsymbol{\Sigma}^{-1}\otimes\boldsymbol{I})\boldsymbol{y} \tag{4.11}$$

as the best linear unbiased estimator of $\boldsymbol{\beta}$ with the following covariance matrix:

$$\mathcal{V}(\hat{\boldsymbol{\beta}})=\left[\boldsymbol{Z}'(\boldsymbol{\Sigma}^{-1}\otimes\boldsymbol{I})\boldsymbol{Z}\right]^{-1}. \tag{4.12}$$

In general, $\hat{\boldsymbol{\beta}}$ is superior to LS applied to each of the L equations separately, but there are two special cases in which these estimation procedures are identical.

The first case is that in which $\boldsymbol{X}_1,\dots,\boldsymbol{X}_L$ are all identical. We can then write $\boldsymbol{X}$ for each of these matrices so that the observation matrix on the exogenous variables in (4.8) and (4.9) takes the form

$$\boldsymbol{Z}=\begin{bmatrix}\boldsymbol{X} & \boldsymbol{0}\dots\boldsymbol{0}\\ \boldsymbol{0} & \boldsymbol{X}\dots\boldsymbol{0}\\ \vdots & \vdots \quad \vdots\\ \boldsymbol{0} & \boldsymbol{0}\dots\boldsymbol{X}\end{bmatrix}=\boldsymbol{I}\otimes\boldsymbol{X}. \tag{4.13}$$

This implies

$$Z'(\Sigma^{-1}\otimes I)Z=(I\otimes X')(\Sigma^{-1}\otimes I)(I\otimes X)=\Sigma^{-1}\otimes X'X$$

and

$$\begin{aligned}[Z'(\Sigma^{-1}\otimes I)Z]^{-1}Z'(\Sigma^{-1}\otimes I)&=[\Sigma\otimes(X'X)^{-1}](I\otimes X')(\Sigma^{-1}\otimes I)\\&=I\otimes(X'X)^{-1}X'.\end{aligned}$$

It is now readily verified from (4.11) that $\hat{\beta}$ consists of L subvectors of the LS form $(X'X)^{-1}X'y_j$. The situation of identical matrices $X_1,\dots,X_L$ occurs relatively frequently in applied econometrics; an example is the reduced form (2.4) for each of the L endogenous variables.

The second case in which (4.11) degenerates into subvectors equal to LS vectors is that of uncorrelated contemporaneous disturbances. Then Σ is diagonal and it is easily verified that $\hat{\beta}$ consists of subvectors of the form $(X_j'X_j)^{-1}X_j'y_j$. See Theil (1971, pp. 311–312) for the case in which Σ is block-diagonal.

Note that the computation of the joint GLS estimator (4.11) requires Σ to be known. This is usually not true and the unknown Σ is then replaced by the sample moment matrix of the LS residuals [see Zellner (1962)]. This approximation is asymptotically (for large n) acceptable under certain conditions; we shall come back to this matter in the opening paragraph of Section 9.

4.3. *Vectorization of matrices*

In eq. (2.3) we wrote Ln equations in matrix form with parameter matrices Γ and B, each consisting of several columns, whereas in (4.8) and (4.9) we wrote Ln equations in matrix form with a "long" parameter vector β. If Z takes the form (4.13), we can write (4.8) in the equivalent form $Y=XB+E$, where Y, B, and E are matrices consisting of L columns of the form y_j, β_j, and ε_j. Thus, the elements of the parameter vector β are then rearranged into the matrix B. On the other hand, there are situations in which it is more attractive to work with vectors rather than matrices that consist of several columns. For example, if $\hat{\beta}$ is an unbiased estimator of the parameter vector β with finite second moments, we obtain the covariance matrix of $\hat{\beta}$ by postmultiplying $\hat{\beta}-\beta$ by its transpose and taking the expectation, but this procedure does not work when the parameters are arranged in a matrix B which consists of several columns. It is then appropriate to rearrange the parameters in vector form. This is a matter of designing an appropriate notation and evaluating the associated algebra.

Let $A=[a_1\dots a_q]$ be a $p\times q$ matrix, a_i being the ith column of A. We define $\operatorname{vec}A=[a_1' \quad a_2'\dots a_q']'$, which is a pq-element column vector consisting of q

subvectors, the first containing the p elements of $\boldsymbol{a}_1$, the second the p elements of $\boldsymbol{a}_2$, and so on. It is readily verified that $\operatorname{vec}(\boldsymbol{A}+\boldsymbol{B}) = \operatorname{vec}\boldsymbol{A} + \operatorname{vec}\boldsymbol{B}$, provided that $\boldsymbol{A}$ and $\boldsymbol{B}$ are of the same order. Also, if the matrix products $\boldsymbol{AB}$ and $\boldsymbol{BC}$ exist,

$$\operatorname{vec}\boldsymbol{AB} = (\boldsymbol{I}\otimes\boldsymbol{A})\operatorname{vec}\boldsymbol{B} = (\boldsymbol{B}'\otimes\boldsymbol{I})\operatorname{vec}\boldsymbol{A},$$
$$\operatorname{vec}\boldsymbol{ABC} = (\boldsymbol{I}\otimes\boldsymbol{AB})\operatorname{vec}\boldsymbol{C} = (\boldsymbol{C}'\otimes\boldsymbol{A})\operatorname{vec}\boldsymbol{B} = (\boldsymbol{C}'\boldsymbol{B}'\otimes\boldsymbol{I})\operatorname{vec}\boldsymbol{A}.$$

For proofs and extensions of these results see Dhrymes (1978, ch. 4).

5. Differential demand and supply systems

The differential approach to microeconomic theory provides interesting comparisons with equation systems such as (2.3) and (4.9). Let $\boldsymbol{g}(\boldsymbol{z})$ be a vector of functions of a vector $\boldsymbol{z}$; the approach uses the total differential of $\boldsymbol{g}(\cdot)$,

$$\mathrm{d}\boldsymbol{g} = \frac{\partial \boldsymbol{g}}{\partial \boldsymbol{z}'}\,\mathrm{d}\boldsymbol{z}, \tag{5.1}$$

and it exploits what is known about $\partial\boldsymbol{g}/\partial\boldsymbol{z}'$. For example, the total differential of consumer demand is $\mathrm{d}\boldsymbol{q} = (\partial\boldsymbol{q}/\partial M)\mathrm{d}M + (\partial\boldsymbol{q}/\partial\boldsymbol{p}')\mathrm{d}\boldsymbol{p}$. Substitution from (3.13) yields:

$$\mathrm{d}\boldsymbol{q} = \frac{\partial \boldsymbol{q}}{\partial M}(\mathrm{d}M - \boldsymbol{q}'\mathrm{d}\boldsymbol{p}) + \lambda\boldsymbol{U}^{-1}\left[\mathrm{d}\boldsymbol{p} - \left(\frac{\partial \boldsymbol{q}'}{\partial M}\mathrm{d}\boldsymbol{p}\right)\boldsymbol{p}\right], \tag{5.2}$$

which shows that the income effect of the price changes is used to deflate the change in money income and, similarly, the general substitution effect to deflate the specific effect. Our first objective is to write the system (5.2) in a more attractive form.

5.1. *A differential consumer demand system*

We introduce the budget share w_i and the marginal share θ_i of good i:

$$w_i = \frac{p_i q_i}{M}, \qquad \theta_i = \frac{\partial(p_i q_i)}{\partial M}, \tag{5.3}$$

and also the Divisia (1925) volume index $\mathrm{d}(\log Q)$ and the Frisch (1932) price index $\mathrm{d}(\log P')$:

$$\mathrm{d}(\log Q) = \sum_{i=1}^{N} w_i \mathrm{d}(\log q_i), \qquad \mathrm{d}(\log P') = \sum_{i=1}^{N} \theta_i \mathrm{d}(\log p_i), \tag{5.4}$$

where log (here and elsewhere) stands for natural logarithm. We prove in the next paragraph that (5.2) can be written in scalar form as

$$w_i \mathrm{d}(\log q_i) = \theta_i \mathrm{d}(\log Q) + \phi \sum_{j=1}^{N} \theta_{ij} \mathrm{d}\left(\log \frac{p_j}{P'}\right), \tag{5.5}$$

where $\mathrm{d}[\log(p_j/P')]$ is an abbreviation of $\mathrm{d}(\log p_j) - \mathrm{d}(\log P')$, while ϕ is the reciprocal of the income elasticity of the marginal utility of income:

$$\phi = \left(\frac{\partial \log \lambda}{\partial \log M}\right)^{-1}, \tag{5.6}$$

and θ_{ij} is an element of the symmetric $N \times N$ matrix

$$\boldsymbol{\Theta} = \frac{\lambda}{\phi M} \boldsymbol{P} \boldsymbol{U}^{-1} \boldsymbol{P}, \tag{5.7}$$

with $\boldsymbol{P}$ defined as the diagonal matrix with the prices $p_1, \dots, p_N$ on the diagonal.

To verify (5.5) we apply (5.1) to $M = \boldsymbol{p}'\boldsymbol{q}$, yielding $\mathrm{d}M = \boldsymbol{q}'\mathrm{d}\boldsymbol{p} + \boldsymbol{p}'\mathrm{d}\boldsymbol{q}$ so that $\mathrm{d}M - \boldsymbol{q}'\mathrm{d}\boldsymbol{p} = M\mathrm{d}(\log Q)$ follows from (5.3) and (5.4). Therefore, premultiplication of (5.2) by $(1/M)\boldsymbol{P}$ gives:

$$\frac{1}{M}\boldsymbol{P}\mathrm{d}\boldsymbol{q} = \boldsymbol{P}\frac{\partial \boldsymbol{q}}{\partial M}\mathrm{d}(\log Q) + \frac{\lambda}{M}\boldsymbol{P}\boldsymbol{U}^{-1}\boldsymbol{P}\left[\boldsymbol{P}^{-1}\mathrm{d}\boldsymbol{p} - \left(\frac{\partial \boldsymbol{q}'}{\partial M}\mathrm{d}\boldsymbol{p}\right)\boldsymbol{\iota}\right], \tag{5.8}$$

where $\boldsymbol{\iota} = \boldsymbol{P}^{-1}\boldsymbol{p}$ is a vector of N unit elements. The ith element of $(1/M)\boldsymbol{P}\mathrm{d}\boldsymbol{q}$ equals $(p_i/M)\mathrm{d}q_i = w_i \mathrm{d}(\log q_i)$, which confirms the left side of (5.5). The vector $\boldsymbol{P}(\partial \boldsymbol{q}/\partial M)$ equals the marginal share vector $\boldsymbol{\theta} = [\theta_i]$, thus confirming the real-income term of (5.5). The jth element of the vector in brackets in (5.8) equals $\mathrm{d}(\log p_j) - \mathrm{d}(\log P')$, which agrees with the substitution term of (5.5). The verification of (5.5) is completed by $(\lambda/M)\boldsymbol{P}\boldsymbol{U}^{-1}\boldsymbol{P} = \phi\boldsymbol{\Theta}$ [see (5.7)]. Note that $\boldsymbol{\Theta}\boldsymbol{\iota} = (\lambda/\phi M)\boldsymbol{P}\boldsymbol{U}^{-1}\boldsymbol{p} = \boldsymbol{P}(\partial \boldsymbol{q}/\partial M)$ [see (3.11) and (5.6)]. Therefore,

$$\boldsymbol{\Theta}\boldsymbol{\iota} = \boldsymbol{\theta}, \qquad \boldsymbol{\iota}'\boldsymbol{\Theta}\boldsymbol{\iota} = \boldsymbol{\iota}'\boldsymbol{\theta} = 1, \tag{5.9}$$

where $\boldsymbol{\iota}'\boldsymbol{\theta} = \sum_i \theta_i = 1$ follows from (2.16). We conclude from $\boldsymbol{\Theta}\boldsymbol{\iota} = \boldsymbol{\theta}$ that the θ_{ij}'s of the ith equation sum to the ith marginal share, and from $\boldsymbol{\iota}'\boldsymbol{\Theta}\boldsymbol{\iota} = 1$ that the θ_{ij}'s of the entire system sum to 1. The latter property is expressed by referring to the θ_{ij}'s as the *normalized price coefficients*.

5.2. A comparison with simultaneous equation systems

The N-equation system (5.5) describes the change in the demand for each good, measured by its contribution to the Divisia index [see (5.4)],[2] as the sum of a real-income component and a substitution component. This system may be compared with the L-equation system (2.1). There is a difference in that the latter system contains in principle more than one endogenous variable in each equation, whereas (5.5) has only one such variable if we assume the $d(\log Q)$ and all price changes are exogenous.[3] Yet, the differential demand system is truly a system because of the cross-equation constraints implied by the symmetry of the normalized price coefficient matrix $\boldsymbol{\Theta}$.

A more important difference results from the utility-maximizing theory behind (5.5), which implies that the coefficients are more directly interpretable than the γ's and β's of (2.1). Writing $[\theta^{ij}] = \boldsymbol{\Theta}^{-1}$ and inverting (5.7), we obtain:

$$\theta^{ij} = \frac{\phi M}{\lambda} \frac{\partial^2 u}{\partial(p_i q_i)\,\partial(p_j q_j)}, \tag{5.10}$$

which shows that θ^{ij} measures (apart from $\phi M/\lambda$ which does not involve i and j) the change in the marginal utility of a dollar spent on i caused by an extra dollar spent on j. Equivalently, the normalized price coefficient matrix $\boldsymbol{\Theta}$ is inversely proportional to the Hessian matrix of the utility function in expenditure terms.

The relation (5.7) between $\boldsymbol{\Theta}$ and $\boldsymbol{U}$ allows us to analyze special preference structures. Suppose that the consumer's tastes can be represented by a utility function which is the sum of N functions, one for each good. Then the marginal utility of each good is independent of the consumption of all other goods, which we express by referring to this case as *preference independence*. The Hessian $\boldsymbol{U}$ is then diagonal and so is $\boldsymbol{\Theta}$ [see (5.7)], while $\boldsymbol{\Theta\iota} = \boldsymbol{\theta}$ in (5.9) is simplified to $\theta_{ii} = \theta_i$. Thus, we can write (5.5) under preference independence as

$$w_i d(\log q_i) = \theta_i d(\log Q) + \phi\theta_i d\left(\log \frac{p_i}{P'}\right), \tag{5.11}$$

which contains only one Frisch-deflated price. The system (5.11) for $i = 1,\dots,N$ contains only N unconstrained coefficients, namely ϕ and $N-1$ unconstrained marginal shares.

The application of differential demand systems to data requires a parameterization which postulates that certain coefficients are constant. Several solutions have

[2] Note that this way of measuring the change in demand permits the exploitation of the symmetry of $\boldsymbol{\Theta}$. When we have $d(\log q_i)$ on the left, the coefficient of the Frisch-deflated price becomes θ_{ij}/w_i, which is an element of an asymmetric matrix.

[3] This assumption may be relaxed; see Theil (1975–76, ch. 9–10) for an analysis of endogenous price changes.

been proposed, but these are beyond the scope of this chapter; see the references quoted in Section 2.4 above and also, for a further comparison with models of the type (2.1), Theil and Clements (1980).

5.3. *An extension to the inputs of a firm: A singularity problem*

Let the p_i's and q_i's be the prices and quantities of N inputs which a firm buys to make a product, the output of which is z. Let $z = g(\boldsymbol{q})$ be the firm's production function, $g(\cdot)$ being three times differentiable. Let the firm's objective be to minimize input expenditure $\boldsymbol{p}'\boldsymbol{q}$ subject to $z = g(\boldsymbol{q})$ for given output z and input prices $\boldsymbol{p}$. Our objective will be to analyze whether this minimum problem yields a differential input demand system similar to (5.5).

As in the consumer's case we construct a Lagrangian function, which now takes the form $\boldsymbol{p}'\boldsymbol{q} - \rho[g(\boldsymbol{q}) - z]$. By equating the derivative of this function with respect to $\boldsymbol{q}$ to zero we obtain a proportionality of $\partial g/\partial \boldsymbol{q}$ to $\boldsymbol{p}$ [compare (2.15)]. This proportionality and the production function provide $N+1$ equations in $N+1$ unknowns: $\boldsymbol{q}$ and ρ. Next we differentiate these equations with respect to z and $\boldsymbol{p}$, and we collect the derivatives in partitioned matrix form. The result is similar to the matrix equation (3.8) of consumption theory, and the Hessian $\boldsymbol{U}$ now becomes the Hessian $\partial^2 g/\partial \boldsymbol{q}\partial \boldsymbol{q}'$ of the production function. We can then proceed as in (3.9) and following text if $\partial^2 g/\partial \boldsymbol{q}\partial \boldsymbol{q}'$ is non-singular, but this is unfortunately not true when the firm operates under constant returns to scale. It is clearly unattractive to make an assumption which excludes this important case. In the account which follows[4] we solve this problem by formulating the production function in logarithmic form.

$$\log z = h(\boldsymbol{q}), \tag{5.12}$$

and using the following $N \times N$ Hessian matrix:

$$\boldsymbol{H} = \left[\frac{\partial^2 h}{\partial(\log q_i)\,\partial(\log q_j)} \right]. \tag{5.13}$$

5.4. *A differential input demand system*

The minimum of $\boldsymbol{p}'\boldsymbol{q}$ subject to (5.12) for given z and $\boldsymbol{p}$ will be a function of z and $\boldsymbol{p}$. We write $C(z, \boldsymbol{p})$ for this minimum: the cost of producing output z at the input

[4]Derivations are omitted; the procedure is identical to that which is outlined above except that it systematically uses logarithms of output, inputs, and input prices. See Laitinen (1980), Laitinen and Theil (1978), and Theil (1977, 1980).

prices $\boldsymbol{p}$. We define

$$\gamma = \frac{\partial \log C}{\partial \log z}, \qquad \frac{1}{\psi} = 1 + \frac{1}{\gamma^2} \frac{\partial^2 \log C}{\partial (\log z)^2}, \tag{5.14}$$

so that γ is the output elasticity of cost and $\psi < 1$ (> 1) when this elasticity increases (decreases) with increasing output; thus, ψ is a curvature measure of the logarithmic cost function. It can be shown that the input demand equations may be written as

$$f_i \mathrm{d}(\log q_i) = \gamma \theta_i \mathrm{d}(\log z) - \psi \sum_{j=1}^{N} \theta_{ij} \mathrm{d}\left(\log \frac{p_j}{P'}\right), \tag{5.15}$$

which should be compared with (5.5). In (5.15), f_i is the factor share of input i (its share in total cost) and θ_i is its marginal share (the share in marginal cost),

$$f_i = \frac{p_i q_i}{C}, \qquad \theta_i = \frac{\partial (p_i q_i)/\partial z}{\partial C/\partial z}, \tag{5.16}$$

which is the input version of (5.3). The Frisch price index on the far right in (5.15) is as shown in (5.4) but with θ_i defined in (5.16). The coefficient θ_{ij} in (5.15) is the (i, j)th element of the symmetric matrix

$$\boldsymbol{\Theta} = \frac{1}{\psi} \boldsymbol{F}(\boldsymbol{F} - \gamma \boldsymbol{H})^{-1} \boldsymbol{F}, \tag{5.17}$$

where $\boldsymbol{H}$ is given in (5.13) and $\boldsymbol{F}$ is the diagonal matrix with the factor shares $f_1, \dots, f_N$ on the diagonal. This $\boldsymbol{\Theta}$ satisfies (5.9) with $\boldsymbol{\theta} = [\theta_i]$ defined in (5.16).

A firm is called *input independent* when the elasticity of its output with respect to each input is independent of all other inputs. It follows from (5.12) and (5.13) that $\boldsymbol{H}$ is then diagonal; hence, $\boldsymbol{\Theta}$ is also diagonal [see (5.17)] and $\boldsymbol{\Theta}\boldsymbol{\iota} = \boldsymbol{\theta}$ becomes $\theta_{ii} = \theta_i$ so that we can simplify (5.15) to

$$f_i \mathrm{d}(\log q_i) = \gamma \theta_i \mathrm{d}(\log z) - \psi \theta_i \mathrm{d}\left(\log \frac{p_i}{P'}\right), \tag{5.18}$$

which is to be compared with the consumer's equation (5.11) under preference independence. The Cobb–Douglas technology is a special case of input independence with $\boldsymbol{H} = \boldsymbol{0}$, implying that $\boldsymbol{F}(\boldsymbol{F} - \gamma \boldsymbol{H})^{-1} \boldsymbol{F}$ in (5.17) equals the diagonal matrix $\boldsymbol{F}$. Since Cobb–Douglas may have constant returns to scale, this illustrates that the logarithmic formulation successfully avoids the singularity problem mentioned in the previous subsection.

5.5. *Allocation systems*

Summation of (5.5) over i yields the identity $\mathrm{d}(\log Q) = \mathrm{d}(\log Q)$, which means that (5.5) is an *allocation system* in the sense that it describes how the change in total expenditure is allocated to the N goods, given the changes in real income and relative prices. To verify this identity, we write (5.5) for $i = 1,\dots,N$ in matrix form as

$$\boldsymbol{W\kappa} = (\boldsymbol{\iota}'\boldsymbol{W\kappa})\boldsymbol{\theta} + \phi\boldsymbol{\Theta}(\boldsymbol{I} - \boldsymbol{\iota\theta}')\boldsymbol{\pi}, \tag{5.19}$$

where $\boldsymbol{W}$ is the diagonal matrix with $w_1,\dots,w_N$ on the diagonal and $\boldsymbol{\pi} = [\mathrm{d}(\log p_i)]$ and $\boldsymbol{\kappa} = [\mathrm{d}(\log q_i)]$ are the vectors logarithmic price and quantity changes so that $\mathrm{d}(\log Q) = \boldsymbol{\iota}'\boldsymbol{W\kappa}$, $\mathrm{d}(\log P') = \boldsymbol{\theta}'\boldsymbol{\pi}$. The proof is completed by premultiplying (5.19) by $\boldsymbol{\iota}'$, which yields $\boldsymbol{\iota}'\boldsymbol{W\kappa} = \boldsymbol{\iota}'\boldsymbol{W\kappa}$ in view of (5.9). Note that the substitution terms of the N demand equations have zero sum.

The input demand system (5.15) is not an allocation system because the firm does not take total input expenditure as given; rather, it minimizes this expenditure for given output z and given input prices $\boldsymbol{p}$. Summation of (5.15) over i yields:

$$\mathrm{d}(\log Q) = \gamma\,\mathrm{d}(\log z), \tag{5.20}$$

where $\mathrm{d}(\log Q) = \sum_i f_i \mathrm{d}(\log q_i) = \boldsymbol{\iota}'\boldsymbol{F\kappa}$ is the Divisia input volume index. Substitution of (5.20) into (5.15) yields:

$$f_i \mathrm{d}(\log q_i) = \theta_i \mathrm{d}(\log Q) - \psi \sum_{j=1}^{N} \theta_{ij} \mathrm{d}\left(\log \frac{p_j}{P'}\right). \tag{5.21}$$

We can interpret (5.20) as specifying the aggregate input change which is required to produce the given change in output, and (5.21) as an allocation system for the individual inputs given the aggregate input change and the changes in the relative input prices. It follows from (5.9) that we can write (5.19) and (5.21) for each i as

$$\boldsymbol{W\kappa} = (\boldsymbol{\iota}'\boldsymbol{W\kappa})\boldsymbol{\Theta\iota} + \phi\boldsymbol{\Theta}(\boldsymbol{I} - \boldsymbol{\iota\iota}'\boldsymbol{\Theta})\boldsymbol{\pi}, \tag{5.22}$$

$$\boldsymbol{F\kappa} = (\boldsymbol{\iota}'\boldsymbol{F\kappa})\boldsymbol{\Theta\iota} - \psi\boldsymbol{\Theta}(\boldsymbol{I} - \boldsymbol{\iota\iota}'\boldsymbol{\Theta})\boldsymbol{\pi}, \tag{5.23}$$

which shows that the normalized price coefficient matrix $\boldsymbol{\Theta}$ and the scalars ϕ and ψ are the only coefficients in the two allocation systems.

5.6. *Extensions*

Let the firm adjust output z by maximizing its profit under competitive conditions, the price y of the product being exogenous from the firm's point of view.

Then marginal cost $\partial C/\partial z$ equals y, while θ_i of (5.16) equals $\partial(p_i q_i)/\partial(yz)$: the additional expenditure on input i resulting from an extra dollar of output revenue. Note that this is much closer to the consumer's θ_i definition (5.3) than is (5.16).

If the firm sells m products with outputs $z_1,\dots,z_m$ at exogenous prices $y_1,\dots,y_m$, total revenue equals $R=\sum_r y_r z_r$ and $g_r = y_r z_r/R$ is the revenue share of product r, while

$$\mathrm{d}(\log Z) = \sum_{r=1}^{m} g_r \mathrm{d}(\log z_r) \tag{5.24}$$

is the Divisia output volume index of the multiproduct firm. There are now m marginal costs, $\partial C/\partial z_r$ for $r=1,\dots,m$, and each input has m marginal shares: θ_i^r defined as $\partial(p_i q_i)/\partial z_r$ divided by $\partial C/\partial z_r$, which becomes $\theta_i^r = \partial(p_i q_i)/\partial(y_r z_r)$ under profit maximization. Multiproduct input demand equations can be formulated so that the substitution term in (5.15) is unchanged, but the output term becomes

$$\gamma \sum_{r=1}^{m} \theta_i^r g_r \mathrm{d}(\log z_r), \tag{5.25}$$

which shows that input i can be of either more or less importance for product r than for product s depending on the values of θ_i^r and θ_i^s.

Maximizing profit by adjusting outputs yields an output supply system which will now be briefly described. The rth supply equation is

$$g_r \mathrm{d}(\log z_r) = \psi^* \sum_{s=1}^{m} \theta_{rs}^* \mathrm{d}\left(\log \frac{y_s}{P'^s}\right), \tag{5.26}$$

which describes the change[5] in the supply of product r in terms of all output price changes, each deflated by the corresponding Frisch input price index:

$$\mathrm{d}(\log P'^r) = \sum_{i=1}^{N} \theta_i^r \mathrm{d}(\log p_i). \tag{5.27}$$

Asterisks are added to the coefficients of (5.26) in order to distinguish output supply from input demand. The coefficient ψ^* is positive, while θ_{rs}^* is a normalized price coefficient defined as

$$\theta_{rs}^* = \frac{1}{\psi^* R} y_r c^{rs} y_s, \tag{5.28}$$

[5]This change is measured by the contribution of product r to the Divisia output volume index (5.24). Note that this is similar to the left variables in (5.5) and (5.15).

where c^{rs} is an element of the inverse of the symmetric $m \times m$ matrix $[\partial^2 C/\partial z_r \partial z_s]$. The similarity between (5.28) and (5.7) should be noted; we shall consider this matter further in Section 6. A multiproduct firm is called *output independent* when its cost function is the sum of m functions, one for each product.[6] Then $[\partial^2 C/\partial z_r \partial z_s]$ and $[\theta^*_{rs}]$ are diagonal [see (5.28)] so that the change in the supply of each product depends only on the change in its own deflated price [see (5.26)]. Note the similarity to preference and input independence [see (5.11) and (5.18)].

6. Definite and semidefinite square matrices

The expression $\boldsymbol{x}'\boldsymbol{A}\boldsymbol{x}$ is a quadratic form in the vector $\boldsymbol{x}$. We met several examples in earlier sections: the second-order term in the Taylor expansion (2.6), $\boldsymbol{\varepsilon}'\boldsymbol{M}\boldsymbol{\varepsilon}$ in the residual sum of squares (2.12), the expression in the exponent in the normal density function (3.3), the denominator $\boldsymbol{p}'\boldsymbol{U}^{-1}\boldsymbol{p}$ in (3.9), and $\boldsymbol{\iota}'\boldsymbol{\Theta}\boldsymbol{\iota}$ in (5.9). A more systematic analysis of quadratic forms is in order.

6.1. *Covariance matrices and Gauss–Markov further considered*

Let $\boldsymbol{r}$ be a random vector with expectation $\mathcal{E}\boldsymbol{r}$ and covariance matrix $\boldsymbol{\Sigma}$. Let $\boldsymbol{w}'\boldsymbol{r}$ be a linear function of $\boldsymbol{r}$ with non-stochastic weight vector $\boldsymbol{w}$ so that $\mathcal{E}(\boldsymbol{w}'\boldsymbol{r}) = \boldsymbol{w}'\mathcal{E}\boldsymbol{r}$. The variance of $\boldsymbol{w}'\boldsymbol{r}$ is the expectation of

$$[\boldsymbol{w}'(\boldsymbol{r}-\mathcal{E}\boldsymbol{r})]^2 = \boldsymbol{w}'(\boldsymbol{r}-\mathcal{E}\boldsymbol{r})(\boldsymbol{r}-\mathcal{E}\boldsymbol{r})'\boldsymbol{w},$$

so that var $\boldsymbol{w}'\boldsymbol{r} = \boldsymbol{w}'\mathcal{V}(\boldsymbol{r})\boldsymbol{w} = \boldsymbol{w}'\boldsymbol{\Sigma}\boldsymbol{w}$. Thus, the variance of any linear function of $\boldsymbol{r}$ equals a quadratic form with the covariance matrix of $\boldsymbol{r}$ as matrix.

If the quadratic form $\boldsymbol{x}'\boldsymbol{A}\boldsymbol{x}$ is positive for any $\boldsymbol{x} \neq \boldsymbol{0}$, $\boldsymbol{A}$ is said to be *positive definite*. An example is a diagonal matrix $\boldsymbol{A}$ with positive diagonal elements. If $\boldsymbol{x}'\boldsymbol{A}\boldsymbol{x} \geqslant 0$ for any $\boldsymbol{x}$, $\boldsymbol{A}$ is called *positive semidefinite*. The covariance matrix $\boldsymbol{\Sigma}$ of any random vector is always positive semidefinite because we just proved that $\boldsymbol{w}'\boldsymbol{\Sigma}\boldsymbol{w}$ is the variance of a linear function and variances are non-negative. This covariance matrix is positive semidefinite but not positive definite if $\boldsymbol{w}'\boldsymbol{\Sigma}\boldsymbol{w} = 0$ holds for some $\boldsymbol{w} \neq \boldsymbol{0}$, i.e. if there exists a non-stochastic linear function of the random vector. For example, consider the input allocation system (5.23) with a

[6] Hall (1973) has shown that the additivity of the cost function in the m outputs is a necessary and sufficient condition in order that the multiproduct firm can be broken up into m single-product firms in the following way: when the latter firms independently maximize profit by adjusting output, they use the same aggregate level of each input and produce the same level of output as the multiproduct firm.

disturbance vector $\boldsymbol{\varepsilon}$ added:

$$\boldsymbol{F\kappa} = (\boldsymbol{\iota'F\kappa})\boldsymbol{\Theta\iota} - \psi\boldsymbol{\Theta}(\boldsymbol{I} - \boldsymbol{\iota\iota'\Theta})\boldsymbol{\pi} + \boldsymbol{\varepsilon}. \tag{6.1}$$

Premultiplication by $\boldsymbol{\iota}'$ and use of (5.9) yields $\boldsymbol{\iota'F\kappa} = \boldsymbol{\iota'F\kappa} + \boldsymbol{\iota'\varepsilon}$, or $\boldsymbol{\iota'\varepsilon} = 0$, which means that the disturbances of the N equations sum to zero with unit probability. This property results from the allocation character of the system (6.1).

We return to the standard linear model described in the discussion following eq. (2.7). The Gauss–Markov theorem states that the LS estimator $\boldsymbol{b}$ in (2.8) is best linear unbiased in the following sense: any other estimator $\tilde{\boldsymbol{\beta}}$ of $\boldsymbol{\beta}$ which is also linear in $\boldsymbol{y}$ and unbiased has the property that $\mathcal{V}(\tilde{\boldsymbol{\beta}}) - \mathcal{V}(\boldsymbol{b})$ is a positive semidefinite matrix. That is,

$$\boldsymbol{w}'\left[\mathcal{V}(\tilde{\boldsymbol{\beta}}) - \mathcal{V}(\boldsymbol{b})\right]\boldsymbol{w} \geqslant 0 \quad \text{for any } \boldsymbol{w},$$

or $\boldsymbol{w}'\mathcal{V}(\tilde{\boldsymbol{\beta}})\boldsymbol{w} \geqslant \boldsymbol{w}'\mathcal{V}(\boldsymbol{b})\boldsymbol{w}$. Since both sides of this inequality are the variance of an estimator of $\boldsymbol{w'\beta}$, the implication is that within the class of linear unbiased estimators LS provides the estimator of any linear function of $\boldsymbol{\beta}$ with the smallest possible variance. This is a stronger result than the statement in the discussion following eq. (2.10); that statement is confined to the estimation of elements rather than general linear functions of $\boldsymbol{\beta}$.

To prove the Gauss–Markov theorem we use the linearity of $\tilde{\boldsymbol{\beta}}$ in $\boldsymbol{y}$ to write $\tilde{\boldsymbol{\beta}} = \boldsymbol{By}$, where $\boldsymbol{B}$ is a $K \times n$ matrix consisting of non-stochastic elements. We define $\boldsymbol{C} = \boldsymbol{B} - (\boldsymbol{X'X})^{-1}\boldsymbol{X}'$ so that $\tilde{\boldsymbol{\beta}} = \boldsymbol{By}$ can be written as

$$\begin{aligned}\left[\boldsymbol{C} + (\boldsymbol{X'X})^{-1}\boldsymbol{X}'\right]\boldsymbol{y} &= \left[\boldsymbol{C} + (\boldsymbol{X'X})^{-1}\boldsymbol{X}'\right](\boldsymbol{X\beta} + \boldsymbol{\varepsilon}) \\ &= (\boldsymbol{CX} + \boldsymbol{I})\boldsymbol{\beta} + \left[\boldsymbol{C} + (\boldsymbol{X'X})^{-1}\boldsymbol{X}'\right]\boldsymbol{\varepsilon}.\end{aligned}$$

The expectation of $\tilde{\boldsymbol{\beta}}$ is thus $(\boldsymbol{CX} + \boldsymbol{I})\boldsymbol{\beta}$, which must be identically equal to $\boldsymbol{\beta}$ in order that the estimator be unbiased. Therefore, $\boldsymbol{CX} = \boldsymbol{0}$ and $\tilde{\boldsymbol{\beta}} = \boldsymbol{\beta} + [\boldsymbol{C} + (\boldsymbol{X'X})^{-1}\boldsymbol{X}']\boldsymbol{\varepsilon}$ so that $\mathcal{V}(\tilde{\boldsymbol{\beta}})$ equals

$$\begin{aligned}&\left[\boldsymbol{C} + (\boldsymbol{X'X})^{-1}\boldsymbol{X}'\right]\mathcal{V}(\boldsymbol{\varepsilon})\left[\boldsymbol{C} + (\boldsymbol{X'X})^{-1}\boldsymbol{X}'\right]' \\ &\quad = \sigma^2\boldsymbol{CC}' + \sigma^2(\boldsymbol{X'X})^{-1} + \sigma^2\boldsymbol{CX}(\boldsymbol{X'X})^{-1} + \sigma^2(\boldsymbol{X'X})^{-1}\boldsymbol{X'C}'.\end{aligned}$$

It thus follows from (2.10) and $\boldsymbol{CX} = \boldsymbol{0}$ that $\mathcal{V}(\tilde{\boldsymbol{\beta}}) - \mathcal{V}(\boldsymbol{b}) = \sigma^2\boldsymbol{CC}'$, which is a positive semidefinite matrix because $\sigma^2\boldsymbol{w'CC'w} = (\sigma\boldsymbol{C'w})'(\sigma\boldsymbol{C'w})$ is the non-negative squared length of the vector $\sigma\boldsymbol{C'w}$.

6.2. Maxima and minima

The matrix $\boldsymbol{A}$ is called *negative semidefinite* if $\boldsymbol{x}'\boldsymbol{A}\boldsymbol{x} \leqslant 0$ holds for any $\boldsymbol{x}$ and *negative definite* if $\boldsymbol{x}'\boldsymbol{A}\boldsymbol{x} < 0$ holds for any $\boldsymbol{x} \neq \boldsymbol{0}$. If $\boldsymbol{A}$ is positive definite, $-\boldsymbol{A}$ is negative definite (similarly for semidefiniteness). If $\boldsymbol{A}$ is positive (negative) definite, all diagonal elements of $\boldsymbol{A}$ are positive (negative). This may be verified by considering $\boldsymbol{x}'\boldsymbol{A}\boldsymbol{x}$ with $\boldsymbol{x}$ specified as a column of the unit matrix of appropriate order. If $\boldsymbol{A}$ is positive (negative) definite, $\boldsymbol{A}$ is non-singular because singularity would imply the existence of an $\boldsymbol{x} \neq \boldsymbol{0}$ so that $\boldsymbol{A}\boldsymbol{x} = \boldsymbol{0}$, which is contradicted by $\boldsymbol{x}'\boldsymbol{A}\boldsymbol{x} > 0$ (< 0). If $\boldsymbol{A}$ is symmetric positive (negative) definite, so is $\boldsymbol{A}^{-1}$, which is verified by considering $\boldsymbol{x}'\boldsymbol{A}\boldsymbol{x}$ with $\boldsymbol{x} = \boldsymbol{A}^{-1}\boldsymbol{y}$ for $\boldsymbol{y} \neq \boldsymbol{0}$.

For the function $g(\cdot)$ of (2.6) to have a stationary value at $z = \bar{z}$ it is necessary and sufficient that the gradient $\partial g/\partial z$ at this point be zero. For this stationary value to be a local maximum (minimum) it is sufficient that the Hessian matrix $\partial^2 g/\partial z \partial z'$ at this point be negative (positive) definite. We can apply this to the supply equation (5.26) which is obtained by adjusting the output vector z so as to maximize profit. We write profit as $y'z - C$, where y is the output price vector and $C =$ cost. The gradient of profit as a function of z is $y - \partial C/\partial z$ (y is independent of z because y is exogenous by assumption) and the Hessian matrix is $-\partial^2 C/\partial z \partial z'$ so that a positive definite matrix $\partial^2 C/\partial z \partial z'$ is a sufficient condition for maximum profit. Since ψ^* and R in (5.28) are positive, the matrix $[\theta^*_{rs}]$ of the supply system (5.26) is positive definite. The diagonal elements of this matrix are therefore positive so that an increase in the price of a product raises its supply.

Similarly, a sufficient conditions for maximum utility is that the Hessian $\boldsymbol{U}$ be negative definite, implying $\phi < 0$ [see (3.9) and (5.6)], and a sufficient condition for minimum cost is that $\boldsymbol{F} - \gamma\boldsymbol{H}$ in (5.17) be positive definite. The result is that $[\theta_{ij}]$ in both (5.5) and (5.15) is also positive definite. Since ϕ and $-\psi$ in these equations are negative, an increase in the Frisch-deflated price of any good (consumer good or input) reduces the demand for this good. For two goods, i and j, a positive (negative) $\theta_{ij} = \theta_{ji}$ implies than an increase in the Frisch-deflated price of either good reduces (raises) the demand for the other; the two goods are then said to be *specific complements* (*substitutes*). Under preference or input independence no good is a specific substitute or complement of any other good [see (5.11) and (5.18)]. The distinction between specific substitutes and complements is from Houthakker (1960); he proposed it for consumer goods, but it can be equally applied to a firm's inputs and also to outputs based on the sign of $\theta^*_{rs} = \theta^*_{sr}$ in (5.26).

The assumption of a definite $\boldsymbol{U}$ or $\boldsymbol{F} - \gamma\boldsymbol{H}$ is more than strictly necessary. In the consumer's case, when utility is maximized subject to the budget constraint $\boldsymbol{p}'\boldsymbol{q} = M$, it is sufficient to assume *constrained* negative definiteness, i.e. $\boldsymbol{x}'\boldsymbol{U}\boldsymbol{x} < 0$ for all $\boldsymbol{x} \neq \boldsymbol{0}$ which satisfy $\boldsymbol{p}'\boldsymbol{x} = 0$. It is easy to construct examples of an indefinite or singular semidefinite matrix $\boldsymbol{U}$ which satisfy this condition. Definiteness obviously implies constrained definiteness; we shall assume that $\boldsymbol{U}$ and $\boldsymbol{F}-\gamma\boldsymbol{H}$ satisfy the stronger conditions so that the above analysis holds true.

6.3. *Block-diagonal definite matrices*

If a matrix is both definite and block-diagonal, the relevant principal submatrices are also definite. For example, if $\boldsymbol{\Sigma}$ of (3.4) is positive definite, then $\boldsymbol{x}_1'\boldsymbol{\Sigma}_1\boldsymbol{x}_1 + \boldsymbol{x}_2'\boldsymbol{\Sigma}_2\boldsymbol{x}_2 > 0$ if either $\boldsymbol{x}_1 \neq \boldsymbol{0}$ or $\boldsymbol{x}_2 \neq \boldsymbol{0}$, which would be violated if either $\boldsymbol{\Sigma}_1$ or $\boldsymbol{\Sigma}_2$ were not definite.

Another example is that of a logarithmic production function (5.12) when the inputs can be grouped into input groups so that the elasticity of output with respect to each input is independent of all inputs belonging to different groups. Then $\boldsymbol{H}$ of (5.13) is block-diagonal and so is $\boldsymbol{\Theta}$ [see (5.17)]. Thus, if i belongs to input group S_g $(g=1,2,\dots)$, the summation over j in the substitution term of (5.15) can be confined to $j \in S_g$; equivalently, no input is then a specific substitute or complement of any input belonging to a different group. Also, summation of the input demand equations over all inputs of a group yields a composite demand equation for the input group which takes a similar form, while an appropriate combination of this composite equation with a demand equation for an individual input yields a conditional demand equation for the input within their group. These developments can also be applied to outputs and consumer goods, but they are beyond the scope of this chapter.

7. Diagonalizations

7.1. *The standard diagonalization of a square matrix*

For some $n \times n$ matrix $\boldsymbol{A}$ we seek a vector $\boldsymbol{x}$ so that $\boldsymbol{A}\boldsymbol{x}$ equals a scalar multiple λ of $\boldsymbol{x}$. This is trivially satisfied by $\boldsymbol{x}=\boldsymbol{0}$, so we impose $\boldsymbol{x}'\boldsymbol{x}=1$ implying $\boldsymbol{x}\neq\boldsymbol{0}$. Since $\boldsymbol{A}\boldsymbol{x}=\lambda\boldsymbol{x}$ is equivalent to $(\boldsymbol{A}-\lambda\boldsymbol{I})\boldsymbol{x}=\boldsymbol{0}$, we thus have

$$(\boldsymbol{A}-\lambda\boldsymbol{I})\boldsymbol{x}=\boldsymbol{0}, \qquad \boldsymbol{x}'\boldsymbol{x}=1, \tag{7.1}$$

so that $\boldsymbol{A}-\lambda\boldsymbol{I}$ is singular. This implies a zero determinant value,

$$|\boldsymbol{A}-\lambda\boldsymbol{I}|=0, \tag{7.2}$$

which is known as the *characteristic equation* of $\boldsymbol{A}$. For example, if $\boldsymbol{A}$ is diagonal with $d_1,\dots,d_n$ on the diagonal, (7.2) states that the product of $d_i-\lambda$ over i vanishes so that each d_i is a solution of the characteristic equation. More generally, the characteristic equation of an $n\times n$ matrix $\boldsymbol{A}$ is a polynomial of degree n and thus yields n solutions $\lambda_1,\dots,\lambda_n$. These λ_i's are the *latent roots* of $\boldsymbol{A}$; the product of the λ_i's equals the determinant of $\boldsymbol{A}$ and the sum of the λ_i's equals the trace of $\boldsymbol{A}$. A vector $\boldsymbol{x}_i$ which satisfies $\boldsymbol{A}\boldsymbol{x}_i=\lambda_i\boldsymbol{x}_i$ and $\boldsymbol{x}_i'\boldsymbol{x}_i=1$ is called a *characteristic vector* of $\boldsymbol{A}$ corresponding to root λ_i.

Even if $\boldsymbol{A}$ consists of real elements, its roots need not be real, but these roots are all real when $\boldsymbol{A}$ is a real symmetric matrix. For suppose, to the contrary, that λ is a complex root and $\boldsymbol{x}+\mathrm{i}\boldsymbol{y}$ is a characteristic vector corresponding to this λ, where $\mathrm{i}=\sqrt{-1}$. Then $\boldsymbol{A}(\boldsymbol{x}+\mathrm{i}\boldsymbol{y})=\lambda(\boldsymbol{x}+\mathrm{i}\boldsymbol{y})$, which we premultiply by $(\boldsymbol{x}-\mathrm{i}\boldsymbol{y})'$:

$$\boldsymbol{x}'\boldsymbol{A}\boldsymbol{x}+\boldsymbol{y}'\boldsymbol{A}\boldsymbol{y}+\mathrm{i}(\boldsymbol{x}'\boldsymbol{A}\boldsymbol{y}-\boldsymbol{y}'\boldsymbol{A}\boldsymbol{x})=\lambda(\boldsymbol{x}'\boldsymbol{x}+\boldsymbol{y}'\boldsymbol{y}). \tag{7.3}$$

But $\boldsymbol{x}'\boldsymbol{A}\boldsymbol{y}=\boldsymbol{y}'\boldsymbol{A}\boldsymbol{x}$ if $\boldsymbol{A}$ is symmetric, so that (7.3) shows that λ is the ratio of two real numbers, $\boldsymbol{x}'\boldsymbol{A}\boldsymbol{x}+\boldsymbol{y}'\boldsymbol{A}\boldsymbol{y}$ and $\boldsymbol{x}'\boldsymbol{x}+\boldsymbol{y}'\boldsymbol{y}$. Roots of asymmetric matrices are considered at the end of this section.

Let λ_i and λ_j be two different roots ($\lambda_i\neq\lambda_j$) of a symmetric matrix $\boldsymbol{A}$ and let $\boldsymbol{x}_i$ and $\boldsymbol{x}_j$ be corresponding characteristic vectors. We premultiply $\boldsymbol{A}\boldsymbol{x}_i=\lambda_i\boldsymbol{x}_i$ by $\boldsymbol{x}_j'$ and $\boldsymbol{A}\boldsymbol{x}_j=\lambda_j\boldsymbol{x}_j$ by $\boldsymbol{x}_i'$ and subtract:

$$\boldsymbol{x}_j'\boldsymbol{A}\boldsymbol{x}_i-\boldsymbol{x}_i'\boldsymbol{A}\boldsymbol{x}_j=(\lambda_i-\lambda_j)\boldsymbol{x}_i'\boldsymbol{x}_j.$$

Since the left side vanishes for a symmetric matrix $\boldsymbol{A}$, we must have $\boldsymbol{x}_i'\boldsymbol{x}_j=0$ because $\lambda_i\neq\lambda_j$. This proves that characteristic vectors of a symmetric matrix are orthogonal when they correspond to different roots. When all roots of a symmetric $n\times n$ matrix $\boldsymbol{A}$ are distinct, we thus have $\boldsymbol{x}_i'\boldsymbol{x}_j=\delta_{ij}$ for all (i,j). This is equivalent to

$$\boldsymbol{X}'\boldsymbol{X}=\boldsymbol{I},\quad \text{where } \boldsymbol{X}=[\boldsymbol{x}_1\quad \boldsymbol{x}_2\dots\boldsymbol{x}_n]. \tag{7.4}$$

Also,

$$\boldsymbol{A}\boldsymbol{X}=[\boldsymbol{A}\boldsymbol{x}_1\dots\boldsymbol{A}\boldsymbol{x}_n]=[\lambda_1\boldsymbol{x}_1\dots\lambda_n\boldsymbol{x}_n],$$

or

$$\boldsymbol{A}\boldsymbol{X}=\boldsymbol{X}\boldsymbol{\Lambda}, \tag{7.5}$$

where $\boldsymbol{\Lambda}$ is the diagonal matrix with $\lambda_1,\dots,\lambda_n$ on the diagonal. Premultiplication of (7.5) by $\boldsymbol{X}'$ yields $\boldsymbol{X}'\boldsymbol{A}\boldsymbol{X}=\boldsymbol{X}'\boldsymbol{X}\boldsymbol{\Lambda}$, or

$$\boldsymbol{X}'\boldsymbol{A}\boldsymbol{X}=\boldsymbol{\Lambda} \tag{7.6}$$

in view of (7.4). Therefore, when we postmultiply a symmetric matrix $\boldsymbol{A}$ by a matrix $\boldsymbol{X}$ consisting of characteristic vectors of $\boldsymbol{A}$ and premultiply by $\boldsymbol{X}'$, we obtain the diagonal matrix containing the latent roots of $\boldsymbol{A}$. This double multiplication amounts to a *diagonalization* of $\boldsymbol{A}$. Also, postmultiplication of (7.5) by $\boldsymbol{X}'$ yields $\boldsymbol{AXX}' = \boldsymbol{X\Lambda X}'$ and hence, since (7.4) implies $\boldsymbol{X}' = \boldsymbol{X}^{-1}$ or $\boldsymbol{XX}' = \boldsymbol{I}$,

$$\boldsymbol{A} = \boldsymbol{X\Lambda X}' = \sum_{i=1}^{n} \lambda_i \boldsymbol{x}_i \boldsymbol{x}_i'. \tag{7.7}$$

In the previous paragraph we assumed that the λ_i's are distinct, but it may be shown that for any symmetric $\boldsymbol{A}$ there exists an $\boldsymbol{X}$ which satisfies (7.4)–(7.7), the columns of $\boldsymbol{X}$ being characteristic vectors of $\boldsymbol{A}$ and Λ being diagonal with the latent roots of $\boldsymbol{A}$ on the diagonal. The only difference is that in the case of multiple roots ($\lambda_i = \lambda_j$ for $i \neq j$) the associated characteristic vectors ($\boldsymbol{x}_i$ and $\boldsymbol{x}_j$) are not unique. Note that even when all λ's are distinct, each $\boldsymbol{x}_i$ may be arbitrarily multiplied by -1 because this affects neither $\boldsymbol{A}\boldsymbol{x}_i = \lambda_i \boldsymbol{x}_i$ nor $\boldsymbol{x}_i'\boldsymbol{x}_j = 0$ for any (i, j); however, this sign indeterminacy will be irrelevant for our purposes.

7.2. *Special cases*

Let $\boldsymbol{A}$ be square and premultiply $\boldsymbol{Ax} = \lambda \boldsymbol{x}$ by $\boldsymbol{A}$ to obtain $\boldsymbol{A}^2\boldsymbol{x} = \lambda \boldsymbol{Ax} = \lambda^2 \boldsymbol{x}$. This shows that $\boldsymbol{A}^2$ has the same characteristic vectors as $\boldsymbol{A}$ and latent roots equal to the squares of those of $\boldsymbol{A}$. In particular, if a matrix is symmetric idempotent, such as $\boldsymbol{M}$ of (2.11), all latent roots are 0 or 1 because these are the only real numbers that do not change when squared. For a symmetric non-singular $\boldsymbol{A}$, premultiply $\boldsymbol{Ax} = \lambda \boldsymbol{x}$ by $(\lambda \boldsymbol{A})^{-1}$ to obtain $\boldsymbol{A}^{-1}\boldsymbol{x} = (1/\lambda)\boldsymbol{x}$. Thus, $\boldsymbol{A}^{-1}$ has the same characteristic vectors as those of $\boldsymbol{A}$ and latent roots equal to the reciprocals of those of $\boldsymbol{A}$. If the symmetric $n \times n$ matrix $\boldsymbol{A}$ is singular and has rank r, (7.2) is satisfied by $\lambda = 0$ and this zero root has multiplicity $n - r$. It thus follows from (7.7) that $\boldsymbol{A}$ can then be written as the sum of r matrices of unit rank, each of the form $\lambda_i \boldsymbol{x}_i \boldsymbol{x}_i'$, with $\lambda_i \neq 0$.

Premultiplication of (7.7) by $\boldsymbol{y}'$ and postmultiplication by $\boldsymbol{y}$ yields $\boldsymbol{y}'\boldsymbol{Ay} = \sum_i \lambda_i c_i^2$, with $c_i = \boldsymbol{y}'\boldsymbol{x}_i$. Since $\boldsymbol{y}'\boldsymbol{Ay}$ is positive (negative) for any $\boldsymbol{y} \neq \boldsymbol{0}$ if $\boldsymbol{A}$ is positive (negative) definite, this shows that all latent roots of a symmetric positive (negative) definite matrix are positive (negative). Similarly, all latent roots of a symmetric positive (negative) semidefinite matrix are non-negative (non-positive).

Let $\boldsymbol{A}_m$ be a symmetric $m \times m$ matrix with roots $\lambda_1, \ldots, \lambda_m$ and characteristic vectors $\boldsymbol{x}_1, \ldots, \boldsymbol{x}_m$; let $\boldsymbol{B}_n$ be a symmetric $n \times n$ matrix with roots $\mu_1, \ldots, \mu_n$ and characteristic vectors $\boldsymbol{y}_1, \ldots, \boldsymbol{y}_n$. Hence, $\boldsymbol{A}_m \otimes \boldsymbol{B}_n$ is of order $mn \times mn$ and has mn latent roots and characteristic vectors. We use $\boldsymbol{A}_m \boldsymbol{x}_i = \lambda_i \boldsymbol{x}_i$ and $\boldsymbol{B}_n \boldsymbol{y}_j = \mu_j \boldsymbol{y}_j$ in

$$(\boldsymbol{A}_m \otimes \boldsymbol{B}_n)(\boldsymbol{x}_i \otimes \boldsymbol{y}_j) = (\boldsymbol{A}_m \boldsymbol{x}_i) \otimes (\boldsymbol{B}_n \boldsymbol{y}_j) = (\lambda_i \boldsymbol{x}_i) \otimes (\mu_j \boldsymbol{y}_j) = \lambda_i \mu_j (\boldsymbol{x}_i \otimes \boldsymbol{y}_j),$$

which shows that $x_i \otimes y_j$ is a characteristic vector of $A_m \otimes B_n$ corresponding to root $\lambda_i \mu_j$. It is easily verified that these characteristic vectors form an orthogonal matrix of order $mn \times mn$:

$$(x_1 \otimes y_1)'(x_1 \otimes y_1) = (x_1' \otimes y_1')(x_1 \otimes y_1) = (x_1' x_1) \otimes (y_1' y_1) = 1,$$
$$(x_1 \otimes y_1)'(x_2 \otimes y_1) = (x_1' x_2) \otimes (y_1' y_1) = 0.$$

Since the determinant of $A_m \otimes B_n$ equals the product of the roots, we have

$$|A_m \otimes B_n| = \prod_{i=1}^{m} \prod_{j=1}^{n} \lambda_i \mu_j = \left(\prod_{i=1}^{m} \lambda_i^n \right) \left(\prod_{j=1}^{n} \mu_j^m \right) = |A_m|^n |B_n|^m.$$

It may similarly be verified that the rank (trace) of $A_m \otimes B_n$ equals the product of the ranks (traces) of A_m and B_n.

7.3. Aitken's theorem

Any symmetric positive definite matrix A can be written as $A = QQ'$, where Q is some non-singular matrix. For example, we can use (7.7) and specify $Q = X\Lambda^{1/2}$, where $\Lambda^{1/2}$ is the diagonal matrix which contains the positive square roots of the latent roots of A on the diagonal. Since the roots of A are all positive, $\Lambda^{1/2}$ is non-singular; X is non-singular in view of (7.4); therefore, $Q = X\Lambda^{1/2}$ is also non-singular.

Consider in particular the disturbance covariance matrix $\sigma^2 V$ in the discussion preceding eq. (2.13). Since $\sigma^2 > 0$ and V is non-singular by assumption, this covariance matrix is symmetric positive definite. Therefore, V^{-1} is also symmetric positive definite and we can write $V^{-1} = QQ'$ for some non-singular Q. We premultiply (2.7) by Q':

$$Q'y = (Q'X)\beta + Q'\varepsilon. \tag{7.8}$$

The disturbance vector $Q'\varepsilon$ has zero expectation and a covariance matrix equal to

$$\sigma^2 Q'VQ = \sigma^2 Q'(QQ')^{-1}Q = \sigma^2 Q'(Q')^{-1}Q^{-1}Q = \sigma^2 I,$$

so that the standard linear model and the Gauss–Markov theorem are now applicable. Thus, we estimate β by applying LS to (7.8).

$$[(Q'X)'Q'X]^{-1}(Q'X)'Q'y = (X'QQ'X)^{-1}X'QQ'y = (X'V^{-1}X)^{-1}X'V^{-1}y,$$

which is the GLS estimator (2.13). The covariance matrix (2.14) is also easily verified.

7.4. *The Cholesky decomposition*

The diagonalization (7.6) uses an orthogonal matrix $\boldsymbol{X}$ [see (7.4)], but it is also possible to use a triangular matrix. For example, consider a diagonal matrix $\boldsymbol{D}$ and an upper triangular matrix $\boldsymbol{C}$ with units in the diagonal,

$$\boldsymbol{C}=\begin{bmatrix}1 & c_{12} & c_{13}\\ 0 & 1 & c_{23}\\ 0 & 0 & 1\end{bmatrix},\qquad \boldsymbol{D}=\begin{bmatrix}d_1 & 0 & 0\\ 0 & d_2 & 0\\ 0 & 0 & d_3\end{bmatrix},$$

yielding

$$\boldsymbol{C}'\boldsymbol{D}\boldsymbol{C}=\begin{bmatrix}d_1 & d_1c_{12} & d_1c_{13}\\ d_1c_{12} & d_1c_{12}^2+d_2 & d_1c_{12}c_{13}+d_2c_{23}\\ d_1c_{13} & d_1c_{12}c_{13}+d_2c_{23} & d_1c_{13}^2+d_2c_{23}^2+d_3\end{bmatrix}.$$

It is readily verified that any 3×3 symmetric positive definite matrix $\boldsymbol{A}=[a_{ij}]$ can be uniquely written as $\boldsymbol{C}'\boldsymbol{D}\boldsymbol{C}$ ($d_1=a_{11}$, $c_{12}=a_{12}/a_{11}$, etc.). This is the so-called *Cholesky decomposition* of a matrix; for applications to demand analysis see Barten and Geyskens (1975) and Theil and Laitinen (1979). Also, note that $\boldsymbol{D}=(\boldsymbol{C}')^{-1}\boldsymbol{A}\boldsymbol{C}^{-1}$ and that $\boldsymbol{C}^{-1}$ is upper triangular with units in the diagonal.

7.5. *Vectors written as diagonal matrices*

On many occasions we want to write a vector in the form of a diagonal matrix. An example is the price vector $\boldsymbol{p}$ which occurs as a diagonal matrix $\boldsymbol{P}$ in (5.8). An alternative notation is $\hat{\boldsymbol{p}}$, with the hat indicating that the vector has become a diagonal matrix. However, such notations are awkward when the vector which we want to write as a diagonal matrix is itself the product of one or several matrices and a vector. For example, in Section 8 we shall meet a nonsingular $N\times N$ matrix $\boldsymbol{X}$ and the vector $\boldsymbol{X}^{-1}\boldsymbol{\iota}$. We write this vector in diagonal matrix form as

$$(\boldsymbol{X}^{-1}\boldsymbol{\iota})_\Delta=\begin{bmatrix}\Sigma_j x^{1j} & 0 & \dots & 0\\ 0 & \Sigma_j x^{2j} & \dots & 0\\ \vdots & \vdots & & \vdots\\ 0 & 0 & \dots & \Sigma_j x^{Nj}\end{bmatrix}, \tag{7.9}$$

where x^{ij} is an element of $\boldsymbol{X}^{-1}$ and all summations are over $j=1,\dots,N$. It is easily verified that

$$(X^{-1}\iota)_{\Delta}\iota = X^{-1}\iota, \qquad \iota'(X^{-1}\iota)_{\Delta} = \iota'(X')^{-1}. \tag{7.10}$$

7.6. *A simultaneous diagonalization of two square matrices*

We extend (7.1) to

$$(A - \lambda B)x = 0, \qquad x'Bx = 1, \tag{7.11}$$

where A and B are symmetric $n \times n$ matrices, B being positive definite. Thus, B^{-1} is symmetric positive definite so that $B^{-1} = QQ'$ for some non-singular Q. It is easily seen that (7.11) is equivalent to

$$(Q'AQ - \lambda I)y = 0, \qquad y'y = 1, \qquad y = Q^{-1}x. \tag{7.12}$$

This shows that (7.11) can be reduced to (7.1) with A interpreted as $Q'AQ$. If A is symmetric, so is $Q'AQ$. Therefore, all results for symmetric matrices described earlier in this section are applicable. In particular, (7.11) has n solutions, $\lambda_1,\dots,\lambda_n$ and $x_1,\dots,x_n$, the x_i's being unique when the λ_i's are distinct. From $y_i'y_j = \delta_{ij}$ and $y_i = Q^{-1}x_i$ we have $x_i'Bx_j = \delta_{ij}$ and hence $X'BX = I$, where X is the $n \times n$ matrix with x_i as the ith column. We write $(A - \lambda B)x = 0$ as $Ax_i = \lambda_i Bx_i$ for the ith solution and as $AX = BX\Lambda$ for all solutions jointly, where Λ is diagonal with $\lambda_1,\dots,\lambda_n$ on the diagonal. Premultiplication of $AX = BX\Lambda$ by X' then yields $X'AX = X'BX\Lambda = \Lambda$. Therefore,

$$X'AX = \Lambda, \qquad X'BX = I, \tag{7.13}$$

which shows that both matrices are simultaneously diagonalized, A being transformed into the latent root matrix Λ, and B into the unit matrix.

It is noteworthy that (7.11) can be interpreted in terms of a constrained extremum problem. Let us seek the maximum of the quadratic form $x'Ax$ for variations in x subject to $x'Bx = 1$. So we construct the Lagrangian function $x'Ax - \mu(x'Bx - 1)$, which we differentiate with respect to x, yielding $2Ax - 2\mu Bx$. By equating this derivative to zero we obtain $Ax = \mu Bx$, which shows that μ must be a root λ_i of (7.11). Next, we premultiply $Ax = \mu Bx$ by x', which gives $x'Ax = \mu x'Bx = \mu$ and shows that the largest root λ_i is the maximum of $x'Ax$ subject to $x'Bx = 1$. Similarly, the smallest root is the minimum of $x'Ax$ subject to $x'Bx = 1$, and all n roots are stationary values of $x'Ax$ subject to $x'Bx = 1$.

7.7. Latent roots of an asymmetric matrix

Some or all latent roots of an asymmetric square matrix $\boldsymbol{A}$ may be complex. If (7.2) yields complex roots, they occur in conjugate complex pairs of the form $a \pm bi$. The absolute value of such a root is defined as $\sqrt{a^2+b^2}$, which equals $|a|$ if $b=0$, i.e. if the root is real. If $\boldsymbol{A}$ is asymmetric, the latent roots of $\boldsymbol{A}$ and $\boldsymbol{A}'$ are still the same but a characteristic vector of $\boldsymbol{A}$ need not be a characteristic vector of $\boldsymbol{A}'$. If $\boldsymbol{A}$ is asymmetric and has multiple roots, it may have fewer characteristic vectors than the number of its rows and columns. For example,

$$\boldsymbol{A}=\begin{bmatrix}1 & 1\\ 0 & 1\end{bmatrix}$$

is an asymmetric 2×2 matrix with a double unit root, but it has only one characteristic vector, $[1 \quad 0]'$. A further analysis of this subject involves the Jordan canonical form, which is beyond the scope of this chapter; see Bellman (1960, ch. 11).

Latent roots of asymmetric matrices play a role in the stability analysis of dynamic equations systems. Consider the reduced form

$$\boldsymbol{y}_t=\boldsymbol{a}+\boldsymbol{A}\boldsymbol{y}_{t-1}+\boldsymbol{A}^*\boldsymbol{x}_t+\boldsymbol{u}_t, \tag{7.14}$$

where $\boldsymbol{y}_t$ is an L-element observation vector on endogenous variables at time t, $\boldsymbol{a}$ is a vector of constant terms, $\boldsymbol{A}$ is a square coefficient matrix, $\boldsymbol{A}^*$ is an $L\times K$ coefficient matrix, $\boldsymbol{x}_t$ is a K-element observation vector on exogenous variables at t, and $\boldsymbol{u}_t$ is a disturbance vector. Although $\boldsymbol{A}$ is square, there is no reason why it should be symmetric so that its latent roots may include conjugate complex pairs. In the next paragraph we shall be interested in the limit of $\boldsymbol{A}^s$ as $s\to\infty$. Recall that $\boldsymbol{A}^2$ has roots equal to the squares of those of $\boldsymbol{A}$; this also holds for the complex roots of an asymmetric $\boldsymbol{A}$. Therefore, $\boldsymbol{A}^s$ has latent roots equal to the sth power of those of $\boldsymbol{A}$. If the roots of $\boldsymbol{A}$ are all less than 1 in absolute value, those of $\boldsymbol{A}^s$ will all converge to zero as $s\to\infty$, which means that the limit of $\boldsymbol{A}^s$ for $s\to\infty$ is a zero matrix. Also, let $\boldsymbol{S}=\boldsymbol{I}+\boldsymbol{A}+\cdots+\boldsymbol{A}^s$; then, by subtracting $\boldsymbol{A}\boldsymbol{S}=\boldsymbol{A}+\boldsymbol{A}^2+\cdots+\boldsymbol{A}^{s+1}$ we obtain $(\boldsymbol{I}-\boldsymbol{A})\boldsymbol{S}=\boldsymbol{I}-\boldsymbol{A}^{s+1}$ so that we have the following result for the limit of $\boldsymbol{S}$ when all roots of $\boldsymbol{A}$ are less than 1 in absolute value:

$$\lim_{s\to\infty}(\boldsymbol{I}+\boldsymbol{A}+\cdots+\boldsymbol{A}^s)=(\boldsymbol{I}-\boldsymbol{A})^{-1}. \tag{7.15}$$

We proceed to apply these results to (7.14) by lagging it by one period, $\boldsymbol{y}_{t-1}=\boldsymbol{a}+\boldsymbol{A}\boldsymbol{y}_{t-2}+\boldsymbol{A}^*\boldsymbol{x}_{t-1}+\boldsymbol{u}_{t-1}$, and substituting this into (7.14):

$$\boldsymbol{y}_t=(\boldsymbol{I}+\boldsymbol{A})\boldsymbol{a}+\boldsymbol{A}^2\boldsymbol{y}_{t-2}+\boldsymbol{A}^*\boldsymbol{x}_t+\boldsymbol{A}\boldsymbol{A}^*\boldsymbol{x}_{t-1}+\boldsymbol{u}_t+\boldsymbol{A}\boldsymbol{u}_{t-1}$$

When we do this s times, we obtain:

$$\begin{aligned} y_t = (I + A + \cdots + A^s)a + A^{s+1}y_{t-s-1} \\ + A^*x_t + AA^*x_{t-1} + \cdots + A^sA^*x_{t-s} \\ + u_t + Au_{t-1} + \cdots + A^s u_{t-s}. \end{aligned} \tag{7.16}$$

If all roots of A are less than 1 in absolute value, so that A^s converges to zero as $s \to \infty$ and (7.15) holds, the limit of (7.16) for $s \to \infty$ becomes

$$y_t = (I - A)^{-1}a + \sum_{s=0}^{\infty} A^sA^*x_{t-s} + \sum_{s=0}^{\infty} A^s \mathbf{u}_{t-s}, \tag{7.17}$$

which is the *final form* of the equation system. This form expresses each current endogenous variable in terms of current and lagged exogenous variables as well as current and lagged disturbances; all lagged endogenous variables are eliminated from the reduced form (7.14) by successive lagged application of (7.14). The coefficient matrices A^sA^* of x_{t-s} for $s = 0,1,2,\ldots$ in (7.17) may be viewed as matrix multipliers; see Goldberger (1959). The behavior of the elements of A^sA^* as $s \to \infty$ is dominated by the root of A with the largest absolute value. If this root is a conjugate complex pair, the behavior of these elements for increasing s is of the damped oscillatory type.

Endogenous variables occur in (7.14) only with a one-period lag. Suppose that Ay_{t-1} in (7.14) is extended to $A_1 y_{t-1} + \cdots + A_\tau y_{t-\tau}$, where τ is the largest lag which occurs in the equation system. It may be shown that the relevant determinantal equation is now

$$|\lambda^\tau(-I) + \lambda^{\tau-1}A_1 + \cdots + A_\tau| = 0. \tag{7.18}$$

When there are L endogenous variables, (7.18) yields $L\tau$ solutions which may include conjugate complex pairs. All these solutions should be less than 1 in absolute value in order that the system be stable, i.e. in order that the coefficient matrix of x_{t-s} in the final form converges to zero as $s \to \infty$. It is readily verified that for $\tau = 1$ this condition refers to the latent roots of A_1, in agreement with the condition underlying (7.17).

8. Principal components and extensions

8.1. Principal components

Consider an $n \times K$ observation matrix Z on K variables. Our objective is to approximate Z by a matrix of unit rank, vc', where v is an n-element vector of values taken by some variable (to be constructed below) and $\boldsymbol{c}$ is a K-element coefficient vector. Thus, the approximation describes each column of $\boldsymbol{Z}$ as proportional to $\boldsymbol{v}$. It is obvious that if the rank of $\boldsymbol{Z}$ exceeds 1, there will be a non-zero $n\times K$ discrepancy matrix $\boldsymbol{Z}-\boldsymbol{v}\boldsymbol{c}'$ no matter how we specify $\boldsymbol{v}$ and $\boldsymbol{c}$; we select $\boldsymbol{v}$ and $\boldsymbol{c}$ by minimizing the sum of the squares of all Kn discrepancies. Also, since $\boldsymbol{v}\boldsymbol{c}'$ remains unchanged when $\boldsymbol{v}$ is multiplied by $k\neq 0$ and $\boldsymbol{c}$ by $1/k$, we shall impose $\boldsymbol{v}'\boldsymbol{v}=1$. It is shown in the next subsection that the solution is $\boldsymbol{v}=\boldsymbol{v}_1$ and $\boldsymbol{c}=\boldsymbol{c}_1$, defined by

$$(\boldsymbol{Z}\boldsymbol{Z}'-\lambda_1\boldsymbol{I})\boldsymbol{v}_1=\boldsymbol{0}, \tag{8.1}$$

$$\boldsymbol{c}_1=\boldsymbol{Z}'\boldsymbol{v}_1, \tag{8.2}$$

where λ_1 is the largest latent root of the symmetric positive semidefinite matrix $\boldsymbol{Z}\boldsymbol{Z}'$. Thus, (8.1) states that $\boldsymbol{v}_1$ is a characteristic vector of $\boldsymbol{Z}\boldsymbol{Z}'$ corresponding to the largest latent root. (We assume that the non-zero roots of $\boldsymbol{Z}\boldsymbol{Z}'$ are distinct.) Note that $\boldsymbol{v}_1$ may be arbitrarily multiplied by -1 in (8.1) but that this changes $\boldsymbol{c}_1$ into $-\boldsymbol{c}_1$ in (8.2) so that the product $\boldsymbol{v}_1\boldsymbol{c}_1'$ remains unchanged.

Our next objective is to approximate the discrepancy matrix $\boldsymbol{Z}-\boldsymbol{v}_1\boldsymbol{c}_1'$ by a matrix of unit rank $\boldsymbol{v}_2\boldsymbol{c}_2'$, so that $\boldsymbol{Z}$ is approximated by $\boldsymbol{v}_1\boldsymbol{c}_1'+\boldsymbol{v}_2\boldsymbol{c}_2'$. The criterion is again the residual sum of squares, which we minimize by varying $\boldsymbol{v}_2$ and $\boldsymbol{c}_2$ subject to the constraints $\boldsymbol{v}_2'\boldsymbol{v}_2=1$ and $\boldsymbol{v}_2'\boldsymbol{v}_1=0$. It is shown in the next subsection that the solution is identical to (8.1) and (8.2) except that the subscript 1 becomes 2 with λ_2 interpreted as the second largest latent root of $\boldsymbol{Z}\boldsymbol{Z}'$. The vectors $\boldsymbol{v}_1$ and $\boldsymbol{v}_2$ are known as the first and second *principal components* of the K variables whose observations are arranged in the $n\times K$ matrix $\boldsymbol{Z}$.

More generally, the ith principal component $\boldsymbol{v}_i$ and the associated coefficient vector $\boldsymbol{c}_i$ are obtained from

$$(\boldsymbol{Z}\boldsymbol{Z}'-\lambda_i\boldsymbol{I})\boldsymbol{v}_i=\boldsymbol{0}, \tag{8.3}$$

$$\boldsymbol{c}_i=\boldsymbol{Z}'\boldsymbol{v}_i, \tag{8.4}$$

where λ_i is the ith largest root of $\boldsymbol{Z}\boldsymbol{Z}'$. This solution is obtained by approximating $\boldsymbol{Z}-\boldsymbol{v}_1\boldsymbol{c}_1'-\cdots-\boldsymbol{v}_{i-1}\boldsymbol{c}_{i-1}'$ by a matrix $\boldsymbol{v}_i\boldsymbol{c}_i'$, the criterion being the sum of the squared discrepancies subject to the unit length condition $\boldsymbol{v}_i'\boldsymbol{v}_i=1$ and the orthogonality conditions $\boldsymbol{v}_i'\boldsymbol{v}_j=0$ for $j=1,\dots,i-1$.

8.2. *Derivations*

It is easily verified that the sum of the squares of all elements of any matrix $\boldsymbol{A}$ (square or rectangular) is equal to $\operatorname{tr}\boldsymbol{A}'\boldsymbol{A}$. Thus, the sum of the squares of the elements of the discrepancy matrix $Z - vc'$ equals

$$\begin{aligned}\operatorname{tr}(Z - vc')'(Z - vc') &= \operatorname{tr} Z'Z - \operatorname{tr} cv'Z - \operatorname{tr} Z'vc' + \operatorname{tr} cv'vc' \\ &= \operatorname{tr} Z'Z - 2v'Zc + (v'v)(c'c),\end{aligned}$$

which can be simplified to

$$\operatorname{tr} Z'Z - 2v'Zc + c'c \tag{8.5}$$

in view of $v'v = 1$. The derivative of (8.5) with respect to c is $-2Z'v + 2c$ so that minimizing (8.5) by varying c for given v yields $c = Z'v$, in agreement with (8.2). By substituting $c = Z'v$ into (8.5) we obtain $\operatorname{tr} Z'Z - v'ZZ'v$; hence, our next step is to maximize $v'ZZ'v$ for variations in v subject to $v'v = 1$. So we construct the Lagrangian function $v'ZZ'v - \mu(v'v - 1)$ and differentiate it with respect to v and equate the derivative to zero. This yields $ZZ'v = \mu v$ so that v must be a characteristic vector of ZZ' corresponding to root μ. This confirms (8.1) if we can prove that μ equals the largest root λ_1 of ZZ'. For this purpose we premultiply $ZZ'v = \mu v$ by v', which gives $v'ZZ'v = \mu v'v = \mu$. Since we seek the maximum of $v'ZZ'v$, this shows that μ must be the largest root of ZZ'.

To verify (8.3) and (8.4) for $i = 2$, we consider

$$\begin{aligned}&\operatorname{tr}(Z - v_1c_1' - v_2c_2')'(Z - v_1c_1' - v_2c_2') \\ &\quad = \operatorname{tr}(Z - v_1c_1')'(Z - v_1c_1') - 2\operatorname{tr}(Z - v_1c_1')'v_2c_2' + \operatorname{tr} c_2v_2'v_2c_2' \\ &\quad = \operatorname{tr}(Z - v_1c_1')'(Z - v_1c_1') - 2c_2'Z'v_2 + c_2'c_2,\end{aligned} \tag{8.6}$$

where the last step is based on $v_1'v_2 = 0$ and $v_2'v_2 = 1$. Minimization of (8.6) with respect to c_2 for given v_2 thus yields $c_2 = Z'v_2$, in agreement with (8.4). Substitution of $c_2 = Z'v_2$ into (8.6) shows that the function to be minimized with respect to v_2 takes the form of a constant [equal to the trace in the last line of (8.6)] minus $v_2'ZZ'v_2$. So we maximize $v_2'ZZ'v_2$ by varying v_2 subject to $v_1'v_2 = 0$ and $v_2'v_2 = 1$, using the Lagrangian function $v_2'ZZ'v_2 - \mu_1v_1'v_2 - \mu_2(v_2'v_2 - 1)$. We take the derivative of this function with respect to v_2 and equate it to zero:

$$2ZZ'v_2 - \mu_1v_1 - 2\mu_2v_2 = \mathbf{0}. \tag{8.7}$$

We premultiply this by v_1', which yields $2v_1'ZZ'v_2 = \mu_1v_1'v_1 = \mu_1$ because $v_1'v_2 = 0$. But $v_1'ZZ'v_2 = 0$ and hence $\mu_1 = 0$ because (8.1) implies $v_2'ZZ'v_1 = \lambda_1v_2'v_1 = 0$. We can thus simplify (8.7) to $ZZ'v_2 = \mu_2v_2$ so that v_2 is a characteristic vector of ZZ' corresponding to root μ_2. This vector must be orthogonal to the characteristic vector v_1 corresponding to the largest root λ_1, while the root μ_2 must be as large as possible because the objective is to maximize $v_2'ZZ'v_2 = \mu_2v_2'v_2 = \mu_2$. Therefore,

μ_2 must be the second largest root λ_2 of $\boldsymbol{ZZ}'$, which completes the proof of (8.3) and (8.4) for $i=2$. The extension to larger values of i is left as an exercise for the reader.

8.3. Further discussion of principal components

If the rank of $\boldsymbol{Z}$ is r, (8.3) yields r principal components corresponding to positive roots $\lambda_1,\dots,\lambda_r$. In what follows we assume that $\boldsymbol{Z}$ has full column rank so that there are K principal components corresponding to K positive roots, $\lambda_1,\dots,\lambda_K$.

By premultiplying (8.3) by $\boldsymbol{Z}'$ and using (8.4) we obtain:

$$(\boldsymbol{Z}'\boldsymbol{Z}-\lambda_i\boldsymbol{I})\boldsymbol{c}_i=\boldsymbol{0}, \tag{8.8}$$

so that the coefficient vector $\boldsymbol{c}_i$ is a characteristic vector of $\boldsymbol{Z}'\boldsymbol{Z}$ corresponding to root λ_i. The vectors $\boldsymbol{c}_1,\dots,\boldsymbol{c}_K$ are orthogonal, but they do not have unit length. To prove this we introduce the matrix $\boldsymbol{V}$ of all principal components and the associated coefficient matrix $\boldsymbol{C}$:

$$\boldsymbol{V}=[\boldsymbol{v}_1\dots\boldsymbol{v}_K], \qquad \boldsymbol{C}=[\boldsymbol{c}_1\dots\boldsymbol{c}_K], \tag{8.9}$$

so that (8.3) and (8.4) for $i=1,\dots,K$ can be written as

$$\boldsymbol{ZZ}'\boldsymbol{V}=\boldsymbol{V}\boldsymbol{\Lambda}, \tag{8.10}$$

$$\boldsymbol{C}=\boldsymbol{Z}'\boldsymbol{V}, \tag{8.11}$$

where $\boldsymbol{\Lambda}$ is the diagonal matrix with $\lambda_1,\dots,\lambda_K$ on the diagonal. By premultiplying (8.10) by $\boldsymbol{V}'$ and using (8.11) and $\boldsymbol{V}'\boldsymbol{V}=\boldsymbol{I}$ we obtain:

$$\boldsymbol{C}'\boldsymbol{C}=\boldsymbol{\Lambda}, \tag{8.12}$$

which shows that $\boldsymbol{c}_1,\dots,\boldsymbol{c}_K$ are orthogonal vectors and that the squared length of $\boldsymbol{c}_i$ equals λ_i.

If the observed variables are measured as deviations from their means, $\boldsymbol{Z}'\boldsymbol{Z}$ in (8.8) equals their sample covariance matrix multiplied by n. Since $\boldsymbol{Z}'\boldsymbol{Z}$ need not be diagonal, the observed variables may be correlated. But the principal components are all uncorrelated because $\boldsymbol{v}_i'\boldsymbol{v}_j=0$ for $i\neq j$. Therefore, these components can be viewed as *uncorrelated linear combinations of correlated variables*.

8.4. The independence transformation in microeconomic theory

The principal component technique can be extended so that two square matrices are simultaneously diagonalized. An attractive way of discussing this extension is in terms of the differential demand and supply equations of Section 5. Recall that under preference independence the demand equation (5.5) takes the form (5.11) with only one relative price. Preference independence amounts to additive utility and is thus quite restrictive. But if the consumer is not preference independent with respect to the N observed goods, we may ask whether it is possible to transform these goods so that the consumer is preference independent with respect to the transformed goods. Similarly, if a firm is not input independent, can we derive transformed inputs so that the firm is input independent with respect to these? An analogous question can be asked for the outputs of a multiproduct firm; below we consider the inputs of a single-product firm in order to fix the attention.

Consider the input allocation equation (5.21) and divide by f_i:

$$\mathrm{d}(\log q_i) = \frac{\theta_i}{f_i}\mathrm{d}(\log Q) - \frac{\psi}{f_i}\sum_{j=1}^{N}\theta_{ij}\mathrm{d}\left(\log\frac{p_j}{P'}\right). \tag{8.13}$$

This shows that θ_i/f_i is the elasticity of the demand for input i with respect to the Divisia input volume index; we shall express this by referring to θ_i/f_i as the *Divisia elasticity* of input i, which is the firm's input version of the consumer's income elasticity of the demand for a good.[7] Also, (8.13) shows that $-\psi\theta_{ij}/f_i$ is the elasticity of input i with respect to the Frisch-deflated price of input j. Under input independence the substitution term is simplified [see (5.15) and (5.18)] so that (8.13) becomes

$$\mathrm{d}(\log q_i) = \frac{\theta_i}{f_i}\mathrm{d}(\log Q) - \frac{\psi\theta_i}{f_i}\mathrm{d}\left(\log\frac{p_i}{P'}\right). \tag{8.14}$$

Hence, all price elasticities vanish except the own-price elasticities; the latter take the form $-\psi\theta_i/f_i$ and are thus proportional to the Divisia elasticities with $-\psi$ as the (negative) proportionality coefficient.

Next consider the input allocation system in the form (5.23):

$$\boldsymbol{F\kappa} = (\boldsymbol{\iota}'\boldsymbol{F\kappa})\boldsymbol{\Theta\iota} - \psi\boldsymbol{\Theta}(\boldsymbol{I} - \boldsymbol{\iota\iota}'\boldsymbol{\Theta})\boldsymbol{\pi}. \tag{8.15}$$

Our objective is to define transformed inputs which diagonalize $\boldsymbol{\Theta}$. We perform this under the condition that total input expenditure and its Divisia decomposition are invariant under the transformation. The derivation is given in Appendix B and the result may be described by means of a simultaneous diagonalization

[7]The consumer's version of θ_i/f_i is θ_i/w_i; it is easily verified [see (5.3)] that θ_i/w_i equals the elasticity of q_i with respect to M.

similar to (7.13):

$$\boldsymbol{X}'\boldsymbol{\Theta}\boldsymbol{X}=\boldsymbol{\Lambda}, \qquad \boldsymbol{X}'\boldsymbol{F}\boldsymbol{X}=\boldsymbol{I}, \tag{8.16}$$

where $\boldsymbol{\Lambda}$ is the diagonal matrix with the roots $\lambda_1,\dots,\lambda_N$ on the diagonal. These roots are the Divisia elasticities of the transformed inputs. The allocation equation for transformed input i takes the form

$$\mathrm{d}(\log q_{\mathrm{T}i})=\lambda_i\mathrm{d}(\log Q)-\psi\lambda_i\mathrm{d}\left(\log\frac{p_{\mathrm{T}i}}{P'}\right), \tag{8.17}$$

where the subscript T stands for "transformed". A comparison of (8.17) and (8.14) shows that the Divisia volume and Frisch price indexes and ψ are all invariant under the transformation.

Recall from (7.11) and (7.13) that any column $\boldsymbol{x}_i$ of the matrix $\boldsymbol{X}$ in (8.16) satisfies

$$(\boldsymbol{\Theta}-\lambda_i\boldsymbol{F})\boldsymbol{x}_i=\boldsymbol{0}. \tag{8.18}$$

We premultiply this by $-\psi\boldsymbol{F}^{-1}$:

$$\left[-\psi\boldsymbol{F}^{-1}\boldsymbol{\Theta}-(-\psi\lambda_i)\boldsymbol{I}\right]\boldsymbol{x}_i=\boldsymbol{0}. \tag{8.19}$$

Since $-\psi\boldsymbol{F}^{-1}\boldsymbol{\Theta}=[-\psi\theta_{ij}/f_i]$ is the price elasticity matrix of the observed inputs [see (8.13)] and $-\psi\lambda_i$ is the own-price elasticity of transformed input i [see (8.17)], (8.19) shows that the latter elasticity is a latent root of the price elasticity matrix $-\psi\boldsymbol{F}^{-1}\boldsymbol{\Theta}$ of the observed inputs. This is an asymmetric matrix, but the λ_i's are nevertheless real. To prove this we premultiply (8.18) by $\boldsymbol{F}^{-1/2}$ and write the result as

$$\left(\boldsymbol{F}^{-1/2}\boldsymbol{\Theta}\boldsymbol{F}^{-1/2}-\lambda_i\boldsymbol{I}\right)\boldsymbol{F}^{1/2}\boldsymbol{x}_i=\boldsymbol{0}. \tag{8.20}$$

Since $\boldsymbol{F}^{-1/2}\boldsymbol{\Theta}\boldsymbol{F}^{-1/2}$ is symmetric positive definite, the λ_i's are all real and positive. Hence, all transformed inputs have positive Divisia elasticities. The diagonalization (8.20) is unique when the λ_i's are distinct. This means that the transformed inputs are identified by their Divisia elasticities.

These elasticities can be used as a tool for the interpretation of the transformed inputs. Another tool is the so-called *composition matrix*

$$\boldsymbol{T}=(\boldsymbol{X}^{-1}\boldsymbol{\iota})_{\Delta}\boldsymbol{X}^{-1}, \tag{8.21}$$

where $(\boldsymbol{X}^{-1}\boldsymbol{\iota})_{\Delta}$ is defined in (7.9). The column sums of $\boldsymbol{T}$ are the factor shares $f_1,\dots,f_N$ of the observed inputs and the row sums are the factor shares of the transformed inputs. Each row of $\boldsymbol{T}$ gives the composition of the factor share of a transformed input in terms of observed inputs; each column of $\boldsymbol{T}$ shows the composition of the factor share of an observed input in terms of transformed inputs. For proofs of these results we refer to Appendix B; below we consider an example illustrating these results, after which a comparison with principal components will follow at the end of this section.

8.5. An example

We consider a two-input translog specification of (5.12):

$$\log z = \text{constant} + \alpha \log K + \beta \log L + \xi\sqrt{\alpha\beta}\,\log K\,\log L, \tag{8.22}$$

where K is capital, L is labor, and α, β, and ξ are constants satisfying $\alpha > 0$, $\beta > 0$, and $-1 < \xi < 1$; units are chosen so that $K = L = 1$ holds at the point of minimum input expenditure. Then it may be shown that the 2×2 price coefficient matrix $-\psi\boldsymbol{\Theta} = [-\psi\theta_{ij}]$ of (8.15) equals

$$-\psi\boldsymbol{\Theta} = \frac{-1}{1-\xi^2}\begin{bmatrix} f_K & \xi\sqrt{f_K f_L} \\ \xi\sqrt{f_K f_L} & f_L \end{bmatrix} \begin{matrix} \text{capital} \\ \text{labor} \end{matrix}, \tag{8.23}$$

where f_K is the factor share of capital and f_L that of labor ($f_K + f_L = 1$). Recall from Section 6 that inputs i and j are called specific complements (substitutes) when $\theta_{ij} = \theta_{ji}$ is positive (negative). Thus, (8.23) combined with $\psi > 0$ shows that capital and labor are specific complements (substitutes) when ξ is positive (negative), i.e. when the elasticity of output with respect to either input is an increasing (decreasing) function of the other input [see (8.22)].

The input independence transformation eliminates all specific substitutability and complementarity relations. The mathematical tool is the simultaneous diagonalization (8.16). It may be verified that, for $-\psi\boldsymbol{\Theta}$ given in (8.23), the matrix

$$\boldsymbol{X} = \frac{1}{\sqrt{2 f_K f_L}}\begin{bmatrix} \sqrt{f_L} & -\sqrt{f_L} \\ \sqrt{f_K} & \sqrt{f_K} \end{bmatrix} \tag{8.24}$$

satisfies $\boldsymbol{X}'\boldsymbol{F}\boldsymbol{X} = \boldsymbol{I}$ and that $\boldsymbol{X}'(\psi\boldsymbol{\Theta})\boldsymbol{X}$ is a diagonal matrix whose diagonal elements are $1/(1-\xi)$ and $1/(1+\xi)$. A comparison with (8.16) and (8.17) shows that the own-price elasticities of the transformed inputs are

$$-\psi\lambda_1 = \frac{-1}{1-\xi}, \qquad -\psi\lambda_2 = \frac{-1}{1+\xi}. \tag{8.25}$$

Multiple roots occur when $\xi = 0$, but this is the uninteresting case in which (8.22) becomes Cobb–Douglas, which is in input independent form and thus requires no transformation.

Substitution of (8.24) into (8.21) yields the composition matrix

$$T = \frac{1}{2}\begin{bmatrix} f_K + \sqrt{f_K f_L} & f_L + \sqrt{f_K f_L} \\ f_K - \sqrt{f_K f_L} & f_L - \sqrt{f_K f_L} \end{bmatrix} \begin{matrix} (\mathrm{T}_1) \\ (\mathrm{T}_2)\,. \end{matrix} \qquad \begin{matrix} \text{(capital)} & \text{(labor)} \end{matrix} \tag{8.26}$$

The column sums are the factor shares of the observed inputs: f_K for capital and f_L for labor. The row sums are the factor shares of the transformed inputs: $\frac{1}{2} + \sqrt{f_K f_L}$ for the input T_1 corresponding to root λ_1 and $\frac{1}{2} - \sqrt{f_K f_L}$ for T_2 corresponding to λ_2. The following is a numerical specification of (8.26), bordered by row and column sums, for $f_K = 0.2$ and $f_L = 0.8$:

0.3	0.6	0.9	(T_1)
-0.1	0.2	0.1	(T_2)
0.2	0.8	1	
(capital)	(labor)		

Both observed inputs are positively represented in T_1, whereas T_2 is a *contrast* between labor and capital. When the firm buys more T_2, its operation becomes more labor-intensive, each dollar spent on T_2 being equivalent to two dollars worth of labor compensated by one dollar worth of capital services which is given up.

8.6. *A principal component interpretation*

We return to (8.8) with $\boldsymbol{Z}'\boldsymbol{Z}$ interpreted as n times the matrix of mean squares and products of the values taken by the observed variables. In many applications of the principal component technique, the observed variables have different dimensions (dollars, dollars per year, gallons, etc.). This causes a problem, because principal components change in value when we change the units in which the observed variables are measured. To solve this problem, statisticians frequently standardize these variables by using their standard deviations as units. This amounts to replacing $\boldsymbol{Z}'\boldsymbol{Z}$ in (8.8) by $\boldsymbol{D}^{-1/2}\boldsymbol{Z}'\boldsymbol{Z}\boldsymbol{D}^{-1/2}$, where $\boldsymbol{D}$ is the diagonal matrix whose diagonal is identical to that of $\boldsymbol{Z}'\boldsymbol{Z}$. Thus, λ_i of (8.8) is now obtained from the characteristic equation

$$|\boldsymbol{D}^{-1/2}\boldsymbol{Z}'\boldsymbol{Z}\boldsymbol{D}^{-1/2} - \lambda_i \boldsymbol{I}| = 0. \tag{8.27}$$

It is of interest to compare this with

$$|F^{-1/2}\Theta F^{-1/2}-\lambda_i I|=0, \tag{8.28}$$

which is the characteristic equation associated with (8.20). In both cases, (8.27) and (8.28), we determine a latent root of a symmetric positive definite matrix ($Z'Z$ or Θ) pre- and postmultiplied by a diagonal matrix. However, the diagonal elements of F are not identical to those of Θ, which is in contrast to D and $Z'Z$ in (8.27). The diagonal elements of F describe the expenditure *levels* of the inputs (measured as fractions of total expenditure), whereas each diagonal element of Θ describes the *change* in the demand for an input caused by a change in its Frisch-deflated price.

Thus, while D in (8.27) is directly obtained from $Z'Z$, the analogous matrix F in (8.28) is unrelated to Θ. Why do we have this unrelated F, which describes expenditure levels, in (8.28) and in the simultaneous diagonalization (8.16)? The answer is that the input independence transformation is subject to the constraint that that total input expenditure and its Divisia decomposition remain invariant. We may view this transformation as a *cost-constrained* principal component transformation. Similarly, when the transformation is applied to the consumer demand system (5.5) or to the output supply system (5.26), it is budget-constrained in the first case and revenue-constrained in the second. Such constraints are more meaningful from an economic point of view than the standardization procedure in (8.27).

9. The modeling of a disturbance covariance matrix

We mentioned in Section 4 that the disturbance covariance matrix Σ which occurs in the GLS estimator (4.11) is normally unknown and that it is then usually replaced by the sample moment matrix of the LS residuals. Although this approximation is acceptable under certain conditions when the sample is sufficiently large, it is less satisfactory when the number of equations, L in (4.7) and (4.8), is also large. The reason is that Σ contains many unknowns when L is large or even moderately large. In fact, the sample moment matrix S of the residuals may be singular so that Σ^{-1} in (4.11) cannot be approximated by S^{-1}. This situation often occurs in applied econometrics, e.g. in the estimation of a fairly large system of demand equations. One way of solving this problem is by modeling the matrix Σ. Below we describe how this can be performed when the equations are behavioral equations of some decision-maker.

9.1. Rational random behavior

Let $\boldsymbol{x}=[x_1 \ldots x_k]'$ be the vector of variables controlled by this decision-maker. We write J for the feasible region of $\boldsymbol{x}$; $\bar{\boldsymbol{x}}$ for the optimal value of $\boldsymbol{x}$ ($\bar{\boldsymbol{x}} \in J$); and $l(\boldsymbol{x}, \bar{\boldsymbol{x}})$ for the loss incurred when the decision is $\boldsymbol{x}$ rather than $\bar{\boldsymbol{x}}$:

$$\begin{aligned} l(\boldsymbol{x}, \bar{\boldsymbol{x}}) &= 0 \quad \text{if } \boldsymbol{x}=\bar{\boldsymbol{x}}, \\ &> 0 \quad \text{if } \boldsymbol{x} \neq \bar{\boldsymbol{x}}. \end{aligned} \tag{9.1}$$

We assume that the optimal decision $\bar{\boldsymbol{x}}$ depends on numerous factors, some of which are unknown, so that $\bar{\boldsymbol{x}}$ is only *theoretically optimal* in the sense that it is optimal under perfect knowledge. The decision-maker can improve on his ignorance by acquiring information. If he does not do this, we describe the decision made as random with a differentiable density function $p_0(\boldsymbol{x})$, to be called the *prior density function*. (The assumption of randomness is justified by the decision-maker's uncertainty as to the factors determining $\bar{\boldsymbol{x}}$.) If he does acquire information, $p_0(\cdot)$ is transformed into some other density function $p(\cdot)$ and the amount of information received is defined as

$$I=\int_J p(\boldsymbol{x}) \log \frac{p(\boldsymbol{x})}{p_0(\boldsymbol{x})} \mathrm{d}x_1 \ldots \mathrm{d}x_k, \tag{9.2}$$

which is a concept from statistical information theory [see Theil (1967)]. We write $c(I)$ for the cost of information and

$$\bar{l}=\int_J l(\boldsymbol{x}, \bar{\boldsymbol{x}}) p(\boldsymbol{x}) \mathrm{d}x_1 \ldots \mathrm{d}x_k \tag{9.3}$$

for the expected loss. If $c(I)$ and $\bar{l}$ are measured in the same unit (dollars or any other unit), the natural solution is the decision distribution with density function $p(\cdot)$ which minimizes $c(I)+\bar{l}$. This $p(\cdot)$ was derived by Barbosa (1975) and the result (see Appendix C) is

$$p(\boldsymbol{x}) \propto p_0(\boldsymbol{x}) \exp\left\{-\frac{l(\boldsymbol{x}, \bar{\boldsymbol{x}})}{c'}\right\} \quad \text{if } \boldsymbol{x} \in J, \tag{9.4}$$

where $\propto$ means "is proportional to", the proportionality coefficient being independent of $\boldsymbol{x}$, and c' is the marginal cost of information $\mathrm{d}c/\mathrm{d}I$ at the solution (9.4).

Behavior generated by the distribution which density function (9.4) is called *rational random behavior*. This distribution is determined by three factors: the prior density function $p_0(\boldsymbol{x})$, the loss function $l(\boldsymbol{x}, \bar{\boldsymbol{x}})$, and the marginal cost of information c'. For example, let $\boldsymbol{x}$ be a scalar and the loss function quadratic: $l(x,\bar{x})=\frac{1}{2}(x-\bar{x})^2$. Then, if $p_0(x)$ equals a constant independent of x for each $x\in J$, (9.4) becomes

$$p(x)\propto\exp\left\{-\frac{1}{2}\frac{(x-\bar{x})^2}{c'}\right\}\quad\text{if } x\in J,\tag{9.5}$$

which is the density function of a truncated normal decision distribution over the interval J.

9.2. *The asymptotics of rational random behavior*

The case of a small marginal cost of information is of particular interest. Imagine that the prices of the goods and services which the decision-maker buys in order to acquire information decline so that c' converges to zero; it is shown in Appendix C that the random decision with density function (9.4) then converges in probability to the theoretically optimal decision $\bar{\boldsymbol{x}}$. Also, if the loss function has a zero gradient and a symmetric positive definite Hessian matrix $\boldsymbol{A}$ at $\boldsymbol{x}=\bar{\boldsymbol{x}}$,

$$\frac{\partial}{\partial\boldsymbol{x}}l(\boldsymbol{x},\bar{\boldsymbol{x}})=\boldsymbol{0}\quad\text{and}\quad\frac{\partial^2}{\partial\boldsymbol{x}\,\partial\boldsymbol{x}'}l(\boldsymbol{x},\bar{\boldsymbol{x}})=\boldsymbol{A}\quad\text{at } \boldsymbol{x}=\bar{\boldsymbol{x}},\tag{9.6}$$

then as $c'\to 0$ the density function $p(\boldsymbol{x})$ of (9.4) converges to

$$\frac{1}{(2\pi)^{k/2}}\left|\frac{1}{c'}\boldsymbol{A}\right|^{1/2}\exp\left\{-\tfrac{1}{2}(\boldsymbol{x}-\bar{\boldsymbol{x}})'\left(\frac{1}{c'}\boldsymbol{A}\right)(\boldsymbol{x}-\bar{\boldsymbol{x}})\right\},\tag{9.7}$$

which is the density function of the multinormal decision distribution [see (3.3)] with mean vector $\bar{\boldsymbol{x}}$ and covariance matrix $c'\boldsymbol{A}^{-1}$. Note that (9.7) is completely determined by c' and two characteristics of the loss function: the theoretically optimal decision $\bar{\boldsymbol{x}}$ at which the loss vanishes [see (9.1)] and the Hessian matrix $\boldsymbol{A}$ of this function at this point. The relationship between the covariance matrix $c'\boldsymbol{A}^{-1}$ and the Hessian matrix $\boldsymbol{A}$ of the loss function enables us to model the disturbance covariance matrix of the decision-maker's behavioral equations; examples will follow in the next subsection.

The prior density function $p_0(\cdot)$ does not occur in the asymptotic result (9.7). This reflects the fact that when information is cheap in the sense that its marginal cost c' is small, the decision-maker acquires information to such an extent that his behavior becomes independent of his prior notions. Thus, whereas we obtained (9.5) under the assumption that $p_0(x)$ is a constant independent of x, this assumption is unnecessary in the asymptotic case $c'\to 0$. Also, (9.5) is the density function of a truncated normal distribution, but the truncation becomes irrelevant as $c' \to 0$. The asymptotic version of (9.5) is the univariate normal density function with mean equal to the theoretically optimal decision $\bar{x}$ and variance equal to the marginal cost of information c'. The declining variance as $c' \to 0$ reflects the attractiveness of a more extensive search for the theoretical optimum when information becomes cheaper.

It is of interest to compare the density function (9.7) of the asymptotic normal decision distribution with the asymptotic normal density of a statistical estimator. In fact, it is not difficult to prove that rational random behavior for small c' is equivalent to large-sample ML estimation of the theoretically optimal decision $\bar{x}$, with a large sample interpreted as a small marginal cost of information. The clue for this equivalence is the similarity of the roles played by the Hessian matrix of the loss function and the information matrix in ML theory.

A second statistical comparison is that with Kadane's (1971) small-σ asymptotics, which consists of an asymptotic series of a multiple σ of the variance of the disturbance of a structural equation. If this equation is a behavioral equation of a decision-maker, Kadane's approach is equivalent to the asymptotic version of rational random behavior when we identify σ with c'.

Another statistical comparison of interest is that with the theorem which states that out of all distributions with range $(-\infty, \infty)$ and a given mean vector and a given covariance matrix, the multinormal distribution has the largest entropy. The link between this theorem and the normal density function (9.7) is the information definition (9.2); both (9.2) and the entropy are measures from information theory. However, note that the normal density (9.7) is not obtained by imposing a given mean vector and covariance matrix *a priori*. The mean vector and covariance matrix (9.7) are determined by the loss function, apart from the scalar c'.

Yet another statistical comparison is with Bayesian inference. There is considerable similarity between the exact (i.e., non-asymptotic) result (9.4) and the Bayesian derivation of the posterior density function of a parameter vector. The occurrence of the prior density function on the right in (9.4) provides one similarity. Another is the presence of c' (which depends on the information acquired) in the exponent of (9.4); this should be compared with the role of the likelihood function (representing the information obtained from the sample) in the Bayesian formula. A third similarity is the disappearance of the prior density function from the asymptotic result (9.7). In Bayesian analysis, too, the posterior density function is dominated by the likelihood function and is no longer affected by the prior density function when the sample is large. All these similarities reflect the fact that rational random behavior and Bayesian inference both describe learning processes based on acquiring information. Nevertheless, the two theories are not equivalent because of the occurrence of the unknown constant $\bar{x}$ in (9.4). The likelihood function in Bayesian analysis involves no unknown constants; this function is determined by the parameter vector, which is viewed as random, and the sample, which is viewed as a set of known constants for the derivation of the posterior density function.

9.3. Applications to demand and supply

When we apply the theory of rational random behavior to the utility-maximizing consumer or the cost-minimizing firm, we must take into consideration that the criterion function is subject to a constraint (a budget or technology constraint). This can be solved by using the constraint to eliminate one of the goods. The consumer's loss function in (9.4) then involves $N-1$ quantities and its derivation from an algebraically specified utility function is straightforward. However, the differential approach provides no such specification so that (9.4) cannot be used; the same holds for the firm because the approach provides no algebraic specification of the production function. But it is possible to use the asymptotic result (9.7) which requires only the theoretically optimal decision and the Hessian matrix of the loss function. The account which follows is therefore based on the asymptotic decision distribution (9.7) rather than (9.4); this also has the advantage of not requiring a specification of the prior density function $p_0(\cdot)$.

Consider the input allocation system in the form (6.1),

$$\boldsymbol{F\kappa} = (\boldsymbol{\iota' F\kappa})\boldsymbol{\Theta\iota} - \psi\boldsymbol{\Theta}(\boldsymbol{I} - \boldsymbol{\iota\iota'\Theta})\boldsymbol{\pi} + \boldsymbol{\varepsilon}, \tag{9.8}$$

or in scalar form, using $\boldsymbol{\Theta\iota} = \boldsymbol{\theta} = [\theta_i]$,

$$f_i \mathrm{d}(\log q_i) = \theta_i \mathrm{d}(\log Q) - \psi \sum_{j=1}^{N} \theta_{ij} \mathrm{d}\left(\log \frac{p_j}{P'}\right) + \varepsilon_i, \tag{9.9}$$

where $[\varepsilon_i] = \boldsymbol{\varepsilon}$. The left variable in (9.9) is the ith decision variable of the firm. The right side, excluding ε_i, is the theoretical optimum of this variable, while ε_i is the random deviation from this optimum which is predicted by the theory of rational random behavior. Since (9.7) implies normality with a mean equal to the theoretically optimal decision, the ε_i's are multinormal with zero mean. Their covariance matrix (see Appendix C) is

$$\mathscr{V}(\boldsymbol{\varepsilon}) = \sigma^2(\boldsymbol{\Theta} - \boldsymbol{\Theta\iota\iota'\Theta}), \tag{9.10}$$

or in scalar form,

$$\operatorname{cov}(\varepsilon_i, \varepsilon_j) = \sigma^2(\theta_{ij} - \theta_i\theta_j), \tag{9.11}$$

where σ^2 is a coefficient which is proportional to the marginal cost of information

c'. The covariance matrix (9.10) equals a scalar multiple of the coefficient matrix of $\boldsymbol{\pi} = [\mathrm{d}(\log p_i)]$ in (9.8) so that the covariance (9.11) is proportional to the substitution effect (specific plus general) of a change in the price of j on the demand for i.

The above result also holds when (9.7) is applied to the consumer; the only modification required is that f_i and ψ in (9.9) become w_i and $-\phi$, respectively [see (5.5)]. Note in particular that the disturbance covariance matrix (9.10) involves only unknown (σ^2) in addition to $\boldsymbol{\Theta}$ which already occurs in the systematic part of the equations. Thus, the implications of rational random behavior are quite strong. We obtain even stronger implications for the demand and supply equations of the multiproduct firm; a brief account follows below.

Recall that when the firm makes m products, the output term of the input demand system (5.15) takes the form (5.25). So, by adding a disturbance ε_i we obtain:

$$f_i \mathrm{d}(\log q_i) = \gamma \sum_{r=1}^{m} \theta_i^r g_r \mathrm{d}(\log z_r) - \psi \sum_{j=1}^{N} \theta_{ij} \mathrm{d}\left(\log \frac{p_j}{P'}\right) + \varepsilon_i. \tag{9.12}$$

Application of (9.7) yields the result that the ε_i's of this input demand system are multinormal with zero means and that their variances and covariances take the same form (9.11) which also holds for the consumer and the single-product firm. Next, by assuming that the firm adjusts its outputs so as to maximize profit, we obtain the output supply system (5.26) which we reproduce with a disturbance ε_r^* added:

$$g_r \mathrm{d}(\log z_r) = \psi^* \sum_{s=1}^{m} \theta_{rs}^* \mathrm{d}\left(\log \frac{y_s}{P'^s}\right) + \varepsilon_r^*. \tag{9.13}$$

By applying (9.7) we find that the ε_r^*'s are multinormal with zero means and the following variance–covariance structure:

$$\mathrm{cov}(\varepsilon_r^*, \varepsilon_s^*) = \frac{\sigma^2 \psi^*}{\gamma \psi} \theta_{rs}^*. \tag{9.14}$$

Since σ^2 already occurs in (9.11) for the input demand disturbances, (9.14) provides no further unknowns. In addition, (9.7) implies that *the input demand disturbances* (the ε_i's) *are stochastically independent of the output supply disturbances* (the ε_r^*'s). This independence has important implications for statistical inference in demand and supply models; it implies that $g_r \mathrm{d}(\log z_r)$ can be viewed as predetermined in the input demand system (9.12). It is also important for the problem of how to organize the firm in terms of its input and output management, but such matters are beyond the scope of this chapter.

10. The Moore–Penrose inverse

A matrix has an inverse only if it is square and nonsingular, but any $m \times n$ matrix $\boldsymbol{A}$ of rank r has a unique Moore–Penrose inverse, written $\boldsymbol{A}^{+}$, which is determined by the following four conditions:

$$\boldsymbol{A}\boldsymbol{A}^{+}\boldsymbol{A} = \boldsymbol{A}, \tag{10.1}$$

$$\boldsymbol{A}^{+}\boldsymbol{A}\boldsymbol{A}^{+} = \boldsymbol{A}^{+}, \tag{10.2}$$

$$\boldsymbol{A}\boldsymbol{A}^{+} \quad \text{and} \quad \boldsymbol{A}^{+}\boldsymbol{A} \text{ are symmetric.} \tag{10.3}$$

It may be verified that these conditions are satisfied by $\boldsymbol{A}^{+} = \boldsymbol{A}^{-1}$ in the special case $m = n = r$. Our first objective is to prove that $\boldsymbol{A}^{+}$ exists and is unique.[8]

10.1. Proof of the existence and uniqueness

The uniqueness of $\boldsymbol{A}^{+}$ is established by assuming that conditions (10.1)–(10.3) have two solutions, $\boldsymbol{A}^{+} = \boldsymbol{B}$ and $\boldsymbol{A}^{+} = \boldsymbol{C}$, and verifying the following 16 steps based on (10.1)–(10.3):

$$\begin{aligned}\boldsymbol{B} &= \boldsymbol{BAB} = \boldsymbol{B}(\boldsymbol{AB})' = \boldsymbol{BB}'\boldsymbol{A}' = \boldsymbol{BB}'(\boldsymbol{ACA})' = \boldsymbol{BB}'\boldsymbol{A}'\boldsymbol{C}'\boldsymbol{A}'\\ &= \boldsymbol{B}(\boldsymbol{AB})'(\boldsymbol{AC})' = \boldsymbol{BABAC} = \boldsymbol{BAC} = \boldsymbol{BACAC} = (\boldsymbol{BA})'(\boldsymbol{CA})'\boldsymbol{C}\\ &= \boldsymbol{A}'\boldsymbol{B}'\boldsymbol{A}'\boldsymbol{C}'\boldsymbol{C} = (\boldsymbol{ABA})'\boldsymbol{C}'\boldsymbol{C} = \boldsymbol{A}'\boldsymbol{C}'\boldsymbol{C} = (\boldsymbol{CA})'\boldsymbol{C} = \boldsymbol{CAC} = \boldsymbol{C}.\end{aligned}$$

Therefore, $\boldsymbol{B} = \boldsymbol{C}$, which proves that $\boldsymbol{A}^{+}$ is unique when it exists.

To prove the existence of $\boldsymbol{A}^{+}$ we consider first a zero matrix $\boldsymbol{A}$ of order $m \times n$; then $\boldsymbol{A}^{+}$ equals the $n \times m$ zero matrix, which may be verified by checking (10.1)–(10.3). Next consider a non-zero matrix $\boldsymbol{A}$ so that its rank r is positive. Then $\boldsymbol{A}'\boldsymbol{A}$ is a symmetric positive semidefinite matrix of order $n \times n$ and rank r, and it is possible to express $\boldsymbol{A}^{+}$ in terms of the positive latent roots of $\boldsymbol{A}'\boldsymbol{A}$ and the characteristic vectors associated with these roots. Write $\boldsymbol{D}$ for the diagonal $r \times r$ matrix which contains the positive roots of $\boldsymbol{A}'\boldsymbol{A}$ on the diagonal and $\boldsymbol{H}$ for an $n \times r$ matrix whose columns are characteristic vectors corresponding to these roots. Then (7.7) applied to $\boldsymbol{A}'\boldsymbol{A}$ yields

$$\boldsymbol{A}'\boldsymbol{A} = \boldsymbol{HDH}', \tag{10.4}$$

[8] There are other generalized inverses besides the Moore–Penrose inverse, most of which are obtained by deleting one or more of the four conditions. For example, using (10.1) and (10.2) but deleting (10.3) yields the reflexive generalized inverse, which in not unique; see Laitinen and Theil (1979) for an application of this inverse to consumption theory. Monographs on applications of generalized inverses to statistics include Albert (1972), Ben-Israel and Greville (1974), Boullion and Odell (1971), Pringle and Rayner (1971), and Rao and Mitra (1971).

and the result for $\boldsymbol{A}^+$ is

$$\boldsymbol{A}^+ = \boldsymbol{H}\boldsymbol{D}^{-1}\boldsymbol{H}'\boldsymbol{A}', \tag{10.5}$$

which is an $n \times m$ matrix of rank r.

To verify (10.5) we introduce an $n \times (n-r)$ matrix $\boldsymbol{K}$ whose columns are characteristic vectors of $\boldsymbol{A}'\boldsymbol{A}$ corresponding to the zero roots:

$$\boldsymbol{A}'\boldsymbol{A}\boldsymbol{K} = \boldsymbol{0}. \tag{10.6}$$

The $n \times n$ matrix $[\boldsymbol{H} \quad \boldsymbol{K}]$ consists of characteristic vectors of $\boldsymbol{A}'\boldsymbol{A}$ corresponding to all roots and is therefore an orthogonal matrix, which can be expressed in two ways. Premultiplying $[\boldsymbol{H} \quad \boldsymbol{K}]$ by its transpose and equating the product to the unit matrix yields

$$\boldsymbol{H}'\boldsymbol{H} = \boldsymbol{I}, \qquad \boldsymbol{K}'\boldsymbol{K} = \boldsymbol{I}, \qquad \boldsymbol{H}'\boldsymbol{K} = \boldsymbol{0}, \tag{10.7}$$

while postmultiplying $[\boldsymbol{H} \quad \boldsymbol{K}]$ by its transpose and equating the product to the unit matrix gives

$$\boldsymbol{H}\boldsymbol{H}' + \boldsymbol{K}\boldsymbol{K}' = \boldsymbol{I}. \tag{10.8}$$

The verification of (10.5) is now a matter of checking conditions (10.1)–(10.3). Premultiplying (10.5) by $\boldsymbol{A}$ yields $\boldsymbol{A}\boldsymbol{A}^+ = \boldsymbol{A}\boldsymbol{H}\boldsymbol{D}^{-1}\boldsymbol{H}'\boldsymbol{A}'$, which is symmetric. Next we postmultiply (10.5) by $\boldsymbol{A}$, $\boldsymbol{A}^+\boldsymbol{A} = \boldsymbol{H}\boldsymbol{D}^{-1}\boldsymbol{H}'\boldsymbol{A}'\boldsymbol{A}$, and hence in view of (10.4) and (10.7), $\boldsymbol{A}^+\boldsymbol{A} = \boldsymbol{H}\boldsymbol{D}^{-1}\boldsymbol{H}'\boldsymbol{H}\boldsymbol{D}\boldsymbol{H}' = \boldsymbol{H}\boldsymbol{H}'$, which is also symmetric. We postmultiply this by (10.5):

$$\boldsymbol{A}^+\boldsymbol{A}\boldsymbol{A}^+ = \boldsymbol{H}\boldsymbol{H}'\boldsymbol{H}\boldsymbol{D}^{-1}\boldsymbol{H}'\boldsymbol{A}' = \boldsymbol{H}\boldsymbol{D}^{-1}\boldsymbol{H}'\boldsymbol{A}' = \boldsymbol{A}^+,$$

which confirms (10.2). Finally, we postmultiply $\boldsymbol{A}\boldsymbol{A}^+ = \boldsymbol{A}\boldsymbol{H}\boldsymbol{D}^{-1}\boldsymbol{H}'\boldsymbol{A}'$ by $\boldsymbol{A}$:

$$\boldsymbol{A}\boldsymbol{A}^+\boldsymbol{A} = \boldsymbol{A}\boldsymbol{H}\boldsymbol{D}^{-1}\boldsymbol{H}'\boldsymbol{A}'\boldsymbol{A} = \boldsymbol{A}\boldsymbol{H}\boldsymbol{D}^{-1}\boldsymbol{H}'\boldsymbol{H}\boldsymbol{D}\boldsymbol{H}' = \boldsymbol{A}\boldsymbol{H}\boldsymbol{H}' = \boldsymbol{A}.$$

To verify the last step, $\boldsymbol{A}\boldsymbol{H}\boldsymbol{H}' = \boldsymbol{A}'$, we premultiply (10.6) by $\boldsymbol{K}'$, which gives $(\boldsymbol{A}\boldsymbol{K})'\boldsymbol{A}\boldsymbol{K} = \boldsymbol{0}$ or $\boldsymbol{A}\boldsymbol{K} = \boldsymbol{0}$. Therefore, $\boldsymbol{A}\boldsymbol{K}\boldsymbol{K}' = \boldsymbol{0}$ so that premultiplication of (10.8) by $\boldsymbol{A}$ yields $\boldsymbol{A}\boldsymbol{H}\boldsymbol{H}' = \boldsymbol{A}$.

10.2. Special cases

If $\boldsymbol{A}$ has full column rank so that $(\boldsymbol{A}'\boldsymbol{A})^{-1}$ exists, $\boldsymbol{A}^+ = (\boldsymbol{A}'\boldsymbol{A})^{-1}\boldsymbol{A}'$, which may either be verified from (10.4) and (10.5) for $r = n$ or by checking (10.1)–(10.3). We may thus write the LS coefficient vector in (2.8) as $\boldsymbol{b} = \boldsymbol{X}^+\boldsymbol{y}$, which may be viewed as an extension of $b = X^{-1}y$ in the special case of a square non-singular X (as many regressors as observations).

If A is a symmetric $n \times n$ matrix of rank r, then

$$A = \sum_{i=1}^{r} \lambda_i x_i x_i', \qquad A^{+} = \sum_{i=1}^{r} \frac{1}{\lambda_i} x_i x_i', \tag{10.9}$$

where $\lambda_1, \ldots, \lambda_r$ are the non-zero latent roots of A and $x_1, \ldots, x_r$ are characteristic vectors associated with these roots. Also, $Ax_i = 0$ and $A^{+}x_i = 0$ for $i = r+1, \ldots, n$, where $x_{r+1}, \ldots, x_n$ are characteristic vectors of A corresponding to the zero roots. Thus, if A is symmetric, A^{+} has characteristic vectors identical to those of A, the same number of zero roots, and non-zero roots equal to the reciprocals of the non-zero roots of A. The verification of these results is again a matter of checking (10.1)–(10.3) and using $x_i'x_j = \delta_{ij}$. Since a symmetric idempotent matrix such as M in (2.11) has only zero and unit roots, it thus follows as a corollary that such a matrix is equal to its own Moore–Penrose inverse.

10.3. *A generalization of Aitken's theorem*

We return to the linear model (2.7), reproduced here:

$$y = X\beta + \varepsilon. \tag{10.10}$$

As before, we assume that X is an $n \times K$ matrix of rank K consisting of non-stochastic elements and that ε has zero expectation, but we now assume that the covariance matrix of ε takes the singular form $\sigma^2 V$, the $n \times n$ matrix V having rank $r < n$. Hence, the Aitken estimator (2.13) does not exist, but is seems reasonable to ask whether

$$\hat{\beta} = \left(X'V^{+}X\right)^{-1} X'V^{+}y \tag{10.11}$$

exists and is a best linear unbiased estimator of β. It will appear that each of these properties (the existence and the best linear unbiasedness) requires a special condition involving both V and X.

The matrix V is comparable to $A'A$ in (10.4) and (10.6) in that both are symmetric positive semidefinite $n \times n$ matrices of rank r. Therefore, we can apply (10.4) and (10.6) to V rather than $A'A$:

$$V = HDH', \tag{10.12}$$

$$VK = 0, \tag{10.13}$$

where $\boldsymbol{D}$ is now the $r \times r$ diagonal matrix with the positive latent roots of $\boldsymbol{V}$ on the diagonal, $\boldsymbol{H}$ is an $n \times r$ matrix whose columns are characteristic vectors of $\boldsymbol{V}$ corresponding to these roots, and $\boldsymbol{K}$ is an $n \times (n-r)$ matrix consisting of characteristic vectors of $\boldsymbol{V}$ that correspond to the zero roots. The results (10.7) and (10.8) are also valid in the present interpretation. In addition, (10.9) and (10.12) imply

$$\boldsymbol{V}^{+} = \boldsymbol{H}\boldsymbol{D}^{-1}\boldsymbol{H}'. \tag{10.14}$$

Our strategy, similar to that of the proof of Aitken's theorem in Section 7, will be to premultiply (10.10) by an appropriate matrix so that the transformed disturbance vector has a scalar covariance matrix. We select $\boldsymbol{D}^{-1/2}\boldsymbol{H}'$, where $\boldsymbol{D}^{-1/2}$ is the diagonal matrix with the reciprocals of the positive square roots of the diagonal elements of $\boldsymbol{D}$ in the diagonal:

$$\boldsymbol{D}^{-1/2}\boldsymbol{H}'\boldsymbol{y} = (\boldsymbol{D}^{-1/2}\boldsymbol{H}'\boldsymbol{X})\boldsymbol{\beta} + \boldsymbol{D}^{-1/2}\boldsymbol{H}'\boldsymbol{\varepsilon}. \tag{10.15}$$

The covariance matrix of $\boldsymbol{D}^{-1/2}\boldsymbol{H}'\boldsymbol{\varepsilon}$ is

$$\mathcal{E}(\boldsymbol{D}^{-1/2}\boldsymbol{H}'\boldsymbol{\varepsilon}\boldsymbol{\varepsilon}'\boldsymbol{H}\boldsymbol{D}^{-1/2}) = \sigma^2\boldsymbol{D}^{-1/2}\boldsymbol{H}'\boldsymbol{V}\boldsymbol{H}\boldsymbol{D}^{-1/2} = \sigma^2\boldsymbol{I},$$

where the last step is based on $\boldsymbol{H}'\boldsymbol{V}\boldsymbol{H} = \boldsymbol{D}$, which is obtained by premultiplying (10.12) by $\boldsymbol{H}'$ and postmultiplying by $\boldsymbol{H}$ and using $\boldsymbol{H}'\boldsymbol{H} = \boldsymbol{I}$ [see (10.7)]. Since $\boldsymbol{D}^{-1/2}\boldsymbol{H}'\boldsymbol{\varepsilon}$ thus has a scalar covariance matrix, let us apply LS to (10.15). Assuming that $\boldsymbol{H}'\boldsymbol{X}$ and hence $\boldsymbol{D}^{-1/2}\boldsymbol{H}'\boldsymbol{X}$ have full column rank, we find the following estimator of $\boldsymbol{\beta}$:

$$(\boldsymbol{D}^{-1/2}\boldsymbol{H}'\boldsymbol{X})^{+}\boldsymbol{D}^{-1/2}\boldsymbol{H}'\boldsymbol{y} = (\boldsymbol{X}'\boldsymbol{H}\boldsymbol{D}^{-1}\boldsymbol{H}'\boldsymbol{X})^{-1}\boldsymbol{X}'\boldsymbol{H}\boldsymbol{D}^{-1}\boldsymbol{H}'\boldsymbol{y}. \tag{10.16}$$

This is indeed identical to (10.11) in view of (10.14).

Two considerations are important for the appraisal of this procedure. First, we assumed that $\boldsymbol{H}'\boldsymbol{X}$ has full column rank; if the rank is smaller, the matrix product in parentheses on the right in (10.16) is singular so that (10.11) does not exist. Therefore, a necessary and sufficient condition for the existence of the estimator (10.11) is that $\boldsymbol{H}'\boldsymbol{X}$ have maximum rank, where $\boldsymbol{H}$ consists of r characteristic vectors of $\boldsymbol{V}$ corresponding to the positive roots. Secondly, we obtained (10.15) by premultiplying (10.10) by $\boldsymbol{D}^{-1/2}\boldsymbol{H}'$, which reduces the number of observations from n to r. We can recover the "missing" $n-r$ observations by premultiplication by $\boldsymbol{K}'$, yielding $\boldsymbol{K}'\boldsymbol{y} = \boldsymbol{K}'\boldsymbol{X}\boldsymbol{\beta} + \boldsymbol{K}'\boldsymbol{\varepsilon}$. The covariance matrix of $\boldsymbol{K}'\boldsymbol{\varepsilon}$ is $\sigma^2\boldsymbol{K}'\boldsymbol{V}\boldsymbol{K} = \boldsymbol{0}$ [see (10.13)] so that $\boldsymbol{K}'\boldsymbol{\varepsilon}$ vanishes with unit probability. Therefore,

$$\boldsymbol{K}'\boldsymbol{y} = \boldsymbol{K}'\boldsymbol{X}\boldsymbol{\beta}, \tag{10.17}$$

which amounts to a linear constraint on $\boldsymbol{\beta}$ unless $\boldsymbol{K}'\boldsymbol{X} = \boldsymbol{0}$.

To clarify this situation, consider the following example for $K=1$, $n=3$, and $r=2$:

$$\boldsymbol{X}=\begin{bmatrix}0\\0\\1\end{bmatrix},\qquad \boldsymbol{V}=\begin{bmatrix}1&0&0\\0&1&0\\0&0&0\end{bmatrix},\qquad \boldsymbol{H}=\begin{bmatrix}1&0\\0&1\\0&0\end{bmatrix},\qquad \boldsymbol{K}=\begin{bmatrix}0\\0\\1\end{bmatrix}. \tag{10.18}$$

Here $\boldsymbol{X}$ has full column rank but $\boldsymbol{H}'\boldsymbol{X}=\boldsymbol{0}$ so that the matrix product in parentheses on the right in (10.16) is singular; in fact, the underlying equation (10.15) does not contain $\boldsymbol{\beta}$ at all when $\boldsymbol{H}'\boldsymbol{X}=\boldsymbol{0}$. Thus, the estimator (10.11) does not exist, but in the case of (10.18) it is nevertheless possible to determine $\boldsymbol{\beta}$ (a scalar in this case) exactly! The reason is that (10.18) implies $\boldsymbol{K}'\boldsymbol{y}=y_3$ and $\boldsymbol{K}'\boldsymbol{X}=1$ so that (10.17) states that y_3 equals the parameter. Ultimately, this results from the zero value of the third diagonal element of $\boldsymbol{V}$ in (10.18) and the non-zero third element of $\boldsymbol{X}$.

Under the assumptions stated in the discussion following eq. (10.10), the estimator (10.11) exists when $\boldsymbol{H}'\boldsymbol{X}$ has full column rank and it is a best linear unbiased estimator of $\boldsymbol{\beta}$ when $\boldsymbol{K}'\boldsymbol{X}=\boldsymbol{0}$ [so that (10.17) is not a real constraint on $\boldsymbol{\beta}$]. A proof of the latter statement follows in the next paragraph. If $\boldsymbol{K}'\boldsymbol{X}$ is a non-zero matrix, (10.17) is a linear constraint on $\boldsymbol{\beta}$ which should be incorporated in the estimation procedure; see Theil (1971, sec. 6.8).

We can write any linear estimator of $\boldsymbol{\beta}$ as

$$\tilde{\boldsymbol{\beta}}=\left[\boldsymbol{A}+(\boldsymbol{X}'\boldsymbol{V}^{+}\boldsymbol{X})^{-1}\boldsymbol{X}'\boldsymbol{V}^{+}\right]\boldsymbol{y}, \tag{10.19}$$

where $\boldsymbol{A}$ is some $K\times n$ matrix consisting of non-stochastic elements. By substituting $\boldsymbol{X\beta}+\boldsymbol{\varepsilon}$ for $\boldsymbol{y}$ in (10.19) and taking the expectation we find that the unbiasedness of $\tilde{\boldsymbol{\beta}}$ requires

$$\boldsymbol{AX}=\boldsymbol{0}, \tag{10.20}$$

so that $\tilde{\boldsymbol{\beta}}-\boldsymbol{\beta}=[\boldsymbol{A}+(\boldsymbol{X}'\boldsymbol{V}^{+}\boldsymbol{X})^{-1}\boldsymbol{X}'\boldsymbol{V}^{+}]\boldsymbol{\varepsilon}$ and the covariance matrix of $\tilde{\boldsymbol{\beta}}$ equals

$$\mathcal{V}(\tilde{\boldsymbol{\beta}})=\sigma^2\left[\boldsymbol{A}+(\boldsymbol{X}'\boldsymbol{V}^{+}\boldsymbol{X})^{-1}\boldsymbol{X}'\boldsymbol{V}^{+}\right]\boldsymbol{V}\left[\boldsymbol{A}'+\boldsymbol{V}^{+}\boldsymbol{X}(\boldsymbol{X}'\boldsymbol{V}^{+}\boldsymbol{X})^{-1}\right]. \tag{10.21}$$

For $\boldsymbol{A}=\boldsymbol{0}$ we have $\tilde{\boldsymbol{\beta}}=\hat{\boldsymbol{\beta}}$ in view of (10.19). Thus, using $\boldsymbol{V}^{+}\boldsymbol{V}\boldsymbol{V}^{+}=\boldsymbol{V}^{+}$ and (10.21), we obtain:

$$\mathcal{V}(\hat{\boldsymbol{\beta}})=\sigma^2(\boldsymbol{X}'\boldsymbol{V}^{+}\boldsymbol{X})^{-1}, \tag{10.22}$$

which is a generalization of (2.14). The excess of (10.21) over (10.22) equals a multiple σ^2 of $\boldsymbol{AVA}'+\boldsymbol{AVV}^{+}\boldsymbol{X}(\boldsymbol{X}'\boldsymbol{V}^{+}\boldsymbol{X})^{-1}+(\boldsymbol{X}'\boldsymbol{V}^{+}\boldsymbol{X})^{-1}\boldsymbol{X}'\boldsymbol{V}^{+}\boldsymbol{VA}'$. But $\boldsymbol{AVV}^{+}\boldsymbol{X}=\boldsymbol{0}$ so that $\mathcal{V}(\tilde{\boldsymbol{\beta}})-\mathcal{V}(\hat{\boldsymbol{\beta}})=\sigma^2\boldsymbol{AVA}'$, which is positive semidefinite and thus establishes that $\hat{\boldsymbol{\beta}}$ is best linear unbiased. To verify that $\boldsymbol{A}\boldsymbol{V}\boldsymbol{V}^{+}\boldsymbol{X}$ is a zero matrix we use (10.12) and (10.14) in

$$\boldsymbol{V}\boldsymbol{V}^{+} = \boldsymbol{H}\boldsymbol{D}\boldsymbol{H}'\boldsymbol{H}\boldsymbol{D}^{-1}\boldsymbol{H}' = \boldsymbol{H}\boldsymbol{H}' = \boldsymbol{I} - \boldsymbol{K}\boldsymbol{K}',$$

where the last two steps are based on (10.7) and (10.8). So, using (10.20) and $\boldsymbol{K}'\boldsymbol{X} = \boldsymbol{0}$ also, we have

$$\boldsymbol{A}\boldsymbol{V}\boldsymbol{V}^{+}\boldsymbol{X} = \boldsymbol{A}\boldsymbol{X} - \boldsymbol{A}\boldsymbol{K}\boldsymbol{K}'\boldsymbol{X} = \boldsymbol{0} - \boldsymbol{0} = \boldsymbol{0}.$$

The matrix $\mathcal{V}(\tilde{\boldsymbol{\beta}}) - \mathcal{V}(\hat{\boldsymbol{\beta}}) = \sigma^2 \boldsymbol{A}\boldsymbol{V}\boldsymbol{A}'$ is obviously zero when we select $\boldsymbol{A} = \boldsymbol{0}$, but it may also be zero for $\boldsymbol{A} \neq \boldsymbol{0}$ when $\boldsymbol{V}$ is singular, which suggests that there is no unique best linear unbiased estimator of $\boldsymbol{\beta}$. This is not true, however; if the estimator (10.11) exists, i.e. if $\boldsymbol{H}'\boldsymbol{X}$ has full column rank, it is the unique best linear unbiased estimator of $\boldsymbol{\beta}$ when $\boldsymbol{K}'\boldsymbol{X} = \boldsymbol{0}$. The reason is that $\boldsymbol{A}\boldsymbol{V}\boldsymbol{A}' = \boldsymbol{0}$ is equivalent to $\mathcal{E}[\boldsymbol{A}\boldsymbol{\varepsilon}(\boldsymbol{A}\boldsymbol{\varepsilon})'] = \boldsymbol{0}$ so that $\boldsymbol{A}\boldsymbol{\varepsilon}$ is a zero vector with unit probability. Using (10.20) also, we obtain $\boldsymbol{A}\boldsymbol{y} = \boldsymbol{A}(\boldsymbol{X}\boldsymbol{\beta} + \boldsymbol{\varepsilon}) = \boldsymbol{0}$, which in conjunction with (10.19) shows that the best linear unbiased estimator of $\boldsymbol{\beta}$ must be of the form (10.11), even though $\boldsymbol{A}$ may be a non-zero matrix.

10.4. *Deleting an equation from an allocation model*

The Moore–Penrose inverse can also be conveniently used to prove that when we estimate an N-equation allocation system such as (6.1), we can simply delete one of the N equations (it does not matter which). The clue is the fact that each equation can be obtained by adding the $N-1$ others. We prove this below for an allocation system which is linear in the parameters. The strategy of the proof will be to start with GLS estimation of $N-1$ equations with a non-singular disturbance covariance matrix, followed by adding the deleted equation (so that the disturbance covariance matrix becomes singular), and then proving that the resulting estimator (10.11) is identical to the original GLS estimator.

We can formulate the problem in the following more general way. Let $\boldsymbol{y} = \boldsymbol{X}\boldsymbol{\beta} + \boldsymbol{\varepsilon}$ have a non-singular covariance matrix $\mathcal{V}(\boldsymbol{\varepsilon}) = \sigma^2\boldsymbol{V}$ of order $n \times n$. We premultiply by a matrix $\boldsymbol{B}$ of order $(n + n') \times n$ and rank n:

$$\boldsymbol{B}\boldsymbol{y} = \boldsymbol{B}\boldsymbol{X}\boldsymbol{\beta} + \boldsymbol{B}\boldsymbol{\varepsilon}. \tag{10.23}$$

For example, take $\boldsymbol{B}' = [\boldsymbol{I} \quad \boldsymbol{C}]$, which means that we add to the original n observations n' linear combinations of these observations. The covariance matrix of $\boldsymbol{B}\boldsymbol{\varepsilon}$ takes the singular form $\sigma^2\boldsymbol{B}\boldsymbol{V}\boldsymbol{B}'$. Thus, the matrix $\boldsymbol{V}$ of the previous subsection becomes $\boldsymbol{B}\boldsymbol{V}\boldsymbol{B}'$ here, while $\boldsymbol{X}$ becomes $\boldsymbol{B}\boldsymbol{X}$. We conclude that condition

$K'X=0$ is now $K'(BX)=0$, where K is a matrix whose n' columns are characteristic vectors of BVB' corresponding to the zero roots: $(BVB')K=0$ and $K'K=I$. Evidently, a sufficient condition for K is $B'K=0$ and $K'K=I$. Such a K can be obtained as a matrix whose columns are characteristic vectors of the idempotent matrix $I-B(B'B)^{-1}B'$ corresponding to the unit roots:

$$\left[I-B(B'B)^{-1}B'\right]K=K.$$

The GLS estimator (10.11) of β in (10.23) is then

$$\left[X'B'(BVB')^{+}BX\right]^{-1}X'B'(BVB')^{+}By. \tag{10.24}$$

This is identical to $(X'V^{-1}X)^{-1}X'V^{-1}y$, and hence to the GLS estimator obtained from the original n observations, because

$$B'(BVB')^{+}B=V^{-1},$$

which follows from $BVB'(BVB')^{+}BVB'=BVB'$ [see (10.1)] premultiplied by $V^{-1}(B'B)^{-1}B'$ and postmultiplied by $B(B'B)^{-1}V^{-1}$. It is unnecessary to check the condition that $H'(BX)$ has full column rank, H being a matrix whose n columns are characteristic vectors of BVB' corresponding to the positive roots. The reason is that the estimator (10.24) would not exist if the condition were not satisfied, whereas we know that (10.24) equals $(X'V^{-1}X)^{-1}X'V^{-1}y$.

Appendix A: Linear independence and related topics

Consider a matrix $V=[v_1 \dots v_n]$ and a linear combination Vc of its n columns. The vectors $v_1,\dots,v_n$ are said to be *linearly independent* if $Vc=0$ implies $c=0$, i.e. if there exists no non-trivial linear combination of the v_i's which is a zero vector. For example, the columns of the 2×2 unit matrix are linearly independent because

$$c_1\begin{bmatrix}1\\0\end{bmatrix}+c_2\begin{bmatrix}0\\1\end{bmatrix}=\begin{bmatrix}c_1\\c_2\end{bmatrix}=\begin{bmatrix}0\\0\end{bmatrix} \quad \text{implies} \quad c_1=c_2=0,$$

but $v_1=[1\quad 0]'$ and $v_2=[2\quad 0]'$ are not linearly independent because $c_1v_1+c_2v_2=0$ if (for example) $c_1=2$ and $c_2=-1$.

For any $m\times n$ matrix A the column rank is defined as the largest number of linearly independent columns, and the row rank as the largest number of linearly independent rows. It can be shown that these two ranks are always equal; we can thus speak about the *rank* r of A, which obviously satisfies $r\leqslant m,n$. If all columns (rows) of $\boldsymbol{A}$ are linearly independent, $\boldsymbol{A}$ is said to have full column (row) rank. For any $\boldsymbol{A}$, the ranks of $\boldsymbol{A}$, $\boldsymbol{A}'$, $\boldsymbol{A}'\boldsymbol{A}$, and $\boldsymbol{A}\boldsymbol{A}'$ are all equal. Also, the rank of $\boldsymbol{AB}$ is at most equal to the rank of $\boldsymbol{A}$ and that of $\boldsymbol{B}$. For example,

$$\begin{bmatrix} 1 & 0 \\ 2 & 0 \end{bmatrix}\begin{bmatrix} 0 \\ 1 \end{bmatrix} = \begin{bmatrix} 0 \\ 0 \end{bmatrix},$$

which illustrates that the rank of $\boldsymbol{AB}$ may be smaller than both that of $\boldsymbol{A}$ and that of $\boldsymbol{B}$. (A zero matrix has zero rank.) If $\boldsymbol{A}$ is square ($n \times n$) and has full rank ($r = n$), it is called *non-singular* and its inverse $\boldsymbol{A}^{-1}$ exists.

For any vector $\boldsymbol{v} = [v_i]$, its *length* is defined as the positive square root of $\boldsymbol{v}'\boldsymbol{v} = \sum_i v_i^2$. If $\boldsymbol{v}'\boldsymbol{v} = 1$, $\boldsymbol{v}$ is said to have *unit length*. The *inner product* of two vectors $\boldsymbol{v} = [v_i]$ and $\boldsymbol{w} = [w_i]$ consisting of the same number of elements is defined as $\boldsymbol{v}'\boldsymbol{w} = \sum_i v_i w_i$. If $\boldsymbol{v}'\boldsymbol{w} = 0$, $\boldsymbol{v}$ and $\boldsymbol{w}$ are called *orthogonal vectors*.

A square matrix $\boldsymbol{X}$ which satisfies $\boldsymbol{X}' = \boldsymbol{X}^{-1}$ is called an *orthogonal matrix*. Premultiplication of $\boldsymbol{X}' = \boldsymbol{X}^{-1}$ by $\boldsymbol{X}$ gives $\boldsymbol{XX}' = \boldsymbol{I}$, which shows that each row of $\boldsymbol{X}$ has unit length and that any two rows of $\boldsymbol{X}$ are orthogonal vectors. Postmultiplication of $\boldsymbol{X}' = \boldsymbol{X}^{-1}$ by $\boldsymbol{X}$ gives $\boldsymbol{X}'\boldsymbol{X} = \boldsymbol{I}$ so that each column of $\boldsymbol{X}$ (each row of $\boldsymbol{X}'$) also has unit length and any two columns of $\boldsymbol{X}$ are also orthogonal vectors.

Appendix B: The independence transformation

The independence transformation is based on three axioms, the first being the invariance of total expenditure. Let a dollar spent on observed input j result in r_{ij} dollars spent on transformed input i, so that the expenditure on i equals $\sum_j r_{ij} p_j q_j$ and the total expenditure on all transformed inputs equals $\sum_j \left(\sum_i r_{ij}\right) p_j q_j$, which must be identical to $\sum_j p_j q_j$ because of the invariance postulated. Therefore, $\sum_i r_{ij} = 1$ for each j, or

$$\boldsymbol{\iota}'\boldsymbol{R} = \boldsymbol{\iota}', \tag{B.1}$$

where $\boldsymbol{R} = [r_{ij}]$. By dividing the expenditure $\sum_j r_{ij} p_j q_j$ on transformed input i by total expenditure C (which is invariant) we obtain the factor share $f_{\mathrm{T}i}$ of this input. Therefore, $f_{\mathrm{T}i} = \sum_j r_{ij} f_j$, or

$$\boldsymbol{F}_{\mathrm{T}}\boldsymbol{\iota} = \boldsymbol{RF\iota}, \tag{B.2}$$

where $\boldsymbol{F}_{\mathrm{T}}$ is the diagonal factor share matrix of the transformed inputs.

The second axiom states that the logarithmic price and quantity changes of the transformed inputs are linear combinations of their observed counterparts, $\boldsymbol{\pi}_{\mathrm{T}} = \boldsymbol{S}_1\boldsymbol{\pi}$ and $\boldsymbol{\kappa}_{\mathrm{T}} = \boldsymbol{S}_2\boldsymbol{\kappa}$, so that the associated Divisia indexes are invariant. The Divisia volume index is $d(\log Q) = \iota' F \kappa$ and its transformed counterpart is $\iota' F_T \kappa_T = \iota' F_T S_2 \kappa = \iota' F(R' S_2)\kappa$ [see (B.2)]. Thus, the invariance of this index requires $R' S_2 = I$ or $S_2 = (R')^{-1}$. We can proceed similarly for the price index $\iota' F \pi$, which yields the same result for S_1, so that the price and quantity transformations use the same matrix, $\pi_T = S\pi$ and $\kappa_T = S\kappa$, where $S = (R')^{-1}$. See remark (3) below for the case of a singular R.

The third axiom diagonalizes Θ. We premultiply (8.15) by R, which yields $RF\kappa = RFR'S\kappa = RFR'\kappa_T$ on the left because $R'S = I$ and $S\kappa = \kappa_T$. When we proceed similarly on the right and use (B.1) also, we obtain:

$$RFR'\kappa_T = (\iota' F \kappa)(R\Theta R')\iota - \psi R\Theta R'[I - \iota\iota'(R\Theta R')]\pi_T, \tag{B.3}$$

which is an allocation system of the same form as (8.15), with logarithmic price and quantity changes π_T and κ_T, provided RFR' on the left equals the diagonal factor share matrix F_T. The new normalized price coefficient matrix is $R\Theta R'$, which occurs in the same three places in (B.3) as Θ does in (8.15). [The matrix $R\Theta R'$ is indeed normalized because $\iota' R\Theta R'\iota = \iota'\Theta\iota = 1$ follows from (B.1).] Therefore, $RFR' = F_T$ and $R\Theta R' =$ diagonal are the conditions under which (B.3) is an input independent allocation system. These are two conditions on R, which must satisfy (B.1) also.

We proceed to prove that

$$R = (X^{-1}\iota)_\Delta X' \tag{B.4}$$

satisfies these three conditions, with X defined in (8.16) and $(X^{-1}\iota)_\Delta$ in (7.9). First, $\iota' R = \iota'$ is true for (B.4) in view of (7.10). Secondly, $RFR' = (X^{-1}\iota)_\Delta X' F X (X^{-1}\iota)_\Delta = (X^{-1}\iota)^2_\Delta$ [see (8.16)] so that

$$RFR' = F_T = (X^{-1}\iota)^2_\Delta = \text{diagonal}. \tag{B.5}$$

Thirdly, using $\Theta = (X')^{-1}\Lambda X^{-1}$ [see (8.16)], we have $R\Theta R' = (X^{-1}\iota)^2_\Delta \Lambda$, which is diagonal. So, using (B.5) also and premultiplying (B.3) by $(RFR')^{-1} = (X^{-1}\iota)^{-2}_\Delta$, we obtain:

$$\kappa_T = (\iota' F\kappa)\Lambda\iota - \psi\Lambda(I - \iota\iota' R\Theta R')\pi_T, \tag{B.6}$$

which is the matrix version of (8.17). The expression which is subtracted in parentheses in the substitution term of (B.6) represents the deflation by the Frisch price index, which is invariant. To prove this we note that the marginal share vector of the transformed inputs equals $R\Theta\iota = R\theta$ in view of the real-income term in (B.3) and $R'\iota = \iota$; the invariance of the Frisch index then follows from $(R\theta)'\pi_T = \theta' R' S\pi = \theta'\pi$.

The expenditure on transformed input i equals $r_{ij}p_jq_j$ dollars insofar as it originates with observed input j. By dividing this amount by total expenditure C we obtain $r_{ij}f_j$, which is thus the factor share of transformed input i insofar as it originates with observed input j. This $r_{ij}f_j$ is an element of the matrix $\boldsymbol{RF}$, to be written $\boldsymbol{T}$:

$$\boldsymbol{T} = \boldsymbol{RF} = (\boldsymbol{X}^{-1}\boldsymbol{\iota})_\Delta \boldsymbol{X}^{-1}, \tag{B.7}$$

where the last step is based on (B.4) and $\boldsymbol{F} = (\boldsymbol{X}')^{-1}\boldsymbol{X}^{-1}$ [see (8.16)]. Postmultiplication of (B.7) by $\boldsymbol{\iota}$ gives $\boldsymbol{T\iota} = \boldsymbol{RF\iota} = \boldsymbol{F}_{\mathrm{T}}\boldsymbol{\iota}$ [see (B.2)]; hence the row sums of $\boldsymbol{T}$ are the factor shares of the transformed inputs. Also, $\boldsymbol{\iota}'\boldsymbol{T} = \boldsymbol{\iota}'\boldsymbol{RF} = \boldsymbol{\iota}'\boldsymbol{F}$, so that the column sums of $\boldsymbol{T}$ are the factor shares of the observed inputs. Note that (B.7) and its row and column sums confirm the results on the composition matrix. Note further that $\boldsymbol{F} = (\boldsymbol{X}')^{-1}\boldsymbol{X}^{-1}$ and $\boldsymbol{\Theta} = (\boldsymbol{X}')^{-1}\boldsymbol{\Lambda}\boldsymbol{X}^{-1}$ [see (8.16)] imply that the price elasticity matrix $-\psi\boldsymbol{F}^{-1}\boldsymbol{\Theta}$ in (8.19) equals $-\psi\boldsymbol{X\Lambda X}^{-1}$. So, using (B.7) also, we have

$$\boldsymbol{T}(-\psi\boldsymbol{F}^{-1}\boldsymbol{\Theta}) = -\psi(\boldsymbol{X}^{-1}\boldsymbol{\iota})_\Delta\boldsymbol{\Lambda}\boldsymbol{X}^{-1} = -\psi\boldsymbol{\Lambda}(\boldsymbol{X}^{-1}\boldsymbol{\iota})_\Delta\boldsymbol{X}^{-1} = -\psi\boldsymbol{\Lambda T}.$$

Combining the first and last member yields $\boldsymbol{t}_i'(-\psi\boldsymbol{F}^{-1}\boldsymbol{\Theta}) = -\psi\lambda_i\boldsymbol{t}_i'$, where $\boldsymbol{t}_i'$ is the ith row of $\boldsymbol{T}$, or

$$\boldsymbol{t}_i'[-\psi\boldsymbol{F}^{-1}\boldsymbol{\Theta} - (-\psi\lambda_i)\boldsymbol{I}] = \boldsymbol{0}.$$

Therefore, each row of the composition matrix is a characteristic row vector of the (asymmetric) price elasticity matrix of the observed inputs.

We conclude with the following remarks.

(1) Although the solution (B.4) satisfies all three conditions, it is not unique. However, it may be shown that this solution is unique up to premultiplication by an arbitrary permutation matrix; such a multiplication affects only the order in which the transformed inputs are listed.

(2) We proved in the second paragraph that the price and quantity transformations take the form $\boldsymbol{\pi}_{\mathrm{T}} = \boldsymbol{S\pi}$ and $\boldsymbol{\kappa}_{\mathrm{T}} = \boldsymbol{S\kappa}$, where $\boldsymbol{S} = (\boldsymbol{R}')^{-1}$. It thus follows from (B.1) that $\boldsymbol{S}^{-1}\boldsymbol{\iota} = \boldsymbol{\iota}$ or $\boldsymbol{S\iota} = \boldsymbol{\iota}$. Therefore, when the prices of the observed inputs change proportionately, $\boldsymbol{\pi}$ being a scalar multiple k of $\boldsymbol{\iota}$, the price of each transformed input changes in the same proportion: $\boldsymbol{\pi}_{\mathrm{T}} = \boldsymbol{S}(k\boldsymbol{\iota}) = k\boldsymbol{S\iota} = k\boldsymbol{\iota}$. The quantities have the same desirable property.

(3) It follows from (B.4) that $\boldsymbol{R}$ is singular when $(\boldsymbol{X}^{-1}\boldsymbol{\iota})_\Delta$ contains a zero diagonal element, and from (B.5) that this implies a zero factor share of one of the transformed inputs. In that case $\boldsymbol{S} = (\boldsymbol{R}')^{-1}$ does not exist. The simplest way to interpret this situation is by means of a perturbation of the firm's technology so

that the ith element of $\boldsymbol{X}^{-1}\boldsymbol{\iota}$ converges from a small non-zero value to zero. It may be shown that $\mathrm{d}(\log p_{\mathrm{T}i})$ then increases beyond bounds. If the increase is toward ∞, transformed input i is priced out of the market; if it is toward $-\infty$, i becomes a free good; in both cases no money is spent on i in the limit. In particular, if (5.12) represents a homothetic technology, $N-1$ elements of $\boldsymbol{X}^{-1}\boldsymbol{\iota}$ are zero and all observed inputs collectively behave as one transformed input with unitary Divisia elasticity; no money is spent on any transformed input whose Divisia elasticity differs from 1. For proofs of these results see Theil (1977).

(4) The independence transformation was first formulated by Brooks (1970) and axiomatically justified by Theil (1975–76, ch. 12) for a finite-change version of the consumer demand system (5.22). The λ_i's are then income elasticities of transformed consumer goods. Rossi (1979a) proved that when all observed goods are specific substitutes, the transformed good with the smallest income elasticity represents all observed goods positively and that all other transformed goods are contrasts between observed goods similar to T_2 in (8.26). The former transformed good serves to satisfy the consumer's wants associated with the observed goods in the least luxurious manner; this result is of particular interest when the transformation is applied to a group of goods which satisfy similar wants such as different brands of the same type of commodity.[9] For an integrated exposition of the independence transformation in consumption and production theory see Theil (1980, ch. 10–11).

Appendix C: Rational random behavior

To verify (9.4) we write $p^*(\boldsymbol{x})=p(\boldsymbol{x})+\delta f(\boldsymbol{x})$ for some density function other that the $p(\boldsymbol{x})$ of (9.4), where δ is independent of $\boldsymbol{x}$ so that $f(\cdot)$ must satisfy

$$\int_J f(\boldsymbol{x})\,\mathrm{d}x_1\ldots\mathrm{d}x_k=0. \tag{C.1}$$

The information I^* and the expected loss $\bar{l}^*$ associated with $p^*(\cdot)$ are

$$I^*=\int_J[p(\boldsymbol{x})+\delta f(\boldsymbol{x})]\log\frac{p(\boldsymbol{x})+\delta f(\boldsymbol{x})}{p_0(\boldsymbol{x})}\,\mathrm{d}x_1\ldots\mathrm{d}x_k, \tag{C.2}$$

$$\bar{l}^*=\bar{l}+\delta\int_J l(\boldsymbol{x},\bar{\boldsymbol{x}})f(\boldsymbol{x})\,\mathrm{d}x_1\ldots\mathrm{d}x_k, \tag{C.3}$$

[9]When $\boldsymbol{\Theta}$ is block-diagonal, so is $\boldsymbol{X}$ in (8.16), which means that the independence transformation can be applied to each block separately. We have a block-diagonal $\boldsymbol{\Theta}$ under block independence. See the end of Section 6 for block independent inputs; the extension to block independent consumer goods is straightforward.

where $\bar{l}$ is the expected loss (9.3) associated with the $p(\cdot)$ of (9.4). We apply a Taylor expansion to (C.2) as a function of δ:

$$I^* = I + k_1\delta + \tfrac{1}{2}k_2\delta^2 + O(\delta^3), \tag{C.4}$$

where I is the information (9.2) associated with (9.4) and

$$k_1 = \int_J f(\boldsymbol{x}) \log \frac{p(\boldsymbol{x})}{p_0(\boldsymbol{x})} \mathrm{d}x_1 \dots \mathrm{d}x_k, \tag{C.5}$$

$$k_2 = \int_J \frac{[f(\boldsymbol{x})]^2}{p(\boldsymbol{x})} \mathrm{d}x_1 \dots \mathrm{d}x_k. \tag{C.6}$$

Next we apply a Taylor expansion to $c(I^*)$, writing $c' = \mathrm{d}c/\mathrm{d}I$ and $c'' = \mathrm{d}^2c/\mathrm{d}I^2$ for the derivatives of $c(\cdot)$ at the I of (9.4):

$$c(I^*) = c(I) + \delta k_1 c' + \tfrac{1}{2}\delta^2(k_2c' + k_1^2c'') + O(\delta^3)$$

and we add this to (C.3):

$$c(I^*) + \bar{l}^* = c(I) + \bar{l} + \delta\left[k_1c' + \int_J l(\boldsymbol{x}, \bar{\boldsymbol{x}}) f(\boldsymbol{x}) \mathrm{d}x_1 \dots \mathrm{d}x_k\right] + \tfrac{1}{2}\delta^2(k_2c' + k_1^2c'') + O(\delta^3) \tag{C.7}$$

For $c(I) + \bar{l}$ to be minimal we require the coefficient of δ in (C.7) to vanish for any $f(\cdot)$ satisfying (C.1) and that of δ^2 to be positive. The latter condition is satisfied when $c' > 0$ and $c'' \geqslant 0$ (a positive nondecreasing marginal cost of information) because (C.6) implies $k_2 > 0$ when $f(\boldsymbol{x}) \neq 0$ for some $\boldsymbol{x}$. It follows from (C.5) that the former condition amounts to a zero value of

$$\int_J \left[c' \log \frac{p(\dot{\boldsymbol{x}})}{p_0(\boldsymbol{x})} + l(\boldsymbol{x}, \bar{\boldsymbol{x}})\right] f(\boldsymbol{x}) \mathrm{d}x_1 \dots \mathrm{d}x_k.$$

This integral vanishes, given (C.1), when the expression in brackets is a constant independent of $\boldsymbol{x}$, which yields (9.4) directly.

To prove the asymptotic results for small c' we take the logarithm of (9.4):

$$\log p(\boldsymbol{x}) = \text{constant} + \log p_0(\boldsymbol{x}) - \frac{l(\boldsymbol{x}, \bar{\boldsymbol{x}})}{c'}, \tag{C.8}$$

and substitute $\bar{\boldsymbol{x}}$ for $\boldsymbol{x}$, using (9.1):

$$\log p(\bar{\boldsymbol{x}}) = \text{constant} + \log p_0(\bar{\boldsymbol{x}}).$$

Since the constants in these equations are equal, subtraction yields

$$\log \frac{p(\bar{x})}{p(x)} = \log \frac{p_0(\bar{x})}{p_0(x)} + \frac{l(x,\bar{x})}{c'}. \tag{C.9}$$

It follows from (9.1) that as $c' \to 0$ the last term increases beyond bounds for any $x \neq \bar{x}$, so that the same holds for $p(\bar{x})/p(x)$ on the left. Hence, as $c' \to 0$ the density $p(x)$ becomes zero for each $x \neq \bar{x}$ and the random decision with density function (9.4) thus converges in probability to $\bar{x}$.

To verify the asymptotic distribution (9.7), we define

$$v = \frac{1}{\sqrt{c'}}(x - \bar{x}), \tag{C.10}$$

so that $l(x, \bar{x}) = l(\bar{x} + \sqrt{c'}\, v, \bar{x})$. We apply a Taylor expansion to $l(x, \bar{x})/c'$, using (9.6):

$$\begin{aligned} \frac{l(x,\bar{x})}{c'} &= \frac{1}{c'}\left[\tfrac{1}{2}(\sqrt{c'}\, v)'A(\sqrt{c'}\, v) + O(c'^{3/2})\right] \\ &= \tfrac{1}{2} v'Av + O(\sqrt{c'}). \end{aligned} \tag{C.11}$$

We assume that $p_0(x)$ is positive and differentiable around $\bar{x}$. Hence, we can apply a Taylor expansion to $\log p_0(x)$ and write it as $\log p_0(\bar{x})$ plus a linear remainder term in $x - \bar{x}$. Therefore, in view of (C.10),

$$\log p_0(x) = \log p_0(\bar{x}) + O(\sqrt{c'}),$$

which in conjunction with (C.8) and (C.11) shows that $\log p(x)$ equals a constant minus $\frac{1}{2}v'Av$ plus two remainder terms which both converge to zero as $c' \to 0$. The result (9.7) is then obtained by substitution from (C.10) for v in $\frac{1}{2}v'Av$.

We obtain (9.11) from (9.7) by using the budget or technology constraint to eliminate one of the decision variables from the criterion function. Let these variables be the quantities bought by the consumer; it was shown by Theil (1975–76, sec. 2.6–2.7) that (9.7) then yields variances and covariances of the form

$$\operatorname{cov}(q_i, q_j) = -k\left(\lambda u^{ij} - \frac{\lambda}{\partial\lambda/\partial M}\frac{\partial q_i}{\partial M}\frac{\partial q_j}{\partial M}\right), \tag{C.12}$$

where $k > 0$ is proportional to the marginal cost of information c'. A comparison of (C.12) with (3.12) shows that $\operatorname{cov}(q_i, q_j)$ is proportional to the substitution component (specific plus general) of $\partial q_i/\partial p_j$. We obtain (9.11) from (C.12) by rearrangements required by the left variable in (5.5).

The results (9.11) and (9.14) for the multiproduct firm and the stochastic independence of the input demand disturbances and the output supply disturbances were derived by Laitinen and Theil (1978). Reference should also be made to Bowman et al. (1979) and to Rossi (1979b, 1979c) for a comparison of rational random behavior and search theory.

References

Aitken, A. C. (1935) "On Least Squares and Linear Combination of Observations", *Proceedings of the Royal Society of Edinburgh*, 55, 42–48.

Albert, A. (1972) *Regression and the Moore–Penrose Pseudoinverse*. New York: Academic Press.

Barbosa, F. de H. (1975) "Rational Random Behavior: Extensions and Applications", Doctoral dissertation, The University of Chicago.

Barten, A. P. (1964) "Consumer Demand Functions under Conditions of Almost Additive Preferences", *Econometrica*, 32, 1–38.

Barten, A. P. (1977) "The Systems of Consumer Demand Functions Approach: A Review", *Econometrica*, 45, 23–51.

Barten, A. P. and E. Geyskens (1975) "The Negativity Condition in Consumer Demand", *European Economic Review*, 6, 227–260.

Bellman, R. (1960) *Introduction to Matrix Analysis*. New York: McGraw-Hill Book Company.

Ben-Israel, A. and T. N. E. Greville (1974) *Generalized Inverses: Theory and Applications*. New York: John Wiley and Sons.

Boullion, T. L. and P. L. Odell (1971) *Generalized Inverse Matrices*. New York: John Wiley and Sons.

Bowman, J. P., K. Laitinen and H. Theil (1979) "New Results on Rational Random Behavior", *Economics Letters*, 2, 201–204.

Brooks, R. B. (1970) "Diagonalizing the Hessian Matrix of the Consumer's Utility Function", Doctoral dissertation, The University of Chicago.

Brown, A. and A. Deaton (1972) "Surveys in Applied Economics: Models of Consumer Behaviour", *Economic Journal*, 82, 1145–1236.

Dhrymes, P. J. (1978) *Mathematics for Econometrics*. New York: Springer-Verlag.

Divisia, F. (1925) "L'indice monétaire et la théorie de la monnaie", *Revue d'Economie Politique*, 39, 980–1008.

Frisch, R. (1932) *New Methods of Measuring Marginal Utility*. Tübingen: J. C. B. Mohr.

Goldberger, A. S. (1959) *Impact Multipliers and Dynamic Properties of the Klein–Goldberger Model*. Amsterdam: North-Holland Publishing Company.

Graybill, F. A. (1969) *Introduction to Matrices with Applications in Statistics*. Belmont, Cal.: Wadsworth Publishing Company.

Hadley, G. (1961) *Linear Algebra*. Reading, Mass.: Addison-Wesley Publishing Company.

Hall, R. E. (1973) "The Specification of Technology with Several Kinds of Output", *Journal of Political Economy*, 81, 878–892.

Houthakker, H. S. (1960) "Additive Preferences", *Econometrica*, 28, 244–257; Errata, 30 (1962) 633.

Kadane, J. B. (1971) "Comparison of *k*-Class Estimators when the Disturbances are Small", *Econometrica*, 39, 723–737.

Laitinen, K. (1980) *A Theory of the Multiproduct Firm*. Amsterdam: North-Holland Publishing Company.

Laitinen, K. and H. Theil (1978) "Supply and Demand of the Multiproduct Firm", *European Economic Review*, 11, 107–154.

Laitinen, K. and H. Theil (1979) "The Antonelli Matrix and the Reciprocal Slutsky Matrix", *Economics Letters*, 3, 153–157.

Phlips, L. (1974) *Applied Consumption Analysis*. Amsterdam: North-Holland Publishing Company.
Pringle, R. M. and A. A. Rayner (1971) *Generalized Inverse Matrices with Applications to Statistics*. London: Charles Griffin and Co.
Rao, C. R. and S. K. Mitra (1971) *Generalized Inverse of Matrices and Its Applications*. New York: John Wiley and Sons.
Rossi, P. E. (1979a) "The Independence Transformation of Specific Substitutes and Specific Complements", *Economics Letters*, 2, 299–301.
Rossi, P. E. (1979b) "The Cost of Search and Rational Random Behavior", *Economics Letters*, 3, 5–8.
Rossi, P. E. (1979c) "Asymptotic Search Behavior Based on the Weibull Distribution", *Economics Letters*, 3, 211–213.
Theil, H. (1967) *Economics and Information Theory*. Amsterdam: North-Holland Publishing Company.
Theil, H. (1971) *Principles of Econometrics*. New York: John Wiley and Sons.
Theil, H. (1975–76) *Theory and Measurement of Consumer Demand*, 2 vols. Amsterdam: North-Holland Publishing Company.
Theil, H. (1977) "The Independent Inputs of Production", *Econometrica*, 45, 1303–1327.
Theil, H. (1980) *The System-Wide Approach to Microeconomics*. Chicago: University of Chicago Press.
Theil, H. and K. W. Clements (1980) "Recent Methodological Advances in Economic Equation Systems", *American Behavioral Scientist*, 23, 789–809.
Theil, H. and K. Laitinen (1979) "Maximum Likelihood Estimation of the Rotterdam Model under Two Different Conditions", *Economics Letters*, 2, 239–244.
Zellner, A. (1962) "An Efficient Method of Estimating Seemingly Unrelated Regressions and Tests for Aggregation Bias", *Journal of the American Statistical Association*, 57, 348–368.

Chapter 2

STATISTICAL THEORY AND ECONOMETRICS

ARNOLD ZELLNER*

University of Chicago

Contents

*Research for this paper was financed by NSF Grant SES 7913414 and by income from the H.G.B. Alexander Endowment Fund, Graduate School of Business, University of Chicago. Part of this work was done while the author was on leave at the National Bureau of Economic Research and the Hoover Institution, Stanford, California.

Handbook of Econometrics, Volume I, Edited by Z. Griliches and M.D. Intriligator

1. Introduction and overview

Econometricians, as well as other scientists, are engaged in learning from their experience and data – a fundamental objective of science. Knowledge so obtained may be desired for its own sake, for example to satisfy our curiosity about aspects of economic behavior and/or for use in solving practical problems, for example to improve economic policymaking. In the process of learning from experience and data, description and generalization both play important roles. Description helps us to understand "what is" and what is to be explained by new or old economic generalizations or theories. Economic generalizations or theories are not only instrumental in obtaining understanding of past data and experience but also are most important in predicting as yet unobserved outcomes, for example next year's rate of inflation or the possible effects of an increase in government spending. Further, the ability to predict by use of economic generalizations or theories is intimately related to the formulation of economic policies and solution of problems involving decision-making.

The methods and procedures by which econometricians and other scientists learn from their data and use such knowledge to predict as yet unobserved data and outcomes and to solve decision problems constitutes the subject-matter of statistical theory. A principal objective of work in statistical theory is to formulate methods and procedures for learning from data, making predictions, and solving decision problems that are generally applicable, work well in applications and are consistent with generally accepted principles of scientific induction and decision-making under uncertainty. Current statistical theories provide a wide range of methods applicable to many problems faced by econometricians and other scientists. In subsequent sections, many theories and methods will be reviewed.

It should be appreciated that probability theory plays a central role in statistical theory. Indeed, it is generally hypothesized that economic and other types of data are generated stochastically, that is by an assumed probabilistic process or model. This hypothesis is a key one which has been found fruitful in econometrics and other sciences. Thus, under this assumption, most operational economic generalizations or theories are probabilistic, and in view of this fact some elements of probability theory and probabilistic models will be reviewed in Section 2.

The use of probability models as a basis for economic generalizations and theories is widespread. If the form of a probability model and the values of its parameters were known, one could use such a model to make probability statements about as yet unobserved outcomes, as for example in connection with games of chance. When the probability model's form and nature are completely known, using it in the way described above is a problem in *direct probability*. That is, with complete knowledge of the probability model, it is usually "direct" to compute probabilities associated with as yet unobserved possible outcomes.

On the other hand, problems usually encountered in science are not those of *direct probability* but those of *inverse probability*. That is, we usually observe data which are assumed to be the outcome or output of some probability process or model, the properties of which are not completely known. The scientist's problem is to infer or learn the properties of the probability model from observed data, a problem in the realm of inverse probability. For example, we may have data on individuals' incomes and wish to determine whether they can be considered as drawn or generated from a normal probability distribution or by some other probability distribution. Questions like these involve considering alternative probability models and using observed data to try to determine from which hypothesized probability model the data probably came, a problem in the area of *statistical analysis of hypotheses*. Further, for any of the probability models considered, there is the problem of using data to determine or estimate the values of parameters appearing in it, a problem of *statistical estimation*. Finally, the problem of using probability models to make predictions about as yet unobserved data arises, a problem of *statistical prediction*. Aspects of these major topics, namely (a) *statistical estimation*, (b) *statistical prediction*, and (c) *statistical analysis of hypotheses*, will be reviewed and discussed.

Different statistical theories can yield different solutions to the problems of statistical estimation, prediction, and analysis of hypotheses. Also, different statistical theories provide different justifications for their associated methods. Thus, it is important to understand alternative statistical theories, and in what follows attention is given to features of several major statistical theories. Selected examples are provided to illustrate differences in results and rationalizations of them provided by alternative statistical theories.[1]

Finally, in a concluding section a number of additional topics are mentioned and some concluding remarks are presented.

2. Elements of probability theory

We commence this section with a discussion of the elements of probability models for observations. Then a brief consideration of several views and definitions of probability is presented. A summary of some properties of axiom systems for probability theory is given followed by a review of selected results from probability theory that are closely related to the formulation of econometric models.

[1]For valuable discussions of many of the statistical topics considered below and references to the statistical literature, see Kruskal and Tanur (1978).

2.1. *Probability models for observations*

As remarked in Section 1, probability models are generally employed in analyzing data in econometrics and other sciences. Lindley (1971, p. 1) explains: "The mathematical model that has been found convenient for most statistical problems contains a *sample space* X of elements x endowed with an appropriate σ-field of sets over which is given a family of probability measures. These measures are indexed by a quantity θ, called a *parameter* belonging to the parameter space Θ. The values x are referred to variously as the *sample*, *observations* or *data*." Thus, a statistical model is often represented by the triplet $(X, \Theta, \mathrm{P}_\theta(x))$, where X denotes the sample space, Θ the parameter space, and $\mathrm{P}_\theta(x)$ the probability measures indexed by the parameter θ belonging to Θ. For many problems, it is possible to describe a probability measure $\mathrm{P}_\theta(x)$ through its *probability density function* (pdf), denoted by $p(x|\theta)$. For example, in the case of n independent, identically distributed normal observations, the sample space is $-\infty < X_i < \infty$, $i = 1,2,\dots,n$, the pdf is $p(\boldsymbol{x}|\boldsymbol{\theta}) = \prod_{i=1}^n f(x_i|\boldsymbol{\theta})$, where $\boldsymbol{x}' = (x_1, x_2, \dots, x_n)$ and $f(x_i|\boldsymbol{\theta}) = (2\pi\sigma^2)^{-1/2}\exp\{-(x_i-\mu)^2/2\sigma^2\}$, with $\boldsymbol{\theta}' = (\mu, \sigma)$ the parameter vector, and Θ: $-\infty < \mu < \infty$ and $0 < \sigma < \infty$ the parameter space.

While the triplet $(X, \Theta, p(x|\theta))$ contains the major elements of many statistical problems, there are additional elements that are often relevant. Some augment the triplet by introducing a decision space D of elements d and a non-negative convex loss function $L(d,\theta)$. For example, in connection with the normal model described at the end of the previous paragraph, $L(d,\theta)$ might be the following "squared error" loss function, $L(d,\mu) = c(\mu - d)^2$, where c is a given positive constant and $d = d(x)$ is some estimate of μ belonging to a decision space D. The problem is then to choose a d from D that is in some sense optimal relative to the loss function $L(d,\mu)$.

An element that is added to the triplet $(X, \Theta, p(x|\theta))$ in the Bayesian approach is a probability measure defined on the σ-field supported by Θ that we assume can be described by its pdf, $\pi(\theta)$. Usually $\pi(\theta)$ is called a prior pdf and its introduction is considered by many to be the distinguishing feature of the Bayesian approach. The prior pdf, $\pi(\theta)$, represents initial or prior information about the possible values of θ available *before* obtaining the observations.

In summary, probabilistic models for observations can be represented by $(X, \Theta, p(x|\theta))$. This representation is often extended to include a loss function, $L(d,\theta)$, a decision space D, and a prior pdf, $\pi(\theta)$. As will be seen in what follows, these elements play very important roles in statistical theories.

Statistical theories indicate how the data x are to be employed to make inferences about the possible value of θ, an estimation problem, how to test hypotheses about possible values of θ, e.g. $\theta = 0$ vs. $\theta \neq 0$, a testing problem, and how to make inferences about as yet unobserved or future data, the problem of prediction. Also, very importantly, the information in the data x can be employed to explore the adequacy of the probability model $(X, \Theta, p(x|\theta))$, a procedure called "model criticism" by Box (1980). Model criticism, involving diagnostic checks of the form of $p(x|\theta)$ and other assumptions, may indicate that the model is adequate or inadequate. If the model is found to be inadequate, then it has to be reformulated. Thus, work with probability models for observations has an important iterative aspect, as emphasized by Box (1976), Box and Tiao (1973), and Zellner (1975, 1979). While some elements of the theory of hypothesis testing are relevant for this process of iterating in on an adequate probability model for the observations, additional research is needed to provide formalizations of the many heuristic procedures employed by applied researchers to iterate in on an adequate model, that is a model that achieves the objectives of an analysis. See Leamer (1978) for a thoughtful discussion of related issues.

2.2. *Definitions of probability*

Above, we have utilized the word "probability" without providing a definition of it. Many views and/or definitions of probability have appeared in the literature. On this matter Savage (1954) has written, "It is unanimously agreed that statistics depends somehow on probability. But, as to what probability is and how it is connected with statistics, there has seldom been such complete disagreement and breakdown of communication since the Tower of Babel. Doubtless, much of the disagreement is merely terminological and would disappear under sufficiently sharp analysis" (p. 2). He distinguishes three main classes of views on the interpretation of probability as follows (p. 3):

> *Objectivistic* views hold that some repetitive events, such as tosses of a penny, prove to be in reasonably close agreement with the mathematical concept of independently repeated random events, all with the same probability. According to such views, evidence for the quality of agreement between the behavior of the repetitive event and the mathematical concept, and for the magnitude of the probability that applies (in case any does), is to be obtained by observation of some repetitions of the event, and from no other source whatsoever.
>
> *Personalistic* views hold that probability measures the confidence that a particular individual has in the truth of a particular proposition, for example, the proposition that it will rain tomorrow. These views postulate that the individual concerned is in some way "reasonable", but they do not deny the possibility that two reasonable individuals faced with the same evidence may have different degrees of confidence in the truth of the same proposition.
>
> *Necessary* views hold that probability measures the extent to which one set of propositions, out of logical necessity and apart from human opinion, confirms the truth of another. They are generally regarded by their holders as

extensions of logic, which tells when one set of propositions necessitates the truth of another.

While Savage's classification scheme probably will not satisfy all students of the subject, it does bring out critical differences of alternative views regarding the meaning of probability. To illustrate further, consider the following definitions of probability, some of which are reviewed by Jeffreys (1967, p. 369 ff).

1. Classical or Axiomatic Definition

If there are n possible alternatives, for m of which a proposition denoted by p is true, then the probability of p is m/n.

2. Venn Limit Definition

If an event occurs a large number of times, then the probability of p is the limit of the ratio of the number of times when p will be true to the whole number of trials, when the number of trials tends to infinity.

3. Hypothetical Infinite Population Definition

An actually infinite number of possible trials is assumed. Then the probability of p is defined as the ratio of the number of cases where p is true to the whole number.

4. Degree of Reasonable Belief Definition

Probability is the degree of confidence that we may reasonably have in a proposition.

5. Value of an Expectation Definition

If for an individual the utility of the uncertain outcome of getting a sum of s dollars or zero dollars is the same as getting a sure payment of one dollar, the probability of the uncertain outcome of getting s dollars is defined to be $u(1)/u(s)$, where $u(\cdot)$ is a utility function. If $u(\cdot)$ can be taken proportional to returns, the probability of receiving s is $1/s$.

Jeffreys notes that Definition 1 appeared in work of De Moivre in 1738 and of J. Neyman in 1937; that R. Mises advocates Definition 2; and that Definition 3 is usually associated with R. A. Fisher. Definition 4 is Jeffreys' definition (1967, p. 20) and close to Keynes' (1921, p. 3). The second part of Definition 5 is involved in Bayes (1763). The first part of Definition 5, embodying utility comparisons, is central in work by Ramsey (1931), Savage (1954), Pratt, Raiffa and Schlaifer (1964), DeGroot (1970), and others.

Definition 1 can be shown to be defective, as it stands, by consideration of particular examples – see Jeffreys (1967, p. 370ff.). For example, if a six-sided die is thrown, by Definition 1 the probability that any particular face will appear is $1/6$. Clearly, this will not be the case if the die is biased. To take account of this possibility, some have altered the definition to read, "If there are n *equally likely* possible alternatives, for m of which p is true, then the probability of p is m/n." If the phrase "equally likely" is interpreted as "equally probable," then the definition is defective since the term to be defined is involved in the definition. Also, Jeffreys (1967) points out in connection with the Venn Limit Definition that, "For continuous distributions there are an infinite number of possible cases, and the definition makes the probability, on the face of it, the ratio of two infinite numbers and therefore meaningless" (p. 371). He states that attempts by Neyman and Cramér to avoid this problem are unsatisfactory.

With respect to Definitions 2 and 3, it must be recognized that they are both non-operational. As Jeffreys (1967) puts it:

> No probability has ever been assessed in practice, or ever will be, by counting an infinite number of trials or finding the limit of a ratio in an infinite series. Unlike the first definition, which gave either an unacceptable assessment or numerous different assessments, these two give none at all. A definite value is got on them *only* by making a hypothesis about what the result would be. The proof even of the existence is impossible. On the limit definition, without some rule restricting the possible orders of occurrence, there might be no limit at all. The existence of the limit is taken as a postulate by Mises, whereas Venn hardly considered it as needing a postulate...the necessary existence of the limit denies the possibility of complete randomness, which would permit the ratio in an infinite series to tend to no limit (p. 373, fn. omitted).

Further, with respect to Definition 3, Jeffreys (1967) writes, "On the infinite population definition, any finite probability is the ratio of two infinite numbers and therefore is indeterminate" (p. 373, fn. omitted). Thus, Definitions 2 and 3 have some unsatisfactory features.

Definition 4, which defines probability in terms of the degree of confidence that we may reasonably have in a proposition, is a primitive concept. It is primitive in the sense that it is not produced by any axiom system; however, it is accepted by some on intuitive grounds. Furthermore, while nothing in the definition requires that probability be measurable, say on a scale from zero to one, Jeffreys (1967, p. 19) does assume measurability [see Keynes (1921) for a critique of this assumption] and explores the consequences of the use of this assumption in a number of applications. By use of this definition, it becomes possible to associate probabilities with hypotheses, e.g. it is considered meaningful to state that the probability that the marginal propensity to consume is between 0.7 and 0.9 is 0.8, a statement that is meaningless in terms of Definitions 1–3. However, the meaningfulness of the metric employed for such statements is a key issue which, as with many measurement problems will probably be resolved by noting how well procedures based on particular metrics perform in practice.

Definition 5, which views probability as a subjective, personal concept, involves the use of a benefit or utility metric. For many, but not all problems, one or the other of these metrics may be considered satisfactory in terms of producing useful results. There may, however, be some scientific and other problems for which a utility or loss (negative utility) function formulation is inadequate.

In summary, several definitions of probability have been briefly reviewed. While the definitions are radically different, it is the case that operations with probabilities, reviewed below, are remarkably similar even though their interpretations differ considerably.

2.3. *Axiom systems for probability theory*

Various axiom systems for probability theory have appeared in the literature. Herein Jeffreys' axiom system is reviewed that was constructed to formalize inductive logic in such a way that it includes deductive logic as a special limiting case. His definition of probability as a degree of reasonable belief, Definition 4 above, allows for the fact that in induction, propositions are usually uncertain and only in the limit may be true or false in a deductive sense. With respect to probability, Jeffreys, along with Keynes (1921), Uspensky (1937), Rényi (1970), and others, emphasizes that all probabilities are conditional on an initial information set, denoted by A. For example, let B represent the proposition that a six will be observed on a single flip of a coin. The degree of reasonable belief or probability that one attaches to B depends on the initial information concerning the shape and other features of the coin and the way in which it is thrown, all of which are included in the initial information set, A. Thus, the probability of B is written $\mathrm{P}(B|A)$, a conditional probability. The probability of B without specifying A is meaningless. Further, failure to specify A clearly and precisely can lead to confusion and meaningless results; for an example, see Jaynes (1980).

Let propositions be denoted by $A, B, C, \ldots$. Then Jeffreys' (1967) first four axioms are:

Axiom 1 (Comparability)

Given A, B is either more, equally, or less probable than C, and no two of these alternatives can be true.

Axiom 2 (Transitivity)

If A, B, C, and D are four propositions and given A, B is more probable than C and C is more probable than D, then given A, B is more probable than D.

Axiom 3 (Deducibility)

All propositions deducible from a proposition A have the same probability given A. All propositions inconsistent with A have the same probability given data A.

Axiom 4

If given A, B_1 and B_2 cannot both be true and if, given A, C_1 and C_2 cannot both be true, and if, given A, B_1 and C_1 are equally probable and B_2 and C_2 are equally probable, then given A, B_1 or B_2 and C_1 or C_2 are equally probable.

Jeffreys states that Axiom 4 is required to prove the addition rule given below. DeGroot (1970, p. 71) introduces a similar axiom.

Axiom 1 permits the comparison of probabilities or degrees of reasonable belief or confidence in alternative propositions. Axiom 2 imposes a transitivity condition on probabilities associated with alternative propositions *based on a common information set A*. The third axiom is needed to insure consistency with deductive logic in cases in which inductive and deductive logic are both applicable. The extreme degrees of probability are certainty and impossibility. As Jeffreys (1967, p. 17) mentions, certainty on data A and impossibility on data A "do not refer to mental certainty of any particular individual, but to the relations of deductive logic..." expressed by B is deducible from A and not-B is deducible from A, or in other words, A entails B in the former case and A entails not-B in the latter. Axiom 4 is needed in what follows to deal with pairs of exclusive propositions relative to a given information set A. Jeffreys' Theorem 1 extends Axiom 4 to relate to more than two pairs of exclusive propositions with the same probabilities on the same data A.

Jeffreys (1967) remarks that it has "...not yet been assumed that probabilities can be expressed by numbers. I do not think that the introduction of numbers is strictly necessary to the further development; but it has the enormous advantage that it permits us to use mathematical technique. Without it, while we might obtain a set of propositions that would have the same meanings, their expression would be much more cumbersome" (pp. 18–19). Thus, Jeffreys recognizes that it is possible to have a "non-numerical" theory of probability but opts for a "numerical" theory in order to take advantage of less cumbersome mathematics and that he believes leads to propositions with about the same meanings.

The following notation and definitions are introduced to facilitate further analysis.

Definitions[2]

(1) $\sim A$ means "not-A", that is, A is false.
(2) $A \cap B$ means "A and B", that is, both A and B are true. The proposition $A \cap B$ is also termed the "intersection" or the "joint assertion" or "conjunction" or "logical product" of A and B.

[2]These are presented in Jeffreys (1967, pp. 17–18) using different notation.

(3) $A \cup B$ means "A or B", that is, at least one of A and B is true. The proposition $A \cup B$ is also referred to as the "union" or "disjunction" or "logical sum" of A and B.
(4) $A \cap B \cap C \cap D$ means "A and B and C and D", that is, A, B, C, and D are all true.
(5) $A \cup B \cup C \cup D$ means "A or B or C or D", that is, at least one of A, B, C, and D is true.
(6) Propositions B_i, $i = 1,2,\dots,n$, are said to be *exclusive* on data A if not more than one of them can be true given A.
(7) Propositions B_i, $i = 1,2,\dots,n$, are said to be *exhaustive* on data A if at least one of them must be true given A.

Note that a set of propositions can be both exclusive and exhaustive. Also, for example, Axiom 4 can be restated using the above notation and concepts as:

Axiom 4

If B_1 and B_2 are exclusive and C_1 and C_2 are exclusive, given data A, and if, given A, B_1 and C_1 are equally probable and B_2 and C_2 are equally probable, then given A, $B_1 \cup B_2$ and $C_1 \cup C_2$ are equally probable.

At this point in the development of his axiom system, Jeffreys introduces numbers associated with or measuring probabilities by the following conventions.

Convention 1

A larger number is assigned to the more probable proposition (and therefore equal numbers to equally probable propositions).

Convention 2

If, given A, B_1 and B_2 are exclusive, then the number assigned on data A to "B_1 or B_2", that is $B_1 \cup B_2$, is the sum of those assigned to B_1 and to B_2.

The following axiom is needed to insure that there are enough numbers available to associate with probabilities.

Axiom 5

The set of possible probabilities on given data, ordered in terms of the relation "more probable than" can be put into a one-one correspondence with a set of real numbers in increasing order.

It is important to realize that the notation $\mathrm{P}(B|A)$ stands for the number associated with the probability of the proposition B on data A. The number expresses or measures the reasonable degree of confidence in B given A, that is, the probability of B given A, but is not identical to it.

The following theorem that Jeffreys derives from Axiom 3 and Convention 2 relates to the numerical assessment of impossible propositions.

Theorem 2

If proposition A entails $\sim B$, then $P(B|A)=0$. Thus, Theorem 2 in conjunction with Convention 1 provides the result that all probability numbers are $\geqslant 0$.

The number associated with certainty is given in the following convention.

Convention 3

If A entails B, then $P(B|A)=1$.

The use of 1 to represent certainty is a pure convention. In some cases it is useful to allow numerical probabilities to range from 0 to ∞ rather than from 0 to 1. On given data, however, it is necessary to use the same numerical value for certainty.

Axiom 6

If $A\cap B$ entails C, then $P(B\cap C|A)=P(B|A)$.

That is, given A throughout, if B is false, then $B\cap C$ is false. If B is true, since $A\cap B$ entails C, C is true and therefore $B\cap C$ is true. Similarly, if $B\cap C$ is true, it entails B and if $B\cap C$ is false, B must be false on data A since if it were true, $B\cap C$ would be true. Thus, it is impossible, given A, that either B or $B\cap C$ should be true without the other. This is an extension of Axiom 3 that results in all equivalent propositions having the same probability on given data.

Theorem 3

If B and C are equivalent in the sense that each entails the other, then each entails $B\cap C$, and the probabilities of B and C on any given data must be equal. Similarly, if $A\cap B$ entails C and $A\cap C$ entails B, $P(B|A)=P(C|A)$, since both are equal to $P(B\cap C|A)$.

A theorem following from Theorem 3 is:

Theorem 4

$P(B|A)=P(B\cap C|A)+P(B\cap\sim C|A)$.

Further, since $P(B\cap\sim C|A)\geqslant 0$, $P(B|A)\geqslant P(B\cap C|A)$. Also, by using $B\cup C$ for B in Theorem 4, it follows that $P(B\cup C|A)\geqslant P(C|A)$.

The *addition rule* for numerical probabilities is given by Theorem 5.

Theorem 5

If B and C are two propositions, not necessarily exclusive on data A, the addition rule is given by

$$P(B|A)+P(C|A)=P(B\cap C|A)+P(B\cup C|A).$$

It follows that

$$\mathrm{P}(B \cup C|A) \leqslant \mathrm{P}(B|A)+\mathrm{P}(C|A),$$

since $\mathrm{P}(B \cap C|A) \geqslant 0$. Further, if B and C are exclusive, then $\mathrm{P}(B \cap C|A) = 0$ and $\mathrm{P}(B \cup C|A) = \mathrm{P}(B|A)+\mathrm{P}(C|A)$.

Theorems 4 and 5 together express upper and lower bounds on the possible values of $\mathrm{P}(B \cup C|A)$ irrespective of exclusiveness, that is

$$\max[\mathrm{P}(B|A),\mathrm{P}(C|A)] \leqslant \mathrm{P}(B \cup C|A) \leqslant \mathrm{P}(B|A)+\mathrm{P}(C|A).$$

Theorem 6

If $B_1, B_2,\dots,B_n$ are a set of equally probable and exclusive alternatives on data A, and if Q and R are unions of two subsets of these alternatives, of numbers m and n, then $\mathrm{P}(Q|A)/\mathrm{P}(R|A) = m/n$. This follows from Convention 2 since $\mathrm{P}(Q|A) = ma$ and $\mathrm{P}(R|A) = na$, where $a = \mathrm{P}(B_i|A)$ for all i.

Theorem 7

Under the conditions of Theorem 6, if $B_1, B_2,\dots,B_n$ are exhaustive on data A, and R denotes their union, then R is entailed by A and by Convention 3, $\mathrm{P}(R|A) = 1$, and it follows that $\mathrm{P}(Q|A) = m/n$.

As Jeffreys notes, Theorem 7 is virtually Laplace's rule stated at the beginning of his *Théorie Analytique*. Since R entails itself and is a possible value of A, it is possible to write, $\mathrm{P}(Q|R) = m/n$, which Jeffreys (1967) interprets as, "...given that a set of alternatives are equally probable, exclusive and exhaustive, the probability that some one of any subset is true is the ratio of the number in that subset to the whole number of possible cases" (p. 23). Also, Theorem 6 is consistent with the possibility that the number of alternatives is infinite, since it requires only that Q and R shall be finite subsets.

Theorems 6 and 7 indicate how to assess the ratios of probabilities and their actual values. Such assessments will always be rational fractions that Jeffreys calls R-probabilities. If all probabilities were R-probabilities, there would be no need for Axiom 5. But, as Jeffreys points out, many propositions are of a form that a magnitude capable of a continuous range of values lies within a specified part of the range and it may not be possible to express them in the required form. He explains how to deal with this problem and puts forward the following theorem:

Theorem 8

Any probability can be expressed by a real number. For a variable z that can assume a continuous set of values, given A, the probability that z's value is less than a given value z_0 is $\mathrm{P}(z < z_0|A) = F(z_0)$, where $F(z_0)$ is referred to as the cumulative probability density function (cdf). If $F(z_0)$ is differentiable, $\mathrm{P}(z_0 < z$

$< z_0 + \mathrm{d}z | A) = f(z_0)\mathrm{d}z + 0(\mathrm{d}z)$, where $f(z_0) = F'(z_0)$ is the probability density function (pdf) and this last expression gives the probability that z lies in the interval z_0 to $z_0 + \mathrm{d}z$.

Theorem 9

If Q is the union of a set of exclusive alternatives, given A, and if R and S are subsets of Q (possibly overlapping), and if the alternatives in Q are all equally probable on data A and also on data $R \cap A$, then

$$\mathrm{P}(R \cap S | A) = \mathrm{P}(R | A)\mathrm{P}(S | R \cap A)/\mathrm{P}(R | R \cap A).$$

Note that if Convention 3 is employed, $\mathrm{P}(R | R \cap A) = 1$, since $R \cap A$ entails R and then Theorem 9 reads

$$\mathrm{P}(R \cap S | A) = \mathrm{P}(R | A)\mathrm{P}(S | R \cap A).$$

In other words, given A throughout, the probability that the true proposition is in the intersection of R and S is equal to the probability that it is in R times the probability that it is in S, given that it is in R. Theorem 9 involves the assumption that the alternatives in Q are equally probable, both given A and also given $R \cap A$. Jeffreys notes that it has not been possible to relax this assumption in proving Theorem 9. However, he regards this theorem as suggestive of the simplest rule that relates probabilities based on different data, here denoted by A and $R \cap A$, and puts forward the following axiom.

Axiom 7

$\mathrm{P}(B \cap C | A) = \mathrm{P}(B | A)\mathrm{P}(C | B \cap A)/\mathrm{P}(B | B \cap A)$.

If Convention 3 is used in Axiom 7, $\mathrm{P}(B | B \cap A) = 1$ and

$$\mathrm{P}(B \cap C | A) = \mathrm{P}(B | A)\mathrm{P}(C | B \cap A),$$

which is the *product rule*. Thus, the product rule relates to probability statements regarding the logical product or intersection $B \cap C$, often also written as BC, while the addition or sum rule relates to probability statements regarding the logical sum or union, $B \cup C$.

Since Axiom 7 is just suggested by Theorem 9, Jeffreys shows that it holds in several extreme cases and concludes that, "The product rule may therefore be taken as general unless it can be shown to lead to contradictions" (p. 26). Also, he states, "When the probabilities...are chances, they satisfy the product rule automatically" (p. 51).[3]

[3]Jeffreys (1967) defines "chance" as follows: "If $q_1, q_2, \ldots, q_n$ are a set of alternatives, mutually exclusive and exhaustive on data r, and if the probabilities of p given any of them and r are the same, each of these probabilities is called the chance of p on data r" (p. 51).

In general, if $\mathrm{P}(C|B \cap A) = \mathrm{P}(C|A)$, B is said to be irrelevant to or independent of C, given A. In this special case, the product rule can be written as $\mathrm{P}(B \cap C|A) = \mathrm{P}(B|A)\mathrm{P}(C|A)$, a form of the product rule that is valid only when B is irrelevant to or independent of C.

Theorem 10

If $q_1, q_2, \dots, q_n$ are a set of alternatives, A, the information already available and x, some additional information, then the ratio

$$\frac{\mathrm{P}(q_r|x \cap A)\mathrm{P}(q_r|q_r \cap A)}{\mathrm{P}(q_r|A)\mathrm{P}(x|q_r \cap A)}$$

is the same for all the q_r.

If we use Convention 3, $\mathrm{P}(q_r|q_r \cap A) = 1$, then

$$\mathrm{P}(q_r|x \cap A) = c\,\mathrm{P}(q_r|A)\mathrm{P}(x|q_r \cap A),$$

where $1/c = \sum_r \mathrm{P}(q_r|A)\mathrm{P}(x|q_r \cap A)$. This is the *principle of inverse probability* or *Bayes' Theorem*, first given in Bayes (1763). The result can also be expressed as

$$\text{Posterior probability} \propto (\text{prior probability})(\text{likelihood function})$$

where $\propto$ denotes "is proportional to", $\mathrm{P}(q_r|x \cap A)$ is the posterior probability, $\mathrm{P}(q_r|A)$ is the prior probability, and $\mathrm{P}(x|q_r \cap A)$ is the likelihood function.

In general terms, Jeffreys describes the use of Bayes' Theorem by stating that if several hypotheses $q_1, q_2, \dots, q_n$ are under consideration and, given background information A there is no reason to prefer any one of them, the prior probabilities, $\mathrm{P}(q_r|A)$, $r = 1, 2, \dots, n$, will be taken equal. Then, the most probable hypothesis after observing the data x, that is, the one with the largest $\mathrm{P}(q_r|x \cap A)$, will be that with the largest value for $\mathrm{P}(x|q_r \cap A)$, the likelihood function. On the other hand, if the data x are equally probable on each hypothesis, the prior views with respect to alternative hypotheses, whatever they were, will be unchanged. Jeffreys (1967) concludes: "The principle will deal with more complicated circumstances also; the immediate point is that it does provide us with what we want, a formal rule in general accord with common sense, that will guide us in our use of experience to decide between hypotheses" (p. 29). Jeffreys (1967, p. 43) also shows that the theory can be utilized to indicate how an inductive inference can approach certainty, though it cannot reach it, and thus explains the usual confidence that most scientists have in inductive inference. These conclusions are viewed as controversial by those who question the appropriateness of introducing prior probabilities and associating probabilities with hypotheses. It appears that these issues can only be settled by close comparative study of the results yielded by various approaches to statistical inference as Anscombe (1961) and

Jaynes (1974), among others, have emphasized. See also Fisher (1959), Savage et al. (1962), Lindley (1971), Barnett (1973), Cox and Hinkley (1974), Rothenberg (1975), and Zellner (1975) for further discussion of these and related issues.

Jeffreys' theory of probability, described above, did not mention utility or benefit since it is primarily a theory of what it is reasonable to believe. However, Jeffreys notes that his theory permits him to define the expectation of a function $u(\tilde{x})$, say a utility function, as follows:

$$E[u(\tilde{x})|A] = \sum_{i=1}^{m} u(x_i)\mathrm{P}(x_i|A),$$

when $\tilde{x}$ is a discrete random variable with possible values $x_1, x_2, \dots, x_m$, or by

$$E[u(\tilde{x})|A] = \int_a^b u(x) f(x|A)\,\mathrm{d}x,$$

when $\tilde{x}$ is a continuous random variable with probability density function $f(x|A)$, $a < x < b$. Thus, utility considerations can be incorporated in Jeffreys' theory. On the other hand, Bayes (1763), Ramsey (1931), Savage (1954), Pratt, Raiffa and Schlaifer (1964), and some others take as the fundamental idea that of expectation of benefit or utility, as pointed out in Definition 5 in Section 2.2. Generally speaking, it is assumed that expectations of benefit or utility can be placed in an order. As Jeffreys (1967) points out, Bayes speaks in terms of monetary stakes, and would say that 1/100 chance of receiving \$100 is as valuable as a certainty of receiving \$1. Bayes' *definition* of a probability of 1/100 would be that it is the probability such that the value of the chance of \$100 is the same as the value of a certain \$1. This requires a postulate that the value of the expectation is proportional to the value to be received. In more modern treatments of the problem, as mentioned in Definition 5 and described below, utility considerations enter into the definition of probability.

Raiffa and Schlaifer (1961) remark that "...when one is forced to compare utility characteristics because one is forced to act, a few basic principles of logically consistent behavior necessarily lead to the introduction of a *weighting function* over Θ [the parameter space]...if this weighting function is normalized it has all the properties of a *probability measure* on Θ..." (p. 25). Raiffa and Schlaifer (1961, pp. 25–27) provide a simple informal proof of this result based on three assumptions for the case that Θ is finite with elements θ_i, $i = 1,2,\dots,r$. The utility characteristic of any decision d is represented by $u = [u(d,\theta_1), u(d,\theta_2),\dots,u(d,\theta_r)] \equiv (u_1, u_2,\dots,u_r)$. Their three assumptions are:

Assumption 1

Let $u = (u_1, u_2,\dots,u_r)$ and $v = (v_1, v_2,\dots,v_r)$ be the utility characteristics of decision functions d_a and d_b, respectively. If $u_i \geqslant v_i$ for all i and if $u_i > v_i$ for some i, then d_a is preferred to d_b.

Assumption 2

Indifference surfaces extend smoothly from boundary to boundary in the r-space, R, of the u_i's, $i = 1, 2, \ldots, r$.

Assumption 3

If d_a, d_b, and d_c are three decision functions such that d_a and d_b are indifferent, then given any p such that $0 \leqslant p \leqslant 1$, a mixed strategy that selects d_a with objective probability p and d_c with objective probability $1 - p$ is indifferent to a strategy which selects d_b with objective probability p and d_c with objective probability $1 - p$.

From these three assumptions that are discussed at length in Luce and Raiffa (1957), Raiffa and Schlaifer (1961) show that "...the decision-maker's indifference surfaces must be parallel hyper-planes with a common normal going into the interior of the first orthant, from which it follows that all utility characteristics $u = (u_1, u_2, \ldots, u_r)$ in R can in fact be ranked by an index which applies a predetermined set of weights $P = (P_1, P_2, \ldots, P_r)$ to their r components" (p. 25). That is $\sum_{i=1}^{r} P_i u_i$ and $\sum_{i=1}^{r} P_i v_i$ can be employed to rank decision functions d_a and d_b with utility characteristics u and v, respectively, where the P_i's are the predetermined set of non-negative weights that can be normalized and have all the properties of a probability measure on Θ. Since the P_i's are intimately related to a person's indifference surfaces, it is clear why some refer to the normalized P_i's as "personal probabilities". For more discussion of this topic see Blackwell and Girshick (1954, ch. 4), Luce and Raiffa (1957, ch. 13), Savage (1954, ch. 1–5), and DeGroot (1970, ch. 6–7). Further, Jeffreys (1967) remarks:

> The difficulty about the separation of propositions into disjunctions of equally possible and exclusive alternatives is avoided by this [Bayes, Ramsey et al.] treatment, but is replaced by difficulties concerning additive expectations [and utility comparisons]. These are hardly practical ones in either case... In my method expectation would be defined in terms of value [or utility] and probability; in theirs [Bayes, Ramsey et al.], probability is defined in terms of values [or utilities] and expectations. The actual propositions [of probability theory] are of course identical (p. 33).

2.4. *Random variables and probability models*

As mentioned in Section 1, econometric and statistical models are usually stochastic, involving random variables. In this section several important probability models are reviewed and some of their properties are indicated.

2.4.1. Random variables

A random variable (rv) will be denoted by $\tilde{x}$. There are discrete, continuous, and mixed rvs. If $\tilde{x}$ is a discrete rv, it can, by assumption, just assume particular values, that is $\tilde{x}=x_j$, $j=0,1,2,\ldots,m$, where m can be finite or infinite and the x_j's are given values, for example $x_0=0$, $x_1=1$, $x_2=2$, and $x_m=m$. These x_j values may represent quantitative characteristics, for example the number of purchases in a given period or qualitative characteristics, for example different occupational categories. If $\tilde{x}$ can assume just two possible values, it is termed a dichotomous rv, if three, a trichotomous rv, if more than three, a polytymous rv. For quantitative discrete rvs the ordering $x_0<x_1<x_2<\cdots<x_m$ is meaningful, while for some qualitative discrete rvs such an ordering is meaningless.

A continuous rv, $\tilde{x}$, such that $a<\tilde{x}<b$, where a and b are given values, possibly with $a=-\infty$ and/or $b=\infty$, can assume a continuum of values in the interval a to b, the range of the rv. A mixed rv, $\tilde{x}$, $a<\tilde{x}<b$, assumes a continuum of values over part of its range, say for $a<\tilde{x}<c$, and discrete values over the remainder of its range, $c\leqslant\tilde{x}<b$. Some econometric models incorporate just continuous or just discrete rvs while others involve mixtures of continuous, discrete, and mixed rvs.

2.4.2. Discrete random variables

For a discrete rv, $\tilde{x}$, that can assume the values, $x_1,x_2,\ldots,x_m$, where the x_j's are distinct and exhaustive, the probability that $\tilde{x}=x_j$, given the initial information A, is

$$\mathrm{P}(\tilde{x}=x_j|A)=p_j,\qquad j=1,2,\ldots,m, \tag{2.1}$$

with

$$p_j\geqslant 0\quad\text{and}\quad\sum_{j=1}^{m}p_j=1. \tag{2.2}$$

The collection of p_j's in (2.1), subject to (2.2), defines the probability mass function (pmf) for the discrete rv, $\tilde{x}$. A plot of the p_j's against the x_j's may be unimodal, bimodal, U-shaped, J-shaped, uniform $(p_1=p_2=\cdots=p_m)$, etc. If it is unimodal, the pmf's modal value is the value of $\tilde{x}$ associated with the largest p_j. Further, the mean of $\tilde{x}$, denoted by $\mu_1'\equiv E\tilde{x}$ is:

$$\mu_1'=E\tilde{x}=\sum_{j=1}^{m}p_jx_j. \tag{2.3}$$

In general, the zeroth and higher order moments about zero are given by

$$\mu_r' = E\tilde{x}^r = \sum_{j=1}^{m} p_j x_j^r, \qquad r = 0,1,2,\dots. \tag{2.4}$$

Moments about the mean, $E\tilde{x}$, called central moments, are given by

$$\mu_r = E(\tilde{x} - E\tilde{x})^r = \sum_{j=1}^{m} p_j (x_j - E\tilde{x})^r, \qquad r = 1,2,\dots. \tag{2.5}$$

Note that $\mu_1 = 0$ and that μ_2, defined as the variance of $\tilde{x}$, $V(\tilde{x})$, is

$$\begin{aligned}\mu_2 \equiv V(\tilde{x}) &= E(\tilde{x} - E\tilde{x})^2 \\ &= \sum_{j=1}^{m} p_j (x_j - E\tilde{x})^2.\end{aligned} \tag{2.6}$$

From the first line of (2.6), $V(\tilde{x}) = E\tilde{x}^2 - 2(E\tilde{x})^2 + (E\tilde{x})^2 = E\tilde{x}^2 - (E\tilde{x})^2 = \mu_2' - (\mu_1')^2$. Similar relations can be obtained that relate higher order central moments to moments about zero. Also, the following unitless measures of skewness are available to characterize unimodal pmfs: sk = (mean − mode)$/\mu_2^{1/2}$, $\beta_1 = \mu_3^2/\mu_2^3$, and $\gamma = \mu_3/\mu_2^{3/2}$. Further, a unitless measure of kurtosis for unimodal pmfs is given by $\beta_2 = \mu_4/\mu_2^2$ that frequently measures the peakedness of a pmf although its value is sensitive to the characteristics of the tails of a pmf.

If $\tilde{x}$ is a discrete rv assuming only non-negative integer values with $P(\tilde{x} = j) = p_j$, $j = 0,1,2,\dots$, then $p(z) = \sum p_j z^j$ is called the probability generating function with the obvious property that $p(1) = 1$, given property (2.6). Further, the αth derivative of $p(z)$, evaluated at $z = 0$, is just $\alpha! p_\alpha$, where $\alpha! = \alpha(\alpha - 1)(\alpha - 2)\dots 1$, and it is in this sense that the probability generating function "generates" the probabilities, the p_j's of a pmf.

If in $p(z) = \sum_{j=0}^{\infty} p_j z^j$, $z = \mathrm{e}^t$, the result is the moment-generating function associated with a pmf, $p_0, p_1, p_2, \dots$, namely

$$p(\mathrm{e}^t) = \sum_{j=0}^{\infty} \mu_j' t^j / j!, \tag{2.7}$$

where the μ_j' are given in (2.4). The expression in (2.7) is obtained by noting that

$$\mathrm{e}^{\alpha t} = \sum_{j=0}^{\infty} (\alpha t)^j / j!$$

and

$$\sum_{\alpha=0}^{\infty} p_\alpha e^{\alpha t} = \sum_{j=0}^{\infty} \sum_{\alpha=0}^{\infty} p_\alpha \alpha^j t^j / j! = \sum_{j=0}^{\infty} \left(\sum_{\alpha=0}^{\infty} p_\alpha \alpha^j \right) t^j / j! = \sum_{\alpha=0}^{\infty} \mu'_j t^j / j!,$$

where $\mu'_j = \sum_{\alpha=0}^{\infty} p_\alpha \alpha^j$. On taking the jth derivative of (2.7) with respect to t and evaluating it at $t = 0$, the result is just μ'_j, and it is in this sense that (2.7) "generates" the moments of a pmf. Upon taking $z = e^{it}$ in $p(z)$, where $i = \sqrt{-1}$, by similar analysis the characteristic function for a pmf can be obtained, namely

$$p(e^{it}) = \sum_{j=0}^{\infty} \mu'_j (it)^j / j!, \tag{2.8}$$

from which the moments, μ'_j, $j = 0, 1, 2, \ldots$, can be obtained by differentiation with respect to t and evaluating the derivatives at $t = 0$. It can be shown by complex Fourier series analysis that a specific pmf function has a unique characteristic function and that a specific characteristic function implies a unique pmf. This is important since, on occasion, manipulating characteristic functions is simpler than manipulating pmfs.

We now turn to consider some specific pmfs for discrete rvs.

2.4.2.1. The binomial process. Consider a dichotomous rv, $\tilde{y}_i$, such that $\tilde{y}_i = 1$ with probability p and $\tilde{y} = 0$ with probability $1 - p$. For example, $\tilde{y}_i = 1$ might denote the appearance of a head on a flip of a coin and $\tilde{y}_i = 0$ the appearance of a tail. Then $E\tilde{y}_i = 1 \cdot p + 0 \cdot (1 - p) = p$ and $V(\tilde{y}_i) = E(\tilde{y}_i - E\tilde{y}_i)^2 = (1 - p)^2 p + (0 - p)^2 (1 - p) = p(1 - p)$. Now consider a sequence of such $\tilde{y}_i$'s, $i = 1, 2, \ldots, n$, such that the value of any member of the sequence provides no information about the values of others, that is, the $\tilde{y}_i$'s are independent rvs. Then *any particular* realization of r ones and $n - r$ zeros has probability $p^r (1 - p)^{n-r}$. On the other hand, the probability of obtaining r ones and $n - r$ zeros is

$$\mathrm{P}(\tilde{r} = r | n, p) = \binom{n}{r} p^r (1 - p)^{n-r}, \tag{2.9}$$

where

$$\binom{n}{r} \equiv n! / r!(n - r)!.$$

Note that the total number of realizations with r ones and $n - r$ zeros is obtained by recognizing that the first one can occur in n ways, the second in $n - 1$ ways, the third in $n - 2$ ways, and the rth in $n - (r - 1)$ ways. Thus, there are $n(n - 1)$

$(n-2)\ldots(n-r+1)$ ways of getting r ones. However, $r(r-1)(r-2)\ldots 2\cdot 1$ of these ways are indistinguishable. Then $n(n-1)(n-2)\ldots(n-r+1)/r! = n!/r!(n-r)!$ is the number of ways of obtaining r ones in n realizations. Since $p^r(1-p)^{n-r}$ is the probability of each one, (2.9) provides the total probability of obtaining r ones in n realizations.

The expression in (2.9) can be identified with coefficients in a binomial expansion,

$$1=(p+q)^n=\sum_{r=0}^{n}\binom{n}{r}p^r q^{n-r}=\sum_{r=0}^{n}\mathrm{P}(\tilde{r}=r|p,n),$$

where $q=1-p$, and hence the name "binomial" distribution. Given the value of p, it is possible to compute various probabilities from (2.9). For example,

$$\Pr(r\leqslant r_0|p,n)=\sum_{r=0}^{r_0}\binom{n}{r}p^r(1-p)^{n-r},$$

where r_0 is a given value of r. Further, moments of r can be evaluated directly from

$$E(\tilde{r}^\alpha|p,n)=\sum_{r=0}^{n}r^\alpha\mathrm{P}(\tilde{r}=r|p,n).$$

By such direct evaluation:

$$\begin{aligned}\mu_1'&=E\tilde{r}=np,\\ \mu_2'&=E\tilde{r}^2=(np)^2+np(1-p),\\ \mu_2&=E(\tilde{r}-E\tilde{r})^2=np(1-p).\end{aligned}\tag{2.10}$$

Further, higher central moments can be evaluated directly or computed from the Ramonovsky recursion formula, $\mu_{\alpha+1}=pq[\alpha n\mu_{\alpha-1}-\mathrm{d}\mu_\alpha/\mathrm{d}q]$, $\alpha=1,2,\ldots$, with $q=1-p$. From these results the skewness measure γ, introduced above, is $\gamma=\mu_3/\mu_2^{3/2}=(q-p)/\sqrt{npq}$, while the kurtosis measure $\beta_2=\mu_4/\mu_2^2=1/npq+3(n-2)/n$ and the "excess" is $\beta_2-3=1/npq-6/n$. For $p=q=1/2$, $\gamma=0$, that is, the binomial pmf is symmetric.

From (2.10), the moments of the proportion of ones, $\tilde{r}/n$, are easily obtained: $E\tilde{r}/n=p$, $E(\tilde{r}/n)^2=p^2+p(1-p)/n$ and $E[\tilde{r}/n-E\tilde{r}/n]^2=p(1-p)/n$. Also note that $E(\tilde{r}/n)^\alpha=(E\tilde{r}^\alpha)/n^\alpha$, $\alpha=1,2,\ldots$.

It is of great interest to determine the form of the binomial pmf when both r and $n-r$ are large and p is fixed, the problem solved in the DeMoivre–Laplace

Theorem. With $m \equiv n - r$,

$$-\log P(\tilde{r} = r | p, n) = \log r! + \log m! - \log n! - r\log p - m\log(1-p).$$

Stirling's formula, $\log n! = (n+1/2)\log n - n + (1/2)\log 2\pi + O(n^{-1})$, can be applied for large r and m to yield

$$-\log P(\tilde{r} = r|n, p) \doteq (1/2)\log(2\pi rm/n) + r\log(r/np) + m\log[m/n(1-p)]$$

or[4]

$$P(\tilde{r} = r | p, n) \doteq [2\pi np(1-p)]^{-1/2}\exp\{-(r-np)^2/2np(1-p)\}, \qquad (2.11)$$

which is a normal approximation[5] to the binomial pmf when r and $m = n - r$ are both large. In (2.11), the mean and variance of $\tilde{r}$ are np and $np(1-p)$, respectively, the same as the exact values for these moments. See, for example, Kenney and Keeping (1951, p. 36ff) and Jeffreys (1967, pp. 61–62) for discussions of the quality of this approximation. If we let $\tilde{f} = \tilde{r}/n$, with the condition underlying (2.11), $\tilde{f}$ has an approximate normal distribution with mean p and variance $p(1-p)/n$. Thus, (2.11) is an important example of a case in which a discrete rv's pmf can be well approximated by a continuous probability density function (pdf).

2.4.2.2. The Poisson process. The Poisson process can be developed as an approximation to the binomial process when n is large and p (or $q = 1 - p$) is small. Such situations are often encountered, for example, in considering the number of children born blind in a large population of mothers, or the number of times the volume of trading on a stock exchange exceeds a large number, etc. For such rare (low p) events from a large number of trials, (2.11) provides a poor approximation to the probabilities of observing a particular number of such rare events and thus another approximation is needed. If n is large but np is of moderate size, say approximately of order 10, the Poisson exponential function can be employed to approximate

$$P(\tilde{r} = r|n, p) = \binom{n}{r} p^r (1-p)^{n-r},$$

[4]Let $r/n = p + \varepsilon/n^{1/2}$, where ε is small, or $r = np + n^{1/2}\varepsilon$ and $n - r = m = n(1-p) - n^{1/2}\varepsilon$. On substituting these expressions for r and m in the logarithmic terms, this produces terms involving $\log[1 + \varepsilon/pn^{1/2}]$ and $\log[1 - \varepsilon/(1-p)n^{1/2}]$. Expanding these as $\log(1+x) = x - x^2/2$ and collecting dominant terms in ε^2, the result is (2.11).

[5]See below for a discussion of the normal distribution.

That is, with $\theta \equiv np$, if θ and r are fixed and $n \to \infty$,

$$\binom{n}{r}\left(\frac{\theta}{n}\right)^r\left(1-\frac{\theta}{n}\right)^{n-r} \to \theta^r \mathrm{e}^{-\theta}/r!, \tag{2.12}$$

which is the Poisson approximation to the probability of r occurrences of the rare event in a large number of trials [see, for example, Kenny and Keeping (1951, p. 44ff) for a discussion of the quality of the approximation]. In the limit as $n \to \infty$, $\mathrm{P}(\tilde{r}=r|\theta)=\theta^r\mathrm{e}^{-\theta}/r!$ is the exact Poisson pmf. Note that $\sum_{r=0}^{\infty}\theta^r\mathrm{e}^{-\theta}/r!=1$ and $E(\tilde{r}|\theta)=\theta$, $E(\tilde{r}^2|\theta)=\theta(\theta+1)$, $E(\tilde{r}-E\tilde{r})^2=\theta$, $E(\tilde{r}-E\tilde{r})^3=\theta$, and $E(\tilde{r}-E\tilde{r})^4 = 3\theta^2+\theta$. It is interesting that the mean, variance, and third central moment are all equal to θ. From these moments, measures of skewness and kurtosis can be evaluated.

2.4.2.3. Other variants of the binomial process. Two interesting variants of the binomial process are the Poisson and Lexis schemes. In the Poisson scheme, the probability that $\tilde{y}_i=1$ is p_i, and not p as in the binomial process. That is, the probability of a one (or "success") varies from trial to trial. As before, the $\tilde{y}_i$'s are assumed independent. Then the expectation of $\tilde{r}$, the number of ones, is

$$E\tilde{r} = E\sum_{i=1}^{n}\tilde{y}_i = \sum_{i=1}^{n}p_i = n\bar{p},$$

where

$$\bar{p} \equiv \sum_{i=1}^{n}p_i/n,$$

and

$$V(\tilde{r}) = E(\tilde{r}-E\tilde{r})^2 = \sum_{i=1}^{n}E(\tilde{y}_i-E\tilde{y}_i)^2 = \sum_{i=1}^{n}p_i(1-p_i) = n\left(\bar{p}\bar{q}-\sigma_p^2\right),$$

where

$$\bar{q}=1-\bar{p} \quad \text{and} \quad \sigma_p^2 \equiv \sum_{i=1}^{n}(p_i-\bar{p})^2/n.$$

Note that $V(\tilde{r})$ is less than the variance of $\tilde{r}$ associated with independent binomial trials with a fixed probability $\bar{p}$ at each trial.

Extensions of the Poisson scheme that are widely used in practice involve the assumption that $p_i = f(\boldsymbol{x}_i, \boldsymbol{\beta})$, where $\boldsymbol{x}_i$ is a given vector of observed variables and

$\boldsymbol{\beta}$ is a vector of parameters. The function $f(\cdot)$ is chosen so that $0<f(\cdot)<1$ for all i. For example, in the *probit model*,

$$p_i=(2\pi)^{-1/2}\int_{-\infty}^{x_i'\boldsymbol{\beta}}\mathrm{e}^{-t^2/2}\,\mathrm{d}t, \tag{2.13}$$

while in the *logit model*,

$$p_i=1/(1+\mathrm{e}^{-x_i'\boldsymbol{\beta}}). \tag{2.14}$$

Then the probability of any particular realization of the $\tilde{y}_i$, $i=1,2,\dots,n$, is given by

$$\prod_{i=1}^{n}p_i^{y_i}(1-p_i)^{1-y_i}, \tag{2.15}$$

where $y_i=0$ or 1 are the observations. By inserting (2.13) or (2.14) in (2.15), the probit and logit models, respectively, for the observations are obtained. Of course, other functions $f(\cdot)$ that satisfy $0<f(\cdot)<1$ for all i can be employed as well, for example $p_i=f(x_i,\beta)=1-\mathrm{e}^{-\beta x_i^2}$, with $\beta>0$ and $0<x_i^2<\infty$, etc.

In the Lexis scheme, m sets of n trials each are considered. The probability of obtaining a one (or a "success") is assumed constant within each set of trials but varies from one set to another. The random number of ones in the jth set is $\tilde{r}_j$, with expectation $E\tilde{r}_j=np_j$. Then, with $\tilde{r}=\sum_{j=1}^{m}\tilde{r}_j$, $E\tilde{r}=n\sum_{j=1}^{m}p_j=nm\bar{p}$, where here $\bar{p}=\sum_{j=1}^{m}p_j/m$. Also, by direct computation,

$$\operatorname{var}(\tilde{r})=n\bar{p}\bar{q}+n(n-1)\sigma_p^2, \tag{2.16}$$

where $\bar{q}=1-\bar{p}$ and $\sigma_p^2=\sum_{j=1}^{m}(p_j-\bar{p})^2/m$. It is seen from (2.16) that var($\tilde{r}$) is larger than from binomial trials with a fixed probability $\bar{p}$ on each trial.

If σ^2 is the variance of $\tilde{r}$, the number of ones or successes in a set of n trials, and if σ_β^2 is the variance calculated on the basis of a binomial process, then the ratio $L=\sigma/\sigma_\beta$ is called the Lexis ratio. The dispersion is said to be subnormal if $L<1$, and supernormal if $L>1$.

The negative binomial process involves observing independent $\tilde{y}_i$'s, as in the binomial process, but with the condition that a preassigned number of ones be observed. Thus r, the number of ones (or successes) to be observed, is fixed and the number of observations or trials, n, is random. Since the last observation is a one with probability p and the probability of getting $r-1$ ones in the first $n-1$ trials is

$$\binom{n-1}{r-1}p^{r-1}(1-p)^{n-r},$$

the desired probability of observing r ones, with r fixed beforehand in n trials, is

$$\mathrm{P}(\tilde{n}=n|r,p)=\binom{n-1}{r-1}p^{r}(1-p)^{n-r}, \tag{2.17}$$

with $n \geqslant r \geqslant 1$, which should be compared with the pmf for the usual binomial process in (2.9).

2.4.2.4. Multinomial process. Let p_j be the probability that a discrete rv, $\tilde{y}_i$, assumes the value j, $j=1,2,\dots,J$. If we observe n independent realizations of $\tilde{y}_i$, $i=1,2,\dots,n$, the probability that r_1 values have $j=1$, r_2 have $j=2,\dots,$ and r_J have $j=J$, with $n=\sum_{i=1}^{J}r_i$, is given by

$$\mathrm{P}(\tilde{\boldsymbol{r}}=\boldsymbol{r}|p)=\frac{n!}{r_1!r_2!\dots r_J!}p_1^{r_1}p_2^{r_2}\cdots p_J^{r_J}, \tag{2.18}$$

with $\tilde{\boldsymbol{r}}'=(\tilde{r}_1,\tilde{r}_2,\dots,\tilde{r}_J)$, $\boldsymbol{r}=(r_1,r_2,\dots,r_J)$, $\boldsymbol{p}'=(p_1,p_2,\dots,p_J)$, $0\leqslant p_j$, and $\sum_{j=1}^{J}p_j=1$. If $J=2$, the multinomial pmf in (2.18) becomes identical to the binomial pmf in (2.9). As with the binomial pmf, for large n and r_j's, we can take the logarithm of both sides of (2.18), use Stirling's approximation, and obtain an approximating multivariate normal distribution [see Kenney and Keeping (1951, pp. 113–114) for analysis of this problem]. Also, as with (2.13) and (2.14), it is possible to develop multivariate probit and logit models.

The pmfs reviewed above are some leading examples of probability models for independent discrete rvs. For further examples, see Johnson and Kotz (1969). When non-independent discrete rvs are considered, it is necessary to take account of the nature of dependencies, as is done in the literature on time series point processes [see, for example, Cox and Lewis (1966) for a discussion of this topic].

2.4.3. *Continuous random variables*

We first describe some properties of models for a single continuous rv, that is, univariate probability density functions (pdfs), and then turn to some models for two or more continuous rvs, that is, bivariate or multivariate pdfs.

Let $\tilde{x}$ denote a continuous rv that can assume a continuum of values in the interval a to b and let $f(x)$ be a non-negative function for $a<x<b$ such that $\Pr(x<\tilde{x}<x+\mathrm{d}x)=f(x)\mathrm{d}x$, where $a<x<b$ and $\int_a^b f(x)\mathrm{d}x=1$. Then $f(x)$ is the normalized pdf for the continuous rv, $\tilde{x}$. In this definition, a may be equal to $-\infty$ and/or $b=\infty$. Further, the cumulative distribution function (cdf) for $\tilde{x}$ is given by $F(x)=\int_b^x f(t)\mathrm{d}t$ with $a<x<b$. Given that $\int_a^b f(t)\mathrm{d}t=1$, $0\leqslant F(x)\leqslant 1$. Further, $\Pr(c<\tilde{x}<d)=F(d)-F(c)$, where $a\leqslant c<d\leqslant b$.

The moments around zero of a continuous rv, $\tilde{x}$, with pdf $f(x)$, are given by

$$\mu_r' = E\tilde{x}^r = \int_a^b x^r f(x)\,\mathrm{d}x, \qquad r = 0,1,2,\ldots, \tag{2.19}$$

with $\mu_1' \equiv$ the mean of $\tilde{x}$. The central moments are given by

$$\mu_r = E(\tilde{x} - E\tilde{x})^r = \int_a^b (\tilde{x} - E\tilde{x})^r f(x)\,\mathrm{d}x, \qquad r = 0,1,2,\ldots, \tag{2.20}$$

with $\mu_2 = \sigma^2$, the variance, and σ, the standard deviation. Note that $\mu_1 = 0$. Also, moments are said to exist when the integrals in (2.19) and (2.20) converge; in cases in which particular integrals in (2.19) and (2.20) fail to converge, the associated moments do not exist or are infinite.[6]

For unimodal pdfs, unitless measures of skewness are sk = (mean − mode)/σ, $\beta_1 = \mu_3^2/\mu_2^3$, and $\gamma_1 = \mu_3/\mu_2^{3/2}$. For symmetric, unimodal pdfs, mean = modal value and thus sk = 0. Since all odd order central moments are equal to zero, given symmetry, $\beta_1 = \gamma_1 = 0$. Measures of kurtosis are given by $\beta_2 = \mu_4/\mu_2^2$ and $\gamma_2 = \beta_2 - 3$, the "excess". For a normal pdf, $\beta_2 = 3$, and $\gamma_2 = 0$. When $\gamma_2 > 0$, a pdf is called leptokurtic, and platykurtic when $\gamma_2 < 0$.

The moment-generating function for $\tilde{x}$ with pdf $f(x)$ is

$$M(t) = \int_a^b f(x)\mathrm{e}^{tx}\,\mathrm{d}x \tag{2.21}$$

from which

$$\left.\frac{\mathrm{d}^r M(t)}{\mathrm{d}t}\right|_{t=0} = E\tilde{x}^r$$

under the assumption that the integral in (2.21) converges for some $t = t_0 > 0$. In (2.21), a may equal $-\infty$ and/or b may equal ∞. Thus, knowing the form of $M(t)$ allows one to compute moments conveniently.

The characteristic function associated with the continuous rv, $\tilde{x}$, with pdf $f(x)$, is given by

$$C(t) = \int_a^b f(x)\mathrm{e}^{\mathrm{i}tx}\,\mathrm{d}x, \tag{2.22}$$

where $\mathrm{i} = \sqrt{-1}$. It is known that the integral (2.22) converges uniformly in t. Thus, the rth derivative of $C(t)$ with respect to t is $\mathrm{i}^r\int_a^b x^r f(x)\mathrm{e}^{\mathrm{i}tx}\mathrm{d}x$. On setting

[6]For analysis regarding the convergence and divergence of integrals, see, for example, Widder (1961, p. 271ff.), or other advanced calculus texts.

$t=0$, $C^{(r)}(0)=\mathrm{i}^r\mu'_r$, which provides a useful expression for evaluating moments when they exist. See Kendall and Stuart (1958, ch. 4) for further discussion and uses of characteristic functions.

For each characteristic function there exists a unique pdf and vice versa. On the other hand, even if moments of all orders exist, it is only under certain conditions that a set of moments uniquely determine a pdf or cdf uniquely. However, as Kendall and Stuart (1958) point out, "...fortunately for statisticians, those conditions are obeyed by all the distributions arising in statistical practice" (p. 86; see also p. 109ff.).

Several examples of univariate pdfs for continuous rvs follow.

2.4.3.1. Uniform. A rv $\tilde{x}$ has a uniform pdf if and only if its pdf is

$$f(x)=\frac{1}{b-a},\qquad a\leqslant x\leqslant b, \tag{2.23}$$

and $f(x)=0$ elsewhere. That (2.23) is a normalized pdf is apparent since $\int_a^b f(x)\mathrm{d}x=1$. By direct evaluation,

$$\mu'_r=\int_a^b\frac{x^r}{b-a}\mathrm{d}x=\frac{1}{b-a}\frac{1}{r+1}(b^{r+1}-a^{r+1}),\qquad r=1,2\dots,$$

and thus $E\tilde{x}=(a+b)/2$ and $E\tilde{x}^2=(b^3-a^3)/3(b-a)=(a^2+ab+b^2)/3$. Also, from $V(\tilde{x})=E\tilde{x}^2-(E\tilde{x})^2$, $V(\tilde{x})=(b-a)^2/12$. Note too that the moment-generating function is

$$\begin{aligned}M(t)&=\int_a^b\frac{\mathrm{e}^{tx}}{b-a}\mathrm{d}x=\frac{\mathrm{e}^{bt}-\mathrm{e}^{at}}{(b-a)t}\\&=1+1/2(a+b)t+1/3(a^2+ab+b^2)2!t^2+\cdots,\end{aligned}$$

where $\mathrm{e}^{bt}=1+bt+(bt)^2/2!+\cdots$ and a similar expression for e^{at} have been employed. Then, for example, $\mu'_1=M'(0)=(a+b)/2$, $\mu'_2=M''(0)=(a^2+ab+b^2)/3$, etc. Finally, observe that (2.23) can be expressed as $\mathrm{d}f/\mathrm{d}x=0$, $a<x<b$. The solution of this differential equation subject to the normalization condition and $f(x)=0$ for $x<a$ and $x>b$ leads to (2.23).

2.4.3.2. Cauchy. A rv $\tilde{x}$ is distributed in the Cauchy form if and only if its pdf has the following form:

$$f(x|\theta,\sigma)=\frac{1}{\pi\sigma}\frac{1}{1+\left(\dfrac{x-\theta}{\sigma}\right)^2},\qquad -\infty<x<\infty,\quad -\infty<\theta<\infty,\quad 0<\sigma<\infty. \tag{2.24}$$

That (2.24) is a normalized pdf can be established by making a change of variable, $z=(x-\theta)/\sigma$, and noting that $\int_{-\infty}^{\infty}(1+z^2)^{-1}\mathrm{d}z=\pi$. Further, note that (2.24) is symmetric about θ, the location parameter which is the modal value and median of the Cauchy pdf. However, θ is not the mean since the mean and higher order moments of the Cauchy pdf do not exist. The non-existence of moments is fundamentally due to the fact that the pdf does not rapidly approach zero as $(x-\theta)^2/\sigma^2$ grows large; that is, the Cauchy pdf has heavy tails. A useful measure of dispersion for such a pdf is the inter-quartile range (IQR), that is, the value of $2c$, with $c>0$, such that $F(\theta+c)-F(\theta-c)=0.5$, where $F(\cdot)$ is the Cauchy cdf. For the Cauchy pdf, IQR $=2\sigma$.

On making a change of variable in (2.24), $z=(x-\theta)/\sigma$, the standardized Cauchy pdf is obtained, namely

$$f(z)=\frac{1}{\pi}\frac{1}{1+z^2},\qquad -\infty<z<\infty, \tag{2.25}$$

which is symmetric about $z=0$, the modal value and median. Further, it is interesting to note that (2.25) can be generated by assuming that the arc tangent of an angle, say ω, ranging from $-\pi/2$ to $\pi/2$ is uniformly distributed, that is, $\omega=\tan^{-1}z$ has a pdf $p(\omega)\mathrm{d}\omega=\mathrm{d}\omega/\pi$, $-\pi/2<\omega<\pi/2$. Since $\mathrm{d}\omega=\mathrm{d}\tan^{-1}z=\mathrm{d}z/(1+z)^2$, this uniform pdf for ω implies (2.25).

2.4.3.3. Normal. A rv $\tilde{x}$ is said to be normally distributed if and only if its pdf is

$$f(x|\theta,\sigma)=(1/\sigma\sqrt{2\pi})\exp\{-(x-\theta)^2/2\sigma^2\},$$
$$-\infty<x<\infty,\quad \infty<\theta<\infty,\quad 0<\sigma<\infty. \tag{2.26}$$

The pdf in (2.26) is the normal pdf that integrates to one and thus is normalized. It is symmetric about θ, a location parameter that is the modal value, median, and mean. The parameter σ is a scale parameter, the standard deviation of the normal pdf as indicated below. Note that from numerical evaluation, $\Pr\{|x-\theta|\leqslant 1.96\sigma\}=0.95$ for the normal pdf (2.26) indicating that its tails are rather thin or, equivalently, that (2.26) decreases very rapidly as $(x-\theta)^2$ grows in value, a fact that accounts for the existence of moments of all orders. Since (2.26) is symmetric about θ, all odd order central moments are zero, that is, $\mu_{2r+1}=0$, $r=0,1,2,\dots$. From $E(\tilde{x}-\theta)=0$, $E\tilde{x}=\theta$, the mean of the normal pdf. As regards even central moments, they satisfy

$$\begin{aligned}\mu_{2r}&=E(\tilde{x}-E\tilde{x})^{2r}=\sigma^{2r}2^r\Gamma(r+1/2)/\sqrt{\pi}\\&=\sigma^{2r}(2r)!/2^r r!,\end{aligned} \tag{2.27}$$

where $\Gamma(r+1/2)$ is the gamma function, $\Gamma(q)$, with argument $q=r+1/2$, that is, $\Gamma(q)=\int_0^\infty u^{q-1}e^{-u}du$, with $0<q<\infty$.[7] From (2.27), $\mu_2=\sigma^2$ and $\mu_4=3\sigma^4$. Thus, the kurtosis measure $\beta_2=\mu_4/\mu_2^2=3$ and $\gamma_2=\beta_2-3=0$ for the normal pdf.

The standardized form of (2.26) may be obtained by making a change of variable, $z=(x-\theta)/\sigma$, to yield

$$f(z)=(1/\sqrt{2\pi})\exp\{-z^2/2\},\qquad -\infty<z<\infty, \tag{2.28}$$

with modal value, median, and mean at $z=0$. Also, $V(\tilde{z})=1$ and $E(\tilde{z}-z)^4=3$ from (2.27).

As was shown in Section 2.4.2.1, the normal pdf can be viewed as a limiting form of the binomial process. Below it will be indicated that central limit theorems show that more general sequences of rvs have limiting normal distributions. These results underline the importance of the normal distribution in theoretical and applied statistics.

2.4.3.4. Student-t. A rv $\tilde{x}$ is distributed in the univariate Student-t (US-t) form if and only if it has the following pdf:

$$f(x|\theta,h,\nu)=c\left(\frac{h}{\nu}\right)^{1/2}\bigg/\left[1+\frac{h}{\nu}(x-\theta)^2\right]^{(\nu+1)/2},$$
$$-\infty<x<\infty,\quad -\infty<\theta<\infty,\quad 0<\nu,h<\infty, \tag{2.29}$$

with $c=\Gamma[(\nu+1)/2]/\sqrt{\pi}\,\Gamma(\nu/2)$, where $\Gamma(\cdot)$ denotes the gamma function. From inspection of (2.29) it is seen that the US-t pdf has a single mode at $x=\theta$ and is symmetric about the modal value. Thus, $x=\theta$ is the median and mean (which exists for $\nu>1$ – see below) of the US-t pdf. As will be seen, the parameter h is intimately linked to the dispersion of the US-t, while the parameter ν, often called the "degrees of freedom" parameter, is involved both in the dispersion as well as the kurtosis of the pdf. Note that if $\nu=1$, the US-t is identical to the Cauchy pdf in (2.24) with $h=1/\sigma^2$. On the other hand, as ν grows in value, the US-t approaches the normal pdf (2.29) with mean θ and variance $1/h$.

In Zellner (1971, p. 367 ff) it is shown that the US-t pdf in (2.29) is a normalized pdf. The odd order moments about θ, $\mu_{2r-1}=E(\tilde{x}-\theta)^{2r-1}$, $r=1,2,\dots$, exist when $\nu>2r-1$ and are all equal to zero given the symmetry of the pdf about θ. In particular, for $\nu>1$, $E(\tilde{x}-\theta)=0$ and $E\tilde{x}=\theta$, the mean which exists, given $\nu>1$. The even order central moments, $\mu_{2r}=E(\tilde{x}-\theta)^{2r}$, $r=1,2,\dots$,

[7]See, for example, Zellner (1971, p. 365) for a derivation of (2.27). From the calculus, $\Gamma(q+1)=q\Gamma(q)$, $\Gamma(1)=1$, and $\Gamma(1/2)=\sqrt{\pi}$. Using these relations, the second line of (2.27) can be derived from the first.

exist given that $\nu > 2r$ and are given by

$$\mu_{2r} = E(\tilde{x} - \theta)^{2r} = \frac{\Gamma(r+\frac{1}{2})\Gamma(\nu/2-r)}{\Gamma(1/2)\Gamma(\nu/2)}\left(\frac{\nu}{h}\right)^{r}, \qquad r = 1,2,\dots, \qquad \nu > 2r. \tag{2.30}$$

From (2.30), the second and fourth central moments are $\mu_2 = E(\tilde{x} - \theta)^2 = \nu/(\nu - 2)h$, $\nu > 2$, and $\mu_4 = E(\tilde{x} - \theta)^4 = 3\nu^2/(\nu - 2)(\nu - 4)h^2$, $\nu > 4$. The kurtosis measure is then $\gamma_2 = \mu_4/\mu_2^2 - 3 = 6/(\nu - 4)$, for $\nu > 4$, and thus the US-t is leptokurtic ($\gamma_2 > 0$). As ν gets large, $\gamma_2 \to 0$, and the US-t approaches a normal form with mean θ and variance $1/h$. When $\nu > 30$, the US-t's form is very close to that of a normal pdf. However, for small ν, the US-t pdf has much heavier tails than a normal pdf with the same mean and variance.

The standardized form of (2.29) is obtained by making the change of variable, $t = \sqrt{h}(x - \theta)$, that yields

$$f(t|\nu) = (c/\sqrt{\nu})/(1 + t^2/\nu)^{(\nu+1)/2}, \qquad -\infty < t < \infty, \tag{2.31}$$

where c has been defined in connection with (2.29). The standardized US-t pdf in (2.31) has its modal value at $t = 0$ which is also the median. The moments of (2.31) may easily be obtained from those of $\tilde{x} - \theta$ presented above.

Finally, it is of interest to note that the US-t pdf in (2.29) can be generated as a "continuous mixture" of normal pdfs, that is,

$$f(x|\theta, h, \nu) = \int_0^\infty f_{\mathrm{N}}(x|\theta, \sigma) f_{\mathrm{IG}}(\sigma|h, \nu)\,\mathrm{d}\sigma, \tag{2.32}$$

where $f_{\mathrm{N}}(x|\theta, \sigma)$ is the normal pdf in (2.26) and $f_{\mathrm{IG}}(\sigma|h, \nu)$ is an inverted gamma pdf "mixing distribution" for σ, $0 < \sigma < \infty$, given by

$$f_{\mathrm{IG}}(\sigma|\nu, h) = \frac{2}{\Gamma(\nu/2)}\left(\frac{\nu}{2h}\right)^{\nu/2} \sigma^{-(\nu+1)} \exp\left\{\frac{-\nu}{2h\sigma^2}\right\}, \tag{2.33}$$

where ν and h are the parameters in the US-t pdf in (2.29).[8] From (2.32), it is seen that $f_{\mathrm{N}}(x|\theta, \sigma)$ is averaged over possible values of σ. This is an example in which the standard deviation σ of a normal pdf can be viewed as random with the pdf shown in (2.33). The fact that (2.32) yields the US-t pdf is a useful interpretation of the US-t pdf. Many well-known pdfs can be generated as continuous mixtures of underlying pdfs.

[8]See Zellner (1971, p. 371ff) for properties of this and other inverted gamma pdfs.

2.4.3.5. Other important univariate pdfs. Among many pdfs that are important in theoretical and applied statistics and econometrics, the following are some leading examples.

The gamma pdf, $f(x|\gamma,\alpha)=x^{\alpha-1}e^{-x/\gamma}/\gamma^{\alpha}\Gamma(\alpha)$, with $0<x<\infty$ and parameters $0<\alpha, \gamma$ is a rich class of pdfs. With a change of variable, $z=x/\gamma$, it can be brought into standardized form, $p(z/\alpha)=z^{\alpha-1}e^{-z}/\Gamma(\alpha)$, $0<z<\infty$. In this form its relation to the gamma function is apparent. If in the non-standardized gamma pdf $\alpha=\nu/2$ and $\gamma=2$, the result is the chi-squared pdf with ν "degrees of freedom", $p(x|\nu)=x^{\nu/2-1}e^{-x/2}/2^{\nu/2}\Gamma(\nu/2)$, with $0<x<\infty$ and $0<\nu<\infty$. If the transformation $x=1/y^2$ is made, the pdf for y is $p(y|\gamma,\alpha)=2e^{-1/\gamma y^2}/y^{2\alpha+1}\Gamma(\alpha)\gamma^{\alpha}$, $0<y<\infty$. The particular inverted gamma pdf in (2.32) can be obtained from $p(y|\gamma,\alpha)$ by setting $\sigma=y$, $\alpha=\nu/2$, and $\gamma=2h/\nu$. Properties of these and other gamma-related densities are discussed in Raiffa and Schlaifer (1961), Zellner (1971), and Johnson and Kotz (1970).

For a continuous rv that has a range $0\leqslant x\leqslant c$, the beta pdf, $f(x|a,b,c)=(x/c)^{a-1}(1-x/c)^{b-1}/cB(a,b)$, where $B(a,b)$ is the beta function[9] with $a,b>0$, is a flexible and useful pdf that can assume a variety of shapes. By a change of variable, $y=x-d$, the range of the beta pdf above can be changed to $y=-d$ to $y=c-d$. Also, by taking $z=x/c$, the standardized form is $f(z|a,b)=z^{a-1}(1-z)^{b-1}/B(a,b)$, with $0\leqslant z\leqslant 1$. There are various pdfs associated with the beta pdf. The inverted beta pdf is obtained from the standardized beta by the change of variable $z=1/(1+u)$, so that $0<u<\infty$ and $f(u|a,b)=u^{b-1}/(1+u)^{a+b}B(a,b)$ is the inverted beta pdf. Another form of the inverted beta pdf is obtained by letting $u=y/c$, with $0<c<\infty$. Then $f(y|a,b,c)=(y/c)^{b-1}/(1+y/c)^{a+b}cB(a,b)$, with $0<y<\infty$. The Fisher–Snedecor F distribution is a special case of this last density with $a=\nu_2/2$, $b=\nu_1/2$, and $c=\nu_2/\nu_1$. The parameters ν_1 and ν_2 are referred to as "degrees of freedom" parameters. Properties of the pdfs mentioned in this paragraph are given in the references cited at the end of the previous paragraph.

The discussion above has emphasized the importance of the normal, Student-t, beta, and gamma distributions. For each of the distributions mentioned above there are often several ways of generating them that are useful, lead to greater insight, and are of value in analysis and applications. For example, generation of the US-t as a special continuous mixture of normal pdfs was mentioned above. A rv with the chi-squared pdf with ν degrees of freedom, say $\tilde{\chi}^2_\nu$, can be considered as the sum of ν squared independent, standardized normal variables, $\tilde{\chi}^2_\nu=\sum_{i=1}^{\nu}\tilde{z}_i^2$, with $\tilde{z}_i=(\tilde{x}_i-\theta)/\sigma$. If $\tilde{\chi}^2_{\nu_1}$ and $\tilde{\chi}^2_{\nu_2}$ are two independent chi-squared variables with ν_1 and ν_2 degrees of freedom, respectively, then $\tilde{F}_{\nu_1,\nu_2}=(\tilde{\chi}^2_{\nu_1}/\nu_1)/(\tilde{\chi}^2_{\nu_2}/\nu_2)$ has an F-pdf with ν_1 and ν_2 degrees of freedom. These are just some of the ways

[9]From the calculus, $B(a,b)=B(b,a)=\Gamma(a)\Gamma(b)/\Gamma(a+b)$. Also, $B(a,b)=\int_0^1 z^{a-1}(1-z)^{b-1}dz$, $a,b>0$.

in which particular pdfs can be generated. Further examples are provided in the references mentioned above.

Above, the reciprocal transformation was employed to produce "inverted" pdfs. Many other transformations can be fruitfully utilized. For example, if the continuous rv $\tilde{y}$ is such that $0 < \tilde{y} < \infty$ and $\tilde{x} = \ln \tilde{y}$, $-\infty < \tilde{x} < \infty$, has a normal pdf with mean θ and variance σ^2, $\tilde{y}$ is said to have a "log-normal" pdf whose form can be obtained from the normal pdf for $\tilde{x}$ by a simple change of variable. The median of the pdf for $\tilde{y} = \mathrm{e}^{\tilde{x}}$ is e^{θ}, while the mean of $\tilde{y}$ is $E\tilde{y} = E\mathrm{e}^{\tilde{x}} = \mathrm{e}^{\theta + \sigma^2/2}$. Thus, there is an interesting dependence of the mean of $\tilde{y}$ on the variance of $\tilde{x}$. This and many other transformations have been analyzed in the literature.

Finally, it should be noted that many of the pdfs mentioned in this section can be obtained as solutions to differential equations. For example, the normal pdf in (2.26) is the solution to $(1/f)\mathrm{d}f/\mathrm{d}x = -(x-\mu)/\sigma^2$. The generalization of this differential equation that yields the Pearson system of pdfs is given by

$$\frac{1}{f}\frac{\mathrm{d}f}{\mathrm{d}x} = \frac{-(x-a)}{(b_0 + b_1 x + b_2 x^2)}. \tag{2.34}$$

The integral of (2.34) is

$$f(x) = A(x - c_1)^{m_1}(c_2 - x)^{m_2}, \tag{2.35}$$

where the value of A is fixed by $\int f(x)\mathrm{d}x = 1$ and c_1 and c_2 are the roots, possibly complex, of $b_0 + b_1 x + b_2 x^2 = 0$. See Jeffreys (1967, p. 74ff.) for a discussion of the solutions to (2.35) that constitute the Pearson system which includes many frequently encountered pdfs. For a discussion of other systems of pdfs, see Kendall and Stuart (1958, p. 167 ff.).

2.4.3.6. Multivariate pdfs for continuous random variables. Consider a random vector $\tilde{\boldsymbol{x}}' = (\tilde{x}_1, \tilde{x}_2, \ldots, \tilde{x}_m)$, with elements $\tilde{x}_i$, $i = 1,2,\ldots,m$, that are scalar continuous rvs. Assume that $\tilde{\boldsymbol{x}} \subset R_x$, the sample space. For example, R_x might be $-\infty < \tilde{x}_i < \infty$, $i = 1,2,\ldots,m$. The pdf for $\tilde{\boldsymbol{x}}$, or equivalently the joint pdf for the elements of $\tilde{\boldsymbol{x}}$, $f(\boldsymbol{x}) = f(x_1, x_2, \ldots, x_m)$, is a non-negative continuous and single-valued function such that $f(\boldsymbol{x})\mathrm{d}\boldsymbol{x} = f(x_1, x_2, \ldots, x_m)\mathrm{d}x_1\mathrm{d}x_2\ldots\mathrm{d}x_m$ is the probability that $\tilde{\boldsymbol{x}}$ is contained in the infinitesimal element of volume $\mathrm{d}\boldsymbol{x} = \mathrm{d}x_1\mathrm{d}x_2\ldots\mathrm{d}x_m$. If

$$\int_{R_x} f(\boldsymbol{x})\mathrm{d}\boldsymbol{x} = \int_{R_x} \cdots \int f(x_1, x_2, \ldots, x_m)\mathrm{d}x_1\mathrm{d}x_2\ldots\mathrm{d}x_m = 1, \tag{2.36}$$

then $f(\boldsymbol{x})$ is a normalized pdf for $\tilde{\boldsymbol{x}}$. When $\tilde{\boldsymbol{x}}$ has just two elements, $m = 2$, $f(\boldsymbol{x})$ is a bivariate pdf, if three elements, $m = 3$, a trivariate pdf, and if $m > 3$, a multivariate pdf.

When R_x is $-\infty < \tilde{x}_i < \infty$, $i = 1,2,\ldots,m$, the cumulative distribution function associated with $f(\boldsymbol{x})$ is $F(\boldsymbol{a})$ given by

$$F(\boldsymbol{a}) = \Pr(\boldsymbol{x} \leqslant \boldsymbol{a}) = \int_{-\infty}^{a_1}\int_{-\infty}^{a_2}\cdots\int_{-\infty}^{a_m} f(x_1, x_2,\ldots,x_m)\,\mathrm{d}x_1\mathrm{d}x_2\ldots\mathrm{d}x_m,$$

where $\boldsymbol{a}' = (a_1, a_2,\ldots,a_m)$ is a given vector and $\Pr(\tilde{\boldsymbol{x}} \leqslant \boldsymbol{a})$ is the probability of the intersection of the events $\tilde{x}_i \leqslant a_i$, $i = 1,2,\ldots,m$.

The mean, assuming that it exists, of an $m \times 1$ random vector $\tilde{\boldsymbol{x}}$ is

$$E\tilde{\boldsymbol{x}} = \begin{pmatrix} E\tilde{x}_1 \\ E\tilde{x}_2 \\ \vdots \\ E\tilde{x}_m \end{pmatrix} = \begin{pmatrix} \theta_1 \\ \theta_2 \\ \vdots \\ \theta_m \end{pmatrix} = \boldsymbol{\theta}, \tag{2.37}$$

where, if $\tilde{\boldsymbol{x}}$ has pdf $f(\boldsymbol{x})$ and $\tilde{\boldsymbol{x}} \subset R_x$,

$$E\tilde{x}_i = \theta_i = \int_{R_x} x_i f(\boldsymbol{x})\,\mathrm{d}\boldsymbol{x}, \qquad i = 1,2,\ldots,m. \tag{2.38}$$

This means that the θ_i's exist and are finite if and only if each integral in (2.38) converges to a finite value.

Second order moments about the mean vector $\boldsymbol{\theta}$ are given by

$$V(\tilde{\boldsymbol{x}}) = E(\boldsymbol{x} - \boldsymbol{\theta})(\boldsymbol{x} - \boldsymbol{\theta})' = \{E(\tilde{x}_i - \theta_i)(\tilde{x}_j - \theta_j)\}, \qquad i, j = 1,2,\ldots,m, \tag{2.39}$$

and the typical element of the symmetric $m \times m$ matrix $V(\tilde{\boldsymbol{x}})$ is given by

$$E(\tilde{x}_i - \theta_i)(\tilde{x}_j - \theta_j) \equiv \sigma_{ij} = \int_{R_x} (x_i - \theta_i)(x_j - \theta_j) f(\boldsymbol{x})\,\mathrm{d}\boldsymbol{x}. \tag{2.40}$$

If, in (2.40), $i = j$, $\sigma_{ii} = E(\tilde{x}_i - \theta_i)^2$ is the variance of $\tilde{x}_i$, $i = 1,2,\ldots,m$, while if $i \neq j$, σ_{ij}, given in (2.40), is the covariance of $\tilde{x}_i$ and $\tilde{x}_j$. Clearly, $\sigma_{ij} = \sigma_{ji}$ and thus the $m \times m$ matrix of variances and covariances,

$$V(\tilde{\boldsymbol{x}}) = \begin{bmatrix} \sigma_{11} & \sigma_{12} & \cdots & \sigma_{1m} \\ \sigma_{21} & \sigma_{22} & \cdots & \sigma_{2m} \\ \vdots & \vdots & \ddots & \vdots \\ \sigma_{m1} & \sigma_{m2} & \cdots & \sigma_{mm} \end{bmatrix} = \Sigma, \tag{2.41}$$

the "covariance matrix" for $\tilde{\boldsymbol{x}}$ is symmetric with $m(m+1)/2$ distinct elements.

The "correlation matrix," denoted by P, associated with $V(\tilde{x})$ is given by

$$P=\begin{bmatrix} 1 & \rho_{12} & \rho_{13} & \cdots & \rho_{1m} \\ \rho_{21} & 1 & \rho_{23} & \cdots & \rho_{2m} \\ \rho_{31} & \rho_{32} & 1 & \cdots & \rho_{3m} \\ \vdots & \vdots & \vdots & \ddots & \vdots \\ \rho_{m1} & \rho_{m2} & \rho_{m3} & \cdots & 1 \end{bmatrix}.$$

where $\rho_{ij}=\sigma_{ij}/\sqrt{\sigma_{ii}\sigma_{jj}}$, $i, j=1,2,3,\dots,m$. Note that P is symmetric and that $P=D^{-1}\Sigma D^{-1}$, with Σ given in (2.41) and D is an $m\times m$ diagonal matrix with typical element $\sigma_{ii}^{1/2}$. In general, mixed central moments are given by $\mu_{l_1,l_2,\dots,l_m}=E(\tilde{x}_1-\theta_1)^{l_1}(\tilde{x}_2-\theta_2)^{l_2}\dots(\tilde{x}_m-\theta_m)^{l_m}$, $l_i=0,1,2,\dots,i=1,2,\dots,m$.

To illustrate linear transformations of the elements of $\tilde{\boldsymbol{x}}-\boldsymbol{\theta}$, consider the $m\times 1$ random vector $\tilde{z}=H(\tilde{\boldsymbol{x}}-\boldsymbol{\theta})$, where H is an $m\times m$ non-stochastic matrix of rank m. Then from the linearity property of the expectation operator, $E\tilde{z}=HE(\tilde{\boldsymbol{x}}-\boldsymbol{\theta})=\mathbf{0}$, since $E\tilde{\boldsymbol{x}}=\theta$ from (2.37). Thus, $\tilde{z}$ has a zero mean vector. By definition from (2.39), $V(\tilde{z})=Ezz'=HE(\tilde{\boldsymbol{x}}-\boldsymbol{\theta})(\tilde{\boldsymbol{x}}-\boldsymbol{\theta})'H'=H\Sigma H'$, the covariance matrix of $\tilde{z}$. Now if Σ is positive definite, there exists a unique H such that $H\Sigma H'=I_m$.[10] If H is so chosen, then $V(\tilde{z})=Ezz'=I_m$; that is, $E\tilde{z}_i^2=1$, $i=1,2,\dots,m$, and $E\tilde{z}_i\tilde{z}_j=0$ for all $i\neq j$. From $H\Sigma H'=I_m$, $\Sigma=H^{-1}(H')^{-1}$ and $\Sigma^{-1}=H'H$. Furthermore, from $z=H(\boldsymbol{x}-\boldsymbol{\theta})$, the Jacobian of the transformation is

$$J=\mathrm{mod}\left|\frac{\partial z}{\partial \boldsymbol{x}}\right|=\mathrm{mod}|H|, \tag{2.42}$$

where "mod" denotes absolute value, and the Jacobian matrix is

$$\frac{\partial z}{\partial \boldsymbol{x}}=\begin{bmatrix} \frac{\partial z_1}{\partial x_1} & \frac{\partial z_1}{\partial x_2} & \cdots & \frac{\partial z_1}{\partial x_m} \\ \frac{\partial z_2}{\partial x_1} & \frac{\partial z_2}{\partial x_2} & \cdots & \frac{\partial z_2}{\partial x_m} \\ \cdot & \cdot & \cdot & \cdot \\ \frac{\partial z_m}{\partial x_1} & \frac{\partial z_m}{\partial x_2} & \cdots & \frac{\partial z_m}{\partial x_m} \end{bmatrix}=H,$$

[10] That is, given that Σ is a positive definite symmetric matrix, Σ can be diagonalized as follows: $P'\Sigma P=D(\lambda_i)$, where P is an $m\times m$ orthogonal matrix and the λ_i are the roots of Σ. Then $D^{-1/2}P'\Sigma PD^{-1/2}=I$ and $H=D^{-1/2}P'$, where $D^{-1/2}=D(\lambda_i^{-1/2})$, an $m\times m$ diagonal matrix with typical element $\lambda_i^{-1/2}$.

and $\mathrm{d}z = J\mathrm{d}\boldsymbol{x}$ shows how the transformation from z to $\boldsymbol{x}$ modifies the infinitesimal unit of volume. Thus, the pdf for z, $f(z)$ gets transformed as follows: $f(z)\mathrm{d}z = Jf[H(\boldsymbol{x}-\boldsymbol{\theta})]\mathrm{d}\boldsymbol{x}$ and $f[H(\boldsymbol{x}-\boldsymbol{\theta})]$ is the pdf for $\boldsymbol{x}$.[11]

Associated with bivariate and multivariate pdfs are marginal and conditional pdfs. For example, in terms of a bivariate pdf, $f(x_1, x_2)$, the marginal pdf for $\tilde{x}_1$ is given by

$$g(x_1) = \int_{R_{x_2}} f(x_1, x_2)\mathrm{d}x_2, \tag{2.43}$$

while the marginal pdf for $\tilde{x}_2$ is

$$h(x_2) = \int_{R_{x_1}} f(x_1, x_2)\mathrm{d}x_1.$$

Note that

$$\int_{R_{x_1}} g(x_1)\mathrm{d}x_1 = \int_{R_{x_2}} h(x_2)\mathrm{d}x_2 = 1$$

given that $f(x_1, x_2)$ satisfies (2.36). Also, $g(x_1)\mathrm{d}x_1$ is the probability that $\tilde{x}_1$ is in the interval $(x_1, x_1+\mathrm{d}x_1)$ for all values of $\tilde{x}_2$ and similarly for $h(x_2)$.

Definition

The rvs $\tilde{x}_1$ and $\tilde{x}_2$ are independent if and only if their joint pdf, $f(x_1, x_2)$, satisfies $f(x_1, x_2) \equiv g(x_1)h(x_2)$.

The conditional pdf for $\tilde{x}_1$ given $\tilde{x}_2$, denoted by $f(x_1|x_2)$, is

$$f(x_1|x_2) = f(x_1, x_2)/h(x_2), \tag{2.44}$$

provided that $h(x_2) > 0$. Similarly, the conditional pdf for $\tilde{x}_2$ given $\tilde{x}_i$, denoted by $f(x_2|x_1)$, if $f(x_2|x_1) = f(x_1, x_2)/g(x_1)$, provided that $g(x_1) > 0$. From (2.44), $f(x_1, x_2) = f(x_1|x_2)h(x_2)$ which, when inserted in (2.43), shows that the marginal pdf

$$g(x_1) = \int_{R_{x_2}} f(x_1|x_2)h(x_2)\mathrm{d}x_2$$

[11]If $z_i = \phi_i(\boldsymbol{x})$, $i = 1,2,\dots,m$, or $z = \boldsymbol{\phi}(\boldsymbol{x})$ represents a set of one-to-one transformations from $\boldsymbol{x}$ to z and if $\tilde{z}$ has pdf $f(z)$, then $f(z)\mathrm{d}z = Jf[\boldsymbol{\phi}(\boldsymbol{x})]\mathrm{d}\boldsymbol{x}$, where $J = \mathrm{mod}|\partial\boldsymbol{\phi}/\partial\boldsymbol{x}|$ is assumed not equal to zero in R_x.

can be interpreted as an average of the conditional pdf $f(x_1|x_2)$ with the marginal pdf $h(x_2)$ serving as the weighting function. Also, from (2.44),

$$\int_{R_{x_1}} f(x_1|x_2)\,\mathrm{d}x_1 = 1$$

since

$$\int_{R_{x_1}} f(x_1, x_2)\,\mathrm{d}x_1 = h(x_2).$$

An instructive interpretation of (2.44) is obtained by writing

$$\begin{aligned} f(x_1,x_2)\,\mathrm{d}x_1\mathrm{d}x_2 &= [h(x_2)\,\mathrm{d}x_2][f(x_1|x_2)\,\mathrm{d}x_1] \\ &= \Pr(x_2 < \tilde{x}_2 < x_2 + \mathrm{d}x_2)\Pr(x_1 < \tilde{x}_1 < x_1 + \mathrm{d}x_1|\tilde{x}_2 = x_2). \end{aligned}$$

Some of these general features of multivariate pdfs will be illustrated below in terms of specific, important pdfs that are frequently employed in econometrics.

(1) Bivariate Normal (BN). A two-element random vector, $\tilde{\mathbf{x}}' = (\tilde{x}_1, \tilde{x}_2)$, has a BN distribution if and only if its pdf is

$$f(x_1, x_2|\boldsymbol{\theta}) = \left(2\pi\sigma_1\sigma_2\sqrt{1-\rho^2}\right)^{-1}\exp\{-Q/2\}, \qquad -\infty < x_1, x_2 < \infty, \tag{2.45}$$

where

$$\begin{aligned} Q = \big[&(x_1-\mu_1)^2/\sigma_1^2 + (x_2-\mu_2)^2/\sigma_2^2 \\ &-2\rho(x_1-\mu_1)(x_2-\mu_2)/\sigma_1\sigma_2\big]/(1-\rho^2) \end{aligned} \tag{2.46}$$

and $\boldsymbol{\theta}' = (\mu_1, \mu_2, \sigma_1, \sigma_2, \rho)$, with $|\rho| < 1$, $-\infty < \mu_i < \infty$, and $0 < \sigma_i < \infty$, $i = 1,2$. Under these conditions, Q is a positive definite quadratic form.

To obtain the standardized form of (2.45), let $z_1 = (x_1 - \mu_1)/\sigma_1$ and $z_2 = (x_2 - \mu_2)/\sigma_2$, with $\mathrm{d}z_1\mathrm{d}z_2 = \mathrm{d}x_1\mathrm{d}x_2/\sigma_1\sigma_2$. Further, let the 2×2 matrix P^{-1} be defined by

$$P^{-1} = \frac{1}{1-\rho^2}\begin{bmatrix} 1 & -\rho \\ -\rho & 1 \end{bmatrix}.$$

Then $Q = \mathbf{z}'P^{-1}\mathbf{z}$, where $\mathbf{z}' = (z_1, z_2)$ and (2.45) becomes

$$f(\mathbf{z}|\rho) = (2\pi)^{-1}|P|^{-1/2}\exp\{-\mathbf{z}'P^{-1}\mathbf{z}/2\}, \tag{2.47}$$

with $-\infty < z_1, z_2 < \infty$, and where $|P|^{-1/2} = 1/(1-\rho^2)^{1/2}$. This is the standardized BN pdf.

Now in (2.47) let $z = Hv$, where $v' = (v_1, v_2)$ and H is a 2×2 non-singular matrix such that $H'P^{-1}H = I_2$. Then $dz = |H|dv$, and, from $H'P^{-1}H = I_2$, $|H'P^{-1}H| = |H|^2|P^{-1}| = 1$ or $|H| = |P|^{1/2}$. Using these results, (2.47) becomes

$$\begin{aligned} f(v) &= (2\pi)^{-1}\exp\{-v'v/2\} \\ &= \left[(2\pi)^{-1/2}\exp\{-v_1^2/2\}\right]\left[(2\pi)^{-1/2}\exp\{-v_2^2/2\}\right]. \end{aligned} \tag{2.48}$$

Thus, $\tilde{v}_1$ and $\tilde{v}_2$ are independent, standardized normal rvs and $\int f(v)dv = 1$, implying that $\int f(x|\theta)dx = 1$. Furthermore, from (2.48), $E\tilde{v} = \mathbf{0}$, so that $E\tilde{z} = HE\tilde{v} = \mathbf{0}$, and from the definition of z, $E\tilde{x}_1 = \mu_1$ and $E\tilde{x}_2 = \mu_2$. Thus, $E\tilde{x} = \mu$, $\mu' = (\mu_1, \mu_2)$, is the mean of $\tilde{x}$. Also, from $E\tilde{v}\tilde{v}' = I_2$ and $\tilde{z} = H\tilde{v}$, $E\tilde{z}\tilde{z}' = HE\tilde{v}\tilde{v}'H' = HH' = P$, since from $H'P^{-1}H = I_2$, $P = HH'$. Therefore,

$$V(\tilde{z}) = E\tilde{z}\tilde{z}' = \begin{bmatrix} E\tilde{z}_1^2 & E\tilde{z}_1\tilde{z}_2 \\ E\tilde{z}_2\tilde{z}_1 & E\tilde{z}_2^2 \end{bmatrix} = \begin{pmatrix} 1 & \rho \\ \rho & 1 \end{pmatrix}, \tag{2.49}$$

where the matrix involving ρ is the inverse of P^{-1}, introduced above. Then (2.49) yields

$$E\tilde{z}_1^2 = E\left(\frac{\tilde{x}_1 - \mu_1}{\sigma_1}\right)^2 = 1 \quad \text{or} \quad E(\tilde{x}_1 - \mu_1)^2 = \sigma_1^2,$$

$$E\tilde{z}_2^2 = E\left(\frac{\tilde{x}_2 - \mu_2}{\sigma_2}\right)^2 = 1 \quad \text{or} \quad E(\tilde{x}_2 - \mu_2)^2 = \sigma_2^2,$$

and

$$E\tilde{z}_1\tilde{z}_2 = E\left(\frac{\tilde{x}_1 - \mu_1}{\sigma_1}\right)\left(\frac{\tilde{x}_2 - \mu_2}{\sigma_2}\right) = \rho \quad \text{or} \quad E(\tilde{x}_1 - \mu_1)(\tilde{x}_2 - \mu_2) = \sigma_1\sigma_2\rho.$$

From these results, $V(\tilde{x}_1) = \sigma_1^2$, $V(\tilde{x}_2) = \sigma_2^2$ and $\rho = E(\tilde{x}_1 - \mu_1)(\tilde{x}_2 - \mu_2)/\sigma_1\sigma_2$ is the correlation coefficient for $\tilde{x}_1$ and $\tilde{x}_2$.

To obtain the marginal and conditional pdfs associated with the BN pdf, write Q in (2.46) as

$$Q = \left[(z_2 - \rho z_1)^2 + (1-\rho^2)z_1^2\right]/(1-\rho^2), \tag{2.50a}$$

$$= \left[(z_1 - \rho z_2)^2 + (1-\rho^2)z_2^2\right]/(1-\rho^2), \tag{2.50b}$$

where $z_i = (x_i - \mu_i)/\sigma_i$, $i = 1,2$. Substituting (2.50a) into (2.45) and noting that $\mathrm{d}z_1\mathrm{d}z_2 = \mathrm{d}x_1\mathrm{d}x_2/\sigma_1\sigma_2$, the pdf for $\tilde{z}_1$ and $\tilde{z}_2$ is

$$f(z_1, z_2|\rho) = f(z_2|z_1, \rho)g(z_1), \tag{2.51}$$

where

$$f(z_2|z_1, \rho) = \left[2\pi(1-\rho^2)\right]^{-1/2}\exp\{-(z_2 - \rho z_1)^2/2(1-\rho^2)\} \tag{2.51a}$$

and

$$g(z_1) = (2\pi)^{-1/2}\exp\{-z_1^2/2\}. \tag{2.51b}$$

That (2.51b) is the marginal pdf for $\tilde{z}_1$ can be established by integrating (2.51) with respect to z_2. Given this result, $f(z_2|z_1, \rho) = f(z_1, z_2|\rho)/g(z_1)$ is the conditional pdf for $\tilde{z}_2$ given $\tilde{z}_1$ and is shown explicitly in (2.51a).

From (2.51b), it is seen that the marginal pdf for $\tilde{z}_1$ is a standardized normal pdf with zero mean and unit variance. Since $\tilde{z}_1 = (\tilde{x}_1 - \mu_1)/\sigma_1$, the marginal pdf for $\tilde{x}_1$ is

$$g(x_1|\mu_1, \sigma_1) = \left(2\pi\sigma_1^2\right)^{-1/2}\exp\{-(x_1 - \mu_1)^2/2\sigma_1^2\},$$

a normal pdf with mean μ_1 and variance σ_1^2.

From (2.51a), the conditional pdf for $\tilde{z}_2$, given $\tilde{z}_1$, is normal with conditional mean ρz_1 and conditional variance $1-\rho^2$. Since $\tilde{z}_2 = (\tilde{x}_2 - \mu_2)/\sigma_2$ the conditional pdf for $\tilde{x}_2$, given $\tilde{x}_1$, is normal, that is,

$$f(x_2|x_1, \boldsymbol{\theta}) = \left[2\pi\sigma_2^2(1-\rho^2)\right]^{-1/2} \times\exp\{-\left[x_2 - \mu_2 - \beta_{2\cdot1}(x_1 - \mu_1)\right]^2/2(1-\rho^2)\sigma_2^2\}, \tag{2.52}$$

where $\boldsymbol{\theta}' = (\rho, \mu_1, \mu_2\beta_{2\cdot1}, \sigma_2)$, with $\beta_{2\cdot1} \equiv \sigma_2\rho/\sigma_1$. From (2.52),

$$E(\tilde{x}_2|\tilde{x}_1 = x_1) = \mu_2 + \beta_{2\cdot1}(x_1 - \mu_1) \tag{2.53}$$

and

$$V(\tilde{x}_2|\tilde{x}_1 = x_1) = \sigma_2^2(1-\rho^2), \tag{2.54}$$

where $E(\tilde{x}_2|\tilde{x}_1 = x_1)$ is the conditional mean of $\tilde{x}_2$, given $\tilde{x}_1$, and $V(\tilde{x}_2|\tilde{x}_1 = x_1)$ is the conditional variance of $\tilde{x}_2$, given $\tilde{x}_1$. Note from (2.53) that the conditional mean of $\tilde{x}_2$ is linear in x_1 with slope or "regression" coefficient $\beta_{2\cdot1} = \sigma_2\rho/\sigma_1$.

The marginal pdf for $\tilde{x}_2$ and the conditional pdf for $\tilde{x}_1$, given $\tilde{x}_2$, may be obtained by substituting (2.50b) into (2.45) and performing the operations in the preceding paragraphs. The results are:

$$f(x_1, x_2|\boldsymbol{\theta}) = f(x_1|x_2, \boldsymbol{\theta}_1)g(x_2|\boldsymbol{\theta}_2), \tag{2.55}$$

with

$$g(x_2|\cdot) = (2\pi\sigma_2^2)^{-1/2}\exp\{-(x_2-\mu_2)^2/2\sigma_2^2\} \tag{2.56}$$

and

$$f(x_1|x_2,\cdot) = \left[2\pi\sigma_1^2(1-\rho^2)\right]^{-1/2} \times\exp\{-[x_1-\mu_1-\beta_{1\cdot 2}(x_2-\mu_2)]^2/2(1-\rho^2)\sigma_1^2\}, \tag{2.57}$$

where $\beta_{1\cdot 2} \equiv \sigma_1\rho/\sigma_2$. From (2.56), it is seen that the marginal pdf for $\tilde{x}_2$ is normal with mean μ_2 and variance σ_2^2, while from (2.57) the conditional pdf for $\tilde{x}_1$, given $\tilde{x}_2$, is normal with

$$E(\tilde{x}_1|\tilde{x}_2 = x_2) = \mu_1 + \beta_{1\cdot 2}(x_2 - \mu_2) \tag{2.58}$$

and

$$V(\tilde{x}_1|\tilde{x}_1 = x_1) = \sigma_1^2(1-\rho^2). \tag{2.59}$$

$\beta_{1\cdot 2} = \sigma_1\rho/\sigma_2$ is the regression coefficient.

From what has been presented above, it is the case that (1) all marginal and conditional pdfs are in the normal form and (2) both conditional means in (2.53) and (2.58) are linear in the conditioning variable. Since $E(\tilde{x}_1|\tilde{x}_2)$ and $E(\tilde{x}_2|\tilde{x}_1)$ define the "regression functions" for a bivariate pdf in general, the bivariate normal pdf is seen to have both regression functions linear. Further, from the definitions of $\beta_{2\cdot 1}$ and $\beta_{1\cdot 2}$, $\beta_{2\cdot 1}\beta_{1\cdot 2} = \rho^2$, the squared correlation coefficient, so that the regression coefficients have the same algebraic signs. Further, if $\rho = 0$, the joint pdf in (2.45) factors into

$$\prod_{i=1}^{2}(2\pi\sigma_i^2)^{-1/2}\exp\{-(x_i-\mu_i)^2/2\sigma_i^2\},$$

showing that with $\rho = 0$, $\tilde{x}_1$ and $\tilde{x}_2$ are independent. Thus, for the BN distribution, $\rho = 0$ implies independence and also, as is true in general, independence implies $\rho = 0$. Note also that with $\rho = 0$, the conditional variances reduce to marginal variances and $\beta_{1\cdot 2} = \beta_{2\cdot 1} = 0$, that is, the regressions in (2.53) and (2.58) have zero slope coefficients.

(2) Multivariate normal (MVN). An m-element random vector, $\tilde{\boldsymbol{x}}' = (\tilde{x}_1, \tilde{x}_2, \ldots, \tilde{x}_m)$, has a MVN distribution if and only if its pdf is

$$f(\boldsymbol{x}|\boldsymbol{\theta}, \Sigma) = (2\pi)^{-m/2}|\Sigma|^{-1/2}\exp\{-(\boldsymbol{x}-\boldsymbol{\theta})'\Sigma^{-1}(\boldsymbol{x}-\boldsymbol{\theta})/2\}, \tag{2.60}$$

with $-\infty < x_i < \infty$, $i = 1,2,\ldots,m$, $\boldsymbol{\theta}' = (\theta_1, \theta_2, \ldots, \theta_m)$, $-\infty < \theta_i < \infty$, $i = 1,2,\ldots,m$, and $\Sigma = \{\sigma_{ij}\}$ is an $m \times m$ positive definite symmetric matrix. When $m = 1$, (2.60) is a univariate normal pdf, and when $m = 2$ it is a bivariate normal pdf.

If H is an $m \times m$ non-singular matrix such that $H'\Sigma^{-1}H = I_m$ and $\boldsymbol{x} - \boldsymbol{\theta} = H\boldsymbol{z}$, then the pdf for $\boldsymbol{z}' = (z_1, z_2, \ldots, z_m)$ is[12]

$$f(\boldsymbol{z}) = (2\pi)^{-m/2}\exp\{-\boldsymbol{z}'\boldsymbol{z}/2\}. \tag{2.61}$$

From (2.61), the $\tilde{z}_i$'s are independent, standardized normal variables and therefore (2.60) and (2.61) integrate to one. In addition, (2.61) implies $E\tilde{\boldsymbol{z}} = H^{-1}E(\tilde{\boldsymbol{x}} - \boldsymbol{\theta}) = \boldsymbol{0}$, or

$$E\tilde{\boldsymbol{x}} = \boldsymbol{\theta}. \tag{2.62}$$

Thus, $\boldsymbol{\theta}$ is the mean of the MVN pdf. Also, (2.61) implies that $E\boldsymbol{z}\boldsymbol{z}' = I_m$, since the elements of $\tilde{\boldsymbol{z}}$ are independent standardized normal rvs. It follows that $H^{-1}E(\tilde{\boldsymbol{x}} - \boldsymbol{\theta})(x-\boldsymbol{\theta})'(H')^{-1} = I_m$; that is, $E(\tilde{\boldsymbol{x}}-\boldsymbol{\theta})(\tilde{\boldsymbol{x}}-\boldsymbol{\theta})' = HH'$, or

$$V(\tilde{\boldsymbol{x}}) = E(\tilde{\boldsymbol{x}}-\boldsymbol{\theta})(\tilde{\boldsymbol{x}}-\boldsymbol{\theta})' = \Sigma, \tag{2.63}$$

since from $H'\Sigma^{-1}H = I_m$, $\Sigma = HH'$. Thus, Σ is the covariance matrix of $\tilde{\boldsymbol{x}}$.

To obtain the marginal and conditional pdfs associated with (2.60), let $G \equiv \Sigma^{-1}$ and partition $\boldsymbol{x} - \boldsymbol{\theta}$ and G correspondingly as

$$\boldsymbol{x} - \boldsymbol{\theta} = \begin{pmatrix} \boldsymbol{x}_1 - \boldsymbol{\theta}_1 \\ \boldsymbol{x}_2 \quad \boldsymbol{\theta}_2 \end{pmatrix} \quad \text{and} \quad G = \begin{pmatrix} G_{11} & G_{12} \\ G_{21} & G_{22} \end{pmatrix}.$$

Then the exponent of (2.60) can be expressed as

$$(\boldsymbol{x}-\boldsymbol{\theta})'G(\boldsymbol{x}-\boldsymbol{\theta}) = Q_1 + Q_2, \tag{2.64}$$

[12] Note that the Jacobian of the transformation from $\boldsymbol{x} - \boldsymbol{\theta}$ to $\boldsymbol{z}$ is $|H|$ and $|\Sigma|^{-1/2} = |H|^{-1}$ from $|H'\Sigma^{-1}H| = |I_m| = 1$. Thus, $|\Sigma|^{-1/2}|H| = 1$.

with

$$Q_1 = \left[x_1 - \theta_1 + G_{11}^{-1}G_{12}(x_2 - \theta_2)\right]'G_{11}\left[x_1 - \theta_1 + G_{11}^{-1}G_{12}(x_2 - \theta_2)\right] \quad (2.65)$$

and

$$Q_2 = (x_2 - \theta_2)'(G_{22} - G_{21}G_{11}^{-1}G_{12})(x_2 - \theta_2). \quad (2.66)$$

From

$$|\Sigma|^{-1} = |G| = |G_{11}||G_{22} - G_{21}G_{11}^{-1}G_{12}|,$$

(2.60) can be expressed as

$$f(x_1, x_2|\theta, G) = f(x_1|x_2, G, \theta)g(x_2|G, \theta_2), \quad (2.67)$$

where

$$f(x_1|x_2, G, \theta) = (2\pi)^{-m_1/2}|G_{11}|^{1/2}\exp\{-Q_1/2\} \quad (2.67a)$$

and

$$g(x_2|G, \theta_2) = (2\pi)^{-m_2/2}|G_{22} - G_{21}G_{11}^{-1}G_{12}|^{1/2}\exp\{-Q_2/2\}, \quad (2.67b)$$

with m_i the number of elements in x_i, $i = 1,2$, $m_1 + m_2 = m$, Q_1, and Q_2 as defined in (2.65) and (2.66), respectively, and $G = \Sigma^{-1}$.

On integrating (2.67) with respect to x_1, the result is $g(x_2|G, \theta_2)$, the marginal pdf for x_2, where $f(x_1|x_2, G, \theta)$ is the conditional pdf for $\tilde{x}_1$, given $\tilde{x}_2$. Both (2.67a) and (2.67b) are normal pdfs. The mean and covariance matrices of the marginal pdf in (2.67b) are[13]

$$E\tilde{x}_2 = \theta_2,$$

$$V(\tilde{x}_2) = (G_{22} - G_{21}G_{11}^{-1}G_{12})^{-1} = \Sigma_{22}.$$

[13] On partitioning $G = \Sigma^{-1}$ and Σ correspondingly as

$$G = \begin{pmatrix} G_{11} & G_{12} \\ G_{21} & G_{22} \end{pmatrix}$$

and

$$\Sigma = \begin{pmatrix} \Sigma_{11} & \Sigma_{12} \\ \Sigma_{21} & \Sigma_{22} \end{pmatrix}$$

then from results on inverting partitioned matrices, $\Sigma_{22} = (G_{22} - G_{21}G_{11}^{-1}G_{12})^{-1}$, $\Sigma_{12}\Sigma_{22}^{-1} = -G_{11}^{-1}G_{12}$, and $G_{11}^{-1} = \Sigma_{11} - \Sigma_{12}\Sigma_{22}^{-1}\Sigma_{21}$.

The mean and covariance matrices of the conditional pdf in (2.67a) are

$$E(\tilde{\boldsymbol{x}}_1|\tilde{\boldsymbol{x}}_2) = \boldsymbol{\theta}_1 - G_{11}^{-1}G_{12}(\boldsymbol{x}_2 - \boldsymbol{\theta}_2)$$

$$= \boldsymbol{\theta}_1 + \Sigma_{12}\Sigma_{22}^{-1}(\boldsymbol{x}_2 - \boldsymbol{\theta}_2), \tag{2.68}$$

$$V(\tilde{\boldsymbol{x}}_1|\tilde{\boldsymbol{x}}_2) = G_{11}^{-1} = \Sigma_{11} - \Sigma_{12}\Sigma_{22}^{-1}\Sigma_{21}. \tag{2.69}$$

Similar operations can be utilized to derive the marginal pdf for $\tilde{\boldsymbol{x}}_1$ and the conditional pdf for $\tilde{\boldsymbol{x}}_2$, given $\tilde{\boldsymbol{x}}_1$. As in the case of (2.67a) and (2.67b), the marginal and conditional pdfs are MVN pdfs. In addition, just as $E(\tilde{x}_1|\tilde{x}_2)$ in (2.68) is linear in x_2, $E(\tilde{\boldsymbol{x}}_2|\tilde{\boldsymbol{x}}_1) = \boldsymbol{\theta}_2 + \Sigma_{21}\Sigma_{11}^{-1}(\boldsymbol{x}_1 - \boldsymbol{\theta}_1)$ is linear in $\boldsymbol{x}_1$. Thus, both conditional expectations, or regression functions, are linear in the conditioning variables.

The conditional expectation of $\tilde{\boldsymbol{x}}_1$, given $\tilde{\boldsymbol{x}}_2$, is called the regression function for $\tilde{\boldsymbol{x}}_1$, given $\tilde{\boldsymbol{x}}_2$. As (2.68) indicates, in the case of the MVN distribution, this regression function is linear and $B' \equiv \Sigma_{12}\Sigma_{22}^{-1}$ is the $m_1 \times m_2$ matrix of partial regression coefficients. If $\boldsymbol{x}_1$ has just one element, then the *vector* of partial regression coefficients is $\boldsymbol{\beta}' = \boldsymbol{\sigma}_{12}'\Sigma_{22}^{-1}$, where $\boldsymbol{\sigma}_{12}'$ is a $1 \times m_2$ vector of covariances of $\tilde{x}_1$, and the elements of $\tilde{\boldsymbol{x}}_2$ are the first row of Σ_{12}. With respect to partial regression coefficients, it is instructive to write

$$(\tilde{\boldsymbol{x}}_1 - \boldsymbol{\theta}_1)' = (\tilde{\boldsymbol{x}}_2 - \boldsymbol{\theta}_2)'B + \tilde{\boldsymbol{u}}', \tag{2.70}$$

where $\tilde{\boldsymbol{u}}'$ is a $1 \times m_1$ random vector with $E\tilde{\boldsymbol{u}}|\tilde{\boldsymbol{x}}_2 = \boldsymbol{0}$ and $E(\tilde{\boldsymbol{x}}_2 - \boldsymbol{\theta}_2)\tilde{\boldsymbol{u}}' = 0$. Then on multiplying (2.70) on the left by $\boldsymbol{x}_2 - \boldsymbol{\theta}_2$ and taking the expectation of both sides, the result is $\Sigma_{21} = \Sigma_{22}B$ or

$$B = \Sigma_{22}^{-1}\Sigma_{21}. \tag{2.71}$$

Note that (2.71) was obtained from (2.70) without assuming normality but, of course, it also holds in the normal case. Without normality, it is not true in general that *both* $E(\tilde{\boldsymbol{x}}_1|\tilde{\boldsymbol{x}}_2 = \boldsymbol{x}_2)$ and $E(\tilde{\boldsymbol{x}}_2|\tilde{\boldsymbol{x}}_1 = \boldsymbol{x}_1)$ are linear in $\boldsymbol{x}_2$ and $\boldsymbol{x}_1$, respectively. For the cases of non-normality, it may be that one of these conditional expectations is linear, but in general both will not be linear except in special cases, for example the multivariate Student-t distribution discussed below.

If in the MVN distribution the elements of $\tilde{\boldsymbol{x}}$ are mutually uncorrelated, that is, $E(\tilde{x}_i - \theta_i)(\tilde{x}_j - \theta_j) = 0$, for all $i \neq j$, then Σ in (2.63) is a diagonal matrix, $\Sigma = D(\sigma_{ii})$ and from (2.60) $f(\boldsymbol{x}|\boldsymbol{\theta}, D) = \prod_{i=1}^{m} g(x_i|\theta_i, \sigma_{ii})$, where $g(x_i|\theta_i, \sigma_{ii})$ is a univariate normal pdf with mean θ_i and variance σ_{ii}. Thus, diagonality of Σ implies that the elements of $\tilde{\boldsymbol{x}}$ are independently distributed and then they are mutually uncorrelated, a result that holds in general. Thus, for the MVN distribution, diagonality of Σ implies independence, and independence of the elements of $\tilde{x}$ implies diagonality of Σ. Also, if Σ is diagonal, $\Sigma_{12}=\Sigma'_{21}=0$ and thus $B=\Sigma_{22}^{-1}\Sigma_{21}=0$; that is, with Σ diagonal, all partial regression coefficients are equal to zero. Further, with Σ diagonal, $V(\tilde{x}_1|\tilde{x}_2)$ in (2.69) specializes to $V(\tilde{x}_1|\tilde{x}_2)=D_{11}(\sigma_{ii})$, an $m_1\times m_1$ diagonal matrix with typical element, σ_{ii}, the variance of $\tilde{x}_i$.

Among many other results for the MVN distribution, the fact that linear combinations of normal variables are normally distributed is of great importance. That is, let $\tilde{w}=C\tilde{x}$, where C is a $q\times m$ given matrix of rank q and thus $\tilde{w}$ is a $q\times 1$ random vector. It can be established that $\tilde{w}$ has a q-dimensional MVN normal distribution with mean vector $E\tilde{w}=C\boldsymbol{\theta}$ and covariance matrix $E(\tilde{w}-E\tilde{w})(\tilde{w}-E\tilde{w})'=CE(\tilde{x}-\boldsymbol{\theta})(\tilde{x}-\boldsymbol{\theta})'C'=C\Sigma C'$, a $q\times q$ positive definite symmetric matrix.

Another important result is that the random variable $\tilde{y}=(\tilde{x}-\boldsymbol{\theta})'\Sigma^{-1}(\tilde{x}-\boldsymbol{\theta})$ has a χ^2 distribution with m degrees of freedom. Let $\tilde{x}-\boldsymbol{\theta}=H\tilde{z}$, where H and $\tilde{z}$ are as defined in connection with (2.61). Then $\tilde{y}=\tilde{z}'\tilde{z}=\sum_{i=1}^{m}\tilde{z}_i^2$ is a sum of independent, squared, standardized normal rvs and thus has a χ^2 pdf with m degrees of freedom. A generalization of this result is that if the random $m\times m$ symmetric matrix $\tilde{A}=\sum_{i=1}^{\nu}\tilde{z}_i\tilde{z}_i'$, where the $\tilde{z}_i$'s are mutually independent normal random vectors, each with zero mean vector and common pds covariance matrix Σ, then the $m(m+1)/2$ distinct elements of $\tilde{A}$ have a central Wishart pdf and those of $\tilde{A}^{-1}$ an inverted Wishart pdf; for $\tilde{A}$ a scalar, its pdf is a χ^2 pdf with ν degrees of freedom, while its reciprocal has an inverted χ^2 pdf.

(3) Multivariate Student-t (MVS). An m-element random vector $\tilde{x}'=(\tilde{x}_1,\tilde{x}_2,\dots,\tilde{x}_m)$, with $-\infty<\tilde{x}_i<\infty$, $i=1,2,\dots,m$, has a MVS distribution if and only if its pdf is

$$f(\boldsymbol{x}|\boldsymbol{\theta},V\nu)=c|V|^{1/2}/[\nu+(\boldsymbol{x}-\boldsymbol{\theta})'V(\boldsymbol{x}-\boldsymbol{\theta})]^{(m+\nu)/2}, \tag{2.72}$$

where $c=\nu^{\nu/2}\Gamma[(\nu+m)/2]/\pi^{m/2}\Gamma(\nu/2)$, $\nu>0$, V is an $m\times m$ pds matrix, and $\boldsymbol{\theta}'=(\theta_1,\theta_2,\dots,\theta_m)$, with $-\infty<\theta_i<\infty$, $i=1,2,\dots,m$. The pdf in (2.72) is an m-dimensional MVS pdf with ν degrees of freedom and is denoted $\mathrm{MVS}_m(\boldsymbol{\theta},V,\nu)$. If $m=1$, (2.72) is the US-t pdf. Further, if $\nu=1$, (2.72) is the multivariate Cauchy pdf and the univariate Cauchy pdf when $m=1$. As ν grows large, the MVS approaches a limiting MVN form with mean $\boldsymbol{\theta}$ and covariance matrix V^{-1}.

From the form of (2.72), it is the case that the pdf is symmetric about $\boldsymbol{\theta}$, the modal value. A standardized form of (2.72) is obtained by making the change of variable $\boldsymbol{x}-\boldsymbol{\theta}=C\boldsymbol{z}$, where C is an $m\times m$ non-singular matrix such that $C'VC=I_m$. Then the pdf for $\boldsymbol{z}$ is

$$f(\boldsymbol{z}|\nu)=c/(\nu+\boldsymbol{z}'\boldsymbol{z})^{(m+\nu)/2}, \tag{2.73}$$

an m-dimensional, standardized MVS density. It can be shown that (2.72) and (2.73) are normalized pdfs, that is, that they integrate to one. Also, the first two moments of $\tilde{z}$ are $E\tilde{z} = \mathbf{0}$ for $\nu > 1$ and $Ezz' = I_m \nu/(\nu - 2)$ for $\nu > 2$, which imply that

$$E\tilde{x} = \boldsymbol{\theta}, \qquad \nu > 1, \tag{2.74}$$

and

$$V(\tilde{x}) = E(x - \boldsymbol{\theta})(x - \boldsymbol{\theta})' = V^{-1}\nu/(\nu - 2), \qquad \nu > 2. \tag{2.75}$$

The conditions $\nu > 1$ and $\nu > 2$ are needed for the existence of moments.

If $\tilde{x}$ is partitioned, $\tilde{x}' = (\tilde{x}_1', \tilde{x}_2')$, marginal and conditional pdfs can be obtained by methods similar to those employed in connection with the MVN. Let V in (2.72) be partitioned, $V = \{V_{ij}\}$, $i, j = 1,2$, to correspond to the partitioning of $\tilde{x}$ into $\tilde{x}_1$ with m_1 elements and $\tilde{x}_2$ with m_2 elements. Then the marginal pdf, for $\tilde{x}_2$, is in the MVS form, that is, $\mathrm{MVS}_{m_2}(\boldsymbol{\theta}_2, V_{2\cdot 1}, \nu)$, where $\boldsymbol{\theta}_2$ is a subvector of $\boldsymbol{\theta}' = (\boldsymbol{\theta}_1', \boldsymbol{\theta}_2')$, partitioned to correspond to the partitioning of $\tilde{x}$, and $V_{2\cdot 1} = V_{22} - V_{12}V_{11}^{-1}V_{12}$. As regards the conditional pdf for $\tilde{x}_1$, given $\tilde{x}_2$, it too is a MVS pdf, namely $\mathrm{MVS}_{m_1}(\delta_{1\cdot 2}, M, \nu')$, with $\nu' = m_2 + \nu$:

$$\boldsymbol{\delta}_{1\cdot 2} = \boldsymbol{\theta}_1 - V_{11}^{-1}V_{12}(x_2 - \boldsymbol{\theta}_2) \tag{2.76}$$

and

$$M = \nu' V_{11} / \left[\nu + (x_2 - \boldsymbol{\theta}_2)' V_{2\cdot 1}(x_2 - \boldsymbol{\theta}_2)\right]. \tag{2.77}$$

For $\nu' > 1$, $\boldsymbol{\delta}_{1\cdot 2}$ in (2.76) is the conditional mean of $\tilde{x}_1$, given $\tilde{x}_2$; note its similarity to the conditional mean for the MVN pdf in (2.68) in that it is linear in x_2. Also, $E\tilde{x}_2|x_1$ is linear in x_1. Thus, the MVS pdf with $\nu > 1$ has all conditional means or regression functions linear. In addition, if V is diagonal, (2.75) indicates that all elements of $\tilde{x}$ are uncorrelated, given $\nu > 2$, and from (2.76), $\boldsymbol{\delta}_{1\cdot 2} = \boldsymbol{\theta}_1$ when V is diagonal. From the form of (2.72), it is clear that diagonality of V, or lack of correlation, does not imply independence for the MVS pdf, in contrast to the MVN case. This feature of the MVS pdf can be understood by recognizing that the MVS can be generated as a continuous mixture of normal pdfs. That is, let $\tilde{x}$ have a MVN pdf with mean vector $\boldsymbol{\theta}$ and covariance matrix $\sigma^2 V^{-1}$, denoted by $f_N(\boldsymbol{\theta}, V^{-1}\sigma^2)$, and consider σ to be random with the inverted gamma pdf in (2.33). Then

$$\mathrm{MVS}_m(\boldsymbol{\theta}, V^{-1}, \nu) = \int_0^\infty f_{\mathrm{N}}(\boldsymbol{\theta}, V^{-1}\sigma^2) f_{\mathrm{IG}}(\sigma|\nu, 1)\,\mathrm{d}\sigma.$$

Thus, if V is diagonal, independence does not result because of the common random σ in $f_N(\boldsymbol{\theta}, V^{-1}\sigma^2)$.

Three other important features of the MVS pdf are: (1) If $\tilde{x}$ is MVS, then $\tilde{w} = L\tilde{x}$, where L is a given $q \times m$ matrix of rank q, is MVS; (2) the quadratic form in (2.72) divided by m, that is, $(\tilde{x} - \boldsymbol{\theta})'V(\tilde{x} - \boldsymbol{\theta})/m$, has an $F_{m,\nu}$ pdf [see, for example, Zellner (1971, p. 385)]; and (3) the quantity $t_i = (\tilde{x}_i - \theta_i)/\sqrt{v^{ii}}$ has a univariate Student-t pdf with ν degrees of freedom where v^{ii} is the i, ith element of V^{-1}.

2.5. *Elements of asymptotic theory*[14]

In previous sections specific pdfs for rvs were described and discussed, the use of which permits one to make probability statements about values of rvs. This section is devoted to a review of some results relevant for situations in which the exact pdfs for rvs are assumed unknown. Use of asymptotic theory provides, among other results, *approximate* pdfs for rvs under relatively weak assumptions. As Cox and Hinkley (1974, p. 247) state: "The numerical accuracy of such an [asymptotic] approximation always needs consideration...." For example, consider a random sample mean, $\bar{X}_n = \sum_{i=1}^n X_i/n$. Without completely specifying the pdfs for the X_i's, central limit theorems (CLTs) yield the result that $\bar{X}_n$ is approximately normally distributed for large finite n given certain assumptions about the X_i. Then it is possible to use this approximate normal distribution to make probability statements regarding possible values of $\bar{X}_n$. Similar results are available relating to other functions of the X_i's, for example, $S_n^2 = \sum_{i=1}^n (X_i - \bar{X}_n)^2/n$, etc. The capability of deducing the "large sample" properties of functions of rvs such as $\bar{X}_n$ or S_n^2 is very useful in econometrics, especially when exact pdfs for the X_i's are not known and/or when the X_i's have known pdfs but functions of the X_i's have complicated or unknown pdfs.

The following are some useful inequalities for rvs that are frequently employed in asymptotic theory and elsewhere.

2.5.1. *Selected inequalities for random variables*

Chebychev's Inequality: Let X be any rv with $EX^2 < \infty$. Then, with ε an arbitrary constant such that $\varepsilon > 0$,

$$\mathrm{P}\{|X| > \varepsilon\} \leqslant EX^2/\varepsilon^2. \tag{2.78}$$

As an example of (2.78), let $X_n = Z_n - \theta$, where Z_n is the random proportion of

[14]For simplicity of notation, in this section rvs are denoted by capital letters, e.g. X, Y, Z, etc. See, for example, Cramér (1946), Loève (1963), Rao (1973), Cox and Hinkley (1974) and the references cited in these works for further consideration of topics in asymptotic theory.

successes in n independent binomial trials with $EZ_n = \theta$, the common probability of success on individual trials, and $EX_n^2 = E(Z_n - \theta)^2 = \theta(1-\theta)/n$, the variance of Z_n. Then $\mathrm{P}\{|Z_n - \theta| > \varepsilon\} \leqslant E(Z_n - \theta)^2/\varepsilon^2 = \theta(1-\theta)/n\varepsilon^2$, which approaches 0 as $n \to \infty$. In this sense, the rv Z_n approaches θ as n grows large, a type of convergence that is defined below as "convergence in probability (i.p.)" and denoted by $\operatorname{plim} Z_n = \theta$ or $Z_n \overset{\text{i.p.}}{\to} \theta$. Note further that if X_n were known to be $N(0,1)$, $\Pr\{|X_n| > 1.96\} = 0.05$ and (2.78) yields $\Pr\{|X_n| > 1.96\} \leqslant 1/(1.96)^2 = 0.260$. Thus, in this case while (2.78) is true, it does not yield a very strong bound. On the other hand, if X_n has a symmetric pdf with mean zero, finite second moment, and heavier tails than the normal pdf, $\mathrm{P}\{|X_n| > 1.96\} > 0.05$ and the bound yielded by (2.78) is closer to $\mathrm{P}\{|X_n| > 1.96\}$.

A useful generalization of (2.78) is

$$\mathrm{P}\{Y > \varepsilon\} \leqslant Eg(Y)/g(\varepsilon), \tag{2.79}$$

where $\varepsilon > 0$ is given, Y is any rv, $g(\cdot)$ is a non-negative, even function, defined on the range of Y such that $g(\cdot)$ is non-decreasing, and $Eg(\cdot) < \infty$. If in (2.79) $Y = |X|$ and $g(|X|) = X^2$ for $|X| \geqslant 0$ and zero otherwise, (2.79) reduces to (2.78). If in (2.79) $Y = |X|$ and $g(|X|) = |X|$ for $|X| \geqslant 0$ and zero otherwise, (2.79) reduces to $P\{|X| > \varepsilon\} \leqslant E|X|/\varepsilon$. Other choices of $g(\cdot)$, for example $g(|X|) = |X|^r$, $r > 0$, produce inequalities involving higher order moments, $P\{|X| > \varepsilon\} \leqslant E|X|^r/\varepsilon^r$, Markov's Inequality that includes (2.78) as a special case.

Some additional inequalities are:[15]

(1) $E|X+Y|^r \leqslant c_r(E|X|^r + E|Y|^r)$, where $c_r = 1$ for $r \leqslant 1$ and $c_r = 2^{r-1}$ for $r \geqslant 1$. Thus, if the rth absolute moments of X and Y exist and are finite, so is the rth absolute moment of $X+Y$.

(2) *Hölder Inequality*: $E|XY| \leqslant [E|X|^r]^{1/r}[E|Y|^s]^{1/s}$, where $r > 1$ and $1/r + 1/s = 1$.

(3) *Minkowski Inequality*: If $r \geqslant 1$, then $[E|X+Y|^r]^{1/r} \leqslant [E|X|^r]^{1/r} + [E|Y|^r]^{1/r}$.

(4) *Schwarz Inequality*: $E|XY| \leqslant [E|X|^2 E|Y|^2]$ or $[E|XY|]^2 \leqslant E|X|^2 E|Y|^2$, which is a special case of Hölder's Inequality with $r = s = 2$.

These inequalities are useful in establishing properties of functions of rvs.

Now various types of convergence of sequences of rvs will be defined.

2.5.2. *Convergence of sequences of random variables*

Consider a sequence of rvs $\{X_n\}$, $n = 1, 2, \ldots$. A specific example of such a sequence is $\bar{X}_1, \bar{X}_2, \bar{X}_3, \ldots, \bar{X}_n, \ldots$, where $\bar{X}_1 = X_1$, $\bar{X}_2 = \sum_{i=1}^2 X_i/2$, $\bar{X}_3 = \sum_{i=1}^3 X_i/3, \ldots, \bar{X}_n = \sum_{i=1}^n X_i/n$, a sequence of means. Since the members of such

[15] For proofs of these and other inequalities, see Loève (1963, p. 154 ff.)

sequences are rvs, the usual mathematical limit of a sequence of non-stochastic quantities does not apply. Thus, it is necessary to define appropriate limits of sequences of rvs. Sequences of rvs can converge, in senses to be defined below, to a non-stochastic constant or to a random variable. The following are the major modes of convergence for sequences of rvs.

1. *Weak convergence or convergence in probability (i.p.).* If for the sequence $\{X_n\}$, $n = 1, 2, \ldots$,

$$\lim_{n \to \infty} \mathrm{P}\{|X_n - c| > \varepsilon\} = 0 \tag{2.80}$$

for every given $\varepsilon > 0$, then the sequence converges weakly or i.p. to the constant c. Alternative ways of writing (2.80) are $X_n \xrightarrow{\text{i.p.}} c$, or $X_n \xrightarrow{P} c$, or $\operatorname{plim} X_n = c$, where "plim" represents the particular limit given in (2.80) and is the notation most frequently employed in econometrics.

2. *Strong convergence or convergence almost surely (a.s.).* If for the sequence $\{X_n\}$, $n = 1, 2, \ldots$,

$$\mathrm{P}\left(\lim_{n \to \infty} X_n = c\right) = 1, \tag{2.81}$$

then the sequence converges strongly or a.s. to c, denoted by $X_n \xrightarrow{\text{a.s.}} c$. An alternative way of expressing (2.81) is

$$\lim_{N \to \infty} \mathrm{P}\left\{\sup_{n \geqslant N} |X_n - c| > \varepsilon\right\} = 0$$

for every given $\varepsilon > 0$.

3. *Convergence in quadratic mean (q.m.).* If for the sequence $\{X_n\}$, $n = 1, 2, \ldots$,

$$\lim_{n \to \infty} E(X_n - c)^2 = 0, \tag{2.82}$$

then the sequence converges in quadratic mean to c, also expressed as $X_n \xrightarrow{\text{q.m.}} c$.

A sequence of rvs $\{X_n\}$, $n = 1, 2, \ldots$, is said to converge to a rv X in the sense of (2.80), (2.81), or (2.82) if and only if the sequence $\{X_n - X\}$, $n = 1, 2, \ldots$, converges to $c = 0$ according to (2.80), (2.81), or (2.82). In the case of (2.80) such convergence is denoted by $X_n \xrightarrow{\text{i.p.}} X$ or $\operatorname{plim}(X_n - X) = 0$; in the case of (2.81) by $X_n \xrightarrow{\text{a.s.}} X$; and in the case of (2.82) by $X_n \xrightarrow{\text{q.m.}} X$.

Rao (1973, p. 110f.) proves the following relations:

(a) Convergence in q.m. (2.82) implies convergence i.p. (2.80).
(b) Convergence a.s. (2.81) implies convergence i.p. (2.80).
(c) If $X_n \overset{\text{q.m.}}{\to} c$ in such a way that $\sum_{n=1}^{\infty} E(X_n - c)^2 < \infty$, then $X_n \overset{\text{a.s.}}{\to} c$.

In connection with a sequence of sample means, $\{\bar{X}_n\}$, $n = 1, 2, \ldots$, with $E\bar{X}_n = \mu$ and $\text{Var}(\bar{X}_n) = \sigma^2/n$, Chebychev's Inequality (2.78) yields $P\{|\bar{X}_n - \mu| > \varepsilon\} \leqslant E(\bar{X}_n - \mu)^2/\varepsilon^2 = \sigma^2/n\varepsilon^2$. Thus, $\lim_{n\to\infty} P\{|\bar{X}_n - \mu| > \varepsilon\} = 0$; that is, $\text{plim}\ \bar{X}_n = \mu$ or $\bar{X}_n \overset{\text{i.p.}}{\to} \mu$. Further, on applying (2.82), $\lim_{n\to\infty} E(\bar{X}_n - \mu)^2 = \lim_{n\to\infty} \sigma^2/n = 0$ and thus $\bar{X}_n \overset{\text{q.m.}}{\to} \mu$.

In the case of a sequence of sample means, and in many other cases, it is valuable to know under what general conditions if and how $\bar{X}_n$ converges to a limiting constant. Laws of large numbers (LLN) provide answers to these questions.

2.5.3. *Laws of large numbers (LLN)*

In this section several LLN are reviewed. Weak laws of large numbers (WLLN) relate to cases in which sequences of rvs converge weakly or in probability while strong laws of large numbers (SLLN) relate to cases in which sequences converge strongly, that is almost surely or with probability 1. In what follows, the sequence of rvs is $\{X_n\}$, $n = 1, 2, \ldots$, and sequences of averages $\bar{X}_n = \sum_{i=1}^{n} X_i/n$, $n = 1, 2, \ldots$, are considered.

Chebychev's WLLN

If $EX_i = \mu_i$, $V(X_i) = E(X_i - \mu_i)^2 = \sigma_i^2$, and $\text{cov}(X_i, X_j) = E(X_i - \mu_i)(X_j - \mu_j) = 0$, $i \neq j$ for all $i, j = 1, 2, \ldots$, then $\lim_{n\to\infty} \bar{\sigma}^2/n = 0$, where $\bar{\sigma}^2 = \sum_{i=1}^{n} \sigma_i^2/n$ implies that

$$\bar{X}_n - \bar{\mu}_n \overset{\text{i.p.}}{\to} 0,$$

or $\text{plim}\,(\bar{X}_n - \bar{\mu}_n) = 0$, with $\bar{\mu}_n = \sum_{i=1}^{n} \mu_i/n$.[16]

As a special case of this WLLN, if $\mu_i = \mu$ and $\sigma_i = \sigma$ for all i and $\text{cov}(X_i, X_j) = 0$, then $\bar{X}_n \overset{\text{i.p.}}{\to} \mu$ or $\text{plim}\ \bar{X}_n = \mu$.

[16]Proof is by use of Chebychev's Inequality (2.78) with $X = \bar{X}_n - \bar{\mu}_n$ since $E(\bar{X}_n - \bar{\mu}_n)^2 = \bar{\sigma}^2/n$. Therefore $P\{|\bar{X}_n - \bar{\mu}_n| > \varepsilon\} \leqslant \bar{\sigma}^2/n\varepsilon$ and thus $\lim_{n\to\infty} P\{|\bar{X}_n - \bar{\mu}_n| > \varepsilon\} = 0$. For proofs and discussion of this and other LLN see, for example, Rao (1973, p. 111 ff.).

Khintchin's WLLN

If $X_1, X_2, \ldots$ are independent and identically distributed (i.i.d.) rvs and $EX_i = \mu < \infty$, then $\bar{X}_n \overset{\text{i.p.}}{\to} \mu$ (or plim $\bar{X}_n = \mu$).

In Khintchin's WLLN, there is no requirement that second moments exist as in Chebychev's WLLN; however, the former WLLN does require that the X_i's be i.i.d. and have a common finite mean. As an example in which Khintchin's WLLN applies but Chebychev's does not, consider the X_i's to be i.i.d. each with a univariate Student-t pdf with degrees of freedom $\nu = 2$. For $\nu = 2$, $EX_i = \mu < \infty$ but the second moment does not exist and thus Chebychev's WLLN cannot be applied but Khintchin's WLLN can. On the other hand, for the X_i's i.i.d. univariate Cauchy, the mean does not exist and thus neither law can be applied. While these exceptions are worth noting, there are many sequences to which these WLLN can be applied. However, special results are needed to handle cases in which the X_i's do not possess moments and/or are correlated.

Kolmogorov's First SLLN

If $X_1, X_2, \ldots$ is a sequence of independent rvs such that $EX_i = \mu_i$ and $V(X_i) = \sigma_i^2$, $i = 1, 2, \ldots$, and if $\sum_{i=1}^{\infty} \sigma_i^2 / i^2 < \infty$, then, with $\bar{\mu}_n = \sum_{i=1}^{n} \mu_i / n$,

$$\bar{X}_n - \bar{\mu}_n \overset{\text{a.s.}}{\to} 0,$$

and the sequence $X_1, X_2, \ldots$ is said to obey the SLLN. Further, if $\mu_i = \mu$ for all i, $\bar{X}_n \overset{\text{a.s.}}{\to} \mu$.

Kolmogorov's Second SLLN

If $X_1, X_2, \ldots$ is a sequence of i.i.d. rvs, then a necessary and sufficient condition that $\bar{X}_n \overset{\text{a.s.}}{\to} \mu$ is that $EX_i = \mu < \infty$ for all i.

Kolmogorov's Second SLLN does not require the existence of the second moments of the independent X_i's as in his first law; however, in the second law the X_i must be independently *and identically* distributed, which is not assumed in his first law. In the first law, if $\mu_i = \mu$ and $\sigma_i^2 = \sigma^2$, the X_i's need not be identically distributed and still, $\bar{X}_n \overset{\text{a.s.}}{\to} \mu$ since $\sigma^2 \sum_{i=1}^{\infty} 1/i^2 < \infty$.

2.5.4. Convergence of sequences of distributions and density functions and central limit theorems (CLTs)

Let $\{F_n\}$, $n = 1, 2, \ldots$, be a sequence of cumulative distribution functions (cdfs) for the rvs $\{X_n\}$, $n = 1, 2, \ldots$, respectively. Then $\{X_n\}$ converges in distribution or in law to a rv X with cdf F if $F_n(t) \to F(t)$ as $n \to \infty$ for every point t such that $F(t)$

is continuous at t. This convergence in distribution or law is denoted by $X_n \xrightarrow{L} X$. The cdf F of the rv X is called the limiting or asymptotic distribution of X_n.

Further, if X_n has pdf $f_n(x)$ and $f_n(x) \to f(x)$ as $n \to \infty$ and if $f(x)$ is a pdf, then $\int |f_n(x) - f(x)| \mathrm{d}x \to 0$ as $n \to \infty$. In addition, if $|f_n(x)| < q(x)$ and $\int q(x) \mathrm{d}x$ exists and is finite, this implies that $f(x)$ is a pdf such that $\int |f_n(x) - f(x)| \mathrm{d}x \to 0$ as $n \to \infty$.

Several additional results that are very useful in practice are:

(1) *Helly–Bray Theorem*: $F_n \to F$ implies that $\int g \mathrm{d}F_n \to \int g \mathrm{d}F$ as $n \to \infty$ for every bounded continuous function g.

For example, the Helly-Bray Theorem can be employed to approximate $Eg^r = \int g^r \mathrm{d}F_n$, $r = 1, 2, \dots$, when g^r satisfies the conditions of the theorem and F's form is known.

(2) With g a continuous function, (a) if $X_n \xrightarrow{L} X$, then $g(X_n) \xrightarrow{L} g(X)$, and (b) if $X_n \xrightarrow{\text{i.p.}} X$, then $g(X_n) \xrightarrow{\text{i.p.}} g(X)$.

As a special case of (b), if $X_n \xrightarrow{\text{i.p.}} c$, a constant, $g(X_n) \xrightarrow{\text{i.p.}} g(c)$.

(3) *Continuity Theorem*. Let $c_n(t)$ be the characteristic function (cf) of X_n. If $X_n \xrightarrow{L} X$, then $c_n(t) \to c(t)$, where $c(t)$ is the cf of X. Also, if $c_n(t) \to c(t)$ and $c(t)$ is continuous at $t = 0$, then $X_n \xrightarrow{L} X$ with the distribution function of X having cf $c(t)$.

By the Continuity Theorem, derivation of the form of $c(t) = \lim_{n \to \infty} c_n(t)$ often permits one to determine the form of the limiting distribution of X_n.

The following convergence results relating to $\{X_n, Y_n\}$, $n = 1, 2, \dots$, a sequence of pairs of rvs are frequently employed:

(a) If $|X_n - Y_n| \xrightarrow{\text{i.p.}} 0$ and $Y_n \xrightarrow{L} Y$, then $X_n \to Y$, that is, the limiting cdf of X_n exists and is the same as that of Y_n.

(b) $X_n \xrightarrow{L} X$ and $Y_n \xrightarrow{\text{i.p.}} 0$, implies that $X_n Y_n \xrightarrow{\text{i.p.}} 0$.

(c) $X_n \xrightarrow{L} X$ and $Y_n \xrightarrow{\text{i.p.}} c$, implies that (i) $X_n + Y_n \xrightarrow{L} X + c$; (ii) $X_n Y_n \xrightarrow{L} cX$; and (iii) $X_n / Y_n \xrightarrow{L} X/c$ if $c \neq 0$.

(d) $X_n - Y_n \xrightarrow{\text{i.p.}} 0$ and $X_n \xrightarrow{L} X$, implies that $Y_n \xrightarrow{L} X$.

The results (a)–(d) can be generalized to apply to cases in which X_n and Y_n are vectors of rvs.

The following lemma relates to the convergence of sequences of random vectors, $\{X_n^{(1)}, X_n^{(2)}, \dots, X_n^{(k)}\}$, $n = 1, 2, \dots$.

Lemma

If for any real $\lambda_1, \lambda_2, \dots, \lambda_k$

$$\lambda_1 X_n^{(1)} + \lambda_2 X_n^{(2)} + \cdots + \lambda_k X_n^{(k)} \xrightarrow{L} \lambda_1 X^{(1)} + \lambda_2 X^{(2)} + \cdots \lambda_k X^{(k)},$$

where $X^{(1)}, X^{(2)}, \dots, X^{(k)}$ have a joint cdf $F(x_1, x_2, \dots, x_k)$, then the limiting joint cdf of the sequence of random vectors exists and is equal to $F(x_1, x_2, \dots, x_k)$.

Central Limit Theorems (CLTs) establish particular limiting cdfs for sequences of rvs. While only CLTs yielding limiting normal cdfs will be reviewed below, it is the case that non-normal limiting cdfs are sometimes encountered.

Lindeberg–Levy CLT

Let $\{X_n\}$, $n = 1, 2, \dots$, be a sequence of i.i.d. rvs such that $EX_n = \mu$ and $V(X_n) = \sigma^2 \neq 0$ exist. Then the cdf of $Y_n = \sqrt{n}(\bar{X}_n - \mu)/\sigma \to \Phi$, where Φ is the normal cdf, $\Phi(y) = (2\pi)^{-1/2}\int_{-\infty}^{y} e^{-t^2/2}\mathrm{d}t$ and $\bar{X}_n = \sum_{i=1}^{n} X_i/n$.

Liapunov CLT

Let $\{X_n\}$, $n = 1, 2, \dots$, be a sequence of independent rvs. Let $EX_n \equiv \mu_n$, $E(X_n - \mu_n)^2 = \sigma_n^2 \neq 0$ and $E(X_n - \mu_n)^3 = \beta_n$ exist for each n. Furthermore, let $B_n = (\sum_{i=1}^{n} \beta_i)^{1/3}$ and $C_n = (\sum_{i=1}^{n} \sigma_i^2)^{1/2}$. Then if $\lim(B_n/C_n) = 0$ as $n \to \infty$, the cdf of $Y_n = \sum_{i=1}^{n}(X_i - \mu_i)/C_n \to \Phi(y)$, a normal cdf.

Lindeberg–Feller CLT

Let $\{X_n\}$, $n = 1, 2, \dots$, be a sequence of independent rvs and G_n be the cdf of X_n. Further, let $EX_n = \mu_n$ and $V(X_n) = \sigma_n^2 \neq 0$ exist. Define $Y_n = \sum_{i=1}^{n}(X_i - \mu_i)/C_n$, where $C_n = \sqrt{n}\,\bar{\sigma}$, with $\bar{\sigma}^2 = \sum_{i=1}^{n} \sigma_i^2/n$. Then the relations

$$\lim_{n \to \infty} \max_{1 \leqslant i \leqslant n} \sigma_i/C_n = 0 \quad \text{and} \quad F_{Y_n} \to \Phi(y)$$

hold if and only if for every $\varepsilon > 0$,

$$\lim_{n \to \infty} \frac{1}{C_n^2} \sum_{i=1}^{n} \int_{|X - \mu_i| > \varepsilon C_n} (X - \mu_i)^2 \mathrm{d}G_i(X) = 0.$$

Multivariate CLT

Let F_n denote the joint cdf of the k-dimensional random vector $\{X_n^{(1)}, X_n^{(2)}, \dots, X_n^{(k)}\}$, $n = 1, 2, \dots$, and $F_{\lambda n}$ the cdf of the linear function $\lambda_1 X_n^{(1)} + \lambda_2 X_n^{(2)} + \cdots + \lambda_k X_n^{(k)}$. A necessary and sufficient condition that F_n tend to a k-variate cdf F is that $F_{\lambda n}$ converges to a limit for each vector λ.

With F_n, $F_{\lambda n}$, and F as defined in the Multivariate CLT, if for each vector λ, $F_{\lambda n} \to F_\lambda$, the cdf of $\lambda_1 X^{(1)} + \lambda_2 X^{(2)} + \cdots + \lambda_k X^{(k)}$, then $F_n \to F$. As an application, consider the random vector $\boldsymbol{U}_n' = (U_{1n}, U_{2n}, \dots, U_{kn})$ with $EU_n = \boldsymbol{\mu}$ and $V(U_n) = \Sigma$, a $k \times k$ matrix. Define $\bar{\boldsymbol{U}}_n = (\bar{U}_{1n}, \bar{U}_{2n}, \dots, \bar{U}_{kn})$, $n = 1, 2, \dots$, with $\bar{U}_{in} = \sum_{j=1}^{n} U_{ij}/n$. Then the asymptotic cdf of $\sqrt{n}(\bar{\boldsymbol{U}}_n - \boldsymbol{\mu})$ is that of a random normal vector with zero mean and covariance matrix Σ.

For use of these and related theorems in proving the asymptotic normality of maximum likelihood estimators and posterior distributions, see, for example, Heyde and Johnstone (1979) and the references in this paper that relate both to cases in which rvs $\{X_n\}$ are i.i.d. and statistically dependent. Finally, for a description of Edgeworth and other asymptotic expansion approaches for approximating finite sample distributions and moments of random variables, see, for example, Kendall and Stuart (1958), Jeffreys (1967, appendix A), Copson (1965), and Phillips (1977a, 1977b). Such asymptotic expansions and numerical integration approaches are useful in checking the quality of approximate asymptotic results and in obtaining more accurate approximations.

3. Estimation theory

Learning the values of parameters appearing in econometric models is important for checking the implications of economic theories and for practical uses of econometric models for prediction and policy-making. Thus, much research in statistical and econometric theory has been concentrated on developing and rationalizing procedures for using data to infer or estimate the values of parameters. It is to a review of the major elements of this work on estimation that we now turn.

3.1. *Point estimation*

Consider a parameter θ contained in Θ, the parameter space. It is often assumed that there exists a true, unknown value of θ, θ_0. Whether the true value θ_0 exists in nature or just in the mind of an investigator is a philosophical issue that will not be discussed. Given θ and its associated parameter space, assume that a sample of data, denoted by $\boldsymbol{x}' = (x_1, x_2, \dots, x_n)$ is randomly drawn or generated and that it has probability density function (pdf), $p(\boldsymbol{x}|\boldsymbol{\theta})$. With this conception of how the data are drawn or generated, $\boldsymbol{x}$ is a particular value of a random vector $\tilde{\boldsymbol{x}}$, with pdf $p(x|\theta)$ and is referred to as a random sample. The problem of point estimation is how to form a function of the sample observations $\boldsymbol{x}$, denoted by $\hat{\theta}(\boldsymbol{x})$, that will, in some sense, be a good approximation to or close to the true, unknown value of θ. The function of the data, $\hat{\theta}(\boldsymbol{x})$, whatever it is, is by definition a *point estimate* of the true, unknown value of θ. A point estimate is a non-random quantity since it depends just on the given, observed data x and, by definition, the function $\hat{\theta}(x)$ does not involve θ. On the other hand, a *point estimator* is the random quantity $\hat{\theta}(\tilde{x})$, where the tilde denotes that $\tilde{x}$ is considered to be random. Also, the random function or estimator, $\hat{\theta}(\tilde{x})$, does not depend on θ and its stochastic properties can be determined before observing the value of $\tilde{x}$, namely x. As a slight generalization, we may be interested in estimating $g(\theta)$, a single-valued function of θ. A special case is $g(\theta)=\theta$. Then $\hat{g}(x)$ is a point estimate of $g(\theta)$, and $\hat{g}(\tilde{x})$ is a point estimator of $g(\theta)$. The problem of point estimation is how to pick the form of $\hat{g}(x)$ so that $\hat{g}(x)$ will be close to the true, unknown value of $g(\theta)$. Various definitions of "closeness" and criteria for choosing the functional forms of estimates have been put forward, and are discussed below.

3.2. *Criteria for point estimation*

There are two general types of criteria that are employed in evaluating properties of point estimates. First, *sampling criteria* involve properties of the sample space and relate to sampling or frequency properties of particular or alternative estimates. The overriding considerations with the use of sampling criteria are properties of estimates in actual or hypothetical repeated samples. Second, *non-sampling criteria* involve judging particular or alternative estimates just on the basis of their properties relative to the given, actually observed data, x. With non-sampling criteria, other as yet unobserved samples of data and long-run frequency properties are considered irrelevant for the estimation of a parameter's value from the actually observed data. The issue of whether to use sampling or non-sampling criteria for constructing and evaluating estimates is a crucial one since these different criteria can lead to different estimates of a parameter's value from the same set of data. However, it is the case, as will be seen, that some non-sampling-based procedures yield estimates that have good sampling properties as well as optimality properties relative to the actually observed data.

3.2.1. *Sampling criteria for point estimation*

According to sampling theory criteria, a particular point estimate, say $\hat{\theta}(x)$ is judged with respect to the properties of $\hat{\theta}(\tilde{x})$, the point estimator. For example, the point estimate, $\hat{\theta}(x)=\sum_{i=1}^{n}x_i/n$, the sample mean is judged by reference to the sampling properties of $\hat{\theta}(\tilde{x})=\sum_{i=1}^{n}\tilde{x}_i/n$, the random sample mean or estimator.

3.2.1.1. "Perfect" estimation criteria. One sampling criterion in point estimation is given by $P[\hat{\theta}(\tilde{x})=\theta]=1$ for whatever the value of θ. Unfortunately, this criterion of perfection cannot be realized since $\tilde{x}$ can assume many different values and thus it is impossible to estimate the value of θ without error. That is, the probability that the random sampling error $\tilde{e} = \hat{\theta}(\tilde{x}) - \theta$ is equal to zero, is zero.

The fact that estimation error is generally unavoidable has led some to introduce the criterion of mean squared error (MSE), $E\tilde{e}^2 = E[\hat{\theta}(\tilde{x}) - \theta]^2$ and to seek the estimator $\hat{\theta}(\tilde{x})$ that minimizes MSE for all possible values of θ. Again, unfortunately, such an estimator does not exist. For example, if we use the "estimator" $\hat{\theta} = 5$, no matter what the sample data are, we will experience a lower MSE when $\theta = 5$ than that associated with any other estimator. Thus, no one estimator can dominate all others in terms of MSE for all possible values of θ. Some other conditions have to be put on the problem in order to obtain a unique, optimal estimator relative to the MSE criterion.

Another sampling criterion for estimation is the highest degree of concentration about the value of the parameter being estimated. That is, we might seek an estimator $\hat{\theta}(\tilde{x})$ such that

$$P[\theta - \lambda_1 < \hat{\theta}(\tilde{x}) < \theta + \lambda_2] \geqslant P[\theta - \lambda_1 < \hat{\theta}_a(\tilde{x}) < \theta + \lambda_2] \tag{3.1}$$

for all θ with λ_1 and λ_2 in the interval 0 to λ and where $\hat{\theta}_a(\tilde{x})$ is any other estimator. A necessary condition for (3.1) to hold is that $E[\hat{\theta}(\tilde{x}) - \theta]^2 \leqslant E[\hat{\theta}_a(\tilde{x}) - \theta]^2$ for all θ. As mentioned above, it is not possible to satisfy this necessary MSE condition and thus the criterion of highest degree of concentration cannot be realized.

Since the strong sampling theory criteria of error-free, minimal MSE, and highest degree of concentration cannot be realized, several weaker criteria for estimators have been put forward. One of these is the criterion of unbiasedness.

3.2.1.2. Unbiasedness.

Definition

An estimator $\hat{\theta}(\tilde{x})$ of a parameter θ is unbiased if and only if $E[\hat{\theta}(\tilde{x})|\theta] = \theta$ for all $\theta \subset \Theta$.

Thus, if an estimator is unbiased, its mean is equal to the value of the parameter being estimated.

Example 3.1

As an example of an unbiased estimator, consider the model, $\tilde{x}_i = \theta + \varepsilon_i$, with $E(\tilde{x}_i|\theta) = \theta$ for $i = 1, 2, \ldots, n$. Then $\hat{\theta}(\tilde{x}) = \sum_{i=1}^n \tilde{x}_i / n$ is an unbiased estimator since $E[\hat{\theta}(\tilde{x})|\theta] = \sum_{i=1}^n E(\tilde{x}_i|\theta)/n = \theta$.

Example 3.2

Consider the multiple regression model, $\tilde{y} = X\beta + \tilde{u}$, where $\tilde{y}' = (\tilde{y}_i, \tilde{y}_2, \dots, \tilde{y}_n)$, X is an $n \times k$ non-stochastic matrix of rank k, $\beta' = (\beta_1, \beta_2, \dots, \beta_k)$, and $\tilde{u}' = (\tilde{u}_1, \tilde{u}_2, \dots, \tilde{u}_n)$, a vector of unobservable random errors or disturbances. Assume that $E(\tilde{y}|X\beta) = X\beta$. Then $\hat{\beta} = (X'X)^{-1}X'\tilde{y}$, the "least squares" estimator of β, is unbiased since $E(\hat{\beta}|X\beta) = (X'X)^{-1}X'E(\tilde{y}|X\beta) = (XX)^{-1}X'X\beta = \beta$ for all β.

While unbiasedness is often regarded as a desirable property of estimators, the following qualifications should be noted. First, there are usually many unbiased estimators for a particular parameter. With respect to Example 3.1, the estimator $\hat{\theta}_w(\tilde{x}) = \sum_{i=1}^n w_i \tilde{x}_i$, with the w_i's given, has mean $E[\hat{\theta}_w(\tilde{x})|\theta] = \sum_{i=1}^n w_i E(x_i|\theta) = \theta \sum_{i=1}^n w_i$, and is unbiased for all w_i's satisfying $\sum_{i=1}^n w_i = 1$. Similarly, with respect to Example 3.2, $\hat{\beta}_c = [(X'X)^{-1}X' + C']\tilde{y}$, where C' is a non-stochastic $k \times n$ matrix and has mean $E(\beta_c|X\beta, C) = \beta + C'X\beta = \beta$ for all C such that $C'X = 0$. Thus, unbiased estimators are not unique.

Secondly, imposing the condition of unbiasedness can lead to unacceptable results in frequently encountered problems. If we wish to estimate a parameter, such as a squared correlation coefficient, that satisfies $0 \leqslant \theta < 1$, an unbiased estimator $\hat{\theta}(\tilde{x})$ must assume negative as well as positive values in order to satisfy $E\hat{\theta}(\tilde{x}) = \theta$ for all θ, $0 \leqslant \theta < 1$. Similarly, an unbiased estimator for a variance parameter τ^2 that is related to two other variances, σ_1^2 and σ_2^2, by $\tau^2 = \sigma_1^2 - \sigma_2^2$, with $\sigma_1^2 \geqslant \sigma_2^2 > 0$, has to assume negative as well as positive values in order to be unbiased for all values of τ^2. Negative estimates of a variance, that is known to be non-negative, are unsatisfactory.

Third, and perhaps of most general importance, the criterion of unbiasedness does not take account of the dispersion or degree of concentration of estimators. Biased estimators can be more closely concentrated about a parameter's value than are unbiased estimators. In this connection, the criterion of MSE can be expressed in general as:[17]

$$\text{MSE}(\hat{\theta}) = V(\hat{\theta}) + (\text{Bias})^2, \tag{3.2}$$

where $V(\hat{\theta}) = E(\hat{\theta} - E\hat{\theta})^2$, the variance of $\hat{\theta}$, and bias $= E\hat{\theta} - \theta$ is the bias of $\hat{\theta}$. Thus, MSE depends on both dispersion, as measured by $V(\hat{\theta})$, and squared bias, and gives them equal weights. In terms of (3.2), the criterion of unbiasedness gives zero weight to $\text{var}(\hat{\theta})$ and unit weight to the bias squared term which is considered unsatisfactory by many.

Fourth, on considering just unbiased estimators for a parameter, denoted by $\hat{\theta}_u(\tilde{x})$, it is clear from (3.2) that MSE $= \text{var}[\hat{\theta}_u(\tilde{x})]$. While the restriction that an

[17]Note MSE $(\hat{\theta}) = E(\hat{\theta} - \theta)^2 = E[(\hat{\theta} - E\hat{\theta}) + (E\hat{\theta} - \theta)]^2 = E(\hat{\theta} - E\hat{\theta})^2 + [E\hat{\theta} - \theta]^2$, since $E(\hat{\theta} - E\hat{\theta})(E\hat{\theta} - \theta) = 0$.

estimator belong to the class of unbiased estimators can be costly in terms of MSE, such a restriction permits estimators to be ordered in terms of their variances and suggests seeking the unbiased estimator with minimal variance. Such an estimator, if it exists, is called a minimum variance unbiased estimator.

3.2.1.3. Criterion of minimum variance unbiased estimation. Some fundamental results on minimum variance unbiased estimators (MVUEs) have been provided by Rao (1973), among others. With reference to the triplet $(X, P_{\theta}, \theta \subset \Theta)$, where θ may be vector-valued, consider estimation of $g(\theta)$ a real-valued function of θ. Let U_g denote the class of unbiased estimators of $g(\theta)$; that is, an estimator $\hat{g}$ belongs to U_g if and only if $E(\hat{g}|\theta) = g(\theta)$ for each $\theta \subset \Theta$. Also, define U_0 as the class of all functions with zero expectation; that is, $f \subset U_0$ if and only if $E(f|\theta) = 0$ for each θ. Then the following results, proved in Rao (1973) are available.

Rao Theorem

A necessary and sufficient condition that an estimator $\hat{g} \subset U_g$, that is, $E(\hat{g}|\theta) = g(\theta)$, has minimum variance at the value $\theta = \theta_0$ is that $\text{cov}(\hat{g}, f|\theta_0) = 0$ for every $f \subset U_0$ such that $\text{var}(f|\theta_0) < \infty$ provided that $\text{var}(\hat{g}|\theta_0) < \infty$.

From the Rao Theorem it follows that: (a) the correlation between a MVUE and any other estimator is non-negative; (b) if there are two unbiased estimators with the same minimum variance, their correlation is equal to one, that is, $\rho(\theta_0) = 1$, and therefore they are the same except for a set of samples of probability measure zero for $\theta = \theta_0$; and (c) if $\hat{g}_1$ and $\hat{g}_2$ are MVUEs of $g_1(\theta)$ and $g_2(\theta)$, respectively, then $b_1\hat{g}_1 + b_2\hat{g}_2$ is a MVUE of $b_1 g_1(\theta) + b_2 g_2(\theta)$, where b_1 and b_2 are fixed constants. Also, if $\hat{g}$ has minimum variance for each θ, $\hat{g}$ is called a *uniformly* MVUE.

As an example of Rao's Theorem, consider Example 3.1 with the $\tilde{x}_i$'s assumed normally and independently distributed, each with mean θ and variance σ^2. Then $f(\tilde{\boldsymbol{x}}) \subset U_0$ implies

$$\int f(\boldsymbol{x}) \exp\left\{-\sum_{i=1}^{n} (x_i - \theta)^2 / 2\sigma^2\right\} \mathrm{d}x = 0.$$

Differentiating this last expression with respect to θ yields

$$\int \bar{x} f(x) \exp\left\{-\sum_{i=1}^{n} (x_i - \theta)^2 / 2\sigma^2\right\} \mathrm{d}x = 0$$

where $\bar{x} = \sum_{i=1}^{n} x_i / n$, the sample mean, and this result implies that $\text{cov}(f, \bar{x}) = 0$, the necessary and sufficient condition for $\bar{x}$ to be a MVUE. By similar analysis, Rao shows that for the multiple regression model in Example 3.2, with $\tilde{\boldsymbol{y}}$ assumed to be normally distributed with mean $X\boldsymbol{\beta}$ and covariance matrix $\sigma^2 I_n$, $\hat{\boldsymbol{\beta}} = (X'X)^{-1}X'y$ and $s^2 = (y - X\hat{\boldsymbol{\beta}})'(y - X\hat{\boldsymbol{\beta}})/(n-k)$ are MVUEs for $\boldsymbol{\beta}$ and σ^2, respectively.

The existence of MVUEs is closely linked to the concepts of sufficiency, sufficient statistics, and completeness, concepts that are important generally. With respect to sufficiency, for some problems certain aspects of the sample data $\boldsymbol{x}$ provide no information about the value of a parameter. For example, in n independent binomial trials with probability of "success" θ on each trial, the order of the occurrence of successes and failures provides no information about the value of θ. The number of successes, $t(\boldsymbol{x}) = \sum_{i=1}^{n} x_i$, where $x_i = 1$, denotes a success and $x_i = 0$ a failure, is a statistic that maps many different sequences of observations into a single value, $t(\boldsymbol{x})$, and thus represents a reduction of the sample data. The basic idea of sufficiency involves a reduction of the dimensionality of the data *without loss of sample information.*

As regards estimation of a scalar parameter, θ, with a sample of size $n \geqslant 2$, consider the joint distribution of a set of r functionally independent statistics, $p_r(t, t_1, t_2, \dots, t_{r-1}|\theta)$, $r = 2, 3,, \dots, n$, where t is a statistic of particular interest. Then

$$p_r(t, t_1, t_2, \dots, t_{r-1}|\theta) = g(t|\theta)h_{r-1}(t_1, t_2, \dots, t_{r-1}|\theta). \tag{3.3}$$

As Kendall and Stuart (1961, pp. 22–23) point out, if $h_{r-1}(t_1, t_2, \dots, t_{r-1}|\theta)$ is independent of θ, then $t_1, t_2, \dots, t_{r-1}$ contribute nothing to our knowledge of the value of θ. Thus, formally t is sufficient for θ if and only if

$$p_r(t, t_1, t_2, \dots, t_{r-1}|\theta) = g(t|\theta)h_{r-1}(t_1, t_2, \dots, t_{r-1}|t), \tag{3.4}$$

where h_{r-1} is independent of θ for $r = 2, 3, \dots, n$, and any choice of $t_1, t_2, \dots, t_{r-1}$. Then t is said to be sufficient for θ. A similar definition applies when θ is a vector of parameters and the t's are vectors. A minimal sufficient statistic is one that achieves the greatest reduction of the sample information without loss of information.[18]

While (3.4) defines sufficiency in general, it does not indicate how sufficient statistics can be found in specific problems. The Factorization Theorem is helpful in this regard.

Factorization Theorem[19]

A necessary and sufficient condition for a statistic $\boldsymbol{t} = \boldsymbol{t}(\boldsymbol{x})$ to be sufficient for a parameter vector $\boldsymbol{\theta}$ appearing in $f(\boldsymbol{x}|\boldsymbol{\theta})$, the probability density function for the

[18]See Kendall and Stuart (1961, pp. 193–194) and Silvey (1970, p. 29) for a discussion of minimal sufficient statistics.

[19]Proofs may be found in Kendall and Stuart (1961, p. 23) and Lehmann (1959, p. 47).

sample data $\boldsymbol{x}$ is that $f(\boldsymbol{x}|\boldsymbol{\theta})$ can be expressed as

$$f(\boldsymbol{x}|\boldsymbol{\theta}) = g(\boldsymbol{t}|\boldsymbol{\theta})h(\boldsymbol{x}), \tag{3.5}$$

where $h(\boldsymbol{x})$ does not depend on $\boldsymbol{\theta}$.

Let us apply the Factorization Theorem to Examples 3.1 and 3.2 under the assumption that in both examples the random observations are normally and independently distributed. For Example 3.1, with the $\tilde{x}_i$'s normally and independently distributed, each with mean θ and variance σ^2,

$$\begin{aligned} f(\boldsymbol{x}|\theta,\sigma) &= (2\pi\sigma^2)^{-n/2}\exp\left\{-\sum_{i=1}^{n}(x_i-\theta)^2/2\sigma^2\right\} \\ &= (2\pi\sigma^2)^{-n/2}\exp\left\{-\left[\nu s^2+n(\bar{x}-\theta)^2\right]/2\sigma^2\right\} = g(\bar{x},s^2|\theta,\sigma^2), \end{aligned}$$

where $n\bar{x} = \sum_{i=1}^{n}x_i$, $\nu s^2 = \sum_{i=1}^{n}(x_i-\bar{x})^2$ and $\nu = n-1$.[20] Thus, $\bar{x}$ and s^2 are sufficient statistics.

In Example 3.2, with $\tilde{\boldsymbol{y}}$ normal with mean $X\boldsymbol{\beta}$ and covariance matrix $\sigma^2 I_n$,

$$\begin{aligned} f(\boldsymbol{y}|\boldsymbol{\beta},\sigma^2) &= (2\pi\sigma^2)^{-n/2}\exp\{-(\boldsymbol{y}-X\boldsymbol{\beta})'(\boldsymbol{y}-X\boldsymbol{\beta})/2\sigma^2\} \\ &= (2\pi\sigma^2)^{-n/2}\exp\{-[\nu s^2+(\hat{\boldsymbol{\beta}}-\boldsymbol{\beta})'X'X(\boldsymbol{\beta}-\hat{\boldsymbol{\beta}})/2\sigma^2\} \\ &= g(\hat{\boldsymbol{\beta}},s^2|\boldsymbol{\beta},\sigma^2), \end{aligned}$$

where $\hat{\boldsymbol{\beta}} = (X'X)^{-1}x'\tilde{\boldsymbol{y}}$, $\nu s^2 = (\boldsymbol{y}-X\hat{\boldsymbol{\beta}})'(\boldsymbol{y}-X\hat{\boldsymbol{\beta}})$, and $\nu = n-k$.[21] Thus, $\hat{\boldsymbol{\beta}}$ and s^2 are sufficient statistics.

The fundamental Rao–Blackwell Theorem provides a link between sufficiency and MVUE.

Rao–Blackwell Theorem[22]

Let t be a sufficient statistic for θ, where both t and θ may be vector-valued and t_1 any other statistic. If g is any function of θ, then

$$E[t_1-g(\theta)]^2 \geqslant E[h(t)-g(\theta)]^2, \tag{3.6}$$

[20] Note that with all summations extending from $i=1$ to $i=n$,

$$\sum(x_i-\theta)^2 = \sum[x_i-\bar{x}-(\theta-\bar{x})]^2 = \sum(x_i-\bar{x})^2+n(\bar{x}-\theta)^2 = \nu s^2+n(\bar{x}-\theta),$$

since $\sum(x_i-\bar{x})=0$.

[21] Note:

$$\begin{aligned} (\boldsymbol{y}-X\boldsymbol{\beta})'(\boldsymbol{y}-X\boldsymbol{\beta}) &= \left[\boldsymbol{y}-X\hat{\boldsymbol{\beta}}-X(\boldsymbol{\beta}-\hat{\boldsymbol{\beta}})\right]'\left[\boldsymbol{y}-X\hat{\boldsymbol{\beta}}-X(\boldsymbol{\beta}-\hat{\boldsymbol{\beta}})\right] \\ &= (\boldsymbol{y}-X\hat{\boldsymbol{\beta}})'(\boldsymbol{y}-X\hat{\boldsymbol{\beta}})+(\hat{\boldsymbol{\beta}}-\boldsymbol{\beta})'X'X(\hat{\boldsymbol{\beta}}-\boldsymbol{\beta}), \end{aligned}$$

since $X'(\boldsymbol{y}-X\hat{\boldsymbol{\beta}})=0$.

[22] See Rao (1973, pp. 320–321) and also Rao (1945) and Blackwell (1947).

where $h(t)=E(t_1|t)$ is independent of θ. Furthermore, $Eh(t)=g(\theta)$, that is, $h(t)$ is unbiased if $Et_1=g(\theta)$.

See Rao (1973, p. 321) for a proof of this theorem. As Rao (1973) notes: "Given any statistic $[t_1]$, we can find a function of the sufficient statistic $[h(t)]$ which is uniformly better in the sense of mean square error or minimum variance (if no bias is imposed)" (p. 321). He also notes that if a complete sufficient statistic exists, that is, one such that no function of it has zero expectation unless it is zero almost everywhere with respect to each of the measures P_θ, then every function of it is a uniformly MVUE of its expected value. In view of the Rao–Blackwell Theorem and assuming the existence of complete sufficient statistics, to find a MVUE it is enough to start with any unbiased estimator and take its conditional expectation given the sufficient statistic, that is, $E(t_1|t)=h(t)$.

Since these results depend strongly on the existence of complete statistics, it is relevant to ask which classes of distribution functions possess sufficient statistics for their parameters. The Pitman–Koopman Theorem [Pitman (1936) and Koopman (1936)] provides an answer to this question.

Pitman–Koopman Theorem

For a parameter θ, scalar or vector-valued, if the range of $f(\boldsymbol{x}|\theta)$ is independent of θ, a distribution will have a sufficient statistic (or statistics) if and only if it is a member of the exponential class of distributions, that is,

$$f(\boldsymbol{x}|\theta)=\exp\{A(\theta)B(\boldsymbol{x})+C(\boldsymbol{x})+D(\theta)\} \tag{3.7}$$

in the scalar case, and

$$f(\boldsymbol{x}|\boldsymbol{\theta})=\exp\left\{\sum_{j=1}^{k}A_j(\boldsymbol{\theta})B_j(\boldsymbol{x})+C(\boldsymbol{x})+D(\boldsymbol{\theta})\right\} \tag{3.8}$$

in the case that $\boldsymbol{\theta}$ is a k-element vector.

The exponential class includes many distributions that arise in practice, for example binomial, normal, geometric, exponential, and Poisson distributions.[23] However, there are many cases in which minimal sufficient statistics are not complete and then it is not possible to use the Rao–Blackwell Theorem to establish the existence of a MVUE. In such cases there may be several different

[23]For example, the binomial pdf,

$$f(x|\theta,n)=\binom{n}{x}\theta^x(1-\theta)^{n-x}=\exp\{x\log[\theta/(1-\theta)]+n\log(1-\theta)\}$$

is in the exponential form (3.7) with $B(x)=x$, $A(\theta)=\log[\theta/(1-\theta)]$, $D(\theta)=n\log(1-\theta)$, and $C(x)=0$.

functions of the minimal sufficient statistic which are unbiased estimators of the parameter and there is no general means of comparing their variances. Silvey (1970, pp. 34–35) presents the following example to illustrate this problem. Suppose that n independent binomial trials, each with probability θ of success, are carried out; then trials are continued until an additional k successes are obtained, this requiring s additional trials.[24] Let the sample be denoted by $x=(x_1, x_2,\dots, x_n, x_{n+1}, x_{n+2},\dots,x_{n+s-1}, 1)$, where $x_i=1$ for a success and $x_i=0$ for a failure. Then $f(x|\theta)=\theta^{r+k}(1-\theta)^{n+s-r-k}$, where $r=\sum_{i=1}^n x_i$ and s depends on x also. The statistic $t=(r,s)$ is sufficient for θ and is also a minimal sufficient statistic. However, t is not complete because if

$$f(t)=\frac{r}{n}-\frac{k-1}{s-1},$$

then

$$Ef(t)=E\frac{r}{n}-E\frac{k-1}{s-1}=\theta-\theta=0 \quad \text{for all } \theta.$$

However, $f(t)\not\equiv 0$, as is required for a complete sufficient statistic, and thus there are problems in applying the Rao–Blackwell Theorem to this and similar problems.

In problems in which it is difficult or impossible to obtain a MVUE, it is useful to consider the Cramér–Rao Inequality that provides a lower bound for the variance of an unbiased estimator.

Cramér – Rao Inequality

Given $(X, p(\boldsymbol{x}|\theta), \theta\subset\Theta)$, with Θ an interval on the real line, then subject to certain regularity conditions, the variance of any unbiased estimator $\hat{g}$ of $g(\theta)$ satisfies the following inequality:

$$\operatorname{var}(\hat{g})\geqslant[g'(\theta)]^2/I_\theta, \tag{3.9}$$

where $I_\theta=E(\partial\log p(x|\theta)/\partial\theta)^2$ was interpreted by R.A. Fisher as the amount of information about θ contained in $\boldsymbol{x}$. If $g(\theta)=\theta$ and $\hat{g}=\hat{\theta}$, then (3.9) becomes

$$\operatorname{var}(\hat{\theta})\geqslant 1/I_\theta. \tag{3.10}$$

[24] This latter part of the process is a negative binomial process.

Proof

Differentiate $E\hat{g}=\int_X \hat{g}p(\boldsymbol{x}|\theta)\,\mathrm{d}\boldsymbol{x}=g(\theta)$ with respect to θ, assuming that it is permissible to differentiate under the integral to obtain:

$$g'(\theta)=\int_X g[\partial\log p(\boldsymbol{x}|\theta)/\partial\theta]p(\boldsymbol{x}|\theta)\,\mathrm{d}\boldsymbol{x}$$

or

$$g'(\theta)=\int_X[\hat{g}-g(\theta)][\partial\log p(\boldsymbol{x}|\theta)/\partial\theta]p(\boldsymbol{x}|\theta)\,\mathrm{d}x, \tag{3.11}$$

since from $\int_X p(\boldsymbol{x}|\theta)\mathrm{d}x=1$, $\int_X[\partial p(\boldsymbol{x}|\theta)/\partial\theta]\mathrm{d}x=0$, and $E[\partial\log p(\boldsymbol{x}|\theta)/\partial\theta]=\int[\partial\log p(\boldsymbol{x}|\theta)/\partial\theta]p(x|\theta)\mathrm{d}\boldsymbol{x}=0$.

On applying the Cauchy–Schwarz integral inequality to (3.11),

$$[g'(\theta)]^2\leqslant\int_X[\hat{g}-g(\theta)]^2p(\boldsymbol{x}|\theta)\,\mathrm{d}\boldsymbol{x}\int_X[\partial\log p(\boldsymbol{x}|\theta)/\partial\theta]^2p(\boldsymbol{x}|\theta)\,\mathrm{d}\boldsymbol{x}$$

or

$$\begin{aligned}\operatorname{var}(\hat{g})&=\int_X[\hat{g}-g(\theta)]^2p(\boldsymbol{x}|\theta)\,\mathrm{d}\boldsymbol{x}\\&\geqslant[g'(\theta)]^2\Big/\int_X[\partial\log p(\boldsymbol{x}|\theta)/\partial\theta]^2p(\boldsymbol{x}|\theta)\,\mathrm{d}\boldsymbol{x}\\&\geqslant[g'(\theta)]^2/I_\theta.\end{aligned}$$

The following lemma provides an alternative expression for I_θ, the Fisher information measure.

Lemma

$I_\theta=E[\partial\log p(\boldsymbol{x}|\theta)/\partial\theta]^2=-E[\partial^2\log p(\boldsymbol{x}|\theta)/\partial\theta^2]$.

Proof

Differentiate $\int_X[\partial\log p(\boldsymbol{x}|\theta)/\partial\theta]p(\boldsymbol{x}|\theta)\,\mathrm{d}\boldsymbol{x}=0$ with respect to θ to obtain:

$$\int_X\{[\partial^2\log p(\boldsymbol{x}|\theta)/\partial\theta^2]\,p(\boldsymbol{x}|\theta)+[\partial\log p(\boldsymbol{x}|\theta)/\partial\theta][\partial p(\boldsymbol{x}|\theta)/\partial\theta]\}\,\mathrm{d}\boldsymbol{x}=0$$

or

$$\int_X \{[\partial^2 \log p(\boldsymbol{x}|\theta)/\partial\theta^2 + [\partial \log p(\boldsymbol{x}|\theta)/\partial\theta]^2\} p(\boldsymbol{x}|\theta)\,\mathrm{d}\boldsymbol{x} = 0$$

or

$$E[\partial \log p(x|\theta)/\partial\theta]^2 = -E[\partial^2 \log p(x|\theta)/\partial\theta^2].$$

In (3.9), $[g'(\theta)]^2/I_\theta$ is called the minimum variance bound (MVB) for the unbiased estimator $\hat{g}$. As Kendall and Stuart (1961, p. 10) point out, the MVB was obtained by application of the Cauchy–Schwarz inequality and thus the necessary and sufficient condition that equality holds in (3.9), that is, that the MVB is attained, is that $\hat{g} - g(\theta)$ is proportional to $\partial \log p(x|\theta)/\partial\theta$ for all sets of observations, that is,

$$\partial \log p(x|\theta)/\partial\theta = A(\theta)[\hat{g} - g(\theta)], \tag{3.12}$$

where $A(\theta)$ may depend on θ but does not depend on x, the observations. From (3.12), $\mathrm{var}[\partial \log p(x|\theta)/\partial\theta] = A^2(\theta)\mathrm{var}(\hat{g})$ and then from the equality form of (3.9),

$$\mathrm{var}\,\hat{g} = g'(\theta)/A(\theta), \tag{3.13}$$

or in terms of (3.10), $\mathrm{var}(\hat{\theta}) = 1/A(\theta)$.

To illustrate use of the MVB in (3.13), consider

$$p(\boldsymbol{x}|\theta,\sigma_0^2) = h(\boldsymbol{x})(2\pi\sigma_0^2/n)^{-1/2}\exp\{-n(\bar{x}-\theta)^2/2\sigma_0^2\},$$

where σ_0^2 is a known value for σ^2 and $\bar{x} = \sum_{i=1}^n x_i/n$. Then $\partial \log p(\boldsymbol{x}|\theta,\sigma_0^2)/\partial\theta = n(\bar{x}-\theta)/\sigma_0^2$. With $\hat{g} = \bar{x}$ and $g(\theta) = \theta$, $A(\theta) = n/\sigma_0^2$. Thus, since $\partial \log p(x|\theta,\sigma_0^2)/\partial\theta$ is proportional to $\bar{x} - \theta$, with the factor of proportionality $A(\theta) = n/\sigma_0^2$, $\bar{x}$ is the MVB unbiased estimator with $\mathrm{var}(\bar{x}) = 1/A(\theta) = \sigma_0^2/n$.

While MVUEs are ingenious constructs and useful in a variety of situations, as Silvey (1970) points out: "There are many situations where either no MVUE exists or where we cannot establish whether or not such an estimator exists" (p. 43). For example, even in the case of a simple binomial parameter, θ, there exists no unbiased estimator of $\eta = \theta/(1-\theta)$, the odds in favor of success. Also, problems arise in obtaining unbiased estimators of the reciprocals and ratios of means and regression coefficients and coefficients of structural econometric models. For these and other problems there is a need for alternative estimation principles.

3.2.1.4. Least squares (LS) and other goodness of fit criteria. With LS and other goodness of fit criteria, a sample estimate of a parameter θ, scalar or vector-valued appearing in a model, is determined so that a given sample of observations is most closely approximated by the estimated or fitted model, a heuristic, non-sampling criterion. The sampling properties of the estimate $\hat{\theta}(\boldsymbol{x})$ so obtained are then shown to be optimal in certain senses to be reviewed below and it is the sampling properties that are usually utilized to justify the LS or goodness of fit estimation procedure.

To illustrate LS and other goodness of fit criteria, let $x_i = \theta + \varepsilon_i$, $i = 1, 2, \dots, n$, be the model for the *actual*, given observations $\boldsymbol{x}' = (x_1, x_2, \dots, x_n)$, where θ is a scalar parameter and the ε_i's are unobserved, non-random errors. To estimate θ, the LS principle involves finding the value of θ that minimizes the sum of squared errors, $\text{SS} = \sum_{i=1}^{n} \varepsilon_i^2 = \sum_{i=1}^{n} (x_i - \theta)^2$. The value of θ that minimizes SS is $\hat{\theta} = \sum_{i=1}^{n} x_i / n$, the sample mean.[25] To this point the x_i's and ε_i's are non-random. Indeed, it is meaningless to attempt to minimize $\widetilde{\text{SS}} = \sum_{i=1}^{n} \tilde{\varepsilon}_i^2 = \sum_{i=1}^{n} (\tilde{x}_i - \theta)^2$, a random function. However, now that we have the LS estimate $\hat{\theta}(\boldsymbol{x})$, the sample mean, it is possible to explore properties of $\hat{\theta}(\tilde{\boldsymbol{x}}) = \sum_{i=1}^{n} \tilde{x}_i / n$, the random sample mean or estimator given various assumptions about the probability model generating the observations, the x_i's. Similarly, with respect to the multiple regression model, $\boldsymbol{y} = X\boldsymbol{\beta} + \boldsymbol{u}$, where $\boldsymbol{y}$ is a *given* $n \times 1$ vector of observations, X is a given $n \times k$ non-stochastic matrix of rank k, $\boldsymbol{\beta}$ is a $k \times 1$ vector of regression parameters with unknown values, and $\boldsymbol{u}$ is an $n \times 1$ vector of unobserved, realized error terms. The sum of squared error terms to be minimized with respect to the value of $\boldsymbol{\beta}$ is $\text{SS} = \boldsymbol{u}'\boldsymbol{u} = (\boldsymbol{y} - X\boldsymbol{\beta})'(\boldsymbol{y} - X\boldsymbol{\beta})$. The minimizing, LS value of $\boldsymbol{\beta}$ is $\hat{\boldsymbol{\beta}} = (X'X)^{-1}X'\boldsymbol{y}$.[26] Note that $\hat{\boldsymbol{\beta}}$ depends on the given sample data and thus is a non-random estimate. Also, $\hat{\boldsymbol{u}} = \boldsymbol{y} - X\hat{\boldsymbol{\beta}}$, the LS residual vector, is an estimate of the unobserved, non-random vector $\boldsymbol{u}$. To determine the sampling properties of $\hat{\boldsymbol{\beta}}$, $\hat{\boldsymbol{u}}$, and other quantities, it is necessary to provide stochastic assumptions about the model for the observations, $\boldsymbol{y} = X\boldsymbol{\beta} + \boldsymbol{u}$.[27] Before considering this problem, it is relevant to ask: Why minimize the sum of squared errors, and not some other function of the errors?

In general, the sum of squared errors, $\text{SS} = \sum_{i=1}^{n} \varepsilon_i^2$, is employed because, as shown in the previous paragraph, it leads in many cases to simple expressions for

[25] Note that $\text{dSS}/\text{d}\theta = -2\sum_{i=1}^{n}(x_i - \theta)$ and $\text{d}^2\text{SS}/\text{d}\theta^2 = 2n > 0$. Thus, the value of θ for which $\text{dSS}/\text{d}\theta = 0$ is $\hat{\theta} = \sum_{i=1}^{n} x_i / n$ and this is a minimizing value since $\text{d}^2\text{SS}/\text{d}\theta^2 > 0$ at $\theta = \hat{\theta}$.

[26] Note $\partial\text{SS}/\partial\beta = -2X'\boldsymbol{y} + 2X'X\boldsymbol{\beta}$ and $\partial^2 SS/\partial\boldsymbol{\beta}^2 = 2X'X$. The value of $\boldsymbol{\beta}$ setting $\partial\text{SS}/\partial\boldsymbol{\beta} = \boldsymbol{0}$ is obtained from $-2X'y + 2X'X\hat{\boldsymbol{\beta}} = \boldsymbol{0}$ or $X'X\hat{\boldsymbol{\beta}} = X'\boldsymbol{y}$, the so-called "normal equations", the solution of which is $\hat{\boldsymbol{\beta}} = (X'X)^{-1}X'y$. Since $\partial^2\text{SS}/\partial\boldsymbol{\beta}^2$ is a positive definite symmetric matrix, $\hat{\boldsymbol{\beta}}$ is a minimizing value of $\boldsymbol{\beta}$.

[27] While these examples involve models linear in the parameters and error terms, it is also possible to use the LS principle in connection with non-linear models, for example $y_i = f(z_i, \theta) + u_i$, $i = 1, 2, \dots, n$, where $f(z_i, \theta)$ is a known function of a vector of given variables, z_i, and a vector of parameters. In this case, the LS principle involves finding the value of θ such that $\sum_{i=1}^{n} [y_i - f(z_i, \theta)]^2$ is minimized.

parameter estimates. Further, as will be shown below, in important problems minimizing SS leads to estimates that are identical to maximum likelihood and Bayesian estimates. However, criteria other than minimizing SS are available, for example minimizing $\text{SAD} = \sum_{i=1}^{n}|\varepsilon_i|$, the sum of absolute deviations that leads to minimum absolute deviation (MAD) estimates or minimizing $\text{WSS} = \sum_{i=1}^{n} w_i \varepsilon_i^2$, the weighted sum of squares, where the w_i's are weights, leads to weighted least squares (WLS) estimates. Basically, the choice among these alternative criteria for generating estimates depends importantly on what is assumed about the probability model for the observations. Given a probability model for the observations and general principles of estimation, it is possible to obtain a unique estimate that is optimal according to the general principle of estimation adopted.

The Gauss–Markov Theorem involves specifying a probability model for the observations and applying a principle of estimation, that of minimum variance linear unbiased estimation to obtain an estimator. A version of the Gauss–Markov Theorem is:

Gauss – Markov Theorem

Assume that the random $n \times 1$ observation vector $\tilde{\boldsymbol{y}}$ is generated by the model $\tilde{\boldsymbol{y}} = X\boldsymbol{\beta} + \tilde{\boldsymbol{u}}$, where X is a non-stochastic, known $n \times k$ matrix of rank k, $\boldsymbol{\beta}$ a $k \times 1$ vector of parameters with unknown values, and $\tilde{\boldsymbol{u}}$ an $n \times 1$ vector of unobserved random errors. Further, assume $E\tilde{\boldsymbol{u}} = \boldsymbol{0}$ and $E\tilde{\boldsymbol{u}}\tilde{\boldsymbol{u}}' = \sigma^2 I_n$, where σ^2 is the common unknown variance of the errors. Then, in the class of unbiased estimators of $\boldsymbol{l}'\boldsymbol{\beta}$, where $\boldsymbol{l}$ is a given $k \times 1$ vector of rank one, the minimum variance linear unbiased estimator[28] of $\boldsymbol{l}'\boldsymbol{\beta}$ is $\boldsymbol{l}'\hat{\boldsymbol{\beta}}$, where $\hat{\boldsymbol{\beta}} = (X'X)^{-1}X'\tilde{\boldsymbol{y}}$, the LS estimator.

Proof

Consider $\tilde{\boldsymbol{\beta}} = [(X'X)^{-1}X' + C']\tilde{\boldsymbol{y}}$, where C' is an arbitrary $k \times n$ matrix. This defines a class of linear (in $\tilde{\boldsymbol{y}}$) estimators. For $\boldsymbol{l}'\tilde{\boldsymbol{\beta}}$ to be an unbiased estimator of $\boldsymbol{l}'\boldsymbol{\beta}$, C must be such that $C'X = 0$ since $E\boldsymbol{l}'\tilde{\boldsymbol{\beta}} = \boldsymbol{l}'\boldsymbol{\beta} + \boldsymbol{l}'C'X\boldsymbol{\beta}$. With the restriction $C'X = 0$ imposed,

$$\operatorname{var}(\boldsymbol{l}'\tilde{\boldsymbol{\beta}}) = E\boldsymbol{l}'(\tilde{\boldsymbol{\beta}} - \boldsymbol{\beta})(\tilde{\boldsymbol{\beta}} - \boldsymbol{\beta})'\boldsymbol{l} = \boldsymbol{l}'\left[(X'X)^{-1} + C'C\right]\boldsymbol{l}\sigma^2,$$

since

$$\begin{aligned} E(\tilde{\boldsymbol{\beta}} - \boldsymbol{\beta})(\tilde{\boldsymbol{\beta}} - \boldsymbol{\beta})' &= \left[(X'X)^{-1}X' + C'\right]E\tilde{\boldsymbol{u}}\tilde{\boldsymbol{u}}'\left[X(X'X)^{-1} + C\right] \\ &= \left[(X'X)^{-1} + C'C\right]\sigma^2 \end{aligned}$$

[28]Some use the term "best linear unbiased estimator" (BLUE) rather than minimum variance linear unbiased estimator" (MVLUE) with "best" referring to the minimal variance property.

on utilizing $E\tilde{\boldsymbol{u}}\tilde{\boldsymbol{u}}' = \sigma^2 I_n$ and $C'X = 0$. Thus,

$$\operatorname{var}(l'\tilde{\boldsymbol{\beta}}) = \boldsymbol{l}'\left[(X'X)^{-1} + C'C\right]\boldsymbol{l}\sigma^2$$

attains a minimum for $C = 0$ which results in $\boldsymbol{l}'\tilde{\boldsymbol{\beta}} = \boldsymbol{l}'\hat{\boldsymbol{\beta}}$, where $\hat{\boldsymbol{\beta}} = (X'X)^{-1}X'\tilde{\boldsymbol{y}}$ is the LS estimator with covariance matrix, $V(\hat{\boldsymbol{\beta}}) = (X'X)^{-1}\sigma^2$.

Thus, the Gauss–Markov (GM) Theorem provides a justification for the LS estimator for the regression coefficient vector $\boldsymbol{\beta}$ under the hypotheses that the regression model $\tilde{y} = X\boldsymbol{\beta} + \tilde{\boldsymbol{u}}$ is properly specified, $E\tilde{y} = X\boldsymbol{\beta}$ and $V(\tilde{y}) = V(\tilde{\boldsymbol{u}}) = \sigma^2 I_n$, that is, that the errors or observations have a common variance and are uncorrelated. Further, the GM Theorem restricts the class of estimators to be linear and unbiased, restrictions that limit the range of candidate estimators and involves the use of the MSE criterion that here is equivalent to variance since only unbiased estimators are considered. As will be shown below, dropping the restrictions of linearity and unbiasedness can lead to biased, non-linear estimators with smaller MSE than that of the LS estimator under frequently encountered conditions.

While the GM Theorem is remarkable in providing a justification for the LS estimator in terms of its properties in repeated (actual or hypothetical) samples, it does not provide direct justification for the LS estimate that is based on a given sample of data. Obviously, good performance on average does not always insure good performance in a single instance.

An expanded version of the GM Theorem in which the assumption that $E\tilde{\boldsymbol{u}}\tilde{\boldsymbol{u}}' = \sigma^2 I_n$ is replaced by $E\tilde{\boldsymbol{u}}\tilde{\boldsymbol{u}}' = V\sigma^2$, with V an $n \times n$ known positive definite symmetric matrix, shows that $\boldsymbol{l}'\tilde{\boldsymbol{\beta}}$ is the MVLUE of $\boldsymbol{l}'\boldsymbol{\beta}$, where $\tilde{\boldsymbol{\beta}} = (X'V^{-1}X)^{-1}X'V^{-1}\tilde{\boldsymbol{y}}$ is the generalized least squares (GLS) estimator. On substituting $\tilde{\boldsymbol{y}} = X\boldsymbol{\beta} + \tilde{\boldsymbol{u}}$ into $\tilde{\boldsymbol{\beta}}$, $\tilde{\boldsymbol{\beta}} = \boldsymbol{\beta} + (X'V^{-1}X)^{-1}X'V^{-1}\tilde{u}$ and thus $E\tilde{\boldsymbol{\beta}} = \boldsymbol{\beta}$ and $V(\tilde{\boldsymbol{\beta}}) = (X'V^{-1}X)^{-1}\sigma^2$. Also, the GLS estimate can be given a weighted least squares interpretation by noting that minimizing the weighted SS, $(y - X\boldsymbol{\beta})'V^{-1}(y - X\boldsymbol{\beta})$ with respect to $\boldsymbol{\beta}$ yields the GLS estimate. Various forms of the GM Theorem are available in the literature for cases in which X is not of full column rank and/or there are linear restrictions on the elements of $\boldsymbol{\beta}$, that is, $A\boldsymbol{\beta} = \boldsymbol{a}$, where A is a $q \times k$ given matrix of rank q and $\boldsymbol{a}$ is a $q \times 1$ given vector.

The parameter σ^2 appears in the GM Theorem and in the covariance matrices of the LS and GLS estimators. The GM Theorem provides no guidance with respect to the estimation of σ^2. Since $\hat{\boldsymbol{u}} = \boldsymbol{y} - X\hat{\boldsymbol{\beta}}$, the $n \times 1$ LS residual vector is an estimate of $\boldsymbol{u}$, the true unobserved error vector, it seems natural to use the average value of the sum of squared residuals as an estimate of σ^2, that is, $\hat{\sigma}^2 = \hat{\boldsymbol{u}}'\hat{\boldsymbol{u}}/n$. As will be seen, $\hat{\sigma}^2$ is the maximum likelihood estimate of σ^2 in the regression model with normally distributed errors. However, when $\hat{\sigma}^2$ is viewed as an estimator, $E\hat{\sigma}^2 = \sigma^2(1-k/n)$, that is, $\hat{\sigma}^2$ is biased downward.[29] The unbiased estimator of σ^2, $s^2 = \hat{\boldsymbol{u}}'\hat{\boldsymbol{u}}/(n-k)$, is widely used even though it has a larger MSE than that of $\hat{\sigma}_m^2 = \hat{\boldsymbol{u}}'\hat{\boldsymbol{u}}/(n-k+2)$, the minimal MSE estimator in the class $\tilde{\sigma}^2 = c\hat{\boldsymbol{u}}'\hat{\boldsymbol{u}}$, where $c>0$ is a constant. That is, with $\tilde{\boldsymbol{u}}$ assumed $N(\boldsymbol{0}, \sigma^2 I_n)$, $\chi_\nu^2 = \hat{\boldsymbol{u}}'\hat{\boldsymbol{u}}/\sigma^2$ has a chi-squared pdf with $\nu = n-k$ degrees of freedom. Then MSE $= E(\tilde{\sigma}^2-\sigma^2)^2 = c^2\sigma^4E(\chi_\nu^2)^2 - 2c\sigma^4E\chi_\nu^2 + \sigma^4$, and the minimizing value of c, c_m, is $c_m = E\chi_\nu^2/E(\chi_\nu^2)^2 = \nu/(\nu^2+2\nu) = 1/(\nu+2)$ and thus $\hat{\sigma}_m^2 = \hat{\boldsymbol{u}}'\hat{\boldsymbol{u}}/(\nu+2)$ is the minimum MSE estimator. The MSE of $\hat{\sigma}_m^2$ is $\text{MSE}(\hat{\sigma}_m^2) = 2\sigma^4/\nu+2)$, while $\text{MSE}(s^2) = 2\sigma^4/\nu$. Thus, $\text{MSE}(\hat{\sigma}_m^2)/\text{MSE}(s^2) = \nu/(\nu+2)$, which for small ν is appreciably below one.[30] This is an example illustrating that a biased estimator, $\hat{\sigma}_m^2$ can have a smaller MSE than an unbiased estimator s^2.

Above, the LS approach was considered in relation to linear models for which minimization of the SS led to solutions, say $\hat{\boldsymbol{\beta}} = (X'X)^{-1}X'\boldsymbol{y}$ or $\tilde{\boldsymbol{\beta}} = (X'V^{-1}X)^{-1}X'V^{-1}\boldsymbol{y}$, that do not depend on parameters with unknown values and hence are estimates that can be viewed as estimators. In a number of generally encountered problems, this is not the case. For example, if in the usual regression model $\tilde{\boldsymbol{y}} = X\boldsymbol{\beta} + \tilde{\boldsymbol{u}}$, $E\tilde{\boldsymbol{u}} = \boldsymbol{0}$, and $E\tilde{\boldsymbol{u}}\tilde{\boldsymbol{u}}' = \sigma^2V(\boldsymbol{\theta})$, where $V(\boldsymbol{\theta})$ is an $n\times n$ matrix with elements depending on some parameters $\boldsymbol{\theta}$ with unknown values, then a problem arises in applying the LS or GLS approach. That is, minimization of $\boldsymbol{u}'V^{-1}(\theta)\boldsymbol{u} = (y - X\boldsymbol{\beta})'V^{-1}(\theta)(y - X\boldsymbol{\beta})$ with respect to $\boldsymbol{\beta}$ yields $\tilde{\boldsymbol{\beta}} = (X'V^{-1}(\theta)X)^{-1}X'V^{-1}(\theta)\boldsymbol{y}$ which is *not* an estimate since $\tilde{\boldsymbol{\beta}}$ depends on $\boldsymbol{\theta}$, a vector of parameters with unknown values. In some problems it is possible to estimate $\boldsymbol{\theta}$ from the LS residuals, $\hat{\boldsymbol{u}} = \boldsymbol{y} - X\hat{\boldsymbol{\beta}}$, where $\hat{\boldsymbol{\beta}} = (X'X)^{-1}X'\boldsymbol{y}$. Let $\hat{\boldsymbol{\theta}}$ be this estimate. Then a "feasible" or "operational" or approximate GLS estimator, $\tilde{\boldsymbol{\beta}}_{\text{a}}$, is defined by $\tilde{\boldsymbol{\beta}}_{\text{a}} = (X'V^{-1}(\hat{\theta})X)^{-1}X'V^{-1}(\hat{\theta})\tilde{y}$. Since $\tilde{\boldsymbol{\beta}}_{\text{a}}$ is just an approximation to $\tilde{\boldsymbol{\beta}}$, it is not exactly a GLS estimate. Often when n is large, the large sample distributions of $\tilde{\boldsymbol{\beta}}$ and $\tilde{\boldsymbol{\beta}}_{\text{a}}$ coincide. However, when n is small, the distributions of $\tilde{\boldsymbol{\beta}}$ and $\tilde{\boldsymbol{\beta}}_{\text{a}}$ are different and further analysis is required to establish the sampling properties of $\tilde{\boldsymbol{\beta}}_{\text{a}}$. To illustrate consider the following example:

Example 3.3

Let $\tilde{\boldsymbol{y}}' = (\tilde{y}_1' \tilde{y}_2')$, $X' = (X_1' X_2')$, and $\tilde{\boldsymbol{u}}' = (\tilde{\boldsymbol{u}}_1'\tilde{\boldsymbol{u}}_2')$ and consider the regression model, $\tilde{\boldsymbol{y}} = X\boldsymbol{\beta} + \tilde{\boldsymbol{u}}$ with $\boldsymbol{\beta}$ a $k\times 1$ vector, $E\tilde{\boldsymbol{u}} = \boldsymbol{0}$, $E\tilde{\boldsymbol{u}}_1\tilde{\boldsymbol{u}}_1' = \sigma_1^2 I_{n_1}$, $E\tilde{\boldsymbol{u}}_2\tilde{\boldsymbol{u}}_2' = \sigma_2^2 I_{n_2}$, and $E\tilde{\boldsymbol{u}}_1\tilde{\boldsymbol{u}}_2' = 0$. For this specification

$$V(\boldsymbol{\theta}) = E\boldsymbol{u}\boldsymbol{u}' - \begin{bmatrix} \sigma_1^2 I_{n_1} & 0 \\ 0' & \sigma_2^2 I_{n_2} \end{bmatrix},$$

[29]From $\hat{\boldsymbol{u}} = \tilde{y} - X\hat{\boldsymbol{\beta}} = [I_n - X(X'X)^{-1}X']\tilde{\boldsymbol{u}}$, $E\hat{\boldsymbol{u}}'\hat{\boldsymbol{u}} = E\tilde{\boldsymbol{u}}'M\tilde{\boldsymbol{u}} = \sigma^2 \operatorname{tr} M = \sigma^2(n-k)$, where $M = I_n - X(X'X)^{-1}X'$ and $\operatorname{tr} M = n-k$. Thus, $E\hat{\sigma}^2 = \sigma^2(n-k)/n$.

[30]The MSE of $\hat{\sigma}^2 = \hat{\boldsymbol{u}}'\hat{\boldsymbol{u}}/n$ is MSE $(\hat{\sigma}^2) = \sigma^4(2/\nu)[(1+k^2/2\nu)/(1+k/\nu)]$ which is smaller than $\text{MSE}(s^2) = 2\sigma^4/\nu$ for $k>2$.

with $\boldsymbol{\theta}' = (\sigma_1^2, \sigma_2^2)$. Then $(y - X\boldsymbol{\beta})'V^{-1}(\boldsymbol{\theta})(y - X\boldsymbol{\beta})$ has minimal value for

$$\begin{aligned}\tilde{\boldsymbol{\beta}} &= \left(X'V^{-1}(\boldsymbol{\theta})X\right)^{-1}X'V^{-1}(\boldsymbol{\theta})y \\ &= \left[X_1{}'X_1/\sigma_1^2 + X_2{}'X_2/\sigma_2^2\right]^{-1}\left(X_1{}'y_1/\sigma^2 + X_2{}'y_2/\sigma^2\right),\end{aligned}$$

which is clearly a function σ_1^2 and σ_2^2. Estimates of σ_1^2 and σ_2^2 are $s_i^2 = (y_i - X_i\hat{\boldsymbol{\beta}}_i)(y_i - X_i\hat{\boldsymbol{\beta}}_i)/\nu_i$, with $\nu_i = n_i - k$ and $\hat{\boldsymbol{\beta}}_i = (X_i'X_i)^{-1}X_i'y_i$ for $i = 1,2$. Then the approximate GLS estimate is

$$\tilde{\boldsymbol{\beta}}_{\mathrm{a}} = \left[X_1'X_1/s_1^2 + X_2'X_2/s_2^2\right]^{-1}\left[X_1'y_1/s_1^2 + X_2'y_2/s_2^2\right].$$

Further, $V(\tilde{\boldsymbol{\beta}}) = [X_1'X_1/\sigma_1^2 + X_2'X_2/\sigma_2^2]^{-1}$ which is often approximated by $V_{\mathrm{a}}(\tilde{\boldsymbol{\beta}}) = [X_1'X_1/s_1^2 + X_2'X_2/s_1^2]^{-1}$. For large sample sizes, these approximations have been shown to be very good. For small samples, some of the properties of these approximations have been studied in the literature. Also, above the unbiased estimates, s_1^2 and s_2^2 have been and usually are inserted for σ_1^2 and σ_2^2. Whether other estimates, for example ML or minimum MSE estimates of σ_1^2 and σ_2^2, would produce better or worse results in estimating $\boldsymbol{\beta}$ requires additional analysis.

The basic problem brought out in the above example is that the GLS approach can lead to a non-operational result. Then approximations to the non-operational result are introduced. Fortunately, in many cases these approximations are rather good in large samples. However, in small samples, the definition of which depends on various properties of the model as well as the sample size, there is no assurance that these approximate GLS estimates will have good sampling properties. Each case must be considered carefully using analytical and/or Monte Carlo techniques. For examples of such studies, see Rao and Griliches (1969), Fomby and Guilkey (1978), Taylor (1978), Revankar (1974), Mehta and Swamy (1976), and Srivastava and Dwivedi (1979). From a practical point of view, it would be desirable to have estimation principles that yield estimators with good small and large sample properties.

3.2.1.5. Maximum likelihood estimation. In maximum likelihood (ML) estimation, a basic element is the likelihood function. Let the sample observations be denoted by $\boldsymbol{x}$ and the joint pdf for the observations be $p(\boldsymbol{x}|\boldsymbol{\theta})$, with $\boldsymbol{x} \subset R_x$ the sample space, and $\boldsymbol{\theta} \subset \Theta$ the parameter space. The likelihood function is $p(\boldsymbol{x}|\boldsymbol{\theta})$ *viewed as a function of* $\boldsymbol{\theta}$ *defined on the parameter space* Θ. To emphasize this point, the likelihood function is denoted by $l(\boldsymbol{\theta}|\boldsymbol{x})$ and is clearly not a pdf for $\boldsymbol{\theta}$. To make this point explicitly, the joint pdf of n independent observations from a normal distribution with mean μ and variance σ^2 is

$$p(\boldsymbol{x}|\mu,\sigma^2)=(2\pi\sigma^2)^{-n/2}\exp\left\{\sum_{i=1}^{n}(x_i-\mu)^2/2\sigma^2\right\}.$$

The likelihood function for this problem is

$$l(\mu,\sigma^2|\boldsymbol{x})=(2\pi\sigma^2)^{-n/2}\exp\left\{-\sum_{i=1}^{n}(x_i-\mu)^2\Big/2\sigma^2\right\},\tag{3.14}$$

with $-\infty<x_i<\infty$, $i=1,2,\dots,n$, the sample space, and $-\infty<\mu<\infty$ and $0<\sigma^2<\infty$, the parameter space. In (3.14), the likelihood function, $l(\mu,\sigma^2|\boldsymbol{x})$, is a function of μ and σ^2 given $\boldsymbol{x}$.

According to the ML estimation principle, estimates of parameters are obtained by maximizing the likelihood function given the data. That is, the ML estimate is the quantity $\hat{\boldsymbol{\theta}}(x)\subset\Theta$ such that $l(\hat{\boldsymbol{\theta}}|\boldsymbol{x})=\max_{\hat{\theta}\subset\Theta}l(\boldsymbol{\theta}|\boldsymbol{x})$. For a very broad range of problems the ML estimate $\hat{\theta}(\boldsymbol{x})$ exists and is unique. In the likelihood approach, $l(\boldsymbol{\theta}|\boldsymbol{x})$ expresses the "plausibility" of various values of $\boldsymbol{\theta}$, and $\hat{\boldsymbol{\theta}}$, the ML estimate, is regarded as the "most plausible" or "most likely" value of $\boldsymbol{\theta}$. This view is a basic non-sampling argument for ML estimation, although the terms "most plausible" or "most likely" cannot be equated with "most probable" since the likelihood function, $l(\boldsymbol{\theta}|\boldsymbol{x})$, is *not* a pdf for $\boldsymbol{\theta}$. As with LS estimates, ML estimates can be viewed as estimators and their properties studied to determine whether they are good or optimal in some senses.

From the likelihood function in (3.14), $\log l(\mu,\sigma^2|\boldsymbol{x})=-n\log\sigma-\sum(x_i-\mu)^2/2\sigma^2+\text{constant}$. The necessary conditions for a maximum are $\partial\log l/\partial\mu=0$ and $\partial\log l/\partial\sigma=0$ which yield $\sum_{i=1}^{n}(x_i-\mu)/2\sigma^2=0$ and $-n/\sigma+\sum_{i=1}^{n}(x_i-\mu)^2/\sigma^3=0$, the solutions of which are $\hat{\mu}=\sum_{i=1}^{n}x_i/n$ and $\hat{\sigma}^2=\sum_{i=1}^{n}(x_i-\hat{\mu})^2/n$ and these values can be shown to be global-maximizing values and hence are ML estimates. In this particular case, $\hat{\mu}$ is the sample mean which, as mentioned above, is a minimum variance unbiased estimator of μ. With respect to $\hat{\sigma}^2=\sum_{i=1}^{n}(\tilde{x}_i-\hat{\mu})^2/n$, it was shown above that this estimator is biased. Thus, with ML estimation, there is no assurance that ML estimators will be unbiased. Indeed, the ML estimator of $\theta=1/\mu$ is $\hat{\theta}=1/\hat{\mu}$, which does not possess a mean. However, as emphasized above, the criterion of unbiasedness is subject to important limitations. Further, note that in this problem the ML estimates are functions of the sufficient statistics, $\hat{\mu}$ and $\sum_{i=1}^{n}(x_i-\hat{\mu})^2$. Given the Factorization Theorem described above, it is the case that ML estimates are always functions of minimal sufficient statistics when they exist. This is not to say, however, that the ML estimator necessarily makes the best possible use of the information contained in minimal sufficient statistics.

For the usual multiple regression model, $\tilde{y} = X\beta + \tilde{u}$, with $\tilde{u}$ assumed normal with mean 0 and covariance matrix, $\sigma^2 I_n$, the joint pdf for the observations is $p(y|X, \beta, \sigma^2) = (2\pi\sigma^2)^{-n/2}\exp\{-(y - X\beta)'(y - X\beta)/2\sigma^2\}$ and the likelihood function, $l(\beta, \sigma^2|X, y)$, is $p(y|X, \beta, \sigma^2)$ viewed as a function of β and σ^2, that is,

$$\begin{aligned} l(\beta, \sigma^2|X, y) &= (2\pi\sigma^2)^{-n/2}\exp\{-(y - X\beta)'(y - X\beta)/2\sigma^2\} \\ &= (2\pi\sigma^2)^{-n/2}\exp\{-[\hat{u}'\hat{u} + (\beta - \hat{\beta})'X'X(\beta - \hat{\beta})]/2\sigma^2\}, \end{aligned}$$

where $\hat{\beta} = (X'X)^{-1}X'y$ and $\hat{u} = y - X\hat{\beta}$. On maximizing $\log l(\beta, \sigma^2|X, y)$ with respect to β and σ^2, the ML estimates are $\hat{\beta} = (X'X)^{-1}X'y$ and $\hat{\sigma}^2 = \hat{u}'\hat{u}/n$. The ML estimate $\hat{\beta}$ is just the LS estimate, $(X'X)^{-1}X'y$, while again the ML estimator for σ^2, $\hat{\sigma}^2$, is biased. Above, the GM Theorem led to the LS estimate $\hat{\beta}$ without a normality assumption. In this connection it is interesting to observe that for all likelihood functions of the form, $\sigma^{-n}f(u'u/\sigma^2)$, where the function $f(\cdot)$ is monotonically decreasing in $u'u/\sigma^2$, minimizing $u'u = (y - X\beta)'(y - X\beta)$ with respect to β produces a ML estimate equal to the LS estimate. Therefore, the normal case $f(u'u/\sigma^2) \propto \exp\{-u'u/2\sigma^2\}$ is just a special case of a class of likelihood functions for which the ML estimate is identical to the LS estimate.[31]

In the case where the error vector in the multiple regression model is normal, with $E\tilde{u} = 0$ and $E\tilde{u}\tilde{u}' = \sigma^2 V(\theta)$, the likelihood function is

$$\begin{aligned} l(\beta, \sigma^2, \theta| y, X) &= (2\pi\sigma^2)^{-n/2}|V(\theta)|^{-1/2} \\ &\quad \times\exp\{-(y - X\beta)'V^{-1}(\theta)(y - X\beta)/2\sigma^2\}. \end{aligned}$$

The quadratic form in the exponential is minimized for any given value of θ by $\tilde{\beta}_\theta = [X'V^{-1}(\theta)X]^{-1}X'V^{-1}(\theta)y$, the GLS quantity. Also, the conditional maximizing value of σ^2, given θ, is $\tilde{\sigma}_\theta^2 = (y - X\tilde{\beta})'V^{-1}(\theta)(y - X\tilde{\beta})/n$. On substitution these conditional maximizing values in the likelihood function, the result is the so-called concentrated log-likelihood function, $\log l_c(\theta| y, X) = \text{constant} - n/2\log\tilde{\sigma}_\theta^2 - \frac{1}{2}\log|V(\theta)|$. By numerical evaluation of this function for various values of θ it is possible to find a maximizing value for θ, say $\tilde{\theta}$, which when substituted into $\tilde{\beta}_\theta$ and $\tilde{\sigma}_\theta^2$ provides ML estimates of *all* of the parameters. This

[31]On the other hand, if the likelihood function is a monotonically decreasing function of $\sum_{i=1}^n |u_i|/\sigma$, then minimizing the sum of the absolute deviations produces ML estimates. Such a likelihood function is encountered when the $\tilde{u}_i$'s are identically and independently distributed, each with a double exponential pdf, $p(u_i) \propto \sigma^{-1}\exp\{-|u_i|/\sigma\}$, $-\infty < u_i < \infty$. Then the joint pdf of the u_i's is

$$\prod_{i=1}^n p(u_i) \propto \sigma^{-n}\exp\left\{-\sum_{i=1}^n |u_i|/\sigma\right\}$$

and minimizing $\sum_{i=1}^n |u_i|$ maximizes the likelihood function.

procedure is useful for computing ML estimates only when there are a few, say one or two, parameters in $\boldsymbol{\theta}$.

A more general procedure, the Newton method, for maximizing log-likelihood functions, $L \equiv \log l(\boldsymbol{\theta}|\boldsymbol{x})$, where $\boldsymbol{\theta}$ is an $m \times 1$ vector of parameters and $\boldsymbol{x}$ a vector of observations, commences with the first-order conditions, $\partial L/\partial\boldsymbol{\theta} = \mathbf{0}$. Given an initial estimate, $\boldsymbol{\theta}^{(0)}$, of the solution $\hat{\boldsymbol{\theta}}$ of $\partial L/\partial\boldsymbol{\theta} = \mathbf{0}$, expand $\partial L/\partial\boldsymbol{\theta}$ in a Taylor's Series about $\boldsymbol{\theta}^{(0)}$, that is,

$$\partial L/\partial\boldsymbol{\theta} = 0 \doteq \partial L/\partial\boldsymbol{\theta} + \left[\partial^2 L/\partial\boldsymbol{\theta}\partial\boldsymbol{\theta}'\right](\hat{\boldsymbol{\theta}} - \boldsymbol{\theta}^{(0)}),$$

where all partial derivatives on the right-side are evaluated at $\boldsymbol{\theta}^{(0)}$. Then,

$$\boldsymbol{\theta}^{(1)} = \boldsymbol{\theta}^{(0)} - \left[\partial^2 L/\partial\boldsymbol{\theta}\partial\boldsymbol{\theta}'\right]^{-1}\left[\partial L/\partial\boldsymbol{\theta}\right], \tag{3.15}$$

with the derivatives evaluated at $\boldsymbol{\theta} = \boldsymbol{\theta}^{(0)}$, is an approximation to the ML estimate, $\hat{\boldsymbol{\theta}}$. By repeating this process, the sequence $\boldsymbol{\theta}^{(1)}, \boldsymbol{\theta}^{(2)}, \dots,$ usually converges to the ML estimate. Other numerical algorithms for solving non-linear optimization problems are described in Goldfeld and Quandt (1972). What is of great importance is that the ML method provides estimates of all parameters of a model in accord with a well-defined criterion and is generally applicable to most econometric estimation problems. Some further properties of ML estimation are summarized in the next paragraph.

First, as pointed out above, ML estimators are not necessarily unbiased in finite samples. However, insistence on the property of unbiasedness in general is not necessarily desirable. The discussion of the criterion of unbiasedness, presented above, and the examples considered are relevant. Second, when an unbiased estimator, $\hat{\theta}$, exists that attains the Cramér–Rao lower bound, it was mentioned above that in such a case $\partial \log l(\theta|\boldsymbol{x})/\partial\theta = a(\theta)(\hat{\theta} - \theta)$ and thus the only solution to $\partial \log l(\theta|\boldsymbol{x})/\partial\theta = 0$ is $\hat{\theta}$. In this case, the ML estimator is identical to the MVB unbiased estimator. Third, in some cases, but not all, when a ML estimator is unbiased, its variance *may* be close to the Cramér–Rao lower bound, a property that has to be checked in individual cases. Fourth, ML estimates have an invariance property, namely if $\hat{\theta}$ is the ML estimate of θ and $\eta = g(\theta)$ is a one-to-one transformation, then $\hat{\eta} = g(\hat{\theta})$ is the ML estimate of η. This property also applies when θ and η are vectors of parameters. Finally, the most important sampling justification for ML estimators is that they usually have very good properties in large samples. That is, under certain regularity conditions,[32] ML estimators are consistent in the sense that the sequence of ML estimators, depending on n, the sample size, $\{\hat{\theta}_{(n)}\}$ converges to θ as $n \to \infty$ in

[32] These regularity conditions are presented and discussed in Kendall and Stuart (1961) and Heyde and Johnstone (1979).

either a weak probability sense, $\hat{\theta}_{(n)} \xrightarrow{\text{i.p.}} \theta$, or a strong probability sense, $\hat{\theta}_{(n)} \xrightarrow{\text{a.s.}} \theta$, depending upon whether a weak or strong law of large numbers is employed in the derivation. Further, in large samples (large n), ML estimators are approximately unbiased and have variances close to the Cramér–Rao lower bound under regularity conditions. Also, as n grows large, the ML estimator's distribution is approximately a normal distribution with mean θ and variance I_θ^{-1}, where I_θ is the Fisher information matrix or Cramér–Rao lower bound for an unbiased estimator based on n independent and identically distributed observations. In the case of a vector ML estimator, $\hat{\boldsymbol{\theta}}_{(n)}$, for large n and under regularity conditions, its approximate distribution is multivariate normal with mean $\boldsymbol{\theta}$ and covariance $I_{\boldsymbol{\theta}}^{-1}$, where $I_{\boldsymbol{\theta}}$ is n times the Fisher information matrix for a single observation. For proofs of these properties that generally assume that observations, vector or scalar, are independently and identically distributed and impose certain conditions on the higher moments or other features of the observations' common distribution, see Cramér (1946, p. 500), Wald (1949), and Anderson (1971). For dependent observations, for example those generated by time series processes, additional assumptions are required to establish the large sample properties of ML estimators; see, for example, Anderson (1971) and Heyde and Johnstone (1979).

A basic issue with respect to the large sample properties of ML estimators is the determination of what constitutes a "large sample". For particular problems, mathematical analysis and/or Monte Carlo experiments can be performed to shed light on this issue. It must be emphasized that not only is the sample size relevant, but also other features of the models, including parameter values, the properties of independent variables, and the distributional properties of error terms. Sometimes the convergence to large sample properties of ML estimators is rapid, while in other cases it can be quite slow. Also, in "irregular" cases, the above large sample properties of ML estimators may not hold. One such simple case is $\tilde{y}_i = \theta_i + \tilde{\varepsilon}_i$, $i = 1,2,\dots,n$, where the $\tilde{\varepsilon}_i$'s are NID$(0,\sigma^2)$. The ML estimate of θ_i is $\hat{\theta}_i = y_i$, and it is clear that $\hat{\theta}_i$ does not converge to θ_i as n grows since there is just one observation for each θ_i. The irregular aspect of this problem is that the number of parameters grows with the sample size, a so-called "incidental parameter" problem. Incidental parameters also appear in the functional form of the errors-in-the-variables model and affect asymptotic properties of ML estimators; see, for example, Neyman and Scott (1948) and Kendall and Stuart (1961). Thus, such "irregular" cases, and also others in which the ranges of observations depend on parameters with unknown values or observations are dependent and generated by non-stationary time series processes and have to be analyzed very carefully since regular large sample ML properties, including consistency, normality, and efficiency, may not hold.

3.2.1.6. Admissibility criterion. The admissibility criterion is a sampling criterion for alternative estimators that involves separating estimators into two classes, namely those that are admissible and those that are inadmissible with respect to estimators' risk properties relative to given loss functions. In this approach, inadmissible estimators are regarded as unacceptable and attention is concentrated on the class of admissible estimators. Since this class usually contains many estimators, additional criteria are required to choose a preferred estimator from the class of admissible estimators.

The basic elements in applying the admissibility criterion are (1) loss functions, (2) risk functions, and (3) comparisons of risk functions associated with alternative estimators. Consider a scalar parameter θ, and $\hat{\theta}$ an estimator for θ. Some examples of loss functions are given below, where the c's are given positive constants:

(1) Quadratic (or squared error): $L(\theta, \hat{\theta}) = c_1(\theta - \hat{\theta})^2$.
(2) Absolute error: $L(\theta, \hat{\theta}) = c_2|\theta - \hat{\theta}|$.
(3) Relative squared error: $L(\theta, \hat{\theta}) = c_3(\theta - \hat{\theta})^2/\theta^2$.
(4) Generalized quadratic: $L(\theta, \hat{\theta}) = h(\theta)(\theta - \theta)^2$.
(5) Exponential: $L(\theta, \hat{\theta}) = c_5[1 - \exp\{-c_6(\theta - \hat{\theta})^2\}]$.

These are but a few of many possible loss functions that can be employed. Note that they all are monotonically increasing functions of the absolute error of estimation, $|e| = |\hat{\theta} - \theta|$. The first three loss functions are unbounded while the fifth is an example of a bounded loss function that attains a maximal value of c_5 as $|\hat{\theta} - \theta| \to \infty$. The relative squared error loss function (3) is a special case of the generalized loss function (4) with $h(\theta) = c_3/\theta^2$. Note too that, as is customary, these loss functions have been scaled so that minimal loss equals zero when $\hat{\theta} - \theta = 0$. Also, negative loss can be interpreted as utility, that is, $U(\theta, \hat{\theta}) \equiv -L(\theta, \hat{\theta})$.

In the case of a vector of parameters, $\boldsymbol{\theta}$, and a vector estimator, $\hat{\boldsymbol{\theta}}$, a quadratic loss function is given by $L(\boldsymbol{\theta}, \hat{\boldsymbol{\theta}}) = (\boldsymbol{\theta} - \hat{\boldsymbol{\theta}})'Q(\boldsymbol{\theta} - \hat{\boldsymbol{\theta}})$, where Q is a given pds matrix. A generalized quadratic loss function is $L(\boldsymbol{\theta}, \hat{\boldsymbol{\theta}}) = h(\boldsymbol{\theta})(\boldsymbol{\theta} - \hat{\boldsymbol{\theta}})'Q(\boldsymbol{\theta} - \hat{\boldsymbol{\theta}})$, where $h(\boldsymbol{\theta})$ is a given function of $\boldsymbol{\theta}$. One example is $h(\boldsymbol{\theta}) = 1/(\boldsymbol{\theta}'\boldsymbol{\theta})^m$, where m is a given non-negative constant.

In a particular estimation problem, the choice of an appropriate loss function is important. Sometimes subject-matter considerations point to a particular form for a loss (or utility) function. The widespread use of quadratic loss functions can perhaps be rationalized by noting that a Taylor's Series expansion of any loss function, $L(e)$, about $e = \hat{\theta} - \theta = 0$, such that $L(0) = 0$ and $L'(0) = 0$, yields $L(e) \doteq L''(0)e^2/2$, an approximate quadratic loss function. This, it must be emphasized, is a *local* approximation which may not be very good for asymmetric loss functions and/or bounded loss functions [see, for example, Zellner and Geisel (1968)].

Given that a loss function, $L(\theta, \hat{\theta})$, has been selected, the next step in applying the admissibility criterion is to evaluate the risk function, denoted by $r_{\hat{\theta}}(\theta)$ and defined by

$$r_{\hat{\theta}}(\theta)=\int_{R_x} L(\hat{\theta},\theta)\, p(x|\theta)\,\mathrm{d}x, \tag{3.16}$$

where $p(x|\theta)$ is the pdf for the observations given θ. It is seen that the risk function in (3.16) is defined for a particular estimator, $\hat{\theta}$, and a particular loss function, $L(\hat{\theta},\theta)$. While it would be desirable to choose an estimator $\hat{\theta}$ so as to minimize $r_{\hat{\theta}}(\theta)$ for all values of θ, unfortunately this is impossible. For example, an "estimator" $\hat{\theta}=5$ will have lower risk when $\theta=5$ than any other estimator and thus no one estimator can minimize risk for all possible values of θ. In view of this fact, all that is possible at this point is to compare the risk functions of alternative estimators, say $\hat{\theta}_1,\hat{\theta}_2,\dots$, with risk functions $r_{\hat{\theta}_1}(\theta)$, $r_{\hat{\theta}_2}(\theta),\dots$, relative to a given loss function. From such a comparison, it may be that $r_{\hat{\theta}_1}(\theta)\le r_{\hat{\theta}_2}(\theta)$ for all θ with the inequality being strict for some θ. In such a case, $\hat{\theta}_2$ is said to be dominated by $\hat{\theta}_1$, and $\hat{\theta}_2$ is termed an *inadmissible* estimator. That $\hat{\theta}_1$ dominates $\hat{\theta}_2$ does not necessarily imply that $\hat{\theta}_1$ is itself admissible. To be admissible, an estimator, say $\hat{\theta}_1$, must have a risk function $r_{\hat{\theta}_1}(\theta)$ such that $r_{\hat{\theta}_1}(\theta)\le r_{\hat{\theta}_a}(\theta)$ for all θ, where $r_{\hat{\theta}_a}(\theta)$ is the risk function associated with any other estimator $\hat{\theta}_a$. Work on proof of admissibility of estimators is given in Brown (1966).

A leading example of the inadmissibility of a maximum likelihood and least squares estimator has been given by Stein (1956) and James and Stein (1961). Let $\tilde{y}_i=\theta_i+\tilde{\varepsilon}_i$, $i=1,2,\dots,n$, with the $\tilde{\varepsilon}_i$'s $\mathrm{NID}(0,\sigma^2)$ and the θ_i's the means of the $\tilde{y}_i$'s, $-\infty<\theta_i<\infty$. Further, let the loss function be $L(\boldsymbol{\theta},\hat{\boldsymbol{\theta}})=(\boldsymbol{\theta}-\hat{\boldsymbol{\theta}})'(\boldsymbol{\theta}-\hat{\boldsymbol{\theta}})$, a quadratic loss function. The likelihood function is $p(\boldsymbol{y}|\boldsymbol{\theta})=(2\pi\sigma^2)^{-n/2}\exp\{-(\boldsymbol{y}-\boldsymbol{\theta})'(\boldsymbol{y}-\boldsymbol{\theta})/2\sigma^2\}$, where $\boldsymbol{y}'=(y_1,y_2,\dots,y_n)$ and $\boldsymbol{\theta}'=(\theta_1,\theta_2,\dots,\theta_n)$. Then the ML estimator is $\hat{\boldsymbol{\theta}}_0=\tilde{\boldsymbol{y}}$, with risk function $r_{\hat{\boldsymbol{\theta}}_0}(\boldsymbol{\theta})=E(\hat{\boldsymbol{\theta}}_0-\boldsymbol{\theta})'(\hat{\boldsymbol{\theta}}_0-\boldsymbol{\theta})=n\sigma^2$. When σ^2 has a known value, say $\sigma^2=1$, James and Stein (1961) put forward the following estimator for $\boldsymbol{\theta}$ when $n\ge 3$,

$$\hat{\boldsymbol{\theta}}_1=\left[1-(n-2)/\boldsymbol{y}'\boldsymbol{y}\right]\boldsymbol{y}, \tag{3.17}$$

that has uniformly lower risk than the ML (and LS) estimator, $\hat{\boldsymbol{\theta}}_0=\boldsymbol{y}$; that is, $r_{\hat{\boldsymbol{\theta}}_1}<r_{\hat{\boldsymbol{\theta}}_0}$ or $E(\hat{\boldsymbol{\theta}}_1-\boldsymbol{\theta})'(\hat{\boldsymbol{\theta}}_1-\boldsymbol{\theta})<E(\hat{\boldsymbol{\theta}}_0-\boldsymbol{\theta})'(\hat{\boldsymbol{\theta}}_0-\boldsymbol{\theta})$ for $0<\theta'\theta<\infty$ [see James and Stein (1961) for a proof]. As James and Stein show, use of $\hat{\boldsymbol{\theta}}_1$ in (3.17) rather than the ML estimator results in a large reduction in risk, particularly in the vicinity of $\boldsymbol{\theta}=\boldsymbol{0}$. They also develop an estimator similar to (3.17) for the case of σ^2 unknown and show that it dominates the ML estimator uniformly. For details , see James and Stein (1961), Zellner and Vandaele (1975), and the references in the latter paper. Also, as shown in Stein (1960), Sclove (1968), and Zellner and Vandaele (1975), Stein's result on the inadmissibility of the ML (and LS) mean estimator carries over to apply to regression estimation problems when the regression coefficients number three or more and an unbounded quadratic loss function is utilized. It is also the case that for certain problems, say estimating the reciprocal of a population mean, the ratio of regression coefficients, and coefficients of simultaneous equation models, ML and other estimators' moments usually or often do not exist, implying that such estimators are inadmissible relative to quadratic and many other unbounded loss functions [see Zellner (1978) and Zellner and Park (1979)]. While use of bounded loss functions will result in bounded risk for these estimators, see Zaman (1981), it is not clear that the ML and other estimators for these problems are admissible.

Another broad class of estimators that are inadmissible are those that are discontinuous functions of the sample data, for example certain "pre-test" estimators. That is, define an estimator by $\hat{\theta}=\hat{\theta}_1$ if $\tilde{t}>a$ and $\hat{\theta}=\hat{\theta}_2$ if $\tilde{t}\leq a$, where $\tilde{t}$ is a test statistic. If $\Pr(\tilde{t}>a)=w$, then the risk of this estimator relative to quadratic loss is $r_{\hat{\theta}}(\theta)=wE(\hat{\theta}_1-\theta)^2+(1-w)E(\hat{\theta}_2-\theta)^2$. As an alternative estimator, consider $\hat{\theta}_3=w\hat{\theta}_1+(1-w)\hat{\theta}_2$ with risk function

$$\begin{aligned}r_{\hat{\theta}_3}(\theta)&=E(\hat{\theta}_3-\theta)^2=E\left[w(\hat{\theta}_1-\theta)+(1-w)(\hat{\theta}_2-\theta)\right]^2\\&=w^2E(\hat{\theta}_1-\theta)^2+(1-w)^2E(\hat{\theta}_2-\theta)^2+2w(1-w)\operatorname{cov}(\hat{\theta}_1,\hat{\theta}_2).\end{aligned}$$

Then $r_{\hat{\theta}}(\theta)-r_{\hat{\theta}_3}(\theta)=w(1-w)E[(\hat{\theta}_1-\theta)-(\hat{\theta}_2-\theta)]^2>0$ and thus the discontinuous estimator $\hat{\theta}$ is inadmissible. For further properties of "preliminary-test" estimators, see Judge and Bock (1978).

Since the class of admissible estimators relative to a specific loss function contains many estimators, further conditions have to be provided in order to choose among them. As seen above, the conditions of the Gauss–Markov Theorem limits the choice to linear and unbiased estimators and thus rules out, for example, the non-linear, biased James–Stein estimator in (3.17) and many others. The limitation to linear and unbiased estimators is not only arbitrary but can lead to poor results in practice [see, for example, Efron and Morris (1975)].

Another criterion for choosing among admissible estimators is the Wald minimax criterion; that is, choose the estimator that minimizes the maximum expected loss. Formally, find $\hat{\theta}$ such that $\max_{\hat{\theta}} r_{\hat{\theta}}(\theta)\leq\max_{\hat{\theta}_a} r_{\hat{\theta}_a}(\theta)$, where $\hat{\theta}_a$ is any other estimator. While this rule provides a unique solution in many problems, its very conservative nature has been criticized; see, for example, Ferguson (1967, p. 58) and Silvey (1970, p. 165). A much less conservative rule is to choose the estimator, when it exists, that minimizes the minimum risk or, equivalently, maximizes the maximum utility. While these rules may have some uses in particular cases, in many others they lead to solutions that are not entirely satisfactory.

To illustrate the use of risk functions, consider Figure 3.1 in which the risk functions associated with three estimators, $\hat{\theta}_1$, $\hat{\theta}_2$, and $\hat{\theta}_3$, have been plotted. As

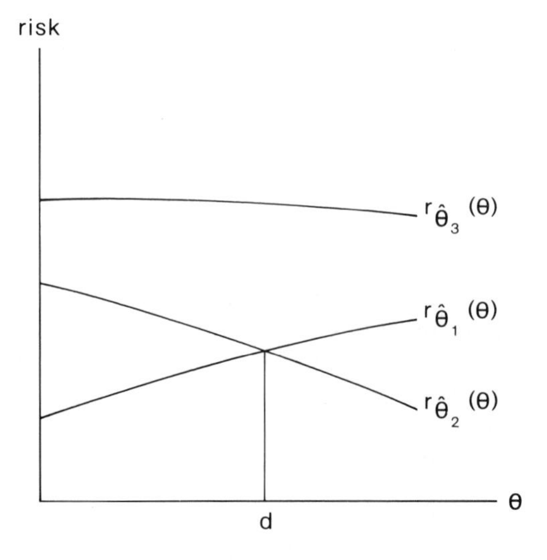

Figure 3.1

drawn, $\hat{\theta}_1$ and $\hat{\theta}_2$ clearly dominate $\hat{\theta}_3$ since $r_{\hat{\theta}_3}$ lies everywhere above the other two risk functions. Thus, $\hat{\theta}_3$ is inadmissible. In choosing between $\hat{\theta}_1$ and $\hat{\theta}_2$, it is clearly important to know whether θ's value is to the right or left of the point of intersection, $\theta = d$. Without this information, choice between $\hat{\theta}_1$ and $\hat{\theta}_2$ is difficult, if not impossible. Further, unless admissibility is proved, there is no assurance that either $\hat{\theta}_1$ or $\hat{\theta}_2$ is admissible. There may be some other estimator, say $\hat{\theta}_4$, that dominates both $\hat{\theta}_1$ and $\hat{\theta}_2$. Given these conditions, there is uncertainty about the choice between $\hat{\theta}_1$ and $\hat{\theta}_2$ and, as stated above, without a proof of admissibility there is no assurance that either is admissible. For a practical illustration of these problems in the context of estimating the parameter ρ in a stationary, normal, first-order autoregressive process, $y_t = \rho y_{t-1} + \varepsilon_t$, see Thornber (1967). He provides estimated risk functions for ML and several other estimators for ρ. These risk functions cross and thus no one estimator uniformly dominates the others. The shapes of the estimated risk functions are also of interest. See also Fomby and Guilkey (1978).

In summary, the criterion of admissibility, a sampling criterion, provides a basis for ruling out some estimators. Indeed, according to this criterion, Stein's results indicate that many ML and LS estimators are inadmissible relative to quadratic loss. In other cases in which estimators do not possess finite moments, they are inadmissible relative to quadratic and other loss functions that require estimators' moments to be finite in order for risk to be finite. Even if just bounded loss functions are considered, there is no assurance that ML and LS estimators are admissible relative to them without explicit proofs that they do indeed possess this property. As regards admissible estimators, they are not in general unique so that the problem of choice among them remains difficult. If information is available about the range of "plausible" or "reasonable" values of parameters, a choice among alternative admissible estimators can sometimes be made. In terms of Figure 3.1, if it is known that $\hat{\theta}_1$ and $\hat{\theta}_2$ are admissible estimators and if it is known that $\theta > d$, then $\hat{\theta}_2$ would be preferred to $\hat{\theta}_1$. Below, in the Bayesian approach, it is shown how such information can be employed in obtaining estimators.

3.2.1.7. Bayesian approach. In the Bayesian approach to estimation, both observations and parameters are considered random. Let $p(\boldsymbol{x}, \boldsymbol{\theta})$ be the joint pdf for an observation vector $\tilde{\boldsymbol{x}} \subset R_x$ and $\boldsymbol{\theta} \subset \Theta$. Then, according to the usual rules for analyzing joint pdfs, the joint pdf can be expressed as

$$\begin{aligned} p(\boldsymbol{x}, \boldsymbol{\theta}) &= p(\boldsymbol{x}|\boldsymbol{\theta})p(\boldsymbol{\theta}) \\ &= p(\boldsymbol{\theta}|\boldsymbol{x})p(\boldsymbol{x}), \end{aligned} \tag{3.18}$$

where the functions $p(\cdot)$ are labelled by their arguments. From (3.18), $p(\boldsymbol{\theta}|\boldsymbol{x}) = p(\boldsymbol{\theta})p(\boldsymbol{x}|\boldsymbol{\theta})/p(\boldsymbol{x})$ or

$$p(\boldsymbol{\theta}|\boldsymbol{x}) \propto p(\boldsymbol{\theta})p(\boldsymbol{x}|\boldsymbol{\theta}), \tag{3.19}$$

where the factor of proportionality in (3.18) is the reciprocal of $\int_\Theta p(\boldsymbol{\theta})p(\boldsymbol{x}|\boldsymbol{\theta})\mathrm{d}\boldsymbol{\theta} = p(\boldsymbol{x})$. The result in (3.19) is Bayes' Theorem with $p(\boldsymbol{\theta}|\boldsymbol{x})$ the posterior pdf for $\boldsymbol{\theta}$, $p(\boldsymbol{\theta})$ the prior pdf for $\boldsymbol{\theta}$, and $p(\boldsymbol{x}|\boldsymbol{\theta})$ the likelihood function. Thus, (3.19) can be expressed as,

$$\text{posterior pdf} \propto (\text{prior pdf}) \times (\text{likelihood function}).$$

As an example of the application of (3.19), consider n independent binomial trials with likelihood function

$$p(r|n, \theta) = \binom{n}{r}\theta^r(1-\theta)^{n-r},$$

with $0 \le \theta \le 1$. As prior pdf for θ, assume that it is given by $p(\theta|a, b) = \theta^{a-1}(1-\theta)^{b-1}/B(a, b)$ a beta pdf with $a, b > 0$ having given values so as to represent the available information regarding possible values of θ. Then the posterior pdf for θ

is given by

$$p(\theta|D) \propto p(\theta|a,b)p(r|n,\theta) \\ \propto \theta^{r+a-1}(1-\theta)^{n-r+b-1}, \tag{3.20}$$

where D denotes the prior and sample information and the factor of proportionality, the normalizing constant is $1/B(r+a, n-r+b)$. It is seen that the posterior pdf in (3.20) is a beta-pdf with parameters $r+a$ and $n-r+b$. The sample information enters the posterior pdf through the likelihood function, while the prior information is introduced via the prior pdf. Note that the complete posterior pdf for θ is available. It can be employed to make probability statements about θ, e.g. $\Pr(c_1 < \theta < c_2|D) = \int_{c_1}^{c_2} p(\theta|D)\mathrm{d}\theta$. Also, the mean and other moments of the posterior pdf are easily evaluated from properties of the beta distribution. Thus, the prior pdf, $p(\theta|a,b)$, has been transformed into a posterior pdf, $p(\theta|D)$, that incorporates both sample and prior information.

As mentioned in Section 2.2, the added element in the Bayesian approach is the prior pdf, $p(\boldsymbol{\theta})$, in (3.19), or $p(\theta|a,b)$ in (3.20). Given a prior pdf, standard mathematical operations yield the posterior pdf as in (3.19). Explicit posterior distributions for parameters of many models encountered in econometrics have been derived and applied in the literature; see, for example, Jeffreys (1967), Lindley (1965), DeGroot (1970), Box and Tiao (1973), Leamer (1978), and Zellner (1971). Further, from (3.19), the marginal pdf for a single element or a subset of the elements of $\boldsymbol{\theta}$ can be obtained by integration. That is, if $\boldsymbol{\theta}' = (\boldsymbol{\theta}_1'\boldsymbol{\theta}_2')$, the marginal posterior pdf for $\boldsymbol{\theta}_1$ is given by

$$p(\boldsymbol{\theta}_1|D) = \int p(\boldsymbol{\theta}_1, \boldsymbol{\theta}_2|D)\mathrm{d}\boldsymbol{\theta}_2 \\ = \int p(\boldsymbol{\theta}_1|\boldsymbol{\theta}_2, D)p(\boldsymbol{\theta}_2|D)\mathrm{d}\boldsymbol{\theta}_2, \tag{3.21}$$

where in the second line the integration over the elements of $\boldsymbol{\theta}_2$ can be interpreted as an averaging of the conditional posterior pdf for $\boldsymbol{\theta}_1$ given $\boldsymbol{\theta}_2$, $p(\boldsymbol{\theta}_1|\boldsymbol{\theta}_2, D)$, with the marginal posterior pdf for $\boldsymbol{\theta}_2$, $p(\boldsymbol{\theta}_2|D)$, serving as the weight function. This integration with respect to the elements of $\boldsymbol{\theta}_2$ is a way of getting rid of parameters that are not of special interest to an investigator, the so-called nuisance parameters. In addition, the conditional posterior pdf, $p(\boldsymbol{\theta}_1|\boldsymbol{\theta}_2, D)$, can be employed to determine how sensitive inferences about $\boldsymbol{\theta}_1$ are to what is assumed about the value of $\boldsymbol{\theta}_2$; that is, $p(\boldsymbol{\theta}_1|\boldsymbol{\theta}_2, D)$ can be computed for various values of $\boldsymbol{\theta}_2$; see, for example, Box and Tiao (1973) and Zellner (1971) for examples of such sensitivity analyses. Finally, as will be explained below, given a loss function point estimates can be obtained.

The prior pdf, $p(\boldsymbol{\theta})$, in (3.19) is formulated to reflect an investigator's prior information, that is, information available about possible values of $\boldsymbol{\theta}$ before observing a current sample. The information represented in a prior distribution may be past sample information and/or non-sample information derived perhaps from economic theory or other sources. The problem of representing such information accurately and adequately is not an easy one even though considerable effort has been devoted to the problem of assessing or determining the forms of prior pdfs [see, for example, Winkler (1980), Kadane et al. (1980), and Zellner (1972, 1980)]. In some cases, particularly when the sample size is moderately large, the posterior properties of pdfs are not very sensitive to minor alterations in the forms of prior pdfs. In terms of the binomial example above, when n and r are moderately large, altering slightly the values of the prior parameters a and b does not change features of the posterior pdf very much.

As regards the often-mentioned issue that different investigators may have different prior pdfs and thus will obtain different posterior distributions from the same likelihood function, this is hardly surprising since they have different initial information. On pooling their initial information, they will obtain similar inferences. Or if it is a matter of comparing the compatibility of prior information with sample information, as explained below predictive and posterior odds techniques can be employed. Given that researchers tend to be individualistic in their thinking, it is not surprising that initial views differ. Generally, the information in data, as reflected in the likelihood, will modify prior views and dominate as the sample size grows large. In fact, for *any* non-degenerate prior, as the sample size grows, the posterior pdf in (3.19) assumes a normal shape centered at the ML estimate with posterior covariance matrix approximately equal to the inverse of the Fisher information matrix evaluated at the ML estimate; see, for example, Jeffreys (1967, p. 193ff.) for details. Jeffreys (1967, p. 194) regards this as a justification of ML estimation in large samples, a non-sampling argument. Thus, in large samples, the information in the sample dominates the posterior pdf in the sense that the prior pdf's influence on the shape of the posterior pdf becomes negligible.

In some cases there may be little or practically no prior information available about the possible values of parameters, as in the early stages of an investigation. In such cases Bayesians employ so-called "non-informative" or "diffuse" prior pdfs. For some work on the formulation of such prior distributions, see Jeffreys (1967), Box and Tiao (1973) Jaynes (1968, 1980), Savage (1961), and Zellner (1971, 1975). In the case of a parameter θ such that $-\infty < \theta < \infty$, Jeffreys recommends using $p(\theta) \propto$ constant, while for a parameter with a semi-infinite range, such as a standard deviation σ satisfying $0 < \sigma < \infty$, he recommends taking $\log \sigma$ uniformly distributed that implies $p(\sigma) \propto 1/\sigma$. It is the case that these are improper priors since they do not integrate to a finite constant and hence are termed "improper".[33] Others, notably Savage (1961) and Box and Tiao (1973), define a "diffuse" prior for $\boldsymbol{\theta}$ as uniform over a very wide, finite interval, that is, $p(\boldsymbol{\theta}) \propto$ constant for $-M < \boldsymbol{\theta} < M$ with M large but finite. In this case the prior is proper but a choice of the value of M is required. An example will be presented to illustrate the use of diffuse prior pdf.

Example 3.4

Consider the normal mean problem $y_i = \mu + \varepsilon_i$, $i = 1,2,\dots,n$, where the ε_i's are NID(0,1). The likelihood function is $(2\pi)^{-n/2}\exp\{-[\nu s^2 + n(\mu - \bar{y})^2]/2\}$, where $\bar{y}$ is the sample mean, $\nu s^2 = \sum_{i=1}^n (y_i - \bar{y})^2$, and $\nu = n-1$. Let the diffuse prior be $p(\mu) \propto$ constant. Then the posterior pdf for μ, $p(\mu|D)$, where D represents the prior, and sample information is

$$p(\mu|D) \propto (2\pi)^{-n/2}\exp\{-[\nu s^2/2\} + n(\mu - \bar{y})^2]/2\} \propto \exp\{-n(\mu - \bar{y})^2/2\},$$

is in the normal form with posterior mean $\bar{y}$, the sample mean and posterior variance $1/n$.[34]

In this example it is seen that the mean and mode of the posterior pdf are equal to the sample mean, $\bar{y}$, the ML estimate. Some have crudely generalized this and similar results to state that with the use of diffuse prior pdfs, Bayesian and non-Bayesian estimation results are equivalent, aside from their differing interpretations. This generalization is not true in general. If a prior pdf is uniform, $p(\boldsymbol{\theta}) \propto$ constant, then the posterior pdf in (3.19) is given by $p(\boldsymbol{\theta}|\boldsymbol{x}) \propto p(\boldsymbol{x}|\boldsymbol{\theta})$, that is, it is proportional to the likelihood function. Thus, the modal value of the posterior pdf will be exactly equal to the ML estimate and in this sense there is an exact correspondence between Bayesian and non-Bayesian results. However, as shown below, the posterior mean of $\boldsymbol{\theta}$ is optimal relative to a quadratic loss function. If a posterior pdf (and likelihood function) is asymmetric, the posterior mean of $\boldsymbol{\theta}$ can be far different from the modal value. Thus, the optimal Bayesian point estimate can be quite different from the ML estimate in finite samples. Asymmetric likelihood functions are frequently encountered in econometric analyses.

As regards point estimation, a part of the Bayesian approach, given a loss function, $L(\boldsymbol{\theta}, \hat{\boldsymbol{\theta}})$, wherein $\boldsymbol{\theta}$ is viewed as random and $\hat{\boldsymbol{\theta}}$ is any non-random estimate, $\hat{\boldsymbol{\theta}} = \hat{\boldsymbol{\theta}}(\boldsymbol{x})$, a non-sampling criterion is to find the value of $\hat{\boldsymbol{\theta}}$ that

[33] Jeffreys (1967) interprets such improper priors as implying that ∞ rather than 1 is being employed to represent the certain event, $\Pr(-\infty < \theta < \infty)$. Then the probability that θ lies in any finite interval, $\Pr(a < \theta < b) = 0$ and $\Pr(a < \theta < b)/\Pr(c < \theta < d)$ being of the form 0/0 is indeterminate.

[34] If the prior pdf $p(\mu) \propto$ constant, $-M < \mu < M$, had been used, the posterior pdf is $p(\mu|D) \propto \exp\{-n(\mu - \bar{y})^2/2\}$ with $-M < \mu < M$. For M large relative to $1/n$, the posterior is very closely normal.

minimizes the posterior expectation of the loss function. Explicitly the problem is as follows:

$$\min_{\hat{\theta}} EL(\theta,\hat{\theta}) = \min_{\hat{\theta}} \int_{\Theta} L(\boldsymbol{\theta},\hat{\boldsymbol{\theta}}) p(\boldsymbol{\theta}|\boldsymbol{x})\,\mathrm{d}\boldsymbol{\theta}, \tag{3.22}$$

where $p(\boldsymbol{\theta}|\boldsymbol{x})$ is the posterior pdf in (3.19). The minimizing value of $\hat{\boldsymbol{\theta}}$, say $\hat{\boldsymbol{\theta}}_{\mathrm{B}}$, is the optimal Bayesian estimate, optimal in a non-sampling sense since the observation vector $\boldsymbol{x}$ is given. In the case of a quadratic loss function, $L(\boldsymbol{\theta},\hat{\boldsymbol{\theta}}) = (\boldsymbol{\theta} - \hat{\boldsymbol{\theta}})'Q(\boldsymbol{\theta} - \hat{\boldsymbol{\theta}})$, where Q is a given pds matrix, $\hat{\boldsymbol{\theta}}_{\mathrm{B}} = \bar{\boldsymbol{\theta}} = E(\boldsymbol{\theta}|\boldsymbol{x})$, the posterior mean vector. That is,

$$\begin{aligned} E(\boldsymbol{\theta}-\hat{\boldsymbol{\theta}})'Q(\boldsymbol{\theta}-\hat{\boldsymbol{\theta}}) &= E\left[(\boldsymbol{\theta}-\bar{\boldsymbol{\theta}})-(\hat{\boldsymbol{\theta}}-\bar{\boldsymbol{\theta}})\right]'Q\left[(\boldsymbol{\theta}-\bar{\boldsymbol{\theta}})-(\hat{\boldsymbol{\theta}}-\hat{\boldsymbol{\theta}})\right] \\ &= E(\boldsymbol{\theta}-\bar{\boldsymbol{\theta}})'Q(\boldsymbol{\theta}-\bar{\boldsymbol{\theta}})+(\hat{\boldsymbol{\theta}}-\bar{\boldsymbol{\theta}})'Q(\hat{\boldsymbol{\theta}}-\bar{\boldsymbol{\theta}}), \end{aligned}$$

since $E(\boldsymbol{\theta} - \bar{\boldsymbol{\theta}}) = 0$. Thus, since Q is pds, taking $\hat{\boldsymbol{\theta}} = \bar{\boldsymbol{\theta}}$ leads to minimal expected loss. For a scalar parameter, θ, and an absolute error loss function, $L(\theta,\hat{\theta}) = c|\theta - \hat{\theta}|$, it can be shown that the median of the posterior pdf for θ is the optimal point estimate in the sense of minimizing posterior expected loss.[35] When the minimization problem in (3.22) cannot be solved analytically, it is often possible to determine the solution by use of numerical integration procedures. Thus, the optimal Bayesian estimate is tailored to be optimal relative to the loss function that is considered appropriate.

Above, the optimal Bayesian estimate is defined as the solution to the minimization problem in (3.22). *Before the data are observed*, it is of interest to consider the sampling properties of the Bayesian estimate $\hat{\boldsymbol{\theta}}_{\mathrm{B}}$. Given a loss function $L(\boldsymbol{\theta},\hat{\boldsymbol{\theta}})$, the risk function, discussed above, is $r_{\hat{\theta}}(\boldsymbol{\theta}) = \int_{R_x} L(\boldsymbol{\theta},\hat{\boldsymbol{\theta}}) p(\boldsymbol{x}|\boldsymbol{\theta})\,\mathrm{d}\boldsymbol{x}$ for $\theta \subset \Theta$. The Bayesian estimator is defined as the solution to the following problem:

$$\min_{\hat{\theta}} Er_{\hat{\theta}}(\boldsymbol{\theta}) = \min_{\hat{\theta}} \int_{\Theta} r_{\hat{\theta}}(\boldsymbol{\theta}) p(\boldsymbol{\theta})\,\mathrm{d}\boldsymbol{\theta}. \tag{3.23}$$

That is, choose the *estimator* $\hat{\boldsymbol{\theta}}$ so as to minimize average risk, $Er_{\hat{\theta}}(\boldsymbol{\theta})$, where the expectation is taken with respect to the prior pdf $p(\boldsymbol{\theta})$. On substituting the integral expression for $r_{\hat{\theta}}(\boldsymbol{\theta})$ in (3.23), the minimand is

$$\int_{\Theta}\int_{R_x} L(\boldsymbol{\theta},\hat{\boldsymbol{\theta}}) p(\boldsymbol{\theta}) p(\boldsymbol{x}|\boldsymbol{\theta})\,\mathrm{d}\boldsymbol{x}\,\mathrm{d}\boldsymbol{\theta} = \int_{\Theta}\int_{R_x} L(\boldsymbol{\theta},\hat{\boldsymbol{\theta}}) p(\boldsymbol{\theta}|\boldsymbol{x}) p(\boldsymbol{x})\,\mathrm{d}\boldsymbol{x}\,\mathrm{d}\boldsymbol{\theta}, \tag{3.24}$$

[35] See, for example, Zellner (1971, p. 25) for a proof. Also, the particular loss structure that yields the modal value of a posterior pdf as an optimal point estimate is described in Blackwell and Girshick (1954, p. 305). This loss structure implies zero loss for very small estimation errors and constant positive loss for errors that are not small.

where $p(\theta)p(x|\theta)=p(x)p(\theta|x)$ from (3.18) has been employed. On interchanging the order of integration in (3.24), the right side becomes

$$\int_{R_x}\left[\int_{\Theta}L(\boldsymbol{\theta},\hat{\boldsymbol{\theta}})p(\boldsymbol{\theta}|\boldsymbol{x})\,\mathrm{d}\boldsymbol{\theta}\right]p(\boldsymbol{x})\,\mathrm{d}\boldsymbol{x}. \tag{3.25}$$

When this multiple integral converges, the quantity $\hat{\boldsymbol{\theta}}_{\mathrm{B}}$ that minimizes the expression in square brackets will minimize the entire expression given that $p(\boldsymbol{x})\geq 0$ for $\boldsymbol{x}\subset R_{\boldsymbol{x}}$.[36] Thus, $\hat{\boldsymbol{\theta}}_{\mathrm{B}}$, the solution to the problem in (3.22), is the Bayesian estimator that minimizes average risk in (3.23).

Some properties of $\hat{\boldsymbol{\theta}}_{\mathrm{B}}$ follow:

(1) $\hat{\boldsymbol{\theta}}_{\mathrm{B}}$ has the optimal non-sampling property in (3.22) and the optimal sampling property in (3.23).

(2) Since $\hat{\boldsymbol{\theta}}_{\mathrm{B}}$ minimizes average risk, it is admissible relative to $L(\boldsymbol{\theta},\hat{\boldsymbol{\theta}})$. This is so because if there were another estimator, say $\hat{\boldsymbol{\theta}}_{\mathrm{A}}$, that uniformly dominates $\hat{\boldsymbol{\theta}}_{\mathrm{B}}$ in terms of risk, it would have lower average risk and this contradicts the fact that $\hat{\boldsymbol{\theta}}_{\mathrm{B}}$ is the estimator with minimal average risk. Hence, no such $\hat{\boldsymbol{\theta}}_{\mathrm{A}}$ exists.

(3) The class of Bayesian estimators is complete in the sense that in the class of all estimators there is no estimator outside the subset of Bayesian estimators that has lower average risk than every member of the subset of Bayesian estimators.

(4) Bayesian estimators are consistent and normally distributed in large samples with mean equal to the ML estimate and covariance matrix equal to the inverse of the estimate information matrix. Further, in large samples the Bayesian estimator (as well as the ML estimator) is "third-order" asymptotically efficient.[37] These results require certain regularity conditions [see, for example, Heyde and Johnstone (1979)].

A key point in establishing these sampling properties of the Bayesian estimator, $\hat{\boldsymbol{\theta}}_{\mathrm{B}}$, is the assumption that the multiple integral in (3.24) converges. It usually does when prior pdfs are proper, although exceptions are possible. One such case occurs in the estimation of the reciprocal of a normal mean, $\theta=1/\mu$, using quadratic loss, $(\theta-\hat{\theta})^2$. The posterior pdf for μ, based on a proper normal prior for μ, is normal. Thus, $\theta=1/\mu$, the reciprocal of a normal variable, possesses no finite moments and the integral defining posterior expected loss does not converge. With more information, say $\theta>0$, this problem becomes amenable to solution. Also, if the loss function is $(\theta-\hat{\theta})^2/\theta^2$, a relative squared error loss

[36]See Blackwell and Girshick (1954), Ferguson (1967), and DeGroot (1970) for consideration of this and the following topics.

[37]See, for example, Takeuchi (1978) and Pfanzagl and Wefelmeyer (1978).

function, there is a well-defined, Bayesian estimator that minimizes average risk [see Zellner (1978) for details]. Also, if the loss function is bounded, solutions exist [see Zaman (1981)]. In terms of the Stein normal vector-mean problem, $\boldsymbol{y} = \boldsymbol{\theta} + \boldsymbol{\varepsilon}$, considered above, if the prior pdf is $p(\boldsymbol{\theta}) \propto$ constant, the posterior pdf is $p(\theta|D) \propto \exp\{-(\boldsymbol{\theta} - \boldsymbol{y})'(\boldsymbol{\theta} - \mathrm{y})/2\}$ that has posterior mean $\boldsymbol{y}$, the inadmissible ML and LS estimator relative to $L(\boldsymbol{\theta}, \hat{\boldsymbol{\theta}}) = (\boldsymbol{\theta} - \hat{\boldsymbol{\theta}})'(\boldsymbol{\theta} - \hat{\boldsymbol{\theta}})$ when $n \geq 3$. However, when $n = 1$ or $n = 2$ the posterior mean is admissible even though it is associated with a posterior pdf based on an improper prior pdf. While inadmissible, the estimate $\hat{\boldsymbol{\theta}} = \boldsymbol{y}$ with $n > 2$ does satisfy the optimality criterion in (3.22). Also, as Hill (1975) points out, if the elements of the mean vector $\boldsymbol{\theta}$ are independently distributed *a priori*, then the joint prior pdf is $p(\boldsymbol{\theta}) = \prod_{i=1}^{n} p_i(\theta_i)$ and the associated posterior pdf is $p(\boldsymbol{\theta}|\boldsymbol{x}) \propto \prod_{i=1}^{n} p_i(\theta_i) p(y_i|\theta_i)$. Thus, the θ_i's are independently distributed *a posteriori* and, using any separable loss function, for example $L(\boldsymbol{\theta}, \hat{\boldsymbol{\theta}}) = \sum_{i=1}^{n}(\theta_i - \hat{\theta}_i)^2$, the Bayesian estimate of θ_i is its posterior mean that just depends on y_i and *not* on the other y_i's.[38] If the priors, $p_i(\theta_i)$, are normal with very large dispersion, Hill shows that the admissible Bayesian estimates, the posterior means of the θ_i's, are not far different from the Bayesian diffuse prior and ML estimates, $\hat{\theta}_i = y_i$, $i = 1, 2, \ldots, n$.

The important point brought out by Hill's (1975) cogent analysis is that if the means in the Stein problem are mean rainfall in Calcutta, mean income in Palo Alto, and mean annual attendance at the Milan opera, these θ_i's are reasonably considered independent *a priori*. Given this property and Hill's analysis, the estimate of a single θ_i, say mean rainfall in Calcutta, will just depend on observed rainfall in Calcutta and not on observed income in Palo Alto and attendance at Milan's opera. Therefore, the Stein–James estimate (3.17) is inappropriate for such data and assumptions. On the other hand, there are many situations in which the θ_i's are dependent *a priori*[39] and for them use of an appropriate prior pdf reflecting such dependence can lead to substantial improvement in estimation and prediction results. Specific prior assumptions leading to a Bayesian estimate close to or equal to the Stein–James estimate in (3.17) are reviewed in Zellner and Vandaele (1975).

In summary, the Bayesian prescription, i.e. choose the estimate that minimizes expected loss, is a general principle that is widely applicable. Its use in finite samples does involve the choice of an appropriate prior pdf for parameters. While this is difficult, particularly in multi-parameter problems, a basic issue is whether it is possible to get sensible estimation results from any point of view without information as to what the probable values of parameters are. Bayesians formally

[38]This implies that the Stein–James estimate in (3.17) is suboptimal for this specification.

[39]Lindley (1962) provides the following model to rationalize dependent θ_i's: $y_i = \theta_i + \varepsilon_i$ and $\theta_i = \theta + v_i$, $i = 1, 2, \ldots, n$, where the ε_i's and v_i's are independent normal error terms and θ is interpreted as a "common effect". Analysis of this model produces estimates of the θ_i's very similar to those in (3.17) [see, for example, Zellner and Vandaele (1975)].

represent such information by use of prior pdfs, while non-Bayesians often use such information informally. Evidence is being accumulated on the relative merits of these alternative approaches to parameter estimation and other inference problems.

3.2.1.8. Robustness criterion. The robustness criterion relates to the sensitivity of point estimation and other inference procedures to departures from specifying assumptions regarding models and prior distributions and to unusual or outlying data. Since specifying assumptions are usually only approximately valid, it is important that the sensitivity of inference techniques to departures from specifying assumptions and to outlying observations be understood and that methods be available that are relatively robust to possible departures and outlying observations. For example, it is well known that least squares estimates can be vitally affected by one or a few outlying data points. Also, in some cases, Bayesian inferences are sensitive to slight changes in the formulation of prior distributions.

In dealing with robustness issues two general approaches have been pursued. In the first, attempts are made to formulate estimation and other inference procedures that retain desirable properties over a range of alternative models and/or in the presence of outlying observations. For example, in estimating a population mean, the sample median is less sensitive to outlying observations that is the sample mean. Such procedures are called "blanket procedures" by Barnett and Lewis (1978, p. 47).[40] The second approach, which may be called a "nesting approach", involves broadening an initial model to accommodate suspected departures from specifying assumptions and/or possible outlying observations and then proceeding with an analysis of the broader model. In both approaches, the nature of alternatives to the initially entertained model must be given careful consideration in order to obtain sensible results in estimation. Mechanical down-weighting of outlying observations does not necessarily lead to satisfactory results. For example, use of the median to estimate the location of a distribution when outlying observations are present may suggest a unimodal distribution when in fact the true distribution is bimodal. In this case, the outlying observations may give some information about the location of a second mode. Or in some cases, outlying observations in regression analysis may indicate that the assumed functional form for the regression equation is incorrect and thus such outlying points should not be carelessly discarded or down-weighted. On the other hand, if outlying observations are in some sense spurious, say the result of transcription errors, then down-weighting them in estimation can lead to more sensible results.

An example of the first approach, the blanket approach, is Huber's (1972) linear order statistics estimators or L-estimators for a location parameter μ based

[40]For examples of this approach, see Tukey (1977), Huber (1964, 1972), and Belsley, Kuh and Welsch (1980).

on a sample of independent observations, $x_1, x_2, \ldots, x_n$. The ordered observations, $x_{(1)} < x_{(2)} < \cdots < x_{(n)}$ are combined with weights c_i to yield the estimate $\tilde{\mu} = \sum_{i=1}^{n} c_i x_{(i)}$, with the c_i's smaller in value for the extreme observations than for the central observations. The L-class of estimators includes various "trimmed-mean" estimates (those that disregard extreme observations and just average central observations), the sample median, and the sample mean ($c_i = 1/n$) as special cases. Judicious choice of the c_i's can lead to estimates that have better sampling properties than the sample mean when the underlying distribution departs from, for example, a normal distribution and in this sense are more robust than the sample mean. Other robust estimates, discussed by Jeffreys (1967, pp. 214–216) and Huber (1972), are maximum likelihood-type estimates, called M-estimates by Huber for estimating a location parameter from n independent observations with log-likelihood function $\sum_{i=1}^{n} \log f(x_i - \mu)$. The necessary condition for a maximum of the likelihood function can be written as

$$\sum_{i=1}^{n} f'(x_i - \hat{\mu})/f(x_i - \hat{\mu}) = \sum_{i=1}^{n} \omega_i (x_i - \hat{\mu}) = 0,$$

where $\hat{\mu}$ is the ML estimate and $\omega_i \equiv f'(x_i - \hat{\mu})/(x_i - \hat{\mu}) f(x_i - \hat{\mu})$, as Jeffreys (1967, p. 214) explains. Or, in Huber's (1972) notation, it may be written as $\sum_{i=1}^{n} \psi(x_i - \hat{\mu}) = 0$ with $\psi(x_i - \hat{\mu}) \equiv (x_i - \hat{\mu})\omega_i$. Thus, $\hat{\mu} = \sum_{i=1}^{n} \omega_i x_i / \sum_{i=1}^{n} \omega_i$ is the form of the estimate with the ω_i's data dependent. Choice of the form of $\psi(\cdot)$ or of the ω_i's depends on the nature of the underlying distribution which usually is not known exactly. If the underlying distribution is normal, then $\omega_i =$ constant and equal weights are appropriate. If f'/f does not increase as fast as $|x_i - \mu|$, then Jeffreys (1967) remarks: "...the appropriate treatment will give reduced weights to the large residuals. If it $[f'/f]$ increases faster, they should receive more weight than the smaller ones" (p. 214). See Jeffreys (1967, pp. 214–215) for an application of this approach, and Huber (1972) for further discussion of the appropriate choice of weights. For independently and identically distributed (i.i.d.) observations $\psi(x_i - \mu) = f'(x_i - \mu)/f(x_i - \mu)$, $i = 1, 2, \ldots, n$, and on integration $\log f(v_i) = \int \psi(v_i) \mathrm{d} v_i +$ constant), where $v_i \equiv x_i - \mu$. Thus, choice of a particular form for $\psi(v_i)$ implies a form of the likelihood function when $\int \psi(v_i) \mathrm{d} v_i$ converges and the i.i.d. assumption is satisfied. Viewed in this way, the M-estimation approach is a "nested approach". However, this interpretation is not generally possible if the observations are not i.i.d.

The second approach, the nesting approach, involves representing suspected departures and/or outlying observations from an initial model by formulating a broader model and analyzing it. There are many examples of this approach in econometrics and statistics. Student-t distributions that include Cauchy and normal distributions as limiting cases can be employed in analyzing regression and other models [see, for example, Jeffreys (1973, p. 68) and Zellner (1976)]. The heavy tails of Student distributions for low degrees of freedom accommodate outlying observations. Also, see Barnett and Lewis (1978) for a review of a number of models for particular kinds of outlying observations. Many production function models including the CES, trans-log, and other generalized production function models include the Cobb–Douglas and other models as special cases. Box–Cox (1964) and other transformations [see, for example, Tukey (1957) and Zellner and Revankar (1969)] can be employed to broaden specifying assumptions and thus to guard against possible specification errors. In regression analysis, it is common practice to consider models for error terms, say autoregressive and/or moving average processes when departures from independence are thought to be present. Such broadened models can of course be analyzed in either sampling theory or Bayesian approaches. With respect to Bayesian considerations of robustness, see, for example, Savage et al. (1963), Box and Tiao (1973), and DeRobertis (1978).

3.2.1.9. Invariance criterion. The invariance criterion, discussed for example in Cox and Hinkley (1974, pp. 41–45) and Arnold (1981, pp. 20–24), relates to properties of estimation and other inference procedures when sample observations and parameters are subjected to certain types of transformations. For example, if in a model for i.i.d. observations the observations are reordered or permuted, the results obtained with most estimation procedures will be unaffected and are thus invariant to such transformations of the data. Further, if the $\tilde{x}_i$'s are i.i.d., each with pdf $f(x-\theta)$, and the $\tilde{x}_i$'s are transformed to $\tilde{y}_i = \tilde{x}_i + a$, where $-\infty < a < \infty$, then each $\tilde{y}_i$ has a pdf $f(y-\theta^*) = f(x-\theta)$, where $\theta^* = \theta + a$. If the parameter spaces for θ and θ^* are identical, say the real line, then the invariance criterion requires that an estimate of θ, $\hat{\theta}(\boldsymbol{x})$, based on the x_i's, be identical to that obtained from the y_i's to estimate θ^*, $\hat{\theta}(\boldsymbol{y})$, and then getting an estimate of θ from $\hat{\theta}(\boldsymbol{y}) = \hat{\theta}^*(\boldsymbol{y}) - a$. That is, an invariant estimate must satisfy $\hat{\theta}(\boldsymbol{x} + a\boldsymbol{\iota}) = \hat{\theta}(\boldsymbol{x}) + a$ for all values of a, where $\boldsymbol{\iota}$ denotes a column of ones. As Cox and Hinkley (1974, p. 43) point out, a crucial point is that there are no external reasons for preferring some values of θ to others. For example, they mention that if $\theta \geq 0$, the invariance condition above would not hold since the parameter space is not invariant under the transformation when $\theta \geq 0$ is imposed.

Conditions under which estimates are invariant in the above sense to more general transformations of the data, say $\tilde{y}_i = c\tilde{x}_i + a$ or $\tilde{y}_i = g\tilde{x}_i$, where g is a member of a class of transformations G, have been analyzed in the literature. Also, Arnold (1981, p. 20 ff.) defines conditions under which an estimation problem and associated loss function are invariant under both transformations of the data, parameters, and loss function and goes on to discuss "best invariant" or "minimum risk invariant" estimators. See also Cox and Hinkley (1974, p. 443) for a discussion of the famous Pitman minimum risk invariant estimate of a location parameter θ in the likelihood function $\prod_{i=1}^{n} f(x_i - \theta)$, $-\infty < \theta < \infty$. The result is that the Pitman estimate is "...the mean of the normalized likelihood function" (p. 444), that is, the mean of a posterior distribution based on a uniform, improper prior for θ.

In the Bayesian approach, invariance of estimation results to transformations of the data and parameters has been considered in Jeffreys (1967), Hartigan (1964), and Zellner (1971). Hartigan, building on Jeffreys' pioneering work, defines various kinds of invariance and provides classes of prior distributions, including Jeffreys' that lead to invariant estimation results.

Requiring that estimation procedures be invariant places restrictions on the forms of estimators. Having invariance with respect to changes in units of measurement and some other types of transformations suggested by the nature of specific problems seems desirable. However, as Cox and Hinkley (1974) state, "...in a decision-making context [as in choice of an estimator relative to a given loss or utility function]...there is sometimes a clash between the invariance principle and other apparently more compelling requirements; there can be a uniform loss of expected utility from following the invariance principle" (p. 45). Thus, for each problem it is important to consider carefully the types of transformations for which invariance is required and their effects on estimation and other inference procedures.

3.2.1.10. Cost criterion. Practically speaking, the cost of applying alternative estimation techniques is of importance. Some estimation procedures involve difficult numerical procedures. Generally, cost–benefit analysis is relevant in choosing among alternative estimation procedures. While this is recognized, it is difficult to generalize about the range of considerations. Each case has to be considered separately. In some cases, cost of computation can be formally introduced in loss functions and these broadened loss functions can be employed to choose among alternative estimation procedures. However, in many cases, cost considerations are dealt with in an informal, heuristic manner.

In this section various approaches and criteria for point estimation have been considered. While point estimation is important, it must be emphasized that a point estimate unaccompanied by a measure of precision is very unsatisfactory. In the sampling theory approach, point estimates are supplemented by their associated standard errors (estimates of standard deviations of estimators). In the Bayesian approach, point estimates are usually accompanied by a measure of the dispersion of posterior distributions, e.g. posterior standard deviations or complete posterior distributions. In the next section attention is directed toward explaining methods for computing intervals or regions that in one of several senses probably include the values of parameters being estimated.

4. Interval estimation: Confidence bounds, intervals, and regions

To provide a quantitative expression of the uncertainty associated with a scalar point estimate $\hat{\theta}$ of a parameter θ, confidence bounds and intervals are available both in the sampling theory and Bayesian approaches. Similarly, for a vector point estimate $\hat{\boldsymbol{\theta}}$ of a parameter vector $\boldsymbol{\theta}$, sampling theory and Bayesian confidence regions for $\boldsymbol{\theta}$ can be computed. As will be seen, the probabilistic interpretations of sampling theory and Bayesian confidence bounds, intervals, and regions are radically different.

4.1. *Confidence bounds*

Let $\theta \subset \Theta$ be a scalar parameter appearing in a probability pdf, $f(\boldsymbol{x}|\theta)$, for a random observation vector $\tilde{\boldsymbol{x}}$. Further, let $\tilde{a}_\alpha = a_\alpha(\tilde{\boldsymbol{x}})$ be a statistic, that is, a function of $\tilde{\boldsymbol{x}}$, such that

$$\mathrm{P}(\tilde{a}_\alpha \geqslant \theta|\theta) = 1-\alpha. \tag{4.1a}$$

In addition, it is required that if $\alpha_1 > \alpha_2$ and if $\tilde{a}_{\alpha_1}$ and $\tilde{a}_{\alpha_2}$ are both defined in accord with (4.1), then $\tilde{a}_{\alpha_1} \leqslant \tilde{a}_{\alpha_2}$, that is, the larger $1-\alpha$, the larger is the upper bound. Then $\tilde{a}_\alpha$ is called a $(1-\alpha)100$ percent upper confidence bound for θ. From (4.1a) the random event $\tilde{a}_\alpha \geqslant \theta$ has probability $1-\alpha$ of occurrence and this is the sense in which $\tilde{a}_\alpha$ is a probabilistic bound for θ. When $\tilde{\boldsymbol{x}}$ is observed, $a_\alpha(\tilde{\boldsymbol{x}})$ can be evaluated with the given sample data $\boldsymbol{x}$. The result is $a_\alpha(\boldsymbol{x})$, a non-stochastic quantity, say $a_\alpha(\boldsymbol{x}) = 1.82$, and the computed upper confidence bound is 1.82. In a similar way a $(1-\alpha)\times 100$ percent lower confidence bound for θ is $\tilde{b}_\alpha = b_\alpha(\tilde{\boldsymbol{x}})$ such that

$$\mathrm{P}(\tilde{b}_\alpha \leqslant \theta|\theta) = 1-\alpha, \tag{4.1b}$$

with $\tilde{b}_{\alpha_1} \geqslant \tilde{b}_{\alpha_2}$ when $\alpha_1 > \alpha_2$; that is, the larger is $1-\alpha$, the smaller is the lower bound.

Bayesian confidence bounds are based on the posterior pdf for θ, $p(\theta|\boldsymbol{x}) \propto \pi(\theta)f(\boldsymbol{x}|\theta)$, where $\pi(\theta)$ is a prior pdf for θ. A $(1-\alpha)\times 100$ percent Bayesian upper bound, $c_\alpha = c_\alpha(\boldsymbol{x})$, is defined by

$$\mathrm{P}(\theta \leqslant c_\alpha|\boldsymbol{x}) = 1-\alpha, \tag{4.2a}$$

where θ is considered random and the sample data $\boldsymbol{x}$ are given. Note that

$$\mathrm{P}(\theta \leqslant c_\alpha|\boldsymbol{x}) = \int_{-\infty}^{c_\alpha} p(\theta|\boldsymbol{x})\,\mathrm{d}\theta$$

is just the posterior cdf evaluated at c_α.

A $(1-\alpha)\times 100$ percent Bayesian lower bound is $d_\alpha = d_\alpha(\boldsymbol{x})$ satisfying

$$P(\theta \geqslant d_\alpha|\boldsymbol{x}) = 1-\alpha, \tag{4.2b}$$

where

$$P(\theta \geqslant d_\alpha|\boldsymbol{x}) = \int_{d_\alpha}^{\infty} p(\theta|\boldsymbol{x})\,\mathrm{d}\theta.$$

The fundamental differences in the probability statements in (4.1) and (4.2) must be emphasized. A sampling theory bound has the interpretation that in repeated samples the bound so computed will be correct in about $1-\alpha$, say $1-\alpha = 0.95$ of the cases. The Bayesian bound states that the random parameter θ will satisfy the bound with posterior probability $1-\alpha$ given the sample and prior information. The following example is a case in which Bayesian and sampling confidence bounds are numerically identical.

Example 4.1

Let n $\tilde{x}_i$'s be i.i.d. $N(\theta,1)$. Then

$$f(\boldsymbol{x}|\theta) = (2\pi)^{-n/2}\exp\{-\textstyle\sum(x_i-\theta)^2/2\} \propto \exp\{-n(\theta-\bar{x})^2/2\},$$

where $\bar{x}$ is the sample mean. Then with the prior $\pi(\theta)\propto$ constant, $-\infty<\theta<\infty$, the posterior pdf is $f(\theta|\boldsymbol{x})\propto\exp\{-n(\theta-\bar{x})^2/2\}$, a normal pdf. Thus, $z=\sqrt{n}(\theta-\bar{x})$ is $N(0,1)$ *a posteriori* and the constant c_α can be found such that $P(z\leqslant c_\alpha|\bar{x}) = 1-\alpha$. $z\leqslant c_\alpha$ is equivalent to $\sqrt{n}(\theta-\bar{x})\leqslant c_\alpha$ or $\theta\leqslant\bar{x}+c_\alpha/\sqrt{n}$. Thus, $P(\theta\leqslant\bar{x}+c_\alpha/\sqrt{n}\,|\bar{x}) = 1-\alpha$ and $\bar{x}+c_\alpha/\sqrt{n}$ is the Bayesian upper confidence bound. Now from a sampling theory point of view, $\bar{x}$ has a normal sampling pdf with mean θ and variance $1/n$. Thus, $z=\sqrt{n}(\bar{x}-\theta)$ is $N(0,1)$, given θ. From $P(z\geqslant -c_\alpha|\theta) = 1-\alpha$ it follows that $P[\sqrt{n}(\bar{x}-\theta)\geqslant -c_\alpha|\theta] = P(\bar{x}+c_\alpha/\sqrt{n}\geqslant\theta|\theta) = 1-\alpha$ and $\bar{x}+c_\alpha/\sqrt{n}$ is the sampling theory upper confidence bound that is numerically identical to the Bayesian bound.

The example indicates that when a uniform prior pdf for the parameter θ is appropriate and when a "pivotal quantity", such as $z=\sqrt{n}(\bar{x}-\theta)$, that has a pdf not involving the parameter θ exists,[41] Bayesian and sampling theory confidence bounds are numerically identical. Other examples involving different pivotal quantities will be presented below. Also, a connection of confidence bounds with construction of tests of hypotheses will be discussed below in the section on hypothesis testing.

[41] Note that the pdf for $z=\sqrt{n}(\bar{x}-\theta)$ in the example is $N(0,1)$ both from the sampling theory and Bayesian points of view.

4.2. Confidence intervals

By use of both a lower and an upper confidence bound for a scalar parameter θ, an interval estimate or a confidence interval is obtained. In the sampling theory approach the random quantities $\tilde{a} = a(\tilde{x})$ and $\tilde{b} = b(\tilde{x})$ such that

$$\mathrm{P}(\tilde{b} < \theta < \tilde{a}|\theta) = 1 - \alpha \tag{4.3}$$

yields a random interval that has probability $1-\alpha$ of including or covering the fixed unknown value of θ. On combining a lower confidence bound, $\tilde{b}_{\alpha_1}$, with an upper confidence bound $\tilde{a}_{\alpha_2}$, with $\alpha_1 + \alpha_2 = \alpha$, (4.3) will be satisfied. Similar considerations apply to combinations of Bayesian lower and upper confidence bounds to obtain a Bayesian confidence interval. In general, since there are many values for α_1 and α_2 satisfying $\alpha_1 + \alpha_2 = \alpha$, confidence intervals with probability content $1-\alpha$ are not unique.

Example 4.2

Consider the standard normal regression model $\tilde{\boldsymbol{y}} = X\boldsymbol{\beta} + \tilde{\boldsymbol{u}}$, where the $n \times 1$ vector $\tilde{\boldsymbol{u}}$ is MVN $(\boldsymbol{0}, \sigma^2 I_n)$. Then $\hat{\boldsymbol{\beta}} = (X'X)^{-1}X'\tilde{\boldsymbol{y}}$ has a pdf that is MVN $[\boldsymbol{\beta}, (X'X)^{-1}\sigma^2]$ and $\nu s^2/\sigma^2$, where $\nu = n-k$ and $\nu s^2 = (\tilde{\boldsymbol{y}} - X\hat{\boldsymbol{\beta}})'(\tilde{\boldsymbol{y}} - X\hat{\boldsymbol{\beta}})$, has a χ^2 pdf with ν degrees of freedom (d.f.). It follows that $t = (\hat{\beta}_i - \beta_i)/s_{\hat{\beta}_i}$ has a univariate Student-t (US-t) pdf with ν d.f., where $\hat{\beta}_i$ and β_i are the ith elements of $\hat{\boldsymbol{\beta}}$ and $\boldsymbol{\beta}$, respectively, and $s^2_{\hat{\beta}_i} = m^{ii}s^2$, with m^{ii} the i,ith element of $(X'X)^{-1}$. Then from tables of the Student-t distribution with ν d.f., a constant $c_\alpha > 0$ can be found such that with given probability $1-\alpha$, $\mathrm{P}(|\tilde{t}| < c_\alpha) = \mathrm{P}(|\hat{\beta}_i - \beta_i|/s_{\hat{\beta}_i} < c_\alpha|\beta_i) = 1-\alpha$. Since $|\hat{\beta}_i - \beta_i|/s_{\hat{\beta}_i} < c_\alpha$ is equivalent to $\hat{\beta}_i - c_\alpha s_{\hat{\beta}_i} < \beta_i < \hat{\beta}_i + cs_{\hat{\beta}_i}$, $\mathrm{P}(\hat{\beta}_i - c_\alpha s_{\hat{\beta}_i} < \beta_i < \hat{\beta}_i + c_\alpha s_{\hat{\beta}_i}|\beta_i) = 1-\alpha$ and $\hat{\beta}_i \pm c_\alpha s_{\hat{\beta}_i}$ is a $(1-\alpha)\times 100$ percent confidence interval for β_i. Note that the interval is random and β_i has a fixed unknown value. With given data $\hat{\beta}_i \pm c_\alpha s_{\hat{\beta}_i}$ can be evaluated to yield, for example 0.56 ± 0.12.

Example 4.3

If the regression model in the previous example is analyzed with a diffuse prior pdf, $p(\boldsymbol{\beta}, \alpha) \propto 1/\sigma$, the posterior pdf is $p(\boldsymbol{\beta}, \sigma|\boldsymbol{y}) \propto \sigma^{-(n+1)}\exp\{-[\nu s^2 + (\boldsymbol{\beta} - \hat{\boldsymbol{\beta}})'X'X(\boldsymbol{\beta} - \hat{\boldsymbol{\beta}})]/2\sigma^2\}$ and on integrating over σ, $0 < \sigma < \infty$, the marginal posterior pdf for $\boldsymbol{\beta}$ is $p(\boldsymbol{\beta}|\boldsymbol{y}) \propto [\nu s^2 + (\boldsymbol{\beta} - \hat{\boldsymbol{\beta}})'X'X(\boldsymbol{\beta} - \hat{\boldsymbol{\beta}})]^{-(\nu+k)/2}$, a pdf in the MVS-$t$ form with $\nu = n-k$, $\hat{\boldsymbol{\beta}} = (X'X)^{-1}X'\boldsymbol{y}$, and $\nu s^2 = (\boldsymbol{y} - X\hat{\boldsymbol{\beta}})'(\boldsymbol{y} - X\hat{\boldsymbol{\beta}})$. Then it follows that $t = (\beta_i - \hat{\beta}_i)/s_{\hat{\beta}_i}$ has a US-t pdf with ν d.f. where β_i and $\hat{\beta}_i$ are the ith elements of $\boldsymbol{\beta}$ and $\hat{\boldsymbol{\beta}}$, respectively, and $s^2_{\hat{\beta}_i} = m^{ii}s^2$, where m^{ii} is the i-ith element of $(X'X)^{-1}$. Thus, c_α can be found such that for given probability $1-\alpha$, $\mathrm{P}(|t| < c_\alpha) = \mathrm{P}(|\beta_i - \hat{\beta}_i|/s_{\hat{\beta}_i} < c_\alpha|\boldsymbol{y}) = 1-\alpha$. Equivalently, $\mathrm{P}(\hat{\beta}_i - c_\alpha s_{\hat{\beta}_i} < \beta_i < \hat{\beta}_i + c_\alpha s_{\hat{\beta}_i}|\boldsymbol{y}) = 1-\alpha$ and $\hat{\beta}_i \pm c_\alpha s_{\hat{\beta}_i}$ is a $(1-\alpha)100$ percent Bayesian confidence interval for β_i in the sense that the posterior probability that the random β_i lies in the fixed interval $\hat{\beta}_i \pm c_\alpha s_{\hat{\beta}_i}$ is $1-\alpha$.

In these two examples $t=(\hat{\beta}_i-\beta_i)/s_{\hat{\beta}_i}$ is a pivotal quantity. Its pdf, $p(t)=c(\nu)/(1+t^2/\nu)^{(\nu+1)/2}$, with $c(\nu)$ a normalizing constant, does not involve $\boldsymbol{\beta}$ and σ. Also, the pdf for t is the same in the sampling theory and Bayesian approaches when the diffuse prior $p(\boldsymbol{\beta},\sigma)\propto 1/\sigma$ is employed and in this case sampling theory and Bayesian confidence intervals are numerically identical. However, if an informative prior pdf were employed, reflecting additional information, the intervals would not be numerically identical. Generally the Bayesian interval, incorporating more information, will be shorter in length for a given confidence level $1-\alpha$. Generally, as $1-\alpha$ is increased in value both sampling theory and Bayesian confidence intervals get broader; that is, to be more confident (higher $1-\alpha$) that the interval probably covers or includes the value of a parameter, it will have to be of greater length.

The intervals discussed in the two examples above, $\hat{\beta}_i\pm c_\alpha s_{\hat{\beta}_i}$, are "central" intervals. As mentioned above, c_{α_1} and c_{α_2} can be found such that $\hat{\beta}_i-c_{\alpha_1}s_{\hat{\beta}_i}$ to $\hat{\beta}_i+c_{\alpha_2}s_{\hat{\beta}_i}$ is a $1-\alpha$ confidence interval for β_i. In this case, the interval is not symmetric with respect to $\hat{\beta}_i$. Similarly, a confidence interval for σ^2 need not be a central interval. That is, the sampling pdf of $z=\nu s^2/\sigma^2$ is χ^2 with ν d.f. Then constants c_{α_1} and c_{α_2} exist such that $\mathrm{P}(c_{\alpha_1}<z<c_{\alpha_2})=\mathrm{P}(c_{\alpha_1}<\nu s^2/\sigma^2<c_{\alpha_2}|\sigma)=1-\alpha$ or, equivalently, $\mathrm{P}(\nu s^2/c_{\alpha_2}<\sigma^2<\nu s^2/c_{\alpha_1}|\sigma)=1-\alpha$. The interval $\nu s^2/c_{\alpha_2}$ to $\nu s^2/c_{\alpha_1}$ is not centered at s^2.[42] Also, there are many ways of selecting c_{α_1} and c_{α_2} such that the probability associated with an interval for σ^2 is $1-\alpha$.

The problem of obtaining a unique confidence interval for a scalar parameter θ can be solved in many cases by applying the criterion that for a given confidence level $1-\alpha$ the interval selected be of shortest, in some sense, length. In a Bayesian context with a posterior pdf for θ, $p(\theta|D)$, where D denotes the sample and prior information, an interval b to a is sought such that $a-b$ is minimized subject to $\int_b^a p(\theta|D)\mathrm{d}\theta=1-\alpha$, where $1-\alpha$ is given. The solution to this constrained minimization problem is to select $a=a_*$ and $b=b_*$ such that $p(a_*|D)=p(b_*|D)$.[43] Then the interval b_* to a_* has probability content $1-\alpha$ and is of minimal length. For a unimodal $p(\theta|D)$, the ordinate of $p(\theta|D)$ in the interval b_* to a_* is everywhere higher than the ordinates outside this interval and thus the interval b_* to a_* is often called a highest posterior density (HPD) interval with probability content $1-\alpha$. If $p(\theta|D)$ is unimodal and symmetric about the modal value $\hat{\theta}_m$, then the HPD interval can be expressed as $\hat{\theta}_m\pm c_\alpha$, a central interval

[42] $\nu s^2/c_{\alpha_2}$ to $\nu s^2/c_{\alpha_1}$ is also a $1-\alpha$ Bayesian interval when the diffuse prior $p(\boldsymbol{\beta},\sigma)\propto 1/\sigma$ is employed, since then $\nu s^2/\sigma^2$ has a χ^2 posterior pdf with ν d.f. In this problem the pivotal quantity is $\nu s^2/\sigma^2$, which has a χ^2 pdf not involving $\boldsymbol{\beta}$ or σ in both the sampling theory and Bayesian approaches.

[43] Write $a-b+\lambda\int_b^a p(\theta|D)\mathrm{d}\theta$, where λ is a Lagrange multiplier. On differentiating this expression partially with respect to a and to b and setting first partial derivatives equal to zero yields $1+\lambda p(a|D)=0$ and $1+\lambda p(b|D)=0$ so that $p(a|D)=p(b|D)$ is necessary for $a-b$ to be minimized subject to the restriction. Under weak conditions, this condition is also sufficient. Also, this interval can be obtained by minimizing expected loss with a loss function of the following type: $L=q(a-b)-1$ if $b\leqslant\theta\leqslant a$ and $L=q(a-b)$ otherwise, with $q>0$ a given constant. This loss function depends on the length of the interval, $a-b$.

with $p(\hat{\theta}_m + c_\alpha | D) = p(\hat{\theta}_m - c_\alpha | D)$ and $\Pr(\hat{\theta}_m - c_\alpha < \theta < \hat{\theta}_m + c_\alpha | D) = 1 - \alpha$. For bimodal and some other types of posterior pdfs, a *single* interval is not very useful in characterizing a range of probable values for θ.

In the sampling theory approach various definitions of interval shortness have been proposed. Since the sampling theory confidence interval is random, its length is random. Attempts to obtain confidence intervals with minimum expected length have not been successful in general. Another criterion is to maximize the probability of coverage, that is, to find $\tilde{b}_\alpha$ and $\tilde{a}_\alpha$ such that $1 - \alpha = \mathrm{P}(\tilde{b}_\alpha \leqslant \theta \leqslant \tilde{a}_\alpha | \theta) \geqslant \mathrm{P}(\tilde{b}_\alpha \leqslant \theta' \leqslant \tilde{a}_\alpha | \theta')$ for every θ and $\theta' \subset \Theta$, where θ is the true value and θ' is some other value. That is, the interval must be at least as likely to cover the true value as any other value. An interval satisfying this criterion is called an unbiased confidence interval of level $1 - \alpha$. Pratt (1961) has shown that in many standard estimation problems there exist $1 - \alpha$ level confidence intervals which have uniformly minimum expected length among all $1 - \alpha$ level unbiased confidence intervals. Also, a concept of shortness related to properties of uniformly most powerful unbiased tests will be discussed below.

In summary, for a scalar parameter θ, or for a function of θ, $g(\theta)$ results are available to compute upper and lower confidence bounds and confidence intervals in both the sampling theory and Bayesian approaches. For some problems, for example $g(\theta) = 1/\theta$, where $\tilde{x}_i = \theta + \tilde{\varepsilon}_i$, with the $\tilde{\varepsilon}_i$'s NID $(0, \sigma^2)$, both the sampling distribution of $1/\bar{x}$ and the posterior pdf for $1/\theta$ can be markedly bimodal and in such cases a single interval is not very useful. Some other pathological cases are discussed in Lindley (1971) and Cox and Hinkley (1974, p. 232 ff.). The relationship of sampling properties of Bayesian and sampling theory intervals is discussed in Pratt (1965).

4.3. *Confidence regions*

A confidence region is a generalization of a confidence interval in the sense that it relates to a vector of parameters rather than a scalar parameter. A sampling theory $1 - \alpha$ confidence region for a vector of parameters $\boldsymbol{\theta} \subset \Theta$ is a nested set of regions in the sample space denoted by $\tilde{\omega}_\alpha = \omega_\alpha(\tilde{\boldsymbol{x}})$, where $\tilde{\boldsymbol{x}}$ is the random observation vector such that for all $\boldsymbol{\theta} \subset \Theta$,

$$\mathrm{P}(\boldsymbol{\theta} \subset \tilde{\omega}_\alpha | \boldsymbol{\theta}) = 1 - \alpha, \tag{4.4}$$

and $\omega_{\alpha_1}(\tilde{\boldsymbol{x}}) \subset \omega_{\alpha_2}(\tilde{\boldsymbol{x}})$ when $\alpha_1 > \alpha_2$. This last condition insures that the confidence region will be larger the larger is $1 - \alpha$. In particular problems, as with confidence intervals, some additional considerations are usually required to determine a unique form for $\tilde{\omega}_\alpha$. If $\tilde{\omega}_\alpha$ is formed so that all parameter values in $\tilde{\omega}_\alpha$ have higher likelihood than those outside, such a region is called a likelihood-based confidence region by Cox and Hinkley (1974, p. 218).

A Bayesian $1-\alpha$ confidence region for a parameter vector $\boldsymbol{\theta}$ is based on the posterior pdf for $\boldsymbol{\theta}$. That is, a nested set of regions $\omega_\alpha = \omega_\alpha(\boldsymbol{x})$, where $\boldsymbol{x}$ is a given vector of observations such that

$$\mathrm{P}(\boldsymbol{\theta} \subset \omega_\alpha(\boldsymbol{x})|\boldsymbol{x}) = 1-\alpha, \tag{4.5}$$

and $\omega_{\alpha_1}(\boldsymbol{x}) \subset \omega_{\alpha_2}(\boldsymbol{x})$ for $\alpha_1 > \alpha_2$ is a $1-\alpha$ Bayesian confidence region for $\boldsymbol{\theta}$. On comparing (4.4) and (4.5) it is the case that they involve different probability statements, (4.4) relating to properties of the random region $\tilde{\omega}_\alpha$ given $\boldsymbol{\theta}$, and (4.5) relating to posterior properties of the random $\boldsymbol{\theta}$ given the region $\omega_\alpha = \omega_\alpha(\boldsymbol{x})$. Two examples will be provided to illustrate these concepts.

Example 4.4

From Example 4.2 the sampling distribution of $\tilde{F} = (\hat{\boldsymbol{\beta}} - \boldsymbol{\beta})'X'X(\hat{\boldsymbol{\beta}} - \boldsymbol{\beta})/ks^2$ is known to be an F pdf with k and ν d.f. From tables of this distribution, c_α can be determined such that $\mathrm{P}(\tilde{F} \leqslant c_\alpha) = 1-\alpha$. Then $F = (\hat{\boldsymbol{\beta}} - \boldsymbol{\beta})'X'X(\hat{\boldsymbol{\beta}} - \boldsymbol{\beta})/ks^2 \leqslant c_\alpha$ defines a set of nested regions (ellipsoids) that constitute a confidence region for $\boldsymbol{\beta}$. In the case where $\boldsymbol{\beta}$ has two elements, $(\hat{\boldsymbol{\beta}} - \boldsymbol{\beta})'X'X(\hat{\boldsymbol{\beta}} - \boldsymbol{\beta})/ks^2 \leqslant c_\alpha$ defines a set of nested ellipses. For $k=1$, the result $(\hat{\beta} - \beta)^2 \Sigma x_i^2/s^2 \leqslant c_\alpha$ is consistent with the confidence interval for a single parameter discussed earlier. The bounding contours of these confidence regions have constant marginal likelihood.

Example 4.5

From the marginal posterior pdf for $\boldsymbol{\beta}$ in Example 4.3, it is the case that $F = (\boldsymbol{\beta} - \hat{\boldsymbol{\beta}})'X'X(\boldsymbol{\beta} - \hat{\boldsymbol{\beta}})/ks^2$ has a posterior pdf in the form of an F pdf with k and ν d.f. Thus, c_α can be obtained such that $\mathrm{P}[(\boldsymbol{\beta} - \hat{\boldsymbol{\beta}})'X'X(\boldsymbol{\beta} - \hat{\boldsymbol{\beta}})/ks^2 \leqslant c_\alpha|D] = 1-\alpha$. Then $(\boldsymbol{\beta} - \hat{\boldsymbol{\beta}})'X'X(\boldsymbol{\beta} - \hat{\boldsymbol{\beta}})/ks^2 \leqslant c_\alpha$ defines a set of nested regions or confidence region for $\boldsymbol{\beta}$. For $\boldsymbol{\beta}$ having two elements, the nested regions are ellipses. Also, it is the case that these confidence regions are highest posterior density regions.

That the regions in Examples 4.4 and 4.5 are identical is due to the fact that there exists a pivotal quantity, namely $F = (\hat{\boldsymbol{\beta}} - \boldsymbol{\beta})'X'X(\hat{\boldsymbol{\beta}} - \boldsymbol{\beta})/ks^2$, that has the same pdf under the sampling theory and Bayesian assumptions and does not involve any parameters with unknown values. These confidence regions relate to the entire coefficient vector $\boldsymbol{\beta}$. Similar results can be obtained for any subvector of $\boldsymbol{\beta}$. Further, there are several types of "simultaneous" confidence intervals for all differences of means or contrasts for various analysis of variance models, including that for independent, normal observations, $\tilde{x}_{ij}$, with $E\tilde{x}_{ij} = \theta + \gamma_i$, $\mathrm{var}(\tilde{x}_{ij}) = \sigma^2$, $i = 1,2,\dots,k$ and $j = 1,2,\dots,n$ and $\sum_{i=1}^k \gamma_i = 0$; see, for example, Arnold (1981, p. 135 ff. and ch. 12) for derivations of simultaneous confidence intervals for regression and analysis of variance models.

For many problems involving a parameter vector $\boldsymbol{\theta}' = (\boldsymbol{\theta}_1', \boldsymbol{\theta}_2')$ if pivotal quantities do not exist, it is difficult to obtain an exact $1-\alpha$ confidence region for $\boldsymbol{\theta}_1$

without additional conditions; see Cox and Hinkley (1974, p. 230 ff.) for analysis of this problem. In a Bayesian approach the marginal posterior pdf for $\boldsymbol{\theta}_1$, $h(\boldsymbol{\theta}_1|D)$, is obtained from the joint posterior pdf $p(\boldsymbol{\theta}_1, \boldsymbol{\theta}_2|D)$ and confidence regions can be based on $h(\boldsymbol{\theta}_1|D)$. Another serious problem arises if the sampling distribution of an estimator $\hat{\boldsymbol{\theta}}$ or the posterior pdf for $\boldsymbol{\theta}$ is multi-modal or has some other unusual features. In such cases sampling theory and Bayesian confidence regions can be misleading. Finally, in large samples, maximum likelihood and other estimators are often approximately normally distributed and the large sample normal distribution can be employed to obtain approximate confidence intervals and regions in a sampling theory approach. Similarly, in large samples, posterior pdfs assume an approximate normal form and approximate Bayesian intervals and regions can be computed from approximate normal posterior distributions. For n large enough, these approximations will be satisfactory.

5. Prediction

Prediction is a most important part of econometrics and other sciences. Indeed, Jeffreys (1967) defines induction to be the process "...of making inferences from past experience to predict future experience" (p. 1). Also, causation has been defined to be confirmed predictability from a law or set of laws by Feigl (1953) and other philosophers. Since induction and causation are directly linked to prediction, and since prediction is intimately involved in economic research,[44] econometric modelling, and policy analysis, it is a topic that deserves considerable emphasis in econometrics and other sciences, a point of view stressed strongly by Geisser (1980).

Prediction usually involves the study of past data, denoted by $\boldsymbol{x}$ and formulation of a probability model for them. For simplicity, assume that the n elements of $\tilde{x}$ are independently and identically distributed and that the model for $\tilde{x}$ is $p(\boldsymbol{x}|\boldsymbol{\theta}) = \prod_{i=1}^{n} f(x_i|\boldsymbol{\theta})$, where $\boldsymbol{\theta}$ is a vector of parameters. Now consider some future or as yet unobserved data. These future or as yet unobserved data are denoted by $\tilde{z}$, a $q \times 1$ random vector. If it is further assumed that the elements of $\tilde{z}$ are generated by the same probability model that generated $\tilde{x}$, then the probability model for $\tilde{z}$ is $g(z|\boldsymbol{\theta}) = \prod_{i=1}^{q} f(z_i|\boldsymbol{\theta})$. If the form of $f(\cdot)$ and the value of $\boldsymbol{\theta}$ were known exactly, then $g(z|\boldsymbol{\theta})$ would be completely determined and could be employed to make various probability statements about the elements of $\tilde{z}$. Unfortunately, the value of $\boldsymbol{\theta}$ is not usually known and this fact makes it important to have prediction techniques that are operational when the value of $\boldsymbol{\theta}$ is uncertain. Also, if there are serious errors in the assumptions underlying the

[44] See, for example, the predictions that Friedman (1957, pp. 214–219) derived from his theory of the consumption function.

probability models for $\tilde{x}$ and for $\tilde{z}$, then predictions of $\tilde{z}$ will usually be adversely affected.[45]

5.1. Sampling theory prediction techniques[46]

With past data x and future or as yet unobserved data $\tilde{z}$, a *point prediction* of the random vector $\tilde{z}$ is defined to be $\hat{z} = \varphi(x)$, where $\varphi(x)' = [\varphi_1(x), \varphi_2(x), \dots, \varphi_q(x)]$ is a function of just x and thus can be evaluated given x. When the value of $\tilde{z}$ is observed, say z_0, then $e_0 = \hat{z} - z_0$ is the observed forecast error vector. In general perfect prediction in the sense $e_0 = 0$ is impossible and thus some other criteria have to be formulated to define good prediction procedures. The parallelism with the problem of defining good estimation procedures, discussed in Section 3, is very close except that here the object of interest, $\tilde{z}$, is random.

In the sampling theory approach, prediction procedures are evaluated in terms of their sampling properties in repeated (actual or hypothetical) samples. That is, the sampling properties of a predictor, $\varphi(\tilde{x})$, are considered in defining good or optimal prediction procedures which involves the choice of an explicit functional form for $\varphi(\tilde{x})$. Note that use of the term "point predictor" or "predictor" implies that the random function $\varphi(\tilde{x})$ is being considered, while use of the term "point prediction" or "prediction" implies that the non-random function $\varphi(x)$ is being considered.

Some properties of predictors are reviewed below with $\tilde{e} = \tilde{z} - \varphi(\tilde{x})$ the random forecast error vector. For brevity of notation, $\varphi(\tilde{x})$ will be denoted by $\tilde{\varphi}$.

(1) *Minimal MSE predictor.* If $l'\tilde{e}$, where l is a given vector of rank one, has minimal MSE, then $\tilde{\varphi}$ is a minimal MSE predictor.
(2) *Unbiasedness.* If $E\tilde{e} = 0$, then $\tilde{\varphi}$ is an unbiased predictor. If $E\tilde{e} \neq 0$, then $\tilde{\varphi}$ is a biased predictor.
(3) *Linearity.* If $\tilde{\varphi} = A\tilde{x}$, where A is a given matrix, then $\tilde{\varphi}$ is a linear predictor.
(4) *Minimum variance linear unbiased (MVLU) predictor.* Consider $l'\tilde{e}$, where l is a given $q \times 1$ vector of rank one and the class of linear, unbiased predictors, $\tilde{\varphi}_u = A_u\tilde{x}$, with A_u, not unique, such that $E(\tilde{z} - A_u\tilde{x}) = 0$. If $\mathrm{var}(l'\tilde{e})$ is minimized by taking $A_u = A_*$, then $\tilde{\varphi}_* = A_*\tilde{x}$ is the MVLU predictor.
(5) *Prediction risk.* If $L(\tilde{e})$ is a convex loss function, then the risk associated with $\tilde{\varphi}$ relative to $L(\tilde{e})$ is $r(0) = EL(\tilde{e})$. For example, if $L(\tilde{e}) = \tilde{e}'Q\tilde{e}$, where Q is a given $q \times q$ positive definite symmetric matrix, the risk associated with

[45] Statistical tests of these assumptions, e.g. the i.i.d. assumption, can be performed.
[46] For further discussion of sampling theory prediction and forecasting techniques for a range of problems, see Granger and Newbold (1977).

$\tilde{\boldsymbol{\varphi}}$ is $E\tilde{\boldsymbol{e}}'Q\tilde{\boldsymbol{e}} = \boldsymbol{\Delta}'Q\boldsymbol{\Delta} + \operatorname{tr} QV(\tilde{\boldsymbol{e}})$, where $\boldsymbol{\Delta} = E\tilde{\boldsymbol{e}}$, the bias vector, and $V(\tilde{\boldsymbol{e}}) = E(\tilde{\boldsymbol{e}} - \boldsymbol{\Delta})(\tilde{\boldsymbol{e}} - \boldsymbol{\Delta})'$, the covariance matrix of $\tilde{\boldsymbol{e}}$. In the case of a scalar forecast error $\tilde{e}$, with $L(\tilde{e}) = \tilde{e}^2$

$$E\tilde{e}^2 = \Delta^2 + \operatorname{var}(\tilde{e}), \tag{5.1}$$

where $E\tilde{e} = \Delta$ is the prediction bias and $\operatorname{var}(\tilde{e})$ is the variance of the forecast error.

(6) *Admissible predictor.* Let $r_1(\boldsymbol{\theta}) = EL(\tilde{\boldsymbol{e}}_1)$ be the risk associated with predictor $\tilde{\boldsymbol{\varphi}}_1$ and $r_a(\boldsymbol{\theta}) = EL(\tilde{\boldsymbol{e}}_a)$ be the risk associated with any other predictor. If there does not exist another predictor such that $r_a(\boldsymbol{\theta}) \leqslant r_1(\boldsymbol{\theta})$ for all $\boldsymbol{\theta}$ in the parameter space, then the predictor $\tilde{\boldsymbol{\varphi}}_1$ is admissible relative to the loss function L. If another predictor exists such that $r_a(\boldsymbol{\theta}) \leqslant r_1(\boldsymbol{\theta})$ for all $\boldsymbol{\theta}$ in the parameter space, with $r_a(\boldsymbol{\theta}) < r_1(\boldsymbol{\theta})$ for some values of $\boldsymbol{\theta}$, then $\tilde{\boldsymbol{\varphi}}_1$ is inadmissible relative to L.

(7) *Robust predictor.* A robust predictor is a predictor that performs well in the presence of model specification errors and/or in the presence of unusual data.

Much of what has been said above with respect to criteria for choice of estimators is also applicable to choice of predictors. Unfortunately, minimal MSE predictors do not in general exist. The unbiasedness property alone does not lead to a unique predictor and insisting on unbiasedness may be costly in terms of prediction MSE. In terms of (5.1), it is clear that a slightly biased predictor with a very small prediction error variance can be better in terms of MSE than an unbiased predictor with a large prediction error variance. Also, as with admissible estimators, there usually are many admissible predictors relative to a given loss function. Imposing the condition that a predictor be linear and unbiased in order to obtain a MVLU predictor can be costly in terms of MSE. For many prediction problems, non-linear biased Stein-like predictors have lower MSE than do MVLU predictors; see, for example, Efron and Morris (1975). Finally, it is desirable that predictors be robust and what has been said above about robust estimators can be adapted to apply to predictors' properties.

To illustrate a close connection between estimation and prediction, consider the standard multiple regression model $\tilde{\boldsymbol{y}} = X\boldsymbol{\beta} + \tilde{\boldsymbol{u}}$, where $\tilde{\boldsymbol{y}}$ is $n \times 1$, X is a given non-stochastic $n \times k$ matrix of rank k, $\boldsymbol{\beta}$ is a $k \times 1$ vector of parameters, and $\tilde{\boldsymbol{u}}$ is an $n \times 1$ disturbance vector with $E\tilde{\boldsymbol{u}} = \boldsymbol{0}$ and $E\tilde{\boldsymbol{u}}\tilde{\boldsymbol{u}}' = \sigma^2 I_n$. Let a future scalar observation $\tilde{z}$ be generated by $\tilde{z} = \boldsymbol{w}'\boldsymbol{\beta} + \tilde{v}$, where $\boldsymbol{w}'$ is a $1 \times k$ given[47] vector of rank one and $\tilde{v}$ is a future disturbance term, uncorrelated with the elements of $\tilde{\boldsymbol{u}}$ with $E\tilde{v} = 0$ and $E\tilde{v}^2 = \sigma^2$. Then a predictor of $\tilde{z}$, denoted by $\hat{z}$, is given by

[47] For some analysis of this problem when $\boldsymbol{w}$ is random, see Feldstein (1971).

$\hat{z} = \boldsymbol{w}'\tilde{\boldsymbol{\beta}}$, where $\tilde{\boldsymbol{\beta}}$ is an estimator for $\boldsymbol{\beta}$. Then $\tilde{e} = \hat{z} - \tilde{z} = \boldsymbol{w}'(\tilde{\boldsymbol{\beta}} - \boldsymbol{\beta}) - \tilde{v}$ and $E\tilde{e} = \boldsymbol{w}'E(\tilde{\boldsymbol{\beta}} - \boldsymbol{\beta})$, so that if $E(\tilde{\boldsymbol{\beta}} - \boldsymbol{\beta}) = \boldsymbol{0}$, that is, $\tilde{\boldsymbol{\beta}}$ is unbiased, $E\tilde{e} = 0$ and the predictor $\hat{z} = \boldsymbol{w}'\tilde{\boldsymbol{\beta}}$ is unbiased. Furthermore, if $\tilde{\boldsymbol{\beta}}$ is a linear estimator, say $\tilde{\boldsymbol{\beta}} = A\tilde{\boldsymbol{y}}$, then $\hat{z}$ is a linear predictor. Further relative to $L(\tilde{e}) = \tilde{e}^2$, the risk of $\hat{z}$ in general is $E\tilde{e}^2 = E[\boldsymbol{w}'(\tilde{\boldsymbol{\beta}} - \boldsymbol{\beta}) - \tilde{v}]\times[(\tilde{\boldsymbol{\beta}} - \boldsymbol{\beta})'\boldsymbol{w} - \tilde{v}] = \boldsymbol{w}'E(\tilde{\boldsymbol{\beta}} - \boldsymbol{\beta})(\tilde{\boldsymbol{\beta}} - \boldsymbol{\beta})'\boldsymbol{w} + \sigma^2$.[48] If in this last expression, $\tilde{\boldsymbol{\beta}} = \hat{\boldsymbol{\beta}} = (X'X)^{-1}X'\tilde{\boldsymbol{y}}$, the MVLU least squares estimator, then $E\tilde{e} = 0$ and $E\tilde{e}^2$ assumes a minimal value in the class of unbiased linear predictors, that is, $\hat{z} = \boldsymbol{w}'A_u\tilde{\boldsymbol{y}}$, where $A_u = (X'X)^{-1}X' + C'$ and C' is such that $C'X = 0$. Finally, if $\tilde{\boldsymbol{\beta}}$ is an estimator such that $\boldsymbol{w}'E(\tilde{\boldsymbol{\beta}} - \boldsymbol{\beta})(\tilde{\boldsymbol{\beta}} - \boldsymbol{\beta})'\boldsymbol{w} \leqslant \boldsymbol{w}'E(\hat{\boldsymbol{\beta}} - \boldsymbol{\beta})(\hat{\boldsymbol{\beta}} - \boldsymbol{\beta})'\boldsymbol{w}$ for all values of $\boldsymbol{\beta}$, then $\hat{z} = \boldsymbol{w}'\hat{\boldsymbol{\beta}}$ is an inadmissible predictor relative to $L(\tilde{e}) = \tilde{e}^2$.

As is seen from the above discussion, much traditional analysis has been carried forward under the assumption that the appropriate loss function is quadratic, for example $L(\tilde{e}) = \tilde{e}^2$ or $L(\tilde{\boldsymbol{e}}) = \tilde{\boldsymbol{e}}'Q\tilde{\boldsymbol{e}}$. While such quadratic loss functions are appropriate or are a reasonable approximation in a number of problems, there are problems for which they are not. For example, Varian (1975) analyzed a problem employing an asymmetric "linex" loss function, $L(\tilde{e}) = b\exp(a\tilde{e}) - c\tilde{e} - b$, with $a = 0.0004$; $b = 2500$, and $c = 0.1$. He considered this asymmetric loss function more appropriate for his problem than $L(\tilde{e}) = \tilde{e}^2$. Also, as pointed out in the estimation section, bounded loss functions may be appropriate for some problems. Use of an appropriate loss function for prediction and other problems is important and solutions can be sensitive to the form of the loss function utilized [see, for example, Varian (1975), Zellner and Geisel (1968), and Zellner (1973)].

Sampling theory prediction bounds, intervals, and regions are available that relate to future observations; for discussion of these topics, see, for example, Christ (1966, pp. 557–564) and Guttman (1970). It must be emphasized that sampling theory prediction intervals or regions are subsets of the sample space that have a specified probability of including the *random* future observations. For example, $\tilde{\varphi} \pm c_\alpha(\tilde{\boldsymbol{x}})$ is a $(1-\alpha)\times 100$ percent central predictive interval for $\tilde{z}$, a future random scalar observation if $\mathrm{P}[\tilde{\varphi} - c_\alpha(\tilde{\boldsymbol{x}}) < \tilde{z} < \tilde{\varphi} + c_\alpha(\tilde{\boldsymbol{x}})] = 1 - \alpha$. Note that $\tilde{\varphi} = \varphi(\tilde{\boldsymbol{x}})$, the predictor, $c_\alpha(\tilde{\boldsymbol{x}})$, and $\tilde{z}$ are all random in this probability statement. For a particular sample, $\boldsymbol{x}$, the computed prediction interval is $\varphi(\boldsymbol{x}) \pm c_\alpha(\boldsymbol{x})$, for example 50.2 ± 1.1, where $\varphi(\boldsymbol{x}) = 50.2$ is the point prediction and $c_\alpha(\boldsymbol{x}) = 1.1$. Other types of prediction intervals, for example random intervals that are constructed so that in repeated sampling the proportion of cases in which $\tilde{z}$ is included in the interval has a specified expected value with a given probability or

[48]Note that $E\boldsymbol{w}'(\tilde{\boldsymbol{\beta}} - \boldsymbol{\beta})\tilde{v} = 0$ if the elements of $\tilde{\boldsymbol{\beta}} - \boldsymbol{\beta}$ and $\tilde{v}$ are uncorrelated as they are under the above assumptions if $\tilde{\boldsymbol{\beta}}$ is a linear estimator. On the other hand, if $\tilde{\boldsymbol{\beta}}$ is a non-linear estimator, sufficient conditions for this result to hold are that the elements of $\tilde{\boldsymbol{u}}$ and $\tilde{v}$ are independently distributed and $E\boldsymbol{w}'(\tilde{\boldsymbol{\beta}} - \boldsymbol{\beta})\tilde{v}$ is finite.

such that the proportion is not less than a specified value with given probability, are called tolerance intervals. See Christ (1966) and Guttman (1970) for further discussion and examples of tolerance intervals. Finally, in many econometric problems, exact prediction intervals and regions are not available and large sample approximate intervals and regions are often employed.

5.2. *Bayesian prediction techniques*

Central in the Bayesian approach to prediction is the predictive pdf for $\tilde{z}$, $p(z|D)$, where $\tilde{z}$ is a vector of future observations and D denotes the sample, $\boldsymbol{x}$, and prior information. To derive the predictive pdf, let $\tilde{z}$ and $\tilde{x}$ be independent[49] with pdfs $g(z|\boldsymbol{\theta})$ and $f(\boldsymbol{x}|\boldsymbol{\theta})$, where $\boldsymbol{\theta} \subset \Theta$ is a vector of parameters with posterior pdf $h(\boldsymbol{\theta}|D) = c\pi(\boldsymbol{\theta})f(\boldsymbol{x}|\boldsymbol{\theta})$, where c is a normalizing constant and $\pi(\boldsymbol{\theta})$ is the prior pdf. Then,

$$p(z|D) = \int_{\Theta} g(z|\boldsymbol{\theta})h(\boldsymbol{\theta}|D)\,\mathrm{d}\boldsymbol{\theta} \tag{5.2}$$

is the predictive pdf for $\tilde{z}$. Note that (5.2) involves an averaging of the conditional pdfs $g(z|\boldsymbol{\theta})$, with $h(\boldsymbol{\theta}|D)$, the posterior pdf for $\boldsymbol{\theta}$, serving as the weight function. Also, $p(z|D)$ incorporates both sample and prior information reflected in $h(\boldsymbol{\theta}|D)$. For examples of explicit predictive pdfs for regression and other models, see Aitchison and Dunsmore (1975), Box and Tiao (1973), Guttman (1970), and Zellner (1971).

If z is partitioned as $z' = (z_1', z_2')$, the marginal predictive pdf for z_1 can be obtained from (5.2) by analytical or numerical integration. Also, a pdf for z_2 given z_1 and/or the distribution of functions of z can be derived from (5.2).

If a point prediction of $\tilde{z}$ is desired, the mean or modal value of (5.2) might be used. If a convex prediction loss function $L(\tilde{z}, \hat{z})$ is available, where $\hat{z} = \hat{z}(D)$ is some point prediction depending on the given sample $\boldsymbol{x}$ and prior information, Bayesians choose $\hat{z}$ so as to minimize expected loss, that is, solve the following problem:

$$\min_{\hat{z}} \int L(z, \hat{z})p(z|D)\,\mathrm{d}z. \tag{5.3}$$

The solution, say $\hat{z} = \hat{z}^*(D)$, is the Bayesian point prediction relative to the loss function $L(\tilde{z}, \hat{z})$. For example, if $L(\tilde{z}, \hat{z}) = (\tilde{z} - \hat{z})'Q(\tilde{z} - \hat{z})$, with Q a given

[49] Independence, assumed here for simplicity, can be relaxed.

positive definite symmetric matrix, the optimal $\hat{z}$ is $\hat{z}^* = E(\tilde{z}|D)$, the mean of the predictive pdf in (5.2).[50] For other loss functions, Bayesian minimum expected loss point predictions can be obtained [see Aitchison and Dunsmore (1975), Litterman (1980), and Varian (1975) for examples]. Prediction intervals and regions can be computed from (5.2) in the same way that posterior intervals and regions are computed for parameters, as described above. These prediction intervals and regions are dependent on the given data D and hence are not viewed as random. For example, in the case of a scalar future observation, $\tilde{z}$, given the predictive pdf for $\tilde{z}$, $p(z|D)$, the probability that $b < \tilde{z} < a$ is just $\int_a^b p(z|D)\mathrm{d}z = 1-\alpha$. If a and b are given, $1-\alpha$ can be calculated. If $1-\alpha$ is given, then a and b are not uniquely determined; however, by requiring that $b-a$ be a minimum subject to a given $1-\alpha$, unique values of a and b can be obtained.

To this point, all results in this subsection are for given data $\boldsymbol{x}$ and given prior information. The sampling properties of Bayesian procedures are of interest, particularly before $\tilde{\boldsymbol{x}}$ is observed and also in characterizing average properties of procedures. In this regard, the solution, $\hat{z}^*$ to the problem in (5.3) can be viewed as a Bayesian predictor, random since it is a function of $\tilde{\boldsymbol{x}}$. For brevity, write a predictor as $\hat{z} = \hat{z}(\tilde{\boldsymbol{x}})$. Then the prediction risk function, $r(\boldsymbol{\theta})$, relative to the loss function $L(\hat{z},\tilde{z})$, is

$$r(\boldsymbol{\theta}) = \int\int L(\hat{z},\tilde{z}) f(\boldsymbol{x}|\boldsymbol{\theta}) g(z|\boldsymbol{\theta})\,\mathrm{d}\boldsymbol{x}\,\mathrm{d}z, \tag{5.4}$$

where the integrations are over the sample spaces of $\tilde{\boldsymbol{x}}$ and $\tilde{z}$. Risk, $r(\boldsymbol{\theta})$, can be computed for alternative predictors, $\hat{z} = \hat{z}(\tilde{\boldsymbol{x}})$. The Bayesian predictor is the one, if it exists, that minimizes average risk, $\mathrm{AR} = \int_\Theta r(\boldsymbol{\theta})\pi(\boldsymbol{\theta})\mathrm{d}\boldsymbol{\theta}$, where $\pi(\boldsymbol{\theta})$ is a prior for $\boldsymbol{\theta}$. If AR is finite, then the Bayesian predictor is admissible and also is given by the solution to the problem in (5.3).

From what has been presented, it is the case that both sampling theory and Bayesian techniques are available for predictive inference. As with estimation, the approaches differ in terms of justifications for procedures and in that the Bayesian approach employs a prior distribution, whereas it is not employed in the sampling theory approach. Further, in both approaches predictive inference has been discussed in terms of *given* models for the observations. Since there is often uncertainty about models' properties, it is important to have testing procedures that help determine the forms of models for observations. In the next section general features of testing procedures are presented.

[50] The proof is very similar to that presented above in connection with Bayesian parameter estimation with a quadratic loss function.

6. Statistical analysis of hypotheses

Statistical procedures for analyzing and testing hypotheses, that is, hypothesized probability models for observations, are important in work to obtain satisfactory econometric models that explain past economic behavior well, predict future behavior reliably, and are dependable for use in analyzing economic policies. In this connection, statistical theory has yielded general procedures for analyzing hypotheses and various justifications for them. In what follows, some basic results in this area will be reviewed.

6.1. *Types of hypotheses*

Relative to the general probability model, $\{X, \Theta, p(x|\boldsymbol{\theta})\}$ hypotheses can relate to the value of $\boldsymbol{\theta}$, or a subvector of $\boldsymbol{\theta}$, and/or to the form of $p(x|\boldsymbol{\theta})$. For example, $\boldsymbol{\theta}=\boldsymbol{0}$ or $\boldsymbol{\theta}=\boldsymbol{c}$, a given vector, are examples of *simple hypotheses*, that is, hypotheses that *completely* specify the parameter vector $\boldsymbol{\theta}$ appearing in $p(x|\boldsymbol{\theta})$. On the other hand, some hypotheses about the value of $\boldsymbol{\theta}$ do not completely specify $p(x|\boldsymbol{\theta})$. For example, $\boldsymbol{\theta}\subset\omega$, a subspace of Θ, does not imply a particular value for $\boldsymbol{\theta}$ and thus is not a simple hypothesis but rather is termed a *composite hypothesis*. In terms of a scalar parameter $\theta\subset\Theta$, where Θ is the entire real line, $\theta=0$ is a simple hypothesis, whereas $\theta<0$ and $\theta>0$ are composite hypotheses. Further, it is often the case that two or more hypotheses are considered. For example $\theta=0$ and $\theta=1$, two simple hypotheses, or $\theta=0$ and $\theta>0$, a simple hypothesis and a composite hypothesis, or $\theta>0$ and $\theta<0$, two composite hypotheses, may be under study. Finally, various forms for $p(x|\boldsymbol{\theta})$ may be hypothesized, for example $p[(x-\mu_1)/\sigma_1]$ normal or $p[(x-\mu_2)/\sigma_2]$ double-exponential are two alternative hypotheses regarding the form of $p(\cdot)$ with the same parameter space Θ: $-\infty<\mu_i<\infty$ and $0<\sigma_i<\infty$, $i=1,2$. In other cases, $p(x|\boldsymbol{\theta})$ and $g(x|\boldsymbol{\varphi})$ may be two alternative hypothesized forms for the pdf for the observations involving parameter vectors $\boldsymbol{\theta}$ and $\boldsymbol{\varphi}$ and their associated parameter spaces. Finally, if the probability model is expanded to include a prior pdf for $\boldsymbol{\theta}$, denoted by $p(\boldsymbol{\theta})$, different $p(\boldsymbol{\theta})$'s can be viewed as hypotheses. For example, for a scalar θ, $p_1(\theta)$ in the form of a normal pdf, with given mean $\bar{\theta}_1$ and given variance σ_1^2, $\theta\sim N(\bar{\theta}_1,\sigma_1^2)$ and $\theta\sim N(\bar{\theta}_2,\sigma_2^2)$, with $\bar{\theta}_2$ and σ_2^2 given, can be viewed as hypotheses.

Whatever the hypothesis or hypotheses, statistical testing theory provides procedures for deciding whether sample observations are consistent or inconsistent with a hypothesis or set of hypotheses. Just as with estimation and prediction procedures, it is desirable that testing procedures have reasonable justifications and work well in practice. It is to these issues that we now turn.

6.2. Sampling theory testing procedures

The Neyman–Pearson (NP) sampling theory testing procedures are widely utilized and described in most statistics and econometrics textbooks.[51] In the NP approach, two hypotheses are considered, which in terms of a scalar parameter $\theta \subset \Theta$ can be described by $\theta \subset \omega$ and $\theta \subset \Theta - \omega$, where ω is a subspace of Θ, and $\Theta - \omega$ denotes the region of Θ not including ω, that is, the complement of ω. For example, $\theta = 0$ and $\theta \neq 0$, with Θ the entire real line, might be two hypotheses under consideration. Usually $\theta \subset \omega$ is called the "null hypothesis" and $\theta \subset \Theta - \omega$, the "alternative hypothesis", nomenclature that suggests an asymmetric view of the hypotheses. In NP theory, the sample space X is partitioned into two regions: (1) the "region of acceptance" or the region in which outcomes are thought to be consistent with the $\theta \subset \omega$, the null hypothesis, and (2) a "region of rejection" or a region, complementary to the "acceptance region", in which outcomes are thought to be inconsistent with the null hypothesis. This "rejection region" is usually called "the critical region" of the test. Example 6.1 illustrates these concepts.

Example 6.1

Let $\tilde{x}_i$, $i = 1,2,\dots,n$, be NID $(\theta,1)$ with Θ: $-\infty < \theta < \infty$, and consider the null hypothesis, H_0: $\theta = \theta_0$ and the alternative hypothesis, H_1: $\theta \neq \theta_0$. Here $\omega \subset \Theta$ is $\theta = \theta_0$ and $\Theta - \omega$ is $\theta \neq \theta_0$. Suppose that we consider the random sample mean $\bar{x} = \sum_{i=1}^{n} \tilde{x}_i / n$. A "region of acceptance" might be $|\bar{x} - \theta_0| \leqslant c$ and a "region of rejection", or critical region $|\bar{x} - \theta_0| > c$, where c is a given constant. Thus, given the value of c, the sample space is partitioned into two regions.

Two major questions raised by Example 6.1 are: Why use $\bar{x}$ in constructing the regions and on what grounds can the value of c be selected? In regard to these questions, NP theory recognizes two types of errors that can be made in testing $\theta \subset \omega$ and $\theta \subset \Theta - \omega$. An error of type I, or of the first kind, is rejecting[52] $\theta \subset \omega$ when it is true, while an error of type II, or of the second kind, is accepting $\theta \subset \omega$ when it is false. The operating characteristics of a NP test are functions that describe probabilities of type I and type II errors. Let $\tilde{t} = \tilde{t}(\tilde{\mathbf{x}})$ be a test statistic, for example $\bar{x}$ in Example 6.1, and let R be the region of rejection, a subset of the sample space. Then $\alpha(\theta) = \mathrm{P}(\tilde{t} \subset R | \theta \subset \omega)$ is the probability of a type I error expressed as a function of θ, which specializes to $\alpha(\theta) = \mathrm{P}(|\bar{x} - \theta_0| \geqslant c | \theta = \theta_0)$ in terms of Example 6.1. The probability of a type II error is given by $\beta(\theta) = \mathrm{P}(\tilde{t} \subset \bar{R} | \theta \subset \Theta - \omega) = 1 - \mathrm{P}(\tilde{t} \subset R | \theta \subset \Theta - \omega)$, where $\bar{R}$ is the region of acceptance, the complement of R. In terms of Example 6.1, $\beta(\theta) = \mathrm{P}(|\bar{x} - \theta_0| \leqslant c | \theta \neq \theta_0) =$

[51] For a detailed account see, for example, Lehmann (1959).

[52] The common terminology "reject" and "accept" will be employed even though "inconsistent with" and "consistent with" the data appear to be preferable.

$1-\mathrm{P}(|\bar{x}-\theta_0|>c|\theta\neq 0)$. The function $1-\beta(\theta)$, the probability of rejecting $\theta\subset\omega$ when $\theta\subset\Theta-\omega$ is true, is called the *power function* of the test.

A test with minimal probabilities of type I and type II errors, that is, minimal $\alpha(\theta)$ and $\beta(\theta)$, would be ideal in the NP framework. Unfortunately, such tests do not exist. What NP do to meet this problem is to look for tests with minimal value for $\beta(\theta)$, the probability of type II error subject to the condition that for all $\theta\subset\omega$, $\alpha(\theta)\leqslant\alpha$, a given value, usually small, say 0.05, the "significance level of the test".[53] By minimizing $\beta(\theta)$, of course, the power of the test, $1-\beta(\theta)$, is maximized. A test meeting these requirements is called a *uniformly most powerful* (UMP) *test*.

Unfortunately, except in special cases, uniformly most powerful tests do not exist. In the case of two simple hypotheses, that is, $\Theta=(\theta_1,\theta_2)$ with ω: $\theta=\theta_1$ and $\Theta-\omega$: $\theta=\theta_2$, and data pdf $p(\boldsymbol{x}|\theta)$, the famous NP lemma[54] indicates that a test based on the rejection region $t(\boldsymbol{x})=p(\boldsymbol{x}|\theta_1)/p(\boldsymbol{x}|\theta_2)\geqslant k_\alpha$, where k_α satisfies $\mathrm{P}[t(\tilde{\boldsymbol{x}})\geqslant k_\alpha|\theta=\theta_1]=\alpha$, with α given, is a UMP test. This is of great interest since in this case $t(\boldsymbol{x})$ is the likelihood ratio and thus the NP lemma provides a justification for use of the likelihood ratio in appraising two simple hypotheses. When composite hypotheses are considered, say $\theta\neq 0$, it is usually the case that UMP tests do not exist. One important exception to this statement is in terms of Example 6.1 testing $\theta=\theta_0$ against $\theta>0$. Then with $\sqrt{n}(\bar{x}-\theta_0)$ as the test statistic and using $\sqrt{n}(\bar{x}-\theta_0)>k_\alpha$ as the region of rejection, where k_α is determined so that $\mathrm{P}(\sqrt{n}(x-\theta_0)>k_\alpha|\theta=\theta_0)=\alpha$, for given α, this test can be shown to be UMP.[55] Similarly, a UMP test of $\theta=\theta_0$ against $\theta<\theta_0$ exists for this problem. However, a UMP test of $\theta=\theta_0$ against $\theta\neq\theta_0$ does not exist. That is, using $\sqrt{n}\,|\bar{x}-\theta_0|\geqslant k_\alpha$ as a region of rejection with k_α such that $\mathrm{P}(\sqrt{n}\,|\bar{x}-\theta_0|>k_\alpha)=\alpha$, given α, is not a UMP test.

Given that UMP tests are not usually available for many testing problems, two further conditions have been utilized to narrow the range of candidate tests. First, only unbiased tests are considered. A test is an unbiased α-level test if its operating characteristics satisfy $\alpha(\theta)\leqslant\alpha$ for all $\theta\subset\omega$ and $1-\beta(\theta)\geqslant\alpha$ for all $\theta\subset\Theta-\omega$. This requirement seems reasonable since it implies $1-\alpha\geqslant\beta(\theta)$, that is, that the probability, $1-\alpha$, of accepting $\theta\subset\omega$ when it is true is greater than or equal to the probability $\beta(\theta)$, $\theta\subset\Theta-\omega$, of accepting it when it is false. Many tests of a null hypothesis, $\theta=\theta_1$, with θ_1 given, against composite hypotheses $\theta\neq\theta_1$ are UMP unbiased tests. In terms of Example 6.1 the test statistic $|\bar{x}-\theta_0|\sqrt{n}$ with rejection region $|\bar{x}-\theta_0|\sqrt{n}\geqslant k_\alpha$ is a UMP unbiased test of $\theta=\theta_0$

[53]Some call α the "size of the test".

[54]See Lehmann (1959, p. 65 ff.) for proof of the NP lemma.

[55]Note that $\bar{x}$ is normal with mean θ_0 and variance $1/n$ under the null hypotheses $\theta=\theta_0$. Thus, $\sqrt{n}(\bar{x}-\theta_0)$ is $N(0,1)$ under $\theta=\theta_0$ and tables of the normal distribution can be utilized to evaluate k_α for given α, say $\alpha=0.10$.

against $\theta \neq \theta_0$. See Lehmann (1959, ch. 4–5) for many examples of UMP unbiased tests.

It is also interesting to note that $|\bar{x}-\theta_0|\sqrt{n} < k_\alpha$ can be written as $\bar{x}-k_\alpha/\sqrt{n} < \theta_0 < \bar{x}+k_\alpha/\sqrt{n}$ and that, given $\theta=\theta_0$, the probability that $\bar{x}\pm k_\alpha/\sqrt{n}$ covers θ_0 is $1-\alpha$. Thus, there is a close *mathematical* relationship between test statistics and confidence intervals, discussed above, and in many cases optimal tests produce optimal intervals (in a shortness sense). However, there is a fundamental difference in that in testing $\theta=\theta_0$, θ_0 is given a specific value, often $\theta_0=0$, which is of special interest. On the other hand, with a confidence interval or interval estimation problem the value of θ_0 is not specified; that is, θ_0 is the true unknown value of θ. Thus, if $\bar{x}\pm k_\alpha/\sqrt{n}$, with $\alpha=0.05$ assumes the value 0.32 ± 0.40, this is a 95 percent confidence interval for θ_0 that extends from -0.08 to 0.72. That the interval includes the value 0 does not necessarily imply that $\theta_0=0$. It may be that $\theta_0\neq0$ and the precision of estimation is low. In terms of testing $\theta=\theta_0=0$, the result 0.32 ± 0.40 implies that the test statistic assumes a value in the region of acceptance, $|\bar{x}|\sqrt{n}<k_\alpha$, and would lead to the conclusion that the data are consistent with $\theta=\theta_0=0$. In NP theory, however, this is an incomplete reporting of results. The power of the test must be considered. For example, if $\theta=\pm0.20$ represent important departures from $\theta=0$, the probabilities of rejecting $\theta=0$ when $\theta=\pm0.20$, that is, $1-\beta(0.2)$ and $1-\beta(-0.2)$, should be reported. Under the above conditions, these probabilities are quite low and thus the test is not very powerful relative to important departures from $\theta=0$. More data are needed to get a more powerful test and more precise estimates.

The above discussion reveals an important dependence of a test's power on the sample size. Generally, for given α, the power increases with n. Thus, to "balance" the probabilities of errors of type I and II as n increases requires some adjustment of α. See DeGroot (1970) for a discussion of this problem.[56]

A second way of delimiting candidate tests is to require that tests be invariant, that is, invariant to a certain group of transformations. See Lehmann (1959, ch. 6–7) for discussion of UMP invariant tests that include the standard t and F tests employed to test hypotheses about regression coefficients that are UMP invariant tests under particular groups of linear transformations. They are also UMP unbiased tests. In a remarkable theorem, Lehmann (1959, p. 229) shows that there exists a unique UMP unbiased test for a given testing problem and that there also exists a UMP almost invariant[57] test with respect to some group of transformations G. Then the latter is also unique and the two sets coincide almost everywhere.

[56]Very few econometrics and statistics texts treat this problem.
[57]See Lehmann (1959, p. 225) for a definition of "almost invariant".

Example 6.2

In terms of the normal regression model of Example 4.2, to test H_0: $\beta_i = \beta_{i0}$, a given value against H_1: $\beta_i \neq 0$ with all other regression coefficients and σ unrestricted, the test statistic $t = (\hat{\beta}_i - \beta_{i0})/s_{\hat{\beta}_i}$ has a univariate Student-t pdf with ν d.f. Then $|t| \geqslant k_\alpha$, where k_α is such that $\Pr\{|t| \geqslant k_\alpha | \beta_i = \beta_{i0}\} = \alpha$, with α, the significance level given, is a rejection region. Such a test is a UMP unbiased and invariant (with respect to a group of linear transformations) α-level test. In a similar fashion, from Example 4.4, the statistic $\tilde{F} = (\hat{\boldsymbol{\beta}} - \boldsymbol{\beta}_0)'X'X(\hat{\boldsymbol{\beta}} - \boldsymbol{\beta}_0)/ks^2$ that has an F pdf with k and ν d.f. under H_0: $\boldsymbol{\beta} = \boldsymbol{\beta}_0$ can be used to test H_0 against H_1: $\boldsymbol{\beta} \neq \boldsymbol{\beta}_0$ with σ^2 unrestricted under both hypotheses. The rejection region is $\tilde{F} \geqslant k_\alpha$, with k_α such that $P(\tilde{F} \geqslant k_\alpha) = \alpha$, a given value. This test is a UMP unbiased and invariant α-level test.

In many testing problems, say those involving hypotheses about the values of parameters of time series models, or of simultaneous equations models, exact tests are generally not available. In these circumstances, approximate large sample tests, for example approximate likelihood ratio (LR), Wald (W), and Lagrange Multiplier (LM) tests, are employed. For example, let $\boldsymbol{\theta}' = (\boldsymbol{\theta}_1', \boldsymbol{\theta}_2')$ be a parameter vector appearing in a model with likelihood function $p(\boldsymbol{x}|\boldsymbol{\theta})$ and let the null hypothesis be H_0: $\boldsymbol{\theta}_1 = \boldsymbol{\theta}_{10}$ and $\boldsymbol{\theta}_2$ unrestricted and H_1: $\boldsymbol{\theta}_1$ and $\boldsymbol{\theta}_2$ both unrestricted. Then $\lambda(\boldsymbol{x})$, the approximate LR, is defined to be

$$\lambda(\boldsymbol{x}) = p(\boldsymbol{x}|\boldsymbol{\theta}_{10}, \hat{\boldsymbol{\theta}}_2)/p(\boldsymbol{x}|\hat{\hat{\boldsymbol{\theta}}}_1, \hat{\hat{\boldsymbol{\theta}}}_2), \tag{6.1}$$

where $\hat{\boldsymbol{\theta}}_2$ is the value of $\boldsymbol{\theta}_2$ that maximizes $p(\boldsymbol{x}|\boldsymbol{\theta}_{10}, \boldsymbol{\theta}_2)$, the restricted likelihood function (LF), while $(\hat{\hat{\boldsymbol{\theta}}}_1, \hat{\hat{\boldsymbol{\theta}}}_2)$ is the value of $(\boldsymbol{\theta}_1, \boldsymbol{\theta}_2)$ that maximizes $p(\boldsymbol{x}|\boldsymbol{\theta}_1, \boldsymbol{\theta}_2)$, the unrestricted LF. Since the numerator of (6.1) is less than or equal to the denominator, given that the numerator is the result of a restricted maximization, $0 < \lambda(\boldsymbol{x}) \leqslant 1$. The larger $\lambda(\boldsymbol{x})$, the "more likely" that the restriction $\boldsymbol{\theta}_1 = \boldsymbol{\theta}_{10}$ is consistent with the data using a relative maximized LF criterion for the meaning of "more likely". In large samples, under regularity conditions and H_0, $-2\log\lambda(\tilde{\boldsymbol{x}}) = \tilde{\chi}^2$ has an *approximate* χ^2 pdf with q d.f., where q is the number of restrictions implied by H_0, here equal to the number of elements of $\boldsymbol{\theta}_1$. Then a rejection region is $\tilde{\chi}^2 \geqslant k_\alpha$, where k_α is such that $P(\tilde{\chi}^2 \geqslant k_\alpha | H_0) \doteq \alpha$, the given significance level.[58] Many hypotheses can be tested approximately in this approach given that regularity conditions needed for $-2\log\lambda(\tilde{\boldsymbol{x}})$ to be approximately distributed as $\tilde{\chi}^2_q$ in large samples under H_0 are satisfied.

In the Wald large sample approach to the test, for example H_0: $\boldsymbol{\theta} = \boldsymbol{\theta}_0$ against H_1: $\boldsymbol{\theta} \neq \boldsymbol{\theta}_0$, let $\hat{\boldsymbol{\theta}}$ be a ML estimator for $\boldsymbol{\theta}$ that, under H_0, is known to be approximately normally distributed in large samples with asymptotic mean $\boldsymbol{\theta}_0$ and

[58] k_α is obtained from the tables for χ^2_q. Since χ^2 is only approximately distributed in the χ^2_q form, the significance level is approximately α.

large sample approximate covariance matrix $\tilde{V} = I_{\hat{\theta}}^{-1}$, where $I_{\hat{\theta}}$ is Fisher's information matrix evaluated at the ML estimate $\hat{\theta}$, then under H_0, the test statistic, $\tilde{W} = (\hat{\theta} - \theta_0)'\hat{V}^{-1}(\hat{\theta} - \theta_0)$, has an approximate large sample χ_q^2 pdf, where q is the number of restrictions implied by H_0.[59] Then a rejection region is $\tilde{W} \geqslant k_\alpha$, where k_α is determined such that $P(\tilde{W} \geqslant k_\alpha) \doteq \alpha$, the given significance level, where k_α is obtained from tables of the χ_q^2 pdf.

In the LM large sample approach to testing restrictions on θ, for example H_0: $\theta_1 = \theta_{10}$ against H_1: $\theta_1 \neq \theta_{10}$ with θ_2 unrestricted under both hypotheses, where $\theta' = (\theta_1', \theta_2')$, use is made of the fact that if H_0 is true, then the restricted and unrestricted estimates of θ will be very close to each other in large samples. If the log LF is regular, then the partial derivatives of this function at the restricted maximizing values will tend to be small. On the other hand, if H_0 is false these partial derivatives will not in general be small. Let $\hat{\theta}_r' = (\theta_{10}, \hat{\theta}_2)$ be the ML estimate for the restricted log $\mathrm{LF}_r = \log p(\boldsymbol{x}|\theta_{10}, \theta_2)$. Then it can be shown that

$$(\partial \log \mathbf{LR}/\partial \theta)'_{\hat{\theta}_r} I_{\hat{\theta}_r}^{-1} (\partial \log \mathbf{LR}/\partial \theta)_{\hat{\theta}_r},$$

where the partial derivatives are evaluated at $\hat{\theta}_r$, the restricted ML estimate, and $I_{\hat{\theta}_r}$ is the information matrix evaluated at $\hat{\theta}_r$ has an approximate χ_q^2 pdf in large samples under H_0 and regularity conditions, and this fact can be employed to construct an approximate α-level test of H_0. The LM test requires just the computation of the restricted ML estimate, $\hat{\theta}_r$, and is thus occasionally much less computationally burdensome than the LR and W tests that require the unrestricted ML estimate. On the other hand, it seems important in applications to view and study both the unrestricted and restricted estimates.

Finally, it is the case that for a given pair of hypotheses, the LR, W, and LM test statistics have the same large sample χ^2 pdf so that in large samples there are no grounds for preferring any one. In small samples, however, their properties are somewhat different and in fact use of large sample test results based on them can give conflicting results [see, for example, Berndt and Savin (1975)]. Fortunately, research is in progress on this problem. Some approximations to the finite sample distributions of these large sample test statistics have been obtained that appear useful; see, for example, Box (1949) and Lawley (1956). Also, Edgeworth expansion techniques to approximate distributions of various test statistics are currently being investigated by several researchers.

6.3. *Bayesian analysis of hypotheses*

Bayesian procedures are available for analyzing various types of hypotheses that yield posterior probabilities and posterior odds ratios associated with alternative

[59]Any consistent, asymptotically efficient estimator can be employed in place of the ML estimator.

hypotheses which incorporate both sample and prior information. Further, given a loss structure, it is possible to choose between or among hypotheses in such a manner so as to minimize expected loss. These procedures, which are discussed in Jeffreys (1967), DeGroot (1970), Leamer (1978), Bernardo et al. (1980), and Zellner (1971), are briefly reviewed below.

With respect to hypotheses relating to a scalar parameter θ, $-\infty<\theta<\infty$, of the form H_1: $\theta>c$ and H_2: $\theta<c$, where c has a given value, e.g. $c=0$, assume that a posterior pdf for θ, $p(\theta|D)$, is available, where D denotes the sample and prior information. Then in what has been called the Laplacian Approach, the posterior probabilities relating to H_1 and to H_2 are given by $\Pr(\theta>c|D)=\int_c^\infty p(\theta|D)\mathrm{d}\theta$ and $\Pr(\theta<c|D)=\int_{-\infty}^c p(\theta|D)\mathrm{d}\theta$, respectively. The posterior odds ratio for H_1 and H_2, denoted by K_{12}, is then $K_{12}=\Pr(\theta>c|D)/\Pr(\theta<c|D)$. Other hypotheses, e.g. $|\theta|<1$ and $|\theta|>1$, can be appraised in a similar fashion. That is, $\Pr(|\theta|<1|D)=\int_{-1}^1 p(\theta|D)\mathrm{d}\theta$ is the posterior probability that $|\theta|<1$ and $1-\Pr(|\theta|<1|D)$ is the posterior probability that $|\theta|>1$.

Example 6.3

Let y_i, $i=1,2,\dots,n$, be independent observations from a normal distribution with mean θ and unit variance. If a diffuse prior for θ, $p(\theta)\propto$ const., is employed, the posterior pdf is $p(\theta|D)\propto\exp\{-n(\theta-\bar{y})^2/2\}$, where $\bar{y}$ is the sample mean; that is, $z=\sqrt{n}(\theta-\bar{y})$ has a $N(0,1)$ posterior pdf. Then for the hypothesis $\theta>0$, $\Pr(\theta>0|D)=\Pr(z>-\sqrt{n}\bar{y}|D)=1-\Phi(-\sqrt{n}\bar{y})$, where $\Phi(\cdot)$ is the cumulative normal pdf. Thus, $\Pr(\theta>0|D)$ can be evaluated from tables $\Phi(\cdot)$.

When a vector of parameters, $\boldsymbol{\theta}$, with $\boldsymbol{\theta}\subset\Theta$ and posterior pdf $p(\boldsymbol{\theta}|D)$, is considered, and the hypotheses are H_A: $\boldsymbol{\theta}\subset\omega$ and H_B: $\boldsymbol{\theta}\subset\Theta-\omega$, where ω is a subspace of Θ, $\Pr(\boldsymbol{\theta}\subset\omega|D)=\int_\omega p(\boldsymbol{\theta}|D)\mathrm{d}\boldsymbol{\theta}$ is the posterior probability associated with H_A while $1-\Pr(\boldsymbol{\theta}\subset\omega|D)$ is that associated with H_B and the posterior odds ratio is the ratio of these posterior probabilities. The above posterior probabilities can be evaluated either analytically, or by the use of tabled values of integrals or by numerical integration. For an example of this type of analysis applied to hypotheses about properties of a second order autoregression, see Zellner (1971, p. 194 ff.).

For a very wide range of different types of hypotheses, the following Jeffreys Approach, based on Bayes Theorem, can be employed in analyzing alternative hypotheses or models for observations. Let $p(y,H)$ be the joint distribution for the data y and an indicator variable H. Then $p(y,H)=p(H)p(y|H)=p(y)p(H|y)$ and $p(H|y)=p(H)p(y|H)/p(y)$. If H can assume values H_1 and H_2, it follows that the posterior odds ratio, K_{12}, is

$$K_{12}=\frac{p(H_1|y)}{p(H_2|y)}=\frac{p(H_1)}{p(H_2)}\cdot\frac{p(y|H_1)}{p(y|H_2)},\tag{6.2}$$

where $p(H_i)$ is the prior probability assigned to H_i, $i=1,2$, $p(H_1)/p(H_2)$ is the prior odds ratio for H_1 versus H_2, and $p(y|H_i)$ is the marginal pdf for y under hypothesis H_i, $i=1,2$. The ratio $p(y|H_1)/p(y|H_2)$ is called the Bayes Factor (BF). In the case that both H_1 and H_2 are simple hypotheses, the BF $p(y|H_1)/p(y|H_2)$ is just the Likelihood Ratio (LR).

Example 6.4

Let $y_i = \theta + \varepsilon_i$, $i=1,2,\dots,n$, with the ε_i's assumed independently drawn from a normal distribution with zero mean and unit variance. Consider two simple hypotheses, H_1: $\theta=0$ and H_2: $\theta=1$, with prior probabilities $p(H_1)=1/2$ and $p(H_2)=1/2$. Then from (6.2),

$$\begin{aligned} K_{12} &= \left(\tfrac{1}{2} \div \tfrac{1}{2}\right) p(y|\theta=0)/p(y|\theta=1) \\ &= \exp\{-y'y/2\}/\exp\{-(y-\iota)'(y-\iota)/2\}, = \exp\{2n\left(\tfrac{1}{2}-\bar{y}\right)\}, \end{aligned}$$

where $\boldsymbol{y}' = (y_1, y_2, \dots, y_n)$, $\boldsymbol{\iota}' = (1,1,\dots,1)$, and $\bar{y}$ is the sample mean. In this case $K_{12} = \text{LR}$ and its value is determined by the value of $2n(\frac{1}{2}-\bar{y})$.

In cases in which non-simple hypotheses are considered, that is, hypotheses that do not involve assigning values to all parameters of a pdf for the data $\boldsymbol{y}$, $p(y|\boldsymbol{\theta}_i, H_i)$, given that a prior pdf for $\boldsymbol{\theta}_i$ is available, $p(\boldsymbol{\theta}_i|H_i)$, it follows that $p(y|H_i) = \int p(y|\boldsymbol{\theta}_i, H_i)p(\boldsymbol{\theta}_i|H_i)\mathrm{d}\boldsymbol{\theta}_i$ and in such cases (6.2) becomes

$$K_{12} = \frac{p(H_1)}{p(H_2)} \cdot \frac{\int p(y|\boldsymbol{\theta}_1, H_1)p(\boldsymbol{\theta}_1|H_1)\mathrm{d}\boldsymbol{\theta}_1}{\int p(y|\boldsymbol{\theta}_2, H_2)p(\boldsymbol{\theta}_2|H_2)\mathrm{d}\boldsymbol{\theta}_2}. \tag{6.3}$$

Thus, in this case, K_{12} is equal to the prior odds ratio, $p(H_1)/p(H_2)$ times a BF that is a ratio of averaged likelihood functions. For discussion and applications of (6.3) to a variety of problems, see, for example, Jeffreys (1967), DeGroot (1970), Leamer (1978), and Zellner (1971).

In (6.2) and (6.3) it is seen that a prior odds ratio gets transformed into a posterior odds ratio. If a loss structure is available, it is possible to choose between or among hypotheses so as to minimize expected loss. To illustrate, consider two mutually exclusive and exhaustive hypotheses, H_1 and H_2, with posterior odds ratio $K_{12} = p_1/(1-p_1)$, where p_1 is the posterior probability for H_1 and $1-p_1$ is the posterior probability for H_2. Suppose that the following two-action, two-state loss structure is relevant:

		State of world	
		H_1	H_2
Acts	A_1: Choose H_1	0	L_{12}
	A_2: Choose H_2	L_{21}	0

The two "states of the world" are: H_1 is in accord with the data or H_2 is in accord with the data; while the two possible actions are: choose H_1 and choose H_2. $L_{12} > 0$ and $L_{21} > 0$ are losses associated with incorrect actions. Then using the posterior probabilities, p_1 and $1 - p_1$, posterior expected losses associated with A_1 and A_2 are:

$$E(L|A_1) = (1 - p_1)L_{12} \quad \text{and} \quad E(L|A_2) = p_1 L_{21},$$

and thus

$$\begin{aligned} \frac{E(L|A_2)}{E(L|A_1)} &= \frac{p_1}{1 - p_1} \frac{L_{21}}{L_{12}} \\ &= K_{12}\left(\frac{L_{21}}{L_{12}}\right). \end{aligned} \tag{6.4}$$

If this ratio of expected losses is larger than one, choosing A_1 minimizes expected loss, while if it is less than one, choosing A_2 leads to minimal expected loss. Note from the second line of (6.4) that both K_{12} and L_{21}/L_{12} affect the decision. In the very special case $L_{21}/L_{12} = 1$, the symmetric loss structure, if $K_{12} > 1$ choose H_1, while if $K_{12} < 1$ choose H_2. The analysis can be generalized to apply to more than two hypotheses. Also, there are intriguing relations between the results provided by the Bayesian approach and sampling theory approaches to testing hypotheses that are discussed in the references cited above.

Finally, given the posterior probabilities associated with the hypotheses, it is possible to use them not only in testing but also in estimation and prediction. That is, if two hypotheses are H_1: $\boldsymbol{\theta} = \boldsymbol{c}$ and H_2: $\boldsymbol{\theta} \neq \boldsymbol{c}$, where $\boldsymbol{c}$ has a given value and p_1 and $1 - p_1$ are the posterior probabilities associated with H_1 and H_2, respectively, then relative to quadratic loss, an optimal estimate is

$$\tilde{\boldsymbol{\theta}} = p_1 \boldsymbol{c} + (1 - p_1)\bar{\boldsymbol{\theta}} = \boldsymbol{c} + (1 - p_1)(\bar{\boldsymbol{\theta}} - \boldsymbol{c}) = \boldsymbol{c} + \frac{1}{1 + K_{12}}(\bar{\boldsymbol{\theta}} - \boldsymbol{c}),$$

where $\bar{\boldsymbol{\theta}}$ is the posterior mean of $\boldsymbol{\theta}$ under H_2 and $K_{12} = p_1/(1 - p_1)$, the posterior odds ratio [see Zellner and Vandaele (1975) for details]. Also, in Geisel (1975) posterior probabilities associated with two different models for the same set of observations are employed to average their predictions and thus to obtain an optimal prediction relative to quadratic loss.

7. Summary and concluding remarks

Research in statistical theory has yielded very useful procedures for learning from data, one of the principal objectives of econometrics and science. In addition, this research has produced a large number of probability models for observations that are widely utilized in econometrics and other sciences. Some of them were reviewed above. Also, techniques for estimation, prediction, and testing were reviewed that enable investigators to solve inference problems in a scientific manner. The importance of utilizing sound, scientific methods in analyzing data and drawing conclusions from them is obvious since such conclusions often have crucial implications for economic policy-making and the progress of economic science. On the other hand, it is a fact that statistical and econometric analysis frequently is a mixture of science and art. In particular, the formulation of appropriate theories and models is largely an art. A challenge for statistical theory is to provide fruitful, formal procedures that are helpful in solving model formulation problems.

While many topics were discussed in this chapter, it is necessary to point to some that were not. These include non-parametric statistics, survey methodology, design of experiments, time series analysis, random parameter models, statistical control theory, sequential and simultaneous testing procedures, empirical Bayes procedures, and fiducial and structural theories of inference. Some of these topics are treated in other parts of this Handbook. Also, readers may refer to Kruskal and Tanur (1978) for good discussions of these topics that provide references to the statistical literature. The annual issues of the ASA/IMS *Current Index to Statistics* are a very useful guide to the current statistical literature.

In the course of this chapter a number of controversial issues were mentioned that deserve further thought and study. First, there is the issue of which concept of probability is most fruitful in econometric work. This is a critical issue since probability statements play a central role in econometric analyses.

Second, there are major controversies concerning the most appropriate approach to statistical inference to employ in econometrics. The two major approaches to statistical inference discussed in this chapter are the sampling theory approach and the Bayesian approach. Examples illustrating both approaches were presented. For further discussion of the issues involved see, for example, Barnett (1973), Bernardo et al. (1980), Cox and Hinkley (1974), Lindley (1971), Rothenberg (1975), Zellner (1975), and the references in these works.

Third, with respect to both sampling theory and Bayesian approaches, while there are many problems for which both approaches yield similar solutions, there are some problems for which solutions differ markedly. Further attention to such problems, some of which are discussed in Bernardo et al. (1980), Cox and Hinkley (1974), Jaynes (1980), Lindley (1971), and the references cited in these works, would be worthwhile.

Fourth, there is controversy regarding the implications of the likelihood principle for econometric and statistical practice. Briefly, the likelihood principle states that if $\boldsymbol{x}$ and $\boldsymbol{y}$ are two data sets such that $p(\boldsymbol{x}|\boldsymbol{\theta}) = cf(\boldsymbol{y}|\boldsymbol{\theta})$, with $\boldsymbol{\theta} \subset \Theta$ and c not depending on $\boldsymbol{\theta}$, then inferences and decisions based on $\boldsymbol{x}$ and on $\boldsymbol{y}$ should be identical. The Bayesian approach satisfies this condition since for a given prior pdf, $\pi(\boldsymbol{\theta})$, the posterior pdfs for $\boldsymbol{\theta}$ based on $p(\boldsymbol{x}|\boldsymbol{\theta})$ and on $cf(\boldsymbol{y}|\boldsymbol{\theta})$ are identical given $p(x|\theta) = cf(y|\theta)$. On the other hand, sampling theory properties and procedures that involve integrations over the sample space, as in the case of unbiasedness, MVU estimation, confidence intervals, and tests of significance violate the likelihood principle. Discussions of this range of issues are provided in Cox and Hinkley (1974, ch. 2) and Lindley (1971, p. 10 ff.) with references to important work by Birnbaum, Barnard, Durbin, Savage, and others.

Fifth, the importance of Bayesian logical consistency and coherence is emphasized by most Bayesians but is disputed by some who argue that these concepts fail to capture all aspects of the art of data analysis. Essentially, what is being criticized here is the Bayesian learning model and/or the precept, "act so as to maximize expected utility (or equivalently minimize expected loss)". If improvements can be made to Bayesian and other learning and decision procedures, they would constitute major research contributions.

Sixth, some object to the introduction of prior distributions in statistical analyses and point to the difficulty in formulating prior distributions in multiparameter problems. Bayesians point to the fact that non-Bayesians utilize prior information informally in assessing the "reasonableness" of estimation results, choosing significance levels, etc. and assert that formal, careful use of prior information provides more satisfactory results in estimation, prediction, and testing.

Seventh, frequentists assert that statistical procedures are to be assessed in terms of their behavior in hypothetical repetitions under the same conditions. Others dispute this assertion by stating that statistical procedures must be justified in terms of the actually observed data and not in terms of hypothetical, fictitious repetitions. This range of issues is very relevant for analyses of non-experimental data, for example macro-economic data.

The above controversial points are just some of the issues that arise in judging alternative approaches to inference in econometrics and statistics. Furthermore, Good has suggested in a 1980 address at the University of Chicago and in Good and Crook (1974) some elements of a Bayes/non-Bayes synthesis that he expects to see emerge in the future. In a somewhat different suggested synthesis, Box (1980) proposes Bayesian estimation procedures for parameters of given models and a form of sampling theory testing procedures for assessing the adequacy of models. While these proposals for syntheses of different approaches to statistical inference are still being debated, they do point toward possible major innovations in statistical theory and practice that will probably be of great value in econometric analyses.

References

Aitchison, J. and I. R. Dunsmore (1975) *Statistical Prediction Analysis*. Cambridge: Cambridge University Press.

Anderson, T. W. (1971) *The Statistical Analysis of Time Series*. New York: John Wiley & Sons, Inc.

Anscombe, F. J. (1961) "Bayesian Statistics", *American Statistician*, 15, 21–24.
Arnold, S. F. (1981) *The Theory of Linear Models and Multivariate Analysis*. New York: John Wiley & Sons, Inc.
Barnett, V. D. (1973) *Comparative Statistical Inference*. New York: John Wiley & Sons, Inc.
Barnett, V. and T. Lewis (1978) *Outliers in Statistical Data*. New York: John Wiley & Sons, Inc.
Bayes, T. (1763) "An Essay Toward Solving a Problem in the Doctrine of Chances", *Philosophical Transactions of the Royal Society (London)*, 53, 370–418; reprinted in *Biometrika*, 45 (1958), 293–315.
Belsley, D. A., E. Kuh and R. E. Welsch (1980) *Regression Diagnostics*. New York: John Wiley & Sons, Inc.
Bernardo, J. M., M. H. DeGroot, D. V. Lindley and A.F.M. Smith (eds.) (1980) *Bayesian Statistics*. Valencia, Spain: University Press.
Berndt, E. and N. E. Savin (1977) "Conflict Among Criteria for Testing Hypotheses in the Multivariate Linear Regression Model", *Econometrica*, 45, 1263–1272.
Blackwell, D. (1947) "Conditional Expectation and Unbiased Sequential Estimation," *Annals of Mathematical Statistics*, 18, 105–110.
Blackwell, D. and M. A. Girschick (1954) *Theory of Games and Statistical Decisions*. New York: John Wiley & Sons, Inc.
Box, G. E. P. (1949) "A General Distribution Theory for a Class of Likelihood Criteria", *Biometrika*, 36, 317–346.
Box, G. E. P. (1976) "Science and Statistics", *Journal of the American Statistical Association*, 71, 791–799.
Box, G. E. P. (1980) "Sampling and Bayes' Inference in Scientific Modelling and Robustness", *Journal of the Royal Statistical Association A*, 143, 383–404.
Box, G. E. P. and D. R. Cox (1964) "An Analysis of Transformations", *Journal of the Royal Statistical Association B*, 26, 211–243.
Box, G. E. P. and G. C. Tiao (1973) *Bayesian Inference in Statistical Analysis*. Reading, Mass.: Addison-Wesley Publishing Co.
Brown, L. (1966) "On the Admissibility of Estimators of One or More Location Parameters", *Annals of Mathematical Statistics*, 37, 1087–1136.
Christ, C. (1966) *Econometric Models and Methods*. New York: John Wiley & Sons, Inc.
Copson, E. T. (1965) *Asymptotic Expansions*. Cambridge: Cambridge University Press.
Cox, D. R. and D. V. Hinkley (1974) *Theoretical Statistics*. London: Chapman and Hall.
Cox, D. R. and P. A. W. Lewis (1966) *The Statistical Analysis of Series of Events*. London: Methuen.
Cramér, H. (1946) *Mathematical Methods of Statistics*. Princeton: Princeton University Press.
DeGroot, M. H. (1970) *Optimal Statistical Decisions*. New York: McGraw-Hill Book Co.
DeRobertis, L. (1978) "The Use of Partial Prior Knowledge in Bayesian Inference", unpublished doctoral dissertation, Department of Statistics, Yale University.
Efron, B. and C. Morris (1975) "Data Analysis Using Stein's Estimator and Its Generalizations", *Journal of the American Statistical Association*, 70, 311–319.
Feigl, H. (1953) "Notes on Causality", in: H. Feigl and M. Brodbeck (eds.), *Readings in the Philosophy of Science*. New York: Appleton-Century-Crofts, Inc., pp. 408–418.
Feldstein, M. S. (1971) "The Error of Forecast in Econometric Models when the Forecast-Period Exogenous Variables are Stochastic", *Econometrica*, 39, 55–60.
Ferguson, T. S. (1967) *Mathematical Statistics: A Decision Theory Approach*. New York: Academic Press, Inc.
Fisher, R. A. (1959) *Statistical Methods and Scientific Inference* (2nd edn.). New York: Hafner Publishing Co.
Fomby, T. B. and D. K. Guilkey (1978) "On Choosing the Optimal Level of Significance for the Durbin–Watson Test and the Bayesian Alternative", *Journal of Econometrics*, 8, 203–214.
Friedman, M. (1957) *A Theory of the Consumption Function*. Princeton: Princeton University Press.
Geisel, M. S. (1975) "Bayesian Comparisons of Simple Macroeconomic Models", in: S. E. Fienberg and A. Zellner (eds.), *Studies in Bayesian Econometrics and Statistics in Honor of Leonard J. Savage*. Amsterdam: North-Holland Publishing Co., pp. 227–256.
Geisser, S. (1980) "A Predictivistic Primer", in: A. Zellner (ed.), *Bayesian Analysis in Econometrics and Statistics: Essays in Honor of Harold Jeffreys*. Amsterdam: North-Holland Publishing Co., pp. 363–381.

Goldfeld, S. M. and R. E. Quandt (1972) *Nonlinear Methods in Econometrics*. Amsterdam: North-Holland Publishing Co.
Good, I. J. and J. F. Crook (1974) "The Bayes/Non-Bayes Compromise and the Multinomial Distribution", *Journal of the American Statistical Association*, 69, 711–720.
Granger, C. W. J. and P. Newbold (1977) *Forecasting Economic Time Series*. New York: Academic Press, Inc.
Guttman, I. (1970) *Statistical Tolerance Regions: Classical and Bayesian*. London: Charles Griffen & Co., Ltd.
Hartigan, J. (1964) "Invariant Prior Distributions," *Annals of Mathematical Statistics*, 35, 836–845.
Heyde, C. C. and I. M. Johnstone (1979) "On Asymptotic Posterior Normality for Stochastic Processes", *Journal of the Royal Statistical Association B*, 41, 184–189.
Hill, B. M. (1975) "On Coherence, Inadmissibility and Inference about Many Parameters in the Theory of Least Squares", in: S. E. Fienberg and A. Zellner (eds.), *Studies in Bayesian Econometrics and Statistics*. Amsterdam: North-Holland Publishing Co., pp. 555–584.
Huber, P. J. (1964) "Robust Estimation of a Location Parameter," *Annals of Mathematical Statistics*, 35, 73–101.
Huber, P. J. (1972) "Robust Statistics: A Review", *Annals of Mathematical Statistics*, 43, 1041–1067.
James, W. and C. Stein (1961) "Estimation with Quadratic Loss", in: *Proceedings of the Fourth Berkeley Symposium on Mathematical Statistics and Probability Theory*, vol. I. Berkeley: University of California Press, pp. 361–397.
Jaynes, E. T. (1968) "Prior Probabilities", *IEEE Transactions on Systems Science and Cybernetics*, SSC-4, 227–241.
Jaynes, E. T. (1974) "Probability Theory", manuscript. St. Louis: Department of Physics, Washington University.
Jaynes, E. T. (1980) "Marginalization and Prior Probabilities", in: A. Zellner (ed.), *Bayesian Analysis in Econometrics and Statistics: Essays in Honor of Harold Jeffreys*. Amsterdam: North-Holland Publishing Co., pp. 43–78.
Jeffreys, H. (1967) *Theory of Probability* (3rd rev. edn.; 1st ed. 1939). London: Oxford University Press.
Jeffreys, H. (1973) *Scientific Inference* (3rd edn.). Cambridge: Cambridge University Press.
Johnson, N. L. and S. Kotz (1969) *Discrete Distributions*. Boston: Houghton Mifflin Publishing Co.
Johnson, N. L. and S. Kotz (1970) *Continuous Univariate Distributions*, vols. 1 and 2. New York: John Wiley & Sons, Inc.
Judge, G. G. and M. E. Bock (1978) *The Statistical Implications of Pre-Test and Stein-Rule Estimators in Econometrics*. Amsterdam: North-Holland Publishing Co.
Kadane, J. B., J. M. Dickey, R. L. Winkler, W. S. Smith and S. C. Peters (1980) "Interactive Elicitation of Opinion for a Normal Linear Model", *Journal of the American Statistical Association*, 75, 845–854.
Kendall, M. G. and A. Stuart (1958) *The Advanced Theory of Statistics*, vol. 1. London: C. Griffen & Co., Ltd.
Kendall, M. G. and A. Stuart (1961) *The Advanced Theory of Statistics*, vol. 2. London: C. Griffen & Co., Ltd.
Kenney, J. F. and E. S. Keeping (1951) *Mathematics of Statistics* (*Part Two*). New York: D. Van Nostrand Company, Inc.
Keynes, J. M. (1921) *Treatise on Probability*. London: Macmillan and Co., Ltd.
Koopman, B. O. (1936) "On Distributions Admitting a Sufficient Statistic", *Transactions of the American Mathematical Society*, 39, 399–409.
Kruskal, W. J. and J. M. Tanur (eds.) (1978), *International Encyclopedia of Statistics*, vols. 1 and 2. New York: The Free Press (Division of Macmillan Publishing Co., Inc.).
Lawley, D. N. (1956) "A General Method for Approximating to the Distribution of Likelihood Ratio Criteria", *Biometrika*, 43, 295–303.
Leamer, E. E. (1978) *Specification Searches*. New York: John Wiley & Sons, Inc.
Lehmann, E. L. (1959) *Testing Statistical Hypotheses*. New York: John Wiley & Sons, Inc.
Lindley, D. V. (1962) "Discussion on Professor Stein's Paper", *Journal of the Royal Statistical Association B*, 24, 285–287.
Lindley, D. V. (1965) *Introduction to Probability and Statistics from a Bayesian Viewpoint. Part 2: Inference*. Cambridge: Cambridge University Press.

Lindley, D. V. (1971) *Bayesian Statistics, A Review*. Philadelphia: Society for Industrial and Applied Mathematics.
Litterman, R. (1980) "A Bayesian Procedure for Forecasting with Vector Autoregressions", manuscript. Department of Economics, MIT; to appear in *Journal of Econometrics*.
Loève, M. (1963) *Probability Theory* (3rd edn.). Princeton: D. Van Nostrand Co., Inc.
Luce, R. D. and H. Raiffa (1957) *Games and Decisions*. New York: John Wiley & Sons, Inc.
Mehta, J. S. and P. A. V. B. Swamy (1976) "Further Evidence on the Relative Efficiencies of Zellner's Seemingly Unrelated Regressions Estimators", *Journal of the American Statistical Association*, 71, 634–639.
Neyman, J. and E. L. Scott (1948) "Consistent Estimates Based on Partially Consistent Observations", *Econometrica*, 16, 1–32.
Pfanzagl, J. and W. Wefelmeyer (1978) "A Third Order Optimum Property of the Maximum Likelihood Estimator", *Journal of Multivariate Analysis*, 8, 1–29.
Phillips, P. C. B. (1977a) "A General Theorem in the Theory of Asymptotic Expansions for Approximations to the Finite Sample Distribution of Econometric Estimators", *Econometrica*, 45, 1517–1534.
Phillips, P. C. B. (1977b) "An Approximation to the Finite Sample Distribution of Zellner's Seemingly Unrelated Regression Estimator", *Journal of Econometrics*, 6, 147–164.
Pitman, E. J. G. (1936), "Sufficient Statistics and Intrinsic Accuracy", *Proceedings of the Cambridge Philosophical Society*, 32, 567–579.
Pratt, J. W. (1961) "Length of Confidence Intervals", *Journal of the American Statistical Association*, 56, 549–567.
Pratt, J. W. (1965) "Bayesian Interpretation of Standard Inference Statements", *Journal of the Royal Statistical Association B*, 27, 169–203.
Pratt, J. W., H. Raiffa and R. Schlaifer (1964) "The Foundations of Decision Under Uncertainty: An Elementary Exposition", *Journal of the American Statistical Association*, 59, 353–375.
Raiffa, H. and R. Schlaifer (1961) *Applied Statistical Decision Theory*. Boston: Graduate School of Business Administration, Harvard University.
Ramsey, F. P. (1931) *The Foundations of Mathematics and Other Essays*. London: Kegan, Paul, Trench, Truber & Co., Ltd.
Rao, C. R. (1945) "Information and Accuracy Attainable in Estimation of Statistical Parameters", *Bulletin of the Calcutta Mathematical Society*, 37, 81–91.
Rao, C. R. (1973) *Linear Statistical Inference and Its Applications*. New York: John Wiley & Sons, Inc.
Rao, P. and Z. Griliches (1969) "Small-Sample Properties of Two-Stage Regression Methods in the Context of Auto-correlated Errors", *Journal of the American Statistical Association*, 64, 253–272.
Rényi, A. (1970) *Foundations of Probability*. San Francisco: Holden-Day, Inc.
Revankar, N. S. (1974) "Some Finite Sample Results in the Context of Two Seemingly Unrelated Regression Equations", *Journal of the American Statistical Association*, 69, 187–190.
Rothenberg, T. J. (1975) "The Bayesian Approach and Alternatives", in: S. E. Fienberg and A. Zellner (eds.), *Studies in Bayesian Econometrics and Statistics*. Amsterdam: North-Holland Publishing Co., pp. 55–67.
Savage, L. J. (1954) *The Foundations of Statistics*. New York: John Wiley & Sons, Inc.
Savage, L. J. (1961) "The Subjective Basis of Statistical Practice", manuscript. University of Michigan, Ann Arbor.
Savage, L. J., et al. (1962) *The Foundations of Statistical Inference*. London: Meuthen.
Savage, L. J., N. Edwards and H. Lindman (1963) "Bayesian Statistical Inference for Psychological Research", *Psychological Review*, 70, 193–242.
Sclove, S. L. (1968) "Improved Estimators for Coefficients in Linear Regression", *Journal of the American Statistical Association*, 63, 596–606.
Silvey, S. D. (1970) *Statistical Inference*. Baltimore, Md.: Penguin Books.
Srivastava, V. K. and T. D. Dwivedi (1979) "Estimation of Seemingly Unrelated Regression Equations: A Brief Survey", *Journal of Econometrics*, 10, 15–32.
Stein, C. (1956) "Inadmissibility of the Usual Estimator for the Mean of a Multivariate Normal Distribution", in: *Proceedings of the Third Berkeley Symposium on Mathematical Statistics and Probability*, vol. I. Berkeley: University of California Press, pp. 197–206.
Stein, C. (1960) "Multiple Regression", in: I. Olkin (ed.), *Contributions to Probability and Statistics: Essays in Honour of Harold Hotelling*. Stanford: Stanford University Press.

Takeuchi, K. (1978) "Asymptotic Higher Order Efficiency of ML Estimators of Parameters in Linear Simultaneous Equations", paper presented at the Kyoto Econometrics Seminar Meeting, University of Kyoto, 27–30 June.

Taylor, W. E. (1978) "The Heteroscedastic Linear Model: Exact Finite Sample Results", *Econometrica*, 46, 663–675.

Thornber, E. H. (1967) "Finite Sample Monte Carlo Studies: An Autoregressive Illustration", *Journal of the American Statistical Association*, 62, 801–818.

Tukey, J. W. (1957) "On the Comparative Anatomy of Transformations", *Annals of Mathematical Statistics*, 28, 602–632.

Tukey, J. W. (1977) *Exploratory Data Analysis*. Reading, Mass.: Addison-Wesley Publishing Co.

Uspensky, J. V. (1937) *Introduction to Mathematical Probability*. New York: McGraw-Hill Book Co., Inc.

Varian, H. R. (1975) "A Bayesian Approach to Real Estate Assessment", in: S. E. Fienberg and A. Zellner (eds.), *Studies in Bayesian Econometrics and Statistics in Honor of Leonard J. Savage*. Amsterdam: North-Holland Publishing Co., pp. 195–208.

Wald, A. (1949) "Note on the Consistency of Maximum Likelihood Estimate", *Annals of Mathematical Statistics*, 20, 595–601.

Widder, D. V. (1961) *Advanced Calculus* (2nd edn.). New York: Prentice-Hall, Inc.

Winkler, R. L. (1980) "Prior Information, Predictive Distributions and Bayesian Model-Building", in: A. Zellner (ed.), *Bayesian Analysis in Econometrics and Statistics: Essays in Honor of Harold Jeffreys*. Amsterdam: North-Holland Publishing Co., pp. 95–109.

Zaman, A. (1981) "Estimators Without Moments: The Case of the Reciprocal of a Normal Mean", *Journal of Econometrics*, 15, 289–298.

Zellner, A. (1971) *An Introduction to Bayesian Inference in Econometrics*. New York: John Wiley & Sons, Inc.

Zellner, A. (1972) "On Assessing Informative Prior Distributions for Regression Coefficients", manuscript. H. G. B. Alexander Research Foundation, Graduate School of Business, University of Chicago.

Zellner, A. (1973) "The Quality of Quantitative Economic Policy-making When Targets and Costs of Change are Misspecified", in: W. Sellekaerts (ed.), *Selected Readings in Econometrics and Economic Theory: Essays in Honor of Jan Tinbergen*, Part II. London: Macmillan Publishing Co., pp. 147–164.

Zellner, A. (1975) "Time Series Analysis and Econometric Model Construction", in: R. P. Gupta (ed.), *Applied Statistics*. Amsterdam: North-Holland Publishing Co., pp. 373–398.

Zellner, A. (1976) "Bayesian and Non-Bayesian Analysis of the Regression Model with Multivariate Student-*t* Error Terms", *Journal of the American Statistical Association*, 71, 400–405.

Zellner, A. (1978) "Estimation of Functions of Population Means and Regression Coefficients Including Structural Coefficients: A Minimum Expected Loss (MELO) Approach", *Journal of Econometrics*, 8, 127–158.

Zellner, A. (1979) "Statistical Analysis of Econometric Models", *Journal of the American Statistical Association*, 74, 628–651.

Zellner, A. (1980) "On Bayesian Regression Analysis with *g*-Prior Distributions", paper presented at the Econometric Society Meeting, Denver, Colorado.

Zellner, A. and M. S. Geisel (1968) "Sensitivity of Control to Uncertainty and Form of the Criterion Function", in: D. G. Watts (ed.), *The Future of Statistics*. New York: Academic Press, Inc., pp. 269–289.

Zellner, A. and S. B. Park (1979) "Minimum Expected Loss (MELO) Estimators for Functions of Parameters and Structural Coefficients of Econometric Models", *Journal of the American Statistical Association*, 74, 185–193.

Zellner, A. and N. S. Revankar (1969) "Generalized Production Functions", *Review of Economic Studies*, 36, 241–250.

Zellner, A. and W. Vandaele (1975) "Bayes–Stein Estimators for *k*-Means, Regression and Simultaneous Equation Models", in: S. E. Fienberg and A. Zellner (eds.), *Studies in Bayesian Econometrics and Statistics*. Amsterdam: North-Holland Publishing Co., pp. 627–653.

PART 2

ECONOMETRIC MODELS

Chapter 3

ECONOMIC AND ECONOMETRIC MODELS

MICHAEL D. INTRILIGATOR*

University of California, Los Angeles

Contents

*The author acknowledges, with appreciation, the helpful suggestions of Ray Fair, Gary Fromm, Zvi Griliches, Elizabeth Landaw, and Henri Theil.

Handbook of Econometrics, Volume I, Edited by Z. Griliches and M.D. Intriligator

1. Introduction and overview

This chapter of the Handbook will present a discussion of models, particularly models used in econometrics.[1] Models play a major role in all econometric studies, whether theoretical or applied. Indeed, defining *econometrics* as the branch of economics concerned with the empirical estimation of economic relationships, models, together with data, represent the basic ingredients of any econometric study. Typically, the theory of the phenomena under investigation is developed into a model which is further refined into an econometric model. This model is then estimated on the basis of data pertaining to the phenomena under investigation using econometric techniques. The estimated model can then be used for various purposes, including structural analysis, forecasting, and policy evaluation.

This chapter provides a discussion of models and economic models in Section 2, and comparative statics in Section 3. Section 4 then presents econometric models, including the structural form, reduced form, and final form. The problem of identification, which is presented in more detail in Chapter 4 of this Handbook, by Cheng Hsiao, is discussed in Section 5. Section 6 provides some examples of specific models, including demand (discussed in more detail in Chapter 30 of this Handbook by Angus Deaton), production (discussed in more detail in Chapter 31 of this Handbook by Dale Jorgenson), macroeconometric models (also discussed in Chapters 33, 34, and 35 of the Handbook by Ray Fair, John Taylor, and Lawrence Klein, respectively), and other econometric models. Section 7 presents a discussion of the uses of econometric models, specifically structural analysis, forecasting (further discussed in Chapter 33 of this Handbook by Ray Fair), and policy evaluation (further discussed in Chapters 34 and 35 of this Handbook by John Taylor and Lawrence Klein, respectively). Section 8 presents a conclusion.

2. Models and economic models

A *model* is a simplified representation of an actual phenomenon, such as an actual system or process. The actual phenomenon is represented by the model in order to explain it, to predict it, and to control it, goals corresponding to the three

[1]This chapter is based to a large extent on material presented in Intriligator (1978, esp. ch. 1, 2, 7, 8, 10, 12, 13, 14, 15, and 16). Other general references on economic and econometric models include Beach (1957), Suits (1963), Christ (1966), Bergstrom (1967), Ball (1968), Kendall (1968), Cramer (1969), Malinvaud (1970), Bridge (1971), Goldberger and Duncan (1973), Maddala (1977), Leamer (1978), Zellner (1979), and Arnold (1981). Other chapters in this Handbook that treat economic and econometric models include Chapter 4 by Hsiao, Chapter 5 by Leamer, Chapter 26 by Lau, Chapter 28 by Maddala, and Chapter 29 by Heckman and Singer.

purposes of econometrics, namely structural analysis, forecasting, and policy evaluation. Sometimes the actual system is called the *real-world system* in order to emphasize the distinction between it and the model system that represents it.

Modeling, that is, the art of model building, is an integral part of most sciences, whether physical or social, because the real-world systems under consideration typically are enormously complex. For example, both the motion of an elementary particle in an accelerator and the determination of national income are real-world phenomena of such complexity that they can be treated only by means of a simplified representation, that is, via a model. To be most useful a model has to strike a reasonable balance between realism and manageability. It should be realistic in incorporating the main elements of the phenomena being represented, specifying the interrelationships among the constituent elements of the system in a way that is sufficiently detailed and explicit so as to ensure that the study of the model will lead to insights concerning the real-world system. It should, however, at the same time be manageable in eliminating extraneous influences and simplifying processes so as to ensure that it yields insights or conclusions not obtainable from direct observation of the real-world system. The art of model building involves balancing the often competing goals of realism and manageability.

Typically the initial models of a phenomena are highly simplified, emphasizing manageability. They may, for example, model the system under study as a "black box", treating only its inputs and outputs without attempting to analyze how the two are related. Later models are typically more elaborate, tracking inputs forward and outputs backward until eventually an analytic model is developed which incorporates all the major interconnections between inputs and outputs in the real-world system. The process of modeling typically involves not only the analysis of interconnections between inputs and outputs but also the treatment of additional or related phenomena and greater disaggregation.

Many different types of models have been used in economics and other social and physical sciences. Among the most important types are verbal/logical models, physical models, geometric models, and algebraic models, involving alternative ways of representing the real-world system.

Verbal/logical models use verbal analogies, sometimes called *paradigms*, to represent phenomena. In economics two of the earliest and still two of the best paradigms were developed by Adam Smith.[2] The first was the pin factory, used by Smith as a model of the concept of division of labor. This concept is applicable at the national and international level, but the participants and processes become so numerous and their interrelations so manifold that the principle could be lost. Smith therefore used the paradigm of the pin factory, where the principle could be readily understood. The second paradigm employed by Smith was that of the

[2]See Smith (1776).

"invisible hand", one of the most important contributions of economics to the study of social processes. Smith observed that in a decentralized economy the price system guides agents to ensure that their individual actions attain a coherent equilibrium for the economy as a whole, promoting the general welfare of society. Again a complex process, in this case that of all economic actions, was represented by a verbal model.

Physical models represent the real-world system by a physical entity. An example is a scale model of a physical object, such as a scaled-down model airframe for an airplane, which is tested in a wind tunnel or a scaled-up model of a protein molecule. Economic systems have also been studied with physical models, including hydraulic models in which flows of fluids represent monetary flows in the economy. The most important physical models of economic phenomena, however, are those relying upon electric circuits, using the modern analog computer.[3]

Geometric models use diagrams to show relationships among variables. Such models have played an important role in the development of economics. For example, the geometric model of price determination in a single isolated market, involving intersecting demand and supply curves, is a fundamental one in microeconomic theory. Similarly the geometric model of the determination of national income, e.g. via the IS–LM diagram, is a fundamental one in macroeconomic theory. Such models are useful in indicating the principal relationships among the major variables representing the phenomena under investigation, but, because of the limited number of dimensions available, it is necessary to restrict geometric models to a relatively few variables. To deal with more variables usually involves use of an algebraic model.

Algebraic models, which are the most important type of models for purposes of econometrics, represent a real-world system by means of algebraic relations which form a system of equations. The system of equations involves certain variables, called *endogenous variables*, which are the jointly dependent variables of the model and which are simultaneously determined by the system of equations. The system usually contains other variables, called *exogenous variables*, which are determined outside the system but which influence it by affecting the values of the endogenous variables. These variables affect the system but are not in turn affected *by* the system. The model also contains *parameters* which are generally estimated on the basis of the relevant data using econometric techniques.

The general algebraic model can be expressed as the following system of g independent and consistent (i.e. mutually compatible) equations in the g endogenous variables, $y_1, y_2, \ldots, y_g$, the k exogenous (or lagged endogenous) variables,

[3]For applications of electronic analog models to economics, see Morehouse, Strotz and Horwitz (1950), Enke (1951), Strotz, McAnulty and Naines (1953), and Tustin (1953).

$x_1, x_2, \ldots, x_k$, and the m parameters, $\delta_1, \delta_2, \ldots, \delta_m$:

$$\begin{aligned} &f^1(y_1, y_2, \ldots, y_g; x_1, x_2, \ldots, x_k; \delta_1, \delta_2, \ldots, \delta_m) = 0, \\ &f^2(y_1, y_2, \ldots, y_g; x_1, x_2, \ldots, x_k; \delta_1, \delta_2, \ldots, \delta_m) = 0, \\ &\vdots \\ &f^g(y_1, y_2, \ldots, y_g; x_1, x_2, \ldots, x_k; \delta_1, \delta_2, \ldots, \delta_m) = 0. \end{aligned} \tag{2.1}$$

In vector notation the general algebraic model can be written

$$f(y, x, \delta) = 0, \tag{2.2}$$

where f is a column vector of g functions, y is a row vector of g endogenous variables, x is a row vector of k exogenous (or lagged endogenous) variables, δ is a row vector of m parameters, and 0 is a column vector of zeros.

Assuming the functions are differentiable and that the Jacobian matrix of first-order partial derivatives is non-singular at a particular point:

$$\left|\frac{\partial f}{\partial y}\right| = \begin{vmatrix} \dfrac{\partial f^1}{\partial y_1} & \dfrac{\partial f^1}{\partial y_2} \cdots & \dfrac{\partial f^1}{\partial y_g} \\ \dfrac{\partial f^2}{\partial y_1} & \dfrac{\partial f^2}{\partial y_2} \cdots & \dfrac{\partial f^2}{\partial y_g} \\ \vdots & & \\ \dfrac{\partial f^g}{\partial y_1} & \dfrac{\partial f^g}{\partial y_2} \cdots & \dfrac{\partial f^g}{\partial y_g} \end{vmatrix} \neq 0 \quad \text{at } (y, x), \tag{2.3}$$

the implicit function theorem implies that at this point it is possible to solve the system of equations (2.2) for the endogenous variables as differentiable functions of the exogenous variables and parameters:[4]

$$y = \phi(x, \delta), \tag{2.4}$$

where ϕ if a column vector of g functions.

A very simple example is the determination of price in an isolated market, where the equations for demand and supply are

$$\begin{aligned} &q - D(p, x, \delta) = 0 \\ &q - S(p, x, \delta) = 0. \end{aligned} \tag{2.5}$$

[4]For discussions of the implicit function theorem, see Rudin (1964), Apostol (1974), and Hoffman (1975).

Here q and p are quantity and price respectively; D and S are demand and supply functions respectively; and x and δ are vectors of exogenous variables and parameters, respectively. The Jacobian condition is

$$\begin{vmatrix} 1 & -\dfrac{\partial D}{\partial p} \\ 1 & -\dfrac{\partial S}{\partial p} \end{vmatrix} \neq 0, \tag{2.6}$$

which is met if the customary slope conditions

$$\frac{\partial D}{\partial p} < 0 \quad \text{and} \quad \frac{\partial S}{\partial p} > 0 \tag{2.7}$$

are satisfied. Then the system of equations (2.4) can be solved for (equilibrium) quantity and price:

$$q = q(x, \delta); \qquad p = p(x, \delta). \tag{2.8}$$

3. Comparative statics

The comparative statics technique is one of the most useful techniques in economic analysis.[5] It involves the comparison of two equilibrium points of a system of equations such as (2.2), describing the phenomena under consideration. The two equilibrium points typically involve equilibrium before and after displacement by a change in one of the parameters of the system of equations.

Consider system (2.2) for which the Jacobian condition in (2.3) is met so the system can be solved for the endogenous variables as in (2.4). Inserting these solutions into (2.2) yields the system of g identities:

$$f[\phi(x, \delta), x, \delta] \equiv 0. \tag{3.1}$$

Now consider the effect of a change in one of the exogenous variables or parameters, say x_j, on the equilibrium values of the variables.[6] Differentiating each of the identities in (3.1) with respect to x_j yields

$$\sum_{h=1}^{g} \frac{\partial f^l}{\partial y_h} \frac{\partial y_h}{\partial x_j} + \frac{\partial f^l}{\partial x_j} = 0 \qquad l = 1, 2, \ldots, g. \tag{3.2}$$

[5]For a general discussion of the theory of comparative statics, see Samuelson (1947), Intriligator (1971), Kalman and Intriligator (1973), Silberberg (1978), and Intriligator (1981).

[6]A similar approach would yield the effect of any parameter, say δ_q, on the equilibrium values of the variables.

Solving for the effect of a change in every x_j, for $j = 1, 2, \ldots, k$, on y_h yields, in matrix notation,

$$\frac{\partial y}{\partial x} = -\left(\frac{\partial f}{\partial y}\right)^{-1}\frac{\partial f}{\partial x}, \tag{3.3}$$

where the three matrices are

$$\frac{\partial y}{\partial x} = \left(\frac{\partial y_h}{\partial x_j}\right); \qquad \frac{\partial f}{\partial y} = \left(\frac{\partial f^l}{\partial y_h}\right); \qquad \frac{\partial f}{\partial x} = \left(\frac{\partial f^l}{\partial x_j}\right). \tag{3.4}$$

Eq. (3.3) expresses the change in the equilibrium levels of each of the endogenous variables as each of the exogenous variables changes. The effect of a change $\mathrm{d}x_j$ in any one exogenous variable on the equilibrium value of any of the endogenous variables $\mathrm{d}y_h$ is then given as

$$\mathrm{d}y_h = \frac{\partial y_h}{\partial x_j}\mathrm{d}x_j, \tag{3.5}$$

where $\partial y_h / \partial x_j$ is the *hj* element of the $\partial y / \partial x$ matrix in (3.3).

Restrictions on the signs or values of the derivatives in $\partial f / \partial y$ and $\partial f / \partial x$ in (3.3) often lead to comparable restrictions on the signs or values of the derivatives in $\partial y / \partial x$. These qualitative restrictions on the effects of exogenous variables on endogenous variables provide some of the most important results in the analysis of economic systems described by an algebraic model.[7]

4. Econometric models

Econometric models are generally algebraic models that are *stochastic* in including random variables (as opposed to *deterministic* models which do not include random variables). The random variables that are included, typically as additive stochastic disturbance terms, account in part for the omission of relevant variables, incorrect specification of the model, errors in measuring variables, etc. The general econometric model with additive stochastic disturbance terms can be written as the *non-linear structural form* system of g equations:

$$f(y, x, \delta) = \varepsilon, \tag{4.1}$$

[7]For a discussion of qualitative economics, involving an analysis of the sign or value restrictions on partial derivatives, see Samuelson (1947) and Quirk and Saposnik (1968). For a specific example of these qualitative restrictions see the discussion of Barten's fundamental matrix equation for consumption theory in Chapter 1 of this Handbook by Theil.

where ε is a vector of stochastic disturbance terms, one for each equation. This form is similar to (2.2) with the addition of disturbance terms in each equation where ε is a vector of stochastic disturbance terms. If the conditions of the implicit function theorem are met these equations can be solved for the endogenous variables as differentiable functions of the exogenous variables and parameters, with the stochastic disturbance terms included as additive error terms. The resulting *non-linear reduced form* is the system of g equations:

$$y=\phi(x,\delta)+u, \tag{4.2}$$

where u is the vector of the stochastic disturbance terms in the reduced form. The corresponding deterministic reduced form of the model is (2.4). From (4.2) it follows that the econometric model uniquely specifies not the endogenous variables but rather the probability distribution of each of the endogenous variables, given the values taken by all exogenous variables and given the values of all parameters of the model. Each equation of the model, other than definitions, equilibrium conditions, and identities, is generally assumed to contain an additive stochastic disturbance term, which is an unobservable random variable with certain assumed properties, e.g. mean, variance, and covariance. The values taken by that variable are not known with certainty; rather, they can be considered random drawings from a probability distribution with certain assumed moments. The inclusion of such stochastic disturbance terms in the econometric model is basic to the use of tools of statistical inference to estimate parameters of the model.

Econometric models are either linear or non-linear. Early econometric models and many current econometric models are linear in that they can be expressed as models that are linear in the parameters. This linearity assumption has been an important one for proving mathematical and statistical theorems concerning econometric models, for estimating parameters, and for using the estimated models for structural analysis, forecasting, and policy evaluation. The linearity assumption has been justified in several ways. First, many economic relationships are by their very nature linear, such as the definitions of expenditure, revenue, cost, and profit. Second, the linearity assumption applies only to parameters, not to variables of the model. Thus, a quadratic cost function, of the form

$$C=a+bq+cq^2, \tag{4.3}$$

where C is cost, q is output, and a, b, and c are parameters, while non-linear in q, is linear in a, b, and c. Third, non-linear models can sometimes be transformed into linear models, such as by a logarithmic transformation. For example, the Cobb–Douglas production function

$$Y=AK^{\alpha}L^{\beta}, \tag{4.4}$$

where Y is output, K is capital, L is labor, and A, a, and β are parameters, can be so transformed into the log-linear form

$$\log Y = a + \alpha \log K + \beta \log L \qquad (a = \log A). \tag{4.5}$$

Fourth, any smooth function can be reasonably approximated in an appropriate range by a linear function, e.g. via a Taylor's theorem approximation. Consider, for example, the general production function

$$Y = F(K, L), \tag{4.6}$$

of which the Cobb–Douglas form (4.4) is one special case. If the function is continuous it can be approximated as a linear function in an appropriate range by taking the linear portion of the Taylor's series expansion. Expanding about the base levels of (K_0, L_0),

$$Y \cong F(K_0, L_0) + \frac{\partial F}{\partial K}(K_0, L_0)(K - K_0) + \frac{\partial F}{\partial L}(K_0, L_0)(L - L_0), \tag{4.7}$$

so that[8]

$$Y \cong a + bK + cL, \tag{4.8}$$

where the parameters can be interpreted as

$$\begin{aligned} a &= F(K_0, L_0) - \frac{\partial F}{\partial K}(K_0, L_0)K_0 - \frac{\partial F}{\partial L}(K_0, L_0)L_0, \\ b &= \frac{\partial F}{\partial K}(K_0, L_0), \\ c &= \frac{\partial F}{\partial L}(K_0, L_0). \end{aligned} \tag{4.9}$$

Finally, linear models are much more convenient and more manageable than

[8]Other approximations are also possible, e.g. expressing the production function as

$$\log Y = \phi(\log K, \log L).$$

Taking a Taylor's series approximation yields

$$\log Y \cong a' + b' \log K + c' \log L,$$

which would approximate any production function as a log-linear Cobb–Douglas production function as in (4.4) and (4.5). See Kmenta (1967). For a more general discussion of transformations see Box and Cox (1964).

non-linear models. Thus, the linearity assumption has frequently been made for econometric models.

Non-linear models, that is, econometric models that are non-linear in the parameters, have become more common in recent years largely due to advances in computer software and numerical analysis that have facilitated the estimation of such models. Techniques and computer software used in the estimation of non-linear econometric models are discussed in Chapters 6 and 12 of this Handbook by Takeshi Amemiya and Richard Quandt, respectively. The parameters of a non-linear model are frequently estimated using successive linear approximations to the model, and the properties of such estimators can be derived asymptotically or approximately. While these properties are valid for large samples the exact small sample properties of estimators for general non-linear econometric models are unknown. Furthermore, some of the properties have been shown to hold only under the assumption of normally distributed stochastic disturbances, and the consequences of model misspecification are generally not known in the non-linear case. A considerable amount of work has been done in recent years on non-linear models, however, as discussed elsewhere in this Handbook.[9]

4.1. *Structural form*

The basic econometric model is the structural form, from which the reduced form and the final form can be obtained. The general *structural form* of the linear (in parameters) stochastic econometric model, assuming there are g endogenous variables, $y_1, y_2, \dots, y_g$, and k exogenous variables, $x_1, x_2, \dots, x_k$, can be written:

$$\begin{aligned} &y_1\gamma_{11} + y_2\gamma_{21} + \cdots + y_g\gamma_{g1} + x_1\beta_{11} + x_2\beta_{21} + \cdots + x_k\beta_{k1} = \varepsilon_1, \\ &y_1\gamma_{12} + y_2\gamma_{22} + \cdots + y_g\gamma_{g2} + x_1\beta_{12} + x_2\beta_{22} + \cdots + x_k\beta_{k2} = \varepsilon_2, \\ &\vdots \\ &y_1\gamma_{1g} + y_2\gamma_{2g} + \cdots + y_g\gamma_{gg} + x_1\beta_{1g} + x_2\beta_{2g} + \cdots + x_k\beta_{kg} = \varepsilon_g. \end{aligned} \tag{4.10}$$

Here the γ's are the coefficients of the endogenous variables, the β's are the coefficients of the exogenous variables, and $\varepsilon_1, \varepsilon_2, \dots, \varepsilon_g$ are g stochastic disturbance terms (random variables). This system of equations can be considered the linear and stochastic version of the system (2.1), where the parameters include not only the coefficients but also those parameters characterizing the stochastic

[9]For discussions of non-linear models see Chapters 6 and 12 of this Handbook by Amemiya and Quandt, respectively. See also Goldfeld and Quandt (1968, 1972), Chow (1973), Jorgenson and Laffont (1974), Goldfeld and Quandt (1976), Belsley (1979, 1980), Gallant and Jorgenson (1979), Fair and Parke (1980), and Gallant and Holly (1980).

disturbance terms. Intercept terms in the equations can be taken into account by specifying one of the exogenous variables, conventionally either the first x_1 or the last x_k, to be identically unity, in which case its coefficients become the intercepts.

Typically, each equation of the structural form (4.10) has an independent meaning and identity, reflecting a behavioral relation (such as a demand function or a consumption function), a technological relation (such as a production function), or some other specific relation suggested by theory for the system under study. Each equation, because it represents one aspect of the structure of the system, is called a *structural equation*, and the set of all structural equations is the *structural form*. Some equations may be deterministic, e.g. definitions, identities, and equilibrium conditions, and for these equations the stochastic disturbance terms are identically zero. In general, however, these equations can be eliminated, reducing both the number of equations and the number of endogenous variables.

The structural form can also be written in *summation notation*, as

$$\sum_{h=1}^{g} y_h \gamma_{hl} + \sum_{j=1}^{k} x_j \beta_{jl} = \varepsilon_l, \qquad l = 1,2,\ldots,g, \tag{4.11}$$

where h is an index of the endogenous variable, l is an index of the equation, and j is an index of the exogenous variable. In *vector–matrix notation* the structural form is written:

$$y\Gamma + xB = \varepsilon, \tag{4.12}$$

which is the linear version of system (4.1), where the coefficient matrices are

$$\Gamma = (\gamma_{hl}) \tag{4.13}$$

and

$$B = (\beta_{jl}). \tag{4.14}$$

Γ is a $g \times g$ matrix of coefficients of endogenous variables, assumed non-singular, and B is a $k \times g$ matrix of coefficients of exogenous variables. Note that the lth columns of Γ and B contain all coefficients in the lth equation of the structural form for $l = 1,2,\ldots,g$. The structural form in vector-matrix notation in (4.12) is the most convenient of the three ways of expressing the structural form, and it will be used in the remainder of this chapter.

There is a trivial indeterminacy in the structural equations in that multiplying all terms in any one of these equations by a non-zero constant does not change the equation. This indeterminacy is eliminated by choosing a *normalization rule*, which is a rule for selecting a particular numerical value for one of the non-zero coefficients in each question. A convenient normalization rule is that which sets all elements along the principal diagonal of the Γ matrix of coefficients of endogenous variables at -1:

$$\gamma_{hh} = -1, \qquad h = 1,2,\dots,g. \tag{4.15}$$

This normalization rule, obtained by dividing all coefficients of equation h by $-\gamma_{hh}$, yields the usual convention of being able to write each equation which specifies one endogenous variable as a function of other endogenous variables, exogenous variables, and a stochastic disturbance term, with a unique such endogenous variable for each equation. Other normalization rules can be used, however, typically involving setting the (non-zero) coefficient of one variable in each equation as 1 or -1 (by dividing by this coefficient or its negative).

Letting i be an index of the observation number, the structural form at the ith observation is

$$y_i\Gamma + x_iB = \varepsilon_i, \qquad i = 1,2,\dots,n. \tag{4.16}$$

Here y_i, x_i, and ε_i are, respectively, the vector of endogenous variables, the vector of exogenous variables, and the vector of stochastic disturbance terms at the ith observation, where i ranges over the sample from 1 to n, n being the sample size (the number of observations). Certain stochastic assumptions are typically made concerning the n stochastic disturbance vectors ε_i. First, they are assumed to have a zero mean:

$$E(\varepsilon_i) = 0, \qquad i = 1,2,\dots,n. \tag{4.17}$$

Second, the covariance matrix of ε_i is assumed to be the same at each observation:

$$\text{cov}(\varepsilon_i) = E(\varepsilon_i'\varepsilon_i) = \Sigma, \qquad i = 1,2,\dots,n, \tag{4.18}$$

where Σ, the positive definite symmetric matrix of variances and covariances, is the same for each i. Third, the ε_i are assumed uncorrelated over the sample

$$E(\varepsilon_i'\varepsilon_j) = 0, \qquad i = 1,2,\dots,n; \quad j = 1,2,\dots,n; \quad i \neq j, \tag{4.19}$$

so that each stochastic disturbance term is uncorrelated with any stochastic disturbance term (including itself) at any other point in the sample. These assumptions are satisfied if, for example, the stochastic disturbance vectors ε_i are independently and identically distributed over the sample, with a zero mean vector and a constant covariance matrix Σ. Sometimes the further assumption of normality is also made, specifying that the ε_i are distributed independently and normally with zero mean vector and $g \times g$ positive definite symmetric covariance matrix Σ:

$$\varepsilon_i \sim N(0, \Sigma), \qquad i = 1, 2, \ldots, n. \tag{4.20}$$

Under these general assumptions (without necessarily assuming normality), while the stochastic disturbance terms are uncorrelated over the sample, they can, by (4.18), be correlated between equations. This latter phenomenon of correlation between stochastic disturbance terms in different equations (due to the fact that there is usually more than one endogenous variable in each equation) is an essential feature of the simultaneous-equation system econometric model and the principal reason why it must be estimated using simultaneous-equation rather than single-equation techniques, as discussed in Chapter 7 of this Handbook by Jerry Hausman.

4.2. *Reduced form*

The structural form (4.10) is a special case of the general system (2.1) (other than the addition of stochastic disturbance terms). The general system could be solved for the endogenous variables if condition (2.3) is met. In the case of the structural form (2.3) is the condition that the matrix Γ of coefficients of endogenous variables be non-singular, which is usually assumed. Then the structural form can be solved for the endogenous variables as explicit (linear, stochastic) functions of all exogenous variables and stochastic disturbance terms – the reduced form. Postmultiplying (4.12) by Γ^{-1} and solving for y yields

$$y = -xB\Gamma^{-1} + \varepsilon\Gamma^{-1}. \tag{4.21}$$

Introducing the $k \times g$ matrix of reduced-form coefficients Π and the $1 \times g$ reduced-form stochastic disturbance vector u, where

$$\Pi \equiv -B\Gamma^{-1}, \qquad u \equiv \varepsilon\Gamma^{-1}, \tag{4.22}$$

the *reduced form* is written

$$y = x\Pi + u. \tag{4.23}$$

This reduced form uniquely determines the probability distributions of the endogenous variables, given the exogenous variables, the coefficients, and the probability distributions of the stochastic disturbance terms.

The matrix of reduced-form coefficients represents the changes in endogenous variables as exogenous variables change:

$$\Pi = \frac{\partial y}{\partial x}, \quad \text{i.e. } \Pi_{jh} = \frac{\partial y_h}{\partial x_j}, \qquad j=1,2,\dots,k; \quad h=1,2,\dots,g. \tag{4.24}$$

Thus, the elements of the matrix of reduced-form coefficients represent the comparative statics results of the model, the jh element of Π measuring the change in the hth endogenous variable as the jth exogenous variable changes, all other predetermined variables and all stochastic disturbance terms being held constant. The estimation of these comparative statics results is an important aspect of structural analysis using the econometric model.

The stochastic assumptions made for the structural form have direct implications for the stochastic disturbance terms of the reduced form. If i is an index of the observation number, the reduced form at the ith observation is

$$y_i = x_i \Pi + u_i, \qquad i=1,2,\dots,n, \tag{4.25}$$

where Π is the same as in (4.22) and the reduced-form stochastic disturbance vector is

$$u_i \equiv \varepsilon_i \Gamma^{-1}. \tag{4.26}$$

This identity is used to obtain conditions on u_i from those assumed for ε_i. From (4.17):

$$E(u_i) = 0, \qquad i=1,2,\dots,n. \tag{4.27}$$

From (4.18):

$$\operatorname{cov}(u_i) = E(u_i' u_i) = (\Gamma^{-1})' E(\varepsilon_i' \varepsilon_i) \Gamma^{-1} = (\Gamma^{-1})' \Sigma \Gamma^{-1} = \Omega, \qquad i=1,2,\dots,n, \tag{4.28}$$

where Ω is the covariance matrix of u_i, which, as is the case of the covariance matrix Σ of ε_i, is constant over the sample. The last equality in (4.28) implies that

$$\Sigma = \Gamma' \Omega \Gamma, \tag{4.29}$$

showing the relationship between the covariance matrix of the structural form Σ and that of the reduced form Ω. Furthermore, from (4.19),

$$E(u_i' u_j) = 0, \qquad i=1,2,\dots,n; \quad j=1,2,\dots,n; \quad i \neq j, \tag{4.30}$$

so the u_i, just as the ε_i, are uncorrelated over the sample. If it is further assumed that the ε_i are independently and normally distributed, as in (4.20), then the u_i are also independently and normally distributed, with zero mean vector and $g \times g$ positive definite symmetric covariance matrix Ω:

$$u_i \sim N(0, \Omega), \qquad i = 1, 2, \dots, n, \tag{4.31}$$

where Ω is given in (4.28) as $(\Gamma^{-1})' \Sigma \Gamma^{-1}$.

Assumptions (4.27), (4.28), and (4.30) summarize the stochastic specification of the reduced-form equations. Under these assumptions the conditions of both the Gauss–Markov Theorem and the Least Squares Consistency Theorem are satisfied for the reduced-form equations, so the least squares estimators

$$\hat{\Pi} = (X'X)^{-1} X'Y, \tag{4.32}$$

where X is the $n \times k$ matrix of data on the k exogenous variables at the n observations and Y is the $n \times g$ matrix of data on the g endogenous variables at the n observations, are the unique best linear unbiased and consistent estimators of the reduced form. The covariance matrix can then be estimated as

$$\hat{\Omega} = \frac{1}{n-k}(Y - X\hat{\Pi})'(Y - X\hat{\Pi}) = \frac{1}{n-k} Y'\left[I - X(X'X)^{-1}X'\right]Y, \tag{4.33}$$

where $I - X(X'X)^{-1}X'$ is the fundamental idempotent matrix of least squares, as introduced in Chapter 1 of this Handbook by Henri Theil. This estimator of the covariance matrix is an unbiased and consistent estimator of Ω.

4.3. *Final form*

Econometric models are either static or dynamic. A *static model* involves no explicit dependence on time, so time is not essential in the model. (Simply adding time subscripts to variables does not convert a static model into a dynamic one.) A *dynamic model* is one in which time plays an essential role, typically by the inclusion of lagged variables or differences of variables over time. Thus, if any equation of the model is a difference equation, then the model is dynamic. (Time also plays an essential role if variables and their rates of change over time are included in the model, such as in a differential equation.)

If the econometric model is dynamic in including lagged endogenous variables, then it is possible to derive another form of the model, the final form.[10] The final

[10] See Theil and Boot (1962).

form expresses the current endogenous variables as functions of base values and all relevant current and lagged exogenous and stochastic disturbance terms. If the structural form involves only one lag, then it can be written[11]

$$y_t\Gamma + \left(y_{t-1} \vdots x_t \right)\left(\frac{B_1}{B_2} \right) = \varepsilon_t \tag{4.34}$$

or

$$y_t\Gamma + y_{t-1}B_1 + x_tB_2 = \varepsilon_t, \tag{4.35}$$

where y_{t-1} is a vector of lagged endogenous variables. The lagged endogenous and exogenous variables are grouped together as the predetermined variables of the system. The B matrix has then been partitioned to conform to the partitioning of the predetermined variables into lagged endogenous and current exogenous variables. The reduced form is then

$$y_t = y_{t-1}\Pi_1 + x_t\Pi_2 + u_t, \tag{4.36}$$

where

$$\Pi_1 \equiv -B_1\Gamma^{-1}; \qquad \Pi_2 \equiv -B_2\Pi^{-1}, \tag{4.37}$$

$$u_t \equiv \varepsilon_t\Gamma^{-1}. \tag{4.38}$$

The final form is obtained by solving this reduced form, which is a difference equation in y_t, iteratively for y_0. In the first iteration

$$\begin{aligned} y_t &= (y_{t-2}\Pi_1 + x_{t-1}\Pi_2 + u_{t-1})\Pi_1 + x_t\Pi_2 + u_t \\ &= y_{t-2}\Pi_1^2 + [x_t\Pi_2 + x_{t-1}\Pi_2\Pi_1] + [u_t + u_{t-1}\Pi_1]. \end{aligned} \tag{4.39}$$

[11] The index has been changed from i in (4.16) to t in (4.34) to emphasize the facts that the variables depend on time and that the model includes lagged variables. It should also be noted that any finite number of lags in both endogenous and exogenous variables can be treated using the approach of this section. With lags up to those of order p, eq. (4.35) generalizes to

$$y_t\Gamma + \sum_{j=1}^{p} y_{t-j}B_{1j} + \sum_{k=0}^{p} x_{t-k}B_{2k} = \varepsilon_t.$$

An infinite number of lags, of the form (for a single endogenous variable)

$$y_t = \alpha + \sum_{k=0}^{\infty} \beta_k x_{t-k} + u_t,$$

is a *distributed lag model*, discussed in Chapters 17–20 of this Handbook by Granger and Watson; Hendry, Pagan and Sargan; Geweke; and Bergstrom, respectively. See also Griliches (1967), Sims (1974), and Dhrymes (1981).

Continuing the iteration back to the base period $t=0$ yields

$$y_t = y_0\Pi_1^t + \sum_{j=0}^{t-1} x_{t-j}\Pi_2\Pi_1^j + \sum_{j=0}^{t-1} u_{t-j}\Pi_1^j. \tag{4.40}$$

This is the *final form*, in which each of the endogenous variables is expressed as a function of base period values, current and lagged exogenous variables, and stochastic disturbance terms. The coefficient of the base period endogenous variable is Π_1^t, and the successive coefficients of the current and lagged exogenous variables

$$\Pi_2, \qquad \Pi_2\Pi_1, \qquad \Pi_2\Pi_1^2, \ldots, \tag{4.41}$$

indicate the influence of the current value of the endogenous variables of successively lagged values of the exogenous variables, starting from the current (non-lagged) values and given as

$$\frac{\partial y_t}{\partial x_{t-j}} = \Pi_2\Pi_1^j, \qquad j=1,2,\ldots,t-1. \tag{4.42}$$

The estimation of these successive coefficients, which can be interpreted as various multipliers, is an important aspect of structural analysis using the econometric model.

5. Identification

The problem of identification is an important issue in econometric model building.[12] Most approaches to the estimation of the structural form start from the estimation of the reduced-form equations, specifically the estimated matrix of reduced-form coefficients $\hat{\Pi}$ in (4.32) and the estimated covariance matrix $\hat{\Omega}$ in (4.33) and use these estimators to obtain estimates of the structural form parameters Γ, B, and Σ in (4.12) and (4.18).

The *problem of identification* is that of using estimates of reduced-form parameters Π and Ω to obtain estimates of structural-form parameters Γ, B, and Σ. Certain information is available from the relations between the structural form and reduced form. In particular, from (4.22) and (4.29) if $\hat{\Pi}$ and $\hat{\Omega}$ are estimates of Π and Ω, respectively, while if $\hat{\Gamma}$, $\hat{B}$, and $\hat{\Sigma}$ are estimates of Γ, B, and Σ,

[12]For an extensive discussion of identification see Chapter 4 of this Handbook by Hsiao. Basic references on identification include Fisher (1966), Rothenberg (1971, 1973), and Bowden (1973). See also Intriligator (1978, ch. 10).

respectively, the estimates must satisfy

$$\hat{B} = \hat{\Pi}\hat{\Gamma} \tag{5.1}$$

and

$$\hat{\Sigma} = \hat{\Gamma}'\hat{\Omega}\hat{\Gamma}. \tag{5.2}$$

These restrictions provide *a posteriori* information, since they follow the estimation of the reduced form. This information, however, is generally not adequate to determine the structural parameters. For example, if the structural form were postmultiplied by any non-singular matrix R:

$$y_i\Gamma R + x_i BR = \varepsilon_i R, \tag{5.3}$$

and this "bogus" system were normalized in the same way as the old one, where the bogus parameters are

$$\bar{\Gamma} = \Gamma R; \qquad \bar{B} = B\Gamma; \qquad \bar{\Sigma} = R'\Sigma R, \tag{5.4}$$

then the reduced form is

$$\bar{\Pi} = -\bar{B}\bar{\Gamma}^{-1} = -BRR^{-1}\Gamma^{-1} = -B\Gamma^{-1} = \Pi, \tag{5.5}$$

$$\bar{\Omega} = (\bar{\Gamma}^{-1})'\bar{\Sigma}\bar{\Gamma}^{-1} = (\Gamma^{-1})'\Sigma\Gamma^{-1} = \Omega. \tag{5.6}$$

Thus, the bogus system has the same reduced-form parameters as the true system. The true and bogus systems are *observationally equivalent* in yielding the same reduced form (more precisely, in implying the same likelihood function for the observed values of the endogenous variables, given the values of the predetermined variables). Thus, the *a posteriori* information in (5.1) and (5.2) cannot distinguish between Γ, B, and Σ, the true parameters, and $\bar{\Gamma}$, $\bar{B}$, and $\bar{\Sigma}$, the bogus parameters. To distinguish the true parameters it is necessary to supplement the *a posteriori* information by *a priori information*, restrictions on the structural parameters imposed prior to the estimation of the reduced form. These restrictions on the structural form, obtained from relevant theory or the results of other studies, have the effect of reducing the class of permissible matrices R in (5.3). If no such restrictions are imposed, or too few are imposed, the system is *not identified*, in which case additional *a priori* information must be imposed in order to identify the structural parameters Γ, B, and Σ. If enough *a priori* information is available, then the system is *identified* in that all structural parameters can be determined from the reduced-form parameters. A structural equation is *underidentified* if there is no way to determine its parameters from the reduced-form parameters. It is *just identified* (or *exactly identified*) if there is a unique way of estimating its parameters from the reduced-form parameters. It is *overidentified* if there is more than one way to calculate its parameters from the reduced-form parameters, leading to restrictions on the reduced-form parameters.

The *a priori* restrictions on the structural-form parameters Γ, B, and Σ usually involve one of three approaches. The first approach is that of zero or linear restrictions, equating some elements of the coefficient matrices *a priori* to zero or, more generally, imposing a set of linear restrictions. The second approach is that of restrictions on the covariance matrix Σ, e.g. via zero restrictions or relative sizes of variances or covariances. A third approach is some mixture of the first two, where certain restrictions, in the form of equalities or inequalities, are imposed on Γ, B, and Σ. An example is that of a *recursive system*, where Γ is a triangular matrix and Σ is a diagonal matrix. Such a system is always just identified, each equation being just identified.[13]

6. Some specific models

This section will present some specific models that have been used in econometrics. It emphasizes systems of equations, as opposed to single equation models.[14]

6.1. Demand models

One of the earliest and most important applications of econometric models is to the estimation of demand relationships.[15] In fact, pioneer empirical analyses of demand, starting in the nineteenth century with the work of Engel and continuing in the early twentieth century with the work of Schultz and Moore, led to later studies of general issues in econometrics.

A complete system of demand equations for n goods consists of the n demand equations:

$$x_j = x_j(p_1, p_2, \ldots, p_n, I, u_j), \qquad j = 1, 2, \ldots, n, \tag{6.1}$$

where x_j is the demand for good j by a single household or a group of households, p_j is the price of good j, I is income, which is the same as the expenditure on the n

[13] For a discussion of recursive systems see Wold (1954, 1960) and Wold (1968).

[14] For a more extensive discussion of various models and a discussion of single equation models see Intriligator (1978, esp. ch. 7, 8, 9, 12, and 13).

[15] For an extensive discussion of demand analysis see Chapter 30 of this Handbook by Deaton. Basic references for econometric studies of consumer demand include Brown and Deaton (1972), Powell (1974), Phlips (1974), Theil (1975/1976), and Barten (1977). See also Intriligator (1978, ch. 7).

goods, and u_j is the stochastic term in the jth demand equation. The n equations determine the quantity demanded of each good, which are the n endogenous variables, as functions of all prices and income, the $n+1$ exogenous variables, and stochastic terms, the latter accounting for omitted variables, misspecification of the equation, and errors in measuring variables. These n equations are the principal results of the theory of the consumer, and their estimation is important in quantifying demand for purposes of structural analysis, forecasting, and policy evaluation.

In order to estimate the system (6.1) it is necessary to specify a particular functional form for the general relationship indicated, and a variety of functional forms has been utilized. Only three functional forms will be considered here, however.

A functional form that has been widely used in demand (and other) studies is the *constant elasticity, log-linear specification.*[16] The n demand functions in (6.1) are specified as

$$x_j = A_j p_1^{\varepsilon_{j1}} p_2^{\varepsilon_{j2}} \dots p_n^{\varepsilon_{jn}} I^{\eta_j} e^{u_j}, \qquad j=1,2,\dots,n, \tag{6.2}$$

so, taking logarithms leads to the log-linear system:

$$\begin{aligned} \ln x_j &= a_j + \varepsilon_{j1} \ln p_1 + \varepsilon_{j2} \ln p_2 + \cdots \varepsilon_{jn} \ln p_n + \eta_j \ln I + u_j, \\ a_j &= \ln A_j, \quad j=1,2,\dots,n. \end{aligned} \tag{6.3}$$

This system is one of constant elasticity, where ε_{jj} are the (own) price elasticities of demand:

$$\varepsilon_{jj} = \frac{\partial \ln x_j}{\partial \ln p_j} = \frac{p_j}{x_j} \frac{\partial x_j}{\partial p_j}, \qquad j=1,2,\dots,n, \tag{6.4}$$

the ε_{jk} for $j \neq k$ are the cross price elasticities of demand:

$$\varepsilon_{jk} = \frac{\partial \ln x_j}{\partial \ln p_k} = \frac{p_k}{x_j} \frac{\partial x_j}{\partial p_k}, \qquad j=1,2,\dots,n; \quad k=1,2,\dots,n, \tag{6.5}$$

[16]Among the studies using the constant elasticity log-linear specification are Wold and Jureen (1953), Stone (1954), and Houthakker (1957, 1965). While this is a frequently used specification of a system of demand equations, such a system is not consistent with the budget constraint and the theoretical restrictions on systems of demand equations discussed in Phlips (1974), Intriligator (1978), and Chapter 1 of this Handbook by Theil. At best it can be treated as a local approximation to the true system of demand equations.

and the η_j are the income elasticities of demand:

$$\eta_j = \frac{\partial \ln x_j}{\partial \ln I} = \frac{I}{x_j}\frac{\partial x_j}{\partial I}, \qquad j = 1,2,\ldots,n. \tag{6.6}$$

The defining characteristic of this specification is that all $n(n+1)$ of these elasticities ($n+1$ for each of the n goods) are constant.

Another functional form used in demand analysis is the *semilogarithmic specification*:[17]

$$x_j = a_j + b_{j1}\ln p_1 + b_{j2}\ln p_2 + \cdots + b_{jn}\ln p_n + c_j \ln I + u_j, \quad j = 1,2,\ldots,n, \tag{6.7}$$

where the coefficients are

$$b_{jk} = \frac{\partial x_j}{\partial \ln p_k} = p_k \frac{\partial x_j}{\partial p_k}, \tag{6.8}$$

so b_{jk}/x_j is the (own or cross) price elasticity of demand.

A third functional form which is widely used in studies of demand is the *linear expenditure system*.[18] This system is

$$x_j = x_j^0 + \frac{\beta_j}{p_j}\left(I - \Sigma p_k x_k^0\right), \quad \text{with } x_j^0 \geqslant 0, \quad j = 1,2,\ldots,n, \tag{6.9}$$

or, in terms of expenditure,

$$p_j x_j = p_j x_j^0 + \beta_j\left(I - \Sigma p_k x_k^0\right), \quad \text{with } x_j^0 \geqslant 0, \quad j = 1,2,\ldots,n. \tag{6.10}$$

It can be interpreted as stating that expenditure on good j is composed of two components, the first being expenditure on a certain base amount x_j^0, which is the amount to which the consumer is committed, and the second being a fraction β_j of the so-called "supernumerary income", given as the income above the "subsistence income" $\Sigma p_k x_k^0$ needed to purchase base amounts of all goods. These two

[17]See Prais and Houthakker (1955). As in the case of the constant elasticity log-linear specification this semilogarithmic specification is not consistent with the theoretical restrictions on systems of demand equations.

[18]Among the many studies using the linear expenditure system are Stone, Brown and Rowe (1965), Pollak and Wales (1969), Stone (1972), Phlips (1974), Deaton (1975), and Barten (1977). The linear expenditure system, unlike the preceding ones, is consistent with the theoretical restrictions on systems of demand equations.

components correspond, respectively, to committed and discretionary expenditure on good j. The parameters that define the system are the n *base quantities*, $x_1^0, x_2^0, \dots, x_n^0$, and the n *marginal budget shares*, $\beta_1, \beta_2, \dots, \beta_n$.

6.2. *Production models*

A second important area of modeling is that of production functions.[19] While many studies treat only the production function itself, a complete system involves the production function and the first-order conditions for profit maximization (under competition) using this production function. Thus, the complete system consists of the $n+1$ equations:

$$\begin{aligned} &y = f(x_1, x_2, \dots, x_n, u) \\ &\frac{1}{w_j}\frac{\partial f}{\partial x_j} = g(v_j), \qquad j = 1, 2, \dots, n, \end{aligned} \tag{6.11}$$

where y is output, $x_1, x_2, \dots, x_n$ are the n inputs, u is a stochastic disturbance term affecting technical efficiency, $f(\cdot)$ is the production function, w_j is the wage of input j relative to the price of output, $\partial f / \partial x_j$ is the marginal product of input j, v_j are stochastic disturbance terms affecting attainment of the first-order conditions, and $g(\cdot)$ is a function expressing how well the firm approximates the first-order conditions, under which g should be unity. These $n+1$ equations determine the output and the n inputs (the endogenous variables), as functions of the wages which, assuming the firm takes prices as given, are the exogenous variables. Estimation of this complete system is generally superior to estimating only the first equation from both an economic and an econometric standpoint. From an economic standpoint, estimating only the first equation reflects only the technology available to the firm, while estimating the complete system reflects the behavior of the firm (profit-maximizing) as well as the technology available to it. From an econometric standpoint estimating the first equation involves simultaneous equations bias, while estimating the complete system can result in consistent estimators. Even estimating the complete system cannot, however, be used to test the hypothesis of profit maximization, which is assumed in (6.11). Furthermore, the system (6.11) assumes that the correct prices and decision rules are known. If they are not known, then the system involves unknown parameters or functional forms, while if they are incorrect, then specification error is introduced in the system.

[19]For an extensive discussion of production (and cost) analysis see Chapter 31 of this Handbook by Jorgenson. Basic references for econometric studies of production functions include Walters (1963, 1968), Frisch (1965), Brown (1967), and Ferguson (1969). See also Intriligator (1978, ch. 8).

As in the case of demand analysis, a variety of functional forms has been utilized in estimating (6.11), but only three will be considered here.

One of the most widely used functional forms for production functions is the same as that used in demand analysis, *the constant elasticity log-linear specification*, also known as the *Cobb–Douglas production function*.[20] This function, already introduced in eqs. (4.4) and (4.5), can be written generally in the form of (6.11) as

$$y = Ax_1^{\alpha_1}x_2^{\alpha_2}\dots x_n^{\alpha_n}e^u,$$
$$\frac{1}{w_j}\frac{\partial y}{\partial x_j} = \frac{\alpha_j y}{w_j x_j} = e^{v_j}, \qquad j = 1,2,\dots,n, \tag{6.12}$$

where disturbances are treated as exponential. Taking logarithms gives the linear system:

$$\ln y = a + \alpha_1 \ln x_1 + \alpha_2 \ln x_2 + \cdots + \alpha_n \ln x_n + u, \qquad a = \ln A,$$
$$\ln y = \ln w_j + \ln x_j - \ln \alpha_j + v_j, \qquad j = 1,2,\dots,n. \tag{6.13}$$

A second widely used specification is the *constant elasticity of substitution (CES) production function*.[21] In the customary two-input case this function is

$$y = A\left[\delta x_1^{-\beta} + (1-\delta)x_2^{-\beta}\right]^{-1/\beta}, \tag{6.14}$$

where $\beta \geqslant -1$, the substitution parameter, is related to the elasticity of substitution σ by

$$\sigma = \frac{1}{1+\beta}. \tag{6.15}$$

This function reduces to the Cobb–Douglas case as $\beta \to 0$ (so $\sigma \to 1$); it reduces to a linear function as $\beta \to -1$ (so $\sigma \to \infty$); and it reduces to the input–output case of fixed coefficients as $\beta \to \infty$ (so $\sigma \to 0$).

A third specification is the *transcendental logarithmic (translog) production function*[22]

$$\ln y = a + \sum_{j=1}^{n} \alpha_j \ln x_j + \tfrac{1}{2}\sum_{i=1}^{n}\sum_{j=1}^{n} \gamma_{ij} \ln x_i \ln x_j, \tag{6.16}$$

[20] See Marschak and Andrews (1944), Douglas (1948), Nerlove (1965), and Zellner, Kmenta and Drèze (1966).

[21] See Arrow, Chenery, Minhas and Solow (1961), Brown and de Cani (1963), and Minhas (1963).

[22] See Christensen, Jorgenson and Lau (1973).

where $\gamma_{ij} = \gamma_{ji}$. This function, which is quadratic in the logarithms of the variables, reduces to the Cobb–Douglas case if $\gamma_{ij} = 0$; otherwise it exhibits non-unitary elasticity of substitution. In general this function is quite flexible in approximating arbitrary production technologies, providing a local approximation to any production frontier. It has also been applied to other frontiers, e.g. to demand functions or to price frontiers.

6.3. *Macroeconometric models*

Macroeconometric models, starting with the work of Tinbergen in the 1930s, represent one of the most important applications of econometrics.[23] Such models generally utilize a Keynesian framework for the determination of national income [usually measured as Gross National Product (GNP) or Gross Domestic Product (GDP)] and its components, consumption, investment, government, and net foreign investment, as well as other macroeconomic variables, such as the distribution of income, prices, wages, interest rates, employment, unemployment, production, and assets. These models are used for all three purposes of econometrics: structural analysis, forecasting, and policy evaluation.

Most macroeconometric models are built around the definition of income, a consumption function, and an investment function:

$$\begin{aligned} &Y = C + I + G,\\ &C = C(Y,\dots,u),\\ &I = I(Y,\dots,v). \end{aligned} \tag{6.17}$$

[23]For further discussions of macroeconometric models see Chapters 33, 34, and 35 of this Handbook, by Fair, Taylor, and Klein, respectively. For surveys of macroeconometric models see Nerlove (1966), Ball (1973), Fromm and Klein (1973), Samuelson (1975), and Klein and Burmeister (1976). For references to econometric models of the United States see Intriligator (1978, footnotes 24–28 and table 12.12 on pp. 454–456). Econometric models have also been developed for other national economics. See the references in Shapiro and Halabuk (1976) and Intriligator (1978). Some examples are Klein et al. (1961), Ball and Burns (1968), Hilton and Heathfield (1970), Ball (1973), Hendry (1974), and Renton (1975) for models of the United Kingdom; Brown (1960) and Helliwell et al. (1969, 1971) for models of Canada; Klein and Shinkai (1963), Ueno (1963), Ichimura et al. (1964), and Kosobud and Minami (1977) for models of Japan; Suits (1964) for a model of Greece; Evans (1969b) for a model of France; Evans (1970) for a model of Israel; Agarwala (1970) for a model of India; and Sylos-Labini (1974) for a model of Italy. In several of these countries these models have been further developed by official agencies, such as the Treasury model of the United Kingdom, the Bank of Japan and Economic Planning Agency models of Japan, and the Bundesbank model for the Federal Republic of Germany. Econometric models have also been built for most centrally planned economies, including the U.S.S.R., the German Democratic Republic, Czechoslovakia, Hungary, and the People's Republic of China. An interesting feature of some of the latter models is the mixture of deterministic and stochastic mechanisms for resource allocation, allowing some discretion for both planners' behavior and market-type mechanisms.

Here Y, national income, is composed of consumption, C, investment, I, and government expenditure, G. Consumption is determined by the consumption function $C(\cdot)$ as a function of income, other relevant variables (e.g. permanent income, assets, measures of income distribution) and a stochastic disturbance term, u. Investment is determined by the investment function $I(\cdot)$ as a function of income, other relevant variables (e.g. lagged income, profits, interest rates), and a stochastic disturbance term, v. Virtually all macroeconometric models involve these basic elements: a definition of national income or a group of such definitions, a consumption function or a group of such functions, and an investment function or a group of such functions. They have, in recent years, however, involved a greater and greater degree of disaggregation of variables and more and more variables as more aspects of the macroeconomy are taken into account.

The early postwar models involved less than ten stochastic equations, an example being the *Klein interwar model* of the U.S. economy over the period 1921–1941, involving three stochastic and three non-stochastic equations in six endogenous and four exogenous variables.[24]

An extremely influential model of the mid-1950s was the *Klein–Goldberger model* of the U.S. economy over the periods 1929–1941 and 1946–1952, which involved 15 stochastic and 5 non-stochastic equations in 20 endogenous and 14 exogenous variables.[25] Among the descendent econometric models of the U.S. economy based in part on or influenced by the Klein–Goldberger model are two models developed in the late 1960s and early 1970s, the Brookings model and the Wharton model.

The Brookings model was, at the time of its development in the 1960s the largest and most ambitious model of the U.S. economy, involving in its "standard" version 176 endogenous and 89 exogenous variables.[26] A major goal in building this model was that of advancing the state of the art in model building both via disaggregation and via the inclusion of sectors not treated in previous models. The resulting model, in representing the detailed structure of the economy, has been used both for structural analysis of cycles and for growth and policy evaluation.

The Wharton model was initiated in 1967 with a quarterly model of 76 endogenous and 42 exogenous variables. Since then later variants have involved the 1972 Wharton Annual and Industry model, with 346 endogenous and 90

[24]See Klein (1950) and Theil and Boot (1962). This model is also discussed in Christ (1966) and Theil (1971).

[25]See Klein and Goldberger (1955), Goldberger (1959), and Adelman and Adelman (1959).

[26]For the Brookings model see Duesenberry, Fromm, Klein and Kuh (1965, 1969), Fromm and Taubman (1968), Fromm (1971), and Fromm and Klein (1975).

exogenous variables, and the 1972 Wharton, Mark III model, with 201 endogenous and 104 exogenous variables.[27] The Wharton models are designed explicitly for developing forecasts of the future of the economy, particularly national income components and unemployment. They are regularly used to forecast ahead eight or more quarters under alternative assumptions regarding the exogenous variables, particularly government monetary and fiscal policy.

More recent macroeconometric models of the U.S., which stem largely from the Brookings and the Wharton models, include the 1968 FMP/MPS model of the Federal Reserve Board/MIT-Penn-SSRC, with 171 endogenous and 119 exogenous variables which emphasizes the monetary and financial sectors; the 1971 Chase Econometrics model, with 150 endogenous and 100 exogenous variables; the 1971 Fair short-run forecasting model with 19 endogenous and 20 exogenous variables; and the 1974 Data Resources Incorporated (DRI) model, with 718 endogenous and 170 exogenous variables in its initial version.[28]

Macroeconometric models have clearly tended to increase in size and scope, involving more variables, more sectors, and the inclusion of related models, such as input–output, financial, and microsimulation models. They have also tended to increase in complexity, including non-linearities. Another trend is the attempt to link various national macroeconometric models into a model of world trade flows.[29] These trends can be understood in terms of a rational response to the falling cost of scale and complexity, given the falling cost of computation and the availability of library programs for econometric routines and computerized data banks. These trends can be expected to continue in the future.

6.4. *Other econometric models*

An important trend in econometric models, particularly in macroeconometric models, has been that of the growing size, scale, and complexity of these models, as noted in the previous section. Another trend in econometric models has been that of applying such models in many other areas of economics, including fields in which such models have traditionally not been applied. The growing knowledge of econometric approaches and availability of library programs for economic

[27] For the Wharton model see Evans and Klein (1967, 1968), Evans (1969a), Evans, Klein and Saito (1972), and Howrey (1972). For the Wharton Annual and Industry model see Preston (1972, 1975). For the Wharton Mark III Quarterly model see McCarthy (1972), Duggal, Klein and McCarthy (1974), and Klein and Young (1980).

[28] For the FMP/MPS model see Rasche and Shapiro (1968), de Leeuw and Gramlich (1968, 1969), Ando and Modigliani (1969), Ando, Modigliani and Rasche (1972), and Muench et al. (1974). For the Chase Econometrics model see Evans (1974). For the Fair model see Fair (1971, 1974, 1976, 1979). For the DRI model see Eckstein, Green and Sinai (1974), Eckstein (1976), and Data Resources, Inc. (1976).

[29] Project LINK represents such an approach. See Ball (1973).

routines and data, including computerized data banks, have aided or accelerated this trend. In fact, econometric models have been developed in virtually all areas of economics, including commodity markets, crime, education, economic history, education, energy, health economics, housing, industrial organization, inflation, international trade and finance, labor economics, monetary economics, transportation, and urban and regional economics. In addition, there have been applications of econometric models to other social sciences, including demography, political science, and sociology.[30]

7. Uses of econometric models

The three principal uses of econometric models are structural analysis, forecasting, and policy evaluation, corresponding to the descriptive, predictive, and prescriptive uses of econometrics. These uses are closely related, the structure determined by structural analysis being used in forecasting, and policy evaluation being based largely on conditional forecasts. These uses will be discussed for both the general non-linear econometric model in (4.1) and (4.2) and the linear econometric model in (4.12) and (4.23).

Writing these models in vector notation and showing the lagged endogenous variables explicitly, the structural form of the general non-linear econometric model (4.1) can be written

$$f(y_t, y_{t-1}, x_t, \delta) = \varepsilon_t, \tag{7.1}$$

where y_t and y_{t-1} are vectors of current and lagged endogenous variables, x_t is a vector of exogenous variables at time t, δ is a vector of parameters, and ε_t is a vector of stochastic disturbance terms at time t. The corresponding reduced form of the general non-linear econometric model in (4.2) is

$$y_t = \phi(y_{t-1}, x_t, \delta) + u_t, \tag{7.2}$$

where u_t is a vector of stochastic disturbance terms for the reduced form at time t. The structural form of the general linear econometric model, also allowing for lagged endogenous variables, is

$$y_t\Gamma + y_{t-1}B_1 + x_tB_2 = \varepsilon_t, \tag{7.3}$$

[30]For discussions of and references to the applications of econometric models in these areas see Intriligator (1978, ch. 9, 13). See especially the bibliography for Chapter 9, which includes references to a wide range of applications of econometrics.

as in (4.35), while the corresponding reduced form of the general linear econometric model is

$$y_t = y_{t-1}\Pi_1 + x_t\Pi_2 + u_t, \tag{7.4}$$

as in (4.36).

7.1. Structural analysis

Structural analysis refers to the use of an estimated econometric model for the quantitative measurement of the underlying interrelationships of the system under consideration. One aspect of structural analysis is the estimation of the parameters of the structural form, particularly the elements of δ in (7.1) and the elements of the Γ, B_1, and B_2 matrices of structural coefficients in (7.3), in order to measure the extent of influence of each of the included variables in any equation of the model.

Another aspect of structural analysis is the estimation of the parameters of the reduced form, particularly the elements of the Π_1 and Π_2 matrices of reduced-form coefficients in (7.4). As indicated in (4.24), these coefficients have the interpretation of comparative statics results of the model, the effects of each of the exogenous variables on each of the endogenous variables of the model. These coefficients are also called *impact multipliers* since they indicate the impact of a change in a current value of an exogenous variable on the current value of an endogenous variable.

A third aspect of structural analysis, for a model with lagged endogenous variables, is the estimation of the final form (4.40), particularly the successive coefficients of the current and lagged exogenous variables in (4.41). These coefficients can also be used to estimate the interim and long-term multipliers of the econometric model. The *τ-period cumulative multiplier* measures the effect on each of the endogenous variables of a change in each of the exogenous variables over τ periods, given as

$$\left.\frac{\partial y}{\partial x}\right|_{\tau} = \sum_{j=0}^{\tau-1} \Pi_2\Pi_1^j = \Pi_2\left(I + \Pi_1 + \Pi_1^2 + \cdots + \Pi_1^{\tau-1}\right), \tag{7.5}$$

where I is the identity matrix.

Setting $\tau = 1$ yields the impact multipliers Π_2, the coefficients of the exogenous variables in the reduced form (4.36). Finite values of τ larger than 1 yield the *cumulative interim multipliers*, indicating the change in each endogenous variable as each exogenous variable experiences a sustained increase over τ periods.

Taking the limit as $\tau \to \infty$ yields the *long-term multipliers*:

$$\left.\frac{\partial y}{\partial x}\right|_{\infty} = \lim_{\tau \to \infty} \Pi_2 \left(I + \Pi_1 + \Pi_1^2 + \cdots + \Pi_1^{\tau} \right) = \Pi_2 (I - \Pi_1)^{-1}, \tag{7.6}$$

assuming the power series converges.[31] These long-term multipliers measure the effect on the endogenous variables of a permanent sustained increase in the exogenous variables.

7.2. *Forecasting*

Forecasting refers to the use of an estimated econometric model to predict quantitative values for certain variables, typically the endogenous variables of the model, outside the sample of data actually observed – typically a prediction for other times or places.[32]

The econometric approach to forecasting is typically based on the reduced form system, which, for the general non-linear case, can be written as in (7.2). A *short-run ex-ante forecast* of values taken by the endogenous variables at time $T+1$, given their values at time T, is then

$$\hat{y}_{T+1} = \phi\left(y_T, \hat{x}_{T+1}, \hat{\delta} \right) + \hat{u}_{T+1}. \tag{7.7}$$

The left-hand side, $\hat{y}_{T+1}$, represents the values forecasted for the endogenous variables at time $T+1$. Three sets of variables enter the $\phi(\cdot)$ function in (7.7). The first set of variables is y_T, the (current) values of the endogenous variables at

[31] The power series in (7.2), called a *Neumann expansion*, converges if $\lim \Pi_1^{\tau} = 0$, or, equivalently, if all characteristic roots of Π_1 have modulus less than unity. The long-term multiplier in (7.2) could have been obtained directly from (4.36) by noting that, in the long run, $y_t = y_{t-1}$, so

$$y_t = y_t \Pi_1 + x_t \Pi_2 + u_t.$$

Solving for y_t:

$$y_t = x_t \Pi_2 (I - \Pi_1)^{-1} + u_t (I - \Pi_1)^{-1},$$

implying (7.6).

[32] For a discussion of forecasting using an econometric model see Chapter 33 of this Handbook by Fair. Basic references on econometric forecasting are Theil (1961, 1966), Zarnowitz (1967), Klein (1971), and Christ (1975). See also Intriligator (1980, ch. 15). For discussions of forecasting based on autoregressive-moving average (ARMA) models see Box and Jenkins (1970), Cooper (1972), Zellner and Palm (1974), Aitchison and Dunsmore (1975), Nicholls et al. (1975), Pindyck and Rubinfeld (1976), Granger and Newbold (1977), Palm (1977), and Nerlove, Grether and Carvalho (1979).

time T, summarizing the systematic dependence on lagged values of the endogenous variables due to constant growth processes, distributed lag phenomena, etc. The second set of variables in $\phi(\cdot)$ in (7.7) is $\hat{x}_{T+1}$, the predicted future values of the exogenous variables. Since the x's are exogenous and hence determined on the basis of factors not explicitly treated in the econometric model, it is reasonable to require that these variables be forecast on the basis of factors other than those treated in the model itself, such as on the basis of extrapolation of past trends, expert opinion, or forecasts from another econometric model (as one example, an econometric model for an industry might use forecasts from a macroeconometric model). The third set of variables in $\phi(\cdot)$ in (7.7) is $\hat{\delta}$, representing the estimated parameters of the econometric model. The final term in (7.7) is $\hat{u}_{T+1}$, the "add factors", which can be interpreted as estimates of future values of the disturbance terms (or, alternatively, as adjustments of intercepts in each of the reduced-form equations). These add factors account for omitted variables, incorrect specification, and errors in measuring variables, which were the reasons for including the stochastic disturbance term in the first place. Their inclusion in short-term *ex-ante* forecasts is appropriate; excluding such terms would be tantamount to ignoring relevant considerations simply because they were omitted from the model. For example, in macroeconometric model forecasting it would be inappropriate to ignore major strikes, external shocks, or new technologies simply because they were not explicitly included in the model. Of course, the add factors are subjective, varying from individual to individual and thus not replicable. Their inclusion thus means that subjective expert opinion is combined with objective factors in generating forecasts. Experience indicates that such judgmental add factors can improve significantly on the accuracy of forecasts made with an econometric model.[33]

In the case of a linear econometric model the short-term *ex-ante* forecast, based on the reduced-form equation (7.4), takes the form

$$\hat{y}_{T+1} = y_T\hat{\Pi}_1 + \hat{x}_{T+1}\hat{\Pi}_2 + \hat{u}_{T+1}. \tag{7.8}$$

[33]The choice of values for the add factors $\hat{u}_{T+1}$ can also be guided by past residuals in estimating the model and past forecast errors, which provide clues to omitted variables, errors in measuring coefficients, and systematic biases in forecasting exogenous variables. For example, one approach is to define add factors so that the computed values of the endogenous variables at the most recent observation, as adjusted by the add factors, are the same as the observed values. See Klein (1971) and Haitovsky, Treyz and Su (1974). It should be noted that professional opinion is by no means unanimous on the subject of add factors. While most of the commercially available macroeconometric models use add factors, Ray Fair has argued against their use. Fair argues, as in Chapter 33 of the Handbook, that the use of add factors means that the information contained in *ex-ante* forecasts has no scientific value. From this point of view he argues that the inclusion of add factors is not appropriate. Others use add factors freely in actual forecasts. A reasonable intermediate position would be to assert that add factors are appropriate in attempting to obtain the most accurate *ex-ante* forecast for a given period, but not appropriate for using predictions from a model for testing and comparison purposes.

Here y_T, $\hat{x}_{T+1}$, and $\hat{u}_{T+1}$ are vectors as before, and $\hat{\Pi}_1$ and $\hat{\Pi}_2$ are matrices of estimated coefficients, summarizing the partial dependence of the forecasted values for the endogenous variables at time $T+1$ on their values at time T and on the values of the forecasted values of the exogenous variables at time $T+1$, respectively. The terms on the right-hand side of (7.8) show explicitly the (linear) dependence on current values of the endogenous variables, on future values of the exogenous variables, and on add factors, representing a combination of objective factors, in $y_T\hat{\Pi}_1$ and $\hat{x}_{T+1}\hat{\Pi}_2$, and subjective factors, in $\hat{u}_{T+1}$.

The econometric forecasts in (7.7) and (7.8) are called "*ex-ante* forecasts" because they are true forecasts, made before the event occurs. By contrast, an *ex-post* forecast, made after the event, would replace predicted values of the exogenous variables by their actual values and would replace the add factors by the zero expected values of the stochastic disturbance terms. Thus, the *short-term ex-post forecast* is, for the non-linear model:

$$\hat{\hat{y}}_{T+1} = \phi\left(y_T, x_{T+1}, \hat{\delta}\right) \tag{7.9}$$

and, for the linear model:

$$\hat{\hat{y}}_{T+1} = y_T\hat{\Pi}_1 + x_{T+1}\hat{\Pi}_2 = \hat{y}_{T+1} + \left(x_{T+1} - \hat{x}_{T+1}\right)\hat{\Pi}_2 - \hat{u}_{T+1}. \tag{7.10}$$

This *ex-post* forecast is useful in focusing on the explicitly estimated parts of the forecast, particularly the estimated coefficient matrices $\hat{\Pi}_1$ and $\hat{\Pi}_2$, eliminating the influence of $\hat{x}_{T+1}$ and $\hat{u}_{T+1}$, which are generally not explicitly estimated. It is possible to replicate *ex-post* forecasts, but not *ex-ante* forecasts. Furthermore, this forecast is optimal given a quadratic loss function.

There are several advantages to the econometric approach to forecasting in (7.7)–(7.10).[34] First, it provides a useful structure in which to consider explicitly various factors, including past values of variables to be forecast, values of other variables, and judgmental factors. Second, it leads to forecasts of variables that are consistent with one another since they must satisfy the requirements of the model, particularly its identities. Third, it leads to forecasts that are explicitly

[34] Both (7.4) and (7.5) refer to short-term forecasts. Long-term forecasts over a forecasting horizon h determine $\hat{y}_{T+h}$ on the basis of a succession of short-term forecasts or, equivalently, on the basis of the final form in (4.40), where the forecast of the endogenous variables at time $T+h$ is

$$\hat{y}_{T+h} = y_T\hat{\Pi}_1^h + \sum_{j=0}^{h-1} \hat{x}_{T+h-j}\hat{\Pi}_2\hat{\Pi}_1^j + \sum_{j=0}^{h-1} \hat{u}_{T+h-j}\hat{\Pi}_1^j.$$

Here y_T is the current value, as in (7.8), the $\hat{x}_{T+h-j}$ are successive expected future values of exogenous variables, and the $\hat{u}_{T+h-j}$ are successive expected future values of stochastic disturbance terms, where the last term on the right can itself be interpreted as the add factor for the long-term forecast.

conditional on current values of endogenous variables, expected future values of exogenous variables, add factors, and estimated coefficient matrices, facilitating analysis of the relative importance of each of these factors and tests of sensitivity. Fourth, and perhaps most important, it has a good record for accuracy and usefulness as compared to other approaches which tend to emphasize one aspect of an econometric forecast but exclude other aspects.

7.3. *Policy evaluation*

Policy evaluation refers to the use of an estimated econometric model to choose among alternative policies.[35] Assume there is a set of policy variables included among the exogenous variables of the model. The structural form (4.35) can then be written, for the non-linear model:

$$f(y_t, y_{t-1}, z_t, r_t, \delta) = \varepsilon_t, \tag{7.11}$$

and, for the linear model:

$$y_t\Gamma + y_{t-1}B_1 + z_tB_2 + r_tB_3 = \varepsilon_t, \tag{7.12}$$

where the vector x_t of exogenous variables has been divided into a vector of (non-policy) exogenous variables z_t and a vector of policy variables r_t, called the *instruments*. The corresponding reduced form is, for the non-linear model:

$$y_t = \phi(y_{t-1}, z_t, r_t, \delta) + u_t, \tag{7.13}$$

and, for the linear model:

$$y_t = y_{t-1}\Pi_1 + z_t\Pi_2 + r_t\Pi_3 + u_t, \tag{7.14}$$

where, in addition to (4.37) and (4.38),

$$\Pi_3 \equiv -B_3\Gamma^{-1}. \tag{7.15}$$

The problem of short-term policy evaluation is that of choosing at time T a particular set of policy variables for time $T+1$, given as r^*_{T+1}, where it is assumed that y_T is known. There are at least three alternative approaches to evaluating

[35] For further discussions of policy evaluation using an econometric model, see Chapters 34 and 35 of this Handbook by Taylor and Klein, respectively. Basic references on econometric policy evaluation are Tinbergen (1955, 1956), Theil (1961, 1964), Suits (1962), Hickman (1965), Fox, Sengupta and Thorbecke (1966), Naylor, Wertz and Wonnacott (1968), Naylor (1971), and Klein (1971, 1977). See also Intriligator (1978, ch. 16).

policy: the *instruments–targets approach*, the *social-welfare-function approach*, and the *simulation approach*.

In the *instruments–targets approach* it is assumed that there is a target for the endogenous variables y^*_{T+1}. The optimal instruments for the non-linear econometric model then solve the equation

$$y^*_{T+1}=\phi\left(y_T,\hat{z}_{T+1},r^*_{T+1},\hat{\delta}\right)+\hat{u}_{T+1}, \tag{7.16}$$

for r^*_{T+1}, where $\hat{z}_{T+1}$ is the vector of expected future values of the exogenous variables, $\hat{\delta}$ is the vector of estimated parameters, and $\hat{u}_{T+1}$ is a vector of add factors. In the linear case it is usually assumed that the number of instruments equals the number of targets, g, so the B_3 matrix is square, as is Π_3.[36] Assuming Π_3 is non-singular, and solving (7.14) for r^*_{T+1} yields

$$r^*_{T+1}=y^*_{T+1}\hat{\Pi}_3^{-1}-y_T\hat{\Pi}_1\hat{\Pi}_3^{-1}-\hat{z}_{T+1}\hat{\Pi}_2\hat{\Pi}_3^{-1}-\hat{u}_{T+1}\hat{\Pi}_3^{-1}, \tag{7.17}$$

giving the optimal value for the instruments as linear functions of the targets, the current values of the endogenous variables, the expected future values of the exogenous variables, and the add factors. This equation indicates the basic interdependence of policies and objectives, with optimal values of each instrument in general depending on all target variables. This approach leads to specific results, but it suffers from three difficulties: it does not allow for tradeoffs among the targets, it assumes that policymakers can specify targets, and it assumes there are enough independent instruments available.

The *social-welfare-function approach* to policy evaluation allows tradeoffs among the endogenous variables by assuming the existence of a social welfare function to be maximized by choice of the instruments subject to the constraints of the model. If $W(y_{T+1},r_{T+1})$ is the social welfare function, dependent on both endogenous variables and policy variables in the next period, the problem is

$$\max_{r_{T+1}} W(y_{T+1},r_{T+1}) \tag{7.18}$$

subject to the constraints of the econometric model. In the case of the non-linear model, W is maximized subject to (7.13), so the problem becomes

$$\max_{r_{T+1}} W\left(\phi\left(y_T,\hat{z}_{T+1},r_{T+1},\hat{\delta}\right)+\hat{u}_{T+1},r_{T+1}\right), \tag{7.19}$$

while in the case of the linear model, W is maximized subject to (7.14), so the

[36]More generally, the targets could be a subset of the endogenous variables, and the number of instruments can exceed (or equal) the number of targets, the difference between the number of instruments and the number of targets being the *policy degrees of freedom*.

problem becomes

$$\max_{r_{T+1}} W\left(y_T\hat{\Pi}_1 + \hat{z}_{T+1}\hat{\Pi}_2 + r_{T+1}\hat{\Pi}_3 + \hat{u}_{T+1}, r_{T+1}\right). \tag{7.20}$$

Frequently the social welfare function is a quadratic loss function, to be minimized. While this approach allows for tradeoffs among the endogenous variables it assumes that policymakers can specify a social welfare function.[37]

The *simulation approach* to policy evaluation does not require either targets or a social welfare function. This approach uses the estimated reduced form to determine alternative combinations of policy variables and endogenous variables for a given set of possible policies. If $r_{T+1}^1, r_{T+1}^2, \dots, r_{T+1}^S$ represent a set of alternative possible policies the simulation approach would determine the endogenous variables implied by each such policy, where, in the non-linear case,

$$\hat{y}_{T+1}^q = \phi\left(y_T, \hat{z}_{T+1}, r_{T+1}^q, \hat{\delta}\right) + \hat{u}_{T+1}, \tag{7.21}$$

and, in the linear case,

$$\hat{y}_{T+1}^q = y_T\hat{\Pi}_1 + \hat{z}_{T+1}\hat{\Pi}_2 + r_{T+1}^q\hat{\Pi}_3 + \hat{u}_{T+1}, \qquad q = 1,2,\dots,S. \tag{7.22}$$

The policymaker would provide the model builder with the alternative policies, and the model builder would, in turn, provide the decisionmaker with their consequences for the endogenous variables. The policymaker would then choose a desired policy and its outcome, r_{T+1}^*, y_{T+1}^*, where r_{T+1}^* is one of the alternative policies available.[38] This approach does not require information on the tastes of the policymaker, such as targets or a social welfare function. Rather, it requires that the policymaker formulate an explicit set of policy alternatives and that an estimated econometric model incorporating the appropriate policy variables be available. Simulation, based in part on communication between policymaker and model builder, represents a valuable approach to policy evaluation that could be used in any policy area in which there exists a relevant estimated econometric model.

[37]See Pindyck (1973), Chow (1975, 1981), Ando and Palash (1976), Klein (1977), and Fair (1978) for an extension of this approach to the problem of optimal control using an econometric model, involving the choice of a time path for policy variables so as to maximize the sum over time (or, in the continuous case, the integral over time) of a social welfare function.

[38]There may, of course, be a problem of simultaneously making a consistent model of expectations of other agents and choosing optimal behavior, as is recognized in the rational expectations literature. In particular, when it becomes necessary to model how agents formulate expectations with respect to alternative policies, some of the structural parameters might themselves change as a result of policy choices, creating severe problems in estimating an econometric model. See Lucas and Sargent (1980) and Sargent (1981).

8. Conclusion

This survey of economic and econometric models indicates that there is a wide range of models and applications. There are many approaches to modeling, and even in the standard linear stochastic algebraic model of econometrics there are many alternative specifications available. These models have been applied in many different areas; in fact, in virtually all areas of economics and in some related social sciences. The models have been used for various purposes, including structural analysis, forecasting, and policy evaluation. Clearly this area is an extremely rich one in which much has been accomplished and much more will be accomplished in the future.

The great diversity of uses and results in the area of economic and econometric models can perhaps be underscored by mentioning some of the issues that have not been treated or treated only briefly, most of which are discussed elsewhere in this Handbook. These issues include, among others:

- Adaptive expectations
- Aggregation
- Asymptotic results
- Autoregressive moving average (ARMA) models
- Bayesian estimation
- Causality
- Certainty equivalence
- Computer simulation
- Disequilibrium models
- Distributed lags
- Dynamic multipliers
- Dynamic simulation
- Errors in variables models
- Exact finite sample results
- Expectations
- Functional forms for relationships
- Identification
- Lag structures
- Latent variables
- Limited dependent variables
- Matrix transition models
- Measurement errors
- Model simplicity/complexity
- Optimal control
- Partial adjustment models
- Path analysis
- Pooling cross-section and time-series data
- Qualitative economics
- Qualitative variables
- Random coefficients model
- Rational expectations
- Residual analysis
- Robust estimation
- Seasonality
- Seemingly unrelated equations
- Sequential hypothesis testing
- Specification error
- Spectral analysis
- Stochastic equilibrium
- Structural change
- Testing
- Time-varying parameters
- Unobserved variables

References

Adelman, I and F. L. Adelman (1959) "The Dynamic Properties of the Klein–Goldberger Model", *Econometrica*, 27, 596–625.

Agarwala, R. (1970) *An Econometric Model of India, 1948–61*. London: Frank Cass & Co.

Aitchison, J. and I. R. Dunsmore (1975) *Statistical Prediction Analysis*. Cambridge: Cambridge University Press.

Ando, A. and F. Modigliani (1969) "Econometric Analyses of Stabilization Policies", *American Economic Review*, 59, 296–314.

Ando, A., F. Modigliani and R. Rasche (1972) "Equations and Definitions of Variables for the FRB–MIT–Penn Econometric Model, November, 1969", in: B. G. Hickman (ed.), *Economic Models of Cyclical Behavior*. National Bureau of Economic Research. New York: Columbia University Press.

Ando, A. and C. Palash (1976) "Some Stabilization Problems of 1971–1975, with an Application of Optimal Control Algorithms", *American Economic Review*, 66, 346–348.

Apostol, T. (1974) *Mathematical Analysis* (2nd edn.). Reading Mass.: Addison-Wesley Publishing Co.

Arnold, S. F. (1981) *The Theory of Linear Models and Multivariable Analysis*. New York: John Wiley & Sons, Inc.

Arrow, K. J., H. B. Chenery, B. S. Minhas and R. M. Solow (1961) "Capital–Labor Substitution and Economic Efficiency", *Review of Economics and Statistics*, 43, 225–235.

Ball, R. J. (1968) "Econometric Model Building," in: *Mathematical Model Building in Economics and Industry*. London: Charles Griffen & Co., Ltd.

Ball, R. J. (ed.) (1973) *The International Linkage of Econometric Models*. Amsterdam: North-Holland Publishing Company.

Ball, R. J. and T. Burns (1968) "An Econometric Approach to Short-Run Analysis of the United Kingdom Economy, 1955–1966", *Operational Research Quarterly*, 19, 225–256.

Barten, A. P. (1977) "The Systems of Consumer Demand Functions Approach: A Review", in: M. D. Intriligator (ed.), *Frontiers of Quantitative Economics*, vol. III. Amsterdam: North-Holland Publishing Co.

Beach, E. F. (1957) *Economic Models: An Exposition*. New York: John Wiley & Sons, Inc.

Belsley, D. A. (1979) "On the Computational Competitiveness of FIML and 3SLS in the Estimation of Nonlinear Simultaneous-Equation Models", *Journal of Econometrics*, 9, 315–342.

Belsley, D. A. (1980) "On the Efficient Computation of the Nonlinear FIML Estimator", *Journal of Econometrics*, 14, 203–225.

Bergstrom, A. R. (1967) *Selected Economic Models and Their Analysis*. New York: American Elsevier Publishing Co., Inc.

Bowden, R. (1973) "The Theory of Parametric Identification", *Econometrica*, 41, 1069–1074.

Box, G. E. P. and D. R. Cox (1964) "An Analysis of Transformations", *Journal of the Royal Statistical Society*, B, 26, 211–243.

Box, G. E. P. and G. M. Jenkins (1970) *Time Series Analysis: Forecasting and Control*. San Francisco: Holden-Day.

Bridge, J. L. (1971) *Applied Econometrics*. Amsterdam: North-Holland Publishing Co.

Brown, J. A. C. and A. S. Deaton (1972) "Surveys in Applied Economics: Models of Consumer Behavior", *Economic Journal*, 82, 1143–1236.

Brown, M. (ed.) (1967) *The Theory and Empirical Analysis of Production*. National Bureau of Economic Research. New York: Columbia University Press.

Brown, M. and J. S. de Cani (1963) "Technological Change and the Distribution of Income", *International Economic Review*, 4, 289–309.

Brown, T. M. (1960) *Specification and Uses of Econometric Models*. London: Macmillan & Co., Ltd.

Chow, G. (1973) "On the Computation of Full Information Maximum Likelihood Estimates for Nonlinear Equation Systems", *Review of Economics and Statistics*, 55, 104–109.

Chow, G. (1975) *Analysis and Control of Dynamic Economic Systems*. New York: John Wiley & Sons, Inc.

Chow, G. (1981) *Econometric Analysis by Control Methods*. New York: John Wiley & Sons, Inc.

Christ, C. (1966) *Econometric Models and Methods*. New York: John Wiley & Sons, Inc.

Christ, C. (1975) "Judging the Performance of Econometric Models of the U.S. Economy", *International Economic Review*, 16, 54–74.
Christensen, L. R., D. W. Jorgenson and L. J. Lau (1973) "Transcendental Logarithmic Production Frontiers", *Review of Economics and Statistics*, 55, 28–45.
Cooper, R. L. (1972) "The Predictive Performance of Quarterly Econometric Models of the United States", in: B. G. Hickman (ed.), *Econometric Models of Cyclical Behavior*. New York: Columbia University Press.
Cramer, J. S. (1969) *Empirical Econometrics*. Amsterdam: North-Holland Publishing Co.
Data Resources, Inc. (1976) *The Data Resources National Economic Information System*. Amsterdam: North-Holland Publishing Co.
Deaton, A. S. (1975) *Models and Projections of Demand in Post-War Britain*. London: Chapman and Hall; New York: Halsted Press.
de Leeuw, F. and E. M. Gramlich (1968) "The Federal Reserve–MIT Econometric Model", *Federal Reserve Bulletin*, 54, 11–40.
de Leeuw, F. and E. M. Gramlich (1969) "The Channels of Monetary Policy: A Further Report on the Federal Reserve–MIT Model", *Journal of Finance*, 24, 265–290.
Douglas, P. H. (1948) "Are There Laws of Production?", *American Economic Review*, 38, 1–49.
Dhrymes, P. J. (1981) *Distributed Lags* (revised edn.). Amsterdam: North-Holland Publishing Company.
Duesenberry, J. S., G. Fromm, L. R. Klein and E. Kuh (eds.) (1965) *The Brookings Quarterly Econometric Model of the United States*. Chicago: Rand-McNally and Company.
Duesenberry, J. S., G. Fromm, L. R. Klein and E. Kuh (eds.) (1969) *The Brookings Model: Some Further Results*. Chicago: Rand-McNally & Company.
Duggal, V. G., L. R. Klein and M. D. McCarthy (1974) "The Wharton Model Mark III: A Modern IS–LM Construct", *International Economic Review*, 15, 572–594.
Eckstein, O. (ed.) (1976) *Parameters and Policies in the U.S. Economy*. Amsterdam: North-Holland Publishing Co.
Eckstein, O., E. W. Green and A. Sinai (1974) "The Data Resources Model: Uses, Structure, and Analysis of the U.S. Economy", *International Economic Review*, 15, 595–615.
Enke, S. (1951) "Equilibrium among Spatially Separated Markets: Solution by Electric Analogue", *Econometrica*, 19, 40–47.
Evans, M. K. (1969a) *Macroeconomic Activity: Theory, Forecasting and Control: An Econometric Approach*. New York: Harper & Row.
Evans, M. K. (1969b) *An Econometric Model of the French Economy: A Short-Term Forecasting Model*. Paris: Organization for Economic Cooperation and Development.
Evans, M. K. (1970) "An Econometric Model of the Israeli Economy 1952–1953", *Econometrica*, 38, 624–660.
Evans, M. K. (1974) "Econometric Models", in: W. F. Butler, R. A. Kavesh and R. B. Platt (eds.), *Methods and Techniques of Business Forecasting*. Englewood Cliffs, N.J.: Prentice-Hall, Inc.
Evans, M. K. and L. R. Klein (1967) *The Wharton Econometric Forecasting Model*. Philadelphia: Economics Research Unit, Wharton School, University of Pennsylvania.
Evans, M. K. and L. R. Klein (1968) *The Wharton Econometric Forecasting Model* (2nd enlarged edn.). Philadelphia: Economics Research Unit, Wharton School, University of Pennsylvania.
Evans, M. K., L. R. Klein and M. Saito (1972) "Short Run Prediction and Long Run Simulation of the Wharton Model", in: B. G. Hickman (ed.), *Econometric Models of Cyclical Behavior*. National Bureau of Economic Research. New York: Columbia University Press.
Fair, R. C. (1971) *A Short-Run Forecasting Model of the United States Economy*. Lexington: Heath Lexington Books.
Fair, R. C. (1974) "An Evaluation of a Short-Run Forecasting Model", *International Economic Review*, 15, 285–303.
Fair, R. C. (1976) *A Model of Macroeconomic Activity, Volume II: The Empirical Model*. Cambridge: Ballinger Publishing Company.
Fair, R. C. (1978) "The Use of Optimal Control Techniques to Measure Economic Performance", *International Economic Review*, 19, 289–309.
Fair, R. C. (1979) "An Analysis of the Accuracy of Four Macroeconometric Models", *Journal of Political Economy*, 87, 701–718.

Fair, R. C. and W. R. Parke (1980) "Full-Information Estimates of a Nonlinear Macroeconometric Model", *Journal of Econometrics*, 13, 269–291.
Ferguson, C. E. (1969) *The Neoclassical Theory of Production and Distribution*. New York: Cambridge University Press.
Fisher, F. M. (1966) *The Identification Problem in Econometrics*. New York: McGraw-Hill Book Company.
Fox, K. A., J. K. Sengupta and E. Thorbecke (1966) *The Theory of Quantitative Economic Policy, with Applications to Economic Growth and Stabilization*. Amsterdam: North-Holland Publishing Co.
Frisch, R. (1965) *Theory of Production*. Dordrecht: Reidel; Chicago: Rand-McNally.
Fromm, G. (ed.) (1971) *Tax Incentives and Capital Spending*. Amsterdam: North-Holland Publishing Co.
Fromm, G. and L. R. Klein (1973) "A Comparison of Eleven Econometric Models of the United States", *American Economic Review*, 63, 385–393.
Fromm, G. and L. R. Klein (eds.) (1975) *The Brookings Model: Perspective and Recent Developments*. Amsterdam: North-Holland Publishing Co.
Fromm, G. and P. Taubman (1968) *Policy Simulations with an Econometric Model*. Washington, D.C.: Brookings Institution.
Gallant, A. R. and D. W. Jorgenson (1979) "Statistical Inference for a System of Simultaneous, Nonlinear, Implicit Equations in the Context of Instrumental Variable Estimation", *Journal of Econometrics*, 11, 275–302.
Gallant, A. R. and A. Holly (1980) "Statistical Inference in an Implicit, Nonlinear, Simultaneous Equation Model in the Context of Maximum Likelihood Estimation", *Econometrica*, 48, 901–929.
Goldberger, A. S. (1959) *Impact Multipliers and Dynamic Properties of the Klein–Goldberger Model*. Amsterdam: North-Holland Publishing Co.
Goldberger, A. S. and O. D. Duncan (eds.) (1973) *Structural Equation Models in the Social Sciences*. New York: Seminar Press.
Goldfeld, S. M. and R. E. Quandt (1968) "Nonlinear Simultaneous Equations: Estimation and Prediction", *International Economic Review*, 9, 113–146.
Goldfeld, S. M. and R. E. Quandt (1972) *Nonlinear Methods in Econometrics*. Amsterdam: North-Holland Publishing Co.
Goldfeld, S. M. and R. E. Quandt (eds.) (1976) *Studies in Nonlinear Estimation*. Cambridge: Ballinger Publishing Co.
Granger, C. W. J. and P. Newbold (1977) *Forecasting Economic Time Series*. New York: Academic Press.
Griliches, Z. (1967) "Distributed Lags: A Survey", *Econometrica*, 35, 16–49.
Haitovsky, Y., G. Treyz and V. Su (1974) *Forecasts with Quarterly Macroeconometric Models*. National Bureau of Economic Research. New York: Columbia University Press.
Helliwell, J. F., L. H. Officer, H. T. Shapiro and J. A. Stewart (1969) *The Structure of RDX1*. Ottawa: Bank of Canada.
Helliwell, J. F., H. T. Shapiro, G. R. Sparks, I. A. Stewart, F. W. Gerbet and D. R. Stevenson (1971) *The Structure of RDX2*. Ottawa: The Bank of Canada.
Hendry, D. F. (1974) "Stochastic Specification in an Aggregate Demand Model of the United Kingdom", *Econometrica*, 42, 559–578.
Hickman, B. G. (ed.) (1965) *Quantitative Planning of Economic Policy*. Washington, D.C.: The Brookings Institution.
Hickman, B. G. (ed.) (1972) *Econometric Models of Cyclical Behavior*. National Bureau of Economic Research. New York: Columbia University Press.
Hilton, K. and D. F. Heathfeld (eds.) (1970) *The Econometric Study of the United Kingdom; Proceedings of the 1969 Southampton Conference on Short-Run Econometric Models of the U.K. Economy*. London: Macmillan.
Hoffman, K. (1975) *Analysis in Euclidean Space*. Englewood Cliffs, N.J.: Prentice-Hall, Inc.
Houthakker, H. S. (1957) "An International Comparison of Household Expenditure Patterns Commemorating the Centenary of Engel's Laws", *Econometrica*, 25, 532–551.
Houthakker, H. S. (1965) "New Evidence on Demand Elasticities", *Econometrica*, 33, 277–288.
Howrey, E. P. (1972) "Dynamic Properties of a Condensed Version of the Wharton Model", in: B. G. Hickman (ed.), *Econometric Models of Cyclical Behavior*. New York: Columbia University Press.

Ichimura, S., L. R. Klein, S. Koizumi, K. Sato and Y. Shinkai (1964) "A Quarterly Econometric Model of Japan, 1952–59", *Osaka Economic Papers*, 12, 19–44.
Intriligator, M. D. (1971) *Mathematical Optimization and Economic Theory*. Englewood Cliffs, N.J.: Prentice-Hall, Inc.
Intriligator, M. D. (1978) *Econometric Models, Techniques, and Applications*. Englewood Cliffs, N.J.: Prentice-Hall, Inc.; Amsterdam: North-Holland Publishing Co.
Intriligator, M. D. (1981) "Mathematical Programming, with Applications to Economics", in: K. J. Arrow and M. D. Intriligator (eds.), *Handbook of Mathematical Economics*, Vol. I. Amsterdam: North-Holland Publishing Co.
Jorgenson, D. W. and J. J. Laffont (1974) "Efficient Estimation of Nonlinear Simultaneous Equations with Additive Disturbances", *Annals of Economic and Social Measurement*, 3, 615–640.
Kalman, P. J. and M. D. Intriligator (1973) "Generalized Comparative Statics, with Applications to Consumer and Producer Theory", *International Economic Review*, 14, 473–486.
Kendall, M. G. (1968) "Introduction to Model Building and its Problems", in: *Mathematical Model Building in Economics and Industry*. London: Charles Griffen & Co., Ltd.
Klein, L. R. (1950) *Economic Fluctuations in the United States, 1921–1941*. Cowles Commission Monograph No. 11. New York: John Wiley & Sons, Inc.
Klein, L. R. (1971) "Forecasting and Policy Evaluation using Large-Scale Econometric Models: The State of the Art", in: M. D. Intriligator (ed.), *Frontiers of Quantitative Economics*. Amsterdam: North-Holland Publishing Co.
Klein, L. R. (1977) "Economic Policy Formation through the Medium of Econometric Models", in: M. D. Intriligator (ed.), *Frontiers of Quantitative Economics*, Vol. III. Amsterdam: North-Holland Publishing Co.
Klein, L. R., R. J. Ball, A. Hazlewood and P. Vandome (1961) *An Econometric Model of the United Kingdom*. Oxford: Basil Blackwell.
Klein, L. R. and E. Burmeister (eds.) (1976) *Econometric Model Performance: Comparative Simulation Studies of the U.S. Economy*. Philadelphia: University of Pennsylvania Press.
Klein, L. R. and A. S. Goldberger (1955) *An Econometric Model of the United States, 1929–1952*. Amsterdam: North-Holland Publishing Co.
Klein, L. R. and Y. Shinkai (1963) "An Econometric Model of Japan, 1930–59", *International Economic Review*, 4, 1–28.
Klein, L. R. and R. M. Young (1980) *An Introduction to Econometric Forecasting Models*. Lexington: Lexington Books.
Kmenta, J. (1967) "On the Estimation of the CES Production Function", *International Economic Review*, 8, 180–189.
Kosobud, R. and R. Minami (eds.) (1977) *Econometric Studies of Japan*. Urbana: University of Illinois Press.
Leamer, E. E. (1978) *Specification Searches*. New York: John Wiley & Sons, Inc.
Lucas, R. E. and T. J. Sargent (eds.) (1980) *Rational Expectations and Econometric Practice*. Minneapolis: University of Minnesota Press.
Maddala, G. S. (1977) *Econometrics*. New York: McGraw-Hill Book Co.
Malinvaud, E. (1970) *Statistical Methods of Econometrics* (2nd rev. edn.). Amsterdam: North-Holland Publishing Co.
Marschak, J. and W. H. Andrews (1944) "Random Simultaneous Equations and the Theory of Production", *Econometrica*, 12, 143–205.
McCarthy, M. D. (1972) *The Wharton Quarterly Econometric Forecasting Model, Mark III*. Philadelphia: Economics Research Unit, University of Pennsylvania.
Minhas, B. S. (1963) *An International Comparison of Factor Costs and Factor Use*. Amsterdam: North-Holland Publishing Co.
Morehouse, N. F., R. H. Strotz and S. J. Horwitz (1950) "An Electro-Analog Method for Investigating Problems in Economic Dynamics: Inventory Oscillations", *Econometrica*, 18, 313–328.
Muench, T., A. Rolneck, N. Wallace and W. Weiler (1974) "Tests for Structural Change and Prediction Intervals for the Reduced Form of Two Structural Models of the U.S.: The FRB–MIT and Michigan Quarterly Models", *Annals of Economic and Social Measurement*, 3, 491–520.
Naylor, T. H. (1971) "Policy Simulation Experiments With Macroeconometric Models: The State of the Art", in: M. D. Intriligator (ed.), *Frontiers of Quantitative Economics*. Amsterdam: North-Holland Publishing Co.

Naylor, T. H., K. Wertz and T. Wonnacott (1968) "Some Methods For Evaluating the Effects of Economic Policies using Simulation Experiments", *Review of the International Statistical Institute*, 36, 184–200.
Nerlove, M. (1965) *Estimation and Identification of Cobb–Douglas Production Functions*. Amsterdam: North-Holland Publishing Co.
Nerlove, M. (1966) "A Tabular Survey of Macroeconometric Models", *International Economic Review*, 7, 127–175.
Nerlove, M., D. M. Grether and J. L. Carvalho (1979) *Analysis of Economic Time Series*. New York: Academic Press.
Nicholls, D. F., A. R. Pagan and R. D. Terrell (1975) "The Estimation and Use of Models with Moving Average Disturbance Terms: A Survey", *International Economic Review*, 16, 113–134.
Palm, F. (1977) "On Univariate Time Series Methods and Simultaneous Equation Econometric Models", *Journal of Econometrics*, 5, 379–388.
Phlips, L. (1974) *Applied Consumption Analysis*. Amsterdam: North-Holland Publishing Co.
Pindyck, R. S. (1973) *Optimal Planning for Economic Stabilization*. Amsterdam: North-Holland Publishing Co.
Pindyck, R. S. and D. L. Rubinfeld (1976) *Econometric Models and Economic Forecasts*. New York: McGraw-Hill Book Co.
Pollak, R. A. and T. J. Wales (1969) "Estimation of the Linear Expenditure System", *Econometrica*, 37, 611–628.
Powell, A. A. (1974) *Empirical Analytics of Demand Systems*. Lexington: Lexington Books.
Prais, S. J. and H. S. Houthakker (1955) *The Analysis of Family Budgets*. New York: Cambridge University Press.
Preston, R. S. (1972) *The Wharton Annual and Industry Forecasting Model*. Philadelphia: Economics Research Unit, Wharton School, University of Pennsylvania.
Preston, R. S. (1975) "The Wharton Long-Term Model: Input–Output Within the Context of a Macro Forecasting Model", *International Economic Review*, 16, 3–19.
Quirk, J. and R. Saposnik (1968) *Introduction to General Equilibrium Theory and Welfare Economics*. New York: McGraw-Hill Book Company.
Rasche, R. H. and H. T. Shapiro (1968) "The FRB–MIT Econometric Model: Its Special Features and Implications for Stabilization Policies", *American Economic Review*, 58, 123–149.
Renton, G. A., Ed. (1975) *Modelling the Economy*. London: Heinemann.
Rothenberg, T. J. (1971) "Identification in Parametric Models", *Econometrica*, 38, 577–591.
Rothenberg, T. J. (1973) *Efficient Estimation with A Priori Information*. Cowles Foundation Monograph 23, New Haven: Yale University Press.
Rudin, W. (1964) *Principles of Mathematical Analysis*. New York: McGraw-Hill Book Company.
Samuelson, P. A. (1947) *Foundations of Economic Analysis*. Cambridge: Harvard University Press.
Samuelson, P. A. (1975) "The Art and Science of Macro Models", in: G. Fromm and L. R. Klein (eds.), *The Brookings Model: Perspective and Recent Developments*. Amsterdam: North-Holland Publishing Co.
Sargent, T. J. (1981) "Interpreting Economic Time Series", *Journal of Political Economy*, 89, 213–248.
Shapiro, H. T. and L. Halabuk (1976) "Macro-Econometric Model Building in Socialist and Non-Socialist Countries: A Comparative Study", *International Economic Review*, 89, 213–248.
Silberberg, E. (1978) *The Structure of Economics: A Mathematical Analysis*. New York: McGraw-Hill Book Company.
Sims, C. A. (1974) "Distributed Lags", in: M. D. Intriligator and D. A. Kendrick (eds.), *Frontiers of Quantitative Economics*, vol. II. Amsterdam: North-Holland Publishing Co.
Smith, A. (1776) *The Wealth of Nations*, Edited by Edwin Cannan (1937). New York: The Modern Library.
Stone, R. (1954) *The Measurement of Consumers' Expenditure and Behavior in the United Kingdom, 1920–1938*. New York: Cambridge University Press.
Stone, R. (1972) *A Computable Model of Economic Growth: A Programme for Growth*, Vol. 1. Cambridge: Chapman and Hall.
Stone, R., A. Brown and D. A. Rowe (1965) "Demand Analysis and Projections in Britain: 1900–1970", in: J. Sandee (ed.), *Europe's Future Consumption*. Amsterdam: North-Holland Publishing Co.

Strotz, R. H., J. C. McAnulty and J. B. Naines, Jr. (1953) "Goodwin's Nonlinear Theory of the Business Cycle: An Electro-Analog Solution", *Econometrica*, 21, 390–411.
Suits, D. B. (1962) "Forecasting and Analysis with an Econometric Model", *American Economic Review*, 52, 104–132.
Suits, D. B. (1963) *The Theory and Application of Econometric Models*. Athens: Center of Economic Research.
Suits, D. B. (1964) *An Econometric Model of the Greek Economy*. Athens: Center of Economic Research.
Sylos-Labini, P. (1974) *Trade Unions, Inflation and Productivity*. Lexington: Lexington Books.
Theil, H. (1961) *Economic Forecasts and Policy* (2nd edn.). Amsterdam: North-Holland Publishing Co.
Theil, H. (1964) *Optimal Decision Rules for Government and Industry*. Chicago: Rand-McNally & Co.; Amsterdam: North-Holland Publishing Co.
Theil, H. (1966) *Applied Economic Forecasting*. Amsterdam: North-Holland Publishing Co.
Theil, H. (1971) *Principles of Econometrics*. New York: John Wiley & Sons, Inc.
Theil, H. (1975/1976) *The Theory and Measurement of Consumer Demand*, Vols. I and II. Amsterdam: North-Holland Publishing Co.
Theil, H. and J. C. G. Boot (1962) "The Final Form of Econometric Equation Systems", *Review of the International Statistical Institute*, 30, 136–152.
Tinbergen, J. (1955) *On the Theory of Economic Policy* (2nd edn.). Amsterdam: North-Holland Publishing Co.
Tinbergen, J. (1956) *Economic Policy: Principles and Design*. Amsterdam: North-Holland Publishing Co.
Tustin, A. (1953) *The Mechanism of Economic Systems*. Cambridge: Harvard University Press.
Ueno, H. (1963) "A Long-Term Model of the Japanese Economy, 1920–1958", *International Economic Review*, 4, 171–193.
Walters, A. A. (1963) "Production and Cost Functions: An Econometric Survey", *Econometrica*, 31, 1–66.
Walters, A. A. (1968) "Econometric Studies of Production and Cost Functions", *Encyclopedia of the Social Sciences*.
Wold, H. (1954) "Causality and Econometrics", *Econometrica*, 22, 162–177.
Wold, H. (1960) "A Generalization of Causal Chain Models", *Econometrica*, 28, 443–463.
Wold, H. (1968) *Econometric Model Building: Essays on the Causal Chain Approach*. Amsterdam: North-Holland Publishing Co.
Wold, H. and L. Jureen (1953) *Demand Analysis*. New York: John Wiley & Sons, Inc.
Zarnowitz, V. (1967) *An Appraisal of Short-Term Economic Forecasts*. New York: National Bureau of Economic Research.
Zellner, A. (1979) "Statistical Analysis of Econometric Models", *Journal of the American Statistical Association*, 74, 628–643, with Comments by D. A. Belsley and E. Kuh (643–645), C. F. Christ (645–646), P. M. Robinson (646–648), T. J. Rothenberg (648–650), and Rejoinder of A. Zellner (650–651).
Zellner, A., J. Kmenta and J. Drèze (1966) "Specification and Estimation of Cobb–Douglas Production Function Models", *Econometrica*, 34, 727–729.
Zellner, A. and F. Palm (1974) "Time Series Analysis and Simultaneous Equation Econometric Models", *Journal of Econometrics*, 2, 17–54.

Strotz, R. H., J. C. McAnulty and J. B. Naines, Jr. (1953) "Goodwin's Nonlinear Theory of the Business Cycle: An Electro-Analog Solution", *Econometrica*, 21, 390–411.
Suits, D. B. (1962) "Forecasting and Analysis with an Econometric Model", *American Economic Review*, 52, 104–132.
Suits, D. B. (1963) *The Theory and Application of Econometric Models*. Athens: Center of Economic Research.
Suits, D. B. (1964) *An Econometric Model of the Greek Economy*. Athens: Center of Economic Research.
Sylos-Labini, P. (1974) *Trade Unions, Inflation and Productivity*. Lexington: Lexington Books.
Theil, H. (1961) *Economic Forecasts and Policy* (2nd edn.). Amsterdam: North-Holland Publishing Co.
Theil, H. (1964) *Optimal Decision Rules for Government and Industry*. Chicago: Rand-McNally & Co.; Amsterdam: North-Holland Publishing Co.
Theil, H. (1966) *Applied Economic Forecasting*. Amsterdam: North-Holland Publishing Co.
Theil, H. (1971) *Principles of Econometrics*. New York: John Wiley & Sons, Inc.
Theil, H. (1975/1976) *The Theory and Measurement of Consumer Demand*, Vols. I and II. Amsterdam: North-Holland Publishing Co.
Theil, H. and J. C. G. Boot (1962) "The Final Form of Econometric Equation Systems", *Review of the International Statistical Institute*, 30, 136–152.
Tinbergen, J. (1955) *On the Theory of Economic Policy* (2nd edn.). Amsterdam: North-Holland Publishing Co.
Tinbergen, J. (1956) *Economic Policy: Principles and Design*. Amsterdam: North-Holland Publishing Co.
Tustin, A. (1953) *The Mechanism of Economic Systems*. Cambridge: Harvard University Press.
Ueno, H. (1963) "A Long-Term Model of the Japanese Economy, 1920–1958", *International Economic Review*, 4, 171–193.
Walters, A. A. (1963) "Production and Cost Functions: An Econometric Survey", *Econometrica*, 31, 1–66.
Walters, A. A. (1968) "Econometric Studies of Production and Cost Functions", *Encyclopedia of the Social Sciences*.
Wold, H. (1954) "Causality and Econometrics", *Econometrica*, 22, 162–177.
Wold, H. (1960) "A Generalization of Causal Chain Models", *Econometrica*, 28, 443–463.
Wold, H. (1968) *Econometric Model Building: Essays on the Causal Chain Approach*. Amsterdam: North-Holland Publishing Co.
Wold, H. and L. Jureen (1953) *Demand Analysis*. New York: John Wiley & Sons, Inc.
Zarnowitz, V. (1967) *An Appraisal of Short-Term Economic Forecasts*. New York: National Bureau of Economic Research.
Zellner, A. (1979) "Statistical Analysis of Econometric Models", *Journal of the American Statistical Association*, 74, 628–643, with Comments by D. A. Belsley and E. Kuh (643–645), C. F. Christ (645–646), P. M. Robinson (646–648), T. J. Rothenberg (648–650), and Rejoinder of A. Zellner (650–651).
Zellner, A., J. Kmenta and J. Dreze (1966) "Specification and Estimation of Cobb–Douglas Production Function Models", *Econometrica*, 34, 784–795.
Zellner, A. and F. Palm (1974) "Time Series Analysis and Simultaneous Equation Econometric Models", *Journal of Econometrics*, 2, 17–54.

Chapter 4

IDENTIFICATION

CHENG HSIAO*

University of Toronto

Contents

*This work was supported by National Science Foundation Grant SES80-07576 and by Social Sciences and Humanities Research Council of Canada Grant 410 80 0080. I am indebted to V. Bencivenga, J. Bossons, M. Deistler, Z. Griliches, F. M. Fisher, J. C. Ham, E. J. Hannan, M. D. Intriligator, J. B. Kadane, E. Leamer, K. Mahjoob, T. Rothenberg, and A. Zellner for helpful comments and discussions. All remaining errors are my own responsibility.

Handbook of Econometrics, Volume I, Edited by Z. Griliches and M.D. Intriligator

1. Introduction

The study of identification has been aptly linked to the design of experiments. In biological and physical sciences an investigator who wishes to make inferences about certain parameters can usually conduct controlled experiments to isolate relations. Presumably, in a well-designed experiment the treatment group and the control group are similar in every aspect except for the treatment. The difference in response may therefore be attributed to the treatment and the parameters of interest are identified.

In economics and other social sciences we are less fortunate. We observe certain facts (which are usually characterized by a set of quantities) and wish to arrange them in a meaningful way. Yet we cannot replace the natural conditions by laboratory conditions. We cannot control variables and isolate relations. The data are produced by an unknown structure, where the effects of a change in this structure are the objects of our study. None of these changes was produced by an investigator as in a laboratory experiment, and often the impact of one factor is confounded with impacts of other factors.

To reduce the complex real-world phenomena into manageable proportions, an economist has to make a theoretical abstraction. The result is a logical model presumably suited to explain the observed phenomena. That is, we assume that there exists an underlying structure which generated the observations of real-world data. However, statistical inference can relate only to characteristics of the distribution of the observed variables. Thus, a meaningful statistical interpretation of the real world through this structure can be achieved only if our assumption that real-world observations are generated by this structure and this structure alone is also true. The problem of whether it is possible to draw inferences from the probability distribution of the observed variables to an underlying theoretical structure is the concern of econometric literature on identification. We now illustrate this concept using the well-known demand and supply example.

Example 1.1

Let

p_t = price at time t,

q_t = quantity at time t.

Linear approximations of the demand and supply functions are

$$q_t = a + bp_t + u_{1t} \quad \text{(demand)}, \tag{1.1}$$

$$q_t = c + dp_t + u_{2t} \quad \text{(supply)}. \tag{1.2}$$

Assume u_1 and u_2 have an independent (over time) bivariate-normal distribution

$$N\left[\begin{pmatrix}0\\0\end{pmatrix},\begin{pmatrix}\sigma_{11} & \sigma_{12}\\ \sigma_{21} & \sigma_{22}\end{pmatrix}\right]. \tag{1.3}$$

Solving for p_t and q_t we obtain the distribution of the observed variables:

$$N\left[\begin{pmatrix}\dfrac{a-c}{d-b}\\[2ex] \dfrac{da-bc}{d-b}\end{pmatrix},\begin{pmatrix}\dfrac{\sigma_{11}+\sigma_{22}-2\sigma_{12}}{(d-b)^2} & \dfrac{d\sigma_{11}-(b+d)\sigma_{12}+b\sigma_{22}}{(d-b)^2}\\[2ex] \dfrac{d\sigma_{11}-(b+d)\sigma_{12}+b\sigma_{22}}{(d-b)^2} & \dfrac{d^2\sigma_{11}+b^2\sigma_{22}-2db\sigma_{12}}{(d-b)^2}\end{pmatrix}\right]. \tag{1.4}$$

There are five functions of parameters that can be estimated. But there are seven parameters of interest. Therefore, in general we can only estimate these functions of the parameters and not any parameter itself.

As is obvious, there are infinitely many possible values of (a, b, c, d, σ_{11}, σ_{22}, and σ_{12}) which could all generate the observed data $\{p_t, q_t\}$. Consequently, without additional information (in the form of *a priori* restrictions), the model specified by (1.1) and (1.2) cannot be estimated and therefore is not useful in confronting economic hypotheses with data. The study of identifiability is undertaken in order to explore the limitations of statistical inference (when working with economic data) or to specify what sort of *a priori* information is needed to make model parameters estimable. It is a fundamental problem concomitant with the existence of a structure. Logically it precedes all problems of estimation or of testing hypotheses.

The general formulation of the identification problems were made by Frisch (1934), Haavelmo (1944), Hurwicz (1950), Koopmans and Reiersøl (1950), Koopmans, Rubin and Leipnik (1950), Marschak (1942), Wald (1950), Working (1925, 1927), and others. An extensive study of the identifiability conditions for the simultaneous equations models under various assumptions about the underlying structures was provided by Fisher (1966). In this chapter I intend to survey the development of the subject since the publication of Fisher's book, although some pre-1966 results will be briefly reviewed for the sake of completeness. Because the purpose of this chapter is expository, I shall draw freely on the work by Anderson (1972), Deistler (1975, 1976), Deistler and Schrader (1979), Drèze (1975), Fisher (1966), Hannan (1969, 1971), Hatanaka (1975), Johnston (1972), Kadane (1975), Kohn (1979), Koopmans and Reiersøl (1950), Preston and Wall (1973), Richmond (1974), Rothenberg (1971), Theil (1971), Wegge (1965), Zellner (1971), etc. without specific acknowledgement in each case.

In Section 2 we define the basic concepts of identification. Section 3 derives some identifiability criteria for contemporaneous simultaneous equation models under linear constraints; Section 4 derives some identifiability criteria for dynamic models. Section 5 discusses criteria for models subject to non-linear continuous differentiable constraints and covariance restrictions with special emphasis on the applications to errors in variables and variance components models. The Bayesian view on identification and concluding remarks are given in Section 6.

2. Basic concepts[1]

It is generally assumed in econometrics that economic variables whose formation an economic theory is designed to explain have the characteristics of random variables. Let y be a set of such observations. A *structure* S is a complete specification of the probability distribution function of y, $P(y)$. The set of all *a priori* possible structures $\mathbb{S}$ is called a *model*. The identification problem consists in making judgements about structures, given the model $\mathbb{S}$ and the observations y. In most applications, y is assumed to be generated by a parametric probability distribution function $P(y|S)=P(y|\alpha)$, where α is an m-dimensional real vector. The probability distribution function P is assumed known, conditional on α, but α is unknown. Hence, a structure is described by a parametric point α, and a model is a set of points $A\subset R^m$. Thus, the problem of distinguishing between structures is reduced to the problem of distinguishing between parameter points. In this framework we have the following definitions.

Definition 2.1

Two structures $S=\alpha$ and $\bar{S}=\bar{\alpha}$ in A are said to be observationally equivalent if $P(y|\alpha)=P(y|\bar{\alpha})$ for all y.

Definition 2.2

The structure $S^0=\alpha^0$ in $\mathbb{S}$ is "globally identified" if there is no other α in A which is observationally equivalent.

Since the set of structures is simply a subset of R^m, it is possible that there may be a number of observationally equivalent structures, but they are isolated from each other. It is natural then to consider the concept of local identification. We define this concept in terms of the distance between two structures.

[1] Professor F. Fisher has pointed out to the author that "overidentification" is part of the general concept of identification and we ought to distinguish collinearity and lack of identification. The concept of "overidentification" has been found relevant for the existence of sampling moments and the efficiency of the estimates in simultaneous-equations models, presumably the topics will be treated in the chapters on estimation and sample distribution. The problem of collinearity is a case of underidentification according to Definition 2.1. So in this chapter we ignore these concepts.

Definition 2.3

A structure $S^0 = \alpha^0$ is "locally identified" if there exists an open neighborhood ω, containing α^0 such that no other α in ω is observationally equivalent to α^0.

On many occasions a structure S may not be identifiable, yet some of its characteristics may still be uniquely determinable. Since the characteristics of a structure S are described by an m-dimensional real vector α, we define this concept of identifiability of a substructure in terms of functions of α.

Definition 2.4

Let $\xi(\alpha)$ be a function of α. $\xi(\alpha)$ is said to be (locally) identifiable if (there exists an open neighborhood ω, such that) all parameter points which are observationally equivalent have the same value for $\xi(\alpha)$ (or lie outside ω).

A special case of $\xi(\alpha)$ will be coordinate functions. For instance, we may be interested in the identifiability of a subset of coordinates α_1 of α. Then the subset of coordinates α_1^0 of α^0 is said to be locally identifiable if there exists an open neighborhood ω, containing α^0, such that all parameter points observationally equivalent to α^0 have the same value for α_1^0 or lie outside ω.

In this chapter, instead of deriving identifiability criteria from the probability law of y [Bowden (1973), Rothenberg (1971)] we shall focus on the first- and second-order moments of y only. If the y are normally distributed, all information is contained in the first- and second-order moments. If the y are not normally distributed, observational information apart from the first and second moments may be available [Reiersøl (1950)]. However, most estimation methods use second-order quantities only; also, if a structure is identifiable with second-order moments, then it is identifiable with a probability law [Deistler and Seifert (1978)]. We shall therefore restrict ourselves to the first and second moments of y (or identifiability in the wide sense) for the sake of simplicity. Thus, we shall view two structures as observationally equivalent if they produce identical first- and second-order moments. Consequently, all the definitions stated above should be modified such that the statements with regard to the probability law of y are replaced by corresponding statements in terms of the first and second moments of y.

3. Contemporaneous simultaneous equation models

3.1. The model

In this section we consider the identification of a contemporaneous simultaneous equation model. We first discuss conditions for two structures to be observationally equivalent. We then derive identifiability criteria by checking conditions which will ensure that no two structures or part of two structures are observationally equivalent. Finally, we illustrate the use of these conditions by considering some simple examples.

For simplicity, let an economic theory specify a set of economic relations of the form

$$By_t + \Gamma x_t = u_t, \qquad t = 1,\dots,T, \tag{3.1}$$

where

y_t is a $G \times 1$ vector of observed endogenous variables;
x_t is a $K \times 1$ vector of observed exogenous variables;
B is a $G \times G$ matrix of coefficients;
Γ is a $G \times K$ matrix of coefficients; and
u_t is a $G \times 1$ vector of unobserved disturbances.

We assume that

Assumption 3.1

B is non-singular.

Assumption 3.2

$\lim_{T\to\infty} \sum_{t=1}^{T} x_t x_t'/T$ exists and is non-singular.

Assumption 3.3

$Eu_t = 0$, $Eu_t x_s' = 0$, and

$$Eu_t u_s' = \begin{cases} \Sigma & \text{if } t = s, \\ 0 & \text{otherwise.} \end{cases}$$

We note that the criteria to be derived can allow for the existence of lagged endogenous variables or autocorrelation in the disturbance term, but not both. We shall indicate how these generalizations can be made in Section 3.4.

3.2. *Observationally equivalent structures*

Suppose u_t has the density $\mathrm{P}(u_t|\Sigma)$, then the joint density of $(u_1,\dots,u_T)$ is

$$\prod_{t=1}^{T} \mathrm{P}(u_t|\Sigma). \tag{3.2}$$

The joint density of $(y_1,\dots,y_T)$ can be derived through the structural relation

(3.1) and the density of u's. Conditional on x_t, we have:

$$\begin{aligned}\mathrm{P}(y_1,\dots,y_T|B,\Gamma,\Sigma,x_1,\dots,x_T)&=|\det B|^T\prod_{t=1}^{T}\mathrm{P}(u_t|\Sigma)\\&=|\det B|^T\prod_{t=1}^{T}\mathrm{P}(By_t+\Gamma x_t|\Sigma).\end{aligned}\tag{3.3}$$

Suppose that we multiply (3.1) through by a $G\times G$ non-singular matrix F. This would involve replacing each equation of the original structure by a linear combination of the equations in that structure. The new structure may be written as

$$(FB)y_t+(F\Gamma)x_t=w_t,\tag{3.4}$$

where $w_t = Fu_t$, so that

$$\mathrm{P}(w_t|F\Sigma F')=|\det F|^{-1}\mathrm{P}(u_t|\Sigma).\tag{3.5}$$

The conditional density for the endogenous variables determined from the new structure is

$$\begin{aligned}|\det FB|^T\cdot\prod_{t=1}^{T}\mathrm{P}(w_t|F\Sigma F')&=|\det F|^T|\det B|^T|\det F|^{-T}\cdot\prod_{t=1}^{T}\mathrm{P}(u_t|\Sigma)\\&=|\det B|^T\cdot\prod_{t=1}^{T}\mathrm{P}(u_t|\Sigma),\end{aligned}\tag{3.6}$$

which is identical to the density (3.3) determined from the original structure. Hence, we say that the two structures (3.1) and (3.4) are observationally equivalent.

A special case of (3.4) occurs when we set $F=B^{-1}$ so that the transformed structure becomes

$$y_t=\Pi x_t+v_t,\tag{3.7}$$

where

$$B\Pi+\Gamma=0,\tag{3.8}$$

$$\begin{aligned}v_t&=B^{-1}u_t,\\Ev_tv_s'&=\begin{cases}B^{-1}\Sigma B^{-1\prime}=V & \text{if } t=s,\\0 & \text{otherwise.}\end{cases}\end{aligned}\tag{3.9}$$

Eq. (3.7) is called the "reduced form" of the "structural system" (3.1). We can alternatively write down the density of y in terms of the reduced form parameters (Π, V) as

$$\prod_{t=1}^{T} \psi(y_t - \Pi x_t | V). \tag{3.10}$$

From (3.4) and (3.6) we know that (3.3) and (3.10) yield identical density functions for the endogenous variables. Thus, if we postulate a set of structural relations (3.1) with reduced form (3.7), then all structures obtained by premultiplying the original structure by an arbitrary non-singular matrix of order G will have this same reduced form, and moreover, all these structures and the reduced forms will be observationally equivalent.[2]

Given that we will focus only on the first and second moments, we may formally state the conditions for two structures to be observationally equivalent in the following lemma.

Lemma 3.2.1

Two structures $S = (B, \Gamma, \Sigma)$ and $\bar{S} = (\bar{B}, \bar{\Gamma}, \bar{\Sigma})$ are observationally equivalent if and only if the following equivalent conditions hold:

(i) $B^{-1}\Gamma = \bar{B}^{-1}\bar{\Gamma}$ and $B^{-1}\Sigma B^{-1\prime} = \bar{B}^{-1}\bar{\Sigma}\bar{B}^{-1\prime}$.
(ii) There exists a non-singular matrix F such that

$$[\bar{B}, \bar{\Gamma}] = F[B, \Gamma] \tag{3.11}$$

and

$$\bar{\Sigma} = F\Sigma F'. \tag{3.12}$$

Proof

(i) follows from (3.1), (3.7), and (3.10). The probability density of the data is assumed to be completely specified by the first and second moments, i.e. the reduced-form parameters (Π, V). If S and $\bar{S}$ are observationally equivalent, they must have identical reduced-form parameter matrix and variance–covariance matrix. Condition (i) is exactly the condition which must be satisfied for the two reduced forms to be equal, and thus (i) is necessary and sufficient.

Now consider (ii). Sufficiency of (ii) is easy to check using (i). To prove its necessity, suppose S and $\bar{S}$ are observationally equivalent. Let $F = \bar{B}B^{-1}$. Assumption 3.1 implies that F is non-singular. Then $\bar{B}^{-1}\bar{\Gamma} = B^{-1}\Gamma$ implies that (3.11) holds. Let $w_t = Fu_t$, we have (3.12).

[2] In other words, we restrict our attention to the class of models which have identifiable reduced forms.

If there were no *a priori* restrictions on the parameters of the model (3.1), any non-singular F will be *admissible* in the sense that the transformed structure satisfies the same restrictions of the model.[3] The situation would indeed be hopeless. An infinite set of structures would have generated any set of sample observations. If economic theories are available to specify a set of *a priori* restrictions on the model, any transformed structure must also satisfy the same *a priori* restrictions if the transformed structure is to belong to the model that has been specified (i.e. the transformation is admissible). For instance, suppose economic theory specifies that (3.1) must obey certain restrictions, then the transformed structure (3.4) will have to obey these same restrictions. *A priori* information on the parameters of the model thus would rule out many structures which would otherwise be observationally equivalent, i.e. they imply restrictions on the elements of F. In this section we shall assume that

Assumption 3.4

All prior information is in the form of linear restrictions on B and Γ.

We ignore the information contained in the variance–covariance matrix for the moment. We shall discuss this situation together with non-linear *a priori* restrictions in Section 5.

The identification problem thus may be stated as

(a) If one considers the transformation matrix F used to obtain the transformed structure (3.4), do the *a priori* restrictions on B and Γ imply sufficient restrictions on the elements of F to make some or all of the coefficients in the original and transformed structures identical (and thus identifiable)?

Since Assumption 3.2 ensures the identifiability of Π (which is consistently estimable by the ordinary least squares method), we may state the identification problem in an alternative but equivalent fashion.

(b) Assuming the elements of Π to be known can one then solve for some or all of the elements of B and Γ uniquely?

3.3. *Identification in terms of trivial transformations*

We first consider the classical identifiability conditions for single equations or a set of equations. These will be expressed in terms of the equivalence conditions (3.11) and (3.12). From Definitions 2.2 and 2.4, we can define the identification

[3] The word "admissible" has a different meaning in statistical decision theory. Its use in these two different contexts should not be confused.

of the gth equation and the complete system as follows:[4]

Definition 3.3.1

The gth equation is identified if and only if all equivalent structures are related by admissible transformations which are trivial with respect to the gth equation. That is, all admissible transformation matrix should have the form

$$F=\begin{pmatrix} & \vdots & \\ 0..0 & f_{gg} & 0..0 \\ & \vdots & \end{pmatrix} \quad g\text{th row.} \qquad (3.13)$$

gth column

Definition 3.3.2

A system of equations is identified if and only if all equations in the system are identified. That is, the admissible transformation matrix has the form

$$F=\begin{pmatrix} f_{11} & 0 & . & . & 0 \\ 0 & f_{22} & 0 & . & 0 \\ . & 0 & . & & \\ \vdots & \vdots & & . & \\ 0 & 0 & & & f_{GG} \end{pmatrix}. \qquad (3.14)$$

When normalization conditions are imposed, such as setting the diagonal elements of B equal to 1 or constraining the variance of each equation to one, we constrain a structure $S^0=\alpha^0$ from a ray to a point; then f_{gg} in Definitions 3.3.1 and 3.3.2 will be equal to 1.

Let $A=[B,\Gamma]$, and a_g' be the gth row of A. We assume that there exists an $(G+K)\times R_g$ matrix ϕ_g' and a $1\times R_g$ vector d_g' whose elements are known constants, such that all prior information about the gth equation of the model including the normalization rule takes the form

$$a_g'\phi_g'=d_g'. \qquad (3.15)$$

[4]Strictly speaking, Definitions 3.3.1 and 3.3.2 are theorems derived from Definitions 2.2 and 2.4. However, the proof of these theorems is trivial and sheds no light on the problem. For simplicity of exposition, we treat them as definitions.

Thus, restrictions stating $\beta_{g2}=0$, $\beta_{g3}=-\beta_{g4}$, and $\beta_{gg}=1$ have the form

$$a_g'\phi_g' = \left(\beta_{g1},\dots,\beta_{gG},\gamma_{g1},\dots,\gamma_{gK}\right)\begin{pmatrix} 0 & 0 & 0 \\ 1 & 0 & . \\ 0 & 1 & . \\ . & 1 & . \\ . & . & . \\ . & . & 1 \\ . & . & . \\ . & . & . \\ 0 & 0 & 0 \end{pmatrix} \begin{matrix} \\ \\ \\ =(0,0,1) \\ \\ \leftarrow g\text{th row} \\ \\ \\ \\ \end{matrix} . \tag{3.16}$$

With the restrictions written in the form of (3.15), we have the following important theorems.

Theorem 3.3.1 (rank condition)

The gth equation is identified if and only if $\operatorname{rank}(A\phi_g')=G$.

Proof

By Lemma 3.2.1 we know that $\bar{a}_g'=f_g'A$, where f_g' denotes the gth row of F. Then

$$\bar{a}_g'\phi_g' = f_g'A\phi_g' = d_g' \tag{3.17}$$

has a solution for $f_g'=e_g'$, where e_g is a vector the elements of which are all zero except for the gth, which is unity. This solution is unique if and only if $A\phi_g'$ has full row rank, i.e. $\operatorname{rank}(A\phi_g')=G$.

Corollary 3.3.1 (order condition)

A necessary condition for the identifiability of the gth equation is that there are at least G linear restrictions including the normalization rule, or $(G-1)$ linear restrictions excluding the normalization rule.

Suppose the prior information about the gth equation is in the form of excluding certain variables from this equation (zero restrictions). Excluding the normalization rule, we may let this prior information take the form

$$a_g'\phi_{g\cdot}' = \underset{1\times(R_g-1)}{0}, \tag{3.18}$$

where $\phi_{g\cdot}$ is an $(R_g-1)\times(G+K)$ matrix. The elements of each row of $\phi_{g\cdot}$ are zero except for the jth which is set to unity in order to restrict the jth element of a_g' to zero.

Corollary 3.3.2

The gth equation is identified if and only if $\operatorname{rank}(A\phi'_{g\cdot}) = G-1$, i.e. the submatrix of A obtained by taking the columns of A with prescribed zeros in the gth row has rank $G-1$.

Proof

Consider two observationally equivalent structures $\bar{A}$ and A. We know that there must exist a non-singular F such that $\bar{A} = FA$ by Lemma 3.2.1. And we wish to show that the gth row of F has zeros everywhere but the gth entry. Without loss of generality, let $g = 1$, so that we are considering the first row of A or $\bar{A}$. Let

$$F = \underbrace{\begin{pmatrix} f_{11} & f'_{1\cdot} \\ f_{\cdot 1} & F_{\cdot\cdot} \end{pmatrix}}_{\substack{1 \quad G-1 \\ \text{columns}}} \begin{matrix} 1 \\ G-1 \end{matrix} \text{ rows,}$$

$$A = \begin{pmatrix} a'_1 \\ A_{\cdot} \end{pmatrix} \begin{matrix} 1 \\ G-1 \end{matrix} \text{ rows.}$$

By Lemma 3.2.1, we know that $\bar{A} = FA$; hence,

$$\bar{a}'_1 = f_{11}a'_1 + f'_{1\cdot}A_{\cdot\cdot}. \tag{3.19}$$

If $\bar{A}$ is equivalent to A,

$$\bar{a}'_1\phi'_{1\cdot} = (f_{11}a'_1 + f'_{1\cdot}A\cdot)\phi'_{1\cdot} = f'_{1\cdot}A_{\cdot}\phi'_{1\cdot} = 0, \tag{3.20}$$

because $a'_1\phi'_{1\cdot} = 0$. Thus, for $f'_{1\cdot} = 0$, $A_{\cdot}\phi'_{1\cdot}$ must have rank $G-1$. But $\operatorname{rank}(A\phi'_{1\cdot}) = \operatorname{rank}(A_{\cdot}\phi'_{1\cdot})$ because

$$A\phi'_1 = \begin{pmatrix} a_1\phi'_{1\cdot} \\ A_{\cdot}\phi'_{1\cdot} \end{pmatrix} = \begin{pmatrix} 0 \\ A_{\cdot}\phi'_1 \end{pmatrix}. \tag{3.21}$$

3.4. *Identification in terms of "linear estimable functions"*

We can also approach the identification problem by considering the reduced form (3.7). Assumptions 3.1–3.3 are sufficient to identify Π. We therefore treat the elements of Π as known and investigate the conditions which will allow some or all of the elements of B and Γ to be uniquely determined.

We approach the problem from the "theory of estimable functions" [Richmond (1974)]. The approach yields rather simple proofs of theorems on the identifiability of individual parameters, thus providing more general results than the conditions for single equation identification described above. The discussion is carried out at the level of the whole system of equations; that is to say, it is not confined to restrictions on the parameters of a particular equation, but rather is also applicable to restrictions on the parameters of different equations (cross-equation restrictions); [Kelly (1975), Mallela (1970), Wegge (1965), etc.]. The cross-equation constraints do exist naturally in a model. For instance, Hansen and Sargent (1980), Revankar (1980), and Wallis (1980) have shown that under the rational expectation hypothesis [Muth (1961)] a model containing an expected value of a variable would imply the existence of cross-equation restrictions.

To implement this more general approach, we first write (3.8) in a more convenient form. Let β_g' $(g=1,\dots,G)$ denote the gth row of B. Let γ_g' $(g=1,\dots,G)$ denote the gth row of Γ. Let $\delta'=(\beta_1',\beta_2',\dots,\beta_G',\gamma_1',\dots,\gamma_G')$ be the $1\times(G+K)G$ constant vector. Then, eq. (3.8) may be written in stacked form as

$$Q\delta=0, \tag{3.22}$$

where Q is the $GK\times G(G+K)$ matrix defined by

$$\underset{GK\times(G+K)G}{Q} = (I_G\otimes\Pi' : I_{GK})$$

$$= \begin{pmatrix} \Pi' & 0 & \cdot & \cdot & \cdot & 0 & I_K & 0 & \cdot & \cdot & \cdot & 0 \\ 0 & \Pi' & \cdot & & \cdot & \cdot & 0 & I_K & & & & \\ \cdot & \cdot & \cdot & & \cdot & \cdot & \cdot & 0 & \cdot & & & \\ \cdot & & \cdot & \cdot & \cdot & \cdot & \cdot & \cdot & & \cdot & & \\ \cdot & & & \cdot & \cdot & \cdot & \cdot & \cdot & & & \cdot & \\ 0 & \cdot & \cdot & 0 & 0 & \Pi' & 0 & 0 & & & & I_K \end{pmatrix}. \tag{3.23}$$

In this equation, $\otimes$ denotes the kronecker product [e.g. see Theil (1972, p. 303)], I_K denotes the $K\times K$ identity matrix, and I_{GK} denotes the $GK\times GK$ identity matrix.

We now assume that additional prior information, including normalization rules, is available in the form of R linear restrictions on the elements of δ. We are given then:

$$\Phi\delta=d, \tag{3.24}$$

where Φ is an $R\times G(G+K)$ matrix and d is an $R\times 1$ vector with known elements. Combining (3.22) and (3.24), and letting

$$W=\begin{pmatrix} Q \\ \Phi \end{pmatrix}, \tag{3.25}$$

we have

$$W\delta=\begin{pmatrix} 0 \\ d \end{pmatrix}=\tilde{d}. \tag{3.26}$$

We note that (3.26) is simply a set of $(GK+R)$ linear equations with $G(G+K)$ unknowns, δ. Thus, the condition for identifying all the parameters of (3.1) is simply that (3.26) has a unique solution in δ. In order for δ to be uniquely solvable, W must have full column rank. Thus, under the assumption that δ is solvable we have the following theorem.[5]

Theorem 3.4.1

The vector δ is identified if and only if $\operatorname{rank}(W)=G(G+K)$.

We can obtain equivalent results in terms of structural parameters. Letting

$$M=(I_G\otimes B : I_G\otimes \Gamma), \tag{3.27}$$

we have

Theorem 3.4.2

The vector δ is identified if and only if $\operatorname{rank}(M\Phi')=G^2$.

Proof

We may write

$$W=\begin{pmatrix} I_G\otimes \Pi' & I_{GK} \\ \Phi_1 & \Phi_2 \end{pmatrix}=\begin{pmatrix} 0 & I_{GK} \\ \Phi M' & \Phi_2 \end{pmatrix}\begin{pmatrix} I_G\otimes B'^{-1} & 0 \\ I_G\otimes \Pi' & I_{GK} \end{pmatrix}. \tag{3.28}$$

Since (by assumption) B is non-singular, the second of the last two partitioned matrices is non-singular, and it follows that

$$\operatorname{rank}(W)=\operatorname{rank}\begin{pmatrix} 0 & I_{GK} \\ \Phi M' & \Phi_2 \end{pmatrix}.$$

[5] See Theorem 3.4.4 for the conditions for solvability of a set of linear equations.

By Theorem 3.4.1, the whole vector δ is identified if and only if $\operatorname{rank}(W)=G(G+K)$, hence if and only if $\operatorname{rank}(M\Phi')=G^2$.

It will sometimes be the case that the whole vector δ is not identified but the *a priori* restrictions would permit a subset of the parameters in δ to be identified. To deal with this case, we now turn to the question of identifiability of individual parameters or a linear combination of the parameters. The problem can be conveniently put in the form of a *linear parametric function*, which is a linear combination of the structural parameters of the form $\xi'\delta$, where ξ' is a $1\times G(G+K)$ vector with known elements. From Definition 2.4 and the identifiability of Π, it follows that

Theorem 3.4.3

A parametric function $\xi'\delta$ is identified, given Π, if it has a unique value for every δ satisfying (3.26).

Finding conditions to identify $\xi'\delta$ is equivalent to finding the conditions for $\xi'\delta$ to be estimable. We note that (3.26) is a set of linear equations. A fundamental theorem with regard to the solvability of linear equations is

Theorem 3.4.4

Exactly one of the following alternatives holds. Either eq. (3.26) has a solution, or there exists c such that

$$c'W=0, \qquad c'\tilde{d}=1. \tag{3.29}$$

The proof of this theorem is given by Gale (1960, p. 41). To see that (3.26) and (3.29) cannot hold simultaneously, we premultiply (3.26) by c', then (3.29) becomes the impossible statement that "zero equals one".

The condition for identifying $\xi'\delta$ follows from this theorem.

Theorem 3.4.5

The parametric function $\xi'\delta$ is identified by eq. (3.26) if and only if there exists η such that $\eta'W=\xi'$, i.e. if and only if $\operatorname{rank}(Q' \vdots \Phi' \vdots \xi)=\operatorname{rank}(Q' \vdots \Phi')$.

Proof

We first prove sufficiency. Let there exist η such that $\eta'W=\xi'$. Then for each δ such that $W\delta=\tilde{d}$, $\xi'\delta=\eta'W\delta=\eta'\tilde{d}$, which is constant and independent of δ.

To prove necessity, let $\xi'\delta$ be unique for each δ such that $W\delta=\tilde{d}$. Let δ_1 be a solution of (3.26). Then for every z such that $Wz=0$, δ_1+z is a solution of (3.26). Hence, for every z such that $Wz=0$, $\xi'(\delta_1+z)=\xi'\delta_1$, i.e. $\xi'z=0$. Thus, the equations $Wz=0$ and $\xi'z=1$ have no solution. By Theorem 3.4.4 there exists η such that $\eta'W=\xi'$.

The Jacobian matrix summarizes the information in the equation system (locally). Theorem 3.4.5 says that if $\xi'\delta$ is identifiable, then knowing the value of $\xi'\delta$ does not add new information to (3.26). An equivalent condition in terms of structural parameters is

Theorem 3.4.6

The parametric function $\xi'\delta$ is identified if and only if there exists κ such that $M\Phi'\kappa = M\xi$, i.e. if and only if $\operatorname{rank}(M\Phi' \vdots M\xi) = \operatorname{rank}(M\Phi')$.

Proof

To prove necessity, suppose $\xi'\delta$ is identified. Then there exists η such that $W'\eta = \xi$. Premultiplying by M, we have

$$\left(MQ' \vdots M\Phi'\right)\eta = M\xi. \tag{3.30}$$

But

$$\begin{aligned} MQ' &= \left(I_G \otimes B \vdots I_G \otimes \Gamma\right)\left(I_G \otimes \Pi' \vdots I_{GK}\right)' \\ &= (I_G \otimes B\Pi) + (I_G \otimes \Gamma) = 0. \end{aligned} \tag{3.31}$$

Hence, there exists η such that $(0 : M\Phi')\eta = M\xi$, i.e. there exists κ such that $M\Phi'\kappa = M\xi$.

To prove sufficiency, first suppose that there exists κ such that $M\Phi'\kappa = M\xi$. Then since $M = (I_G \otimes B)(I_{G^2} : I_G \otimes (-\Pi))$ and since B is non-singular, there exists κ such that

$$\left[I_{G^2} \vdots I_G \otimes (-\Pi)\right]\Phi'\kappa = \left[I_{G^2} \vdots I_G \otimes (-\Pi)\right]\xi. \tag{3.32}$$

But if (3.32) has a solution, then the equations

$$z'\left[I_{G^2} \vdots I_G \otimes (-\Pi)\right]\Phi' = 0 \quad \text{and} \quad z'\left[I_{G^2} \vdots I_G \otimes (-\Pi)\right]\xi = 1 \tag{3.33}$$

have no solution.

Partitioning $\Phi = (\Phi_1 \vdots \Phi_2)$ and $\xi = (\xi_1 \vdots \xi_2)$, the equations

$$z'\Phi_1' - z'(I_G \otimes \Pi)\Phi_2' = 0 \quad \text{and} \quad z'\xi_1 - z'(I_G \otimes \Pi)\xi_2 = 1 \tag{3.34}$$

have no solution. Now suppose $\xi'\delta$ is not identified. Then the equations $(Q' : \Phi')\eta = \xi$ have no solution. Hence there exists a solution to

$$c'\left(Q' \vdots \Phi'\right) = 0; \qquad c'\xi = 1. \tag{3.35}$$

Partitioning $c' = (z', w')$, these equations may be written as

$$\begin{aligned} &z'(I_G \otimes \Pi) + w' = 0, \\ &z'\Phi_1' + w'\Phi_2' = 0, \\ &z'\xi_1 + w'\xi_2 = 0. \end{aligned} \tag{3.36}$$

Substituting the first into the second and third of these equations, we obtain the result that there must exist a solution to (3.34). But this is a contradiction. Hence, if there exists κ such that $M\Phi'\kappa = M\xi$, then $\xi'\delta$ is identified.

We can easily obtain the conditions for the identifiability of individual parameters from Theorems 3.4.5 and 3.4.6 by letting $\xi = e_j$, where e_j is a $G(G+K)\times 1$ vector whose elements are all zero except for the jth which is unity.

Lemma 3.4.1

The jth element of δ, $\delta_j = e_j'\delta$, is identified if and only if the following equivalent conditions hold:

(i) $\operatorname{rank}(W) = \operatorname{rank}(W_j) + 1$, where W_j is the matrix obtained by deleting the jth column from W;
(ii) $\operatorname{rank}(M\Phi' \vdots m_j) = \operatorname{rank}(M\Phi')$, where m_j is the jth column of M.

If all the restrictions are confined to restrictions on the parameters of a particular equation, including the normalization rule, we can obtain the conventional rank condition using Lemma 3.4.1.

Theorem 3.4.7

The gth equation of (3.1) is identified if and only if the following equivalent conditions hold:

(i) $\operatorname{rank}\begin{pmatrix} \Pi' \quad I_K \\ \cdots\cdots \\ \phi_g \end{pmatrix} = G + K;$

(ii) $\operatorname{rank}(A\phi_g') = G,$

where ϕ_g denotes the $R_g \times (G+K)$ matrix whose elements are known constants such that all prior information on the gth equation takes the form (3.15).

We note that in the derivation of the above theorems, the basic assumption is that Π is identifiable. Under Assumption 3.2, $\operatorname{plim}(\sum_{t=1}^T x_t u_t / T) = 0$ is sufficient to ensure it. Consequently, we can relax the restrictive assumptions about our model in two ways. The first is to allow the model to contain lagged endogenous variables and keep the assumption that u_t are serially uncorrelated. In this case, the lagged endogenous variables are included in the x_t vector and treated as predetermined. The second relaxation will be to allow serial correlation in u_t, but to keep the assumption that x_t are exogenous. It would of course be desirable in general to permit both relaxations. However, we cannot use the theorems of this section and allow both lagged endogenous variables and autocorrelation in the disturbance term to appear since then $\operatorname{plim}(\sum_{t=1}^{T} x_t u_t / T) \neq 0$. The more general case, with both autocorrelated disturbances and lagged endogenous variables, needs a different treatment. We take this up in the next section after a brief illustration of the conditions of identifiability already discussed.

3.5. *Examples*

To illustrate the application of the conditions for identifiability we consider a two-equation system:

$$\begin{aligned} &\beta_{11} y_{1t} + \beta_{12} y_{2t} + \gamma_{11} x_{1t} + \gamma_{12} x_{2t} = u_{1t}, \\ &\beta_{21} y_{1t} + \beta_{22} y_{2t} + \gamma_{21} x_{1t} + \gamma_{22} x_{2t} = u_{2t}. \end{aligned} \tag{3.37}$$

Without further restrictions neither equation is identified since we can premultiply (3.37) by any 2×2 non-singular constant matrix F, and the new equation will be observationally equivalent to (3.37). We consider the identifiability of the parameters of this model under various *a priori* restrictions. We normalize the equations by letting $\beta_{11} = 1$ and $\beta_{22} = 1$.

Case 1: $\gamma_{21} = 0$.

The constraint matrices in (3.15) for the first and second equations are

$$\phi_1' = \begin{pmatrix} 1 \\ 0 \\ 0 \\ 0 \end{pmatrix}, \qquad d_1' = (1),$$

and

$$\phi_2' = \begin{pmatrix} 0 & 0 \\ 1 & 0 \\ 0 & 1 \\ 0 & 0 \end{pmatrix}, \qquad d_2' = (1,0).$$

We thus have

$$\operatorname{rank}(A\phi_1') = \operatorname{rank}\begin{pmatrix} 1 \\ \beta_{21} \end{pmatrix} = 1,$$

$$\operatorname{rank}(A\phi_2') = \operatorname{rank}\begin{pmatrix} \beta_{12} & \gamma_{11} \\ 1 & 0 \end{pmatrix} = 2,$$

so that from Theorem 3.3.1 the first equation is not identified, but the second equation is.

Case 2: $\gamma_{12}=0$ and $\gamma_{11}+\gamma_{21}=0$.

Since we have a cross-equations constraint, we put the model in stack form (as in Section 3.4) and then write the constraint matrices of (3.24) as

$$\Phi=\begin{pmatrix}1&0&0&0&0&0&0&0\\0&0&0&1&0&0&0&0\\0&0&0&0&0&1&0&0\\0&0&0&0&1&0&-1&0\end{pmatrix},\qquad d=\begin{bmatrix}1\\1\\0\\0\end{bmatrix}.$$

In this case $\delta'=(\beta_{11},\beta_{12},\beta_{21},\beta_{22},\gamma_{11},\gamma_{12},\gamma_{21},\gamma_{22})$ and

$$W=\begin{pmatrix}\pi_{11}&\pi_{21}&0&0&1&0&0&0\\\pi_{12}&\pi_{22}&0&0&0&1&0&0\\0&0&\pi_{11}&\pi_{21}&0&0&1&0\\0&0&\pi_{12}&\pi_{22}&0&0&0&1\\1&0&0&0&0&0&0&0\\0&0&0&1&0&0&0&0\\0&0&0&0&0&1&0&0\\0&0&0&0&1&0&-1&0\end{pmatrix}.$$

Since $\operatorname{rank}(W)=8$, then from Theorem 3.4.1 we know both equations are identified.

Case 3: $\beta_{21}=0$ and $\gamma_{21}=0$.

The first equation is not identified, but the second equation is because $\operatorname{rank}(A\phi_1')=1$ and $\operatorname{rank}(A\phi_2')=2$. However, the individual coefficient γ_{11} is also identified by Lemma 3.4.1 since

$$M\Phi'=\begin{pmatrix}\beta_{11}&\beta_{12}&0&0&\gamma_{11}&\gamma_{12}&0&0\\\beta_{21}&\beta_{22}&0&0&\gamma_{21}&\gamma_{22}&0&0\\0&0&\beta_{11}&\beta_{12}&0&0&\gamma_{11}&\gamma_{12}\\0&0&\beta_{21}&\beta_{22}&0&0&\gamma_{21}&\gamma_{22}\end{pmatrix}\begin{pmatrix}1&0&0\\0&0&0\\0&0&0\\0&1&0\\0&0&0\\0&0&0\\0&0&1\\0&0&0\end{pmatrix}$$

$$=\begin{pmatrix}\beta_{11}&0&0\\\beta_{21}&0&0\\0&\beta_{12}&\gamma_{11}\\0&\beta_{22}&\gamma_{21}\end{pmatrix},$$

and (since γ_{11} is the fifth element of δ):

$$m_5 = \begin{pmatrix} \gamma_{11} \\ \gamma_{21} \\ 0 \\ 0 \end{pmatrix}.$$

Noting that $\beta_{11} = \beta_{22} = 1$ and $\beta_{21} = \gamma_{21} = 0$, we thus have

$$(M\Phi' : m_5) = \begin{pmatrix} 1 & 0 & 0 & \vdots & \gamma_{11} \\ 0 & 0 & 0 & \vdots & 0 \\ 0 & \beta_{12} & \gamma_{11} & \vdots & 0 \\ 0 & 1 & 0 & \vdots & 0 \end{pmatrix}$$

so that $\operatorname{rank}(M\Phi' : m_5) = 3 = \operatorname{rank}(M\Phi')$.

4. Dynamic models with serially correlated residuals[6]

4.1. The model

In this section we consider identification criteria for models of the form

$$\sum_{s=0}^{p} B_s y_{t-s} + \sum_{s=0}^{q} \Gamma_s x_{t-s} = u_t, \tag{4.1}$$

where y_t, x_t, and u_t are $G \times 1$, $K \times 1$, and $G \times 1$ vectors of observed jointly dependent variables, exogenous variables, and unobserved disturbance terms, respectively; B_s and Γ_s are $G \times G$ and $G \times K$ parameter matrices. Rewriting (4.1) in terms of the lag operator L, $Ly_t = y_{t-1}$, we have

$$B(L) y_t + \Gamma(L) x_t = u_t, \tag{4.2}$$

where $B(L)$ and $\Gamma(L)$ are polynomials in the lag operator L, with

$$\begin{aligned} B(L) &= B_0 + B_1 L + \cdots + B_p L^p, \\ \Gamma(L) &= \Gamma_0 + \Gamma_1 L + \cdots + \Gamma_q L^q. \end{aligned} \tag{4.3}$$

[6]Readers who are not familiar with polynomial matrices are advised to read the Appendix before proceeding with this section.

We assume that

Assumption 4.1

The maximum order of lags for the jointly dependent variables and exogenous variables, p and q, are known *a priori*.

Assumption 4.2

$\det|B_0| \neq 0$.

Assumption 4.3

The roots of the polynomial equations $|B_0 + B_1 L + \cdots + B_p L^p| = 0$ lie outside the unit circle.

Assumption 4.4

x_t is stationary in the wide sense, ergodic, has finite second moments, and is independent of the $\{u_t\}$ sequences.[7] Furthermore, we assume that there is no set of vectors d'_s for which

$$\sum_{s=0}^{\infty} d'_s x_{t-s} = 0,$$

where d'_s is a K-component row vector and $d_0 \neq 0$.

Assumption 4.5

$\mathrm{E}u_t = 0$, and $\{u_t\}$ is jointly stationary in the wide sense. We shall denote the sequence of the second-order moments,

$$C_u(\tau) = \mathrm{E}u_t u'_{t-\tau}, \qquad \tau = -\infty, \dots, -1, 0, 1, \dots, \infty, \tag{4.4}$$

by $\{C_u\}$. We also assume that $C_u(0)$ is non-singular.

Assumptions 4.1–4.3 are conventional. In fact, Assumption 4.3 implies Assumption 4.2. However, for ease of comparison between contemporaneous and dynamic models, we state them separately. Below we shall relax Assumption 4.1. Assumption 4.5 will also be modified to take into account additional information in the disturbance term. For the moment we assume we have no such information. Assumption 4.4 is a dynamic analogy to Assumptions 3.2 and 3.3 made for contemporaneous models. It is made to ensure the identifiability of the transfer

[7]A stochastic process is stationary in the wide sense if the mean $Ex_t = Ex_s$ is a constant and the covariance $E[x_t - Ex_t][x_s - Ex_s]'$ depends only on their distance apart in time $(t-s)$. A stationary process is said to be ergodic if the sample mean of every function of finite observations tends to its expected value in mean square, at least when the expected value of the square of the function exists.

function. [The transfer function is the dynamic equivalent of the matrix Π in the contemporaneous model. It is defined in Section 4.2, eq. (4.8) below.] This assumption rules out the occurrence of multicollinearity; it also implies no deterministic right-hand-side variables so that the prediction errors about these variables will not be zero. In particular, it means that among the K components of x_t none is a constant, a trend, or a seasonal variation, which are expressed as a solution of a linear difference equation with constant coefficients. The assumption is not as restrictive as it appears. The model can be generalized to take care of these cases as Hannan (1971) and Hatanaka (1975) have shown.

We know that in the contemporaneous model (3.1) two structures are observationally equivalent if and only if they are connected by a $G\times G$ non-singular constant matrix (Lemma 3.2.1). However, in the dynamic model (4.1) we may premultiply by a non-singular $G\times G$ polynomial matrix $F(L)$ to yield observationally equivalent structures. This may be illustrated by the following example of a two-equation system [Koopmans, Rubin and Leipnik (1950, p. 109)] for which the identification conditions discussed in the previous section are incomplete.

Example 4.1.1

Let

$$\begin{bmatrix} 1 & 0 \\ \beta_{21,0} & 1 \end{bmatrix}\begin{bmatrix} y_{1t} \\ y_{2t} \end{bmatrix}+\begin{bmatrix} \beta_{11,1} & \beta_{12,1} \\ 0 & 0 \end{bmatrix}\begin{bmatrix} y_{1,t-1} \\ y_{2,t-1} \end{bmatrix}=\begin{bmatrix} u_{1t} \\ u_{2t} \end{bmatrix}. \tag{4.5}$$

If u_t is serially uncorrelated, the only admissible transformation is an identity matrix, and hence both equations are identified. However, if serial correlation in u_t is allowed, then

$$F(L)=I+\begin{bmatrix} 0 & f_{12,1} \\ 0 & 0 \end{bmatrix}L \tag{4.6}$$

is an admissible transformation, and thus observational information cannot distinguish the transformed first equation from the true first equation.

As one can see from this example, the major complication in the identification of (4.1) lies in the possibility of premultiplication by a non-singular $G\times G$ polynomial matrix $F(L)$ while retaining observational equivalence. This additional complexity does not arise in the case of lagged dependent variables alone, since if u_t is serially uncorrelated, the left factor $F(L)$ can be at most a non-singular constant matrix; otherwise, serial correlation will be introduced. The problem arises when both lagged dependent variables and serially correlated residuals are allowed. In what follows we first state formally the requirements for two structures to be observationally equivalent, and then illustrate how conditions may be imposed so that $F(L)$ is again restricted to a constant matrix.

4.2. Observationally equivalent structures

We now derive conditions for two structures to be observationally equivalent when both lagged endogenous variables and serial correlation in the disturbance terms are allowed. Premultiplying (4.2) by $B(L)^{-1}$ we obtain the model in transfer function form:

$$y_t = \Pi(L)x_t + D(L)u_t, \tag{4.7}$$

where

$$\Pi(L) = -B(L)^{-1}\Gamma(L), \tag{4.8}$$

$$D(L) = B(L)^{-1} = D_0 + D_1 L + D_2 L^2 + \cdots. \tag{4.9}$$

Conditional on x_{t-s}, $s \geqslant 0$, the mean of y_t is

$$\bar{y}_t = E(y_t | x_{t-s}, s \geqslant 0) = \Pi(L)x_t. \tag{4.10}$$

The second-order moments of $\{y_t\}$ conditional on x_t are given by the sequence

$$\begin{aligned} E[y_t - \bar{y}_t][y_{t+\tau} - \bar{y}_{t+\tau}]' &= E\left[B(L)^{-1}u_t\right]\left[B(L)^{-1}u_{t+\tau}\right]' \\ &= \sum_{i=0}^{\infty}\sum_{j=0}^{\infty} D_i C_u(j-\tau-i)D_j', \end{aligned}$$

$$\tau = -\infty,\dots,-1,0,1,\dots,\infty, \tag{4.11}$$

which will be denoted by $B^{-1} * C_u * B^{-1'}$. The convolution notation $*$ will be used also for the autocovariance sequence of any other transformation of $\{u_t\}$.[8]

Lemma 4.2.1

Two structures $S = [B(L), \Gamma(L), \{C_u\}]$ and $\bar{S} = [\bar{B}(L), \bar{\Gamma}(L), \{\bar{C}_u\}]$ are observationally equivalent if and only if the following equivalent conditions hold:

(i)

$$B(L)^{-1}\Gamma(L) = \bar{B}(L)^{-1}\bar{\Gamma}(L), \tag{4.12}$$

$$B^{-1} * C_u * B^{-1'} = \bar{B}^{-1} * \bar{C}_u * \bar{B}^{-1'}. \tag{4.13}$$

[8] A more convenient way of representing the sequence of the covariance of y (4.11) is to use the spectral representation $B(e^{i\lambda})^{-1}f_u B'(e^{-i\lambda})^{-1}$, where $f_u = (1/2\pi)\Sigma_{-\infty}^{\infty}C_u(\tau)e^{-i\tau\lambda}$ and $B(e^{i\lambda}) = B_0 + B_1 e^{i\lambda} + \cdots + B_p e^{ip\lambda}$. In fact we can derive the identifiability criteria more elegantly using the spectral approach. However, the spectral approach will require a more advanced reasoning in statistics than is required in this chapter. If readers find (4.11), (4.13), and (4.15) confusing, they may be ignored. They are presented for the completeness of the argument. They are not used in the actual derivation of the identifiability criteria.

(ii) There exists an admissible $G \times G$ non-singular matrix $F(L)$ with elements that are rational functions of L such that

$$[\bar{B}(L) \quad \bar{\Gamma}(L)] = F(L)[B(L) \quad \Gamma(L)], \tag{4.14}$$

$$\{\bar{C}_u\} = \{F * C_u * F'\}. \tag{4.15}$$

Proof

(i) Follows from (4.8), (4.10), and (4.11).

(ii) Sufficiency is easy to check using (i). To prove necessity, we suppose that S and $\bar{S}$ are observationally equivalent. Let

$$F(L) = \bar{B}(L)B(L)^{-1}. \tag{4.16}$$

From (4.12) and (4.13) we have (4.14) and (4.15). Owing to (4.16) and Assumption 4.2, $F(L)$ must be rational and non-singular.

Although $\{C_u\}$ can be derived from (4.11) once $B(L)$ is known, Assumption 4.5 provides us with no information about the factorization of (4.11) between $\{C_u\}$ and $B(L)$ if $B(L)$ is unknown. Therefore, we shall focus only on the prior information of $[B(L) \quad \Gamma(L)]$ which will enable us to identify them from the transfer function $\Pi(L)$ defined in (4.8). Just as in the case of the contemporaneous model where the reduced-form parameters are assumed identifiable to begin with, here we approach the problem as if the transfer function, $\Pi(L)$, is known.

Lemma 4.2.2

Under Assumptions 4.1–4.5 the transfer function, $\Pi(L)$, is identifiable.

Proof

Supposing two transfer functions are observationally equivalent, we have

$$\Pi(L)x_t = \bar{\Pi}(L)x_t. \tag{4.17}$$

That is,

$$[\Pi(L) - \bar{\Pi}(L)]x_t = 0. \tag{4.18}$$

This implies either that $\Pi(L) = \bar{\Pi}(L)$ or that there exist d_s such that $\sum_{s=0}^{\infty} d_s' x_{t-s} = 0$. However, Assumption 4.4 excludes the existence of such d_s. Therefore $\Pi(L) = \bar{\Pi}(L)$.

4.3. Linear restrictions on the coefficients

Given $\Pi(L)$, the model (4.2) is completely identified if it is possible to factor $\Pi(L)$ into the two unique polynomial matrices $B(L)$ and $\Gamma(L)$. In order to obtain a unique decomposition of $\Pi(L)$ in the form of $-B(L)^{-1}\Gamma(L)$, given (4.8), we must eliminate the possibility that there is a common (left) factor in $B(L)$ and $\Gamma(L)$ which cancels when $B(L)^{-1}\Gamma(L)$ is formed. This is usually done in three steps [Hannan (1969, 1971)]:

(i) eliminate the redundancy specification;[9]
(ii) restrict the admissible transformations to constant matrices; and
(iii) restrict the constant transformation matrix to an identity matrix.

The first step is achieved by imposing

Condition 4.3.1

$[B(L) \quad \Gamma(L)]$ are relatively left prime.

By "relatively left prime" we mean that the greatest common left divisor is a unimodular matrix, i.e. its determinant is a non-zero constant. [For this and some useful definitions and theorems on polynomial matrices see Gantmacher (1959), MacDuffee (1956), or the Appendix.] When a left common divisor is not unimodular, it means that there is redundancy in the specification. The following is an example of an equation with a redundancy.

Example 4.3.1

Let

$$\begin{bmatrix} (1+\rho L)(1+\beta_{11,1}L) & 0 \\ 0 & (1+\beta_{22,1}L) \end{bmatrix} \begin{bmatrix} y_{1t} \\ y_{2t} \end{bmatrix} + \begin{bmatrix} (1+\rho L)\gamma_{11} \\ 0 \end{bmatrix} x_t = \begin{bmatrix} u_{1t} \\ u_{2t} \end{bmatrix} \tag{4.19}$$

The first equation has a common factor $(1+\rho L)$. However, $[B(L) \quad \Gamma(L)]$ in this case can be factored into

$$\begin{aligned}[B(L) \quad \Gamma(L)] &= \begin{bmatrix} 1+\rho L & 0 \\ 0 & 1 \end{bmatrix} \begin{bmatrix} 1+\beta_{11,1}L & 0 & \gamma_{11} \\ 0 & 1+\beta_{22,1}L & 0 \end{bmatrix} \\ &= F(L)[\bar{B}(L) \quad \bar{\Gamma}(L)], \end{aligned} \tag{4.20}$$

where the common left devisor $F(L)$ has a non-constant determinant $(1+\rho L)$.

[9]By "redundancy" we mean that there are common (polynomial) factors appearing in $B(L)$ and $\Gamma(L)$.

By ruling out such a redundancy, we help rule out the transformation matrix $F(L)$ which has the form like that of (4.20). Also, from a practical point of view this imposition of the relatively left prime condition may not be unreasonable, since if common factors are allowed there may be infinitely many structures which are observationally equivalent. Redundant specification also poses serious estimation problems [Box and Jenkins (1970) Hannan, Dunsmuir and Deistler (1980)]. Condition 4.3.1 puts restrictions on the observationally equivalent structures in the sense that

Lemma 4.3.1

Under Condition 4.3.1, two structures $S = [B(L), \Gamma(L), \{C_u\}]$ and $\bar{S} = [\bar{B}(L), \bar{\Gamma}(L), \{\bar{C}_u\}]$ are observationally equivalent if and only if there exists a unimodular matrix $F(L)$ such that (4.14) and (4.15) hold.

Having removed non-unimodular factors from the model (4.1) by imposing Condition 4.3.1, the second step is to restrict $F(L)$ to a constant matrix F. This is achieved by imposing

Condition 4.3.2

$\text{Rank}[B_p \quad \Gamma_q] = G$.

To see that this condition constrains F to be a constant matrix, let $F(L) = F_0 + F_1 L$ be a unimodular matrix. Condition (4.14) states that

$$\begin{aligned}[\bar{B}(L) \quad \bar{\Gamma}(L)] &= F(L)[B(L) \quad \Gamma(L)] \\ &= \big[F_0 B_0 + (F_0 B_1 + F_1 B_0)L + \cdots + (F_0 B_p + F_1 B_{p-1})L^p \\ &\quad + F_1 B_p L^{p+1} \vdots F_0 \Gamma_0 + \cdots + (F_0 \Gamma_q + F_1 \Gamma_{q-1})L^q + F_1 \Gamma_q L^{q+1}\big]. \end{aligned} \tag{4.21}$$

We notice that as long as there exists a non-zero F_1 such that

$$F_1[B_p \quad \Gamma_q] = [0 \vdots 0], \tag{4.22}$$

then the order condition p and q will still be preserved. On the other hand, if $[B_p \quad \Gamma_q]$ has rank G, application of $F(L)$ will increase the length of lag, and thus the two structures will not be equivalent unless $F_1 = 0$. Condition 4.3.2 is imposed to eliminate the possibility that F_j may not be equal to 0.

After imposing conditions to restrict the non-singular transformation matrix $F(L)$ to a constant matrix F, we then look at conditions which will restrict F to an identity matrix and thus make the dynamic equations identified. From Section

3 we know that one such condition is

Condition 4.3.3

Let

$$A = \begin{bmatrix} B_0 & B_1 & \dots & B_p & \Gamma_0 & \Gamma_1 & \dots & \Gamma_q \end{bmatrix}. \tag{4.23}$$

Let at least $(G-1)$ zeros be prescribed in each row of A and let one element in each row of B_0 be prescribed as unity. Let the rank of each submatrix of A obtained by taking the columns of A with prescribed zeros in a certain row be $G-1$.

Then we have

Theorem 4.3.1

Under Conditions 4.3.1–4.3.3 the model (4.1) with Assumptions 4.1–4.5 is identifiable.[10]

In the contemporaneous case discussed in Section 3, $p = q = 0$, and hence Condition 4.3.2 is automatically satisfied. For dynamic models this condition is sufficient,[11] but not necessary since even if it is not satisfied, it may not be possible to find a non-trivial unimodular factor which preserves other *a priori* restrictions [Hannan (1971)]. Condition 4.3.2 may be replaced by other conditions which serve the same purpose of constraining $F(L)$ to a constant matrix F [Hannan (1971)]. For instance, in Theorem 4.3.1 we may replace Condition 4.3.2 by

Condition 4.3.2′

The maximum degree of each column of $[B(L) \quad \Gamma(L)]$ is known *a priori* and there exist G columns of $[B(L) \quad \Gamma(L)]$ such that the matrix of coefficient vectors corresponding to the maximum degrees of these G columns has rank G.

Similarly, instead of zero restrictions we may impose linear restrictions, e.g. the gth row of A may satisfy

$$a_g' \phi_g' = d_g, \tag{4.24}$$

where ϕ_g and d_g are an $R_g \times [G(p+1) + K(q+1)]$ matrix and an $R_g \times 1$ vector of known elements, respectively. Then we may replace Condition 4.3.3 in Theorem

[10]Here we refer to the identification of a complete model. If only certain rows in A satisfy Condition 4.3.3, then the theorem should be modified to refer to the identification of those rows of A. Similarly for Condition 4.3.3′ below.

[11]For instance, Condition 4.3.2 implied Condition 6.2.1 of Fisher (1966).

4.3.1 by

Condition 4.3.3′

For each row of A, $\operatorname{rank}(A\phi_g') = G$.

On the other hand, there may be cross-equation linear constraints. We can derive a similar condition for restricting F to an identity matrix by stacking the coefficients of A in one row, $\operatorname{vec}(A) = (a_1', a_2', \ldots, a_G')$, and state the restrictions as

$$\operatorname{vec}(A)\Phi' = d, \tag{4.25}$$

where Φ' and d are a $[G^2(p+1)+GK(q+1)]\times R$ matrix and an $R\times 1$ vector with known elements. We then replace Condition 4.3.3 in Theorem 4.3.1 by

Condition 4.3.3″

$\operatorname{Rank}\{[I_G \otimes A]\Phi'] = G^2$.

For further details, see Deistler (1976).

It is possible to identify a dynamic model by relaxing some of these assumptions and conditions. For example, when the prior information is in the form of excluding certain variables from an equation, the identification of (4.1) may be achieved without specifying *a priori* the maximum order of lags (Assumption 4.1), but instead allowing the lags to be empirically determined. The restrictive Condition 4.3.2 or its variant may also be dropped. These relaxations are permitted by the following theorem.

Theorem 4.3.2

The model (4.1) with Assumptions 4.2–4.5 under Condition 4.3.1 is identified if and only if the following condition holds.

Condition 4.3.3‴

Let $A(L) = [B(L) \quad \Gamma(L)]$. Let at least $(G-1)$ zeros be prescribed in each row of $A(L)$ and let one element in each row of B_0 be prescribed as unity. For each row, the matrix consisting of the columns of $A(L)$ having prescribed null elements in that row is of $\operatorname{rank}(G-1)$.

Proof

The sufficiency follows from the fact that under Condition 4.3.3‴ $F(L)$ must be diagonal. Condition 4.3.1 restricts $F(L)$ to be unimodular, i.e the determinant is a constant, and hence the diagonal elements of $F(L)$ must be constants. The normalization rules then further restrict F equal to an identity matrix.

To prove the necessity, suppose the structure S does not satisfy Condition 4.3.3‴. Without loss of generality we suppose that the matrix consisting of the columns of $A(L)$ having prescribed null elements in the first row, $\hat{A}(L)$ say, has rank less than $(G-1)$. Let $b(L)$ be the first column of $B(L)$. Then there exists a $G\times 1$ vector polynomial $f(L)=(1, f_{.1}(L)')'$, with $f_{.1}(L)\not\equiv 0$, $c=f(0)'b(0)\neq 0$, and $f(L)'\hat{A}(L)=0$. Let $\chi(L)$ be the greatest common divisor of the elements of $f(L)'A(L)$, with $\chi(0)=c$. Let

$$F(L)=\begin{pmatrix} \chi(L)^{-1}f(L)' \\ \cdots\cdots\cdots \\ 0 \;\vdots\; I_{G-1} \end{pmatrix}. \tag{4.26}$$

Then the structure $[F(L)\quad A(L), \{C_u\}]$ is observationally equivalent to S and belongs to the model. Because $F(L)\not\equiv I_G$ by construction, S is not identifiable.

Although the elimination of redundancy is intuitively appealing, in practice it may be difficult to determine since the elements of $[B(L)\quad \Gamma(L)]$ are unknown. Hatanaka (1975) has suggested that we treat a model as identified when the structures that are observationally equivalent (in the wide sense) and consistent with the a priori specifications are unique apart from the redundancy in each equation.[12] In this the identification of a model is achieved when the only admissible transformation $F(L)$ is a $G\times G$ non-singular diagonal matrix of rational functions of L. Condition 4.3.3$'''$ is then necessary and sufficient for the identification of model (4.1) with Assumptions 4.2–4.5. For other identification conditions without the relatively left prime assumption, see Deistler and Schrader (1979).

4.4. *Additional information about the disturbances*

Assumption 4.5 provides us with no information about the identification of $[B(L)\quad \Gamma(L)]$. However, if $[B(L)\quad \Gamma(L)]$ are identified through conditions stated in Section 4.3, then by Wold's decomposition theorem [Hannan (1970) and Rozanov (1967)] we can obtain a unique representation of u_t from the covariance of y (4.11) as

$$u_t=\sum_{s=0}^{\infty}\theta_s v_{t-s}, \tag{4.27}$$

where $\theta_0=I_G$, $\Sigma_s\|\theta\|^2<\infty$, and v_t are independently, identically distributed

[12] It should be noted that although in general we do not know the true order, it is essential to have a prior bound on the order. Deistler and Hannan (1980) and Hannan, Dunsmuir and Deistler (1980) have shown that the maximum likelihood estimator of the transfer function will converge to the true value provided the assumed order is not smaller than the true order, but some pathology arises that the numerical problem of finding the unique maximum of the likelihood can be difficult.

random variables with mean 0 and covariance matrix Ω. We did not pursue such a factorization in Section 4.3 because it does not affect the estimation of $[B(L) \quad \Gamma(L)]$.

However, if we have some information about u_t, then this will aid identification; also, making use of it will improve the efficiency of estimates [e.g., see Espasa and Sargan (1977) and Hannan and Nicholls (1972)]. In this section we illustrate how we may use this additional information to identify the model. For additional discussion, see Deistler (1975), Hannan (1971, 1976), Hatanaka (1976), Kohn (1979), Sargan (1961, 1979), etc.

Suppose instead of Assumption 4.5, we assume

Assumption 4.5′

Let u_t be an rth order moving average process

$$u_t = \sum_{s=0}^{r} \theta_s v_{t-s}, \tag{4.28}$$

where $\det|\theta_0| \neq 0$, and v_t are independently distributed random variables having zero mean and non-singular covariance matrix Ω. The roots of the determinant equation

$$|\theta_0 + \theta_1 L + \cdots + \theta_r L^r| = 0 \tag{4.29}$$

lie outside the unit circle.

Then

$$C_u(\tau) = Eu_t u'_{t-\tau} = \begin{cases} \sum_{s=\tau}^{r} \theta_s \Omega \theta'_{s-\tau} & \text{for } 0 \leqslant \tau \leqslant r, \\ 0 & \text{for } \tau > r, \end{cases} \tag{4.30}$$

and

$$C_u(-\tau) = C'_u(\tau). \tag{4.31}$$

Hence

Lemma 4.4.1

$[\theta(L) \quad \Omega]$ and $[\bar{\theta}(L) \quad \bar{\Omega}]$ are observationally equivalent if and only if there exists a $G \times G$ non-singular constant matrix H such that

$$\bar{\theta}(L) = \theta(L) H, \tag{4.32}$$

and

$$\Omega = H \bar{\Omega} H'. \tag{4.33}$$

Proof

Sufficiency is easy to check using (4.31) and (4.32).

To prove necessity, let $\theta(L)v_t$ and $\bar{\theta}(L)\bar{v}_t$ be two rth-order moving average processes which yield identical $\{C_u\}$. Define

$$H(L)=\theta(L)^{-1}\bar{\theta}(L) = \sum_{s=0}^{\infty} H_s L^s. \tag{4.34}$$

We then obtain:

$$Ev_t v'_{t-s} = C_v(s) = 0 = H_s C_{\bar{v}}(0) H'_0 = H_s \bar{\Omega} H'_0 \quad \text{for all } s>0. \tag{4.35}$$

H_0 is non-singular because θ_0 and $\bar{\theta}_0$ are non-singular. $\bar{\Omega}$ is non-singular by assumption, hence $H_s=0$ for all $s>0$, and thus $H\colon= H_0$ is constant and (4.32) and (4.33) hold.

Combining Lemma 4.2.1 and Lemma 4.4.1 we obtain the following proposition about observationally equivalent structures with regard to model (4.1).

Lemma 4.4.2

Two structures, $S=[B(L)\quad \Gamma(L)\quad \theta(L)\quad \Omega]$ and $\bar{S}=[\bar{B}(L)\quad \bar{\Gamma}(L)\quad \bar{\theta}(L)\quad \bar{\Omega}]$, are observationally equivalent if and only if there exists an admissible $G\times G$ non-singular matrix $F(L)$, with elements that are rational functions of L, and a $G\times G$ constant non-singular matrix H such that

$$[\bar{B}(L)\quad \bar{\Gamma}(L)\quad \bar{\theta}(L)] = F(L)[B(L)\quad \Gamma(L)\quad \theta(L)H] \tag{4.36}$$

and

$$\Omega = H\bar{\Omega}H'. \tag{4.37}$$

By the same reasoning as previously stated in Section 4.3 we have:

Theorem 4.4.1

Let

$$A=(B_0,\dots,B_p,\Gamma_0,\dots,\Gamma_q,\theta_0,\dots,\theta_r). \tag{4.38}$$

Define the model by the conditions: (i) at least $(G-1)$ zero elements are prescribed in each row of A; (ii) the diagonal elements of B_0 are unity; and (iii) $\theta_0 = B_0$.

Sufficient conditions for a structure belonging to the model to be identified are:

Condition 4.4.1

$[B(L) \quad \Gamma(L) \quad \theta(L)]$ are relatively left prime.

Condition 4.4.2

$[B_p \quad \Gamma_q \quad \theta_r]$ has rank G.

Condition 4.4.3

The rank of each submatrix of A obtained by taking the columns of A with prescribed zeros in a certain row is $(G-1)$.

The normalization rule $\theta_0 = B_0$ is imposed to restrict H to an identity matrix. In some cases we may wish to replace the restrictions $\theta_0 = B_0$ by $\theta_0 = I_G$ to facilitate computation. Then we can redefine A as $(B_0,\dots,B_p,\Gamma_0,\dots,\Gamma_q)$ and replace Conditions 4.4.2 and 4.4.3 by those similar to Conditions 4.3.2 and 4.3.3 to keep the system identified. Or we may rely on [Kohn (1979)]:

Theorem 4.4.2

Under Condition 4.4.1 and $\theta_0 = I_G$, when $B(L)$ is lower triangular with diagonal elements of B_0 equal to unity and the maximum degree of the off-diagonal elements $b_{gl}(L)$ less than the maximum degree of the diagonal elements $b_{ll}(L)$ for $g > l$, the system is identified.

When identities are present, this is equivalent to knowing that certain elements of the variance–covariance matrix of u_t are zero. Suppose that the last $(G-G_1)$ equations $(0 < G_1 < G)$ in (4.1) are identities. Then we can write

$$B(L)=\begin{pmatrix} B_{11}(L) & \vdots & B_{12}(L) \\ \cdots & \cdots & \cdots \\ B_{21}(L) & \vdots & B_{22}(L)\end{pmatrix}, \qquad \Gamma(L)=\begin{pmatrix}\Gamma_1(L)\\ \cdots \\ \Gamma_2(L)\end{pmatrix},$$

$$u_t=\begin{pmatrix}u_t^1\\ \cdots \\ 0\end{pmatrix},\qquad y_t=\begin{pmatrix}y_t^1\\ \cdots \\ y_t^2\end{pmatrix}, \tag{4.39}$$

where $B_{11}(L)$ is $G_1\times G_1$, $B_{12}(L)$ is $G_1\times(G-G_1)$, $\Gamma_1(L)$ is $G_1\times K$, u_t^1 is $G_1\times 1$, and y_t^1 is $G_1\times 1$; $B_{21}(L)$, $B_{22}(L)$, and $\Gamma_2(L)$ are generally assumed known since identities are usually definitional equations.

We eliminate identities by substitution and obtain:

$$\tilde{B}(L)y_t^1+\tilde{\Gamma}(L)x_t=u_t^1, \tag{4.40}$$

where $\tilde{B}(L)=B_{11}(L)-B_{12}(L)B_{22}^{-1}(L)B_{21}(L)$ and $\tilde{\Gamma}(L)=\Gamma_1(L)-B_{12}(L)B_{22}^{-1}(L)\Gamma_2(L)$. We can then restate the above theorems in terms of (4.40). But (4.40) is usually not a suitable form for identification because it will be difficult to state prior knowledge on $B(L)$ and $\Gamma(L)$ in terms of $\tilde{B}(L)$ and $\tilde{\Gamma}(L)$. However, Conditions 4.3.1–4.3.3 (Theorem 4.3.1) remain sufficient for identification of model (4.1) even if identities are present. This should be intuitively obvious since additional information about the disturbance cannot hinder identification of a model. Further prior information can only help by ruling out certain transformations which would otherwise be admissible. For a more thorough discussion, see Deistler (1975), Hannan (1971), and Kohn (1979). For examples, see Phillips and Wickens (1978, ch. 7).

5. Non-linear *a priori* constraints and covariance restrictions

5.1. Some useful theorems

In every problem so far considered we have been confined to linear models with linear *a priori* constraints only. Hence, we have either a unique solution (globally identified) or else an infinite number of solutions forming a continuous variety in the m-dimensional space of the elements of $\alpha \subset A$. In this section we turn to a special case of non-linearity in which the model itself is non-linear in parameters. When non-linear restrictions are imposed, some of the equations defining the prior and observational restrictions are not linear in the elements of α. It is therefore possible that these equations would admit more than one solution; one of them may correspond to the true structure. Without additional information to distinguish them we may only speak of local identifiability.

Suppose the characteristics of y may be representable by the n population moments ω_i, $i=1,\dots,n$, which are continuously differentiable functions of the m-dimensional structural parameters, so $\omega_i = \psi_i(\alpha)$, $i=1,\dots,n$. Thus, one may say that the probability density for y depends on the structural parameter α only through an n-dimensional function $[\psi_i(\alpha)]$.

Let the structural parameter α also satisfy a set of continuously differentiable constraint equations $\phi_j(\alpha)=0$ $(j=1,\dots,R)$. Then α^0 is globally identifiable if and only if the equations

$$\begin{aligned} \omega_i &= \psi_i(\alpha), \quad i=1,\dots,n, \\ 0 &= \phi_i(\alpha), \quad j=1,\dots,R, \end{aligned} \tag{5.1}$$

have a unique solution α^0; it is locally identifiable if all the solutions of (5.1) are isolated from each other. The fundamental theorem used in this analysis is the implicit function theorem.[13]

[13] Here we only look at conditions implied by the first derivative matrix because conditions implied by higher order derivatives are difficult to formulate. In fact, a model may suffer from first order lack of identifications but is still locally identified. However, in this case although the usual estimators are consistent, they are not asymptotically normally distributed. For details, see Sargan (1980).

Theorem 5.1.1

If α^0 is a regular point, a necessary and sufficient condition that α^0 be a locally isolated solution is that the $(n+R)\times m$ Jacobian matrix formed by taking partial derivatives of (5.1) with respect to α,

$$J(\alpha)=\begin{bmatrix}\bar{\psi}(\alpha)\\ \cdots\\ \Phi(\alpha)\end{bmatrix}, \tag{5.2}$$

where

$$\underset{n\times m}{\bar{\psi}(\alpha)}=\left[\frac{\partial\psi_i(\alpha)}{\partial\alpha_j}\right],\qquad \underset{R\times m}{\Phi(\alpha)}=\left[\frac{\partial\phi_i(\alpha)}{\partial\alpha_j}\right],$$

has rank m at α^0.

α^0 is a regular point if and only if for all α in some sufficiently small neighborhood of α^0, $\operatorname{rank}(J(\alpha))=\operatorname{rank}(J(\alpha^0))$. The regularity assumption can usually be assumed since when the elements of $J(\alpha)$ are analytic, the set of irregular points has Lebesque measure zero [Fisher (1966, p. 167)]. The proof of this theorem is given by Fisher (1966, pp. 163–165).

The case of global identification when the ψ_i and ϕ_i are non-linear is more difficult. An overly strong sufficient condition using the result of Gale and Nikaido (1965) is given by Fisher (1966) and Rothenberg (1971).

Richmond (1976) has provided a useful theorem with regard to the identifiability of a continuously differentiable function of α, namely $\xi(\alpha)$. It includes the identifiability of a structural parameter as a special case by letting $\xi(\alpha)=e_j'\alpha$, where e_j' is an $1\times m$ vector whose elements are all zero except for the jth which is unity.

Theorem 5.1.2

Suppose α^0 is a regular point with respect to (5.1) and $\xi(\alpha)$. Then $\xi(\alpha)$ is locally identifiable if and only if

$$\operatorname{rank}\begin{pmatrix}J(\alpha)\\ \cdots\cdots\\ \nabla\xi(\alpha)\end{pmatrix}=\operatorname{rank}(J(\alpha)), \tag{5.3}$$

where

$$\nabla\xi(\alpha)=\left(\frac{\partial\xi(\alpha)}{\partial\alpha_1},\dots,\frac{\partial\xi(\alpha)}{\partial\alpha_m}\right). \tag{5.4}$$

Theorem 5.1.2 is a non-linear generalization of Theorem 3.4.5. Together with Theorem 5.1.1 we can derive local identifiability conditions for α or a subset of α, and hence theorems stated in previous sections can easily be modified to take account of the non-linear constraints. For instance, for the dynamic simultaneous equations model (4.1) with Assumptions 4.1–4.5, instead of assuming that the coefficients are subject to linear constraints we may assume that they are subject to R continuously differentiable cross-equation constraints:

$$\Phi(\alpha)=\begin{pmatrix}\phi_1(\alpha)\\ \phi_2(\alpha)\\ \vdots\\ \phi_R(\alpha)\end{pmatrix}=0. \tag{5.5}$$

Using Theorem 5.1 we obtain the following theorem with regard to the local identifiability of (4.1).

Theorem 5.1.3

Sufficient conditions for the local identifiability of the model (4.1) with Assumptions 4.1–4.5 at $\alpha=\alpha^0$ are conditions 4.3.1, 4.3.2, and the condition that the rank of

$$\left[I_G\otimes(B_0,B_1,\dots,B_p,\Gamma_1,\dots,\Gamma_q)\right]\left(\frac{\partial\Phi'}{\partial\alpha}\right)_{\alpha=\alpha^0} \tag{5.6}$$

be G^2, where $\alpha'=\operatorname{vec}(B_0,\dots,B_p,\Gamma_1,\dots,\Gamma_q)$ is a $1\times[G(p+1)+K(q+1)]G$ vector.

In the case where ψ_i and ϕ_i are linear, the theorems are globally valid. In general, identification conditions for different problems can be derived by putting them in the framework of Theorems 5.1.1 and 5.1.2. For instance, the derivation of identification conditions using constraints on the coefficients matrix (Section 3.4) may be viewed as a linear application of these theorems. In the following sections we illustrate the use of these theorems by considering errors-in-variables models and covariance restrictions.

5.2. *Contemporaneous error-shock models*

Models with errors-in-variables may arise either because of inaccurate measurements of economic variables, or because an economist uses a theoretical construct which cannot be observed in the real world. Examples of the latter are Friedman's

(1958) permanent income hypothesis or Muth's (1960) rational expectation hypothesis. We do not intend to survey the subject here [for recent surveys, see Goldberger (1972) and Griliches (1974)]. We shall only illustrate how Theorems 5.1.1 and 5.1.2 may be used to find identifiability conditions for models containing random measurement errors.

The classical errors-in-variables models were mainly discussed in a single equation context [e.g. see Madansky (1959) and Malinvaud (1970)]. The occurrence of random measurement errors in the explanatory variables would make standard estimation methods invalid. To circumvent this incidental parameters problem, statisticians have made assumptions about the distribution of random measurement errors, whereas econometricians have postulated prior information about unobservables from economic theory (e.g. that they are linear functions of other observables). The former leads to weighted regression and the latter to instrumental variable methods. Although these different emphases led to different approaches towards the problem, researchers in both fields have been aware of the close association between methods developed in the statistical literature and in the econometric literature [e.g. see Anderson (1976)].

Recently, Goldberger (1972), by way of an example, has shown that in a multiple equations model we may identify the parameters of a model containing unobservables without bringing in additional *a priori* information. We illustrate this connection between errors-in-variables and simultaneous equations by considering a contemporaneous simultaneous equations model. For additional details, see Geraci (1976) and Hsiao (1976).

Let an economic theory specify a set of economic relations between the $G\times 1$ endogenous variables y^* and $K\times 1$ exogenous variables x^* by

$$By_t^* + \Gamma x_t^* = u_t, \tag{5.7}$$

where u is a $G\times 1$ vector of unobservable stochastic term; B and Γ are $G\times G$ and $G\times K$ constant matrices, respectively. We assume that:

Assumption 5.2.1

B is non-singular.

Assumption 5.2.2

The process of the exogenous variables x_t^* is stationary in the wide sense with finite second-order moments. $C_{x^*}(0) = Ex_t^*x_t^{*\prime}$ is non-singular.

Assumption 5.2.3

$Eu_t = 0$, $Eu_tx_s^{*\prime} = 0$, and

$$Eu_tu_s' = \begin{cases} \Sigma & \text{if } t = s, \\ 0 & \text{otherwise.} \end{cases}$$

Suppose, further, that y^* and x^* are unobservable and instead we observe

$$y_t = y_t^* + \zeta_t, \qquad x_t = x_t^* + \varepsilon_t. \tag{5.8}$$

The ζ_t and ε_t are $G\times 1$ and $K\times 1$ vector processes which satisfy:

Assumption 5.2.4

$E\zeta_t = 0$, $E\zeta_t y_s^{*\prime} = 0$,

$$E\zeta_t\zeta_s' = \begin{cases} \Xi & \text{if } t = s, \\ 0 & \text{if } t \neq s, \end{cases}$$

$E\varepsilon_t = 0$, $E\varepsilon_t\zeta_s' = 0$, $E\varepsilon_t x_s^{*\prime} = 0$, $E\varepsilon_t y_s^{*\prime} = 0$, and

$$E\varepsilon_t\varepsilon_s' = \begin{cases} \Lambda & \text{if } t = s, \\ 0 & \text{otherwise.} \end{cases}$$

The variance–covariance matrices of the measurement error processes Ξ and Λ may be singular. Diagonal elements of Ξ and Λ may be zero if some variables are measured without error. Off-diagonal elements may be zero to reflect a lack of correlation among components of the corresponding measurement error process.[14]

Rewriting (5.7) in terms of observables y and x, we have

$$By_t + \Gamma x_t = u_t + B\zeta_t + \Gamma\varepsilon_t. \tag{5.9}$$

Since

$$E(u_t + B\zeta_t + \Gamma\varepsilon_t)x_t' = \Gamma\Lambda, \tag{5.10}$$

the ordinary least squares regression of y_t on x_t will not yield a consistent estimate of the reduced-form parameters $\Pi = -B^{-1}\Gamma$.

Suppose that, in addition to Assumptions 5.2.1–5.2.4, we also postulate:

Assumption 5.2.5

x_t^* are serially uncorrelated.

Then all the (second-order) observational information is contained in

$$C_x(0) = Ex_tx_t' = C_{x^*}(0) + \Lambda = C_x, \tag{5.11}$$

$$C_y(0) = Ey_ty_t' = B^{-1}\Gamma C_x \Gamma B'^{-1} + B^{-1}\Sigma B'^{-1} + \Xi, \tag{5.12}$$

$$C_{yx}(0) = Ey_tx_t' = -B^{-1}\Gamma(C_x - \Lambda). \tag{5.13}$$

[14]When $\Lambda = 0$, we have a model without measurement error, i.e. (5.7) is basically the same as (3.1).

We note that when Σ and Ξ are unrestricted, it is not possible to identify them separately from (5.12) even if we are given B and Γ. In fact, the identification of them is not a theoretically interesting problem since they do not affect the identification and estimation of B and Γ in (5.9). We shall therefore ignore measurement errors in y and treat $u_t + B\zeta_t$ as an amalgamated disturbance term of the model with constant variance–covariance matrix Σ^*, hence (5.12) may be rewritten as

$$C_y(0) = B^{-1}\Gamma C_x \Gamma' B'^{-1} + \Sigma^*. \tag{5.12'}$$

We can rewrite the observational information (5.13) in stacked form as in Section 3.4. Let $\delta' = (\beta_1', \beta_2', \dots, \beta_G', \gamma_1', \dots, \gamma_G')$ be a $1 \times G(G+K)$ vector. Then

$$\left[I_G \otimes C_{yx}' : I_G \otimes (C_x - \Lambda) \right] \delta = 0. \tag{5.14}$$

We assume that additional prior information, including normalization rules, is in the form of linear restrictions on the elements of δ:

$$\Phi \delta = d, \tag{5.15}$$

where Φ and d are an $R \times G(G+K)$ matrix and an $R \times 1$ vector with known elements, respectively.

When δ is identified, a unique solution for Σ^* can be obtained from (5.12′). Therefore, the identification problem becomes simply a question of the uniqueness of solutions to systems of equations (5.14) and (5.15) with respect to δ and unknown elements of Λ. Using Theorem 5.1.1, we obtain:

Theorem 5.2.1

The structure $S = (B, \Gamma, \Sigma^*, \Lambda)$ is identifiable locally if and only if the Jacobian formed by taking partial derivatives of (5.14) and (5.15) with respect to δ and unknown elements of Λ has rank $G(G+K)+l$ around its true value, where l is the number of unknowns in Λ.

Let λ' denote the $1 \times l$ row vector consisting of unknown elements of Λ.[15] Let $\mu' = (\delta', \lambda')$. Using Theorem 5.1.2 we can obtain conditions for identifying

[15] Suppose

$$\Lambda = \begin{bmatrix} \lambda_{11} & \lambda_{12} & 0 & . & 0 \\ \lambda_{21} & \lambda_{22} & 0 & . & 0 \\ 0 & 0 & 0 & & . \\ . & . & . & . & . \\ 0 & . & . & . & 0 \end{bmatrix},$$

then $\lambda' = (\lambda_{11}, \lambda_{12}, \lambda_{22})$.

individual parameters. Let $\mu_j = e_j'\mu$, where e_j is now a $[G(G+K)+l]\times 1$ vector whose elements are all zero except for the jth which is unity. We have:

Theorem 5.2.2

The jth element of μ, $\mu_j = e_j'\mu$, is (locally) identified if and only if $\operatorname{rank}(J' : e_j) = \operatorname{rank}(J')$.

Although Theorems 5.2.1 and 5.2.2 are local conditions, using an instrumental variable argument we can show that when they hold, δ is identified globally. Theorem 5.2.1 or 5.2.2 is possibly a local one only in the sense that we wish to identify both δ and λ. Our prior information may be incomplete such that additional elements of γ's may be zero, thus making it impossible to evaluate λ.

We notice that the Jacobian of (5.14) and (5.15) is of the form

$$J = \begin{bmatrix} I_G \otimes C_{yx}' & \vdots & I_G \otimes (C_x - \Lambda) & \vdots & U \\ \cdots & \cdots & \cdots & \vdots & \cdots \\ & \Phi & & \vdots & 0 \end{bmatrix}, \tag{5.16}$$

where U is a $(GK \times l)$ matrix whose elements are either zero or functions of $(\gamma_1', \dots, \gamma_G')$.[16] Let $\tilde{l}$ rows of U contain non-zero elements $(\tilde{l} \geq l)$.[17] Let $(I_G \tilde{\otimes} C_{yx}' : I_G \tilde{\otimes} (C_x - \Lambda))$ be the $(GK - \tilde{l}) \times G(G+K)$ matrix formed by deleting the rows corresponding to the $\tilde{l}$ non-zero rows of U. Let

$$\tilde{J} = \begin{bmatrix} I_G \tilde{\otimes} C_{yx}' & \vdots & I_G \tilde{\otimes} (C_x - \Lambda) \\ \cdots & \cdots & \cdots \\ & \Phi & \end{bmatrix}. \tag{5.17}$$

Then $\operatorname{rank}(J) \geq \operatorname{rank}(\tilde{J}) + \operatorname{rank}(U)$. Using Theorem 5.2.2, we know if $\operatorname{rank}(\tilde{J})$ has the appropriate rank, δ is identified. Since $\operatorname{rank}(\tilde{J})$ does not depend on Λ, we have the following sufficient identification condition for δ.

Theorem 5.2.3

The parameter vector δ is locally identifiable if $\operatorname{rank}(\tilde{J}) = G(G+K)$.

[16] The i, jth element of $BC_{yx} + \Gamma(C_x - \Lambda) = 0$ [eq. (5.13)] has the form

$$\sum_{g=1}^{G} \beta_{ig} C_{y_g x_j} + \sum_{k=1}^{K} \gamma_{ik} \left(C_{x_k x_j} - \lambda_{kj} \right).$$

Thus, each non-zero element of U has the form $-\gamma_{ik}$.

[17] The possible inequality arises because the same measurement error variances and covariances may appear in different equations.

If all the prior information including the normalization rule is in the form of within-equation constraints, we may rewrite (5.15) as

$$\phi_g \delta_g = d_g, \qquad g = 1,\dots,G, \tag{5.18}$$

where $\delta_g' = (\beta_g', \gamma_g')$ is the gth row of (B, Γ), and ϕ_g and d_g are respectively an $R_g \times (G+K)$ matrix and an $R_g \times 1$ vector with known elements. Let λ_g' be a $1 \times l_g$ vector of unknown measurement error variances and covariances associated with those exogenous variables which appear in the gth equation. Let $a_g' = (\delta_g', \lambda_g')$ and let U_g be the $K \times l_g$ matrix obtained by taking partial derivatives of $(C_x - \Lambda)\gamma_g$ with respect to λ_g'. Using Theorem 5.2.2 we have

Theorem 5.2.4

A necessary and sufficient condition for a_g to be locally identifiable is that the Jacobian matrix

$$J_g = \begin{bmatrix} C_{yx}' & (C_x - \Lambda) & U_g \\ \multicolumn{2}{c}{\phi_g} & 0 \end{bmatrix} \tag{5.19}$$

has rank$(G + K + l_g)$.

We note that this condition is necessary only in the sense that we are considering the identification of λ_g using the information of the gth equation alone. The same measurement error variances and covariances may appear in other equations and hence may be identifiable by those restrictions of other equations. Consequently, either we will have to rely on Theorems 5.2.1 and 5.2.3 to identify δ_g and λ, or if Theorem 5.2.4 is to be interpreted as a necessary condition, we must exclude from consideration those measurement error variances and covariances which can be identified by the restrictions of other equations.

Suppose all the prior information is in the form of exclusion restrictions. Let there be G_Δ and K_* jointly dependent and exogenous variables appearing in the gth equation, and $G_{\Delta\Delta}$ and K_{**} jointly dependent variables and exogenous variables excluded from this equation; then $R_g = G_{\Delta\Delta} + K_{**}. + 1$. Theorem 5.2.4 says that $R_g + K$ has to be greater than $G + K + l_g$. Excluding one restriction for the normalization rule we have $K_{**} \geq G_\Delta + l_g - 1$. If Λ is diagonal (i.e. components of the measurement error ε_t are uncorrelated), we obtain the following order condition.

Corollary 5.2.1

A necessary condition for the identification of the gth row of (B, Γ) through exclusion restrictions is that the number of exogenous variables excluded from this equation be at least as great as the number of included jointly dependent variables plus the number of inaccurately measured included exogenous variables less one.

We illustrate the use of these conditions by considering the following two equation model for which $G = 2$ and $K = 3$. Let the structural relations be:

$$\begin{aligned} \beta_{11}y_t + \beta_{12}y_{2t} &= \gamma_{11}x^*_{1t} + \gamma_{12}x^*_{2t} + \gamma_{13}x^*_{3t} + u_{1t}, \\ \beta_{21}y_{1t} + \beta_{22}y_{2t} &= \gamma_{21}x^*_{1t} + \gamma_{22}x^*_{21} + \gamma_{23}x^*_{3t} + u_{2t}, \end{aligned} \tag{5.20}$$

where the normalization rules are $\beta_{11} = \beta_{22} = 1$, and the prior restrictions are $\gamma_{12} = \gamma_{13} = \gamma_{21} = 0$.

Example 5.2.1

Suppose that x^*_{1t} is the single unobservable variable. That is,

$$x_{1t} = x^*_{1t} + \varepsilon_{1t}, \qquad x_{2t} = x^*_{2t}, \qquad x_{3t} = x^*_{3t},$$

and

$$\Lambda = \begin{bmatrix} \lambda_{11} & 0 & 0 \\ 0 & 0 & 0 \\ 0 & 0 & 0 \end{bmatrix}.$$

Since $\gamma_{21} = 0$, the Jacobian of (5.16) is

$$J =$$

$$\left[\begin{array}{cccccccccc|c}
C_{y_1x_1} & C_{y_2x_1} & 0 & 0 & C_{x_1x_1}-\lambda_{11} & C_{x_1x_2} & C_{x_1x_3} & 0 & 0 & 0 & -\gamma_{11} \\
C_{y_1x_2} & C_{y_2x_2} & 0 & 0 & C_{x_1x_2} & C_{x_2x_2} & C_{x_2x_3} & 0 & 0 & 0 & 0 \\
C_{y_1x_3} & C_{y_2x_3} & 0 & 0 & C_{x_1x_3} & C_{x_2x_3} & C_{x_3x_3} & 0 & 0 & 0 & 0 \\
0 & 0 & C_{y_1x_1} & C_{y_2x_1} & 0 & 0 & 0 & C_{x_1x_1}-\lambda_{11} & C_{x_1x_2} & C_{x_1x_3} & 0 \\
0 & 0 & C_{y_1x_2} & C_{y_2x_2} & 0 & 0 & 0 & C_{x_1x_2} & C_{x_2x_2} & C_{x_2x_3} & 0 \\
0 & 0 & C_{y_1x_3} & C_{y_2x_3} & 0 & 0 & 0 & C_{x_1x_3} & C_{x_2x_3} & C_{x_3x_3} & 0 \\
\hline
1 & 0 & 0 & 0 & 0 & 0 & 0 & 0 & 0 & 0 & 0 \\
0 & 0 & 0 & 1 & 0 & 0 & 0 & 0 & 0 & 0 & 0 \\
0 & 0 & 0 & 0 & 0 & 1 & 0 & 0 & 0 & 0 & 0 \\
0 & 0 & 0 & 0 & 0 & 0 & 1 & 0 & 0 & 0 & 0 \\
0 & 0 & 0 & 0 & 0 & 0 & 0 & 1 & 0 & 0 & 0
\end{array}\right], \tag{5.21}$$

where $C_{y_i x_j}$ denotes the covariance between y_i and x_j. Since only the first element $(-\gamma_{11})$ of U is non-zero, the Jacobian $\tilde{J}$ of (5.17) is that of (5.21) without the first row. Theorem 5.2.3 says that the parameter vector δ is identifiable if rank$(\tilde{J})=10$. This is possible if C_x is non-singular and either $C_{x_1x_2}\neq 0$ or $C_{x_1x_3}\neq 0$ or both. Furthermore, if $\gamma_{11}\neq 0$, then both δ and λ_{11} are (globally) identifiable.

Example 5.2.2

Suppose instead that x^*_{2t}, not x^*_{1t}, were unobserved such that $x_{1t}=x^*_{1t}$, $x_{2t}=x^*_{2t}+\varepsilon_{2t}$, and $x_{3t}=x^*_{2t}$, then

$$\Lambda=\begin{pmatrix}0&0&0\\0&\lambda_{22}&0\\0&0&0\end{pmatrix}. \tag{5.22}$$

The Jacobian of (5.16) is now of the form

$$J=$$

$$\left[\begin{array}{cccccccccc:c}
C_{y_1x_1} & C_{y_2x_1} & 0 & 0 & C_{x_1x_1} & C_{x_1x_2} & C_{x_1x_3} & 0 & 0 & 0 & 0\\
C_{y_1x_2} & C_{y_2x_2} & 0 & 0 & C_{x_1x_2} & C_{x_2x_2}-\lambda_{22} & C_{x_2x_3} & 0 & 0 & 0 & 0\\
C_{y_1x_3} & C_{y_2x_3} & 0 & 0 & C_{x_1x_3} & C_{x_2x_3} & C_{x_3x_3} & 0 & 0 & 0 & 0\\
0 & 0 & C_{y_1x_1} & C_{y_2x_1} & 0 & 0 & 0 & C_{x_1x_2} & C_{x_2x_1} & C_{x_2x_3} & 0\\
0 & 0 & C_{y_1x_2} & C_{y_2x_2} & 0 & 0 & 0 & C_{x_1x_2} & C_{x_2x_2}-\lambda_{22} & C_{x_2x_3} & -\gamma_{22}\\
0 & 0 & C_{y_1x_3} & C_{y_2x_3} & 0 & 0 & 0 & C_{x_1x_3} & C_{x_2x_3} & C_{x_3x_3} & 0\\
\hdashline
1 & 0 & 0 & 0 & 0 & 0 & 0 & 0 & 0 & 0 & 0\\
0 & 0 & 0 & 1 & 0 & 0 & 0 & 0 & 0 & 0 & 0\\
0 & 0 & 0 & 0 & 0 & 1 & 0 & 0 & 0 & 0 & 0\\
0 & 0 & 0 & 0 & 0 & 0 & 1 & 0 & 0 & 0 & 0\\
0 & 0 & 0 & 0 & 0 & 0 & 0 & 1 & 0 & 0 & 0
\end{array}\right], \tag{5.23}$$

The Jacobian $\tilde{J}$ (5.17) is obtained by deleting the fifth row of (5.23). As one can see, the rank of $\tilde{J}$ is at most 9, so that by Theorem 5.2.1 not all the parameter vector δ is identifiable. However, from (5.19), J_1 is of the form

$$J_1=\begin{bmatrix}
C_{y_1x_1} & C_{y_2x_1} & C_{x_1x_1} & C_{x_1x_2} & C_{x_1x_3}\\
C_{y_1x_2} & C_{y_2x_2} & C_{x_2x_1} & C_{x_2x_2}-\lambda_{22} & C_{x_2x_3}\\
C_{y_1x_3} & C_{y_2x_3} & C_{x_3x_1} & C_{x_3x_2} & C_{x_3x_3}\\
1 & 0 & 0 & 0 & 0\\
0 & 0 & 0 & 1 & 0\\
0 & 0 & 0 & 0 & 1
\end{bmatrix}, \tag{5.24}$$

which is of rank five by our assumption. By Theorem 5.2.4 the first equation is identified.

The error-in-variables model discussed above may be viewed as a special case of the factor analysis model (with known factor loading matrix) discussed in psychometric literature. For a general discussion on the relation between unobservable models and factor analysis models, see Griliches (1974) and Malinvaud (1971). For a discussion of dynamic unobservable models, see Hsiao (1977, 1979), and Maravall and Aigner (1977).

5.3. *Covariance restrictions*

So far we have not explicitly considered the use of *a priori* information on the variance–covariance matrix to identify a model. If there is enough prior information on the coefficient matrix, we may identify the model without making use of it. If there is not enough prior information on the coefficient matrix, the information on the covariance matrix may help us achieve identification.

Residuals are commonly attributed to the effects of omitted variables. In the analysis of panel or group data it is not unreasonable to decompose the residuals into two components, those associated with the ith individual unit alone and those which are independently identically distributed over cross-sectional unit i and time t. Anderson and Hsiao (1980), Balestra and Nerlove (1966), Hausman and Taylor (1980a), Maddala (1971) and others have dealt with the single-equation version of this variance components model. Chamberlain (1976, 1977a, 1977b), Chamberlain and Griliches (1975), Taubman (1976), etc. have dealt with the simultaneous-equations-model version of it.

For a simultaneous equation model (3.1) it is natural not only to allow the correlation of the same individual at different time (or different members of a group), but also across equations. A convenient way to model both types of correlation is to postulate latent variables connecting the residuals. Let y_{git} denote the value of the variable y_g for the ith individual (or group) at time t (or tth member). We may assume that[18]

$$u_{git} = \lambda_g h_{it} + \nu_{git}, \tag{5.25}$$

where the ν's are uncorrelated across equations (g) and across i and t. The correlations across equations are all generated by the common omitted variable h. The group structure of the residuals can be generated by assigning h a variance

[18]Note that the notations γ, ξ_i, and ϵ_{it} used in Section 5.3 have a different meaning from the same notations used in Section 5.2.

components structure:

$$h_{it} = \zeta_i + \varepsilon_{it}, \tag{5.26}$$

where ζ_i is invariant over t but is independently, identically distributed across i (groups) with mean zero and variance σ_ζ^2, and ε_{it} is independently, identically distributed across i and t with mean zero and variance σ_ε^2.

Since h is unobservable, we can have

$$\lambda_g^2\left(\sigma_\zeta^2 + \sigma_\varepsilon^2\right) = c\lambda_g^2\left(\frac{1}{c}\sigma_\zeta^2 + \frac{1}{c}\sigma_\varepsilon^2\right)$$

and so its scale is indeterminate. We normalize h by letting $\sigma_\zeta^2 = 1$. Then

$$\begin{aligned} E\boldsymbol{u}_{it}\boldsymbol{u}'_{it} &= \left(1+\sigma_\varepsilon^2\right)\lambda\lambda' + \operatorname{diag}\left(\sigma_1^2,\ldots,\sigma_G^2\right) \\ &= \Sigma, \end{aligned} \tag{5.27}$$

$$E\boldsymbol{u}_{it}\boldsymbol{u}'_{is} = \lambda\lambda' = \Sigma_w \quad \text{if } t \neq s, \tag{5.28}$$

$$E\boldsymbol{u}_{it}\boldsymbol{u}'_{js} = 0 \qquad \text{if } i \neq j, \tag{5.29}$$

where $\boldsymbol{u}_{it} = (u_{1it},\ldots,u_{Git})'$, $\lambda = (\lambda_1,\ldots,\lambda_G)$, and $\operatorname{diag}(\sigma_1^2,\ldots,\sigma_G^2)$ denotes a $G\times G$ diagonal matrix with $\sigma_1^2,\ldots,\sigma_G^2$ on the diagonal.

Under the assumption that ζ_i, ε_{it}, and ν_{git} are normally distributed, or if we limit our attention to second-order moments, all the information with regard to the distribution of y is contained in

$$C_{y_{tt}} = -B^{-1}\Gamma C_{x_{tt}}\Gamma' B'^{-1} + B^{-1}\Sigma B'^{-1}, \tag{5.30}$$

$$C_{y_{t,s}} = -B^{-1}\Gamma C_{x_{t,s}}\Gamma' B'^{-1} + B^{-1}\Sigma_w B'^{-1}, \tag{5.31}$$

$$C_{yx} = -B^{-1}\Gamma C_x, \tag{5.32}$$

where $C_{y_{tt}} = E\boldsymbol{y}_{it}\boldsymbol{y}'_{it}$, $C_{y_{ts}} = E\boldsymbol{y}_{it}\boldsymbol{y}'_{is}$, $C_{x_{tt}} = E\boldsymbol{x}_{it}\boldsymbol{x}'_{it}$, $C_{x_{ts}} = E\boldsymbol{x}_{it}\boldsymbol{x}'_{is}$, $C_{yx} = E\boldsymbol{y}_{it}\boldsymbol{x}'_{js}$, $C_x = E\boldsymbol{x}_{it}\boldsymbol{x}'_{js}$, $\boldsymbol{y}_{it} = (y_{1it},\ldots,y_{Git})'$, and $\boldsymbol{x}_{it} = (x_{1it},\ldots,x_{Kit})'$. Suppose that the $1\times(G+K)G$ vector $\delta' = (\beta'_1,\ldots,\beta'_G,\gamma'_1,\ldots,\gamma'_G)$ is also subject to R *a priori* constraints:

$$\Phi(\delta) = d. \tag{5.33}$$

Then by Theorem 5.1.1 a necessary and sufficient condition for the local identification of B, Γ, σ_ε^2, $\lambda_1,\ldots,\lambda_G$, and $\sigma_1^2,\ldots,\sigma_G^2$ is that the rank of the Jacobian formed by taking partial derivatives of (5.30)–(5.33) with respect to the unknowns is equal to $G(G+K)+2G+1$.

Since a widely used structure for longitudinal micro data is the triangular structure [e.g. Chamberlain (1976, 1977a, 1977b), and Chamberlain and Griliches (1975)], we shall consider a model with B being lower triangular. If the λ_g's are all zero, the residuals are *uncorrelated* across equations. A structure of this form is called a *recursive* structure [Anderson (1972), Fisher (1966), and Wold (1960)] and it is identifiable. To see this, we note that

Lemma 5.3.1

If B is lower block triangular, the admissible transformation matrix F is lower block triangular.

Proof

Suppose we can partition the G joint dependent variables, y, into two sets, y^1 and y^2, where y^1 is $G_1 \times 1$ and y^2 is $G_2 \times 1$. Then we can also partition B, Γ, and the transformation matrix F correspondingly as

$$(B, \Gamma) = \underset{\substack{G_1 \quad G_2 \quad K \\ \text{columns}}}{\begin{pmatrix} B_{11} & B_{12} & \Gamma_1 \\ B_{21} & B_{22} & \Gamma_2 \end{pmatrix}} \begin{matrix} G_1 \\ G_2 \end{matrix} \text{ rows;} \qquad F = \underset{\substack{G_1 \quad G_2 \\ \text{columns}}}{\begin{pmatrix} F_{11} & F_{12} \\ F_{21} & F_{22} \end{pmatrix}} \begin{matrix} G_1 \\ G_2 \end{matrix} \text{ rows,} \tag{5.34}$$

where B_{11}, B_{22}, F_{11}, and F_{22} are square matrices. If y^1 does not depend on y^2, the $G_1 \times G_2$ matrix B_{12} is a zero matrix. By Lemma 3.2.1 all equivalent structures must satisfy

$$\begin{pmatrix} \bar{B}_{11} & 0 & \bar{\Gamma}_1 \\ \bar{B}_{21} & \bar{B}_{22} & \bar{\Gamma}_2 \end{pmatrix} = \begin{pmatrix} F_{11} & F_{12} \\ F_{21} & F_{22} \end{pmatrix} \begin{pmatrix} B_{11} & 0 & \Gamma_1 \\ B_{21} & B_{22} & \Gamma_2 \end{pmatrix}$$

$$= \begin{pmatrix} F_{11}B_{11} + F_{12}B_{12} & F_{12}B_{22} & F_{11}\Gamma_1 + F_{12}\Gamma_2 \\ F_{21}B_{11} + F_{22}B_{21} & F_{22}B_{22} & F_{21}\Gamma_1 + F_{22}\Gamma_2 \end{pmatrix}. \tag{5.35}$$

Thus, $F_{12}B_{22} = 0$. Since B is non-singular, so is B_{22} and we have $F_{12} = 0$.

Lemma 5.3.2

If B is lower block triangular, and the disturbance from an equation in one block is uncorrelated with the disturbance from any equation in another block (i.e. Σ block diagonal), the admissible transformation matrix F is block diagonal.

Proof

Corresponding to the partition in Lemma 5.3.1, we can partition Σ as

$$\Sigma = \begin{pmatrix} \Sigma_{11} & \Sigma_{12} \\ \Sigma_{21} & \Sigma_{22} \end{pmatrix} \begin{matrix} G_1 \\ G_2 \end{matrix} . \tag{5.36}$$
$$\qquad \begin{matrix} G_1 & G_2 \end{matrix}$$

The assumption is that Σ_{12} and Σ_{21} are zero matrices. Lemma 5.3.1 says that F is lower block triangular. By Lemma 3.2.1 we know that all equivalent structures are related by

$$\begin{aligned} \bar{\Sigma} &= \begin{pmatrix} \bar{\Sigma}_{11} & 0 \\ 0 & \bar{\Sigma}_{22} \end{pmatrix} \\ &= \begin{pmatrix} F_{11} & 0 \\ F_{21} & F_{22} \end{pmatrix} \begin{pmatrix} \Sigma_{11} & 0 \\ 0 & \Sigma_{22} \end{pmatrix} \begin{pmatrix} F_{11}' & F_{21}' \\ 0' & F_{22}' \end{pmatrix} \\ &= \begin{pmatrix} F_{11}\Sigma_{11}F_{11}' & F_{11}\Sigma_{11}F_{21}' \\ F_{21}\Sigma_{11}F_{11}' & F_{21}\Sigma_{11}F_{21}' + F_{22}\Sigma_{22}F_{22}' \end{pmatrix}. \end{aligned} \tag{5.37}$$

Since $|F_{11}| \neq 0$ and $|\Sigma_{11}| \neq 0$, we have $|F_{11}\Sigma_{11}| \neq 0$ which implies $F_{21} = 0$.

Lemma 5.3.1 says that if an equation is identified within the first subset (of G_1 equations), it is identifiable within the whole system. But it is still not possible to consider the block G_2 equations separately. In order for the two blocks to be considered separately, we need Lemma 5.3.2. Models that satisfy Lemma 5.3.2 are called *block recursive* models.

Theorem 5.3.1

If B is lower triangular and Σ is diagonal, the entire system (3.1) is identified.

Proof

If B and $\bar{B}$ are lower triangular, it implies that F is lower block triangular for every pair G_1 and G_2 such that $G_1 + G_2 = G$. Therefore F is lower triangular. Furthermore, B and $\bar{B}$ triangular, and Σ and $\bar{\Sigma}$ block diagonal imply that $F_{12} = 0$ and $F_{21} = 0$. So B triangular and Σ diagonal imply that F is diagonal.

Intuitively, the (block) triangularity of B means that movements in any element of u_g lead, through effects on the endogenous variables, to movements in the elements of $u_{g'}$, if and only if $g' \geq g$. They do not affect the elements of $y_1, \dots, y_{g-1}$. Similarly, the (block) diagonality of Σ means that movements in elements of u_g are not systematically associated with movements in the elements of $u_{g'}$, for $g \neq g'$. The two together thus allow us to treat endogenous variables with respect to higher numbered equations as predetermined. [For additional discussion on recursive, block recursive systems, and zero restrictions on the covariance matrix, see Anderson (1972), Fisher (1966), Hausman and Taylor (1980b), Kelly (1972), Maddala and Patil (1976), etc.]

If some of the λ_g's are different from zero, the residuals for different i (groups) are still uncorrelated by our assumption, but the residuals for the same i (within the group) are correlated. To concentrate on this non-standard aspect of the covariance structure, we assume a structure not carrying any exogenous variables (i.e. $\Gamma \equiv 0$). Thus, we have

$$By_{it} = \boldsymbol{u}_{it}, \tag{5.38}$$

where B is lower triangular with unity as the diagonal elements. Then the information about y is contained in

$$BC_{y_{tt}}B' = \Sigma, \tag{5.39}$$

$$BC_{y_{ts}}B' = \Sigma_w. \tag{5.40}$$

We note that Σ is symmetric and we have $G(G+1)/2$ independent equations from (5.39). But Σ_w is of rank 1, therefore we can derive at most G independent equations from (5.40). Excluding the normalization conditions, we still have $G(G-1)/2$ unknowns in B, $2G+1$ unknowns of $(\lambda_1,\dots,\lambda_G)$, $(\sigma_1^2,\dots,\sigma_G^2)$, and σ_ε^2. We have one less equation than the number of unknowns. In order for the Jacobian matrix formed by (5.39), (5.40), and *a priori* restrictions to be non-singular, we need at least one additional a priori restriction.

Lemma 5.3.3

A necessary condition for the identification of (5.38) under exclusion restrictions is that at least one $\beta_{gl} = 0$ for $g > l$.

Theorem 5.3.2

A necessary and sufficient condition for the local identification of (5.38) under exclusion restrictions on B is that one $\beta_{gl} = 0$, $g > l$, provided that

$$\operatorname{rank}\begin{pmatrix} \bar{\Psi} \\ e_j \end{pmatrix} = \frac{G(G-1)}{2} + 2G + 1, \tag{5.41}$$

where $\bar{\Psi}$ denotes the $G(G+1)/2 + G$ partial derivative equations of (5.39) and (5.40) with respect to $\alpha' = (\beta_{21}, \beta_{31}, \beta_{32}, \dots, \beta_{G,G-1}, \lambda_1, \dots, \lambda_G, \sigma_1^2, \dots, \sigma_G^2, \sigma_\varepsilon^2)$, and

e_j is a $G(G-1)/2+2G+1$ vector whose elements are all zero except for the $(g-1)(g-2)/2+l$th element which is unity.

We illustrate the use of this theorem by considering the following simple example.

Example 5.3.1

Suppose that

$$\begin{aligned} y_{1it} &= \lambda_1 h_{it} + \nu_{1it}, \\ \beta_{21} y_{1it} + y_{2it} &= \lambda_2 h_{it} + \nu_{2it}, \\ \beta_{31} y_{1it} + \beta_{32} y_{2it} + y_{3it} &= \lambda_3 h_{it} + \nu_{3it}, \end{aligned} \tag{5.42}$$

where y_1 is years of schooling, y_2 is a late (post-schooling) test score, y_3 is earnings, and h is an unobserved measure of early (pre-school) ability. The model is formally equivalent to the one estimated by Chamberlain and Griliches (1975) using data on brothers (t) within a family (i).

The information with regard to y is contained in

$$\sigma_{y_1}^2 = Ey_{1it}^2 = \lambda_1^2(1+\sigma_\varepsilon^2)+\sigma_1^2, \tag{5.43}$$

$$\sigma_{y_2}^2 = Ey_{2it}^2 = (\lambda_2 - \beta_{21}\lambda_1)^2(1+\sigma_\varepsilon^2)+\beta_{21}^2\sigma_1^2+\sigma_2^2, \tag{5.44}$$

$$\begin{aligned} \sigma_{y_3}^2 = Ey_{3it}^2 &= [\lambda_3 - \beta_{31}\lambda_1 - \beta_{32}(\lambda_2 - \beta_{21}\lambda_1)]^2(1+\sigma_\varepsilon^2) \\ &\quad + (\beta_{31} - \beta_{32}\beta_{21})^2\sigma_1^2 + \beta_{32}^2\sigma_2^2 + \sigma_3^2, \end{aligned} \tag{5.45}$$

$$\begin{aligned} \sigma_{y_1 y_2} &= Ey_{1it}y_{2it} \\ &= [\lambda_1\lambda_2 - \beta_{21}\lambda_1^2](1+\sigma_\varepsilon^2) - \beta_{21}\sigma_1^2, \end{aligned} \tag{5.46}$$

$$\begin{aligned} \sigma_{y_1 y_3} = Ey_{1it}y_{3it} &= [(\lambda_3 - \lambda_2\beta_{32}) + \lambda_1(\beta_{32}\beta_{21} - \beta_{31})]\lambda_1(1+\sigma_\varepsilon^2) \\ &\quad + (\beta_{32}\beta_{21} - \beta_{31})\sigma_1^2, \end{aligned} \tag{5.47}$$

$$\begin{aligned} \sigma_{y_2 y_3} = Ey_{2it}y_{3it} &= [\lambda_3(\lambda_2 - \beta_{21}\lambda_1) + \lambda_1^2\beta_{31}\beta_{21} - \lambda_1\lambda_2\beta_{31} \\ &\qquad - \beta_{32}(\lambda_2 - \beta_{21}\lambda_1)^2] \\ &\quad \times (1+\sigma_\varepsilon^2) + (\beta_{31}\beta_{21} - \beta_{32}\beta_{21}^2)\sigma_1^2 - \beta_{32}\sigma_2^2, \end{aligned} \tag{5.48}$$

$$\sigma_{y_{1t} y_{1s}} = Ey_{1it}y_{1is} = \lambda_1^2, \tag{5.49}$$

$$\sigma_{y_{1t} y_{2s}} = Ey_{1it}y_{2is} = \lambda_1(\lambda_2 - \beta_{21}\lambda_1), \tag{5.50}$$

$$\sigma_{y_{1t} y_{3s}} = Ey_{1it}y_{3is} = \lambda_1[\lambda_3 - \beta_{31}\lambda_1 - \beta_{32}(\lambda_2 - \beta_{21}\lambda_1)], \tag{5.51}$$

$$\sigma_{y_{2t}y_{2s}} = Ey_{2it}y_{2is} = (\lambda_2 - \beta_{21}\lambda_1)^2, \tag{5.52}$$

$$\sigma_{y_{2t}y_{3s}} = Ey_{2it}y_{3is} = (\lambda_2 - \beta_{21}\lambda_1)[\lambda_3 - \beta_{31}\lambda_1 - \beta_{32}(\lambda_2 - \beta_{21}\lambda_1)], \tag{5.53}$$

$$\sigma_{y_{3t}y_{3s}} = Ey_{3it}y_{3is} = [\lambda_3 - \beta_{31}\lambda_1 - \beta_{32}(\lambda_2 - \beta_{21}\lambda_1)]^2. \tag{5.54}$$

We note that $\sigma_{y_{2t}y_{2s}} = \sigma^2_{y_{1t}y_{2s}}/\sigma_{y_{1t}y_{1s}}$, $\sigma_{y_{2t}y_{3s}} = \sigma_{y_{1t}y_{3s}} \cdot \sigma_{y_{1t}y_{2s}}/\sigma_{y_{1t}y_{1s}}$, and $\sigma_{y_{3t}y_{3s}} = \sigma^2_{y_{1t}y_{3s}}/\sigma_{y_{1t}y_{1s}}$. Therefore, only (5.43)–(5.51) contain independent information about y. The identification of (5.42) is equivalent to whether the unknown parameters may be solved uniquely from (5.43)–(5.51). However, there are ten unknowns, and only nine independent equations. We need an additional constraint. Suppose $\beta_{gl} = 0$ for $g > l$ and the rank condition (Theorem 5.3.2) is satisfied; then the model is locally identified. Under the assumption that σ_1^2, σ_2^2, and $\sigma_3^2 \neq 0$ when $\beta_{21} = 0$, the rank condition is equivalent to $\lambda_1\lambda_2 \neq 0$. When $\beta_{31} = 0$, the rank condition is equivalent to $\lambda_1\lambda_3\{\lambda_1\sigma_2^2 + \beta_{21}\lambda_2\sigma_1^2\} \neq 0$. When $\beta_{32} = 0$, the rank condition is equivalent to $\lambda_1\lambda_2\lambda_3 \neq 0$.

This is a particular example of using the information in the covariance matrix to identify a structure. We can further generalize the analysis to derive identification conditions for models involving multi-factors or the multiple-indicator, multiple-cause models. In a sense, they are all applications of Theorems 5.1.1 and 5.1.2. For details see Chamberlain (1977a) and Jöreskog and Goldberger (1975).

6. Bayesian theory of identification and concluding remarks

The identification problem is "a general and fundamental problem arising, in many fields of inquiry, as a concomitant of the scientific procedure that postulates the existence of a structure" [Koopmans and Reiersøl (1950)]. In this chapter we restricted our attention to a particular class of models, namely contemporaneous and dynamic simultaneous equation models. The basic characteristics of the identification problems in these models are approached from the point of view of observationally equivalent structures. Conditions for their identifiability are derived by giving an unambiguous definition to parameters to be estimated.

The analysis so far is based entirely on the classical approach to statistical inference where we assume that we have exact knowledge on some of the parameters and no knowledge at all on others. It was shown that if enough exact prior information was available on the structural parameters, a model may be identifiable. More often we possess only vague information. In such cases the Bayesian approach which permits replacing deterministic constraints by probabilistic constraints – that is, by a prior density on parameters – would seem more appealing. The unresolved question is whether Bayesian theory requires a different definition of identification from the classical one.

Bayesians and non-Bayesians both attempt to extract the information contained in the sample through the use of the likelihood function. However, what is unsettled is whether the prior has anything to do with the identification of a structure. Forms of priors put forward and used by Box and Tiao (1973), Jeffreys (1961), Raiffa and Schlaifer (1961), etc. depend on the form of the likelihood function. With this consideration in mind, Zellner (1971) defines the Bayesian observationally equivalent structures which involve both likelihood functions and prior distributions. Explicitly, suppose that we have two structures: S, with its parameter vector α and prior probability density function $p(\alpha)$, and $\overline{S}$, with its parameter vector $\overline{\alpha}$ and prior probability density function $p(\overline{\alpha})$. If both structures, along with their associated prior information, lead to exactly the same marginal probability distribution for a set of observations, y,

$$p(y|S)=\int p(y|\alpha)p(\alpha)\,\mathrm{d}\alpha=\int p(y|\overline{\alpha})p(\overline{\alpha})\,\mathrm{d}\overline{\alpha}=p(y|\overline{S}), \tag{6.1}$$

we shall say that S and its associated prior information and $\overline{S}$ and its associated prior information are observationally equivalent. In this sense, if diffuse priors are employed, a structure that is unidentified in classical analysis will be unidentified in Bayesian analysis as well [e.g. see Zellner (1971, p. 255)]. But a structure may be unidentified in the classical view, yet be identifiable from a Bayesian view, if priors on α and $\overline{\alpha}$ are informative. Alternatively, even if priors are informative, if (6.1) is satisfied, there can still be an identification problem. Theorems on the identification of a structure under this approach remain to be worked out.

On the other hand, Kadane (1975) takes the Savage view that "identification is a property of the likelihood function, and is the same whether considered classically or from the Bayesian approach".[19] In this view, the definition of observationally equivalent structures for a Bayesian is the same as the definition (Definition.2.1) for a classical statistician; it has nothing to do with the prior density of the parameter vector. Thus, the conditions required to identify a model derived in previous sections using exact prior information remain valid as sufficient conditions.

Under the Kadane–Savage view, when a model is identified, the experiment is *informative* with regard to unknown parameters, that is, the prior will be different from the posterior. Clearly, the concepts of identification and informativeness are different.[20] If you are opinionated in the sense that you believe that only one

[19]If prior information about parameters' values is used in designing a survey or an experiment, then the likelihood function will be linked to the prior probability density function and it will not be possible to consider them separately. I owe this point to Professor A. Zellner.

[20]E. Leamer takes a different view from this. For instance, see Leamer (1978, ch. 5.3).

particular value, α^0, of α holds, then no experiment is informative about α^0, yet the model is identified since no other α will satisfy these exact constraints except that $\alpha = \alpha^0$. On the other hand, if we are less than certain about the unknown parameters we may characterize identification in terms of informativeness of an experiment.

Theorem 6.1

A structure S with its (unknown) parameter vector α and prior $p(\alpha)$ such that $0 < p(\alpha) < 1$ is identified if and only if $p(\alpha|y) \neq p(\alpha)$.

Proof

There is no loss of generality in assuming a two-point prior, α^* and $\bar{\alpha}$, such that $0 < p(\alpha^*) = 1 - p(\bar{\alpha}) < 1$. First, suppose that S is not identified, by Definition 2.1:

$$p(y|\alpha^*) = p(y|\bar{\alpha}) \quad \text{for all } y. \tag{6.2}$$

Then

$$p(\alpha^*|y) = \frac{p(\alpha^*)p(y|\alpha^*)}{p(\alpha^*)p(y|\alpha^*) + p(\bar{\alpha})p(y|\bar{\alpha})} = p(\alpha^*). \tag{6.3}$$

Similarly, $p(\bar{\alpha}|y) = p(\bar{\alpha})$. So the experiment is not *informative*.

Suppose now that S is identified. Let α^* and $\bar{\alpha}$ be distinguished points in model $\mathbb{S}$. From Definition 2.2:

$$p(y|\alpha^*) \neq p(y|\bar{\alpha}) \quad \text{for at least one } y. \tag{6.4}$$

Let $0 < p(\alpha^*) = 1 - p(\bar{\alpha}) < 1$. Now suppose

$$p(\alpha^*|y) = \frac{p(\alpha^*)p(y|\alpha^*)}{p(\alpha^*)p(y|\alpha^*) + p(\bar{\alpha})p(y|\bar{\alpha})} = p(\alpha^*), \tag{6.5}$$

$$p(\bar{\alpha}|y) = \frac{p(\bar{\alpha})p(y|\bar{\alpha})}{p(\alpha^*)p(y|\alpha^*) + p(\bar{\alpha})p(y|\bar{\alpha})} = p(\bar{\alpha}). \tag{6.6}$$

Eqs. (6.5) and (6.6) hold if and only if

$$\frac{p(\alpha^*)p(y|\alpha^*)}{p(\bar{\alpha})p(y|\bar{\alpha})} = \frac{p(\alpha^*)}{p(\bar{\alpha})} \quad \text{for all } y. \tag{6.7}$$

That is, $p(y|\alpha^*) = p(y|\bar{\alpha})$ for all y, which contradicts (6.4). Therefore (6.5) and (6.6) cannot hold, so the experiment is *informative* with respect to this distinguished two-point prior.

Although Theorem 6.1 can be viewed both as a consequence and conditions of identifiability, it is not easy to check in practice. In particular, if economic knowledge is subject to uncertainties of various kinds, the prior information will usually be described by means of exact constraints on some of the parameters, and a probability distribution function defined on some or all of the remaining parameters. Unless the exact constraints satisfy conditions derived in previous sections, to check identifiability by the derivation of the posterior is usually extremely complicated for models considered in this paper.

Drèze (1974) has indicated a valuable approach to derivation of the necessary conditions for identification when stochastic prior information is also utilized. Consider the model (3.1). From Theorem 6.1 we know that a necessary condition is that the posterior of (B, Γ, Σ) is informative. Let $p(B, \Gamma, \Sigma)$ be the prior density on the structural parameters. Under the assumption that for given B the transformation from Σ to V is non-singular, one may then perform an integrand transformation from the parameter space (B, Γ, Σ) to the parameter space (B, Π, V) to define the alternative prior density $p(B, \Pi, V)$. The prior density $p(B, \Pi, V)$ then may be factorized into a conditional density on B, given (Π, V), and a marginal density on (Π, V), that is,

$$p(B, \Pi, V) = p(B|\Pi, V)p(\Pi, V). \tag{6.8}$$

Given this factorization, we have the following important theorem:

Theorem 6.2

$p(y|B, \Gamma, \Sigma, x) \equiv p(y|\Pi, V, x)$ implies

$$p(B, \Pi, V|y, x) = p(B|\Pi, V)p(\Pi, V|y, x), \tag{6.9}$$

whenever the posterior density exists.

Proof

By Bayes' theorem

$$\begin{aligned} p(B, \Pi, V|y, x) &= \frac{p(B, \Pi, V)p(y|B, \Pi, V, x)}{\int p(B, \Pi, V)p(y|B, \Pi, V, x)\,\mathrm{d}B\,\mathrm{d}\Pi\,\mathrm{d}V} \\ &= \frac{p(B|\Pi, V)p(\Pi, V)p(y|\Pi, V, x)}{\int p(\Pi, V)p(y|\Pi, V, x)\left[\int p(B|\Pi, V)\,\mathrm{d}B\right]\mathrm{d}\Pi\,\mathrm{d}V} \\ &= p(B|\Pi, V)p(\Pi, V|y, x). \end{aligned} \tag{6.10}$$

Theorem 6.2 says that the conditional posterior density of B, given (Π, V), is proportional to the prior density for B, given (Π, V). Since $p(\Pi, V|y, x)$ is informative under the usual assumption, a necessary condition for the existence of a well-defined posterior (B, Π, V) is the existence of a well-defined conditional prior density $p(B|\Pi, V)$. Typically, one may expect the prior information to be truly informative about a subset of the structural parameters $\alpha = (B, \Gamma, \Sigma)$ only. One may then wish to know whether the prior density on such a proper subset of structural parameters implies a proper prior density on B, given the reduced-form parameters (Π, V). For instance, suppose that prior information is described by R exact constraints and $\tilde{R}$ inexact constraints. Let the R exact prior constraints be expressed by $\Phi(\alpha) = 0$ and the $\tilde{R}$ inexact constraints be in the form of informative priors on a subset of the vector α, say α_1. If a proper prior density on α_1 conditional on $\Phi(\alpha)$ $(=0)$ and (Π, V) implies a proper prior density on B, conditional on $\Phi(=0)$ and (Π, V), then the posterior (B, Γ, Σ) is well defined. Drèze (1974) has derived a rank condition for this transformation to be well defined. This rank condition can be viewed as a natural extension of the rank condition found in Section 3 and reasonably easy to check. Unfortunately, the informativeness in the posterior is only a necessary condition for the model to be identified, hence the rank condition is only necessary and not sufficient. A well-defined prior density will satisfy the rank condition, yet the model may be unidentified in the classical sense. A development of easily checked sufficient identification conditions incorporating stochastic prior information appears to be one of the fruitful areas for future research.

The Bayesian approach is more general in the sense that it allows one to make use of vague (or inexact) prior information. However, there is still a fundamental question of whether there exists enough prior information to identify a structure. Liu (1960) and Sims (1980) have argued that most econometric models are hopelessly underidentified as opposed to the commonly assumed case that there is more prior information than is necessary to identify a structure (i.e. overidentification). They maintain that identification of a structure is often achieved not because we have sufficient prior information but because we want it to be so. For instance, to identify a structure by zero restrictions often ends up in limiting the number of variables included in each equation and in the system as a whole. The complexity of modern economic society makes this approach much more likely to be incorrect. There are always more variables "really" in each equation than are assumed in the practical approximations. Nor are the equations stated in the model a complete set of equations. Other, outside equations always exist, either relating variables in the model alone or relating them to unincluded variables. This means that the number of equations in the system has been understated and variables which are supposed to be exogenous are in fact endogenous. The necessary order condition for identification is then not satisfied. Thus, the only obvious thing to do is to proceed from the reduced form rather than haphazardly adding false restrictions to achieve identification.

This is indeed a powerful and disturbing argument. An economist who confronts models which are otherwise unidentified should resist the temptation to arbitrarily specify models in such a way as to produce identifiability of relevant characteristics. Scientific honesty demands that the specifications of a model be based on prior knowledge of the phenomena studied, but not on the desire for identifiability of characteristics in which the researcher happens to be interested. Theoretical statisticians or econometricians can only gradually increase the repertoire of models and methods available to applied workers. It is the responsibility of each empirical researcher to make the best choice based on available knowledge of statistical properties of different procedures under alternative circumstances, his *a priori* knowledge and beliefs about the workings of the economic sector studied, and the particular purpose of his study. A study of identification is useful if it is used to guide the economist's strategy for collecting richer data or developing theoretical specifications to overcome the lack of identifiability.

However, since the Liu–Sims attack cannot easily be dismissed, it appears that the following two lines of research will be useful. One is to study identification conditions of a far more general kind than those considered in this chapter. The other is to evaluate the cost of approximate specification. Fisher (1961) and Sargan (1980) have studied the properties and significance tests of some estimators for models under approximate specification or when the models are almost unidentified. Chow (1980), realizing that the specification error will almost always be present, is more concerned with total error, due to both specification and sampling, in using an estimated model for forecasting. He has followed Akaike (1974) and suggested using the information criterion to evaluate the cost of approximate specification. A thorough investigation of the theorems of identification and the cost of approximate specification from a decision theoretical framework would seem highly desirable.

Appendix

In this appendix we state some definitions and theorems with regard to polynomial matrices. The main references are Gantmacher (1959) and MacDuffee (1956).

Definition A.1

A polynomial matrix is a rectangular matrix $A(\lambda)$ whose elements are polynomials in λ:

$$A(\lambda)=\left(a_{ij}(\lambda)\right)=\left(a_{ij,0}+a_{ij,1}\lambda+\cdots+a_{ij,l}\lambda^{l}\right),$$

$$i=1,\dots,m;\quad j=1,\dots,n,\tag{A.1}$$

where l is the largest of the degrees of the polynomials $a_{ij}(\lambda)$.

Setting $A_k = (a_{ij,k})$ we may represent the polynomial matrix $A(\lambda)$ in the form of a polynomial in λ with matrix coefficients:

$$A(\lambda) = A_0 + A_1\lambda + \cdots + A_l\lambda^l. \tag{A.2}$$

Corresponding to the elementary row (or column) operations of a constant matrix we can define the elementary row (or column) operations of a polynomial matrix:

(1) Multiplication of any row (or column), for example the ith, by a number $c \neq 0$.
(2) Addition to any row (or column), for example the ith, of any other row or column, for example the jth, multiplied by any arbitrary polynomial $b(\lambda)$.
(3) Interchange of any two rows (or columns), for example the ith and the jth.

The elementary row operations, (1), (2), and (3), are equivalent to a multiplication of the polynomial matrix $A(\lambda)$ on the left by the following square matrices of order m, respectively:[21]

$$T_1 = \begin{pmatrix} 1 & 0 & & \overset{(i)}{} & & & 0 \\ 0 & \cdot & & & & & \\ \cdot & & \cdot & & & & \\ \cdot & & & c & & & \\ \cdot & & & & \cdot & & \\ \cdot & & & & & \cdot & 0 \\ 0 & \cdot & \cdot & & & 0 & 1 \end{pmatrix}, \quad T_2 = \begin{pmatrix} 1 & 0 & \overset{(i)}{\cdot} & \cdot & \overset{(j)}{\cdot} & \cdot & 0 \\ 0 & \cdot & & \cdot & \cdot & & \\ \cdot & & 1 & \cdot & b(\lambda) & & \cdot \\ & & & \cdot & \cdot & & \cdot \\ & & & & \cdot & \cdot & 0 \\ 0 & \cdot & \cdot & & \cdot & 0 & 1 \end{pmatrix},$$

$$T_3 = \begin{pmatrix} 1 & 0 & \overset{(i)}{\cdot} & \cdot & \overset{(j)}{\cdot} & \cdot & \cdot & 0 \\ 0 & \cdot & & & & & & \cdot \\ \cdot & \cdot & 0 & \cdot & 1 & & & \cdot \\ \cdot & & \cdot & \cdot & \cdot & & & \cdot \\ \cdot & & 1 & \cdot & 0 & & & \cdot \\ & & & & & \cdot & & \\ \cdot & & & & & & \cdot & 0 \\ 0 & \cdot & \cdot & \cdot & \cdot & \cdot & 0 & 1 \end{pmatrix}. \tag{A.1}$$

The elementary column operations, (1), (2), and (3), are equivalent to a multiplication of the polynomial matrix $A(\lambda)$ on the right by the following

[21] In these matrices all the elements that are not shown are 1 on the main diagonal and 0 elsewhere.

square matrices of order n, respectively:

$$T_1' = \begin{pmatrix} 1 & 0 & \cdot & \cdot & \cdot & 0 \\ 0 & \cdot & \cdot & & & \cdot \\ \cdot & & C & & & \cdot \\ \cdot & & & \cdot & & \cdot \\ \cdot & & & \cdot & & 0 \\ 0 & \cdot & \cdot & \cdot & 0 & 1 \end{pmatrix}, \quad T_2' = \begin{pmatrix} 1 & 0 & \cdot & \cdot & \cdot & \cdot & \cdot & 0 \\ 0 & \cdot & & & & & & \cdot \\ \cdot & \cdot & 1 & & & & & \cdot \\ \cdot & \cdot & \cdot & \cdot & & & & \cdot \\ \cdot & & b(\lambda) & \cdot & \cdot & & & \cdot \\ \cdot & & & & & & \cdot & 0 \\ 0 & \cdot & \cdot & \cdot & \cdot & \cdot & 0 & 1 \end{pmatrix} \begin{matrix} \\ \\ (i) \\ \\ (j) \\ \\ \\ \end{matrix},$$

$$T_3' = \begin{pmatrix} 1 & \cdot & \cdot & \cdot & \cdot & \cdot & 0 \\ \cdot & \cdot & & & & & \cdot \\ \cdot & & 0 & \cdot & 1 & & \cdot \\ \cdot & & \cdot & \cdot & \cdot & & \cdot \\ \cdot & & 1 & \cdot & 0 & & \cdot \\ \cdot & & & & & \cdot & \cdot \\ 0 & \cdot & \cdot & \cdot & \cdot & 0 & 1 \end{pmatrix} \begin{matrix} \\ \\ (i) \\ \\ (j) \\ \\ \\ \end{matrix}. \tag{A.2}$$

(In T_1', C stands in column (i).)

The determinant of every elementary matrix (A.1) and (A.2) does not depend on λ and is different from zero. Therefore each left (right) elementary operation has an inverse operation which is also a left (right) elementary operation.

A unimodular matrix is a square polynomial matrix whose determinant is a non-zero constant. Hence

Lemma A.1

Every unimodular matrix can be represented in the form of a product of a finite number of elementary matrices.

If three polynomial matrices are in the relation

$$A(\lambda) = C(\lambda)D(\lambda), \tag{A.3}$$

then C is called a *left divisor* of A, and A is called a *right multiple* of C. A *greatest common left divisor* (g.c.l.d.) C of two matrices A and B is a common left divisor which is a right multiple of every common left divisor of A and B. A *least common right multiple* of two matrices C and F is a common right multiple which is a left divisor of every common right multiple of C and F.

Lemma A.2

If A and B have a non-singular greatest common left divisor C, every g.c.l.d. of A and B is of the form CU, where U is unimodular.

Proof

If F is another g.c.l.d., then $C = FP$, $F = CQ$, $C = CQP$, and $QP = I$.

If A and B are relatively left prime, then there is no g.c.l.d. that is not unimodular. To determine whether A and B are relatively left prime, we can use [Gantmacher (1959, pp. 135–136)]

Theorem A.1

An arbitrary rectangular polynomial matrix $m \times n$ $(n \geq m)$ can always be brought into the form

$$H(\lambda) = \begin{pmatrix} h_{11}(\lambda) & 0 & \cdot & \cdot & \cdot & \cdot & \cdot & 0 \\ h_{21}(\lambda) & h_{22}(\lambda) & \cdot & \cdot & & & & \cdot \\ \cdot & & \cdot & & & & & \cdot \\ \cdot & & & \cdot & \cdot & & & \cdot \\ h_{m1}(\lambda) & h_{m2}(\lambda) & \cdot & \cdot & h_{mm}(\lambda) & 0 & \cdot & 0 \end{pmatrix} = (H_I(\lambda):0) \tag{A.4}$$

by means of right elementary operations, where the polynomials $h_{i1}(\lambda)$, $h_{i2}(\lambda),\dots,h_{i,i-1}(\lambda)$ are of degree less than that of $h_{ii}(\lambda)$, provided $h_{ii}(\lambda) \neq 0$, and all are identically equal to zero if $h_{ii}(\lambda) = \text{constant} \neq 0$.

Theorem A.2

Let $A(\lambda) = [A_I(\lambda) \quad A_{II}(\lambda)]$, where $A_I(\lambda)$ and $A_{II}(\lambda)$ are $m \times m$ and $m \times (n-m)$ polynomial matrices, respectively. If $A_I(\lambda)$ is non-singular, then $H_I(\lambda)$ of (A.4) is a g.c.l.d. of $A_I(\lambda)$ and $A_{II}(\lambda)$. Any matrix $\tilde{H}_I(\lambda)$ that is a right multiple of $H_I(\lambda)$ by a unimodular matrix $U(\lambda)$ is also a g.c.l.d; conversely, any g.c.l.d. $\tilde{H}_I(\lambda)$ satisfies $H_I(\lambda) = H_I(\lambda)U(\lambda)$ for unimodular $U(\lambda)$ [MacDuffee (1956)].

Corollary A.1

If A and B are relatively left prime, then the g.c.l.d. is I.

Proof

If A and B are relatively left prime, then there is no g.c.l.d. that is not unimodular. Theorem A.1 says that the determinant of H_I must be a constant matrix. Since

$$h_{11}(\lambda)\dots h_{mm}(\lambda) = \text{constant} \neq 0$$

implies

$$h_{ii}(\lambda) = \text{constant} \neq 0 \qquad (i = 1,\dots,m),$$

by an elementary operation this can be reduced to an identity matrix.

We can use these results to find the g.c.l.d. of a dynamic model. For instance, in Example 4.3.1 we assumed that

$$[B(L) \quad \Gamma(L)] = \begin{bmatrix} (1+\rho L)(1+\beta_{11,1}L) & 0 & (1+\rho L)\gamma_{11} \\ \beta_{21,0} & (1+\beta_{22,1}L) & 0 \end{bmatrix}. \tag{A.5}$$

Through the elementary column operation we can bring (A.5) into the form

$$[B(L) \quad \Gamma(L)]U(L) = [H_I(L) \vdots 0], \tag{A.6}$$

where

$$H_I(L) = \begin{bmatrix} 1+\rho L & 0 \\ 0 & 1 \end{bmatrix},$$

$$U(L) = \frac{1}{\gamma_{11}\beta_{21,0}} \begin{bmatrix} 0 & \gamma_{11} & -\gamma_{11}(1+\beta_{22,1}L) \\ 0 & 0 & \gamma_{11}\beta_{21,0} \\ \beta_{21,0} & -(1+\beta_{22,1}L) & (1+\beta_{22,1}L)^2 \end{bmatrix}$$

and where $\det|U(L)| = \gamma_{11}\beta_{21,0}$, which is a non-zero constant, but $\det|H_I(L)| = (1+\rho L)$ is not.

References

Akaike, H. (1974) "A New Look at the Statistical Model Identification", *IEEE Transactions on Automatic Control*, AC-19, 716–723.

Anderson, T. W. (1972) "Lecture Notes in Econometrics", mimeograph.

Anderson, T. W. (1976) "Estimation of Linear Functional Relationships: Approximate Distributions and Connections with Simultaneous Equations in Econometrics", *The Journal of the Royal Statistical Society*, B 38, 1–36.

Anderson, T. W. and C. Hsiao (1980) "Formulation and Estimation of Dynamic Models Using Panel Data", Technical Report, Institute for Mathematical Studies in the Social Sciences, Stanford University (forthcoming in the *Journal of Econometrics*).

Balestra, P. and M. Nerlove (1966) "Pooling Cross-Section and Time-Series Data in the Estimation of a Dynamic Model: The Demand for Natural Gas", *Econometrica*, 34, 585–612.

Bowden, R. (1973) "The Theory of Parametric Identification", *Econometrica*, 41, 1069–1074.

Box, G. E. P. and G. M. Jenkins (1970) *Time Series Analysis, Forecasting and Control*. San Francisco: Holden Day.

Box, G. E. P. and G. C. Tiao (1973) *Bayesian Inference in Statistical Analysis*. Reading, Mass.: Addison-Wesley.

Chamberlain, G. (1976) "Identification in Variance Components Models", Discussion Paper No. 486, Harvard Institute of Economic Research, Harvard University.

Chamberlain, G. (1977a) "An Instrumental Variable Interpretation of Identification in Variance-Components and MIMIC Models", in: P. Taubman (ed.), *Kinometrics: Determinants of Socioeconomic Success With and Between Families*. Amsterdam: North-Holland Publishing Co., pp. 235–254.

Chamberlain, G. (1977b) "Are Brothers as Good as Twins?", in: P. Taubman (ed.), *Kinometrics: Determinants of Socioeconomic Success Within and Between Families*. Amsterdam: North-Holland Publishing Co., pp. 287–297.

Chamberlain, G. and Z. Griliches (1975) "Unobservables with a Variance-Components Structure: Ability, Schooling, and the Economic Success of Brothers", *International Economic Review*, 16, 422–449.

Chow, G. (1980) "Evaluation of Econometric Models by Decomposition and Aggregation", Econometric Research Program Research Memorandum No. 235, Princeton University.

Deistler, M. (1975) "z-Transform and Identification of Linear Econometric Models with Autocorrelated Errors", *Metrika*, 22, 13–25.

Deistler, M. (1976) "The Identifiability of Linear Econometric Models with Autocorrelated Errors", *International Economic Review*, 17, 26–45.

Deistler, M. (1978) "The Structural Identifiability of Linear Models with Autocorrelated Errors in the Case of Affine Cross-Equation Restriction", *Journal of Econometrics*, 8, 23–31.

Deistler, M. and E. J. Hannan (1980) "Some Properties of the Parameterization of ARMA Systems with Unknown Order", mimeograph.

Deistler, M. and J. Schrader (1976) "Identifiability of the Transfer Function of Linear systems", Research Memo, Institute of Econometrics, University of Bonn, West Germany.

Deistler, M. and J. Schrader (1979) "Linear Models with Autocorrelated Errors: Structural Identifiability in the Absence of Minimality Assumptions", *Econometrica*, 47, 495–504.

Deistler, M. and H. Seifert (1978) "Identifiability and Consistent Estimability in Econometric Models", *Econometrica*, 46, 969–980.

Drèze, J. (1975) "Bayesian Theory of Identification in Simultaneous Equations Models", in: S. E. Fienberg and A. Zellner (eds.), *Studies in Bayesian Econometrics and Statistics*. Amsterdam: North-Holland Publishing Co.

Espasa, A. and J. D. Sargan (1977) "The Spectral Estimation of Simultaneous Equation Systems with Lagged Endogenous Variables", *International Economic Review*, 18, 583–606.

Fisher, F. M. (1961) "On the Cost of Approximate Specification in Simultaneous Equation Estimation", *Econometrica*, 29, 139–170.

Fisher, F. M. (1966) *The Identification Problem in Econometrics*. New York: McGraw-Hill.

Friedman, M. (1958) *A Theory of the Consumption Function*. Princeton: Princeton University Press.

Frisch, R. (1934) *Statistical Confluence Analysis by Means of Complete Regression Systems*, Publication no. 5. Oslo: Universitetes Økonomiske Institutt.

Gale, D. (1960) *The Theory of Linear Economic Models*. New York: McGraw-Hill.

Gale, D. and H. Nikaido (1965) "The Jacobian Matrix and Global Univalence of Mappings", *Mathematische Annalen*, 159.

Gantmacher, F. R. (1959) *Matrix Theory*, vol. I. New York: Chelsea.

Geraci, V. J. (1976) "Identification of Simultaneous Equation Models with Measurement Error", *Journal of Econometrics*, 4, 262–283.

Goldberger, A. S. (1972) "Structural Equation Methods in the Social Sciences", *Econometrica*, 40, 979–1001.

Griliches, Z. (1974) "Errors in Variables and Other Unobservables", *Econometrica*, 42, 971–998.

Haavelmo, T. (1944) "The Probability Approach in Econometrics", *Econometrica*, 12, Suppl. 1–118.

Hannan, E. J. (1969) "The Identification of Vector Mixed Autoregressive Moving Average Systems", *Biometrika*, 56, 223–225.

Hannan, E. J. (1970) *Multiple Time Series*. New York: John Wiley & Sons.

Hannan, E. J. (1971) "The Identification Problem for Multiple Equation Systems with Moving Average Errors", *Econometrica*, 39, 751–765.

Hannan, E. J. (1976) "The Identification and Parameterization of ARMAX and State Space Forms", *Econometrica*, 44, 713–723.

Hannan, E. J., W. Dunsmuir and M. Deistler (1980) "Estimation of Vector ARMAX Models", forthcoming in the *Journal of Multivariate Analysis*.

Hannan, E. J. and D. F. Nicholls (1972) "The Estimation of Mixed Regression, Autoregression, Moving Average and Distributed Lag Models", *Econometrica*, 40, 529–547.

Hansen, L. P. and T. J. Sargent (1980) "Formulating and Estimating Dynamic Linear Rational Expectations Models", *Journal of Economic Dynamics and Control*, 2, 7–46.

Hatanaka, M. (1975) "On the Global Identification of the Dynamic Simultaneous Equations Model with Stationary Disturbances" *International Economic Review*, 16, 545–554.
Hatanaka, M. (1976) "Several Efficient Two-Step Estimators for the Dynamic Simultaneous Equation Model with Autocorrelated Disturbances", *Journal of Econometrics*, 4, 189–204.
Hausman, J. A. and W. E. Taylor (1980a) "Panel Data and Unobservable Individual Effects", forthcoming in *Econometrica*.
Hausman, J. A. and W. E. Taylor (1980b) "Identification, Estimation and Testing in Simultaneous Equations Models with Disturbance Covariance Restrictions", unpublished manuscript.
Hsiao, C. (1976) "Identification and Estimation of Simultaneous Equation Models with Measurement Error", *International Economic Review*, 17, 319–339.
Hsiao, C. (1977) "Identification for Linear Dynamic Simultaneous Error-Shock Model", *International Economic Review*, 18, 181–194.
Hsiao, C. (1979) "Measurement Error in a Dynamic Simultaneous Equations Model with Stationary Disturbances", *Econometrica*, 47, 475–494.
Hurwicz, L. (1950) "Generalization of the Concept of Identification", in: *Statistical Inference in Dynamic Economic Models*, Cowles Commission Monograph 10, New York: John Wiley & Sons.
Jeffreys, H. (1961) *Theory of Probability* (3rd edn.). Oxford: Clarendon Press.
Johnston, J. (1972) *Econometric Methods* (2nd edn.). New York: McGraw-Hill.
Jöreskog, K. and A. S. Goldberger (1975) "Estimation of a Model with Multiple-Indicators and Multiple-Causes of a Single Latent Variable", *Journal of the American Statistical Association*, 70, 631–639.
Kadane, J. B. (1975) "The Role of Identification in Bayesian Theory", in: S. E. Fienberg and A. Zellner (eds.), *Studies in Bayesian Econometrics and Statistics*. Amsterdam: North-Holland Publishing Co.
Kelly, J. S. (1972) "Proportionate Variances and Identification Problem", *Econometrica*, 40, 1147–1150.
Kelly, J. S. (1975) "Linear Cross-Equation Constraints and the Identification Problem", *Econometrica*, 43, 125–140.
Kohn, R. (1979) "Identification Results for ARMAX Structures", *Econometrica*, 47, 1295–1304.
Koopmans, T. C. and O. Reiersøl (1950) "The Identification of Structural Characteristics", *Annals of Mathematical Statistics*, 21, 165–181.
Koopmans, T. C., H. Rubin and R. B. Leipnik (1950) "Measuring the Equation Systems of Dynamic Economics", in: *Statistical Inference in Dynamic Economic Models*, Cowles Commission Monograph 10. New York: John Wiley & Sons.
Leamer, E. (1978) *Specification Searches*. New York: John Wiley & Sons.
Liu, T. C. (1960) "Underidentification, Structural Estimation, and Forecasting", *Econometrica*, 28, 855–865.
MacDuffee, C. C. (1956) *The Theory of Matrices*. New York: Chelsea.
Madalla, G. S. (1971) "The Use of Variance-Components Models in Pooling Time-Series and Cross-Section Data", *Econometrica*, 39, 341–358.
Madansky, A. (1959) "The Fitting of Straightlines when both Variables are Subject to Error", *Journal of the American Statistical Association*, 54, 173–205.
Malinvaud, E. (1970) *Statistical Methods of Econometrics* (2nd edn.). Amsterdam: North-Holland Publishing Co.
Mallela, P. (1970) "Dimension Conditions of Identifiability", *International Economic Review*, 11, 378–387.
Mallela, P. and G. H. Patil (1976) "On the Identification with Covariance Restrictions: A Note", *International Economic Review*, 17, 741–750.
Maravall A. and D. J. Aigner (1977) "Identification of the Dynamic Shock-Error Model: The Case of Dynamic Regression" in: D. J. Aigner and A. S. Goldberger (eds.), *Latent Variables in Social-Economic Models*. Amsterdam: North-Holland Publishing Co.
Marschak, J. (1942) "Economic Interdependence and Statistical Analysis" in: *Studies in Mathematical Economics and Econometrics*. Chicago: University of Chicago Press, pp. 135–150.
Muth, J. F. (1961) "Rational Expectations and the Theory of Price Movements", *Econometrica*, 29, 315–335.
Phillips, P. C. B. and M. R. Wickens (1978) *Exercises in Econometrics*. Oxford: Philip Alan.

Preston, A. J. and K. D. Wall (1973) "An Extended Identification Problem for State Space Representations of Econometric Models", Discussion Paper no. 6, Programme of Research into Econometric Methods, University of London.
Raiffa, H. and R. Schlaifer (1961) *Applied Statistical Decision Theory*. Cambridge, Mass: Harvard University Press.
Reiersøl, O. (1950) "Identifiability of a Linear Relation Between Variables Which are Subject to Error", *Econometrica*, 18, 375–389.
Revankar, N. S. (1980) "Testing of the Rational Expectations Hypothesis", *Econometrica*, 48, 1347–1363.
Richmond, J. (1974) "Identifiability in Linear Models", *Econometrica*, 42, 731–736.
Richmond, J. (1976) "Aggregation and Identification", *International Economic Review*, 17, 47–56.
Rothenberg, T. J. (1971) "Identification in Parametric Models", *Econometrica*, 39, 577–592.
Rozanov, Y. S. (1967) *Stationary Random Processes*. San Francisco: Holden-Day.
Sargan, J. D. (1961) "The Maximum Likelihood Estimation of Economic Relationships with Autoregressive Residuals", *Econometrica*, 29, 414–426.
Sargan, J. D. (1979) "The Identification and Estimation of Sets of Simultaneous Stochastic Equations", mimeograph.
Sargan, J. D. (1980) "Identification and Lack of Identification", mimeograph.
Sims, C. A. (1980) "Macroeconomics and Reality", *Econometrica*, 48, 1–48.
Taubman, P. (1976) "Earnings, Education, Genetics, and Environment", *Journal of Human Resources*, 11, 447–461.
Theil, H. (1971) *Principles of Econometrics*. New York: John Wiley & Sons.
Wald, A. (1950) "Note on the Identification of Economic Relations", in: *Statistical Inference in Dynamic Economic Models*, Cowles Commission Monograph, 10. New York: John Wiley & Sons.
Wallis, K. F. (1980) "Econometric Implications of the Rational Expectations Hypothesis", *Econometrica*, 48, 49–73.
Wegge, L. L. (1965) "Identifiability Criteria for a System of Equations as a Whole", *Australian Journal of Statistics*, 3, 67–77.
Wold, H. O. (1960) "A Generalization of Causal Chain Model", *Econometrica*, 28, 443–463.
Working, E. J. (1925) "The Statistical Determination of Demand Curves", *Quarterly Journal of Economics*, 39, 503–543.
Working, E. J. (1927) "What Do Statistical 'Demand Curves' Show?", *Quarterly Journal of Economics*, 41, 212–235.
Zellner, A. (1971) *An Introduction to Bayesian Inference in Econometrics*. New York: John Wiley & Sons.

Chapter 5

MODEL CHOICE AND SPECIFICATION ANALYSIS

EDWARD E. LEAMER*

University of California, Los Angeles

Contents

*Helpful comments from David Belsley, Zvi Griliches, Michael Intriligator, and Peter Schmidt are gratefully acknowledged. Work was supported by NSF grant SOC78-09477.

Handbook of Econometrics, Volume I, Edited by Z. Griliches and M.D. Intriligator

1. Introduction

The data banks of the National Bureau of Economic Research contain time-series data on 2000 macroeconomic variables. Even if observations were available since the birth of Christ, the degrees of freedom in a model explaining gross national product in terms of all these variables would not turn positive for another two decades. If annual observations were restricted to the 30-year period from 1950 to 1979, the degrees of freedom deficit would be 1970. A researcher who sought to sublimate the harsh reality of the degrees of freedom deficit and who restricted himself to exactly five explanatory variables could select from a menu of

$$\binom{2000}{5} = 2.65 \times 10^{14}$$

equations to be estimated, which, at the cost of ten cents per regression, would consume a research budget of twenty-six trillion dollars.

What is going on? Although it is safe to say that economists have not tried anything like 10^{14} regressions to explain GNP, I rather think a reasonably large number like 1000 is likely an underestimate. Does this make any sense at all? Can our profession use data to make progress? The answer is not necessarily "yes". But what does seem clear is that until the complex phenomenon of specification searches is well understood, the answer to this question cannot unambiguously be in the affirmative.

This chapter contains a summary of the statistical theories of model selection. Sections 2 and 3 include most of the traditional model selection problems. Section 2 deals with alternative models which spring from *a priori* judgment. It is maintained that a constrained model might be either "true" or "approximately true". Classical hypotheses testing is discussed, as well as Bayesian treatments with complete and incomplete prior distributions. Subsections deal also with measuring the "multicollinearity problem" and with inference given zero degrees of freedom. Models in Section 3 arise not from the judgment of the investigator as in Section 2, but from his purpose, which is measured by a formal loss function. Quadratic loss functions are considered, both with and without fixed costs. A brief comment is made about "ridge regression". The main conclusion of Section 3 is that quadratic loss does not imply a model selection problem.

Sections 4, 5, and 6 discuss problems which are not as well known. The problem of selecting the best proxy variable is treated in Section 4. Akaike's Information Criterion is discussed in Section 5, although it is pointed out that, except for subtle conceptual differences, his problem reduces to estimation with quadratic loss. Section 6 deals with methods for discounting evidence when models are discovered after having previewed the data. Finally, Section 7 contains material on "stepwise regression", "cross-validation", and "goodness-of-fit" tests.

A uniform notation is used in this chapter. The T observations of the dependent variable are collected in the $T\times 1$ vector $\boldsymbol{Y}$, and the T observations of a potential set of k explanatory variables are collected in the $T\times k$ matrix $\boldsymbol{X}$. The hypothesis that the vector $\boldsymbol{Y}$ is normally distributed with mean $\boldsymbol{X\beta}$ and variance matrix $\sigma^2\boldsymbol{I}$ will be indicated by H:

$$H\colon \boldsymbol{Y}\sim N(\boldsymbol{X\beta},\sigma^2\boldsymbol{I}).$$

A subset of explanatory variables will be denoted by the $T\times k_J$ matrix $\boldsymbol{X}_J$, where J is a subset of the first k integers, each integer selecting an included variable. The excluded variables will be indexed by the set $\bar{J}$, and the hypothesis that the variables in $\bar{J}$ have no effect is

$$H_J\colon \boldsymbol{Y}\sim N(\boldsymbol{X}_J\boldsymbol{\beta}_J,\sigma^2\boldsymbol{I}).$$

The least-squares estimate of $\boldsymbol{\beta}$ is a solution to the normal equations $\boldsymbol{X}'\boldsymbol{Xb}=\boldsymbol{X}'\boldsymbol{Y}$. Except where indicated, $\boldsymbol{X}'\boldsymbol{X}$ is assumed to be invertible:

$$\boldsymbol{b}=(\boldsymbol{X}'\boldsymbol{X})^{-1}\boldsymbol{X}'\boldsymbol{Y}. \tag{1.1}$$

The corresponding residual operator and residual sum-of-squares are

$$\boldsymbol{M}=\boldsymbol{I}-\boldsymbol{X}(\boldsymbol{X}'\boldsymbol{X})^{-1}\boldsymbol{X}', \tag{1.2}$$

$$ESS=(\boldsymbol{Y}-\boldsymbol{Xb})'(\boldsymbol{Y}-\boldsymbol{Xb})=\boldsymbol{Y}'\boldsymbol{MY}. \tag{1.3}$$

The same three concepts for the restricted model are

$$\hat{\boldsymbol{\beta}}_J=(\boldsymbol{X}_J'\boldsymbol{X}_J)^{-1}\boldsymbol{X}_J'\boldsymbol{Y}, \tag{1.4}$$

$$\boldsymbol{M}_J=\boldsymbol{I}-\boldsymbol{X}_J(\boldsymbol{X}_J'\boldsymbol{X}_J)^{-1}\boldsymbol{X}_J', \tag{1.5}$$

$$ESS_J=(\boldsymbol{Y}-\boldsymbol{X}\hat{\boldsymbol{\beta}}_J)'(\boldsymbol{Y}-\boldsymbol{X}\hat{\boldsymbol{\beta}}_J)=\boldsymbol{Y}'\boldsymbol{M}_J\boldsymbol{Y}. \tag{1.6}$$

The reduction in the error sum-of-squares which results when the variables $\bar{J}$ are added to the model J is

$$ESS_J-ESS=\boldsymbol{b}_{\bar{J}}'\boldsymbol{X}_{\bar{J}}'\boldsymbol{M}_J\boldsymbol{X}_{\bar{J}}\boldsymbol{b}_{\bar{J}}.$$

2. Model selection with prior distributions

One very important source of model selection problems is the existence of *a priori* opinion that constraints are "likely". Statistical testing is then designed either to determine if a set of constraints is "true" or to determine if a set of constraints is "approximately true". The solutions to these two problems, which might be supposed to be essentially the same, in fact diverge in two important respects. (1) The first problem leads clearly to a significance level which is a decreasing function of sample size, whereas the second problem selects a relatively constant significance level. (2) The first problem has a set of alternative models which is determined entirely from *a priori* knowledge, whereas the second problem can have a data-dependent set of hypotheses.

The problem of testing to see if constraints are "true" is discussed in Section 2.1 under the heading "hypothesis testing searches", and the problem of testing to see if constraints are "approximately true" is discussed in Section 2.2 under the heading "interpretive searches".

Before proceeding it may be useful to reveal my opinion that "hypothesis testing searches" are very rare, if they exist at all. An hypothesis testing search occurs when the subjective prior probability distribution allocates positive probability to a restriction. For example, when estimating a simple consumption function which relates consumption linearly to income and an interest rate, many economists would treat the interest rate variable as doubtful. The method by which this opinion is injected into a data analysis is usually a formal test of the hypothesis that the interest rate coefficient is zero. If the t-statistic on this coefficient is sufficiently high, the interest rate variable is retained; otherwise it is omitted. The opinion on which this procedure rests could be characterized by a subjective prior probability distribution which allocates positive probability to the hypothesis that the interest-rate coefficient is exactly zero. A Bayesian analysis would then determine if this atom of probability becomes larger or smaller when the data evidence is conditioned upon. But the same statistical procedure could be justified by a continuous prior distribution which, although concentrating mass in the neighborhood of zero, allocates zero probability to the origin. In that event, the posterior as well as the prior probability of the sharp hypothesis is zero.

Although the subjective logic of Bayesian inference allows for any kind of prior distribution, I can say that I know of no case in economics when I would assign positive probability to a point (or, more accurately, to a zero volume subset in the interior of the parameter space). Even cooked-up examples can be questioned. What is your prior probability that the coin in my pocket produces when flipped a binomial sequence with probability precisely equal to 0.5? Even if the binomial assumption is accepted, I doubt that a physical event could lead to a probability precisely equal to 0.5. In the case of the interest-rate coefficient described above, ask yourself what chance a 95 percent confidence interval has of covering zero if the sample size is enormous (and the confidence interval tiny). If you say infinitesimal, then you have assigned at most an infinitesimal prior probability to the sharp hypothesis, and you should be doing data-interpretation, not hypothesis testing.

2.1. Hypothesis testing searches

This subsection deals with the testing of a set of M alternative hypotheses of the form $\boldsymbol{R}_i\boldsymbol{\beta}=\boldsymbol{0}$, $i=1,\dots,M$. It is assumed that the hypotheses have truth value in the sense that the prior probability is non-zero, $\mathrm{P}(\boldsymbol{R}_i\boldsymbol{\beta}=\boldsymbol{0})>0$. Familiarity with the concepts of significance level and power is assumed and the discussion focuses first on the issue of how to select the significance level when the hypotheses have a simple structure. The clear conclusion is that the significance level should be a decreasing function of sample size.

Neyman and Pearson (1928) are credited with the notion of a power function of a test and, by implication, the need to consider specific alternative models when testing an hypothesis. The study of power functions is, unfortunately, limited in value, since although it can rule in favor of uniformly most powerful tests with given significance levels, it cannot select between two tests with different significance levels. Neyman's (1958) advice notwithstanding, in practice most researchers set the significance level equal to 0.05 or to 0.01, or they use these numbers to judge the size of a P-value. The step from goodness-of-fit testing, which considers only significance levels, to classical hypothesis testing, which includes in principle a study of power functions, is thereby rendered small.

The Bayesian solution to the hypothesis testing problem is provided by Jeffreys (1961). The posterior odds in favor of the "null" hypothesis H_0 versus an "alternative" H_1 given the data Y is

$$\frac{\mathrm{P}(H_0|Y)}{\mathrm{P}(H_1|Y)}=\left\{\frac{\mathrm{P}(Y|H_0)}{\mathrm{P}(Y|H_1)}\right\}\cdot\left\{\frac{\mathrm{P}(H_0)}{\mathrm{P}(H_1)}\right\}.$$

In words, the posterior odds ratio is the prior odds ratio times the "Bayes factor", $B(Y)=\mathrm{P}(Y|H_0)/\mathrm{P}(Y|H_1)$. This Bayes factor is the usual likelihood ratio for testing if the data were more likely to come from the distribution $\mathrm{P}(Y|H_0)$ than from the distribution $\mathrm{P}(Y|H_1)$. If there were a loss function, a Bayesian would select the hypothesis that yields lowest expected loss, but without a loss function it is appropriate to ask only if the data favor H_0 relative to H_1. Since a Bayes factor in excess of one favors the null hypothesis, the inequality $B(Y)\leqslant 1$ implicitly defines the region of rejection of the null hypothesis and thereby selects the significance level and the power. Equivalently, the loss function can be taken to penalize an error by an amount independent of what the error is, and the prior probabilities can be assumed to be equal.

In order to contrast Jeffreys' solution from Neyman and Pearson's solution, consider testing the null hypothesis that the mean of a sample of size n, $\overline{Y}_n$, is distributed normally with mean 0 and variance n^{-1} versus the alternative that $\overline{Y}_n$ is normal with mean μ_a and variance n^{-1}, where $\mu_a > 0$. Classical hypothesis testing at the 0.05 level of significance rejects the null hypothesis if $\overline{Y}_n > 1.6n^{-1/2}$, whereas the Bayes rejection region defined by $B(Y) \leqslant 1$ is $\overline{Y}_n \geqslant \mu_a/2$. The classical rejection region does not depend on μ_a, and, somewhat surprisingly, the treatment of the data does not depend on whether $\mu_a = 1$ or $\mu_a = 10^{10}$. Also, as sample size grows, the classical rejection region gets smaller and smaller, whereas the Bayesian rejection region is constant. Thus, the classical significance level is fixed at 0.05 and the power $P(\overline{Y}_n < 1.6n^{-1/2} | \mu = \mu_a)$ goes to one as the sample size grows. The Bayes rule, in contrast, has the probabilities of type one and type two errors equal for all sample sizes $P(\overline{Y}_n > \mu_a/2 | \mu = 0)/P(\overline{Y}_n < \mu_a/2 | \mu = \mu_a) = 1$. The choice between these two treatments is ultimately a matter of personal preference; as for myself, I much prefer Jeffreys' solution. The only sensible alternative is a minimax rule, which in this case is the same as the Bayes rule. Minimax rules, such as those proposed by Arrow (1960), generally also have the significance level a decreasing function of sample size.

Jeffreys' Bayesian logic is not so compelling for the testing of composite hypotheses because a prior distribution is required to define the "predictive distribution", $P(Y|H_i)$, because prior distributions are usually difficult to select, and because the Bayesian answer in this case is very sensitive to the choice of a prior. Suppose, for example, that the null hypothesis H_0 is that the mean of a sample of size n, $\overline{Y}_n$, is distributed normally with mean 0 and variance n^{-1}. Take the alternative H_1 to be that $\overline{Y}_n$ is normally distributed with unknown mean μ and variance n^{-1}. In order to form the predictive distribution $P(\overline{Y}_n|H_1)$ it is necessary to assume a prior distribution for μ, say normal with mean m^* and variance $(n^*)^{-1}$. Then the marginal distribution $P(\overline{Y}_n|H_1)$ is also normal with mean m^* and variance $(n^*)^{-1}+(n)^{-1}$. The Bayes factor in favor of H_0 relative to H_1 therefore becomes

$$
\begin{aligned}
B(\overline{Y}_n) &= P(\overline{Y}_n|H_0)/P(\overline{Y}_n|H_1) \\
&= \left[n^*(n+n^*)^{-1}\right]^{-1/2} \exp\{-\tfrac{1}{2}\overline{Y}_n^2 n\}/\exp\{-\tfrac{1}{2}(\overline{Y}_n - m^*)^2 \\
&\quad \times (n^{-1}+n^{*-1})^{-1}\} \\
&= \left[n^*(n+n^*)^{-1}\right]^{-1/2} \exp\{-\tfrac{1}{2}(n^2\overline{Y}_n^2 + 2nn^*\overline{Y}m^* - nn^*m^{*2}) \\
&\quad /(n+n^*)\},
\end{aligned} \tag{2.1}
$$

and the corresponding region of rejection defined by $B \leqslant 1$ is

$$(n+n^*)^{-1}(n^2\overline{Y}^2+2nn^*\overline{Y}m^*-nn^*m^{*2})+\log n^*(n+n^*)^{-1}\geqslant 0. \tag{2.2}$$

It should be observed that the region of rejection is not one-sided if $m^* \neq 0$ and $n^* < \infty$. Furthermore, both the Bayes factor and the region of rejection depend importantly on the prior parameters n^* and m^*. If, for example, the prior is located at the origin, $m^* = 0$, then given the data $\overline{Y}_n$, the Bayes factor $B(\overline{Y}_n; n^*)$ varies from infinity if $n^* = 0$ to one as n^* approaches infinity. The minimum value, $B(\overline{Y}_n; n^*) = (n\overline{Y}^2)^{1/2}\exp\{-(n\overline{Y}^2-1)/2\}$, is attained at $n^* = n/(n\overline{Y}^2-1)$ if $n\overline{Y}^2-1$ is positive. Otherwise $B(\overline{Y}_n; n^*)$ is strictly greater than one. The region of rejection varies from the whole line to the region $\overline{Y}^2 \geqslant n^{-1}$. This Bayesian logic for selecting the significance level of a test is therefore hardly useful at all if you have much trouble selecting your prior distribution.

The Bayesian logic is very useful, however, in emphasizing the fact that the significance level should be a decreasing function of sample size. As the sample size grows, the Bayes factor becomes

$$\lim_{n\to\infty} B(\overline{Y}_n) = (n/n^*)^{1/2}\exp\{-\tfrac{1}{2}\overline{Y}^2 n\},$$

with corresponding region of rejection $n\overline{Y}^2 \geqslant \frac{1}{2}\log n/n^*$. This contrasts with the usual classical region of rejection $n\overline{Y}^2 \geqslant c$, where c is a constant independent of sample size chosen such that $\mathrm{P}(n\overline{Y}^2 \geqslant c|\mu = 0) = 0.05$. The important point which needs to be made is that two researchers who study the same or similar problems but who use samples with different sample sizes should use the *same* significance level only if they have *different* prior distributions. In order to maintain comparability, it is necessary for each to report results based on the same prior. Jeffreys, for example, proposes a particular "diffuse" prior which leads to the critical t-values reported in Table 2.1, together with some of my own built on a somewhat different limiting argument. It seems to me better to use a table such as this built on a somewhat arbitrary prior distribution than to use an arbitrarily selected significance level, since in the former case at least you know what you are doing. Incidentally, it is surprising to me that the t-values in this table increment so slowly.

Various Bayes factors for the linear regression model are reported in Zellner (1971), Lempers (1971), Gaver and Geisel (1974), and in Leamer (1978). The Bayes factor in favor of model J relative to model J^* is $\mathrm{P}(Y|H_J)/\mathrm{P}(Y|H_{J^*})$. A marginal likelihood of model J is given by the following result.

Theorem 1 (Marginal likelihood)

Suppose that the observable $(T\times 1)$ vector $\boldsymbol{Y}$ has mean vector $\boldsymbol{X}_J\boldsymbol{\beta}_J$ and variance matrix $h_J^{-1}\boldsymbol{I}_T$, where $\boldsymbol{X}_J$ is a $T\times k_J$ observable matrix of explanatory variables, $\boldsymbol{\beta}_J$

Table 2.1
Selected critical values of the t-statistic
(T obs. from a normal distribution with unknown mean and unknown variance)

	Degrees of freedom $(T-1)$:						
	5	10	50	100	1000	10,000	100,000
Classical value, 0.05 level	2.57	2.23	2.01	1.98	1.96	1.96	1.96
Classical value, 0.01 level	4.03	3.17	2.67	2.60	2.58	2.58	2.58
Jeffreys' values[a]	1.84	1.84	2.14	2.28	2.72	3.11	3.46
Leamer's values[b]	1.32	1.56	2.00	2.16	2.63	3.04	3.39

[a]Jeffreys (1961, p. 439):

$$t^2=(T-1)\left[\left\{\frac{\pi(T-1)}{2}\right\}^{1/(T-2)}-1\right].$$

[b]Leamer (1978, p. 116): $t^2=(T-1)(T^{1/T}-1)$ [same as Schwarz (1978)].

is a $k\times 1$ unobservable vector of regression coefficients, and h_J is an unobservable scalar precision parameter. Suppose also that conditional on h_J, $\boldsymbol{\beta}_J$ is normal with mean $\boldsymbol{b}_J^*$ and covariance $h_J^{-1}\boldsymbol{N}_J^{*-1}$, and suppose that h_J has a gamma distribution with location and scale parameters s^{*2} and ν^*. Then the marginal data density [Leamer (1978, p. 110)] is

$$f(\boldsymbol{Y}|H_J)=c(\nu^*,T)\left|\frac{\boldsymbol{M}_J^*}{s^{*2}}\right|^{1/2}\left(\nu^*+\frac{Q_J}{s^{*2}}\right)^{-(\nu^*+T)/2}, \tag{2.3}$$

where

$$\boldsymbol{M}_J^*=\boldsymbol{I}_T-\boldsymbol{X}_J(\boldsymbol{N}_J^*+\boldsymbol{X}_J'\boldsymbol{X}_J)^{-1}\boldsymbol{X}_J',$$

$$c(\nu^*,T)=(\nu^*)^{\nu^*/2}\left(\frac{\nu^*}{2}+\frac{T}{2}-1\right)!\Big/\pi^{T/2}\left(\frac{\nu^*}{2}-1\right)!,$$

$$Q_J\doteq(\boldsymbol{Y}-\boldsymbol{X}_J\hat{\boldsymbol{\beta}}_J)'(\boldsymbol{Y}-\boldsymbol{X}_J\hat{\boldsymbol{\beta}}_J)+(\hat{\boldsymbol{\beta}}_J-\boldsymbol{b}_J^*)'$$
$$\times\left(\boldsymbol{N}_J^{*-1}+(\boldsymbol{X}_J'\boldsymbol{X}_J)^{-1}\right)^{-1}(\hat{\boldsymbol{\beta}}_J-\boldsymbol{b}_J^*),$$

$$\hat{\boldsymbol{\beta}}_J=(\boldsymbol{X}_J'\boldsymbol{X}_J)^{-1}\boldsymbol{X}_J'\boldsymbol{Y}_J.$$

A large Q_J implies a small value for $f(\boldsymbol{Y}|H_J)$, which means that the data $\boldsymbol{Y}$ were not likely to have come from the distribution $f(\boldsymbol{Y}|H_J)$. It may be observed that Q_J increases with the error sum-of-squares $(\boldsymbol{Y}-\boldsymbol{X}_J\hat{\boldsymbol{\beta}}_J)'(\boldsymbol{Y}-\boldsymbol{X}_J\hat{\boldsymbol{\beta}}_J)$ and with the discrepancy between the prior mean $\boldsymbol{b}_J^*$ and the least-squares estimate $\hat{\boldsymbol{\beta}}_J$.

As the sample size grows, the marginal likelihood (2.3) converges to

$$f(\boldsymbol{Y}|H_J)=c_JT^{-k_J/2}(ESS_J)^{-T/2},$$

where c_J is a constant depending on the precise choice of prior distribution.[1] Leamer (1978) argues somewhat unconvincingly that the term c_J can be treated as if it were the same for all models. This leads to the model selection criterion appropriate for one form of diffuse prior:

$$\max_J T^{-k_J/2}(ESS_J)^{-T/2}, \tag{2.4}$$

which is actually the formula used to produce my critical t-values in Table 2.1. Schwarz (1978) also proposes criterion (2.4) and uses the same logic to produce it.

This Bayesian treatment of hypothesis testing generalizes straightforwardly to all other settings, at least conceptually. The ith composite hypothesis $\boldsymbol{Y} \sim f_i(\boldsymbol{Y}|\boldsymbol{\theta}_i)$, $\boldsymbol{\theta}_i \in \Theta_i$, is mixed into the point hypothesis $\boldsymbol{Y} \sim f_i(\boldsymbol{Y}) \equiv \int f_i(Y|\theta_i) f_i(\theta_i)\,\mathrm{d}\theta_i$, where $f_i(\boldsymbol{\theta}_i)$ is the prior distribution. The Bayes factor in favor of hypothesis i relative to hypothesis j is simply $f_i(\boldsymbol{Y})/f_j(\boldsymbol{Y})$. Numerical difficulties can arise if computation of the "marginal" likelihood $\int f_i(\boldsymbol{Y}|\boldsymbol{\theta}_i) f(\boldsymbol{\theta}_i)\,\mathrm{d}\boldsymbol{\theta}_i$ is unmanageable. Reporting difficulties can arise if the choice of prior, $f_i(\boldsymbol{\theta}_i)$, is arguable.

This relatively settled solution contrasts greatly with the classical treatment. The sample space $\mathcal{Y}$ must be partitioned into a set of acceptance regions A_i, $\mathcal{Y} = \bigcup_i A_i$, $A_i \cap A_j = \varnothing$ for $i \neq j$, such that if the data fall in A_i then hypothesis i is "accepted". From the Bayesian perspective, these regions can be implicitly defined by the process of maximizing the marginal likelihood, $A_i = \{\boldsymbol{Y}|f_i(\boldsymbol{Y}) > f_j(\boldsymbol{Y}), j \neq i\}$. But, classically, one is free to choose any partition. Once a partition is chosen, one is obligated to study its error properties: the type I error $\mathrm{P}(\boldsymbol{Y} \notin A_i|H_i, \theta_i)$ and the type II error $\mathrm{P}(\boldsymbol{Y} \in A_i|H_j, \theta_j)$, the first error being the probability that H_i is not accepted when in fact it is the true model, and the second being the probability that H_i is accepted when in fact it is not the true model. A partition, $\bigcup_i A_i = \mathcal{Y}$, is ruled inadmissible if there exists another partition, $\bigcup_i A_i^* = \mathcal{Y}$ such that

$$\mathrm{P}(\boldsymbol{Y} \notin A_i|H_i, \theta_i) \geqslant \mathbf{P}(\boldsymbol{Y} \notin A_i^*|H_i, \theta_i) \quad \text{for all } i, \theta_i,$$

[1]It is assumed here that $\lim(\boldsymbol{X}'\boldsymbol{X}/T) = \boldsymbol{\Sigma} \neq \boldsymbol{0}$ and $\lim(ESS_J/T) = \sigma_J^2$. Using $\lim(Q_J/ESS_J) = 1$,

$$\begin{aligned}\lim\left[\nu^* s^{*2} + T(ESS/T)\right]^{-T/2} &\div T(ESS/T)^{-T/2} \\ &= \lim\left[\nu^* s^{*2}/(TESS/T)+1\right]^{-T/2} = \exp\left(-\nu^* s^{*2}/2(ESS/T)\right)\end{aligned}$$

and

$$\lim\left(|\boldsymbol{M}_J^*|\cdot T^{k_J}\right) = \lim\left(|\boldsymbol{N}_J^*|T^{k_J}/|\boldsymbol{N}_J^* + \boldsymbol{X}_J'\boldsymbol{X}_J|\right) = |\boldsymbol{N}_J^*|/|\boldsymbol{\Sigma}_{JJ}|$$

we may derive the exact expression

$$c_J = c(\nu^*, T)\cdot s^{*\nu^*}|\boldsymbol{N}_J^*|^{-1/2}|\boldsymbol{\Sigma}_{JJ}|^{1/2}\left(\sigma_J^2\right)^{-\nu^*/2}\cdot T^{-\nu^*/2}\cdot\exp\left(-\nu^* s^{*2}/2\sigma_J^2\right).$$

and

$$\mathrm{P}(Y \in A_i|H_j, \theta_j) \geqslant \mathrm{P}(Y \in A_i^*|H_j, \theta_j) \quad \text{for all } i, j, \theta_j,$$

with at least one strict inequality. Otherwise, a partition is ruled admissible.

The criterion of admissibility can rule out certain obviously silly procedures; for example, when testing H_1: $x \sim N(0,1)$ versus H_2: $x \sim N(1,1)$, the partition with $A_1 = \{x|x \geqslant 1\}$ is inadmissible. (Let $A_1^* = \{x|x \leqslant 1\}$.) But otherwise the criterion is rarely useful. As a result, there is a never-ending sequence of proposals for alternative ways of partitioning the sample space (just as there is a never-ending sequence of alternative prior distributions!).

The most commonly used test discriminates between hypothesis H_J and hypothesis H and is based on the following result.

Theorem 2 (F distribution)

Conditional on hypothesis H_J the following random variable has an F-distribution with $k - k_J$ and $T - k$ degrees of freedom:

$$F = \frac{(ESS_J - ESS)/(k - k_J)}{ESS/(T-k)}. \tag{2.5}$$

The traditional region of acceptance of H_J is

$$A_J = \{Y|F \leqslant F_{T-k}^{k-k_J}(\alpha)\}, \tag{2.6}$$

where $F_{T-k}^{k-k_J}(\alpha)$ is the upper αth percentile of the F distribution with $k - k_J$ and $T-k$ degrees of freedom. What is remarkable about Theorem 2 is that the random variable F has a distribution independent of $(\boldsymbol{\beta}_J, \sigma^2)$; in particular, $\mathrm{P}(\boldsymbol{Y} \notin A_J|H_J, \boldsymbol{\beta}_J, \sigma^2) = \alpha$. Nonetheless, the probability of a type II error, $\mathrm{P}(\boldsymbol{Y} \in A_J|H, \boldsymbol{\beta}, \sigma^2)$, does depend on $(\boldsymbol{\beta}, \sigma^2)$. For that reason, the substantial interpretive clarity associated with tests with uniquely defined error probabilities is not achieved here. The usefulness of the uniquely defined type I error attained by the F-test is limited to settings which emphasize type I error to the neglect of type II error.

When hypotheses are not nested, it is not sensible to set up the partition such that $\mathrm{P}(Y \notin A_J|H_j, \boldsymbol{\beta}_J, \sigma_J^2)$ is independent of $(\boldsymbol{\beta}_J, \sigma^2)$. For example, a test of model J against model J' could use the partition defined by (2.6), in which case $\mathrm{P}(\boldsymbol{Y} \notin A_J|H_J, \boldsymbol{\beta}_J, \sigma_J^2)$ would be independent of $\boldsymbol{\beta}_J, \sigma_J^2$. But this treats the two hypotheses asymmetrically for no apparent reason. The most common procedure instead is to select the model with the maximum $\bar{R}^2$:

$$\bar{R}_J^2 = 1 - cESS_J/(T - k_J), \tag{2.7}$$

where $c=(T-1)/\Sigma_t(Y_t-\overline{Y})^2$. Equivalently, the model is selected which minimizes the estimated residual variance

$$s_J^2 = ESS_J/(T-k_J). \tag{2.8}$$

Theil (1971) gives this rule some support by showing that if model J is true then $E(s_J^2)\leqslant E(s_{J'}^2)$ for all J'. Error probabilities for this rule have been studied by Schmidt (1973).

Other ways to partition the sample space embed non-nested hypotheses into a general model. Models J and J' can each separately be tested against the composite model. This partitions the sample space into four regions, the extra pair of regions defined when both J and J' are rejected and when both J and J' are accepted.[2] Although it is possible to attach meaning to these outcomes if a more general viewpoint is taken, from the standpoint of the problem at hand these are nonsensical outcomes.

The specification error tests of Wu (1973) and Hausman (1978), interpreted by Hausman and Taylor (1980), can also be discussed in this section. The constrained estimator (4) has bias $(\boldsymbol{X}_J'\boldsymbol{X}_J)^{-1}\boldsymbol{X}_J'\boldsymbol{X}_{\bar{J}}\boldsymbol{\beta}_{\bar{J}}\equiv \boldsymbol{E}_J\boldsymbol{\beta}_{\bar{J}}$ and is therefore unbiased (and consistent) if $\boldsymbol{E}_J\boldsymbol{\beta}_{\bar{J}}=\boldsymbol{0}$. The hypothesis that misspecification is inconsequential, $\boldsymbol{E}_J\boldsymbol{\beta}_{\bar{J}}=\boldsymbol{0}$, differs from the more traditional hypothesis, $\boldsymbol{\beta}_{\bar{J}}=\boldsymbol{0}$, if $\boldsymbol{E}_J$ has rank less than $k_{\bar{J}}$, in particular if $k_J<k_{\bar{J}}$. A classical test of $\boldsymbol{E}_J\boldsymbol{\beta}_{\bar{J}}=\boldsymbol{0}$ can therefore differ from a classical test of $\boldsymbol{\beta}_{\bar{J}}=\boldsymbol{0}$. But if your prior has the feature $\Pr(\boldsymbol{\beta}_{\bar{J}}=\boldsymbol{0}|\boldsymbol{E}_J\boldsymbol{\beta}_{\bar{J}}=\boldsymbol{0})=1$, then Bayesian tests of the two hypotheses are identical. Because $\boldsymbol{E}_J\boldsymbol{\beta}_{\bar{J}}$ is not likely to be a set of special linear combinations of the coefficients, it is quite likely that the only mass point of your prior in the subspace $\boldsymbol{E}_J\boldsymbol{\beta}_{\bar{J}}=\boldsymbol{0}$ is at $\boldsymbol{\beta}_{\bar{J}}=\boldsymbol{0}$. As a consequence, the special specification error hypothesis becomes uninteresting.

The hypotheses so far discussed all involve linear constraints on parameters in univariate normal linear regression models. In more complex settings the most common test statistic is the likelihood ratio

$$LR=\max_{\boldsymbol{\theta}\in\Omega} f(\boldsymbol{Y}|\boldsymbol{\theta})\Big/\max_{\boldsymbol{\theta}\in\Omega_0} f(\boldsymbol{Y}|\boldsymbol{\theta}),$$

where $\boldsymbol{\theta}$ is a vector of parameters assumed to come from some set Ω, and the null hypothesis is that $\boldsymbol{\theta}\in\Omega_0\subset\Omega$. Recently, two alternatives have become popular in the econometric theory literature: the Wald test and the Lagrange multiplier test. These amount to alternative partitions of the sample space and are fully discussed in Chapter 13 of this Handbook by Engle.

[2]See the discussion in Gaver and Geisel (1974) and references including the methods of Cox (1961, 1962) and Quandt (1974).

2.2. *Interpretive searches*

Although it is rare to have the prior probability of an exact constraint be non-zero, it is fairly common to assign substantial prior probability to the hypothesis that a constraint is "approximately true". You may think that various incidental variables have coefficients which, though certainly not zero, are quite likely to be close to zero. You may realize that the functional form is certainly not linear, but at the same time you may expect the departures from linearity to be small. Similar variables are very unlikely to have exactly the same coefficients, but nonetheless are quite likely to have coefficients roughly the same size.

In all these cases it is desirable when interpreting a given data set to use all prior information which may be available, especially so when the data are weak where the prior is strong. In practice, most researchers test sequentially the set of *a priori* likely constraints. If the data do not cast doubt on a constraint, it is retained; otherwise, it is discarded. Section 2.2.1. comments on these *ad hoc* methods by comparing them with formal Bayesian procedures. Section 2.2.2. deals with sensitivity analyses appropriate when the prior is "ambiguous". Various measures of multi-collinearity are discussed in Section 2.2.3. Finally, the dilemma of the degrees-of-freedom deficit is commented on in Section 2.2.4.

2.2.1. *Interpretive searches with complete prior distributions*

A Bayesian with a complete prior distribution and a complete sampling distribution straightforwardly computes his posterior distribution. Although no model selection issues seem to arise, the mean of his posterior distribution is a weighted average of least-squares estimates from various constrained models. For that reason, he can get an idea where his posterior distribution is located by looking at particular constrained estimates and, conversely, the choice of which constrained estimates he looks at can be used to infer his prior distribution. Suppose in particular that the prior for $\boldsymbol{\beta}$ is normal with mean zero and precision matrix $\boldsymbol{D}^* = \text{diag}\{d_1, d_2, \dots, d_k\}$. Then, the posterior mean is

$$\boldsymbol{b}^{**} = E(\boldsymbol{\beta}|\boldsymbol{Y}, \boldsymbol{X}, \sigma^2) = (\boldsymbol{H} + \boldsymbol{D}^*)^{-1}\boldsymbol{X}'\boldsymbol{Y}, \tag{2.9}$$

where $\boldsymbol{H} = \sigma^{-2}\boldsymbol{X}'\boldsymbol{X}$. The following two theorems from Leamer and Chamberlain (1976) link $\boldsymbol{b}^{**}$ to model selection strategies.

Theorem 3 (The 2^k regressions)

The posterior mean (2.9) can be written as

$$\boldsymbol{b}^{**} = (\boldsymbol{H} + \boldsymbol{D}^*)^{-1}\boldsymbol{H}\boldsymbol{b} = \sum_J w_J \hat{\boldsymbol{\beta}}_J,$$

where J indexes the 2^k subsets of the first k integers, $\hat{\boldsymbol{\beta}}_J$ is the least-squares estimate subject to the constraints $\beta_i = 0$ for $i \in J$, and

$$w_J \propto \left(\prod_{i \in J} d_i\right)|\sigma^{-2}\boldsymbol{X}_{\bar{J}}'\boldsymbol{X}_{\bar{J}}|, \qquad \sum_J w_J = 1.$$

Theorem 4 (Principal Component Regression)

If $\boldsymbol{D}^* = d\boldsymbol{I}$, then the posterior mean (14) can be written as

$$\boldsymbol{b}^{**} = (\boldsymbol{H} + d\boldsymbol{I})^{-1}\boldsymbol{H}\boldsymbol{b} = \sum_{j=0}^{k} w_j(d/\sigma^2)\boldsymbol{c}_j,$$

where $\boldsymbol{c}_j$ is the jth "principal component regression" estimate, formed by imposing the j principal component constraints with the smallest eigenvalues, and

$$\sum_{j=0}^{k} w_j(d/\sigma^2) = 1.$$

Theorem 3 establishes conditions under which a posterior mean is a weighted average of the 2^k regressions. If the prior is located at the origin and if the coefficients are *a priori* independent, then a Bayesian can determine the location of his posterior by computation of the 2^k regressions. Conversely, a researcher who selects a particular coordinate system in which to omit variables (or equivalently selects k linearly independent constraints) and then proceeds to compute regressions on all subsets of variables (or equivalently uses all subsets of the k constraints) thereby reveals a prior located at the origin with coefficients independently distributed in the prior. It is, perhaps, worth emphasizing that this solves the puzzle of how to choose a parameterization. For example, if a variable y depends on x and lagged x, $y_t = \beta_1 x_t + \beta_2 x_{t-1}$, it is not likely that β_1 and β_2 are independent *a priori* since if I tell you something about β_2, it is likely to alter your opinions about β_1. But if the model is written as $y_t = (\beta_1 + \beta_2)(x_t + x_{t-1})/2 + (\beta_1 - \beta_2)(x_t - x_{t-1})/2$, it is likely that the long-run effect $(\beta_1 + \beta_2)$ is independent of the difference in the effects $(\beta_1 - \beta_2)$. In that event, computation of the 2^k regressions should be done in the second parameterization, not the first.

Theorem 4 makes use of the extra information that the prior variances are all the same. If the prior distribution is completely specified, as is assumed in this section, there is always a choice of parameterization such that the parameters are independent and identically distributed. In practice, it may not be easy to select such a parameterization. This is especially so when the explanatory variables are measured in different units, although logarithmic transformations can be useful in that event. When this difficulty is overcome Theorem 4 links principal component regression selection strategies with a full Bayesian treatment. The usual arbitrariness of the normalization in principal component regression is resolved by using a parameterization such that the prior is spherical. Furthermore, the principal component restrictions should be imposed as ordered by their eigenvalues, not by their t-values as has been suggested by Massy (1965).

2.2.2. *Model selection and incomplete priors*

The Bayesian logic is to my mind compellingly attractive, and it is something of a paradox that economic data are not routinely analyzed with Bayesian methods. It seems clear to me that the principal resistance to Bayesian methods is expressed in the incredulous grin which greets Bayesians when they make statements like: "We need to begin with a multivariate prior distribution for the parameter vector $\boldsymbol{\beta}$." Because prior distributions are not precisely measurable, or because potential readers may differ in their opinions, an analysis built on a particular distribution is of little interest. Instead, a researcher should report as completely as possible the mapping implied by the given data from priors into posteriors. In a slogan, "the mapping is the message".

In fact, many researchers attempt now to report this mapping. Often different least-squares equations with different lists of variables are included to give the reader a sense of the sensitivity of the inferences to choice of model. Sometimes a researcher will report that the inferences are essentially the same for a certain family of specifications.

The reporting of sensitivity analyses should be greatly encouraged. As readers we have a right to know if an inference "holds up" to minor and/or major changes in the model. Actually, results are often ignored until a thorough sensitivity analysis is completed, usually by other researchers. Sensitivity analyses are not now routinely reported largely because we do not have particularly useful tools for studying sensitivity, nor do we have economical ways of reporting. I believe the Bayesian logic ultimately will have its greatest practical impact in its solutions to this problem.

A Bayesian sensitivity analysis supposes that there is a class Π of prior distributions. This may be a personal class of distributions, containing all possible measurements of my uniquely maintained prior, or it may be a public class of distributions, containing alternative priors which readers are likely to maintain. In either case, it is necessary to characterize completely the family of posterior distributions corresponding to the family of priors. For a finite universe of elemental events e_i with corresponding probabilities $\mathrm{P}(e_i)=\pi_i$, the set Π may be generated by a set of inequalities:

$$\Pi=\{\boldsymbol{\pi}|\boldsymbol{G}\boldsymbol{\pi}\geqslant\boldsymbol{a}\},$$

which can involve constraints on the probabilities of sets, $\sum_i \pi_i \delta_i$, where δ_i is the indicator function, or constraints on the expectations of random variables $\sum_i \pi_i X(e_i)$, where X is the random variable. Given the data Y, a set A of interest with indicator function δ, the likelihood function $\mathrm{P}(Y|e_i) \equiv f_i$, and a particular prior, $\pi \in \Pi$, then the prior and posterior probabilities are

$$\mathrm{P}_\pi(A) = \sum_i \pi_i \delta_i,$$

$$\mathrm{P}_\pi(A|Y) = \sum_i f_i \pi_i \delta_i \Big/ \sum_i f_i \pi_i.$$

Prior, upper and lower, probabilities are then

$$\mathrm{P}_*(A) = \min_{\pi \in \Pi} \mathrm{P}_\pi(A),$$

$$\mathrm{P}^*(A) = \max_{\pi \in \Pi} \mathrm{P}_\pi(A),$$

with posterior bounds defined analogously. The interval $\mathrm{P}^*(A) - \mathrm{P}_*(A)$ is called the "confusion" by Dempster (1967). If Π is a public class of distribution $\mathrm{P}^* - \mathrm{P}_*$ might better be called the "disagreement".[3] Similarly, upper and lower prior expectations of the random variable X are

$$E_*(X) = \min_{\pi \in \Pi} \sum_i \pi_i X(e_i),$$

$$E^*(X) = \max_{\pi \in \Pi} \sum_i \pi_i X(e_i),$$

with posterior bounds defined analogously.

Although indeterminate probabilities have been used in many settings, reviewed by Shafer (1978) and DeRobertis (1979), as far as I know their use for

[3] In Shafer's view (1976, p. 23), which I share, the Bayesian theory is incapable of representing ignorance: "It does not allow one to withhold belief from a proposition without according that belief to the negation of the proposition." Lower probabilities do not necessarily have this restriction, $\mathrm{P}_*(A) + \mathrm{P}_*(\sim A) \neq 1$, and are accordingly called "non-additive". Shafer's review (1978) includes references to Bernoulli, Good, Huber, Smith and Dempster but excludes Keynes (1921). Keynes elevates indeterminate probabilities to the level of primitive concepts and, in some cases, takes only a partial ordering of probabilities as given. Except for some fairly trivial calculations on some relationships among bounded probabilities, Keynes' *Treatise* is devoid of practical advice. Braithwaite, in the editorial forward, reports accurately at the time (but greatly wrong as a prediction) that "this leads to intolerable difficulties without any compensating advantages". Jeffreys, in the preface to his third edition, began the rumor that Keynes had recanted in his (1933) review of Ramsey and had accepted the view that probabilities are both aleatory and additive. Hicks (1979), who clearly prefers Keynes to Jeffreys, finds no recantation in Keynes (1933) and refers to Keynes' 1937 *Quarterly Journal of Economics* article as evidence that he had not changed his mind.

analyzing the regression model is confined to Leamer and Chamberlain (1976), Chamberlain and Leamer (1976), and Leamer (1978). In each of these papers the prior location is taken as given and a study is made of the sensitivity of the posterior mean (or modes) to changes in the prior covariance matrix. One important sensitivity result is the following [Chamberlain and Leamer (1976)].

Theorem 5

The posterior mean $\boldsymbol{b}^{**}=(\boldsymbol{X}'\boldsymbol{X}+\boldsymbol{N}^{*})^{-1}\boldsymbol{X}'\boldsymbol{Y}$, regardless of the choice of the prior precision matrix N^*, lies in the ellipsoid

$$(\boldsymbol{b}^{**}-\boldsymbol{b}/2)'\boldsymbol{X}'\boldsymbol{X}(\boldsymbol{b}^{**}-\boldsymbol{b}/2)\leqslant \boldsymbol{b}'\boldsymbol{X}'\boldsymbol{X}\boldsymbol{b}/4. \tag{2.10}$$

Conversely, any point in this ellipsoid is a posterior mean for some $\boldsymbol{N}^*$.

The "skin" of this ellipsoid is the set of all constrained least-squares estimates subject to constraints of the form $\boldsymbol{R\beta}=\boldsymbol{0}$ [Leamer (1978, p. 127)]. Leamer (1977) offers a computer program which computes extreme values of $\boldsymbol{\psi}'\boldsymbol{b}^{**}$ for a given $\boldsymbol{\psi}$ over the ellipsoid (2.10) and constrained also to a classical confidence ellipsoid of a given confidence level. This amounts to finding upper and lower expectations within a class of priors located at the origin with the further restriction that the prior cannot imply an estimate greatly at odds with the data evidence. Leamer (1982) also generalizes Theorem 4 to the case $\boldsymbol{A}\leqslant \boldsymbol{N}^{*-1}\leqslant \boldsymbol{B}$, where A and B are lower and upper variance matrices and $\boldsymbol{A}\leqslant \boldsymbol{N}^{*-1}$ mean $\boldsymbol{N}^{*-1}-\boldsymbol{A}$ is positive definite.

Other results of this form can be obtained by making other assumptions about the class Π of prior distributions. Theorems 3 and 4, which were used above to link model selection methods with Bayesian procedures built on completely specified priors, can also be used to define sets of posterior means for families of prior distributions. Take the class Π to be the family of distributions for $\boldsymbol{\beta}$ located at the origin with β_i independent of β_j, $i\neq j$. Then, Theorem 3 implies that the upper and lower posterior modes of $\boldsymbol{X}'\boldsymbol{\beta}$ occur at one of the 2^k regressions. If the class Π includes all distributions uniform on the spheres $\boldsymbol{\beta}'\boldsymbol{\beta}=c$ and located at the origin, the set of posterior modes is a curve called by Dickey (1975) the "curve decolletage", by Hoerl and Kennard (1970a) the "ridge trace", and by Leamer (1973) the "information contract curve". This curve is connected to principal component regression by Theorem 3.

2.2.3. *The multi-collinearity problem and model selection*

There is no pair of words that is more misused both in econometrics texts and in the applied literature than the pair "multi-collinearity problem". That many of our explanatory variables are highly collinear is a fact of life. And it is completely clear that there are experimental designs $\boldsymbol{X}'\boldsymbol{X}$ which would be much preferred to the designs the natural experiment has provided us. But a complaint about the apparent malevolence of nature is not at all constructive, and the *ad hoc* cures for a bad design, such as stepwise regression or ridge regression, can be disastrously inappropriate. Better that we should rightly accept the fact that our non-experiments are sometimes not very informative about parameters of interest.

Most proposed measures of the collinearity problem have the very serious defect that they depend on the coordinate system of the parameter space. For example, a researcher might use annual data and regress a variable y on the current and lagged values of x. A month later, with a faded memory, he might recompute the regression but use as explanatory variables current x and the difference between current and lagged x. Initially he might report that his estimates suffer from the collinearity problem because x and x_{-1} are highly correlated. Later, he finds x and $x - x_{-1}$ uncorrelated and detects no collinearity problem. Can whimsy alone cure the problem?

To give a more precise example, consider the two-variable linear model $y = \beta_1 x_1 + \beta_2 x_2 + u$ and suppose that the regression of x_2 on x_1 yields the result $x_2 = r x_1 + e$, where e by construction is orthogonal to x_1. Substitute this auxiliary relationship into the original one to obtain the model

$$y = \beta_1 x_1 + \beta_2 (r x_1 + e) + u = (\beta_1 + \beta_2 r) x_1 + \beta_2 e + u \equiv \theta_1 z_1 + \theta_2 z_2 + u,$$

where $\theta_1 = (\beta_1 + \beta_2 r)$, $\theta_2 = \beta_2$, $z_1 = x_1$, and $z_2 = x_2 - r x_1$. A researcher who used the variables x_1 and x_2 and the parameters β_1 and β_2 might report that β_2 is estimated inaccurately because of the collinearity problem. But a researcher who happened to stumble on the model with variables z_1 and z_2 and parameters θ_1 and θ_2 would report that there is no collinearity problem because z_1 and z_2 are orthogonal (x_1 and e are orthogonal by construction). This researcher would nonetheless report that $\theta_2 (= \beta_2)$ is estimated inaccurately, not because of collinearity, but because z_2 does not vary adequately.

What the foregoing example aptly illustrates is that collinearity as a cause of weak evidence is indistinguishable from inadequate variability as a cause of weak evidence. In light of that fact, it is surprising that all econometrics texts have sections dealing with the "collinearity problem" but none has a section on the "inadequate variability problem". The reason for this is that there *is* something special about collinearity. It not only causes large standard errors for the coefficients but also causes very difficult interpretation problems when there is prior information about one or more of the parameters. For example, collinear data may imply weak evidence about β_1 and β_2 separately but strong evidence about the linear combination $\beta_1 + \beta_2$. The interpretation problem is how to use the sample information about $\beta_1 + \beta_2$ to draw inferences about β_1 and β_2 in a context where there is prior information about β_1 and/or β_2. Because classical inference is not concerned with pooling samples with prior information, classical econometrics texts ought not to have special sections devoted to collinearity as distinct from inadequate variability. Researchers do routinely pool prior information with sample information and do confront the interpretation problem, usually without the support of formal statistical theory. Because they do, most text writers using a classical framework feel compelled to write a section on the collinearity problem, which usually turns out confused and lame. An exception is Kmenta (1971, p. 391) who accurately writes: "that a high degree of multicollinearity is simply a feature of the sample that contributes to the unreliability of the estimated coefficients, but has no relevance for the conclusions drawn as a result of this unreliability".

This view of collinearity, which seems to me to be entirely straightforward, is nonetheless not widely held and I have one more communication device which I trust will be decisive. Consider again the two-variable model $y = \beta_1 x_1 + \beta_2 x_2 + u$ and suppose that you have been commissioned to estimate β_1. Suppose, further, that the computer program you are using prints out estimates and standard errors, but neither the covariance between β_1 and β_2 nor the correlation between x_1 and x_2. Another program which does compute covariances is available for \$100. Are you willing to buy it? That is, given the estimate and standard error of β_1, are you willing to bear a cost to find out if the standard error of β_1 is big because of the correlation between x_1 and x_2? The correct answer is yes if there is another source of information about β_1 or β_2 which you wish to use to estimate β_1. Otherwise, the answer is no. The best you can do is to use the given estimate of β_1 and the given standard error. Thus, the interesting aspects of the collinearity problem arise in an information pooling problem. Otherwise, collinearity can be ignored.

Measurements of the collinearity problem generally fall into one of four categories, each of which measures different aspects of the problem.[4]

(1) Measures of the "quality of the design" $\boldsymbol{X}'\boldsymbol{X}$.
(2) Measures of the usefulness of other information.
(3) Measures of the inappropriate use of multivariate priors.
(4) Measures of the sensitivity of inferences to choice of prior distribution.

The last three of these measures deal with information pooling problems. The first makes no reference to pooling and has the defects just described.

The first set of measures indicates the distance between some ideal design matrix, say $\boldsymbol{V}$, and the actual design, $\boldsymbol{X}'\boldsymbol{X}$. Apparently thinking that an ideal design is proportional to the identity matrix $\boldsymbol{V} = v\boldsymbol{I}$, many theorists, including

[4] One other way of measuring collinearity, proposed by Farrar and Glauber (1967), is to test if the explanatory variables are drawn independently. This proposal has not met with much enthusiasm since, if the design is badly collinear, it is quite irrelevant to issues of inference from the given data that another draw from the population of design matrices might not be so bad.

Raduchel (1971) and Belsley, Kuh and Welch (1980), have proposed the condition number of $X'X$ as a measure of the collinearity problem. The condition number is the square root of the ratio of the largest to the smallest eigenvalues. But there is always a parameterization in which the $X'X$ matrix is the identity, and all eigenvalues are equal to one. In fact, the condition number can be made to take on any value greater than or equal to one by suitable choice of parameterization. Aside from the fact that the condition number depends on the parameterization, even if it did not it would be nothing more than a complaint and would not point clearly to any specific remedial action. And, if a complaint is absolutely required, it is much more direct merely to report the standard error of the parameter of interest, and to observe that the standard error would have been smaller if the design were different – in particular, if there were less collinearity or more variability, the two being indistinguishable.

Category (2), in contrast, includes measures which do point to specific remedial action because they identify the value of specific additional information. For example, Leamer (1978, p. 197) suggests the ratio of the conditional standard error of β_1 given β_2 divided by the unconditional standard error as a measure of the incentive to gather information about β_2 if interest centers on β_1. If the data are orthogonal, that is, $X'X$ is diagonal, then this measure is equal to one. Otherwise it is a number less than one. The usefulness of this kind of measure is limited to settings in which it is possible to imagine that additional information (data-based or subjective) can be gathered.

Measures in category (3) have also been proposed by Leamer (1973), who contrasts Bayesian methods with *ad hoc* methods of pooling prior information and sample information in multivariate settings. When the design is orthogonal, and the prior precision matrix is diagonal, the informal use of prior information is not altogether misleading, but when the design is collinear or the prior covariances are not zero, the pooling of prior and sample information can result in surprising estimates. In particular, the estimates of the issue $\psi'\beta$ may not lie between the prior mean and the least-squares estimate. Thus, collinearity creates an incentive for careful pooling.

A category measure of collinearity has been proposed by Leamer (1973). If there were a one-dimensional experiment to measure the issue $\psi'\beta$ with the family of priors Π located at the origin, then the posterior confusion, the difference between the upper and lower expectations, is just $|\psi'b|$, where b is the least-squares estimate. Because the regression experiment in fact is k-dimensional, the confusion is increased to $|\max\psi'b^{**} - \min\psi'b^{**}|$, where b^{**} is constrained to the ellipsoid (4). The percentage increase in the confusion due to the dimensionality, $|\max\psi'b^{**} - \min\psi'b^{**}|/|\psi'b|$ has been proposed as a collinearity measure and is shown by Chamberlain and Leamer (1976) to be equal to $\sqrt{\chi^2/z^2}$, where χ^2 is the chi-squared statistic for testing $\beta = 0$ and z^2 is the square of the normal statistic for testing $\psi'\beta = 0$.

2.2.4. The degrees-of-freedom deficit

If it is admitted that the degree of a polynomial could be as high as k, then it would usually be admitted that it could be $k+1$ as well. A theory which allows k lagged explanatory variables would ordinarily allow $k+1$. In fact, I know of no setting in economics in which the list of explanatory variables can be said to be finite. Lists of variables in practice are finite, not because of theoretical belief, but only because of the apparent inferential hazards of degrees-of-freedom deficits. Actually, it is standard practice to increase the dimensionality of the parameter space as the number of observations increases, thereby revealing that an analysis at any given sample size is founded on a parameterization which is misleadingly abbreviated.

The method of abbreviation usually is based on prior information: variables which are not likely to be very important are excluded, unless the researcher is "wealthy" enough to be able to "afford" the luxury of "spending" some of the data evidence on these incidental issues. Since prior information is at the foundation of the method, it is unsurprising that a Bayesian has no special problem in dealing with k in excess of T. In particular, the posterior mean given by eq. (2.8) makes no reference to the invertibility of $\boldsymbol{X}'\boldsymbol{X}$. The usual practice of restricting attention to subsets of variables with $k_J \leqslant T$ can be justified by Theorem 1 which assigns zero weights to any subsets such that $|\boldsymbol{X}_J'\boldsymbol{X}_J|$ is zero.[5]

3. Model selection with loss functions

The formal model selection theories presented in Section 2 assume a world in which thinking, information gathering, and computer processing are all errorless and costless. In such a world, exogeneity issues aside, you would explain GNP in terms of all 2000 variables on the NBER data files and millions of other variables, as well, including the size of the polar ice cap. Mortals would find the collection of the data and the assessment of priors for such a study to be unacceptably burdensome and would simplify the model before and after having collected the data.

Pre-simplification and post-simplification will refer to decisions made respectively before and after data observation. A statistical decision-theoretic solution to the pre-simplification problem would require us to identify the size of the polar ice cap as a possible variable and to reflect upon its probable importance in the

[5]The degrees-of-freedom deficit does cause special problems for estimating the residual variance σ^2. It is necessary to make inferences about σ^2 to pick a point on the contract curve, as well as to describe fully the posterior uncertainty. But if σ^2 is assigned the Jeffreys' diffuse prior, and if $\boldsymbol{\beta}$ is *a priori* independent of σ^2, then the posterior distribution for $\boldsymbol{\beta}$ has a non-integrable singularity on the subspace $\boldsymbol{X}'\boldsymbol{Y}=\boldsymbol{X}'\boldsymbol{X}\boldsymbol{\beta}$. Raiffa and Schlaifer's (1961) conjugate prior does not have the same feature.

equation in terms of both its regression coefficient and its variability. We might decide not to observe this variable, and thereby to save observation and processing costs, but we would have suffered in the process the intolerable costs of thinking consciously about the millions of variables which might influence GNP. I know of no solution to this dilemma. In practice, one selects "intuitively" a "horizon" within which to optimize. There is no formal way to assure that a given pre-simplification is optimal, and a data analysis must therefore remain an art.

Useful formal theories of post-simplification can be constructed, however. It is most convenient to do so in the context of a model which has not been pre-simplified. For that reason, in this section we continue to assume that the researcher faces no costs for complexity until after the model has been estimated. Statistical analysis of pre-simplified models is further discussed in Section 6 which deals with data-instigated hypotheses.

In order to simplify a model it is necessary to identify the purposes for which the model is intended. An ideal model for forecasting will differ from an ideal model for policy evaluation or for teaching purposes. For scientific purposes, simplicity is an important objective of a statistical analysis because simple models can be communicated, understood, and remembered easily. Simplicity thereby greatly facilitates the accumulation of knowledge, both publicly and personally. The word simplicity is properly defined by the benefits it conveys: A simple model is one that can be easily communicated, understood, and remembered. Because these concepts do not lend themselves to general mathematical descriptions, statistical theory usually, and statistical practice often, have sought parsimonious models, with parsimony precisely measured by the number of uncertain parameters in the model. (An input–output model is a simple model but not a parsimonious one.)

Actually, most of the statistical theory which deals with parsimonious models has not sought to identify simple models. Instead, the goal has been an "estimable" model. A model is estimable if it leads to accurate estimates of the parameters of interest. For example, variables may be excluded from a regression equation if the constrained estimators are more accurate than the unconstrained estimators. "Overfitting" is the name of the disease which is thought to be remedied by the omission of variables. In fact, statistical decision theory makes clear that inference ought always to be based on the complete model, and the search for "estimable" models has the appeal but also the pointlessness of the search for the fountain of youth. I do not mean that "overfitting" is not an error. But "overfitting" can be completely controlled by using a proper prior distribution. Actually, I would say that overfitting occurs when your prior is more accurately approximated by setting parameters to zero than by assigning them the improper uniform prior.

The framework within which we will be operating in Section 3 is the following. The problem is to estimate $\boldsymbol{\beta}$ given the data $\boldsymbol{Y}$ with estimator $\hat{\boldsymbol{\beta}}(\boldsymbol{Y})$. The loss incurred in selecting an inaccurate estimate is

$$\text{loss} = L(\boldsymbol{\beta}, \hat{\boldsymbol{\beta}}).$$

The risk function is the expected loss conditional on $\boldsymbol{\beta}$:

$$R(\boldsymbol{\beta}, \hat{\boldsymbol{\beta}}) = E(L(\boldsymbol{\beta}, \hat{\boldsymbol{\beta}})|\boldsymbol{\beta}).$$

The estimator $\hat{\boldsymbol{\beta}}$ is said to be *inadmissible* if there exists another estimator $\hat{\boldsymbol{\beta}}^*$ with uniformly smaller risk:

$$R(\boldsymbol{\beta}, \hat{\boldsymbol{\beta}}^*) \leqslant R(\boldsymbol{\beta}, \hat{\boldsymbol{\beta}}) \quad \text{for all } \boldsymbol{\beta},$$

with strict inequality for at least one $\boldsymbol{\beta}$. Otherwise, $\hat{\boldsymbol{\beta}}$ is admissible. A Bayes estimator is found by minimizing expected posterior loss:

$$\min_{\hat{\beta}} E(L(\boldsymbol{\beta}, \hat{\boldsymbol{\beta}})|\boldsymbol{Y})$$

or, equivalently, by minimizing expected prior risk (*Bayes risk*):

$$\min_{\hat{\beta}(Y)} E[R(\boldsymbol{\beta}, \hat{\boldsymbol{\beta}})].$$

When loss is quadratic, the Bayes estimator is the posterior mean (2.9) and is admissible, e.g. Leamer (1978, p. 141). Because this posterior mean will not have any zero elements, a Bayesian treatment of estimation with quadratic loss creates a *prima facie* case against model selection procedures for this problem.

3.1. Model selection with quadratic loss

A huge literature has been built on the supposition that quadratic loss implies a model selection problem. The expected squared difference between an estimator $\hat{\theta}$ and the true value θ, the mean-squared-error, can be written as the variance plus the square of the bias:

$$\begin{aligned} E[(\hat{\theta}-\theta)^2|\theta] &= E([\hat{\theta}-E(\hat{\theta}|\theta)]^2|\theta)+(E(\hat{\theta}|\theta)-\theta)^2 \\ &= \operatorname{var}\hat{\theta}+\operatorname{bias}^2\hat{\theta}. \end{aligned}$$

This seems to suggest that a constrained least-squares estimator might be better than the unconstrained estimator, since although the constrained estimator is biased it also has smaller variance. Of course, the constrained estimator will do better if the constraint is true but will do worse if the constraint is badly violated. The choice between alternative estimators is therefore ultimately a choice between alternative prior distributions, a subject discussed in Section 2. If the prior is fairly diffuse, even if loss is quadratic, the estimator ought to be least-squares, inadmissibility results notwithstanding (see Section 3.4).

The generalization of the mean-squared-error to the case of a vector of parameters is the matrix

$$\begin{aligned} \boldsymbol{F} &= E\left((\hat{\boldsymbol{\theta}}-\boldsymbol{\theta})(\hat{\boldsymbol{\theta}}-\boldsymbol{\theta})'|\boldsymbol{\theta}\right) \\ &= \mathrm{var}(\hat{\boldsymbol{\theta}}) + (\mathrm{bias}\,\hat{\boldsymbol{\theta}})(\mathrm{bias}\,\hat{\boldsymbol{\theta}})', \end{aligned}$$

where $\mathrm{bias}\,\hat{\boldsymbol{\theta}} = E(\hat{\boldsymbol{\theta}}|\boldsymbol{\theta}) - \boldsymbol{\theta}$. The mean-squared-error of the linear combination $\boldsymbol{\psi}'\boldsymbol{\theta}$ is $E((\boldsymbol{\psi}'\boldsymbol{\theta}-\boldsymbol{\psi}'\hat{\boldsymbol{\theta}})^2|\boldsymbol{\theta}) = \boldsymbol{\psi}'\boldsymbol{F}\boldsymbol{\psi}$. An estimator $\hat{\boldsymbol{\theta}}_a$ which has mean-squared-error matrix $\boldsymbol{F}_a$ will be said to *dominate in the mean-squared-error sense* an estimator $\hat{\boldsymbol{\theta}}_b$ with mean-squared-error F_b if $\boldsymbol{F}_b - \boldsymbol{F}_a$ is positive definite for all $\boldsymbol{\theta}$, that is, if $\boldsymbol{\psi}'\boldsymbol{F}_b\boldsymbol{\psi} > \boldsymbol{\psi}'\boldsymbol{F}_a\boldsymbol{\psi}$ for all $\boldsymbol{\psi}$ and all $\boldsymbol{\theta}$.

Because the least-squares estimator $\boldsymbol{b}$ is unbiased, the mean-squared-error is just the variance

$$\mathrm{MSE}(\boldsymbol{b},\boldsymbol{\beta}) = \sigma^2(\boldsymbol{X}'\boldsymbol{X})^{-1}.$$

The mean-squared-error of the constrained estimator (1.4) with $\hat{\boldsymbol{\beta}}_{\bar{J}} = 0$ is

$$\mathrm{MSE}(\hat{\boldsymbol{\beta}},\boldsymbol{\beta}) = \sigma^2\begin{bmatrix}(\boldsymbol{X}_J'\boldsymbol{X}_J)^{-1} & 0\\ 0 & 0\end{bmatrix} + \begin{bmatrix}\boldsymbol{E}\\ -\boldsymbol{I}\end{bmatrix}\boldsymbol{\beta}_{\bar{J}}\boldsymbol{\beta}_{\bar{J}}'\begin{bmatrix}\boldsymbol{E}\\ -\boldsymbol{I}\end{bmatrix}',$$

where $\boldsymbol{E} = (\boldsymbol{X}_J'\boldsymbol{X}_J)^{-1}\boldsymbol{X}_J'\boldsymbol{X}_{\bar{J}}$.

By appeal to the partitioned inverse rule, the differences in the mean-squared-error matrices can be written as

$$\begin{aligned} \mathrm{MSE}(\boldsymbol{b},\boldsymbol{\beta}) - \mathrm{MSE}(\hat{\boldsymbol{\beta}},\boldsymbol{\beta}) &= \sigma^2\begin{bmatrix}\boldsymbol{E}\boldsymbol{C}^{-1}\boldsymbol{E}' - \boldsymbol{E} & \boldsymbol{C}^{-1}\\ -\boldsymbol{C}^{-1}\boldsymbol{E} & \boldsymbol{C}^{-1}\end{bmatrix} - \begin{bmatrix}\boldsymbol{E}\\ -\boldsymbol{I}\end{bmatrix}\boldsymbol{\beta}_{\bar{J}}\boldsymbol{\beta}_{\bar{J}}'\begin{bmatrix}\boldsymbol{E}\\ -\boldsymbol{I}\end{bmatrix} \\ &= \begin{bmatrix}\boldsymbol{E}\\ -\boldsymbol{I}\end{bmatrix}\left[\sigma^2\boldsymbol{C}^{-1} - \boldsymbol{\beta}_{\bar{J}}\boldsymbol{\beta}_{\bar{J}}'\right]\begin{bmatrix}\boldsymbol{E}\\ -\boldsymbol{I}\end{bmatrix}', \end{aligned} \tag{3.1}$$

where $\sigma^2\boldsymbol{C}^{-1} = \sigma^2(\boldsymbol{X}_{\bar{J}}'\boldsymbol{M}_J\boldsymbol{X}_{\bar{J}})^{-1} = \mathrm{var}(\boldsymbol{b}_{\bar{J}})$.

By inspection of eq. (3.1), we obtain the basic dominance result:

Theorem 6 (Mean-squared-error dominance)

If $\mathrm{var}(\boldsymbol{b}_{\bar{J}}) - \boldsymbol{\beta}_{\bar{J}}\boldsymbol{\beta}_{\bar{J}}'$ is positive definite, then $\mathrm{MSE}(\boldsymbol{b},\boldsymbol{\beta}) - \mathrm{MSE}(\hat{\boldsymbol{\beta}},\boldsymbol{\beta})$ is positive definite.

In words, if $\boldsymbol{\beta}_{\bar{J}}$ is small compared to the least-squares sampling variance $V(\boldsymbol{b}_{\bar{J}})$, it is better to estimate with $\boldsymbol{\beta}_{\bar{J}}$ set to zero. When $\boldsymbol{\beta}_{\bar{J}}$ is a scalar, the dominance condition can be written in terms of the "true t" as in Wallace (1964):

$$\text{MSE}(b,\beta)-\text{MSE}(\hat{\beta},\beta) \text{ pos. def.} \Leftrightarrow \tau^2 > 1,$$

where

$$\tau^2 = \beta_{\bar{J}}^2/\text{var}(b_{\bar{J}}).$$

More generally, if the "true squared t" is larger than one for all linear combinations of the omitted variables, then unconstrained least-squares estimates have a smaller mean-squared-error than constrained least-squares estimates. The answer to the question "Which is the better estimator?" is then only the answer to the question "What is your prior distribution?", or more particularly the question "Do you think $\boldsymbol{\beta}_{\bar{J}}$ is small relative to var($\boldsymbol{b}_{\bar{J}}$)?" Thus, the problem of model selection with quadratic loss is turned back into a problem of model selection with prior information and the quadratic loss function actually becomes incidental or unnecessary except that it selects one feature of the posterior distribution, namely the mean, for special attention.

This dark cloud of subjectivity is pierced by the sunshine of ingenuity when it is noted that the "true t", τ^2, can be estimated. That suggests selecting $\hat{\boldsymbol{\beta}}$ if the t value for testing $\beta_{\bar{J}} = 0$ is less than one and otherwise selecting $\boldsymbol{b}$, or, equivalently, selecting the model with the higher $\bar{R}^2$ [Edwards (1969)]. The bootstraps turn out to be a bit too loose for the researcher actually to get himself off the ground and this "pre-test" estimator has smaller mean-squared-error than least-squares for some values of $\boldsymbol{\beta}$ but not for others [Wallace and Ashar (1972) and Feldstein (1973)].

Since *a priori* information is clearly necessary to choose between $\boldsymbol{\psi}'\hat{\boldsymbol{\beta}}$ and $\boldsymbol{\psi}'\boldsymbol{b}$ as estimators of $\boldsymbol{\psi}'\boldsymbol{\beta}$, it is useful here to form the decision-theoretic estimator of $\boldsymbol{\psi}'\boldsymbol{\beta}$. Given the data $\boldsymbol{Y}$ and therefore the posterior distribution of $\boldsymbol{\beta}$, $f(\boldsymbol{\beta}|\boldsymbol{Y})$, a Bayesian chooses to estimate $\boldsymbol{\psi}'\boldsymbol{\beta}$ by $\hat{\boldsymbol{\theta}}$ so as to minimize $E((\boldsymbol{\psi}'\boldsymbol{\beta}-\hat{\boldsymbol{\theta}})^2|\boldsymbol{Y})$, which quite simply produces the posterior mean

$$\hat{\theta} = E(\boldsymbol{\psi}'\boldsymbol{\beta}|\boldsymbol{Y}) = \boldsymbol{\psi}'E(\boldsymbol{\beta}|\boldsymbol{Y}). \tag{3.2}$$

Because this posterior mean will, with probability one, contain no zero elements, there is no model selection problem implied by quadratic loss. If, as is suggested above, there is prior information that $\beta_{\bar{J}}$ is small, and there is a diffuse prior for all the other parameters, then the Bayes estimator of $\boldsymbol{\psi}'\boldsymbol{\beta}$ will fall somewhere between $\boldsymbol{\psi}'\boldsymbol{b}$ and $\boldsymbol{\psi}'\hat{\boldsymbol{\beta}}$, but will never equal $\boldsymbol{\psi}'\hat{\boldsymbol{\beta}}$.

Non-Bayesians cannot afford to ignore these observations merely because they resist the notion of a prior distribution. The class of Bayesian estimators forms an essentially complete class of admissible decision rules, and estimators which are "far from" Bayes' estimators are consequently inadmissible. In particular the "pre-test" estimator of $\boldsymbol{\psi}'\boldsymbol{\beta}$ is a discontinuous function of the data $\boldsymbol{Y}$, since an infinitesimal change in $\boldsymbol{Y}$ which shifts the hypothesis from acceptance to rejection causes a discrete jump in the estimate from $\boldsymbol{\psi}'\hat{\boldsymbol{\beta}}$ to $\boldsymbol{\psi}'\boldsymbol{b}$. The Bayes decisions (3.2) are in contrast necessarily continuous functions of $\boldsymbol{Y}$, and this test-and-estimate procedure has been shown by Cohen (1965) to be inadmissible when σ^2 is known.

The erroneous connection of model selection procedures and quadratic loss is most enticing if the loss function is

$$L(\boldsymbol{\beta},\hat{\boldsymbol{\beta}}) = (\boldsymbol{\beta}-\hat{\boldsymbol{\beta}})'\boldsymbol{X}'\boldsymbol{X}(\boldsymbol{\beta}-\hat{\boldsymbol{\beta}}), \tag{3.3}$$

with a dependence on the sample design which can be justified by supposing that we wish to estimate the mean values of $\boldsymbol{Y}$ at the sampled points $\boldsymbol{X}$, say $\tilde{\boldsymbol{Y}} = \boldsymbol{X}\boldsymbol{\beta}$, with loss

$$(\tilde{\boldsymbol{Y}}-\hat{\boldsymbol{Y}})'(\tilde{\boldsymbol{Y}}-\hat{\boldsymbol{Y}}) = (\boldsymbol{X}\boldsymbol{\beta}-\hat{\boldsymbol{Y}})'(\boldsymbol{X}\boldsymbol{\beta}-\hat{\boldsymbol{Y}}) = (\boldsymbol{\beta}-\hat{\boldsymbol{\beta}})'\boldsymbol{X}'\boldsymbol{X}(\boldsymbol{\beta}-\hat{\boldsymbol{\beta}}).$$

If least-squares is used to estimate $\boldsymbol{\beta}$, then the expected loss is

$$\begin{aligned} E((\boldsymbol{\beta}-\boldsymbol{b})'\boldsymbol{X}'\boldsymbol{X}(\boldsymbol{\beta}-\boldsymbol{b})|\boldsymbol{\beta},\sigma^2) &= E(\operatorname{tr}(\boldsymbol{\beta}-\boldsymbol{b})(\boldsymbol{\beta}-\boldsymbol{b})'\boldsymbol{X}'\boldsymbol{X}|\boldsymbol{\beta},\sigma^2) \\ &= \sigma^2\operatorname{tr}(\boldsymbol{X}'\boldsymbol{X})^{-1}\boldsymbol{X}'\boldsymbol{X} = k\sigma^2, \end{aligned} \tag{3.4}$$

which is an increasing function of the number of parameters k.

Alternatively, partition $\boldsymbol{\beta}' = (\boldsymbol{\beta}_J', \boldsymbol{\beta}_{\bar{J}}')$, where $\boldsymbol{\beta}_J$ has k_J elements, and estimate $\boldsymbol{\beta}$ by constrained least-squares with $\boldsymbol{\beta}_{\bar{J}} = \boldsymbol{0}$. The expected loss is a function of $\boldsymbol{\beta}$ and is (using the mean-squared-errors matrix above)

$$\begin{aligned} R(\boldsymbol{\beta},\hat{\boldsymbol{\beta}}) &= E[(\boldsymbol{\beta}-\hat{\boldsymbol{\beta}})'\boldsymbol{X}'\boldsymbol{X}(\boldsymbol{\beta}-\hat{\boldsymbol{\beta}})|\boldsymbol{\beta},\sigma^2) = \operatorname{tr}\{E[(\boldsymbol{\beta}-\hat{\boldsymbol{\beta}})(\boldsymbol{\beta}-\hat{\boldsymbol{\beta}})']\boldsymbol{X}'\boldsymbol{X}\} \\ &= \sigma^2 k_J + \boldsymbol{\beta}_{\bar{J}}'\begin{bmatrix}\boldsymbol{E}\\ -\boldsymbol{I}\end{bmatrix}\boldsymbol{X}'\boldsymbol{X}\begin{bmatrix}\boldsymbol{E}\\ -\boldsymbol{I}\end{bmatrix}'\boldsymbol{\beta}_{\bar{J}} \\ &= \sigma^2 k_J + \boldsymbol{\beta}_{\bar{J}}'\left[\boldsymbol{X}_{\bar{J}}'\boldsymbol{X}_{\bar{J}} - \boldsymbol{X}_{\bar{J}}'\boldsymbol{X}_J(\boldsymbol{X}_J'\boldsymbol{X}_J)^{-1}\boldsymbol{X}_J'\boldsymbol{X}_{\bar{J}}\right]\boldsymbol{\beta}_{\bar{J}}. \end{aligned} \tag{3.5}$$

The term $\sigma^2 k_I$ has been called a penalty for complexity and the second term a penalty for misspecification.

It is natural to select model J over the complete model if the expected loss (3.5) is less than the expected loss (3.4). But since (3.5) depends on $\boldsymbol{\beta}_{\bar{J}}$ this rule is non-operational. One way of making it operational is to estimate $\boldsymbol{\beta}_{\bar{J}}$ in (3.5) and to choose the model with the smallest estimated risk. A consistent estimate of $\boldsymbol{\beta}_{\bar{J}}$ is least-squares $\boldsymbol{b}_{\bar{J}} = (\boldsymbol{X}_{\bar{J}}'\boldsymbol{M}_J\boldsymbol{X}_{\bar{J}})^{-1}\boldsymbol{X}_{\bar{J}}'\boldsymbol{M}_J\boldsymbol{Y}$. Substituting this into (3.5) yields the estimated risk:

$$\hat{R}_J = \sigma^2 k_J + \boldsymbol{Y}'\boldsymbol{M}_J\boldsymbol{X}_{\bar{J}}(\boldsymbol{X}_{\bar{J}}'\boldsymbol{M}_J\boldsymbol{X}_{\bar{J}})^{-1}\boldsymbol{X}_{\bar{J}}'\boldsymbol{M}_J\boldsymbol{Y}$$
$$= \sigma^2 k_J + ESS_J - ESS. \tag{3.6}$$

Model selection rules based on estimated risk similar to $\hat{R}_J$ have been proposed by Allen (1971), Sawa (1978), and Amemiya (1980).

Although this is a consistent estimate of the risk, it is also biased. An unbiased estimator can be found by noting that

$$\begin{aligned} E(ESS_J) &= E(\boldsymbol{Y}'\boldsymbol{M}_J\boldsymbol{Y}) = \operatorname{tr}\boldsymbol{M}_J E(\boldsymbol{YY}') \\ &= \operatorname{tr}\boldsymbol{M}_J(\boldsymbol{X\beta\beta}'\boldsymbol{X}' + \sigma^2\boldsymbol{I}) \\ &= \boldsymbol{\beta}'\boldsymbol{X}'\boldsymbol{M}_J\boldsymbol{X\beta} + \sigma^2 \operatorname{tr}\boldsymbol{M}_J \\ &= \boldsymbol{\beta}_{\bar{J}}'\boldsymbol{X}_{\bar{J}}'\boldsymbol{M}_J\boldsymbol{X}_{\bar{J}}\boldsymbol{\beta}_{\bar{J}} + \sigma^2(T - k_J). \end{aligned}$$

Thus, an unbiased estimator of $R(\beta, \hat{\beta})$ is

$$\begin{aligned} \hat{R}_J &= \sigma^2 k_J + ESS_J - \sigma^2(T - k_J) \\ &= ESS_J - \sigma^2(T - 2k_J). \end{aligned}$$

Dividing this by σ^2 yields Mallow's (1973) criterion "C_p",

$$C_J = \frac{ESS_J}{\sigma^2} - (T - 2k_J), \tag{3.7}$$

which surfaced in the published literature in Gorman and Toman (1966) and Hocking and Leslie (1967). If σ^2 in the formula is replaced by the unbiased estimator $\hat{\sigma}^2 = (ESS)/(T-k)$, then C_p for the complete model ($k_J = k$) is just k. Models with C_p less than k are therefore "revealed to yield smaller prediction error".

Wallace's mean-squared-error tests and Mallow's C_p statistics are terribly appealing but they suffer from one substantial defect: neither can assure that the model which is most estimable will be selected. Errors in the selection process which are a necessary feature of any statistical analysis may mean that the mean-squared-errors of the two-stage estimator are larger than unconstrained least-squares. It depends on the value of $\boldsymbol{\beta}$. The only way really to solve the problem is therefore to apply the Bayesian logic with a prior distribution whose location and dispersion determine where you especially want to do better than least-squares and where you do not care too much. But a full Bayesian treatment of this problem quickly makes clear that there is no reason to set estimates to zero, and quadratic loss implies an estimation problem, not a model selection problem.

3.2. *Simplification searches: Model selection with fixed costs*

Although quadratic loss does not imply a model-selection problem, it is easy enough to find a loss function which does. For example, merely add to the quadratic loss a fixed cost dependent on the model but not on the true parameters. If future prediction of $\boldsymbol{Y}$ at the T sampled values of X is the goal, and if loss is quadratic with a fixed cost, then the expected loss can be written as

$$L(\boldsymbol{\beta},\hat{\boldsymbol{\beta}}) = (\boldsymbol{\beta}-\hat{\boldsymbol{\beta}})'\boldsymbol{X}'\boldsymbol{X}(\boldsymbol{\beta}-\hat{\boldsymbol{\beta}}) + T\sigma^2 + c\dim(\hat{\boldsymbol{\beta}}), \tag{3.8}$$

where c is a given constant and $\dim(\hat{\boldsymbol{\beta}})$ is the number of non-zero elements of $\hat{\boldsymbol{\beta}}$, and where $T\sigma^2$ measures the expected contribution to the loss due to the residual error. Loss (3.8) differs from (3.3) in that it includes the term $T\sigma^2$ and also the complexity penalty $c\dim(\hat{\boldsymbol{\beta}})$. The posterior expected value of this loss function is

$$\begin{aligned} E\big(L(\boldsymbol{\beta},\hat{\boldsymbol{\beta}})|\boldsymbol{Y},\sigma^2\big) &= E\{(\boldsymbol{\beta}-\bar{\boldsymbol{\beta}})'\boldsymbol{X}'\boldsymbol{X}(\boldsymbol{\beta}-\bar{\boldsymbol{\beta}})|\boldsymbol{Y}\} \\ &\quad + (\bar{\boldsymbol{\beta}}-\hat{\boldsymbol{\beta}})'\boldsymbol{X}'\boldsymbol{X}(\bar{\boldsymbol{\beta}}-\hat{\boldsymbol{\beta}}) + T\sigma^2 + c\dim(\hat{\boldsymbol{\beta}}) \\ &= \operatorname{tr}[\boldsymbol{X}'\boldsymbol{X}\cdot\operatorname{var}(\boldsymbol{\beta}|\boldsymbol{Y})] + (\bar{\boldsymbol{\beta}}-\hat{\boldsymbol{\beta}})'\boldsymbol{X}'\boldsymbol{X}(\bar{\boldsymbol{\beta}}-\hat{\boldsymbol{\beta}}) \\ &\quad + T\sigma^2 + c\dim(\hat{\boldsymbol{\beta}}), \end{aligned}$$

where $\bar{\boldsymbol{\beta}} = E(\boldsymbol{\beta}|\boldsymbol{Y})$. If $\hat{\boldsymbol{\beta}}$ is partitioned as above, and if $\hat{\boldsymbol{\beta}}_{\bar{J}}$ is set at zero, then the second term in this equation is minimized by setting

$$\hat{\boldsymbol{\beta}}_J = \bar{\boldsymbol{\beta}}_J + (\boldsymbol{X}_J'\boldsymbol{X}_J)^{-1}\boldsymbol{X}_J'\boldsymbol{X}_{\bar{J}}\bar{\boldsymbol{\beta}}_{\bar{J}}, \tag{3.9}$$

with the quadratic form becoming

$$(\bar{\boldsymbol{\beta}}-\hat{\boldsymbol{\beta}})'\boldsymbol{X}'\boldsymbol{X}(\bar{\boldsymbol{\beta}}-\hat{\boldsymbol{\beta}}) = \bar{\boldsymbol{\beta}}_{\bar{J}}'\boldsymbol{X}_{\bar{J}}'\boldsymbol{M}_J\boldsymbol{X}_{\bar{J}}\bar{\boldsymbol{\beta}}_{\bar{J}}.$$

Finally, in order to distinguish model selection with prior distributions from model selection with loss functions, let the prior be diffuse, so that $\bar{\boldsymbol{\beta}}$ is just the least-squares estimate with $\bar{\boldsymbol{\beta}}_{\bar{J}} = \boldsymbol{b}_{\bar{J}}$ and $V(\boldsymbol{\beta}|\boldsymbol{Y},\sigma^2) = \sigma^2(\boldsymbol{X}'\boldsymbol{X})^{-1}$. Then the expected loss becomes

$$\begin{aligned} E_J\big(L(\boldsymbol{\beta},\hat{\boldsymbol{\beta}})|\boldsymbol{Y},\sigma^2\big) &= k\sigma^2 + \boldsymbol{Y}'\boldsymbol{M}_J\boldsymbol{X}_{\bar{J}}(\boldsymbol{X}_{\bar{J}}'\boldsymbol{M}_J\boldsymbol{X}_{\bar{J}})^{-1}\boldsymbol{X}_{\bar{J}}'\boldsymbol{M}_J\boldsymbol{Y} + T\sigma^2 + ck_J \\ &= (k+T)\sigma^2 + ck_J + ESS_J - ESS. \end{aligned}$$

Thus, the best model is the one that solves the problem

$$\min_J (ck_J + ESS_J), \tag{3.10}$$

where k_J is the number of parameters and ESS_J is the error-sum-of-squares formed by regressing $\boldsymbol{Y}$ on these k_J explanatory variables. This can be compared with Mallows' C_p statistic (3.7) which has $c = 2\sigma^2$, and the estimated risk (3.6) with $c = \sigma^2$.

The exact interpretation attaching to the complexity penalty ck_J is a matter of considerable importance, but will likely remain elusive. Most of us simplify models to facilitate communication and innovation. The features of a model which impair communication and innovation are certain to vary with the subject-matter and the number of parameters is at best an incomplete measure of complexity. Moreover, the trade-off between complexity and prediction error (the scale of c) is likewise elusive. Leamer (1978) advocates avoiding these issues by reporting the percentage increase in the expected loss due to the restriction, exclusive of the complexity penalty,

$$\frac{L_J - L}{L} = \frac{ESS_J - ESS}{(k+T)\sigma^2}.$$

Lindley (1968), which is the source of these ideas, studies the conditional prediction problem in which the complexity penalty is the cost of observing "future" explanatory variables preliminary to forming the forecast. Lindley (1968) also studies the choice of variables for control and contrasts the solution of the prediction problem with the solution of the control problem, the latter depending and the former not depending on the posterior variance of $\boldsymbol{\beta}$.

It is important to notice that this solution (3.10) is applicable for prediction *only* at the observed values of the explanatory variables X. Essentially the same solution applies if the future explanatory variables are unknown and are treated as a sample out of a multivariate normal distribution, as in Lindley (1968). It should also be observed that this solution forces the included variables to play partly the role of the excluded variables, as is evident from eq. (3.9). Leamer (1978) has argued that the model which results is not simple in the sense of being easily communicated, and he recommends that $\hat{\boldsymbol{\beta}}_J$ should be set equal to $\bar{\boldsymbol{\beta}}_J$ rather than the adjusted coefficients (3.9). Using this restriction and the logic which produces (3.10), we obtain the expected loss $E_J(L(\boldsymbol{\beta}, \hat{\boldsymbol{\beta}}^*)|\boldsymbol{Y}, \sigma^2) = (k+T)\sigma^2 + \boldsymbol{b}_{\bar{J}}'\boldsymbol{X}_{\bar{J}}'\boldsymbol{X}_{\bar{J}}\boldsymbol{b}_{\bar{J}} + ck_J$, with the best model being the one that minimizes

$$\min_J (ck_J + \boldsymbol{b}_{\bar{J}}'\boldsymbol{X}_{\bar{J}}'\boldsymbol{X}_{\bar{J}}\boldsymbol{b}_{\bar{J}}), \tag{3.11}$$

the second term being a multivariate generalization of a "beta coefficient",

$b_k^2 \operatorname{var}(x_k)/\operatorname{var}(Y)$. This contrasts with (3.10) which minimizes $ck_J + \boldsymbol{b}'_{\bar{J}}\boldsymbol{X}'_{\bar{J}}\boldsymbol{M}_J\boldsymbol{X}_{\bar{J}}\boldsymbol{b}_{\bar{J}}$, where $\boldsymbol{X}'_{\bar{J}}\boldsymbol{X}'_{\bar{J}}$ is the unconditional variance of the excluded variables and $\boldsymbol{X}'_{\bar{J}}\boldsymbol{M}_J\boldsymbol{X}_{\bar{J}}$ is the variance conditional on the included variables.

3.3. *Ridge regression*

Hoerl and Kennard (1970) proposed to estimate $\boldsymbol{\beta}$ by the formula

$$\hat{\boldsymbol{\beta}}^{\mathrm{r}}(c) = (\boldsymbol{X}'\boldsymbol{X} + c\boldsymbol{I})^{-1}\boldsymbol{X}'\boldsymbol{Y}, \tag{3.12}$$

which they called the *ridge estimator*. Although this procedure is thoroughly discussed in Chapter 10 of this Handbook by Judge and Bock, it is useful here to note first that the ridge estimator is connected to all-subsets regression by Theorem 1 and to principal component regression by Theorem 2. In particular, the ridge estimator is a weighted average of regressions on all subsets and is also a weighted average of the principal component regressions. Secondly, the ridge estimator is proposed as a solution to the problem discussed in Section 3.1, estimation with quadratic loss, and it suffers from the same defect as the pre-test estimator, namely the risk is lower than least-squares risk for some values of $\boldsymbol{\beta}$ but higher for others. Whether you want to use ridge regression therefore depends on prior information. This is an aggravating example of the importance of packaging for the marketing of professional ideas. Hoerl and Kennard (1970) themselves observed that (3.12) has a Bayesian interpretation as the posterior mean with a spherical prior. What they proposed in effect is that you ought to act as if you have a spherical prior located at the origin, even when you do not. For reasons which escape me, some who resist Bayesian methods as being too "subjective", are nonetheless receptive to the use of spherical priors even when the true prior is something altogether different! Smith and Campbell (1980) may signal the beginning of a backlash.

3.4. *Inadmissibility*

The argument against model selection procedures when the loss function is quadratic rests primarily on the fact that methods which select discretely from points in the model space are inadmissible. But when there are three or more coefficients the unconstrained least-squares estimator itself is inadmissible and there exist known estimators which dominate least-squares. These are fully discussed in Chapter 10 of this Handbook by Judge and Bock. What these estimators have in common is an arbitrary location toward which the ordinary least-squares estimate is shrunk. The only way I know to choose this location is by appeal to prior information. Thus, in the context of a decision problem with quadratic loss a convincing argument can be made against least-squares, but the sensible choice of another estimator still rests on prior information. Moreover, I cannot think of a setting in which an economist has a quadratic loss function.

4. Proxy searches: Model selection with measurement errors

In practice, many model selection exercises are aimed at selecting one from a set of alternative proxy variables which measure a common hypothetical construct. However, the large and growing literature on errors-in-variables problems, reviewed in Chapter 23 of this Handbook by Aigner et al., rarely if ever touches on model selection issues. In this section I point out a few model selection problems which may arise when there are multiple proxy variables. The sources of these problems are the same as discussed in Sections 2 and 3: prior information and loss functions. The purpose of this section is not to provide solutions but only to alert the reader to an important set of issues.

The model which will serve as a basis for our comments is

$$\begin{aligned} y_t &= \beta\chi_t + \gamma z_t + u_t, \\ x_{1t} &= \delta_1\chi_t + \varepsilon_{1t}, \\ x_{2t} &= \delta_2\chi_t + \varepsilon_{2t}, \end{aligned}$$

where $(y_t, z_t, x_{1t}, x_{2t})$, $t = 1,\dots,T$, are observable variables, each with its sample mean removed, and $(\chi_t, u_t, \varepsilon_{1t}, \varepsilon_{2t})$, $t = 1,\dots,T$, are random vectors drawn independently from a normal distribution with mean vector $(wz_t, 0, 0, 0)$ and diagonal covariance matrix $\operatorname{diag}\{\sigma_\chi^2, \sigma_u^2, \sigma_1^2, \sigma_2^2\}$. In words, x_1 and x_2 are alternative proxy variables for the unobservable χ. In settings like this, researchers look for proxy variables which provide the "correct" estimates and high R^2's. The purpose of the following is to demonstrate the appropriateness of these informal techniques.

We first consider likelihood ratio tests for the hypothesis that x_1 is a better proxy than x_2. To make it as simple as possible, consider the hypothesis $\sigma_1^2 = 0$ versus $\sigma_2^2 = 0$. If $\sigma_1^2 = 0$, the sampling distribution can be written as

$$\begin{aligned} &\sigma_u^{-T/2}\left\{\exp - \sum_t (y_t - \beta x_{1t} - \gamma z_t)^2 / 2\sigma_u^2\right\}. \\ &\left(\delta_1^2\sigma_\chi^2\right)^{-T/2} \exp\left\{-\sum_t x_{1t}^2 / 2\delta_1^2\sigma_\chi^2\right\}. \\ &\left(\sigma_2^2\right)^{-T/2}\left\{\exp - \sum_t (x_{2t} - \delta_2 x_{1t})^2 / 2\sigma_2^2\right\}. \end{aligned}$$

Maximizing this with respect to the parameters produces the likelihood statistic:

$$L_1 = [ESS_1 \cdot \boldsymbol{x}_1'\boldsymbol{x}_1 \cdot \boldsymbol{x}_2'\boldsymbol{M}_1\boldsymbol{x}_2]^{-T/2},$$

where ESS_1 is the error sum-of-squares formed by regressing y on x_1 and z, and $\boldsymbol{M}_1 = \boldsymbol{I} - \boldsymbol{x}_1(\boldsymbol{x}_1'\boldsymbol{x}_1)^{-1}\boldsymbol{x}_1'$. With L_2 defined analogously, the likelihood ratio for testing $\sigma_1^2 = 0$ versus $\sigma_2^2 = 0$ is

$$L_1/L_2 = (ESS_2/ESS_1)^{T/2},$$

and the better proxy is the one that produces the higher R^2.

If it is known that both δ_1 and δ_2 are equal to one, then the likelihood ratio becomes

$$L_1/L_2 = [ESS_2 \cdot x_2'x_2/ESS_1 \cdot x_1'x_1]^{T/2}$$

and the variable with the lower variance is thereby favored since high variance suggests great measurement error. If, in addition, the values of β and γ are known, the likelihood ratio becomes

$$L_1/L_2 = (ESS_2 + Q_2(\beta,\gamma))(\boldsymbol{x}_2'\boldsymbol{x}_2)/(ESS_1 + Q_1(\beta,\gamma))(\boldsymbol{x}_1'\boldsymbol{x}_1), \tag{4.1}$$

where Q_i measures the difference between (β,γ) and the least-squares estimates

$$Q_i(\beta,\gamma) = (\hat{\beta}_i - \beta, \hat{\gamma}_i - \gamma)\begin{bmatrix} \boldsymbol{x}_i'\boldsymbol{x}_i & \boldsymbol{x}_i'z \\ z'\boldsymbol{x}_i & z'z \end{bmatrix}\begin{bmatrix} \hat{\beta}_i - \beta \\ \hat{\gamma}_i - \beta \end{bmatrix},$$

where

$$\begin{bmatrix} \hat{\beta}_i \\ \hat{\gamma}_i \end{bmatrix} = \begin{bmatrix} \boldsymbol{x}_i'\boldsymbol{x}_i & \boldsymbol{x}_i'z \\ z'\boldsymbol{x}_i & z'z \end{bmatrix}^{-1}\begin{bmatrix} \boldsymbol{x}_i'y \\ z'y \end{bmatrix}.$$

Thus, eq. (4.1) reveals that a good proxy yields a high R^2, generates estimates which are close to *a priori* estimates, and has a low variance.

5. Model selection without a true model

The preceding sections have taken as given the rather far-fetched assumption that the "true" model is necessarily one of a given class of alternatives. The word "true" can be given either an objectivist or a subjectivist definition. The data may be thought actually to have been drawn independently from some unknown distribution, $g(Y_i)$, which happens not to be in the maintained class of distributions $f(Y_i|\boldsymbol{\theta}), \boldsymbol{\theta} \in H$. Alternatively, my beliefs about the exchangeable sequence Y can be described by saying that I act as if each observation Y_i were drawn from a distribution $g(Y_i|\boldsymbol{\gamma})$, with uncertain parameter $\boldsymbol{\gamma}$ having the prior distribution $g(\boldsymbol{\gamma})$; thus, my true marginal density is $g(\boldsymbol{Y}) = \int_{\gamma}[\Pi_i g(Y_i|\boldsymbol{\gamma})]g(\boldsymbol{\gamma})\mathrm{d}\boldsymbol{\gamma}$. For a variety of reasons I choose to approximate $g(\boldsymbol{Y})$ by $f(\boldsymbol{Y}) = \int_{\theta}[\Pi_i f(Y_i|\boldsymbol{\theta})]f(\boldsymbol{\theta})\mathrm{d}\boldsymbol{\theta}$. That is, I act instead as if the observations were drawn from the distribution $f(Y_i|\boldsymbol{\theta})$, with uncertain parameter $\boldsymbol{\theta}$ having the prior distribution $f(\boldsymbol{\theta})$.

A well-known result in the objectivist tradition is that the method of maximum likelihood asymptotically produces an estimated density closest to the true density in the "information sense". This follows from the fact that $T^{-1}\Sigma_i \log f(Y_i|\boldsymbol{\theta})$ estimates $E(\log f(Y_i|\boldsymbol{\theta})) = \int_{Y_i} \log f(Y_i|\boldsymbol{\theta})g(Y_i)\mathrm{d}Y_i$ and from the fact that the difference between f and g "in the information sense" is

$$E\log[g(Y_i)/f(Y_i|\boldsymbol{\theta})] = E[\log g(Y_i)] - E[\log f(Y_i|\boldsymbol{\theta})].$$

Thus, maximizing the log-likelihood is equivalent asymptotically to maximizing $\mathrm{E}[\log f(Y_i|\boldsymbol{\theta})]$, which in turn is equivalent to minimizing the expected logarithmic difference between the two densities.

A more active role in searching for a model closely approximating the true density has been suggested by Akaike (1973, 1974) and explored by Sawa (1978). The considerable appeal of this framework is that none of the hypothesized linear regression models need be considered to be the true model. Although this might imply a model selection problem, Leamer (1979) shows that maximizing information is essentially the same as estimating with quadratic loss and consequently there is no proper penalty for complexity inherent in the approach. In fact, when σ^2 is known, the "Akaike Information Criterion" is the same as Mallows' C_p statistic and suffers from exactly the same defects.

In discussing the information criterion it is important to keep in mind the distinction between parameters of the data distribution and the decision arguments of the loss function. In the more familiar estimation problem the parameters of the data distribution are the regression coefficient vector $\boldsymbol{\beta}$ and the residual variance σ^2; the decision arguments of the loss function are the corresponding estimators, $\hat{\boldsymbol{\beta}}$ and $\hat{\sigma}^2$. In the Akaike–Sawa information framework, there are no data parameters corresponding to the decisions $\hat{\boldsymbol{\beta}}$ and $\hat{\sigma}^2$. Instead, the data $\boldsymbol{Y}$ are assumed to be distributed multivariate normal, for example, with mean vector $\boldsymbol{\mu}$ and variance–covariance matrix $\boldsymbol{\Omega}$. Inferences are drawn only about the parameters $\boldsymbol{\mu}$ and $\boldsymbol{\Omega}$; the choice of regression model is a decision problem which logically occurs after the inference problem has been solved. The decision problem is to approximate the uncertain mean vector $\boldsymbol{\mu}$ with a vector $\boldsymbol{X}\hat{\boldsymbol{\beta}}$, where $\boldsymbol{X}$ is an observed matrix and $\hat{\boldsymbol{\beta}}$ is a vector of free decision parameters. To make this distinction as clear as possible we may pose the rhetorical question: "What is the prior distribution for the coefficient vector β?" The answer is that there is no prior for β. Subjective probability distributions apply only to the uncertain state of nature (μ, Ω), and not to the decisions (β, σ^2).

Solutions are well known to the inference problem with Y multivariate normal with uncertain mean vector μ and uncertain covariance matrix Ω [Degroot (1970, p. 183)]. The decision problem is then to select a matrix X, a vector β, and a scalar σ^2 such that the regression model, $Y = X\beta + u$, is a good approximation to reality, in particular so that the difference between the true normal density $f_N(Y|\mu, \Omega)$ and the approximating density $f_N(Y|X\beta, \sigma^2 I)$ is minimized, where difference is measured in terms of the Kullback–Leibler information criterion. Expressed in terms of a loss function this becomes

$$L(\mu, \Omega; X, \beta, \sigma^2) = \frac{2}{n} E\left[\log\{ f_N(Y|\mu, \Omega)/f_N(Y|X\beta, \sigma^2 I)\}\right].$$

In the event that $\Omega = \omega^2 I$, this function becomes

$$L(\mu, \omega^2; X, \beta, \sigma^2) = c(\mu, \omega^2) + \log \sigma^2 + \frac{\omega^2}{\sigma^2} + \frac{1}{n}\frac{(\mu - X\beta)'(\mu - X\beta)}{\sigma^2},$$

where $c(\mu, \omega^2)$ is a loss which is independent of the decision (X, β, σ^2). The loss function is written to emphasize the fact that the first pair of arguments (μ, ω^2) are parameters and the second triple of arguments are decision variables. Inspection of this loss function reveals that the problem reduces to estimating μ with quadratic loss with the further restriction that the estimate $\hat{\mu}$ must be of the form $X\hat{\beta}$. Akaike (1973, 1974) has suggested an estimate of the approximation loss $L(\mu, \Omega; X, \beta, \sigma^2)$ equal to the maximum log-likelihood minus the number of parameters, and suggests selecting the model with lowest estimated loss. This is conceptually equivalent to selecting the model with smallest estimated risk, i.e. eq. (3.6) or (3.7). Just as in Section 3.1, an estimation problem is incorrectly interpreted as a model selection problem and the resultant estimator is almost certainly inadmissible.

Finally, it should be observed that the fundamental reason why the information criterion does not imply anything especially different from maximum likelihood methods is that it uses the logarithmic "scoring rule" which is implicit in maximum likelihood estimation. Alternative measures of the distance between two densities could produce dramatically different results.

6. Data-instigated models

In response to a question from Dr. Watson concerning the likely perpetrators of the crime, Sherlock Holmes replied "No data yet... It is a capital mistake to theorize before you have all the evidence. It biases the judgments" [Doyle (1888)]. Were Doyle trained as a theoretical statistician, he might have had Watson poised to reveal various facts about the crime, with Holmes admonishing: "No theories yet... It is a capital mistake to view the facts before you have all the theories. It biases the judgments."

Each of these quotations has a certain appeal. The first warns against placing excessive confidence in the completeness of any set of theories and suggests that over-confidence is a consequence of excessive theorizing before the facts are examined. The second quotation, on the other hand, points to the problem which data-instigated theories necessarily entail. Theories which are constructed to explain the given facts, cannot at the same time be said to be supported by these facts.

To give an example reviewed by Keynes (1921, ch. 25), De Morgan argues that a random choice of inclinations to the ecliptic of the orbits of the planets is highly unlikely to produce a set of inclinations with sum as small or smaller than those of our solar system. De Morgan derives from this an enormous presumption that "there was a necessary cause in the formation of the solar system...". D'Alembert in 1768 observed that the same conclusion could be drawn regardless of the set of inclinations, since any particular set of inclinations is highly unlikely to have been drawn randomly.

Keynes (1921, p. 338) points out that the solution to this dilemma is "simply" to find the correct prior probability of the data-instigated model: "If a theory is first proposed and is then confirmed by the examination of statistics, we are inclined to attach more weight to it than to a theory which is constructed in order to suit the statistics. But the fact that the theory which precedes the statistics is more likely than the other to be supported by general considerations – for it has not, presumably, been adopted for no reason at all – constitutes the only valid ground for this preference."

In order to make Keynes' observation as clear as possible, consider the two sequences of digits: A: 1, 2, 3, 4, 5; and B: 2, 8, 9, 1, 4. Ask yourself how probable is it that the next digit in each sequence is a six. Does it affect your opinion if you notice that the first and second pairs of digits of the B sequence add to ten? Does it affect your opinion if A and B came from IQ tests which included questions of the form: "Which digit is the sixth digit in the sequence?" What if A and B are the first five digits of six-digit license plates?

My own informal thinking would lead me initially to suppose that a six is highly likely (probability 0.9?) for sequence A but not very likely for sequence B (probability 0.1?). My opinion is little affected by the observation that pairs of digits add to ten, until I am told that these are sequences from IQ tests. Then I think six is very likely under both A and B, with probabilities 0.99 and 0.9, respectively. On the other hand, if these were license plates, I would expect a six with probability 0.1 for both sequences.

Both these sequences instigate hypotheses which were not explicitly identified before the data were observed. The preceding discussion is meant to suggest that the inferences you make in such circumstances depend critically on the prior probability you apply to the data-instigated hypothesis. In order to interpret the given evidence it is therefore necessary only to have the correct prior probabilities. The problem with data-instigated hypotheses is that prior probabilities have to be computed after having seen the data. Most of us are subject to what a psychologist [Fischoff (1975)] has called "the silly certainty of hindsight": Once an event is known to occur (Napoleon lost the battle of Waterloo) we tend to think it was an inevitable consequence of events which preceded it (Napoleon was suffering from a head cold). Before the battle, it is fair to say the outcome was very much in doubt, even given the fact that Napoleon had a cold. And before we knew the orbits of the planets, it is fair to say that it is unlikely that a "necessary cause" would select orbits in roughly the same plane.

The solution to this problem then reduces to policing the assignment of probabilities to data-instigated hypotheses. Leamer (1974) has provided a framework for doing this in the context of the linear regression model by mimicking the following sequential decision problem. Suppose that the true model has two explanatory variables $y_t = \beta_0 + \beta_1 x_t + \beta_2 z_t + u_t$, but suppose that it is costly to observe z. If it is known also that z_t obeys the auxiliary relationship $z_t = r_0 + r_1 x_t + \varepsilon_t$, then a regression of y on x alone will yield a useful estimate of β_1 if β_2 is zero or if r_1 is zero, since, conditional on x, $y_t = (\beta_0 + \beta_2 r_0) + (\beta_1 + \beta_2 r_1)x_t + (u_t + \beta_2 \varepsilon_t)$. Even if neither parameter is identically zero, it may be uneconomic to suffer the costs of observing z. However, once $\boldsymbol{Y}$ and $\boldsymbol{x}$ are observed, you may change your mind about observing z, possibly because the sample correlation between $\boldsymbol{Y}$ and $\boldsymbol{x}$ is too low or of the wrong sign.

This formal decision theory problem requires a supermind, capable of specifying a complete model and the relevant prior distribution. But the principal reason most of us use pre-simplified models is to avoid the unlimited cost of a full assessment. Although a simplified model cannot therefore result from formal decision theory, we nonetheless can act as if our models were so derived. The reason for doing so is that it implies constraints on the probabilities of data-instigated models, and consequently a very appealing system of discounting evidence built on these models. In particular, for the model described above, when y is regressed on x alone, the researcher is required to assess a prior for the experimental bias parameter $\beta_2 r_1$. This he must be able to do if he thinks he is getting evidence about β_1, even if z is not formally identified. If the regression coefficient is thought to be "almost" unbiased, then $\beta_2 r_1$ is "almost" zero. Next, when peculiarities in the data force a reconsideration of the model and z is added to the list of explanatory variables, then the prior for β_2 must be consistent with the originally assumed prior for $\beta_2 r_1$. Given r_1, this requirement will locate the prior for β_1 at zero and will constrain the prior variance. Consequently, when y is regressed on x and z the data will have to overcome the *a priori* prejudice that β_2 is small.

Parenthetically the word "instigate" is used here advisedly to mean that the data suggest a hypothesis already known to the researcher. This contrasts with data-initiated models which I am comfortable assuming do not exist.

7. Miscellaneous topics

7.1. Stepwise regression

"Stepwise regression" refers to any of a large number of subject-free sequential methods of selecting variables as originated by Ezekial and Fox (1930) and discussed in Draper and Smith (1966), in Efroymson (1960), and in Beale (1970). Variables are either added to or deleted from the regression equation depending on how they affect the residual sum-of-squares, for example to increase Mallows' C_p statistic. It is to the credit of economists that they rarely turn over the task of selecting variables to the computer. Aside from lacking a clear logical foundation, stepwise procedures almost always seek sets of variables which are relatively uncorrelated. Economists have avoided stepwise methods because they do not think nature is pleasant enough to guarantee orthogonal explanatory variables, and they realize that, if the true model does not have such a favorable design, then omitting correlated variables can have an obvious and disastrous effect on the estimates of the parameters.

7.2. Cross-validation

Cross-validation is a method for choosing models which is intended to respond to the following argument: "It is not difficult for a statistician to find a model which accurately explains a given data set. A polynomial of sufficiently high degree is flexible enough to wander from observation to observation. And there are always variables or other changes in a model which can be employed to improve the apparent fit. Specification searches are routinely employed for precisely this purpose. However, the usefulness of a model derives not from the fact that it is an adequate description of the process that generated a given data set, but rather because it works in other settings. A statistician can argue that model A will work better than model B in some new setting only by making a sequence of implausible assumptions. For all of these reasons, we should insist that a model survives to a new relevant environment, and we should reserve judgment as to which model is best until post-sample studies are performed."

A possible remedy is cross-validation in which part of the data is saved to check the adequacy of the model selection. References include Herzberg (1969), Stone (1974, 1977, 1978), Geisser (1975), and Geisser and Eddy (1979). Because cross-validation is explicitly designed to avoid assumptions, it is not possible legitimately to criticize the various proposals. Until the cross-validation tools are used and evaluated in a substantial number of real prediction situations, we are limited to making aesthetic judgments as we view the tools on display in the museums of the statistics journals. My own aesthetic senses have been rather sharply honed by my many visits to the main, probabilistic hall of the museum, and I do not find the cross-validation room very appealing. Moreover, all the lofty intentions notwithstanding, cross-validation ends up choosing models in essentially the same way as traditional tools.

The two most common cross-validatory methods for selecting a model involve splitting the data set and one-at-a-time deletions. A "test" of the hypothesis $\boldsymbol{\beta}_{\bar{J}}=\boldsymbol{0}$ can be built by splitting the data into a part used for estimating and a part used for prediction, say $\boldsymbol{Y}'=(\boldsymbol{Y}_1',\boldsymbol{Y}_2')$. The first subset of the data is used to estimate both the restricted and the unrestricted models, and to predict the second subset. The model is selected which yields the better predictions, measured as the sum-of-squared deviations between the actual $\boldsymbol{Y}_2$ and the predicted $\hat{\boldsymbol{Y}}_2$. The data may be treated symmetrically by predicting also $\boldsymbol{Y}_1$ given $\boldsymbol{Y}_2$. Consider in particular the full model with

$$\boldsymbol{X}=\begin{bmatrix}\boldsymbol{X}_1\\ \boldsymbol{X}_2\end{bmatrix}$$

and with least squares estimates based on the first and second parts of the data

$$\boldsymbol{b}_1=(\boldsymbol{X}_1'\boldsymbol{X}_1)^{-1}\boldsymbol{X}_1'\boldsymbol{Y}_1, \qquad \boldsymbol{b}_2=(\boldsymbol{X}_2'\boldsymbol{X}_2)^{-1}\boldsymbol{X}_2'\boldsymbol{Y}_2.$$

The penalty, which we seek to minimize by choice of model, is

$$P_1=(\boldsymbol{Y}_1-\boldsymbol{X}_1\boldsymbol{b}_2)'(\boldsymbol{Y}_1-\boldsymbol{X}_1\boldsymbol{b}_2)+(\boldsymbol{Y}_2-\boldsymbol{X}_2\boldsymbol{b}_1)'(\boldsymbol{Y}_2-\boldsymbol{X}_2\boldsymbol{b}_1), \tag{7.1}$$

where $\boldsymbol{X}_1\boldsymbol{b}_2$ is the prediction of $\boldsymbol{Y}_1$ based on $\boldsymbol{Y}_2$ and $\boldsymbol{X}_2\boldsymbol{b}_1$ is the prediction of $\boldsymbol{Y}_2$ based on $\boldsymbol{Y}_1$.

Since the criterion (7.1) is meant to be a method of choosing models when there is no theory of model selection, it is difficult to comment on it theoretically. It seems sensible. If it works, it works. But we can ask how it contrasts with the methods already discussed, which are a consequence of a certain list of assumptions. In fact, criterion (7.1) looks something like the usual error sum-of-squares.

If $\boldsymbol{X}_1 = \boldsymbol{X}_2$, then

$$\begin{aligned}
&\boldsymbol{b} = (\boldsymbol{X}'\boldsymbol{X})^{-1}\boldsymbol{X}'\boldsymbol{Y} = (\boldsymbol{b}_1 + \boldsymbol{b}_2)/2,\\
&(\boldsymbol{Y}_i - \boldsymbol{X}_i\boldsymbol{b}_i)'\boldsymbol{X}_i = \boldsymbol{0},\\
&\boldsymbol{Y}_i - \boldsymbol{X}_i\boldsymbol{b}_j = \boldsymbol{Y}_i - \boldsymbol{X}_i\boldsymbol{b} + \boldsymbol{X}_i(\boldsymbol{b}_i - \boldsymbol{b}_j)/2,\\
&\boldsymbol{Y}_i - \boldsymbol{X}_i\boldsymbol{b} = \boldsymbol{Y}_i - \boldsymbol{X}_i\boldsymbol{b}_i + \boldsymbol{X}_i(\boldsymbol{b}_i - \boldsymbol{b}_j)/2,
\end{aligned}$$

and

$$\begin{aligned}
(\boldsymbol{Y}_1 - \boldsymbol{X}_1\boldsymbol{b}_2)'(\boldsymbol{Y}_1 - \boldsymbol{X}_1\boldsymbol{b}_2) &= (\boldsymbol{Y}_1 - \boldsymbol{X}_1\boldsymbol{b})'(\boldsymbol{Y}_1 - \boldsymbol{X}_1\boldsymbol{b})\\
&\quad + (\boldsymbol{b}_1 - \boldsymbol{b}_2)'\boldsymbol{X}_1'\boldsymbol{X}_1(\boldsymbol{b}_1 - \boldsymbol{b}_2)/4\\
&\quad + (\boldsymbol{Y}_1 - \boldsymbol{X}_1\boldsymbol{b})'\boldsymbol{X}_1(\boldsymbol{b}_1 - \boldsymbol{b}_2)\\
&= (\boldsymbol{Y}_1 - \boldsymbol{X}_1\boldsymbol{b})'(\boldsymbol{Y}_1 - \boldsymbol{X}_1\boldsymbol{b})\\
&\quad + 3(\boldsymbol{b}_1 - \boldsymbol{b}_2)'\boldsymbol{X}_1'\boldsymbol{X}_1(\boldsymbol{b}_1 - \boldsymbol{b}_2)/4.
\end{aligned}$$

Thus,

$$\begin{aligned}
P_1 &= (Y - Xb)'(Y - Xb) + 3(b_1 - b_2)'X_1'X_1(b_1 - b_2)/2\\
&= ESS + 3(b_1 - b_2)'X_1'X_1(b_1 - b_2)/2.
\end{aligned}$$

That is, the cross-validation index is the usual error sum-of-squares plus a penality for coefficient instability. The complete neglect of coefficient instability evidenced by the traditional least-squares methods is certainly a mistake, but whether cross-validation is the proper treatment is very much in doubt. There are many formal statistical models of parameter drift that could be used instead. These methods yield estimates of the speed of parameter drift and pick the model which most closely tracks the data, allowing for parameter drift. In contrast, the cross-validation approach seeks a model with no drift at all, which seems inappropriate in our unstable world.

Cross-validation can also be done by deleting observations one at a time. Let $\boldsymbol{\delta}_i$ be a T-dimensional vector with a one in location i and zeros everywhere else. Then the prediction error of observation i is the same as the coefficient of the dummy variable $\boldsymbol{\delta}_i$ when $\boldsymbol{Y}$ is regressed on $\boldsymbol{X}$ and the dummy; this estimated coefficient is $(\boldsymbol{\delta}_i'\boldsymbol{M}\boldsymbol{\delta}_i)^{-1}\boldsymbol{\delta}_i\boldsymbol{M}\boldsymbol{Y} = e_i/M_{ii}$, where $\boldsymbol{M} = \boldsymbol{I} - \boldsymbol{X}(\boldsymbol{X}'\boldsymbol{X})^{-1}\boldsymbol{X}'$ and $\boldsymbol{e} = \boldsymbol{M}\boldsymbol{Y}$ is the vector of residuals when $\boldsymbol{Y}$ is regressed on $\boldsymbol{X}$. The sum-of-squared prediction errors is then the cross-validation penalty

$$P_2 = \sum_{i=1}^{T} e_i^2/M_{ii}^2 = \left(\sum_i e_i^2 M_{ii}^{-2} \Big/ \sum_i M_{ii}^{-2}\right)\sum_i M_{ii}^{-2}, \tag{7.2}$$

which is called SSPE by Schmidt (1974) and PRESS by Allen (1974). It is discussed by Hocking (1972, 1976) and studied by Stone (1974). The penalty P_2 is just a weighted error sum-of-squares times the sum-of-squares of the inverse diagonal elements of the matrix $\boldsymbol{M}$. It should be noted that in the nested case with the subset $\boldsymbol{X}_J$ tested against the full model $\boldsymbol{X}$, $\boldsymbol{\delta}_i'\boldsymbol{M}\boldsymbol{\delta}_i$ is strictly less than $\boldsymbol{\delta}_i'\boldsymbol{M}_J\boldsymbol{\delta}_i$, and $\sum_i(\boldsymbol{M}_J)_{ii}^{-2} < \sum_i \boldsymbol{M}_{ii}^{-2}$. For this reason the cross-validation penalty (7.2) can be interpreted as a weighted error sum-of-squares times a complexity penalty. The complexity penalty $\sum_i \boldsymbol{M}_{ii}^{-2}$ is very similar to the complexity penalty implicit in the adjusted R^2, which is a transformation of the unweighted error sum-of-squares, ESS, times $(\operatorname{tr}\boldsymbol{M})^{-1} = (\sum_i \boldsymbol{M}_{ii})^{-1} = (T-k)^{-1}$. As will be shown below, the cross-validation penalty (7.2) can be regarded to be inferior to the adjusted R^2 rule because it does not consistently select the true model.

The preceding paragraph has evaluated predictions in terms of their absolute errors and has ignored the fact that a prediction might have a large standard error. This is sensibly and easily corrected. The computed variance attaching to the dummy variable $\boldsymbol{\delta}_i$ is $\sigma^2(\boldsymbol{\delta}_i'\boldsymbol{M}\boldsymbol{\delta}_i)^{-1} = \sigma^2 M_{ii}^{-1}$, and the squared prediction error scaled by its variance is $(e_i^2/M_{ii}^2)/\sigma^2 M_{ii}^{-1} = e_i^2/\sigma^2 M_{ii}$. The sum-of-squares of standardized prediction errors is then the cross-validation penalty proposed by Schmidt (1974):

$$P_3 = \sum_i \left(e_i^2/M_{ii}\right).$$

Yet another possibility is offered by Stone (1974):

$$P_4 = \sum_i e_i^2 \Big/ \sum_i M_{ii}^2.$$

What P_2, P_3, P_4, and $\overline{R}^2$ all have in common is that they select the model which minimizes

$$\min_J \boldsymbol{e}_J'\boldsymbol{D}_J\boldsymbol{e}_J,$$

where $\boldsymbol{D}_J$ is a diagonal matrix. As in Schmidt (1974), we can compare the mean of $\boldsymbol{e}_J'\boldsymbol{D}_J\boldsymbol{e}_J$ for the model, J_1, with the mean for an alternative model, J_2. For any model J,

$$\begin{aligned} E(\boldsymbol{e}_J'\boldsymbol{D}_J\boldsymbol{e}_J) &= \operatorname{tr}\boldsymbol{D}_J E(\boldsymbol{e}_J\boldsymbol{e}_J') \\ &= \operatorname{tr}\boldsymbol{D}_J\boldsymbol{M}_J\boldsymbol{X}\boldsymbol{\beta}\boldsymbol{\beta}\boldsymbol{X}\boldsymbol{M}_J + \sigma^2 \operatorname{tr}\boldsymbol{D}_J\boldsymbol{M}_J, \end{aligned} \tag{7.3}$$

where $\boldsymbol{M}_J = \boldsymbol{I} - \boldsymbol{X}_J(\boldsymbol{X}_J'\boldsymbol{X}_J)^{-1}\boldsymbol{X}_J'$. If model J_1 is the true model, then $\boldsymbol{X}\boldsymbol{\beta} = \boldsymbol{X}_{J_1}\boldsymbol{\beta}_{J_1}$,

$\boldsymbol{M}_{J_1}\boldsymbol{X}_{J_1} = \boldsymbol{0}$, and

$$\mathrm{E}(\boldsymbol{e}'_{J_1}\boldsymbol{D}_{J_1}\boldsymbol{e}_{J_1}) = \sigma^2 \operatorname{tr} \boldsymbol{D}_{J_1}\boldsymbol{M}_{J_1}. \tag{7.4}$$

Since the first term in expression (7.3) is positive, a sufficient condition for (7.4) to be less than or equal to (7.3) is

$$\operatorname{tr} \boldsymbol{D}_{J_1}\boldsymbol{M}_{J_1} = \operatorname{tr} \boldsymbol{D}_{J_2}\boldsymbol{M}_{J_2}. \tag{7.5}$$

One choice of $\boldsymbol{D}_J$ with this property is $\boldsymbol{D}_J = \boldsymbol{I}/\operatorname{tr}\boldsymbol{M}_J$, which generates Theil's $\bar{R}^2$ criterion. Another choice is $\boldsymbol{D}_J = \operatorname{diag}\{M_{ii}^{-1}\}$, the basis for P_3. Neither P_2 with $\boldsymbol{D}_J = \operatorname{diag}\{M_{ii}^{-2}\}$, nor P_4 with $\boldsymbol{D}_J = \boldsymbol{I}/\sum_i M_{ii}^2$, has the property (7.5) and it is not the case that the average value is necessarily less for the true model than for the alternative.

In summary, cross-validation, though clearly well intentioned, seems on close inspection not likely to fulfill its promises. In the case of criterion P_1, it selects stable models when models which allow parameter drift might be better. In the case of P_2, it reduces to the suggestion that the usual least-squares residuals should be weighted. This is hardly a major departure from selecting the model with maximum adjusted R^2; nor is it a desirable departure, if one of the models is the true model.

7.3. *Goodness-of-fit tests*

Goodness-of-fit tests are intended to establish the adequacy of a given model without reference to specific alternatives. The most common of these is the chi-square test due to Karl Pearson in 1900. If e_i is the probability of observing an event in class i, and o_i is the observed relative frequency, then

$$\chi^2 = \sum_{i=2}^{k} (e_i - o_i)^2 \Big/ \sum_i e_i^2$$

asymptotically has a chi-square distribution with $k-1$ degrees of freedom. For example, a test of the assumption that observations x_t, $t = 1,\dots,n$, come from a standard normal distribution begins with a partitioning of the line into a set of intervals $I_i = \{x | c_i < x \leqslant c_{i+1}\}$ for $i = 1,\dots,k$ with $c_1 = -\infty$ and $c_{k+1} = \infty$. The probability of drawing an observation from interval i is $e_i = F(c_{i+1}) - F(c_i)$, where F is the cumulative of the standard normal. The hypothesis that observations are drawn from a standard normal distribution is rejected if the expected relative frequencies e_i differ substantially from the observed relative frequencies o_i.

The problem with this kind of a test is that the null hypothesis is virtually impossible and will surely be rejected if the sample size is large enough. The procedure therefore degenerates into an elaborate exercise to measure the effective sample size. Approximate hypotheses studied by Hodges and Lehman (1954) are not rejectable at the outset and do not suffer this logical defect. Perhaps more importantly, once having rejected the null hypothesis, in the absence of a null defined alternative, it is hard to know where to turn.

Goodness-of-fit tests are rare in economics but seem to be increasing in popularity with the increased interest in "diagnostics". Ramsey's (1969, 1974) work especially bears mentioning for the wrinkle of discriminating among alternative hypotheses on the basis of goodness-of-fit tests, that is, selecting the model which passes a battery of goodness-of-fit tests.

8. Conclusion

There is little question that the absence of completely defined models impinges seriously on the usefulness of data in economics. On pessimistic days I doubt that economists have learned anything from the mountains of computer print-outs that fill their offices. On especially pessimistic days, I doubt that they ever will. But there are optimistic days as well. There have been great advances in the last decades. A conceptual framework within which to discuss the model selection issues is emerging, largely because econometricians are learning about statistical decision theory. A large number of results have been obtained and many seem likely to be useful in the long run.

References

Akaike, H. (1973) "Information Theory and an Extension of the Maximum Likelihood Principle", in: B. N. Petrov and F. Csaki (eds.), *Proceedings of the Second International Symposium on Information Theory*. Budapest: Akademiai Kaido, pp. 267–281.

Akaike, H. (1974) "A New Look at Statistical Model Identification", *IEEE Transactions on Automatic Control*, AC-19, 716–723.

Akaike, H. (1978) "A Bayesian Analysis of the Minimum AIC Procedure", *Annals of the Institute of Mathematical Statistics*, 30, 9–14.

Akaike, H. (1979) "A Bayesian Extension of the Minimum AIC Procedure of Autogressive Model Fitting", *Biometrika*, 66, 237–242.

Allen, D. M. (1971) "Mean Square Error of Prediction as a Criterion for Selecting Variables", *Technometrics*, 13, 469–475.

Allen, D. M. (1974) "The Relationship Between Variable Selection and Data Augmentation and a Method of Prediction", *Technometrics*, 16, 125–127.

Amemiya, T. (1980) "Selection of Regressors", *International Economic Review*, 21, 331–354.

Ames, Edward and Stanley Reiter (1961) "Distributions of Correlation Coefficients in Economic Time Series", *Journal of the American Statistical Association*, 56, 637–656.

Anderson, T. W. (1951) "Estimating Linear Restrictions on Regression Coefficients for Multivariate Normal Distributions", *Annals of Mathematical Statistics*, 22, 327–351.

Anderson, T. W. (1958) *An Introduction to Multivariate Statistical Analysis*. New York: John Wiley & Sons.
Anderson, T. W. (1962) "The Choice of the Degree of a Polynomial Regression as a Multiple Decision Problem", *Annals of Mathematical Statistics*, 33, 255–265.
Ando, A. and G. M. Kaufman (1966) "Evaluation of an Ad Hoc Procedure for Estimating Parameters of Some Linear Models", *Review of Economics and Statistics*, 48, 334–340.
Anscombe, F. J. and J. W. Tukey (1963) "The Examination and Analysis of Residuals", *Technometrics*, 5, 141–160.
Anscombe, F. J. (1963) "Tests of Goodness of Fit", *Journal of the Royal Statistical Society*, Ser. B 25, 81–94.
Arrow, K. J. (1960) "Decision Theory and the Choice of Significance for the *t*-Test", in: I. Olkin et al., *Contributions to Probability and Statistics. Essays in Honor of Harold Hotelling*. Stanford: Stanford University Press.
Atkinson, A. C. (1970) "A Method for Discriminating Between Models", *Journal of the Royal Statistical Society*, Ser. B 32, 323–353.
Beale, E. M. L. (1970) "A Note on Procedures for Variable Selection in Multiple Regression", *Technometrics*, 12, 909–914.
Belsley, D., E. Kuh and R. Welsch (1980) *Regression Diagnostics*. New York: John Wiley & Sons.
Chamberlain, G. and E. Leamer (1976) "Matrix Weighted Averages and Posterior Bounds", *Journal of the Royal Statistical Society*, Ser. B 38, 73–84.
Chipman, J. S. (1964) "On Least Squares with Insufficient Observations", *Journal of the American Statistical Association*, 59, 1078–1111.
Chow, G. C. (1960) "Tests of Equality Between Sets of Coefficients in Two Linear Regressions", *Econometrica*, 28, 591–605.
Chow, G. C. (1979) "A Comparison of the Information and Posterior Probability Criteria for Model Selection", unpublished.
Cohen, A. (1965) "Estimates of Linear Combinations of the Parameters in the Mean Vector of a Multivariate Distribution", *Annals of Mathematical Statistics*, 36, 78–87.
Cohen, A. (1965) "A Hybrid Problem on the Exponential Family", *Annals of Mathematical Statistics*, 36, 1185–1206.
Cover, T. M. (1969) "Hypothesis Testing with Finite Statistics", *Annals of Mathematical Statistics*, 40, 828–835.
Cox, D. R. (1958) "Some Problems Connected with Statistical Inference", *Annals of Mathematical Statistics*, 29, 352–372.
Cox, D. R. (1961) "Tests of Separate Families of Hypotheses", *Proceedings of the Berkeley Symposium on Mathematical Statistical Probability*, vol. 1. Berkeley: University of California Press, pp. 105–123.
Cox, D. R. (1962) "Further Results on Tests of Separate Hypotheses", *Journal of the Royal Statistical Society*, Ser. B 24, 406–424.
Degroot, M. H. (1970) *Optimal Statistical Decisions*. New York: McGraw-Hill.
Dempster, A. P. (1967) "Upper and Lower Probabilities Induced by Multivalued Maps", *Annals of Mathematical Statistics*, 38, 325–339.
Dempster, A. P. (1968) "A Generalization of Bayesian Inference", *Journal of the Royal Statistical Society*, Ser. B 30, 205–248.
Dempster, A. P. (1971) "Model Searching and Estimation in the Logic of Inference" (with discussion), in: V. P. Godambe and D. A. Sprott (eds.), *Foundations of Statistical Inference*. Toronto: Holt, Rinehart and Winston, pp. 56–81.
Dempster, A. (1973) "Alternatives to Least Squares in Multiple Regression", in: D. G. Kabe and R. P. Gupta (eds.), *Multivariate Statistical Inference*. Amsterdam: North-Holland Publishing Co., pp. 25–40.
De Robertis, Lorraine (1979) "The Use of Partial Prior Knowledge in Bayesian Inference", unpublished Ph.D. dissertation, Yale University.
Dhrymes, P., et al. (1972) "Criteria for Evaluation of Econometric Models", *Annals of Economic and Social Measurement*, 1, 259–290.
Dickey, J. M. (1973) "Scientific Reporting and Personal Probabilities: Student's Hypothesis", *Journal of the Royal Statistical Society*, Ser. B 35, 285–305.
Dickey, J. M. (1975) "Bayesian Alternatives to the *F*-test and Least-Squares Estimates in the Normal Linear Model", in: S. E. Fienberg and A. Zellner (eds.), *Bayesian Studies in Econometrics and Statistics*. Amsterdam: North-Holland Publishing Co.

Doyle, A. C. (1888) "A Study in Scarlet", Reprinted in W. S. Baring-Gould (1967) *The Annotated Sherlock Holmes*. New York: Clarkson N. Potter, Inc.

Draper, N. R. and H. Smith (1966) *Applied Regression Analysis*. New York: John Wiley & Sons.

Durbin, J. (1953) "A Note on Regression When There is Extraneous Information About One of the Coefficients", *Journal of the American Statistical Association*, 48, 799–808.

Edwards, J. B. (1969) "The Relationship Between the F-test and $\bar{R}^2$", *American Statistician*, 23, 28.

Efroymson, M. A. (1960) "Multiple Regression Analysis", in: A. Ralston and H. S. Wilf, (eds.), *Mathematical Methods for Digital Computers*. New York: John Wiley & Sons.

Ezekial, Mordecai (1930) and K. A. Fox (1960), *Methods of Correlation and Regression Analysis: Linear and Curvilinear*. New York: John Wiley & Sons.

Farrar, D. E. and R. R. Glauber (1967) "Multicollinearity in Regression Analysis: The Problem Revisited", *Review of Economics and Statistics*, 49, 92–107.

Feldstein, M. (1973) "Multicollinearity and the Mean Square Error Criterion", *Econometrica*, 41, 337–346.

Ferguson, T. (1967) *Mathematical Statistics: A Decision Theoretic Approach*. New York: Academic Press.

Fienberg, S. E. and A. Zellner (eds.) (1975) *Studies in Bayesian Econometrics and Statistics*. Amsterdam: North-Holland Publishing Co.

Fisher, R. A. (1925), *Statistical Methods for Research Workers*. New York: Hafner.

Fisher, W. D. (1962) "Estimation in the Linear Decision Model", *International Economic Review*, 3, 1–29.

Fischoff, B. (1975) "The Silly Certainty of Hindsight", *Psychology Today*, 32, 72–77.

Frisch, R. (1934) *Statistical Confluence Analysis by Means of Complete Regression Systems*. Oslo: University Institute of Economics.

Furnival, George M. (1971) "All Possible Regressions With Less Computation", *Technometrics*, 13, 403–412.

Furnival, George M. and R. W. Wilson, Jr. (1974) "Regression by Leaps and Bounds", *Technometrics*, 16, 499–511.

Garside, M. J. (1965) "The Best Subset in Multiple Regression Analysis", *Applied Statistics*, 14, 196–201.

Gaver, K. M. and M. S. Geisel (1974) "Discriminating Among Alternative Models: Bayesian and Non-Bayesian Methods", in: P. Zarembka (ed.), *Frontiers in Econometrics*. New York: Academic Press, pp. 49–77.

Geisser, S. (1975) "The Predictive Sample Reuse Method with Applications", *Journal of the American Statistical Association*, 70, 320–328.

Geisser, S. and Eddy, W. F. (1979) "A Predictive Approach to Model Selection", *Journal of the American Statistical Association*, 74, 153–160.

Gorman, J. W. and R. J. Toman (1966) "Selection of Variables for Fitting Equations to Data", *Technometrics*, 8, 27–51.

Haitovsky, Y. (1969) "A Note on the Maximization of R^2", *American Statistician*, 23, 20–21.

Halpern, E. F. (1973) "Polynomial Regression from a Bayesian Approach", *Journal of the American Statistical Association*, 68, 137–143.

Hausman, J. A. (1978) "Specification Tests in Econometrics", *Econometrica*, 46, 1251–1272.

Hausman, J. and W. E. Taylor (1980) "Comparing Specification Tests and Classical Tests", M.I.T. Working Paper no. 266.

Hellman, M. E. and T. M. Cover (1970) "Learning With Finite Memory", *Annals of Mathematical Statistics*, 41, 765–782.

Herzberg, P. A. (1969) "The Parameters of Cross Validation", *Psychometrika*, 34, no. 2, part 2.

Hillier, G. (1977) "Constrained Linear Regression: Comparing the 'Best Linear' Estimator with Other Techniques", unpublished Monarch University Working Paper no. 21.

Hocking, R. R. and R. N. Leslie (1967) "Selection of the Best Subset in Regression Analysis", *Technometrics*, 9, 531–540.

Hocking, R. R. (1972) "Criteria for the Selection of a Subset Regression: Which One Should Be Used?", *Technometrics*, 14, 967–970.

Hocking, R. R. (1976) "The Analysis and Selection of Variables in Linear Regression", *Biometrics*, 32, 1–49.

Hodges, J. L., Jr. and E. L. Lehmann (1954) "Testing the Approximate Validity of Statistical

Hypotheses", *Journal of the Royal Statistical Society*, Ser. B 16, 261–268.
Hoerl, A. E. and R. W. Kennard (1970a) "Ridge Regression: Biased Estimation for Nonorthogonal Problems", *Technometrics*, 12, 55–67.
Hoerl, A. E. and R. W. Kennard (1970b) "Ridge Regression: Applications to Nonorthogonal Problems", *Technometrics*, 12, 69–82.
Hotelling, H. (1940) "The Selection of Variates for Use in Prediction With Some Comments on the Problem of Nuisance Parameters", *Annals of Mathematical Statistics*, 11, 271–283.
Huntsberger, D. V. (1955) "A Generalization of a Preliminary Testing Procedure for Pooling Data", *Annals of Mathematical Statistics*, 26, 734–743.
James, W. and C. Stein (1961) "Estimation with Quadratic Loss", in: *Proceedings of the Fourth Berkeley Symposium on Mathematical Statistics and Probability*, vol. 1, pp. 361–379.
Jeffreys, H. (1957) *Scientific Inference* (2nd edn.). Cambridge: Cambridge University Press.
Jeffreys, H. (1961) *Theory of Probability* (3rd edn.). London: Oxford University Press.
Kennard, Robert (1971) "A Note on the C_p Statistic", *Technometrics*, 13, 899–900.
Kennedy, W. J. and T. A. Bancroft (1971) "Model Building for Predicting in Regression Based Upon Repeated Significance Tests", *Annals of Mathematical Statistics*, 42, 1273–1284.
Keynes, J. M. (1921) *A Treatise on Probability*. New York: Harper and Row.
Keynes, J. M. (1933, 1951) in: Geoffrey Keynes (ed.), *Essays in Biography*. New York: Horizon Press (paperback by Norton, 1963).
Kiefer, J. and J. Wolfowitz (1956) "Consistency of Maximum Likelihood Estimator in the Presence of Infinitely Many Nuisance Parameters", *Annals of Mathematical Statistics*, 27, 887–906.
Klein, G. E. (1968) "Selection Regression Programs", *Review of Economics and Statistics*, 50, 288–290.
Kmenta, J. (1971) *Elements of Econometrics*. New York: John Wiley & Sons, Inc.
Kruskal, W. H. (1978) "Tests of Significance", in: W. H. Kruskal and J. M. Tanur (eds.), *International Encyclopedia of Statistics*, vol. 2. New York: The Free Press, pp. 944–958.
Lamotte, L. R. and R. R. Hocking (1970) "Computational Efficiency in the Selection of Regression Variables", *Technometrics*, 12, 83–93.
Larson, H. J. and T. A. Bancroft (1963a) "Biases in Prediction by Regression for Certain Incompletely Specified Models", *Biometrika*, 50, 391–402.
Larson, H. J. and T. A. Bancroft (1963b) "Sequential Model Building for Prediction in Regression Analysis", *Annals of Mathematical Statistics*, 34, 462–479.
Leamer, E. E. (1973) "Multicollinearity: A Bayesian Interpretation", *Review of Economics and Statistics*, 55, 371–380.
Leamer, E. E. (1974) "False Models and Post-Data Model Construction", *Journal of the American Statistical Association*, 69, 122–131.
Leamer, E. E. (1975) "A Result on the Sign of Restricted Least Squares Estimates", *Journal of Econometrics*, 3, 387–390.
Leamer, E. E. and G. Chamberlain (1976) "A Bayesian Interpretation of Pretesting", *Journal of the Royal Statistical Society*, Ser. B 38, 85–94.
Leamer, E. E. (1977) "SEARCH, A Linear Regression Computer Package", mimeo.
Leamer, E. E. (1978) "Regression-Selection Strategies and Revealed Priors", *Journal of the American Statistical Association*, 73, 580–587.
Leamer, E. E. (1979) "Information Criteria for Choice of Regression Models: A Comment", *Econometrica*, 47, 507–510.
Leamer, E. E. (1979) *Specification Searches*. New York: John Wiley & Sons.
Leamer, E. E. (1982) "Sets of Posterior Means With Bounded Variance Priors", *Econometrica*, 50, 725–736.
Lempers, F. B. (1971) *Posterior Probabilities of Alternative Linear Models*. Rotterdam: Rotterdam University Press.
Lindley, D. V. (1968) "The Choice of Variables in Multiple Regression", *Journal of the Royal Statistical Society*, Ser. B 31, 31–66.
Lindley, D. V. (1971a) *Bayesian Statistics, A Review*. Regional Conference Series in Applied Mathematics, S.I.A.M.
Mallows, C. L. (1973) "Some Comments on C_p", *Technometrics*, 15, 661–675.
Massy, W. F. (1965) "Principal Components Regression in Exploratory Statistical Research", *Journal of the American Statistical Association*, 60, 234–256.
Mayer, T. (1975) "Selecting Economic Hypotheses by Goodness of Fit", *Economics Journal*, 85, 877–882.

Neyman, Jerzy (1958) "The Use of the Concept of Power in Agricultural Experimentation", *Journal of Indian Society of Agricultural Statistics*, 2, 9–17.

Neyman, Jerzy and E. S. Pearson (1928) "On the Use and Interpretation of Certain Test Criteria for Purposes of Statistical Inference", *Biometrika*, 20A, 175–240, 263–294.

Neyman, Jerzy, and E. S. Pearson (1933) "On the Problem of the Most Efficient Tests of Statistical Hypotheses", *Royal Society of London, Philosphical Transactions*, Series A 231, 289–337.

Pearson, Karl (1900) "On the Criterion that a Given System of Deviations From the Probable in the Case of a Correlated System of Variables is Such that it Can Be Reasonably Supposed to Have Arisen From Random Sampling", *Philosophical Magazine*, 5th Ser., 50, 157–175.

Pereira, B. de B. (1977) "Discriminating Among Several Models: A Bibliography", *International Statistical Review*, 45, 163–172.

Quandt, R. E. (1974) "A Comparison of Methods for Testing Nonnested Hypotheses", *Review of Economics and Statistics*, 56, 92–99.

Raduchel, W. J. (1971) "Multicollinearity Once Again", Harvard Institute of Economic Research Paper no. 205.

Raiffa, H. and R. Schlaifer (1961) *Applied Statistical Decision Theory*. Cambridge, Mass.: Harvard University Press.

Ramsey, J. B. (1969) "Tests for Specification Errors in Classical Linear Least-Squares Regression Analysis", *Journal of the Royal Statistical Society*, Ser. B 31, 350–371.

Ramsey, J. B. (1974) "Classical Model Selection Through Specification Error Tests", pp. 13–47 in: P. Zarembka (ed.), *Frontiers in Econometrics*, Academic Press.

Sawa, T. (1978) "Information Criteria for Discriminating Among Alternative Regression Models", *Econometrica*, 46, 1273–1291.

Schatzoff, M., R. Tsao and S. Fienberg (1968) "Efficient Calculations of All Possible Regressions", *Technometrics*, 10, 769–780.

Schmidt, P. (1973) "Calculating the Power of the Minimum Standard Error Choice Criterion", *International Economic Review*, 14, 253–255.

Schmidt, P. (1974) "Choosing Among Alternative Linear Regression Models", *Atlantic Economic Journal*, 1, 7–13.

Schwarz, G. (1978), "Estimating the Dimension of a Model", *Annals of Statistics*, 6, 461–464.

Sclove, S. L., C. Morris and R. Radhakrishnan (1972) "Non-optimality of Preliminary-Test Estimators for the Mean of a Multivariate Normal Distribution", *Annals of Mathematical Statistics*, 43, 1481–1490.

Sen, P. K. (1979) "Asymptotic Properties of Maximum Likelihood Estimators Based on Conditional Specifications", *Annals of Statistics*, 7, 1019–1033.

Shafer, Glenn (1976) *A Mathematical Theory of Evidence*. Princeton: Princeton University Press.

Shafer, Glenn (1978) "Non-additive Probabilities in the Work of Bernoulli and Lambert", *Archives for History of Exact Sciences*, 19, 309–370.

Smith, Gary and Frank Campbell (1980) "A Critique of Some Ridge Regression Methods", *Journal of the American Statistical Association*, 75, 74–103.

Stein, C. (1956) "Inadmissibility of the Usual Estimator for the Mean of a Multivariate Normal Distribution", in: *Proceedings of the Third Berkeley Symposium on Mathematical Statistical Probability*, vol. 1, pp. 197–206.

Stone, M. (1974) "Cross-validatory Choice and Assessment of Statistical Predictions", *Journal of the Royal Statistical Society*, Ser. B 36, 111–147.

Stone, M. (1977) "An Asymptotic Equivalence of Choice of Model by Cross-Validation and Akaike's Criterion", *Journal of the Royal Statistical Society*, Ser. B 39, 44–47.

Stone, M. (1978) "Cross-Validation: A Review", *Mathematische Operationsforschung und Statistik: Series Statistics*, 9, 127–139.

Stone, Charles (1979) "Admissible Selection of an Accurate and Parsimonious Normal Linear Regression Model", unpublished discussion paper.

Strawderman, W. E. and A. Cohen (1971) "Admissibility of Estimators of the Mean Vector of a Multivariate Normal Distribution With Quadratic Loss", *Annals of Mathematical Statistics*, 42, 270–296.

Theil, H. (1957) "Specification Errors and the Estimation of Economic Relationships", *Review of the International Statistical Institute*, 25, 41–51.

Theil, H. (1971) *Principles of Econometrics*. New York: John Wiley & Sons.

Theil, H. and A. S. Goldberger (1961) "On Pure and Mixed Statistical Estimation in Economics", *International Economic Review*, 2, 65–78.

Thompson, M. L. (1978) "Selection of Variables in Multiple Regression: Part I. A Review and Evaluation", *International Statistical Review*, 46, 1–19; "Part II. Chosen Procedures, Computations and Examples", 46, 129–146.

Toro-Vizcarrondo, C. and T. D. Wallace (1968) "A Test of the Mean Square Error Criterion for Restrictions in Linear Regressions", *Journal of the American Statistical Association*, 63, 558–572.

Wallace, T. D. (1964) "Efficiencies for Stepwise Regressions", *Journal of the American Statistical Association*, 59, 1179–1182.

Wallace, T. D. and V. G. Ashar (1972) "Sequential Methods of Model Construction", *Review of Economics and Statistics*, 54, 172–178.

Watson, S. R. (1974) "On Bayesian Inference With Incompletely Specified Prior Distributions", *Biometrika*, 61, 193–196.

Wu, D. (1973) "Alternative Tests of Independence Between Stochastic Regressors and Disturbances", *Econometrica*, 41, 733–750.

Zellner, A. (1971) *An Introduction to Bayesian Inference in Econometrics*. New York: John Wiley & Sons.

PART 3

ESTIMATION AND COMPUTATION

Chapter 6

NON-LINEAR REGRESSION MODELS

TAKESHI AMEMIYA*

Stanford University

Contents

*This work was supported by National Science Foundation Grant SES-7912965 at the Institute for Mathematical Studies in the Social Sciences, Stanford University. The author is indebted to the following people for valuable comments: R. C. Fair, A. R. Gallant, Z. Griliches, M. D. Intriligator, T. E. MaCurdy, J. L. Powell, R. E. Quandt, N. E. Savin, and H. White.

Handbook of Econometrics, Volume I, Edited by Z. Griliches and M.D. Intriligator

1. Introduction

This is a survey of non-linear regression models, with an emphasis on the theory of estimation and hypothesis testing rather than computation and applications, although there will be some discussion of the last two topics. For a general discussion of computation the reader is referred to Chapter 12 of this Handbook by Quandt. My aim is to present the gist of major results; therefore, I will sometimes omit proofs and less significant assumptions. For those, the reader must consult the original sources.

The advent of advanced computer technology has made it possible for the econometrician to estimate an increasing number of non-linear regression models in recent years. Non-linearity arises in many diverse ways in econometric applications. Perhaps the simplest and best known case of non-linearity in econometrics is that which arises as the observed variables in a linear regression model are transformed to take account of the first-order autoregression of the error terms. Another well-known case is the distributed-lag model in which the coefficients on the lagged exogenous variables are specified to decrease with lags in a certain non-linear fashion, such as geometrically declining coefficients. In both of these cases, non-linearity appears only in parameters but not in variables.

More general non-linear models are used in the estimation of production functions and demand functions. Even a simple Cobb–Douglas production function cannot be transformed into linearity if the error term is added rather than multiplied [see Bodkin and Klein (1967)]. CES [Arrow, Chenery, Minhas and Solow (1961)] and VES [Revankar (1971)] production functions are more highly non-linear. In the estimation of expenditure functions, a number of highly non-linear functions have been proposed (some of these are used in the supply side as well)–Translog [Christensen, Jorgenson and Lau (1975)], Generalized Leontief [Diewert (1974)], S-Branch [Brown and Heien (1972)], and Quadratic [Howe, Pollack and Wales (1979)], to name a few. Some of these and other papers with applications will be mentioned in various relevant parts of this chapter.

The non-linear regression models I will consider in this chapter can be written in their most general form as

$$f_i(y_t, x_t, \alpha_i) = u_{it}; \qquad i = 1,2,\dots,n; \quad t = 1,2,\dots,T, \tag{1.1}$$

where y_t, x_t, and α_i are vectors of endogenous variables, exogenous variables, and parameters, respectively, and u_{it} are unobservable error terms with zero mean. Eqs. (1.1), with all generality, constitute the non-linear simultaneous equations model, which is analyzed in Section 5. I devote most of the discussion in the chapter to this section because this area has been only recently developed and therefore there is little account of it in general references.

Many simpler models arising as special cases of (1.1) are considered in other sections. In Section 2 I take up the simplest of these, which I will call the standard non-linear regression model, defined by

$$y_t = f(x_t, \beta) + u_t, \qquad t = 1,2,\dots,T, \tag{1.2}$$

where $\{u_t\}$ are scalar i.i.d. (independent and identically distributed) random variables with zero mean and constant variance. Since this is the model which has been most extensively analyzed in the literature, I will also devote a lot of space to the analysis of this model. Section 3 considers the non-i.i.d. case of the above model, and Section 4 treats its multivariate generalization.

Now, I should mention what will not be discussed. I will not discuss the maximum likelihood estimation of non-linear models unless the model is written in the regression form (1.1). Many non-linear models are discussed elsewhere in this Handbook; see, for example, the chapters by Dhrymes, McFadden, and Maddala. The reader is advised to recognize a close connection between the non-linear least squares estimator analyzed in this chapter and the maximum likelihood estimator studied in the other chapters; essentially the same techniques are used to derive the asymptotic properties of the two estimators and analogous computer algorithms can be used to compute both.

I will not discuss splines and other methods of function approximation, since space is limited and these techniques have not been as frequently used in econometrics as they have in engineering applications. A good introduction to the econometric applications of spline functions can be found in Poirier (1976).

Above I mentioned the linear model with the transformation to reduce the autocorrelation of the error terms and the distributed-lag model. I will not specifically study these models because they are very large topics by themselves and are best dealt with separately. (See the chapter by Hendry, Pagan, and Sargan in this Handbook). There are a few other important topics which, although non-linearity is involved, woud best be studied within another context, e.g. non-linear error-in-variable models and non-linear time-series models. Regarding these two topics, I recommend Wolter and Fuller (1978) and Priestley (1978). Finally, I conclude this introduction by citing general references on non-linear regression models. Malinvaud (1970b) devotes one long chapter to non-linear regression models in which he discusses the asymptotic properties of the non-linear least squares estimator in a multivariate model. There are three references which are especially good in the discussion of computation algorithms, confidence regions, and worked out examples: Draper and Smith (1966), Bard (1974), and Judge, Griffiths, Hill and Lee (1980). Several chapters in Goldfeld and Quandt (1972) are devoted to the discussion of non-linear regression models. Their Chapter 1 presents an excellent review of optimization techniques which can be used in the computation of both the non-linear least squares and the maximum likelihood estimators. Chapter 2 discusses the construction of confidence regions in the non-linear regression model and the asymptotic properties of the maximum likelihood estimator (but not of the non-linear least squares estimator). Chapter 5 considers the Cobb–Douglas production function with both multiplicative and additive errors, and Chapter 8 considers non-linear (only in variables) simultaneous equations models. There are two noteworthy survey articles: Gallant (1975a), with emphasis on testing and computation, and Bunke, Henscheke, Strüby and Wisotzki (1977), which is more theoretically oriented. None of the above-mentioned references, however, discusses the estimation of simultaneous equations models non-linear both in variables and parameters.

2. Single equation–i.i.d. case

2.1. *Model*

In this section I consider the standard non-linear regression model

$$y_t = f(x_t, \beta_0) + u_t, \qquad t = 1,2,\dots,T, \tag{2.1}$$

where y_t is a scalar endogenous variable, x_t is a vector of exogenous variables, β_0 is a K-vector of unknown parameters, and $\{u_t\}$ are unobservable scalar i.i.d. random variables with $Eu_t = 0$ and $Vu_t = \sigma_0^2$, another unknown parameter. Note that, unlike the linear model where $f(x_t, \beta_0) = x_t'\beta_0$, the dimensions of the vectors x_t and β_0 are not necessarily the same. We will assume that f is twice continuously differentiable. As for the other assumptions on f, I will mention them as they are required for obtaining various results in the course of the subsequent discussion.

Econometric examples of (2.1) include the Cobb–Douglas production function with an additive error,

$$Q_t = \beta_1 K_t^{\beta_2} L_t^{\beta_3} + u_t, \tag{2.2}$$

and the CES (constant elasticity of substitution) production function:

$$Q_t = \beta_1 \left[\beta_2 K_t^{-\beta_3} + (1-\beta_2) L_t^{-\beta_3} \right]^{-\beta_4/\beta_3} + u_t. \tag{2.3}$$

Sometimes I will write (2.1) in vector notation as

$$y = f(\beta_0) + u, \tag{2.4}$$

where y, $f(\beta_0)$, and u are T-vectors whose tth element is equal to y_t, $f(x_t, \beta_0)$, and u_t, respectively. I will also use the symbol $f_t(\beta_0)$ to denote $f(x_t, \beta_0)$.

The non-linear least squares (NLLS) estimator, denoted $\hat{\beta}$, is defined as the value of β that minimizes the sum of squared residuals

$$S_T(\beta) = \sum_{t=1}^{T} [y_t - f(x_t, \beta)]^2. \tag{2.5}$$

It is important to distinguish between the β that appears in (2.5), which is the argument of the function $f(x_t, \cdot)$, and β_0, which is a fixed true value. In what follows, I will discuss the properties of $\hat{\beta}$, the method of computation, and statistical inference based on $\hat{\beta}$.

2.2. *Asymptotic properties*

2.2.1. *Consistency*

The consistency of the NLLS estimator is rigorously proved in Jennrich (1969) and Malinvaud (1970a). The former proves strong consistency ($\hat{\beta}$ converging to β_0 almost surely) and the latter weak consistency ($\hat{\beta}$ converging to β_0 in probability). Weak consistency is more common in the econometric literature and is often called by the simpler name of consistency. The main reason why strong consistency, rather than weak consistency, is proved is that the former implies the latter and is often easier to prove. I will mainly follow Jennrich's proof but translate his result into weak consistency.

The consistency of $\hat{\beta}$ is proved by proving that $\operatorname{plim} T^{-1}S_T(\beta)$ is minimized at the true value β_0. Strong consistency is proved by showing the same holds for the almost sure limit of $T^{-1}S_T(\beta)$ instead. This method of proof can be used to prove the consistency of any other type of estimator which is obtained by either minimizing or maximizing a random function over the parameter space. For example, I used the same method to prove the strong consistency of the maximum likelihood estimator (MLE) of the Tobit model in Amemiya (1973b).

This method of proof is intuitively appealing because it seems obvious that if $T^{-1}S_T(\beta)$ is close to $\operatorname{plim} T^{-1}S_T(\beta)$ and if the latter is minimized at β_0, then $\hat{\beta}$, which minimizes the former, should be close to β_0. However, we need the following three assumptions in order for the proof to work:

The parameter space B is compact (closed and bounded) and β_0 is its interior point. (2.6)

$S_T(\beta)$ is continuous in β. (2.7)

$\operatorname{plim} T^{-1}S_T(\beta)$ exists, is non-stochastic, and its convergence is uniform in β. (2.8)

The meaning of (2.8) is as follows. Define $S(\beta)=\operatorname{plim} T^{-1}S_T(\beta)$. Then, given $\varepsilon, \delta>0$, there exists T_0, independent of β, such that for all $T \geqq T_0$ and for all β, $P[|T^{-1}S_T(\beta)-S(\beta)|>\varepsilon]<\delta$.

It is easy to construct examples in which the violation of any single assumption above leads to the inconsistency of $\hat{\beta}$. [See Amemiya (1980).]

I will now give a sketch of the proof of the consistency and indicate what additional assumptions are needed as I go along. From (2.1) and (2.5), we get

$$\frac{1}{T}S_T(\beta)=\frac{1}{T}\sum u_t^2+\frac{2}{T}\sum[f_t(\beta_0)-f_t(\beta)]u_t+\frac{1}{T}\sum[f_t(\beta_0)-f_t(\beta)]^2$$
$$\equiv A_1+A_2+A_3, \tag{2.9}$$

where $\sum$ means $\sum_{t=1}^{T}$ unless otherwise noted. First, plim $A_1=\sigma_0^2$ by a law of large numbers [see, for example, Kolmogorov Theorem 2, p. 115, in Rao (1973)]. Secondly, for fixed β_0 and β, plim $A_2=0$ follows from the convergence of $T^{-1}\sum[f_t(\beta_0)-f_t(\beta)]^2$ by Chebyshev's inequality:

$$\mathrm{P}\left[\left\{T^{-1}\sum[f_t(\beta_0)-f_t(\beta)]u_t\right\}^2>\varepsilon^2\right]<\frac{\sigma^2}{\varepsilon^2T^2}\sum[f_t(\beta_0)-f_t(\beta)]^2. \tag{2.10}$$

Since the uniform convergence of A_2 follows from the uniform convergence of the right-hand side of (2.10), it suffices to assume

$$\frac{1}{T}\sum f_t(\beta_1)f_t(\beta_2) \text{ converges uniformly in } \beta_1,\beta_2\in B. \tag{2.11}$$

Having thus disposed of A_1 and A_2, we need only to assume that $\lim A_3$ is uniquely minimized at β_0; namely,

$$\lim\frac{1}{T}\sum[f_t(\beta_0)-f_t(\beta)]^2\neq 0 \quad \text{if } \beta\neq\beta_0. \tag{2.12}$$

To sum up, the non-linear least squares estimator $\hat{\beta}$ of the model (2.1) is consistent if (2.6), (2.11), and (2.12) are satisfied. I will comment on the significance and the plausibility of these three assumptions.

The assumption of a compact parameter space (2.6) is convenient but can be rather easily removed. The trick is to dispose of the region outside a certain compact subset of the parameter space by assuming that in that region $T^{-1}\sum[f_t(\beta_0)-f_t(\beta)]^2$ is sufficiently large. This is done by Malinvaud (1970a). An essentially similar argument appears also in Wald (1949) in the proof of the consistency of the maximum likelihood estimator.

It would be nice if assumption (2.11) could be paraphrased into separate assumptions on the functional form of f and on the properties of the exogenous sequence $\{x_t\}$, which are easily verifiable. Several authors have attempted to obtain such assumptions. Jennrich (1969) observes that if f is bounded and continuous, (2.11) is implied by the assumption that the empirical distribution function of $\{x_t\}$ converges to a distribution function. He also notes that another way to satisfy (2.11) is to assume that $\{x_t\}$ are i.i.d. with a distribution function F, and f is bounded uniformly in β by a function which is square integrable with respect to F. Malinvaud (1970a) generalizes the first idea of Jennrich by introducing the concept of weak convergence of measure, whereas Gallant (1977) generalizes the second idea of Jennrich by considering the notion of Cesaro summability. However, it seems to me that the best procedure is to leave (2.11) as it is and try to verify it directly.

The assumption (2.12) is comparable to the familiar assumption in the linear model that $\lim T^{-1}X'X$ exists and is positive definite. It can be easily proved that in the linear model the above assumption is not necessary for the consistency of least squares and it is sufficient to assume $(X'X)^{-1} \to 0$. This observation suggests that assumption (2.12) can be relaxed in an analogous way. One such result can be found in Wu (1981).

2.2.2. *Asymptotic normality*

The asymptotic normality of the NLLS estimator $\hat{\beta}$ is rigorously proved in Jennrich (1969). Again, I will give a sketch of the proof, explaining the required assumptions as I go along, rather than reproducing Jennrich's result in a theorem–proof format.

The asymptotic normality of the NLLS estimator, as in the case of the MLE, can be derived from the following Taylor expansion:

$$\left.\frac{\partial S_T}{\partial \beta}\right|_{\hat{\beta}} = \left.\frac{\partial S_T}{\partial \beta}\right|_{\beta_0} + \left.\frac{\partial^2 S_T}{\partial \beta \partial \beta'}\right|_{\beta^*} (\hat{\beta} - \beta_0), \tag{2.13}$$

where $\partial^2 S_T / \partial \beta \partial \beta'$ is a $K \times K$ matrix of second-order derivatives and β^* lies between $\hat{\beta}$ and β_0. To be able to write down (2.13), we must assume that f_t is twice continuously differentiable with respect to β. Since the left-hand side of (2.13) is zero (because $\hat{\beta}$ minimizes S_T), from (2.13) we obtain:

$$\sqrt{T}(\hat{\beta} - \beta_0) = -\left[\left.\frac{1}{T} \frac{\partial^2 S_T}{\partial \beta \partial \beta'}\right|_{\beta^*} \right]^{-1} \left.\frac{1}{\sqrt{T}} \frac{\partial S_T}{\partial \beta}\right|_{\beta_0}. \tag{2.14}$$

Thus, we are done if we can show that (i) the limit distribution of $\sqrt{T}^{-1}(\partial S_T / \partial \beta)_{\beta_0}$ is normal and (ii) $T^{-1}(\partial^2 S_T / \partial \beta \partial \beta')_{\beta^*}$ converges in probability to a non-singular matrix. We will consider these two statements in turn.

The proof of statement (i) is straightforward. Differentiating (2.5) with respect to β, we obtain:

$$\frac{\partial S_T}{\partial \beta} = -2\sum [y_t - f_t(\beta)]\frac{\partial f_t}{\partial \beta}. \tag{2.15}$$

Evaluating (2.15) at β_0 and dividing it by $\sqrt{T}$, we have:

$$\frac{1}{\sqrt{T}}\left.\frac{\partial S_T}{\partial \beta}\right|_{\beta_0} = -\frac{2}{\sqrt{T}}\sum u_t \left.\frac{\partial f_t}{\partial \beta}\right|_{\beta_0}. \tag{2.16}$$

But it is easy to find the conditions for the asymptotic normality of (2.16) because the summand in the right-hand side is a weighted average of an i.i.d. sequence–the kind encountered in the least squares estimation of a linear model. Therefore, if we assume

$$\lim \frac{1}{T}\sum \left.\frac{\partial f_t}{\partial \beta}\right|_{\beta_0} \left.\frac{\partial f_t}{\partial \beta'}\right|_{\beta_0} (\equiv C) \text{ exists and is non-singular,} \tag{2.17}$$

then

$$\frac{1}{\sqrt{T}}\left.\frac{\partial S_t}{\partial \beta}\right|_{\beta_0} \to N(0, 4\sigma_0^2 C). \tag{2.18}$$

This result can be straightforwardly obtained from the Lindberg–Feller central limit theorem [Rao (1973, p. 128)] or, more directly, from of Anderson (1971, Theorem 2.6.1, p. 23).

Proving (ii) poses a more difficult problem. Write an element of the matrix $T^{-1}(\partial^2 S_T/\partial\beta\partial\beta')_{\beta^*}$ as $h_T(\beta^*)$. One might think that $\operatorname{plim} h_T(\beta^*) = \operatorname{plim} h_T(\beta_0)$ follows from the well-known theorem which says that the probability limit of a continuous function is the function of the probability limit, but the theorem does not apply because h_T is in general a function of an increasing number of random variables $y_1, y_2, \dots, y_T$. But, by a slight modification of lemma 4, p. 1003, of Amemiya (1973b), we can show that if $h_T(\beta)$ converges almost surely to a certain non-stochastic function $h(\beta)$ uniformly in β, then $\operatorname{plim} h_T(\beta^*) = h(\operatorname{plim}\beta^*) = h(\beta_0)$. Differentiating (2.15) again with respect to β and dividing by T yields

$$\frac{1}{T}\frac{\partial^2 S_T}{\partial\beta\partial\beta'} = \frac{2}{T}\sum \frac{\partial f_t}{\partial \beta}\frac{\partial f_t}{\partial \beta'} - \frac{2}{T}\sum [f_t(\beta_0) - f_t(\beta)]\frac{\partial^2 f_t}{\partial\beta\partial\beta'} - \frac{2}{T}\sum u_t \frac{\partial^2 f_t}{\partial\beta\partial\beta'}. \tag{2.19}$$

We must show that each of the three terms in the right-hand side of (2.19) converges almost surely to a non-stochastic function uniformly in β. For this purpose the following assumptions will suffice:

$$\frac{1}{T}\sum \frac{\partial f_t}{\partial \beta}\frac{\partial f_t}{\partial \beta'} \text{ converges uniformly in } \beta \text{ in an open neighborhood of } \beta_0, \tag{2.20}$$

and

$$\frac{1}{T}\sum \left[\frac{\partial^2 f_t}{\partial \beta_i \partial \beta_j}\right]^2 \text{ converges uniformly in } \beta \text{ in an open neighborhood of } \beta_0. \tag{2.21}$$

Then, we obtain;

$$\operatorname{plim}\frac{1}{T}\left.\frac{\partial^2 S_T}{\partial \beta \partial \beta'}\right|_{\beta^*} = 2C. \tag{2.22}$$

Finally, from (2.14), (2.18), and (2.22) we obtain:

$$\sqrt{T}\left(\hat{\beta}-\beta_0\right) \to N\left(0, \sigma_0^2 C^{-1}\right). \tag{2.23}$$

The assumptions we needed in proving (2.23) were (2.17), (2.20), and (2.21) as well as the assumption that $\hat{\beta}$ is consistent.

It is worth pointing out that in the process of proving (2.23) we have in effect shown that we have, asymptotically,

$$\hat{\beta}-\beta_0 \cong (G'G)^{-1}G'u, \tag{2.24}$$

where I have put $G = (\partial f/\partial \beta')_{\beta_0}$, a $T \times K$ matrix. Note that (2.24) exactly holds in the linear case. The practical consequence of the approximation (2.24) is that all the results for the linear regression model are asymptotically valid for the non-linear regression model if we treat G as the regressor matrix. In particular, we can use the usual t and F statistics with an approximate precision, as I will explain more fully in Sections 2.4 and 2.5 below. Since the matrix G depends on the unknown parameters, we must in practice evaluate it at $\hat{\beta}$.

2.3. *Computation*

Since there is in general no explicit formula for the NLLS estimator $\hat{\beta}$, the minimization of (2.5) must usually be carried out by some iterative method. There are two general types of iteration methods: general optimization methods applied to the non-linear least squares problem in particular, and procedures which are specifically designed to cope with the present problem. In this chapter I will discuss two representative methods – the Newton–Raphson iteration which belongs to the first type and the Gauss–Newton iteration which belongs to the second type – and a few major variants of each method. These cover a majority of the iterative methods currently used in econometric applications. Although not discussed here, I should mention another method sometimes used in econometric applications, namely the so-called conjugate gradient method of Powell (1964), which does not require the calculation of derivatives and is based on a different principle from the Newton methods. Much more detailed discussion of these and other methods can be found in Chapter 12 of this Handbook and in Goldfeld and Quandt (1972, ch. 1).

2.3.1. Newton–Raphson iteration

The Newton–Raphson method is based on the following quadratic approximation of a minimand (it also works for a maximand):

$$S_T(\beta) \cong S_T(\hat{\beta}_1) + \left.\frac{\partial S_T}{\partial \beta'}\right|_{\hat{\beta}_1}(\beta - \hat{\beta}_1) + \tfrac{1}{2}(\beta - \hat{\beta}_1)' \left.\frac{\partial^2 S_T}{\partial \beta \partial \beta'}\right|_{\hat{\beta}_1}(\beta - \hat{\beta}_1), \tag{2.25}$$

where $\hat{\beta}_1$ is the initial estimate [obtained by a pure guess or by a method such as the one proposed by Hartley and Booker (1965) described below]. The second-round estimator $\hat{\beta}_2$ of the iteration is obtained by minimizing the right-hand side of (2.25). Therefore,

$$\hat{\beta}_2 = \hat{\beta}_1 - \left[\left.\frac{\partial^2 S_T}{\partial \beta \partial \beta'}\right|_{\hat{\beta}_1}\right]^{-1} \left.\frac{\partial S_T}{\partial \beta}\right|_{\hat{\beta}_1}. \tag{2.26}$$

The iteration is to be repeated until the sequence $\{\hat{\beta}_n\}$ thus obtained converges to the desired degree of accuracy.

Inserting (2.26) into (2.25) and writing $n+1$ and n for 2 and 1, we obtain:

$$S_T(\hat{\beta}_{n+1}) \simeq S_T(\hat{\beta}_n) - \tfrac{1}{2}(\hat{\beta}_{n+1} - \hat{\beta}_n)' \left.\frac{\partial^2 S_T}{\partial \beta \partial \beta'}\right|_{\hat{\beta}_n}(\hat{\beta}_{n+1} - \hat{\beta}_n). \tag{2.27}$$

The above equation shows two weaknesses of the Newton–Raphson iteration. (i) Even if (2.27) holds exactly, $S_T(\hat{\beta}_{n+1}) < S_T(\hat{\beta}_n)$ is not guaranteed unless $(\partial^2 S_T/\partial \beta \partial \beta')_{\hat{\beta}_n}$ is a positive definite matrix. (ii) Even if the matrix is positive definite, $\hat{\beta}_{n+1}-\hat{\beta}_n$ may be too large or too small – if it is too large, it overshoots the target, and if it is too small, the speed of convergence is slow.

The first weakness may be alleviated if we modify (2.26) as

$$\hat{\beta}_{n+1}=\hat{\beta}_n-\left[\left.\frac{\partial^2 S_T}{\partial\beta\partial\beta'}\right|_{\hat{\beta}_n}+\alpha_n I\right]^{-1}\left.\frac{\partial S_T}{\partial\beta}\right|_{\hat{\beta}_n}, \tag{2.28}$$

where I is the identity matrix and α_n is a scalar to be appropriately chosen by the researcher subject to the condition that $(\partial^2 S_T/\partial\beta\partial\beta')_{\hat{\beta}_n}+\alpha_n I$ is positive definite. This modification was proposed by Goldfeld, Quandt and Trotter (1966) and is called *quadratic hill-climbing* (since they were considering maximization). See the same article or Goldfeld and Quandt (1972, ch. 1) for a discussion of how to choose α_n and the convergence properties of the method.

The second weakness may be remedied by the modification:

$$\hat{\beta}_{n+1}=\hat{\beta}_n-\lambda_n\left[\left.\frac{\partial^2 S_T}{\partial\beta\partial\beta'}\right|_{\hat{\beta}_n}\right]^{-1}\left.\frac{\partial S_T}{\partial\beta}\right|_{\hat{\beta}_n}, \tag{2.29}$$

where the scalar λ_n is to be appropriately determined. See Fletcher and Powell (1963) for a method to determine λ_n by a cubic interpolation of $S_T(\beta)$ along the current search direction. [This method is called the DFP iteration since Fletcher and Powell refined the method originally proposed by Davidon (1959).] Also, see Berndt, Hall, Hall and Hausman (1974) for another method to choose λ_n.

Ordinarily, the iteration (2.26) is to be repeated until convergence takes place. However, if $\hat{\beta}_1$ is a consistent estimator of β_0 such that $\sqrt{T}(\hat{\beta}_1-\beta_0)$ has a proper limit distribution, the second-round estimator $\hat{\beta}_2$ has the same asymptotic distribution as $\hat{\beta}$. In this case, a further iteration does not bring any improvement so far as the asymptotic distribution is concerned. This is shown below.

By a Taylor expansion of $(\partial S_T/\partial\beta)_{\hat{\beta}_1}$ around β_0, we obtain:

$$\left.\frac{\partial S_T}{\partial\beta}\right|_{\hat{\beta}_1}=\left.\frac{\partial S_T}{\partial\beta}\right|_{\beta_0}+\left.\frac{\partial^2 S_T}{\partial\beta\partial\beta'}\right|_{\beta^*}(\hat{\beta}_1-\beta_0), \tag{2.30}$$

where β^* lies between $\hat{\beta}_1$ and β_0. Inserting (2.30) into (2.26) yields

$$\begin{aligned}\sqrt{T}(\hat{\beta}_2-\beta_0)=&\left\{I-\left[\left.\frac{\partial^2 S_T}{\partial\beta\partial\beta'}\right|_{\hat{\beta}_1}\right]^{-1}\left.\frac{\partial^2 S_T}{\partial\beta\partial\beta'}\right|_{\beta^*}\right\}\sqrt{T}(\hat{\beta}_1-\beta_0)\\&-\left[\frac{1}{T}\left.\frac{\partial^2 S_T}{\partial\beta\partial\beta'}\right|_{\hat{\beta}_1}\right]^{-1}\frac{1}{\sqrt{T}}\left.\frac{\partial S_T}{\partial\beta}\right|_{\beta_0}.\end{aligned} \tag{2.31}$$

But, under the assumptions of Section 2.2 from which we proved the asymptotic normality of $\hat{\beta}$, we have

$$\operatorname{plim}\frac{1}{T}\left.\frac{\partial^2 S_T}{\partial\beta\partial\beta'}\right|_{\hat{\beta}_1}=\operatorname{plim}\frac{1}{T}\left.\frac{\partial^2 S_T}{\partial\beta\partial\beta'}\right|_{\beta^*}=\operatorname{plim}\frac{1}{T}\left.\frac{\partial^2 S_T}{\partial\beta\partial\beta'}\right|_{\beta_0}. \tag{2.32}$$

Therefore,

$$\begin{aligned}\sqrt{T}\left(\hat{\beta}_2-\beta_0\right) &\overset{\text{LD}}{=} -\left[\operatorname{plim}\frac{1}{T}\left.\frac{\partial^2 S_T}{\partial\beta\partial\beta'}\right|_{\beta_0}\right]^{-1}\frac{1}{\sqrt{T}}\left.\frac{\partial S_T}{\partial\beta}\right|_{\beta_0}\\ &\overset{\text{LD}}{=} \sqrt{T}\left(\hat{\beta}-\beta_0\right),\end{aligned} \tag{2.33}$$

where $\overset{\text{LD}}{=}$ means that both sides of the equation have the same non-degenerate limit distribution.

To start an iteration, we need an initial estimate. Since there may be more than one local minima in S_T, it is helpful to use the starting value as close to the true value as possible. Thus, it would be desirable to have available an easily computable good estimator, such as $\hat{\beta}_1$; all the better if it is consistent so that we can take advantage of the result of the preceding paragraph. Surprisingly, I know only one such estimator – the one proposed by Hartley and Booker (1965). Their initial estimator is obtained as follows. Let us assume for simplicity $mK=T$ for some integer m and partition the set of integers $(1,2,\dots,T)$ into K non-overlapping consecutive subsets $\Psi_1,\Psi_2,\dots,\Psi_K$, each of which contains m elements. If we define $\bar{y}_{(i)}=m^{-1}\Sigma_{t\in\Psi_i}y_t$ and $\bar{f}_{(i)}(\beta)=m^{-1}\Sigma_{t\in\Psi_i}f_t(\beta)$, $i=1,2,\dots,K$, the Harley–Booker estimator is defined as the value of β that satisfies K equations:

$$\bar{y}_{(i)}=\bar{f}_{(i)}(\beta),\qquad i=1,2,\dots,K. \tag{2.34}$$

Since (2.34) cannot generally be solved explicitly for β, one still needs an iteration to solve it. Hartley and Booker propose the minimization of $\Sigma_{i=1}^{m}[\bar{y}_{(i)}-\bar{f}_{(i)}(\beta)]^2$ by an iterative method, such as one of the methods being discussed in this section. This minimization is at least simpler than the original minimization of (2.5) because the knowledge that the minimand is zero at $\beta=\beta_0$ is useful. However, if there are multiple solutions to (2.34), an iteration may lead to the wrong solution.

Hartley and Booker proved the consistency of their estimator. Jennrich (1969) gave a counterexample to their consistency proof; however, their proof can easily be modified to take account of Jennrich's counter-example. A more serious weakness of the Hartley–Booker proof is that their assumptions are too restrictive: one can easily construct a benign example for which their assumptions are violated and yet their estimator is consistent.

Gallant (1975a) suggested a simpler variation of the Hartley–Booker idea: just select K observations appropriately and solve them for β. This estimator is simpler to compute, but inconsistent. Nevertheless, one may obtain a good starting value by this method, as Gallant's example shows.

2.3.2. Gauss–Newton iteration

This is the method specifically designed to calculate the NLLS estimator. Expanding $f_t(\beta)$ in a Taylor series around the initial estimate $\hat{\beta}_1$, we get:

$$f_t(\beta) \cong f_t(\hat{\beta}_1) + \left.\frac{\partial f_t}{\partial \beta'}\right|_{\hat{\beta}_1} (\beta - \hat{\beta}_1). \tag{2.35}$$

Substituting the right-hand side of (2.35) for $f_t(\beta)$ in (2.5) yields

$$S_T(\beta) \cong \sum \left[y_t - f_t(\hat{\beta}_1) - \left.\frac{\partial f_t}{\partial \beta'}\right|_{\hat{\beta}_1} (\beta - \hat{\beta}_1) \right]^2. \tag{2.36}$$

The second-round estimator $\hat{\beta}_2$ of the Gauss–Newton iteration is obtained by minimizing the right-hand side of (2.36) with respect to β. Thus,

$$\hat{\beta}_2 = \hat{\beta}_1 - \left[\sum \left.\frac{\partial f_t}{\partial \beta}\right|_{\hat{\beta}_1} \left.\frac{\partial f_t}{\partial \beta'}\right|_{\hat{\beta}_1} \right]^{-1} \sum [y_t - f_t(\hat{\beta}_1)] \left.\frac{\partial f_t}{\partial \beta}\right|_{\hat{\beta}_1}. \tag{2.37}$$

The iteration is to be repeated until convergence is obtained. By an argument similar to the one I used in proving (2.33), we can prove that the asymptotic distribution of $\hat{\beta}_2$ defined in (2.37) is the same as that of $\hat{\beta}$ if we use a consistent estimator (such as the Hartley–Booker estimator) to start this iteration. An advantage of the Gauss–Newton iteration over the Newton–Raphson iteration is that the former requires only the first derivatives of f_t.

The Gauss–Newton iteration may be alternatively motivated as follows. Evaluating the approximation (2.35) at β_0 and inserting it into eq. (2.1) yields

$$y_t - f_t(\hat{\beta}_1) + \left.\frac{\partial f_t}{\partial \beta'}\right|_{\hat{\beta}_1} \hat{\beta}_1 \cong \left.\frac{\partial f_t}{\partial \beta'}\right|_{\hat{\beta}_1} \beta_0 + u_t. \tag{2.38}$$

Then, the second-round estimator $\hat{\beta}_2$ can be obtained as the least squares estimator of β_0 applied to the linear regression equation (2.38), where the whole left-hand side is treated as the dependent variable and $(\partial f_t / \partial \beta')_{\hat{\beta}_1}$ as the vector of independent variables. Eq. (2.38) reminds us of the point raised above: namely, the non-linear regression model asymptotically behaves like the linear regression model if we treat $(\partial f / \partial \beta')_{\hat{\beta}}$ as the regressor matrix.

The Gauss–Newton iteration suffers from weaknesses similar to those of the Newton–Raphson iteration: namely, the possibility of a total or near singularity of the matrix to be inverted in (2.37), and the possibility of too much or too little change from $\hat{\beta}_n$ to $\hat{\beta}_{n+1}$.

In order to deal with the first weakness, Marquardt (1963) proposed a modification:

$$\hat{\beta}_{n+1} = \hat{\beta}_n - \left[\sum \frac{\partial f_t}{\partial \beta}\bigg|_{\hat{\beta}_n} \frac{\partial f_t}{\partial \beta'}\bigg|_{\hat{\beta}_n} + \alpha_n I \right]^{-1} \sum [y_t - f_t(\hat{\beta}_n)] \frac{\partial f_t}{\partial \beta}\bigg|_{\hat{\beta}_n}, \tag{2.39}$$

where α_n is a positive scalar to be appropriately determined by a rule based on the past behavior of the algorithm.

In order to deal with the second weakness, Hartley (1961) proposed the following modification. First, calculate

$$\Delta_n = -\left[\sum \frac{\partial f_t}{\partial \beta}\bigg|_{\hat{\beta}_n} \frac{\partial f_t}{\partial \beta'}\bigg|_{\hat{\beta}_n} \right]^{-1} \sum [y_t - f_t(\hat{\beta}_n)] \frac{\partial f_t}{\partial \beta}\bigg|_{\hat{\beta}_n}, \tag{2.40}$$

and, secondly, choose λ_n so as to minimize

$$S_T(\hat{\beta}_n + \lambda_n \Delta_n), \qquad 0 \leqq \lambda_n \leqq 1. \tag{2.41}$$

Hartley proves that under general conditions his iteration converges to a stationary point: that is, a root of the normal equation $\partial S_T / \partial \beta = 0$. He also proves (not so surprisingly) that if the iteration is started at a point sufficiently close to $\hat{\beta}$, it converges to $\hat{\beta}$. See Tornheim (1963) for an alternative proof of the convergence of the Hartley iteration. Some useful comments on Marquardt's and Hartley's algorithms can be found in Gallant (1975a). The methods of determining λ_n in the Newton–Raphson iteration (2.29) mentioned above can be also applied to the determination of λ_n in (2.41).

Jennrich (1969) proves that if the Gauss–Newton iteration is started at a point sufficiently close to the true value β_0, and if the sample size T is sufficiently large, the iteration converges to β_0. This is called the asymptotic stability of the iteration. The following is a brief sketch of Jennrich's proof. Rewrite the Gauss-Newton iteration (2.37) as (I have also changed 1 to n and 2 to $n+1$ in the subscript)

$$\hat{\beta}_{n+1} = h(\hat{\beta}_n), \tag{2.42}$$

where h is a vector-valued function implicitly defined by (2.37). By a Taylor expansion:

$$\hat{\beta}_{n+1} - \hat{\beta}_n = \left.\frac{\partial h}{\partial \beta'}\right|_{\beta^*_{n-1}} (\hat{\beta}_n - \hat{\beta}_{n-1}), \tag{2.43}$$

where β^*_{n-1} lies between $\hat{\beta}_n$ and $\hat{\beta}_{n-1}$. If we define $A_n = (\partial h/\partial \beta')_{\beta^*_n}$ and denote the largest characteristic root of $A'_n A_n$ by λ_n, we can show that $A_n \to 0$ almost surely for all n as $T \to \infty$ and hence

$$\lambda_n \to 0 \text{ almost surely for all } n \text{ as } T \to \infty. \tag{2.44}$$

But (2.44) implies two facts. First, the iteration converges to a stationary point, and secondly, this stationary point must lie sufficiently close to the starting value $\hat{\beta}_1$ since

$$(\hat{\beta}_n - \hat{\beta}_1)'(\hat{\beta}_n - \hat{\beta}_1) \leqq \delta'\delta(1 + \lambda_1 + \lambda_1\lambda_2 + \cdots + \lambda_1\lambda_2 \cdots \lambda_{n-2}), \tag{2.45}$$

where $\delta = \hat{\beta}_2 - \hat{\beta}_1$. Therefore, this stationary point must be $\hat{\beta}$ if $\hat{\beta}_1$ is within a neighborhood of β_0 and if $\hat{\beta}$ is the unique stationary point in the same neighborhood.

In closing this section I will mention several empirical papers in which the above-mentioned and related iterative methods are used. Bodkin and Klein (1967) estimated the Cobb–Douglas (2.2) and the CES (2.3) production functions by the Newton–Raphson method. Charatsis (1971) estimated the CES production function by a modification of the Gauss–Newton method similar to that of Hartley (1961) and showed that in 64 samples out of 74, it converged in six iterations. Mizon (1977), in a paper the major aim of which was to choose among nine production functions, including the Cobb–Douglas and CES, used the conjugate gradient method of Powell (1964). Mizon's article is a useful compendium on the econometric application of various statistical techniques such as sequential testing, Cox's test of separate families of hypotheses [Cox (1961, 1962)], the Akaike Information Criterion [Akaike (1973)], the Box–Cox transformation [Box and Cox (1964)], and comparison of the likelihood ratio, Wald, and Lagrange multiplier tests (see the end of Section 2.4 below). Sargent (1978) estimates a rational expectations model (which gives rise to non-linear constraints among parameters) by the DFP algorithm mentioned above.

2.4. *Tests of hypotheses*

In this section I consider tests of hypotheses on the regression parameters β. It is useful to classify situations into four cases depending on the nature of the

Table 2.1 Four cases of hypotheses tests

Hypothesis \ Error distribution	Normal	Non-normal
Linear	I	II
Non-linear	III	IV

hypotheses and the distribution of the error term as depicted in Table 2.1. I will discuss the t and F tests in Case I and the likelihood ratio, Wald, and Rao tests in Case IV. I will not discuss Cases II and III because the results in Case IV are *a fortiori* valid in Cases II and III.

2.4.1. Linear hypotheses under normality

Partition the parameter vector as $\beta' = (\beta'_{(1)}, \beta'_{(2)})$, where $\beta_{(1)}$ is a K_1-vector and $\beta_{(2)}$ is a K_2-vector. By a linear hypothesis I mean a hypothesis which specifies that $\beta_{(2)}$ is equal to a certain known value $\overline{\beta}_{(2)}$. Student's t test is applicable if $K_2 = 1$ and the F test if $K_2 > 1$.

The hypothesis of the form $Q\beta = c$, where Q is a known $K_2 \times K$ matrix and c is a known K_2-vector, can be transformed into a hypothesis of the form described above and therefore need not be separately considered. Assuming Q is full rank, we can find a $K_1 \times K$ matrix R such that $(R', Q') \equiv A'$ is non-singular. If we define $\alpha = A\beta$ and partition $\alpha' = (\alpha'_{(1)}, \alpha'_{(2)})$, the hypothesis $Q\beta = c$ is equivalent to the hypothesis $\alpha_{(2)} = c$.

As noted after eq. (2.24), all the results of the linear regression model can be extended to the non-linear model by treating $G = (\partial f/\partial \beta')_{\beta_0}$ as the regressor matrix if the assumptions of Section 2.2 are satisfied. Since β_0 is unknown, we must use $\hat{G} = (\partial f/\partial \beta')_{\hat{\beta}}$ in practice. We will generalize the t and F statistics of the linear model by this principle. If $K_2 = 1$, we have approximately

$$\frac{\sqrt{T-K}\left(\hat{\beta}_{(2)} - \overline{\beta}_{(2)}\right)}{\sqrt{S_T(\hat{\beta})\hat{c}}} \sim t(T-K), \tag{2.46}$$

where $\hat{c}$ is the last diagonal element (if $\beta_{(2)}$ is the ith element of β, the ith diagonal element) of $(\hat{G}'\hat{G})^{-1}$ and $t(T-K)$ denotes Student's t distribution with $T-K$ degrees of freedom. For the case $K_2 \geqq 1$, we have asymptotically under the null hypothesis:

$$\frac{(T-K)\left(\hat{\beta}_{(2)} - \overline{\beta}_{(2)}\right)'\left[J'(\hat{G}'\hat{G})^{-1}J\right]^{-1}\left(\hat{\beta}_{(2)} - \overline{\beta}_{(2)}\right)}{K_2 S_T(\hat{\beta})} \sim F(K_2, T-K), \tag{2.47}$$

where $J' = (0, I)$, 0 being the $K_2 \times K_1$ matrix of zeros and I being the identity matrix of size K_1, and $F(K_2, T-K)$ denotes the F distribution with K_2 and $T-K$ degrees of freedom.

Gallant (1975a) examined the accuracy of the approximation (2.46) by a Monte Carlo experiment using the model

$$f(x_t, \beta) = \beta_1 x_{1t} + \beta_2 x_{2t} + \beta_4 e^{\beta_3 x_{3t}}. \tag{2.48}$$

For each of the four parameters, the empirical distribution of the left-hand side of (2.46) matched the distribution of $t(T-K)$ reasonably well, although, as we would suspect, the performance was the poorest for $\hat{\beta}_3$.

In testing $\beta_{(2)} = \bar{\beta}_{(2)}$ when $K_2 \geqq 1$, we may alternatively use the asymptotic approximation (under the null hypothesis):

$$\frac{(T-K)[S_T(\tilde{\beta}) - S_T(\hat{\beta})]}{K_2 S_T(\hat{\beta})} \sim F(K_2, T-K), \tag{2.49}$$

where $\tilde{\beta}$ is the constrained non-linear least squares estimator obtained by minimizing $S_T(\beta)$ subject to $\beta_{(2)} = \bar{\beta}_{(2)}$. Although, as is well known, the statistics (2.47) and (2.49) are identical in the linear model, they are different in the non-linear model.

The study of Gallant (1975c) sheds some light on the choice between (2.47) and (2.49). He obtained the asymptotic distribution of the statistics (2.47) and (2.49) under the alternative hypothesis as follows. Regarding $S_T(\hat{\beta})$, which appears in both formulae, we have asymptotically:

$$S_T(\hat{\beta}) \cong u'\left[I - G(G'G)^{-1}G'\right]u, \tag{2.50}$$

where $G = (\partial f/\partial \beta')_{\beta_0}$ as before. Define $G_1 = (\partial f/\partial \beta'_{(1)})_{\beta_0}$. Then, Gallant shows (asymptotically) that

$$S_T(\tilde{\beta}) \cong (u+\delta)'\left[I - G_1(G_1'G_1)^{-1}G_1'\right](u+\delta), \tag{2.51}$$

where $\delta = f(\beta_0) - f(\beta^*_{(1)}, \bar{\beta}_{(2)})$ in which $\beta^*_{(1)}$ is the value of $\beta_{(1)}$ that minimizes $\| f(\beta_0) - f(\beta_{(1)}, \bar{\beta}_{(2)})\|^2$. ($\|x\|^2 \equiv x'x$ for any vector x.) He also shows

$$\begin{aligned}(\hat{\beta}_{(2)} - \bar{\beta}_{(2)})'\left[J'(\hat{G}'\hat{G})^{-1}J\right]^{-1}(\hat{\beta}_{(2)} - \bar{\beta}_{(2)}) \\ \cong \left[J'(G'G)^{-1}G'u + \beta_{(2)0} - \bar{\beta}_{(2)}\right]'\left[J'(G'G)^{-1}J\right]^{-1} \\ \times \left[J'(G'G)^{-1}G'u + \beta_{(2)0} - \bar{\beta}_{(2)}\right],\end{aligned} \tag{2.52}$$

where $\beta_{(2)0}$ is the true value of $\beta_{(2)}$.[1] The asymptotic distribution of the statistic (2.47) under the alternative hypothesis can now be derived from (2.50) and (2.52) and, similarly, that of (2.49) from (2.50) and (2.51).

Gallant (1975c) conducted a Monte Carlo study using the model (2.48) to compare the above two tests in testing $\beta_1 = 0$ against $\beta_1 \neq 0$ and $\beta_3 = -1$ against $\beta_3 \neq -1$. His results show that (i) the asymptotic approximation under the alternative hypothesis matches the empirical distribution reasonably well for both statistics but works a little better for the statistic (2.49) and (ii) the power of (2.49) tends to be higher than that of (2.47).[2] Gallant (1975a) observes that (2.49) is easier to calculate than (2.47) except when $K_2 = 1$. All these observations indicate a preference for (2.49) over (2.47). See Gallant (1975b) for a tabulation of the power function of the test based on $S_T(\beta^+)/S_T(\hat{\beta})$, which is equivalent to the test based on (2.49).

2.4.2. *Non-linear hypotheses under non-normality*

Now I consider the test of a non-linear hypothesis

$$h(\beta) = 0, \tag{2.53}$$

where h is a q-vector valued non-linear function such that $q < K$.

If β are the parameters that characterize a concentrated likelihood function $L(\beta)$, where L may or may not be derived from the normal distribution, we can test the hypothesis (2.53) using one of the following well-known test statistics: the likelihood ratio test (LRT), Wald's test [Wald (1943)], or Rao's test [Rao (1947)]:

$$\text{LRT} = 2\left[\log L(\hat{\beta}) - \log L(\tilde{\beta})\right], \tag{2.54}$$

$$\text{Wald} = -h(\hat{\beta})'\left\{ \left.\frac{\partial h}{\partial \beta'}\right|_{\hat{\beta}} \left[\left.\frac{\partial^2 \log L}{\partial \beta \partial \beta'}\right|_{\hat{\beta}} \right]^{-1} \left.\frac{\partial h'}{\partial \beta}\right|_{\hat{\beta}} \right\}^{-1} h(\hat{\beta}), \tag{2.55}$$

and

$$\text{Rao} = -\left.\frac{\partial \log L}{\partial \beta'}\right|_{\tilde{\beta}} \left[\left.\frac{\partial^2 \log L}{\partial \beta \partial \beta'}\right|_{\tilde{\beta}} \right]^{-1} \left.\frac{\partial \log L}{\partial \beta}\right|_{\tilde{\beta}}, \tag{2.56}$$

[1] In deriving the asymptotic approximations (2.51) and (2.52), Gallant assumes that the "distance" between the null and alternative hypotheses is sufficiently small. More precisely, he assumes that there exists a sequence of hypothesized values $\{\bar{\beta}_{(2)}^T\}$ and hence a sequence $\{\beta_{(1)}^{*T}\}$ such that $\sqrt{T}(\beta_{(2)0} - \bar{\beta}_{(2)}^T)$ and $\sqrt{T}(\beta_{(1)0} - \beta_{(1)}^{*T})$ converge to constant vectors as T goes to infinity.

[2] Actually, the powers of the two tests calculated either from the approximation or from the empirical distribution are identical in testing $\beta_1 = 0$. They differ only in the test of $\beta_3 = -1$.

where $\hat{\beta}$ is the unconstrained maximum likelihood estimator and $\tilde{\beta}$ is the constrained maximum likelihood estimator obtained maximizing $L(\beta)$ subject to (2.53).[3] By a slight modification of the proof of Rao (1973) (a modification is necessary since Rao deals with a likelihood function rather than a concentrated likelihood function), it can be shown that all the three test statistics have the same limit distribution – $\chi^2(q)$, chi-square with q degrees of freedom. For more discussion of these tests, see Chapter 13 of this Handbook by Engle.

Gallant and Holly (1980) obtained the asymptotic distribution of the three statistics under an alternative hypothesis in a non-linear simultaneous equations model. Translated into the present simpler model, their results can be stated as follows. As in Gallant (1975c) (see footnote 1), they assume that the "distance" between the null hypothesis and the alternative hypothesis is small: or, more precisely, that there exists a sequence of true values $\{\beta_0^T\}$ such that $\delta \equiv \lim\sqrt{T}(\beta_0^T - \beta_0)$ is finite and $h(\beta_0)=0$. Then, statistics (2.54), (2.55), and (2.56) converge to $\chi^2(q,\lambda)$, chi-square with q degrees of freedom and the noncentrality parameter λ,[4] where

$$\lambda = \frac{1}{\sigma_0^2 T}\delta' \left.\frac{\partial h'}{\partial \beta}\right|_{\beta_0} \left\{ \left.\frac{\partial h}{\partial \beta}\right|_{\beta_0} \left[\left.\frac{\partial^2 \log L}{\partial\beta\partial\beta'}\right|_{\beta_0}\right]^{-1} \left.\frac{\partial h'}{\partial \beta}\right|_{\beta_0} \right\}^{-1} \left.\frac{\partial h}{\partial \beta'}\right|_{\beta_0} \delta. \tag{2.57}$$

If we assume the normality of u in the non-linear regression model (2.1), we can write (2.54), (2.55), and (2.56) as[5]

$$\text{LRT} = T\left[\log T^{-1}S_T(\tilde{\beta}) - \log T^{-1}S_T(\hat{\beta})\right], \tag{2.58}$$

$$\text{Wald} = \frac{h(\hat{\beta})'\left[\left.\frac{\partial h}{\partial \beta'}\right|_{\hat{\beta}} (\hat{G}'\hat{G})^{-1} \left.\frac{\partial h'}{\partial \beta}\right|_{\hat{\beta}} \right]^{-1} h(\hat{\beta})}{S_T(\hat{\beta})/T}, \tag{2.59}$$

and

$$\text{Rao} = \frac{[y - f(\tilde{\beta})]'\tilde{G}(\tilde{G}'\tilde{G})^{-1}\tilde{G}'[y - f(\tilde{\beta})]}{S_T(\tilde{\beta})/T}, \tag{2.60}$$

where $G = (\partial f/\partial \beta')_{\tilde{\beta}}$. Since (2.58)–(2.60) are special cases of (2.54)–(2.56), all these statistics are asymptotically distributed as $\chi^2(q)$ if u are normal. However,

[3] See Silvey (1959) for an interpretation of Rao's test as a test on Lagrange multipliers.

[4] If ξ is distributed as a q-vector $N(0, V)$, then $(\xi + \mu)'V^{-1}(\xi + \mu) \sim \chi^2(q, \mu'V^{-1}\mu)$.

[5] In the following derivation I have omitted some terms whose probability limit is zero in evaluating $\sqrt{T}^{-1}(\partial \log L/\partial\beta')$ and $T^{-1}(\partial^2 \log L/\partial\beta\partial\beta')$.

using a proof similar to Rao's, we can show that the statistics (2.58), (2.59), and (2.60) are asymptotically distributed as $\chi^2(q)$ even if u are not normal. Thus, these statistics can be used to test a non-linear hypothesis under a non-normal situation.

In the linear regression model we can show Wald $\geqq$ LRT $\geqq$ Rao [see Berndt and Savin (1977)]. Although the inequalities do not exactly hold for the non-linear model, Mizon (1977) found Wald $\geqq$ LRT most of the time in his samples.

2.5. *Confidence regions*

Confidence regions on the parameter vector β or its subset can be constructed using any of the test statistics considered in the preceding section. In this section I discuss some of these as well as other methods of constructing confidence regions.

A $100\times(1-\alpha)$ percent confidence interval on an element of β can be obtained from (2.46) as

$$\frac{\sqrt{T-K}\,|\hat{\beta}_{(2)}-\beta_{(2)}|}{\sqrt{S_T(\hat{\beta})\hat{c}}}<t_{\alpha/2}(T-K), \tag{2.61}$$

where $t_{\alpha/2}(T-K)$ is the $\alpha/2$ critical value of $t(T-K)$.

A confidence region $-100\times(1-\alpha)$ percent throughout this section – on the whole vector β can be constructed using either (2.47) or (2.49). If we use (2.47) we obtain:

$$\frac{(T-K)(\hat{\beta}-\beta)'\hat{G}'\hat{G}(\hat{\beta}-\beta)}{KS_T(\hat{\beta})}<F_\alpha(K,T-K), \tag{2.62}$$

and if we use (2.49) we obtain:

$$\frac{(T-K)[S_T(\beta)-S_T(\hat{\beta})]}{KS_T(\hat{\beta})}<F_\alpha(K,T-K). \tag{2.63}$$

Goldfeld and Quandt (1972, p. 53) give a striking example in which the two regions defined by (2.62) and (2.63) differ markedly, even though both statistics have the same asymptotic distribution – $F(K,T-K)$. I have not come across any reference discussing the comparative merits of the two methods.

Beale (1960) shows that the confidence region based on (2.63) gives an accurate result – that is, the distribution of the left-hand side of (2.63) is close to $F(K,T-K)$ – if the "non-linearity" of the model is small. He defines a measure of non-linearity as

$$\hat{N}=\sum_{i=1}^{m}\sum_{t=1}^{T}\left[f_t(b_i)-f_t(\hat{\beta})-\left.\frac{\partial f_t}{\partial \beta'}\right|_{\hat{\beta}}(b_i-\hat{\beta})\right]^2\cdot K(T-K)^{-1}s_T(\hat{\beta})$$

$$\times\left\{\sum_{i=1}^{m}\left[\sum_{t=1}^{T}\left[f_t(\beta_i)-f_t(\hat{\beta})\right]^2\right]^2\right\}^{-1}, \tag{2.64}$$

where $b_1, b_2, \ldots, b_m$ are m arbitrarily chosen K-vectors of constants, and states that (2.63) gives a good result if $\hat{N}F_\alpha(K,T-K)<0.01$, but unsatisfactory if $\hat{N}F_\alpha(K,T-K)>1$. Guttman and Meeter (1965), on the basis of their experience in applying Beale's measure of non-linearity to real data, observe that $\hat{N}$ is a useful measure if the degree of "true non-linearity" (which can be measured by the population counterpart of Beale's measure) is small. Also, see Bates and Watts (1980) for a further development.

The standard confidence ellipsoid in the linear regression model can be written as

$$\frac{(T-K)(y-X\beta)'X(X'X)^{-1}X'(y-X\beta)}{K(y-X\beta)'\left[I-X(X'X)^{-1}X'\right](y-X\beta)}<F_\alpha(K,T-K). \tag{2.65}$$

Note that β actually drops out of the denominator of (2.65), which makes the computation of the confidence region simple in this case. In analogy to (2.65), Hartley (1964) proposed the following confidence region:

$$\frac{(T-K)(y-f)'Z(Z'Z)^{-1}Z'(y-f)}{K(y-f)'\left[I-Z(Z'Z)^{-1}Z'\right](y-f)}<F_\alpha(K,T-K), \tag{2.66}$$

where Z is an appropriately chosen $T\times K$ matrix of constants with rank K. The computation of (2.66) is more difficult than that of (2.65) because β appears in both the numerator and denominator of (2.66). In a simple model where $f_t(\beta)=\beta_1+\beta_3 e^{\beta_2 x_t}$, Hartley suggests choosing Z such that its tth row is equal to $(1, x_t, x_t^2)$. This suggestion may be extended to a general recommendation that we should choose the column vectors of Z to be those independent variables which we believe best approximate G. Although the distribution of the left-hand side of (2.66) is exactly $F(K,T-K)$ for any Z under the null hypothesis, its power depends crucially on the choice of Z.

3. Single equation–non-i.i.d. case

3.1. Autocorrelated errors

In this section we consider the non-linear regression model (2.1) where $\{u_t\}$ follow a general stationary process

$$u_t = \sum_{j=0}^{\infty} \gamma_j \varepsilon_{t-j}, \tag{3.1}$$

where $\{\varepsilon_t\}$ are i.i.d. with $E\varepsilon_t = 0$ and $V\varepsilon_t = \sigma^2$, and the γ's satisfy the condition

$$\sum_{j=0}^{\infty} \gamma_j^2 < \infty, \tag{3.2}$$

and where

$$\text{the spectral density } g(\omega) \text{ of } \{u_t\} \text{ is continuous.} \tag{3.3}$$

I will add whatever assumptions are needed in the course of the subsequent discussion. The variance–covariance matrix Euu' will be denoted by Σ.

I will indicate how to prove the consistency and the asymptotic normality of the non-linear least squares estimator $\hat{\beta}$ in the present model, given the above assumptions as well as the assumptions of Section 2.2. Changing the assumption of independence to autocorrelation poses no more difficulties in the non-linear model than in the linear model.

To prove consistency, we consider (2.9) as before. Since A_1 does not depend on β and A_3 does not depend on u_t, we need to be concerned with only A_2. Since A_2 involves the vector product $f'u$ and since $E(f'u)^2 = f'\Sigma f \leqq f'f\lambda_l(\Sigma)$, where $\lambda_l(\Sigma)$ is the largest characteristic root of Σ, assumption (2.11) implies plim $A_2 = 0$ by Chebyshev's inequality, provided that the characteristic roots of Σ are bounded from above. But this last condition is implied by assumption (3.3).

To prove the asymptotic normality in the present case, we need only prove the asymptotic normality of (2.16) which, just as in the linear model, follows from theorem 10.2.11, page 585, of Anderson (1971) if we assume

$$\sum_{j=0}^{\infty} |\gamma_j| < \infty \tag{3.4}$$

in addition to all the other assumptions. Thus,

$$\sqrt{T}\left(\hat{\beta} - \beta_0\right) \to N\left[0, \sigma_0^2 \lim T^{-1}(G'G)^{-1}G'\Sigma G(G'G)^{-1}\right], \tag{3.5}$$

which indicates that the linear approximation (2.24) works for the autocorrelated model as well. Again it is safe to say that all the results of the linear model are asymptotically valid in the non-linear model. This suggests, for example, that the Durbin–Watson test will be approximately valid in the non-linear model, though this has not been rigorously demonstrated.

Now, let us consider the non-linear analogue of the generalized least squares estimator, which I will call the non-linear generalized least squares (NLGLS) estimator.

Hannan (1971) investigated the asymptotic properties of the class of estimators, denoted by $\beta(\Lambda)$, obtained by minimizing $(y-f)'\Lambda^{-1}(y-f)$ for some Λ, which is the variance–covariance matrix of a stationary process with bounded (both from above and from below) characteristic roots. This class contains the NLLS estimator, $\hat{\beta} \equiv \hat{\beta}(I)$, and the NLGLS estimator, $\hat{\beta}(\Sigma)$.

Hannan actually minimized an approximation of $(y-f)'\Lambda^{-1}(y-f)$ expressed in the frequency domain; therefore, his estimator is analogous to his spectral estimator proposed for the linear regression model [Hannan (1963)]. If we define the periodograms

$$\begin{aligned} I_y(\omega) &= \frac{1}{2\pi T}\left|\sum_t y_t \mathrm{e}^{it\omega}\right|^2, \\ I_f(\omega) &= \frac{1}{2\pi T}\left|\sum_t f_t \mathrm{e}^{it\omega}\right|^2, \\ I_{yf}(\omega) &= \frac{1}{2\pi T}\sum_t y_t \mathrm{e}^{it\omega}\sum_t f_t \mathrm{e}^{-it\omega}, \\ \omega &= 0, \frac{2\pi}{T}, \frac{4\pi}{T}, \dots, \frac{2\pi(T-1)}{T}, \end{aligned} \tag{3.6}$$

we have approximately:

$$(y-f)'\Lambda^{-1}(y-f) \cong \sum_\omega \left[I_y(\omega)+I_f(\omega)-2\mathrm{Re}I_{yf}(\omega)\right]\phi(\omega), \tag{3.7}$$

where $\phi(\omega)$ is the spectral density associated with Λ. This approximation is based on an approximation of Λ by a circular matrix. [See Amemiya and Fuller (1967, p. 527).]

Hannan proves the strong consistency of his non-linear spectral estimator obtained by minimizing the right-hand side of (3.7) under the assumptions (2.6),

(2.12), and the new assumption

$$\frac{1}{T}\sum_t f_t(c_1)f_{(t+s)}(c_2) \text{ converges uniformly in } c_1, c_2 \in B \text{ for every integer } s. \tag{3.8}$$

Note that this is a generalization of the assumption (2.11). However, the assumption (3.8) is merely sufficient and not necessary. Hannan shows that in the model

$$y_t = \alpha_1 + \alpha_2 \cos\beta_0 t + \alpha_3 \sin\beta_0 t + u_t, \tag{3.9}$$

assumption (3.8) does not hold and yet $\hat{\beta}$ is strongly consistent if we assume (3.4) and $0 < \beta_0 < \pi$. In fact, $T(\hat{\beta} - \beta_0)$ converges to zero almost surely in this case.

In proving the asymptotic normality of his estimator, Hannan needs to generalize (2.20) and (2.21) as follows:

$$\frac{1}{T}\sum \left.\frac{\partial f_t}{\partial \beta}\right|_{c_1} \left.\frac{\partial f_{(t+s)}}{\partial \beta'}\right|_{c_2} \text{ converges uniformly in } c_1 \text{ and } c_2$$
$$\text{in an open neighborhood of } \beta_0 \tag{3.10}$$

and

$$\frac{1}{T}\sum \left.\frac{\partial^2 f_t}{\partial \beta_i \partial \beta_j}\right|_{c_1} \left.\frac{\partial^2 f_{(t+s)}}{\partial \beta_i \partial \beta_j}\right|_{c_2} \text{ converges uniformly in } c_1 \text{ and } c_2$$
$$\text{in an open neighborhood of } \beta_0. \tag{3.11}$$

He also needs an assumption comparable to (2.17), namely

$$\lim \frac{1}{T} G'\Lambda^{-1}G(\equiv A) \text{ exists and is non-singular.} \tag{3.12}$$

Using (3.10), (3.11), and (3.12), as well as the assumptions needed for consistency, Hannan proves

$$\sqrt{T}\left[\hat{\beta}(\Lambda) - \beta_0\right] \to N(0, A^{-1}BA^{-1}), \tag{3.13}$$

where $B = \lim T^{-1}G'\Lambda^{-1}\Sigma\Lambda^{-1}G$. If we define a matrix function F by

$$\lim \frac{1}{T}\frac{\partial f_t}{\partial \beta}\frac{\partial f_{t+s}}{\partial \beta'} = \int_{-\pi}^{\pi} \mathrm{e}^{\mathrm{i}s\omega}\,\mathrm{d}F(\omega), \tag{3.14}$$

we can write $A = (2\pi)^{-1}\int_{-\pi}^{\pi} g(\omega)\phi(\omega)^2\,\mathrm{d}F(\omega)$ and $B = (2\pi)^{-1}\int_{-\pi}^{\pi}\phi(\omega)\,\mathrm{d}F(\omega)$.

In the model (3.9), assumptions (3.10) and (3.11) are not satisfied; nevertheless, Hannan shows that the asymptotic normality holds if one assumes (3.4) and $0 < \beta_0 < \pi$. In fact, $\sqrt{T}\,T(\hat{\beta} - \beta_0) \to$ normal in this case.

An interesting practical case is where $\phi(\omega) = \hat{g}(\omega)^{-1}$, where $\hat{g}(\omega)$ is a consistent estimator of $g(\omega)$. I will denote this estimator by $\hat{\beta}(\hat{\Sigma})$. Hannan proves that $\hat{\beta}(\hat{\Sigma})$ and $\hat{\beta}(\Sigma)$ have the same asymptotic distribution if $g(\omega)$ is a rational spectral density.

Gallant and Goebel (1976) propose a NLGLS estimator of the autocorrelated model which is constructed in the time domain, unlike Hannan's spectral estimator. In their method, they try to take account of the autocorrelation of $\{u_t\}$ by fitting the least squares residuals $\hat{u}_t$ to an autoregressive model of a finite order. Thus, their estimator is a non-linear analogue of the generalized least squares estimator analyzed in Amemiya (1973a).

The Gallant–Goebel estimator is calculated in the following steps. (1) Obtain the NLLS estimator $\hat{\beta}$. (2) Calculate $\hat{u} = y - f(\hat{\beta})$. (3) Assume that $\{u_t\}$ follow an autoregressive model of a finite order and estimate the coefficients by the least squares regression of $\hat{u}_t$ on $\hat{u}_{t-1}, \hat{u}_{t-2}, \dots$. (4) Let Σ be the variance–covariance matrix of u obtained under the assumption of an autoregressive model. Then we can find a lower triangular matrix R such that $\Sigma^{-1} = R'R$, where R depends on the coefficients of the autoregressive model.[6] Calculate $\hat{R}$ using the estimates of the coefficients obtained in Step (3) above. (5) Finally, minimize $[\hat{R}(y - f)]'[\hat{R}(y - f)]$ to obtain the Gallant–Goebel estimator.

Gallant and Goebel conducted a Monte Carlo study of the model $y_t = \beta_1 e^{\beta_2 x_t} + u_t$ to compare the performance of the four estimators – the NLLS, the Gallant–Goebel AR1 (based on the assumption of a first-order autoregressive model), the Gallant–Goebel AR2, and Hannan's $\hat{\beta}(\hat{\Sigma})$ – when the true distribution of $\{u_t\}$ is i.i.d., AR1, AR2, or MA4 (a fourth-order moving average process). Their major findings were as follows. (1) The Gallant–Goebel AR2 was not much better than the AR1 version. (2) The Gallant–Goebel estimators performed far better than the NLLS estimator and a little better than Hannan's $\hat{\beta}(\hat{\Sigma})$, even when the true model was MA4 – the situation most favorable to Hannan. They think the reason for this is that in many situations an autoregressive model produces a better approximation of the true autocovariance function than the circular approximation upon which Hannan's spectral estimator is based. They

[6] If we assume a first-order autoregressive model, for example, we obtain:

$$R = \frac{1}{\alpha_\varepsilon^2}\begin{bmatrix} \sqrt{1-\alpha^2} & 0 & \cdot & \cdot & \cdot & 0 \\ -\alpha & 1 & 0 & & & \\ 0 & -\alpha & 1 & & & \\ \cdot & & \cdot & \cdot & & \\ \cdot & & & \cdot & \cdot & \\ 0 & & & & -\alpha & 1 \end{bmatrix}.$$

illustrate this point by approximating the autocovariance function of the U.S. wholesale price index by the two methods. (3) The empirical distribution of the t statistic based on the Gallant-Goebel estimators was reasonably close to the theoretical distribution obtained under the pretense that the assumed model was the true model.

For an application of a non-linear regression model with an autocorrelated error, see Glasbey (1979) who estimated a growth curve for the weight of a steer:

$$y_t = \frac{1}{\beta_1}\left[1+\beta_2 \exp(-\beta_3-\beta_4 t)\right]^{1/\beta_2} + u_t, \tag{3.15}$$

where $\{u_t\}$ follow a first-order autoregressive process, using the maximum likelihood estimator assuming normality.

3.2. *Heteroscedastic errors*

White (1980a) considers a model which differs from the standard non-linear regression model in that $\{x_t\}$ are regarded as vector random variables distributed independently of $\{u_t\}$ and that $\{x_t, u_t\}$ are serially independent but not identically distributed. White is especially interested in the "stratified sample" case where $Vu_t = \sigma_1^2$ for $1 \leqq t \leqq T_1$, $Vu_t = \sigma_2^2$ for $T_1 < t \leqq T_2, \dots$.

For his model White first considers the non-linear weighted least squares estimator which minimizes $\sum W_t(y_t - f_t)^2 \equiv Q_T(\beta)$, where the weights $\{W_t\}$ are bounded constants.

A major difference between his proof of consistency and the one employed in Section 2.2 is that he must account for the possibility that $\operatorname{plim} T^{-1}Q_T(\beta)$ may not exist due to the heteroscedasticity of $\{u_t\}$. [See White (1980a, p. 728) for an example of this.] Therefore, instead of proving that $\operatorname{plim} T^{-1}Q_T(\beta)$ attains the minimum at β_0 as done in Section 2.2, White proves that $\operatorname{plim} T^{-1}[Q_T(\beta) - EQ_T(\beta)] = 0$ and there exists T_0 such that for any neighborhood $N(\beta_0)$

$$\inf_{T>T_0}\left[\min_{\beta\in\bar{N}} EQ_T(\beta) - EQ_T(\beta_0)\right] > 0,$$

from which consistency follows.

Another difference in his proof, which is necessitated by his assumption that $\{x_t\}$ are random variables, is that instead of using assumption (2.11), he appeals to Hoadley's (1971) strong law of large numbers, which essentially states that if $\{X_t(\beta)\}$ are independent random variables such that $|X_t(\beta)| \leq X_t^*$ for all β and $E|X_t^*|^{1+\delta} < \infty$ for some $\delta > 0$, then $\sup_\beta T^{-1}|\sum_{t=1}^T (X_t - EX_t)|$ converges to zero almost surely.

White's proof of asymptotic normality is a modification of the proof given in Section 2.2 and uses a multivariate version of Liapounov's central limit theorem due to Hoadley (1971).

White shows that the results for non-stochastic weights, W_t, hold also for stochastic weights, $\hat{W}_t$, provided that $\hat{W}_t$ converges to W_t uniformly in t almost surely. The last assumption is satisfied, for example, if $\{\hat{\sigma}_t^{-2}\}$ are used as stochastic weights in the stratified sample case mentioned above.

Just and Pope (1978) consider a non-linear regression model where the variance of the error term is a non-linear function of parameters:

$$y_t = f_t(\beta) + u_t, \qquad V_t = h_t(\alpha). \tag{3.16}$$

This is a non-linear analogue of the model considered by Hildreth and Houck (1968). Generalizing Hildreth and Houck's estimator, Just and Pope propose regressing $\hat{u}_t^2$ on $h_t(\alpha)$ by NLLS.

4. Multivariate models

In this section I consider the multivariate non-linear regression model

$$y_{it} = f_i(x_{it}, \beta_{i0}) + u_{it}, \qquad i = 1,2,\ldots,N, \quad t = 1,2,\ldots,T. \tag{4.1}$$

Sometimes I will write $f_i(x_{it}, \beta_{i0})$ more simply as $f_{it}(\beta_{i0})$ or just f_{it}. Defining N-vectors $y_t = (y_{1t}, y_{2t},\ldots,y_{Nt})'$, $f_t = (f_{1t}, f_{2t},\ldots,f_{Nt})'$, and $u_t = (u_{1t}, u_{2t},\ldots,u_{Nt})'$ we can write (4.1) as

$$y_t = f_t(\theta_0) + u_t, \qquad t = 1,2,\ldots,T, \tag{4.2}$$

where I have written the vector of all the unknown regression parameters as θ_0, allowing for the possibility that there are constraints among $\{\beta_{i0}\}$. Thus, if there is no constraint, $\theta_0 = (\beta'_{10}, \beta'_{20},\ldots,\beta'_{N0})'$. We assume that $\{u_t\}$ are i.i.d. vector random variables with $Eu_t = 0$ and $Eu_t u'_t = \Sigma$.

The reader will immediately notice that this model is a non-linear analogue of the well-known seemingly unrelated regressions (SUR) model proposed by Zellner (1962). The estimation of the parameters also can be carried out using the same iterative method (the Zellner iteration) used in the linear SUR model. The Zellner iteration in the non-linear SUR model (4.2) can be defined as follows. Let $\hat{\theta}(\Lambda)$ be the estimator obtained by minimizing

$$\sum_{t=1}^{T} (y_t - f_t)'\Lambda^{-1}(y_t - f_t) \tag{4.3}$$

for some matrix Λ. Let $\hat{\theta}_n$ be the nth-round estimator of the Zellner iteration. Then,

$$\begin{aligned} \hat{\theta}_1 &= \hat{\theta}(I), \\ \hat{\theta}_{n+1} &= \hat{\theta}(\hat{\Lambda}_n), \qquad n = 1,2,\ldots, \end{aligned} \tag{4.4}$$

where

$$\hat{\Lambda}_n = \frac{1}{T}\sum\left[y_t - f_t(\hat{\theta}_n)\right]\left[y_t - f_t(\hat{\theta}_n)\right]'. \tag{4.5}$$

Note that in each step of the Zellner iteration we must use some other iterative method to minimize the minimand (4.3). For this minimization, Malinvaud (1970b) suggests a multivariate version of the Gauss–Newton iteration, Beauchamp and Cornell (1966) recommend a multivariate version of Hartley's iteration, and Gallant (1975d) shows how to transform the problem to a univariate problem so that standard univariate programs found at most computing centers can be used.

The consistency and the asymptotic normality of $\hat{\theta}(\Lambda)$ for a fixed Λ can be proved by a straightforward modification of the proofs of Section 2.2. Gallant (1975d) proves that $\hat{\theta}(\hat{\Sigma})$ has the same asymptotic distribution as $\hat{\theta}(\Sigma)$ if $\hat{\Sigma}$ is a consistent estimator of Σ. In particular, his result means that the second-round estimator $\hat{\theta}_2$ of the Zellner iteration (4.4) has the same asymptotic distribution as $\hat{\theta}(\Sigma)$ – a result analogous to the linear case. Gallant also generalizes another well-known result in the linear SUR model and proves that the asymptotic distributions of $\hat{\theta}(I)$ and $\hat{\theta}(\hat{\Sigma})$ are the same if $\{x_{it}\}$ are the same for all i and $f_i(x_{it}, \beta_i)$ has the same functional form for all i, provided that there are no constraints among the $\{\beta_i\}$.[7]

By a Monte Carlo study of a two-equation model, Gallant (1975d) finds that an estimate of the variance of the estimator calculated from the asymptotic formula tends to underestimate the true variance and recommends certain corrections.

If $u \sim N(0, \Sigma)$, the concentrated log-likelihood function can be written as

$$-\log\det\sum(y_t - f_t)(y_t - f_t)'. \tag{4.6}$$

However, it is possible to define the estimator $\hat{\theta}$ as the value of θ that maximizes (4.6) without the normality assumption. Since we do not assume the normality of u_t here, we will call $\hat{\theta}$ the quasi maximum likelihood estimator. Phillips (1976) proves that the Zellner iteration (4.4) converges to the quasi MLE $\hat{\theta}$ if T is

[7]In the linear SUR model, the least squares and the generalized least squares estimators are identically equal for every finite sample if the same conditions are met.

sufficiently large and if $\hat{\theta}_1$ (or any other initial estimate) is sufficiently close to $\hat{\theta}$. Therefore, Phillips' proof is of the same nature as Jennrich's proof (1969) of the asymptotic stability of the Gauss–Newton iteration, discussed in Section 2.3.

Since (4.3) is minimized at each step of the Zellner iteration (4.4), the iteration can be also defined implicitly by the normal equation

$$0=\sum_t\left\{\left.\frac{\partial f_t'}{\partial\theta}\right|_{\hat{\theta}_{n+1}}\left[\sum_t[y_t-f_t(\hat{\theta}_n)][y_t-f_t(\hat{\theta}_n)]'\right]^{-1}[y_t-f_t(\hat{\theta}_n)]\right\}$$
$$\equiv H_T(\hat{\theta}_{n+1},\hat{\theta}_n). \tag{4.7}$$

But, since

$$\frac{\partial}{\partial\theta}\log\det\sum_t(y_t-f_t)(y_t-f_t)'$$
$$=-2\sum_t\left\{\frac{\partial f_t'}{\partial\theta}\left[\sum_t(y_t-f_t)(y_t-f_t)'\right]^{-1}(y_t-f_t)\right\}, \tag{4.8}$$

the quasi MLE $\hat{\theta}$ is a stationary point of the iteration (4.7): that is,

$$H_T(\hat{\theta},\hat{\theta})=0. \tag{4.9}$$

Phillips proves that (4.7) defines a unique function $\hat{\theta}_{n+1}=h(\hat{\theta}_n)$ by showing a mapping $(a,b)\to(z,w)$ defined by $z=H_T(a,b)$ and $w=b$ has a Jacobian which is a P-matrix (every principal minor is positive) and therefore is one-to-one by a theorem of Gale and Nikaido (1965). The rest of this proof of the asymptotic stability proceeds similarly to the arguments following (2.42) in Section 2.3.

In the discussion above we have assumed that $\{u_t\}$ are i.i.d. When $\{u_t\}$ follow a vector stationary process, we have a multivariate generalization of the model of Section 3.1. Robinson (1972) considered such a model and extended the results of Hannan (1971) discussed above.

There are many empirical applications of the Zellner iteration in multivariate non-linear models, especially in the analysis of expenditures. I will mention a few representative papers. In all of these works, the dependent variables are expenditure shares, which means that the sum of the dependent variables across commodity groups $(i=1,2,\dots,N)$ is unity. Therefore, (quasi) maximum likelihood estimation is applied to $N-1$ equations after dropping one equation. [See Theil (1971, page 275) regarding when this procedure is justified.]

Deaton (1974) estimates various expenditure models (such as the Rotterdam model, the linear expenditure system, and the direct addilog system), some of which lead to multivariate $(N=9)$ non-linear regression models with linear constraints on the parameters, using Marquardt's modification of the

Gauss–Newton iteration to minimize (4.3) at each step of the Zellner iteration. Darrough (1977) estimates Diewert's (1974) Generalized Leontief model with $N=4$ and Berndt, Darrough and Diewert (1977) compare Translog [Christensen, Jorgenson and Lau (1975)], Generalized Leontief, and Generalized Cobb–Douglas [Diewert (1973)] models with $N=2$. Both of these studies used some modification of the Gauss–Newton iteration to minimize (4.3). They are not explicit on this point except that they refer to Berndt, Hall, Hall and Hausman (1974).

In the expenditure analysis it is important to test various hypotheses on the form of the utility function (such as symmetry, homotheticity, and additivity). Also, if, as in the study of Deaton, different models can be nested in a single family, the choice of models can be done by a standard testing procedure. The above-mentioned studies use the likelihood ratio test to test these hypotheses. Deaton discusses the adequacy of the asymptotic chi-square distribution of the likelihood ratio test in these models and suggests the use of a simple multiplicative correction [Anderson (1958, pp. 207–210)].

Although it is not a Zellner iteration, I should also mention MacKinnon (1976), who estimated the S-branch utility model of Brown and Heien (1972) for the case of $N=6$ by maximum likelihood estimation using the quadratic hill-climbing method, Powell's conjugate gradient method, and the DFP method.

5. Simultaneous equations models

5.1. *Non-linear two-stage least squares estimator*

In this section we consider the non-linear regression equation

$$y_t = f(Y_t, X_{1t}, \alpha_0) + u_t, \qquad t = 1,2,\dots,T, \tag{5.1}$$

where y_t is a scalar endogenous variable, Y_t is a vector of endogenous variables, X_{1t} is a vector of exogenous variables, α_0 is a K-vector of unknown parameters, and $\{u_t\}$ are scalar i.i.d. random variables with $Eu_t = 0$ and $Vu_t = \sigma^2$. The model does not specify the distribution of Y_t. Eq. (5.1) may be one of many structural equations which simultaneously define the distribution of y_t and Y_t, but here we are not concerned with the other equations. I will sometimes write $f(Y_t, X_{1t}, \alpha_0)$ simply as $f_t(\alpha_0)$ or as f_t. Also, I will define T-vectors y, f, and u in the same way I defined them in Section 2.1, and matrices Y and X_1, whose tth rows are Y_t' and X_{1t}', respectively.

The non-linear least squares estimator of α_0 in this model generally yields an inconsistent estimator essentially for the same reason that the least squares estimator is inconsistent in a linear simultaneous equations model. We can see this by considering (2.9) and noting that plim $A_2 \neq 0$ in general because f_t may be correlated with u_t in the model (5.1) due to a possible dependence of Y_t on u_t. In this section I will consider how we can generalize the two-stage least squares (2SLS) method to the non-linear model (5.1) so that we can obtain a consistent estimator.

5.1.1. *Non-linear only in parameters*

The case where the non-linearity of f occurs only in α_0 does not pose much of a problem. Such a case will occur, for example, if the variables of a linear structural equation are transformed to take account of the autocorrelation of the error terms. The model which is nonlinear only in parameters can be written in vector form as

$$y = Y\gamma(\theta_0) + X_1\beta(\theta_0) + u. \tag{5.2}$$

We can generalize the two-stage least squares estimator in this model using either Theil's interpretation of 2SLS [Theil (1953)] or the instrumental variables (I.V.) interpretation. Suppose the reduced form for Y is given by

$$Y = X\Pi + V. \tag{5.3}$$

If we use Theil's interpretation, we should replace Y by $\hat{Y} = X(X'X)^{-1}X'Y$ in the right-hand side of (5.2) and then apply the non-linear least squares estimation to (5.2). If we use the I.V. interpretation, we should premultiply (5.2) by $X(X'X)^{-1}X'$ and then apply the non-linear least squares estimation. Clearly, we get the same estimator of θ_0 in either method. I will call the estimator thus obtained the non-linear two-stage least squares (NL2S) estimator of the model (5.2). In Amemiya (1974) it was proved that this NL2S estimator is asymptotically efficient in the model of (5.2) and (5.3); that is to say, it has the same asymptotic distribution as the maximum likelihood estimator of the same model (called the limited information maximum likelihood estimator). For an application of this estimator, see Zellner, Huang and Chau (1965).

5.1.2. *Non-linear only in variables*

Next, I consider the case where f is non-linear only in variables. Let $F_t(Y_t, X_t)$ be a vector-valued function and let F be the matrix whose tth row is equal to F_t'. Then, the present case can be written as

$$y = F\gamma_0 + X_1\beta_0 + u. \tag{5.4}$$

We will assume that the reduced form for F_t is not linear in X_t, for then the model is reduced to a linear model. Eq. (5.3) may or may not hold. This model is more problematical than the model (5.2). Here, the estimator obtained according to Theil's interpretation is no longer consistent and the one based on the I.V. interpretation is consistent but no longer asymptotically efficient.

The following simple example illustrates why the application of Theil's interpretation does not work in the present situation.[8] Suppose a structural equation is

$$y_t = \gamma z_t^2 + u_t \tag{5.5}$$

and is the reduced form for z_t is

$$z_t = x_t + v_t. \tag{5.6}$$

Note that I have simplified the matter by assuming that the reduced form coefficient is known. Inserting (5.6) into (5.5) yields

$$y_t = \gamma x_t^2 + \gamma\sigma_v^2 + \left(u_t + 2\gamma x_t v_t + \gamma v_t^2 - \gamma\sigma_v^2\right), \tag{5.7}$$

where the composite error term in parentheses has zero mean. Since the application of Theil's interpretation to (5.5) means regressing y_t on x_t^2 without a constant term, (5.7) clearly demonstrates that the resulting estimator is inconsistent.

That the estimator obtained according to the I.V. interpretation in the model (5.4) is not fully efficient can be seen by noting that the reduced form for F_t is not linear in X_t. This suggests that one may hope to obtain a more efficient estimator by premultiplying (5.4) by $W(W'W)^{-1}W'$, where the tth row of W, denoted W_t', is a vector-valued function of X_t such that the linear dependence of F_t on W_t is greater than that of F_t on X_t. Thus, in the present situation it is useful to consider the class of NL2S estimators with varying values for W. The elements of W_t will be called instrumental variables.

Goldfeld and Quandt (1968) were the first to consider NL2S estimators in simultaneous equations models non-linear in variables. One of the two models they analyzed is given by[9]

$$\log y_{1t} = \gamma_1 \log y_{2t} + \beta_1 + \beta_2 x_t + u_{1t}, \tag{5.8}$$

$$y_{2t} = \gamma_2 y_{1t} + \beta_3 x_t + u_{2t}. \tag{5.9}$$

These equations constitute a whole simultaneous equations model, but I will

[8] This statement should not be construed as a criticism of Theil's interpretation. I know of at least six more interpretations of 2SLS: a certain interpretation works better than others when one tries to generalize 2SLS in a certain way. Thus, the more interpretations, the better.

[9] The subscript "0" indicating the true value is suppressed to simplify the notation.

consider only the estimation of the parameters of (5.8) for the time being. Note that in this model y_{2t} does not have a linear reduced form like (5.3). Goldfeld and Quandt compared four estimators of (5.8) by a Monte Carlo study: (1) the least squares, (2) NL2S where $W_t' = (1, x_t)$, (3) NL2S where $W_t' = (1, x_t, x_t^2)$, and (4) maximum likelihood.[10] The ranking of the estimators turned out to be (2), (4), (3), and (1). However, the top ranking of (2) was later disclaimed by Goldfeld and Quandt (1972) as a computational error, since they found this estimator to be inconsistent. This was also pointed out by Edgerton (1972). In fact, the consistency of NL2S requires that the rank of W must be at least equal to the number of regression parameters to be estimated, as I will explain below. Goldfeld and Quandt (1968) also tried the Theil interpretation and discarded it since it gave poor results, as they expected.

Kelejian (1971) points out the consistency of the NL2S estimator using W of a sufficient rank and recommends that powers of the exogenous variables as well as the original exogenous variables be used as the instrumental variables to form W. Kelejian also shows the inconsistency of the estimator obtained by Theil's interpretation. Edgerton (1972) also noted the consistency of NL2S. See Strickland and Weiss (1976) and Rice and Smith (1977) for applications of the NL2S estimator using Kelejian's recommendation. The former estimates a three-equations model and the latter an eleven-equations model. Rice and Smith use other statistical techniques in conjunction with NL2S – a correction for the autocorrelation of the error terms and the use of principal components in the definition of W.

5.1.3. Non-linear in parameters and variables

Amemiya (1974) considered a general model (5.1) and defined the class of NL2S estimators as the value of α that minimizes

$$S_T(\alpha|W) = (y - f)'W(W'W)^{-1}W'(y - f), \tag{5.10}$$

where W is some matrix of constants with rank at least equal to K. The minimization can be done using the iterative methods discussed in Section 2.3. The advantage of this definition is two-fold. First, it contains 2SLS and the NL2S estimators defined in the preceding two subsections as special cases. Second, and more importantly, the definition of the estimator as the solution of a minimization problem makes it possible to prove the consistency and the asymptotic

[10] Goldfeld and Quandt state that they generated u's according to the normal distribution. But, their model is not well-defined unless the domain of u's is somehow restricted. Thus, I must interpret the distribution they used as the truncated normal distribution. This means that the maximum likelihood derived under normality is not the genuine maximum likelihood.

normality of the NL2S estimator by standard techniques similar to the ones used in Section 2.2.

I will give an intuitive proof of the consistency by writing the expression corresponding to (2.9). [The method of proof used in Amemiya (1974) is slightly different.] Using a Taylor expansion of $f(\alpha)$ around α_0, we obtain the approximation

$$\frac{1}{T}S_T(\alpha|W) \cong \frac{1}{T}u'P_W u + \frac{2}{T}(\alpha_0 - \alpha)'G'P_W u + \frac{1}{T}(\alpha - \alpha_0)'G'P_W G(\alpha - \alpha_0), \tag{5.11}$$

where $P_W = W(W'W)^{-1}W'$ and $G = (\partial f/\partial \alpha')_{\alpha_0}$. It is apparent from (5.11) that the consistency is attained if $\operatorname{plim} T^{-1}G'P_W u = 0$ and $\operatorname{plim} T^{-1}G'P_W G$ exists and has rank K. (Note that this implies that W must have at least rank K, which I assumed at the outset.) Intuitively it means that W should be chosen in such a way that the multiplication by P_W eliminates all the part of G that is correlated with u but retains enough of it to make $\operatorname{plim} T^{-1}G'P_W G$ full rank.

The asymptotic normality of NL2S can be proved by a method analogous to the one used in Section 2.2 using formula (2.14). In the present case we have

$$\left.\frac{\partial S_T}{\partial \alpha}\right|_{\alpha_0} = -2G'P_W u \tag{5.12}$$

and

$$\operatorname{plim}\frac{1}{T}\left.\frac{\partial^2 S_T}{\partial \alpha \partial \alpha'}\right|_{\alpha_0} = 2\operatorname{plim}\frac{1}{T}G'P_W G. \tag{5.13}$$

Hence, if we denote the NL2S estimator by $\hat{\alpha}$:

$$\sqrt{T}(\hat{\alpha} - \alpha_0) \to N\left[0, \sigma^2 \operatorname{plim} T(G'P_W G)^{-1}\right]. \tag{5.14}$$

Remember that I stated that the multiplication of G by P_W should retain enough of G to make $\operatorname{plim} T^{-1}G'P_W G$ full rank. The variance formula in (5.14) suggests that the more retained, the higher the efficiency.

Amemiya (1975) considered, among other things, the optimal choice of W. It is easy to show that $\operatorname{plim} T(G'P_W G)^{-1}$ is minimized in the matrix sense (i.e. $A > B$ means $A - B$ is a positive definite matrix) when one chooses $W = EG$. I will call the resulting estimator the best non-linear two-stage least squares estimator (BNL2S). Its asymptotic covariance matrix is given by

$$V_B = \sigma^2 \operatorname{plim} T(EG'EG)^{-1}. \tag{5.14}$$

However, BNL2S is not a practical estimator because of the following two problems: (1) it is often difficult to find an explicit expression for EG, and (2) EG generally depends on the unknown parameter vector α_0. The second problem is less serious since α_0 may be replaced by any consistent member of the NL2S class using some W, such as that recommended by Kelejian (1971).

Given the first problem, the following procedure recommended by Amemiya (1976) seems the best practical way to approximate BNL2S. (1) Compute $\hat{\alpha}$, a member of the NL2S class. (2) Evaluate G at $\hat{\alpha}$ – call it $\hat{G}$. (3) Treat $\hat{G}$ as the dependent variables of regressions and search for the optimal set of independent variables, denoted W_0, that best predict $\hat{G}$. (4) Set $W = W_0$. (If we wanted to be more elaborate, we could search for a different set of independent variables for each column of $\hat{G}$, say W_i for the ith column $\hat{g}_i$, and set $W = [P_{W_1}\hat{g}_1, P_{W_2}\hat{g}_2, \dots, P_{W_k}\hat{g}_k]$.)

Kelejian (1974) proposed another way to approximate EG. He proposed this method for the model which is non-linear only in variables, but it could also work for certain fully non-linear cases. Let the tth row of G be G_t' – that is, $G_t' = (\partial f_t / \partial \alpha')_{\alpha_0}$. Then, since G_t is a function of Y_t and α_0, it is also a function of u_t and α_0; therefore, write $G_t(u_t, \alpha_0)$. Kelejian's suggestion is to generate u_t independently n times by simulation and approximate EG_t by $n^{-1}\sum_{i=1}^{n} G_t(u_{ti}, \hat{\alpha})$ when $\hat{\alpha}$ is some consistent estimator of α_0. Kelejian also points out that $G_t(0, \hat{\alpha})$ is also a possible approximation for EG_t; although it is computationally simpler, it is likely to be a worse approximation than that given above.

Before Amemiya (1974), Tsurumi (1970) had actually used what amounted to the Gauss–Newton iteration to minimize (5.10). I overlooked this reference at the time of writing the 1974 article. In his article Tsurumi estimates the CES production function (2.3) by first linearizing the function around certain initial estimates of the parameters as in (2.38) and then proceeding as if he had the model which is nonlinear only in variables – the model of Section 5.1.2 above. The only difference between Tsurumi's linearized function and (2.38) is that in Tsurumi's case $(\partial f_t / \partial \beta')_{\hat{\beta}_1}$ is endogenous, since it contains L and K which are assumed to be endogenous variables in his model. Tsurumi tried two methods – the method according to Theil's interpretation and the method according to the I.V. interpretation described in Section 5.1.2 above. In each method Tsurumi carried out the iteration until convergence occurred. Thus, the method according to the I.V. interpretation is actually the Gauss–Newton iteration to obtain NL2S. In the end Tsurumi discards the estimates obtained by the method according to Theil's interpretation on the grounds that the convergence was slower and the estimates it produced were sometimes untenable. Tsurumi does not discuss the asymptotic properties of estimators derived by either method.

Hurd (1974) used a modification of NL2S to estimate the wage rate as a non-linear function of manhours, unemployment rate, and price level. His regression model has the form $y = Z(\lambda)\alpha + u$, where many of the right-hand side variables are transformed as $z_t - \lambda z_{t-1}$. He proposes transforming exogenous variables similarly, and minimizing $[y - Z(\lambda)\theta]'X(\lambda)[X(\lambda)'X(\lambda)]^{-1}X(\lambda)'[y - Z(\lambda)\theta]$ with respect to λ and θ proves the consistency of the resulting NL2S estimator by a modification of the proof of Amemiya (1974).

5.1.4. *Tests of hypotheses*

I will consider the test of the hypothesis of the form

$$h(\alpha) = 0, \tag{5.15}$$

where h is a q-vector-valued non-linear function. Since we have not specified the distribution of all the endogenous variables of the model, we could not use the three tests defined in (2.54), (2.55), and (2.56), even if we assumed the normality of u. But we can use the following two test statistics: (1) the Wald test statistic analogous to (2.59), and (2) the difference between the constrained and the unconstrained sums of squared residuals (I will call it the SSRD test). Both statistics are based on the NL2S estimator $\hat{\alpha}$. I will give the asymptotic distribution of the two statistics under the alternative hypothesis derived by Gallant and Jorgenson (1979). The normality of u is assumed.

I will assume that (5.15) can be alternatively written as

$$\alpha = r(\theta), \tag{5.16}$$

where θ is a $(K - q)$-vector of freely varying parameters,[11] and write the asymptotic distribution in terms of both (5.15) and (5.16).

First, I will give a list of the symbols used in this section:

$$\begin{aligned}
&\hat{\alpha} \text{ minimizes } S_T(\alpha|W) \text{ without constraint,}\\
&\tilde{\alpha} \text{ minimizes } S_T(\alpha|W) \text{ subject to } h(\alpha) = 0,\\
&\tilde{\theta} \text{ minimizes } S_T[r(\theta)|W]; \quad \text{therefore } \tilde{\alpha} = r(\tilde{\theta}),\\
&\text{plim } \hat{\alpha} = \alpha_0; \quad \text{plim } \textit{is always taken under the model } (5.1),\\
&\text{plim } \tilde{\alpha} = \alpha^*,\\
&\text{plim } \tilde{\theta} = \theta^*; \quad \text{therefore } \alpha^* = r(\theta^*),\\
&E\left.\frac{\partial f}{\partial \alpha'}\right|_{\alpha_0} = \bar{G},\\
&\left.\frac{\partial f}{\partial \alpha'}\right|_{\hat{\alpha}} = \hat{G},
\end{aligned} \tag{5.17}$$

[11]See Gallant and Jorgenson (1979, p. 299) for the conditions under which this is possible.

$$\left.\frac{\partial h}{\partial \alpha'}\right|_{\alpha_0} = H,$$

$$\left.\frac{\partial h}{\partial \alpha'}\right|_{\hat{\alpha}} = \hat{H},$$

$$\left.\frac{\partial r}{\partial \theta'}\right|_{\theta^*} = R,$$

$P = W(W'W)^{-1}W'$; I called this P_W before;

$$P_1 = P\overline{G}(\overline{G}'P\overline{G})^{-1}\overline{G}'P,$$

$$P_2 = P\overline{G}R(R'\overline{G}'P\overline{G}R)^{-1}R'\overline{G}'P.$$

The aforementioned two statistics are defined as

$$\text{Wald} = \frac{T}{S_T(\hat{\alpha})}h(\hat{\alpha})'\left[\hat{H}(\hat{G}'P\hat{G})^{-1}\hat{H}'\right]^{-1}h(\hat{\alpha}) \tag{5.18}$$

and

$$\text{SSRD} = \frac{T}{S_T(\hat{\alpha})}\left[S_T(\tilde{\alpha}) - S_T(\hat{\alpha})\right]. \tag{5.19}$$

Under the assumption that the alternative hypothesis is "close" to the null hypothesis or, more rigorously,

$$\alpha_0 - \alpha^* = 0\left(\frac{1}{\sqrt{T}}\right), \tag{5.20}$$

Gallant and Jorgenson (1979) obtain the following limit distribution of the two statistics:

$$\text{Wald, SSRD} \sim \chi^2\left\{q, \frac{1}{\sigma^2}h(\alpha_0)'\left[H(\overline{G}'P\overline{G})^{-1}H'\right]^{-1}h(\alpha_0)\right\}. \tag{5.21}$$

Differentiating both sides of (5.15) with respect to θ, we obtain:

$$HR \cong 0, \tag{5.22}$$

which implies

$$\overline{G}'(P_1 - P_2)\overline{G} = H'\left[H(\overline{G}'P\overline{G})^{-1}H'\right]^{-1}H. \tag{5.23}$$

Using (5.23) we can write (5.21) alternatively as

$$\text{Wald, SSRD} \sim \chi^2\left[q, \frac{1}{\sigma^2}(\alpha_0 - \alpha^*)'\bar{G}'(P_1 - P_2)\bar{G}(\alpha_0 - \alpha^*)\right]. \tag{5.24}$$

As an application of the SSRD test, Gallant and Jorgenson tested the hypothesis of homogeneity of degree zero of the equation for durables in the two-equation translog expenditure model of Jorgenson and Lau (1978). In this model, the ratios of prices to total expenditures in the right-hand side of the regression equations, as well as the expenditure shares in the left-hand side, are regarded as endogenous variables; therefore, a non-linear simultaneous equation estimator such as NL2S must be used to yield consistent estimates.

5.2. *Other single equation estimators*

This section is based mainly on Amemiya (1975). In the preceding section we assumed the model (5.1) without specifying the model for Y_t or assuming the normality of u_t and derived the asymptotic distribution of the class of NL2S estimators and the optimal member of the class – BNL2S. In this section we specify the model for Y_t and assume that all the error terms are normally distributed; and, under these assumptions, we derive two estimators which are asymptotically more efficient than BNL2S. These two estimators take advantage of the added assumptions and consequently their asymptotic properties crucially depend on the validity of the assumptions. Thus, we are aiming at a higher efficiency, but possibly sacrificing robustness.

We now assume, in addition to (5.1),

$$Y_t' = X_t'\Pi + V_t', \tag{5.25}$$

where V_t is a vector of random variables, X_t is a vector of known constants, and Π is a matrix of unknown parameters. We assume that (u_t, V_t') are independent drawings from a multivariate normal distribution with zero mean and variance–covariance matrix

$$\Sigma = \begin{bmatrix} \sigma^2 & \sigma_{12}' \\ \sigma_{12} & \Sigma_{22} \end{bmatrix}. \tag{5.26}$$

We will define X and V as matrices whose tth rows are X_t' and V_t', respectively. Since u and V are jointly normal, we can write

$$u = V\Sigma_{22}^{-1}\sigma_{12} + \varepsilon, \tag{5.27}$$

where ε is independent of V and distributed as $N(0, \sigma^{*2}I)$, where $\sigma^{*2} = \sigma^2 - \sigma_{12}'\Sigma_{22}^{-1}\sigma_{12}$.

The model defined above may be regarded either as a simplified non-linear simultaneous equations model in which both the non-linearity and the simultaneity appear only in the first equation, or as the model that represents the "limited information" of the investigator. In the latter interpretation X_t are not necessarily the original exogenous variables of the system, some of which appear in the arguments of f, but, rather, are the variables a linear combination of which the investigator believes will explain Y_t effectively.

5.2.1. *Modified NL2S*

In Section 5.1.3 we learned that the consistency of NL2S follows from the fact that the projection matrix, P_W, removes the stochastic part of G, but that the "larger" $P_W G$, the more efficient the estimator. This suggests that the projection matrix $M_V \equiv I - V(V'V)^{-1}V'$ should perform better than P_W since it precisely eliminates the part of G that depends on V. Thus, if V were known (which is the same as if Π were known), we could define the estimator that minimizes

$$(y-f)'M_V(y-f). \tag{5.28}$$

In order to verify the consistency of the estimator, write T^{-1} times the minimand in the same way as in (5.11). Then we get the same expression as the right-hand side of (5.11), except that P_W is replaced by M_V. Thus, the first condition for consistency is plim $T^{-1}G'M_V u = 0$. But because of (5.27) this condition is equivalent to

$$\operatorname{plim}\frac{1}{T}G'M_V\varepsilon = 0. \tag{5.29}$$

The above is generally satisfied if V and ε are independent because G depends only on V. Note that here I have used the joint normality of u and V.[12] Concerning the asymptotic efficiency, one can easily prove

$$\operatorname{plim}\frac{1}{T}G'M_V G \geqslant \operatorname{plim}\frac{1}{T}G'P_W G \tag{5.30}$$

for any constant W, so that this estimator is asymptotically more efficient than any member of the class of NL2S estimators.

[12] However, normality is not necessary. For example, it is sufficient to assume that $u = V\alpha + \varepsilon$ for some α with V and ε being independent.

However, since this estimator is not practical, I define the modified non-linear two-stage least squares estimator (MNL2S) as the value of α that minimizes (5.28) after replacing V with $Y - X\hat{\Pi}$, where $\hat{\Pi} = (X'X)^{-1}X'Y$. Thus, MNL2S minimizes

$$(y-f)'\left[I - M_X Y(Y'M_X Y)^{-1}Y'M_X\right](y-f), \tag{5.31}$$

where $M_X = I - X(X'X)^{-1}X'$. Its asymptotic covariance matrix is given in Amemiya (1975) as

$$V_M = \operatorname{plim} T(G'M_V G)^{-1}\left[\sigma^{*2}G'M_V G + (\sigma^2 - \sigma^{*2})G'P_X G\right](G'M_V G)^{-1}, \tag{5.32}$$

where $P_X = X(X'X)^{-1}X'$. Amemiya (1975) proved

$$V_M < V_B. \tag{5.33}$$

It is interesting to note that if f is linear in α and Y, MNL2S is reduced to the usual 2SLS.

5.2.2. *Non-linear limited information maximum likelihood estimator (NLLI)*

Since the Jacobian of the transformation from (u, V) to (y, Y) is unity in our model, the log-likelihood function assuming normality can be written, apart from a constant, as

$$L^{**} = -\frac{T}{2}\log|\Sigma| - \tfrac{1}{2}\operatorname{tr}\Sigma^{-1}Q, \tag{5.34}$$

where

$$Q = \begin{bmatrix} u'u & u'V \\ V'u & V'V \end{bmatrix},$$

u and V representing $y - f$ and $Y - X\Pi$, respectively. Solving $\partial L^{**}/\partial\Sigma = 0$ for Σ yields

$$\Sigma = \frac{1}{T}Q. \tag{5.35}$$

Putting (5.35) into (5.34), we obtain a concentrated likelihood function

$$L^{*} = -\frac{T}{2}\left(\log u'u + \log|V'M_u V|\right). \tag{5.36}$$

Solving $\partial L^*/\partial \Pi = 0$ for Π, we obtain:

$$\Pi = (X'M_u X)^{-1} X'M_u Y. \tag{5.37}$$

Putting (5.37) into (5.36) yields a further concentrated likelihood function:

$$L = -\frac{T}{2}\left(\log u'u + \log|Y'M_u Y - Y'M_u X(X'M_u X)^{-1}X'M_u Y|\right), \tag{5.38}$$

which depends only on α. Interpreting our model as that which represents the limited information of the researcher, I call the value of α that minimizes (5.38) the NLLI estimator. Its asymptotic covariance matrix is given by

$$V_L = \operatorname{plim} T\left[\frac{1}{\sigma^{*2}}G'M_V G - \left(\frac{1}{\sigma^{*2}} - \frac{1}{\sigma^2}\right)G'M_X G\right]^{-1}. \tag{5.39}$$

It is proved in Amemiya (1975) that

$$V_L < V_M. \tag{5.40}$$

The maximization of (5.38) may be done by standard iterative procedures such as those discussed in Section 2.3. Another iterative method may be defined as follows. Rewrite (5.36) equivalently as

$$L^* = -\frac{T}{2}\left(\log u'M_V u + \log|V'V|\right) \tag{5.41}$$

and iterate back and forth between (5.37) and (5.41). That is, obtain $\hat{\Pi} = (X'X)^{-1}X'Y$ and $\hat{V} = Y - X\hat{\Pi}$, maximize (5.41) with respect to α after replacing V with $\hat{V}$ (at this point one gets the MNL2S estimator of α), call this estimator $\hat{\alpha}$ and define $\hat{u} = y - f(\hat{\alpha})$, insert it into (5.37) to obtain another estimator of Π, and repeat the procedure until convergence.

5.2.3. *Comparison*

In this and the preceding sections we have discussed four estimators: (1) NL2S (as a class), (2) BNL2S, (3) MNL2S, and (4) NLLI. If we denote NL2S($W = X$) by SNL2S (the first S stands for standard), we have in the linear case

$$\text{SNL2S} \equiv \text{BNL2S} \equiv \text{MNL2S} \cong \text{NLLI}, \tag{5.42}$$

where $\equiv$ means exact identity and $\cong$ means asymptotic equivalence. In the non-linear model defined by (5.1) and (5.25) with the normality assumption we have obtained the following ranking in terms of the asymptotic covariance matrix:

$$\text{SNL2S} \prec \text{BNL2S} \prec \text{MNL2S} \prec \text{NLLI}, \tag{5.43}$$

where $\prec$ means "is worse than". However, it is important to remember that if the normality assumption is removed the last two estimators may lose consistency, whereas the first two estimators retain it.[13]

Quandt (1975) conducted a Monte Carlo experiment to study the comparative performance of several estimators, including those discussed above, in the following model:

$$y_{1t} = \alpha_1 \log y_{2t} + \alpha_2 x_{1t} + \alpha_3 + u_{1t} \tag{5.44}$$

$$y_{2t} = (x_{3t})^{\alpha_4} y_{1t} + \alpha_5 x_{2t} + u_{2t}, \tag{5.45}$$

where u_{1t} and u_{2t} are jointly normal.[14]

Let us suppose that we want to estimate the parameters of (5.44), so that (5.44) is a specific example of (5.1). Then, there are two ways to define the reduced-form equation for y_{2t} – the equation comparable to (5.25). How we define this equation determines SNL2S and NLFI. One way, which I think is the way Quandt adopted, is to define

$$y_{2t} = \pi_1 + \pi_2 x_{1t} + \pi_3 x_{2t} + \pi_4 x_{3t} + v_t, \tag{5.46}$$

and the other way is to define

$$y_{2t} = \pi_1 + \pi_2 x_{1t} + \pi_3 x_{2t} + \pi_4 x_{3t} + \pi_5 x_{1t}^2 + \pi_6 x_{2t}^2 + \pi_7 x_{3t}^2 + \pi_8 x_{1t} x_{2t} + \cdots + v_t, \tag{5.47}$$

where the right-hand side of (5.47) minus v_t is a linear function of the x's and the transformation of the x's which the researcher believes best explains y_{2t}, subject to the condition that the number of parameters should not be so large.

We will assume that (5.46) is a specific example of (5.25); thus, SNL2S uses $(1, x_{1t}, x_{2t}, x_{3t})$ as W_t' and NLLI is the maximum likelihood estimator applied to (5.44) and (5.46) assuming the joint normality of u_{1t} and v_t. Quandt compares the performance of ordinary least squares (OLS), SNL2S, the NL2S estimators where the W is chosen to be various sets of powers of the x's, MNL2S, NLLI, and the full information maximum likelihood estimator (NLFI), which is the maximum

[13] In the simple model defined by (5.5) and (5.6), the necessary and sufficient condition for the consistency of either MNL2S or NLLI can be shown to be $Ev^2uEv^2 = Ev^3Evu$.

[14] The same qualification as footnote 10 applies to this model as well.

likelihood estimator applied to the original equations (5.44) and (5.45). Quandt concludes that NLFI is the best, OLS is the worst, and the rest are more or less similar, although, to a certain extent, the asymptotic ranking (5.43) is preserved.

5.3. Non-linear simultaneous equations

So far I have considered the estimation of the parameters of a single equation in the system of simultaneous equations. Now, I will consider the estimation of all the parameters of the system.

The equations of the model are

$$f_i(y_t, x_t, \alpha_i) = u_{it}; \qquad i = 1,2,\dots,n; \quad t = 1,2,\dots,T, \tag{5.48}$$

where y_t is an n-vector of endogenous variables, x_t is a vector of exogenous variables, and α_i is a K_i-vector of unknown parameters. I will assume that the n-vector $u_t = (u_{1t}, u_{2t}, \dots, u_{nt})'$ is an i.i.d. vector random variable with zero mean and variance–covariance matrix Σ. Not all of the elements of the vectors y_t and x_t may actually appear in the arguments of each f_i. I assume that each equation has its own vector of parameters α_i and that there are no constraints among the α_i's, but the results I state subsequently can easily be modified if we can express each α_i parametrically as $\alpha_i(\theta)$, where the number of elements in θ is fewer than $\sum_{i=1}^{n} K_i$.

Strictly speaking, (5.48) is not a complete model by itself because there is no guarantee that a unique solution for y_t exists for every possible value of u_{it} unless some stringent assumptions are made on the form of f_i. Therefore, we will assume either that f_i satisfies such assumptions or that if there is more than one solution for y_t there is some additional mechanism by which a unique solution is chosen.

I have already mentioned two simple examples of (5.48): the model of Goldfeld and Quandt (1968), defined by (5.8) and (5.9), and the model of Quandt (1975), defined by (5.44) and (5.45). The first model is shown to possess two solutions occurring in two different regions; therefore, the model is not complete unless we specify some mechanism by which one of the two solutions is chosen. Goldfeld and Quandt conduct a Monte Carlo study in which they analyze how the performance of several estimators is affected by various mechanisms for choosing solutions such as always choosing solutions from one region or mixing solutions from two regions. [See Kelejian (1975) for a further study on this issue.] Quandt (1975) shows that in the second model above there is a one-to-one correspondence between y and u if it is assumed that $\alpha_1 < 0$ and $x_{3t} > 0$ for all t.

I will not discuss the problem of identification in the model (5.48). There are not many useful results in the literature beyond the basic discussion of Fisher (1966) as summarized in Goldfeld and Quandt (1972, p. 221 ff). I will merely point out that non-linearity generally helps rather than hampers identification, so that, for example, the number of excluded exogenous variables in a given equation need not be greater than or equal to the number of parameters of the same equation in a non-linear model. I should also point out that I have actually given one sufficient condition for identifiability – that plim $T^{-1}(G'P_WG)$ in the right-hand side of (5.14) be non-singular.

To facilitate the discussion of the subsequent sections I will give a list of symbols:

$$
\begin{aligned}
&\alpha = (\alpha_1', \alpha_2', \dots, \alpha_n')', \\
&\Lambda = \Sigma \otimes I, \quad \text{where} \otimes \text{is the Kronecker product}, \\
&f_{it} = f_i(y_t, x_t, \alpha_i), \\
&f_t = \text{an } n\text{-vector whose } i\text{th element is } f_{it}, \\
&f_{(i)} = \text{a } T\text{-vector whose } t\text{th element is } f_{it}, \\
&\boldsymbol{f} = (f_{(1)}', f_{(2)}', \dots, f_{(n)}'), \quad \text{an } nT\text{-vector}, \\
&F = (f_{(1)}, f_{(2)}, \dots, f_{(n)}), \quad \text{a } T \times n\text{-matrix}, \\
&g_{it} = \frac{\partial f_{it}}{\partial \alpha_i}, \quad \text{a } K_i\text{-vector}, \\
&G_i = \frac{\partial f_{(i)}}{\partial \alpha_i'}, \quad \text{a } T \times K_i \text{ matrix whose } t\text{th row is } g_{it}', \\
&\boldsymbol{G} = \operatorname{diag}\{G_1, G_2, \dots, G_n\}, \quad \text{an } nT \times \left(\sum_{i=1}^n K_i\right) \text{ block diagonal matrix.}
\end{aligned}
\tag{5.49}
$$

5.4. *Non-linear three-stage least squares estimator*

Before starting the main discussion I wish to point out that all the results of Sections 5.1 and 5.2 are valid if we change (5.1) to

$$f(y_t, Y_t, X_{1t}, \alpha_0) = u_t. \tag{5.50}$$

Consequently, the minimand (5.10) which defines the class of NL2S estimators should be changed to

$$f'W(W'W)^{-1}W'f. \tag{5.51}$$

The asymptotic normality result (5.14) needs not be changed. The significance of the above modification is that a NL2S estimator can be applied to each equation of (5.48).[15]

As a natural extension of the class of the NL2S estimators defined in Section 5.1.3, Jorgenson and Laffont (1974) defined the class of non-linear three-stage least squares estimators (NL3S) as the value of α that minimizes

$$f'\left[\hat{\Sigma}^{-1}\otimes W(W'W)^{-1}W'\right]f, \tag{5.52}$$

where $\hat{\Sigma}$ is a consistent estimate of Σ. For example,

$$\hat{\Sigma}=\frac{1}{T}\sum_{t=1}^{T}f_t(\hat{\alpha})f_t(\hat{\alpha})', \tag{5.53}$$

where $\hat{\alpha}$ is the NL2S estimator obtained from each equation. The above definition is analogous to the definition of the linear 3SLS as a generalization of the linear 2SLS. The consistency and the asymptotic normality of the estimator defined above are proved in Jorgenson and Laffont (1974) and Gallant (1977).

The consistency of the NL2S and NL3S estimators of the parameters of the model (5.48) can be proved with minimal assumptions on u_{it} – namely, those stated after (5.48). This robustness makes the estimators attractive. Another important strength of the estimators is that they retain their consistency regardless of whether or not (5.48) yields a unique solution for y_t and, in the case of multiple solutions, regardless of what additional mechanism chooses a unique solution. See MaCurdy (1980) for an interesting discussion of this point.

Amemiya (1977) defined the class of the NL3S estimators slightly more generally as the value of α that minimizes

$$f'\hat{\Lambda}^{-1}S(S'\hat{\Lambda}^{-1}S)^{-1}S'\hat{\Lambda}^{-1}f, \tag{5.54}$$

where $\hat{\Lambda}$ is a consistent estimate of Λ and S is a matrix of constants with nT rows and with the rank of at least $\sum_{i=1}^{n}K_i$. This definition is reduced to the Jorgenson–Laffont definition if $S=\text{diag}(W,W,\dots,W)$. Its asymptotic variance-covariance matrix is given by

$$V_3=\text{plim}\,T\left[G'\Lambda^{-1}S(S'\Lambda^{-1}S)^{-1}S'\Lambda^{-1}G\right]^{-1}. \tag{5.55}$$

Its lower bound is equal to

$$V_{B3}=\lim T\left[EG'\Lambda^{-1}EG\right]^{-1}, \tag{5.56}$$

[15]Another advantage of this modification is that the Box–Cox transformation model $(y_t^{\lambda}-1)/\lambda=\beta'x_t+u_t$ [see Box and Cox (1964)] can be regarded as a special case of (5.52). See Amemiya and Powell (1980) for the application of NL2S to the Box–Cox model.

which is attained when one chooses $S = EG$. I will call this estimator the BNL3S estimator (B for best).

We can also attain the lower bound (5.56) using the Jorgenson–Laffont definition, but that is possible if and only if the space spanned by the column vectors of W contains the union of the spaces spanned by the column vectors of EG_i for $i = 1,2,\dots,n$. This necessitates including many columns in W, which is likely to increase the finite sample variance of the estimator although it has no effect asymptotically. This is a disadvantage of the Jorgenson–Laffont definition compared to my definition.

Noting that BNL3S is not practical just as BNL2S, Amemiya (1976) suggests the following approximation. (1) Compute $\hat{\alpha}_i$, an SNL2S estimator of α_i, $i = 1,2,\dots,n$. (2) Evaluate G_i at $\hat{\alpha}_i$ – call it $\hat{G}_i$. (3) Treat $\hat{G}_i$ as the dependent variables of the regression and search for the optimal set of independent variables W_i that best predict $\hat{G}_i$. (4) Choose $S = \text{diag}\{P_1\hat{G}_1, P_2\hat{G}_2,\dots, P_n\hat{G}_n\}$, where $P_i = W_i(W_i'W_i)^{-1}W_i'$.

In Section 5.1.4 I discussed tests of hypotheses based on NL2S developed by Gallant and Jorgenson (1979). These tests can easily be generalized to the tests based on NL3S, as shown by the same authors.

Let $\hat{\alpha}$ be the NL3S estimator here and let $\hat{\Sigma}$ be a consistent estimate of Σ. Also, let $S_T(\alpha)$ refer to the NL3S minimand—and let $\tilde{\alpha}$ refer to the constrained NL3S subject to the condition (5.15) or (5.16). Then, the Wald and SSRD test statistics can now be defined as

$$\text{Wald} = h(\hat{\alpha})'\left\{\hat{H}'\left[\hat{\boldsymbol{G}}'(\hat{\Sigma}^{-1}\otimes P)\hat{\boldsymbol{G}}\right]^{-1}\hat{H}\right\}^{-1}h(\hat{\alpha}), \tag{5.57}$$

and

$$\text{SSRD} = S_T(\tilde{\alpha}) - S_T(\hat{\alpha}), \tag{5.58}$$

where $\hat{\boldsymbol{G}}$ is $\boldsymbol{G}$ evaluated at $\hat{\alpha}$ and $P = W(W'W)^{-1}W'$. Note that (5.57) and (5.58) are similar to (5.18) and (5.19). The limit distribution of the two statistics under the alternative hypothesis (if it is "close" to the null hypothesis as before) is given by

$$\text{Wald, SSRD} \sim \chi^2\left[q, h(\alpha_0)'\left\{H'\left[\overline{\boldsymbol{G}}'(\Sigma^{-1}\otimes P)\overline{\boldsymbol{G}}\right]H\right\}^{-1}h(\alpha_0)\right], \tag{5.59}$$

or, alternatively, using (5.22):

$$\text{Wald, SSRD} \sim \chi^2\left[q, (\alpha_0 - \alpha^*)'\overline{\boldsymbol{G}}'(P_1 - P_2)\overline{\boldsymbol{G}}\,(\alpha_0 - \alpha^*)\right], \tag{5.60}$$

where

$$P_1 = (\Sigma^{-1}\otimes P)\overline{\boldsymbol{G}}\left[\overline{\boldsymbol{G}}'(\Sigma^{-1}\otimes P)\overline{\boldsymbol{G}}\right]^{-1}\overline{\boldsymbol{G}}'(\Sigma^{-1}\otimes P)$$

and

$$P_2 = (\Sigma^{-1}\otimes P)\overline{G}R\left[R'\overline{G}'(\Sigma^{-1}\otimes P)\overline{G}R\right]^{-1}R'\overline{G}'(\Sigma^{-1}\otimes P).$$

As an application of the SSRD test, Gallant and Jorgenson tested the hypothesis of symmetry of the matrix of parameters in the three-equation translog expenditure model of Jorgenson and Lau (1975).

The other applications of the NL3S estimators include Jorgenson and Lau [1978], which was previously mentioned in Section 5.1.4, and Haessel [1976], who estimated a system of demand equations, nonlinear only in parameters, by both NL2S and NL3S estimators.

5.5. *Non-linear full information (NLFI) maximum likelihood estimator*

In this section we consider the maximum likelihood estimator of model (5.48) under the normality assumption of u_{it}. To do so we must assume that (5.48) defines a one-to-one correspondence between y_t and u_t. This assumption enables us to write down the likelihood function in the usual way as the product of the density of u_t and the Jacobian. Unfortunately, this is a rather stringent assumption, which considerably limits the usefulness of the NLFI estimator in practice. We have already noted that Goldfeld and Quandt's model defined by (5.8) and (5.9) does not satisfy this assumption. This example illustrates two types of problems which confront NLFI:

(1) Since there is no solution for y for some values of u, the domain of u must be restricted, which implies that the normality assumption cannot hold.
(2) Since there are two solutions for y for some values of u, one must specify a mechanism for choosing a unique solution in order to write down the likelihood function. One should note that the NL2S and the NL3S estimators are free from both of these problems.

Assuming $u_t \sim N(0,\Sigma)$, we can write the log-likelihood function of the model (5.48) as

$$L^* = -\frac{T}{2}\log|\Sigma| + \sum_{t=1}^{T}\log\left\|\frac{\partial f_t}{\partial y_t'}\right\| - \tfrac{1}{2}\sum_{t=1}^{T} f_t'\Sigma^{-1}f_t. \tag{5.61}$$

Solving $\partial L^*/\partial\Sigma = 0$ for Σ, we get:

$$\Sigma = \frac{1}{T}\sum_{t=1}^{T} f_t f_t'. \tag{5.62}$$

Inserting (5.62) into (5.61) yields the concentrated log-likelihood function

$$L = \Sigma\log\left\|\frac{\partial f_t}{\partial y_t'}\right\| - \frac{T}{2}\log|\frac{1}{T}\Sigma f_t f_t'|. \tag{5.63}$$

The NLFI maximum likelihood estimator of α is defined as the value of α that maximizes (5.63).

It is shown in Amemiya (1977) that the NLFI estimator is consistent if the true distribution of u_t is normal but is generally inconsistent if u_t is not normal.[16] This result is contrary to the result in the linear case where the FIML estimator derived from the normality assumption retains its consistency even if the true distribution is not normal. It is also shown in the same article that NLFI is asymptotically more efficient than BNL3S in general if the true distribution of u_t is normal. (On the other hand, NL3S is more robust than NLFI because NL3S is consistent even if the distribution of u_t is not normal.) This result is also contrary to the result in the linear case where FIML and 3SLS have the same asymptotic distribution. In the subsequent subsections I will further discuss these results as well as some other problems related to the NLFI estimator. I will not discuss hypotheses testing here since the discussion of Section 2.4.2 is applicable to the present model.

5.5.1. *Consistency*

Differentiating (5.63) with respect to α_i, we obtain:

$$\frac{\partial L}{\partial \alpha_i} = \sum_{t=1}^{T} \frac{\partial g_{it}}{\partial u_{it}} - T \sum_{t=1}^{T} g_{it} f_t' (\Sigma f_t f_t')_i^{-1}, \tag{5.64}$$

where $(\cdot)_i^{-1}$ denotes the ith column of the inverse of the matrix within the parentheses. The consistency of NLFI is equivalent to the condition:

$$\lim E \frac{1}{T} \frac{\partial L}{\partial \alpha_i}\bigg|_{\alpha_0} = 0, \tag{5.65}$$

and hence to the condition:

$$\lim \frac{1}{T} \sum_{t=1}^{T} E \frac{\partial g_{it}}{\partial u_{it}} = \lim \frac{1}{T} \sum_{t=1}^{T} E g_{it} u_t' \sigma^i, \tag{5.66}$$

where σ^i is the ith column of Σ^{-1}. Now, (5.66) could hold even if each term of a summation is different from the corresponding term of the other, but that event is extremely unlikely. Therefore, we can say that the consistency of NLFI is

[16] This result is completely separate from and in no way contradicts the quite likely fact that the maximum likelihood estimator of a non-linear model derived under the assumption of a certain regular non-normal distribution is consistent if the true distribution is the same as the assumed distribution.

essentially equivalent to the condition:

$$E\frac{\partial g_{it}}{\partial u_{it}} = Eg_{it}u_t'\sigma^i. \tag{5.67}$$

It is interesting to note that condition (5.67) holds if u_t is normal because of the following lemma.[17]

Lemma

Suppose $u = (u_1, u_2, \dots, u_n)'$ is distributed as $N(0, \Sigma)$, where Σ is positive definite. If $\partial h(u)/\partial u_i$ is continuous, $E|\partial h/\partial u_i| < \infty$, and $E|hu_i| < \infty$, then

$$E\frac{\partial h}{\partial u_i} = Ehu'\sigma^i, \tag{5.68}$$

where σ^i is the ith column of Σ^{-1}.

Another interesting question is: Does (5.67) hold even if u_t is not normal? To throw some light on this question, let us suppose that the joint density of u_t is $\psi(u_t)$, which may or may not be normal and that (5.65) holds where E is taken with respect to $\psi(u_t)$. If g_{it} and ψ satisfy certain conditions,[18] we have by integration by parts:

$$\int \frac{\partial g_i}{\partial u_i}\psi(u)\,\mathrm{d}u_i = -\int g_i \frac{\partial \psi}{\partial u_i}\,\mathrm{d}u_i, \tag{5.69}$$

where I have suppressed the subscript t from g and u for simplicity. Integrating both sides of (5.69) with respect to the elements of u other than u_i (this is not always permissible but we assume again that certain conditions on g and ψ enable us to do so)[19] and combining the resulting equation with (5.69), we obtain:

$$\int \dots \int \left[\frac{\partial \psi}{\partial u_i} + u'\sigma^i\psi(u)\right] g_i\,\mathrm{d}u = 0. \tag{5.70}$$

[17]This lemma, stated slightly differently, is originally due to Stein (1973). In Amemiya (1977), I rediscovered this lemma being unaware of Stein's work. But, there was an error in the conditions of the lemma as presented in my article, which was pointed out by Phillips (1981). A correct proof is given in Amemiya (1981).

[18]If we treat $\partial g_i/\partial u_i$ and ψ as functions of u_i alone, it is sufficient to assume: (1) $\partial g_i/\partial u_i$ is continuous, (2) $\int_{-\infty}^{\infty}|\partial g_i/\partial u_i|\psi \mathrm{d}u_i < \infty$, and (3) $\psi(u_i)$ is monotone decreasing for $|u_i| > M$ for some M and $\lim_{|u_i|\to\infty}\psi(u_i) = 0$.

[19]It is sufficient if all the conditions of footnote 18 hold uniformly with respect to all the other elements of u.

Now, the question is: Does (5.70) hold for a density $\psi(u)$ other than normal? The term within the square brackets in (5.70) is clearly zero if u is normal. Moreover, we can say "if and only if" in the preceding sentence, provided that we restrict our attention to the class of continuously differentiable densities, ψ, as proved by White (1980b). However, $\partial\psi/\partial u_i + u'\sigma^i\psi(u) = 0$ is not a necessary condition for (5.70) to hold, as we have noted in footnote 13 regarding a simple example of (5.5) and (5.6). This was first noted by Phillips (1981), who gives another interesting example. His model is defined by

$$\log y_{1t} + \alpha_1 x_t = u_{1t}, \tag{5.71}$$

$$y_{2t} + \alpha_2 y_{1t} = u_{2t}. \tag{5.72}$$

In this example $g_1 = 1$ and $g_2 = e^{u_1 - \alpha_1}$; therefore, (5.70) clearly holds for $i = 1$ for any density ψ, and Phillips found a class of densities for which (5.70) holds for the case $i = 2$.

What is the significance of these examples? It is that given g_i we can sometimes find a class of non-normal densities ψ for which (5.70) holds. When g_i are simple, as in these examples, we can find a fairly broad class of non-normal densities for which (5.70) holds. However, if g_i is a more complicated non-linear function of the exogenous variables and the parameters $\{\alpha_i\}$ as well as of u, (5.70) can be made to hold only when we specify a density which depends on the exogenous variables and the parameters of the model. In such a case, normality can be regarded, for all practical purposes, as a necessary and sufficient condition for the consistency of NLFI.

5.5.2. *Comparison between NLFI and NL3S*

Amemiya (1977) showed that the asymptotic equivalence of NLFI and BNL3S occurs if and (almost) only if f_{it} can be written in the form

$$f_i(y_t, x_t, \alpha_i) = A_i(\alpha_i)'z(y_t, x_t) + B_i(\alpha_i, x_t), \tag{5.73}$$

where z is an n-vector of surrogate variables.

Another instructive way to compare NLFI and BNL3S is to compare certain iterative methods to obtain the two estimators.

By equating the right-hand side of (5.64) to zero and rearranging terms we can obtain the following iteration to obtain NLFI:

$$\hat{\alpha}_{(2)} = \hat{\alpha}_{(1)} - (\tilde{G}'\hat{\Lambda}^{-1}G)^{-1}\tilde{G}'\hat{\Lambda}^{-1}f, \tag{5.74}$$

where

$$\tilde{G}_i' = G_i' - \frac{1}{T}\sum_{t=1}^{T}\frac{\partial g_{it}}{\partial u_t'}F' \tag{5.75}$$

and $\tilde{\boldsymbol{G}} = \text{diag}(\tilde{G}_1, \tilde{G}_2, \dots, \tilde{G}_n)$ and where all the variables that appear in the second term of the right-hand side of (5.74) are evaluated at $\hat{\alpha}_{(1)}$.

The Gauss–Newton iteration to obtain BNL3S can be written as

$$\hat{\alpha}_{(2)} = \hat{\alpha}_{(1)} - (\overline{\boldsymbol{G}}'\hat{\Lambda}^{-1}\boldsymbol{G})^{-1}\overline{\boldsymbol{G}}'\hat{\Lambda}^{-1}\boldsymbol{f}, \tag{5.76}$$

where $\overline{G}_i' = EG_i'$ and $\overline{\boldsymbol{G}} = \text{diag}(\overline{G}_1, \overline{G}_2, \dots, \overline{G}_n)$ as before.

Thus, we see that the only difference between (5.74) and (5.76) is in the respective "instrumental variables" used in the formulae. Note that $\tilde{G}_i$ defined in (5.75) can work as a proper set of "instrumental variables" (that is, the variables uncorrelated with u_t) only if u_t satisfies the condition of the aforementioned lemma, whereas $\overline{G}_i$ is always a proper set of instrumental variables, which implies that BNL3S is more robust than NLFI. If u_t is normal, however, $\tilde{G}_i$ catches more of the part of G_i uncorrelated with u_t than $\overline{G}_i$ does, which implies that NLFI is more efficient than BNL3S under normality.

Note that (5.74) is a generalization of the formula expounded by Hausman (1975) for the linear case. Unlike in the linear case, however, the iteration defined by (5.74) does not have the property that $\hat{\alpha}_{(2)}$ is asymptotically equivalent to NLFI when $\hat{\alpha}_{(1)}$ is consistent. Therefore, its main value may be pedagogical, and it may not be recommendable in practice.

5.5.3. Computation of NLFI

The discussion of the computation of NLFI preceded the theoretical discussion of the statistical properties of NLFI by more than ten years. The first paper on computation was by Eisenpress and Greenstadt (1966) who proposed a modified Newton–Raphson iteration. Their modification is the kind that combines both (2.28) and (2.29). Chow (1973) essentially differs from the above two authors in that he obtained simpler formulae by assuming that different parameters appear in different equations as in (5.48). I have already mentioned the iterative method considered by Amemiya (1977) mainly for a pedagogical purpose. Dagenais (1978) modified my iteration to speed up the convergence and compared it with a Newton–Raphson method due to Chow and Fair (1973) and the DFP iteration mentioned in Section 2.3.1 in certain examples of non-linear models. Results are inconclusive. Belsley (1979) compared the computation speed of the DFP iteration in computing NLFI and NL3S in five models of varying degrees of complexity and found that NL3S was three to ten times faster. Nevertheless, Belsley shows that the computation of NLFI is quite feasible and can be improved by using a more suitable algorithm and by using the approximation of the Jacobian due to Fair – see eq. (5.79) below.

5.5.4. *Other related papers*

Fair and Parke (1980) estimated Fair's (1976) macro model (97 equations, 29 of which are stochastic, with 182 parameters including 12 first-order autoregressive coefficients), which is non-linear in variables as well as in parameters (this latter non-linearity caused by the transformation to take account of the first-order autogression of the errors), by OLS, SNL2S, the Jorgenson–Laffont NL3S, and NLFI. The latter two estimators are calculated by a derivative-free algorithm due to Parke. This algorithm of NLFI uses the approximation of the Jacobian:

$$\sum_{t=1}^{T} \log|J_t| = \frac{T}{N}\left(\log|J_{t_1}| + \log|J_{t_2}| + \cdots + \log|J_{t_N}|\right), \tag{5.79}$$

where $J_t = \partial f_t / \partial y_t'$, N is a small integer, and $t_1, t_2, \ldots, t_N$ are equally spaced between 1 and T. Fair finds that in terms of predictive accuracy there is not much difference among different estimators, but in terms of policy response OLS is set apart from the rest.

Bianchi and Calzolari (1980) propose a method by which one can calculate the mean squared prediction error matrix of a vector predictor based on any estimator of the non-linear simultaneous equations model. Suppose the structural equations can be written as $f(y_p, x_p, \alpha) = u_p$ at the prediction period p and we can solve it for y_p as $y_p = g(x_p, \alpha, u_p)$. Define the predictor $\hat{y}_p$ based on the estimator $\hat{\alpha}$ by $\hat{y}_p = g(x_p, \hat{\alpha}, 0)$. (Note that y_p is an n-vector.) Then we have

$$\begin{aligned} E(y_p - \hat{y}_p)(y_p - \hat{y}_p)' &\\ = E\|g(x_p, \alpha, u_p) - g(x_p, \alpha, 0)\|^2 &+ E\|g(x_p, \alpha, 0) - g(x_p, \hat{\alpha}, 0)\|^2 \\ \equiv A_1 + A_2. & \end{aligned} \tag{5.80}$$

The authors suggest that A_1 be evaluated by simulation. As for A_2, we can easily obtain its asymptotic value from the knowledge of the asymptotic distribution of $\hat{\alpha}$.

Hatanaka (1978) considers a simultaneous equations model non-linear only in variables. Such a model can be written as $F(Y, X)\Gamma + XB = U$. Define $\hat{Y}$ by $F(\hat{Y}, X)\hat{\Gamma} + X\hat{B} = 0$, where $\hat{\Gamma}$ and $\hat{B}$ are the OLS estimates. Then, Hatanaka proposes using $F(\hat{Y}, X)$ as the instruments to calculate 3SLS. He proposes the method-of-scoring iteration to calculate NLFI where the iteration is started at the aforementioned 3SLS. He also proves the consistency and the asymptotic normality of NLFI and obtains its asymptotic covariance matrix [which can be also obtained from Amemiya (1977) by an appropriate reduction].

References

Akaike, H. (1973) "Information Theory and an Extension of the Maximum Likelihood Principle", in: B. N. Petrov and F. Csaki (eds.), *Second International Symposium on Information Theory*. Budapest: Akademiai Kiado, pp. 267–281.

Amemiya, T. (1973a) "Generalized Least Squares with an Estimated Autocovariance Matrix", *Econometrica*, 41, 723–732.

Amemiya, T. (1973b) "Regression Analysis When the Dependent Variable Is Truncated Normal", *Econometrica*, 41, 997–1016.

Amemiya, T. (1974) "The Nonlinear Two-Stage Least-Squares Estimator", *Journal of Econometrics*, 2, 105–110.

Amemiya, T. (1975) "The Nonlinear Limited-Information Maximum-Likelihood Estimator and the Modified Nonlinear Two-Stage Least-Squares Estimator", *Journal of Econometrics*, 3, 375–386.

Amemiya, T. (1976) "Estimation in Nonlinear Simultaneous Equation Models", Paper presented at Institut National de La Statistique et Des Etudes Economiques, Paris, March 10 and published in French in: E. Malinvaud (ed.), *Cahiers Du Seminaire D'econometrie*, no. 19 (1978).

Amemiya, T. (1977) "The Maximum Likelihood and the Nonlinear Three-Stage Least Squares Estimator in the General Nonlinear Simultaneous Equation Model", *Econometrica*, 45, 955–968.

Amemiya, T. (1980) "Lecture Notes in Advanced Theory of Econometrics", Department of Economics, Stanford University.

Amemiya, T. (1981) "Correction to a Lemma", forthcoming in *Econometrica*.

Amemiya, T. and W. A. Fuller (1967) "A Comparative Study of Alternative Estimators in a Distributed Lag Model", *Econometrica*, 35, 509–529.

Amemiya, T. and J. L. Powell (1980) "A Comparison of the Box-Cox Maximum Likelihood Estimator and the Nonlinear Two Stage Least Squares Estimator", Technical Report No. 322, August, Institute for Mathematical Studies in the Social Sciences, Stanford University.

Anderson, T. W. (1958) *An Introduction to Multivariate Statistical Analysis*. New York: John Wiley & Sons.

Anderson, T. W. (1971) *The Statistical Analysis of Time Series*. New York: John Wiley & Sons.

Arrow, K. J., H. B. Chenery, B. S. Minhas and R. M. Solow (1961) "Capital–Labor Substitution and Economic Efficiency", *Review of Economics and Statististics*, 43, 225–250.

Bard, Y. (1974) *Nonlinear Parameter Estimation*. New York: Academic Press.

Bates, D. M. and D. G. Watts (1980) "Relative Curvature Measures of Nonlinearity", *Journal of the Royal Statistical Society*, Ser. B, 42, 1–25 (with discussion).

Beale, E. M. L. (1960) "Confidence Regions in Non-Linear Estimation", *Journal of the Royal Statistical Society*, Ser. B, 22, 41–88 (with discussion).

Beauchamp, J. J. and R. G. Cornell (1966) "Simultaneous Nonlinear Estimation", *Technometrics*, 8, 319–326.

Belsley, D. A. (1979) "On the Computational Competitiveness of Full-Information Maximum-Likelihood and Three-Stage Least-Squares in the Estimation of Nonlinear Simultaneous-Equations Models", *Journal of Econometrics*, 9, 315–342.

Berndt, E. R., W. E. Diewert and M. N. Darrough (1977) "Flexible Functional Forms and Expenditure Distributions: An Application to Canadian Consumer Demand Functions", *International Economic Review*, 18, 651–676.

Berndt, E. R., B. H. Hall, R. E. Hall and J. A. Hausman (1974) "Estimation and Inference in Nonlinear Structural Models", *Annals of Econometric and Social Measurement*, 3, 653–666.

Berndt, E. R. and N. E. Savin (1977) "Conflict Among Criteria for Testing Hypotheses in the Multivariate Linear Regression Model", *Econometrica*, 45, 1263–1278.

Bianchi, C. and G. Calzolari (1980) "The One-Period Forecast Error in Non-linear Econometric Models", *International Economic Review*, 21, 201–208.
Bodkin, R. G. and L. R. Klein (1967) "Nonlinear Estimation of Aggregate Production Functions", *Review of Economics and Statistics*, 49, 28–44.
Box, G. E. P. and D. R. Cox (1964) "An Analysis of Transformations", *Journal of Royal Statistical Society*, Ser. B, 26, 211–252 (with discussion).
Brown, M. and D. Heien (1972) "The S-Branch Utility Tree: A Generalization of the Linear Expenditure System", *Econometrica*, 40, 737–747.
Bunke, H., K. Henscheke, R. Strüby and C. Wisotzki (1977) "Parameter Estimation in Nonlinear Regression Models", *Mathematische Operationsforschung und Statistik*, Series Statistics, 8, 23–40.
Charatsis, E. G. (1971) "A Computer Program for Estimation of the Constant Elasticity of Substitution Production Function", *Applied Statistics*, 20, 286–296.
Chow, G. C. (1973) "On the Computation of Full-Information Maximum Likelihood Estimates for Nonlinear Equation Systems", *Review of Economics and Statistics*, 55, 104–109.
Chow, G. C. and R. C. Fair (1973) "Maximum Likelihood Estimation of Linear Equation Systems with Auto-Regressive Residuals", *Annals of Economic and Social Measurement*, 2, 17–28.
Christensen, L. R., D. W. Jorgenson and L. J. Lau (1975) "Transcendental Logarithmic Utility Functions", *American Economic Review*, 65, 367–383.
Cox, D. R. (1961) "Tests of Separate Families of Hypotheses", in: J. Neyman (ed.), *Proceedings of the Fourth Berkeley Symposium on Mathematical Statistics and Probability*, vol. I. Berkeley: University of California Press, pp. 105–123.
Cox, D. R. (1962) "Further Results on Tests of Separate Families of Hypotheses", *Journal of the Royal Statistical Society*, Ser. B, 24, 406–424.
Dagenais, M. G. (1978) "The Computation of FIML Estimates as Iterative Generalized Least Squares Estimates in Linear and Nonlinear Simultaneous Equations Models", *Econometrica*, 46, 1351–1362.
Darrough, M. N. (1977) "A Model of Consumption and Leisure in an Intertemporal Framework: A Systematic Treatment Using Japanese Data", *International Economic Review*, 18, 677–696.
Davidon, W. C. (1959) "Variable Metric Method for Minimization", AEC Research Development Report, ANL-5990.
Deaton, A. S. (1974) "The Analysis of Consumer Demand in the United Kingdom, 1900–1970", *Econometrica*, 42, 341–368.
Diewert, W. E. (1973) "Separability and a Generalization of the Cobb–Douglas Cost, Production, and Indirect Utility Functions", Technical Report 86, January, Institute for Mathematical Studies in the Social Sciences, Stanford University.
Diewert, W. E. (1974) "Applications of Duality Theory", in: M. Intriligator and D. Kendrick (eds.), *Frontiers of Quantitative Economics*, vol. II. Amsterdam: North-Holland Publishing Co., pp. 106–171.
Draper, N. R. and H. Smith (1966) *Applied Regression Analysis*. New York: John Wiley & Sons.
Edgerton, D. L. (1972) "Some Properties of Two Stage Least Squares as Applied to Non-Linear Models", *International Economic Review*, 13, 26–32.
Eisenpress, H. and J. Greenstadt (1966) "The Estimation of Nonlinear Econometric Systems", *Econometrica*, 34, 851–861.
Fair, R. C. (1976) *A Model of Macroeconomic Activity, Vol. II: The Empirical Model*. Cambridge, Mass.: Ballinger.
Fair, R. C. and W. R. Parke (1980) "Full-Information Estimates of a Non-linear Macroecometric Model", *Journal of Econometrics*, 13, 269–291.
Fisher, F. M. (1966) *The Identification Problem in Econometrics*. New York: McGraw-Hill.
Fletcher, R. and M. J. D. Powell (1963) "A Rapidly Convergent Descent Method for Minimization", *Computer Journal*, 6, 163–168.
Gale, D. and H. Nikaido (1965) "The Jacobian Matrix and Global Univalence of Mappings", *Mathematische Annalen*, 159, 81–93.
Gallant, A. R. (1975a) "Nonlinear Regression", *The American Statistician*, 29, 73–81.
Gallant, A. R. (1975b) "The Power of the Likelihood Ratio Test of Location in Nonlinear Regression Models", *Journal of the American Statistical Association*, 70, 198–203.
Gallant, A. R. (1975c) "Testing a Subset of the Parameters of a Nonlinear Regression Model", *Journal of the American Statistical Association*, 70, 927–932.

Gallant, A. R. (1975d) "Seemingly Unrelated Nonlinear Regression", *Journal of Econometrics*, 3, 35–50.

Gallant, A. R. (1977) "Three-Stage Least-Squares Estimation for a System of Simultaneous, Nonlinear, Implicit Equations", *Journal of Econometrics*, 5, 71–88.

Gallant, A. R. and J. J. Goebel (1976) "Nonlinear Regression with Autocorrelated Errors", *Journal of the American Statistical Association*, 71, 961–967.

Gallant, A. R. and A. Holly (1980) "Statistical Inference in an Implicit, Nonlinear, Simultaneous Equation Model in the Context of Maximum Likelihood Estimation", *Econometrica*, 48, 697–720.

Gallant, A. R. and D. W. Jorgenson (1979) "Statistical Inference for a System of Simultaneous, Non-Linear, Implicit Equations in the Context of Instrumental Variable Estimation", *Journal of Econometrics*, 11, 275–302.

Glasbey, C. A. (1979) "Correlated Residuals in Non-Linear Regression Applied to Growth Data", *Applied Statistics*, 28, 251–259.

Guttman, I. and D. A. Meeter (1965) "On Beale's Measures of Non-Linearity", *Technometrics*, 7, 623–637.

Goldfeld, S. M. and R. E. Quandt (1968) "Nonlinear Simultaneous Equations: Estimation and Prediction", *International Economic Review*, 9, 113–136.

Goldfeld, S. M. and R. E. Quandt (1972) *Nonlinear Methods in Econometrics*. Amsterdam: North-Holland Publishing Co.

Goldfeld, S. M., R. E. Quandt and H. F. Trotter (1966) "Maximization by Quadratic Hill-Climbing", *Econometrica*, 34, 541–551.

Haessel, W. (1976) "Demand for Agricultural Commodities in Ghana: An Application of Nonlinear Two-Stage Least Squares with Prior Information", *American Journal of Agricultural Economics*, 58, 341–345.

Hannan, E. J. (1963) "Regression for Time Series", in: M. Rosenblatt (ed.), *Time Series Analysis*. New York: John Wiley & Sons, pp. 17–37.

Hannan, E. J. (1971) "Non-Linear Time Series Regression", *Journal of Applied Probability*, 8, 767–780.

Hartley, H. O. (1961) "The Modified Gauss-Newton Method for the Fitting of Non-Linear Regression Functions By Least Squares", *Technometrics*, 3, 269–280.

Hartley, H. O. (1964) "Exact Confidence Regions for the Parameters in Non-Linear Regression Laws", *Biometrika*, 51, 347–353.

Hartley, H. O. and A. Booker (1965) "Non-Linear Least Squares Estimation", *Annals of Mathematical Statistics*, 36, 638–650.

Hatanaka, M. (1978) "On the Efficient Estimation Methods for the Macro-Economic Models Nonlinear in Variables", *Journal of Econometrics*, 8, 323–356.

Hausman, J. A. (1975) "An Instrumental Variable Approach to Full Information Estimators for Linear and Certain Nonlinear Econometric Models", *Econometrica*, 43, 727–738.

Hildreth, C. and J. P. Houck (1968) "Some Estimators for a Linear Model with Random Coefficients", *Journal of the American Statistical Association*, 63, 584–595.

Hoadley, B. (1971) "Asymptotic Properties of Maximum Likelihood Estimators for the Independent Not Identically Distributed Case", *Annals of Mathematical Statistics*, 42, 1977–1991.

Howe, H., R. A. Pollack and T. J. Wales (1979) "Theory and Time Series Estimation of the Quadratic Expenditure System", *Econometrica*, 47, 1231–1248.

Hurd, M. D. (1974) "Wage Changes, Desired Manhours and Unemployment", Memorandum No. 155 (Revised), October, Center for Research in Economic Growth, Stanford University.

Jennrich, R. I. (1969) "Asymptotic Properties of Non-linear Least Squares Estimation", *Annals of Mathematical Statistics*, 40, 633–643.

Jorgenson, D. W. and J. Laffont (1974) "Efficient Estimation of Nonlinear Simultaneous Equations with Additive Disturbances", *Annals of Economic and Social Measurement*, 3, 615–640.

Jorgenson, D. W. and L. J. Lau (1975) "The Structure of Consumer Preferences", *Annals of Economic and Social Measurement*, 4, 49–101.

Jorgenson, D. W. and L. J. Lau (1978) "Testing the Integrability of Consumer Demand Functions, United States, 1947–1971", mimeo.

Judge, G. G., W. E. Griffiths, R. C. Hill and T. C. Lee (1980) *The Theory and Practice of Econometrics*. New York: John Wiley & Sons.

Just, R. E. and R. D. Pope (1978) "Stochastic Specification of Production Functions and Economic Implications", *Journal of Econometrics*, 7, 67–86.
Kelejian, H. H. (1971) "Two-Stage Least Squares and Econometric Systems Linear in Parameters but Nonlinear in the Endogenous Variables", *Journal of the American Statistical Association*, 66, 373–374.
Kelejian, H. H. (1974) "Efficient Instrumental Variable Estimation of Large Scale Nonlinear Econometric Models", mimeo.
Kelejian, H. H. (1975) "Nonlinear Systems and Non-Unique Solutions: Some Results Concerning Estimation", mimeo., May (Revised).
MacKinnon, J. G. (1976) "Estimating the Linear Expenditure System and Its Generalizations", in: S. M. Goldfeld and R. E. Quandt (eds.), *Studies in Nonlinear Estimation*. Cambridge, Mass.: Ballinger, pp. 143–166.
MaCurdy, T. E. (1980) "An Intertemporal Analysis of Taxation and Work Disincentives", Working Papers in Economics no. E-80-4, The Hoover Institution, Stanford University.
Malinvaud, E. (1970a) "The Consistency of Nonlinear Regressions", *Annals of Mathematical Statistics*, 41, 956–969.
Malinvaud, E. (1970b) *Statistical Methods of Econometrics* (2nd rev. edn.). Amsterdam: North-Holland Publishing Co.
Marquardt, D. W. (1963) "An Algorithm for Least Squares Estimation of Nonlinear Parameters", *Journal of the Society for Industrial and Applied Mathematics*, 11, 431–441.
Mizon, G. E. (1977) "Inference Procedures in Nonlinear Models: An Application in a UK Industrial Cross Section Study of Factor Substitution and Returns to Scale", *Econometrica*, 45, 1221–1242.
Phillips, P. C. B. (1976) "The Iterated Minimum Distance Estimator and the Quasi-Maximum Likelihood Estimator", *Econometrica*, 44, 449–460.
Phillips, P. C. B. (1981) "On the Consistency of Non-Linear FIML", mimeo.
Poirier, D. J. (1976) *The Econometrics of Structural Change*. Amsterdam: North-Holland Publishing Co.
Powell, M. J. D. (1964) "A Efficient Method for Finding the Minimum of a Function of Several Variables Without Calculating Derivatives", *Computer Journal*, 7, 115–162.
Priestley, M. B. (1978) "Non-Linear Models in Time Series Analysis", *The Statistician*, 27, 159–176.
Quandt, R. E. (1975) "A Note on Amemiya's Nonlinear Two-Stage Least Squares Estimators", Research Memorandum no. 178, May, Econometric Research Program, Princeton University.
Rao, C. R. (1947) "Large Sample Tests of Statistical Hypotheses Concerning Several Parameters with Applications to Problems of Estimation", *Proceedings of Cambridge Philosophical Society*, 44, 50–57.
Rao, C. R. (1973) *Linear Statistical Inference and Its Applications* (2nd edn.). New York: John Wiley & Sons.
Revankar, N. S. (1971) "A Class of Variable Elasticity of Substitution Production Function", *Econometrica*, 39, 61–71.
Rice, P. and V. K. Smith (1977) "An Econometric Model of the Petroleum Industry", *Journal of Econometrics*, 6. 263–288.
Robinson, P. M. (1972) "Non-Linear Regression for Multiple Time-Series", *Journal of Applied Probability*, 9, 758–768.
Sargent, T. J. (1978) "Estimation of Dynamic Labor Demand Schedules Under Rational Expectations", *Journal of Political Economy*, 86, 1009–1044.
Silvey, S. D. (1959) "The Lagrangian Multiplier Test", *Annals of Mathematical Statistics*, 30, 389–407.
Stein, C. (1973) "Estimation of the Mean of a Multivariate Normal Distribution", Technical Report no. 48, June 26, Department of Statistics, Stanford University.
Strickland, A. D. and L. W. Weiss (1976) "Advertising, Concentration, and Price-Cost Margins", *Journal of Political Economy*, 84, 1109–1121.
Theil, H. (1953) "Repeated Least-Squares Applied to Complete Equation Systems", mimeo. The Hague: Central Planning Bureau.
Theil, H. (1971) *Principles of Econometrics*. New York: John Wiley & Sons.
Tornheim, L. (1963) "Convergence in Nonlinear Regression", *Technometrics*, 5, 513–514.
Tsurumi, H. (1970) "Nonlinear Two-Stage Least Squares Estimation of CES Production Functions Applied to the Canadian Manufacturing Industries", *Review of Economics and Statistics*, 52, 200–207.

Wald, A. (1943) "Tests of Statistical Hypotheses Concerning Several Parameters When the Number of Observations is Large", *Transaction of American Mathematical Society*, 54, 426–482.

Wald, A. (1949) "Note on the Consistency of the Maximum Likelihood Estimate", *Annals of Mathematical Statistics*, 60, 595–601.

White, H. (1980a) "Nonlinear Regression on Cross-Section Data", *Econometrica*, 48, 721–746.

White, H. (1980b) "A Note on Normality and the Consistency of the Nonlinear Simultaneous Equations Maximum Likelihood Estimator", May, mimeo.

Wolter, K. M. and W. A. Fuller (1978) "Estimation of Nonlinear Errors-in-Variables Models", mimeo.

Wu, C. F. (1981) "Asymptotic Theory of Nonlinear Least Squares Estimation", *Annals of Statistics*, 9, 501–513.

Zellner, A. (1962) "An Efficient Method of Estimating Seemingly Unrelated Regressions and Tests for Aggregation Bias", *Journal of the American Statistical Association*, 57, 348–368.

Zellner, A., D. S. Huang and L. C. Chau (1965) "Further Analysis of the Short-Run Consumption Function with Emphasis on the Role of Liquid Assets", *Econometrica*, 33, 571–581.

Chapter 7

SPECIFICATION AND ESTIMATION OF SIMULTANEOUS EQUATION MODELS

JERRY A. HAUSMAN*

Massachusetts Institute of Technology

Contents

*I would like to thank A. Deaton, F. Fisher, W. Newey, D. Poirier, P. Ruud, W. Taylor, and the editors for helpful comments. The NSF provided research support.

Handbook of Econometrics, Volume I, Edited by Z. Griliches and M.D. Intriligator

1. Introduction

The simultaneous equation model is perhaps the most remarkable development in econometrics. Many of the models used in the statistical analysis of economic data arose from previous work in statistics. Econometric research has, of course, led to further developments and applications of these statistical models. But in the case of the simultaneous equation problem, econometrics has provided unique insight. And this insight arises from economic theory in terms of the operations of markets and the simultaneous determination of economic variables through an equilibrium model. Consider a linear regression specification which relates the quantity purchased of a commodity to its price at time t:

$$q_t = \beta_0 + \beta_1 p_t + Z_t \gamma + \varepsilon_t, \tag{1.1}$$

where Z_t is a $k \times 1$ vector of other variables thought to affect the relationship. What economic meaning can be given to the statistical specification of eq. (1.1)? More explicitly, is eq. (1.1) a demand curve or a supply curve or should we examine the least squares estimate $\hat{\beta}_1$ to decide upon our answer?

The econometricians' answer is that *both* quantity and price are simultaneously determined by the actions of the market so that to understand the quantity and price relationship we need to treat the two variables as jointly endogenous. Thus, eq. (1.1) considered in isolation is not sufficient to determine the economic meaning of the statistical relationship. Instead, we must consider a more complete model in which both quantity and price are determined simultaneously by the operation of economic markets. With joint endogeneity, least squares becomes the wrong statistical technique to use in estimation of the unknown coefficients of eq. (1.1). The implications of joint endogeneity of variables give the simultaneous equation model its unique characteristics which distinguishes it from most other regression type models in statistics. Developments in both identification and estimation of simultaneous equation models arise from the jointly endogenous feature of economic variables when they are treated from either a theoretical or statistical viewpoint.

The seminal papers from which the simultaneous equation model developed established the importance of joint endogeneity for statistical analysis of economic relationships. Haavelmo (1943, 1944) realized that in the presence of jointly endogenous variables that a joint probability distribution was necessary to analyze the data. He also distinguished the essence of the identification problem which arises in the attempt to determine what meaning can be given to specifications like eq. (1.1):

> Clearly no more complete description of the interconnections between certain random variables can be given than that which is formulated in their joint probability law. *If*, therefore, *two different formulations* of an economic theory lead to identically *the same joint probability law* of the observed random variables involved, we can *not distinguish between them on the basis of observations* [Haavelmo (1944, p. 88), his emphasis].[1]

We now return to our demand and supply example to see the importance of the concepts of identification and joint endogeneity for our statistical model. Let us specify eq. (1.1) to be the demand curve where we limit Z_t to a single variable, consumer income, for simplicity. We then specify the supply curve as

$$P_t = \alpha_0 + \alpha_1 q_t + W_t \delta + \eta_t. \tag{1.2}$$

Again to simplify we let W_t consist of a single variable, say the wage rate. We assume that both Z_t and W_t are determined exogenously to our model in the sense that both are (asymptotically) uncorrelated with the stochastic residuals ε_t and η_t. But in general neither p_t nor q_t can be taken as exogenous even in the particular equation in which they appear on the right-hand side because even if ε_t and η_t are independent, p_t is not uncorrelated with ε_t and likewise for q_t and η_t. The variables q_t and p_t are jointly endogeneous and require non-standard statistical treatment.

Before even considering questions of estimation, the problem of identification must be settled first. We have agreed to call eq. (1.1) the demand curve and eq. (1.2) the supply curve, but can we actually distinguish them statistically? Without the presence of Z_t and W_t our position is hopeless, because we would simply observe a scatter of points near the intersection (equilibrium position) of the demand and supply curves. The scatter would arise only because of the stochastic disturbance; and no statistical estimation procedure could establish the position of the demand or supply curve in price–quantity space from the observation of a single point perturbed by stochastic disturbances. But a change in Z_t in eq. (1.1), independent of ε_t and η_t, causes the demand curve to shift and permits points on the supply curve to be established. Likewise, shifts in W_t, again independent of ε_t and η_t, shift the supply curve so that points on the demand curve can be established. It is interesting to note that exogenous shifts in variables in the *other*

[1]Of course, Haavelmo's research had many antecedents. Working (1927) gave an early account of the identification problem. Joint endogeneity (although not caused by simultaneous determination of economic variables) might be said to have arisen first in the errors in variables problem in regression. Adcock (1878) is the first reference that I know to the errors in variables problem.

equation lead to identification of the equation in question. This finding is the basis for the previous remark than an equation cannot be considered in isolation but that a more complete model is required. Koopmans (1949), Koopmans and Reiersol (1950), and Koopmans, Rubin and Leipnik (1950) established conditions for identification in linear simultaneous models. We will further consider the identification problem in Section 3.

We now turn to statistical estimation. What are the properties of our estimates if we use least squares on eq. (1.1)? Let us assume that we measure all variables in deviations from their means so that β_0 and α_0 are eliminated. The least squares estimate of β_1 will be biased because of the correlation of p_t with ε_t and it has a probability limit

$$\operatorname{plim}\hat{\beta}_{1,\mathrm{OLS}}=\beta_1+\operatorname{plim}\left[\left(\frac{1}{T}p'Q_Z p\right)^{-1}\left(\frac{1}{T}p'Q_Z\varepsilon\right)\right], \tag{1.3}$$

where $Q_Z=I-Z(Z'Z)^{-1}Z'$.[2] The second term in eq. (1.3) is not zero because solving eqs. (1.1) and (1.2) in terms of the exogenous variables and the residuals yields

$$p_t=\frac{1}{1-\alpha_1\beta_1}(\alpha_1 Z_t\gamma+W_t\delta+\alpha_1\varepsilon_t+\eta_t). \tag{1.4}$$

Since ε_t appears in the reduced-form equation for p_t, its presence leads to both bias and inconsistency in $\hat{\beta}_1$. The direction of the bias depends on the magnitude of α_1 as well as the covariance between ε_t and η_t. Thus, least squares is an inappropriate estimator, in general, for the simultaneous equation model specification. But it turns out that an intimate relationship exists between identification and estimation in simultaneous equation models: an identified model can always be consistently estimated. We use W_t, the identifying variable from the other equation in our model, as an instrumental variable to find

$$\hat{\beta}_{1,\mathrm{IV}}=(W'Q_Z p)^{-1}W'Q_Z q. \tag{1.5}$$

We now have a consistent estimator because

$$\hat{\beta}_{1,\mathrm{IV}}=\beta+\operatorname{plim}\left[\left(\frac{1}{T}W'Q_Z p\right)^{-1}\left(\frac{1}{T}W'Q_Z\varepsilon\right)\right]. \tag{1.6}$$

So long as the first term in brackets has a finite plim, consistent estimation occurs

[2] Haavelmo (1944) was first to point out that least squares estimates of the coefficients of a structural equation are inconsistent. For a simple example he derived the plim of the inconsistent estimation. His argument against least squares is based on the correct point that the conditional expectation of the residuals given the right-hand-side variables is not zero, i.e. some of the right-hand-side variables are jointly endogenous.

because the second term has a zero plim under the assumption that W and Z are exogenous. Thus, just the variable that led to identification of the demand curve also provides us with the means to estimate consistently its parameters. This point is the basis for Haavelmo's discussion of the simultaneous equation problem.

I shall use the concept of instrumental variables to organize this survey. The most important set of identification conditions, namely coefficient restrictions, involves determining whether a sufficient number of instruments are available [cf. Fisher (1966)]. Furthermore, it has recently been proven that the other type of identification restrictions used in linear simultaneous equation models, namely covariance restrictions, are also most easily understood in terms of instrumental variables [Hausman and Taylor (1980a)]. In terms of estimation almost all consistent estimators are either instrumental variables estimators or asymptotic approximations to them. The original maximum likelihood estimator (FIML) proposed for the simultaneous equation model is an instrumental variable estimator [Hausman (1975)]; other estimators rely on asymptotic approximations to the basic likelihood equations [Hendry (1976)].[3] Estimation is considered in Section 4.

Two other interrelated notions that we consider are the endogeneity–exogeneity concept and tests of specification. We have emphasized joint endogeneity as the principle behind simultaneous equations models. Yet both identification and estimation rest on an exogeneity assumption, as our example indicates. We attempt to explore this assumption from a statistical point of view. The question naturally arises of whether the key exogeneity assumptions are testable, perhaps using recently developed techniques on causality from time-series analysis. The answer is no. If we have a surplus of exogenous variables, then a subset may be tested for endogeneity. But these tests depend on the maintained assumption of exogeneity in other variables. Specification tests look at questions of endogeneity and also at coefficient restrictions. In empirical work they are not used as often as they should be. In Section 5 we consider exogeneity tests and specification tests in reference to the simultaneous equation model.

Finally, in Section 6 we briefly consider the non-linear simultaneous equation model. In the general case it appears that identification ceases to be a problem. Consistent estimation by instrumental variables or by minimum distance estimators is possible [Amemiya (1974b)]. Yet at the present time problems which arise

[3]Another possible classification of estimators arises from a minimum distance (minimum chi square) interpretation. Malinvaud (1970) and Rothenberg (1973) use this approach. We consider this approach in Section 4. The reason that I prefer the instrumental variable approach is because it carries over to the case of non-linear simultaneous equations. As results in Section 6 demonstrate, the attraction of maximum likelihood estimation is reduced in the non-linear case because consistent estimation usually requires correct specification of the unknown stochastic distributions. Instrumental variable estimation does not require knowledge of the distributions. Furthermore, maximum likelihood is a particular application of instrumental variables, but not vice versa. Minimum distance estimation is severely limited in the non-linear case by the non-existence of a convenient reduced-form expression.

with estimation by maximum likelihood estimation have not been completely resolved.

2. Model specification

Three basic specifications have been used in the interpretation of linear simultaneous equation models: the structural form, the reduced form, and the recursive form. The structural form has stochastic equations and sometimes accounting identities which correspond to the basic economic theory underlying the model. It typically contains all the economic knowledge that we are able to include in the model. The demand and supply example of the previous section is a simple example of a structural model. The major difference between structural models and more traditional linear models in statistics is the presence of jointly endogenous variables. The reduced-form model can be obtained from the structural model by a non-singular linear transformation.[4] The joint endogeneity is eliminated from the model by the reduced-form transformation as each endogenous variable can be written as a linear function of only exogenous variables. Thus, the reduced-form specification is similar to the well-known multivariate least squares regression specification (although non-linear parameter constraints are typically present). The question might well be asked: Why then do we need the structural form?[5] Estimation and prediction might well proceed with the reduced form. The traditional answer is that the change in one structural equation will change the entire (restricted) reduced form. I do not find the answer particularly persuasive because we could re-estimate the reduced form after a structural change occurs.

Can a case be made for structural estimation? First, structural models provide a crucial inductive method to increase our knowledge about economic relationships and to test hypotheses about economic behavior. Almost all economic theory is concerned with structural models so that the unresolved questions of economics will usually be set within a structural framework. Also, when one considers a reduced form it contains all the current and lagged exogenous and endogenous (or predetermined) variables in the model on the right-hand side. We often will not have enough observations to estimate such a model in unrestricted form.[6]

[4]It is important to note that while the reduced form follows in a straightforward manner from the structural model in the linear case, usually no simple reduced-form specification exists in the non-linear case.

[5]Transformation from a structural form to a reduced form with regression properties cannot in general be accomplished in the *non-linear* simultaneous equation model. Important differences arise in identification and estimation from the presence of non-linearities.

[6]Since all restrictions arise from the structural model, specification and estimation of reduced-form models would presumably be done on unrestricted models.

Arguments have been put forth [Liu (1960) and Sims (1980)] that in reality structural models would also contain all endogenous and exogenous variables without identification being possible. Thus, they do not escape the problems inherent in reduced-form estimation. Models are always at best approximations to reality. It seems that whichever approach is taken, either structural specifications or reduced-form specifications, restrictions on either the structural form or reduced form will be necessary for estimation to be possible.[7] Economic theory provides some guidance about restrictions on the structural form which in part can be tested, while we have little theory to guide us about which variables to omit from a reduced-form specification. Lastly, while we can always go from the structural form to the reduced form, the reverse transformation is impossible to perform when the unrestricted reduced form is used. It is unclear how much we can learn about economic behavior by specification and estimation of a reduced form unless it is derived from a structural form.[8] This point may have important implications for the analysis of economic policy. If a structural parameter were to change, perhaps because policymakers change their behavior, analysis via the structural form may be useful. Possible analysis via the unrestricted reduced form seem quite limited here.

The last specification, the recursive form, can also be derived from the structural form via a non-singular linear transformation. In unrestricted form the recursive form can be estimated by least squares type techniques. However, interpretation of the resulting parameter estimates is not straightforward. Moreover, severe restrictions have to be placed on the distribution of the stochastic disturbances for least squares to yield consistent estimates and for the parameters to be readily interpretable. Without the restrictions the recursive form has little to add to the analysis beyond the structural form and the reduced form. Very little research is currently done on recursive-form specifications of simultaneous equation models. In previous work Wold (1964) and others have argued that the structural form is naturally recursive so that least square type techniques are appropriate because the structural specification takes the recursive form. But the necessary assumptions for the recursive specification are usually beyond acceptance. However, the use of a block recursive specification is sometimes made so that analysis of a large econometric model may be simplified [Fisher (1966, ch. 4) and Koopmans (1950)].

[7]Fisher (1961) considers the effects on identification and estimation when the restrictions are very close to true but not exactly met.

[8]For forecasting purposes we have many unresolved questions about what type of model might be more appropriate. For further discussion, see Fair (Chapter 33 in this Handbook). To the extent the unrestricted reduced form is used to test the structural specification, it is an underutilized tool of econometric methodology. We discuss such tests in Section 5. Use of the reduced form for tests of dynamic specification may also be quite important.

The general specification of a linear structural model is

$$YB + Z\Gamma = U, \tag{2.1}$$

where Y is the $T \times M$ matrix of jointly endogenous variables, Z is the $T \times K$ matrix of predetermined variables which contains both exogenous and lagged endogenous variables, and U is a $T \times M$ matrix of the structural disturbances of the system. The matrices B and Γ consist of unknown parameters to be estimated as well as known values (usually zeros) which arise from *a priori* economic knowledge. We assume that all identities have been substituted out of the system which typically may lead to an arithmetic combination of some of the original economic variables. We also assume that all variables are accurately measured because a consideration of errors in variables in simultaneous equation models would lead us too far afield.[9]

We now consider some assumptions that permit statistical analysis of the structural equation model of eq. (2.1).

Assumption 2.1

B is a non-singular matrix.

The non-singularity assumption allows us to solve for the endogenous variables in terms of the predetermined variables, i.e. to solve for the reduced form. We can write the reduced form after postmultiplication of eq. (2.1) by B^{-1} as

$$Y = -Z\Gamma B^{-1} + UB^{-1} = Z\Pi + V. \tag{2.2}$$

If contrary to assumption, B were singular, then the model does not provide a complete theory of the determination of the endogenous variables. Also, eq. (2.2) demonstrates that a small structural disturbance could lead to an infinite change in some of the dependent variables. Such an event is contrary to most economic theory.

Assumption 2.2.

Z has full column rank equal to k.

We rule out linear dependence so that the reduced form has a unique interpretation in terms of its unknown coefficients.

Assumption 2.3

The rows of U are independent and identically distributed. U has mean zero and non-singular covariance matrix $\Sigma \otimes I_T$. (Thus, the tth row of U, denoted U_t, has mean zero and covariance matrix Σ.)

[9]Some recent work in this area is found in Goldberger (1970), Geraci (1977, 1978), Hausman (1977), and Hsiao (1976). Also, see Aigner, Hsiao, Kapteyn and Wansbeek (Chapter 23 in this Handbook).

We assume independence of the structural disturbances across time and allow only for contemporaneous covariance in the case of time-series model specifications. Non-independence across time is briefly treated in subsequent sections, but is left mainly to other chapters in this Handbook which deal with time-series problems. For cross-section model specifications we are assuming independence of the structural disturbances across individuals who would be indexed by t. Note that with these three assumptions, all information contained in the structural model is also contained in the reduced-form model. In particular, the reduced-form model determines the conditional distribution (on Z) of the endogenous variable since V has mean zero and non-singular covariance matrix $\Omega\otimes I_T$, where $\Omega = B^{-1\prime}\Sigma B^{-1}$. We now consider a more precise statistical definition of jointly endogenous and predetermined variables. We separate the exogenous variables into two sets, truly exogenous variables R and lagged endogenous variables Y_0 which occur before the start of the sample period. The variables Y_0 are treated as initial conditions for the analysis that follows which is done conditionally on Y_0. We consider the joint probability distribution of U, R, and Y_0 which by eq. (2.2) determines the joint distribution of Y:

$$G(U,R,Y_0) = G_1(U|R,Y_0)G_2(R,Y_0). \tag{2.3}$$

A decomposition of a joint distribution into conditional and marginal distributions is always possible, but the importance of the exogenous variable assumption arises from

Assumption 2.4

$$G_1(U|R,Y_0) = G(U). \tag{2.4}$$

Thus, the conditioning information adds no knowledge to the joint distribution of U which is assumed independent of all past, current, and future realizations of the exogenous variables. As Assumption 2.4 makes clear, the distribution of exogenous variables is independent of the structural population parameters. This assumption corresponds to the Koopmans et al. (1950, p. 56) definition: "[exogenous variables] are defined as variables that influence the endogenous variables but are not themselves influenced by the endogenous variables." In particular note that an implication of eq. (2.4) is $EU = E(U|R,Y_0) = 0$.

We now turn to the other component of predetermined variables, namely lagged endogenous variables. Clearly, the conditioning argument is nonsensical here because knowledge of a realization of a lagged endogenous variable together with the right-hand-side variables from the previous period certainly imparts information about the distribution of the stochastic disturbances in the previous period. But an implication of our assumptions is that if we consider the marginal distribution of U_t, which corresponds to a row of the U matrix, this marginal distribution equals the conditional distribution given the knowledge of all *past* realization of the endogenous variables

$$G_t(U_t) = G_t(U_t|R, Y_0, Y(-)), \tag{2.5}$$

where $Y(-)$ denotes lagged endogenous variables. Lagged endogenous variables can thus be treated along with R and Y_0 as predetermined at a particular point in time because they are not affected by the realization U_t. Thus, the assumption of temporal independence allows an important simplification since $Y(-)$ can be solved for in terms of exogenous variables and lagged stochastic disturbances assumed independent of current stochastic disturbances.

We have attempted a definition of variables which corresponds to the framework of Koopmans (1950). A time-series perspective on these issues is also present in Zellner and Palm (1975) and Wallis (1977). Further understanding may be gained from the following considerations. Given our assumptions, suppose we want to estimate the model by maximum likelihood. By Assumption 2.4 we need to choose a parametric form for G. Suppose for the moment that all predetermined variables are exogenous with no lags. Let θ denote all unknown parameters. We have the joint density:

$$f(Y, Z, \theta) = g(U, Z, \theta)\left[\frac{\partial u}{\partial y}\right] = g_1(u, \theta_1|Z)\left[\frac{\partial u}{\partial y}\right] g_2(Z, \theta_2). \tag{2.6}$$

The Jacobian of the transformation for the linear case is $|B|^T$ which is non-zero by Assumption 2.1. So long as the unknown parameter vector θ can be separated into two parts so that θ_1 and θ_2 are separate from an estimation point of view, then the exogenous variables can be taken as fixed numbers for purposes of estimation of θ_1 by ML.[10] Thus, $f(Y, Z, \theta)$ can be understood as two consecutive experiments (by nature). The first experiment chooses Z as a function of θ_2 through $g_2(Z, \theta_2)$. This first experiment does not give any information regarding the parameters of interest, θ_1. Given the realized values of Z, the second experiment yields information on θ_1 only. The additional knowledge of $g_2(Z, \theta_2)$ is irrelevant since all information about θ_1 arises from the second experiment. This setup corresponds to R. A. Fisher's (1935, 1956) definition of *ancillarity*. Here, it is certainly the case that inference on θ_1 depends on the first experiment since inference is typically done conditional upon its outcome. The first experiment affects the precision of our inference about θ_1, but not the direct inference itself.[11] Furthermore, as I emphasized in Hausman (1975), it is the presence of the

[10]Note that inference regarding the parameters may well depend on the distribution g_2 given the conditional form of equation (2.6).

[11]Of course, this statement does not imply that the precision about θ_1 can be improved by further analysis of $g_2(Z, \theta_2)$.

Jacobian term that distinguishes the jointly endogenous variables in simultaneous equation estimation by ML and removes estimation from the least squares framework.

When lagged endogenous variables are included, we shall assume that Y_0 is fixed in repeated samples. The Jacobian of the transformation then becomes

$$\begin{bmatrix} \frac{\partial u_1}{\partial y_1} & \frac{\partial u_2}{\partial y_1} & \cdots & \frac{\partial u_T}{\partial y_1} \\ 0 & \frac{\partial u_2}{\partial y_2} & \cdots & \frac{\partial u_T}{\partial y_2} \\ 0 & 0 & \cdot & \cdot \\ & 0 & & \vdots \\ \vdots & \vdots & \ddots & \\ 0 & 0 & & \frac{\partial u_T}{\partial y_T} \end{bmatrix} = \prod_{t=1}^{T} \frac{\partial u_t}{\partial y_t}. \tag{2.7}$$

Again, knowledge of previous realizations of endogenous variables do not alter the distribution of the U_t. But note that if the U_t's are not assumed independent, lagged endogenous variables are no longer predetermined in a statistical sense but instead become jointly endogenous. The factorization present in eq. (2.6) thus seems to be a useful way of characterizing statistical exogeneity in that the joint distribution can be separated into a conditional and marginal distribution with unknown parameters also separated into two subvectors, using the notion of ancillarity. The further extension to predetermined variables then can be made when the assumption of independence of stochastic disturbances is made.[12]

The recursive form can be derived from the structural form of eq. (1.1) by finding a matrix P which simultaneously diagonalizes Σ and triangularizes B. Since Σ is symmetric positive definite and B non-singular, at least one such matrix P exists.[13] Postmultiplication yields

$$YBP + Z\Gamma P = UP \quad \text{or} \quad YC + ZD = W. \tag{2.8}$$

[12] Engle et al. (1981) take a somewhat different approach to the definition of exogeneity. It is important to note that a given variable may be jointly endogenous with respect to one equation in a structural system but predetermined with respect to another equation. Examples of such "relative recursivity" are presented in the next section. In these cases the factorization of (2.5) need not be for the entire structural system, but instead it is for a given equation being estimated which corresponds to a subvector of θ_1.

[13] P can be found by the following method. Take the reduced-form system of eq. (2.2), $Y - Z\Pi = V$. Then take the Cholesky factorization $\Omega^{-1} = RR'$, where R is a lower triangular matrix. Thus, $YR - Z\Pi R = VR$ so that var(VR) $= I_M$. Then rescale so that $r_{jj} = 1$ for the conventional normalization. Therefore, $P = B^{-1}R\Lambda$, where Λ is the diagonal rescaling matrix.

C is now lower triangular and $\text{var}(W) = \Lambda^2$, a diagonal matrix upon choice of renormalization. It was once claimed that the recursive form has special advantages in estimation. However, in fact it has no particular advantage over either the structural form or reduced forms from which it can be derived. Furthermore, the parameter matrices C and D are mixtures of structural slope coefficients and covariance coefficients as can be seen by the construction of P. Given our usual almost complete lack of knowledge regarding Σ, the recursive-form coefficients are even more difficult to interpret than are the reduced-form coefficients Π. Of course, if the structural specification took the special form of eq. (2.8) without any needed transformation, i.e. P is the identity matrix, then special characteristics do occur. Basically, the simultaneous equation problem disappears because no special problems of identification or estimation beyond the usual least squares case occur, as we demonstrate in the next section. However, the specification of B as triangular and Σ as diagonal seems unacceptable in most model specifications. While the recursive form offers few advantages and is now not often used, it does provide a useful reference point for special cases of the structural form. We will see this distinction as we now turn to identification of simultaneous equation models.

3. Identification

Identification in parametric statistical models has an extremely simple form. Suppose we intend to estimate the unknown parameter θ from eq. (2.6). The identification assumption, which is a regularity condition for the consistency of maximum likelihood, states there cannot exist $\theta' \neq \theta$ such that $f(Y, Z, \theta') = f(Y, Z, \theta)$ for all Y and Z. In the linear simultaneous equation model, where we estimate the conditional form of eq. (2.6), the parameter vector of interest $\theta_1 = (B, \Gamma, \Sigma)$. The identification assumption then determines whether the *a priori* structural assumptions are sufficient to guarantee the uniqueness of (B, Γ, Σ). Often we may similarly be interested in a subset of the parameters, say B_1 and Γ_1, the first columns of B and Γ, which correspond to the parameters of the first equation. We would then partition θ_1 and discuss identification of the subset of parameters in terms of the non-existence of alternative subvectors which lead to the same density function. A considerable body of literature has arisen on this topic. Koopmans (1949) and Koopmans, Rubin and Leipnik (1950) solved the identification problem for the case of linear restrictions on B and Γ. Fisher (1966) reinterpreted the Koopmans' conditions and also considered linear restrictions on the elements of Σ. Wegge (1965) also considered covariance restrictions. Recently, Hausman and Taylor (1980) have provided convenient sufficient conditions for the case of restrictions on Σ.[14]

[14]Hsiao (Chapter 4 in this Handbook) also discusses identification of simultaneous equation models. He discusses the case of errors in variables also.

We briefly repeat our assumptions of Section 2: B is non-singular, Z has full column rank, and the rows of U are i.i.d. As we mentioned above, since the reduced-form model determines the conditional distribution of the endogenous variables, all the parameters (Π, Ω) are identified. Identification of the structural parameters thus is equivalent to the question of whether any other structural model can have the same reduced form model. Define $A = [B \quad \Gamma]$. Then a structural model is not identified if there exists a non-singular linear transformation H such that $A' = AH$ and $U' = UH$ is i.i.d. with mean zero and covariance matrix Σ. The possible existence of such a matrix H is closely tied with Fisher's (1966) approach to identification via admissible transformations.

We first consider the case of *a priori* restrictions on B and Γ while Σ is left unrestricted. It is important to note that we use no information which might arise from possible knowledge of the form of the probability distribution $G(U)$. If we required the matrix H to have the property that $G(U) = G(UH)$, then in many cases the necessary and sufficient conditions for identification would be very much weaker, except in the case when $G(\cdot)$ is M-variate normal.[15] We want to interpret the identification conditions on B and Γ in instrumental variable form since we emphasize this approach to estimation. For the case of restrictions on B and Γ we limit attention to the case of linear restrictions. These restrictions arise from *a priori* economic theory and usually correspond to omission of elements Y and Z from a given equation. We impose the exclusion restrictions and a normalization $B_{ii} = 1$ $(i = 1, \dots, M)$ and rewrite the equations in regression form as

$$y_i = Y_i\beta_i + Z_i\gamma_i + u_i = X_i\delta_i + u_i \qquad (i = 1, \dots, M), \tag{3.1}$$

where $X_i = [Y_i \quad Z_i]$, $\delta_i' = [\beta_i' \quad \gamma_i']$ and eq. (3.1) contains r_i jointly endogenous variables on its right-hand side and s_i predetermined variables for a total of $k_i = r_i + s_i$ right-hand-side variables. In the current case, without loss of generality, we are concerned with the identification of the unknown parameter vector $(\beta_1, \gamma_1, \sigma_{11})$ in the first structural equation. Thus, the identification problem is to derive necessary and sufficient conditions so that the equations $\Pi = -\Gamma B^{-1}$ and $\Omega = (B')^{-1}\Sigma B^{-1}$ yield a unique solution for $[B_1, \Gamma_1, \sigma_{11}]$ given (Π, Ω) and prior information on $[B, \Gamma, \Sigma]$, where B_1 is the first column of B and Γ_1 is the first column of Γ.

For the case of linear restrictions on (B_1, Γ_1) we write the prior information as

$$\Phi \begin{bmatrix} B_1 \\ \Gamma_1 \end{bmatrix} = \phi, \tag{3.2}$$

[15] This situation is analogous to the classical bivariate errors in variables problem, e.g. Kendall and Stuart (1967), where Reiersol (1950) demonstrated that the model is identified for all bivariate distributions *except* the normal distribution.

when Φ is a $g\times(M+K)$ matrix of known constants and ϕ is a known g vector. Since at present we take Σ as entirely unknown, the only restrictions which arise from the reduced-form equations on (B_1,Γ_1) take the form $\Pi B_1 = -\Gamma_1$ together with eq. (3.2). Therefore (B_1,Γ_1) is identified if and only if

$$\begin{bmatrix} \Pi & I \\ \multicolumn{2}{c}{\Phi} \end{bmatrix}\begin{bmatrix} B_1 \\ \Gamma_1 \end{bmatrix} = \begin{bmatrix} 0 \\ \phi \end{bmatrix} \tag{3.3}$$

has a unique solution for (B_1,Γ_1). A necessary and sufficient condition is that

$$\operatorname{rank}\begin{bmatrix} \Pi & I \\ \multicolumn{2}{c}{\Phi} \end{bmatrix} = M+K. \tag{3.4}$$

Equation (3.4) is the rank condition for identification and is proven in Hood and Koopmans (1953), Fisher (1966), and most econometrics textbooks. The necessary order condition is that $g \geqslant M$ so that eq. (3.3) has at least $M+K$ rows. Then for our normalized equation (3.1), we see that (β_1,γ_1) are identified if and only if

$$\Pi_{12}\tilde{\beta}_1 = 0 \tag{3.5}$$

has a unique solution for $\tilde{\beta}_1 = (-1,\beta_1)$, where Π_{12} is the submatrix of Π which relates the endogenous variables included in the first equation (y_1, Y_1) with the excluded predetermined variables. The order condition is then $k_1 = r_1 + s_1 \leqslant K$ or $r_1 \leqslant K - s_1$. That is, the number of included jointly endogenous right-hand-side variables must be no greater in number than the excluded predetermined variables. These excluded predetermined variables are used to form the instruments for consistent estimation, as our example in Section 1 demonstrated. We discuss instruments further in the next section on estimation. Letting W_1 be the matrix of instruments, the rank condition takes the form that identification is present if and only if $(W_1'X_1)$ is non-singular.

How do conditions change if Σ_1 is known? We then have the additional equation to $\Pi B_1 = -\Gamma_1$ that $\Omega B_1 = (B')^{-1}\Sigma_1$. For identification of the first equation the only useful restriction is $\sigma_{11}=0$, so that the first structural disturbance is identically zero. We then have the result that (B_1,Γ_1) are identified, using only prior restrictions on (B_1,Γ_1) and $\sigma_{11}=0$ if and only if

$$\operatorname{rank}\begin{bmatrix} \Pi & I \\ \multicolumn{2}{c}{\Phi} \\ \Omega & 0 \end{bmatrix} = M+K. \tag{3.6}$$

This result is equivalent to the generalized rank condition of Fisher (1966, ch. 3).

Partition $\Omega = [\Omega_1 \; \vdots \; \Omega_2]$ as we did Π and we find the necessary condition that $\text{rank}(\Omega_1) \geqslant (r_1 - 1) \dot{-} (K - s_1)$. If the rank of Σ and thus Ω is $M-1$, the order condition is fulfilled even without coefficient restrictions. This rather peculiar result for the non-stochastic situation arises because if $\sigma_{11} = 0$, $\text{plim}(1/T)Y_i'U_1 = 0$ for $i = 1, \dots, M$, so that every jointly endogenous variable is predetermined in the first equation and can be used as an instrument so long as the rank condition is satisfied. The case of variance restrictions, $\sigma_{ii} = 0$, is not of very much importance in econometric applications.

Lastly, we consider covariance restrictions. It turns out that covariance restrictions can yield identification in one of two ways: an otherwise jointly endogenous variable can be made predetermined, somewhat like the last example, or an estimated residual from an otherwise identified equation can serve as an instrument. We report results from Hausman and Taylor (1980a) where proofs are given. Besides linear restrictions on (B_1, Γ_1) and zero restrictions on Σ_1 we also use exclusion (zero) restrictions on certain other elements of B. We begin with two definitions:

Definition

For a $G \times G$ matrix B, a *chain product* corresponding to the ith row and the jth column is a product of no more than $G-1$ elements of B of the form $\beta_{ia}\beta_{ab}\beta_{bc} \cdots \beta_{fj}$, where all indices are distinct. The set of all such chain products is denoted $B_{[i,j]}$.

Definition

Equations (i, j) are *relatively triangular* if and only if $B_{[i,j]} = \{0\}$. Equations (i, j) relatively triangular does not imply that equations (j, i) are relatively triangular.

It turns out to be the case that the relative triangularity of equations (i, j) is equivalent to a zero in the (i, j)th position of $(B')^{-1}$. The relative triangularity of equations (i, j) is a necessary condition for y_j to be uncorrelated with u_i and thus to be predetermined in the ith equation. We now need to consider zero restrictions on (B, Σ_1) which are useful for identification.

We have the result that $((B')^{-1}\Sigma_1)_j = 0$ if and only if equations $(j, 1)$ are relatively triangular and u_1 is uncorrelated with u_k for equations $(k, 1)$ which are not relatively triangular. This condition is less restrictive than $\sigma_{1i} = 0$ for $i = 2, \dots, M$. We now give our final definition:

Definition

Equations $(1, j)$ are *relatively recursive* if and only if $((B')^{-1}\Sigma_1)_j = 0$.

Then y_j is uncorrelated with u_1 (because v_j is uncorrelated with u_1) and can be considered predetermined in the first equation along the lines of eq. (2.4).

Relative recursion occurs if and only if y_j is predetermined in the first equation. Thus, we have demonstrated our first method by which zero restrictions on (B_1, Σ_1) are useful for identification. They cause the jointly endogenous variable y_j to be predetermined in the first equation so that it can be used to form the instrument matrix W_1. Writing the zero restrictions as $\Psi(B')^{-1}\Sigma_1 = 0$ we note that this equation yields $\Psi\Omega B_1 = 0$. Then Hausman and Taylor (1980, p. 27) prove that a necessary and sufficient condition for the identification of (B_1, Γ_1) using linear restrictions on (B_1, Γ_1) and zero restrictions on (B, Σ_1) is

$$\operatorname{rank}\begin{bmatrix} \Pi & \vdots & I \\ & \Phi & \\ \Psi\Omega & \vdots & 0 \end{bmatrix} = M + K. \tag{3.7}$$

The necessary order condition takes the form that $(B_1, \Gamma_1, \Sigma_1)$ is identified given Φ and Ψ so long as the number of unconstrained coefficients k_1 does not exceed the number of instrumental variables which can include all predetermined variables for the first equation. The necessary condition for the instruments is

$$\operatorname{rank}\begin{bmatrix} \Pi_{12} \\ \Psi\Omega_1 \end{bmatrix} = r_1 - 1.$$

Thus, we see that identification via covariance restrictions is closely connected with the existence of instruments for estimation.

A particular example of the importance of covariance restrictions for identification is the recursive system specification. For this specification B is assumed to be lower triangular and Σ is specified to be diagonal [Wold (1964)]. Then all equations (i, j) are relatively triangular for i greater than j, $i, j = 1,\dots,M$. Then y_j is uncorrelated with u_i because $((B')^{-1}\Sigma_i)_j = 0$ and v_j is uncorrelated with u_i. Therefore y_j is predetermined in the ith equation. Two implications arise from this result. The rank condition of eq. (3.7) is satisfied. Furthermore, since all right-hand-side variables are predetermined, least squares on each equation is the appropriate estimator. But, as we discussed above, the assumptions necessary for a recursive system are unacceptable, especially that Σ is diagonal. If B is specified to be lower triangular, but no assumptions on Σ are made beyond the usual conditions, the system is called triangular. Interesting differences in estimation still arise from the non-triangular specification, but no special issues in identification arise. We consider estimation of triangular systems in the next section [eq. (4.20)].

The previous discussion of (relative) recursiveness raises the important question of when can a variable, say y_j, be treated as predetermined in a particular equation? An analogous concern is whether y_j can be used as an instrumental variable for estimation. As the discussion demonstrated, in the presence of covariance restrictions y_j will not necessarily be predetermined with respect to all equations as we assumed the z's are. For y_j to be predetermined with respect to equation i, it is necessary that $\text{plim}(1/T)v_j'u_i = 0$. Two important consequences of this condition are as follow. (1) We cannot generally say that y_j is predetermined (exogenous) with respect to another variable y_i apart from the specification of the complete system of equations. Bivariate relations which attempt to test for "causality" cannot be used to decide whether y_j can be used as an instrumental variable in equation i. We discuss this point further in Section 5. (2) For any economic variable to be predetermined its reduced-form residual must be uncorrelated with u_i. This event may often seem unlikely in many models. But as a model approximation, it is sometimes assumed that a variable is relatively recursive with respect to an equation because there is no structural feedback from variable y_i to the predetermined variable y_j. But note that covariance restrictions are also necessary, as our definition of relative recursivity demonstrates.[16] Therefore it is the relationship of v_j and u_j which is critical; conditions on both B and Σ are required for a variable to be predetermined with respect to another equation.

The second form of covariance identification occurs when $\sigma_{1j} = 0$ but equation j is otherwise identified. Consistent estimates of the residuals $\hat{u}_j$ can then be used for identification. We thus need to consider the more general case of linear restrictions on all of (B, Γ, Σ). The proposition is that if $\sigma_{1j} = 0$ and B_j is either known or estimable, then $Y\hat{B}_j$ is predetermined in the first structural equation where $\hat{\boldsymbol{B}}_j \equiv (1 - \hat{\beta}_j)$. Then either $Y\hat{B}_j$ or the residual $\hat{u}_j$ can be used interchangeably as an instrumental variable in the first equation. The rank condition arises from the proposition that if y_1 is predetermined in the jth equation, then $\hat{u}_j$ cannot be used as an instrument for the first equation. Likewise, if y_k is predetermined in the jth equation, then $\hat{u}_j$ cannot be used as an instrument for y_k in the first equation. Otherwise, estimated residuals can be used to form instruments. We give an example in the next section.

To summarize: four forms of linear restrictions can lead to identification of $(B_1, \Gamma_1, \Sigma_1)$, all of which have straightforward interpretations via instrumental variable interpretation. (1) Coefficient restrictions of the Koopmans type which state that the number of included right-hand-side jointly endogenous variables must be no greater in number than the excluded predetermined variables which can serve as instruments. (2) The Fisher condition $\sigma_{11} = 0$ so that no simultaneous equation problem occurs. (3) The Hausman–Taylor condition so that jointly

[16] Fisher (1966, ch. 4) discusses "block recursive" systems where it is assumed that B is block triangular and Σ is block diagonal. Then endogenous variables from a block of B are predetermined with respect to the equations of a higher numbered block because the condition of relative recursiveness is met.

endogenous variables become predetermined in the first equation. (4) Estimated residuals from another equation can serve as instruments in the first equation.[17] This list exhausts all linear identification restrictions for the linear simultaneous model which have been considered in the literature.

4. Estimation

Estimation of simultaneous equation models is the central focus of this chapter. We have considered questions of specification and identification. We now assume that sufficient restrictions are present for identification and that all restrictions are linear restrictions on the parameters of a given equation (B_i, Γ_i). Below, we will consider covariance restrictions on the elements of Σ. The early work by Haavelmo (1944) and the Cowles Commission, i.e. Koopmans (1950) and Hood and Koopmans (1953), emphasized maximum likelihood estimation of simultaneous equation models. The probability distribution of Assumption 2.4 was taken to be multivariate normal, $G(U) = N(0, \Sigma \otimes I_T)$. At that time it was realized [Koopmans et al. (1950, section 3)] that consistency of maximum likelihood was maintained even if $G(U)$ was not normal so long as Assumption 2.3 concerning the moments of U was satisfied. Still, in the pre-computer age of estimation, maximum likelihood estimation represented a laborious task. Not until the two-stage least squares (2SLS) procedure of Basmann (1957) and Theil (1958) was invented did consistent estimation of simultaneous equation models require only the same order of magnitude of computation as least squares for regression specifications. 2SLS is an example of instrumental variable (IV) estimation. Sargan (1958) introduced instrumental variable estimation for simultaneous equation models. It is interesting to note that IV estimation for errors in variables models was known at the time of the Cowles Commission studies, i.e. Geary (1949) and Reiersol (1945). But application of IV estimation to simultaneous equation models did not occur until after a further decade had passed.

4.1. *Single equation estimation*

First we consider estimation of a single equation, say the demand equation from our initial example, eq. (1.1). We denote it as the first equation and rewrite eq. (3.1) as

$$y_1 = X_1\delta_1 + u_1 \quad \text{for } X_1 = [Y_1 \quad Z_1]. \tag{4.1}$$

[17]Leamer (1981) considers the use of covariance restrictions to estimate an interval which contains the true parameter of the endogenous variables in a two-equation system. The technique is very similar to that sometimes used in errors in variables problems.

Least squares is inconsistent because

$$\operatorname{plim}(\hat{\delta}_{1,\mathrm{OLS}} - \delta_1) = \operatorname{plim}\left[(X_1'X_1)^{-1}X_1'u_1\right] \neq 0. \tag{4.2}$$

To evaluate eq. (4.2) we need two other assumptions:

Assumption 4.1

The matrix $M = \operatorname{plim}(1/T)Z'Z$ is non-singular as T goes to infinity.

Assumption 4.2

If lagged endogenous variables are present as predetermined variables, all roots of the characteristic equation $|\alpha^n B + \alpha^{n-1}H_1 + \cdots + \alpha H_{n-1} + H_n| = 0$ lie within the unit circle where $\tilde{Z}_t$ are exogenous variables and $y_t B + \tilde{Z}_t G + \sum_{\tau=1}^{n} y_{t-\tau}H_\tau = u_t$, i.e. the system is stable.

We can now evaluate eq. (4.2):[18]

$$\operatorname{plim}(\hat{\delta}_{1,\mathrm{OLS}} - \delta_1) = \operatorname{plim}\begin{bmatrix} \hat{\beta}_{1,\mathrm{OLS}} - \beta_1 \\ \hat{\gamma}_{1,\mathrm{OLS}} - \gamma_1 \end{bmatrix}$$

$$= \begin{bmatrix} (R_{11} - S_{11}M_{11}^{-1}S_{11}')^{-1} \\ -M_{11}^{-1}S_{11}'(R_{11} - S_{11}M_{11}^{-1}S_{11}')^{-1} \end{bmatrix}\left[(B')^{-1}\Sigma_1\right]_1, \tag{4.3}$$

where $R_{11} = \Pi_1'M\Pi_1 + \Omega_{11}$ for Π_1 corresponding to Y_1, M_{11} is the submatrix of M associated with Z_1, and $S_{11} = \Pi_1'M[I_{s_1} \;\; 0]'$. Thus, both subvectors of δ_1 are inconsistently estimated because of the correlation of Y_1 with u_1. Note that the matrix $[(B')^{-1}\Sigma_1]_1$ is the covariance of Y_1 and u_1 as expected given the covariance restriction results on relatively recursive variables which were discussed in the previous section.

Instrumental variable estimation provides a consistent estimator for δ_1. For consistent IV estimation we require a $T \times K_1$ matrix W_1 of instruments to estimate

$$\hat{\delta}_{1,\mathrm{IV}} = (W_1'X_1)^{-1}W_1'y_1. \tag{4.4}$$

We see the first requirement of the instrument matrix: it must be correlated with X_1. The second requirement follows from

$$\operatorname{plim}(\hat{\delta}_{1,\mathrm{IV}} - \delta_1) = \operatorname{plim}\left[(W_1'X_1)^{-1}W_1'u_1\right]. \tag{4.5}$$

[18]The notation $[(B')^{-1}\Sigma]_1$ selects the endogenous variables from the equation analogously to our notation used in the definition of relatively recursive in the previous section.

So long as plim $T(W_1'X_1)^{-1}$ exists and is finite, we require that $\operatorname{plim}(1/T)W_1'u_1 = 0$. That is, the instruments must not be contemporaneously correlated with the stochastic disturbances. In general, finite sample bias will not be zero due to the stochastic nature of Y_1.

Where do instruments arise in the linear simultaneous equation problem? The reduced form of eq. (2.2) determines the conditional distribution of Y_1. Therefore, the predetermined variables Z provide a source of instrumental variables since Y_1 is a stochastic linear function of Z_t, while the predetermined variables are uncorrelated with u_t by assumption. Therefore, we consider linear combinations of the predetermined variables $W_1 = ZA_1$, where A_1 is a $K \times K_1$ matrix of rank K_1 to form the matrix of instruments. (Note the distinction between the instruments W_1, which have column rank K_1, and the instrumental variables Z, which have column rank $\geqslant K_1$ and are used to form W_1.) A_1 can either be known or estimated as $\hat{A}_1$. To determine the first-order asymptotic approximation to the distribution of $\hat{\delta}_{1,\mathrm{IV}}$, we consider only the case of Z to be exogenous. For lagged endogenous variables, the results are identical but the issue of appropriate central limit theorems in the time-series case arises [see, for example, Mann and Wald (1943) and Fuller (1976)]. Then,

$$\sqrt{T}\left(\hat{\delta}_{1,\mathrm{IV}} - \delta_1\right) = \left(\frac{1}{T}A_1'Z'X_1\right)^{-1}\left(\frac{1}{\sqrt{T}}A_1'Z'u_1\right). \tag{4.6}$$

The first matrix on the right-hand side of eq. (4.6) has plim equal to $A_1'MD_1$, which is non-singular where $D_1 = [\Pi_1 \;\vdots\; I_1]$ with I_1 a selection matrix which chooses Z_1. The vector $(1/\sqrt{T})A_1'Z'u_1$ forms a sequence of independent and non-identically distributed random variables. We can apply either the Liapounoff version of the central limit theorem or the slightly weaker Lindberg–Feller version to claim that the vector converges in distribution to a normal random vector with distribution $N(0, \sigma_{11}A_1'MA_1)$ [Rao (1973, p. 128)].[19] Then using the rules on products of random variables where we have a finite plim and the other converges in distribution [Rao (1973, p. 122)], we find the asymptotic distribution

$$\sqrt{T}\left(\hat{\delta}_{1,\mathrm{IV}} - \delta_1\right) \overset{\mathrm{A}}{\sim} N\left(0, \sigma_{11}\left[(A_1'MD_1)^{-1}A_1'MA_1(D_1'MA_1)^{-1}\right]\right). \tag{4.7}$$

If A_1 is replaced by an estimate $\hat{A}_1$ which has plim A_1, we obtain identical asymptotic results again because of the product formulae for random variables with limiting distributions.

Given the formula for the asymptotic covariance matrix for $\hat{\delta}_{1,\mathrm{IV}}$, we would like to find the best choice of A_1 to form the matrix of instruments W_1. That is, we

[19]Assumptions need to be made either about third moments of the random variable or about limiting behavior of the sums to apply these central limit theorems.

want to choose A_1 to minimize, in a matrix sense, the asymptotic covariance.[20] It turns out that the best A_1 is not unique but any optimum choice must satisfy the condition that plim $\hat{A}_1 = D_1$. We can show an optimum choice is $\hat{A}_1 = (Z'Z)^{-1}Z'X_1$. For this choice of $\hat{A}_1$ we calculate the asymptotic covariance matrix from eq. (4.7):

$$V(\hat{\delta}_{1,\mathrm{IV}}) = \sigma_{11}\operatorname{plim}\Big[\big(X_1'Z(Z'Z)^{-1}MD_1\big)^{-1}\big(X_1'Z(Z'Z)^{-1}M(Z'Z)^{-1}Z'X_1\big) \times \big(D_1'M(Z'Z)^{-1}Z'X_1\big)^{-1}\Big]$$
$$= \sigma_{11}[D_1'MD_1]^{-1}, \tag{4.8}$$

since plim $\hat{A}_1 = D_1$. We now compare the asymptotic covariance matrix for another choice of A_1, say E_1, such that $E_1 \neq D_1$ (or plim $\hat{E}_1 \neq D_1$). For $\tilde{W}_1 = ZE_1$ we need to compare the term within brackets in eq. (4.7) to that of eq. (4.8). We use as our standard of comparison all possible linear combinations $g'\delta_1$, where g' is a k_1 vector and say $\hat{A}_1$ is best if the quadratic form

$$g'\big[V(\hat{\delta}_{1,A_1}) - V(\hat{\delta}_{1,E_1})\big]g \leqslant 0 \quad \text{for all } g. \tag{4.9}$$

Equivalently, we can use the inverse covariance matrices to show the reverse inequality.

Factor $M = NN'$ by the symmetric factorization and define $h = N^{-1}Z'X_1g$. Thus, the plim $h'h = g'[D_1'MD_1]g$, the inner term of which comes from eq. (4.8). For the comparison estimator $g'[(E_1'MD_1)(E_1'ME_1)^{-1}(D_1'ME_1)]g =$ plim $h'G(G'G)^{-1}G'h$, where $G = NE_1$. Therefore the difference of the inverse covariance matrices is

$$\operatorname{plim} h'\big[I - G(G'G)^{-1}G'\big]h \geqslant 0, \tag{4.10}$$

since by the generalized Pythagorean theorem, $h'h \geqslant h'P_Gh$, where $P_G = G(G'G)^{-1}G'$, the orthogonal projection. That is, P_G is an orthogonal projection so all its characteristic roots are either one or zero [Rao (1973, p. 72)]. Somewhat analogously to the Gauss–Markov theorem, we have shown that among all instrument matrices W_1 formed by linear combinations of the predetermined variables, that the best choice is a matrix which has probability limit equal to $D_1 = [\Pi_1 \quad I_1]$. For $\hat{A}_1$ then, $W_1 = Z\hat{A}_1 = Z[\hat{\Pi}_1 \quad I_1]$.

[20] When the terminology "asymptotic covariance matrix" is used, we mean, more precisely, the covariance matrix of the asymptotic distribution. Minimization of the asymptotic covariance matrix means that the matrix difference between it and the asymptotic covariance matrix of comparison estimator is negative semi-definite. Equivalently, for any vector P, $P'\hat{\delta}_{1,\mathrm{IV}}$ has minimum variance in the class of instrumental variable estimators for the A_1 which minimizes the asymptotic covariance.

It is now straightforward to demonstrate that the two-stage least squares estimator (2SLS) is numerically identical to the optimal IV estimator with $\hat{A}_1 = (Z'Z)^{-1}Z'X_1$. 2SLS "purges" all jointly endogenous variables in eq. (4.1) and replaces them with their conditional expectations estimated from the unrestricted reduced form: $\hat{Y}_1 = Z\hat{\Pi}_1 = Z(Z'Z)^{-1}Z'Y_1 = P_Z Y_1$. The second stage consists of replacing the jointly endogenous variables in eq. (4.1):

$$y_1 = \hat{Y}_1\beta_1 + Z_1\gamma_1 + \hat{V}_1\beta_1 + u_1 = \hat{X}_1\delta_1 + \tilde{u}_1, \tag{4.11}$$

where $\hat{V}_1 = Y_1 - \hat{Y}_1$ and is orthogonal to $\hat{Y}_1$ and Z_1. Then least squares is done:

$$\hat{\delta}_{1,2SLS} = \left(\hat{X}_1'\hat{X}_1\right)^{-1}\hat{X}_1'y_1. \tag{4.12}$$

2SLS is thus identical to the IV estimator using $\hat{A}_1$ because $(\hat{X}_1'\hat{X}_1) = (X_1'P_Z P_Z X_1) = (\hat{X}_1'X_1) = (\hat{A}_1'Z'X_1)$.[21] The two estimators remain identical only so long as *all* predetermined variables are used to form $\hat{X}_1$. In fact, in purging Y_1 to form $\hat{Y}_1$, Z_1 must be included among the regressors. If not, $\hat{V}_1$ is not necessarily orthogonal to Z_1 and inconsistency may result. This mistake has been made by numerous researchers. However, if the IV estimator of eq. (4.4) is used, this problem does not occur since W_1 is a linear combination of predetermined variables. Thus, the IV estimator will continue to be consistent for any matrix A_1, but will have a larger asymptotic covariance than the IV estimator with $W_1 = ZD_1$. Also, consistent IV estimation is computable for the case $K > T > K_1$. As we previously discussed in Section 2, in many cases the number of predetermined variables K will exceed T, the number of observations. 2SLS is no longer computable because the unrestricted reduced-form estimates are not computable. But IV estimation can still be used in this case.[22]

Another member of the IV estimator class for single equation models is the limited information maximum likelihood (LIML) estimator which arises when a normality assumption is made for the distribution $G(U)$. We will defer consideration of LIML until we discuss maximum likelihood estimation. However, a related class of estimators to LIML, referred to as k-class estimators, deserve a brief mention [Theil (1961)]. They can also be derived as IV estimators, and can be considered analogous to the 2SLS estimators. Define the instrument $W_1 = [\hat{Y}_1 \quad Z_1]$, where we now use $\hat{Y}_1 = Y_1 - \kappa\hat{V}_1$ for κ, a scalar. Clearly 2SLS has $\kappa = 1$

[21]Another possible interpretation of 2SLS is as a minimum distance estimator where $\min_{\delta_1}(y_1 - X_1\delta_1)'Z(Z'Z)^{-1}Z'(y_1 - X_1\delta_1)$. We thus project the model into the subspace spanned by the columns of the predetermined variables to obtain orthogonality.

[22]An extensive literature exists on which IV estimator should be used when $T > K$. The problem is difficult to resolve because it is small sample in nature, while our current optimality conditions depend on asymptotic approximations as T grows large. Swamy (1980) reviews the "undersized sample" problem.

and OLS has $\kappa = 0$. Consistency requires plim $\kappa = 1$ which also is the condition for them to be an IV estimator because $\operatorname{plim}(1/T)[((1-\kappa)I + \kappa P_Z)Y_1 \quad Z_1]'u_1 = 0$ requires that plim $\kappa = 1$. However, to have the same asymptotic distribution as the optimal IV estimators, κ must satisfy the stricter condition plim $\sqrt{T}(\kappa - 1) = 0$. Consider the difference in the estimators when plim $\kappa = 1$:

$$\operatorname{plim}\sqrt{T}\left(\hat{\delta}_{1,\kappa} - \hat{\delta}_{1,2\mathrm{SLS}}\right) = Q_{11}^{-1}\operatorname{plim}\left(\frac{1}{\sqrt{T}}\left[Y_1'(I - P_Z)(1-\kappa) \quad 0\right]u_1\right), \tag{4.13}$$

where $Q_{11} = \operatorname{plim}(1/T)\hat{X}_1'\hat{X}_1$. The term containing κ can be written as

$$\operatorname{plim}\left(\sqrt{T}(1-\kappa)\right)\left(\frac{1}{T}\hat{V}_1'u_1\right) = \operatorname{plim}\left(\sqrt{T}(1-\kappa)\right)\left((B')^{-1}\Sigma_1\right)_1,$$

which yields the requirement plim $\sqrt{T}(\kappa - 1) = 0$ for asymptotic equivalence. Thus, to be an IV estimator we need $\operatorname{plim}(\kappa - 1) = 0$; to be an optimal IV estimator, we need plim $\sqrt{T}(\kappa - 1) = 0$. Nagar (1959), Sawa (1973), and Fuller (1977) have considered the choice of κ with the Nagar $\kappa = 1 + (K - k_1)/T$ sometimes used because it eliminates bias in the first term of the asymptotic expansion.

4.2. System estimation

We now consider estimation of the entire system of equations rather than only a single equation. Under correct specification of the other equations, estimates of the coefficients of the first equation will then have a smaller asymptotic covariance matrix so long as Σ_1 has $\sigma_{1j} \neq 0$ for some $j \neq 1$ and the jth equation is overidentified. The term "overidentification" refers to the case where there are more than enough restrictions so that the rank conditions of Section 2 are satisfied even if one or more prior restrictions are disregarded.[23] Again we will only consider linear restrictions for a single equation so that a necessary condition for equation j to be overidentified is that the number of right-hand-side variables included (after the normalization) is strictly less than the number of predetermined variables, $r_j + s_j = k_j < K$. However, the gain of the reduction in the

[23] Alternative notations of overidentification exist. When only exclusion restrictions are present, in terms of the order condition overidentification can be defined when $K > k_1 = r_1 + s_1$ [Hood and Koopmans (1953, p. 139) and Theil (1971, p. 449)]. The precise definition of overidentification is that at least two sets of exact identifying restrictions exist which are not identical and the deletion of any restriction loses identification if they are the only restrictions. Overidentification can also be defined in terms of restrictions on the reduced form [Malinvaud (1970, p. 663)]. However, Malinvaud's definition can lead to problems in that an equation defined to be overidentified may not, in fact, be identified.

asymptotic covariance matrix brings with it an increased risk of inconsistent estimation. We emphasized in single equation estimation that IV estimators required only that the two properties of instruments W_1 be satisfied for consistent estimation. With system estimation misspecification of any equation in the system will generally lead to inconsistent estimation of *all* equations in the system.[24]

In discussing system estimation, it is often easier to stack the system of equations. Each equation has the form

$$y_i = X_i\delta_i + u_i \qquad (i=1,2,\dots,M). \tag{4.14}$$

The stacked form of the complete system of equations is then

$$y = X\delta + u, \tag{4.15}$$

with

$$y=\begin{bmatrix} y_1 \\ \vdots \\ y_M \end{bmatrix}; \quad X=\begin{bmatrix} X_1 & & 0 \\ & \ddots & \\ 0 & & X_M \end{bmatrix}; \quad \delta=\begin{bmatrix} \delta_1 \\ \vdots \\ \delta_M \end{bmatrix}; \quad u=\begin{bmatrix} u_1 \\ \vdots \\ u_M \end{bmatrix}.$$

The important point to notice is that $V(u)=\Sigma\otimes I_T$ so that a multivariate least squares type approach to IV estimation is called for to account for the non-diagonal covariance matrix.

The general system IV estimator may be defined as

$$\hat{\delta}_{\mathrm{IV}} = (W'X)^{-1}W'y. \tag{4.16}$$

The particular form that W takes for system estimation is $W'=\hat{X}'(\hat{\Sigma}\otimes I_T)^{-1}$, where $\hat{X}=\operatorname{diag}(\hat{X}_1, \hat{X}_2,\dots,\hat{X}_M)$ and $\hat{\Sigma}$ is a consistent estimate of Σ. Note that each $\hat{X}_i = ZA_i$ must satisfy the two properties for instruments as well as the additional property that $\operatorname{plim}(1/T)\hat{X}_i'u_j=0$ for all i and j, i.e. $\hat{X}_i$ must be asymptotically uncorrelated with *all* the disturbance vectors in the system instead of only u_i as in single equation estimation. If $\hat{\Sigma}$ is replaced by another matrix with plim not equal to Σ, consistent estimation still follows, but a larger asymptotic covariance matrix for the estimator results. Derivation of the asymptotic distribution is very similar to the single equation case:

$$\sqrt{T}\left(\hat{\delta}_{\mathrm{IV}}-\delta\right)=\left(\frac{1}{T}W'X\right)^{-1}\left(\frac{1}{\sqrt{T}}W'u\right). \tag{4.17}$$

[24]This potential problem is mitigated somewhat by specification tests proposed in the next section.

The first matrix on the right-hand side of eq. (4.17) has plim equal to $\tilde{A}'N\tilde{D}$, where $\tilde{A} = \text{diag}(A_1,\dots,A_M)$, $N = \Sigma^{-1}\otimes M$, and $\tilde{D} = \text{diag}(D_1,\dots,D_M)$. The second term has an asymptotic normal distribution with mean zero and covariance $\tilde{A}'N\tilde{A}$. Therefore, the asymptotic distribution for the IV estimator is

$$\sqrt{T}(\hat{\delta}_{IV} - \delta) \overset{A}{\sim} N\left(0, \left[(\tilde{A}'N\tilde{D})^{-1}(\tilde{A}'N\tilde{A})(\tilde{D}'N\tilde{A})^{-1}\right]\right). \tag{4.18}$$

Again, A_i can be replaced by $\hat{A}_i$ which has plim A_i with no change in the limiting distribution.

Choice of the optimal matrix $\tilde{A}$ follows exactly as in the single equation case and is not repeated here. The best choice of $\tilde{A}$ has each A_i satisfy the condition that plim $A_i = D_i = [\Pi_i \quad I_i]$. The asymptotic covariance matrix of the optimal system IV estimator is then

$$V(\hat{\delta}_{IV}) = [\tilde{D}'N\tilde{D}]^{-1} = \text{plim}\left(\frac{1}{T}\hat{X}'(\Sigma^{-1}\otimes I)\hat{X}\right)^{-1}. \tag{4.19}$$

We now consider IV estimators which have an optimal A_i:

(1) Three-Stage Least Squares (3SLS) [Zellner and Theil, (1962)]. The 3SLS estimator takes $\hat{A}_i = (Z'Z)^{-1}Z'X_i$. Its estimate of Σ is $\hat{S}$ derived from the residuals of the structural equations estimated by 2SLS. For 3SLS W takes the particularly simple form $W' = X'(\hat{S}^{-1}\otimes Z(Z'Z)^{-1}Z')$. Note that if $\hat{S}$ is replaced by the identity matrix I_M we have 2SLS done on each equation. The term 3SLS again arises because in the Zellner–Theil formulation Y_i was replaced by $\hat{Y}_i$ in each equation and "seemingly unrelated regression" was done on the system.[25] Iterated 3SLS has been considered [Dhrymes (1973)] where $\hat{S}$ is updated at each iteration. The asymptotic distribution is not changed by the iterative procedure.

(2) Iterated Instrumental Variables. The 3SLS estimator requires $T > K$ just like the 2SLS estimator. Brundy and Jorgenson (1971) and Dhrymes (1971) propose an estimator which only requires that $T > r_i + s_i$ for all $i = 1,\dots,M$. The procedure first estimates $\hat{\delta}_i$ for each equation by an IV estimator. These consistent, but inefficient, estimates are used to form $\hat{S}$, a consistent estimate of Σ. System instruments are formed with $\overline{W}' = \hat{X}'(\hat{S}\otimes I_T)^{-1}$, where $\hat{X}_i = [-Z(\hat{\Gamma}\hat{B})_i^{-1} \quad Z_i]$,

[25]Again differences can arise between the "repeated least squares" form and the IV form. The optimal IV estimator requires only a consistent estimator of the D_i. However, the repeated least squares form requires an estimate of Π_i at least as efficient as $\Pi_i = P_z Y_i$. In particular, if 2SLS estimates are used to form an estimate of Π_i for the "purged" variable Y_i, then a 3SLS-like estimator no longer has an asymptotic covariance matrix as small as that of the optimal IV estimator. Also, if 3SLS is done as a seemingly unrelated regression, the terms which arise from the first stage residuals are ignored.

where $(\hat{\Gamma}\hat{B}^{-1})_i$ is formed from the consistent estimates $\hat{\delta}_i = [\hat{\beta}_i \hat{\gamma}_i]'$ and the prior (zero) restrictions. Then $\hat{\delta}_{\mathrm{IV}} = (\overline{W}'X)^{-1}\overline{W}'y$. This estimator has the identical asymptotic distribution to the optimal IV estimator since $\text{plim} - (\hat{\Gamma}\hat{B}^{-1}) = \Pi$ (so long as B is non-singular). However, since $T < K$ is a "small sample" problem, it is unclear how much the asymptotic argument can be relied on. Small sample approximations will be required to evaluate the IV estimators better. Also, knowledge about the effect of the initial consistent estimator on the small sample properties of $\hat{\delta}_{\mathrm{IV}}$ remains to be established.

Dhrymes (1971) proposed to iterate the process by replacing $(\hat{\Gamma}\hat{B}^{-1})$ at each iteration and $\hat{S}$ with the new estimates. Hausman (1975) demonstrated that if the iterative process converged, then it would yield the maximum likelihood (FIML) estimates as is demonstrated when FIML estimation is considered, so long as $T > K + M$. Lyttkens (1970) considered iteration with S replaced by I_M as did Brundy and Jorgenson. While this estimator is properly regarded as a full information system estimator, since all equations must be specified in structural form, the asymptotic distribution is the same as the (system) 2SLS estimator.

(3) System k-class [Srivastava (1971) and Savin (1973)]. This estimator is a straightforward generalization of the single equation case. Replace Y by $\hat{\hat{Y}}' = [(I - \tilde{\kappa})Y' + \tilde{\kappa}\hat{\Pi}'Z']$ for a matrix $\tilde{\kappa}$. Then a system IV type estimator is used. Again consistency requires $\text{plim}\, \tilde{\kappa} = I$, while asymptotic efficiency requires $\text{plim}\, \sqrt{T}(\tilde{\kappa} - I) = 0$. The proof of these requirements is the same as in the single equation case.

An interesting special case of system estimation arises when the system is triangular [Lahiri and Schmidt (1978)]. This specification occurs when B is lower triangular after the a priori restrictions have been applied.[26] If Σ is also specified to be diagonal, we then have the recursive specification [Wold (1964)]. All right-hand-side variables are predetermined so least squares on each equation is the optimal estimator. But if Σ is not diagonal, least squares is inconsistent. If Σ were known, the system could be transformed and generalized least squares (GLS) used. With Σ unknown, it can be estimated and 3SLS provides an optimal estimator in the sense of having an asymptotic distribution identical to the optimal IV estimator so long as the system is identified (apart from restrictions on Σ). But a relevant question is whether GLS with a consistently estimated covariance matrix also has an identical asymptotic distribution. The answer is no, although the estimator is consistent, because

$$\text{plim}\sqrt{T}\left(\hat{\delta}_{\mathrm{IV}} - \hat{\delta}_{\mathrm{GLS}}\right) = [\tilde{D}N\tilde{D}]^{-1}\text{plim}\frac{1}{\sqrt{T}}\left[\tilde{A}'(I\otimes Z')(\hat{\Sigma}\otimes I)^{-1}u - X'(\hat{S}\otimes I)^{-1}u\right] \neq 0. \tag{4.20}$$

[26] This specification arises in the path analysis model often used in other social sciences [see Goldberger (1972)].

While $\text{plim}(1/T)Z'u=0$, $\text{plim}(1/T)X'u\neq 0$ so that eq. (4.20) would equal zero only if $\text{plim}\sqrt{T}(\hat{\Sigma}-\hat{S})=0$. Thus, to apply GLS to a triangular system an efficient estimate of Σ is required or the estimator will have a larger asymptotic covariance matrix than the optimal IV estimator.

4.3. *Reduced-form estimation*

So far we have directed our attention to estimation of the structural parameters (B,Γ,Σ). Estimation of the reduced-form parameters (Π,Ω) can be important to derive forecasts or for optimal control procedures. If the entire system is just identified so that the appropriate rank conditions of eqs. (3.4) and (3.6) are equalities, then Π is unrestricted. This result follows from the first block of eq. (3.3), $\Pi B_1+\Gamma_1=0$, which can be uniquely solved for (B_1,Γ_1) for any Π.[27] On the other hand, when any structural equation is overidentified, then Π is subject to restrictions. In the case of linear equation restrictions, the total number of restrictions on Π, called the degree of system overidentification, is determined by the expression $\sum_i(K-k_i)$.[28]

Again let us stack the reduced-form equation (2.2) into a $MT\times 1$ vector system:

$$y=(I\otimes Z)\pi+v, \tag{4.21}$$

where $\pi'=(\pi_{11},\pi_{21},\dots,\pi_{KM})$ and $v'=(v_{11},v_{21},\dots,v_{TM})$. Note that $V(v)=\Omega\otimes I$, so that eq. (4.21) represents a multivariate least squares problem. If π is unrestricted, then the GLS estimator and the OLS estimator are identical

$$\begin{aligned}\hat{\pi}&=\left[(I\otimes Z')(\Omega^{-1}\otimes I)(I\otimes Z)\right]^{-1}(I\otimes Z')(\Omega^{-1}\otimes I)y\\&=\left[\Omega^{-1}\otimes Z'Z\right]^{-1}(\Omega^{-1}\otimes Z')y=\left[I\otimes(Z'Z)^{-1}Z'\right]y.\end{aligned} \tag{4.22}$$

In the case of overidentification with restrictions on π, then non-linear GLS is used with an estimated $\hat{\Omega}$ [Zellner (1962)]. This approach is very close to the minimum distance methods of Malinvaud (1970) and Rothenberg (1973, ch. 4). Let $\tilde{\pi}_{ij}=-(\Gamma B^{-1})_{ij}$ be subject to the *a priori* restrictions, while $\hat{\pi}_{ij}$ represents the unrestricted OLS estimates. Then put these elements into vectors and the

[27]This result also leads to the structural estimation model of indirect least squares (ILS) for just identified systems. Estimates of the reduced-form parameters, Π, are used to solve for (B_1,Γ_1). ILS has been superseded by the IV estimators. The shortcoming of ILS is its inapplicability to overidentified models. A generalization of ILS to this case is proposed by Khazzom (1976).

[28]If any equation is underidentified so the rank is less than $M+K$, then this equation adds no restrictions to Π and is omitted from the sum.

minimum distance estimation is

$$\min_{\tilde{\pi}} (\hat{\pi} - \tilde{\pi})'(Z'Z \otimes \hat{\Omega}^{-1})(\hat{\pi} - \tilde{\pi}), \tag{4.23}$$

where $\hat{\Omega} = (1/T)y'(I - P_{\bar{Z}})y$ for $\bar{Z} = (I \otimes Z)$. Solution of eq. (4.23) represents a non-linear problem because of the restrictions on $\tilde{\pi}$.[29]

Other estimators of Π are possible. An obvious suggestion would be to estimate Π from the structural parameter estimates, $\hat{\Pi}_{\text{IV}} = -\hat{\Gamma}_{\text{IV}}\hat{B}_{\text{IV}} = h(\hat{A}_{\text{IV}})$. Since h is differentiable (totally) if $\hat{\Gamma}_{\text{IV}}$ and $\hat{B}_{\text{IV}}$ are optimal system IV estimators, e.g. 3SLS, we might expect $\hat{\Pi}_{\text{IV}}$ to have good asymptotic properties [Rao (1973, ch. 6)]. The asymptotic covariance matrix for an efficient system IV estimator, e.g. 3SLS, is[30]

$$V(\hat{\pi}) = \left((\hat{B}')^{-1} \otimes I\right)\tilde{A}\left(X'(\hat{\Sigma}^{-1} \otimes P_Z)X\right)^{-1}\tilde{A}'\left((\hat{B})^{-1} \otimes I\right). \tag{4.23a}$$

In fact, the $\hat{\tilde{\pi}}$ from eq. (4.23) and $\hat{\Pi}_{\text{IV}}$ are asymptotically equivalent, $\operatorname{plim}\sqrt{T}(\hat{\tilde{\pi}} - \hat{\pi}_{\text{IV}}) = 0$ [Rothenberg (1973, ch. 4)]. This result follows because both estimators are asymptotically equivalent to FIML which we will examine momentarily. However, if non-optimal system IV estimators are used to form $\hat{\Pi}$, then no optimal asymptotic results hold. Dhrymes (1973b) demonstrated that $\hat{\Pi}_{\text{2SLS}}$ is not necessarily better than unrestricted least squares estimation of Π, even though the overidentifying restrictions have been imposed.

4.4. *Maximum likelihood estimation*

ML estimation formed the original approach to estimation of linear simultaneous equation models by Haavelmo (1944) and Koopmans et al. (1950). The likelihood function follows from an additional assumption:

Assumption 4.3

The structural disturbances follow a non-singular normal distribution $U \sim N(0, \Sigma \otimes I_T)$.

[29]Brown (1960) considered setting $\Omega = I$ which leads to a loss of asymptotic efficiency.
[30]If an efficient single equation estimator is used, e.g. 2SLS, the asymptotic covariance matrix is

$$V(\Pi_{2S}) = \left[(B_{2S}')^{-1} \otimes I\right)\tilde{A}_{2S}\left(X'(I \otimes P_Z)X\right)^{-1}\left(X'(\Sigma \otimes P_Z)X\right)\left(X'(I \otimes P_Z)X\right)^{-1}\tilde{A}_{2S}'\left((B_{2S})^{-1} \otimes I\right).$$

Note this covariance matrix is larger than that of eq. (4.23a) unless all equations are just identified. A derivation of this $V(\Pi_{2S})$ is found in Goldberger et al. (1961).

This happens after all identities have been substituted out. We can then write the log-likelihood function:

$$L(B,\Gamma,\Sigma) = C + \frac{T}{2}\log\det(\Sigma)^{-1} + T\log|\det(B)|$$
$$-\frac{T}{2}\operatorname{tr}\left(\frac{1}{T}\Sigma^{-1}(YB+Z\Gamma)'(YB+Z\Gamma)\right), \tag{4.24}$$

where the constant C is disregarded in maximization. The presence of the Jacobian term $\log|\det(B)|$ of the transformation from U to Y represents the essential feature of the ML approach to simultaneous equation estimation [see eqs. (2.5) and (2.6)]. In its absence the LF becomes identical to that of the multivariate least squares specification. In fact, for the triangular specification $\det(B)=1$, and we have already seen how this case represents a special situation for estimation.

Finding the maximum of eq. (4.23) involves non-linear optimization if the system is overidentified. We impose the *a priori* restrictions and solve for the first-order conditions only with respect to unknown elements of B and Γ:

$$\text{(i)}\quad \frac{\partial L}{\partial B^u}:\quad \left[T(B')^{-1} - Y'(YB+Z\Gamma)\Sigma^{-1}\right]^u = 0,$$
$$\text{(ii)}\quad \frac{\partial L}{\partial \Gamma^u}:\quad \left[-Z'(YB+Z\Gamma)\Sigma^{-1}\right]^u = 0, \tag{4.25}$$
$$\text{(iii)}\quad \frac{\partial L}{\partial (\Sigma^{-1})^u}:\quad \tfrac{1}{2}\left[T\Sigma - (YB+Z\Gamma)'(YB+Z\Gamma)\right]^u = 0.$$

Here the $[\cdot]^u$ notation stands for the equations corresponding to unknown elements, i.e. u is a selection operator which chooses unrestricted elements of a matrix. If we postmultiply within the brackets of eq. (4.25, i) by Σ/T, we find that $[(B')^{-1}\Sigma - (1/T)Y'U)]^u = 0$. We know that the plim of this equation must be zero for ML to be consistent. In fact, the plim *is* zero by our earlier calculation that $\operatorname{plim}(1/T)Y'U = (B')^{-1}\Sigma$. Therefore, it is the presence of the Jacobian term $\det(B)$ in the likelihood function which "corrects" for the correlation of the jointly endogenous variables and the structural disturbances which is the essential feature of the simultaneous equation specification. Hausman (1975) combines eqs. (4.25, i) and $(B')^{-1}U'U = Y'U + (B')^{-1}\Gamma'Z'U$ to obtain $[(B')^{-1}\Gamma'Z'(YB + Z\Gamma)\Sigma^{-1}]^u = 0$ to derive the first-order conditions with respect to the unknown elements of B and Γ:

$$\left[\begin{pmatrix} -Z' \\ (B')^{-1}\Gamma'Z' \end{pmatrix}(YB+Z\Gamma)\Sigma^{-1}\right]^u = 0. \tag{4.26}$$

We impose the usual normalization $\beta_{ii}=1$ and now stack eq. (4.26) into the form of eq. (4.15) where δ_i contains only unknown elements of β_i and γ_i. Thus, for each equation i and predetermined variable z_j the top line of eq. (4.26) corresponds to

$$z_j'\left[(\Sigma^{-1})_i\otimes I_T\right]y = z_j'\left[(\Sigma^{-1})_i\otimes I_T\right]X\delta \quad \text{for } i=1,\dots,M \text{ and } j=1,\dots,K \tag{4.27}$$

if γ_{ji} is unknown. Likewise, the second block of eq. (4.26) for equation i when β_{ji} is unknown leads to

$$\hat{y}_j'\left[(\Sigma^{-1})_i\otimes I_T\right]y = \hat{y}_j'\left[(\Sigma^{-1})_i\otimes I_T\right]X\delta \quad \text{for } i=1,\dots,M \text{ and } j=1,\dots,K, \tag{4.28}$$

where $\hat{y}_j=(-Z\hat{\Gamma}\hat{B}^{-1})_j$. To stack the equations, we take the elements of eqs. (4.27) and (4.28) and rearrange them so that they are in the order $X_i=[Y_i \quad Z_i]$ for each equation $i=1,\dots,M$ to find the IV form of the first-order conditions for FIML:

$$\hat{\delta}_{\text{FIML}}=(\overline{W}'X)^{-1}\overline{W}'y, \tag{4.29}$$

where the instrument matrix $\overline{W}'=\hat{X}'(\hat{\Sigma}\otimes I_T)^{-1}$, with $\hat{X}=\text{diag}(\hat{X}_1,\dots,\hat{X}_M)$, $\hat{X}_i=[-Z(\hat{\Gamma}\hat{B}^{-1})_i \quad Z_i]$, and $\hat{\Sigma}=T^{-1}(Y\hat{B}+Z\hat{\Gamma})'(Y\hat{B}+Z\hat{\Gamma})$. Eq. (4.29) is non-linear because $\hat{X}_i$ contains unknown elements of δ. The equations can be solved by an iterative process called Durbin's method by Malinvaud (1970, p. 686), where for the kth iteration

$$\hat{\delta}_{k+1}=(W_k'X)^{-1}W_k'y, \tag{4.30}$$

where $\hat{X}^k$ and $\hat{\Sigma}^k$ are formed from the $k-1$ values of $\hat{\delta}$.[31] FIML is a member of the class of optimal system IV estimators since $\text{plim}\,\hat{A}_i=\text{plim}[(-\hat{\Gamma}\hat{B}^{-1})_i \quad I_i]=(\Pi_i \quad I_i)=D_i$ and $\text{plim}\,\hat{\Sigma}=\Sigma$ using eqs. (4.18) and (4.19) we have its asymptotic distribution:

$$\sqrt{T}\left(\hat{\delta}_{\text{FIML}}-\delta\right)\overset{A}{\sim}N\left[0,(\tilde{D}'N\tilde{D})^{-1}\right]=N\left[0,\text{plim}\left(\frac{1}{T}\hat{X}'(\Sigma^{-1}\otimes I)X\right)^{-1}\right]. \tag{4.31}$$

We can now calculate the Fisher information matrix to demonstrate that the FIML estimator of eq. (4.31) attains the asymptotic Cramer–Rao bound. Two asymptotically equivalent approaches are possible: the covariance matrix of the

[31] It is unlikely that eq. (4.30) provides a good iterative process because it lacks desirable properties. See Hausman (1974) and Quandt (Chapter 12 in this Handbook).

gradients (in stacked form) from eq. (4.25) can be calculated or the Hessian of the likelihood function can be calculated. Here we take the latter approach. The information matrix takes the form (corresponding to unknown coefficients)

$$J = -\lim_{T\to\infty} E\begin{bmatrix} L_{\delta\delta} & L_{\delta\Sigma^{-1}} \\ L_{\Sigma^{-1}\delta} & L_{\Sigma^{-1}\Sigma^{-1}} \end{bmatrix}$$

$$= \begin{bmatrix} \tilde{B}R\tilde{B}' + \operatorname{plim}\frac{1}{T}X'(\Sigma^{-1}\otimes I)X & -\frac{1}{2}\tilde{B}(I\otimes\Sigma)(I\otimes I+R) \\ J_{21} & \frac{1}{2}(I\otimes\Sigma)R(I\otimes\Sigma) \end{bmatrix}, \tag{4.31a}$$

where the non-zero off-diagonal block arises from $\operatorname{plim}(1/T)Y'U = \operatorname{plim}(1/T)V'U = (B')^{-1}\Sigma$, with terms $\tilde{B} = \operatorname{diag}(\tilde{B}_1,\dots,\tilde{B}_M)$ for $B_i = [(B'^{-1})_i \quad 0_i]$, where the first matrix corresponds to Y_i and the second matrix corresponds to Z_i. The matrix $R = (E_{ji})$ or

$$R = \begin{bmatrix} E_{11} & E_{21}\dots E_{M1} \\ E_{12} & \\ E_{1M} & E_{2M}\dots E_{MM} \end{bmatrix}, \tag{4.31b}$$

where E_{nk} is an $M\times M$ matrix with a one in the nkth position and zero elsewhere. The first term in J_{11} and J_{12} both arise from the presence of the Jacobian in the likelihood function. We invert J and look at J^{11} which corresponds to the lower bound for the slope parameters δ.[32] Now $J^{11} = (J_{11} - J_{12}J_{22}^{-1}J_{21})^{-1}$, with

$$J_{12}J_{22}^{-1}J_{21} = \tfrac{1}{2}\tilde{B}(I\otimes\Sigma)(\tilde{I}+R)(I\otimes\Sigma^{-1})R(I\otimes\Sigma^{-1})(\tilde{I}+R)(I\otimes\Sigma)\tfrac{1}{2}\tilde{B}',$$

where $\tilde{I} = (I\otimes I)$. $J_{11} = \tilde{B}R\tilde{B}' + \operatorname{plim}((1/T)X'(\Sigma^{-1}\otimes I)X)$. But since $X_i = [Z\Pi_i + V_i \quad Z_i]$, we find

$$\operatorname{plim}\frac{1}{T}X'(\Sigma^{-1}\otimes I)X = \tilde{B}(\Sigma^{-1}\otimes\Sigma)\hat{B}' + \operatorname{plim}\frac{1}{T}\tilde{D}'\tilde{Z}'(\Sigma^{-1}\otimes I)\tilde{Z}\tilde{D}$$

for $\tilde{Z} = I\otimes Z$ using $V_i = (UB^{-1})_i$ so that $J^{11} = (\tilde{D}'N\tilde{D})^{-1} = \operatorname{plim}((1/T)\hat{X}'(\Sigma\otimes I)^{-1}\hat{X})^{-1}$ in eq. (4.31). This calculation provides an alternative demonstration of

[32]Whitney Newey has helped this derivation. In the derivation the following properties of R are used: $R^{-1} = R$, $(I\otimes A)R(I\otimes A^{-1})R = A^{-1}\otimes A$ for symmetric A, and $(A^{-1}\otimes I)R(I\otimes A) = R$. I did not find that J_{22} has a Kronecker product form as claimed by Rothenberg (1973, p. 87) in his derivation of the information matrix for the unrestricted reduced-form estimator. A permutation of the columns of Σ for σ_{jk} and σ_{kj} is required to achieve the Kronecker form. Note that no use has been made of the symmetry of Σ.

the optimality of our class of IV estimators since their asymptotic covariance matrices equal J^{11}. Thus, FIML is another IV estimator for the linear simultaneous equation model with properties which can be derived easily from the IV formulae. It is asymptotically efficient (BAN) among the class of CUAN estimators.

Another approach to FIML estimation is via the reduced form. The appropriate log-likelihood function is

$$L(\Pi,\Omega)=C+\frac{T}{2}\log\det(\Omega)^{-1}-\frac{T}{2}\operatorname{tr}\left(\frac{1}{T}\Omega^{-1}(Y-Z\Pi)'(Y-Z\Pi)\right). \quad (4.32)$$

By the invariance theorem of ML we know that the FIML estimates from eqs. (4.24) and (4.32) are numerically identical: $\hat{\Pi}_{\text{FIML}}=-\hat{\Gamma}_{\text{FIML}}\hat{B}^{-1}_{\text{FIML}}$ and likewise for $\hat{\Omega}_{\text{FIML}}=(\hat{B}'_{\text{FIML}})^{-1}\hat{\Sigma}_{\text{FIML}}(\hat{B}_{\text{FIML}})^{-1}$. However, FIML on the structural form is easier to perform computationally because of the way in which the constraints affect only one coefficient at a time, while eq. (4.32) is subject to cross-equation constraints. But the invariance theorem also allows us to claim that $\hat{\Pi}_{3\text{SLS}}$ and $\hat{\Omega}_{3\text{SLS}}$ are also asymptotically efficient.

We now consider limited information maximum likelihood (LIML) [Anderson and Rubin (1949)]. LIML estimates the parameters of a single equation. Although not originally derived in this manner, LIML can be usefully considered as a special case of FIML where all but the first equation are just identified. Therefore we write the system of equations as[33]

$$\begin{aligned} y_1&=X_1\delta_1+u_1\\ y_i&=Z\Pi_i+v_i \quad \text{for } i=2,\dots,M. \end{aligned} \quad (4.33)$$

All Π_j are unrestricted. Note that the specification of eq. (4.33) corresponds to our notion of limited information estimation – the structural specification of only one equation is required along with a list of all predetermined variables in the system. Before proceeding to derive the LIML estimator from FIML on eq. (4.33), we demonstrate the 3SLS on the system is numerically identical to 2SLS on the first equation [Narayanan (1969)]. We stack the system

$$\begin{bmatrix} y_1\\ y_2\end{bmatrix}=\begin{bmatrix} X_1 & 0\\ 0 & I\otimes Z\end{bmatrix}\begin{bmatrix}\delta_1\\ \Pi_2\end{bmatrix}+\begin{bmatrix}u_1\\ v_2\end{bmatrix}, \quad (4.34)$$

where the subscript 2 denotes equations $i=2,\dots,M$. The covariance matrix of the

[33]We only consider LIML estimation for a single equation. The approach can be easily extended to LIML estimation on a subset of equations. This interpretation of LIML is given in words by Hood and Koopmans but not mathematically derived.

disturbances is

$$V\begin{pmatrix} u_1 \\ v_2 \end{pmatrix} = \begin{bmatrix} \sigma_{11} & (\Sigma B^{-1})_{12} \\ (\Sigma B^{-1})_{21} & \Omega_{22} \end{bmatrix} \otimes I = \begin{bmatrix} \sigma_{11} & \psi_{12} \\ \psi_{21} & \Omega_{22} \end{bmatrix} \otimes I = \Psi \otimes I. \tag{4.35}$$

We apply 3SLS to eq. (4.34) to find after use of partitioned inversion and the fact that $P_Z Z = Z$ [Court (1973)]:

$$\begin{aligned} \begin{bmatrix} \hat{\delta} \\ \hat{\Pi}_2 \end{bmatrix} &= \begin{bmatrix} X_1'(\psi^{11} \otimes P_Z) X_1 & X_1'(\psi^{12} \otimes Z) \\ (\psi^{21} \otimes Z') X_1 & \psi^{22} \otimes (Z'Z) \end{bmatrix}^{-1} \\ &\quad \times \begin{bmatrix} X_1'(\psi^{11} \otimes P_Z) y_1 + X_1'(\psi^{12} \otimes P_Z) y_2 \\ (\psi^{21} \otimes Z') y_1 + (\psi^{22} \otimes Z') y_2 \end{bmatrix} \\ &= \begin{bmatrix} (X_1'(\sigma^{11} \otimes P_Z) X_1)^{-1} & (X_1'(\sigma^{11} \otimes P_Z) X_1)^{-1} X_1'(\sigma^{11} \psi_{12} \otimes Z(Z'Z)^{-1}) \\ m_{21} & \Omega_{22} \otimes (Z'Z)^{-1} - (\psi_{21} \otimes I)(G - GH'(HGH')^{-1} HG)(\psi_{12} \otimes I) \end{bmatrix} \\ &\quad \times \begin{bmatrix} w_1 \\ w_2 \end{bmatrix}, \end{aligned} \tag{4.36}$$

where m_{21} equals the upper right-hand block transposed, $G = \sigma^{11} \otimes (Z'Z)^{-1}$, and $H = X_1'Z$ and w_1 and w_2 correspond to the vectors directly above. We multiply eq. (4.36) using the fact that $\psi^{11}\psi_{12}\Omega_{22}^{-1} + \psi^{12} = 0$ and $\psi^{11} + \sigma^{11}\psi_{12}\psi^{21} = \sigma^{11}$ to find

$$\hat{\delta} = (X_1'(\sigma^{11} \otimes P_Z) X_1)^{-1} X_1'(\sigma^{11} \otimes P_Z) y_1 = (\hat{X}_1' X_1)^{-1} \hat{X}_1' y_1. \tag{4.37}$$

Thus, the 3SLS and 2SLS estimators for the first equation coincide.[34] We likewise now apply FIML to the system of eqs. (4.25) to derive the LIML estimator.

For the limited information specification of eq. (4.34) the basic first-order equations of eq. (4.26) for FIML take the form

$$\left[\begin{pmatrix} -Z' \\ (B')^{-1}\Gamma'Z' \end{pmatrix} \begin{pmatrix} YB_1 + Z\Gamma_1 \\ Y_2 - Z\Pi_2 \end{pmatrix} \Psi^{-1} \right]^u = 0, \tag{4.38}$$

[34] While the estimator of δ remains the same, the "3SLS" estimates of Π_2 would be more efficient than the OLS estimates [Court (1973)].

where the subscripts denote the separate sets of equations and where B and Γ correspond to the LIML specification of eq. (4.33). Now we stack the equations as we did in eq. (4.29) for FIML [see also eq. (4.36)]. To solve the system of equations we also need to solve for $\hat{\Pi}$ as in eq. (4.36) using the FIML first-order conditions:

$$\hat{\Pi}_2=\left[I\otimes(Z'Z)^{-1}Z'\right]y_2-\left[\hat{\psi}_{21}\hat{\sigma}^{11}\otimes(Z'Z)^{-1}Z'\right](y_1-X_1\hat{\delta}). \tag{4.39}$$

We thus see that LIML adjusts its estimate of $\hat{\Pi}$ away from the OLS reduced-form estimates on each iteration using the estimated structural parameters δ as well as estimates of the covariance matrix terms $\hat{\psi}_{21}$ and $\hat{\sigma}^{11}$. The iteration for $\hat{\delta}$ is just as it was for FIML so that $\hat{X}_1=[Z\hat{\Pi}_1 \quad Z_1]$ rather than $\hat{Y}_1=P_ZY_1$ as in the 3SLS–2SLS estimator of eq. (4.36). Also, the IV form is used as in eq. (4.37): $\hat{\delta}=(\hat{X}_1'X_1)^{-1}\hat{X}_1'y_1$. Here, then, is the IV interpretation of LIML. At each iteration, $\hat{\delta}_k$ is updated using estimates of $\hat{\Pi}_{k-1}$. Then, using $\hat{\delta}_k$, $\hat{\psi}_{21,k}$, and $\hat{\sigma}_k^{11}$ eq. (4.39) is used for a new estimate $\hat{\Pi}_k$. As usual with ML, at convergence the estimates are all mutually consistent with each other. The asymptotic distribution of the LIML estimator is then straightforward to find (without regard to a certain characteristic root) because LIML is a member of our class of optimal IV estimators for single equation estimation. The asymptotic distribution of LIML follows directly from eq. (4.6) so that

$$\sqrt{T}(\hat{\delta}_{\text{LIML}}-\delta)\overset{\text{A}}{\sim}N\left[0,\sigma_{11}(D_1'MD_1)^{-1}\right]=N\left[0,\sigma_{11}\text{plim}\left(\frac{1}{T}\hat{X}_1'\hat{X}_1\right)^{-1}\right], \tag{4.40}$$

since plim $\hat{\Pi}=\Pi$. Again the regularity conditions for ML are satisfied so the Fisher information for δ_1 is σ_{11}^{-1} plim$((1/T)\hat{X}_1'\hat{X}_1)$. I find this approach to LIML a satisfactory way to understand the estimator.

We now derive LIML as a k-class estimator where κ is derived as a characteristic root [Hendry (1976)]. First note in eq. (4.35) that

$$\hat{\sigma}_{11}=\frac{1}{T}(Y\hat{B}_1+Z\hat{\Gamma}_1)'(Y\hat{B}_1+Z\hat{\Gamma}_1)=\frac{1}{T}(X\hat{A}_1)'(X\hat{A}_1)$$

and that

$$\hat{\psi}_{21}=\frac{1}{T}(Y_2-Z\hat{\Pi}_2)'(Y\hat{B}_1+Z\hat{\Gamma}_1)=\frac{1}{T}(Y_2-Z\hat{\Pi}_2)'(X\hat{A}_1)$$

from eq. (4.38). Then from eq. (4.39) we find after premultiplication by $\hat{A}_1'X'Z$ that

$$\begin{aligned}\hat{A}_1'X'Z\hat{\Pi}_2&=\hat{A}_1'X'P_ZY_2-(\hat{A}_1'X'X\hat{A}_1)^{-1}\hat{A}_1'X'P_ZX\hat{A}_1\hat{A}_1'X'(Y_2-Z\hat{\Pi}_2)\\&=\hat{A}_1'X'P_ZY_2-\hat{\lambda}\hat{A}_1'X'(Y_2-Z\hat{\Pi}_2),\end{aligned} \tag{4.41}$$

where $\hat{\lambda} = \hat{A}_1' X' P_Z X \hat{A}_1 / \hat{A}_1' X' X \hat{A}_1$. From the derivation, $\hat{\lambda}$ satisfies the first-order conditions for LIML (FIML). To choose the particular $\hat{\lambda}$ from the possible values which satisfy (4.41), we choose the minimum $\hat{\lambda}$ since it leads to a maximum of the likelihood function as demonstrated in Hood and Koopmans (1953, p. 170, eq. (6.23)):

$$L(A_1) = -\tfrac{1}{2}\log\left(\frac{A_1' X' X A_1}{A_1' X'(I - P_Z) X A_1}\right) + C. \tag{4.42}$$

Of course, the solution to eq. (4.41) requires a non-linear iterative process because $\hat{A}_1$ and $\hat{\Pi}_2$ are functions of $\hat{\lambda}$ which depends on $\hat{A}_1$.

From eq. (4.42) the likelihood function is a monotonic transformation of $1 - \lambda = A_1' X'(I - P_Z) X A_1 / A_1' X' X A_1$, so that λ can be solved for as the smallest characteristic root of $|X_1' P_Z X_1 - \lambda X_1' X_1|$,[35] and A_1 is the corresponding characteristic vector given a normalization $\beta_{11} = 1$. The characteristic equation for λ and A_1 is

$$X_1' P_Z X \hat{A}_1 = \hat{\lambda} X_1' X \hat{A}_1, \tag{4.42a}$$

which can be used to find

$$\hat{\delta}_{\text{LIML}} = \left(\hat{\hat{X}}_1' X_1\right)^{-1} \hat{\hat{X}}_1' y_1, \tag{4.43}$$

where $\hat{\hat{X}}_1 = [\hat{\hat{Y}}_1 \quad Z_1] = [((1 - \hat{\kappa})I + \hat{\kappa} P_Z) Y_1 \quad Z_1]$, with $\hat{\kappa} = 1/(1 - \hat{\lambda})$. Eq. (4.43) is the k-class formula for LIML, since

$$\kappa = 1/(1 - \lambda) = A_1' X' X A_1 / A_1' X'(I - P_Z) X A_1 = A_1' X' X A_1 / B_1' Y'(I - P_Z) Y B_1$$

is monotonic in λ, and A_1 corresponds to the minimum λ. We thus have the traditional LIML formula

$$\kappa = 1/(1 - \hat{\lambda}) = \hat{B}_1' Y'(I - P_{Z_1}) Y \hat{B}_1 / \hat{B}_1' Y'(I - P_Z) Y \hat{B}_1 \tag{4.43a}$$

[Koopmans and Hood (1953, p. 167) and Theil (1971, p. 504)]. Eq. (4.43a) can be derived by noting that the subvector of A_1 which corresponds to Z_1 appears only in the numerator of the expression for $1/(1 - \lambda)$. Therefore, κ in eq. (4.43a) is minimized by "partialling out" the Z_1's and solving for B_1 which gives the minimum. Taking plims of eq. (4.41) yields plim $\hat{\kappa}$ = plim $1/(1 - \hat{\lambda}) = 1$. However,

[35] The result that λ is the characteristic root which gives the minimum of the ratio of the quadratic forms from the characteristic equation is a standard result in matrix algebra, e.g. Gantmacher (1959, pp. 310ff.).

the proof of plim $\sqrt{T}(\hat{\kappa}-1) = -\text{plim } \sqrt{T}(\hat{\lambda}/(1-\hat{\lambda})) = 0$ is more difficult to prove directly.[36] But we do not need to demonstrate this proposition since we have already derived the properties of LIML as an optimal IV estimator (which in fact does provide a proof).

One last approach is to treat the limited information specification of eq. (4.34) as a triangular system [Pagan (1979)]. The Jacobian term in the likelihood function det(B) has the form

$$\det(B) = \det\begin{bmatrix} & B_1 & \\ 0 & & I \end{bmatrix} = 1, \tag{4.44}$$

since B is upper triangular and its determinant equals the product of the diagonal elements, with the normalization $\beta_{11} = 1$. Thus, our previous discussion regarding triangular systems holds true for limited information specifications. Thus, maximization of the "multivariate least squares type" likelihood function

$$L(B_1, \Gamma_1, \Pi, \Psi) = C + \frac{T}{2}\log\det(\Psi)^{-1} - \frac{T}{2}\operatorname{tr}\left[\frac{1}{T}\Psi^{-1}\begin{pmatrix} YB_1 + Z\Gamma_1 \\ Y_2 - Z\Pi_2 \end{pmatrix}'\begin{pmatrix} YB_1 + Z\Gamma_1 \\ Y_2 - Z\Pi_2 \end{pmatrix}\right] \tag{4.45}$$

yields LIML. It seems interesting to note that limited information specifications can be put into triangular form. However, some limited computational experiments seem to indicate that estimation of LIML in this form is an inferior method of computation compared to calculation of the characteristic value and the use of equation (4.43).[37] Nevertheless, this approach does re-emphasize the point made by eqs. (4.38) and (4.39) of the role of the covariance matrix of the disturbances in LIML estimation.

4.5. *Estimation with covariance restrictions*

So far we have only considered the case where Σ is unrestricted. And this situation is by far the most common. Yet we saw in Section 3 that identification can occur via covariance restrictions. We now give a brief example of how covariance restrictions provide extra instruments for estimation and therefore lead to more efficient estimates. Rothenberg and Leenders (1964) first noted that

[36] Malinvaud (1970, p. 702) proves this result by noting that the root $T(\kappa-1)$ is distributed as a χ^2 random variable with $K - r_1$ degrees of freedom. Therefore plim$\sqrt{T}(\kappa-1) = 0$. We prove this result by a somewhat easier method in the next section.

[37] I began the iteration from the 2SLS estimates of the first equation and OLS estimates of the other equations.

3SLS is not asymptotically equivalent to FIML when the covariance matrix Σ is known. Rothenberg (1973) demonstrates that more efficient estimates can be obtained by FIML in the case of Σ restricted to a diagonal matrix. Our example shows why this proposition holds true. A general treatment of the problem is left to future work [Hausman and Taylor (1981)].

For our example we use the system specification of eq. (2.1) together with the assumptions we have previously made. In particular we assume that each equation is identified by linear coefficient restrictions and that U is distributed as multivariate normal so that we can consider FIML estimation. We shall consider covariance restrictions so that Σ is assumed to be diagonal.[38] When we consider the first-order conditions for FIML in eq. (4.25), note that we know all the elements of Σ *a priori* except the diagonal elements σ_{ii}. Using the identity $\Sigma\Sigma^{-1} = I_M$ we transform equation (4.25,i) by postmultiplication:

$$\left(\left[T(B')^{-1}\Sigma - Y'(YB+Z\Gamma)\right]\Sigma^{-1}\right)^u = 0. \tag{4.46}$$

We consider the first term in eq. (4.46) and look at the ijth element of the matrix product:

$$\left[T(B')^{-1}\Sigma\right]_{ij} = T\beta^{ji}\sigma_{jj} = \left(v_i' - \sum_{\substack{k=1\\k\neq j}}^{M}\beta^{ki}u_k'\right)u_j, \tag{4.47}$$

where β^{ji} is the jith element of the inverse matrix B^{-1}. We now insert eq. (4.47) back into the bracket terms in eq. (4.46) to find

$$\begin{aligned}\left[T(B')^{-1}\Sigma - Y'(YB+Z\Gamma)\right]_{ij} &= \left(v_i' - \sum_{k\neq j}\beta^{ki}u_k' - (Z\Pi_i)' - v_i'\right)u_j\\ &= -\left(Z\Pi_i + \sum_{k\neq j}\beta^{ki}u_k\right)'u_j.\end{aligned} \tag{4.48}$$

We thus see the differences that covariance restrictions makes in the first-order conditions. In the previous case of no restrictions on Σ in eq. (4.26), the instruments used were $W' = \hat{X}'(\hat{\Sigma}\otimes I_T)^{-1}$, with $\hat{X}_i = [-Z(\hat{\Gamma}\hat{B}^{-1})_i \quad Z_i]$ and $\hat{\Sigma} = T^{-1}(Y\hat{B}+Z\hat{\Gamma})'(Y\hat{B}+Z\hat{\Gamma})$. Two differences are present here. First, rather than setting $\hat{Y}_i = Z(\hat{\Gamma}\hat{B}^{-1})_i$, eq. (4.48) uses $\hat{Y}_i = -Z(\hat{\Gamma}\hat{B}^{-1})_i + \hat{\tilde{v}}_i$, where each term in $\hat{\tilde{v}}_i$ contains all the terms in the reduced-form disturbance $\hat{v}_i = (\hat{U}\hat{B}^{-1})_i$ *except* the

[38] Malinvaud (1970, pp. 678–679) derives the concentrated log-likelihood function for this case. Rothenberg (1973, p. 78) derives the Cramer–Rao bound.

term $\beta^{ii}u_i$ which is, of course, correlated with the structural disturbance in the ith equation. If Y_i were not jointly endogenous, so that it was relatively recursive, we know given the results in Section 3 that all the $[(B')^{-1}\Sigma]_{ij} = 0$ so eq. (4.46) would lead to $\hat{Y}_i = Y_i$. FIML does the next best thing. $Y_i = Y\Pi_i + v_i$ and FIML uses all of that part of v_i which satisfies the condition of $\hat{Y}_i$ being a valid instrument. FIML uses the covariance restrictions to form a better prediction of Y_i. The second difference is that to estimate Σ, we use $\hat{\Sigma}_{ii} = [T^{-1}(Y\hat{B} + Z\hat{\Gamma})'(Y\hat{B} + Z\hat{\Gamma})]_{ii}$ and set all $\hat{\Sigma}_{ij} = 0$. Thus, FIML both imposes the *a priori* restrictions on Σ and forms better instruments, in the sense of being more highly correlated with Y_i when covariance restrictions are present.

Compared to FIML, how good are the system IV estimators when restrictions on Σ are present? They can easily impose the restrictions on Σ but do not use the reduced-form disturbances v_i to form better instruments. For example, 3SLS would continue to use $\hat{Y}_i = Z\hat{\Pi}_i$, where $\hat{\Pi}_i$ is the unrestricted estimator of the reduced-form coefficients. Thus, the IV estimators are inefficient relative to FIML when covariance restrictions are present. Hausman and Taylor (1981) define "augmented" IV estimators, AIV and A3SLS, which are asymptotically more efficient than regular IV or 3SLS estimators. These augmented estimators take into account that part of v_i which is asymptotically uncorrelated with u_i when forming Y_i. And the AIV estimator, if iterated until convergence, yields the FIML estimator. However, the non-iterated AIV and A3SLS seem asymptotically less efficient that FIML. I conjecture that an additional term will be found through linearization of the likelihood function which will make an estimator much like AIV asymptotically efficient. For now, we require an iterative procedure if we want to achieve the asymptotic Cramer–Rao lower bound. Lastly, Hausman and Taylor (1981) extend FIML and the augmented estimators to the case when only some $\sigma_{ij} = 0$, rather than all. To answer partially Rothenberg's (1973, p. 73) question about the likely value of covariance restrictions, we use the IV setup of FIML in this case. In a time-series context with slowly trending variables, $Y_i = Z\Pi_i$ often does a good job of "predicting" Y_i, i.e. the R^2's are high. Thus, covariance restrictions would usually lead to small gains in asymptotic efficiency. On the other hand, in cross-section simultaneous equation models on individual or firm data, the covariance restrictions might well lead to substantial gains in asymptotic efficiency. In cross-section models the R^2's are often low so that the residuals could play an important part in providing better instruments, if the restrictions are indeed true.

4.6. *Other considerations*

We have looked at most of the simultaneous equation estimators. We now turn to three other problems all involved with estimation.

(1) Among the equivalent classes of IV estimators, might some be better than others? We have proven first-order asymptotic equivalence but we might consider higher orders of approximation. R. A. Fisher claimed that ML was best here; and Rao (1961, 1962) and Efron (1975) have defined and considered "second-order efficiency" properties of ML. As we iterate from an efficient estimate to ML, we know that contact with the likelihood function is increased at each iteration. The unanswered question might be to ask what that means for estimation or hypothesis testing. Pfanzagl (1975) gives a partial answer where he demonstrates that (higher order) bias corrected ML estimates may lead to more powerful tests in regular situations. In fact, Pfanzagl (1975) and Akahira and Takeuchi (1981) define second-order efficiency on the basis of estimators which yield the most powerful one-sided test. Pfanzagl and Wefelmeyer (1978) and Akahira and Takeuchi (1981) also go on to consider "third-order efficiency" which is defined in terms of the concentration of the distribution of the estimator in convex sets which contain the true parameter. The latter authors prove that LIML and FIML, when bias corrected, are superior to 3SLS and 2SLS on third-order efficiency grounds. But the issues raised are far from settled, especially the acceptance of the definition of third-order efficiency. We know that FIML and LIML both impose the overidentifying restrictions in forming the instruments and thus differ from 3SLS and 2SLS. The consequence of these additional restrictions still remain to be worked out fully. One suggestion which does arise from the second-order literature is the improvement which follows from asymptotic bias corrections to ML estimators. Investigation of these corrections in some simple cases might be worthwhile.[39]

(2) Even when first-order approximations are used, we do not yet know the best method to compute the asymptotic covariance matrices. For instance, the FIML estimated inverse information matrix $(\hat{X}'(\Sigma\otimes I)^{-1}\hat{X})^{-1}$ is typically larger than the 3SLS estimated covariance matrix which does not impose restrictions in forming $\hat{X}$. And the difference can be substantial, as simple examples demonstrate [Hausman (1974)]. Furthermore, recently Efron and Hinckley (1978) and Hinckley (1980) have given non-simultaneous equation ML estimator examples which tend to show that the estimated inverse information matrix may not be that accurate. Better approximations may be required; the question of second-order approximations arises. It might be the case that the next order of approximation to the Hessian, using the Newton–Raphson matrix of second derivatives at convergence, might be superior to use of the inverse information matrix.[40]

[39]However, this issue is far from settled in the statistics literature, even for simple models. See the article by Berkson (1980) together with the discussion which follows the paper.

[40]This statement only pertains to estimation of the covariance matrix, *not* to efficient methods of calculation of ML estimates.

(3) Lastly, we have the related question of the accuracy of the asymptotic approximations in finite samples. We know that finite sample moments of the estimators, either structural estimators or reduced-form estimators, no matter how large the sample is, need not exist [Basmann (1961) and Sargan (1978)].[41] Existence of finite sample moments for the single equation instrumental variable type estimators usually depends on the degree of overidentification [Phillips (1980)]. Whether non-existence of moments should be troublesome depends on the use that we plan for the estimates. Certainly, for hypothesis testing the problem may not seem too important.

Next, we might want to assess the accuracy of the asymptotic normal approximations. Exact finite sample distributions are known for some cases [Basmann (1975) and Phillips (1980)] but they are quite complicated to compute. They have not received much use to date. Much current research is being done on the quality of the asymptotic approximations [Phillips (1981a) and Phillips (Chapter 8 in this Handbook)].[42] From the current results we know that the first order normal approximations are not generally thick enough in the tails. Moreover, the actual distributions are also usually asymmetric. Thus, we might well expect more refined asymptotic approximations to be useful. I feel that their most important use might be to let us choose among estimators. Some recent work of Anderson et al. (1982) uses approximations to demonstrate the mean square error superiority of LIML over 2SLS when the equation is overidentified to a sufficient degree. We certainly want the best estimators possible. But I expect that in the near future hypothesis testing will continue to be based on the asymptotic approximations. However, the outstanding current unresolved problem seems to be whether even better asymptotic approximations or, rather, direct approximations to the finite sample distributions, will be the more useful approach in the long run for the development of convenient and useful approximations for econometric estimators and tests.[43]

5. Specification tests

Given a specification of a linear simultaneous equation system as in eq. (2.1), the opportunity exists to test both coefficient restrictions and asymptotic orthogonality assumptions. The first type of test, which considers overidentifying restrictions, was developed in the early work at the Cowles Commission by

[41] Estimators have recently been proposed which are guaranteed to have finite sample moments. See Zellner (1978) for structural form estimators and Maasoumi (1978) for reduced form estimators.

[42] The small sigma approach of Kadane (1971) is also of relevance here.

[43] Phillips (1981a) discusses this issue.

Anderson–Rubin (1949) and Koopmans–Hood (1953). We shall refer to these tests as classical specification tests. It is perhaps surprising that these classical tests are not used more in applied work. One problem, already referred to, is that for large macro models, sufficient observations may not be present to estimate the unrestricted reduced form which is required to apply the tests.[44] But perhaps more importantly, a rejection of the overidentifying restrictions gives no clear indication of what action to follow next. Since we saw in Section 3 that M restrictions in each equation are required for just identification, multiple comparison procedures of the Scheffe (1958) form seem difficult to apply to help decide which restrictions are not correct.[45] Lastly, the tests are inapplicable to non-linear simultaneous equation systems, which we turn to in the next section, where a closed-form reduced-form specification does not usually exist. The asymptotic orthogonality tests, applied to simultaneous equation specifications by Wu (1973) and Hausman (1978), deal with the structural form rather than the reduced form. Thus, a maintained set of just identifying assumptions are required but specific overidentifying assumptions may be tested. Neither sample size nor non-linearity restrict use of the tests, to the extent that the asymptotic distribution theory provides a reliable guide to the behavior of the test statistics.

We first consider the complete systems test which takes all overidentifying restrictions in the system into account. Define $K - r_i - s_i = K - k_i$ to be the degree of overidentification in the ith equation, where each restriction is assumed to affect overidentification by our previous definition in Section 4. Then let $p = MK - \sum_{k=1}^{M} k_i$ be the degree of overidentification in the entire system.[46] A joint test of all overidentifying restrictions then can be constructed by comparing the unrestricted estimates of the reduced-form coefficients Π_u to be restricted estimates of the coefficients Π_R. Since the latter estimates are subject to p overidentifying restrictions, we can perform a test of the restrictions. Various asymptotically equivalent tests can be defined. First, we might consider a likelihood ratio (LR) test based on the likelihood function of eq. (4.32). The restricted estimates can be based on any of the optimal system IV estimations, although the FIML estimates seem best given that we base the test on a comparison of

[44]Wegge (1978) offers a sequential test procedure of beginning with a just identified system and adding restrictions to help alleviate this problem. He bases his procedure on the indirect least squares estimator.

[45]But conditional on the *just*-identifying restrictions, multiple comparison procedures could be applied to the overidentifying restrictions of the reduced-form coefficients. However, these tests may be quite difficult to interpret.

[46]As always when restrictions are tested, one must be careful not to count redundant restrictions in determining the appropriate degrees of freedom. The precise criterion to use here is the number of non-redundant restrictions on the reduced-form coefficients. Such a determination may be quite difficult in actual situations.

likelihood values. Regularity conditions for LR tests are satisfied given a normality assumption so that after simplification:[47]

$$W_p = -2\log\left(\frac{L(\hat{\Pi}_R,\hat{\Omega}_R)}{L(\hat{\Pi}_u,\hat{\Omega}_u)}\right) = T\log\left(\frac{\det\hat{\Omega}_R}{\det\hat{\Omega}_u}\right). \tag{5.1}$$

Under the null hypothesis of correct specification, W_p is distributed as central χ^2_p. We could also directly compare $\hat{\Pi}_u$ with $\hat{\Pi}_R$ via Wald or Lagrange multiplier (LM) which are asymptotically equivalent to LR tests [Silvey (1970, ch. 7)]. Let us now consider the stacked form of reduced-form coefficients, π, as they were defined in eq. (4.21). If the overidentifying restrictions are put into the form $h(\pi)=0$, we then use a first-order expansion around $\hat{\pi}_u$, $h(\pi)\simeq h(\hat{\pi}_u)+\hat{H}_u(\pi-\hat{\pi}_u)$, where $\hat{H}_u$ is evaluated at $\hat{\pi}_u$. Then a possible test of the overidentifying restrictions is [Silvey (1970, p. 116) and Byron (1974)]

$$\hat{W}_p = h(\hat{\pi}_u)'\left[\hat{H}_u\hat{\Psi}_u^{-1}\hat{H}_u'\right]h(\hat{\pi}_u), \tag{5.2}$$

where $\hat{\Psi}_u = \bar{Z}'(\hat{\Omega}_u^{-1}\otimes I)\bar{Z}$, and where $\bar{Z}=I\otimes Z$. $\hat{W}_p$ has the same asymptotic distribution under the null hypothesis as does W_p. Another asymptotically equivalent test, which is perhaps the computationally most convenient, is to note that $\hat{\pi}_R \simeq \hat{\pi}_u + \hat{\Psi}_u^{-1}\hat{H}_u'[\hat{H}_u\hat{\Psi}_u^{-1}\hat{H}_u']^{-1}h(\hat{\pi}_u)$ [Silvey (1970, p. 119), Malinvaud (1970, p. 360), and Maasoumi (1978)] so that we find the LM-type test statistic:

$$\tilde{\tilde{W}}_p = (\hat{\pi}_R-\hat{\pi}_u)'\hat{\Psi}_u(\hat{\pi}_R-\hat{\pi}_u), \tag{5.3}$$

where again under the null hypothesis $\tilde{\tilde{W}}_p$ is distributed as central χ^2_p.[48] Besides the problems previously mentioned, even the most convenient test $\tilde{\tilde{W}}_p$ is difficult to interpret since it is based on estimated π's. Since almost all our economic theory and most of our empirical knowledge centers about the structural parameters (B,Γ), we usually find it difficult to determine what is wrong with the structural equation specification when we base tests on the estimated reduced-form coefficients.

We now turn to single equation tests of overidentification.[49] Anderson and Rubin (1949) based their version of this test on the asymptotic distribution of the

[47] Viewed from a minimum distance perspective, this test continues to hold even if the stochastic disturbances are not normal, asymptotically. See Berndt, Hall, Hall and Hausman (1974).

[48] Note we are not quite using a true LM statistic which would require $\hat{\Omega}_R$ rather than $\hat{\Omega}_u$ in forming $\hat{\Psi}$. Yet another asymptotically equivalent test of the overidentifying restrictions is to use the 3SLS estimates to form $\tilde{\tilde{W}}_p = \hat{u}'(S^{-1}\otimes Z(Z'Z)^{-1}Z')\hat{u}$. However, the test of eq. (5.3) may be preferred since it allows comparison of the restricted and unrestricted coefficient estimates.

[49] We do not discuss the test of just identification proposed in the single equation context. It can be found in Koopmans and Hood (1953, pp. 183–184). It is based on the two smallest characteristic roots of the LIML determinental equation.

smallest characteristic root derived from LIML [eq. (4.42)]. The test is based on $T(\hat{\kappa}-1)$ which under the null hypothesis is distributed as χ^2 with $p_1 = K - k_1$ degrees of freedom. An asymptotically equivalent test for overidentification of, say, the first equation, can be applied with any optimal single equation IV estimator, e.g. 2SLS which is asymptotically equivalent to LIML. First estimate $\hat{\sigma}_{11}$ from an optimal IV estimator. Define the vector $\tilde{Y}_1 = [y_1 \quad Y_1]$ and estimate $M_{11} = \tilde{Y}_1' Q_z \tilde{Y}_1$ for $Q_z = I - Z(Z'Z)^{-1}Z'$. Then define $\tilde{\beta}_{1,\mathrm{IV}} = [1 \quad \hat{\beta}_{1,\mathrm{IV}}]$. The test statistic is then [Malinvaud (1970, p. 712)]

$$w_{p_1} = T\left[\hat{\sigma}_{11}/\left(\tilde{\beta}_{1,\mathrm{IV}} M_{11} \tilde{\beta}_{1,\mathrm{IV}}'\right) - 1\right], \tag{5.4}$$

where w_{p_1} is distributed as $\chi^2_{p_1}$, and where $p_1 = K - k_1$ under the null hypothesis.[50] Eq. (5.4) is straightforward to derive from the LIML likelihood function for the restricted and unrestricted reduced-form estimates as in eq. (5.1). Since $\sigma_{11} = (B'\Omega B)_{11}$, while M_{11} is the submatrix of Ω containing jointly endogenous variables in the first equation, we see that the ratio in eq. (5.4) resembles the ratio of the restricted to unrestricted reduced form covariance estimates of eq. (5.1).

An asymptotically equivalent test for misspecification which is easier to compute can be derived by using $\hat{\lambda}$ rather than $\hat{\kappa}$ from eq. (4.43a) since they are monotonic transformations of each other. After a normalization we find $\hat{\lambda} = (y_1 - X_1\hat{\delta}_1)'P_Z(y_1 - X_1\hat{\delta}_1)/\hat{\sigma}_{11}$. We now derive the limiting distribution of $\hat{\lambda}$. The idea is to form $\hat{q}_1 = P_Z\hat{u}_1$; under the null hypothesis of no misspecification $\hat{q}$ should be near zero given the orthogonality of the instruments. Suppose we estimate $\hat{\delta}_1$ by 2SLS; LIML estimates will yield the same asymptotic distribution. Note that $\hat{u}_1 = u_1 - X_1(\hat{\delta}_1 - \delta_1)$ so that $\hat{q}_1 = P_z\hat{u}_1 = P_z u_1 - P_z X_1(X_1'P_z X_1)^{-1}X_1'P_z u_1$. We then form the statistic:

$$\begin{aligned} \tilde{w}_{p_1} &= \hat{q}_1'\hat{q}_1/\hat{\sigma}_{11} = \hat{\lambda} = \hat{u}_1' P_z \hat{u}_1/\hat{\sigma}_{11} \\ & u_1' A u_1/\hat{\sigma}_{11} \overset{\mathrm{A}}{\sim} \chi^2_{p_1}, \end{aligned} \tag{5.4a}$$

where $A = [P_z - P_z X_1(X_1'P_z X_1)^{-1}X_1'P_z]$, which is an orthogonal projection since it equals $P_z - P_{\hat{X}_1}$, where $\hat{X}_1 = P_z X_1$ which lies in the subspace spanned by the columns of Z. The degrees of freedom follow since the dimension of P_z is K, while the dimension of $P_{\hat{X}_1}$ is k_1. Computationally this statistic is easily calculated. Regress $\hat{u}_1$ on all the instruments and calculate $\tilde{w}_{p_1} = TR^2$, where R^2 is the coefficient of multiple correlation. Of course, this result demonstrates the Lagrange multiplier characteristic of the test. Both the derivation and computation of this

[50] Note that the expression in eq. (5.4) is just $T(\kappa - 1)$ in our earlier notation except that the LIML estimates have been replaced by asymptotically equivalent estimates and we have used κ from eq. (4.42).

test statistic based on $\hat{\lambda}$ seem a good deal easier than the Anderson-Rubin (1949) test for $T(\hat{\kappa}-1)$ which is the traditional overidentification test statistic. A problem of interpretation exists for these limited information tests as it did with the system tests of overidentification since we cannot compare structural parameter estimates under the null and alternative hypotheses.

An alternative test of overidentification which does permit such a comparison is given by Hausman and Taylor (1980b).[51] This test is closely related to the asymptotic orthogonality tests of Wu (1973) and Hausman (1978). We again consider the first equation, $y_1 = X_1\delta_1 + u_1$. As the maintained hypothesis, we assume that the r_1 instruments which correspond to Z_1 plus an additional s_1 instruments are valid, $\tilde{Z}_1$. We then test the validity of the additional $K - r_1 - s_1 = p_1$ instruments. Note that we then have a test of overidentification of these instruments, Z_1^*, because we are partly testing the exclusion restrictions on them in the first equation. Our test is based on a comparison of the structural coefficient estimates, $\hat{\delta}_1$ and $\hat{\delta}_1^*$, where the latter estimator is IV with both W_1 and W_1^* as instruments. Define $Z = [\tilde{Z}_1 \vdots Z_1^*]$. Then we define

$$\hat{q} = \hat{\delta}_1 - \hat{\delta}_1^* = \left[(X_1'P_{\tilde{Z}_1}X_1)^{-1}X_1'P_{Z_1} - (X_1'P_ZX_1)^{-1}X_1'P_Z\right]y_1. \tag{5.5}$$

Then the statistic

$$\tilde{w}_{p_1} = \frac{1}{\hat{\sigma}_{11}}\hat{q}'\left[(X_1'P_{\tilde{Z}_1}X_1)^{-1} - (X_1'P_ZX_1)^{-1}\right]^+\hat{q} \tag{5.6}$$

is distributed as χ^2 with degrees of freedom $\min[p_1, k_1]$ under the null hypothesis where $[\cdot]^+$ denotes any generalized inverse. A subset of the overidentifying restrictions can be tested in the same way. Note that we have explicitly stated a sufficient number of instruments to just identify the equation under the maintained hypothesis. But we can then compare differences in the structural estimates about which our economic knowledge is typically much greater than the reduced-form coefficients.

A somewhat similar system test is given by Hausman (1978). He proposes comparing full information and limited information estimators, e.g. 3SLS and 2SLS, to test the specification of the rest of the system if any equation is overidentified.[52] The test is again based on the structural parameter estimates,

[51]W. Taylor has pointed out that Anderson–Rubin type tests of overidentification are only necessary but not sufficient as tests for the predeterminedness of the overidentifying instruments. The following type of test is appropriate for a test of the legitimacy of the overidentifying instruments.

[52]We assume here that Σ is unrestricted. If $\sigma_{ij} = 0$, then equation j is dropped from the system for this test.

$\hat{q} = \hat{\delta}_{1,FI} - \hat{\delta}_{1,LI}$. The test statistic is

$$\tilde{\hat{w}}_{p\dagger} = \hat{q}'\left[V(\hat{\delta}_{1,LI}) - V(\hat{\delta}_{1,FI})\right]^{+}\hat{q} \tag{5.7}$$

which, under the null hypothesis of correct specification, is distributed as χ^2 with degrees of freedom $\min[k_1, p - p_1]$. Again the interpretation of the structural coefficients should be more fruitful in an applied problem.

The last type of specification test considered are the so-called causality tests of Granger (1969) and Sims (1972). We do not attempt a discussion of the somewhat difficult notion of causality here; nor do we consider the possible usefulness of causality tests as tests of proper dynamic specification. Instead, we consider the tests within the context of a simultaneous equation estimation, in particular as tests of whether variables can be tested as predetermined.[53] Wu (1973) and Hausman (1978) have proposed tests of whether right-hand-side variables can be treated as predetermined. We know that this condition is one of the two conditions which allow a variable to be used as a valid instrument. The test of eq. (5.6) on the validity of instruments is an example of Wu–Hausman type tests. Recently, Sims (1977) and Geweke (1978) have used the Granger–Sims causality tests as "exogeneity tests." We want to see what Granger–Sims tests can tell us about the validity of instruments.

The example often used in the causality literature is the case of a bivariate time series:

$$\begin{pmatrix} a_{11}(L) & a_{12}(L) \\ a_{21}(L) & a_{22}(L) \end{pmatrix}\begin{pmatrix} x_1 \\ x_2 \end{pmatrix} = \begin{pmatrix} u_1 \\ u_2 \end{pmatrix}, \tag{5.8}$$

where $a_{ij}(L) = \sum_{k=0}^{\infty} a_{ijk} L^k$ is a polynomial in the lag operator $L^k x_{1,t} = x_{1,t-k}$. We make the usual assumption that u is i.i.d. with mean zero and covariance matrix Σ. The system is assumed stable so we can write

$$x_2 = b(L)x_1 + v, \tag{5.9}$$

where $b(L) = -a_{21}(L)/a_{22}(L)$ and $v = u_2/a_{22}(L)$. What are the conditions required for x_1 to be predetermined? From Section 3 we know that for x_1 to be relatively recursive, we require that $a_{12}(0) = 0$ and $\sigma_{12} = 0$. The stronger condition of exogeneity requires that all $a_{12}(L) = 0$ [Koopmans (1950) and Zellner and Palm (1975)]. The Granger–Sims causal ordering requirement, sometimes referred to as exogeneity, is that the projection of x_{2t} on the entire x_1 process has zero

[53] The following section is closely related to Wu (1979). A discussion of Granger causality for dynamic rational expectation models and its relationship to the validity of instruments is contained in Hansen and Sargent (1980).

coefficients for future x_1's covariance $E(x_1u_2')=0$. Thus, the Granger–Sims requirement is that $a_{12k}/a_{22k}=\alpha\sigma_{12}/\sigma_{22}$ for all $k\geqslant 0$ and a scalar α, which is a weaker requirement than simultaneous equation type exogeneity.

The question at issue is whether under these three assumptions x_1 is a valid instrument for eq. (5.9), i.e. does least squares provide consistent estimates for $b(L)$?[54] We know that the first two definitions provide consistent estimates. Since under both assumptions $\text{plim}(1/T)X_1'V=0$, eq. (4.5) guarantees consistent estimation. But the causal ordering requirement is neither necessary nor sufficient for consistent estimation of $b(L)$. Granger–Sims' causality may not hold, but x_1 can still be predetermined: $a_{120}=0$ but some $a_{12k}\neq 0$ for $k\geqslant 1$. Remember that a valid instrument need only be predetermined, not exogenous, which is a much stronger requirement. Neither is Granger–Sims causality sufficient for consistent estimation since in the case of a static model $a_{12k}=a_{22k}=0$ for all $k\geqslant 1$ but $a_{120}\neq 0$ so that x_1 is jointly endogenous. Thus, we conclude that a Granger–Sims causality test cannot answer the question of validity of instruments. Furthermore, the necessary assumption for its use, $\sigma_{12}=0$, places it together with recursive type specifications. In Sections 2 and 3 we argued that system recursive specifications are of very limited usefulness compared with structural specifications. Relative recursivity of a variable with respect to a given equation seems to be a more useful notion since it determines whether a variable can be treated as predetermined *in a given equation*. But the specification of the equation is crucial to the question of relative recursitivity since $\text{plim}(1/T)v_j'u_i$ is necessary for y_j to be predetermined with respect to equation i. Consideration of y_i and y_j in isolation cannot answer whether a variable can be treated as predetermined since u_i has not been properly specified. Other important uses of causality tests may exist, but they cannot answer the crucial question that arises in the specification of structural simultaneous equation models.[55]

6. Non-linear specifications

Specification and estimation of non-linear simultaneous equation models has many similarities to the linear case. But important differences arise also. In this section we give an overview without the detailed investigation that we attempted in the linear case. The subject is much less settled than in the linear case. First

[54] Of course, we can always estimate a recursive form of eq. (5.8). But as we argued in Section 2, we then no longer have the coefficients which correspond to our structural specification.

[55] Zellner (1979a) examines the usefulness of causality tests from a different perspective but he also concludes that additional structure is needed for them to be useful. Engle, Hendry and Richard (1981) conclude that (Granger) causality is "irrelevant" given the assumptions which were made in Section 2. They also give another example when Granger causality is neither necessary nor sufficient for predeterminedness.

consider the specification of a single non-linear simultaneous equation with additive errors:

$$f_i(y, z, \alpha) = u_i, \qquad i = 1, \dots, M, \tag{6.1}$$

where as before y and z are jointly endogenous and predetermined variables and α is the vector of unknown coefficients. The stochastic disturbance u_i has mean zero and covariance $V(u_i) = \sigma_{ii} I_T$. The contemporaneous covariance matrix is Σ so that $E(uu') = \Sigma \otimes I_T$. Non-linear simultaneous equation specifications can arise in a number of ways.

(1) Non-linear functional forms, f_i, can lead to jointly endogenous variables entering the specification non-linearly. For instance, a factor demand equation which arises from a two-factor production function might well have the specification:

$$q_{it} = \beta_{i1} p_{1t} + \beta_{i2} p_{2t} + \beta_{i3} p_{1t} p_{2t} + \beta_{i4} p_{1t}^2 + \beta_{i5} p_{2t}^2 + Z_{it} \gamma_i + u_i, \tag{6.1a}$$

for $i = 1, 2$, where the factor prices, p_{1t} and p_{2t}, are jointly endogenous. Of course, more complicated functions f_i can arise which do not have a polynomial form.

(2) Non-linear identities can transform an otherwise linear specification into a non-linear specification which is linear in parameters form like eq. (6.1a) but is non-linear in variables. An example arises when price and quantity are used to determine expenditure.

(3) Essentially linear specifications may become non-linear due to non-linear *a priori* restrictions among the parameters or by induced transformations to account for serial correlation in the stochastic disturbances. These latter, "non-linear in parameter" specifications are much closer to our previous linear case. We first turn to this latter situation.

Assume that the ith equation can be written in the form

$$y_i = Y_i \beta(\alpha_i) + Z_i \gamma(\alpha_i) + u_i = X_i \delta(\alpha_i) + u_i, \qquad i = 1, \dots, M. \tag{6.2}$$

Eq. (6.2), the non-linear in parameters specification, is non-linear only because the structural parameters depend on the α_i in a non-linear manner. As before we have a reduced-form specification which is now non-linear in parameters:

$$Y = Z\Pi(\alpha) + V, \tag{6.3}$$

where $\alpha = (\alpha_1', \dots, \alpha_M')$. A sufficient additional assumption for identification of eq.

(6.2) arises due to the non-linearity induced by the function $\delta(\alpha_i)$:

Assumption 6.1

The inverse mapping of $\delta(\alpha_i)$ is one-to-one and continuous in the neighborhood of the true value $\delta^* = \delta(\alpha_i^*)$. Also, $\delta(\alpha_i)$ has bounded third derivatives in the neighborhood of α_i^* and the matrix $\partial\delta/\partial\alpha_i$ has full column rank.

This assumption basically assures a *unique* solution locally; therefore, the notion of identification holds only locally rather than globally as in linear specifications.[56]

The optimal single equation instrumental variable estimator can be transformed for this non-linear in parameters specification. Consider the IV estimator for the first equation:

$$\min_{\alpha_1}(\hat{\delta}_1 - \delta(\alpha_1))'\left[(\hat{D}_1'Z'Z\hat{A}_1)(\hat{A}_1'Z'Z\hat{A}_1)^{-1}(\hat{A}_1'Z'Z\hat{D}_1)\right](\hat{\delta}_1 - \delta(\alpha_1)), \quad (6.4)$$

where $\hat{\delta}_1$ has been estimated via the instruments $W_1 = Z\hat{A}_1$. Note that the term in brackets arises from the estimated covariance matrix from eq. (4.7). Here the optimal instruments remain the same as before with the requirement that plim $\hat{A}_1 = D_1$, so that $\hat{A}_1 = (Z'Z)^{-1}Z'X_1$ then yields

$$\min_{\alpha}(\hat{\delta}_1 - \delta(\alpha_1))'\left[\hat{A}_1'(Z'Z)\hat{A}_1\right](\hat{\delta}_1 - \delta(\alpha_1)). \quad (6.5)$$

A linear approximation to the constraints around the true value, $\hat{\delta}_1 = \delta(\alpha_1^*) + (\partial\delta(\alpha_1^*)/\partial\alpha_1)\cdot(\hat{\alpha}_1 - \alpha_1^*)$ allows straightforward proof that the estimator $\hat{\alpha}_{1,\mathrm{IV}}$ has an asymptotic normal distribution so that

$$\sqrt{T}(\hat{\alpha}_{1,\mathrm{IV}} - \alpha_1) \overset{A}{\sim} N\left(0, \operatorname{plim}\hat{\sigma}_{11}\left(\frac{\partial\delta(\hat{\alpha}_{\mathrm{IV}})'}{\partial\alpha_1}\left[\hat{A}_1'Z'Z\hat{A}_1\right]\frac{\partial\delta(\hat{\alpha}_{\mathrm{IV}})}{\partial\alpha_1}\right)^{-1}\right). \quad (6.6)$$

An important special case of this estimator is the autoregressive specification where in the original equation $u_{1t} = \rho_1 u_{1,t-1} + \eta_{1t}$. The specification of eq. (4.1) then becomes

$$y_1 = \rho_1 y_{1l} + (X_1 - \rho_1 X_{1l})\delta + \eta_1, \quad (6.7)$$

where $X_{1l} = X_{1,t-1}$. The optimal instruments now depend on a consistent estimate of ρ_1, $\hat{W}_1 = (Z\hat{\Pi}_1 - \hat{\rho}_1 Y_{1l} \quad Z_1 - \hat{\rho}_1 Z_{1l})$. Similarly autoregressive LIML and auto-

[56] Fisher (1966) and Rothenberg (1970) discuss local identification conditions.

regressive 2SLS have been considered. An important computational simplification can often be made by using the fact that the instruments need only be based on initial consistent estimates with no effect on the asymptotic distribution.[57]

The non-linear in parameters specification (with across-equation constraints) can also be made for the full system specification:[58]

$$YB(\alpha)+Z\Gamma(\alpha)=XA(\alpha)=U. \tag{6.8}$$

For the stacked specification $y=X\delta(\alpha)+u$, optimal IV estimates take the form

$$\min_{\alpha}\left(\hat{\delta}-\delta(\alpha)\right)'\left(\hat{X}'(\hat{\Sigma}\otimes I)^{-1}\hat{X}\right)\left(\hat{\delta}-\delta(\alpha)\right), \tag{6.9}$$

where $\hat{X}=\operatorname{diag}(\hat{X}_1,\dots,\hat{X}_M)$ with $\hat{X}_i=Z\hat{A}_i$ and $\operatorname{plim}\hat{A}_i=D_i=[\Pi_i \quad I_i]$. Hausman (1975) demonstrates that this estimator is asymptotically equivalent to the FIML estimator given a normality assumption on the U. Autoregressive specifications can also be considered in this formulation; Hendry (1976) gives a review.

We now turn to IV estimation of the general non-linear simultaneous equation specification with additive disturbances. We forgo a discussion of regularity conditions; Amemiya (1977, and Chapter 6 in this Handbook) outlines the appropriate assumptions. The general method of application is quite straightforward. We linearize the equations around an initial parameter estimate, α^1, and apply an instrumental variable estimator. A new estimate α^2 is obtained and iteration continues via minimization of a sum of squares type objective function. The approach is identical to the Gauss–Newton approach to non-linear least squares problems, e.g. Seber (1966, ch. 10) or Malinvaud (1970, ch. 9).[59] If we expand the first equation from our specification of eq. (6.1) around some α^1, we find

$$f_1(y,Z,\alpha^1)\doteq f_1(y,Z,\alpha)+g_1^1(\alpha-\alpha^1)=g_1^1(\alpha-\alpha^1)+u_1, \tag{6.10}$$

where g_1^1 is the matrix of first derivatives $g_1^1=\partial f_1(y,Z,\alpha^1)/\partial\alpha$. The simultaneous equation problem exists because $\operatorname{plim}(1/T)g_1'u_1\neq 0$ since g_1 contains the jointly

[57] However, iteration is still required to find the estimate of ρ_1 as a parameter in the equation. See Sargan (1959, 1961, 1964), and Fair (1970) for these estimators. Hendry (1976) provides a convenient summary. An asymptotically optimal IV estimator which does not require iteration is given by Hatanaka (1976).

[58] This specification has recently received considerable attention in rational expectation models [Wallis (1980)].

[59] Goldfeld and Quandt (1968) and Kelejian (1971) first applied this methodology to specifications which were linear in the parameters, but non-linear in the variables. Amemiya (1974b) proposed the use of the Gauss–Newton linearization technique (although he took a minimum distance approach to the problem). Hansen and Singleton (1982) apply the IV technique to non-linear rational expectations models.

endogenous variables. Suppose we have a $T \times K_1$ matrix of instruments, W_1. We use eq. (4.4) to apply the instruments to eq. (6.10) to find the new estimate of α:

$$\alpha^2 - \alpha^1 = (W_1'g_1')^{-1}W_1'f_1(y, Z, \alpha^1). \tag{6.11}$$

We now expand around α^2 and the iteration continues.[60] An alternative, but equivalent, approach is to use the matrix of instruments W_1 to solve $W_1'f_1(y, Z, \alpha^1) = 0$. The Gauss–Newton method applied to this equation again leads to eq. (6.11). At convergence, we apply eqs. (4.6) and (4.7) to derive the asymptotic distribution of the non-linear IV estimator $\hat{\alpha}_{\text{IV}}$:[61]

$$\sqrt{T}(\hat{\alpha}_{\text{IV}} - \alpha) \overset{A}{\sim} N\left(0, \operatorname{plim} \hat{\sigma}_{11}\left[\frac{1}{T}(W_1'g_1(\hat{\alpha}_{\text{IV}}))^{-1}(W_1'W_1)(g_1'(\hat{\alpha}_{\text{IV}})W_1)^{-1}\right]\right). \tag{6.12}$$

Where do we find the instruments? In the linear case they arose from the reduced form. The crucial difference between the non-linear and linear simultaneous equation specification is the absence of a reduced-form specification which allows the additive separation of jointly endogenous variables into a function of the predetermined variables and stochastic disturbances, $Y = Z\Pi + V$ in the linear case. In the non-linear case the reduced-form specification has the form $y_t = h(z_t, \alpha, u_t)$, where y_t, z_t, and u_t are all $M \times 1$ vectors for some function h. Analogously to the linear specification we might set $u_t = 0$ and use $\hat{y}_t = h(z_t, \hat{\alpha}, 0)$ for some $\hat{\alpha}$ and perhaps even iterate on the instruments. But the function h does not usually exist in convenient closed form so the alternative [Kelejian (1971)] suggestion is to form W_1 from linear and non-linear combinations of the predetermined variables which serve as instrumental variables. No best choice of instruments exists; but as in the linear case we want to come as close to $Eg_1(\alpha)$ as possible while still retaining the instrument property of W_1. Given the wide choice of instrumental variables, what about the identification problem since in the linear case we stressed the equivalence of identifiability with the existence of sufficient instruments? Speaking somewhat loosely, the identification problem no longer exists in its previous form. The reduced form is no longer linear so that non-linear functions of predetermined variables can be used to form instruments,

[60] It is important to do true instrumental variable estimation here not "repeated least squares". True IV estimation leads to consistent estimates while the analogue of 2SLS is often inconsistent. Inconsistency occurs when a non-linear function of a predicted endogenous variable is used in the second stage of the estimation procedure. At MIT this latter estimator is called the "forbidden regression". However, the IV procedure of eq. (6.11) is guaranteed to be consistent so long as the orthogonality conditions are satisfied.

[61] We have no guarantee that this procedure converges. Nonlinear optimization routines exist which do guarantee convergence given regularity conditions; see Quandt (Chapter 12 in this Handbook) or Berndt, Hall, Hall and Hausman (1974).

W_1. Therefore, a wide variety of non-linear functions might be used. Analogously, the jointly endogenous variables y are no longer normally distributed even if u is so we are no longer restricted to linear combinations of z in forming W_1. We thus usually have more instrumental variables than are needed.[62]

Let us therefore consider formation of W_1 to allow for the presence of greater than K_1 instrumental variables. The situation is analogous to the overidentified linear case where the predetermined variables Z provide greater than K_1 instrumental variables. There we formed $W_1 = Z\hat{A}_1 = Z(Z'Z)^{-1}Z'X_1$ which also had a minimum distance interpretation. Here we use linear and non-linear combinations of Z to form a $T\times\tilde{k}_1$ matrix of instrumental variables $\tilde{W}_1$. We now use the minimum distance formulation for a $T\times\tilde{k}_1$ matrix of instrumental variables $\tilde{W}_1$, $\operatorname{rank}(\tilde{W}_1)\geqslant k_1$, to find the k_1 linear combinations of $\tilde{W}_1$ most highly correlated with g_1 by minimization of[63]

$$\min_{\alpha} J = \tfrac{1}{2} f_1'(y,z,\alpha)\tilde{W}_1\left(\tilde{W}_1'\tilde{W}_1\right)^{-1}\tilde{W}_1' f_1(y,z,\alpha). \tag{6.13}$$

The solution of eq. (6.13), called the non-linear 2SLS (NL2SLS) estimator by Amemiya (1974), has the asymptotic distribution

$$\sqrt{T}\left(\hat{\alpha}_{\text{NL2SLS}}-\alpha\right) \overset{A}{\sim} N\left(0, \operatorname{plim}\sigma_{11}\left[\frac{1}{T}g_1'(\alpha)\tilde{W}_1\left(\tilde{W}_1'\tilde{W}_1\right)^{-1}\tilde{W}_1'g_1(\alpha)\right]^{-1}\right), \tag{6.14}$$

which follows from an application of a Taylor expansion in α of the first-order conditions of eq. (6.13):

$$f_1'(\alpha)\tilde{W}_1\left(\tilde{W}_1'\tilde{W}_1\right)^{-1}\tilde{W}_1'g_1(\alpha) = 0 \tag{6.15}$$

Note also that the asymptotic covariance matrix in eq. (6.14) coincides with that in eq. (6.12) when $W_1 = \tilde{W}_1(\tilde{W}_1'\tilde{W}_1)^{-1}\tilde{W}_1'g_1(\hat{\alpha}_{\text{IV}})$. Thus, the "minimum distance" NL2SLS estimator is an IV estimator on the linearized equation (6.10). The particular choice of instrumental W_1 corresponds to the optimal choice of instruments in the linear case. Thus, the asymptotic distribution follows directly from the IV theory applied to the linearized equation. Inclusion of additional instruments in $\tilde{W}_1$ will decrease the asymptotic covariance of $\hat{\alpha}_{\text{NL2SLS}}$ but the finite sample properties are unknown.

We now turn to estimation of the complete system of eq. (6.1) written in stacked form as $f(y,Z,\alpha)=u$, where u has mean zero and covariance matrix

[62] In certain special cases we may still be close to the linear situation. An example is given by Fisher (1966, ch. 5).

[63] This function follows directly from the proof of the optimal IV estimator in the linear case.

$\Sigma\otimes I_T$. The systems IV estimator matrix in the linear case was based on W_i each of which was a linear function of the predetermined variables, $W_i=[Z\Pi_i \quad Z_i]$. Here we might well want to allow for different combinations of instrumental variables in each equation. Therefore, we consider the NL3SLS estimator:

$$\min_{\alpha} J=\tfrac{1}{2}f'(y,z,\alpha)(S^{-1}\otimes I)H\left(H'(S^{-1}\otimes I)H\right)^{-1}H'(S^{-1}\otimes I)f(y,z,\alpha), \tag{6.16}$$

where $H=\operatorname{diag}(H_1,\dots,H_M)$, $H_i=\tilde{W}_i$ from above, and S is a consistent estimator of Σ.[64] The asymptotic distribution of the estimator is derived via a first-order Taylor expansion of the first-order conditions to find

$$\sqrt{T}\,(\hat{\alpha}_{\text{NL3SLS}}-\alpha)\overset{A}{\sim}N\left(0,\operatorname{plim}\left[\frac{1}{T}g'(\alpha)(S^{-1}\otimes I)H\left(H'(S^{-1}\otimes I)H\right)^{-1}\right.\right.$$
$$\left.\left.\times H'(S^{-1}\otimes I)g(\alpha)\right]^{-1}\right). \tag{6.17}$$

In the case where we restrict the number of instrumental variables in H_i to be k_i [as we did in eq. (6.11)] we can simplify the covariance matrix in eq. (6.17). But again it seems best to include a large number of instrumental variables to increase the asymptotic efficiency of our estimator.[65]

Maximum likelihood estimation

As in the linear case, the likelihood function follows from Assumption 4.3 that the structural disturbances follow a non-singular normal distribution. We rewrite the specification of eq. (6.1) in non-stacked form

$$F(y,Z,\alpha)=U, \tag{6.18}$$

where U is a $T\times M$ matrix with distribution $U\sim N(0,\Sigma\otimes I_T)$. The log-likelihood

[64] This first estimator was first defined by Berndt et al. (1974). An alternative definition of NL3SLS is given by Jorgenson and Laffont (1974) who take the case $H_i=H_j$ for all i and j. Their criterion function is then

$$G=\tfrac{1}{2}f'\left(S^{-1}\otimes H_i(H_i'H_i)^{-1}H_i'\right)\quad\text{since } H=I\otimes H_i,$$

which is analogous to the linear 3SLS estimator.

[65] Amemiya (1977) explores the best chcice of instruments for eq. (6.16). However, for practical usage we still need more guidance since he concentrates on forming instruments from the expectation of g which would necessitate involved computations given the non-linear structure of the problem and the absence of an additive reduced-form specification. Bowden and Turkington (1981) do a study for a specific example. Their results emphasize the importance of having the instrument close to the expectation of g.

function then is

$$L(\alpha,\Sigma)=C+\frac{T}{2}\log\det(\Sigma)^{-1}+\sum_{t=1}^{T}\log|J_t|-\frac{T}{2}\operatorname{tr}\left[\frac{1}{T}\Sigma^{-1}F'F\right], \tag{6.19}$$

where the Jacobian $J_t=\det(\partial U_t/\partial y_t)$. The important difference between the log-likelihood of eq. (6.19) and the log-likelihood function for the linear specification of eq. (4.24) is the non-constancy of the Jacobian term. The first-order conditions for a maximum are

$$\text{(i)}\quad \frac{\partial L}{\partial \alpha}:\quad \left[\sum_{t=1}^{T}\frac{\partial g_t}{\partial U_t}-G'F\Sigma^{-1}\right]^u=0,$$

$$\text{(ii)}\quad \frac{\partial L}{\partial \Sigma^{-1}}:\quad [T\Sigma-F'F]^u=0, \tag{6.20}$$

where as before g is the vector of first derivatives and the $M\times T$ matrix G corresponds also the derivatives of F with respect to α. To put ML into IV form, we consider eq. (6.1) in structural form, $f(y,z,\alpha)=U$ and define the instruments $\overline{W}'=\overline{G}'(\hat{\Sigma}\otimes I_T)^{-1}$ with $\hat{\Sigma}=T^{-1}\hat{F}'\hat{F}, \overline{G}=\operatorname{diag}(\overline{G}_1,\dots,\overline{G}_M)$, for

$$G_i'=-\frac{1}{T}\sum_t\frac{\partial g_{it}}{\partial U_t'}F'+G_i'$$

which arises from eq. (6.20, i) by using the same substitution used in the linear case [eq. (4.26)] to get:

$$\left[\left(T^{-1}\sum_t\frac{\partial g_t}{\partial U_t'}F'F-GF\right)\Sigma^{-1}\right]^u=0. \tag{6.21}$$

Then all derivatives and $\hat{\Sigma}$ are evaluated at α^1 to derive the FIML IV iteration corresponding to eq. (6.11):[66]

$$\alpha^2-\alpha^1=(\overline{W}'G)^{-1}\overline{W}'f. \tag{6.22}$$

The first term in the instruments corrects for the endogeneity present in the G_i. In the linear case the first term of the instruments took the form $(B')^{-1}U'U=(Y-Z\Pi)'U=(Y-\hat{Y})'U$, where the last term stands for the reduced form prediction $\hat{Y}'=-(\hat{B}^{-1})'\hat{\Gamma}'Z'$ which forms the basis for construction of optimal instruments. In the non-linear case the lack of an additively separable reduced form precludes formation of optimal instruments so that the NL3SLS estimator to eq. (6.16) has a larger asymptotic covariance matrix than does FIML.

[66]As usual, this form of the FIML first-order conditions does not provide the best numerical method to employ. For alternative methods, see Quandt (Chapter 12 in this Handbook) and Belsley (1980). Hatanaka (1978) presents an interesting two-step estimator for the special case of a specification which is linear in the parameters but non-linear in the variables.

Lastly, we consider consistency issues. For consistency the large sample expectation of the first-order condition must equal zero so that we require

$$\lim_{T\to\infty}\frac{1}{T}\sum_t E\frac{\partial g_t}{\partial U_t}=\lim_{T\to\infty}\frac{1}{T}EG'F\Sigma^{-1}. \tag{6.23}$$

Amemiya (1977) demonstrates that this condition is satisfied for U distributed as multivariate normal. However, it is not satisfied in the general case of $EU=0$, $V(U)=\Sigma\otimes I_T$. Thus, FIML in the non-linear case may not be consistent if U is not normal. In the linear case, since the IV form FIML did not depend on normality, FIML was consistent even without normality. But in the non-linear case, the instruments, $\overline{G}_i$, contain the non-constant term, $\partial g_{it}/\partial u_t$, which arises from the Jacobian and maintains the distributional assumption in the first-order conditions. As in the previous paragraph, because of the additive separability of the reduced form we have $E\hat{Y}=Z\hat{\Pi}+\hat{V}=Y$ for the linear case when we form instruments regardless of the distribution of U and V. The reduced form in the non-linear case $y=h(z,\alpha,u)$ does not have the separable property. Phillips (1981b) has recently shown that consistency does not of necessity require normality.[67] But for the general situation it appears that FIML is likely to be inconsistent without normality. That is why the IV approach to estimation of simultaneous equations seems more useful than the approximate ML approach of Hendry (1976) since it need not be based on the assumption of normality.

Since the preceding IV estimators, NL2SLS and NL3SLS, do not require normality, a specification test seems in order. For instance, comparison of $\hat{\alpha}_{\text{NL3SLS}}$ with $\hat{\alpha}_{\text{FIML}}$ using the result of Hausman (1978) permits a test of normality since both estimators are consistent under the null hypothesis of normality, while $\hat{\alpha}_{\text{NL3SLS}}$ is asymptotically inefficient. Under the alternative hypothesis, $\hat{\alpha}_{\text{NL3SLS}}$ remains consistent while $\hat{\alpha}_{\text{FIML}}$ usually loses consistency. Fair and Parke (1980) employed such a test but lack of sufficient degrees of freedom seemed to cause problems in the asymptotic approximations employed. Note that the specification tests discussed in Section 5 form the basis for testing the structural specification due to the absence of a convenient reduced-form specification. Further development of tests for the misspecification of non-linear simultaneous equation models clearly seems an important next step in their use.[68]

[67]A class of distributions does exist for each non-linear model under which FIML will be consistent. But the generality of the class depends on the particular model specification. Phillips also points out that Amemiya's regularity conditions are not sufficient for consistency. Strengthened assumptions can be made, however.

[68]Tests of overidentification are possible for non-linear systems. The non-linear analogues of eqs. (5.4a) and $\overset{=}{W}_p$ of footnote 48 can be used. However, these tests seem less interesting than in the linear case due to the absence of a benchmark unrestricted reduced form to base the test on.

References

Adcock, R. (1878) "A Problem in Least Squares", *The Analyst*, 5, 53–54.

Akahira T. and K. Takeuchi (1981) *Asymptotic Efficiency of Statistical Estimators.* New York: Springer-Verlag.

Amemiya, T. (1974a) "The Nonlinear Two-Stage Least Squares Estimator", *Journal of Econometrics*, 2, 105–110.

Amemiya, T. (1978) "The Maximum Likelihood and the NL3SLS Estimator in the General Nonlinear Simultaneous Equation Model", *Econometrica*, 45, 955–958.

Anderson, T. W. and H. Rubin (1949) "Estimation of the parameters of a Single Equation in a Complete System of Stochastic Equations", *Annals of Mathematical Statistics*, 20, no. 1, 46–63.

Anderson, T. W., N. Kunitomo and T. Sawa (1982) "Evaluation of the Distribution Function of the LIML Estimation", *Econometrica*, 50, 1009–1027.

Basmann, R. L. (1957) "A Generalized Classical Method of Linear Estimation of Coefficients in a Structural Equation", *Econometrica*, 25, 77–83.

Basmann, R. L. (1961) "A Note on the Exact Finite Sample Frequency Functions of a Generalized Classical Linear Estimator in Two Leading Overidentified Cases", *Journal of the American Statistical Association*, 56, 619–636.

Basmann, R. L. (1974) "Exact Finite Sample Distributions and Test Statistics" in M. D. Intriligator and D. A. Kendrick (eds.), *Frontiers of Quantitative Economics*, vol. II. Amsterdam: North-Holland Publishing Co.

Belsley, D. (1980) "On the Efficient Computation of the Non-linear FIML Estimator", *Journal of Econometrics*, 14, 203–225.

Berkson, J. A. (1980) "Minimum Chi-Square, not Maximum Likelihood!", *Annals of Statistics*, 8, 457–487.

Berndt, E., B. Hall, R. Hall and J. Hausman (1974) "Estimation and Inference in Nonlinear Structural Models", *Annals of Economic and Social Measurement*, 3, 653–666.

Bowden, R. and D. Turkingtron (1981) "A Comparative Study of Instrumental Variable Estimators for Nonlinear Simultaneous Models", *Journal of American Statistical Society*, 76, 985–996.

Brown, T. M. (1960) "Simultaneous Least Squares: A Distribution Free Method of Equation System Structure Estimation", *International Economic Review*, 1, 173–191.

Brundy, J. M. and D. W. Jorgenson (1971) "Efficient Estimation of Simultaneous Equations by Instrumental Variables", *Review of Economics and Statistics*, 53, 207–224.

Byron, R. P. (1974) "Testing Structural Specification Using the Unrestricted Reduced Form", *Econometrica*, 42, 869–884.

Court, R. H. (1973) "Efficient Estimation of the Reduced Form from Econometric Models", *Review of Economic Studies*, 40, 411–418.

Dhrymes, P. (1971) "A Simplified Estimator for Large-Scale Econometric Models", *The Australian Journal of Statistics*, 13, 168–175.

Dhrymes, P. J. (1973a) "Small Sample and Asymptotic Relations Between Maximum Likelihood and Three-Stage Least Squares", *Econometrica*, 41, 357–364.

Dhrymes, P. (1973b) "Restricted and Unrestricted Reduced Forms", *Econometrica*, 41, 119–134.

Dhrymes, P. and J. Taylor (1976) "On an Efficient Two-Step Estimate for Dynamic Simultaneous Equation Systems with Autoregressive Errors", *International Economic Review*, 17, 362–376.

Drèze, Jacques H. (1976) "Bayesian Limited Information Analysis of the Simultaneous Equations Model", *Econometrica*, 44, 1045–1075.

Efron, B. (1975) "Defining the Curvature of a Statistical Problem", *Arrivals of Statistics*, 3, 1189–1242.

Efron, B. and D. Hinkley (1978) "Assessing the Accuracy of the Maximum Likelihood Estimator", *Biometrika*, 65, 457–487.

Engle, R., D. Hendry and J. Richard (1981) "Exogeneity", unpublished, mimeo.

Fair, R. C. (1970) "The Estimation of Simultaneous Equation Models with Lagged Endogenous Variables and First Order Serially Correlated Errors", *Econometrica*, 38, 507–516.

Fair, R. and W. Parke (1980) "Full Information Estimates of a Nonlinear Macroeconometric Model", *Journal of Econometrics*, 269–291.

Fisher, F. M. (1961) "On the Cost of Approximate Specification in Simultaneous Equation Estimation", *Econometrica*, 29, 349–381.

Fisher, F. M. (1966) *The Identification Problem in Econometrics.* New York: McGraw-Hill.
Fisher, R. A. (1935) "The Logic of Inductive Inference", *Journal of the Royal Statistical Society*, 98, 39–82.
Fisher, R. A. (1956) *Statistical Methods and Scientific Inference.* Edinburgh: Oliver and Boyd.
Fuller, W. (1976) *Introduction to Statistical Time Series*, New York: John Wiley & Sons.
Fuller, W. (1977) "Some Properties of a Modification of the Limited Information Estimator", *Econometrica*, 45, 939–954.
Gallant, A. Ronald (1977) "Three-Stage Least Squares Estimation for a System of Simultaneous, Nonlinear, Implicit Equations", *Journal of Econometrics*, 5, 71–88.
Gantmacher, F. R. (1959) *The Theory of Matrices.* New York: Chelsea Publishing Company.
Geary, R. C. (1949) "Determination of Linear Relations between Systematic Parts of Variables with Errors of Observation, the Variances of which are Unknown", *Econometrica*, 17, 30–59.
Geraci, V. (1976) "Identification of Simultaneous Equation Models with Measurement Error", *Journal of Econometrics*, 4, 263–283.
Geraci, V. (1977) "Estimation of Simultaneous Equation Models with Measurement Error", *Econometrica*, 45, 1243–1256.
Geweke, J. (1978) "Testing the Exogeneity Specification in the Complete Dynamic Simultaneous Equation Model", *Journal of Econometrics*, 7, 163–186.
Goldberger, A. S. (1972) "Structural Equation Models in the Social Sciences", *Econometrica*, 40, 979–1002.
Goldberger, A., A. L. Nagar and H. S. Odeh (1961) "The Covariance Matrices of Reduced Form Coefficients and of Forecasts in a Structural Model", *Econometrica*, 29, 556–573.
Goldfeld, S. M. and R. E. Quandt (1968) "Nonlinear Simultaneous Equations: Estimation and Prediction", *International Economic Review*, 9, 113–136.
Granger, C. W. J. (1969) "Investigating Causal Relations by Econometric Models and Cross-Spectral Methods", *Econometrica*, 37, 424–438.
Haavelmo, T. (1943) "The Statistical Implications of a System of Simultaneous Equations", *Econometrica*, 11, 1–12.
Haavelmo, T. (1944) "The Probability Approach in Econometrics", *Econometrica*, 12, Suppl., 1–118.
Hansen, L. P. and T. J. Sargent (1980) "Formulating and Estimating Dynamic Linear Rational Expectation Model", *Journal of Economic Dynamic and Control*, 2, 7–46.
Hansen, L. P. and K. Singleton (1982) "Generalized Instrumental Variables Estimation of Nonlinear Rational Expectations Models", *Econometrica*, 50, 1269–1286.
Hatanaka, M. (1976) "Several Efficient Two-Step Estimators for the Dynamic Simultaneous Equations Model with Autoregressive Disturbances", *Journal of Econometrics*, 4, 189–204.
Hatanaka, M. (1978) "On the Efficient Estimation Methods for the Macro-Economic Models Nonlinear in Parameters", *Journal of Econometrics*, 18, 323–356.
Hausman, J. A. (1974) "Full Information Instrumental Variable Estimation of Simultaneous Equation Systems", *Annals of Economic and Social Measurement*, 3, 641–652.
Hausman, J. A. (1975) "An Instrumental Variable Approach to Full Information Estimators for Linear and Certain Non-Linear Econometric Models", *Econometrica*, 43, 727–738.
Hausman, J. A. (1977) "Errors in Variables in Simultaneous Equation Models", *Journal of Econometrics*, 5, 389–401.
Hausman, J. A. (1978) "Specification Tests in Econometrics", *Econometrica*, 46, 1251–1272.
Hausman, J. A. and W. B. Taylor (1980a) "Identification in Simultaneous Equation Systems with Covariance Restrictions", MIT, mimeo.
Hausman, J. A. and W. B. Taylor (1980b) "The Relationship Between Specification Tests and Classical Tests", MIT, mimeo.
Hausman, J. A. and W. B. Taylor (1981) "Efficient Estimation with Covariance Restrictions in Simultaneous Equations Models", mimeo.
Hendry, D. F. (1976) "The Structure of Simultaneous Equations Estimators", *Journal of Econometrics*, 4, 51–88.
Hinkley, D. V. (1980) "Fisher's Development of Conditional Inference, in: D. V. Hinkley (ed.), *Papers on the Work of R. A. Fisher*, New York.
Hood, W. C. and T. C. Koopmans (eds.) (1953) *Studies in Econometric Method*, Cowles Commission Monograph 14. New York: John Wiley & Sons.
Hsiao, C. (1976) "Identification and Estimation of Simultaneous Equation Models with Measurement Error", *International Economic Review*, 17, 319–339.

Jorgenson, D. W. and J. Laffont (1974) "Efficient Estimation of Nonlinear Simultaneous Equations with Additive Disturbances", *Annals of Economic and Social Measurement*, 3, 615–640.

Kadane, J. B. (1971) "Comparison of K-class Estimators when the Disturbances are Small", *Econometrica*, 39, 727–737.

Kelejian, H. H. (1971) "2SLS and Econometric Systems Linear in Parameters but Nonlinear in the Endogenous Variables", *Journal of the American Statistical Association*, 66, 373–378.

Kendall, M. and A. Stuart (1967) *The Advanced Theory of Statistics*, vol. II. London: Griffen.

Khazzom, J. Daniel (1976) "An Indirect Least Squares Estimator for Overidentified Equations", *Econometrica*, 44, 741–750.

Koopmans, T. C. (1949) "Identification Problems in Economic Model Construction", *Econometrica*, 17, 125–144.

Koopmans, T. C. (ed.) (1950) *Statistical Inference in Dynamic Economic Models*, Cowles Commission Monograph 10. New York: John Wiley & Sons.

Koopmans, T. C. and W. C. Hood (1953) "The Estimation of Simultaneous Linear Economic Relationships", in: W. C. Hood and T. C. Koopmans (eds.), *Studies in Econometric Method*, Cowles Commission Monograph 14. New York: John Wiley & Sons.

Koopmans, T. C. and O. Reiersol (1950) "The Identification of Structural Characteristics", *The Annals of Mathematical Statistics*.

Koopmans, T. C., H. Rubin and R. B. Leipnik (1950) "Measuring the Equation Systems of Dynamic Economics", in: Koopmans, T. C. (ed.), *Statistical Inference in Dynamic Economic Models*, Cowles Commission Monograph 10. New York: John Wiley & Sons.

Lahiri, Kajal and Peter Schmidt (1978) "On the Estimation of Triangular Structural Systems", *Econometrica*, 46, 1217–1222.

Leamer, E. (1981) "Is It a Demand Curve, or Is It a Supply Curve?: Partial Identification Through Inequality Constraints", *Review of Economics and Statistics*, 63, 319–327.

Liu, T. C. (1960) "Underidentification, Structural Estimation, and Forecasting", *Econometrica*, 28, 855–871.

Lyttkens, E. (1970) "Symmetric and Asymmetric Estimation Methods", in: E. Mosback and H. Wold (eds.), *Interdependent Systems*. Amsterdam: North-Holland Publishing Co.

Maasoumi, E. (1978) "A Modified Stein-like Estimator for the Reduced Form Coefficients of Simultaneous Equations", *Econometrica*, 46, 695–704.

Malinvaud, E. (1970) *Statistical Methods of Econometrics* (2nd edn.). Amsterdam: North-Holland Publishing Co.

Mann, H. B. and A. Wald (1943) "On the Statistical Treatment of Linear Stochastic Difference Equations", *Econometrica*, 11, 173–220.

Nagar, A. L. (1959) "The Bias and Moment Matrix of the General k-class Estimators of the Parameters in Simultaneous Equations", *Econometrica*, 27, 575–595.

Narayanan, R. (1969) "Computation of Zellner–Theil's Three Stage Least Squares Estimates", *Econometrica*, 37, 298–306.

Pagan, A. (1979) "Some Consequences of Viewing LIML as an Iterated Aitken Estimator", *Economics Letters*, 3, 369–372.

Pfanzagl, J. (1975) "On Asymptotically Complete Classes", in: *Proceedings Summer Institute of Statistical Inference for Stochastic Processes*, vol. 2, pp. 1–43.

Pfanzagl, J. and W. Wefelmeyer (1978) "A Third Order Optimum Property of Maximum Likelihood Estimators", *Journal of Multivariate Analysis*, 8, 1–29.

Phillips, P. C. B. (1980) "The Exact Distribution of Instrumental Variable Estimators in an Equation Containing $n+1$ Endogenous Variables", *Econometrica*, 48, 861–878.

Phillips, P. C. B. (1981a) "A New Approach to Small Sample Theory", Yale, mimeo.

Phillips, P. C. B. (1981b) "On the Consistency of Nonlinear FIML", mimeo.

Rao, C. R. (1961) "Asymptotic Efficiency and Limiting Information", in: *Proceedings of the Fourth Berkeley Symposium*, University of California Press, pp. 531–545.

Rao, C. R. (1962) "Efficient Estimates and Optimum Inference Procedures in Large Samples", *Journal of the Royal Statistical Society*, Ser. B, 24, 46–72.

Rao, C. R. (1973) *Statistical Inference and Its Applications*. New York: John Wiley & Sons.

Reiersol, O. (1945) "Confluence Analyses by Means of Instrumental Sets of Variables", *Arkiv for Mathematik, Astronomi och Fysik*, 32, 1–119.

Reiersol, O. (1950) "Identifiability of a Linear Relation Between Variables Which are Subject to Error", *Econometrica*, 18, 375–389.

Rothenberg, T. J. (1971) "Identification in Parametric Models", *Econometrica*, 39, 577–595.
Rothenberg, T. J. (1973) *Efficient Estimation with a priori Information*, Cowles Foundation Monograph 23. New Haven: Yale University Press.
Rothenberg, T. J. and C. T. Leenders (1964) "Efficient Estimation of Simultaneous Equations Systems", *Econometrica*, 32, 57–76.
Sargan, J. D. (1958) "On the Estimation of Economic Relationships by Means of Instrumental Variables", *Econometrica*, 26, 393–415.
Sargan, J. D. (1959) "The Estimation of Relationships with Autocorrelated Residuals by the Use of Instrumental Variables", *Journal of the Royal Statistical Society*, B21, 91–105.
Sargan, J. D. (1961) "The Maximum Likelihood Estimation of Economic Relationships with Autocorrelated Residuals", *Econometrica*, 29, 414–426.
Sargan, J. D. (1964) "Three-Stage Least-Squares and Full Maximum Likelihood Estimates", *Econometrica*, 32, 77–81.
Sargan, J. D. (1978) "On the Existence of the Moments of the 3SLS Estimator", *Econometrica*, 46, 1329–1350.
Savin, N. E. (1973) "Systems *k*-class Estimators", *Econometrica*, 41, 1125–1136.
Sawa, T. (1973) "Almost Unbiased Estimators in Simultaneous Equations Systems", *International Economic Review*, 14, 97–106.
Scheffe, H. (1959) *Analysis of Variance*. New York: John Wiley & Sons.
Seber, G. A. F. (1966) *The Linear Hypothesis*. London.
Silvey, D. S. (1970) *Statistical Inference*. New York: John Wiley & Sons.
Sims, C. A. (1972) "Money, Income and Causality", *American Economic Review*, 62, 540–532.
Sims, C. A. (1977) "Exogeneity and Causal Ordering in Macroeconomic Models", in: C. A. Sims (ed.), *New Methods in Business Cycle Research*. Minneapolis.
Sims, C. A. (1980) "Macroeconomics and Reality", *Econometrica*, 48, 1–48.
Srivastava, V. K. (1971) "Three-Stage Least-Squares and Generalized Double K-class Estimators: A Mathematical Relationship", *International Economic Review*, 12, 312–316.
Swamy, P. (1980) "A Comparison of Estimators for Undersized Samples", *Journal of Econometrics*, 14, 161–181.
Theil, H. (1958) *Economic Forecasts and Policy*. Amsterdam: North-Holland Publishing Co.
Theil, H. (1961) *Economic Forecasts and Policy* (2nd edn.). Amsterdam: North-Holland Publishing Co.
Theil, H. (1971) *Principles of Econometrics*. New York: John Wiley & Sons.
Wallis, K. F. (1977) "Multiple Time Series Analysis and the Final Form of Econometric Models", *Econometrica*, 45, 1481–1497.
Wallis, K. F. (1978) "Econometric Implications of the Rational Expectations Hypothesis", *Econometrica*, 48, 49–73.
Wallis, K. F. (1980) "Econometric Implications of the Rational Expectations Hypothesis", *Econometrica*, 48, 49–74.
Wegge, L. (1965) "Identifiability Criteria for a System of Equations as a Whole", *The Australian Journal of Statistics*.
Wegge, L. (1978) "Constrained Indirect Least Squares Estimators", *Econometrica*, 46, 435–450.
Wold, H. (ed.) (1964) *Econometric Model Building: Essays on the Causal Chain Approach*. Amsterdam: North-Holland Publishing Co.
Working, E. (1927) "What do 'Statistical Demand Curves' Show", *Quarterly Journal of Economics*.
Wu, De-Min (1973) "Alternative Tests of Independence Between Stochastic Regressors and Disturbances", *Econometrica*, 41, 733–750.
Zellner, A. (1962) "An Efficient Method of Estimating Seemingly Unrelated Regressions and Tests for Aggregation Bias", *Journal of the American Statistical Association*, 57, 348–368.
Zellner, A. (1978) "Estimation of Functions of Population Means and Regression Coefficients including Structural Coefficients: A Minimum Expected Loss (MELO) Approach", *Journal of Econometrics*, 8, 127–158.
Zellner, A. (1979a) "Causality and Econometrics", in: K. Brunner and A. H. Meltzer (eds.), *Three Aspects of Policy and Policymaking*. Amsterdam: North-Holland Publishing Co.
Zellner, A. and F. Palm (1975) "Time Series Analysis and Simultaneous Equation Econometric Models", *Journal of Econometrics*, 2, 17–54.
Zellner, A. and H. Theil (1962) "Three-Stage Least-Squares: Simultaneous Estimation of Simultaneous Equations", *Econometrica*, 30, 54–78.

Chapter 8

EXACT SMALL SAMPLE THEORY IN THE SIMULTANEOUS EQUATIONS MODEL

P. C. B. PHILLIPS*

Yale University

Contents

*The present chapter is an abridgement of a longer work that contains *inter alia* a fuller exposition and detailed proofs of results that are surveyed herein. Readers who may benefit from this greater degree of detail may wish to consult the longer work itself in Phillips (1982e).

My warmest thanks go to Deborah Blood, Jerry Hausmann, Esfandiar Maasoumi, and Peter Reiss for their comments on a preliminary draft, to Glena Ames and Lydia Zimmerman for skill and effort in preparing the typescript under a tight schedule, and to the National Science Foundation for research support under grant number SES 8007571.

Handbook of Econometrics, Volume I, Edited by Z. Griliches and M.D. Intriligator

Little experience is sufficient to show that the traditional machinery of statistical processes is wholly unsuited to the needs of practical research. Not only does it take a cannon to shoot a sparrow, but it misses the sparrow! The elaborate mechanism built on the theory of infinitely large samples is not accurate enough for simple laboratory data. Only by systematically tackling small sample problems on their merits does it seem possible to apply accurate tests to practical data. Such at least has been the aim of this book. [From the Preface to the First Edition of R. A. Fisher (1925).]

1. Introduction

Statistical procedures of estimation and inference are most frequently justified in econometric work on the basis of certain desirable asymptotic properties. One estimation procedure may, for example, be selected over another because it is known to provide consistent and asymptotically efficient parameter estimates under certain stochastic environments. Or, a statistical test may be preferred because it is known to be asymptotically most powerful for certain local alternative hypotheses.[1] Empirical investigators have, in particular, relied heavily on asymptotic theory to guide their choice of estimator, provide standard errors of their estimates and construct critical regions for their statistical tests. Such a heavy reliance on asymptotic theory can and does lead to serious problems of bias and low levels of inferential accuracy when sample sizes are small and asymptotic formulae poorly represent sampling behavior. This has been acknowledged in mathematical statistics since the seminal work of R. A. Fisher,[2] who recognized very early the limitations of asymptotic machinery, as the above quotation attests, and who provided the first systematic study of the exact small sample distributions of important and commonly used statistics.

The first step towards a small sample distribution theory in econometrics was taken during the 1960s with the derivation of exact density functions for the two stage least squares (2SLS) and ordinary least squares (OLS) estimators in simple simultaneous equations models (SEMs). Without doubt, the mainspring for this research was the pioneering work of Basmann (1961), Bergstrom (1962), and Kabe (1963, 1964). In turn, their work reflected earlier influential investigations in econometrics: by Haavelmo (1947) who constructed exact confidence regions for structural parameter estimates from corresponding results on OLS reduced form coefficient estimates; and by the Cowles Commission researchers, notably Anderson and Rubin (1949), who also constructed confidence regions for structural coefficients based on a small sample theory, and Hurwicz (1950) who effectively studied and illustrated the small sample bias of the OLS estimator in a first order autoregression.

[1]The nature of local alternative hypotheses is discussed in Chapter 13 of this Handbook by Engle.

[2]See, for example, Fisher (1921, 1922, 1924, 1928a, 1928b, 1935) and the treatment of exact sampling distributions by Cramér (1946).

The mission of these early researchers is not significantly different from our own today: ultimately to relieve the empirical worker from the reliance he has otherwise to place on asymptotic theory in estimation and inference. Ideally, we would like to know and be able to compute the exact sampling distributions relevant to our statistical procedures under a variety of stochastic environments. Such knowledge would enable us to make a better assessment of the relative merits of competing estimators and to appropriately correct (from their asymptotic values) the size or critical region of statistical tests. We would also be able to measure the effect on these sampling distributions of certain departures in the underlying stochastic environment from normally distributed errors. The early researchers clearly recognized these goals, although the specialized nature of their results created an impression[3] that there would be no substantial payoff to their research in terms of applied econometric practice. However, their findings have recently given way to general theories and a powerful technical machinery which will make it easier to transmit results and methods to the applied econometrician in the precise setting of the model and the data set with which he is working. Moreover, improvements in computing now make it feasible to incorporate into existing regression software subroutines which will provide the essential vehicle for this transmission. Two parallel current developments in the subject are an integral part of this process. The first of these is concerned with the derivation of direct approximations to the sampling distributions of interest in an applied study. These approximations can then be utilized in the decisions that have to be made by an investigator concerning, for instance, the choice of an estimator or the specification of a critical region in a statistical test. The second relevant development involves advancements in the mathematical task of extracting the form of exact sampling distributions in econometrics. In the context of simultaneous equations, the literature published during the 1960s and 1970s concentrated heavily on the sampling distributions of estimators and test statistics in single structural equations involving only two or at most three endogenous variables. Recent theoretical work has now extended this to the general single equation case.

The aim of the present chapter is to acquaint the reader with the main strands of thought in the literature leading up to these recent advancements. Our discussion will attempt to foster an awareness of the methods that have been used or that are currently being developed to solve problems in distribution theory, and we will consider their suitability and scope in transmitting results to empirical researchers. In the exposition we will endeavor to make the material accessible to readers with a working knowledge of econometrics at the level of the leading textbooks. A cursory look through the journal literature in this area may give the impression that the range of mathematical techniques employed is quite diverse, with the method and final form of the solution to one problem being very different from the next. This diversity is often more apparent than real and it is

[3]The discussions of the review article by Basmann (1974) in Intriligator and Kendrick (1974) illustrate this impression in a striking way. The achievements in the field are applauded, but the reader is left with a skeptical view of the usefulness of the results.

hoped that the approach we take to the subject in the present review will make the methods more coherent and the form of the solutions easier to relate.

Our review will not be fully comprehensive in coverage but will report the principal findings of the various research schools in the area. Additionally, our focus will be directed explicitly towards the SEM and we will emphasize exact distribution theory in this context. Corresponding results from asymptotic theory are surveyed in Chapter 7 of this Handbook by Hausman; and the refinements of asymptotic theory that are provided by Edgeworth expansions together with their application to the statistical analysis of second-order efficiency are reviewed in Chapter 15 of this Handbook by Rothenberg. In addition, and largely in parallel to the analytical research that we will review, are the experimental investigations involving Monte Carlo methods. These latter investigations have continued traditions established in the 1950s and 1960s with an attempt to improve certain features of the design and efficiency of the experiments, together with the means by which the results of the experiments are characterized. These methods are described in Chapter 16 of this Handbook by Hendry. An alternative approach to the utilization of soft quantitative information of the Monte Carlo variety is based on constructive functional approximants of the relevant sampling distributions themselves and will be discussed in Section 4 of this chapter.

The plan of the chapter is as follows. Section 2 provides a general framework for the distribution problem and details formulae that are frequently useful in the derivation of sampling distributions and moments. This section also provides a brief account of the genesis of the Edgeworth, Nagar, and saddlepoint approximations, all of which have recently attracted substantial attention in the literature. In addition, we discuss the Wishart distribution and some related issues which are central to modern multivariate analysis and on which much of the current development of exact small sample theory depends. Section 3 deals with the exact theory of single equation estimators, commencing with a general discussion of the standardizing transformations, which provide research economy in the derivation of exact distribution theory in this context and which simplify the presentation of final results without loss of generality. This section then provides an analysis of known distributional results for the most common estimators, starting with certain leading cases and working up to the most general cases for which results are available. We also cover what is presently known about the exact small sample behavior of structural variance estimators, test statistics, systems methods, reduced-form coefficient estimators, and estimation under misspecification. Section 4 outlines the essential features of a new approach to small sample theory that seems promising for future research. The concluding remarks are given in Section 5 and include some reflections on the limitations of traditional asymptotic methods in econometric modeling.

Finally, we should remark that our treatment of the material in this chapter is necessarily of a summary nature, as dictated by practical requirements of space. A more complete exposition of the research in this area and its attendant algebraic detail is given in Phillips (1982e). This longer work will be referenced for a fuller discussion of the material whenever it is appropriate in the present chapter.

2. Simple mechanics of distribution theory

2.1. Primitive exact relations and useful inversion formulae

To set up a general framework we assume a model which uniquely determines the joint probability distribution of a vector of n endogenous variables at each point in time ($t=1,\dots,T$), namely $\{y_1,\dots,y_T\}$, conditional on certain fixed exogenous variables $\{x_1,\dots,x_T\}$ and possibly on certain initial values $\{y_{-k},\dots,y_0\}$. This distribution can be completely represented by its distribution function (d.f.), $\mathrm{df}(y|x, y_{-1};\theta)$ or its probability density function (p.d.f.), $\mathrm{pdf}(y|x, y_-;\theta)$, both of which depend on an unknown vector of parameters θ and where we have set $y'=(y_1',\dots,y_T')$, $x'=(x_1',\dots,x_T')$, and $y_-'=(y_{-k}',\dots,y_0')$. In the models we will be discussing in this chapter the relevant distributions will not be conditional on initial values, and we will suppress the vector y_- in these representations. However, in other contexts, especially certain time-series models, it may become necessary to revert to the more general conditional representation. We will also frequently suppress the conditioning x and parameter θ in the representation $\mathrm{pdf}(y|x;\theta)$, when the meaning is clear from the context. Estimation of θ or a subvector of θ or the use of a test statistic based on an estimator of θ leads in all cases to a function of the available data. Therefore we write in general $\theta_T=\theta_T(y,x)$. This function will determine the numerical value of the estimate or test statistic.

The small sample distribution problem with which we are faced is to find the distribution of θ_T from our knowledge of the distribution of the endogenous variables and the form of the function which defines θ_T. We can write down directly a general expression for the distribution function of θ_T as

$$\begin{aligned} \mathrm{df}(r) &= \mathrm{P}(\theta_T \leqslant r) = \int_{y\in\Theta(r)} \mathrm{pdf}(y)\,dy, \\ \Theta(r) &= \{y: \theta_T(y,x)\leqslant r\}. \end{aligned} \tag{2.1}$$

This is an nT-dimensional integral over the domain of values $\Theta(r)$ for which $\theta_T\leqslant r$.

The distribution of θ_T is also uniquely determined by its characteristic function (c.f.), which we write as

$$\mathrm{cf}(s) = E(\mathrm{e}^{\mathrm{i}s\theta_T}) = \int \mathrm{e}^{\mathrm{i}s\theta_T(y,x)}\mathrm{pdf}(y)\,\mathrm{d}y, \tag{2.2}$$

where the integration is now over the entire y-space. By inversion, the p.d.f. of θ_T is given by

$$\mathrm{pdf}(r) = \frac{1}{2\pi}\int_{-\infty}^{\infty} \mathrm{e}^{-\mathrm{i}sr}\mathrm{cf}(s)\,\mathrm{d}s, \tag{2.3}$$

and this inversion formula is valid provided $\mathrm{cf}(s)$ is absolutely integrable in the Lebesgue sense [see, for example, Feller (1971, p. 509)]. The following two inversion formulae give the d.f. of θ_T directly from (2.2):

$$\mathrm{df}(r)-\mathrm{df}(0)=\frac{1}{2\pi}\int_{-\infty}^{\infty}\frac{1-\mathrm{e}^{-isr}}{is}\mathrm{cf}(s)\,\mathrm{d}s \tag{2.4}$$

and

$$\mathrm{df}(r)=\frac{1}{2}+\frac{1}{2\pi}\int_{0}^{\infty}\frac{\mathrm{e}^{isr}\mathrm{cf}(-s)-\mathrm{e}^{-isr}\mathrm{cf}(s)}{is}\mathrm{d}s. \tag{2.5}$$

The first of these formulae is valid whenever the integrand on the right-hand side of (2.4) is integrable [otherwise a symmetric limit is taken in defining the improper integral – see, for example, Cramér (1946, pp. 93–94)]. It is useful in computing first differences in $\mathrm{df}(r)$ or the proportion of the distribution that lies in an interval (a, b) because, by subtraction, we have

$$\mathrm{df}(b)-\mathrm{df}(a)=\frac{1}{2\pi}\int_{-\infty}^{\infty}\frac{\mathrm{e}^{-isa}-\mathrm{e}^{-isb}}{is}\mathrm{cf}(s)\,\mathrm{d}s. \tag{2.6}$$

The second formula (2.5) gives the d.f. directly and was established by Gil-Pelaez (1951).

When the above inversion formulae based on the characteristic function cannot be completed analytically, the integrals may be evaluated by numerical integration. For this purpose, the Gil-Pelaez formula (2.5) or variants thereof have most frequently been used. A general discussion of the problem, which provides bounds on the integration and truncation errors, is given by Davies (1973). Methods which are directly applicable in the case of ratios of quadratic forms are given by Imhof (1961) and Pan Jie Jian (1968). The methods provided in the latter two articles have often been used in econometric studies to compute exact probabilities in cases such as the serial correlation coefficient [see, for example, Phillips (1977a)] and the Durbin–Watson statistic [see Durbin and Watson (1971)].

2.2. *Approach via sample moments of the data*

Most econometric estimators and test statistics we work with are relatively simple functions of the sample moments of the data (y, x). Frequently, these functions are rational functions of the first and second sample moments of the data. More specifically, these moments are usually well-defined linear combinations and matrix quadratic forms in the observations of the endogenous variables and with the weights being determined by the exogenous series. Inspection of the relevant formulae makes this clear: for example, the usual two-step estimators in the linear model and the instrumental variable (IV) family in the SEM. In the case of limited information and full information maximum likelihood (LIML, FIML), these estimators are determined as implicit functions of the sample moments of the data through a system of implicit equations. In all of these cases, we can proceed to write $\theta_T = \theta_T(y, x)$ in the alternative form $\theta_T = \theta_T^*(m)$, where m is a vector of the relevant sample moments.

In many econometric problems we can write down directly the p.d.f. of the sample moments, i.e. pdf(m), using established results from multivariate distribution theory. This permits a convenient resolution of the distribution of θ_T. In particular, we achieve a useful reduction in the dimension of the integration involved in the primitive forms (2.1) and (2.2). Thus, the analytic integration required in the representation

$$\mathrm{pdf}(m) = \int_{a \in \mathcal{A}} \mathrm{pdf}(y) \left| \frac{\partial y}{\partial(m, a)} \right| \mathrm{d}a \tag{2.7}$$

has already been reduced. In (2.7) a is a vector of auxiliary variates defined over the space $\mathcal{A}$ and is such that the transformation $y \to (m, a)$ is $1:1$.

The next step in reducing the distribution to the density of θ_T is to select a suitable additional set of auxiliary variates b for which the transformation $m \to (\theta_T, b)$ is $1:1$. Upon changing variates, the density of θ_T is given by the integral

$$\mathrm{pdf}(r) = \int_{b \in \mathcal{B}} \mathrm{pdf}(m) \left| \frac{\partial m}{\partial(r, b)} \right| \mathrm{d}b, \tag{2.8}$$

where $\mathcal{B}$ is the space of definition of b. The simplicity of the representation (2.8) often belies the major analytic difficulties that are involved in the practical execution of this step.[4] These difficulties center on the selection of a suitable set of auxiliary variates b for which the integration in (2.8) can be performed analytically. In part, this process depends on the convenience of the space, $\mathcal{B}$, over which the variates b are to be integrated, and whether or not the final integral has a recognizable form in terms of presently known functions or infinite series.

All of the presently known exact small sample distributions of single equation estimators in the SEM can be obtained by following the above steps. When reduced, the final integral (2.8) is most frequently expressed in terms of infinite

[4]See, for example, Sargan (1976a, Appendix B) and Phillips (1980a). These issues will be taken up further in Section 3.5.

series involving some of the special functions of applied mathematics, which themselves admit series representations. These special functions are often referred to as higher transcendental functions. An excellent introduction to them is provided in the books by Whittaker and Watson (1927), Rainville (1963), and Lebedev (1972); and a comprehensive treatment is contained in the three volumes by Erdéyli (1953). At least in the simpler cases, these series representations can be used for numerical computations of the densities.

2.3. Asymptotic expansions and approximations

An alternative to searching for an exact mathematical solution to the problem of integration in (2.8) is to take the density pdf(m) of the sample moments as a starting point in the derivation of a suitable approximation to the distribution of θ_T. Two of the most popular methods in current use are the Edgeworth and saddlepoint approximations. For a full account of the genesis of these methods and the constructive algebra leading to their respective asymptotic expansions, the reader may refer to Phillips (1982e). For our present purpose, the following intuitive ideas may help to briefly explain the principles that underlie these methods.

Let us suppose, for the sake of convenience, that the vector of sample moments m is already appropriately centered about its mean value or limit in probability. Let us also assume that $\sqrt{T}m \xrightarrow{\mathfrak{D}} N(0, \mathfrak{V})$ as $T \to \infty$, where $\xrightarrow{\mathfrak{D}}$ denotes "tends in distribution". Then, if $\theta_T = f(m)$ is a continuously differentiable function to the second order, we can readily deduce from a Taylor series representation of $f(m)$ in a neighborhood of $m = 0$ that $\sqrt{T}\{f(m) - f(0)\} \xrightarrow{\mathfrak{D}} N(0, \mathfrak{W})$, where $\mathfrak{W} = (\partial f(0)/\partial m')\mathfrak{V}\partial f'(0)/\partial m$. In this example, the asymptotic behavior of the statistic $\sqrt{T}\{f(m) - f(0)\}$ is determined by that of the linear function $\sqrt{T}(\partial f(0)/\partial m')m$ of the basic sample moments. Of course, as $T \to \infty$, $m \to 0$ in probability, so that the behavior of $f(m)$ in the immediate locality of $m = 0$ becomes increasingly important in influencing the distribution of this statistic as T becomes large.

The simple idea that underlies the principle of the Edgeworth approximation is to bridge the gap between the small sample distribution (with T finite) and the asymptotic distribution by means of correction terms which capture higher order features of the behavior of $f(m)$ in the locality of $m = 0$. We thereby hope to improve the approximation to the sampling distribution of $f(m)$ that is provided by the crude asymptotic. Put another way, the statistic $\sqrt{T}\{f(m) - f(0)\}$ is approximated by a polynomial representation in m of higher order than the linear representation used in deducing the asymptotic result. In this sense, Edgeworth approximations provide refinements of the associated limit theorems which give us the asymptotic distributions of our commonly used statistics. The reader may usefully consult Cramér (1946, 1972), Wallace (1958), Bhattacharya and Rao

(1976), and the review by Phillips (1980b) for further discussion, references, and historical background.

The concept of using a polynomial approximation of θ_T in terms of the elements of m to produce an approximate distribution for θ_T can also be used to approximate the moments of θ_T, where these exist, or to produce pseudo-moments (of an approximating distribution) where they do not.[5] The idea underlies the work by Nagar (1959) in which such approximate moments and pseudo-moments were developed for k-class estimators in the SEM. In popular parlance these moment approximations are called Nagar approximations to the moments. The constructive process by which they are derived in the general case is given in Phillips (1982e).

An alternative approach to the development of asymptotic series approximations for probability densities is the saddlepoint (SP) method. This is a powerful technique for approximating integrals in asymptotic analysis and has long been used in applied mathematics. A highly readable account of the technique and a geometric interpretation of it are given in De Bruijn (1958). The method was first used systematically in mathematical statistics in two pathbreaking papers by Daniels (1954, 1956) and has recently been the subject of considerable renewed interest.[6]

The conventional approach to the SP method has its starting point in inversion formulae for the probability density like those discussed in Section 2.1. The inversion formula can commonly be rewritten as a complex integral and yields the p.d.f. of θ_T from knowledge of the Laplace transform (or moment-generating function). Cauchy's theorem in complex function theory [see, for example, Miller (1960)] tells us that we may well be able to deform the path of integration to a large extent without changing the value of the integral. The general idea behind the SP method is to employ an allowable deformation of the given contour, which is along the imaginary axis, in such a way that the major contribution to the value of the integral comes from the neighborhood of a point at which the contour actually crosses a saddlepoint of the modulus of the integrand (or at least its dominant factor). In crude terms, this is rather akin to a mountaineer attempting to cross a mountain range by means of a pass, in order to control the maximum

[5] This process involves a stochastic approximation to the statistic θ_T by means of polynomials in the elements of m which are grouped into terms of like powers of $T^{-1/2}$. The approximating statistic then yields the "moment" approximations for θ_T. Similar "moment" approximations are obtained by developing alternative stochastic approximations in terms of another parameter. Kadane (1971) derived such alternative approximations by using an expansion of θ_T (in the case of the k-class estimator) in terms of increasing powers of σ, where σ^2 is a scalar multiple of the covariance matrix of the errors in the model and the asymptotics apply as $\sigma \to 0$. Anderson (1977) has recently discussed the relationship between these alternative parameter sequences in the context of the SEM.

[6] See, for example, Phillips (1978), Holly and Phillips (1979), Daniels (1980), Durbin (1980a, 1980b), and Barndorff-Nielson and Cox (1979).

altitude he has to climb. This particular physical analogy is developed at some length by De Bruijn (1958).

A new and elegant approach to the extraction of SP approximations has recently been developed by Durbin (1980a). This method applies in cases where we wish to approximate the p.d.f. of a sufficient statistic and has the great advantage that we need only know the p.d.f. of the underlying data pdf$(y;\theta)$ and the limiting mean information matrix $\lim_{T\to\infty}E\{-T^{-1}\partial^2\ln[\text{pdf}(y;\theta)]/\partial\theta\partial\theta'\}$ in order to construct the approximation. This is, in any event, the information we need to extract the maximum likelihood estimator of θ and write down its asymptotic covariance matrix. Durbin's approach is based on two simple but compelling steps. The first is the fundamental factorization relation for sufficient statistics, which yields a powerful representation of the required p.d.f. for a parametric family of densities. The second utilizes the Edgeworth expansion of the required p.d.f. but at a parametric value (of θ) for which this expansion has its best asymptotic accuracy. This parametric recentering of the Edgeworth expansion increases the rate of convergence in the asymptotic series and thereby can be expected to provide greater accuracy at least for large enough T. Algebraic details, further discussion and examples of the method are given in Phillips (1982e).

2.4. *The Wishart distribution and related issues*

If $X=[x_1,\dots,x_T]$ is an $n\times T$ matrix variate (i.e. matrix of random variates) whose columns are independent $N(0,\Omega)$ then the $n\times n$ symmetric matrix $A=XX'=\sum_{t=1}^{T}x_t x_t'$ has a Wishart distribution with p.d.f. given by

$$\text{pdf}(A)=\frac{1}{(2)^{nT/2}\Gamma_n\left(\frac{T}{2}\right)(\det\Omega)^{T/2}}\text{etr}\left(-\tfrac{1}{2}\Omega^{-1}A\right)(\det A)^{(T-n-1)/2}. \tag{2.9}$$

Since A is symmetric $n\times n$, this density has $N=\frac{1}{2}n(n+1)$ independent arguments and is supported on the subset (a natural cone) of N dimensional Euclidean space for which A is positive definite (which we write as $A>0$). It is a simple and useful convention to use the matrix A as the argument of the density in (2.9), although in transforming the distribution we must recognize the correct number of independent arguments.

In (2.9) above $\Gamma_n(z)$ is the multivariate gamma function defined by the integral

$$\Gamma_n(z)=\int_{S>0}\text{etr}(-S)(\det S)^{z-(1/2)(n+1)}\,\text{d}S.$$

This integral is a (matrix variate) Laplace transform [see, for example, Herz

(1955) and Constantine (1963)] which converges absolutely for $\text{Re}(z) > \frac{1}{2}(n-1)$ and the domain of integration is the set of all positive definite matrices. It can be evaluated in terms of univariate gamma functions as

$$\Gamma_n(z) = \pi^{(1/4)n(n-1)} \prod_{i=1}^{n} \Gamma\left(z - \tfrac{1}{2}(i-1)\right)$$

[see James (1964)]. In (2.9) we also use the abbreviated operator representation $\text{etr}(\cdot) = \exp\{\text{tr}(\cdot)\}$.

The parameters of the Wishart distribution (2.9) are: (i) the order of the symmetric matrix A, namely n; (ii) the degrees of freedom T, of the component variates x_t in the summation $A = XX' = \sum_{t=1}^{T} x_t x_t'$; and (iii) the covariance matrix, Ω, of the normally distributed columns x_t in X. A common notation for the Wishart distribution (2.9) is then $\mathcal{W}_n(T, \Omega)$ [see, for example, Rao (1973, p. 534)]. This distribution is said to be central (in the same sense as the central χ^2 distribution) since the component variates x_t have common mean $E(x_t) = 0$. In fact, when $n = 1$, $\Omega = 1$, and $A = a$ is a scalar, the density (2.9) reduces to $(2)^{-T/2}\Gamma(T/2)^{-1}a^{T/2-1}e^{-(1/2)a}$, the density of a central χ^2 with T degrees of freedom.

If the component variates x_t in the summation are not restricted to have a common mean of zero but are instead independently distributed as $N(m_t, \Omega)$, then the joint distribution of the matrix $A = XX' = \sum_{t=1}^{T} x_t x_t'$ is said to be (non-central) Wishart with non-centrality matrix $\overline{M} = MM'$, where $M = [m_1, \dots, m_T]$. This is frequently denoted $\mathcal{W}_n(T, \Omega, \overline{M})$, although M is sometimes used in place of $\overline{M}$ [as in Rao (1973), for example]. The latter is a more appropriate parameter in the matrix case as a convenient generalization of the non-centrality parameter that is used in the case of the non-central χ^2 distribution – a special case of $\mathcal{W}_n(T, \Omega, \overline{M})$ in which $n = 1$, $\Omega = 1$, and $\overline{M} = \sum_{t=1}^{T} m_t^2$.

The p.d.f. of the non-central Wishart matrix $A = XX' = \sum_{t=1}^{T} x_t x_t'$, where the x_t are independent $N(m_t, \Omega)$, $M = [m_1, \dots, m_T] = E(X)$, and $\overline{M} = MM'$ is given by

$$\text{pdf}(A) = \frac{\text{etr}\left(-\frac{1}{2}\Omega^{-1}\overline{M}\right)}{2^{nT/2}\Gamma_n\left(\frac{T}{2}\right)(\det \Omega)^{T/2}} \, {}_0F_1\left(\frac{T}{2}; \frac{1}{4}\Omega^{-1}\overline{M}\Omega^{-1}A\right) \times \text{etr}\left(-\tfrac{1}{2}\Omega^{-1}A\right)(\det A)^{(T-n-1)/2}. \tag{2.10}$$

In (2.10) the function ${}_0F_1(;)$ is a matrix argument hypergeometric function, closely related to the Bessel function of matrix argument discussed by Herz (1955). Herz extended the classical hypergeometric functions of scalar argument [see, for example, Erdéyli (1953)] to matrix argument functions by using multidimensional Laplace transforms and inverse transforms. Constantine (1963) discovered that hypergeometric functions ${}_pF_q$ of a matrix argument have a general series representation in terms of zonal polynomials as follows:

$$ {}_pF_q(a_1,\dots,a_p;\, b_1,\dots,b_q;\, S) = \sum_{j=0}^{\infty} \sum_J \frac{(a_1)_J \cdots (a_p)_J}{(b_1)_J \cdots (b_q)_J} \frac{C_J(S)}{j!}. \tag{2.11} $$

In (2.11) J indicates a partition of the integer j into not more than n parts, where S is an $n \times n$ matrix. A partition J of weight r is a set of r positive integers $\{j_1,\dots,j_r\}$ such that $\sum_{i=1}^{r} j_i = j$. For example $\{2, 1\}$ and $\{1, 1, 1\}$ are partitions of 3 and are conventionally written (21) and (1^3). The coefficients $(a)_J$ and $(b)_J$ in (2.11) are multivariate hypergeometric coefficients defined by

$$ (a)_J = \prod_{i=1}^{n} \left(a - \tfrac{1}{2}(i-1)\right)_{j_i} \quad \text{for } J = \{j_1,\dots,j_n\}, $$

and where

$$ (\lambda)_j = \lambda(\lambda+1)\dots(\lambda+j-1) = \Gamma(\lambda+j)/\Gamma(\lambda). $$

The factor $C_J(S)$ in (2.11) is a zonal polynomial and can be represented as a symmetric homogeneous polynomial of degree j of the latent roots of S. General formulae for these polynomials are presently known only for the case $m = 2$ or when the partition of j has only one part, $J = (j)$ [see James (1964)]. Tabulations are available for low values of j and are reported in James (1964). These can be conveniently expressed in terms of the elementary symmetric functions of the latent roots of S [Constantine (1963)] or in terms of the quantities:

s_m = sum of the mth powers of the latent roots of S.

Thus, the first few zonal polynomials take the form:

degree j	partition J	zonal polynomial $C_J(S)$
1	1	s_1
2	1^2	$\frac{2}{3}(s_1^2 - s_2)$
	2	$\frac{1}{3}(s_1^2 + 2s_2)$
3	1^3	$\frac{1}{3}(s_1^3 - 3s_1s_2 + 2s_3)$
	21	$\frac{3}{5}(s_1^3 + s_1s_2 - 2s_3)$
	3	$\frac{1}{15}(s_1^3 + 6s_1s_2 + 8s_3)$

[see, for example, Johnson and Kotz (1972, p. 171)]. Algorithms for the extraction of the coefficients in these polynomials have been written [see James (1968) and McLaren (1976)] and a complete computer program for their evaluation has recently been developed and made available by Nagel (1981). This is an important development and will in due course enhance what is at present our very limited ability to numerically compute and readily interpret multiple infinite series such as (2.11). However, certain special cases of (2.11) are already recognizable in terms of simpler functions: when $n=1$ we have the classical hypergeometric functions

$$ {}_pF_q(a_1,\dots,a_p;\, b_1,\dots,b_q;\, s) = \sum_{j=0}^{\infty} \frac{(a_1)_j\dots(a_p)_j s^j}{(b_1)_j\dots(b_q)_j j!} $$

[see, for example, Lebedev (1965, ch. 9)]; and when $p=q=0$ we have

$$ {}_0F_0(S) = \sum_{j=0}^{\infty} \sum_J C_J(S)/j! = \operatorname{etr}(S), $$

which generalizes the exponential series and which is proved in James (1961); and when $p=1$ and $q=0$ we have

$$ {}_1F_0(a;S) = \sum_{j=0}^{\infty} \sum_J \frac{(a)_J}{j!} C_J(S) = (\det(I-S))^{-a}, $$

which generalizes the binomial series [Constantine (1963)]. The series ${}_0F_1(;)$ in the non-central Wishart density (2.10) generalizes the classical Bessel function. [The reader may recall that the non-central χ^2 density can be expressed in terms of the modified Bessel function of the first kind – see, for example, Johnson and Kotz (1970, p. 133).] In particular, when $n=1$, $\Omega=1$, $\overline{M}=\lambda$, and $A=a$ is a scalar, we have

$$ \begin{aligned} \operatorname{pdf}(a) &= \frac{\exp\{-\frac{1}{2}(a+\lambda)\}}{2^{T/2}\Gamma(T/2)} a^{T/2-1} \sum_{j=0}^{\infty} \frac{(\frac{1}{4}\lambda a)^j}{(T/2)_j j!} \\ &= \frac{\exp\{-\frac{1}{2}(a+\lambda)\}}{2^{T/2}} \sum_{j=0}^{\infty} \frac{\lambda^j a^{T/2+j-1}}{\Gamma(T/2+j)\, j! 2^{2j}}. \end{aligned} \tag{2.12} $$

This is the usual form of the p.d.f. of a non-central χ^2 variate.

3. Exact theory in the simultaneous equations model

3.1. The model and notation

We write the structural form of a system of G contemporaneous simultaneous stochastic equations as

$$YB + ZC = U, \tag{3.1}$$

and its reduced form as

$$Y = Z\Pi + V, \tag{3.2}$$

where $Y' = [y_1, \dots, y_T]$ is a $G \times T$ matrix of T observations of G endogenous variables, $Z' = [z_1, \dots, z_T]$ is a $K \times T$ matrix of T observations of K non-random exogenous variables, and $U' = [u_1, \dots, u_T]$ is a $G \times T$ matrix of the structural disturbances of the system. The coefficient matrices B $(G \times G)$ and C $(K \times G)$ comprise parameters that are to be estimated from the data and about which some *a priori* economic knowledge is assumed; usually this takes the form of simple (and frequently zero exclusion type) restrictions upon certain of the coefficients together with conventional normalization restrictions. As is usual in this contemporaneous version of the SEM (see Chapter 4 and Chapter 7 in this Handbook by Hsiao and Hausman, respectively), it is also assumed that the u_t $(t = 1, \dots, T)$ are serially independent random vectors distributed with zero mean vector and (non-singular) covariance matrix Σ. The coefficient matrix B is assumed to be non-singular and these conditions imply that the rows, v_t', of V in (3.2) are independent random vectors with zero mean vector and covariance matrix $\Omega = B'^{-1}\Sigma B^{-1}$. To permit the development of a distribution theory for finite sample sizes we will, unless otherwise explicitly stated, extend these conventional assumptions by requiring v_t $(t = 1, \dots, T)$ to be i.i.d. $N(0, \Omega)$. Extensions to non-normal errors are possible [see Phillips (1980b), Satchell (1981), and Knight (1981)] but involve further complications.

We will frequently be working with a single structural equation of (3.1) which we write in the following explicit form that already incorporates exclusion type restrictions:

$$y_1 = Y_2\beta + Z_1\gamma + u \tag{3.3}$$

or

$$y_1 = W_1\delta + u, \qquad W_1 = \left[Y_2 \vdots Z_1\right], \qquad \delta' = (\beta', \gamma'), \tag{3.4}$$

where y_1 $(T \times 1)$ and Y_2 $(T \times n)$ contain T observations of $n + 1$ included endogenous variables, Z_1 is a $T \times K_1$ matrix of included exogenous variables, and u is the vector of random disturbances on this equation. Thus, (3.3) explicitly represents one column of the full model (3.1). The reduced form of (3.3) is written

$$\left[y_1 \vdots Y_2\right] = \left[Z_1 \vdots Z_2\right] \begin{bmatrix} \pi_{11} & \Pi_{12} \\ \pi_{21} & \Pi_{22} \end{bmatrix} + \left[v_1 \vdots V_2\right] \tag{3.5}$$

or

$$X = Z\Pi^s + V^s, \qquad X = \left[y_1 \vdots Y_2\right], \qquad Z = \left[Z_1 \vdots Z_2\right], \tag{3.5'}$$

where Z_2 is a $T \times K_2$ matrix of exogenous variables excluded from (3.3). To simplify notation the selection superscripts in (3.5′) will be omitted in what follows. The system (3.5) represents $n+1$ columns of the complete reduced form (containing $G \geqslant n+1$ columns) given in (3.2). The total number of exogenous variables in (3.5) is $K = K_1 + K_2$ and the observation matrix Z is assumed to have full rank, K. We also assume that $K_2 \geqslant n$ and the submatrix Π_{22} $(K_2 \times n)$ in (3.4) has full rank $(=n)$ so that the structural equation is identified. Note that (3.3) can be obtained by postmultiplication of (3.5) by $(1, -\beta')'$ which yields the relations

$$\pi_{11} - \Pi_{12}\beta = \gamma, \qquad \pi_{21} - \Pi_{22}\beta = 0. \tag{3.6}$$

We will sometimes use the parameter $N = K_2 - n$ to measure the degree by which the structural relation (3.3) is overidentified.

3.2. *Generic statistical forms of common single equation estimators*

As argued in Section 2.2, most econometric estimators and test statistics can be expressed as simple functions of the sample moments of the data. In the case of the commonly used single equation estimators applied to (3.3) we obtain relatively simple generic statistical expressions for these estimators in terms of the elements of moment matrices which have Wishart distributions of various degrees of freedom and with various non-centrality parameter matrices. This approach enables us to characterize the distribution problem in a simple but powerful way for each case. It has the advantage that the characterization clarifies those cases for which the estimator distributions will have the same mathematical forms but for different values of certain key parameters and it provides a convenient first base for the mathematics of extracting the exact distributions. Historically the approach was first used by Kabe (1963, 1964) in the econometrics context and has since been systematically employed by most authors working in this field. An excellent recent discussion is given by Mariano (1982).

We will start by examining the IV estimator, δ_{IV}, of the coefficient vector $\delta' = (\beta', \gamma')$ in (3.3)–(3.4) based on the instrument matrix H. δ_{IV} minimizes the quantity

$$(y - W_1\delta)'H(H'H)^{-1}H'(y - W_1\delta), \tag{3.7}$$

and writing

$$P_D = D(D'D)^{-1}D', \qquad Q_D = I - P_D, \tag{3.8}$$

we obtain by stepwise minimization of (3.7) the following explicit expressions for the IV estimators of the subvectors β and γ:

$$\gamma_{IV} = (Z_1'P_HZ_1)^{-1}Z_1'P_H(y_1 - Y_2\beta_{IV}), \tag{3.9}$$

$$\begin{aligned}\beta_{IV} = &\left\{Y_2'\left[P_H - P_HZ_1(Z_1'P_HZ_1)^{-1}Z_1'P_H\right]Y_2\right\}^{-1} \\ &\times\left\{Y_2'\left[P_H - P_HZ_1(Z_1'P_HZ_1)^{-1}Z_1'P_H\right]y_1\right\}.\end{aligned} \tag{3.10}$$

In the usual case where H includes Z_1 as a subset of its instruments and $P_HZ_1 = Z_1$ we have the simple formulae:

$$\gamma_{IV} = (Z_1'Z_1)^{-1}Z_1'(y_1 - Y_2\beta_{IV}), \tag{3.11}$$

$$\beta_{IV} = \left[Y_2'(P_H - P_{Z_1})Y_2\right]^{-1}\left[Y_2'(P_H - P_{Z_1})y_1\right]. \tag{3.12}$$

We define the moment matrix

$$\begin{aligned}A(P_H) &= \begin{bmatrix} a_{11}(P_H) & a_{21}'(P_H) \\ a_{21}(P_H) & A_{22}(P_H)\end{bmatrix} = \begin{bmatrix} y_1'(P_H - P_{Z_1})y_1 & y_1'(P_H - P_{Z_1})Y_2 \\ Y_2'(P_H - P_{Z_1})y_1 & Y_2'(P_H - P_{Z_1})Y_2\end{bmatrix} \\ &= X'(P_H - P_{Z_1})X.\end{aligned} \tag{3.13}$$

The generic statistical form for the estimator β_{IV} in (3.12) is then

$$\beta_{IV} = A_{22}^{-1}(P_H)a_{21}(P_H). \tag{3.14}$$

This specializes to the cases of OLS and 2SLS where we have, respectively,

$$\beta_{OLS} = \left[Y_2'Q_{Z_1}Y_2\right]^{-1}\left[Y_2'Q_{Z_1}y_1\right] = A_{22}^{-1}(I)a_{21}(I), \tag{3.15}$$

$$\beta_{2SLS} = \left[Y_2'(P_Z - P_{Z_1})Y_2\right]^{-1}\left[Y_2'(P_Z - P_{Z_1})y_1\right] = A_{22}^{-1}(P_Z)a_{21}(P_Z). \tag{3.16}$$

In a similar way we find that the k-class estimator $\beta_{(k)}$ of β has the generic form

$$\begin{aligned}\beta_{(k)} &= \{Y_2'[k(P_Z - P_{Z_1}) + (1-k)Q_{Z_1}]Y_2\}^{-1} \\ &\quad \times \{Y_2'[k(P_Z - P_{Z_1}) + (1-k)Q_{Z_1}]y_1\} \\ &= [kA_{22}(P_Z) + (1-k)A_{22}(I)]^{-1}[ka_{21}(P_Z) + (1-k)a_{21}(I)]. \end{aligned} \tag{3.17}$$

The LIML estimator, β_{LIML}, of β minimizes the ratio

$$\frac{\beta_\Delta' A(I)\beta_\Delta}{\beta_\Delta'[A(I) - A(P_Z)]\beta_\Delta} = 1 + \frac{\beta_\Delta' A(P_Z)\beta_\Delta}{\beta_\Delta'[A(I) - A(P_Z)]\beta_\Delta} = 1 + \frac{\beta_\Delta' W\beta_\Delta}{\beta_\Delta' S\beta_\Delta}, \quad \text{say,} \tag{3.18}$$

where $\beta_\Delta' = (1, -\beta')$ and β_{LIML} satisfies the system

$$\{A(I) - \lambda[A(I) - A(P_Z)]\}\beta_\Delta = 0, \tag{3.19}$$

where λ is the minimum of the variance ratio in (3.18). Thus, β_{LIML} is given by the generic form

$$\beta_{\mathrm{LIML}} = [\lambda A_{22}(P_Z) + (1-\lambda)A_{22}(I)]^{-1}[\lambda a_{21}(P_Z) + (1-\lambda)a_{21}(I)], \tag{3.20}$$

that is, the k-class estimator (3.17) with $k = \lambda$.

The above formulae show that the main single equation estimators depend in a very similar way on the elements of an underlying moment matrix of the basic form (3.13) with some differences in the projection matrices relevant to the various cases. The starting point in the derivation of the p.d.f. of these estimators of β is to write down the joint distribution of the matrix A in (3.13). To obtain the p.d.f. of the estimator we then transform variates so that we are working directly with the relevant function $A_{22}^{-1}a_{21}$. The final step in the derivation is to integrate over the space of the auxiliary variates, as prescribed in the general case of (2.8) above, which in this case amounts essentially to (a_{11}, A_{22}). This leaves us with the required density function of the estimator.

The mathematical process outlined in the previous section is simplified, without loss of generality, by the implementation of standardizing transformations. These transformations were first used and discussed by Basmann (1963, 1974). They reduce the sample second moment matrix of the exogenous variables to the identity matrix (orthonormalization) and transform the covariance matrix of the endogenous variables to the identity matrix (canonical form). Such transformations help to reduce the parameter space to an essential set and identify the critical parameter functions which influence the shape of the distributions.[7] They are fully discussed in Phillips (1982e) and are briefly reviewed in the following section.

3.3. *The standardizing transformations*

We first partition the covariance matrix Ω conformably with $[y_1 \vdots Y_2]$ as

$$\Omega = \begin{bmatrix} \omega_{11} & \omega_{21}' \\ \omega_{21} & \Omega_{22} \end{bmatrix}. \tag{3.21}$$

Then the following result [proved in Phillips (1982e)] summarizes the effect of the standardizing transformations on the model.

Theorem 3.3.1

There exist transformations of the variables and parameters of the model given by (3.3) and (3.5) which transform it into one in which

$$T^{-1}Z'Z = I_K \quad \text{and} \quad \Omega = I_{n+1}. \tag{3.22}$$

Under these transformations (3.3) and (3.5) can be written in the form

$$y_1^* = Y_2^*\beta^* + \bar{Z}_1\bar{\gamma}^* + u^* \tag{3.23}$$

and

$$\left[y_1^* \vdots Y_2^*\right] = \bar{Z}\bar{\Pi}^* + \bar{V}, \tag{3.24}$$

where $T^{-1}\bar{Z}'\bar{Z} = I_K$ and the rows of $[y_1^* \vdots Y_2^*]$ are uncorrelated with covariance matrix given by I_{n+1}. Explicit formulae for the new coefficients in (3.23) are

$$\beta^* = \left(\omega_{11} - \omega_{21}'\Omega_{22}^{-1}\omega_{21}\right)^{-1/2}\Omega_{22}^{1/2}\left(\beta - \Omega_{22}^{-1}\omega_{21}\right) \tag{3.25}$$

and

$$\bar{\gamma}^* = \left(\frac{Z_1'Z_1}{T}\right)^{1/2}\left(\omega_{11} - \omega_{21}'\Omega_{22}^{-1}\omega_{21}\right)^{-1/2}\gamma. \tag{3.26}$$

[7]As argued recently by Mariano (1982), these reductions also provide important guidelines for the design of Monte Carlo experiments (at least in the context of SEMs) by indicating the canonical parameter space which is instrumental in influencing the shape of the relevant small sample distributions and from which a representative sample of points can be taken to help reduce the usual specificity of simulation findings.

These transformations preserve the number of excluded exogenous variables in the structural equation and the rank condition for its identifiability. □

It turns out that the commonly used econometric estimators of the standardized coefficients β^* and $\bar{\gamma}^*$ in (3.23) are related to the unstandardized coefficient estimators by the same relations which define the standard coefficients, namely (3.25) and (3.26). Thus, we have the following results for the 2SLS estimator [see Phillips (1982e) once again for proofs].

Theorem 3.3.2

The 2SLS estimator, β_{2SLS}, of the coefficients of the endogenous variables in (3.3) are invariant under the transformation by which the exogenous variables are orthornormalized. The 2SLS estimator, γ_{2SLS}, is not, in general, invariant under this transformation. The new exogenous variable coefficients are related to the original coefficients under the transformation $\bar{\gamma} = J_{11}\gamma$ and to the estimators by the corresponding equation $\bar{\gamma}_{2SLS} = J_{11}\gamma_{2SLS}$, where $J_{11} = (Z_1'Z_1/T)^{1/2}$. □

Theorem 3.3.3

The 2SLS estimators of β^* and $\bar{\gamma}^*$ in the standardized model (3.23) are related to the corresponding estimators of β and γ in the unstandardized model (3.3) by the equations:

$$\beta^*_{2SLS} = \left(\omega_{11} - \omega_{21}'\Omega_{22}^{-1}\omega_{21}\right)^{-1/2}\Omega_{22}^{1/2}\left(\beta_{2SLS} - \Omega_{22}^{-1}\omega_{21}\right) \tag{3.27}$$

and

$$\bar{\gamma}^*_{2SLS} = \left(\frac{Z_1'Z_1}{T\left(\omega_{11} - \omega_{21}'\Omega_{22}^{-1}\omega_{21}\right)}\right)^{1/2}\gamma_{2SLS}. \tag{3.28}$$

□

Results that correspond to these for 2SLS can be derived similarly for other estimators such as IV and LIML [see Phillips (1982e) for details].

The canonical transformation induces a change in the coordinates by which the variables are measured and therefore (deliberately) affects their covariance structure. Some further properties of the transformed structural equation (3.23) are worth examining. Let us first write (3.23) in individual observation form as

$$y^*_{1t} = y^{*\prime}_{2t}\beta^* + \bar{z}_{1t}'\bar{\gamma}^* + u^*_t. \tag{3.29}$$

Then, by simple manipulations we find that

$$\operatorname{cov}(y^*_{2t}, u^*_t) = -\beta^*, \tag{3.30}$$

$$\operatorname{var}(u^*_t) = 1 + \beta^{*\prime}\beta^*, \tag{3.31}$$

and

$$\operatorname{corr}(y_{2t}^*, u_t^*) = -\beta^*/(1+\beta^{*\prime}\beta^*)^{1/2}. \tag{3.32}$$

These relations show that the transformed coefficient vector, β^*, in the standardized model contains the key parameters which determine the correlation pattern between the included variables and the errors. In particular, when the elements of β^* become large the included endogenous variables and the error on the equation become more highly correlated. In these conditions, estimators of the IV type will normally require larger samples of data to effectively purge the included variables of their correlation with the errors. We may therefore expect these estimators to display greater dispersion in small samples and slower convergence to their asymptotic distributions under these conditions than otherwise. These intuitively based conjectures have recently been substantiated by the extensive computations of exact densities by Anderson and Sawa (1979)[8] and the graphical analyses by Phillips (1980a, 1982a) in the general case.

The vector of correlations corresponding to (3.32) in the unstandardized model is given by

$$\operatorname{corr}(y_{2t}, u_t) = \frac{\Omega_{22}^{1/2}(\Omega_{22}^{-1}\omega_{21} - \beta)}{(\omega_{11} - 2\beta'\omega_{21} + \beta'\Omega_{22}\beta)^{1/2}} = \frac{-\beta^*}{(1+\beta^{*\prime}\beta^*)^{1/2}}, \tag{3.33}$$

so that for a fixed reduced-form error covariance matrix, Ω, similar conditions persist as the elements of β grow large. Moreover, as we see from (3.33), the transformed structural coefficient β^* is itself determined by the correlation pattern between regressors and error in the unstandardized model. The latter (like β^*) can therefore be regarded as one of the critical sets of parameters that influence the shape of the distribution of the common estimators of the coefficient β.

3.4. *The analysis of leading cases*

There are two special categories of models in which the exact density functions of the common SEM estimators can be extracted with relative ease. In the first category are the just identified structural models in which the commonly used consistent estimators all reduce to indirect least squares (ILS) and take the form

$$\beta_{\text{ILS}} = [\bar{Z}_2'Y_2]^{-1}[\bar{Z}_2'y_1] \tag{3.34}$$

[8] See also the useful discussion and graphical plots in Anderson (1982).

of a matrix ratio of normal variates. In the two endogenous variable case (where $n=1$) this reduces to a simple ratio of normal variates whose p.d.f. was first derived by Fieiller (1932) and in the present case takes the form[9]

$$\mathrm{pdf}(r)=\frac{\exp\left\{-\frac{\mu^2}{2}(1+\beta^2)\right\}}{\pi(1+r^2)}\,{}_1F_1\left(1,\frac{1}{2};\frac{\mu^2}{2}\frac{(1+\beta r)^2}{1+r^2}\right), \tag{3.35}$$

where $\mu^2=T\Pi_{22}'\Pi_{22}$ is the scalar concentration parameter.[10] In the general case of $n+1$ included endogenous variables the density (3.35) is replaced by a multivariate analogue in which the ${}_1F_1$ function has a matrix argument [see (3.46) below]. The category of estimators that take the generic form of a matrix ratio of normal variates, as in (3.34), also include the general IV estimator in the overidentified case provided the instruments are non-stochastic: that is, if $\beta_{\mathrm{IV}}=[W'Y_2]^{-1}[W'y_1]$ and the matrix W is non-stochastic, as distinct from its usual stochastic form in the case of estimators like 2SLS in overidentified equations. This latter case has been discussed by Mariano (1977). A further application of matrix ratios of normal variates related to (3.34) occurs in random coefficient SEMs where the reduced-form errors are a matrix quotient of the form $A^{-1}a$ where both a and the columns of A are normally distributed. Existing theoretical work in this area has proceeded essentially under the hypothesis that $\det A$ is non-random [see Kelejian (1974)] and can be generalized by extending (3.35) to the multivariate case in much the same way as the exact distribution theory of (3.34), which we will detail in Section 3.5 below.

The second category of special models that facilitate the development of an exact distribution theory are often described as leading cases of the fully parameterized SEM.[11] In these leading cases, certain of the critical parameters are set equal to zero and the distribution theory is developed under this null hypothesis. In the most typical case, this hypothesis prescribes an absence of simultaneity and a specialized reduced form which ensures that the sample moments of the data on which the estimator depends have central rather than (as is typically the case) non-central distributions.[12] The adjective "leading" is used advisedly since the distributions that arise from this analysis typically provide the leading term in the multiple series representation of the true density that applies when the null

[9] This density is given, for example, in Mariano and McDonald (1979).

[10] This parameter is so called because as $\mu^2\to\infty$ the commonly used single equation estimators all tend in probability to the true parameter. Thus, the distributions of these estimators all "concentrate" as $\mu^2\to\infty$, even if the sample size T remains fixed. See Basmann (1963) and Mariano (1975) for further discussion of this point.

[11] See Basmann (1963) and Kabe (1963, 1964).

[12] Some other specialized SEM models in which the distributions of commonly used estimators depend only on central Wishart matrices are discussed by Wegge (1971).

hypothesis itself no longer holds. As such the leading term provides important information about the shape of the distribution by defining a primitive member of the class to which the true density belongs in the more general case. In the discussion that follows, we will illustrate the use of this technique in the case of IV and LIML estimators.[13]

We set $\beta = 0$ in the structural equation (3.3) and $\Pi_{22} = 0$ in the reduced form so that y_1 and y_2 (taken to be a vector of observations on the included endogenous variable now that $n = 1$) are determined by the system[14]

$$y_1 = Z_1\gamma + u, \qquad y_2 = Z_1\Pi_{12} + v_2. \tag{3.36}$$

The IV estimator of β is

$$\beta_{\mathrm{IV}} = (y_2'Z_3Z_3'y_2)^{-1}(y_2'Z_3Z_3'y_1). \tag{3.37}$$

under the assumption that standardizing transformations have already been performed. Let Z_3 be $T \times K_3$ with $K_3 \geqslant 1$ so that the total number of instruments is $K_1 + K_3$. Simple manipulations now confirm that the p.d.f. of β_{IV} is given by [see Phillips (1982e)]

$$\mathrm{pdf}(r) = \left[B\left(\frac{1}{2}, \frac{K_3}{2}\right)\right]^{-1}(1+r^2)^{-(K_3+1)/2}, \tag{3.38}$$

where $B(\frac{1}{2}, K_3/2)$ is the beta function. This density specializes to the case of 2SLS when $K_3 = K_2$ and OLS when $K_3 = T - K_1$. [In the latter case we may use (3.15) and write $Q_{Z_1} = I - T^{-1}Z_1Z_1' = C_1C_1'$, where C_1 is a $T \times (T - K_1)$ matrix whose columns are the orthogonal latent vectors of Q_{Z_1} corresponding to unit latent roots.] The density (3.38) shows that integral moments of the distribution exist up to order $K_3 - 1$: that is, in the case of 2SLS, $K_2 - 1$ (or the degree of overidentification) and, in the case of OLS, $T - K_1 - 1$.

The result corresponding to (3.38) for the case of the LIML estimator is [see Phillips (1982e) for the derivation]

$$\mathrm{pdf}(r) = \left[\pi(1+r^2)\right]^{-1}, \qquad -\infty < r < \infty. \tag{3.39}$$

[13]An example of this type of analysis for structural variance estimators is given in Section 3.7.

[14]In what follows it will often not be essential that both $\beta = 0$ and $\Pi_{22} = 0$ for the development of the "leading case" theory. What is essential is that $\Pi_{22} = 0$, so that the structural coefficients are, in fact, unidentifiable. Note that the reduced-form equations take the form

$$y_1 = Z_1\pi_{11} + v_1, \qquad Y_2 = Z_1\Pi_{12} + V_2,$$

when $\Pi_{22} = 0$. The first of these equations corresponds to (3.36) in the text when $\beta = 0$.

Thus, the exact sampling distribution of the β_{LIML} is Cauchy in this leading case. In fact, (3.39) provides the leading term in the series expansion of the density of LIML derived by Mariano and Sawa (1972) in the general case where $\beta \neq 0$ and $\Pi_{22} \neq 0$. We may also deduce from (3.39) that β_{LIML} has no finite moments of integral order, as was shown by Mariano and Sawa (1972) and Sargan (1970). This analytic property of the exact distribution of β_{LIML} is associated with the fact that the distribution displays thicker tails than that of β_{IV} when $K_3 > 1$. Thus, the probability of extreme outliers is in general greater for β_{LIML} than for β_{IV}. This and other properties of the distributions of the two estimators will be considered in greater detail in Sections 3.5 and 3.6.

3.5. *The exact distribution of the IV estimator in the general single equation case*

In the general case of a structural equation such as (3.3) with $n+1$ endogenous variables and an arbitrary number of degrees of overidentification, we can write the IV estimator β_{IV} of β in the form

$$\beta_{\mathrm{IV}} = (Y_2'Z_3Z_3'Y_2)^{-1}(Y_2'Z_3Z_3'y_1), \tag{3.40}$$

where the standardizing transformations are assumed to have been carried out. This is the case where $H = [Z_1 \vdots Z_3]$ is a matrix of $K_1 + K_3$ instruments used in the estimation of the equation. To find the p.d.f. of β_{IV} we start with the density of the matrix:

$$A = \begin{bmatrix} a_{11} & a_{21}' \\ a_{21} & A_{22} \end{bmatrix} = T^{-1}\begin{bmatrix} y_1'Z_3Z_3'y_1 & y_1'Z_3Z_3'Y_2 \\ Y_2'Z_3Z_3'y_1 & Y_2'Z_3Z_3'Y_2 \end{bmatrix}.$$

In general this will be non-central Wishart with a p.d.f. of the form

$$\mathrm{pdf}(A) = \frac{\mathrm{etr}\left(-\frac{1}{2}MM'\right)}{2^{(1/2)K_3(n+1)}\Gamma_{n+1}\left(\frac{K_3}{2}\right)}\,{}_0F_1\left(\frac{K_3}{2};\frac{1}{4}MM'A\right)$$
$$\times\,\mathrm{etr}\left(-\tfrac{1}{2}A\right)(\det A)^{(1/2)(K_3-n-2)}$$

[see (2.10) above] where $M = E(T^{-1/2}X'Z_3) = T^{-1/2}\Pi'Z'Z_3$.

We now introduce a matrix S which selects those columns of Z_2 which appear in Z_3, so that $Z_3 = Z_2S$. Then, using the orthogonality of the exogenous variables, we have

$$M' = T^{-1/2}Z_3'Z\Pi = T^{1/2}[OS']\Pi = T^{1/2}S'[\pi_{21}\Pi_{22}] = T^{1/2}S'\Pi_{22}[\beta, I]$$

in view of the relations (3.6) given above. Writing $\Pi_{22}'SS'\Pi_{22}$ as $\bar{\Pi}_{22}'\bar{\Pi}_{22}$, where $\bar{\Pi}_{22}$ is an $n\times n$ matrix (which is non-singular since the structural equation (3.3) is assumed to be identified), we find that

$$\operatorname{etr}\left(-\tfrac{1}{2}MM'\right)=\operatorname{etr}\left\{-\frac{T}{2}(I+\beta\beta')\bar{\Pi}_{22}'\bar{\Pi}_{22}\right\}.$$

Moreover, since the non-zero latent roots of $MM'A$ are the latent roots of

$$T\bar{\Pi}_{22}[\beta,I]A\begin{bmatrix}\beta'\\ I\end{bmatrix}\bar{\Pi}_{22}',$$

(3.41) becomes

$$\frac{\operatorname{etr}\left\{-\frac{T}{2}(I+\beta\beta')\bar{\Pi}_{22}'\bar{\Pi}_{22}\right\}}{2^{(1/2)K_3(n+1)}\Gamma_{n+1}\left(\frac{K_3}{2}\right)}\,{}_0F_1\left(\frac{K_3}{2};\frac{T}{4}\bar{\Pi}_{22}[\beta,I]A\begin{bmatrix}\beta'\\ I\end{bmatrix}\bar{\Pi}_{22}'\right)$$
$$\times\operatorname{etr}\left(-\tfrac{1}{2}A\right)(\det A)^{(1/2)(K_3-n-2)}.$$

We now transform variables from the matrix variate A to $w=a_{11}-r'A_{22}r$, $r=A_{22}^{-1}a_{21}$, and $A_{22}=A_{22}$. The Jacobian of the transformation is $\det A_{22}$ and we have

$$\begin{aligned}
&\operatorname{pdf}(w,r,A_{22})\\
&=\frac{\operatorname{etr}\left\{-\frac{T}{2}(I+\beta\beta')\bar{\Pi}_{22}'\bar{\Pi}_{22}\right\}}{2^{(1/2)K_3(n+1)}\Gamma_{n+1}\left(\frac{K_3}{2}\right)}\\
&\quad\times{}_0F_1\left(\frac{K_3}{2};\frac{T}{4}\{w\bar{\Pi}_{22}\beta\beta'\bar{\Pi}_{22}'+\bar{\Pi}_{22}(1+\beta r')A_{22}(I+r\beta')\bar{\Pi}_{22}'\}\right)\\
&\quad\times\exp\left(-\tfrac{1}{2}(w+r'A_{22}r)\right)\operatorname{etr}\left(-\tfrac{1}{2}A_{22}\right)\{w\det A_{22}\}^{(1/2)(K_3-n-2)}\det A_{22}\\
&=\frac{\operatorname{etr}\left\{-\frac{T}{2}(I+\beta\beta')\bar{\Pi}_{22}'\bar{\Pi}_{22}\right\}}{2^{(1/2)K_3(n+1)}\Gamma_{n+1}\left(\frac{K_3}{2}\right)}\\
&\quad\times{}_0F_1\left(\frac{K_3}{2};\left\{\frac{T}{4}w\bar{\Pi}_{22}\beta\beta'\bar{\Pi}_{22}'+\bar{\Pi}_{22}(I+\beta r')A_{22}(I+r\beta')\bar{\Pi}_{22}'\right\}\right)\\
&\quad\times\exp\left(-\tfrac{1}{2}w\right)\operatorname{etr}\left(-\tfrac{1}{2}(I+rr')A_{22}\right)w^{(1/2)(K_3-n-2)}(\det A_{22})^{(1/2)(K_3-n)}.
\end{aligned}$$

Define $L = K_3 - n$ and introduce the new matrix variate $B = (I + rr')^{1/2} A_{22} (I + rr')^{1/2}$. The Jacobian of this transformation is $[\det(I + rr')]^{-(n+1)/2}$ and we have

$$\mathrm{pdf}(w, r, B) = \frac{\mathrm{etr}\left\{-\frac{T}{2}(I + \beta\beta')\bar{\Pi}_{22}'\bar{\Pi}_{22}\right\}}{2^{(1/2)(L+n)(n+1)}\Gamma_{n+1}\left(\frac{L+n}{2}\right)[\det(I + rr')]^{(L+n+1)/2}}$$
$$\times {}_0F_1\left[\frac{L+n}{2}; \frac{T}{4}\{w\bar{\Pi}_{22}\beta\beta'\bar{\Pi}_{22}' + \bar{\Pi}_{22}(I + \beta r')(I + rr')^{-1/2}\right.$$
$$\left.\times B(I + rr')^{-1/2}(I + \beta r')\bar{\Pi}_{22}'\}\right]\exp(-\tfrac{1}{2}w)$$
$$\times \mathrm{etr}(-\tfrac{1}{2}B)w^{L/2-1}(\det B)^{L/2}. \tag{3.42}$$

As in the general scheme of development outlined in Section 2.2 we need to integrate out the auxiliary variates (w, B) in (3.42) in order to find the analytic form of the density of β_{IV}. This problem was the main obstacle in the development of an exact distribution theory for single equation estimators in the general case for over a decade following the work of Richardson (1968) and Sawa (1969) that dealt explicitly with the two endogenous variable case $(n = 1)$. In this latter case the ${}_0F_1$ function in (3.42) can be replaced by a power series in the argument:

$$\frac{T}{4}\{w\bar{\Pi}_{22}\beta\beta'\bar{\Pi}_{22}' + \bar{\Pi}_{22}(I + \beta r')(I + rr')^{-1/2}B(I + rr')^{-1/2}(I + r\beta')\bar{\Pi}_{22}'\} \tag{3.43}$$

which, when $n = 1$, is a scalar. Powers of this variable may now be expanded in binomial series and inspection of (3.42) shows that terms of this double series may then be integrated simply as gamma functions. When $n > 1$, (3.43) is a matrix and the series development of the ${}_0F_1$ function is in terms of zonal polynomials of this matrix. In the absence of an algebra to develop a binomial type expansion for zonal polynomials of the sum of two matrices, integration of the auxiliary variables (w, B) in (3.42) appeared impossible. However, a solution to this difficulty was found by Phillips (1980a). The idea behind the method developed in this article is to use an alternative representation of the ${}_0F_1$ function in which the argument matrix (3.43) is thrown up into an exponent. The two elements of the binomial matrix sum (3.43) can then effectively be separated and integrated out. (We will not give the full argument here but refer the reader to the article for details.)[15] In short, the process leads to the following analytic form for the exact

[15]An alternative approach to the extraction of the exact density of β_{IV} from (3.42) is given in Phillips (1980a, appendix B) and directly involves the algebra of expanding the zonal polynomial of a sum of two matrices into a sum of more basic polynomials in the constituent matrices. This algebra was developed by Davis (1980a, 1980b) and has recently been extended by Chikuse (1981) to matrix multinomial expansions of zonal polynomials of the sum of several matrices.

finite sample density of β_{IV}:

$$\operatorname{pdf}(r)=\frac{\operatorname{etr}\left\{-\frac{T}{2}(I+\beta\beta')\bar{\Pi}'_{22}\bar{\Pi}_{22}\right\}\Gamma_n\left(\frac{L+n+1}{2}\right)}{\pi^{n/2}\left[\det(I+rr')\right]^{(L+n+1)/2}}$$
$$\times\sum_{j=0}^{\infty}\frac{\left(\frac{L}{2}\right)_j}{j!\Gamma_n\left(\frac{L+n}{2}+j\right)}$$
$$\times\left[\left(\frac{T}{2}\beta'\bar{\Pi}'_{22}\left(\operatorname{adj}\frac{\partial}{\partial W}\right)\bar{\Pi}_{22}\beta\right)^j(\det(I+W))^{(L-1)/2+j}\right.$$
$$\times{}_1F_1\left(\frac{L+n+1}{2},\frac{L+n}{2}+j;\frac{T}{2}(I+W)\bar{\Pi}_{22}\right.$$
$$\left.\left.\times(I+\beta r')(I+rr')^{-1}(I+r\beta')\bar{\Pi}_{22}\right)\right]_{W=0}. \tag{3.44}$$

In (3.44) $L=K_3-n$ is the number of surplus instruments used in the estimation of β. That is, K_1+K_3 instruments are used and at least K_1+n are needed to perform the estimation by the traditional IV procedure. Thus, when $K_3=K_2$ and $L=K_2-n$, (3.44) gives the p.d.f. of the 2SLS estimator of β; and when $K_1+K_3=T$, so that $K_3=T-K_1$ and $L=T-K_1-n$, (3.44) gives the p.d.f. of the OLS estimator of β.

The matrix $W(n\times n)$ in (3.44) contains auxiliary variables that are useful in reducing the integral from which (3.44) is derived and $\operatorname{adj}(\partial/\partial W)$ denotes the adjoint of the matrix differential operator $\partial/\partial W$. We note that when $n=1$, W is a scalar, $\operatorname{adj}(\partial/\partial W)=1$, and (3.44) becomes

$$\operatorname{pdf}(r)=\frac{\exp\left\{-\frac{\mu^2}{2}(1+\beta^2)\right\}\Gamma\left(\frac{L+2}{2}\right)}{\pi^{1/2}\Gamma\left(\frac{L+1}{2}\right)(1+r^2)^{(L+2)/2}}$$
$$\times\sum_{j=0}^{\infty}\frac{\left(\frac{L}{2}\right)_j}{j!\left(\frac{L+1}{2}\right)_j}\left(\frac{\mu^2}{2}\beta^2\right)^j{}_1F_1\left(\frac{L+2}{2},\frac{L+1}{2}+j;\frac{\mu^2}{2}\frac{(1+\beta r)^2}{1+r^2}\right), \tag{3.45}$$

in which $\mu^2=T\bar{\Pi}_{22}^2=T\Pi'_{22}\Pi_{22}$ is the scalar concentration parameter [recall (3.35) and footnote 10]. The density (3.45) was first derived for 2SLS ($L=K_2-1$) and OLS ($L=T-K_1-1$) by Richardson (1968) and Sawa (1969).

When $L=0$ in (3.44) the series corresponding to the suffix j terminates at the first term and we have

$$\text{pdf}(r)=\frac{\operatorname{etr}\left\{-\frac{T}{2}(I+\beta\beta')\bar{\Pi}_{22}'\bar{\Pi}_{22}\right\}\Gamma_n\left(\frac{n+1}{2}\right)}{\pi^{n/2}\Gamma_n\left(\frac{n}{2}\right)[\det(I+rr')]^{(n+1)/2}}$$
$$\times {}_1F_1\left(\frac{n+1}{2};\frac{n}{2};\frac{T}{2}\bar{\Pi}_{22}(I+\beta r')(I+rr')^{-1}(I+r\beta')\bar{\Pi}_{22}'\right). \tag{3.46}$$

That is, a single term involving a matrix argument hypergeometric function as obtained by Sargan (1976a) in this special case.

While (3.44) gives us a general representation of the exact joint density function of instrumental variable estimators in simultaneous equation models, this type of series representation of the density is not as easy to interpret as we would like. It can be said that the leading term in the density reveals the order to which finite sample moments of the estimator exist [cf. Basmann (1974)]. In the present case, we see that when $L=0$ the leading term involves $[\det(I+rr')]^{-(n+1)/2}=(1+r'r)^{-(n+1)/2}$, which is proportional to the multivariate Cauchy density [see Johnson and Kotz (1972)]; when $L>0$ the term involves $[\det(I+rr')]^{-(L+n+1)/2}=(1+r'r)^{-(L+n+1)/2}$, which is similar to a multivariate t-density. These expressions enable us to verify directly Basmann's conjecture [Basmann (1961, 1963)] that integer moments of the 2SLS estimator ($L=K_2-n$) exist up to the degree of overidentification. In other respects, the analytic form of (3.44) is not by itself very revealing. Moreover, series representations such as (3.44) and (3.46) cannot as yet be implemented for numerical calculations as easily as might be expected. The formulae rely on the matrix argument ${}_1F_1$ function and numerical evaluation depends on available tabulations and computer algorithms for the zonal polynomials that appear in the series representation of such matrix argument functions [see (2.11)]. This is an area in which important developments are currently taking place [some discussion and references are given in Section 2 following (2.11)]. Unfortunately, the availability of tabulations and algorithms for zonal-type polynomials[16] will cover only part of the computational difficulty. As noted by Muirhead (1978), the series that involve these polynomials often converge very slowly. This problem arises particularly when the polynomials have large arguments (large latent roots) and it becomes necessary to work deeply into the higher terms of the series in order to achieve convergence. This in turn raises additional

[16]This is a generic term that I am using to denote zonal polynomials and more general polynomials of this class but which may involve several argument matrices, as in the work of Davis (1980a, 1980b) and Chikuse (1981).

problems of underflow and overflow in the computer evaluations of the coefficients in the series and the polynomials themselves. To take as a simple example the case of the exact density of the IV estimator in the two endogenous variable case, the author has found that in a crude summation of the double infinite series for the density a thousand or more terms seem to be necessary to achieve adequate convergence when the true coefficient [that is, β in (3.45)] is greater than 5 and the concentration parameter, μ^2, is greater than 10. These are not in any way unrealistic values and the problems increase with the size of the coefficient and concentration parameter. When the density is expressed as a single series involving the ${}_1F_1$ function of a scalar argument, as in (3.45), these considerations necessitate the computation of the ${}_1F_1$ function for scalar arguments greater than 225. Use of the conventional asymptotic expansion of the ${}_1F_1$ function [which is normally recommended when the argument is greater than 10, see Slater (1965)] fails here because one of the parameters of the ${}_1F_1$ function grows as we enter more deeply into the series and the series itself no longer converges. Undoubtedly, the additional problems encountered in this example quickly become much worse as the dimension of the argument matrices in the special functions and the zonal polynomials increases and as we need to make use of the more general zonal-type polynomials (see footnote 16).

For direct computational work in the case of the IV estimator when there are more than two endogenous variables in the structural equation, the problems reported in the previous section were overcome in Phillips (1980a) by extracting an asymptotic expansion of the exact joint density of the vector coefficient estimator. This involves the use of a multidimensional version of Laplace's method of approximating integrals [see, for example, Bleistein and Handelsman (1976)]. Marginal density expansions were obtained by similar techniques in Phillips (1982a). These results give us direct and readily computable formulae for the joint and marginal densities of the coefficient estimator. The leading terms of these expansions of the joint and marginal densities have an error of $O(T^{-1})$, where T is the sample size and in the univariate (two endogenous variable) case the resulting approximation can be otherwise obtained by the saddlepoint technique as in Holly and Phillips (1979). The latter article demonstrates that the approximation gives high accuracy for some plausible values of the parameters throughout a wide domain of the distribution, including the tails.

The main conclusions about the shape and sensitivity of the p.d.f. of β_{IV} and its components which emerge from the computational work in these articles confirm the results of earlier numerical investigations dealing with the two endogenous variable case by Sawa (1969) and Anderson and Sawa (1979) and the recent experimental investigations by Richardson and Rohr (1982). A full discussion of the two endogenous variable case will be taken up in Section 3.6. In what follows we report briefly the principal results which apply in the multi-endogenous variable cases investigated by Phillips (1980a, 1982a).

(1) For comparable parameter values the marginal distributions of β_{IV} appear to concentrate more slowly as $T \to \infty$ when the number of endogenous variables $(n+1)$ in the equation increases.

(2) The marginal densities are particularly sensitive to the degree of correlation in the concentration parameter matrix $\overline{M} = T\overline{\Pi}'_{22}\overline{\Pi}_{22}$ in (3.44) Setting, for example,

$$\overline{M} = \mu^2 \begin{bmatrix} 1 & \rho \\ \rho & 1 \end{bmatrix}$$

in the $n+1=3$ endogenous variable case, the location, dispersion, and skewness of the marginal distributions all seem to be sensitive to ρ. Since $\overline{M}$ approaches singularity as $|\rho| \to 1$ when the equation becomes unidentifiable [Π_{22} in (3.5) and hence $\overline{\Pi}_{22}$ must be of full rank $= n$ for identifiability of the equation] we would expect the dispersion of the marginal distributions of the structural estimator β_{IV} to increase with $|\rho|$. This phenomenon is, in fact, observed in the graphical plots recorded by Phillips (1980a, 1982a) for different values of ρ. The central tendencies of the marginal distributions also seem to be sensitive to the relative signs of ρ and the elements of the true coefficient vector β. We give the following example. When the coefficients β_i and ρ all have the same sign the common set of exogenous variables are compatible as instruments for Y_2 in the regression and the marginal distributions appear to be adequately centered (for small values of L and moderate μ^2); but when β_i and ρ take opposite signs the exogenous variables are less compatible as instruments for the columns of Y_2 and the marginal distributions become less well centered about the true coefficients.

(3) The effect of increasing the number of endogenous variables, *ceteris paribus*, in a structural equation is a decrease in the precision of estimation. This accords with well-known results for the classical regression model.

(4) The marginal distribution of β_{IV} displays more bias in finite samples as L, the number of additional instruments used for the n right-hand-side endogenous variables, increases in value. When L becomes small the distribution is more centrally located about the true value of the parameter but also has greater dispersion than when L is large.

3.6. *The case of two endogenous variables ($n = 1$)*

As seen in (3.45) the general form of the joint density (3.44) can be specialized to yield results which apply in the two endogenous variable case. These results were first established independently by Richardson (1968) and Sawa (1969) for 2SLS and OLS [to which (3.45) applies], by Mariano and Sawa (1972) for LIML, and by Anderson and Sawa (1973) for k-class estimators. Moreover, as demonstrated by Richardson and Wu (1970) and by Anderson (1976) the exact p.d.f.s for 2SLS and LIML directly apply after appropriate changes in notation to the OLS and orthogonal regression estimators of the slope coefficient in the errors in variables model.

Details of the argument leading to the exact density of the 2SLS (or OLS) estimator can be outlined in a few simple steps arising from (3.42) [see Phillips (1982e) for details]. The final result is expression (3.45), obtained above as a specialized case of the more general result in Section 3.5. Expression (3.45) gives the density of β_{2SLS} when $L = K_2 - 1$ and the density of β_{OLS} when $L = T - K_1 - 1$. An alternative method of deriving the density of β_{2SLS} (or β_{OLS}) is given in Phillips (1980b, appendix A), where the Fourier inversion [of the form (2.3)] that yields the density is performed by contour integration.

Similar methods can be used to derive the exact densities of the LIML and k-class estimators, β_{LIML} and $\beta_{(k)}$. In the case of LIML the analysis proceeds as for the leading case but now the joint density of sample moments is non-central [see Phillips (1982e) for details]. This joint density is the product of independent Wishart densities with different degrees of freedom (K_2 and $T - K$, respectively) and a non-centrality parameter matrix closely related to that which applies in the case of the IV estimator analyzed in Section 3.5. The parameterization of the joint density of the sample moments upon which β_{LIML} depends clarifies the key parameters that ultimately influence the shape of the LIML density. These are the (two) degrees of freedom, the non-centrality matrix, and the true coefficient vector. For an equation with two endogenous variables the relevant parameters of the LIML density are then: K_2, $T - K$, μ^2, and β. The mathematical form of the density was first derived for this case by Mariano and Sawa (1972).[17] The parameterization of the LIML density is different from that of the IV density given above. In particular, the relevant parameters of (3.45) are L, μ^2, and β; or in the case of 2SLS, K_2, μ^2, and β. We may note that the IV density depends on the sample size T only through the concentration parameter μ^2, as distinct from the LIML density which depends on the sample size through the degrees of freedom, $T - K$, of one of the underlying Wishart matrices as well as the concentration parameter.

Similar considerations apply with respect to the distribution of the k-class estimator, $\beta_{(k)}$. We see from (3.17) that for $k \neq 0, 1$ the p.d.f. of $\beta_{(k)}$ depends on the joint density of two underlying Wishart matrices. The relevant parameters of the p.d.f. of $\beta_{(k)}$ are then: K_2, $T - K$, k, μ^2, and β. The mathematical form of this

[17]See Mariano and McDonald (1979) for a small correction.

density for $0 \leqslant k \leqslant 1$ was found by Anderson and Sawa (1973) as a fourth-order infinite series.

Extensive computations are now available for at least some of the exact densities (and associated distribution functions) discussed in this section. Most of this work is due to a series of substantial contributions by T. W. Anderson, T. Sawa, and their associates. An excellent account of their work is contained in Anderson (1982). We summarize below the main features that emerge from their numerical tabulations of the relevant distributions, all of which refer to the two endogenous variable case.

(1) The distribution of β_{2SLS} is asymmetric about the true parameter value, except when $\beta = 0$ [the latter special case is also evident directly from expression (3.45) above]. The asymmetry and skewness of the distribution increase as both β and K_2 increase. For example, when $\beta = 1$, $\mu^2 = 100$, and $K_2 = 30$ the median of the distribution is -1.6 (asymptotic) standard deviations from the true parameter value, whereas at $K_3 = 3$ the median is -0.14 standard deviations from β. As K_2 becomes small the distribution becomes better located about β (as the numbers just given illustrate) but displays greater dispersion. Thus, at $\beta = 1$, $\mu^2 = 100$, and $K_2 = 30$ the interquartile range (measured again in terms of asymptotic standard deviations) is 1.031, whereas at $\beta = 1$, $\mu^2 = 100$, and $K_2 = 3$ the interquartile range is 1.321. Table 3.1 table illustrates how these effects are magnified as β increases:[18]

Table 3.1
Median (MDN) and interquartile range (IQR) of $\beta_{2SLS} - \beta$ in terms of asymptotic standard deviations ($\mu^2 = 100$)

K_2 \ β		1	2	5
3	MDN	-0.140	-0.177	-0.194
	IQR	1.321	1.310	1.304
30	MDN	-1.599	-2.021	-2.215
	IQR	1.031	0.924	0.860

(2) The rate at which the distribution of β_{2SLS} (appropriately centered and standardized) approaches normality depends critically on the values of β and K_2. If either (or both) of these parameters are large, then the approach to normality is quite slow. At $\beta = 1$ and $K_2 = 3$, for example, the value of μ^2 must be at least 100 to hold the maximum error on the asymptotic normal approximation to 0.05; but

[18] The numbers in Tables 3.1 and 3.2 have been selected from the extensive tabulations in Anderson and Sawa (1977, 1979) which are recommended to the reader for careful study. My thanks to Professors Anderson and Sawa for their permission to quote from their tables.

when $K_2 = 10$, μ^2 must be at least 3000 to ensure the same maximum error on the asymptotic distribution.

(3) Since the exact distribution of β_{LIML} involves a triple infinite series, Anderson and Sawa (1977, 1979) tabulated the distribution of a closely related estimator known as LIMLK. This estimator represents what the LIML estimator would be if the covariance matrix of the reduced-form errors were known. In terms of (3.18), β_{LIMLK} minimizes the ratio $\beta_\Delta' W \beta_\Delta / \beta_\Delta' \Omega \beta_\Delta$, where Ω is the reduced-form error covariance matrix and satisfies the system $(W - \lambda_m \Omega)\beta_\Delta = 0$, where λ_m is the smallest latent root of $\Omega^{-1}W$. The exact distribution of β_{LIMLK} can be obtained from the non-central Wishart distribution of W. Anderson and Sawa (1975) give this distribution in the form of a double infinite series that is more amenable to numerical computation than the exact distribution of LIML. In a sampling experiment Anderson et al. (1980) investigated the difference between the LIML and LIMLK distributions and found this difference to be very small except for large values of K_2. Anderson (1977) also showed that expansions of the two distributions are equivalent up to terms of $O(\mu^{-3})$. These considerations led Anderson and Sawa to take LIMLK and a proxy for LIML in analyzing the small sample properties of the latter and in the comparison with 2SLS. They found the central location of LIMLK to be superior to that of 2SLS. In fact, LIMLK is median unbiased for all β and K_2. Moreover, its distribution (appropriately centered and standardized) approaches normality much faster than that of 2SLS. However, LIMLK displays greater dispersion in general than 2SLS and its distribution function approaches unity quite slowly. These latter properties result from the fact that LIMLK, like LIML, has no integral moments regardless of the sample size and its distribution can therefore be expected to have thicker tails than those of 2SLS. Table 3.2 [selected computations from Anderson and Sawa (1979)] illustrates these effects in relation to the corresponding results for 2SLS in Table 3.1.[19]

Table 3.2
Median and interquartile range of $\beta_{\text{LIMLK}} - \beta$
in terms of asymptotic standard deviations ($\mu^2 = 100$)

K_2 \ β		1	2	5
3	MDN	0	0	0
	IQR	1.360	1.357	1.356
30	MDN	0	0	0
	IQR	1.450	1.394	1.363

[19]We note that since β_{LIMLK} depends only on the non-central Wishart matrix W with degrees of freedom K_2, the distribution of β_{LIMLK} depends on the sample size T only through the concentration parameter μ^2, unlike the distribution of β_{LIML}.

These features of the exact small sample distributions of 2SLS and LIMLK give rise to the following two conclusions reported by Anderson (1982): (a) the distribution of β_{2SLS} may be badly located and skewed unless β and K_2 are small or μ^2 is very large; and (b) the approach to the asymptotic normal distribution is slow for 2SLS and rapid for LIMLK and, apparently, LIML. Thus, in many cases the asymptotic normal theory may be a fairly adequate approximation to the actual distribution of LIML but a less than adequate approximation to the distribution of 2SLS.

These conclusions clearly suggest the use of caution in the application of asymptotic theory and thereby agree with the results of many other studies. One additional point is worthy of mention. The above exact results and reported numerical experience refer to the standardized model as discussed in Section 3.3. When we referred to the true coefficient β above, we therefore meant the true standardized coefficient [as given by β^* in expression (3.25) of Theorem 3.3.1]. But we note that the correlation between the included endogenous regressor, y_{2t}, and the structural error, u_t, in the unstandardized model is a simple function of β^*, namely $\text{corr}(y_t, u_t) = -\beta^*/(1+\beta^{*2})^{1/2}$ as given by (3.33) in the general case. Thus, as the modulus of the standardized coefficient, $|\beta^*|$, increases, the correlation between y_{2t} and u_t increases. We therefore need, *ceteris paribus*, a larger sample of data to effectively purge y_{2t} of its correlation with u_t in estimation by 2SLS (or more generally IV). This correlation is explicitly taken into account when we estimate by LIMLK (or LIML), since we directly utilize the reduced-form error covariance matrix (or an estimate of it) in this procedure. Thus, it may not be too surprising that the finite sample distribution of β_{2SLS} displays a far greater sensitivity to the value (particularly large values) of β^* than does the distribution of LIML, as the computations in Tables 3.1 and 3.2 illustrate.

3.7. *Structural variance estimators*

In Sections 3.3–3.6 our attention has focused on the distribution of structural coefficient estimators. Structural variance estimators are also of importance, both as measures of residual variation and as components in commonly used test statistics (such as coefficient significance tests of the t ratio and asymptotic χ^2 variety where the metric relies on an estimate of the structural equation error variance). Basmann (1974) has pointed to an additional role that structural variance estimators may play by indicating the demands for accuracy which a model such as (3.3) and (3.5) may place on the measurement of the data.

Structural error variance estimators typically rely on the residuals from an estimated structural equation and their distributions rely, in turn, on those of the structural coefficient estimators. The following quadratic forms define three alternative classes of estimator for the structural variance, σ^2, of the errors in (3.3):

$$G_1(\beta)=\beta_\Delta' XQ_{Z_1}X\beta_\Delta=\beta_\Delta' A(I)\beta_\Delta, \tag{3.47}$$

$$G_2(\beta)=\beta_\Delta' X'Q_Z X\beta_\Delta=\beta_\Delta'[A(I)-A(P_Z)]\beta_\Delta, \tag{3.48}$$

$$Q(\beta)=G_1(\beta)-G_2(\beta)=\beta_\Delta' X'(P_Z-P_{Z_1})X\beta_\Delta=\beta_\Delta' A(P_Z)\beta_\Delta.^{20} \tag{3.49}$$

Corresponding to (3.47) we have the most common structural error variance estimators, namely

$$\begin{aligned}\sigma_{\mathrm{IV}}&=T^{-1}G_1(\beta_{\mathrm{IV}})=T^{-1}(y_1-Y_2\beta_{\mathrm{IV}})'Q_{Z_1}(y_1-Y_2\beta_{\mathrm{IV}})\\&=T^{-1}(y_1-Y_2\beta_{\mathrm{IV}}-Z_1\gamma_{\mathrm{IV}})'(y_1-Y_2\beta_{\mathrm{IV}}-Z_1\gamma_{\mathrm{IV}})\end{aligned} \tag{3.50}$$

and

$$\begin{aligned}\sigma_{\mathrm{LIML}}&=T^{-1}G_1(\beta_{\mathrm{LIML}})=T^{-1}(y_1-Y_2\beta_{\mathrm{LIML}})'Q_{Z_1}(y_1-Y_2\beta_{\mathrm{LIML}})\\&=T^{-1}(y_1-Y_2\beta_{\mathrm{LIML}}-Z_1\gamma_{\mathrm{LIML}})'(y_1-Y_2\beta_{\mathrm{LIML}}-Z_1\gamma_{\mathrm{LIML}})\end{aligned} \tag{3.51}$$

for estimation by IV and LIML, respectively.

The exact distributions of σ_{IV} and σ_{LIML} [as well as those of the related estimators based on $G_2(\beta)$ and $Q(\beta)$] can be extracted in steps that reduce the dimensionality of the problem from the essential sample moments and coefficient estimators in (3.50)–(3.51) to the required marginal densities. The analytic derivations relate very closely to those that apply in the distribution theory for the relevant coefficient estimator. The published work on the distribution of structural variance estimators refers to the two endogenous variable case and, in conjunction with other important unpublished material, has been well reviewed by Basmann (1974, pp. 252–254). The essential contributions in this area are by Basmann and Richardson (1969, 1973), who found the exact p.d.f.s of 2SLS variance estimators based on (3.47)–(3.49), and by McDonald (1972), who found the exact p.d.f.s of the LIML variance estimators for the same trinity of quadratic forms. Some of their principal findings can otherwise be obtained by examination of the leading case considered above in Section 3.4.

[20] In the case of estimation by IV (with instrument matrix H) it will sometimes be more appropriate to consider the following quadratic form instead of (3.49):

$$Q(\beta)=\beta'_\Delta X'(P_H-P_{Z_1})X\beta_\Delta=\beta_\Delta' A(P_H)\beta_\Delta.$$

In particular, the exact density of σ_{IV} in standardized form and in the leading case is given by [see Phillips (1982e) for derivations]

$$\operatorname{pdf}(\sigma_{\mathrm{IV}}) = \frac{TB\left(\frac{1}{2},\frac{T-K_1+K_3}{2}\right)e^{-T\sigma_{\mathrm{IV}}/2}}{2\Gamma\left(\frac{T-K_1}{2}\right)B\left(\frac{1}{2},\frac{K_3}{2}\right)}\sum_{j=0}^{\infty}\frac{\left(\frac{1}{2}\right)_j\left(\frac{T\sigma_{\mathrm{IV}}}{2}\right)^{(T-K_1)/2+j-1}}{\left(\frac{T-K_1+K_3+1}{2}\right)_j j!}. \tag{3.52}$$

Expression (3.52) gives the density of σ_{IV}, where K_1+K_3 is the number of instrumental variables used in the estimation of the equation. When $K_3=1$ this corresponds to the just identified case and also to LIML. The latter follows because in the leading case the density of β_{LIML} is Cauchy, as shown in (3.39), and this is just the special case of (3.38) when $K_3=1$.

Analysis of (3.52) shows that moments, $E(\sigma_{\mathrm{IV}}^h)$, of σ_{IV} are finite provided $h<K_3/2$. In the case of 2SLS, where $K_3=K_2$, this corresponds to the results reported by Basmann (1974). And by setting $K_3=1$, we deduce that σ_{LIML} has no finite moments of integral order, as shown by McDonald (1972). In this connection we may also note that since $\beta_{2\mathrm{SLS}}$ minimizes $Q(\beta)$ in (3.49) and since $Q(\beta)$ is proportional to a $\chi_{K_2}^2$ variate when β takes on the true value of the coefficient vector, the structural variance estimator $\hat{\sigma}_{2\mathrm{SLS}}=Q(\beta_{2\mathrm{SLS}})/(K_2-n)$ possesses finite moments of all orders. However, $\hat{\sigma}_{2\mathrm{SLS}}$ (unlike σ_{IV} and σ_{LIML}) is in general an inconsistent estimator of the structural error variance. In fact, Basmann and Richardson (1973) show that the statistic $\hat{\sigma}_{2\mathrm{SLS}}$ has a limiting $\chi_{K_2-1}^2$ distribution as the concentration parameter $\mu^2\to\infty$. The errors involved in this asymptotic distribution of $\hat{\sigma}_{2\mathrm{SLS}}$ were analyzed for the two endogenous variable case by Ebbeler and McDonald (1973) who found that the errors increased with the size of β^2 and with the number of excluded exogenous variables, K_2.

3.8. Test statistics

The finite sample distributions of certain test statistics as well as structural coefficient and variance estimators have also received attention in the literature. As with the classical linear regression model, knowledge of the distribution of test criteria allows us in principle to construct exact confidence intervals and to carry out significance tests for which the size is exact (or the critical region appropriate for a given test size). However, an important practical difference arises in this context between the classical regression model and the SEM. In the former, the usual t-ratio, F, and χ^2 tests are parameterized only by degrees of freedom which are available upon simple data, parameter, and restriction counts. In the latter, however, most finite sample distributions depend on a number of parameters some of which figure prominently in the parameterization of the structural and reduced forms. This is evident from the results reported above in sections that pertain to estimators rather than test criteria. It prevents, in particular, our using those results directly to mount significance tests on the coefficients in just the same way as the presence of an unknown error variance prevents our using a normal theory as the basis of an exact significance test for a coefficient in a linear regression. Whereas this problem is simply overcome in the regression model by the use of the t-ratio, it is not so simply resolved in the SEM. Unless we can assume values for the unknown parameters upon which the relevant distribution depends, an exact theory will in most cases be beyond reach.

Two different approaches have been adopted in the literature to assist in resolving this difficulty. The first of these is to develop an Edgeworth expansion of the distribution of the test statistic, then to replace the unknown parameters that arise in the coefficients of this expansion by (consistent) estimates of them. Thus, if a symmetric confidence interval for a parameter based on crude asymptotic theory is corrected by the $O(T^{-1})$ terms in the Edgeworth expansion and the parameters, say ψ, that arise in these terms are replaced by consistent estimates $\hat{\psi}$ for which $\hat{\psi} = \psi + O_p(T^{-1/2})$ the order of magnitude of the error in the Edgeworth correction will be maintained. This approach forms the basis of the work by Sargan (1975, 1976a, 1980) and will be considered more closely in Chapter 15 of this Handbook by Rothenberg.

The second approach to the problem, at least in a sampling theoretic framework, is to use test criteria whose distributions are more parsimoniously parameterized and therefore more useful in providing revisions to asymptotic significance levels and critical regions. The published work in this area is less general than the literature which deals with Edgeworth corrections and the associated distribution theory is more limited than that which has been developed for structural estimators. Nevertheless, some interesting and important results have been obtained which we will now briefly review.

3.8.1. Anderson–Rubin exact confidence intervals and tests

A small sample theory of interval estimation and testing in structural equations such as (3.3) was developed by Anderson and Rubin (1949).[21] Their procedure is applicable when the confidence region or null hypothesis concerns the full vector of endogenous variable coefficients in the structural equation. Thus, if we consider the hypothesis H_0: $\beta = \beta_0$ in (3.3), we may define $y^* = y_1 - Y_2\beta_0$ and rewrite the structural equation under the null hypothesis as $y^* = Z_1\gamma + u$. On the

[21]As pointed out by Anderson and Rubin (1949, p. 61) their method was independently suggested by Bartlett (1948).

other hand, when H_0 is not true, y^* will [in view of the reduced form (3.5)] be a linear function of both Z_1 and Z_2. Thus, H_0 may be tested by a conventional F-test of the hypothesis that the coefficient vector of Z_2 is zero in the regression of y^* on Z_1 and Z_2. The statistic for this test takes the usual form of

$$F = \frac{T-K}{K_2}\frac{y^{*\prime}(Q_{Z_1}-Q_Z)y^*}{y^{*\prime}Q_Z y^*} \tag{3.53}$$

and has an $F_{K_2,T-K}$ distribution under H_0. When H_0 is false, the denominator of (3.53) is still proportional to a χ^2_{T-K} variate while the numerator becomes non-central $\chi^2_{K_2}$ with the non-centrality dependent upon the vector of coefficient inaccuracy under the null, $\beta-\beta_0$, and a subset of the reduced-form parameters. Thus (3.53) is non-central $F_{K_2,T-K}$ under the alternative hypothesis, $\beta \neq \beta_0$. This test can readily be extended to accommodate hypotheses that involve exogenous variable coefficients and even (under suitable conditions) coefficients from several structural equations. The common requirement in each version of the test is that all structural coefficients pertaining to endogenous variables be specified under the null. This requirement ensures that the model can be rewritten, as above, as a multiple (or multivariate) regression when the null hypothesis holds. The test based on (3.53) is consistent and its power function was considered briefly by Revankar and Mallela (1972). Confidence regions follow from (3.53) in the usual way as the set of all β satisfying the inequality

$$\frac{(y_1-Y_2\beta)'(Q_{Z_1}-Q_Z)(y_1-Y_2\beta)}{(y_1-Y_2\beta)'Q_Z(y-Y_2\beta)} \leqslant \frac{K_2}{T-K}F_{K_2,T-K}(\alpha) \tag{3.54}$$

at the $100(1-\alpha)$ percent confidence level.

3.8.2. *An exact structural t-statistic*

Richardson and Rohr (1971) studied a structural t-statistic [introduced by Dhrymes (1969)] that can be used to test hypotheses and construct confidence intervals for individual structural coefficients in SEMs. They found the exact distribution of this statistic for an equation with two endogenous variables and showed: (a) that this distribution is, indeed, Student's t-distribution when the true standardized structural coefficient (β) is zero; and (b) that this distribution tends to the t-distribution as $\mu^2 \to \infty$. However, their numerical computations of this exact distribution and its first three moments indicate that the exact distribution will often be poorly approximated by the t-distribution unless β is very small or μ^2 very large. The exact density of their statistic is in fact highly skewed even for large μ^2 and small β. Exact probabilities for intervals symmetric about the origin are lower than those for the t-distribution (unless $\beta = 0$), so that confidence levels will be overstated and levels of significance will be understated if the t-distribution is used as an approximation in constructing confidence intervals or in two-sided tests.

Their analysis can be illustrated by considering the IV estimator, β_{IV}, in the two endogenous variable case and for the leading null hypothesis $\beta = 0$, $\Pi_{22} = 0$ of Section 3.4. The Richardson-Rohr structural t-statistic is given by

$$t = (y_2' Z_3 Z_3 y_2)^{1/2} \beta_{\mathrm{IV}}/s, \tag{3.55}$$

where $s^2 = Q(\beta_{\mathrm{IV}})/(K_3 - 1)$. Simple manipulations show that this has a Student's t-distribution with $K_3 - 1$ degrees of freedom. In the 2SLS case that is considered by Richardson and Rohr (1971), $K_3 - 1 = K_2 - 1 =$ degree of overidentification of the structural equation and it is assumed that $K_2 - 1 \geqslant 1$.

An interesting experimental investigation that bears on this test has been reported by Maddala (1974). Maddala studied the power functions of the Dhrymes-Richardson-Rohr (DRR) statistic, the Anderson-Rubin (AR) statistic, and the conventional t-ratio statistic (corresponding to what would be justified if the equation were a classical linear regression and estimation were by OLS). For the model and parameter values used by Maddala, he found that the DRR test had very low power in comparison with the AR and conventional test. This outcome is partially explained by Maddala in terms of the different structural variance estimators that are used in the various test statistics. He argues, in particular, that the DRR statistic involves a variance estimator based on $Q(\beta)$ in (3.49). This estimator relies on linear forms in the data such as $Z_3'X$ and does not involve the sample second moments $X'X$ directly, as do the more conventional estimators σ_{IV} and σ_{LIML} in (3.50)–(3.51). To this extent they neglect useful sample information about the error variance and this is reflected in the observed low power of the DRR test in comparison with the conventional tests.

3.8.3. *Identifiability test statistics*

The structural equation (3.3) may be written in the alternative form

$$y_1 = Y_2\beta + Z_1\gamma_1 + Z_2\gamma_2 + u, \tag{3.56}$$

under what is known [compare Basmann (1960)] as the "identifiability hypothesis":

$$H_0\colon \gamma_2 = 0. \tag{3.57}$$

It is usually assumed that $K_2 \geqslant n + 1$ so that attention is focused on the overidentifying restrictions in (3.57). Several tests of these restrictions have been suggested in the literature and are referred to under the name of *identifiability test statistics*. The most common of these arise naturally in 2SLS and LIML estimation. Their construction relies on the quadratic forms (3.47)–(3.49) studied in connection with structural variance estimators. Explicitly we have

$$\phi_{2SLS}=\frac{G_1(\beta_{2SLS})-G_2(\beta_{2SLS})}{G_2(\beta_{2SLS})}=\frac{Q(\beta_{2SLS})}{G_2(\beta_{2SLS})}, \tag{3.58}$$

$$\phi_{LIML}=\frac{G_1(\beta_{LIML})-G_2(\beta_{LIML})}{G_2(\beta_{LIML})}=\frac{Q(\beta_{LIML})}{G_2(\beta_{LIML})}. \tag{3.59}$$

If the identifiability hypothesis (3.57) is valid, the limiting distributions of $T\phi_{2SLS}$ and $T\phi_{LIML}$ are both $\chi^2_{K_2-n}$ as $T\to\infty$ under fairly general conditions. These asymptotic results were derived by Anderson and Rubin (1950), Hood and Koopmans (1953), and Basmann (1960) and are reported in Chapter 7 of this Handbook by Hausman. They lead to the common critical region (of rejection) in a large sample test of identifying restrictions:

$$T\phi\geqslant\chi^2_{K_2-n}(\alpha), \tag{3.60}$$

where α is the chosen significance level and ϕ denotes either ϕ_{2SLS} or ϕ_{LIML}.

As an approximate finite sample test, Anderson and Rubin (1949) suggested the alternative critical region:

$$\frac{T-K}{K_2}\phi_{LIML}\geqslant F_{K_2,T-K}(\alpha). \tag{3.61}$$

This may be justified on the argument that for fixed β in (3.59) the ratio $(T-K)\phi/K_2$ is indeed distributed as $F_{K_2,T-K}$ [compare (3.53) above]. Basmann (1960) criticized this suggestion on the grounds that as $T\to\infty$

$$T\phi_{2SLS},\,T\phi_{LIML}\overset{\mathfrak{D}}{\to}\chi^2_{K_2-n},$$

whereas

$$K_2F_{K_2,T-K}\overset{\mathfrak{D}}{\to}\chi^2_{K_2}.$$

He also argued that these considerations suggested an adjustment to the numerator degrees of freedom in the F-ratio and, as a result, the alternative critical regions:

$$\left.\begin{aligned} &\frac{T-K}{K_2-n}\phi_{2SLS} \geqq \\ &\frac{T-K}{K_2-n}\phi_{LIML} \geqq \end{aligned}\right\} F_{K_2-n,T-K}(\alpha), \qquad \begin{aligned}(3.62)\\(3.63)\end{aligned}$$

as approximate finite sample tests of identifying restrictions. An experimental study was performed by Basmann to determine the adequacy of the new tests (*inter alia*) and his results give very favorable support to the alternative critical region for the 2SLS statistic in (3.62). An extensive series of experiments recently conducted by Rhodes and Westbrook (1982) adds support to this evidence, indicating that the adequacy of the critical regions (3.62)–(3.63) depends on the degree of overidentification of the equation under the identifiability hypothesis. In particular, the adequacy of these regions in finite samples deteriorates as $K_2 - n$ increases.

The exact finite sample distribution of ϕ_{2SLS} in (3.58) was found by Basmann (1965) for the case in which $n=1$ and $K_2-n=1$ and by Richardson (1969) for $n=1$ and an arbitrary degree of overidentification K_2-1. McDonald (1972) discovered the exact p.d.f. of ϕ_{LIML} and showed that it has finite moments of order less than $T-K$. Since ϕ_{2SLS} has finite integer moments of order less than $(T-K)/2$ [from the results of Richardson (1969)] it follows that the tails of the distribution of ϕ_{2SLS} will be thicker than those of ϕ_{LIML}. We notice that this feature of the finite sample distributions of the two statistics is consonant with the observed inequality between the statistics, namely $\phi_{LIML} \leqslant \phi_{2SLS}$, that arises from the construction of the LIML estimator [see (3.18) and (3.59)].

Some of the results in the preceding section have been generalized to the case of an equation containing $n+1$ endogenous variables in an important contribution by Rhodes (1981). Rhodes found the exact distribution of ϕ_{LIML} and showed that it has finite moments of order less than $(n+1)(T-K)/2$. The exact distribution depends on the non-centrality parameter matrix which we will denote by $\overline{M}$, as in Section 3.5 above. In our notation:

$$\overline{M} = \tfrac{1}{2}\Sigma^{-1}E(X')(P_Z - P_{Z_1})E(X). \tag{3.64}$$

After standardizing transformations have been carried out and when the null hypothesis is correct, this becomes:

$$\overline{M} = T\begin{bmatrix}\beta' \\ \cdots \\ I_n\end{bmatrix}\Pi_{22}'\Pi_{22}\left[\beta \;\vdots\; I_n\right]. \tag{3.65}$$

Thus, under the null $\overline{M}$ has one zero latent root and generally n non-zero roots.

When the null is false, the simpler form of $\overline{M}$ in (3.65) no longer holds and (3.64) normally has rank $n+1$ rather than n. Thus, the true power functions of tests such as (3.60), (3.61), or (3.63) depend on the values of these non-zero latent roots. Rhodes (1981) investigates the actual size and power of these tests for a selected set of latent roots of $\overline{M}$ and finds that when the non-zero roots are small (less than 10) the true size of each test is very poorly represented by the nominal level of significance. To relate these results to those of Basmann (1960) reported above, Rhodes calculated the non-zero latent roots of the relevant non-centrality matrix for Basmann's experiment and found the roots to be large, explaining in part why (3.63) proved to be quite accurate in those experiments.

Since the exact distribution of ϕ_{LIML} is not amenable to computation, some steps have been taken to provide improvements on the critical regions (3.60), (3.61), and (3.63). McDonald (1974) obtained an approximate F distribution for ϕ_{LIML} by selecting parameters for the former in order that the first two moments of the distributions would be the same. Rhodes (1981) developed an alternative critical region for the test by considering the conditional distribution of ϕ_{LIML} given the other roots of the LIML determinantal equation. In particular, this conditional distribution has a simple asymptotic form as the largest n latent roots of $\overline{M}$ tend to infinity and can be used for the computation of a new critical region for a test based on ϕ_{LIML} and for power function evaluations. It has the advantage (over the conventional asymptotic and other tests we have discussed) of incorporating more sample information, and preliminary experimental results in Rhodes (1981) indicate that it may provide a more accurate critical region for the identifiability test.

3.9. Systems estimators and reduced-form coefficients

In comparison with the analytic results reviewed in previous sections for single equation estimators and test statistics, much less is known about the distribution of full systems estimators, reduced-form coefficient estimators, and their associated test statistics. Most progress has in fact been made in the application of small sample asymptotics by the use of Edgeworth expansions. Here the theory and constructive process detailed in Phillips (1982e) are directly applicable and machine programmable for both structural and reduced-form coefficient estimators. We will consider the analytic results for the two groups of coefficients separately below.

3.9.1. Structural coefficient estimators

Some manageable formulae for the first correction term of $O(T^{-1/2})$ in the Edgeworth expansion have been obtained by Sargan (1976a, appendix C) for

3SLS and FIML systems estimators. But no work is presently available to shed light on the adequacy of these approximations. What we know of their performance in the case of single equation estimators[22] suggests that their adequacy (at least for 3SLS estimation) will deteriorate as certain equations in the system become heavily overidentified. It also seems clear that the size of the system will have an important bearing in this respect, given other relevant factors such as the sample size, reduced-form parameter values, and features of the exogenous series. Some evidence which relates to this issue is available in Phillips (1977c), who developed formulae for the Edgeworth expansion of two-stage Aitken estimators of the parameters in a linear multivariable system subject to general linear cross-equation restrictions.[23] These formulae show that to terms of $O(T^{-1})$ the finite sample distribution is a rescaled version of the exact distribution of the Aitken estimator. This scaling factor depends on the moments of the estimated error covariance matrix and the sample second moments of the exogenous variables. As the number of equations in the system increases, the scale generally changes in such a way that the dispersion of the distribution increases. This corresponds with exact results obtained by Kataoka (1974) for a somewhat simpler version of this model and squares with the intuition that as the precision of our error covariance estimator decreases (through reductions in the effective degrees of freedom) the sampling dispersion of the resulting two-stage coefficient estimator increases. These results for the multivariate linear model furnish interesting conjectures for systems estimation in the SEM. Finally in this connection, we may mention that Nagar-type approximating moments may be deduced from the Edgeworth formulae [see Phillips (1982e)]. Such approximating moments, or pseudo-moments (where this term is appropriate), were derived independently for the 3SLS structural coefficient estimator by Mikhail (1969) in doctoral dissertation work at the London School of Economics.

In addition to the approximate distribution theory discussed above some progress on a leading case analysis for systems estimation along the lines of Section 3.4 is possible. The principles may be illustrated by considering FIML applied to a two-equation system of the form (3.1) with

$$B = \begin{bmatrix} 1 & b_{21} \\ b_{12} & 1 \end{bmatrix}, \tag{3.66}$$

and overidentifying restrictions imposed on each column of the exogenous variable coefficient matrix C. We may consider the null hypothesis in which

[22] See Anderson and Sawa (1979), Holly and Phillips (1979), and Richardson and Rohr (1981). An attempt to tackle this problem by asymptotic expansions in which the degree of overidentification grows large is given by Morimune (1981).

[23] Recent work in the same framework has been published by Maekawa (1980) for t ratio type test statistics.

$C=0$ and hence $\Pi=0$ in the reduced form (3.2). In this case, it is shown in Phillips (1982e) that the joint density of the unrestricted coefficient estimates $(b_{12}^{\text{FIML}}, b_{21}^{\text{FIML}})$ is bivariate Cauchy. This result confirms that the FIML estimator of the structural coefficients (b_{12}, b_{21}) has no finite integral moments. Sargan (1970) originally established the latter result by using the fact that the FIML estimator is independent of the normalization of the structural equation. In the context of (3.66) this argument takes the form that if the implied normalization $(b_{11}, b_{12})=(1, b_{12})$ were changed to $(b_{11}, 1)$, then the FIML estimates under the alternative normalizations would satisfy the reciprocal relationship $b_{11}^{\text{FIML}} = 1/b_{12}^{\text{FIML}}$. Thus, the FIML estimate of a structural coefficient can be interpreted as the reciprocal of another FIML estimate under a different normalization. This fact would normally imply that the distribution of such an estimator has no integral moments. As in the case of LIML (see Section 3.4) this property of the exact distribution of FIML estimates of the structural coefficients (b_{12}, b_{21}) means that the probability of extreme outliers is generally higher for FIML than for other structural coefficient estimators.

Moments of the 3SLS structural coefficient estimator have been investigated by Sargan (1978). Some difficulty occurs in the treatment of the 2SLS estimated error covariance matrix, Σ_{2SLS}, arising in the conventional 3SLS formula. Sargan, therefore, considers two cases. The first case treats Σ_{2SLS} as non-random or, more generally, allows Σ_{2SLS} to be random but bounds the ratio of its largest and smallest latent roots. For this case, Sargan demonstrates that the 3SLS estimator of the coefficients in any equation has finite moments of integral order up to (and including) the degree of overidentification for that equation. Thus, for this case it is proved that 2SLS and 3SLS estimators have finite moments to the same integral order. The second case considered by Sargan allows Σ_{2SLS} to be the conventional estimator of the error covariance matrix. Here it is proved that moments of the 3SLS estimator will be finite provided the order of the moment is less than $(N+1)/2$, where N is the degree of overidentification. Thus, the mean and variance of the 3SLS estimator will certainly be finite if the degree of overidentification is two and four, respectively. These are sufficient conditions and Sargan conjectures that the earlier result for Σ_{2SLS} non-random also applies in this second case where Σ_{2SLS} is the conventional error covariance matrix estimator.

3.9.2. *Reduced-form coefficients*

In an important article, McCarthy (1972) initiated the analytic study of the finite sample properties of restricted reduced-form (RRF) coefficient estimators and associated predictions. The RRF incorporates additional information that is embodied in overidentifying restrictions on the structural equations of the system. To the extent that RRF estimators utilize this information, they were thought for many years to possess higher asymptotic efficiency and, as a result, smaller variances than unrestricted reduced-form (URF) coefficient estimators.[24] McCarthy demonstrated that if there are overidentifying restrictions on the structural equations the solved reduced-form coefficients from 2SLS will in general possess no integral moments. This property influences the probability in small samples of outliers in the 2SLS reduced-form coefficients and associated predictions. It warns us that RRF estimators may give rise to occasional very poor forecasts and cautions against the use of quadratic loss criteria in estimator evaluations and in Monte Carlo work. Since the publication of McCarthy's article, generalizations of these results have been made to estimators other than 2SLS and various reduced-form estimator modifications have been suggested which attempt to improve on small sample performance. Much of this work is contained in a fundamental paper by Sargan (1976b) and in the doctoral dissertation research of Maasoumi (1977). Regrettably, a good deal of their work has not yet been published. However, with the permission of these authors some of their unpublished results will briefly be reported within the general discussion that follows.

An intuitive explanation for the McCarthy result and its generalizations arises from the transformation which takes the structural system (3.1) into its reduced form (3.2), namely $\Pi = -CB^{-1}$. For structural coefficient estimators $(\hat{B}, \hat{C})$ we deduce the reduced-form estimator $\hat{\Pi} = -\hat{C}(\operatorname{adj}\hat{B})/(\det\hat{B})$. Now if there exists a value of $(\hat{B}, \hat{C})$ for which $\det\hat{B} = 0$ while $\hat{C}(\operatorname{adj}\hat{B}) \neq 0$ and $\operatorname{pdf}(\hat{B}, \hat{C}) > 0$, then at least some elements of the reduced-form estimator $\hat{\Pi}$ will have no integral moments. This follows because the integral that defines the first moment, namely $\int \hat{\Pi}\operatorname{pdf}(\hat{B}, \hat{C})\mathrm{d}\hat{B}\mathrm{d}\hat{C}$, does not converge under the stated conditions (just as the integral $\int_a^b |x-a|^{-s}\mathrm{d}x$ diverges for all values of s in the interval $1 \leq s < \infty$). This intuitive argument underlies the mathematical proof of the 2SLS result by McCarthy (1972) and forms the basis of the following general result first given (in a modified form) by Sargan (1976b) and proved in Phillips (1982e).

Theorem 3.9.1

If $\hat{\beta} = \psi(p)/\phi(p)$, where p is a random n-vector and $\hat{\beta}$ is a scalar function of p and there exists a p_0 in the domain of definition of p such that:

(i) $\psi(p)$ is continuous at p_0 with $\psi(p_0) \neq 0$,
(ii) $\phi(p)$ has continuous first derivatives at p_0, denoted by the vector ϕ_p, for which $\phi_p'\phi_p > 0$ and $\phi(p_0) = 0$,
(iii) p has a continuous p.d.f. with $\operatorname{pdf}(p_0) > 0$, then $\hat{\beta}$ has no integral moments. □

This theorem applies readily to a wide variety of reduced-form estimators of the type $\hat{\Pi} = -\hat{C}(\operatorname{adj}\hat{B})/\det(\hat{B})$ considered above. Its conclusion gives some

[24] Dhrymes (1973) showed that this ranking in terms of asymptotic efficiency does not hold for RRF estimators, such as 2SLS, which are not fully efficient.

general analytical support for the presumption that reduced-form coefficients extracted from OLS, 2SLS, and 3SLS structural estimators have no moments in overidentified models. The particular two-equation case studied by McCarthy (1972) also follows directly. Sargan (1976b) extends McCarthy's analysis to the general case. His results confirm that for most overidentified models 2SLS and 3SLS reduced-form coefficient estimators possess no integral moments. Exceptions do occur when the model is of a recursive type in which $\det \hat{B} = \text{const.}$ Another important exception occurs for just identified models. Here the structural coefficient estimators for 2SLS, 3SLS, and FIML are all equal to indirect least squares and the corresponding reduced-form coefficients are equal to the OLS estimates, which have moments of all orders under normality.

We can also consider estimation of the reduced form by FIML. If θ is the vector of unconstrained elements in the structural coefficient matrices (B, C) of (3.1) then we may write $\Pi = \Pi(\theta)$ in (3.2) and θ_{FIML} is obtained by minimizing

$$\ln\left[\det\{T^{-1}(Y - Z\Pi(\theta))'(Y - Z\Pi(\theta))\}\right]. \tag{3.67}$$

From the criterion function (3.67) it is clear that low probabilistic weight will be attached to events in θ space which imply large values of Π since the latter will normally imply large values for the criterion (3.67). This will not be the case as the columns of Z become highly collinear or more generally when the complete data matrix $T^{-1}W'W$ is close to singularity. Thus, we might expect the FIML reduced-form $\Pi_{\text{FIML}} = \Pi(\theta_{\text{FIML}})$ to possess finite moments provided T is large in relation to the number of variables, $n + K$, in the system. In fact, Sargan (1976b) proves that Π_{FIML} has finite moments of integral order up to $T - n - K$.

The fact that many reduced-form estimators possess no integral moments has led to the suggestion of improved estimators which combine URF and RRF estimators in such a way that the tail behavior of the combined estimator is improved. A fundamental contribution in this area is due to Maasoumi (1978).[25] Maasoumi develops a new reduced-form estimator which combines the corresponding restricted 3SLS and the unrestricted OLS estimators. The new estimator incorporates the outcome of an asymptotic χ^2 test of the model's overidentifying restrictions and thereby opens up a middle road of methodology that lies between completely unrestricted and fully restricted estimation. Specifically, Maasoumi proposes the following estimator:

$$\Pi^* = \lambda\Pi_{3\text{SLS}} + (1 - \lambda)\Pi_{\text{OLS}}, \tag{3.68}$$

[25] More recent work by Maasoumi (1981) dealing with generic reduced forms that allow for reduced-form estimation in the light of intrinsically uncertain structural information is also pertinent to this discussion. Nagar pseudo-moment expansions for 3SLS reduced-form coefficients have also been developed in Maasoumi (1977).

where

$$\lambda = \begin{cases} 1 & \text{if } \phi \leq C_\alpha, \\ \left(\dfrac{C_\alpha}{\phi}\right)^{1/2} \quad \text{or} \quad \left(\dfrac{C_\alpha}{\phi}\right) & \text{if } \phi > C_\alpha. \end{cases} \tag{3.69}$$

The weight coefficient λ depends on the outcome of a test of the overidentifying restrictions based on the statistic [see, for example, Malinvaud (1980, p. 378)]:

$$\phi = \operatorname{tr}\{W^{-1}(\Pi_{\text{OLS}} - \Pi_{\text{3SLS}})'Z'Z(\Pi_{\text{OLS}} - \Pi_{\text{3SLS}})\}, \tag{3.70}$$

where $W = T^{-1}Y'(I - P_Z)Y$ is the usual consistent estimator of the reduced-form error covariance matrix and C_α in (3.69) is the (asymptotic) critical value of the test corresponding to a chosen significance level α.

The combined estimator Π^* can be additionally motivated by appealing to the principles of Stein–James estimation,[26] mixed regression,[27] and minimum expected loss[28] (MELO) methodologies, all of which lead to estimators which can be expressed as combinations of restricted and unrestricted estimators. To highlight the similarity of (3.68) with Stein-like procedures [and in particular the positive rule estimator proposed by Sclove (1968, 1971)] we may write Π^* as

$$\Pi^* = \Pi_{\text{3SLS}} + I_{(C_\alpha,\infty)}\left(1 - \frac{C_\alpha}{\phi}\right)(\Pi_{\text{OLS}} - \Pi_{\text{3SLS}}), \tag{3.71}$$

where $I_{(\cdot)}$ is an indicator function equal to unity when ϕ is in the indicated range and equal to zero otherwise. This estimator differs from the traditional Stein-like variety in that it takes the unrestricted estimate Π_{OLS} as the point of attraction rather than simply the origin.

The finite sample and asymptotic properties of the combined estimator Π^* are investigated in Maasoumi (1978).[29] It is shown that Π^* has finite integral moments to the order $T-n-K$ (as for the FIML reduced-form discussed earlier in this section) and that the limiting distribution of $\sqrt{T}(\Pi^* - \Pi)$ is close to that of $\sqrt{T}(\Pi_{\text{3SLS}} - \Pi)$ for conventional choices of the significance level C_α. Thus, Π^* has close to asymptotic equivalence with Π_{3SLS} and has apparently superior small sample properties in terms of outlier elimination. Practical implementation of the method is as straightforward as 3SLS. What remains problematic is the selection

[26]See, for example, James and Stein (1961), Zellner and Vandaele (1975), and Chapter 10 in this Handbook by Judge and Bock.
[27]See Goldberger (1973).
[28]See Zellner (1978) and Zellner and Park (1979).
[29]The finite sample properties of Stein-like improved estimators in the context of the linear regression model have been studied by Ullah (1974, 1980).

of the critical level, C_α. The statistic ϕ in (3.70) has a limiting χ^2_N distribution, where N is the total number of overidentifying restrictions. Even in moderately sized models, N may be quite large and strict application of the test based on (3.70) at conventional significance levels usually leads to a rejection of the restrictions. Thus, frequent occurrence of $\Pi^* = \Pi_{\text{OLS}}$ in practical situations might be expected and this might raise the very genuine objection to the combined estimator that it will frequently result in the extreme alternative of unrestricted reduced-form estimation by OLS. This criticism should be tempered by the knowledge that the critical value, C_α, will often be a very poor (asymptotically based) indicator of the correct finite sample critical value for a test with a chosen size of α. Monte Carlo results by Basmann (1960), Byron (1974), Maddala (1974), Basmann, Richardson and Rohr (1974), Maasoumi (1977), Laitinen (1978), Meisner (1979), Hausman and McFadden (1981), and Rhodes and Westbrook (1982) all indicate that many conventional asymptotic tests of restrictions lead to an unduly high rate of rejection (that is often severe) in small sample situations. This evidence suggests that conventional asymptotic tests are often not sufficiently reliable to justify the extreme alternative of completely unrestricted reduced-form estimation. It would therefore seem wise in the light of this evidence to set the size of the test at a level much lower than usual so that the implied (asymptotic) critical value, C_α, is larger and the probability of a test rejection reduced. The problem of the most appropriate selection of C_α for a given model, data set, and limited knowledge about the exact distribution of ϕ clearly warrants substantially more attention than it has received. Mechanical correctors to the asymptotic critical region (C_α, ∞) can be based on Edgeworth expansions along the lines of Section 2.3 and this is an area of extensive current research in mathematical statistics. However, little is known at present concerning the adequacy of such corrections.

In addition to the above work on reduced forms, attention has also been given in the literature to the partially restricted reduced-form (PRRF) suggested by Amemiya (1966) and Kakwani and Court (1972). The PRRF coefficients can be obtained equation by equation from relationships such as

$$\begin{bmatrix} \pi_{11} \\ \pi_{21} \end{bmatrix} = \begin{bmatrix} \Pi_{12} \\ \Pi_{22} \end{bmatrix} \beta + \begin{bmatrix} \gamma \\ 0 \end{bmatrix} = \left[\begin{array}{c|c} \pi_{12} & I \\ \hline \pi_{22} & 0 \end{array}\right] \begin{bmatrix} \beta \\ \gamma \end{bmatrix} \tag{3.72}$$

[deduced from (3.6) above] which relate the reduced-form coefficients of one (relevant) equation to those of other equations in terms of the identifying restrictions. The PRRF estimator of the coefficients in the first reduced-form equation [given by the left-hand side of (3.72)] is then

$$\begin{bmatrix} \hat{\pi}_{11} \\ \hat{\pi}_{21} \end{bmatrix} = \left[(Z'Z)^{-1}Z'Y \,\middle|\, \begin{array}{c} I \\ \hline 0 \end{array} \right] \begin{bmatrix} \beta_{2\text{SLS}} \\ \gamma_{2\text{SLS}} \end{bmatrix}, \tag{3.73}$$

where $(Z'Z)^{-1}Z'Y_2$ is the OLS estimator of the coefficients in the reduced-form equations for the variables Y_2 appearing in the structural equation (3.3), and β_{2SLS} and γ_{2SLS} are the usual structural 2SLS estimators. The small sample properties of the estimator (3.73) and associated forecasts have been studied by Knight (1977), Nagar and Sahay (1978), Swamy and Mehta (1980), and Sahay (1981). Knight proved that this reduced-form estimator has finite moments of all orders. Nagar and Sahay found expressions for the exact bias and mean squared error of forecasts based on (3.73) in the two endogenous variable case; and Sahay (1981) has extended part of this work to the case of an equation with three endogenous variables. Their conclusions suggest that the mean squared error of forecasts based on (3.73) will be smaller than that based on unrestricted reduced-form estimation by OLS unless $\|\beta\|$ is large. These authors work with the standardized model studied in Section 3.3. Their conclusion accords with the intuitive ideas discussed in the paragraph following (3.32) that when β has large elements 2SLS estimators may display considerable dispersion in view of the correlation between the included endogenous variables and the structural error [which in the standardized model is measured by $-\beta(1+\beta'\beta)^{-1/2}$ as in (3.32)].

3.10. *Improved estimation of structural coefficients*

The idea of developing alternative estimators which may improve on certain aspects of the small behavior behavior of the more conventional estimators has been applied to structural as well as reduced-form coefficient estimation. Here, many of the ideas are based on simple corrections to the usual formulae that are suggested by an analysis of the higher order terms [frequently terms up to $O(T^{-1})$] in series expansions of the distributions and (pseudo-) moments of the conventional estimators.[30] While these corrections are for the most part mechanical, there is an accumulating body of evidence which supports the view that their use will result in actual improvements in estimation as measured by the probability of concentration about true parameter values. In this respect, Morimune (1981) has provided a useful review of the performance characteristics of the main improved estimators.

One of the first suggestions appears in Nagar (1959) who provided (pseudo-) moment expansions for the first two moments of the k-class estimator and deduced values of k which removed the bias to $O(T^{-1})$ and minimized the determinant of the mean squared error moment matrix to $O(T^{-2})$. Zellner (1978) and Zellner and Park (1979) also developed an optimal member of the k-class

[30]A similar analysis of higher order terms (in Edgeworth expansions) of the distributions of conventional test statistics can be performed. Much work has already been done on this topic in mathematical statistics leading to some general results on the higher order efficiency of tests based on maximum likelihood estimators. See, for example, Pfanzagl and Wefelmeyer (1978, 1979).

family in terms of the minimum expected loss (MELO) criterion whereby the posterior expectation of a weighted quadratic loss function is minimized with respect to the structural coefficients. Both Nagar and Zellner–Park reported applications of their improved estimators in the context of small macroeconometric models. Zellner and Park found in their application that the (asymptotic) standard errors of the MELO estimates were consistently smaller and often much smaller than their 2SLS counterparts.

Alternative estimators constructed by taking linear combinations of 2SLS with OLS and 2SLS with LIML were proposed by Sawa (1973a, 1973b) and Morimune (1978),[31] respectively. The weights in these combined estimators were selected so as to remove the bias (or pseudo-bias when this is appropriate) in the estimator up to terms of $O(\sigma^2)$, where σ^2 is a scalar multiple of the covariance matrix of the errors in the model. That is, the improvements were based on the use of small-σ asymptotic expansions (see footnote 5 in Section 2). Sawa (1973b) numerically computed the first two exact moments of the combined 2SLS–OLS estimator but no clear conclusion concerning its superiority over 2SLS emerged from these computations. Morimune (1978) examined the (asymptotic) mean squared error[32] of the 2SLS–LIML combination and demonstrated its superiority over LIML according to this criterion. In the context of systems estimation related work has been done by Maasoumi (1980) on a ridge-like modification to the 3SLS estimator.

Fuller (1977) introduced modifications to the LIML and fixed k-class estimators which ensure that the new estimators possess finite moments. The modifications add weight to the denominators in the matrix ratios that define the unmodified estimators. Their generic form, in the notation of Section 3.2 above, is as follows:

$$\beta_M = \left[A_{22}(I) - l\{A_{22}(I) - A_{22}(P_Z)\}\right]^{-1}\left[a_{21}(I) - l\{a_{21}(I) - a_{21}(P_Z)\}\right], \tag{3.74}$$

where

$$l = a\lambda + b, \qquad a, b = \text{const.}, \tag{3.75}$$

and λ is the smallest latent root of the equation:

$$\det\left[A(I) - \lambda\{A(I) - A(P_Z)\}\right] = 0 \tag{3.76}$$

as in (3.19) above. The estimator β_M in (3.74) specializes: to LIML for $a = 1$,

[31]See also Morimune and Kunitomo (1980).

[32]That is, the mean squared error of the asymptotic expansion of the distribution up to a certain order.

$b=0$; to the fixed k-class for $a=0$, $b=\text{const.}$; to 2SLS for $a=0$, $b=1$; and to OLS for $a=0$, $b=0$.

Fuller's proposal is to set $a=1$ and $b=-\alpha/(T-K)$ for some fixed real number $\alpha>0$. For these values of a and b in (3.75), β_M is a direct modification of β_{LIML} [compare (3.20) above]. Fuller shows that when estimators of the class (3.74) are restricted to have the same bias to $O(T^{-1})$ the modified LIML estimator dominates the class according to mean squared error through to terms of $O(T^{-2})$. This result can be regarded in the light of the second-order efficiency of the maximum likelihood estimator, as has been mentioned by Rothenberg (1980). Asymptotic expansions of the densities of estimators in the class (3.74) have been explicitly derived and compared by Kunitomo (1981) who gives the same conclusion concerning the second-order efficiency of LIML. Additional small sample comparisons of estimators in the class (3.74) have been made by Morimune (1981) who suggests a variant of Fuller's modified LIML that displays superior concentration when the degree of equation overidentification is large. The numerical computations in this paper report the empirical distribution functions of the various estimators from Monte Carlo experiments with extensive replications. They indicate that the modifications to LIML thin out the tails of the distribution, as we would expect. They also confirm results recorded above (in Section 3.6) concerning the good locational properties of the LIML estimator and evaluate the performance of some new asymptotic approximations based on a large degree of equation overidentification.

3.11. Supplementary results on moments

In Sections 3.4–3.9 questions of existence of moments were dealt with in our discussion of the respective exact distributions. In most of these cases, direct formulae for the moments that exist can be extracted using term-by-term integration of the series expressions for the p.d.f.s. Direct results concerning the existence of moments and formulae for them have also appeared in the literature. The following result, which was first established in its complete form by Kinal (1980), encapsulates our knowledge concerning the existence of moments for k-class estimators.

Theorem 3.10.1 (Existence of moments for k-class estimators, $0\leq k\leq 1$)

Integral moments of the estimator $\beta_{(k)}$ given by (3.17) for non-stochastic k in the structural equation (3.3) exist up to order M where

$$M=\begin{cases} T-K_1-n & \text{for} \quad 0\leq k<1, \\ K_2-n & \text{for} \quad k=1. \end{cases} \qquad \square$$

Earlier work was done by Mariano (1973) who covered the 2SLS ($k = 1$) case for even-order moments and by Hatanaka (1973) who gave sufficient conditions for existence. Sawa (1972) dealt with the two endogenous variable case, established the above result, and further demonstrated that $\beta_{(k)}$ has no integral moments when $k > 1$. Sawa also gave exact formulae for the first two moments when $0 \leq k \leq 1$ and developed asymptotic expansions for them in terms of the reciprocal of the concentration parameter, namely $1/\mu^2$. Similar formulae were derived by Takeuchi (1970) for OLS, 2SLS, and IV estimators in the two endogenous variable case.[33] Ullah and Nagar (1974) gave analytic formulae for the mean of the 2SLS estimator and their results were used by Sahay (1979) in finding an expression for the mean of the 2SLS structural equation residuals. Extending this work to the general single equation case (with n and K_2 arbitrary), Hillier and Srivastava (1981) and Kinal (1982) have derived exact formulae for the bias and mean squared error of the OLS and 2SLS estimator of a single endogenous variable coefficient. This generalizes the work of Sawa (1972). Unfortunately, the presence of zonal-type polynomials in the final formulae prevents their use for numerical computations in the general single equation case, at least with present-day tabulations and algorithmic machinery (see the discussion of this point in Section 3.5 above).

Before leaving this topic it may be worth mentioning that moments are useful to the extent that they shed light on the distribution itself. In particular, they provide summary information about the location, dispersion, and shape of a distribution. However, as many of the cases that are analyzed in Sections 3.5 and 3.6 attest, an important feature of many exact distributions in econometrics is their asymmetry. Obviously, moment analyses of higher order than the second are necessary to inform on such aspects of the shape of a distribution. In some cases, of course, such higher order moments may not exist. When they do, the formulae will often be as complicated as the series expressions for the p.d.f.s themselves. Considerations of research strategy therefore indicate that it may well be wise to direct most of our attention to the distributions, their numerical evaluation, and their approximation rather than that of the moments.[34]

Finally, we remark that direct results concerning the moments of estimated coefficients of the exogenous variables in a structural equation can be deduced from the relevant formulae given in Section 3.3 above and the results for the coefficients of the included endogenous variables. Thus, in the case of the IV

[33]As reported by Maasoumi and Phillips (1982a) there appear to be errors in his expression arising out of his formulae (2-7) and (2-8) which appear to confuse even- and odd-order moments.

[34]The issues raised in this section have an obvious bearing on Monte Carlo experimentation, where it is customary to work with summary measures defined in terms of low order moments. Caution in the use of such methods has been advised by several authors, for example Basmann (1961) and Maasoumi and Phillips (1982a). Problems of accurately estimating high order moments by Monte Carlo replications (and the demands this may place on the experimental design) are apposite here but seem not to have been discussed in the literature in this field.

estimator we have

$$\gamma_{IV} = T^{-1}Z_1'y_1 - T^{-1}Z_1'Y_2\beta_{IV} \tag{3.77}$$

(assuming standardizing transformations are performed). Then $Z_1'X = Z_1'[y_1 \vdots Y_2]$ is statistically independent of β_{IV} (under error normality of course) and moments of γ_{IV} in (3.77) are defined to the same order as those of β_{IV} (see Section 3.5). Formulae for these moments can also be deduced from those that apply for β_{IV} and $Z_1'X$. Similar considerations enable us to treat the general k-class estimator of the exogenous variable coefficients.

3.12. *Misspecification*

Earlier results in this section have all been obtained on the presumption that the model has been correctly specified. When this is not so, the sampling distributions undergo modifications contingent upon the nature and extent of misspecification and earlier conclusions about estimator and test statistic performance in small samples no longer necessarily apply. Fisher (1961, 1966, 1967) carried out an asymptotic analysis of estimator performance in the presence of specification error consisting of incorrectly excluded variables in structural equations such as (3.3). An exact small sample theory can also be developed for this problem using the approach of Sections 3.5 and 3.6. We illustrate by considering OLS and 2SLS estimation of the (incorrectly specified) structural equation (3.3) when the true equation includes additional exogenous variables and is of the form

$$y_1 = Y_2\beta + Z_1\gamma_1 + Z_4\gamma_4 + u. \tag{3.78}$$

We write the reduced form as

$$\begin{aligned} \left[y_1 \vdots Y_2\right] = X &= \left[Z_1 \vdots Z_4 \vdots Z_5\right]\begin{bmatrix} \pi_{11} & \Pi_{12} \\ \pi_{41} & \Pi_{42} \\ \pi_{51} & \Pi_{52} \end{bmatrix} + \left[v_1 \vdots V_2\right] \\ &= \left[Z_1 \vdots Z_4 \vdots Z_5\right]\begin{bmatrix} \Pi_1 \\ \cdots \\ \Pi_4 \\ \cdots \\ \Pi_5 \end{bmatrix} + V, \end{aligned} \tag{3.79}$$

where the coefficients satisfy

$$\pi_{11} - \Pi_{12}\beta = \gamma_1, \qquad \pi_{41} - \Pi_{42}\beta = \gamma_4, \qquad \pi_{51} - \Pi_{52}\beta = 0. \tag{3.80}$$

We define $Z_2 = [Z_4 \vdots Z_5]$ and then

$$M = E\left[T^{-1/2}X'Z_2\right] = T^{-1/2}\Pi'Z'Z_2 = T^{1/2}\left[\Pi_4' \vdots \Pi_5'\right]. \tag{3.81}$$

The non-centrality parameter matrix is

$$MM' = T\left[\Pi_4'\Pi_4 + \Pi_5'\Pi_5\right] = TD, \quad \text{say.} \tag{3.82}$$

We may note that this reduces to

$$\begin{bmatrix}\beta' \\ I\end{bmatrix}(\Pi_{42}'\Pi_{42} + \Pi_{52}'\Pi_{52})\left[\beta \vdots I\right] = \begin{bmatrix}\beta' \\ I\end{bmatrix}\Pi_{22}'\Pi_{22}\left[\beta \vdots I\right] \tag{3.83}$$

when $\gamma_4 = 0$ and the eq. (3.3) is correctly specified.

As in Section 3.5, OLS and 2SLS estimators of β in (3.78) depend on the matrix

$$A = T^{-1}X'FF'X, \tag{3.84}$$

with $FF' = Z_2Z_2'$ in the case of 2SLS and $FF' = I - T^{-1}Z_1Z_1'$ for OLS, where F is a $T \times f$ matrix of rank f and $F'F = TI_f$. Formulae for the exact densities of β_{OLS} and β_{2SLS} in the general case are then obtained by arguments which closely follow those of Section 3.5, as shown by Maasoumi and Phillips (1982b). For the two endogenous variable case we obtain [see Phillips (1982e) for derivations]:

$$\begin{aligned}\text{pdf}(r) = {} & \frac{\operatorname{etr}\left(\frac{-TD}{2}\right)}{B\left(\frac{1}{2},\frac{f}{2}\right)(1+r^2)^{(f+1)/2}} \sum_{k=0}^{\infty} \frac{\left\{\det\left(\frac{TD}{2}\right)\right\}^k \left(\frac{f+1}{2}\right)_k}{\left(\frac{f}{2}\right)_{2k} k!} \\ & \times \sum_{h=0}^{\infty} \frac{\left(\frac{T}{2}\right)^h \left(\frac{f+1}{2}+k\right)_h \left(r^2 d_{11} + 2rd_{12} + d_{22}\right)^h}{\left(\frac{f}{2}+2k\right)_h h!(1+r^2)^{k+h}} \\ & \times {}_1F_1\left(\frac{f-1}{2}+k,\ \frac{f}{2}+h+2k;\ \frac{Tdu}{2}\right).\end{aligned} \tag{3.85}$$

This expression gives the exact density under misspecification of β_{OLS} when $f = T - K_1$ and of β_{2SLS} when $f = K_2$. The density reduces to (3.45) when the structural equation is correctly specified ($\gamma_4 = 0$) as can be shown by rearrangement of the series.

Formula (3.85) was derived by Rhodes and Westbrook (1981)[35] and formed the basis of the computational work reported in their paper. These numerical computations provide valuable evidence concerning the practical consequences of misspecification. Two principal results emerge from their study: misspecification can substantially increase the concentration of the distribution of both OLS and 2SLS; and in some cases it may also reduce the bias (as well as the dispersion) of both estimators. These results led Rhodes and Westbrook to conclude that, when a structural equation is misspecified by incorrectly excluded variables, OLS may indeed be a superior technique of estimation to 2SLS.

The same general conclusion was reached by Hale, Mariano and Ramage (1980) who examined exact and approximate asymptotic expressions for the bias and mean squared error (MSE) of k-class estimators (for k non-stochastic in the interval $0 \leq k \leq 1$). Their results, which also refer to the two endogenous variable case, show that OLS is relatively insensitive to specification error and that when errors of specification are a more serious problem than simultaneity, OLS is preferable to 2SLS. Moreover, the entire k-class is dominated in terms of MSE under misspecification by either OLS or 2SLS.

Similar analysis of the effect of misspecification upon the LIML estimator in the two endogenous variable case has been performed by Mariano and Ramage (1978). Some extensions of this work, involving asymptotic expansions and moment approximations to the general single equation case, are contained in Mariano and Ramage (1979). Exact formulae for the p.d.f.s of OLS and 2SLS estimators in the general single equation case under misspecification are given by Maasoumi and Phillips (1982b).[36-39]

[35] Their stated result in theorem 2.1 contains a small error in that $|T_n/4|^k$ in their formula (2.11) should be replaced by $|T_n/2|^k$.

[36] In addition, Knight (1981) has shown how, in the two endogenous variable case, expressions for the exact moments of k-class estimators under misspecification can be extracted from the corresponding expressions that apply in correctly specified situations.

[37] Related work on the effect of multicollinearity on the shape of the distributions of OLS and 2SLS estimators has been done by Mariano, McDonald and Tishler (1979).

[38] Hale (1979) has also studied the effects of misspecification on the two-stage Aitken estimator (2SAE) and OLS estimator in a two-equation seemingly unrelated regression model. Hale's main conclusion is that the distribution of 2SAE appears to be more affected by misspecification than that of OLS.

[39] Analysis of the effects of distributional shape are also possible. Knight (1981) has, in particular, found expressions for the first two exact moments of the k-class estimator in the two endogenous variable case when the reduced-form errors follow a non-normal distribution of the Edgeworth type. Phillips (1980b) indicated generalizations of existing results for asymptotic expansions of coefficient estimators and test statistics under non-normal errors of this type. Explicit formulae for such asymptotic expansions have in fact been derived by Satchell (1981) for the distribution of the serial correlation coefficient.

4. A new approach to small sample theory

4.1. Intuitive ideas

This section outlines the elements of a new approach to small sample theory that is developed in Phillips (1982c). The idea that underlies the method in this article is very simple. It is motivated by the observation that, in spite of the complex analytic forms of many of the exact p.d.f.s presently known for econometric statistics (such as those in Section 3), when we do turn around and obtain numerical tabulations or graphical plots of the densities we typically end up with well-behaved, continuous functions that tend to zero at the limits of their domain of definition. The form of these p.d.f.s strongly suggests that we should be able to get excellent approximations to them in the class of much simpler functions and certainly without the use of multiple infinite series. We need to deal with approximating functions (or approximants as they are often called) that are capable of capturing the stylized form of a density: in particular, we want the approximant to be able to go straight for long periods in a direction almost parallel to the horizontal axis and yet still be able to bend, quite sharply if necessary, to trace out the body of the distribution wherever it is located. One class of functions that seems particularly promising in this respect, as well as being simple in form, are rational functions. Even low degree rational functions can go straight for long periods and then bend quite sharply. In this, of course, they are very different from low degree polynomials whose graphs typically display a distinct roly-poly character.

The possibility of finding rational functions which provide good global approximations to a general class of p.d.f.s is considered in Phillips (1982c). The technique developed there is based on the idea of working from local Taylor series approximations at certain points of the distribution towards a global approximation which performs well in the whole domain over which the distribution is defined and yet retains the good performance of the Taylor series approximations in the immediate locality of the points of expansion. This is, in part, achieved by the use of multiple-point Padé approximants. These Padé approximants[40] are rational functions constructed so as to preserve the local Taylor series behavior of the true p.d.f. (or d.f.) to as high an order as possible. The points selected for local expansion will often be simply the origin (in the central body of the distribution) and the tails. These local expansions can, in fact, be obtained from information about the characteristic function of the distribution so that direct knowledge of the local behavior of the true p.d.f. is not necessary for the successful application

[40]Padé approximants have a long tradition in mathematics and have recently been successfully applied to a large number of problems in applied mathematics and mathematical physics. References to this literature may be found in Phillips (1982c).

of the technique. Local information may also be based on estimates obtained from the empirical d.f. arising in Monte Carlo simulations. Supplementary information about the distribution such as (i) its analytic form in leading cases (compare Section 3.4), (ii) knowledge of its moments where these exist, (iii) leading terms in its Edgeworth expansion (see Section 2.3), or even (iv) the crude asymptotic distribution, may all be utilized in the construction of the approximant. The final step in the method is to modify the Padé approximant so that it does display appropriate global behavior. This may involve the removal of unwanted zeros and poles which occur in the bridging region between the points of local expansion and possible modifications to ensure non-negativity in the approximant.

4.2. *Rational approximation*

Phillips (1982c) considers a general class of continuous marginal densities and defines the following family of potential rational function approximants:

$$R_{m,n}(r) = s(r)\frac{P_m(r)}{Q_n(r)} = s(r)\frac{a_0 + a_1 r + \dots + a_m r^m}{b_0 + b_1 r + \dots + b_n r^n}, \tag{4.1}$$

where m and n are even integers with $m \leq n$ and $s(r)$ is a real continuous function satisfying $s(r) > 0$ and $s(r) \to 0$ as $r \to \pm\infty$.

The coefficient function $s(r)$ in (4.1) is a vehicle by which additional information about the true density can be readily embodied in the approximant. This can be soft quantitative information, for example of the type that $\mathrm{pdf}(r) > 0$ and $\mathrm{pdf}(r) \to 0$ as $r \to \pm\infty$ [already explicit in $s(r)$]; or hard quantitative information, for example of the type (i) that $\mathrm{pdf}(r)$ has moments up to a certain order or (ii) that $\mathrm{pdf}(r)$ takes an especially simple form in an important and relevant leading case or (iii) that $\mathrm{pdf}(r)$ has a known Edgeworth series expansion up to a certain order (suitably modified to ensure that it is everywhere positive and still tends to zero at infinity).

Practical considerations frequently suggest a specialization of (4.1) to the family of rational fractions in which numerator and denominator polynomials have the same degrees (i.e. $m = n$).[41] In addition, a normalization condition is imposed on the coefficients of the polynomials in (4.1) to eliminate the redundancy that results from the multiplication of $P_n(r)$ and $Q_n(r)$ by an arbitrary constant. In density function approximation this can be simply achieved by setting $b_0 = 1$, which also ensures that the rational approximant is well behaved as

[41] This is discussed at some length in Phillips (1982c, 1982d).

r passes through the origin. In distribution function approximation it is convenient to set $a_n = b_n = 1$ and then, after an appropriate choice of $s(r)$ as a primitive distribution, the resulting approximant $R_{nn}(r)$ embodies desirable tail behavior as $r \to \pm\infty$.

A theory of goodness of approximation to continuous p.d.f.s based on rational approximants of the type (4.1) is developed in Phillips (1982c, 1982d). This theory uses the uniform norm

$$\| \operatorname{pdf}(r) - R_{nn}(r) \| = \sup_{r \in (-\infty, \infty)} |\operatorname{pdf}(r) - R_{nn}(r)| \tag{4.2}$$

to measure the error in the approximation. Under this error norm it is shown that best uniform approximants within the family (4.1) exist and are unique for a general class of continuous p.d.f.s. Setting $m = n$ and defining $\gamma' = \{a_0, \dots, a_n; b_1, \dots, b_n\}$ in (4.1) means that there exists a vector γ^* and a corresponding rational fraction $R'_{nn}(r)$ for which

$$\| \operatorname{pdf}(r) - R'_{nn}(r) \| = \inf_{\gamma} \| \operatorname{pdf}(r) - R_{nn}(r) \| \tag{4.3}$$

given some continuous density pdf(r); and, moreover, the rational fraction $R'_{nn}(r)$ with the property (4.3) is unique. As $n \to \infty$ $R'_{nn}(r)$ converges uniformly to pdf(r). Hence, an arbitrarily good approximation is possible within this family of rational functions.

Practical implementation of rational approximation requires the degree of $R_{nn}(r)$ to be prescribed, the coefficient function $s(r)$ to be selected, and the parameters of the polynomials to be specified. The problem is one of constructive functional approximation to a given distribution within the family of approximants (4.1) Operational guidelines for this constructive process are laid out in Phillips (1982c) and the final solution in any particular case will rely intimately on the information that is available about the true distribution. Typically, we will want the approximant to embody as much analytic and reliable experimental information about the distribution as possible. This will directly affect the choice of $s(r)$ and the prescribed degree of $R_{nn}(r)$. Leading case analyses such as those in Section 3.4 will often lead to a suitable choice of $s(r)$. Knowledge of the local behavior of the distribution in the body and in the tails can be used to determine the polynomial coefficients in $R_{nn}(r)$ which will then magnify or attenuate as appropriate the leading case distribution. Local information about the distribution may take the form of Taylor expansions at certain points or estimates of the function values obtained from Monte Carlo simulations. In cases where numerical or Monte Carlo integration is possible, a selected set of points within the main body and in the tails of the distribution can be used for these evaluations, which can then assist in determining the parameters of $R_{nn}(r)$. This has the advantage of keeping the number of numerical integrations within economic limits and at the same time marrying the information from these integrations with other useful knowledge about the distribution. Constructive functional approximants of this type will have Bayesian applications in the problem of reducing multidimensional posterior distributions to manageable and readily interpreted marginal posteriors. They also provide a convenient avenue for combining Monte Carlo experimental evidence and analytic knowledge in a simple useable form.

The procedure outlined above was successfully applied in the context of approximating the exact density (3.45) of the 2SLS estimator by the rational family $R_{nm}(r)$ in (4.8) with $n = m = 4$ and $s(r)$ set equal to the leading density given in (3.38). The results obtained were very encouraging even for extremely small values of the concentration parameter μ^2, when other approximations such as those based on Edgeworth and saddlepoint methods produced very poor results. In particular, the modified two-point Padé approximant (using information at only the origin and infinity) yielded two decimal place accuracy to the exact distribution over the whole real axis and an unmodified seven-point Padé approximant gave three decimal place accuracy with a maximum error of 0.0008. Further development of these approximants seems likely on the basis of these results to produce a method that is flexible and general enough to be widely used and sufficiently accurate to be relied upon in empirical work.

4.3. *Curve fitting or constructive functional approximation?*

The above discussion and reported application present a favorable picture of the strengths and potential of this new approach. An important contributory factor in this optimistic view is the flexible mathematical apparatus that underlies constructive functional approximation in the class defined by (4.1). As much analytic knowledge as is available about a distribution can be embodied in $R_{mn}(r)$ through the dual vehicles of the coefficient function $s(r)$ and the rational coefficients $\{a_0,\dots,a_m, b_1,\dots,b_n\}$. Thus, Edgeworth expansions and saddlepoint approximations are just subcases of (4.1). For if these expansions are known to yield good approximants in certain problems they themselves may be used to construct $s(r)$. Simple modifications to the Edgeworth expansion will ensure that $s(r)$ is everywhere positive, continuous, and still tends to zero as $|r| \to \infty$. Additional information about the distribution can then be incorporated in the rational coefficients and in adjustments to $s(r)$ that ensure the same tail behavior as the true distribution, where this is known by separate analytic investigation. Other choices of $s(r)$ that stem directly from analytic knowledge of the true distribution are also possible, as the example cited demonstrates. Moreover, experimental data about the distribution can be utilized in the choice of the rational coefficients by least squares or generalized least squares fitting to the empirical distribution in place of (or in addition to) exact solutions. Thus, constructive approximants belonging to the family (4.1) can successfully embrace a wide range of different types of analytic and experimental information in a functional form that is useful for both descriptive and inferential purposes.

In a more limited mode of thinking, rational approximation *per se* is an exercise in interpolation or curve fitting. Of course, rational approximants do provide an excellent medium for such exercises, as the extensive examples in Hastings (1955) amply demonstrate, and they are already a standard technique in computer function algorithms because of their accuracy and economy [see Hart (1968)]. But such an interpretation in the present context would be myopic, ignoring as it does the extent to which the family (4.1) can build on the strengths of alternative, less flexible approximation methods and simultaneously blend analytic and experimental information from many diverse sources. It is this constructive mode of analysis that underlies the new approach and distinguishes it from more mechanical methods of asymptotic analysis and exercises in curve fitting.

5. Concluding remarks

This review began with some remarks taken from the first edition of R. A. Fisher's (1925) influential manual for practising statisticians. Fisher's keen awareness of the limitations of asymptotic theory, his emphasis on statistical tools which are appropriate in the analysis of small samples of data, and his own research on the exact sampling distributions of variance ratios and correlation coefficients contributed in significant ways to the growth of what is now an extensive literature in mathematical statistics on small sample distribution theory. The challenge of developing such a theory in models that are of interest to econometricians has produced the corpus of knowledge that forms the subject-matter of this review. Questions of the relevance of this research and its operational payoff in terms of empirical practice are as much a topic of debate in econometrics as they were (and still are to a lesser degree) in mathematical statistics.

In contrast to small sample theory, the power of asymptotic theory lies unmistakedly in the generality with which its conclusions hold, extending over a wide domain of models and assumptions that now allow for very general forms of dependent random processes, non-linear functional forms, and model misspecifications. However, the generality of this theory and the apparent robustness of many of its conclusions should not necessarily be presumed to be strengths. For the process by which asymptotic machinery works inevitably washes out sensitivities that are present and important in finite samples. Thus, generality and robustness in asymptotic theory are achieved at the price of insensitivity with respect to such ingredients as the distributional characteristics of a model's random elements and the values of many of its parameters. These ingredients do influence finite sample behaviour, often in vital ways (as the numerical work reported in Sections 3.5 and 3.6 substantiates). But their effects fade out in very large samples and are totally lost in an asymptotic theory.

There is a second major weakness in the operation of traditional asymptotic machinery. In econometrics, models are most frequently interpreted as approximate data-generating mechanisms that are useful to the extent that they assist in the explanation of observed data, in making predictions, and in other relevant empirical goals. In such exercises, the fact that an investigator can marshall only a finite sample of data is itself a critical factor. For, in the absence of experimentation, the design of a model for empirical use is always tailored by the data that is available and the extent to which the latter may limit the goals of the investigation. As the size and coverage of the sample increase, there is a natural tendency to model an always more complex phenomenon in increasing degrees of sophistication. This process of sophistication arises from the knowledge that a model may be an adequate descriptive and inferential tool over a certain span of data but may be far less adequate as the data evolves further and as the horizons of the investigation widen. When a model grows in complexity with increases in the availability of data, traditional asymptotic methods of statistical analysis inevitably become self-defeating. To extend Fisher's own metaphor, by the time the cannon has rumbled into place, and been loaded and sighted to fire, the sparrow will have flown away.[42]

The central issue is, and will remain, how relevant asymptotic results are in the context of a given model, data set, and plausible hypotheses concerning the stochastic environment. Refinements of asymptotic theory such as those in Section 2.3 can shed light on this issue but are far from conclusive. The ultimate criterion of evaluation will inevitably be the relevant sampling distributions themselves. The recent advancements we have reviewed in the mathematical task of representing the analytic form of these distributions already play an important role in the evaluation of asymptotic theory as the results of Section 3 demonstrate. Moreover, these exact mathematical results and the construction of reliable functional approximants to them, such as those given in Section 4, suggest an alternative basis for estimator choice and a means by which the accuracy of inferential procedures may be improved. It is the challenge of future researchers to extend these results to more sophisticated models and to transmit the machinery in an operational form to the practitioner by the development of appropriate computer software.

[42]Some further reflections on the problems inherent in asymptotic theory are given in Phillips (1982b).

References

Amemiya, T. (1966) "On the Use of Principal Components of Independent Variables in Two-Stage Least-Squares Estimation", *International Economic Review*, 7, 283–303.

Anderson, T. W. (1976) "Estimation of Linear Functional Relationships: Approximate Distribution and Connections with Simultaneous Equations in Econometrics", *Journal of the Royal Statistical Society*, B38, 1–36.

Anderson, T. W. (1977) "Asymptotic Expansions of the Distributions of Estimates in Simultaneous Equations for Alternative Parameter Sequences", *Econometrica*, 45, 509–518.

Anderson, T. W. (1982) "Some Recent Developments on the Distributions of Single-Equation Estimators", in: W. Hildebrand (ed.), *Advances in Econometrics*. Amsterdam: North-Holland Publishing Co. (forthcoming).

Anderson, T. W., N. Kunitomo and T. Sawa (1980) "Evaluation of the Distribution Function of the Limited Information Maximum Likelihood Estimator", Technical Report No. 319, The Economic Series, Stanford University.

Anderson, T. W., K. Morimune and T. Sawa (1978) "The Numerical Values of Some Key Parameters in Econometric Models", Stanford University, IMSSS TR no. 270.

Anderson, T. W. and H. Rubin (1949), "Estimation of the Parameters of a Single Equation in a Complete System of Stochastic Equations", *Annals of Mathematical Statistics*, 20, 46–63.

Anderson, T. W. and H. Rubin (1950), "The Asymptotic Properties of Estimates of the Parameters of a Single Equation in a Complete System of Stochastic Equations", *Annals of Mathematical Statistics*, 21, 570–582.

Anderson, T. W. and T. Sawa (1973) "Distributions of Estimates of Coefficients of a Single Equation in a Simultaneous System and Their Asymptotic Expansions", *Econometrica*, 41, 683–714.

Anderson, T. W. and T. Sawa (1975) "Distribution of a Maximum Likelihood Estimate of a Slope Coefficient: The LIML Estimate for Known Covariance Matrix", Technical Report no. 174, IMSSS, Stanford University.

Anderson, T. W. and T. Sawa (1977) "Numerical Evaluation of the Exact and Approximate Distribution Functions of the Two Stage Least Squares Estimate", Stanford Economic Series Technical Report no. 239.

Anderson, T. W. and T. Sawa (1979) "Evaluation of the Distribution Function of the Two-Stage Least Squares Estimate", *Econometrica*, 47, 163–182.

Bartlett, M. S. (1948) "A Note on the Statistical Estimation of Demand and Supply Relations from Time Series", *Econometrica*, 16, 323–329.

Barndorff-Nielson, O. and D. R. Cox (1979) "Edgeworth and Saddle-point Approximations with Statistical Applications", *Journal of the Royal Statistical Society*, Ser. B, 41, 279–312.

Basmann, R. L. (1960) "On Finite Sample Distributions of Generalized Classical Linear Identifiability Test Statistics", *Journal of the American Statistical Association*, 55, 650–659.

Basmann, R. L. (1961) "Note on the Exact Finite Sample Frequency Functions of Generalized Classical Linear Estimators in Two Leading Overidentified Cases", *Journal of the American Statistical Association*, 56, 619–636.

Basmann, R. L. (1963) "A Note on the Exact Finite Sample Frequency Functions of Generalized Classical Linear Estimators in a Leading Three Equation Case", *Journal of the American Statistical Association*, 58, 161–171.

Basmann, R. L. (1965) "On the Application of the Identifiability Test Statistic in Predictive Testing of Explanatory Economic Models", *Econometric Annual of the Indian Economic Journal*, 13, 387–423.

Basmann, R. L. (1974) "Exact Finite Sample Distribution for Some Econometric Estimators and Test Statistics: A Survey and Appraisal", M. D. Intriligator and D. A. Kendricks (eds.), in: *Frontiers of Quantitative Economics*, vol. 2. Amsterdam: North-Holland Publishing Co., ch. 4, pp. 209–271.

Basmann, R. L. and R. H. Richardson (1969) "The Finite Sample Distribution of a Structural Variance Estimator", Research Papers in Theoretical and Applied Economics, 24, University of Kansas, mimeographed.

Basmann, R. L. and D. H. Richardson (1973), "The Exact Finite Sample Distribution of a Non-Consistent Structural Variance Estimator", *Econometrica*, 41, 41–58.

Basmann, R. L., D. H. Richardson and R. J. Rohr (1974) "An Experimental Study of Structural Estimators and Test Statistics Associated with Dynamical Econometric Models", *Econometrica*, 42, 717–730.

Bergstrom, A. R. (1962) "The Exact Sampling Distributions of Least Squares and Maximum Likelihood Estimators of the Marginal Propensity to Consume", *Econometrica*, 30, 480–490.

Bhattacharya, R. N. and R. R. Rao (1976) *Normal Approximation and Asymptotic Expansions*. New York: John Wiley & Sons.

Bleistein, N. and R. A. Handelsman (1976) *Asymptotic Expansions of Integrals*. New York: Holt, Rinehart and Winston.

Byron, R. P. (1974) "Testing Structural Specification Using the Unrestricted Reduced Form", *Econometrica*, 42, 869–883.

Chikuse, Y. (1981) "Invariant Polynomials with Matrix Arguments and Their Applications", in: R. P. Gupta (ed.), *Multivariate Statistical Analysis*. Amsterdam: North-Holland Publishing Co.

Constantine, A. G. (1963) "Some Noncentral Distribution Problems in Multivariate Analysis", *Annals of Mathematical Statistics*, 34, 1270–1285.

Cramér, H. (1946) *Mathematical Methods of Statistics*. Princeton: Princeton University Press.

Cramér, H. (1972) "On the History of Certain Expansions Used in Mathematical Statistics", *Biometrika*, 59, 204–207.

Daniels, H. E. (1954) "Saddlepoint Approximations in Statistics", *Annals of Mathematical Statistics*, 25, 631–650.

Daniels, H. E. (1956) "The Approximate Distribution of Serial Correlation Coefficients", *Biometrika*, 43, 169–185.

Daniels, H. E. (1980) "Exact Saddlepoint Approximations", *Biometrika*, 67, 59–63.

Davies, R. B. (1973) "Numerical Inversion of a Characteristic Function", *Biometrika*, 60, 415–417.

Davis, A. W. (1980a) "Invariant Polynomials with Two Matrix Arguments Extending the Zonal Polynomials", in: P. R. Krishnaiah (ed.), *Multivariate Analysis*. Amsterdam: North-Holland Publishing Co.

Davis, A. W. (1980b) "Invariant Polynomials with Two Matrix Arguments Extending the Zonal Polynomials: Applications to Multivariate Distribution Theory", *Annals of the Institute of Statistical Mathematics*, forthcoming.

De Bruijn, N. G. (1958) *Asymptotic Methods in Analysis*. Amsterdam: North-Holland Publishing Co.

Dhrymes, P. J. (1969) "Alternative Asymptotic Tests of Significance and Related Aspects of 2SLS and 3SLS Estimated Parameters", *Review of Economic Studies*, 36, 213–226.

Drymes, P. J. (1973) "Restricted and Unrestricted Reduced Froms: Asymptotic Distribution and Relative Efficiency", *Econometrica*, 41, 119–134.

Durbin, J. (1980a) "Approximations for Densities of Sufficient Estimators", *Biometrika*, 67, 311–333.

Durbin, J. (1980b) "The Approximate Distribution of Partial Serial Correlation Coefficients Calculated from Residuals from Regression on Fourier Series ", *Biometrika*, 67, 335–349.

Durbin, J. and G. S. Watson (1971) "Testing for Serial Correlation in Least Squares Regression, III", *Biometrika*, 58, 1–19.

Ebbeler, D. H. and J. B. McDonald (1973) "An Analysis of the Properties of the Exact Finite Sample Distribution of a Nonconsistent GCL Structural Variance Estimator", *Econometrica*, 41, 59–65.

Erdéyli, A. (1953) *Higher Transcendental Functions*, vol. 1. New York: McGraw-Hill.

Feller, W. (1971) *An Introduction to Probability Theory and Its Applications*, vol. II. Wiley: New York.

Fieiller, E. C. (1932) "The Distribution of the Index in a Normal Bivariate Population", *Biometrika*, 24, 428–440.

Fisher, F. M. (1961) "On the Cost of Approximate Specification in Simultaneous Equation Estimation", *Econometrica*, 29, 139–170.

Fisher, F. M. (1966) "The Relative Sensitivity to Specification Error of Different k-Class Estimators", *Journal of the American Statistical Association*, 61, 345–356.

Fisher, F. M. (1967) "Approximate Specification and the Choice of a k-Class Estimator", *Journal of the American Statistical Association*, 62, 1265–1276.

Fisher, R. A. (1921) "On the Probable Error of a Coefficient of Correlation Deduced From a Small Sample", *Metron*, 1, 1–32.

Fisher, R. A. (1922) "The Goodness of Fit of Regression Formulae and the Distribution of Regression Coefficients", *Journal of the Royal Statistical Society*, 85, 597–612.

Fisher, R. A. (1924) "The Distribution of the Partial Correlation Coefficient", *Metron*, 3, 329–332.
Fisher, R. A. (1925) *Statistical Methods for Research Workers*. Edinburgh: Oliver and Boyd,
Fisher, R. A. (1928a) "On a Distribution Yielding the Error Functions of Several Well Known Statistics", in: *Proceedings of the International Congress of Mathematics*. Toronto, pp. 805–813.
Fisher, R. A. (1928b) "The General Sampling Distribution of the Multiple Correlation Coefficient", *Proceedings of the Royal Statistical Society*, 121, 654–673.
Fisher, R. A. (1935) "The Mathematical Distributions Used in the Common Tests of Significance", *Econometrica*, 3, 353–365.
Fuller, W. A. (1977) "Some Properties of a Modification of the Limited Information Estimator", *Econometrica*, 45, 939–953.
Gil-Pelaez, J. (1951) "Note on the Inversion Theorem", *Biometrika*, 38, 481–482.
Goldberger, A. (1973) "Efficient Estimation in Overidentified Models: An Interpretive Analysis", in: A. S. Goldberger and O. D. Duncan (eds.), *Structural Equation Models in the Social Sciences*. Seminar Press.
Haavelmo, T. (1947) "Methods of Measuring the Marginal Propensity to Consume ", *Journal of the American Statistical Association*, 42, 105–122.
Hale, C. (1979) "Misspecification in Seemingly Unrelated Regression Equations", Kent University, mimeo.
Hale, C., R. S. Mariano and J. G. Ramage (1980), "Finite Sample Analysis of Misspecification in Simultaneous Equation Models", *Journal of the American Statistical Association*, 75, 418–427.
Hart, J. F. (1968) *Computer Approximations*. New York: John Wiley & Sons.
Hastings, C. (1955) *Approximations for Digital Computers*. Princeton: Princeton University Press.
Hatanaka, M. (1973) "On the Existence and the Approximation Formulae for the Moments of the k-Class Estimators", *The Economic Studies Quarterly*, 24, 1–15.
Hausman, J. A. and D. McFadden (1981) "Specification Tests for the Multinomial Logit Model", Discussion Paper no. 292, Department of Economics, Massachusetts Institute of Technology.
Herz, C. S. (1955) "Bessel Functions of Matrix Argument", *Annals of Mathematics*, 61, 474–523.
Hillier, G. H. and V. R. Srivastava (1981) "The Exact Bias and Mean Square Error of the k-Class Estimators for the Coefficient of an Endogenous Variable in a General Structural Equation", Monash University, mimeo.
Holly, A. and P. C. B. Phillips (1979) "A Saddlepoint Approximation to the Distribution of the k-Class Estimator of a Coefficient in a Simultaneous System", *Econometrica*, 47, 1527–1547.
Hood, Wm. C. and T. C. Koopmans (1953) "The Estimation of Simultaneous Linear Economic Relationships", in: Wm. C. Hood and T. C. Koopmans (eds.), *Studies in Econometric Method*, Cowles Commission. New York: John Wiley & Sons.
Hurwicz, L. (1950) "Least Squares Bias in Time Series", in: T. C. Koopmans (ed.), *Statistical Inference in Dynamic Economic Models*. New York: John Wiley & Sons.
Imhof, J. P. (1961) "Computing the Distribution of Quadratic Forms in Normal Variables", *Biometrika*, 48, 419–426.
Intriligator, M. D. and D. A. Kendrick (eds.) (1974) *Frontiers of Quantitative Economics*, vol. 11. Amsterdam: North-Holland Publishing Co.
James, A. T. (1961) "Zonal Polynomials of the Real Positive Definite Symmetric Matrices", *Annals of Mathematics*, 74, 456–469.
James, A. T. (1964) "Distribution of Matrix Variates and Latent Roots Derived from Normal Samples", *Annals of Mathematical Statistics*, 35, 475.
James, A. T. (1968) "Circulation of Zonal Polynomial Coefficients by Use of the Laplace–Beltrami operator", *Annals of Mathematical Statistics*, 39, 1711–1718.
James, W. and C. Stein (1961) "Estimation with Quadratic Loss", in: *Proceedings of the Fourth Berkeley Symposium on Mathematical Statistics and Probability*, vol. 1. Berkeley: University of California Press, pp. 361–379.
Johnson, N. L. and S. Kotz (1970) *Continuous Univariate Distributions*–2. Boston: Houghton Mifflin.
Johnson, N. L. and S. Kotz (1972) *Distributions in Statistics: Continuous Multivariate Distributions*. New York: John Wiley & Sons.
Kabe, D. G. (1963) "A Note on the Exact Distributions of the GCL Estimators in Two-Leading Overidentified Cases", *Journal of the American Statistical Association*, 58, 535–537.

Kabe, D. G. (1964) "On the Exact Distributions of the GCL Estimators in a Leading Three-Equation Case", *Journal of the American Statistical Association*, 58, 535–537.

Kadane, J. (1971) "Comparison of *k*-Class Estimators When the Disturbances are Small", *Econometrica*, 39, 723–737.

Kakwani, N. C. and R. H. Court (1972) "Reduced-Form Coefficient Estimation and Forecasting from a Simultaneous Equation Model", *Australian Journal of Statistics*, 14, 143–160.

Kataoka, Y. (1974) "The Exact Finite Sample Distributions of Joint Least Squares Estimators for Seemingly Unrelated Regression Equations", *Economic Studies Quarterly*, 25, 36–44.

Kelejian, H. H. (1974) "Random Parameters in a Simultaneous Equation Framework: Identification and Estimation", *Econometrica*, 42, 517–527.

Kinal, T. W. (1980) "The Existence of Moments of *k*-Class Estimators", *Econometrica*, 48, 241–249.

Kinal, T. W. (1982) "On the Comparison of Ordinary and Two Stage Least Squares Estimators", SUNY, Albany, mimeo.

Knight, J. L. (1977) "On the Existence of Moments of the Partially Restricted Reduced-Form Estimators from a Simultaneous-Equation Model", *Journal of Econometrics*, 5, 315–321.

Knight, J. L. (1981) "Non-Normality of Disturbances and the *k*-Class Structural Estimator", School of Economics, The University of New South Wales, mimeo.

Kunitomo, N. (1981) "On a Third Order Optimum Property of the LIML Estimator When the Sample Size is Large", Discussion Paper no. 502, Department of Economics, Northwestern University.

Laitinen, K. (1978) "Why is Demand Homogeneity Rejected so Often?", *Economic Letters*, 1, 187–191.

Lebedev, N. N. (1972) *Special Functions and Their Application*. New York: Dover.

Maasoumi, E. (1977) "A Study of Improved Methods of Estimating Reduced Form Coefficients Based Upon 3SLS", unpublished Ph.D. Thesis, London School of Economics.

Maasoumi, E. (1978) "A Modified Stein-Like Estimator for the Reduced Form Coefficients of Simultaneous Equations", *Econometrica*, 46, 695–703.

Maasoumi, E. (1980) "A Ridge-like Method for Simultaneous Estimation of Simultaneous Equations ", *Journal of Econometrics*, 12, 161–176.

Maasoumi, E. (1981) "Uncertain Structural Models and Generic Reduced Form Estimation", Iowa University, mimeo.

Maasoumi, E. and P. Phillips (1982a) "On the Behaviour of Instrumental Variable Estimators", *Journal of Econometrics* (forthcoming).

Maasoumi, E. and P. Phillips (1982b) "Misspecification in the General Single Equation Case", Yale Univerisity, mimeo.

Maddala, G. S. (1974) "Some Small Sample Evidence on Tests of Significance in Simultaneous Equations Models", *Econometrica*, 42, 841–851.

Maekawa, K. (1980) "An Asymptotic Expansion of the Distribution of the Test Statistics for Linear Restrictions in Zellner's SUR Model", *The Hiroshima Economic Review*, 4, 81–97.

Malinvaud, E. (1980) *Statistical Methods of Econometrics*. Amsterdam: North-Holland Publishing Co.

Mariano, R. S. (1973) "Approximations to the Distribution Functions of the Ordinary Least-Squares and Two-stage Least Squares Estimates in the Case of Two Included Endogenous Variables", *Econometrica*, 41, 67–77.

Mariano, R. S. (1975) "Some Large-Concentration-Parameter Asymptotics for the *k*-Class Estimators", *Journal of Econometrics* 3, 171–177.

Mariano, R. S. (1977) "Finite-Sample Properties of Instrumental Variable Estimators of Structural Coefficients", *Econometrica*, 45, 487–496.

Mariano, R. S. (1982) "Analytical Small-Sample Distribution Theory in Econometrics: The Simultaneous-Equations Case", *International Economic Review* (forthcoming).

Mariano, R. S. and J. McDonald (1979) "A Note on the Distribution Functions of LIML and 2SLS Coefficient Estimators in the Exactly Identified Case", *Journal of the American Statistical Association*, 74, 847–848.

Mariano, R. S., J. McDonald and A. Tishler (1979) "On the Effects of Multicollinearity upon the Properties of Structural Coefficient Estimators", mimeo.

Mariano, R. S. and J. G. Ramage (1978) "Ordinary Least Squares versus other Single Equation Estimators: A Return Bout under Misspecification in Simultaneous Systems", University of Pennsylvania, Department of Economics Discussion Paper no. 400.

Mariano, R. S. and J. G. Ramage (1979), "Large Sample Asymptotic Expansions for General Linear Simultaneous Systems under Misspecification", University of Pennsylvania, mimeo.

Mariano, R. S. and T. Sawa (1972) "The Exact Finite-Sample Distribution of the Limited Information Maximum Likelihood Estimator in the Case of Two Included Endogenous Variables", *Journal of the American Statistical Association*, 67, 159–163.

McCarthy, M. D. (1972) "A Note on the Forecasting Properties of 2SLS Restricted Reduced Forms", *International Economic Review*, 13, 757–761.

McDonald, J. B. (1972) "The Exact Finite Sample Distribution Function of the Limited-Information Maximum Likelihood Identifiability Test Statistic", *Econometrica*, 40, 1109–1119.

McDonald, J. B. (1974) "An Approximation of the Distribution Function of the LIML Identifiability Test Statistic Using the Method of Moments", *Journal of Statistical Computation and Simulation*, 3, 53–66.

McLaren, M. L. (1976) "Coefficients of the Zonal Polynomials", *Applied Statistics*, 25, 82–87.

Meisner, J. F. (1979) "The Sad Fate of the Asymptotic Slutsky Symmetry Test for Large Systems", *Economic Letters*, 2, 231–233.

Mikhail, W. M. (1969) "A Study of the Finite Sample Properties of Some Economic Estimators", unpublished Ph.D. Thesis, London School of Economics.

Miller, K. S. (1960) *Advanced Complex Calculus*. New York: Harper.

Morimune, K. (1978) "Improving the Limited Information Maximum Likelihood Estimator When the Disturbances are Small", *Journal of the American Statistical Association*, 73, 867–871.

Morimune, K. (1981) "Approximate Distributions of the k-Class Estimators when the Degree of Overidentifiability is Large Compared with the Sample Size", Discussion Paper no. 159, Institute of Economic Research, Kyoto University.

Morimune, K. and N. Kunitomo (1980) "Improving the Maximum Likelihood Estimate in Linear Functional Relationships for Alternative Parameter Sequences", *Journal of the American Statistical Association*, 75, 230–237.

Muirhead, R. J. (1978) "Latent Roots and Matrix Variates: A Review of Some Asymptotic Results", *The Annals of Statistics*, 6, 5–33.

Nagar, A. L. (1959) "The Bias and Moment Matrix of the General k-Class Estimators of the Parameters in Structural Equations", *Econometrica*, 27, 575–595.

Nagar, A. L. and S. N. Sahay (1978) "The Bias and Mean Squared Error of Forecasts from Partially Restricted Reduced Form", *Journal of Econometrics*, 7, 227–243.

Nagel, P. J. A. (1981) "Programs for the Evaluation of Zonal Polynomials", *American Statistician*, 35, 53.

Pan Jie-Jian (1968) "Distribution of the Noncircular Serial Correlation Coefficients", *American Mathematical Society and Institute of Mathematical Statistics, Selected Translations in Probability and Statistics*, 7, 281–291.

Pfanzagl, J. and W. Wefelmeyer (1978) "A Third-Order Optimum Property of the Maximum Likelihood Estimator", *Journal of Multivariate Analysis*, 8, 1–29.

Pfanzagl, J. and W. Wefelmeyer (1979) "Addendum to a Third-Order Optimum Property of the Maximum Likelihood Estimator", *Journal of Multivariate Analysis*, 9, 179–182.

Phillips, P. C. B. (1977a) "Approximations to Some Finite Sample Distributions Associated with a First Order Stochastic Difference Equation", *Econometrica*, 45, 463–486.

Phillips, P. C. B. (1977b) "A General Theorem in the Theory of Asymptotic Expansions as Approximations to the Finite Sample Distributions of Econometric Estimators", *Econometrica*, 45, 1517–1534.

Phillips, P. C. B. (1977c) "An Approximation to the Finite Sample Distribution of Zellner's Seemingly Unrelated Regression Estimator", *Journal of Econometrics*, 6, 147–164.

Phillips, P. C. B. (1978) "Edgeworth and Saddlepoint Approximations in the First-Order Noncircular Autoregression", *Biometrika*, 65, 91–98.

Phillips, P. C. B. (1980a) "The Exact Finite Sample Density of Instrumental Variable Estimators in an Equation with $n+1$ Endogenous Variables", *Econometrica*, 48, no. 4, 861–878.

Phillips, P. C. B. (1980b) "Finite Sample Theory and the Distributions of Alternative Estimators of the Marginal Propensity to Consume", *Review of Economic Studies*, 47, no. 1, 183–224.

Phillips, P. C. B. (1982a) "Marginal Densities of Instrumental Variable Estimators in the General Single Equation Case", *Advances in Econometrics* (forthcoming).

Phillips, P. C. B. (1982b) "Comments on the Unification of Asymptotic Theory for Nonlinear Econometric Models", *Econometric Reviews* (forthcoming).
Phillips, P. C. B. (1982c) "A New Approach to Small Sample Theory", Cowles Foundation, Yale University, mimeo.
Phillips, P. C. B. (1982d) "Best Uniform and Modified Padé Approximation of Probability Densities in Econometrics", in: W. Hildenbrand (ed.), *Advances in Econometrics*. Cambridge: Cambridge University Press.
Phillips, P. C. B. (1982e) "Small Sample Distribution Theory in Econometric Models of Simultaneous Equations", Cowles Foundation Discussion Paper no. 617, Yale University.
Phillips, P. C. B. and M. R. Wickens (1978) *Exercises in Econometrics*, vols. I and II. Ballinger & Philip Allan.
Rainville, E. D. (1963) *Special Functions*. New York: Macmillan.
Rao, C. R. (1973) *Linear Statistical Inference and its Applications*. New York: John Wiley & Sons.
Revankar, N. S. and P. Mallela (1972) "The Power Function of an F-Test in the Context of a Structural Equation", *Econometrica*, 40, 913–916.
Rhodes, G. F. (1981) "Exact Density Functions and Approximate Critical Regions for Likelihood Ratio Identifiability Test Statistics", *Econometrica*, 49, 1035–1056.
Rhodes, G. F. and M. D. Westbrook (1982) "Simultaneous Equations Estimators, Identifiability Test Statistics, and Structural Forms", *Advances in Econometrics* (forthcoming).
Rhodes, G. F. and M. D. Westbrook (1981) "A Study of Estimator Densities and Performance Under Misspecification", *Journal of Econometrics* (forthcoming).
Richardson, D. H. (1968) "The Exact Distribution of a Structural Coefficient Estimator", *Journal of the American Statistical Association*, 63, 1214–1226.
Richardson, D. H. (1969) "On the Distribution of the Identifiability Test Statistic", in: J. P. Quirk and A. M. Zarley (eds.), *Papers in Quantitative Economics*. Kansas: Lawrence University Press.
Richardson, D. H. and R. J. Rohr (1971) "The Distribution of a Structural t-Statistic for the Case of Two Included Endogenous Variables", *Journal of the American Statistical Association*, 66, 375–382.
Richardson, D. H. and R. J. Rohr (1981) "Experimental Evidence on Approximations to Two-Stage Least-Squares Distributions", Brown University, mimeo.
Richardson, D. H. and R. J. Rohr (1982), "An Experimental Study of Two Stage Least Squares Distributions in Equations Containing Three Endogenous Variables", *Advances in Econometrics* (forthcoming).
Richardson, D. H. and D. Wu (1969) "A Note on the Comparison of Ordinary and Two-Stage Least Squares Estimators", Research Papers in Theoretical and Applied Economics (University of Kansas), Paper no. 25.
Richardson, D. H. and D. Wu (1970) "Least Squares and Grouping Method Estimators in the Errors in Variables Model", *Journal of the American Statistical Association*, 65, 724–748.
Rothenberg, T. J. (1977) "Edgeworth Expansions for Multivariate Test Statistics", University of California, Berkeley Discussion Paper, unpublished.
Rothenberg, T. J. (1978) "Second-Order Efficiency of Estimators and Tests in Simultaneous Equations Models", University of California, Berkeley, mimeo.
Rothenberg, T. J. (1980) "Comparing Alternative Asymptotically Equivalent Tests", presented to World Congress of Econometric Society, Aix-en-Provence. 1980.
Sahay, S. N. (1979) "The Exact Mean of the Two-Stage Least Squares Residuals in a Structural Equation Having Three Endogenous Variables", Discussion Paper no. 69/79, Department of Economics, University of Bristol.
Sahay, S. N. (1981) "The Bias of the Forecast from Partially Restricted Reduced Form When There are Three Endogenous Variables in the Structural Equation", Discussion Paper no. 81, University of Bristol.
Sargan, J. D. (1970) "The Finite Sample Distribution of FIML Estimators", Paper Presented to The World Congress of the Econometric Society, Cambridge.
Sargan, J. D. (1975) "Gram–Charlier Approximations Applied to t-Ratios of k-Class Estimators", *Econometrica*, 43, 326–346.
Sargan, J. D. (1976a) "Econometric Estimators and the Edgeworth Approximation", *Econometrica*, 44, 421–428, and (1977) "Erratum", *Econometrica*, 15, 272.

Sargan, J. D. (1976b) "Existence of the Moments of Estimated Reduced Form Coefficients", London School of Economics Discussion Paper no. A6.

Sargan, J. D. (1978) "On the Existence of the Moments of 3SLS Estimators", *Econometrica*, 46, 1329–1350.

Sargan, J. D. (1980) "Some Approximation to the Distribution of Econometric Criteria which are Asymptotically Distributed as Chi-squared", *Econometrica*, 48, 1107–1138.

Sargan, J. D. and S. E. Satchell (1981) "The Validity of Edgeworth Expansions for Autoregressive Models", in preparation.

Satchell, S. E. (1981) "Edgeworth Approximations in Linear Dynamic Models", unpublished Ph.D. dissertation, London School of Economics.

Sawa, T. (1969) "The Exact Finite Sampling Distribution of Ordinary Least Squares and Two Stage Least Squares Estimator", *Journal of the American Statistical Association*, 64, 923–936.

Sawa, T. (1972) "Finite-Sample Properties of the *k*-Class Estimators", *Econometrica*, 40, 653–680.

Sawa, T. (1973a) "Almost Unbiased Estimator in Simultaneous Equations Systems", *International Economic Review*, 14, 97–106.

Sawa, T. (1973b) "The Mean Square Error of a Combined Estimator and Numerical Comparison with the TSLS Estimator", *Journal of Econometrics*, 1, 115–132.

Sclove, S. (1968) "Improved Estimators for Coefficients in Linear Regression", *Journal of the American Statistical Association*, 63, 596–606.

Sclove, S. (1971) "Improved Estimation of Parameters in Multivariate Regression", *Sankya*, Ser. A, 61–66.

Slater, L. J. (1965) "Confluent Hypergeometric Functions", in: M. Abramowitz and I. A. Stegun (eds.), *Handbook of Mathematical Functions*. New York: Dover.

Srinivasan, T. N. (1970) "Approximations to Finite Sample Moments of Estimators Whose Exact Sampling Distributions are Unknown", *Econometrica*, 38, 533–541.

Swamy, P. A. V. B. and J. S. Mehta (1980) "On the Existence of Moments of Partially Restricted Reduced Form Coefficients", *Journal of Econometrics*, 14, 183–194.

Takeuchi, K. (1970) "Exact Sampling Moments of Ordinary Least Squares, Instrumental Variable and Two Stage Least Squares Estimators", *International Economic Review*, 11, 1–12.

Ullah, A. (1974) "On the Sampling Distribution of Improved Estimators for Coefficients in Linear Regression", *Journal of Econometrics*, 2, 143–150.

Ullah, A. (1980) "The Exact, Large-Sample and Small-Disturbance Conditions of Dominance of Biased Estimators in Linear Models", *Economic Letters*, 6, 339–344.

Ullah, A. and A. L. Nagar (1974) "The Exact Mean of the Two Stage Least Squares Estimator of the Structural Parameters in an Equation Having Three Endogenous Variables", *Econometrica*, 42, 749–758.

Wallace, D. L. (1958) "Asymptotic Approximations to Distributions", *Annals of Mathematical Statistics*, 29, 635–654.

Wegge, L. L. (1971) "The Finite Sampling Distribution of Least Squares Estimators with Stochastic Regressors", *Econometrica*, 38, 241–251.

Whittaker, E. T. and G. N. Watson (1927) *Modern Analysis*. Cambridge.

Widder, D. V. (1941) *The Laplace Transform*. Princeton: Princeton University Press.

Widder, D. V. (1961) *Advanced Calculus*. Prentice-Hall.

Zellner, A. (1978) "Estimation of Functions of Population Means and Regression Coefficients Including Structural Coefficients: A Minimum Expected Loss (MELO) Approach", *Journal of Econometrics*, 8, 127–158.

Zellner, A. and W. Vandaele (1975) "Bayes–Stein Estimators for *k*-Means, Regression and Simultaneous Equation Models", in: S. E. Feinberg and A. Zellner (eds.), *Studies in Bayesian Econometrics and Statistics in Honor of Leonard J. Savage*. Amsterdam: North-Holland Publishing Co.

Zellner, A. and S. B. Park (1979) "Minimum Expected Loss (MELO) Estimators for Functions of Parameters and Structural Coefficients of Econometric Models", *Journal of the American Statistical Association*, 74, 183–185.

Chapter 9

BAYESIAN ANALYSIS OF SIMULTANEOUS EQUATION SYSTEMS

JACQUES H. DRÈZE and JEAN-FRANÇOIS RICHARD*

Université Catholique de Louvain

Contents

*The authors thank David F. Hendry, Teun Kloek, Hans Tompa, Herman van Dijk, and Arnold Zellner for helpful comments on a preliminary version. They are particularly grateful to Luc Bauwens for his assistance with computations and his detailed comments on the manuscript.

Handbook of Econometrics, Volume I, Edited by Z. Griliches and M.D. Intriligator

1. Introduction and summary

1.1. The simultaneous equation model

The standard specification of the static *simultaneous equation model* (SEM) in econometrics is [see, for example, Goldberger (1964) or Theil (1971)]

$$YB + Z\Gamma = U, \tag{1.1}$$

where

Y is a $T\times m$ matrix of observed endogenous variables,
Z is a $T\times n$ matrix of observed exogenous[1] variables,
B is an $m\times m$ non-singular matrix of unknown coefficients,
Γ is an $n\times m$ matrix of unknown coefficients, and
U is a $T\times m$ matrix of unobserved disturbances.

U and Z are uncorrelated and the T rows of U are assumed identically independently normally distributed, each with zero expectation and positive definite symmetric (PDS) covariance matrix Σ; hence the matrix U has the matricvariate normal density (as defined in Appendix A):

$$p(U) = f_{\mathrm{MN}}^{T\times m}(U|0, \Sigma\otimes I_T). \tag{1.2}$$

The m equations (1.1) are called *structural equations*. Solving these explicitly for Y yields the *reduced form*:

$$Y = Z\Pi + V, \qquad p(V) = f_{\mathrm{MN}}^{T\times m}(V|0, \Omega\otimes I_T), \tag{1.3}$$

where

$$\Pi = -\Gamma B^{-1}, \qquad \Omega = B'^{-1}\Sigma B^{-1}. \tag{1.4}$$

The data density $p(Y|\Pi,\Omega)$ – see footnote 1 – is

$$p(Y|\Pi,\Omega) = f_{\mathrm{MN}}^{T\times m}(Y|Z\Pi, \Omega\otimes I_T) = \|B\|^T f_{\mathrm{MN}}^{T\times m}(YB + Z\Gamma|0, \Sigma\otimes I_T). \tag{1.5}$$

[1]Strictly speaking the variables in Z should be "weakly exogenous" in the terminology of Engle et al. (1983). The analysis applies to dynamic simultaneous equation models where Z includes lagged variables, whether exogenous or endogenous. In that case, however, the derivation of a *joint* predictive density for several successive periods of time raises problems, some of which are discussed in Richard (1979). Finally, even though our analysis is conditional on Z, we shall systematically omit Z from the list of conditioning variables, for notational convenience.

Given Y, (1.5) yields two alternative expressions for the likelihood function, namely

$$L(\Pi,\Omega|Y)\propto|\Omega|^{-(1/2)T}\exp\left[-\tfrac{1}{2}\operatorname{tr}\Omega^{-1}(Y-Z\Pi)'(Y-Z\Pi)\right], \tag{1.6}$$

with $(\Pi,\Omega)\in R^{nm}\times\mathcal{C}^{m}$, where $\mathcal{C}^{m}$ denotes the space of all real PDS matrices of size m; and

$$L(B,\Gamma,\Sigma|Y)\propto\|B\|^{T}|\Sigma|^{-(1/2)T} \times\exp\left[-\tfrac{1}{2}\operatorname{tr}\Sigma^{-1}(YB+Z\Gamma)'(YB+Z\Gamma)\right], \tag{1.7}$$

with $(B,\Gamma,\Sigma)\in\mathcal{B}^{m}\times R^{nm}\times\mathcal{C}^{m}$, where $\mathcal{B}^{m}$ denotes the space of all real non-singular matrices of size m.

In this model, the reduced-form parameters (Π,Ω) are identified, but the structural parameters (B,Γ,Σ) are not. The likelihood function $L(B,\Gamma,\Sigma|Y)$ is constant over every m^2-dimensional subspace defined by given values of Π and Ω in (1.5). (Every such subspace consists of "observationally equivalent" parameter points and defines an "equivalence class" in the parameter space.) Identification of (B,Γ,Σ) is achieved by imposing on the structural parameters a set of exact *a priori* conditions (including the normalization rule):

$$\Psi_k(B,\Gamma,\Sigma)=0,\qquad k=1\dots K, \tag{1.8}$$

such that (1.4) *and* (1.8) can be solved uniquely for (B,Γ,Σ) as functions of (Π,Ω). When a unique solution exists for almost all (Π,Ω) in $R^{nm}\times\mathcal{C}^{m}$, the structural model is *just-identified* and (Π,Ω) are "unconstrained". When a solution exists only for some, but not for almost all (Π,Ω) in $R^{nm}\times\mathcal{C}^{m}$, the structural model is *overidentified* and (Π,Ω) are constrained. When (1.4) and (1.8) admit multiple solutions the structural model is *underidentified*. Note that over-identification and underidentification are not mutually exclusive. The simplest example of identifying conditions is obtained when (1.8) is separable across equations and predetermines m or more coefficients in each row of $(B'\quad\Gamma')$.

Maximum likelihood estimation of the structural parameters calls for maximizing the likelihood function in (1.7) subject to the constraints (1.8). This estimation method, known as *full information maximum likelihood* (FIML), is conceptually straightforward, but relies on numerical procedures and may be computationally demanding. Also, it may be sensitive to specification errors in the overidentified case. In particular, estimators of (Π,Ω) are then subject to exact *a priori* restrictions, some of which may be in conflict with the data.

An alternative, which is less efficient but also less demanding computationally and less sensitive to some classes of specification errors, is *limited information maximum likelihood* (LIML). It consists in estimating each structural equation separately by maximizing the likelihood function subject only to those constraints in (1.8) which involve solely the coefficients of the equation being estimated. In the standard case where the constraints (1.8) are separable across equations, the LIML and FIML estimators of the coefficients of a structural equation coincide when the other equations in the system are not overidentified.

Exact finite sample distributions of maximum likelihood estimators have been obtained only for LIML analysis of an equation containing two endogenous variables [see, for example, Mariano (1980)]. LIML and FIML estimators can be numerically approximated respectively by two-stage (2SLS) and three-stage (3SLS) least squares estimators with no loss of asymptotic efficiency [see, for example, Theil (1971) and Hendry (1976)]. In the special case of just-identified models, all these methods are equivalent.

1.2. *Bayesian inference and identification*

A Bayesian analysis of the SEM proceeds along the same lines as any other Bayesian analysis. Thus, if the analyst has chosen to work in a given parameter space, he defines a prior density[2] on that space and applies Bayes theorem to revise this prior density in the light of available data. The resulting posterior density is then used to solve problems of decision and inference. Predictive densities for future observations can also be derived.

Thus, let $p(y|\theta)$ and $p(\theta)$ denote respectively a data density with parameter θ and a prior density. The predictive density $p(y)$ and the posterior density $p(\theta|y)$ are then given by

$$p(y)=\int p(y|\theta)p(\theta)\,\mathrm{d}\theta, \tag{1.9}$$

$$p(\theta|y)=\frac{p(y|\theta)p(\theta)}{p(y)}\propto L(\theta|y)p(\theta), \quad (\text{provided } p(y)\neq 0), \tag{1.10}$$

where $L(\theta|y)$ is the likelihood function.[3] The product $L(\theta|y)p(\theta)$ defines a *kernel* of the posterior density [see, for example, Raiffa and Schlaifer (1961, section 2.1.2)].

The operation (1.10) is well-defined, *whether θ is identified or not*; hence the remark by Lindley (1971, p. 46) that "unidentifiability causes no real difficulty in

[2] It is assumed, mainly for convenience, that all the relevant probability distributions are continuous with respect to an appropriate measure, typically the Lebesgue measure; they are therefore represented by density functions. By abuse of language, we still use the word "density" rather than "measure" when the function is not integrable.

[3] In (1.10) the proportionality factor may depend on y, since the posterior density is conditional on y.

the Bayesian approach".[4] In particular, we may analyze an underidentified model, with a prior density which substitutes stochastic prior information for exact *a priori* restrictions. And we may substitute stochastic prior information for over-identifying *a priori* restrictions, whenever the underlying economic theory is less than fully compelling.

Note, however, an obvious implication of (1.10). If θ_a and θ_b are two observationally equivalent points in the parameter space, so that $p(y|\theta_a) \equiv p(y|\theta_b)$, then

$$\frac{p(\theta_a|y)}{p(\theta_b|y)} = \frac{p(\theta_a)}{p(\theta_b)}, \quad \text{identically in } y. \tag{1.11}$$

Over equivalence classes of the parameter space, the prior density is not revised through observations. If, in particular, θ can be partitioned into $\theta = (\rho, \lambda)$ in such a way that $p(y|\rho, \lambda) \equiv p(y|\lambda)$, identically in ρ and λ, then

$$p(\rho|\lambda, y) \equiv p(\rho|\lambda), \quad \text{identically in } \lambda \text{ and } y. \tag{1.12}$$

The *conditional* prior density $p(\rho|\lambda)$ is not revised through observations; however, the *marginal* prior density $p(\rho)$ will be revised unless ρ and λ are *a priori* independent. See Section 3 for details.

1.3. Bayesian treatment of exact restrictions

Incorporating in a Bayesian analysis exact prior restrictions such as (1.8) raises conditionalization paradoxes [see, for example, Kolmogorov (1950)] which we briefly illustrate by means of a simple bivariate example. Let $p(\theta_1, \theta_2)$ be a bivariate uniform density on the open unit square and let $D = \{(\theta_1, \theta_2)|\theta_1 = \theta_2\}$ be a diagonal of that square. There is no unique way to derive from $p(\theta_1, \theta_2)$ a density for θ_1 conditionally on $(\theta_1, \theta_2) \in D$. Let, for example, $\lambda = \theta_1 - \theta_2$ and $\mu = \theta_1/\theta_2$; then

$$p(\theta_1|\lambda = 0) = 1, \qquad \forall \theta_1 \in]0,1[, \tag{1.13}$$

$$p(\theta_1|\mu = 1) = 2\theta_1, \qquad \forall \theta_1 \in]0,1[. \tag{1.14}$$

In (1.13) D is implicitly considered as the limit of the infinite sequence $\{D_{1n}\}$, where $D_{1n} = \{(\theta_1, \theta_2)| -(1/n) < \theta_1 - \theta_2 < 1/n\}$; in (1.14) it is considered instead

[4] The reader may usefully be reminded at once that: "identification is a property of the likelihood function, and is the same whether considered classically or from the Bayesian approach" [Kadane (1975, p. 175)].

as the limit of the infinite sequence $\{D_{2n}\}$, where

$$D_{2n} = \{(\theta_1, \theta_2) \mid 1 - (1/n) < \theta_1/\theta_2 < 1 + 1/n\}.$$

In order to avoid such paradoxes we shall write explicitly all the exact prior restrictions,[5] and assign probabilities only for parameter vectors defined over a space of positive Lebesgue measure. In the above example, this approach calls for selecting among (1.13) and (1.14) – or among other similar expressions – the density which seems best suited for inference on θ_1, without attempting to derive it from an underlying joint density. If needed, inferences on θ_2 can always be drawn through the integrand transformation $\theta_2 = \theta_1$.

1.4. *Bayesian analysis of the reduced form*

We shall discuss in Section 7 numerical methods for evaluating key characteristics (integrating constant, moments, fractiles, etc.) of posterior and predictive density functions. However, for models with many parameters, like most simultaneous equation models, analytical methods remain indispensable to evaluate these densities – either fully, or conditionally on a few parameters amenable to numerical treatment, or approximately to construct importance functions for Monte Carlo integration. The classes of prior densities permitting analytical evaluation of the posterior density are limited. In most Bayesian analyses they consist essentially of the so-called *non-informative* and *natural-conjugate* families. Loosely speaking, a natural-conjugate prior density conveys the same information as a hypothetical previous sample, whereas a non-informative prior density conveys as little information as possible. (Typically, a non-informative prior is a limiting member of the natural-conjugate family.)

In the simultaneous equation model, the *unrestricted reduced form* is a traditional multivariate regression model, which has been studied extensively [see, for example, Zellner (1971, ch. VIII)]. The natural-conjugate density for that model has the Normal–Wishart form. It follows that the $mn \times mn$ marginal covariance matrix of the column expansion of Π is restricted to a Kronecker product, say $W \otimes M^{-1}$, where $W \in \mathcal{C}^m$ and $M \in \mathcal{C}^n$ are matrices of prior parameters.[6] This restriction is harmless if we wish the prior density to be non-informative about Π ($M = 0$), or to be informative about the parameters of a single reduced-form equation (W is zero except for a single diagonal element). It is a severe restriction

[5]It is only when such explicitations are not critical for the sake of the argument that we shall use notations such as $p(B, \Gamma, \Sigma)$, even though (B, Γ, Σ) is subject to the restrictions (1.8).

[6]In the natural-conjugate framework, this property reflects the fact that the covariance matrix of the column expansion of $\hat{\Pi}$, the unrestricted ordinary least squares estimator of Π, is given by $\Omega \otimes (Z'Z)^{-1}$.

in other cases since it implies that all columns of Π should have the same covariance matrix, up to a proportionality factor; see Section 4 for details. As discussed in Section 6.4, there exist generalizations of the Normal–Wishart density which are more flexible in that respect and are also natural conjugate for the *seemingly unrelated regression* model (SUR) or for reduced forms subject only to exclusion restrictions. However, the evaluation of these densities requires some application of numerical methods.

1.5. Bayesian analysis of the structural form

A "natural-conjugate" approach to inference about the structural parameters is fraught with even more difficulties. First, if the model is underidentified, a natural-conjugate prior density is necessarily improper. Second, the restrictive covariance structure obtained for the reduced-form prior applies also to the conditional prior density $p(\Gamma, \Sigma | B)$. Third, a full natural-conjugate prior density for the structural parameters does not have the Normal–Wishart form, due to the presence of the additional factor $\|B\|^T$ in the likelihood (1.7).

Two alternative approaches have been developed, corresponding respectively to limited information analysis and to full information analysis. The limited information analysis relies on prior information (both exact and stochastic) about the parameters of a single structural equation; call them θ. A suitable non-informative prior density is then defined on the reduced-form parameters (Π, Ω), conditionally on θ. Bayes' theorem yields a posterior density in the form $p(\theta | Y)p(\Pi, \Omega | \theta, Y)$. For a class of prior densities on θ (including the Normal–gamma and Student–gamma families), the posterior marginal density for the regression coefficients in the equation of interest is a poly-t density, i.e. a density whose kernel is a product, or ratio of products, of Student-t kernels[7] [see Drèze (1977)]. Properties of poly-t densities are discussed in Section 7. In the cases reviewed here, evaluation of posterior and predictive densities requires at most unidimensional numerical integration and an efficient software package is available. A Bayesian limited information analysis is thus fully operational. It is presented in Section 5, together with an application. In order to provide a simple preview of the main results with a minimum of technicalities, we treat first a special case in Section 2. That section is self-contained, except for proofs.

[7]"Poly-t densities are defined by the simple property that their kernel is a product, or ratio of products, of Student-t kernels. They are obtained as posterior marginal densities for regression coefficients, under a variety of specifications for the prior density and the data generating process. No analytical expressions are available for the integrating constant or moments of poly-t densities, and the family is not suited for tabulations. Yet, it may (by contemporary standards) be regarded as 'tractable', because it lends itself to numerical analysis by integration in a number of dimensions that does *not* depend upon the number of variables but rather upon the number of Student-t kernels in the product, or in the numerator (minus one)" [Drèze (1977, p. 330)].

The full information analysis is more complex. One approach uses the *extended natural-conjugate* prior density corresponding to the seemingly unrelated regressions model. That prior density includes as a special case the possibility of specifying m independent Student densities on the coefficients of the m structural equations. The posterior density is obtained analytically, either for (B, Γ) marginally with respect to Σ, or for (B, Σ) marginally with respect to Γ. Neither of these expressions has yet been found amenable to further analytical treatment, except in the special case of two-equation models, reviewed in Section 6.5.

The only general procedure is to integrate the posterior density numerically by a Monte Carlo technique. This approach gives additional flexibility in the specification of prior densities. It has been developed by Kloek and van Dijck (1978); see Sections 6.6.2 for an application and 7.3.2 for computational aspects.

The posterior density for the coefficients of a single equation, defined marginally with respect to Σ but conditionally on all other parameters of the model, belongs to the class of poly-t densities whenever the corresponding conditional prior density belongs to that class (e.g. is a multivariate Student density). When attention is focused on the coefficients of a single equation, this conditional full information approach and the limited information approach define two extreme ways of handling prior information on the other coefficients of the model.

Full information methods are discussed in Section 6, together with applications. A table at the end of the chapter contains references to the main formulae for prior and posterior densities on the coefficients of one equation. Finally, the formulae pertaining to the Normal–Wishart density, which play a central role in these analyses, are collected in Appendix A.

1.6. Summary

In summary, Bayesian limited information analysis of general models and Bayesian full information analysis of two-equation models have now been developed to a point where:

(i) they allow a flexible specification of the prior density, including well-defined non-informative prior measures;

(ii) they yield exact finite sample posterior and predictive densities with known properties; and

(iii) they can be evaluated through numerical methods (either exact or involving integration in a few dimensions), using an efficient integrated software package.

The treatment of stochastic prior information is illustrated in Sections 2.4, 5.5, and 6.6.1. The use of posterior densities for policy analysis is illustrated in Sections 6.6.2 and 6.6.3.

Remarks (i) and (ii) apply also to full information analysis of general models. But the numerical evaluation (by Monte Carlo) is more demanding, and no integrated software package is available yet. Remark (iii) also applies to the analysis of one structural equation at a time, conditionally on point estimates for the other parameters. But avenues for further developments are open.

These advances must be weighted against some lasting drawbacks, in particular:

(i) The computations remain demanding – as could be expected of exact finite sample results in a complex model;
(ii) the specification of the prior densities requires careful thought to avoid misrepresentation of the prior information; and
(iii) the treatment of identities, non-linear restrictions, reparameterizations, etc. is more complicated than under a maximum likelihood approach, to the extent that integrand transformations are more complicated than functional transformations.

1.7. Bibliographical note

The Bayesian approach to simultaneous equations analysis was reviewed earlier by Rothenberg (1975). The intellectual origins of the subject go back to unpublished papers by Drèze (1962), Rothenberg (1963), and Zellner (1965). These three papers contain ideas developed more fully in ensuing work by the same authors – see the list of references. These ideas also influenced other researchers at Louvain, Rotterdam (where Rothenberg had worked in 1962–1964), Madison, and Chicago. Much of the more definitive work reviewed here was carried out at these places, notably by Morales, Richard and Tompa, Harkema, Kloek and van Dijk, and Chetty, Kmenta, Tiao and Vandaele; see the list of references.

Rothenberg's 1963 paper contained an application to Haavelmo's model, studied in greater details by Chetty (1968). The next applications came from Harkema (1971), Morales (1971), and Richard (1973). See also Drèze (1976), Kaufman (1975), and Maddala (1976) for the analysis of Tintner's model of supply and demand for meat; and Zellner, Kmenta and Drèze (1966) as well as Zellner and Richard (1973) for an application to Cobb–Douglas production functions.

2. A special case

2.1. Limited information maximum likelihood estimation

In order to give a simple preview of the more general analysis to follow, we shall discuss first a special case, namely limited information analysis of a single, identified equation. As usual, maximum likelihood analysis provides a natural reference point for the Bayesian approach. Accordingly, we retrace some steps of LIML analysis and bring out the analogies.

To single out one equation – labelled equation 1 – we partition the system (1.1) as

$$Y[\beta \quad B_2]+Z[\gamma \quad \Gamma_2]=[u \quad U_2], \tag{2.1}$$

where β is $m \times 1$, γ is $n \times 1$, and u is $T \times 1$. Let σ^2 denote the first diagonal element of Σ. The equation of interest is thus

$$Y\beta + Z\gamma = u, \tag{2.2}$$

where u is normally distributed with expectation 0 and covariance matrix $\sigma^2 I_T$. Compatibility with (1.4) requires

$$\Pi\beta + \gamma = 0, \qquad \beta'\Omega\beta = \sigma^2. \tag{2.3}$$

Restricting attention to exclusion restrictions, we consider the partitioning

$$[Y_\Delta \quad Y_{\Delta\Delta}]\begin{bmatrix}\beta_\Delta \\ 0\end{bmatrix}+[Z_* \quad Z_{**}]\begin{bmatrix}\gamma_* \\ 0\end{bmatrix}=u, \tag{2.4}$$

where Y_Δ is a $T \times m_1$ matrix consisting of T observations on the m_1 endogenous variables included in the first equation; Z_* is a $T \times n_1$ matrix of T observations on the n_1 exogenous variables so included; and (β_Δ, γ_*) are the associated parameters. The partitioning in (2.4) is achieved through appropriate labelling of the variables (as distinct from the equations). Π and Ω are partitioned conformably, in an obvious notation, and (2.3) is then rewritten as

$$\begin{pmatrix}\Pi_{*\Delta} & \Pi_{*\Delta\Delta} \\ \Pi_{**\Delta} & \Pi_{**\Delta\Delta}\end{pmatrix}\begin{pmatrix}\beta_\Delta \\ 0\end{pmatrix}+\begin{pmatrix}\gamma_* \\ 0\end{pmatrix}=0, \qquad (\beta_\Delta' \quad 0)\Omega\begin{pmatrix}\beta_\Delta \\ 0\end{pmatrix}=\sigma^2. \tag{2.5}$$

Under the rank condition

$$\operatorname{rank} \Pi_{**\Delta} = m_1 - 1, \tag{2.6}$$

$(\beta_\Delta, \gamma_*, \sigma)$ is identified *up to a scale factor*. LIML analysis recognizes this feature and estimates $(\beta_\Delta, \gamma_*, \sigma)$ up to an arbitrary scale factor, to be selected freely. In particular, if the scale normalization consists in setting one element of β equal to unity, the statistical analysis is invariant with respect to the choice of that particular element. (This is in contrast with 2SLS analysis, whereby a choice of normalization is made first and affects distinctly the statistical analysis to follow.)

In order to proceed in "limited information" spirit, i.e. without explicit consideration for the restrictions pertaining to the remaining structural equations, it is convenient to partition the reduced form (1.3) as

$$[Y_1 \quad Y_2] = Z[\Pi_1 \quad \Pi_2] + [V_1 \quad V_2], \tag{2.7}$$

where Y_1 is $T \times 1$, Π_1 is $n \times 1$, and V_1 is $T \times 1$. Here Y_1 is any column of Y_Δ (the labelling inside Y_Δ is arbitrary).

One may then exhibit the joint density of the disturbances $[u \quad V_2]$, i.e. of the T realizations of one structural disturbance (u) and $m-1$ reduced-form disturbances (V_2). Since $(u \quad V_2) = VL$, where L is the triangular matrix

$$L = \left[\begin{array}{c|c} \beta & \begin{matrix} 0' \\ I_{m-1} \end{matrix} \end{array} \right], \tag{2.8}$$

we have

$$p(u \quad V_2) = f_{\mathrm{MN}}^{T \times m}((u \quad V_2) | 0, \Omega^* \otimes I_T), \tag{2.9}$$

with $\Omega^* = L'\Omega L$. It follows that the likelihood function (1.7) may be rewritten as

$$\begin{aligned} L(\beta_\Delta, \gamma_*, \Pi_2, \Omega^* | Y) \propto & |L|^T |\Omega^*|^{-(1/2)T} \\ & \times \exp\left[-\tfrac{1}{2} \operatorname{tr} \Omega^{*-1} [Y_\Delta \beta_\Delta + Z_* \gamma_* \quad Y_2 - Z\Pi_2]' \right. \\ & \left. \times [Y_\Delta \beta_\Delta + Z_* \gamma_* \quad Y_2 - Z\Pi_2] \right]. \end{aligned} \tag{2.10}$$

Maximizing (2.10) with respect to Π_2 and Ω^* yields the concentrated likelihood

$$L(\beta_\Delta, \gamma_* | Y) \propto \left\{ \frac{\beta_\Delta' W_{\Delta\Delta} \beta_\Delta}{(Y_\Delta \beta_\Delta + Z_* \gamma_*)'(Y_\Delta \beta_\Delta + Z_* \gamma_*)} \right\}^{(1/2)T}, \tag{2.11}$$

where $W_{\Delta\Delta} = Y_\Delta'[I - Z(Z'Z)^{-1}Z']Y_\Delta$. This expression is homogeneous of degree 0 in the parameters (β_Δ, γ_*). It can therefore be maximized with respect to any $(m_1 + n_1 - 1)$ elements of (β_Δ, γ_*) conditionally on a normalization rule, in order to obtain the LIML estimators. Since γ_* is unrestricted, (2.11) reaches its maximum at $\gamma_* = (Z_*'Z_*)^{-1}Z_*'Y_\Delta \beta_\Delta$ and the resulting concentrated likelihood function is

$$L(\beta_\Delta | Y) \propto \left\{ \frac{\beta_\Delta' W_{\Delta\Delta} \beta_\Delta}{\beta_\Delta' W_{\Delta\Delta}^* \beta_\Delta} \right\}^{(1/2)T}, \tag{2.12}$$

where $W_{\Delta\Delta}^* = Y_\Delta'[I - Z_*(Z_*'Z_*)^{-1}Z_*']Y_\Delta$. (2.12) is homogeneous of degree 0 in β_Δ.

Let $\hat{\Pi} = (Z'Z)^{-1}Z'Y$ be partitioned conformably with Π in (2.5). We can then verify that rank $\hat{\Pi}_{**\Delta} = m_1 - 1$, the sampling analogue of (2.6), is equivalent to rank$(W^*_{\Delta\Delta} - W_{\Delta\Delta}) = m_1 - 1$. The properties of LIML estimators are discussed for example in Anderson and Rubin (1949) or Theil (1971).

2.2. *A Bayesian analogue*

To develop a Bayesian analogue of LIML, we may start from either (1.6) or (2.10) and combine the likelihood with a prior density which is non-informative with respect to all the parameters except $(\beta_\Delta, \gamma_*, \sigma^2)$. The expression (2.10) provides a natural starting point to the extent that $(\beta_\Delta, \gamma_*, \sigma^2)$ appear explicitly as arguments of that likelihood function. Also, (Π_2, Ω^*) are not subject to exact prior restrictions. On the other hand, the parameterization of (2.10) includes an element of arbitrariness, to the extent that the matrix L defined in (2.8) is itself arbitrary. In particular, the selection of $m-1$ columns from Π (from Y) destroys the symmmetry among all the endogenous variables included in the equation of interest. Thus, if Q is any $m(m-1)$ matrix such that $P = [\beta \quad Q]$ is non-singular, then the parameterization (Φ_2, Λ), where

$$\Phi_2 = \Pi Q, \qquad \Lambda = P'\Omega P, \tag{2.13}$$

is just as meaningful as the parameterization (Π_2, Ω^*).

This element of arbitrariness is not a difficulty in LIML analysis. Indeed, (2.11) is invariant with respect to Q in the sense that reformulating (2.10) in terms of (Φ_2, Λ) and maximizing with respect to (Φ_2, Λ) still yields (2.11) independently of Q.[8] Similarly, in a Bayesian analysis resting on a *proper* prior density $p(\beta_\Delta, \gamma_*, \Pi_2, \Omega^*)$, the prior information so defined in $(\beta_\Delta, \gamma_*, \Pi_2, \Omega^*)$-space could be translated into *the same* prior information in $(\beta_\Delta, \gamma_*, \Phi_2, \Lambda)$-space, by an integrand transformation with a well-defined Jacobian. But this element of arbitrariness raises a difficulty in defining a *non-informative* prior density on (Π_2, Ω^*) because such a density would generally fail to be equally non-informative about (Φ_2, Λ). Fortunately, it is possible to select a particular prior measure (improper density) which can be claimed "non-informative" in a sense that is *both natural and invariant with respect to Q*. It is shown in Section 5 that the density

$$p(\beta_\Delta, \gamma_*, \Pi_2, \Omega^*) \propto p(\beta_\Delta, \gamma_*) \cdot |\Omega^*_{22.1}|^{(1/2)n} |\Omega^*|^{(1/2)(m+1)} \tag{2.14}$$

has the desired property. This density can be interpreted as the product of three

[8] The LIML estimators of (Π, Ω) are also invariant with respect to Q. Indeed, under general conditions, the ML estimator of a function of the parameters is that function of the ML estimators of the parameters.

terms:

(i) $p(\beta_\Delta, \gamma_*)$, a prior density which is left unspecified at this stage;
(ii) $p(\Pi_2|\Omega^*) \propto |\Omega^*_{22.1}|^{-(1/2)n}$, the limiting form of an $(m-1)\times n$ multivariate normal density for Π_2 with covariance matrix $\Omega^*_{22.1}\otimes M^{-1}$, where $M \simeq 0$; and
(iii) $p(\Omega^*) \propto |\Omega^*|^{-(1/2)(m+1)}$, the limiting form of a Wishart density with ν degrees of freedom, where $\nu \simeq 0$.

Combining (2.10) and (2.14) yields (see footnote 1)

$$\begin{aligned} p(\beta_\Delta, \gamma_*, \Pi_2, \Omega^* | Y) \propto p(\beta_\Delta, \gamma_*) \cdot |\Omega^*_{22.1}|^{-(1/2)n} |\Omega^*|^{-(1/2)(m+1)} \\ \times |L|^T |\Omega^*|^{-(1/2)T} \exp\big[-\tfrac{1}{2} \operatorname{tr} \Omega^{*-1} \\ \times [Y_\Delta\beta_\Delta + Z_*\gamma_* \quad Y_2 - Z\Pi_2]' \\ \times [Y_\Delta\beta_\Delta + Z_*\gamma_* \quad Y_2 - Z\Pi_2]\big]. \end{aligned} \tag{2.15}$$

As shown in Section 5, this expression can be integrated analytically with respect to (Π_2, Ω^*) to yield

$$p(\beta_\Delta, \gamma_* | Y) \propto p(\beta_\Delta, \gamma_*) \cdot \left\{ \frac{\beta_\Delta' W_{\Delta\Delta} \beta_\Delta}{(Y_\Delta\beta_\Delta + Z_*\gamma_*)'(Y_\Delta\beta_\Delta + Z_*\gamma_*)} \right\}^{(1/2)(T-m+1)}. \tag{2.16}$$

The posterior density (2.16) is invariant with respect to Q in the sense that if the analysis is conducted in terms of (Φ_2, Λ) instead of (Π_2, Ω^*) the *same* posterior density is obtained for (β_Δ, γ_*): both the prior density (2.14) and the likelihood function are invariant with respect to Q.

Expression (2.16) differs from (2.11) in three respects: (i) it is a density and not a likelihood function; (ii) it contains as an additional term the prior density $p(\beta_\Delta, \gamma_*)$; and (iii) the ratio of quadratic forms in (β_Δ, γ_*) is raised to the power $(T-m+1)/2$ instead of $T/2$. The first two differences are self-explanatory. The last one reflects the difference in exponents (degrees of freedom) between conditionalization – as in (2.11) – and marginalization – as in (2.16) – of a Wishart density.[9]

It is also shown in Section 5 that a prior density of the form

$$p(\beta_\Delta, \gamma_*, \Pi_2, \Omega^*) \propto \sigma^{-(m_1+n_1)} |\Omega^*_{22.1}|^{-(1/2)n} |\Omega^*|^{-(1/2)(m+1)} \tag{2.17}$$

[9]Indeed, the stepwise maximization leading to (2.11) is algebraically equivalent to conditionalization of a joint density.

is invariant with respect to Q and leads to the posterior densities:

$$p(\beta_\Delta,\gamma_*|Y)\propto\{\beta_\Delta' W_{\Delta\Delta}\beta_\Delta\}^{(1/2)(T-m+1)}(Y_\Delta\beta_\Delta+Z_*\gamma_*)'$$
$$\times(Y_\Delta\beta_\Delta+Z_*\gamma_*)\}^{-(1/2)(T-m+m_1+n_1+1)}, \tag{2.18}$$
$$p(\beta_\Delta|Y)\propto\{\beta_\Delta' W_{\Delta\Delta}\beta_\Delta\}^{(1/2)(T-m+1)}\{\beta_\Delta' W_{\Delta\Delta}^*\beta_\Delta\}^{-(1/2)(T-m+m_1+1)}. \tag{2.19}$$

The prior density (2.17) differs from (2.14) in that $p(\beta_\Delta,\gamma_*)$ is replaced by

(i′) $p(\beta_\Delta,\gamma_*|\Omega^*)\propto(\sigma^2)^{-(1/2)(m_1+n_1)}$, the limiting form of a normal density in m_1+n_1 variables with covariance matrix σ^2N^{-1}, where $N\simeq 0$.

We now turn to a discussion of the normalization issue – Section 2.3.1 – and show that (2.16) with $p(\beta_\Delta,\gamma_*)$ suitably chosen, (2.18), and (2.19) define poly-t densities – Section 2.3.2.

2.3. *The normalization issue*

2.3.1. *Invariance*

We noted above that the concentrated likelihood functions (2.11)–(2.12) are homogeneous of degree 0 in (β_Δ,γ_*). Thus, if β_1 is any element of β_Δ, such that $\beta_1\neq 0$ with probability one, define:

$$\alpha'=(\alpha_1'\quad\alpha_2')=\frac{1}{\beta_1}(\beta_2\ldots\beta_{m_1}\quad\gamma_1\ldots\gamma_{n_1})\in R^{(m_1-1)+n_1}, \tag{2.20}$$

and also, for ease of notation:

$$\bar{\alpha}'=(1\quad\alpha')=(\bar{\alpha}_1'\quad\alpha_2'). \tag{2.21}$$

It is obvious that, in the parameterization $(\beta_1,\alpha,\Pi_2,\Omega^*)$, β_1 is not identified. In particular, (2.11) may be rewritten as

$$L(\beta_1,\alpha|Y)\propto\left\{\frac{\bar{\alpha}_1' W_{\Delta\Delta}\bar{\alpha}_1}{(Y_\Delta\bar{\alpha}_1+Z_*\alpha_2)'(Y_\Delta\bar{\alpha}_1+Z_*\alpha_2)}\right\}^{(1/2)T}, \tag{2.22}$$

an expression which no longer depends on β_1. It follows from (1.12) that

$$p(\beta_1|\alpha,Y)\equiv p(\beta_1|\alpha),\quad\text{identically in }\alpha\text{ and }(Y,Z). \tag{2.23}$$

Similar identities hold conditionally on Π_2 or Ω^*. The whole analysis can, without loss of information, be conducted in terms of the parameters (α,Π_2,Ω^*).

Taking into account the Jacobian $|\beta_1|^{m_1+n_1}$ of the integrand transformation from (β_Δ, γ_*) to (β_1, α), the posterior marginal densities (2.16), (2.18), and (2.19) become, respectively:

$$p(\alpha|Y) \propto p(\alpha)\cdot\left\{\frac{\bar{\alpha}_1' W_{\Delta\Delta}\bar{\alpha}_1}{(Y_\Delta\bar{\alpha}_1 + Z_*\alpha_2)'(Y_\Delta\bar{\alpha}_1 + Z_*\alpha_2)}\right\}^{(1/2)(T-m+1)}, \tag{2.24}$$

$$p(\alpha|Y) \propto (\bar{\alpha}_1' W_{\Delta\Delta}\bar{\alpha}_1)^{(1/2)(T-m+1)}$$
$$\times\{(Y_\Delta\bar{\alpha}_1 + Z_*\alpha_2)'(Y_\Delta\bar{\alpha}_1 + Z_*\alpha_2)\}^{-(1/2)(T-m+m_1+n_1+1)}, \tag{2.25}$$

$$p(\alpha_1|Y) \propto (\bar{\alpha}_1' W_{\Delta\Delta}\bar{\alpha}_1)^{(1/2)(T-m+1)}(\bar{\alpha}_1' W_{\Delta\Delta}^*\bar{\alpha}_1)^{-(1/2)(T-m+m_1+1)}. \tag{2.26}$$

It is shown in Section 5.6 that the (functional forms of the) *densities* (2.25) *and* (2.26) *are invariant with respect to the normalization rule*, i.e. with respect to the labelling of variables under Y_Δ. The density (2.24) is invariant with respect to the normalization rule if and only if the prior density $p(\alpha)$ has that property.

In conclusion, our approach to normalization consists in writing a prior density $p(\beta_\Delta, \gamma_*|\cdot)$ as $p(\beta_1|\alpha,\cdot)p(\alpha|\cdot)$ and then working with the parameters α alone. However, for convenience we shall often write our formulae in terms of full parameter sets such as B, β, or β_Δ and refer to "β normalized by $\beta_1 = 1$" as a substitute for α_1. The more explicit notation would be heavier, especially when we shall discuss full information approaches.

2.3.2. *Poly-t densities*

Using (2.21), it is easily recognized that (2.25) and (2.26) define 1–1 poly-*t* densities. This is done in Section 5.4 – see in particular formulae (5.37) and (5.38) – where it is shown that $\bar{\alpha}_1' W_{\Delta\Delta}\bar{\alpha}_1 = s_1^2 + (\alpha_1 - a_1)' H_1(\alpha_1 - a_1)$. The statistics $W_{\Delta\Delta}$ and (s_1^2, a_1, H_1) are in one-to-one correspondence. Thus, $\bar{\alpha}_1' W_{\Delta\Delta}\bar{\alpha}_1$ is proportional to the quadratic form appearing in the kernel of a Student density for α_1 centered at a_1 with the covariance matrix proportional to $s_1^2 H_1^{-1}$. A similar argument applies to the other factors in the right-hand side of (2.25) and (2.26) which, therefore, define 1–1 poly-*t* densities respectively in $(m_1 + n_1 - 1)$ and $(m_1 - 1)$ variables – the elements of (β_Δ, γ_*) and β_Δ after normalization. These densities are well defined and integrable, they possess finite fractiles but no moments of order 1 or higher.[10]

When $p(\alpha)$ is a Student density, (2.24) defines a 2–1 poly-*t* density on $(m_1 + n_1 - 1)$ normalized elements of (β_Δ, γ_*). Invariance with respect to the normalization requires $p(\alpha)$ to be a Cauchy density (Student density with one

[10]LIML estimators do not possess finite sampling moments.

Table 2.1
Limited information analysis of the demand equation[a]

	β_{22}	γ_{23}	γ_{25}
LIML estimators	-0.6876	$0.7254E-03$	$0.3456E+02$
	(0.1223)	(0.0730)	(0.0514)
Prior (2.17)	-0.7118	$0.7370E-03$	$0.3550E+02$
	(0.1520)	(0.0853)	(0.0621)
Prior (2.14)–(2.29)	-0.6484	$0.6860E-03$	$0.3361E+02$
	(0.1518)	(0.0853)	(0.0655)
Prior (2.31)–(2.29)	-0.6005	$0.6655E-03$	$0.3150E+02$
	(0.1116)	(0.0657)	(0.0482)

[a]In order to ensure the existence of moments, the marginal posterior densities of β_{22}, γ_{23}, and γ_{25} have been truncated over the intervals $(-1.5,0.0)$, $(0.30E-03, 1.10E-03)$, and $(0.0,65.0)$, respectively. The probability of falling outside these intervals is in all cases less than 0.01.

degree of freedom). In that case, (2.24) is integrable, but possesses no moments of order 1 or higher.

2.4. *An application*

As an illustration, we consider the two-equation supply–demand model of the retail beef market in Belgium discussed in Morales (1971) and Richard (1973). The model is just identified and the structural coefficients are labelled as follows:

$$(B' \quad \Gamma') = \begin{pmatrix} 1 & -\beta_{12} & 0 & -\gamma_{14} & -\gamma_{15} \\ 1 & -\beta_{22} & -\gamma_{23} & 0 & -\gamma_{25} \end{pmatrix}. \tag{2.27}$$

The corresponding variables in $(Y \quad Z)$ are successively: quantity (kg) consumed per capita, a meat price index, national income (Belgian francs) per capita, the cattle stock (number of heads at the beginning of each period) per capita, and a constant term. Price and income have been deflated by an index of consumer prices. Sixteen annual observations (1950–1965) are used. We only consider here the second (demand) equation. The LIML[11] estimators of its coefficients are reported in the first row of Table 2.1, together with their asymptotic sampling standard deviations.

We first analyze the demand equation under the non-informative prior density (2.17). The corresponding posterior density of β_{22} is given by (2.26). It is integrable and its graph is given in Figure 2.1 (curve 1). However, the difference of exponents between the denominator and the numerator being equal to one, the

[11]Since the model is just-identified, the LIML and 2SLS estimators of each equation coincide.

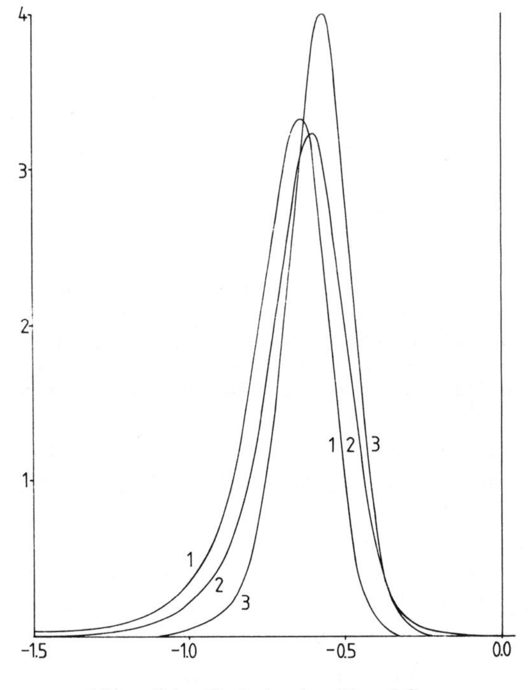

Figure 2.1. Posterior densities of β_{22}.

posterior moments do not exist. The posterior means and standard deviations reported in Table 2.1 are truncated moments.

Prior information on the income coefficient γ_{23} is available from a subsample of a budget study undertaken by the Belgian National Institute of Statistics and is summarized in the following mean and variance:

$$E(\gamma_{23}) = 0.415 \times 10^{-3}, \qquad V(\gamma_{23}) = 0.1136 \times 10^{-7}. \tag{2.28}$$

This leads to the following Student prior density:[12]

$$p(\gamma_{23}) \propto \left[1 + 22.007(10^3 \cdot \gamma_{23} - 0.415)^2\right]^{-3.5}, \tag{2.29}$$

together with (2.14). The corresponding posterior means and standard deviations as derived from (2.24) are reported in the third row of Table 2.1. The graph of the posterior density of β_{22} is given in Figure 2.1 (curve 2). The standard errors in

[12]The choice of the degree of freedom parameter will be discussed in Section 5 where additional results are reported.

rows 2 and 3 reveal that the asymptotic LIML variances underestimate by 30–60 percent the exact Bayesian results.

However, since we are non-informative on all coefficients but one, we might wish to follow the principle underlying the specification of (2.17), replacing, therefore, the prior density (2.14) by

$$p(\beta_\Delta,\gamma_*,\Pi_2,\Omega^*)\propto p(\gamma_{23})\sigma^{-(m_1+n_1-1)}|\Omega^*_{22.1}|^{-(1/2)n}|\Omega^*|^{-(1/2)(m+1)} \tag{2.30}$$

which leads to the posterior density

$$p(\alpha|Y)\propto p(\gamma_{23})(\bar{\alpha}_1'W_{\Delta\Delta}\bar{\alpha}_1)^{(1/2)(T-m+1)} \times\{(Y_\Delta\bar{\alpha}_1+Z_*\alpha_2)'(Y_\Delta\bar{\alpha}_1+Z_*\alpha_2)\}^{-(1/2)(T-m+m_1+n_1)}. \tag{2.31}$$

The corresponding posterior means and standard deviations are reported in the fourth row of Table 2.1. The graph of the posterior density of β_{22} is given in Figure 2.1 (curve 3).

Note that the larger the exponent of the quadratic form $(Y_\Delta\bar{\alpha}_1+Z_*\alpha_2)'(Y_\Delta\bar{\alpha}_1+Z_*\alpha_2)$ in the posterior $p(\alpha|Y)$, the more weight is given the OLS values, say $\hat{\alpha}$, which are respectively -0.3864, $0.5405E-03$, and $0.2304E+02$. The posterior variances also go down as the exponent increases; those in row 4 seem artificially low, in comparison with rows 2 and 3.

3. Identification

3.1. Classical concepts

The parameters of the model (1.1)–(1.2) are (B,Γ,Σ) in $\mathfrak{B}^m\times R^{nm}\times\mathcal{C}^m$. Let (1.8) denote the set of exact *a priori* restrictions imposed on these parameters, including the normalization rule. The parameter space is then

$$\mathfrak{S}=\{(B,\Gamma,\Sigma)\in\mathfrak{B}^m\times R^{nm}\times\mathcal{C}^m|\Psi_k(B,\Gamma,\Sigma)=0,\qquad k=1\dots K\}. \tag{3.1}$$

Consider the transformation from (B,Γ,Σ) to (B,Π,Ω) as given by (1.4). The restrictions (3.1) may be expressed equivalently as

$$\Psi_k(B,-\Pi B,B'\Omega B)=_{\text{def}}\Xi_k(B,\Pi,\Omega)=0,\qquad k=1\dots K, \tag{3.2}$$

so that the image space of $\mathfrak{S}$ by the transformation (1.4) is

$$\mathfrak{T}=\{(B,\Pi,\Omega)\in\mathfrak{B}^m\times R^{nm}\times\mathcal{C}^m|\Xi_k(B,\Pi,\Omega)=0,\qquad k=1\dots K\}. \tag{3.3}$$

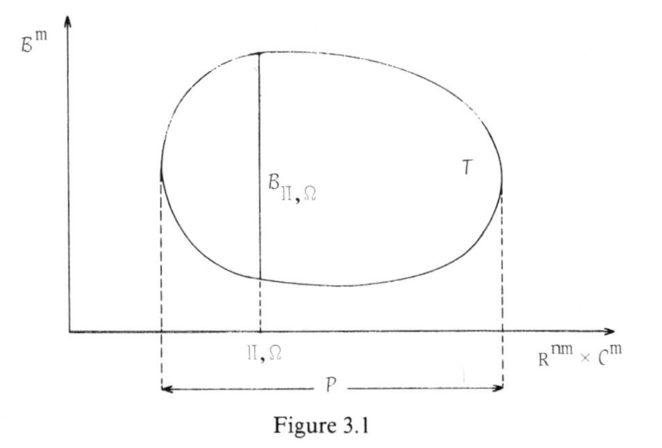

Figure 3.1

The transformation (1.4) is a mapping of $\mathcal{S}$ onto $\mathcal{T}$. Conditionally on B, it is also linear and one-to-one on the appropriate subspaces. The projection of $\mathcal{T}$ on the space of (Π, Ω) is

$$\mathcal{P} = \{(\Pi, \Omega) \in R^{nm} \times \mathcal{C}^m | \exists B \in \mathcal{B}^m : (B, \Pi, \Omega) \in \mathcal{T}\}. \tag{3.4}$$

The section of $\mathcal{T}$ at given (Π, Ω) is

$$\mathcal{B}_{\Pi,\Omega} = \{B \in \mathcal{B}^m | (B, \Pi, \Omega) \in \mathcal{T}\}. \tag{3.5}$$

These concepts are illustrated in Figure 3.1.

The model is *identified at* (Π, Ω) in $\mathcal{P}$ if and only if $\mathcal{B}_{\Pi,\Omega}$ is a singleton; it is *identified* if and only if $\mathcal{B}_{\Pi,\Omega}$ is a singleton for almost all (Π, Ω) in $\mathcal{P}$; otherwise, it is *underidentified*, and the set $\mathcal{B}_{\Pi,\Omega}$ defines, for each (Π, Ω), an equivalence class of the parameter space. The model is *overidentified* if and only if $\mathcal{P}$ is a proper subset of $R^{nm} \times \mathcal{C}^m$. (Thus, Figure 3.1 corresponds to a model which is both overidentified and underidentified.)

3.2. *Posterior densities and identification*

In a formal Bayesian analysis (see footnote 5) one defines a prior density on $\mathcal{S}$, or on $\mathcal{T}$, or on any parameter space which is related to $\mathcal{S}$ and $\mathcal{T}$ by one-to-one mappings. Because B is non-singular the corresponding integrand transformations have non-vanishing Jacobians. In particular, the prior density $p(B, \Pi, \Omega)$ can be factorized as

$$p(B, \Pi, \Omega) = p(\Pi, \Omega) \cdot p(B | \Pi, \Omega), \tag{3.6}$$

where the support of $p(\Pi, \Omega)$ is $\mathcal{P}$ and the support of $p(B|\Pi, \Omega)$ is $\mathcal{B}_{\Pi,\Omega}$. If the model is identified at (Π, Ω), then $p(B|\Pi, \Omega)$ has all its mass concentrated at a single point. Otherwise, $p(B|\Pi, \Omega)$ is a density on $\mathcal{B}_{\Pi,\Omega}$.

The posterior density is obtained by application of Bayes theorem:

$$p(B, \Pi, \Omega | Y) = \frac{1}{p(Y)} p(Y|B, \Pi, \Omega) p(B, \Pi, \Omega), \tag{3.7}$$

where, in the continuous case,

$$p(Y) = \int_{\mathcal{T}} p(Y|B, \Pi, \Omega) p(B, \Pi, \Omega) \, \mathrm{d}B \, \mathrm{d}\Pi \, \mathrm{d}\Omega. \tag{3.8}$$

In view of (1.6), the data density satisfies

$$p(Y|B, \Pi, \Omega) \equiv p(Y|\Pi, \Omega), \quad \text{identically in } B, \Pi, \text{ and } \Omega. \tag{3.9}$$

Consequently, using (1.12) with $\rho = B$ and $\lambda = (\Pi, \Omega)$, we have[13]

Proposition 3.1

$$p(B|\Pi, \Omega, Y) \equiv p(B|\Pi, \Omega), \quad \text{identically in } \Pi, \Omega, \text{ and } Y. \tag{3.10}$$

□

The *conditional* density $p(B|\Pi, \Omega)$ is not revised through observations; but unless $p(B|\Pi, \Omega) \equiv p(B)$ the *marginal* density $p(B)$ is revised, because

$$\begin{aligned} p(B|Y) &= \int_{\mathcal{P}} p(B|\Pi, \Omega) p(\Pi, \Omega | Y) \, \mathrm{d}\Pi \, \mathrm{d}\Omega \\ &\not\equiv \int_{\mathcal{P}} p(B|\Pi, \Omega) p(\Pi, \Omega) \, \mathrm{d}\Pi \, \mathrm{d}\Omega = p(B). \end{aligned} \tag{3.11}$$

Note that (3.10) and (3.11) remain valid under any parameterization, say $\theta = (\theta_a, \theta_b)$, where θ_a and θ_b are related through one-to-one mappings respectively to (Π, Ω) and B.

3.3. Prior densities and identification

The Bayesian analysis of the observations can be conducted in the reduced-form parameter space $\mathcal{P}$ only. To the extent that prior information based on economic

[13] This proposition appeared in Drèze (1962, 1975); it is discussed in Zellner (1971, p. 257) and generalized in Kadane (1975, theorem 5).

theories is subject to revision through observations, it might sometimes seem desirable to avoid overidentification, and to embody such prior information in a marginal prior density $p(\Pi, \Omega)$. However, when the prior information is provided as $p(B, \Gamma, \Sigma)$, it may not be convenient to perform the integrand transformation from (B, Γ, Σ) to (B, Π, Ω) *conditionally on* (1.8); and/or to separate $p(B, \Pi, \Omega)$ into the marginal density $p(\Pi, \Omega)$ and the conditional density $p(B|\Pi, \Omega)$. As we shall see below, these transformations are not necessary to obtain the posterior density $p(B, \Gamma, \Sigma | Y, Z)$.[14] But it is advisable to check, before engaging in detailed computations, that enough prior information has been introduced, so that the joint posterior density will at least be proper. When the model is identified by the exact restrictions (1.8), then the posterior density will be proper under almost any prior density of interest. Whether the model is so identified can be verified by means of operational conditions extensively discussed in the econometric literature [see, for example, Goldberger (1964)].

When the model is not identified by the exact restrictions (1.8), then the posterior density will not be proper, unless the prior density entails a proper conditional prior density on the equivalence classes of the parameter space – for instance in the form $p(B|\Pi, \Omega)$. Whether this property holds can be verified through natural generalizations of the operational conditions for identifiability just mentioned.

Let, for example, the conditions (1.8) assign predetermined values to a subset of (B, Γ) and let the prior density consist of a product of independent proper densities, each defined on some coefficients from a given structural equation, times a non-informative density on the remaining parameters. Then, in order for $p(B|\Pi, \Omega)$ to be a proper density, it is necessary that at least m coefficients in each structural equation be *either predetermined or arguments of a proper prior density* (generalized order condition); it is sufficient that the corresponding submatrix from Π have full rank [generalized rank condition (2.6)].

More general cases are discussed in Drèze (1975, Section 2.3).

3.4. *Choice of models and identification*

Consider two simultaneous equation models, say M_1 with parameter $(B_1, \Gamma_1, \Sigma_1)$ and M_2 with parameter $(B_2, \Gamma_2, \Sigma_2)$. The variables Y and Z are taken to be the same for both models which, therefore, differ only by the set of exact *a priori* restrictions, say

$$\Psi_k^i(B, \Gamma, \Sigma) = 0; \qquad k = 1 \dots K_i, \tag{3.12}$$

[14] See, however, Harkema (1971) for an example of explicit derivation of the complete marginal prior density $p(\Pi, \Omega)$ in a class of underidentified models.

with associated parameter spaces $\mathcal{S}^i$, $\mathcal{T}^i$, $\mathcal{P}^i$ and $\mathcal{B}^i_{\Pi,\Omega}$ $(i=1,2)$. When $\mathcal{P}^1=\mathcal{P}^2$, we could say that the models M_1 and M_2 are not identified in relation to one another. A Bayesian generalization of this concept, introduced in Zellner (1971, p. 254), takes into account the prior densities $p(\Pi,\Omega|M_i)$ associated with models M_i $(i=1,2)$. The predictive densities, conditional on M_i, are

$$p(Y|M_i)=\int_{\mathcal{P}^i} p(Y|\Pi,\Omega)\,p(\Pi,\Omega|M_i)\,\mathrm{d}\Pi\,\mathrm{d}\Omega, \tag{3.13}$$

since, by definition of M_1 and M_2, $p(Y|\Pi,\Omega)\equiv p(Y|\Pi,\Omega,M_i)$, $i=1,2$. Then M_1 and M_2 and their associated prior densities are *not identified in relation to one another*[15] if and only if

$$p(Y|M_1)\equiv p(Y|M_2). \tag{3.14}$$

In such a case, the prior odds for M_1 and M_2 are not revised through observations, and the posterior odds satisfy

$$\frac{p(M_1|Y)}{p(M_2|Y)}=\frac{p(M_1)}{p(M_2)},\quad \text{identically in } Y. \tag{3.15}$$

4. Reduced-form analytics

4.1. Natural-conjugate prior densities

The analytics in this section are *not* presented because we regard natural-conjugate analysis of the reduced form as useful for applications. They are presented for easy reference in subsequent sections and as a review of basic tools in Bayesian econometrics.

Provided rank $Z=n<T$ (and $n+m<T$ to validate other results below), the likelihood function (1.6) can be expressed in terms of the least-squares estimates:

$$\hat{\Pi}=(Z'Z)^{-1}Z'Y,\qquad (T-n)\hat{\Omega}=W=Y'\left[I-Z(Z'Z)^{-1}Z'\right]Y,$$
$$M=Z'Z, \tag{4.1}$$

$$L(\Pi,\Omega|Y)\propto|\Omega|^{-(1/2)T}\exp\left[-\tfrac{1}{2}\operatorname{tr}\Omega^{-1}\left[(\Pi-\hat{\Pi})'M(\Pi-\hat{\Pi})+W\right]\right]. \tag{4.2}$$

[15] See also Florens et al. (1974) where the authors discuss a closely related concept of D-identification.

The right-hand side of (4.2) is also a kernel of a Normal–Inverted–Wishart density on (Π, Ω). The likelihood function may therefore be rewritten as

$$L(\Pi,\Omega|Y) \propto f_{\mathrm{MN}}^{T\times m}(\Pi|\hat{\Pi},\Omega\otimes M^{-1}) f_{\mathrm{iW}}(\Omega|W,T-n-m-1). \tag{4.3}$$

Thus a natural-conjugate prior density for (Π, Ω) is given by the product of a conditional matricvariate normal prior density:

$$p(\Pi|\Omega) = f_{\mathrm{MN}}^{n\times m}(\Pi|\Pi_0,\Omega\otimes M_0^{-1}), \quad \text{with } (\Pi_0, M_0) \in R^{nm}\times \mathcal{C}^n, \tag{4.4}$$

and a marginal Inverted–Wishart prior density:

$$p(\Omega) = f_{\mathrm{iW}}^{m}(\Omega|W_0,\nu_0), \quad \text{with } W_0 \in \mathcal{C}^m \text{ and } \nu_0 > m-1. \tag{4.5}$$

The prior expectations of Π and Ω are

$$E(\Pi) = E(\Pi|\Omega) = \Pi_0, \qquad E(\Omega) = \frac{1}{\nu_0 - m - 1} W_0 \qquad (\nu_0 > m+1). \tag{4.6}$$

The conditional and marginal covariance matrices of $\operatorname{vec}\Pi$, the column expansion of Π, are

$$V(\Pi|\Omega) = \Omega\otimes M_0^{-1}, \qquad V(\Pi) = \frac{1}{\nu_0 - m - 1} W_0\otimes M_0^{-1}, \tag{4.7}$$

where advantage has been taken of the fact that Π and Ω are *linearly* independent even though they are clearly non-independent (implications of this feature for applied work are discussed below). The (hyper) parameters (Π_0, M_0, W_0) and ν_0 should reflect the *a priori* information. The marginal prior density on Π implied by (4.4) and (4.5) is a matricvariate-t density:

$$\begin{aligned} p(\Pi) &= f_{\mathrm{M}t}^{n\times m}(\Pi|\Pi_0,W_0,M_0,\nu_0) \\ &\propto |W_0 + (\Pi-\Pi_0)'M_0(\Pi-\Pi_0)|^{-(1/2)(\nu_0+n)}. \end{aligned} \tag{4.8}$$

As usual in a natural-conjugate framework, both the prior and the posterior densities have the same functional form, so that

$$p(\Pi|\Omega,Y) = f_{\mathrm{MN}}^{n\times m}(\Pi|\Pi_*,\Omega\otimes M_*^{-1}), \tag{4.9}$$

$$p(\Omega|Y) = f_{\mathrm{iW}}^{m}(\Omega|W_*,\nu_*), \tag{4.10}$$

$$p(\Pi|Y) = f_{\mathrm{M}t}^{n\times m}(\Pi|\Pi_*,W_*,M_*,\nu_*), \tag{4.11}$$

where[16]

$$M_* = M_0 + M, \qquad \Pi_* = M_*^{-1}(M_0\Pi_0 + M\hat{\Pi}), \qquad \nu_* = \nu_0 + T \tag{4.12a}$$

and

$$\begin{aligned} W_* &= (W_0 + \Pi_0' M_0 \Pi_0) + (W + \hat{\Pi}' M \hat{\Pi}) - \Pi_*' M_* \Pi_* \\ &= W_0 + W + (\hat{\Pi} - \Pi_0)'[M_0^{-1} + M^{-1}]^{-1}(\hat{\Pi} - \Pi_0) \\ &= W_0 + (Y - X\Pi_0)'(I_T - XM_*^{-1}X')(Y - X\Pi_0). \end{aligned} \tag{4.12b}$$

As initially noted by Rothenberg (1963), the natural-conjugate family has a very restrictive feature. In (4.7), the $nm \times nm$ covariance matrix of vec Π takes the form of Kronecker product with, therefore $\frac{1}{2}m(m+1)+\frac{1}{2}n(n+1)$ instead of $\frac{1}{2}nm(nm+1)$ free (hyper) parameters. This implies in particular that the covariance matrices between pairs of columns of Π are all proportional to M_0^{-1}.

The only relevant situations where this feature is harmless are:

(i) $M_0 = 0$: The prior density is non-informative about Π (and may or may not be informative about Ω).
(ii) $w_{ij}^0 = 0$ for all i, j except when $i = j = 1$: The prior density is informative about the parameters of a single reduced-form equation (here taken to be the first, through appropriate labelling of variables).

In Section 6.4 we shall define an *extended natural-conjugate* prior density for (Π, Ω), where the (conditional) covariance matrix of vec Π is no longer constrained to the form of a Kronecker product. The cost of this generalization is the need to rely, partially at least, on numerical methods – a typical trade-off in the Bayesian framework.

4.2. *Further results*

We have already mentioned that, under a natural-conjugate prior density, Π and Ω are not independent. The practical consequences of this undesirable feature will

[16] As shown for example in Raiffa and Schlaifer (1961), these formulae derive from the standard convolution rules for the sufficient statistics of successive independent samples. Let $(Y_0 \;\; Z_0)$ be a matrix of observations for a (hypothetical) sample and $(Y_* \;\; Z_*)$ be the pooled samples. Then

$$M_* = Z_*'Z_* = Z_0'Z_0 + Z'Z = M_0 + M,$$

$$M_*\Pi_* = Z_*'Y_* = Z_0'Y_0 + Z'Y = M_0\Pi_0 + M\hat{\Pi},$$

$$W_* + \Pi_*'M_*\Pi_* = Y_*'Y_* = Y_0'Y_0 + Y'Y = (W_0 + \Pi_0'M_0\Pi_0) + (W + \hat{\Pi}'M\hat{\Pi}).$$

be illustrated in Section 6.6. We could assume instead prior independence between Π and Ω and, in particular, consider the prior density

$$p(\Pi,\Omega)=f_{\mathrm{M}t}^{n\times m}(\Pi|\Pi_0,Q_0,M_0,\mu_0)f_{\mathrm{iW}}^{m}(\Omega|W_0,\lambda_0). \tag{4.13}$$

When $Q_0=W_0$ and $\mu_0=\lambda_0=\nu_0$, the first and second moments of (Π,Ω) under (4.13) are the same as those obtained under the natural-conjugate prior densities (4.4) and (4.5). A kernel of the posterior density is given by the product of the prior density (4.13) and the likelihood function (4.3). Marginalization with respect to Ω yields the posterior density

$$p(\Pi|Y)\propto f_{\mathrm{M}t}^{n\times m}(\Pi|\Pi_0,Q_0,M_0,\mu_0)f_{\mathrm{M}t}^{n\times m}\big(\Pi|\hat{\Pi},W+W_0,M,\lambda_0+T-n\big). \tag{4.14}$$

As such this density is not amenable to further analytical treatment. However, if we partition the reduced form as in (2.7), the posterior density of Π_1, the vector of coefficients of the first reduced-form equation, *conditionally* on Π_2, is a 2–0 poly-*t* density,[17] i.e. a product of two kernels of Student densities. As discussed in Section 7.2, such densities can be evaluated numerically by means of one-dimensional numerical integrations, independently of n, the number of coefficients in Π_1. Furthermore, as we shall outline in Section 7.3, this property can be exploited in order to construct so-called importance functions for the Monte Carlo numerical integration of the joint posterior density (4.14).

If we are interested only in Π_1, an obvious alternative to the above conditional analysis amounts to using a prior density which is non-informative on (Π_2,Ω), following thereby the limited information techniques outlined in Section 2. Following (2.14), let

$$p(\Pi,\Omega)\propto p(\Pi_1)\cdot|\Omega_{22.1}|^{-(1/2)n}|\Omega|^{-(1/2)(m+1)}, \tag{4.15}$$

where, for the purpose of comparison with (4.8), $p(\Pi_1)$ is taken to be the Student density:

$$p(\Pi_1)\propto f_t^{n}\left(\Pi_1|\Pi_1^0,\frac{1}{w_{11}^0}M_0,\nu_0-m+1\right). \tag{4.16}$$

The posterior density of Π_1 is then obtained by a direct transposition of formula

[17] To see this, factorize the two kernels in the right-hand side of (4.14) according to formulae (A.33) and (A.34) in Appendix A.

(2.16):[18]

$$p(\Pi_1) \propto f_t^n\left(\Pi_1|\Pi_1^0, \frac{1}{w_{11}^0}M_0, \nu_0 - m + 1\right) f_t^n\left(\Pi_1|\hat{\Pi}_1, \frac{1}{w_{11}}M, T - n - m + 1\right), \tag{4.17}$$

and is, therefore, again a 2–0 poly-t density.

It will be shown in Section 6.4 that 2–0 poly-t densities are also obtained under extended natural-conjugate prior densities.

As noted by Zellner (1971, p. 101) the density (4.17) can be approximated by the normal density $f_N^n(\Pi_1|\overline{\Pi}_1, \overline{M})$, with parameters

$$\overline{M} = (\nu_0 - m + 1)\frac{M_0}{w_{11}^0} + (T - n - m + 1)\frac{M}{w_{11}} \tag{4.18}$$

and

$$\overline{\Pi}_1 = \overline{M}^{-1}\left[(\nu_0 - m + 1)\frac{M_0}{w_{11}^0}\Pi_1^0 + (T - n - m + 1)\frac{M}{w_{11}}\hat{\Pi}_1\right]. \tag{4.19}$$

However, this approximation should not be used indiscriminately. For example, the density (4.17) could be bimodal or strongly asymmetrical, in which cases the normal approximation would be misleading.

Two forms of "non-informative" prior densities have been used for (Π, Ω):

(i) If we start from the natural-conjugate prior density (4.4) and (4.5) and let M_0, W_0, and ν_0 tend to zero, we obtain:

$$p(\Pi, \Omega) \propto |\Omega|^{-(1/2)(m+n+1)}, \tag{4.20}$$

which is found for example in Drèze (1976). Note that (4.20) includes a factor $|\Omega|^{-(1/2)n}$ arising from the integrating constant of (4.4).

(ii) If we start instead from the independent prior density (4.13) and let M_0, W_0, μ_0, and λ_0 tend to zero, we obtain:

$$p(\Pi, \Omega) \propto |\Omega|^{-(1/2)(m+1)}, \tag{4.21}$$

which is advocated [for example by Zellner (1971)] on the basis of Jeffreys' invariance principles.

[18] We just replace β_Δ' by $(1 \quad 0')$ and γ_* by Π_1.

We are not aware of any compelling argument to prefer either specification over the other. In the posterior densities (4.10)–(4.11), the choice affects only ν_*, which is equal to $T+n$ under (4.20) and T under (4.21). A conservative attitude would favor (4.21) which results in larger posterior variances. As shown in Section 5, however, the prior densities used under limited information analysis are closer to (4.20) than to (4.21).

5. Limited information analysis

5.1. Introduction

As indicated in Sections 1.1 and 2.1, classical limited information analysis estimates the parameters of a single structural equation, taking into account the *a priori* restrictions imposed on these parameters, but ignoring the restrictions imposed on the parameters of the remaining structural equations. The specification of these additional equations is used only to complete the list of exogenous variables. Stated alternatively, limited information analysis ignores all the over-identifying restrictions pertaining to these remaining equations, and uses only the restrictions on the reduced-form parameters implied by overidentifying restrictions pertaining to the single structural equation of interest.[19]

A Bayesian analogue of this approach similarly uses the exact *a priori* restrictions pertaining only to the parameters of a single structural equation, and ignores the remaining restrictions. In addition, it rests on a fully specified prior density for the parameters of the structural equation of interest, together with a non-informative prior density on the remaining parameters of the system. There is some arbitrariness in the choice of a parameterization for the rest of the system. It seems accordingly desirable to specify the prior density in such a way that it be invariant with respect to this choice. This problem is discussed in Section 5.2. The reader who is not interested in the technicalities associated with the definition of non-informative prior densities may turn immediately to Sections 5.3 and 5.4 which are essentially self-contained. The illustration of Section 2.4 is extended in Section 5.5. Section 5.6 deals with normalization and Section 5.7 with generalizations.

5.2. Parameterization and invariance

Using the notation of Section 2, we study a single structural equation

$$Y\beta + Z\gamma = u, \tag{2.2}$$

[19]See, for example, Goldberger (1964, p. 346), Maddala (1977, p. 231), Malinvaud (1978, p. 759), Theil (1971, p. 503), or Zellner (1971, pp. 264–265).

where u is normally distributed with expectation zero and covariance $\sigma^2 I_T$. Conditionally on $(\beta, \gamma, \sigma^2)$ the reduced-form parameters (Π, Ω) satisfy the $n+1$ linear equalities:

$$\Pi\beta + \gamma = 0, \qquad \beta'\Omega\beta = \sigma^2. \tag{2.3}$$

Define, for notational convenience, $\theta = (\beta, \gamma, \sigma^2)$ – see footnote 5 – and let Θ denote the set of admissible values for θ. Conditionally on θ, the space of reduced-form parameters is

$$\mathcal{P}_\theta = \{\Pi \in R^{nm} | \Pi\beta + \gamma = 0\} \times \{\Omega \in \mathcal{C}^m | \beta'\Omega\beta = \sigma^2\} =_{\text{def}} \mathcal{P}_\theta^\Pi \times \mathcal{P}_\theta^\Omega. \tag{5.1}$$

In a limited information framework we want to define a "non-informative" prior measure on $\mathcal{P}_\theta$. If we ignore the restrictions (2.3), then (Π, Ω) will vary freely over $R^{nm} \times \mathcal{C}^m$, and we could use the non-informative prior measures (4.20) or (4.21).

Given (2.3), however, (Π, Ω) are not variation free. In order to overcome this difficulty (see also the remark at the end of this section), we shall define a family of one-to-one mappings from $\mathcal{P}_\theta = \mathcal{P}_\theta^\Pi \times \mathcal{P}_\theta^\Omega$ onto $R^{n(m-1)} \times (R^{m-1} \times \mathcal{C}^{m-1})$. These mappings are indexed by an (essentially) arbitrary $m(m-1)$ matrix Q. Each mapping (each Q) defines a variation-free reparameterization of the reduced form. The corresponding expressions for the likelihood function are given in Lemma 5.2. We then define in Lemma 5.3 a condition under which non-informative prior measures on the new parameter space are invariant with respect to Q. This condition defines uniquely our non-informative prior measure on $\mathcal{P}_\theta$. Finally, two corollaries relate this result to the special case of Section 2 and to prior measures appearing in earlier work.

We first note that $\mathcal{P}_\theta^\Pi$ is an $n(m-1)$-dimensional linear subspace of R^{nm}. Let Q be an arbitrary $m(m-1)$ matrix of rank $m-1$, such that

$$P = (\beta \quad Q) \tag{5.2}$$

is non-singular almost everywhere[20] and let

$$\Pi P =_{\text{def}} \Phi = (\Pi\beta \quad \Pi Q) = (-\gamma \quad \Phi_2). \tag{5.3}$$

Conditionally on θ, (5.3) defines a one-to-one mapping from $\mathcal{P}_\theta^\Pi$ onto $R^{n(m-1)}$, the space of Φ_2, with reciprocal

$$\Pi = \Phi P^{-1} = (-\gamma \quad \Phi_2) P^{-1}. \tag{5.4}$$

[20] The notation Q_β might be advisable in order to stress this condition.

$\mathcal{P}_\theta^\Omega$ is not a linear subspace of $\mathcal{C}^m$. However, it is isomorphic to $R^{m-1} \times \mathcal{C}^{m-1}$, as evidenced by the following property:

Lemma 5.1

Let Λ be an $m \times m$ matrix partitioned as

$$\Lambda = \begin{pmatrix} \Lambda_{11} & \Lambda_{12} \\ \Lambda_{21} & \Lambda_{22} \end{pmatrix}, \tag{5.5}$$

where Λ_{ij} is $m_i \times m_j$, and define

$$\Delta_{12} = \Lambda_{11}^{-1}\Lambda_{12}, \qquad \Lambda_{22.1} = \Lambda_{22} - \Lambda_{21}\Lambda_{11}^{-1}\Lambda_{12}. \tag{5.6}$$

Then $\Lambda \in \mathcal{C}^m$ if and only if

$$(\Lambda_{11}, \Delta_{12}, \Lambda_{22.1}) \in \mathcal{C}^{m_1} \times R^{m_1 m_2} \times \mathcal{C}^{m_2}. \tag{5.7}$$

Proof

We rewrite Λ as

$$\Lambda = \begin{pmatrix} I_{m_1} & 0 \\ \Delta_{12}' & I_{m_2} \end{pmatrix} \begin{pmatrix} \Lambda_{11} & 0 \\ 0 & \Lambda_{22.1} \end{pmatrix} \begin{pmatrix} I_{m_1} & \Delta_{12} \\ 0 & I_{m_2} \end{pmatrix}. \tag{5.8}$$

□

We shall apply Lemma 5.1 with $m_1 = 1$ and

$$\Lambda = P'\Omega P = \begin{pmatrix} \sigma^2 & \beta'\Omega Q \\ Q'\Omega\beta & Q'\Omega Q \end{pmatrix}. \tag{5.9}$$

Conditionally on θ, (5.9) defines a one-to-one mapping from $\mathcal{P}_\theta^\Omega$ onto $R^{m-1} \times \mathcal{C}^{m-1}$, namely

$$\Delta_{12} = \sigma^{-2}\beta'\Omega Q, \tag{5.10}$$

$$\Lambda_{22.1} = Q'\Omega Q - \sigma^{-2}Q'\Omega\beta\beta'\Omega Q, \tag{5.11}$$

with reciprocal

$$\Omega = P'^{-1}\begin{pmatrix} 1 & 0 \\ \Delta_{12}' & I_{m-1} \end{pmatrix} \begin{pmatrix} \sigma^2 & 0 \\ 0 & \Lambda_{22.1} \end{pmatrix} \begin{pmatrix} 1 & \Delta_{12} \\ 0 & I_{m-1} \end{pmatrix} P^{-1}. \tag{5.12}$$

The mappings (5.3), (5.10), and (5.11) define the parameterization $(\theta, \Phi_2, \Delta_{12}, \Lambda_{22.1})$, with parameter space $\Theta \times R^{n(m-1)} \times R^{m-1} \times \mathcal{C}^{m-1}$. These parameters are *variation free*. Conditionally on θ, $(\Phi_2, \Delta_{12}, \Lambda_{22.1})$ are in one-to-one correspondence with $(\Pi, \Omega) \in \mathcal{P}_\theta$. Note, however, that the definition of these parameters depends on the choice of Q.

We can now rewrite the likelihood function in terms of the new parameters.

Lemma 5.2

The likelihood function (1.6) may be rewritten and factorized as

$$L(\theta, \Phi_2, \Delta_{12}, \Lambda_{22.1} | Y) = L_1(\theta | Y) \cdot L_2(\theta, \Phi_2, \Delta_{12}, \Lambda_{22.1} | Y), \tag{5.13}$$

where

$$L_1(\theta | Y) \propto f_N^T(Y\beta + Z\gamma | 0, \sigma^2 I_T), \tag{5.14}$$

$$L_2(\theta, \Phi_2, \Delta_{12}, \Lambda_{22.1} | Y)$$

$$\propto \|P\|^T f_{MN}^{T \times (m-1)}\left(YQ | (Y\beta + Z\gamma \quad Z)\begin{pmatrix} \Delta_{12} \\ \Phi_2 \end{pmatrix}, \Lambda_{22.1} \otimes I_T \right). \tag{5.15}$$

Proof

Combining formulae (1.6), (5.4), (5.10), and (5.11), we have successively:

$$\begin{aligned} &\operatorname{tr} \Omega^{-1}(Y - Z\Pi)'(Y - Z\Pi) \\ &\quad = \operatorname{tr} \Lambda^{-1}(Y\beta + Z\gamma \quad YQ - Z\Phi_2)'(Y\beta + Z\gamma \quad YQ - Z\Phi_2) \\ &\quad = \sigma^{-2}(Y\beta + Z\gamma)'(Y\beta + Z\gamma) + \operatorname{tr} \Lambda_{22.1}^{-1} \\ &\qquad \times [YQ - Z\Phi_2 - (Y\beta + Z\gamma)\Delta_{12}]'[YQ - Z\Phi_2 - (Y\beta + Z\gamma)\Delta_{12}]. \end{aligned}$$

Also, $|\Omega| = \|P\|^{-2}|\Lambda| = \sigma^2 |P|^{-2} |\Lambda_{22.1}|$. □

Our prior density will be similarly expressed as

$$p(\theta, \Phi_2, \Delta_{12}, \Lambda_{22.1}) = p(\theta) \cdot p(\Phi_2, \Delta_{12}, \Lambda_{22.1} | \theta). \tag{5.16}$$

We want $p(\Phi_2, \Delta_{12}, \Lambda_{22.1} | \theta)$ to be both non-informative and invariant with respect to Q. In the Normal–Wishart and Student–Wishart families, a non-informative measure will be of the form

$$p(\Phi_2, \Delta_{12}, \Lambda_{22.1} | \theta) \propto |\Lambda_{22.1}|^{-(1/2)(\nu + m + n + 1)}, \qquad \nu \in R. \tag{5.17}$$

The following lemma yields the desired invariance.

Lemma 5.3

The prior measure (5.17) is invariant with respect to Q if and only if $\nu = 0$.

Proof

See Appendix B. □

Note that for $\nu = 0$, the exponent in (5.17) may be written as $(m-1)+(n+1)+1$, where $(m-1)$ and $(n+1)$ represent respectively the number of "dependent" variables and the number of "regressors" in the partial model (5.15), θ being given. The prior measure (5.17) could therefore also be justified on the basis of Jeffreys' invariance principles. Yet, as mentioned at the end of Section 4.2, these principles may lead to different choices of exponents, depending on whether or not the number of regressors is taken into account. If it is not, the resulting prior measure is no longer invariant with respect to Q, as evidenced by Lemma 5.3.

Lemma 5.3 can now be applied to the prior measures introduced in Section 2. In Section 2.1, Q was specified by (2.8) as[21]

$$Q = \begin{pmatrix} 0' \\ I_{m-1} \end{pmatrix}, \tag{5.18}$$

in which case $\Phi_2 = \Pi_2$ and $\Lambda = \Omega^*$, the covariance matrix of $(u \quad V_2)$. Δ_{12} and $\Lambda_{22.1}$ are then given by

$$\Delta_{12} = \sigma^{-2}\Omega^*_{12} \tag{5.19}$$

and

$$\Lambda_{22.1} = \Omega^*_{22.1} = \Omega^*_{22} - \Omega^*_{21}\Omega^{*-1}_{11}\Omega^*_{12}, \tag{5.20}$$

and represent respectively the regression coefficients of V_2 on u and the covariance matrix of any row of V_2 given u.

Corollary 5.4

The prior measures

$$p(\beta_\Delta, \gamma_*, \Pi_2, \Omega^*) \propto p(\beta_\Delta, \gamma_*)|\Omega^*_{22.1}|^{-(1/2)n}|\Omega^*|^{-(1/2)(m+1)}, \tag{2.14}$$

$$p(\beta_\Delta, \gamma_*, \Pi_2, \Omega^*) \propto \sigma^{-(m_1+n_1)}|\Omega^*_{22.1}|^{-(1/2)n}|\Omega^*|^{-(1/2)(m+1)}, \tag{2.17}$$

are invariant with respect to Q. Under (2.14), the "marginal" prior measure $p(\theta)$

[21] A similar choice appears in Zellner (1971, section 9.5).

is given by

$$p(\theta) \propto p(\beta_\Delta, \gamma_*)(\sigma^2)^{(1/2)(m-3)}. \tag{5.21}$$

Under (2.17), it is given instead by

$$p(\theta) \propto (\sigma^2)^{(1/2)(m-m_1-n_1-3)}. \tag{5.22}$$

Proof

We make the integrand transformation from Ω^* $(=\Lambda)$ to $(\sigma^2, \Delta_{12}, \Lambda_{22.1})$, where Δ_2 and $\Lambda_{22.1}$ are given in (5.19) and (5.20). The Jacobian of this transformation is $(\sigma^2)^{m-1}$. After transformation, (2.14) and (2.17) factorize into the product of (5.17) with (5.21) and (5.22), respectively. The proof follows from the invariance of (5.17). □

Remark

Earlier work on limited information Bayesian procedures often relied on prior measures expressed in terms of (θ, Π, Ω). This raises problems of interpretation, due to the restrictions (2.3). The following corollary helps to relate such measures to the invariant measures (5.17).

Corollary 5.5

Prior measures of the form

$$p(\beta, \gamma, \Phi_2, \Lambda) \propto f_0(\beta, \gamma, \sigma^2)|\Lambda|^{-(1/2)(m+n+1)}, \tag{5.23}$$

where $f_0(\beta, \gamma, \sigma^2)$ is an arbitrary measure, are invariant with respect to Q. After reparameterization they factor into the product of (5.17) and

$$p(\beta, \gamma, \sigma^2) \propto (\sigma^2)^{-(1/2)(n-m+3)} f_0(\beta, \gamma, \sigma^2). \tag{5.24}$$

Proof

As in the proof of Corollary 5.4, make the integrand transformation from Λ to $(\sigma^2, \Delta_{12}, \Lambda_{22.1})$. □

Corollary 5.5 suggests that the marginal density $p(\beta, \gamma, \sigma^2)$ should include the additional factor $(\sigma^2)^{-(1/2)(n-m+3)}$ as in (5.24), thereby removing an ambiguity in the interpretation of (5.23) which has hampered previous presentations, as in Drèze (1976).

5.3. Posterior conditional densities and moments

We now derive posterior densities under the invariant prior measures (5.17). Let

$$S = P'WP = \begin{pmatrix} \beta'W\beta & \beta'WQ \\ Q'W\beta & Q'WQ \end{pmatrix} = \begin{pmatrix} s_{11} & S_{12} \\ S_{21} & S_{22} \end{pmatrix}, \tag{5.25}$$

where, following (4.1), $W = Y'[I - Z(Z'Z)^{-1}Z']Y$.

Lemma 5.6

Under the prior measure

$$p(\theta, \Phi_2, \Delta_{12}, \Lambda_{22.1}) \propto p(\theta)|\Lambda_{22.1}|^{-(1/2)(m+n+1)} \tag{5.26}$$

the joint posterior density $p(\theta, \Phi_2, \Delta_{12}, \Lambda_{22.1}|Y)$ may be written as the product of[22]

$$p(\theta|Y) \propto p(\theta)|\beta'W\beta|^{(1/2)(T-m+1)} f_N^T(Y\beta + Z\gamma|0, \sigma^2 I_T), \tag{5.27}$$

$$p(\Lambda_{22.1}|\theta, Y) = f_{iW}^{m-1}(\Lambda_{22.1}|S_{22.1}, T), \tag{5.28a}$$

$$p(\Delta_{12}|\Lambda_{22.1}, \theta, Y) = f_{MN}^{1\times(m-1)}\left(\Delta_{12}|(\beta'W\beta)^{-1}\beta'WQ, (\beta'W\beta)^{-1}\Lambda_{22.1}\right), \tag{5.28b}$$

$$p(\Phi_2|\Delta_{12}, \Lambda_{22.1}, \theta, Y) = f_{MN}^{n\times(m-1)}\left(\Phi_2|\hat{\Pi}Q - (\hat{\Pi}\beta + \gamma)\Delta_{12}, \Lambda_{22.1}\otimes M^{-1}\right). \tag{5.28c}$$

Proof

See Appendix B. □

Digressing briefly, we use Lemma 5.6 at once to validate results quoted in Section 2.2.

Corollary 5.7

Under the prior measure (2.14), the posterior density of (β_Δ, γ_*) is given by

$$p(\beta_\Delta, \gamma_*|Y) \propto p(\beta_\Delta, \gamma_*)\left\{\frac{\beta_\Delta' W_{\Delta\Delta}\beta_\Delta}{(Y_\Delta\beta_\Delta + Z_*\gamma_*)'(Y_\Delta\beta_\Delta + Z_*\gamma_*)}\right\}^{(1/2)(T-m+1)}. \tag{2.16}$$

[22]The density (5.28b) could indifferently be written as $f_N^{m-1}(\Delta_{12}'|S_{21}s_{11}^{-1}, s_{11}^{-1}\Lambda_{22.1})$. The matrix normal notation generalizes to subsystems, as discussed in Section 5.5.2 below.

Under the prior measure (2.17) it is given by

$$p(\beta_\Delta, \gamma_*|Y) \propto (\beta_\Delta' W_{\Delta\Delta}\beta_\Delta)^{(1/2)(T-m+1)}$$
$$\times\left[(Y_\Delta\beta_\Delta + Z_*\gamma_*)'(Y_\Delta\beta_\Delta + Z_*\gamma_*)\right]^{-(1/2)(T-m+m_1+n_1+1)}. \tag{2.18}$$

Proof

Under (2.14), we combine formulae (5.21) and (5.27), obtaining:

$$p(\theta|Y) \propto p(\beta_\Delta, \gamma_*)(\beta_\Delta' W_{\Delta\Delta}\beta_\Delta)^{(1/2)(T-m+1)}(\sigma^2)^{-(1/2)(T-m+3)}$$
$$\times\exp\left[\left(-\frac{1}{2\sigma^2}\right)(Y_\Delta\beta_\Delta + Z_*\gamma_*)'(Y_\Delta\beta_\Delta + Z_*\gamma_*)\right],$$

from which (2.16) follows. Under (2.18) we simply replace (5.21) by (5.22). □

The posterior density (5.27) is discussed extensively in Section 5.4. As for (5.28), it enables us to derive the posterior moments for (Π, Ω) and the predictive moments for Y, conditionally on θ.

Corollary 5.8

Under the prior measure (5.26), the posterior mean of Π, the posterior covariance matrix of the column expansion of Π, and the posterior mean of Ω, are given conditionally on θ by

$$E(\Pi|\theta, Y) = \hat{\Pi} - (\hat{\Pi}\beta + \gamma)(\beta' W\beta)^{-1}\beta' W, \tag{5.29}$$

$$V(\operatorname{vec}\Pi|\theta, Y) = \frac{1}{T-m}\left[W - W\beta(\beta' W\beta)^{-1}\beta' W\right]$$
$$\otimes\left[M^{-1} + (\hat{\Pi}\beta + \gamma)(\beta' W\beta)^{-1}(\hat{\Pi}\beta + \gamma)'\right], \tag{5.30}$$

$$E(\Omega|\theta, Y) = W\beta(\beta' W\beta)^{-1}\sigma^2(\beta' W\beta)^{-1}\beta' W$$
$$+\frac{1}{T-m}\left[1 + \sigma^2(\beta' W\beta)^{-1}\right]\left[W - W\beta(\beta' W\beta)^{-1}\beta' W\right]. \tag{5.31}$$

Proof

See Appendix B. □

All these expressions are homogeneous of degree zero in (β, γ, σ) and are therefore invariant with respect to the normalization rule – a property discussed in Section 5.6.

As a result of (2.3), the conditional moments of (Π, Ω) satisfy:

$$E(\Pi\beta + \gamma|\theta, Y) \equiv 0, \tag{5.32}$$

$$V(\Pi\beta + \gamma|\theta, Y) \equiv 0, \tag{5.33}$$

$$E(\beta'\Omega\beta|\theta, Y) \equiv \sigma^2. \tag{5.34}$$

Predictive moments are by-products of formulae (5.29)–(5.31). Let (Z_f, Y_f) denote a future sample. We then have

$$\begin{aligned} &E(Y_f|Z_f, \theta, Y) = Z_f E(\Pi|\theta, Y), \\ &V(\operatorname{vec} Y_f|Z_f, \theta, Y) = E(\Omega|\theta, Y) \otimes I + V[\operatorname{vec}(Z_f\Pi)|\theta, Y], \end{aligned} \tag{5.35}$$

where $V[\operatorname{vec}(Z_f\Pi)|\theta, Y]$ is obtained from (5.30) by pre- and postmultiplying the last term respectively by Z_f and Z_f'.

5.4. *Posterior marginal densities*

We now concentrate on the analysis of the posterior density of θ, as given in (5.27). This density is of direct interest for inference on θ and it is also required to marginalize the expression derived in Corollary 5.8. The form of $p(\theta|Y)$ indicates that, conditionally on β, we can apply the usual analysis of univariate regression models, as discussed for example in Zellner (1971), whereby $p(\theta)$ is either a Normal–gamma density – as in (5.40) below – or a Student–gamma density – as in (5.49) below. More general cases must be treated by means of numerical integration, using the techniques described in Section 7, and will not be discussed. Also, we only consider exclusion restrictions, since more general linear restrictions can be handled through suitable linear transformations of the coefficients and the variables.

The notations are those of Section 2 and we have $\beta' = (\beta_\Delta' \;\; 0')$ and $\gamma' = (\gamma_*' \;\; 0')$, where $\beta_\Delta \in R^{m_1}$ and $\gamma_* \in R^{n_1}$. The data matrices are partitioned conformably as in (2.4). As mentioned in Section 3.3, we need not impose the condition $m_1 + n_1 \leqslant n$ provided $p(\theta)$ is sufficiently informative. Normalization is dealt with by introducing the transformation from $(\beta_\Delta, \gamma_*, \sigma^2)$ to $(\beta_1, \alpha, \sigma_1^2)$, where

$$\alpha' = (\alpha_1' \;\; \alpha_2') = \frac{1}{\beta_1}(\beta_2 \ldots \beta_{m_1} \;\; \gamma_1 \ldots \gamma_{n_1}) \in R^{l_1}, \tag{2.20}$$

with $l_1 = (m_1 - 1) + n_1$ and

$$\sigma_1^2 = \left(\frac{\sigma}{\beta_1}\right)^2. \tag{5.36}$$

β_1 is now unidentified and we shall draw inference only on α and σ_1^2. The quadratic form $\beta_\Delta' W_{\Delta\Delta} \beta_\Delta$ may be rewritten as

$$\beta_\Delta' W_{\Delta\Delta} \beta_\Delta = s_1^2 + (\alpha_1 - a_1)' H_1 (\alpha_1 - a_1), \tag{5.37}$$

where the statistics $(a_1, H_1, s_1^2) \in R^{m_1 - 1} \times \mathcal{C}^{m_1 - 1} \times \mathcal{C}^1$ – see Lemma 5.1 – are defined through the identity

$$W_{\Delta\Delta} = \begin{pmatrix} s_1^2 + a_1' H_1 a_1 & -a_1' H_1 \\ -H_1 a_1 & H_1 \end{pmatrix}. \tag{5.38}$$

Following (5.27), the posterior density of (α, σ_1^2) is given by

$$\begin{aligned} p(\alpha, \sigma_1^2 | Y) &\propto p(\alpha, \sigma_1^2) \left[s_1^2 + (\alpha_1 - a_1)' H_1 (\alpha_1 - a_1) \right]^{(1/2)(T-m+1)} \\ &\quad \times f_N^T(y | -X\alpha, \sigma_1^2 I_T) \\ &\propto p(\alpha, \sigma_1^2) \left[f_t^{m_1 - 1}(\alpha_1 | a_1, s_1^{-2} H_1, \mu_1) \right]^{-1} f_N^T(y | -X\alpha, \sigma_1^2 I_T), \end{aligned} \tag{5.39}$$

where $\mu_1 = T - m - m_1 + 2$ and X is a $T \times l_1$ matrix consisting of the columns of $(Y \quad Z)$ corresponding to the elements of α.

Lemma 5.9

Under the prior density

$$p(\alpha, \sigma_1^2) = f_N^{l_1}(\alpha | a_0, \sigma_1^2 H_0^{-1}) f_{i\gamma}(\sigma_1^2 | s_0^2, \mu_0), \tag{5.40}$$

the posterior density of (α, σ_1^2) is given by

$$\begin{aligned} p(\alpha, \sigma_1^2 | Y) \propto &\left[f_t^{m_1 - 1}(\alpha_1 | a_1, s_1^{-2} H_1, \mu_1) \right]^{-1} \\ &\times f_N^{l_1}(\alpha | a_*, \sigma_1^2 H_*^{-1}) f_{i\gamma}(\sigma_1^2 | s_*^2, \mu_*), \end{aligned} \tag{5.41}$$

where

$$H_* = H_0 + X'X, \qquad a_* = H_*^{-1}(H_0 a_0 - X'y), \qquad \mu_* = \mu_0 + T, \tag{5.42}$$

$$s_*^2 = s_0^2 + (y + X a_0)'(I_T - X H_*^{-1} X')(y + X a_0), \tag{5.43}$$

and the posterior density of α is the 1–1 poly-t density

$$p(\alpha | Y) \propto \left[f_t^{m_1 - 1}(\alpha_1 | a_1, s_1^{-2} H_1, \mu_1) \right]^{-1} f_t^{l_1}(\alpha | a_*, s_*^{-2} H_*, \mu_*). \tag{5.44}$$

Proof

Up to an additional Student kernel in α_1, these expressions may be found for example in Zellner (1971, ch. 4). □

It is instructive to decompose (5.41) into conditional and marginal densities.

Corollary 5.10

Under the prior density (5.40), we have

$$p(\alpha_2|\alpha_1, Y) \propto f_t^{n_1}\Big(\alpha_2|a^*_{2.1}, \big[s^2_* + (\alpha_1 - a^*_1)'H^*_{11.2}(\alpha_1 - a^*_1)\big]^{-1}H^*_{22}, \times \mu_* + (m_1 - 1)\Big), \tag{5.45}$$

$$p(\sigma^2_1|\alpha_1, Y) \propto f_{i\gamma}\big(\sigma^2_1|s^2_* + (\alpha_1 - a^*_1)'H^*_{11.2}(\alpha_1 - a^*_1), \mu_* + (m_1 - 1)\big), \tag{5.46}$$

$$p(\alpha_1|Y) \propto \big[f_t^{m_1-1}(\alpha_1|a_1, s_1^{-2}H_1, \mu_1)\big]^{-1} f_t^{m_1-1}(\alpha_1|a^*_1, s_*^{-2}H^*_{11.2}, \mu_*), \tag{5.47}$$

where a_* and H_* have been partitioned conformably with α, and

$$a^*_{2.1} = a^*_2 - H^{*-1}_{22}H^*_{21}a^*_1, \qquad H^*_{11.2} = H^*_{11} - H^*_{12}H^{*-1}_{22}H^*_{21}. \tag{5.48}$$

Proof

These results follow from (5.41), by well-known properties of the Student and Inverted–gamma densities, and in particular by the identity

$$\begin{aligned} &s^2_* + (\alpha - a_*)'H_*(\alpha - a_*) \\ &\equiv \big[s^2_* + (\alpha_1 - a^*_1)'H^*_{11.2}(\alpha_1 - a^*_1)\big] + (\alpha_2 - a^*_{2.1})'H^*_{22}(\alpha_2 - a^*_{2.1}). \end{aligned}$$ □

The posterior density of α_1, being defined through a ratio of Student kernels, is a so-called 1–1 poly-t density. Furthermore, following Corollaries 5.8 and 5.10, the conditional posterior expectations of (Π, Ω) and covariance matrix of vec Π depend on quadratic forms in α_1 which are precisely those characterizing the Student kernels in $p(\alpha_1|Y)$. As discussed in Section 7, this makes the Bayesian limited information analysis of a single equation fully operational under the prior density (5.40).

Specifying a sensible prior density for σ^2_1 is generally a difficult task. Consequently, model-builders often use diffuse specifications. In such cases, it is advisable to assume prior independence between α and σ^2_1, as we shall illustrate in

Section 5.5. The prior density

$$p(\alpha,\sigma_1^2)=f_t^{l_1}(\alpha|a_0,G_0,\mu_0)f_{i\gamma}(\sigma_1^2|s_0^2,\lambda_0) \tag{5.49}$$

satisfies this independence requirement. Note that when $G_0=s_0^{-2}H_0$ and $\lambda_0=\mu_0$, the prior densities (5.40) and (5.49) imply the same first *and* second moments for (α,σ_1^2); in particular, α and σ_1^2 are still *linearly* independent under (5.40).

Lemma 5.11

Under the prior density (5.49), the posterior densities $p(\sigma_1^2|\alpha,Y)$ and $p(\alpha|Y)$ are given by

$$p(\sigma_1^2|\alpha,Y)\propto f_{i\gamma}(\gamma^2|s_0^2+s^2+(\alpha-\hat{a})'H(\alpha-\hat{a}),\lambda_*), \tag{5.50}$$

$$p(\alpha|Y)\propto\left[f_t^{m_1-1}(\alpha_1|a_1,s_1^{-2}H_1,\mu_1)\right]^{-1}f_t^{l_1}(\alpha|a_0,G_0,\mu_0)$$
$$\times f_t^{l_1}\left(\alpha|\hat{a},(s_0^2+s^2)^{-1}H,\lambda_*-m_1-n_1+1\right), \tag{5.51}$$

where

$$H=X'X,\qquad \hat{a}=-(X'X)^{-1}X'y,\qquad s^2=y'\left[I_T-X(X'X)^{-1}X'\right]y,$$
$$\lambda_*=\lambda_0+T. \tag{5.52}$$

Proof

See, for example, Zellner (1971, ch. 4). □

The posterior density $p(\alpha|Y)$ is now a so-called 2–1 poly-t density. As discussed in Section 7, this preserves the tractability of the approach.

5.5. *An application*

To illustrate, we review the application described in Section 2.4 under both the Normal–gamma prior density (5.40) and the Student–gamma density (5.49). In all cases under consideration, the (hyper) parameters in (5.40) and (5.49) are chosen in such a way that the prior density $p(\alpha)$ is given by (2.29).[23] This leaves

[23] That is, $\lambda_0=\mu_0=6$, and

$$a_0=0.415E-03\begin{pmatrix}0\\1\\0\end{pmatrix},$$

$$G_0=s_0^{-2}H_0=0.22007E+08\begin{pmatrix}0&0&0\\0&1&0\\0&0&0\end{pmatrix}.$$

Table 5.1
Limited information analysis of the demand equation

		β_{22}	γ_{23}	γ_{25}
Prior (5.40)	A_1	-0.6136	$0.6508E-03$	$0.3302E+02$
		(0.1831)	(0.0962)	(0.0821)
	A_2	-0.3992	$0.5108E-03$	$0.2518E+02$
		(0.5122)	(0.1896)	(0.2561)
	A_3	0.2115	$0.3578E-03$	$0.0797E+02$
		(1.001)	(0.1900)	(0.5812)
Prior (5.49)	B_1	-0.6346	$0.6808E-03$	$0.3307E+02$
		(0.1292)	(0.0763)	(0.0558)
	B_2	-0.8047	$0.7552E-03$	$0.4090E+02$
		(0.2575)	(0.1405)	(0.1143)
	B_3	-0.8538	$0.7706E-03$	$0.4343E+02$
		(0.9325)	(0.4003)	(0.4490)

only s_0^2 to be specified. Three different values are considered, namely 4, 16, and 64. The corresponding prior means and standard deviations of σ_{22} are 1, 4, and 16. For comparison, the FIML estimator of σ_{22} is 0.832, a value which should be unknown when $p(\sigma_1^2)$ is selected. Hopefully, the second and third values of s_0^2 should appear most implausible to a careful analyst. They have been chosen in order to exemplify the dangers of a careless assessment of $p(\sigma_1^2)$, especially under the Normal–gamma specification (5.40). The runs corresponding to the successive values of s_0^2 are labelled $A_1 \to A_3$ under (5.40) and $B_1 \to B_3$ under (5.49). The posterior means and standard deviations of α are reported in Table 5.1. The graphs of the posterior densities of β_{22} are given in Figures 5.1 and 5.2.

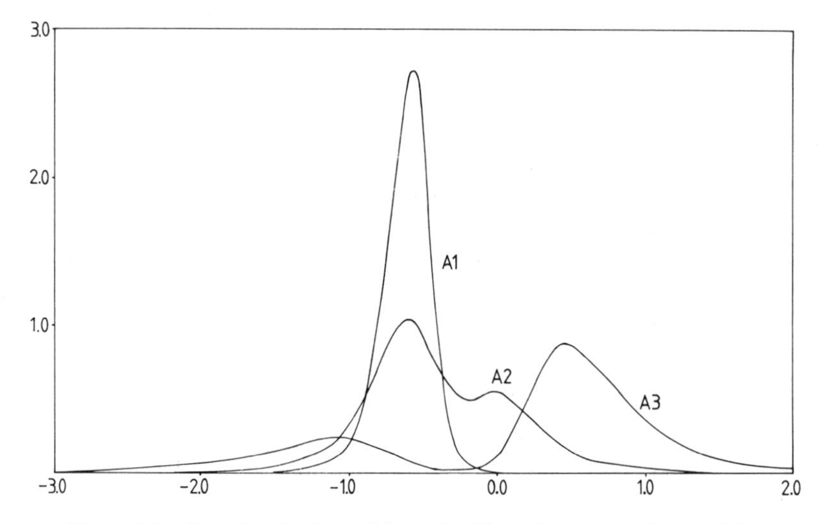

Figure 5.1. Posterior densities of β_{22} under Normal–gamma prior densities.

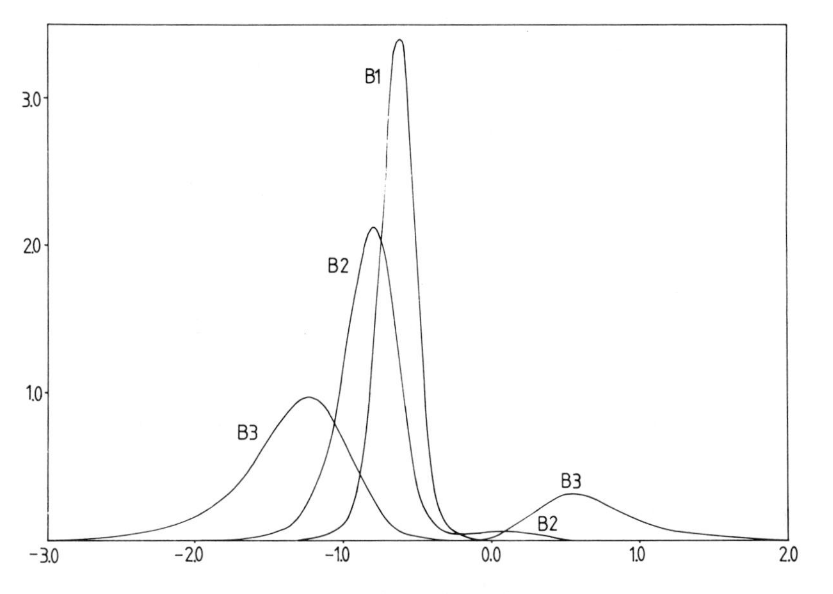

Figure 5.2. Posterior densities of β_{22} under Student–gamma prior densities.

As evidenced by Figures 5.1 and 5.2, the results are more sensitive to the choice of $p(\sigma_1^2)$ under (5.40) than they are under (5.49). It is shown in Richard (1973) that the positive local modes of β_{22} under A_2, B_2, A_3, and B_3 reflect doubtful rank identification, due to a strong positive correlation between national income and cattle stock.

5.6. *Normalization and invariance*

We mentioned in Section 2.1 that the LIML estimators of (β_Δ, γ_*) are invariant with respect to normalization. It is therefore natural to analyze which of the posterior densities derived in the previous sections are invariant with respect to normalization (with respect to the labelling of the endogenous variables). Clearly, this question is particularly relevant when the prior density is "non-informative", as in (2.25) and (2.26) [see Fisher (1976)].

Assuming β_1 and β_{m_1} to be non-zero with probability one, let $(\beta_1 \quad \alpha)$ and $(\beta_{m_1} \quad \zeta)$ denote two reparameterizations associated respectively with the normalization rules "$\beta_1 = 1$" and "$\beta_{m_1} = 1$" (see the concluding sentences in Section 2.3):

$$\alpha' = (\alpha_1' \quad \alpha_2') = \frac{1}{\beta_1}(\beta_2 \ldots \beta_{m_1} \quad \gamma_1 \ldots \gamma_{n_1}), \qquad \bar{\alpha}_1' = (1 \quad \alpha_1'), \quad \text{(2.20)-(2.21)}$$

$$\zeta' = (\zeta_1' \quad \zeta_2') = \frac{1}{\beta_{m_1}}(\beta_1 \ldots \beta_{m_1 - 1} \quad \gamma_1 \ldots \gamma_{n_1}), \qquad \bar{\zeta}_1' = (\zeta_1' \quad 1). \tag{5.53}$$

A density $p(\alpha|\cdot)$ is said to be invariant with respect to normalization if and only if its functional form is invariant with respect to the integrand transformation from α to ζ.

Lemma 5.12

The posterior densities (2.25) and (2.26) are invariant with respect to normalization.

Proof

By definition $\beta_{m_1}(\bar{\zeta}_1' \quad \zeta_2') = \beta_1(\bar{\alpha}_1' \quad \alpha_2') = (\beta_\Delta' \quad \gamma_*')$. Let ζ_{11} denote the first element of ζ_1, i.e. $\zeta_{11} = \beta_1/\beta_{m_1}$. The integrand transformation from α to ζ has Jacobian $|\zeta_{11}^{-1}|^{m_1+n_1}$ (ζ and α contain only m_1+n_1-1 elements but ζ_{11} is the reciprocal of the last element of α_1). Applying this transformation to (2.25) yields, therefore,

$$p(\zeta|Y) \propto |\zeta_{11}^{-1}|^{m_1+n_1}\left(\zeta_{11}^{-2}\cdot\bar{\zeta}_1'W_{\Delta\Delta}\bar{\zeta}_1\right)^{(1/2)(T-m+1)}$$
$$\times\left\{\zeta_{11}^{-2}\cdot(Y_\Delta\bar{\zeta}_1+Z_*\zeta_2)'(Y_\Delta\bar{\zeta}_1+Z_*\zeta_2)\right\}^{-(1/2)(T-m+m_1+n_1+1)},$$

and $|\zeta_{11}|$ cancels out. The same reasoning holds for (2.26) since the integrand transformation from α_1 to ζ_1 has Jacobian $|\zeta_{11}^{-1}|^{n_1}$. □

Lemma 5.13

The posterior density (2.24) is invariant with respect to normalization if and only if the prior density $p(\alpha)$ has that property.

Proof

We repeat the argument in the proof of Lemma 5.12. The Jacobian of the transformation from α to ζ is absorbed in $p(\alpha)$ and the additional factor in (2.24) is homogeneous of degree zero in α and, therefore, in ζ. □

Note that the Student density

$$p(\alpha) = f_t^{l_1}(\alpha|a_0, G_0, \mu_0) \tag{5.54}$$

is invariant with respect to normalization if and only if $\mu_0 = 1$, i.e. if and only if $p(\alpha)$ is Cauchy. A Cauchy density does not possess finite moments of order 1 or higher, but it possesses finite fractiles. The corresponding posterior density, as given in (2.24), has the same property since the ratio of quadratic forms in the right-hand term of (2.24) is homogeneous of degree zero.

We can apply the same line of reasoning to the posterior densities (5.44) and (5.51).

Lemma 5.14

The posterior density (5.44) is invariant with respect to normalization if and only if $\mu_0 = 2 - m$. The posterior density (5.51) is invariant with respect to normalization if and only if $\mu_0 + \lambda_0 = 2 - m$. □

These results may seem strange at first sight. Actually, they are not surprising once it is recognized that prior densities like (5.40) or (5.49) are not equivalent to hypothetical samples. It would take more complex densities, in the form of (5.41) or (5.50)–(5.51), to represent the information contained in a previous sample (where equations 2 to m are just identified). The conditions for invariance with respect to normalization of these more general densities admit a natural interpretation.

A stronger form of invariance is achieved by imposing *a priori exchangeability* among all the elements of (β_Δ, γ_*), i.e. by requiring that (5.54) be of the form

$$p(\alpha) \propto \left\{ \bar{\alpha}'\left(I_{l_1+1} - c_0 \iota\iota'\right)\bar{\alpha} \right\}^{-(1/2)(l_1+1)}, \tag{5.55}$$

where ι is an $(l_1+1)\times 1$ vector with every element equal to one and c_0 is a scalar (hyper) parameter $(c_0 \leqslant 1/(l_1+1))$ [see Lindley and Smith (1972)].

5.7. *Two generalizations*

We conclude this section by outlining two generalizations of the limited information analysis. Details may be found in Richard (1979), Drèze (1976, section 5) and, for a wider class of linear models, in Florens et al. (1979). Another generalization to non-linear equations is presented in Ter Berg and Harkema (1980).

5.7.1. *Informative prior densities*

The non-informative conditional prior measure $p(\Phi_2, \Delta_{12}, \Lambda_{22.1}|\theta)$ as given in (5.17), could be replaced by an informative prior density, provided it be natural-conjugate for the partial likelihood function (5.15).

As pointed out in Section 4, such a prior density is still restrictive; this is the price we have to pay in order to preserve the operational features of limited information procedures. The alternative is to use the computationally more costly full information procedures discussed in Section 6.

Given (5.15), a natural-conjugate prior density $p(\Phi_2, \Delta_{12}, \Lambda_{22.1}|\theta)$ is defined as the product of a matricvariate normal density on (Φ_2, Δ_{12}), conditionally on $(\Lambda_{22.1}, \theta)$ and an Inverted–Wishart density on $\Lambda_{22.1}$, conditionally on θ. The joint density is fully characterized by a scalar $\nu_0 > m-1$ and a matrix $\Psi_0 \in C^{m+n}$.

Following Lemma 5.1, ψ_0 is in one-to-one correspondence with $(W_0, \Pi_0, M_0) \in C^m \times R^{nm} \times C^n$ with

$$\Psi_0 = \begin{pmatrix} W_0 + \Pi_0' M_0 \Pi_0 & \Pi_0' M_0 \\ M_0 \Pi_0 & M_0 \end{pmatrix}. \tag{5.56}$$

The corresponding conditional prior moments of (Π, Ω) are given by formulae (5.29)–(5.31) if we substitute (M_0, Π_0, W_0, ν_0) for $(M, \hat{\Pi}, W, T)$; they will typically be used for the elicitation of the prior density $p(\Phi_2, \Delta_{12}, \Lambda_{22.1}|\theta)$. These moments, being conditional on θ, we must proceed numerically, trying successive values of Ψ_0 and computing the corresponding marginal moments of (Π, Ω) until a satisfactory assessment is reached.

As usual within a natural-conjugate framework, the (hyper) parameters of the posterior density $p(\Phi_2, \Delta_{12}, \Lambda_{22.1}|\theta, Y)$ are obtained by the convolution

$$\Psi_* = \Psi_0 + \begin{pmatrix} Y'Y & Y'Z \\ Z'Y & Z'Z \end{pmatrix}, \qquad \nu_* = \nu_0 + T, \tag{5.57}$$

from which (W_*, Π_*, M_*) can be obtained through the decomposition in (5.56). The conditional posterior moments of (Π, Ω) are also given by formulae (5.29)–(5.31), upon substituting (M_*, Π_*, W_*, ν_*) for $(M, \hat{\Pi}, W, T)$.

If, in addition, the conditional prior density $p(\sigma_1^2|\alpha)$ is given by

$$p(\sigma_1^2|\alpha) = f_{i\gamma}(\sigma_1^2|\beta' W_0 \beta, \nu_0 - m + 1), \tag{5.58}$$

i.e. if the prior density $p(\sigma_1^2, \Delta_{12}, \Lambda_{22.1}|\theta)$ derives from an underlying Inverted–Wishart prior density on Ω,

$$p(\Omega) = f_{iW}(\Omega | W_0, \nu_0), \tag{5.59}$$

then the results derived above carry over to the informative case. In particular, the functional form of the posterior densities (2.24) and (5.51) is preserved, the data moment matrix being replaced by Ψ_*.

This short discussion indicates that a generalization of the Bayesian limited information analysis, whereby the joint prior density is now proper, can be achieved at little additional computational cost.

5.7.2. *Subsystem analysis*

Our analysis can be extended from a single equation to a subset of equations from a simultaneous equation model. Let such a subset be

$$YB_1 + Z\Gamma_1 = U_1, \tag{5.60}$$

$$p(U_1) = f_{MN}^{T \times p_1}(0, \Sigma_{11} \otimes I_T), \tag{5.61}$$

where B_1 and Γ_1 are respectively $m \times p_1$ and $n \times p_1$ matrices of unknown coefficients. Let $\theta = (B_1, \Gamma_1, \Sigma_{11})$. Most formulae derived in the single equation case remain valid except that dimensions must be readjusted (essentially $m-1$ becomes $m - p_1$). In particular, under the non-informative prior measure (5.17) the posterior density of θ is given by

$$p(\theta|Y) \propto p(\theta)|B_1'WB_1|^{(1/2)(T-m+p_1)} f_{\mathrm{MN}}^{T\times p_1}(YB_1 + Z\Gamma_1|0, \Sigma_{11}\otimes I_T). \quad (5.62)$$

Comparing (5.62) and (1.7), it should be clear that:

(i) the analysis of the posterior density of θ is the subject-matter of full information analysis, to which we turn in Section 6; and
(ii) formula (5.62) enables us to combine full information analysis of a subsystem with limited information analysis of the remaining equations.

6. Full information analysis

6.1. *Introduction*

The Bayesian full information analysis of simultaneous equation models raises difficult problems as regards both the specification of the prior densities and the (numerical) analysis of the posterior densities. These difficulties are introduced in Section 6.2 by means of a special case. The more general case is discussed in Section 6.3 which is also more technical. In Section 6.4 we apply this analysis to the seemingly unrelated regression model of Zellner (1962). An alternative approach is presented in Section 6.5, with special emphasis on the two-equation model, for which a general analysis is fully operational. Three applications are reviewed in Section 6.6. The more technical details and proofs are regrouped in Appendix B.

6.2. *A special case*

The starting point of our analysis is the likelihood function (1.7) which we rewrite as

$$L(B, \Gamma, \Sigma|Y) \propto \|B\|^T |\Sigma|^{-(1/2)T} \exp\left[-\tfrac{1}{2}\operatorname{tr}\Sigma^{-1}A'X'XA\right], \quad (6.1)$$

where $X = (Y \quad Z)$ is the $T\times(m+n)$ matrix of observations and $A' = (B' \quad \Gamma')$. We only discuss the case where the elements of A are subject to linear restrictions. For convenience – see footnote 5 – we assume that B is normalized on the diagonal

($\beta_{ii} = 1$, i: $1 \to m$) and that the other constraints on A take the form of exclusion restrictions. Let $l_i = (m_i - 1) + n_i$ represent the number of unrestricted coefficients in the ith equation, δ_i be the $l_i \times 1$ vector of these coefficients, $X_{(i)} = [Y_{(i)} \quad Z_{(i)}]$ be the $T \times l_i$ submatrix consisting of the corresponding columns in X, and y_i be the ith column of Y.[24] The ith equation of the system becomes

$$y_i + X_{(i)}\delta_i = u_i. \tag{6.2}$$

The A matrix contains $l = \sum_{i=1}^{m} l_i$ unrestricted coefficients. Let $\delta' = (\delta_1' \dots \delta_m')$ be the $1 \times l$ vector of these coefficients.

The expression (6.1) suggests integrating out Σ first.

Lemma 6.1

Under the prior density

$$p(\delta, \Sigma) = p(\delta) f_{\mathrm{iW}}^{m}(\Sigma | S_0, \nu_0) \tag{6.3}$$

the posterior density of (δ, Σ) is

$$p(\Sigma | \delta, Y) = f_{\mathrm{iW}}^{m}(\Sigma | S_0 + A'X'XA, \nu_0 + T), \tag{6.4}$$

$$p(\delta | Y) \propto p(\delta) \|B\|^{T} |S_0 + A'X'XA|^{-(1/2)(\nu_0 + T)}. \tag{6.5}$$

Proof

We multiply together the likelihood function (6.1) and the prior density (6.3). The result follows by properties of the Inverted–Wishart density. □

The density (6.5) is the starting point of the analysis by Kloek and van Dijk (1978) – see also van Dijk and Kloek (1980) – who evaluate the posterior density of δ numerically by Monte Carlo procedures – see Section 7. One can also use (partially) analytical procedures at the cost of imposing constraints on the class of admissible prior densities. The two approaches are in fact complementary, as we shall argue in Section 7.

In order to illustrate the more analytical approach, we assume that the prior density of δ takes the form of a product of m independent Student densities:

$$p(\delta) = \prod_{i=1}^{m} f_t^{l_i}(\delta_i | d_i^0, H_i^0, \nu_i^0). \tag{6.6}$$

[24] Note the customary change of notation between single equation and full system analysis; thus, δ_1 and β_{11} in this section were denoted α and β_1 in Sections 2 and 5; also $[y_1 \quad Y_{(1)}]$ and $Z_{(1)}$ were denoted Y_Δ and Z_*, respectively.

Lemma 6.2

Under the prior densities (6.3)–(6.6) and provided S_0 is either a constant or of the form $B'W_0B$ with W_0 constant,[25] the posterior density of δ_1 conditionally on $(\delta_2 \dots \delta_m)$ is a 2–1 poly-*t* density.

Proof

See Appendix B. By comparison with formulae such as (4.17), the additional quadratic form in the kernel of the density originates from the factor $\|B\|^T$ in (6.5). □

Lemma 6.2 is of interest for the following reasons.

(1) It reveals that, under a Student (conditional) prior density for δ_1, the *full information conditional* posterior density $p(\delta_1 | \delta_2 \dots \delta_m, Y)$ has the same analytical form as the *limited information marginal* posterior density $p(\delta_1 | Y)$, as given for example in (5.51). A comparison between these two densities would reveal whether or not inferences on the coefficients of a single equation are sensitive to the prior information on the coefficients of the other equations.

(2) As we shall argue in Section 7, Lemma 6.2 can be used to construct approximate posterior densities from which δ can be drawn at random. [This is not true for the exact posterior density (6.5) since its integrating constant is not known.] These so-called "importance functions" are critical for the application of Monte Carlo numerical integration procedures.

In addition, the prior density (6.3)–(6.6) lends itself to the definition of a non-informative prior density for full information analysis of a model identified by exact *a priori* restrictions. Indeed, a natural requirement to impose on such a prior density is invariance with respect to normalization (as defined in Section 5.6). When $\nu_0 = 0$ in (6.3) and $\nu_i^0 = 1$ for each i in (6.6), the prior density

$$p(\delta, \Sigma) \propto |\Sigma|^{-(1/2)(m+1)} \prod_{i=1}^{m} f_t^{l_i}(\delta_i | d_i^0, H_i^0, 1) \tag{6.7}$$

has the desired property. Also, the additional property of exchangeability may be introduced by appropriate choice of (d_i^0, H_i^0) [see formula (5.55)].

6.3. Extended natural-conjugate prior densities

We now turn to a class of *extended natural-conjugate* prior densities analyzed by Drèze and Morales (1976). The form of these densities is again suggested by an

[25] For example, if $p(\Omega) = f_{iW}^m(\Omega | W_0, \nu_0)$, then $p(\Sigma | B) = f_{iW}^m(\Sigma | B'W_0B, \nu_0)$. Note also that Lemma 6.2 remains true if (6.6) is generalized to $p(\delta) = f_t^{l_1}(\delta_1 | d_1^0, H_1^0, \nu_1^0) \cdot p(\delta_2 \dots \delta_m)$.

appropriate reformulation of the likelihood function. Let

$$\Delta = \begin{pmatrix} \delta_1 & 0\dots0 \\ 0 & \delta_2\dots0 \\ & \vdots\ \ \vdots \\ 0 & 0\dots\delta_m \end{pmatrix}, \qquad \Xi = \left[X_{(1)}X_{(2)}\dots X_{(m)}\right]. \tag{6.8}$$

Δ and Ξ are respectively $l \times m$ and $T \times l$ matrices. The likelihood function (6.1) may be rewritten as

$$L(\Delta,\Sigma|Y) \propto \|B\|^T|\Sigma|^{-(1/2)T}\exp\left[-\tfrac{1}{2}\operatorname{tr}\Sigma^{-1}(Y+\Xi\Delta)'(Y+\Xi\Delta)\right] \tag{6.9}$$

or equivalently

$$L(\Delta,\Sigma|Y) \propto \|B\|^T|\Sigma|^{-(1/2)T}\exp\left[-\tfrac{1}{2}\operatorname{tr}\Sigma^{-1}\left[S+(\Delta-\hat{\Delta})'\Xi'\Xi(\Delta-\hat{\Delta})\right]\right], \tag{6.10}$$

where $\hat{\Delta}$ denotes an arbitrary solution to the normal equations

$$\Xi'\Xi\hat{\Delta} = -\Xi'Y, \tag{6.11}$$

and

$$S = (Y+\Xi\hat{\Delta})'(Y+\Xi\hat{\Delta}). \tag{6.12}$$

Note that Ξ typically includes columns of Y, in which case S is a singular matrix. Similarly, columns of X may repeat themselves in Ξ which is then not of full column rank.

An *extended natural-conjugate* (ENC) prior density for (Δ,Σ) is then defined as[26]

$$p(\Delta,\Sigma) \propto \|B\|^{\tau_0}|\Sigma|^{-(1/2)(\nu_0+m+1)} \times\exp\left[-\tfrac{1}{2}\operatorname{tr}\Sigma^{-1}\left[S_0+(\Delta-D_0)'\mathfrak{M}_0(\Delta-D_0)\right]\right], \tag{6.13}$$

where $\mathfrak{M}_0$ and S_0 are respectively $l \times l$ and $m \times m$ PSDS matrices. This prior density has the same properties as the posterior density discussed below; when

[26]Formula (6.13) is equivalent to formula (2.3) in Drèze and Morales (1976) when the scalar θ appearing there is set equal to $\nu_0 - m - 1$.

$\nu_0 > 0$, it factorizes into the product of the densities:

$$p(\Sigma|\Delta) = f_{\mathrm{iW}}^{m}(\Sigma|S_0 + (\Delta - D_0)'\mathfrak{M}_0(\Delta - D_0), \nu_0), \tag{6.14}$$

$$p(\Delta) \propto \|B\|^{\tau_0}|S_0 + (\Delta - D_0)'\mathfrak{M}_0(\Delta - D_0)|^{-(1/2)\nu_0}. \tag{6.15}$$

Lemma 6.3

Under the prior density (6.13), the posterior density of (Δ, Σ) is

$$p(\Sigma|\Delta, Y) = f_{\mathrm{iW}}^{m}(\Sigma|S_* + (\Delta - D_*)'\mathfrak{M}_*(\Delta - D_*), \nu_*), \tag{6.16}$$

$$p(\Delta|Y) \propto \|B\|^{\tau_*}|S_* + (\Delta - D_*)'\mathfrak{M}_*(\Delta - D_*)|^{-(1/2)\nu_*}, \tag{6.17}$$

where

$$\mathfrak{M}_* = \mathfrak{M}_0 + \Xi'\Xi, \qquad \nu_* = \nu_0 + T, \qquad \tau_* = \tau_0 + T, \tag{6.18}$$

$$S_* = S_0 + D_0'\mathfrak{M}_0 D_0 + Y'Y - D_*'\mathfrak{M}_* D_*, \tag{6.19}$$

and D_* is an arbitrary solution of the equation system

$$\mathfrak{M}_* D_* = \mathfrak{M}_0 D_0 - \Xi'Y. \tag{6.20}$$

Proof

By application of formulae (4.9)–(4.12). □

Morales (1971, section 3.2) has shown that the *posterior modes* of the density (6.17) can be computed by the same algorithms as FIML estimators. Other properties of the densities (6.15) and (6.17) have been derived in Drèze and Morales (1976) and will be briefly reviewed here. Since the same arguments apply to both densities, the subscripts 0 and * will be omitted in the rest of this section.

Lemma 6.4

Let B and Δ in the density (6.15) be partitioned as

$$\Delta = \begin{pmatrix} \Delta_1 & 0 \\ 0 & \Delta_2 \end{pmatrix}, \qquad B = (B_1 \quad B_2), \tag{6.21}$$

where Δ_i and B_i are respectively $l \times m_i$ and $m \times m_i$ matrices. The matrices S, D, and $\mathfrak{M}$ are partitioned conformably. Then, under (6.15) we have

$$p(\Delta_1|\Delta_2) \propto |B_1'\left[I_m - B_2(B_2'B_2)^{-1}B_2'\right]B_1|^{(1/2)\tau} \times |\bar{S}_{11} + (\Delta_1 - \bar{D}_{11})'R_{11}(\Delta_1 - \bar{D}_{11})|^{-(1/2)\nu}, \tag{6.22}$$

where $\bar{S}_{11}$, $\bar{D}_{11}$, and R_{11} are obtained by application of Lemma B.1 in Appendix B.

Proof

By application of Lemma B.1 together with the identity (B.9) applied here with $\beta = B_1$. □

Lemma 6.6 applies in particular to $\Delta_1' = \delta_1' = (\beta_1' \quad \gamma_1')$, the $1 \times l_1$ vector of unconstrained coefficients in the first equation.

Corollary 6.5

Let S and $\mathfrak{M}_{11}$ be PDS matrices. The conditional density $p(\delta_1 | \Delta_2)$, as given in (6.22), is then a 1–1 poly-t density. The same holds for $p(\beta_1 | \Delta_2)$.

Proof

See Appendix B. □

In the special case where $m = 2$ and $|B| = 1$, the first factor in the right-hand side of (6.22) drops out and under the conditions of Corollary 6.5, the conditional density $p(\delta_1 | \Delta_2)$ is a Student density. This enables us to obtain an *analytical expression* for the *marginal* density $p(\Delta_2)$ which, provided $\mathfrak{M}$ is PDS, is a 2–1 poly-t density [see Drèze and Morales (1976)].

The existence of moments of Δ can be ascertained from the following lemma.

Lemma 6.6

Let $\mathfrak{M} = (\mathfrak{M}_{ij})$ be SPDS, $i, j = 1, m$. Let $p(\Delta)$ be the density (6.15) with S and $\mathfrak{M}_{ii}$ PDS matrices. A *sufficient* condition for the existence of moments of order r is

$$\nu - m + 1 - \mu - \tau > r, \tag{6.23}$$

where $\mu = \sup\{l_i;\ i = 1, m\}$.[27]

Proof

See Appendix B. □

The import of the results obtained under the ENC prior density (6.13) is limited for two reasons.

(1) The benefit of using the ENC prior density instead of the more easily interpretable prior densities (6.3) and (6.6) lies essentially in the fact that the posterior density $p(\delta_1 | \Delta_2, Y)$ is a 1–1 poly-t density (Corollary 6.5) instead of a

[27]Similarly, a sufficient condition for the existence of the conditional moments of order r in (6.22) is $\nu - l_i - \tau > r$.

2–1 poly-t density (Lemma 6.2). However – see Section 7 – this no longer represents a major advantage in computations.

(2) In contrast with the prior density (6.6), the ENC prior density (6.15) must be elicited numerically, except in the special case where $\tau_0 = 0$ and

$$D_0 = \begin{pmatrix} d_1^0 & 0 \dots 0 \\ 0 & d_2^0 \dots 0 \\ \vdots & \vdots \quad \vdots \\ 0 & 0 \dots d_m^0 \end{pmatrix}, \qquad S_0 = \begin{pmatrix} S_{11}^0 & 0 \dots 0 \\ 0 & S_{22}^0 \dots 0 \\ \vdots & \vdots \quad \vdots \\ 0 & 0 \dots S_{mm}^0 \end{pmatrix},$$

$$\mathfrak{M}_0 = \begin{pmatrix} \mathfrak{M}_{11}^0 & 0 \dots 0 \\ 0 & \mathfrak{M}_{22}^0 \dots 0 \\ \vdots & \vdots \quad \vdots \\ 0 & 0 \dots \mathfrak{M}_{mm}^0 \end{pmatrix}. \tag{6.24}$$

The ENC prior density takes then the form of a product of m independent Student densities with a common exponent ν_0:

$$p(\Delta) = \prod_{i=1}^{m} f_t^{l_i}\left(\delta_i | d_i^0, \frac{1}{s_{ii}^0}, \mathfrak{M}_{ii}^0, \nu_0 - l_i \right). \tag{6.25}$$

The common exponent restriction might not be too severe for most practical purposes, except in the non-informative case of formula (6.7), which does not fit into the ENC framework unless l_i has the same value for all i. Note, however, that the prior density (6.6) may still be completed by an arbitrary Inverted–Wishart density for Σ, as shown in Lemma 6.1, while in the ENC framework $p(\Delta)$ and $p(\Sigma|\Delta)$ depend on the same coefficient matrix. As illustrated in Section 6.5, the ENC framework imposes therefore strong restrictions on the prior density of Σ.

However, we shall argue in Section 7 that it is possible within an ENC framework to design general procedures for constructing importance functions. This property, which does not hold with other priors, may prove important for Monte Carlo applications.

6.4. *Seemingly unrelated regression models*

The same analysis applies to the *seemingly unrelated regression model* of Zellner (1962). The likelihood function may be written as (6.9) except that $B = I_m$ and Ξ

contains only exogenous variables. The ENC prior for that model is therefore of the form (6.13) with $\tau_0 = 0$. The analysis simplifies since B is now an identity matrix.

As a special case, consider the reduced-form model (1.5). Let $\Pi_i(\hat{\Pi}_i)$ denote the ith column of $\Pi(\hat{\Pi})$. Since the regressors are now common to all equations, the likelihood function (4.2) may be rewritten as

$$L(\Pi,\Omega|Y) \propto |\Omega|^{-(1/2)T} \times \exp\left[-\tfrac{1}{2}\operatorname{tr}\Omega^{-1}\left[W + (\Delta_\Pi - \hat{\Delta}_\Pi)'(\iota\iota' \otimes Z'Z)(\Delta_\Pi - \hat{\Delta}_\Pi)\right]\right], \tag{6.26}$$

where ι is the $m \times 1$ vector with all elements equal to 1 and

$$\Delta_\Pi = \begin{pmatrix} \Pi_1 & 0 & \dots & 0 \\ 0 & \Pi_2 & \dots & 0 \\ \vdots & \vdots & \vdots & \\ 0 & 0 & \dots & \Pi_m \end{pmatrix}, \qquad \Delta_\Pi \begin{pmatrix} \hat{\Pi}_1 & 0 & \dots & 0 \\ 0 & \hat{\Pi}_2 & \dots & 0 \\ \vdots & \vdots & & \vdots \\ 0 & 0 & \dots & \hat{\Pi}_m \end{pmatrix}. \tag{6.27}$$

An ENC prior density for (Π,Ω) is given by

$$p(\Pi,\Omega) \propto |\Omega|^{-(1/2)(\nu_0+m+1)} \times \exp\left[-\tfrac{1}{2}\operatorname{tr}\Omega^{-1}\left[W_0 + (\Delta_\Pi - \Delta_\Pi^0)'\mathfrak{M}_0(\Delta_\Pi - \Delta_\Pi^0)\right]\right], \tag{6.28}$$

where $W_0 \in \mathcal{C}^m$, $\mathfrak{M}_0 \in \mathcal{C}^l$, and Δ_Π^0 has the same block-diagonal structure as $\hat{\Delta}_\Pi$. Comparing (6.28) with (4.4) it is seen that the extension consists in replacing the "restrictive" expression $W_0 \otimes M_0^{-1}$ by the general matrix $\mathfrak{M}_0$. The posterior density $p(\Pi,\Omega|Y)$ has the same functional form as the prior density (6.28) with parameters $(W_*, \Delta_\Pi^*, \mathfrak{M}_*, \nu_*)$, as obtained from formulae (6.18)–(6.20). Note, however, that Δ_Π^* will not have the same block-diagonal structure as Δ_Π^0.

6.5. *Two-equation models*

Alternatives to the full information procedures described so far start from a reformulation of the likelihood function (1.6). Let $H = \Sigma^{-1}$ denote the sample precision matrix and let

$$y = \begin{pmatrix} y_1 \\ y_2 \\ \vdots \\ y_m \end{pmatrix}, \qquad \mathfrak{X} = \begin{pmatrix} X_{(1)} & 0 & \dots & 0 \\ 0 & X_{(2)} & \dots & 0 \\ \vdots & \vdots & & \vdots \\ 0 & 0 & \dots & X_{(m)} \end{pmatrix}, \tag{6.29}$$

be respectively $mT \times 1$ and $mT \times l$ data matrices. The likelihood function may be written as

$$L(\delta, H|y) \propto \|B\|^T |H|^{(1/2)/T} \exp\left[-\tfrac{1}{2}(y + \mathcal{X}\delta)'(H \otimes I_T)(y + \mathcal{X}\delta)\right] \quad (6.30)$$
$$\propto \|B\|^T f_N^m(y| - \mathcal{X}\delta, H^{-1} \otimes I_T).$$

Lemma 6.7

Under the prior density

$$p(\delta, H) = p(\delta)p(H), \quad (6.31)$$

the conditional posterior density of δ, given H, is

$$p(\delta|H) \propto \|B\|^T p(\delta) f_N^l(\delta|\hat{d}, \mathcal{X}'(H \otimes I_T)\mathcal{X}), \quad (6.32)$$

where

$$\hat{d} = [\mathcal{X}'(H \otimes I_T)x]^{-1}\mathcal{X}'(H \otimes I_T)y. \quad (6.33)$$

Proof

See, for example, Zellner (1971, ch. 3) (the generalization to a non-scalar covariance matrix is obvious).[28] □

The algebra simplifies somewhat if we are willing to assume prior dependence between δ and H in the form which is usual within a natural-conjugate framework. Note in particular that the ENC prior density (6.13) may be rewritten as

$$p(\delta, H) \propto \|B\|^{\tau_0} |H|^{(1/2)(\nu_0 - m - 1)}$$
$$\times \exp\left[-\tfrac{1}{2}\left[\operatorname{tr} HS_0 + (\delta - d_0)'H \square \mathfrak{M}_0(\delta - d_0)\right]\right], \quad (6.34)$$

[28] Partition δ' into $(\beta' \quad \gamma')$, where β regroups the unrestricted elements of B and γ those of Γ, and consider the case where $p(\gamma|\beta)$ is a Student density. Conditionally on (β, H), the posterior density of γ is the product of a Student kernel and a Normal density function. Its integrating constant and moments can be obtained by means of the procedures described in Richard and Tompa (1980). Conditionally on β, the transformation from (γ, H^{-1}) to (Π, Ω) is linear, since $\Gamma = -\Pi B$ and $H^{-1} = B'\Omega B$. It follows that, conditionally on (β, H), we can easily obtain analytical expressions for the moments of (Π, Ω) and for the predictive moments; these expressions will have to be marginalized numerically with respect to β and H at the final stage of the analysis.

where $H\square\mathfrak{M}_0$ denotes the $l\times l$ matrix:

$$H\square\mathfrak{M}_0=\begin{pmatrix} h_{11}\mathfrak{M}_{11}^0 & h_{12}\mathfrak{M}_{12}^0 & \cdots & h_{1m}\mathfrak{M}_{1m}^0 \\ h_{21}\mathfrak{M}_{21}^0 & h_{22}\mathfrak{M}_{22}^0 & \cdots & h_{2m}\mathfrak{M}_{2m}^0 \\ \vdots & \vdots & & \vdots \\ h_{m1}\mathfrak{M}_{m1}^0 & h_{m2}\mathfrak{M}_{m2}^0 & \cdots & h_{mm}\mathfrak{M}_{mm}^0 \end{pmatrix}. \tag{6.35}$$

Lemma 6.8

Under the ENC prior density (6.34), the conditional posterior density of δ, given H, is

$$p(\delta|H)\propto\|B\|^T f_N^l(\delta|d_*,H\square\mathfrak{M}_*), \tag{6.36}$$

where $\mathfrak{M}_*$ is defined in (6.18) and d_* is a solution of the equation system

$$(H\square\mathfrak{M}_*)d_*=(H\square\mathfrak{M}_0)d_0+\mathfrak{X}'(H\otimes I_T)\mathfrak{X}. \tag{6.37}$$

Proof

(6.36) is related to (6.16)–(6.17) in the same way that (6.34) is related to (6.13). Note that $\mathfrak{X}'(H\otimes I_T)\mathfrak{X}=H\square\Xi'\Xi$. □

Partition δ' into (β',γ'), where β regroups the unrestricted elements of B and γ those of Γ. The posterior density $p(\gamma|\beta,H)$ is now a normal density. Since the emphasis is shifted to analytical integration of γ, it is no longer necessary to impose on $p(\beta,H)$ all the restrictions which are implicit in the ENC prior density (6.34). This extension of the natural-conjugate framework has been developed for two-equation models in Richard (1973) under the more general prior density:

$$\begin{aligned} p(\delta,h_{11},h_{22},\rho)\propto{}& g(\beta,\rho)h_{11}^{(1/2)\nu_1^0}h_{22}^{(1/2)\nu_2^0}\\ &\times\exp\left[-\tfrac{1}{2}\left[\operatorname{tr}HS_0+(\delta-d_0)'H\square\mathfrak{M}_0(\delta-d_0)\right]\right], \end{aligned} \tag{6.38}$$

where $\rho=\sigma_{12}(\sigma_{11}\sigma_{22})^{-1/2}=-h_{12}(h_{11}h_{22})^{-1/2}$ and $g(\beta,\rho)$ is any suitable positive function of the simultaneity parameters $(\beta,\rho)\in R^2\times(-1,1)$. Through the specification of $\mathfrak{M}_0$ – it can for example contain zeros in the rows and columns corresponding to β – and of the function $g(\beta,\rho)$, we can incorporate within (6.38) an arbitrary prior density for β and ρ. We note that the ENC prior density (6.34) is a special case of (6.38) with $\nu_1^0=\nu_2^0=\nu_0-2$ and

$$g(\beta,\rho)\equiv\|B\|^{\tau_0}(1-\rho^2)^{(1/2)(\nu_0-3)}. \tag{6.39}$$

This last expression illustrates the restrictiveness of the ENC prior density, as regards Σ in particular. It can be shown that, under the prior density (6.38), the joint posterior density factorizes into

$$p(\beta,\gamma,h_{11},h_{22},\rho|Y)=p(\gamma|\beta,h_{11},h_{22},\rho,Y)\times p(h_{11}|\beta,\rho,Y)p(h_{22}|\beta,\rho,Y)p(\beta,\rho|Y), \tag{6.40}$$

where, as a result of Lemma 6.8, $p(\gamma|\beta,h_{11},h_{22},\rho,Y)$ is a multivariate normal density. Furthermore, the posterior density $p(h_{ii}|\beta,\rho,Y)$, $i=1$ or 2, takes the form of an infinite sum of gamma density functions. Consequently, integration with respect to (γ,h_{11},h_{22}) can be carried out analytically and we are left with trivariate numerical integration on (β_1,β_2,ρ), the simultaneity parameters. The computational burden implied by the presence of infinite sums in the integrands can be drastically cut by the use of hypergeometric series, which can be computed efficiently and with high numerical precision by means of finite stable recurrence relationships. It is also fairly easy to obtain analytical expressions for the moments of Γ, Π, and Ω conditionally on (β,ρ) and next to marginalize them numerically with respect to β and ρ. It follows that a Bayesian full information analysis of two-equation models under the prior density (6.38) is fully operational.

Note finally that the same analysis applies to the two-equation seemingly unrelated regression model of Zellner (1962), in which case we are simply left with univariate numerical integration on $\rho\in]-1,1[$.

6.6. *Applications*

We conclude this section by discussing three applications.

6.6.1. *The Belgian beef market*

The two-equation model (2.27) has been analyzed under full information by Morales (1971) and Richard (1973). The prior density $p(\gamma_{23})$, as given in (2.29), was completed by independent prior information on γ_{14}, taken from cross-section data and summarized by the density

$$p(\gamma_{14})\propto\left[1+0.00161(\gamma_{14}-110)^2\right]^{-3.5}, \tag{6.44}$$

with mean and variance

$$E(\gamma_{14})=110,\qquad V(\gamma_{14})=155. \tag{6.45}$$

Whereas Morales used an ENC prior density, Richard used the joint prior density (6.38), together with

$$g(\beta_1, \beta_2, \rho) \equiv 1, \qquad s_{12}^0 = m_{12}^0 = 0, \qquad \nu_1^0 = \nu_2^0 = 5, \tag{6.46}$$

$$\frac{m_{11}^0}{s_{11}^0} = 0.00161, \qquad \frac{m_{22}^0}{s_{22}^0} = 0.2207E+08. \tag{6.47}$$

This choice ensures that the joint prior density (6.38) is compatible with the marginal prior densities (2.29) and (6.44). s_{11}^0 and s_{22}^0 are then chosen in a way which reflects prior information on σ_{11} and σ_{22}, if any. Three different sets of values were considered, namely

$$s_{22}^{0i} = 2s_{11}^{0i} = 4^i, \qquad i = 1,2,3, \tag{6.48}$$

which imply the following conditional prior means for σ_{11} and σ_{22}:[29]

$$E(\sigma_{22}|\rho = 0) = 2E(\sigma_{11}|\rho = 0) = 4^{i-1}, \qquad i = 1,2,3. \tag{6.49}$$

The corresponding runs are labelled A_1, A_2, and A_3. As in Section 5.5, the runs A_2 and A_3 imply implausibly large values for the prior means of σ_{11} and σ_{22} but were considered mainly for the purpose of illustrating the risks inherent in the specification of (extended) natural-conjugate prior densities. The posterior means and standard deviations of the coefficients of the demand equation are reported in Table 6.1. The graphs of the posterior densities of β_{22} are given in Figure 6.1. Detailed comments on these and other results may be found in Richard (1973). In short, the following remarks deserve mention, even though they might be somewhat specific to the application under review.

(1) Prior information on Σ has a marked influence on the posterior density of B (see, in particular, the long left-hand tails in Figure 6.1). The sample information tends to concentrate the posterior distribution of (Π, Ω) near the maximum likelihood estimators – see formulae (4.10) to (4.12) – and the posterior densities of (B, Γ, Σ) are left to adjust through the identities $B'\Omega B = \Sigma$ and $\Pi B + \Gamma = 0$. Given a non-informative prior density on B, the prior information on Σ affects more markedly the posterior density of B than that of Ω.

(2) As usual within a natural-conjugate framework, more weight is given to the prior information on $(\gamma_{14}, \gamma_{23})$ as one increases s_{11}^0 and s_{22}^0, leaving the prior

[29] Under (6.38) the marginal moments of σ_{11} and σ_{22} have to be computed numerically. This contrasts with the limited information analysis where, taking advantage of the diffuse prior density on Ω, we can derive these moments analytically.

Table 6.1
Full information analysis of the demand equation

		β_{22}	γ_{23}	γ_{25}
Prior (6.38)	A_1	−0.6423	0.6607*E*-03	0.3446*E* +02
		(0.1557)	(0.0818)	(0.0721)
	A_2	−0.6968	0.6041*E*-03	0.4071*E* +02
		(0.2226)	(0.0858)	(0.1207)
	a_3	−1.110	0.5624*E*-03	0.7091*E* +02
		(0.475)	(0.0853)	(0.3002)

moments of the γ's unchanged (and, therefore, increasing proportionally m_{11}^0 and m_{22}^0).

(3) By introducing prior information on the γ's, one solves the problems raised by the strong multicollinearity between Z_3 and Z_4. In contrast with the limited information results reported in Section 5.5, positive values of β_{22} are now overwhelmingly rejected.

6.6.2. *Klein's model I*

Van Dijk and Kloek (1980) have analyzed Klein's model I [Klein (1950)] under several full information prior densities by means of nine-dimensional Monte Carlo numerical integration procedures – see Section 7.3.2. Their results illustrate the flexibility of the Bayesian approach in dealing with prior information of a

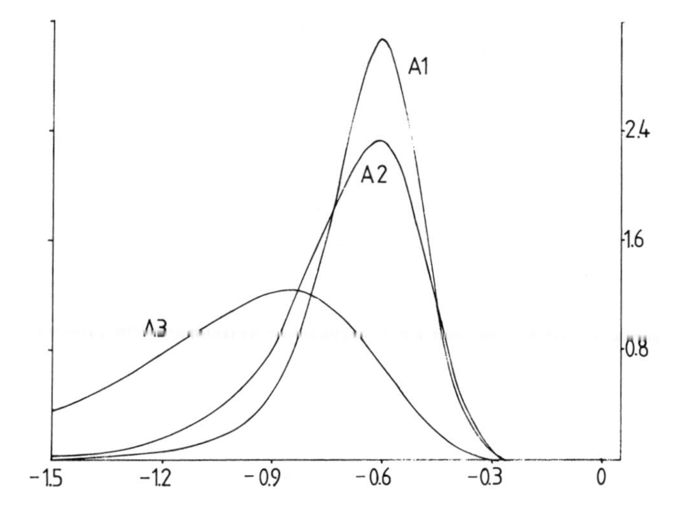

Figure 6.1. Posterior densities of β_{22}.

complex nature. An interesting feature of the model is that the FIML estimators of two structural coefficients and, therefore, of several multipliers have "wrong" signs (not significantly, however).

Several prior densities have been considered, all of which are non-informative on Σ and on the intercepts of the structural equations. The prior densities on the nine remaining coefficients are uniform on the unit region $]0,1[^9$ minus the region where $\|B\| < 0.01$ (this restriction ensures the existence of prior moments for the multipliers). Depending on the runs, additional constraints were imposed on the parameters, namely:

(i) stability of the system,
(ii) long-run coefficients in the structural equations in the unit interval,
(iii) short-run multipliers less than five in absolute value and with the correct signs (ten for the capital lagged multiplier),
(iv) same constraints on the long-run multipliers, and
(v) period of oscillation between three and ten years.

Two runs are analyzed in greater detail and are labelled respectively run 2, subject to constraint (i), and run 8, subject to constraints (i)–(v). Posterior densities and moments are obtained for the structural coefficients, the short-run and the long-run multipliers. In Table 6.2 we reproduce the FIML estimators and the posterior means and standard deviations for the short-run and the long-run multipliers of government wage expenditures (W_2), business taxes (T), and government non-wage expenditures (G) on net national income (Y). Graphs of the corresponding posterior densities are given in Figure 6.2.

These results are illustrative, since the authors' main purpose is "to demonstrate that the integrations can be done and how they are done". However, they are representative of the output of a Bayesian analysis. Indeed, for such essential characteristics as multipliers, period of oscillations, etc. complete posterior densi-

Table 6.2
Net national income multipliers in Klein's model I

		W_2	T	G
Short-run	FIML	1.50	−0.36	0.62
	Prior 2	2.30	−1.32	1.65
		(0.16)	(0.20)	(0.19)
	Prior 8	2.49	−1.81	2.06
		(0.26)	(0.34)	(0.30)
Long-run	FIML	2.57	−1.30	1.96
	Prior 2	2.87	−1.73	2.38
		(0.16)	(0.23)	(0.14)
	Prior 8	2.63	−1.67	2.25
		(0.16)	(0.20)	(0.13)

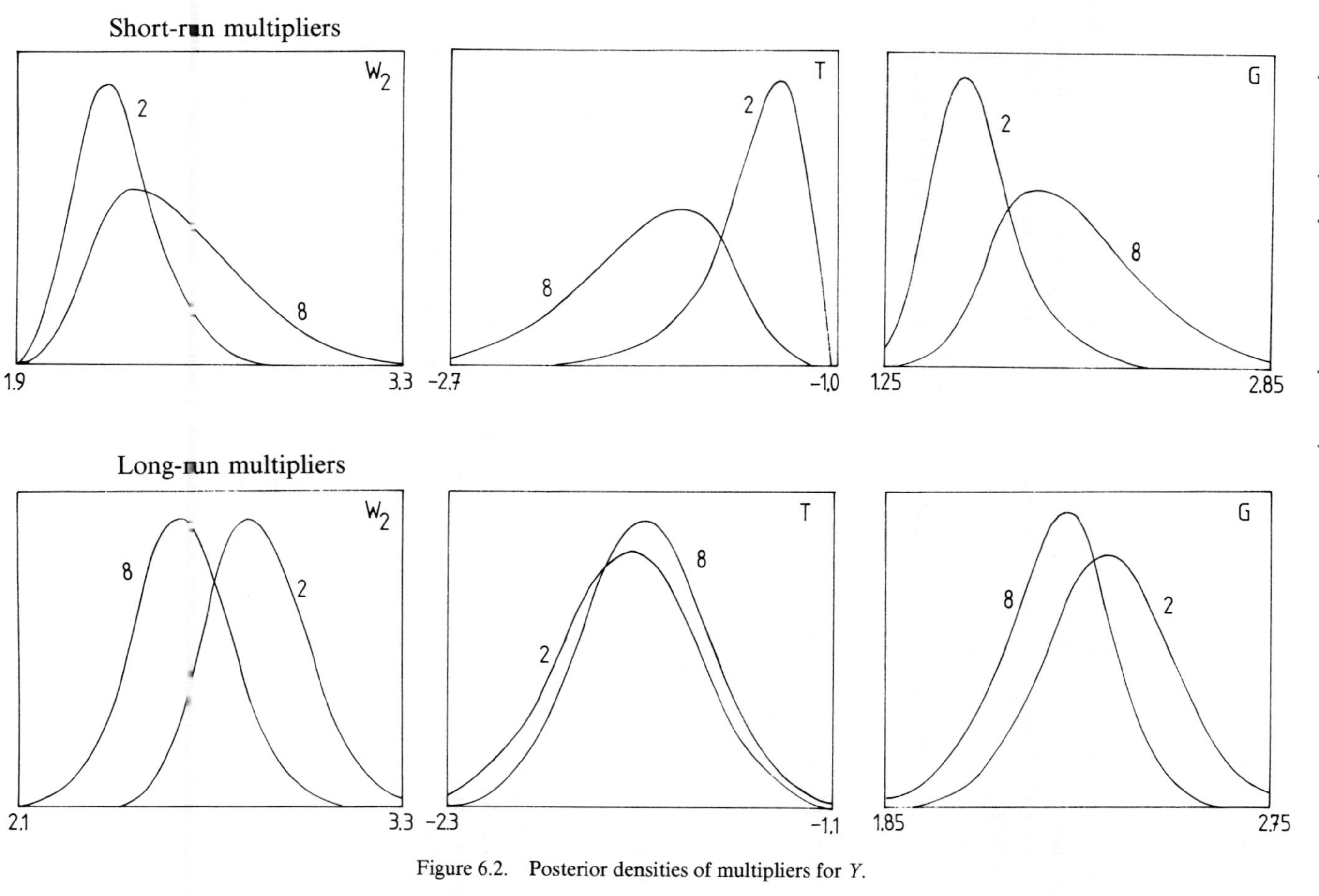

Figure 6.2. Posterior densities of multipliers for Y.

ties are more relevant than point estimates and *asymptotic* standard errors. This feature is also central to our third illustration.

6.6.3. *Belgian foreign trade, wages, and employment*

In a paper concerned with substantive policy issues rather than with methodology, Drèze and Modigliani (1981) estimate short-run and medium-run elasticities of employment with respect to real wages for Belgium. Their theoretical model, treating the balance of payments as a binding constraint, leads to the short-run formula

$$\eta_{L(W/P)} = \left\{ \left[\eta_{P_XC} + (1-\eta_{MX})\eta_{XC} \right] \eta_{CW} - \eta_{MP}\eta_{PW} \right\} \eta_{LY} / \eta_{MY}(1-\eta_{PW}), \tag{6.50}$$

where η = elasticity symbol,
L = employment,
W = nominal wages,
P = price level,
C = index of factor costs,
P_X = export prices,
X = export quantities,
M = import quantities, and
Y = national income.

The estimates of four elasticities $(\eta_{MX}, \eta_{CW}, \eta_{LY}, \eta_{PW})$ are borrowed from previous models of the Belgian economy. The analysis is conditional on point estimates for the first three of these and uses a posterior density for η_{PW} subjectively assigned on the basis of published estimates. A three-equation model of Belgium's foreign trade, due to Bauwens and d'Alcantara (1981), is used to estimate the remaining elasticities, namely η_{MP} and η_{MY} (import quantities equation), η_{P_XC} (export prices equation), and η_{XC} (export quantities equation). Eleven annual observations (1966–1976) are used. The system is treated as block-recursive between the import equation and the two export equations. These last two structural equations involve seven exogenous variables and the following (restricted) coefficient matrices:

$$B' = \begin{pmatrix} 1 & 0 \\ \beta & 1 \end{pmatrix}, \qquad \Gamma' = \begin{pmatrix} \alpha & 0 & 0 & 0 & 0 & 0 & 0 \\ 0 & 1-\alpha & \gamma_1 & \gamma_2 & \gamma_3 & \gamma_4 & \gamma_5 \end{pmatrix},$$

with $\eta_{P_XC} = \alpha$ and $\eta_{XC} = \alpha(\beta+\gamma_4) - (1-\alpha)\gamma_5$.

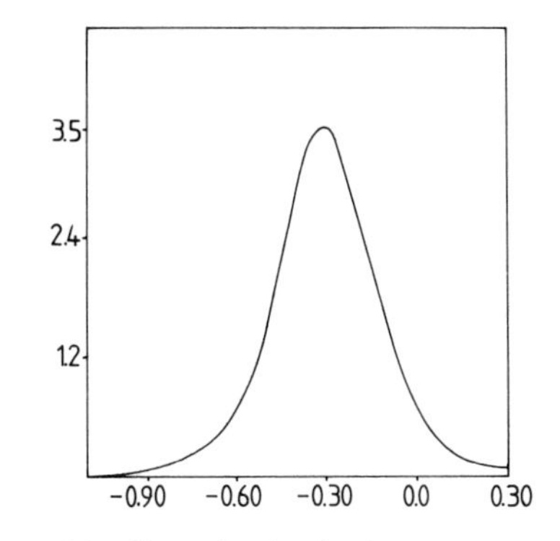

Figure 6.3. Posterior density for (6.52), short run.

A full information Bayesian analysis of these two equations is conducted, using the "non-informative" prior density[30]

$$p(\alpha,\beta,\gamma,\Sigma)\propto|\Sigma|^{-3/2},\qquad \alpha\in\left]0,1\right[,\quad \beta\in R,\quad \gamma\in R^5,\quad \Sigma\in\mathcal{C}^2. \tag{6.51}$$

Σ is integrated out analytically as in (6.5), whereby the posterior density $p(\beta,\gamma|\alpha,$ data) is obtained in the form of a Student density ($|B|=1$). Since

$$\eta_{P_XC}+(1-\eta_{MX})\eta_{XC}=\alpha+(1-\eta_{MX})[\alpha(\beta+\gamma_4)-(1-\alpha)\gamma_5], \tag{6.52}$$

a posterior density for this quantity, conditionally on α and η_{MX}, is obtained as a univariate Student density and marginalized numerically with respect to α (for a given point estimate of η_{MX}). The result is given in Figure 6.3, after multiplication by $\eta_{CW}=0.33$.

To evaluate the posterior density of $\eta_{L(W/P)}$ in (6.50), three densities are used, namely:

(i) the subjective density for η_{PW}, expressed as a normal density;
(ii) the posterior joint density of η_{MP} and η_{MY}, obtained from Bayesian single equation analysis of the import quantities equation, under a non-informative prior density; this joint density is bivariate Student, but approximated by the Normal density with the same mean vector and covariance matrix;

[30] The "degrees of freedom" parameter $-\frac{3}{2}=-\frac{1}{2}(m+1)$ corresponds to (4.21) and has the "conservative" bias mentioned at the end of Section 4. With $T=11$ and $n=7$, the difference between (4.21) and (4.20) is significant. Our computations with (4.21) give approximately the same results as those initially reported by Drèze and Modigliani, who used (4.20) but adjusted subjectively the posterior variances. In the light of our discussion of invariance with respect to normalization in Sections 5.6 and 6.2, we could add to (6.51) a Cauchy prior on (β,γ).

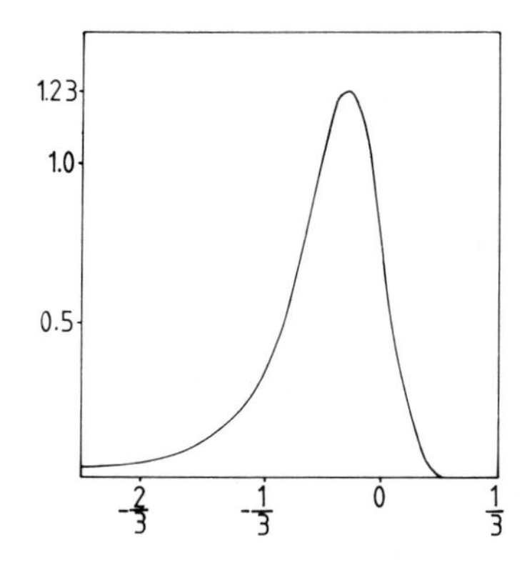

Figure 6.4. Posterior density for $\eta_{L(W/P)}$, short run.

and

(iii) the posterior density of $[\eta_{P_X C}+(1-\eta_{MX})\eta_{XC}]\eta_{CW}$ in Figure 6.3, approximated by the normal density with identical mean and variance.

A combination of analytical and numerical procedures then yields the marginal density of $\eta_{L(W/P)}$ in Figure 6.4.

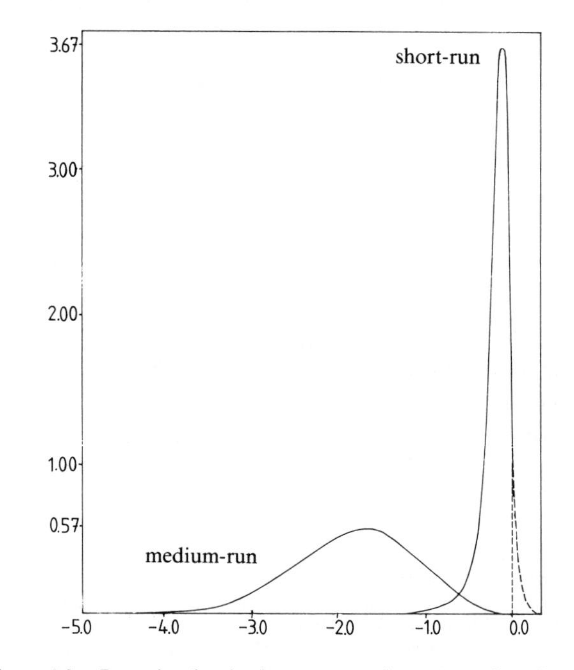

Figure 6.5. Posterior density for $\eta_{L(W/P)}$, short run and medium run.

The questions of central interest to the authors receive rather unequivocal answers in the form

$$\Pr(\eta_{L(W/P)} \geqslant 0) = 0.07, \qquad \Pr(\eta_{L(W/P)} \geqslant -1) \simeq 1. \tag{6.53}$$

A similar analysis of the medium-run elasticity attempts to measure the more elusive impact of wages on capacity adjustments and capital/labor substitution. The correspondingly more imprecise results are presented side by side with the short-run posterior density in Figure 6.5. This time,

$$\Pr(\eta_{L(W/P)} \geqslant -1) = 0.15. \tag{6.54}$$

7. Numerics

7.1. Introduction

In the course of this chapter we have been confronted mainly with two types of numerical problems: (i) the evaluation of poly-t densities, whose kernels are products, or ratios of products, of multivariate Student kernels; this problem is discussed in Section 7.2; and (ii) the computation of multidimensional integrals by numerical procedures; this problem is discussed in Section 7.3. These two sections can be read independently of each other. Other numerical problems may arise in specific contexts (computation of hypergeometric functions, computation of orthogonal polynomials, etc.); these are not discussed here.

7.2. Evaluation of poly-t densities

Let $y \in R^k$ be a random vector and $\bar{y}' = (1 \quad y')$. Let still $\mathcal{C}_k(\bar{\mathcal{C}}_k)$ denote the cone of $k \times k$ positive (semi) definite symmetric matrices.

Definition 7.1 [Drèze (1977)]

(i) The kernel of an m–0 (product form) poly-t density is defined as

$$\phi_m(y|\mathcal{S}) = \prod_{j=1}^{m} (\bar{y}'S_j\bar{y})^{-(1/2)\nu_j}, \tag{7.1}$$

where $\mathcal{S} = ((S_j, \nu_j); j = 1 \dots m)$ and

$$(S_j, \nu_j) \in \bar{\mathcal{C}}_{k+1} \times R_+; \qquad j = 1 \dots m, \tag{7.2a}$$

$$\sum_{j=1}^{m} S_j \in \mathcal{C}_{k+1}, \qquad \nu = \sum_{j=1}^{m} \nu_j - k > 0. \tag{7.2b}$$

(ii) An $m-n$ (ratio form) poly-t density is essentially[31] defined as the ratio of an $m-0$ poly-t density to an $n-0$ poly-t density.

In this section we shall restrict our attention to $m-1$ poly-t densities with kernels

$$\phi_{m,1}(y|\mathbb{S}_0,\mathbb{S}) = (\bar{y}'S_0\bar{y})^{(1/2)\nu_0}\phi_m(y|\mathbb{S}), \tag{7.3}$$

where, in addition to (7.2),

$$(S_0,\nu_0) \in \bar{\mathcal{C}}_{k+1} \times R_+, \tag{7.4a}$$

$$\rho = \nu - \nu_0 = \sum_{j=1}^{m} \nu_j - \nu_0 - k > 0. \tag{7.4b}$$

The numerical evaluation of $m-1$ poly-t densities has been discussed in Richard and Tompa (1980). In this section we only report their two main results and refer the interested reader to the paper for more details and proofs.

(1) $m-0$ poly-t densities can be analyzed by means of $(m-1)$-dimensional numerical integrations on an auxiliary vector $c \in (0,1)^{m-1}$, *irrespective of* k, the number of components in y. This key result is based on an integral identity due to Dickey (1968). For $m \leqslant 3$, which covers all the cases encountered in this paper, iterative Simpson rules, as described in Section 7.3.1, have proved very efficient, especially since the distribution of c may exhibit considerable skewness.

(2) Under the additional assumption that $\frac{1}{2}\nu_0$ is an integer, the integrating constant and the moments of a $1-1$ poly-t density can be computed by means of *finite analytical* recurrence relationships.[32] Non-integer values of $\frac{1}{2}\nu_0$ (usually half-integer for Bayesian applications) are treated by interpolation between neighboring cases with integer values. This requires only a few additional iterations with the recurrence formulae.

There now exists a computer program "poly-t distribution" (PTD) written in FORTRAN IV and based on the ideas discussed above. It can handle 1–0, 1–1, 2–0, 2–1, 3–0, and 3–1 poly-t distributions and computes the integrating constant and the first- and second-order moments of y, together with its covariance matrix. Optionally, the program also computes the skewness and the fractiles for the components of y, together with plots of the marginal density functions. [It is clear from the definition of poly-t densities, as given in (7.1) and (7.3), that the marginal density of a single component of y can be evaluated at any given point by integrating a $(k-1)$-dimensional poly-t density of the same type.] Test runs

[31] The constraints (7.2b) need not be satisfied by the $n-0$ poly-t density in the denominator, as illustrated by our definition of the $m-1$ poly-t density in (7.3)–(7.4).

[32] These recurrence formulae are obtained by application of the formula for integration by parts.

and applications have shown the results to be reliable and no problems of numerical instability have been encountered. A user's manual is available from the authors. The program PTD has also been incorporated in a Bayesian Regression software (BRP) which is designed to evaluate posterior density functions and their moments for a class of models which are discussed in Drèze (1977). This class covers the single equation analysis, both marginally in the limited information framework and conditionally (Lemma 6.2 and Corollary 6.5) in the full information framework.

7.3. *Numerical integration*

At various stages of our discussion we have been confronted with the necessity of evaluating multiple integrals by numerical methods. More generally, such methods are required whenever the restrictions embodied in analytically convenient prior specifications are thought to be unacceptable. We draw a distinction based on the dimensionality of the integrals to be evaluated.

(1) Some of the Bayesian procedures described here call for low-dimensional numerical integrations, say not more than three-dimensional. This class of problems includes the limited information procedures described in Section 5, taking into account our discussion of poly-t densities and the full information procedures for two-equations model. Such low-dimensional integrals can easily be computed with *predetermined* accuracy by means of the so-called product rules described in Section 7.3.1. These rules are of general application in the sense that they do not depend on the function to be evaluated. (Evidently, the choice of one rule rather than another may very much depend on the function.)

(2) In cases of higher-dimensional numerical integrations the Monte Carlo procedures described in Section 7.3.2 will generally prove far more efficient. However, they suffer from two drawbacks: their numerical accuracy is often difficult to determine; and they must be adapted to specific features of the functions to be integrated – features that are typically unknown.

7.3.1. *Product rules*

A univariate integral may be evaluated by a weighted mean of the values of the integrand at certain values x_i of the abscissa:[33]

$$\int_a^b f(x)\,\mathrm{d}x = \sum_{i=1}^{n} w_i f(x_i) + R, \tag{7.5}$$

[33]We shall limit our attention to integrals over intervals since the real line is isomorphic to]0, 1[.

where the error R depends on n, on the w's, the x's, and of course on f. Two rules have been applied successfully in our field, namely the iterative three-point Newton–Cotes formula, also known as the iterative Simpson's rule, and the n-point Gaussian rule. Other rules may be found for example in Davis and Rabinowitz (1967).

(1) Simpson's rule uses the function values at the two end points (x_1 and x_3) and at the mid point (x_2) of the range of integration

$$\int_a^b f(x)\,\mathrm{d}x \simeq \frac{b-a}{6}\left[f(x_1)+4f(x_2)+f(x_3)\right]. \tag{7.6}$$

The error is then given by

$$R = -\frac{1}{90}\left(\frac{b-a}{2}\right)^5 f^{(4)}(\xi), \tag{7.7}$$

where $f^{(4)}(\xi)$ represents the fourth derivative evaluated at an intermediate point ξ. Simpson's rule amounts therefore to approximating f by a cubic parabola.

In the iterative Simpson's rule, the integral is evaluated by Simpson's rule and then a second time by applying Simpson's rule to the two halves of the range. If the sum of these two integrals equals the integral over the first range to the required precision, the value is accepted; if not, the same procedure is applied to each half of the range successively. Evidently, the stopping rule will not be used before the function has been evaluated at a minimum number of points. Properties of the iterative Simpson's rule are discussed for example in Tompa (1973). In particular: "...the iterative Simpson process is local in the sense that it chooses a large number of abscissae in regions where it is necessary to do so but uses comparatively few where a small number suffices; of course the abscissae are still chosen among a set of predetermined values" [Tompa (1973), p. 14)].

(2) In the Gaussian n-point formula, both the abscissae x_i and the weights w_i have been determined in such a way that the rule is exact for polynomials up to the order $2n-1$. There exists, therefore, no n-point rule of higher order than the Gaussian rule. The abscissae are irrational – they are in fact the zeros of the nth degree Legendre polynomials – and do not contain the end points of the range of integration so that, in contrast with the iterative Simpson's rule, the values of the integrand computed for a given value of n cannot be used for any other value of n.

Tompa (1967) has proposed a scheme of automatic Gaussian integration which he has found to be on the whole more efficient than iterative Simpson integration. Still, the latter has proved more efficient for evaluating integrals of the form met in the evaluation of 1–1 poly-t densities.

Multiple integrals can be evaluated by product rules which are defined by a recursive application of the corresponding univariate rules. For example, an integral of the form

$$I=\int\int_{D}f(x,y)\,\mathrm{d}x\,\mathrm{d}y \tag{7.8}$$

can be evaluated by integrating for example first over y and then over x:

$$I=\int_{a_1}^{b_1}\left[\int_{a_2}^{b_2}f(x,y)\,\mathrm{d}y\right]\mathrm{d}x, \tag{7.9}$$

the bounds of integration being appropriatedly defined. (Note that a_2 and b_2 may be functions of x since the domain D need not be a rectangle.) When the procedure is applied to density functions, the marginal density of x is obtained without further computing effort since it is given by the inner integral in (7.9). In fact, provided D is a rectangle and the same intermediate points are used (in which case f is evaluated over a grid of points), both marginal densities are obtained as by-products of the integral (7.8). The same holds for higher dimensions. The major disadvantage of product rules lies in the fact that the number of points which are required in order to achieve a predetermined accuracy grows more or less exponentially with the dimensionality of the integrals. This limits their usefulness to dimensions up to, say, 3 or 4.

7.3.2. *Monte Carlo procedures*

For higher dimensional problems, Kloek and van Dijk (1978) and van Dijk and Kloek (1980) rightly advocate the use of Monte Carlo procedures. The critical step in the application of such procedures lies in the construction of an *importance function*, i.e. of an approximation of the density function to be evaluated, from which vectors can be drawn at random (this requires knowledge of the integrating constant). Let f be the (multivariate) function to be integrated over the domain D, and g be an importance function. One may then write the integral as

$$I=\int_{D}f(x)\,\mathrm{d}x=\int_{D}\left[\frac{f(x)}{g(x)}\right]g(x)\,\mathrm{d}x. \tag{7.10}$$

Let then $x_1, x_2\dots x_n$ be a sequence of n independent random drawings from the density g. The integral (7.10) is evaluated by

$$\hat{I}_n=\frac{1}{n}\sum_{i=1}^{n}\frac{f(x_i)}{g(x_i)}. \tag{7.11}$$

To evaluate the number of drawings required in order to achieve a given level of accuracy, let σ^2 be the (unknown) variance of f/g. Under the usual assumptions for the application of central limit theorems, the random variable $(\sqrt{n}/\sigma)(\hat{I}_n - I)$ converges in distribution towards a standardized normal random variable. Let c_α be the α-percent value of the standardized normal distribution. If we wish to guarantee a relative error less than ε with probability $1-\alpha$, the number of independent drawings should be

$$n \geqslant \left(\frac{c_\alpha}{\varepsilon}\right)^2 \cdot \left(\frac{\sigma}{I}\right)^2. \tag{7.12}$$

In fact many of the expressions to be evaluated (moments, fractiles, etc.) are ratios of integrals. As discussed in Kloek and van Dijk (1978), the number of drawings may then often be reduced, provided the covariance between the approximations of the numerator and of the denominator is positive. In any case, an importance function should be chosen in such a way that the variation coefficient σ/I be as small as possible. This is typically a difficult task, especially when little is known about the properties of the function to be integrated. Van Dijk and Kloek (1980) have made several pathbreaking advances as regards the design of importance functions and have applied them successfully to the analysis of Klein's model I, an application which requires nine-dimensional numerical integration (see Section 6.6.2). The prospects are therefore promising. But more work is needed before Monte Carlo methods can be claimed to be operational – which would for example require the automatized derivation of importance functions. It is our belief that the analytical methods discussed in Sections 6.2–6.3 will be helpful in that respect. Indeed:

(i) the properties of posterior distributions are more easily investigated under (extended) natural-conjugate prior densities than they are under more general priors;
(ii) we could therefore also consider approximating any given prior density by an (extended) natural-conjugate prior and using the corresponding importance function for the more general case.

Although these comments are presently highly tentative, we can illustrate them by proposing a general procedure for constructing an implicit importance function for the posterior density (6.17). The proposition combines a suggestion made by van Dijk and Kloek (1980) with the results of Corollary 6.5 and works as follows.

(1) We draw first a random matrix $\tilde{\Delta}$ from the normal approximation to the posterior density (6.17) which, as discussed in Morales (1971), can be constructed by application of the usual full information maximum likelihood estimation procedures.

(2) We then draw a complete set of δ's from the conditional densities $D(\delta_{\cdot}|\Delta_{\cdot})$, as given in Corollary 6.5, with the conditioning variables being fixed at the values obtained in step (1).

The procedure is repeated n times. It is evidently required that we should be able to draw random vectors from a 1–1 poly-t density. Preliminary investigations indicate that this should be feasible. The analysis of the properties of such procedures belongs in a list of priorities for further research. They suggest an avenue to make the application of Bayesian full information procedures operational for simultaneous equation models with more than two equations.

Appendix A: Elements of multivariate analysis

We have regrouped here a set of formulae which we have used in the course of this chapter. For more details and for proofs, see for example the appendices in Zellner (1971) and Richard (1979).

A.1. Notations

$\mathcal{C}^q(\bar{\mathcal{C}}^q)$ denotes the set of $q \times q$ symmetric positive (semi) definite matrices. Vec A is the column expansion of the $p \times q$ matrix A:

$$\operatorname{vec} A = \begin{pmatrix} a_1 \\ \vdots \\ a_q \end{pmatrix}, \tag{A.1}$$

where a_i is the ith column of A. Let X be a $p \times q$ random matrix. $E(X)$ and $V(X)$ represent respectively the expectation matrix of X and the covariance matrix of vec X. The following partitionings will be used:

$$X = (X_1 \quad X_2) = \begin{pmatrix} X_{(1)} \\ X_{(2)} \end{pmatrix} = \begin{pmatrix} X_{11} & X_{12} \\ X_{21} & X_{22} \end{pmatrix}, \tag{A.2}$$

where X_{ij} is $p_i \times q_j$. When X is square and non-singular, we can partition conformably X^{-1} and we have

$$X^{-1} = \begin{pmatrix} X^{11} & X^{12} \\ X^{21} & X^{22} \end{pmatrix} = \begin{pmatrix} X_{11}^{-1} + X_{11}^{-1}X_{12}X_{22.1}^{-1}X_{21}X_{11}^{-1} & -X_{11}^{-1}X_{12}X_{22.1}^{-1} \\ -X_{22.1}^{-1}X_{21}X_{11}^{-1} & X_{22.1}^{-1} \end{pmatrix}, \tag{A.3}$$

where $X_{22.1} = X_{22} - X_{21}X_{11}^{-1}X_{12}$. The Moore–Penrose inverse of X is denoted X^{+}. A remarkable property of symmetric semi definite matrices is that formula (A.3) still applies with Moore–Penrose inverses (since we have in particular $X_{21}X_{11}^{+}X_{11} = X_{21}$). The same is *not* true with other matrices.

A.2.

The $p\times 1$ random vector x has a multivariate Normal distribution if and only if its density function is given by

$$p(x)=f_{\mathrm{N}}^{p}(x|\bar{x},\Omega)=_{\mathrm{def}}\left[(2\pi)^{p}|\Omega|\right]^{-1/2}\exp\left[-\tfrac{1}{2}(x-\bar{x})'\Omega^{-1}(x-\bar{x})\right], \tag{A.4}$$

where $\bar{x}\in R^{p}$ and $\Omega\in\mathcal{C}^{p}$. The properties of the multivariate Normal distribution need not be reiterated here since they are special cases ($q=1$) of the formulae we give next.

A.3.

The $p\times q$ random matrix X has a matricvariate Normal distribution if and only if its density function $p(X)$ is given by

$$\begin{aligned} p(X)&=f_{\mathrm{MN}}^{p\times q}(X|\bar{X},\Omega\otimes P)\\ &=_{\mathrm{def}}\left[(2\pi)^{pq}|\Omega|^{p}|P|^{q}\right]^{-1/2}\exp\left[-\tfrac{1}{2}\operatorname{tr}\Omega^{-1}(X-\bar{X})'P^{-1}(X-\bar{X})\right],\end{aligned} \tag{A.5}$$

where $\bar{X}\in R^{p\times q}$, $P\in\mathcal{C}^{p}$, and $\Omega\in\mathcal{C}^{q}$. We then have

$$E(X)=\bar{X},\qquad V(X)=\Omega\otimes P, \tag{A.6}$$

$$p(AX)=f_{\mathrm{MN}}^{l\times q}(AX|A\bar{X},\Omega\otimes APA'),\qquad A_{l\times p},\ \operatorname{rank}A=l\leqslant p, \tag{A.7}$$

$$p(XB)=f_{\mathrm{MN}}^{p\times l}(XB|\bar{X}B,B'\Omega B\otimes P),\qquad B_{q\times l},\ \operatorname{rank}B=l\leqslant q, \tag{A.8}$$

$$p(X_{(1)})=f_{\mathrm{MN}}^{p_1\times q}(X_{(1)}|\bar{X}_{(1)},\Omega\otimes P_{11}), \tag{A.9}$$

$$p(X_{(2)}|X_{(1)})=f_{\mathrm{MN}}^{p_2\times q}(X_{(2)}|\bar{X}_{(2.1)}+P_{21}P_{11}^{-1}X_{(1)},\Omega\otimes P_{22.1}), \tag{A.10}$$

where

$$\bar{X}_{(2.1)}=\bar{X}_{(2)}-P_{21}P_{11}^{-1}\bar{X}_{(1)}=\bar{X}_{(2)}+(P^{22})^{-1}P^{21}\bar{X}_{(1)}, \tag{A.11}$$

$$\left(\bar{X}_{(1)},P_{11},\bar{X}_{(2.1)},P_{21}P_{11}^{-1},P_{22.1}\right)\in\left(R^{p_1\times q}\times\mathcal{C}^{p_1}\right)\times\left(R^{p_2\times q}\times R^{p_2\times p_1}\times\mathcal{C}^{p_2}\right), \tag{A.12}$$

$$p(X_1)=f_{\mathrm{MN}}^{p\times q_1}(X_1|\bar{X}_1,\Omega_{11}\otimes P), \tag{A.13}$$

$$p(X_2|X_1)=f_{\mathrm{MN}}^{p\times q_2}(X_2|\bar{X}_{2.1}+X_1\Omega_{11}^{-1}\Omega_{12},\Omega_{22.1}\otimes P), \tag{A.14}$$

where

$$\bar{X}_{2.1}=\bar{X}_2-\bar{X}_1\Omega_{11}^{-1}\Omega_{12}=\bar{X}_2+\bar{X}_1\Omega^{12}(\Omega^{22})^{-1}, \tag{A.15}$$

$$(\bar{X}_1,\Omega_{11},\bar{X}_{2.1},\Omega_{11}^{-1}\Omega_{12},\Omega_{22.1})\in(R^{p\times q_1}\times\mathcal{C}^{q_1})\times(R^{p\times q_2}\times R^{q_1\times q_2}\times\mathcal{C}^{q_2}). \tag{A.16}$$

A.4.

The random variable σ^2 has an Inverted–gamma distribution if and only if its density function is given by

$$p(\sigma^2)=f_{\mathrm{i}\gamma}(\sigma^2|s^2,\nu) \tag{A.17}$$

$$=_{\mathrm{def}}\left[\Gamma\left(\frac{\nu}{2}\right)\right]^{-1}\left(\frac{s^2}{2}\right)^{\nu/2}(\sigma^2)^{-(1/2)(\nu+2)}\exp\left(-\frac{1}{2}\frac{s^2}{\sigma^2}\right),$$

where $s^2>0$ and $\nu>0$. It is a special case ($q=1$) of the Inverted–Wishart density.

A.5.

The $q\times q$ random matrix $\Omega\in\mathcal{C}^q$ has an Inverted–Wishart distribution if and only if its density function is given by

$$p(\Omega)=f_{\mathrm{iW}}^{q}(\Omega|Q,\nu) \tag{A.18}$$

$$=_{\mathrm{def}}\left[2^{(1/2)\nu q}\pi^{(1/4)q(q-1)}\prod_{i=1}^{q}\Gamma\left(\frac{\nu+1-i}{2}\right)\right]^{-1}$$

$$\times|Q|^{(1/2)\nu}|\Omega|^{-(1/2)(\nu+q+1)}\exp\left(-\tfrac{1}{2}\mathrm{tr}\,\Omega^{-1}Q\right),$$

where $Q\in\mathcal{C}^q$ and $\nu>q-1$. We then have

$$E(\Omega)=\frac{1}{\nu-q-1}Q\qquad(\nu>q+1), \tag{A.19}$$

$$p(\Omega_{11},\Omega_{11}^{-1}\Omega_{12},\Omega_{22.1})=p(\Omega_{11})p(\Omega_{11}^{-1}\Omega_{12}|\Omega_{22.1})p(\Omega_{22.1}), \tag{A.20}$$

with

$$p(\Omega_{11}) = f_{iW}^{q_1}(\Omega_{11}|Q_{11}, \nu - q_2), \tag{A.21}$$

$$p(\Omega_{11}^{-1}\Omega_{12}|\Omega_{22.1}) = f_{MN}^{q_1 \times q_2}(\Omega_{11}^{-1}\Omega_{12}|Q_{11}^{-1}Q_{12}, \Omega_{22.1} \otimes Q_{11}^{-1}), \tag{A.22}$$

$$p(\Omega_{22.1}) = f_{iW}^{q_2}(\Omega_{22.1}|Q_{22.1}, \nu), \tag{A.23}$$

$$E(\Omega|\Omega_{11}) = \frac{1}{\nu - q_2 - 1}(1 + \operatorname{tr}\Omega_{11}Q_{11}^{-1})\begin{pmatrix} 0 & 0 \\ 0 & Q_{22.1} \end{pmatrix} + \begin{pmatrix} I_{q_1} \\ Q_{21}Q_{11}^{-1} \end{pmatrix}\Omega_{11}\begin{pmatrix} I_{q_1} & Q_{11}^{-1}Q_{12} \end{pmatrix}. \tag{A.24}$$

If, furthermore, $p(X|\Omega) = f_{MN}^{p \times q}(X|\bar{X}, \Omega \otimes H^{-1})$ and $X_{2.1} = X_2 - X_1\Omega_{11}^{-1}\Omega_{12}$, then (X_1, Ω_{11}) and $(X_{2.1}, \Omega_{11}^{-1}\Omega_{12}, \Omega_{22.1})$ are independent and we have in particular

$$p(X_{2.1}|\Omega_{11}^{-1}\Omega_{12}, \Omega_{22.1}) = f_{MN}^{p \times q_2}(X_{2.1}|\bar{X}_{2.1}, \Omega_{22.1} \otimes H^{-1}), \tag{A.25}$$

$$E(X|X_1, \Omega_{11}) = E(X|X_1) = \bar{X} + (X_1 - \bar{X}_1)\begin{pmatrix} I_{q_1} & Q_{11}^{-1}Q_{12} \end{pmatrix}, \tag{A.26}$$

$$V(X|X_1, \Omega_{11}) = V(X|X_1) = \frac{1}{\nu - q_2 - 1}\begin{pmatrix} 0 & 0 \\ 0 & Q_{22.1} \end{pmatrix} \otimes [H^{-1} + (X_1 - \bar{X}_1)Q_{11}^{-1}(X_1 - \bar{X}_1)']. \tag{A.27}$$

A.6.

The $p \times 1$ random vector x has a multivariate Student distribution if and only if its density function is given by

$$p(x) = f_t^p(x|\bar{x}, H, \nu) =_{\text{def}} \left[\pi^{-(1/2)p}\Gamma\left(\frac{\nu + p}{2}\right)/\Gamma\left(\frac{\nu}{2}\right)\right] \times |H|^{1/2}[1 + (x - \bar{x})'H(x - \bar{x})]^{-(1/2)(\nu + p)}, \tag{A.28}$$

where $\bar{x} \in R^p$, $H \in \mathcal{C}^p$, and $\nu > 0$. Its properties are obtained as special cases $(q = 1)$ of the formulae which are given next.

A.7.

The $p \times q$ random matrix X has a matricvariate Student distribution if and only if its density function is given by

$$p(X) = f_{\mathrm{M}t}^{p\times q}(X|\bar{X}, Q, H, \nu)$$

$$=_{\mathrm{def}} \left\{ \pi^{(1/2)pq} \prod_{i=1}^{q} \frac{\Gamma\left(\frac{\nu+1-i}{2}\right)}{\Gamma\left(\frac{\nu+p+1-i}{2}\right)} \right\}^{-1} \tag{A.29}$$

$$\times |Q|^{(1/2)\nu} |H|^{(1/2)q} |Q + (X-\bar{X})'H(X-\bar{X})|^{-(1/2)(\nu+p)},$$

where $\bar{X} \in R^{p\times q}$, $H \in \mathcal{C}^p$, $Q \in \mathcal{C}^q$, and $\nu > q-1$. We then have

$$E(X) = \bar{X}\,(\nu > q), \qquad V(X) = \frac{1}{\nu - q - 1} Q \otimes H^{-1} \qquad (\nu > q+1), \tag{A.30}$$

$$p(X_{(1)}) = f_{\mathrm{M}t}^{p_1\times q}\left(X_{(1)}|\bar{X}_{(1)}, Q, H_{11.2}, \nu\right), \tag{A.31}$$

$$p(X_{(2)}|X_{(1)}) = f_{\mathrm{M}t}^{p_2\times q}\Big(X_{(2)}|\bar{X}_{(2)} - H_{22}^{-1}H_{21}\left(X_{(1)} - \bar{X}_{(1)}\right), \tag{A.32}$$

$$Q + \left(X_{(1)} - \bar{X}_{(1)}\right)'H_{11.2}\left(X_{(1)} - \bar{X}_{(1)}\right), H_{22}, \nu + p_1\Big),$$

$$p(X_1) = f_{\mathrm{M}t}^{p\times q_1}\left(X_1|\bar{X}_1, Q_{11}, H, \nu - q_2\right), \tag{A.33}$$

$$p(X_2|X_1) = f_{\mathrm{M}t}^{p\times q_2}\Big(X_2|\bar{X}_2 + \left(X_1 - \bar{X}_1\right)Q_{11}^{-1}Q_{12}, Q_{22.1}, \tag{A.34}$$

$$\left[H^{-1} + \left(X_1 - \bar{X}_1\right)Q_{11}^{-1}\left(X_1 - \bar{X}_1\right)'\right]^{-1}, \nu\Big).$$

If $p(X|\Omega) = f_{\mathrm{MN}}^{p\times q}(X|\bar{X}, \Omega \otimes H^{-1})$ and $p(\Omega) = f_{\mathrm{iW}}^{q}(\Omega|Q, \nu)$, then

$$p(X) = f_{\mathrm{M}t}^{p\times q}(X|\bar{X}, Q, H, \nu), \tag{A.35}$$

$$p(\Omega|X) = f_{\mathrm{iW}}^{q}(\Omega|Q + (X-\bar{X})'H(X-\bar{X}), \nu + p). \tag{A.36}$$

Appendix B: Proofs

B.1. Proof of Lemma 5.3

Let $P_{\mathrm{a}} = (\beta \quad Q_{\mathrm{a}})$ and $P_{\mathrm{b}} = (\beta \quad Q_{\mathrm{b}})$, where the indices a and b are used to denote two alternative choices of Q. Consider the sequence of transformations

$(\Phi^{a}, \sigma^{2}, \Delta^{a}_{12}, \Lambda^{a}_{22.1}) \xrightarrow{1} (\Phi^{a}, \Lambda^{a}) \xrightarrow{2} (\Phi^{b}, \Lambda^{b}) \xrightarrow{3} (\Phi^{b}, \sigma^{2}, \Delta^{b}_{12}, \Lambda^{b}_{22.1})$, where $\Phi^{\cdot}, \Lambda^{\cdot}, \Delta^{\cdot}_{12}, \Lambda^{\cdot}_{22.1}$ are obtained respectively from (5.3), (5.9), (5.10), and (5.11). The transformations 1 and 3 have Jacobians $(\sigma^{2})^{-(m-1)}$ and $(\sigma^{2})^{m-1}$ which cancel out.

Furthermore,

$$\Lambda^{b} = \left(P_{a}^{-1}P_{b}\right)'\Lambda^{a}\left(P_{a}^{-1}P_{b}\right), \tag{B.1}$$

$$\Phi^{b} = \Phi^{a}\left(P_{a}^{-1}P_{b}\right). \tag{B.2}$$

The Jacobian of the transformation 2 is therefore $\|P_{a}^{-1}P_{b}\|^{-(m+n+1)}$. Also,

$$\|\Lambda^{b}_{22.1}| = \sigma^{-2}|\Lambda^{b}| = \sigma^{-2}|P_{a}^{-1}P_{b}|^{2}|\Lambda^{a}| = |P_{a}^{-1}P_{b}|^{2}|\Lambda^{a}_{22.1}|. \tag{B.3}$$

Consequently, if $p(\Phi^{a}_{2}, \Delta^{a}_{12}, \Lambda^{a}_{22.1}|\theta)$ is of the form (5.17), then

$$p\left(\Phi^{b}_{2}, \Delta^{b}_{12}, \Lambda^{b}_{22.1}|\theta\right) \propto \|P_{a}^{-1}P_{b}\|^{\nu}|\Lambda^{b}_{22.1}|^{-(1/2)(\nu+m+n+1)}.\square \tag{B.4}$$

B.2. *Proof of Lemma 5.6*

Our first step is to combine the conditional prior measure $|\Lambda_{22.1}|^{-(1/2)(m+n+1)}$ and the partial likelihood $L_{2}(\theta, \Phi_{2}, \Delta_{12}, \Lambda_{22.1}|Y)$, as defined in (5.15). With $\hat{\Pi}$ as defined in (4.1), the quadratic form in the exponential part of this product is then rewritten as:

$$\begin{aligned}&\{[YQ-(Y\beta+Z\gamma)\Delta_{12}]-Z\Phi_{2}\}'\{[YQ-(Y\beta+Z\gamma)\Delta_{12}]-Z\Phi_{2}\} \\ &= \{\Phi_{2}-[\hat{\Pi}Q-(\hat{\Pi}\beta+\gamma)\Delta_{12}]\}'Z'Z\{\Phi_{2}-[\hat{\Pi}Q-(\hat{\Pi}\beta+\gamma)\Delta_{12}]\} \\ &\quad+(\Delta_{12}-s_{11}^{-1}S_{12})'s_{11}(\Delta_{12}-s_{11}^{-1}S_{12})+S_{22.1}.\end{aligned} \tag{B.5}$$

This decomposition is familiar in stepwise regression whereby, for given Δ_{12}, $YQ-(Y\beta+Z\gamma)\Delta_{12}$ is regressed on Z and Δ_{12} is estimated by regressing $[I-Z(Z'Z)^{-1}Z']YQ$ on $[I-Z(Z'Z)^{-1}Z']Y\beta$. Formula (5.28) follows immediately. Note that the integrating constants of these densities include respectively the factors $|S_{22.1}|^{(1/2)T}$, $|s_{11}^{-1}\Lambda_{22.1}|^{-1/2}$, and $|\Lambda_{22.1}|^{-(1/2)n}$. This justifies the "degrees of freedom" parameter T in (5.28a) and the posterior density

$$p(\theta|Y) \propto p(\theta)\|P\|^{T}(s_{11})^{-(1/2)(m-1)}|S_{22.1}|^{-(1/2)T}\cdot L_{1}(\theta|Y), \tag{B.6}$$

which reduces to (5.27) upon noting from (5.25) that

$$|S_{22.1}| = s_{11}^{-1}|S| = s_{11}^{-1}|W|\,|P|^{2}.\square \tag{B.7}$$

B.3. Proof of Corollary 5.8

Let x and y represent two random vectors. We have, in a self-explanatory notation:

$$\begin{aligned} E(x) &= E_y[E(x|y)], \\ V(x) &= E_y[V(x|y)] + V_y[E(x|y)], \\ \operatorname{cov}(x, y) &= \operatorname{cov}_y[E(x|y), y]. \end{aligned}$$

Corollary 5.8 is then proved by applying these formulae to the conditional moments of $\Lambda_{22.1}$, Δ_{12}, and Φ_2, as obtained from the densities in (5.28). For example, (5.28c) and (5.28b) yield first

$$\begin{aligned} E(\Phi_2|\theta, Y) &= \hat{\Pi}Q - (\hat{\Pi}\beta + \gamma)E(\Delta_{12}|\theta, Y) \\ &= \hat{\Pi}Q - (\hat{\Pi}\beta + \gamma)(\beta' W\beta)^{-1}\beta' WQ. \end{aligned} \tag{B.8}$$

Applying (5.3) and rewriting γ as $(\hat{\Pi}\beta + \gamma) - \hat{\Pi}\beta$ yields formula (5.29). The derivation of (5.30) and (5.31) goes along the same lines but is somewhat more tedious. A direct proof based on properties of matricvariate Normal and Inverted–Wishart densities may be found in Richard (1979). □

B.4. Proof of Lemma 6.2

Let $B = (\beta_1 \quad B_2)$. We then have

$$\begin{aligned} \|B\| = |B'B|^{1/2} &= \begin{vmatrix} \beta_1'\beta_1 & \beta_1' B_2 \\ B_2'\beta_1 & B_2' B_2 \end{vmatrix}^{1/2} \\ &= \left[\beta_1'\{I_m - B_2(B_2'B_2)^{-1}B_2'\}\beta_1\right]^{1/2} |B_2'B_2|^{1/2}. \end{aligned} \tag{B.9}$$

The factor $\|B\|^T$ in (6.5) is thereby decomposed into the product of a Student kernel in the unconstrained elements of β_1, and a factor which no longer depends on β_1.

Let $A = (\alpha_1 \quad A_2)$. A similar argument – see formula (B.11) below – enables us to factorize the determinant $|S_0 + A'X'XA|$ into the product of a Student kernel in δ_1, the vector of unconstrained elements of α_1, and a factor which no longer depends on δ_1.

The third Student kernel arises from the marginal prior density $f_t^{l_1}(\delta_1 | d_1^0, H_1^0, \nu_1^0)$. □

B.5. Lemma B.1

Let S and $\mathfrak{M}$ be respectively $m \times m$ and $l \times l$ PSDS matrices and X be an arbitrary $l \times m$ matrix, partitioned conformably as

$$X = (X_1 \quad X_2) = \begin{pmatrix} X_{11} & X_{12} \\ X_{21} & X_{22} \end{pmatrix}, \qquad S = \begin{pmatrix} S_{11} & S_{12} \\ S_{21} & S_{22} \end{pmatrix},$$

$$\mathfrak{M} = \begin{pmatrix} \mathfrak{M}_{11} & \mathfrak{M}_{12} \\ \mathfrak{M}_{21} & \mathfrak{M}_{22} \end{pmatrix}. \tag{B.10}$$

Then

$$|S + X'\mathfrak{M}X| = |S_{22} + X_2'\mathfrak{M}X_2| \cdot |\bar{S}_{11} + (X_{11} - \bar{X}_{11})'R_{11}(X_{11} - \bar{X}_{11})| \tag{B.11}$$

where

$$\bar{S}_{11} = S_{11.2} + (X_{21} - X_{22}S_{22}^{+}S_{21})'R_{22.1}(X_{21} - X_{22}S_{22}^{+}S_{21}), \tag{B.12}$$

$$\bar{X}_{11} = X_{12}S_{22}^{+}S_{21} - R_{11}^{+}R_{12}(X_{21} - X_{22}S_{22}^{+}S_{21}), \tag{B.13}$$

$$R = \mathfrak{M} - \mathfrak{M}X_2(S_{22} + X_2'\mathfrak{M}X_2)^{+}X_2'\mathfrak{M}. \tag{B.14}$$

The superscript $+$ denotes Moore–Penrose inverses and the following notation has been used: $A_{ii.j} = A_{ii} - A_{ij}A_{jj}^{+}A_{ji}$.

Proof

We first establish two preliminary results:

(1) Since $S \in \mathcal{C}^m$ we have $S_{12}S_{22}^{+}S_{22} = S_{12}$ and

$$\begin{pmatrix} I_{m_1} & -S_{12}S_{22}^{+} \\ 0 & I_{m_2} \end{pmatrix} S \begin{pmatrix} I_{m_1} & 0 \\ -S_{22}^{+}S_{21} & I_{m_2} \end{pmatrix} = \begin{pmatrix} S_{11.2} & 0 \\ 0 & S_{22} \end{pmatrix}, \tag{B.15}$$

from which

$$\begin{aligned} |S + X'\mathfrak{M}X| &= \begin{vmatrix} S_{11.2} + X_{1.2}'\mathfrak{M}X_{1.2} & X_{1.2}'\mathfrak{M}X_2 \\ X_2'\mathfrak{M}X_{1.2} & S_{22} + X_2'\mathfrak{M}X_2 \end{vmatrix} \\ &= |S_{22} + X_2'\mathfrak{M}X_2| \cdot |S_{11.2} + X_{1.2}'RX_{1.2}|, \end{aligned} \tag{B.16}$$

with

$$X_{1.2} = X_1 - X_2 S_{22}^{+} S_{21}, \tag{B.17}$$

$$R = \mathfrak{M} - \mathfrak{M} X_2 \left(S_{22} + X_2' \mathfrak{M} X_2 \right)^{+} X_2' \mathfrak{M}. \tag{B.18}$$

(2) Similarly we have

$$\begin{pmatrix} I_{l_1} & 0 \\ -\mathfrak{M}_{21}\mathfrak{M}_{11}^{+} & I_{l_2} \end{pmatrix} \mathfrak{M} \begin{pmatrix} I_{l_1} & -\mathfrak{M}_{11}^{+}\mathfrak{M}_{12} \\ 0 & I_{l_2} \end{pmatrix} = \begin{pmatrix} \mathfrak{M}_{11} & 0 \\ 0 & \mathfrak{M}_{22.1} \end{pmatrix}, \tag{B.19}$$

from which

$$S + X'\mathfrak{M}X = S + X_{(2)}'\mathfrak{M}_{22.1}X_{(2)} + X_{(1.2)}'\mathfrak{M}_{11}X_{(1.2)}, \tag{B.20}$$

with

$$X_{(1.2)} = X_{(1)} + \mathfrak{M}_{11}^{+}\mathfrak{M}_{12}X_{(2)}. \tag{B.21}$$

Formulae (B.11)–(B.14) are then obtained by applying formula (B.20) to the last term in formula (B.16). □

B.6. Proof of Corollary 6.5

Given (B.18) we first have

$$R_{11} = \mathfrak{M}_{11} - \left(\mathfrak{M}_{11} \quad \mathfrak{M}_{12} \right) X_2 \left(S_{22} + X_2'\mathfrak{M}X_2 \right)^{+} X_2' \begin{pmatrix} \mathfrak{M}_{11} \\ \mathfrak{M}_{21} \end{pmatrix} \tag{B.22}$$

and, together with formula (A.3) applied to Moore–Penrose inverses,

$$R_{11}^{+} = \mathfrak{M}_{11}^{+} + \mathfrak{M}_{11}^{+} \left(\mathfrak{M}_{11} \quad \mathfrak{M}_{12} \right) X_2 \left(S_{22} + X_{22}'\mathfrak{M}_{22.1}X_{22} \right)^{+} X_2' \begin{pmatrix} \mathfrak{M}_{11} \\ \mathfrak{M}_{21} \end{pmatrix} \mathfrak{M}_{11}^{+}. \tag{B.23}$$

It follows that if $\mathfrak{M}_{11}$ and S_{22} are PDS matrices then R_{11} is also PDS. This condition, together with $\bar{S}_{11} \geqslant S_{11.2} > 0$, insures that the last term in (6.22) is a proper Student density. Since the first term in (6.22) does not depend on γ_1, integration with respect to γ_1 can be carried out analytically by property of the Student density. □

B.7. Proof of Lemma 6.6

By application of Lemma 6.4 to the case where $\Delta_1 = \delta_1$ (see also Corollary 6.5) we first have

$$
\begin{aligned}
&|S+(\Delta-D)'\mathfrak{M}(\Delta-D)|^{-(1/2)\nu} \\
&\propto \left| S_{22} + \begin{pmatrix} -D_{12} \\ \Delta_2 - D_{22} \end{pmatrix}' \mathfrak{M} \begin{pmatrix} -D_{12} \\ \Delta_2 - D_{22} \end{pmatrix} \right|^{-(1/2)\nu} |R_{11}|^{-1/2} \\
&\times |\bar{S}_{11}|^{-(1/2)(\nu-l_1)} f_t^{l_1}(\delta_1 | \bar{D}_{11}, \bar{S}_{11}^{-1}, R_{11}, \nu - l_1).
\end{aligned}
\tag{B.24}
$$

Consider next the matrix:

$$
\left(\begin{array}{c|c}
\mathfrak{M}_{11} & (\mathfrak{M}_{11} \quad \mathfrak{M}_{12}) \begin{pmatrix} -D_{12} \\ \Delta_2 - D_{22} \end{pmatrix} \\
\hline
\begin{pmatrix} -D_{12} \\ \Delta_2 - D_{22} \end{pmatrix}' \begin{pmatrix} \mathfrak{M}_{11} \\ \mathfrak{M}_{21} \end{pmatrix} & S_{22} + \begin{pmatrix} -D_{12} \\ \Delta_2 - D_{22} \end{pmatrix}' \mathfrak{M} \begin{pmatrix} -D_{12} \\ \Delta_2 - D_{22} \end{pmatrix}
\end{array}\right),
\tag{B.25}
$$

from which

$$
\begin{aligned}
&|\mathfrak{M}_{11}| \cdot |S_{22} + (\Delta_2 - D_{22})'\mathfrak{M}_{22.1}(\Delta_2 - D_{22})| \\
&= |R_{11}| \left| S_{22} + \begin{pmatrix} -D_{12} \\ \Delta_2 - D_{22} \end{pmatrix}' \mathfrak{M} \begin{pmatrix} -D_{12} \\ \Delta_2 - D_{22} \end{pmatrix} \right|
\end{aligned}
\tag{B.26}
$$

and, under the conditions of Corollary 6.5,

$$
|R_{11}|^{-1} \leqslant |S_{22}|^{-1} |\mathfrak{M}_{11}|^{-1} \left| S_{22} + \begin{pmatrix} -D_{12} \\ \Delta_2 - D_{22} \end{pmatrix}' \mathfrak{M} \begin{pmatrix} -D_{12} \\ \Delta_2 - D_{22} \end{pmatrix} \right|
\tag{B.27}
$$

together with $\bar{S}_{11} \geqslant S_{11.2} > 0$. It follows then from (B.24) that

$$
\begin{aligned}
&|S+(\Delta-D)'\mathfrak{M}(\Delta-D)|^{-(1/2)\nu} \\
&\leqslant f_t^{l_1}(\delta_1 | \bar{D}_{11}, \bar{S}_{11}^{-1}, R_{11}, \nu - l_1) \\
&\quad \times |\mathfrak{M}_{11}|^{-1/2} |S_{22}|^{-1/2} |S_{11.2}|^{-(1/2)(\nu - l_1)} \\
&\quad \times \left| S_{22} + \begin{pmatrix} -D_{12} \\ \Delta_2 - D_{22} \end{pmatrix}' \mathfrak{M} \begin{pmatrix} -D_{12} \\ \Delta_2 - D_{22} \end{pmatrix} \right|^{-(1/2)(\nu-1)}.
\end{aligned}
\tag{B.28}
$$

Reference table

Sections and models	Prior density	Posterior density	Characteristics	Properties[a]	References to applications
5. *Limited information analysis*					
Normal – gamma prior					
General	(5.17)–(5.40)	(5.44)	1–1 poly-*t*		5.5
Non-informative	(2.17)	(2.25)	1–1 poly-*t*	Invariance with respect to normalization, no moments exist	2.4
2.4					
Student – gamma prior					
General	(5.17)–(5.49)	(5.51)	2–1 poly-*t*		5.5
Non-informative	(5.17)–(5.49), $\mu_0 = 1$	(5.51)	2–1 poly-*t*	Invariance with respect to normalization, no moments exist	2.4
Non-informative exchangeable	(5.17)–(5.49)–(5.55)	(5.51)	2–1 poly-*t*	Invariance with respect to normalization, no moments exist	
6. *Full information analysis*					
Extended natural – conjugate prior	(6.13)	(6.22)	1–1 poly-*t*	Conditional on Δ_2	
Student – Wishart prior					
General	(6.3)–(6.6)	Lemma 6.2	2–1 poly-*t*	Conditional on Δ_2	
Non-informative	(6.7)	Lemma 6.2	2–1 poly-*t*	Conditional on Δ_2, invariance with respect to normalization	6.6.3
Two-equation models	(6.38)	(6.40)		Requires trivariate numerical integrations	6.6.1
4. *Reduced-form analysis*					
Normal – gamma prior					
General	(4.4)–(4.5)	(4.11)	Student		
Non-informative	(4.20), $M_0 = W_0 = 0$, $\nu_0 = 0$	(4.11)	Student		
Student – gamma prior					
General	(4.13)	(4.14)	2–0 poly-*t*	Conditional on Π_2	
Limited information	(4.15)–(4.16)	(4.17)	2–0 poly-*t*		
Non-informative	(4.21), $M_0 = W_0 = 0$, $\nu_0 = 0$	(4.14)	Student		
Extended natural – conjugate prior	(6.28)	(6.28), (6.18)–(6.20)	Student	Conditional on Π_2	

[a] Unless otherwise specified, the posterior densities are marginal densities for the coefficients of one equation. All the cases mentioned in the table are covered by the Bayesian Regression Program (BRP) and documented in Bauwens et al. (1981).

Since

$$\left| S_{22} + \begin{pmatrix} -D_{12} \\ \Delta_2 - D_{22} \end{pmatrix}' \mathfrak{M} \begin{pmatrix} -D_{12} \\ \Delta_2 - D_{22} \end{pmatrix} \right| = \text{say} \, |\bar{S}_{22} + (\Delta_2 - \bar{D}_{22})' \mathfrak{M}_{22} (\Delta_2 - \bar{D}_{22})|$$

has the same properties as $|S + (\Delta - D)'\mathfrak{M}(\Delta - D)|$, it may be decomposed similarly and a recursive argument yields

$$|S + (\Delta - D)'\mathfrak{M}(\Delta - D)|^{-(1/2)\nu} \leqslant \kappa(\mathfrak{M}, S) \prod_{j=1}^{m} f_t^{l_j}(\delta_j | \bar{d}_j, H_j, \nu - l_j - j + 1), \tag{B.29}$$

where $(\bar{d}_j, H_j;\ j = 1, m)$ are functions of $(\delta_{j+1} \ldots \delta_m)$ obtained by a recursive application of formulae (B.12)–(B.14). Well-known results on the existence of moments for Student densities yield the result. □

References

Anderson, T. W. and H. Rubin (1949) "Estimation of the Parameters of a Single Equation in a Complete System of Stochastic Equations", *Annals of Mathematical Statistics*, 20, 46–68.

Bauwens, L. and G. d'Alcantara (1981) "An Export Model for the Belgian Industry", CORE Discussion Paper 8105, Université Catholique de Louvain.

Bauwens, L., J.-P. Bulteau, P. Gille, L. Longrée, M. Lubrano and H. Tompa (1981) *Bayesian Regression Program (BRP) User's Manual*, CORE Computing Report 81-A-01.

Chetty, V. K. (1968) "Bayesian Analysis of Haavelmo's Models", *Econometrica*, 36, 582–602.

Davis, P. J. and P. Rabinowitz (1967) *Numerical Integration*. Waltham, Mass.: Blaisdell.

Dickey, J. (1968) "Three Multidimensional–Integral Identities with Bayesian Applications", *The Annals of Mathematical Statistics*, 39, 1615–1628.

Drèze, J. H. (1962) "The Bayesian Approach to Simultaneous Equations Estimation", ONR Research Memorandum 67, The Technological Institute, Northwestern University.

Drèze, J. H. (1975) "Bayesian Theory of Identification in Simultaneous Equations Models", in: S. E. Fienberg and A. Zellner (eds.), *Studies in Bayesian Econometrics and Statistics*. Amsterdam: North-Holland Publishing Co.

Drèze, J. H. (1976) "Bayesian Limited Information Analysis of the Simultaneous Equations Model", *Econometrica*, 44, 1045–1075.

Drèze, J. H. (1977) "Bayesian Regression Analysis Using Poly-*t* Densities", *Journal of Econometrics*, 6, 329–354.

Drèze, J. H. and F. Modigliani (1981) "The Trade-off Between Real Wages and Employment in an Open Economy (Belgium)", *European Economic Review*, 15, 1–40.

Drèze, J. H. and J. A. Morales (1976) "Bayesian Full Information Analysis of Simultaneous Equations", *Journal of the American Statistical Association*, 71, 919–923.

Engle, R. F., D. F. Hendry and J. F. Richard (1983) "Exogeneity", forthcoming in *Econometrica.*

Fisher, W. (1976) "Normalization in Point Estimation", *Journal of Econometrics*, 4, 243–252.

Florens, J. P., M. Mouchart and J. F. Richard (1974) "Bayesian Inference in Error-In-Variables Models", *Journal of Multivariate Analysis*, 4, 419–452.

Florens, J. P., M. Mouchart and J. F. Richard (1979) "Specification and Inference in Linear Models", CORE Discussion Paper 7943, Université Catholique de Louvain.

Gantmacher, F. (1960) *Matrix Theory*, vol. 1. New York: Chelsea.
Goldberger, A. S. (1964) *Econometric Theory*. New York: John Wiley & Sons.
Harkema, R. (1971) *Simultaneous Equations. A Bayesian Approach*. Rotterdam: Universitaire Pers.
Hendry, D. F. (1976) "The Structure of Simultaneous Equations Estimators", *Journal of Econometrics*, 4, 51–88.
Kadane, J. (1975) "The Role of Identification in Bayesian Theory", in: S. E. Fienberg and A. Zellner (eds.), *Studies in Bayesian Econometrics and Statistics*. Amsterdam: North-Holland Publishing Co.
Kaufman, G. M. (1975) "Posterior Inference for Structural Parameters Using Cross-Section and Time Series Data", in: S. E. Fienberg and A. Zellner (eds.), *Studies in Bayesian Econometrics and Statistics*, Amsterdam: North-Holland Publishing Co.
Klein, L. R. (1950) *Economic Fluctuations in the United States, 1921–1941*. New York: John Wiley & Sons.
Kloek, T. and H. K. van Dijk (1978) "Bayesian Estimates of Equation Systems Parameters: An Application of Integration by Monte Carlo", *Econometrica*. 46, 1–19.
Kolmogorov, A. N. (1950) *Foundations of the Theory of Probability*. New York: Chelsea.
Lindley, D. V. (1971) *Bayesian Statistics: A Review*. Philadelphia: SIAM.
Lindley, D. V. and A. F. M. Smith (1972) "Bayes' Estimates for the Linear Model", *Journal of the Royal Statistical Society*, Ser. B, 34, 1–41.
Maddala, G. S. (1976) "Weak Priors and Sharp Posteriors in Simultaneous Equation Models", *Econometrica*, 44, 345–351.
Maddala, G. S. (1977) *Econometrics*. New York: McGraw-Hill.
Malinvaud, E. (1978) *Méthodes statistiques de l'économétrie* (3rd edn.). Paris: Dunod.
Mariano, R. S. (1980) "Analytical Small-Sample Distribution Theory in Econometrics: The Simultaneous-Equations Case", CORE Discussion Paper 8026, Université Catholique de Louvain.
Morales, J. A. (1971) *Bayesian Full Information Structural Analysis*. Berlin: Springer-Verlag.
Raiffa, H. and R. Schlaifer (1961) *Applied Statistical Decision Theory*. Cambridge, Mass.: Massachusetts Institute of Technology Press.
Richard, J. F. (1973) *Posterior and Predictive Densities for Simultaneous Equation Models*. Berlin: Springer-Verlag.
Richard, J. F. (1979) "Exogeneity, Inference and Prediction in So-Called Incomplete Dynamic Simultaneous Equation Models", CORE Discussion Paper 7922, Université Catholique de Louvain.
Richard, J. F. and H. Tompa (1980) "On the Evaluation of Poly-*t* Density Functions", *Journal of Econometrics*, 12, 335–351.
Rothenberg, T. (1963) "A Bayesian Analysis of Simultaneous Equation Systems", Econometric Institute Report 6315, Erasmus Universiteit, Rotterdam.
Rothenberg, T. (1973) *Efficient Estimation with A Priori Information*, Cowles Foundation Monograph 23. Yale University Press.
Rothenberg, T. (1975) "Bayesian Analysis of Simultaneous Equations Models", in: S. E. Fienberg and A. Zellner (eds.), *Studies in Bayesian Econometrics and Statistics*. Amsterdam: North-Holland Publishing Co.
Ter Berg, P. and R. Harkema (1980) "Bayesian Limited Information Analysis of Nonlinear Simultaneous Equations Systems", Econometric Institute Report 8023/E, Erasmus Universiteit, Rotterdam.
Theil, H. (1971) *Principles of Econometrics*. New York: John Wiley & Sons.
Tiao, G. C. and A. Zellner (1964) "On the Bayesian Estimation of Multivariate Regression", *Journal of the Royal Statistical Society*, Ser. B. 277–285.
Tompa, H. (1967) "Gaussian Numerical Integration of a Function Depending on a Parameter", *The Computer Journal*, 10, 204–205.
Tompa, H. (1973) "The Iterative Simpson Method of Numerical Integration", CORE Discussion Paper 7336, Université Catholique de Louvain.
van Dijk, H. K. and T. Kloek (1980) "Further Experience in Bayesian Analysis Using Monte Carlo Integration", Econometric Institute Report 8012/E, Erasmus Universiteit, Rotterdam.
Zellner, A. (1962) "An Efficient Method of Estimating Seemingly Unrelated Regressions and Tests for Aggregation Bias", *Journal of the American Statistical Association*, 57, 348–368.
Zellner, A. (1965) "Bayesian and Non-Bayesian Analysis of Simultaneous Equation Models", Paper presented at the First World Congress of the Econometric Society, Rome, September.

Zellner, A. (1971) *An Introduction to Bayesian Inference in Econometrics*. New York: John Wiley & Sons.
Zellner, A. and J. F. Richard (1973) "Use of Prior Information in the Analysis and Estimation of Cobb–Douglas Production Function Models", *International Economic Review*, 14, 107–119.
Zellner, A. and W. Vandaele (1975) "Bayes–Stein Estimators for k-Means, Regression and Simultaneous Equation Models", in: S. E. Feinberg and A. Zellner (eds.), *Studies in Bayesian Econometrics and Statistics*. Amsterdam: North-Holland Publishing Co.
Zellner, A., J. Kmenta and J. H. Drèze (1966) "Specification and Estimation of Cobb–Douglas Production Function Models", *Econometrica*, 34, 784–795.

Chapter 10

BIASED ESTIMATION

G. G. JUDGE and M. E. BOCK*

University of Illinois and Purdue University

Contents

*This work was facilitated by National Science Foundation Grants. Useful comments by Arnold Zellner are gratefully acknowledged.

Handbook of Econometrics, Volume I, Edited by Z. Griliches and M.D. Intriligator

1. Introduction

Much of the literature concerned with estimation and inference from a sample of data deals with a situation when the statistical model is correctly specified. Consequently, in econometric practice it is customary to assume that the statistical model employed for purposes of estimation and inference is consistent with the sampling process whereby the sample observations were generated. In this happy event, statistical theory provides techniques for obtaining point and interval estimates of the population parameters and for hypothesis testing.

Under this scenario for the traditional linear statistical model with normal, independent, and identically distributed errors it is conventional to make use of the maximum likelihood–least squares rule when estimating the unknown location parameters. From the sampling theory point of view this approach is justified since it leads to a minimum variance among linear unbiased estimators and under squared error loss, the least squares estimator is minimax. From the Bayesian point of view, under a uniform–non-informative prior for the coefficients, the property of minimum posterior mean squared error is achieved. All in all this is a fairly impressive set of credentials and doubtless this goes a long way toward explaining the popularity of the least squares estimator, which is really best in a class of one. These results also suggest that if improvement in estimator performance is to be achieved, one must go outside of the traditional sampling theory rules and consider a range of alternative estimators that are biased and possibly nonlinear.

Despite its popularity the statistical implications of remaining in the linear unbiased family of rules may in many cases be rather severe. One indication of the possibly questionable stature of the least squares rule occurred when Stein (1955) showed, under conditions normally fulfilled in practice, that there were other minimax estimators. Following Stein's result, James and Stein (1961) exhibited an estimator which under squared error loss dominates the least squares estimator and thus demonstrates its inadmissibility. This result means that the unbiased least squares rule may have an inferior mean square error when compared to other biased estimators.

Another trouble spot for the conventional least squares estimator arises in case of a false statistical model. Just as few economic variables are free of measurement error and few economic relations are non-stochastic, few statistical models are correctly specified and many of these specification errors imply a biased outcome when the least squares rule is used. For example, consider the problem of an investigator who has a single data set and wants to estimate the parameters of a linear model which are known to lie in a high dimensional parameter space $\underline{\theta}_1$. The researcher may suspect the relationship may be characterized by a lower dimensional parameter space $\theta_2 \subset \theta_1$. Under this uncertainty if the θ_1 dimensional parameter space is estimated by least squares the result, from the possibly overspecified model, will be unbiased but have large variance and thus may make a poor showing in terms of mean square error. Alternatively, the $\underline{\theta}_2$ dimensional parameter space may incorrectly specify the statistical model and thus if estimated by least squares will be biased and this bias may or may not outweigh the reduction in variance if evaluated in a mean square error context.

Although uncertainty concerning the proper column dimension of the matrix of explanatory variables is the rule, in many cases prior information exists about the individual parameters and/or relationships among the unknown parameters. Ignoring this information and using only sample information and the least squares rule may lead to a loss of precision, while taking the information into account may lead to a more precise though biased estimator. Intuitively it would seem any estimator that does not take account of existing non-sample information should lead to suboptimal rules.

Furthermore, since most economic data are passively generated and thus do not come from an experimental design situation where the investigator has a good degree of control, the data may be nearly collinear and this means that approximate linear relations may hold among the columns of the explanatory variables that appear in the design matrix X. When this happens the least squares estimates are unstable, the $X'X$ matrix is often nearly singular and small changes in the observations may result in large changes in the estimates of the unknown coefficients. Ridge and minimax general ridge estimators have been suggested as alternatives to the least squares rule when handling data with these characteristics.

In the linear statistical model when the errors are long tailed and the conventional normally distributed constant variance error specification is not appropriate, the least squares rule loses some of its inferential reach. Under this scenario it is necessary to consider biased alternatives which are conceptually different from, for example the Stein and ridge approaches noted above. In this chapter we do no more than identify the problem, since it will be discussed in full elsewhere in this Handbook.

To cope with some of the problems noted above and to avoid the statistical consequences of remaining with the conventional estimator, researchers have proposed and evaluated a range of alternatives to least squares. Useful summaries of some of the results to date include papers by Dempster (1973), Mayer and Willke (1973), Gunst and Mason (1977), and Draper and Van Nostrand (1979).

In laying out the statistical implications of a range of biased alternatives to the least squares rule the chapter is organized as follows: In Section 2 conventional linear statistical models, estimators, and a hypothesis testing framework are presented and the sampling theory and Bayes bases for gauging estimator performance are specified. In Section 3 sampling theory and Bayes estimators which permit sample information and various types of non-sample information to be jointly considered, are specified and appropriately evaluated. In Section 4 testing frameworks are specified for evaluating the compatibility of the sample information and the various types of non-sample information and the corresponding pretest estimators are derived, compared, and evaluated. In Section 5 the inadmissibility of the least squares estimator is discussed and a range of Stein-rule estimators are considered for alternative loss functions and design matrices. In Section 6 alternatives to least squares are considered for the stochastic regressor case. In Section 7 the problem of nearly collinear data is discussed and the ridge-type and general minimax estimators which have been suggested to cope with this age old problem, are compared and evaluated. Finally, in Section 8 some comments are made about the statistical implications of these biased alternatives for econometric theory and practice.

2. Conventional statistical models, estimators, tests, and measures of estimator performance

We are concerned with the sampling performance of a family of biased estimators for the following linear statistical model:

$$y = X\boldsymbol{\beta} + \boldsymbol{e}, \tag{2.1}$$

where y is a $(T \times 1)$ vector of observations, X is a known $(T \times K)$ design matrix of rank K, $\boldsymbol{\beta}$ is a $(K \times 1)$ fixed vector of unknown parameters, $\boldsymbol{e}$ is a $(T \times 1)$ vector of unobservable normal random variables with mean vector zero and finite covariance matrix $E[\boldsymbol{ee}'] = \sigma^2\phi$, with σ^2 unknown, and ϕ is a known symmetric positive definite matrix. We assume throughout that the random variables which comprise $\boldsymbol{e}$ are independently and identically distributed, i.e. $E[\boldsymbol{ee}'] = \sigma^2 I_T$ or can be transformed to this specification since ϕ is known. In almost all cases we will assume $\boldsymbol{e}$ is a *normal* random vector.

2.1. *Conventional estimators and tests*

Given that y is generated by the linear statistical model (2.1) the least squares basis for estimating the unknown coefficients is given by the linear rule

$$\boldsymbol{b} = (X'X)^{-1}X'y, \tag{2.2}$$

which is best linear unbiased. If it is assumed that $\boldsymbol{e}$ is multivariate normal then (2.2) is the maximum likelihood estimator and is a minimax estimator no longer limited to the class of linear estimators. Furthermore, if $\boldsymbol{e}$ is normal then $\boldsymbol{b}$ has minimum risk $E[(\boldsymbol{b}-\boldsymbol{\beta})'(\boldsymbol{b}-\boldsymbol{\beta})]$ among the unbiased (not necessarily linear) estimators of $\boldsymbol{\beta}$.

The assumption that $\boldsymbol{y}$ is a normally distributed vector implies that the random vector $(\boldsymbol{b}-\boldsymbol{\beta})$ is normally distributed with mean vector zero and covariance

$$E\left[(\boldsymbol{b}-\boldsymbol{\beta})'(\boldsymbol{b}-\boldsymbol{\beta})\right]=\sigma^2(X'X)^{-1}. \tag{2.3}$$

Therefore, the quadratic form $(\boldsymbol{b}-\boldsymbol{\beta})'X'X(\boldsymbol{b}-\boldsymbol{\beta})/\sigma^2$ is distributed as a central chi-square random variable with K degrees of freedom.

A best quadratic unbiased estimator of the unknown scalar σ^2 is given by

$$\begin{aligned}\hat{\sigma}^2&=(\boldsymbol{y}-X\boldsymbol{b})'(\boldsymbol{y}-X\boldsymbol{b})/(T-K)=\boldsymbol{y}'\left(I_T-X(X'X)^{-1}X'\right)\boldsymbol{y}/(T-K)\\&=\boldsymbol{y}'M\boldsymbol{y}/(T-K)=\boldsymbol{e}'M\boldsymbol{e}/(T-K),\end{aligned} \tag{2.4}$$

where M is an idempotent matrix of rank $(T-K)$. If we leave the class of unbiased quadratic estimators of σ^2, the minimum variance quadratic estimator, with smallest mean square error, is $\tilde{\sigma}^2=\boldsymbol{y}'M\boldsymbol{y}/(T-K+2)$. Since e is a normally distributed vector with mean vector zero and covariance $\sigma^2 I_T$, the quadratic form

$$(T-K)\hat{\sigma}^2/\sigma^2=\boldsymbol{e}'M\boldsymbol{e}/\sigma^2 \tag{2.5}$$

is distributed as a central chi-square random variable with $(T-K)$ degrees of freedom.

Let us represent the hypotheses we have about the K dimensional unknown parameters in the form of the following linear hypotheses:

$$\boldsymbol{\beta}=\boldsymbol{r} \tag{2.6}$$

or

$$\boldsymbol{\delta}=0,$$

where $\boldsymbol{\delta}=\boldsymbol{\beta}-\boldsymbol{r}$ is a $(K\times 1)$ vector representing specification errors and $\boldsymbol{r}$ is a K dimensional known vector. Given this formulation it is conventional to use likelihood ratio procedures to test the null hypothesis H_0: $\boldsymbol{\beta}=\boldsymbol{r}$ against the alternative hypothesis $H_{\mathrm{A}}: \boldsymbol{\beta}\neq\boldsymbol{r}$, by using the test statistic

$$u=(\boldsymbol{b}-\boldsymbol{r})'X'X(\boldsymbol{b}-\boldsymbol{r})/K\hat{\sigma}. \tag{2.7}$$

If the hypotheses are correct and indeed $\boldsymbol{r}=\boldsymbol{\beta}$, the test statistic u is a central F random variable with K and $(T-K)$ degrees of freedom, i.e. $u\sim F_{(K,T-K)}$. If the linear hypotheses are incorrect, u is distributed as a non-central F random variable with K and $(T-K)$ degrees of freedom and non-centrality parameter

$$\lambda = (\boldsymbol{\beta}-\boldsymbol{r})'X'X(\boldsymbol{\beta}-\boldsymbol{r})/2\sigma^2 = \boldsymbol{\delta}'X'X\boldsymbol{\delta}/2\sigma^2. \tag{2.8}$$

The traditional test procedure for H_0 against H_A is to reject the linear hypotheses H_0 if the value of the test statistic u is greater than some specified value c. The value of c is determined for a given significance level α by

$$\int_c^{\infty} \mathrm{d}F_{(K,T-K)} = \mathrm{P}\left[F_{(K,T-K)} \geq c\right] = \alpha. \tag{2.9}$$

The above test mechanism leads to an estimator that will be specified and evaluated in Section 4.

For some of the estimators to be discussed in the coming sections, it is convenient for expository purposes to reparameterize the linear statistical model (2.1) in one of the following two forms:

$$\boldsymbol{y} = X\boldsymbol{\beta} + \boldsymbol{e} = XS^{-1/2}S^{1/2}\boldsymbol{\beta} + \boldsymbol{e} = Z\boldsymbol{\theta} + \boldsymbol{e}, \tag{2.10a}$$

where S is a positive definite symmetric matrix with $S^{1/2}S^{1/2} = S = X'X$, $\boldsymbol{\theta} = S^{1/2}\boldsymbol{\beta}$, $Z = XS^{-1/2}$, and $Z'Z = I_K$. Under this reparameterization a best linear unbiased estimator of $\boldsymbol{\theta}$ is $\boldsymbol{w} = Z'\boldsymbol{y}$ with covariance $\Sigma_w = \sigma^2 I_K$. Note also we may write (2.10a) as

$$Z'\boldsymbol{y} = \boldsymbol{\theta} + Z'\boldsymbol{e}, \tag{2.10b}$$

where $\boldsymbol{z} = Z'\boldsymbol{y}$ has a K variate normal distribution with mean vector $\boldsymbol{\theta}$ and covariance $\sigma^2 I_K$. This formulation is equivalent to the K mean statistical model usually analyzed in the statistical literature. Although (2.10b) is a convenient form for analysis purposes we will remain in this chapter with the linear statistical (regression) form since this is the one most commonly dealt with in econometrics. The common nature of the two problems should be realized in interpreting the results to be developed. Alternatively consider the following canonical form:

$$\boldsymbol{y} = X\boldsymbol{\beta} + \boldsymbol{e} = XTT^{-1}\boldsymbol{\beta} + \boldsymbol{e}, \tag{2.11}$$

where T is a non-singular matrix chosen so that the columns of XT are orthogonal. One choice of T is to choose an orthogonal matrix P whose columns are orthonormal characteristic vectors of $X'X$. Consequently, $PP' = I$ and

$$\boldsymbol{y} = X\boldsymbol{\beta} + \boldsymbol{e} = XPP'\boldsymbol{\beta} + \boldsymbol{e} = H\boldsymbol{\alpha} + \boldsymbol{e}. \tag{2.12}$$

The columns of H are orthogonal since $H'H = \Lambda$, which is a diagonal matrix with elements $\lambda_1 > \lambda_2 > \ldots > \lambda_K$, that are the characteristic roots of $X'X$. The best linear unbiased estimator of $\boldsymbol{\alpha}$ is $\hat{\boldsymbol{\alpha}}=\Lambda^{-1}{}'H'y$, with covariance $\sigma^2\Lambda^{-1}$. The variance of $\hat{\alpha}_1$, $i=1,2,\dots,K$, is σ^2/λ_i.

2.2. *Measures of performance*

Finally, let us consider the basis for gauging the performance of a range of alternative estimators. We can, as we did with the estimators considered above, require the property of unbiasedness, and in this context $\boldsymbol{b}$ is the only unbiased estimate of $\boldsymbol{\beta}$ based on sufficient statistics. But why the concept of unbiasedness? If the information from sample observations is to be used for decision purposes why not make use of statistical decision theory which is based on the analyses of losses due to incorrect decisions? This is in fact the approach we use in this chapter as a basis for comparing estimators as we go outside of traditional rules and enter the family of non-linear biased estimators.

Although there are many forms for representing the loss or risk functions we will to a large extent be concerned with estimation alternatives under a squared error loss measure. However, the estimators we consider are in general robust under a range of loss functions.

Assume that $\boldsymbol{y}$ is a $(T\times 1)$ random vector. If $\boldsymbol{\delta}(y)$ is some estimator of the K dimensional parameter vector $\boldsymbol{\beta}$, then the weighted squared error or weighted quadratic loss function is

$$L(\boldsymbol{\beta},\boldsymbol{\delta}(y))=(\boldsymbol{\delta}(y)-\boldsymbol{\beta})'Q(\boldsymbol{\delta}(y)-\boldsymbol{\beta}), \tag{2.13}$$

and has risk

$$\rho(\boldsymbol{\beta},\boldsymbol{\delta})=E[(\boldsymbol{\delta}-\boldsymbol{\beta})'Q(\boldsymbol{\delta}-\boldsymbol{\beta})], \tag{2.14}$$

where Q is a known positive definite weight matrix. If $Q=I_K$ under this criterion, the unbiased estimator with minimum risk is the unbiased estimator with minimum variance. If we make use of the condition that $\boldsymbol{\delta}(y)$ be both linear in $\boldsymbol{y}$ and unbiased, this leads to the Gauss–Markoff criterion and the minimum risk or best linear unbiased estimator is $\boldsymbol{\delta}(y)=\boldsymbol{y}$ if $E[y]=\boldsymbol{\beta}$.

Reparameterizing the statistical model and transforming from one parameter space to another in many cases changes the measure of goodness used to judge performance. For example, if interest centers on statistical model (2.1) and sampling performance in the $\boldsymbol{\beta}$ space (estimation problem), specifying an unweighted loss function in the $\boldsymbol{\theta}$ space (2.10), results in a weighted function in the $\boldsymbol{\beta}$ space, i.e.

$$\begin{aligned}(\hat{\boldsymbol{\theta}}-\boldsymbol{\theta})'(\hat{\boldsymbol{\theta}}-\boldsymbol{\theta})&=(S^{1/2}\hat{\boldsymbol{\beta}}-S^{1/2}\boldsymbol{\beta})'(S^{1/2}\hat{\boldsymbol{\beta}}-S^{1/2}\boldsymbol{\beta})\\&=(\hat{\boldsymbol{\beta}}-\boldsymbol{\beta})'S(\hat{\boldsymbol{\beta}}-\boldsymbol{\beta})=(\hat{\boldsymbol{\beta}}-\boldsymbol{\beta})'X'X(\hat{\boldsymbol{\beta}}-\boldsymbol{\beta}).\end{aligned} \tag{2.15}$$

Therefore, while the reparametrized model (2.10) is appropriate for analyzing the conditional mean forecasting problem of estimating $X\boldsymbol{\beta}$ by $X\boldsymbol{b}$, it is not appropriate for analyzing the performance of $\boldsymbol{b}$ as an estimate of $\boldsymbol{\beta}$ unless one is interested in the particular weight matrix $(X'X)$.

Alternatively, an unweighted squared error loss risk in the $\boldsymbol{\beta}$ space results in a weighted risk function in the $\boldsymbol{\theta}$ space, i.e.

$$\begin{aligned} E[(\boldsymbol{b}-\boldsymbol{\beta})(\boldsymbol{b}-\boldsymbol{\beta})] &= E[(S^{-1/2}\hat{\boldsymbol{\theta}}-S^{-1/2}\boldsymbol{\theta})'(S^{-1/2}\hat{\boldsymbol{\theta}}-S^{-1/2}\boldsymbol{\theta})] \\ &= E[(\hat{\boldsymbol{\theta}}-\boldsymbol{\theta})'S^{-1}(\hat{\boldsymbol{\theta}}-\boldsymbol{\theta})] \\ &= E[(\hat{\boldsymbol{\theta}}-\boldsymbol{\theta})'(X'X)^{-1}(\hat{\boldsymbol{\theta}}-\boldsymbol{\theta})]. \end{aligned} \tag{2.16}$$

In some of the evaluations to follow it will be convenient or analytically more tractable to consider the weighted risk function in the $\boldsymbol{\theta}$ space instead of the unweighted counterpart in the $\boldsymbol{\beta}$ space.

Finally, let us note for the canonical form (2.12) that the orthogonal transformation preserves the distance measure, i.e.

$$\begin{aligned} (\hat{\boldsymbol{\alpha}}-\boldsymbol{\alpha})'(\hat{\boldsymbol{\alpha}}-\boldsymbol{\alpha}) &= (P'\boldsymbol{b}-P'\boldsymbol{\beta})'(P'\boldsymbol{b}-P'\boldsymbol{\beta}) \\ &= (\boldsymbol{b}-\boldsymbol{\beta})'PP'(\boldsymbol{b}-\boldsymbol{\beta}) = (\boldsymbol{b}-\boldsymbol{\beta})'(\boldsymbol{b}-\boldsymbol{\beta}). \end{aligned} \tag{2.17}$$

The minimum mean square error criterion is another basis we will use for comparing the sampling performance of estimators. This generalized mean square error or risk measure for some estimator $\hat{\boldsymbol{\beta}}$ of $\boldsymbol{\beta}$ may be defined as

$$\begin{aligned} MSE[\boldsymbol{\beta},\hat{\boldsymbol{\beta}}] &= E[(\hat{\boldsymbol{\beta}}-\boldsymbol{\beta})(\hat{\boldsymbol{\beta}}-\boldsymbol{\beta})'] \\ &= (\text{bias}\,\hat{\boldsymbol{\beta}})(\text{bias}\,\hat{\boldsymbol{\beta}})'+\text{cov}\,\hat{\boldsymbol{\beta}}. \end{aligned} \tag{2.18}$$

Under this measure the diagonal elements are mean square errors and the trace of (2.18) is the squared error risk, when $Q=I_K$. In using the mean square error criterion an estimator $\hat{\boldsymbol{\beta}}$ is equal or superior to another estimator $\tilde{\boldsymbol{\beta}}$ if, for all $\boldsymbol{\beta}$,

$$\Delta = E[(\tilde{\boldsymbol{\beta}}-\boldsymbol{\beta})(\tilde{\boldsymbol{\beta}}-\boldsymbol{\beta})'] - E[(\hat{\boldsymbol{\beta}}-\boldsymbol{\beta})(\hat{\boldsymbol{\beta}}-\boldsymbol{\beta})'] \tag{2.19}$$

is a positive semidefinite matrix. This implies $\boldsymbol{l}'\Delta\boldsymbol{l}\geq \boldsymbol{0}$ for any K dimensional real vector $\boldsymbol{l}$.

2.3. *Bayes estimation*

The assumption that the vector $\boldsymbol{\beta}$ is itself a random vector with a known distribution leads, when combined with previously developed measures of performance, to a well-defined estimator for $\boldsymbol{\beta}$. In this approach one chooses, optimally, a Bayes estimator, $\hat{\boldsymbol{\beta}}_\tau$, which minimizes for all $\hat{\boldsymbol{\beta}}$ the expected value of $\rho(\boldsymbol{\beta}, \hat{\boldsymbol{\beta}})$, where the expectation is taken over $\boldsymbol{\beta}$ with respect to its known distribution τ. The Bayes risk for $\hat{\boldsymbol{\beta}}$ is

$$E_\tau[\rho(\boldsymbol{\beta}, \hat{\boldsymbol{\beta}}_\tau)] = \inf_{\hat{\boldsymbol{\beta}}} E_\tau[\rho(\boldsymbol{\beta}, \hat{\boldsymbol{\beta}})]. \tag{2.20}$$

In particular, for a weighted quadratic loss, such as (2.13),

$$\hat{\boldsymbol{\beta}}_\tau = E[\boldsymbol{\beta}|y], \tag{2.21}$$

the mean of the conditional distribution of $\boldsymbol{\beta}$ given the sample data.

3. Some possibly biased alternatives

Under the standard linear normal statistical model and a sampling theory framework, when only sample information is used, the least squares estimator gives a minimum variance among unbiased estimators. In the Bayesian framework for inference, if a non-informative–uniform prior is used in conjunction with the sample information, the minimum posterior mean square error property is achieved via the least squares rule. One problem with least squares in either framework is that it does not take into account the often existing prior information or relationships among the coefficients. A Bayesian might even say that the non-informative prior which leads to least squares should be replaced by a proper distribution which reflects in a realistic way the existing non-sample information.

To mitigate the impact of ignoring this non-sample information, and to patch up their basis of estimation and inference so that it makes use of all of the information at hand, sampling theorists have developed procedures for combining sample and various types of non-sample information. When the non-sample information is added and certain of these rules are used, although we gain in precision, biased estimators result if the prior information specification is incorrect. In other cases biased estimators result even if the prior specification is correct. Thus, we are led, in comparing the estimators, to a bias–variance dichotomy for measuring performance and some of the sampling theory estimators which make use of non-sample information show, for example, superior mean square error over much of the relevant parameter space. Alternatively, there are other conventionally used biased sampling theory alternatives for which this result does not hold. In the remainder of this section we review the sampling properties of these possibly biased estimators and evaluate their performance under a squared error loss measure.

3.1. *Exact non-sample information*

Let us assume that in addition to the sample information contained in (2.1) there also exists information about the K dimensional unknown vector $\boldsymbol{\beta}$, in the form of J independent linear equality restrictions or hypotheses, where $J \leq K$. This information may be specified as

$$R\boldsymbol{\beta} = \boldsymbol{r}, \tag{3.1}$$

where R is a $(J \times K)$ known matrix of rank J which expresses the structure of the outside information as it relates to the individual parameters or their linear combinations and $\boldsymbol{r}$ is a $(J \times 1)$ vector of known elements. The restrictions may also be written as $\boldsymbol{\delta} = \boldsymbol{0}$, where $\boldsymbol{\delta} = R\boldsymbol{\beta} - \boldsymbol{r}$ and the $(J \times 1)$ vector $\boldsymbol{\delta}$ represents the specification errors in the prior information. Under this scenario the maximum likelihood estimator which includes this non-sample information is

$$\boldsymbol{b}^* = \boldsymbol{b} + S^{-1}R'[RS^{-1}R']^{-1}(\boldsymbol{r} - R\boldsymbol{b}), \tag{3.2}$$

and is the solution to the problem minimizing the quadratic form $(\boldsymbol{y} - X\boldsymbol{\beta})'(\boldsymbol{y} - X\boldsymbol{\beta})$ subject to $R\boldsymbol{\beta} = \boldsymbol{r}$, where $S = X'X$. Thus, $\boldsymbol{b}^*$ is multinormally distributed, with mean

$$\begin{aligned} E[\boldsymbol{b}^*] &= \boldsymbol{\beta} - S^{-1}R'[RS^{-1}R']^{-1}(R\boldsymbol{\beta} - \boldsymbol{r}) \\ &= \boldsymbol{\beta} - S^{-1}R'[RS^{-1}R']^{-1}\boldsymbol{\delta}, \end{aligned} \tag{3.3}$$

covariance matrix

$$E[(\boldsymbol{b}^* - E[\boldsymbol{b}^*])(\boldsymbol{b}^* - E[\boldsymbol{b}^*])'] = \sigma^2[S^{-1} - C], \tag{3.4}$$

where $C = S^{-1}R'[RS^{-1}R']^{-1}RS^{-1}$. The mean square error or risk matrix

$$\begin{aligned} E[(\boldsymbol{b}^* - \boldsymbol{\beta})(\boldsymbol{b}^* - \boldsymbol{\beta})'] = \sigma^2[S^{-1} - C] + S^{-1}R'[RS^{-1}R']^{-1} \\ \times \boldsymbol{\delta}\boldsymbol{\delta}'[RS^{-1}R']^{-1}RS^{-1} \end{aligned} \tag{3.5}$$

and weighted quadratic risk is

$$\begin{aligned} E[(\boldsymbol{b}^* - \boldsymbol{\beta})'Q(\boldsymbol{b}^* - \boldsymbol{\beta})] &= \rho(\boldsymbol{\beta}, \boldsymbol{b}^*) \\ &= \sigma^2 \operatorname{tr}(S^{-1} - C)Q + \boldsymbol{\delta}'[RS^{-1}R']^{-1}RS^{-1} \\ &\quad \times QS^{-1}R'[RS^{-1}R']^{-1}\boldsymbol{\delta}. \end{aligned} \tag{3.6}$$

These results imply that if $\boldsymbol{\delta} = R\boldsymbol{\beta} - \boldsymbol{r} = \boldsymbol{0}$, then $\boldsymbol{b}^*$ is best linear unbiased within the class of unbiased estimators which are linear functions of $\boldsymbol{y}$ and $\boldsymbol{r}$. If $\boldsymbol{\delta} \neq \boldsymbol{0}$, then $\boldsymbol{b}^*$ is biased (3.3) and has mean square or quadratic risk (3.5) and (3.6).

Under the general mean square error or risk criterion, in comparing the restricted and unrestricted maximum likelihood estimators $\boldsymbol{b}^*$ and $\boldsymbol{b}$, Toro-Vizcorrondo and Wallace (1968) show that $E(\boldsymbol{b}-\boldsymbol{\beta})(\boldsymbol{b}-\boldsymbol{\beta})'] - E[(\boldsymbol{b}^*-\boldsymbol{\beta})(\boldsymbol{b}^*-\boldsymbol{\beta})'] = \Delta$, where Δ is a positive semi-definite if and only if

$$\frac{\boldsymbol{\delta}'(RS^{-1}R')^{-1}\boldsymbol{\delta}}{\sigma^2} < 1 \quad \text{or} \quad \frac{\boldsymbol{\delta}'(RS^{-1}R')^{-1}\boldsymbol{\delta}}{2\sigma^2} \geq \frac{1}{2}. \tag{3.7}$$

Under the weighted quadratic risk criterion, $\boldsymbol{b}^*$ has smaller risk than b, i.e. $E[\boldsymbol{b}^*-\boldsymbol{\beta})'Q(\boldsymbol{b}^*-\boldsymbol{\beta})] \leq E[(\boldsymbol{b}-\boldsymbol{\beta})'\mathrm{Q}(\boldsymbol{b}-\boldsymbol{\beta})]$, if and only if $2^{-1}\sigma^{-2}\boldsymbol{\delta}'[RS^{-1}R']^{-1} RS^{-1}QS^{-1}R'[RS^{-1}R']^{-1}\boldsymbol{\delta} \leq \operatorname{tr} CQ2^{-1}$. If the weight matrix under the quadratic risk criterion is $X'X$, i.e. the conditional mean forecasting problem, then the restricted maximum likelihood estimator has risk

$$E[(\boldsymbol{b}^*-\boldsymbol{\beta})'X'X(\boldsymbol{b}^*-\boldsymbol{\beta})] = \sigma^2(K-J) + \boldsymbol{\delta}'(RS^{-1}R')^{-1}\boldsymbol{\delta}$$

and

$$E[(\boldsymbol{b}-\boldsymbol{\beta})'X'X(\boldsymbol{b}-\boldsymbol{\beta})] - E[(\boldsymbol{b}^*-\boldsymbol{\beta})'X'X(\boldsymbol{b}^*-\boldsymbol{\beta})] \geq 0$$

if

$$\frac{\boldsymbol{\delta}'[RS^{-1}R']^{-1}\boldsymbol{\delta}}{2\sigma^2} \leq \frac{J}{2}. \tag{3.8}$$

Therefore, as the prior information errors grow, i.e. the length of the vector $\boldsymbol{\delta}$ increases, the risk of the restricted maximum likelihood estimator increases without bound. Consequently, this biased estimator alternative permits us the possibility of being precisely wrong.

3.2. *Stochastic non-sample information*

Assume the following stochastic prior information exists about $\boldsymbol{\beta}$:

$$\boldsymbol{r} = R\boldsymbol{\beta} + \boldsymbol{v}, \tag{3.9}$$

where $\boldsymbol{r}$ and R are defined in conjunction with (3.1) and $\boldsymbol{v}$ is a $(J \times 1)$ unobservable, normally distributed random vector with mean $\boldsymbol{\delta}$ and covariance $\sigma^2\Phi$, with Φ

known. Following Theil and Goldberger (1961) and Theil (1963) we may combine the sample information (2.1) with the stochastic prior information (3.9) in the linear statistical model

$$\begin{bmatrix} \boldsymbol{y} \\ \boldsymbol{r} \end{bmatrix} = \begin{bmatrix} X \\ R \end{bmatrix} \boldsymbol{\beta} + \begin{bmatrix} \boldsymbol{e} \\ \boldsymbol{v} \end{bmatrix}, \tag{3.10}$$

where $(\boldsymbol{e}', \boldsymbol{v}')'$ is multivariate normal with mean $(\boldsymbol{0}', \boldsymbol{\delta})$ and covariance

$$\sigma^2 \begin{bmatrix} I_T & 0 \\ o & \Phi \end{bmatrix}.$$

When σ^2 is known the mixed (Aitken) estimator for (3.10) is

$$\begin{aligned} \boldsymbol{b}^{**} &= (S + R'\Phi^{-1}R)^{-1}(X'\boldsymbol{y} + R\Phi^{-1}\boldsymbol{r}) \\ &= \left[S^{-1} - S^{-1}R'(RS^{-1}R' + \Phi)^{-1}RS^{-1}\right](X'\boldsymbol{y} + R'\Phi^{-1}\boldsymbol{r}), \end{aligned} \tag{3.11}$$

with mean

$$E[\boldsymbol{b}^{**}] = \boldsymbol{\beta} + [S + R'\Phi^{-1}R]^{-1}R'\Phi^{-1}\boldsymbol{\delta}, \tag{3.12}$$

covariance

$$E[(\boldsymbol{b}^{**} - E(\boldsymbol{b}^{**}))(\boldsymbol{b}^{**} - E(\boldsymbol{b}^{**}))'] = \sigma^2[S + R'\Phi^{-1}R]^{-1} \equiv \sigma^2 W^{-1}, \tag{3.13}$$

and mean square error or risk matrix

$$E[(\boldsymbol{b}^{**} - \boldsymbol{\beta})(\boldsymbol{b}^{**} - \boldsymbol{\beta})'] = \sigma^2 W^{-1} + W^{-1}R'\Phi^{-1}\boldsymbol{\delta}\boldsymbol{\delta}'\Phi^{-1}RW^{-1}. \tag{3.14}$$

The difference between the covariance matrix of the unrestricted maximum likelihood estimator $\boldsymbol{b}$ and the stochastic restricted estimator $\boldsymbol{b}^{**}$ is

$$\begin{aligned} &E[(\boldsymbol{b} - \boldsymbol{\beta})(\boldsymbol{b} - \boldsymbol{\beta})'] - E[(\boldsymbol{b}^{**} - E[\boldsymbol{b}^{**}])(\boldsymbol{b}^{**} - E[\boldsymbol{b}^{**}])'] \\ &\quad = \sigma^2 S^{-1}R'[RS^{-1}R^1 + \Phi]^{-1}RS^{-1}, \end{aligned} \tag{3.15}$$

a positive semi-definite matrix. Furthermore, the difference between the mean square error matrices for $\boldsymbol{b}$ and $\boldsymbol{b}^{**}$ is

$$\begin{aligned} MSE_{\boldsymbol{b}} - MSE_{\boldsymbol{b}^{**}} &= \sigma^2 S^{-1} - \sigma^2 W^{-1} - W^{-1}R'\Phi^{-1}\boldsymbol{\delta}\boldsymbol{\delta}'\Phi^{-1}RW^{-1} \\ &= W^{-1}R'\Phi^{-1}\left[\sigma^2(RS^{-1}R' + \Phi) - \boldsymbol{\delta}\boldsymbol{\delta}'\right]\Phi^{-1}RW^{-1}. \end{aligned} \tag{3.16}$$

This means that (3.16) is positive semi-definite if and only if $\sigma^2(RS^{-1}R'+\Phi)-\boldsymbol{\delta}\boldsymbol{\delta}'$ is positive semi-definite. Consequently, under the generalized mean square error criterion the stochastic restricted estimator $\boldsymbol{b}^{**}$ is superior to the least squares estimator $\boldsymbol{b}$ in the part of the parameter space where

$$\frac{\boldsymbol{\delta}'(RS^{-1}R'+\Phi)^{-1}\boldsymbol{\delta}}{2\sigma^2}<\frac{1}{2}. \tag{3.17}$$

In terms of quadratic risk [Judge and Bock (1978, pp. 41–43)], $\rho(\boldsymbol{\beta},\boldsymbol{b}^{**})\leq\rho(\boldsymbol{\beta},\boldsymbol{b})$ for those values of the parameter space where

$$\begin{aligned}&\boldsymbol{\delta}'[RS^{-1}R'+\Phi]^{-1}RS^{-2}R'[RS^{-1}R'+\Phi]^{-1}\boldsymbol{\delta}\sigma^{-2}\\&\quad<\operatorname{tr}\left([RS^{-1}R'+\Phi]^{-1}RS^{-2}R'\right).\end{aligned} \tag{3.18}$$

Similar results are obtained for the $X'X$ weighted risk function case.

One weakness of the mixed estimator lies in assuming the random vector $\boldsymbol{v}$, representing the uncertainty of the prior information, has a zero mean vector, an assumption necessary for the estimator to be unbiased. Since the frequency and subjective interpretations of probability are different, the argument that prior judgements are equivalent to prior unbiased estimates seems unsatisfactory. As shown by Swamy and Mehta (1977) the estimator (3.11) may be less efficient than the least squares estimator based only on sample data if the stochastic prior information (3.9) is misspecified. The requirement that the prior covariance matrix Φ be known must in most cases be unreasonably demanding. In addition, the fixed $\boldsymbol{\beta}$ and random $\boldsymbol{r}$ and $\boldsymbol{v}$ in eq. (3.9) does not fit any Bayesian axiom system and it would appear that no set of principles have been set down which would justify this specification.

3.3. *Inequality non-sample information*

Assume now the non-sample information about the unknown parameters exists in the form of linear inequality constraints which may be represented as

$$R\boldsymbol{\beta}\geq\boldsymbol{r} \tag{3.19}$$

or

$$R\boldsymbol{\beta}+\boldsymbol{\delta}=\boldsymbol{r},$$

where $\boldsymbol{\delta}$ is a $(J\times 1)$ unknown vector. The estimation problem which results from combining both sample (3.1) and inequality constraint information (3.19) may be specified as a quadratic programming problem for which a number of solution algorithms exists.

In order to give bias and risk evaluations we consider the orthonormal statistical model $y = Z\boldsymbol{\theta} + \boldsymbol{e}$ (2.10a) and the case where the information design matrix R has the form $[I_J \quad 0]$. In fact, without loss of generality, for expository purposes we consider the problem of estimating the ith unknown parameter θ, when non-sample information exists in the form $\theta \geq r$, where r is a known scalar. Since for any sample of data either the maximum likelihood estimator $\hat{\theta}$ violates the constraint or it does not, the inequality restricted estimator may be expressed as

$$\begin{aligned}\theta^+ &= I_{(-\infty,r)}(\hat{\theta})r + I_{[r,\infty)}(\hat{\theta})\hat{\theta}\\ &= I_{(-\infty,\delta/\sigma)}((\hat{\theta}-\theta)/\sigma)r + I_{[\delta/\sigma,\infty)}((\hat{\theta}-\theta)/\sigma)\hat{\theta}\\ &= \hat{\theta} + I_{(-\infty,\delta/\sigma)}(\omega)\delta - I_{(-\infty,\delta/\sigma)}(\omega)\omega\sigma,\end{aligned} \tag{3.20}$$

where $I_{(\cdot)}(\cdot)$ is an indicator function that takes on the value 1 if the argument takes on a value within the subscripted interval and zero otherwise, $\omega = (\hat{\theta}-\theta)/\sigma$ is a standard normal random variable, and $\delta = r - \theta$.

3.3.1. Bias

Making use of (3.20) and corollaries 1 and 2 from Judge and Yancey (1978) the mean of θ^+ is, when $\delta < 0$ and the direction of the inequality is correct,

$$E[\theta^+] = \theta + (\delta/2)P\left(\chi^2_{(1)} \geq \delta^2/\sigma^2\right) + \left(\sigma/\sqrt{2\pi}\right)P\left(\chi^2_{(2)} \geq \delta^2/\sigma^2\right). \tag{3.21}$$

When $r \geq \theta$ and thus $\delta \geq 0$, the inequality is not correct and the mean of the θ^+ is

$$E[\theta^+] = \theta + \delta - (\delta/2)P\left(\chi^2_{(1)} \geq \delta^2/\sigma^2\right) + \left(\sigma/\sqrt{2\pi}\right)P\left(\chi^2_{(2)} \geq \delta^2/\sigma^2\right). \tag{3.22}$$

These results imply that if the direction of non-sample information is correct, as $\delta \to -\infty$, then $E[\theta^+] \to \theta$. As $\delta \to 0$, then $E[\theta^+] \to \theta + \sigma/\sqrt{2\pi}$. If the direction of the inequality is incorrect and as $\delta \to \infty$, the $E[\theta^+] \to \theta + \delta = r$, the mean of the restricted least squares estimator. The bias characteristics of the inequality estimator are presented in Figure 3.1.

3.3.2. Risk

The risk–mean square error of the inequality estimator θ^+ is

$$\begin{aligned}\rho(\theta,\theta^+) &= E\left[(\theta^+ - \theta)^2\right]\\ &= E\left[(\hat{\theta}-\theta)^2\right] + E\left[I_{(-\infty,\delta/\sigma)}(\omega)\delta^2\right] - \sigma^2\left[I_{(-\infty,\delta/\sigma)}(\omega)\omega^2\right].\end{aligned} \tag{3.23}$$

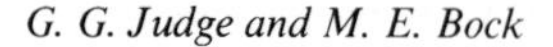

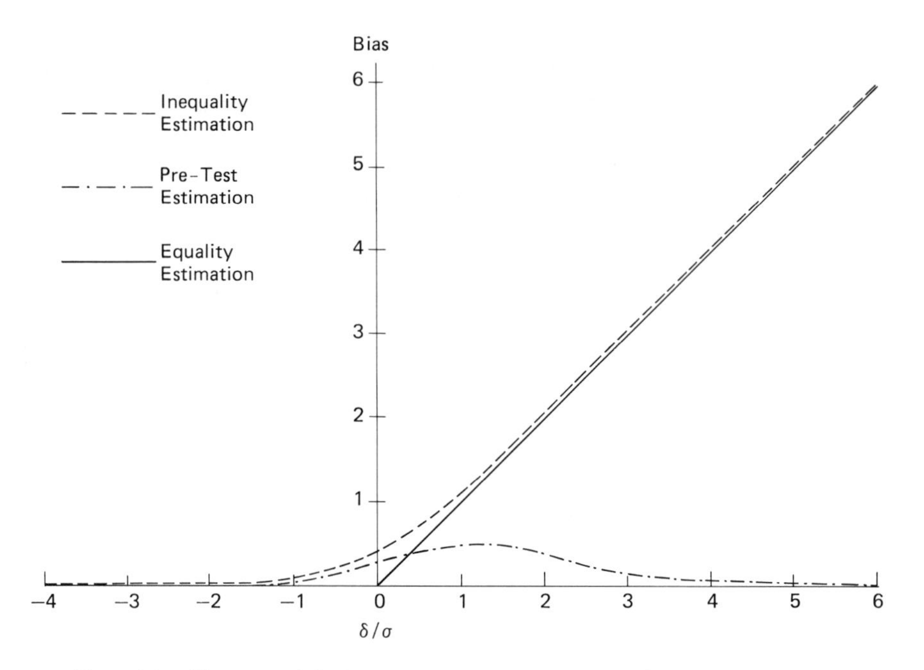

Figure 3.1. The mean of the inequality restricted, estimator θ^+ as a function of δ/σ.

Using corollary 3 from Judge and Yancey (1978) the risk function may be expressed, when $\delta < 0$, as

$$\rho(\theta,\theta^+) = \sigma^2 + (\delta^2/2)P\left(\chi^2_{(1)} \geq \delta^2/\sigma^2\right) - (\sigma^2/2)P\left(\chi^2_{(3)} \geq \delta^2/\sigma^2\right), \quad (3.24a)$$

and when $\delta \geq 0$, as

$$\rho(\theta,\theta^+) = \delta^2 + (\sigma^2/2)P\left(\chi^2_{(3)} \geq \delta^2/\sigma^2\right) - (\delta^2/2)P\left(\chi^2_{(1)} \geq \delta^2/\sigma^2\right). \quad (3.24b)$$

These results imply that if the direction of the inequality is correct and $\delta < 0$, then (i) as $\delta \to -\infty$ the $\rho(\theta,\theta^+) \to \sigma^2$ and (ii) as $\delta \to 0$ the $\rho(\theta,\theta^+) \to \sigma^2/2$. Consequently, if $\delta < 0$ the inequality estimator is minimax. If the direction of the inequality is not correct and $\delta \to \infty$, then $\rho(\theta,\theta^+) - \delta^2 \to 0$, where δ^2 is the risk of the restricted least squares estimator. The risk characteristics of the inequality restricted estimator are presented in Figure 3.2.

When the linear statistical model is orthogonal and the information design matrix R is diagonal these results generalize directly to the K mean or K dimensional linear model problem. The results also hold for a general linear statistical model and diagonal R under weighted squared error loss when $X'X$ is the weight matrix.

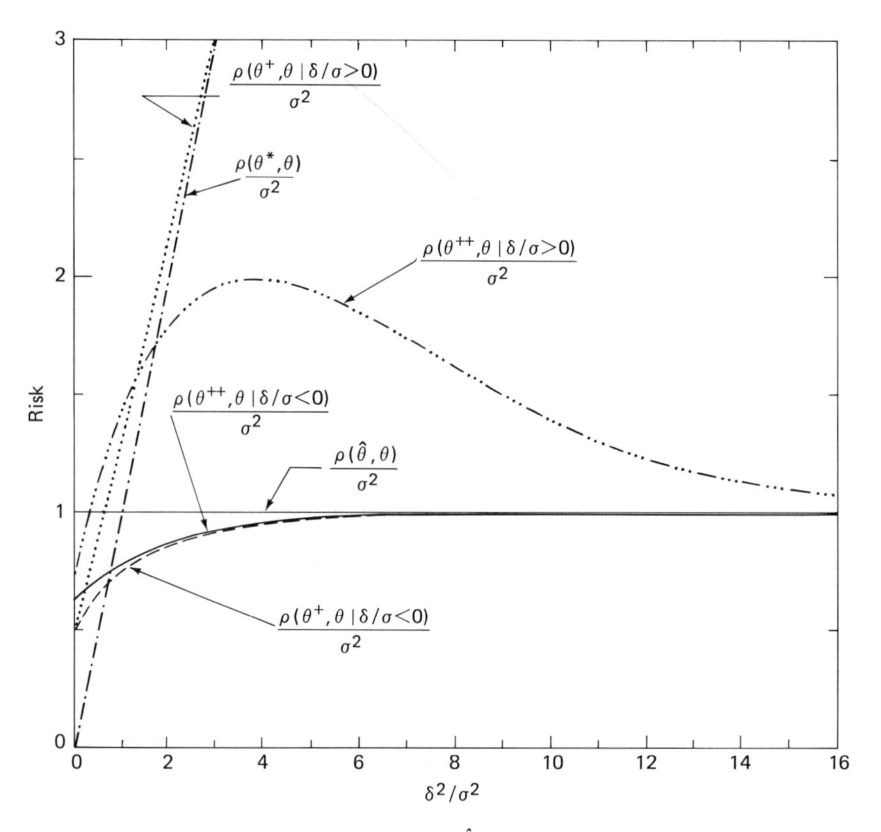

Figure 3.2. Risks for the maximum likelihood $\hat{\theta}$ restricted θ^*, inequality restricted θ^+ and pre-test inequality restricted θ^{++} estimators as a function of δ^2/σ^2.

3.4. *Parameter distribution information (prior)*

The assumption that the parameter vector $\boldsymbol{\beta}$ is stochastic and that its distribution, τ, called the prior distribution for $\boldsymbol{\beta}$, is known, leads within the context of an appropriate loss function to the selection of a Bayes' estimator $\hat{\boldsymbol{\beta}}_\tau$ which may be biased. As noted in Section 1, the least squares estimator $\boldsymbol{b}$ may be regarded as the Bayes estimator with respect to a diffuse prior distribution and a quadratic loss function and this particular Bayes estimator is unbiased.

Alternatively, consider the use of an informative proper prior, such as the specification of the natural conjugate prior distribution τ for $\boldsymbol{\beta}$ which is normal with mean $\bar{\boldsymbol{\beta}}$ and covariance $\sigma^2 A^{-1}$. In the case of quadratic loss, this formulation results in the following optimal point estimate which minimizes posterior expected loss:

$$\hat{\boldsymbol{\beta}}_{\tau} = (A + X'X)^{-1}(A\bar{\boldsymbol{\beta}} + X'\boldsymbol{y})$$
$$= (A + X'X)^{-1}(A\bar{\boldsymbol{\beta}} + X'X\boldsymbol{b}). \tag{3.25}$$

This Bayes estimator has sampling bias

$$E[\hat{\boldsymbol{\beta}}_{\tau}] - \boldsymbol{\beta} = (A + X'X)^{-1}A(\bar{\boldsymbol{\beta}} - \boldsymbol{\beta}). \tag{3.26}$$

A comparison of the sampling properties of $\boldsymbol{b}$ and $\boldsymbol{\beta}_{\tau}$, under the *MSE* (risk matrix) criteria, has been made by Giles and Rayner (1979). In particular they examine some of the conditions under which any linear combination of the elements of the natural conjugate Bayes estimator of $\boldsymbol{\beta}$ has a smaller mean squared error than has the corresponding linear combination of the elements of $\boldsymbol{b}$. Their principal result is given in the form of an inequality that involves the sample observations, the prior parameters, and the unknown parameters of the statistical model. It should be noted that under this measure of performance both estimators are admissible.

Zellner (1980) has proposed a prior for $\boldsymbol{\beta}$ which is the same form as the natural conjugate prior, where $A = g(X'X)$ and g is a positive constant. The posterior mean of this resulting Bayes estimation $\hat{\boldsymbol{\beta}}_g$ is

$$\hat{\boldsymbol{\beta}}_g = (\boldsymbol{b} + g\bar{\boldsymbol{\beta}})/(1+g) = \boldsymbol{b} - g(\boldsymbol{b} - \bar{\boldsymbol{\beta}})/(1+g), \tag{3.27}$$

with sampling mean

$$E[\hat{\boldsymbol{\beta}}_g] = (\boldsymbol{\beta} + g\bar{\boldsymbol{\beta}})/(1+g) \tag{3.27a}$$

and bias

$$E[\hat{\boldsymbol{\beta}}_g] - \boldsymbol{\beta} = g(\bar{\boldsymbol{\beta}} - \boldsymbol{\beta})/(1+g), \tag{3.27b}$$

which approaches zero as $g \to 0$.

Under a squared error loss measure Zellner notes that the Bayes estimator $\boldsymbol{\beta}_g$ has risk

$$\rho(\boldsymbol{\beta}, \hat{\boldsymbol{\beta}}_g) = K_0\sigma^2(1 + g^2(\bar{\boldsymbol{\beta}} - \boldsymbol{\beta})'(\bar{\boldsymbol{\beta}} - \boldsymbol{\beta})/K_0\sigma^2)(1+g)^2 \tag{3.28}$$

and average risk $K_0\sigma^2/(1+g)$, given σ^2, where $K_0 = \operatorname{tr}(X'X)^{-1}$. The risk of $\hat{\boldsymbol{\beta}}_g$ will be superior to that of b if $(1 + g^2(\bar{\boldsymbol{\beta}} - \boldsymbol{\beta})'(\bar{\boldsymbol{\beta}} - \boldsymbol{\beta})/K_0\sigma^2)/(1+g)^2 < 1$. Furthermore, there may be a considerable reduction in average risk if one uses $\hat{\boldsymbol{\beta}}_g$

instead of $\boldsymbol{b}$. Therefore, the g prior may provide a useful Bayesian analysis of the linear statistical model when there is information about the $\boldsymbol{\beta}$ vector but little information about the prior covariance.

3.5. Some remarks

We have discussed in this section three sampling theory estimators that are biased if the non-sample information is incorrect. Both the exact and stochastic restricted estimators win in terms of precision, but may lose, possibly heavily, in terms of bias. If we could be sure of the direction of the inequality non-sample information, and in economics there are many cases when this may be true, the inequality estimator, although biased, wins in terms of precision and mean squared error and thus has appealing sampling properties.

Rothenberg (1973) has studied inequality restrictions under very general assumptions regarding the form of the inequalities and has suggested an alternative class of estimators which are biased. In particular he has shown that in the case of the linear regression model, if the true parameter vector β is known to be constrained in a convex, proper subset of the parameter space, then the restricted least squares estimator dominates the unconstrained least squares estimator under a mean squared prediction error loss criterion. However, if the sample is not normally distributed, the constrained estimator does not necessarily dominate its unconstrained counterpart under a generalized mean square error loss criterion.

4. Pre-test–variable selection estimators

The previous subsections are informative relative to the sampling performance of the equality, stochastic, and inequality restricted least squares estimators. One problem with these results is that the researcher is seldom certain about the correctness of the non-sample information and thus may have only a vague notion about $\boldsymbol{\theta}$ or the $\boldsymbol{\delta}'\boldsymbol{\delta}/2\sigma^2$ specification error parameter space. Therefore, the results are of little help in choosing between the restricted and unrestricted estimators or, more to the point, choosing the biased estimator with a minimum risk.

Since there may be reasons to doubt the compatibility of the sample and non-sample information or uncertainty about the dimensions of the design matrix X or Z, some biased estimators result when the investigator performs a preliminary test of significance (chooses a criterion) and on the basis of the test (criterion) makes a choice between the unbiased estimator and a possibly biased one. To see the possible significance of these biased alternatives let us consider the equality and inequality pre-test estimators and one or more conventional variable selection procedures. For expository purposes we stay in the orthonormal linear statistical model world where $Z'Z = I_K$ and continue to assume $R = I_K$.

4.1. Conventional equality by pre-test estimator

Using likelihood ratio test procedures we may test the null hypotheses $H_0:\boldsymbol{\theta} = \boldsymbol{r}$ against the hypothesis $\boldsymbol{\theta} \neq \boldsymbol{r}$, by using the test statistic

$$u = (\hat{\boldsymbol{\theta}} - \boldsymbol{r})'(\hat{\boldsymbol{\theta}} - \boldsymbol{r})/K\hat{\sigma}^2, \tag{4.1}$$

which is distributed as a central F random variable with K and $(T-K)$ degrees of freedom if the hypotheses (restrictions) are correct. Of course if the restrictions are incorrect $E[\hat{\boldsymbol{\theta}} - \boldsymbol{r}] = (\boldsymbol{\theta} - \boldsymbol{r}) = \boldsymbol{\delta} \neq 0$, and u (4.1) is distributed as a non-central F with non-centrality parameter $\lambda = (\boldsymbol{\theta} - \boldsymbol{r})'(\boldsymbol{\theta} - \boldsymbol{r})/2\sigma^2 = \boldsymbol{\delta}'\boldsymbol{\delta}/2\sigma^2$. As a test mechanism the null hypothesis is rejected if $u \geq F^{\alpha}_{(K,T-K)} = c$, where c is determined for a given level of the test α by $\int_c^{\infty} \mathrm{d}F_{(K,T-K)} = \mathrm{P}[F_{(K,T-K)} \geq c] = \alpha$. This means that by accepting the null hypothesis we use the restricted least squares estimator $\boldsymbol{\theta}^*$ as our estimate of $\boldsymbol{\theta}$, and by rejecting the null hypothesis $\boldsymbol{\theta} - \boldsymbol{r} = \boldsymbol{\delta} = \boldsymbol{0}$ we use the unrestricted least squares estimator $\hat{\boldsymbol{\theta}}$. Thus, the estimate that results is dependent upon a preliminary test of significance and this means the estimator used by many applied workers is of the form

$$\hat{\hat{\boldsymbol{\theta}}} = \begin{bmatrix} \boldsymbol{\theta}^* & \text{if } u < c, \\ \hat{\boldsymbol{\theta}} & \text{if } u \geq c. \end{bmatrix} \tag{4.2}$$

Alternatively the estimator may be written as

$$\begin{aligned} \hat{\hat{\boldsymbol{\theta}}} &= I_{(0,c)}(u)\boldsymbol{\theta}^* + I_{[c,\infty)}(u)\hat{\boldsymbol{\theta}} \\ &= \hat{\boldsymbol{\theta}} - I_{(0,c)}(u)(\hat{\boldsymbol{\theta}} - \boldsymbol{\theta}^*) = \hat{\boldsymbol{\theta}} - I_{(0,c)}(u)(\hat{\boldsymbol{\theta}} - \underline{r}), \end{aligned} \tag{4.3}$$

This specification means that in a repeated sampling context the data, the linear hypotheses, and the selected level of statistical significance all determine the combination of the two estimators that is chosen.

4.1.1. Bias

From (4.3) the mean of the pre-test estimator is

$$E\left[\hat{\hat{\boldsymbol{\theta}}}\right] = \boldsymbol{\theta} - E\left[I_{(0,c)}(u)(\hat{\boldsymbol{\theta}} - \boldsymbol{r})\right], \tag{4.4}$$

which by theorem 3.1 in Judge and Bock (1978, p. 71) may be expressed as

$$E\left[\hat{\hat{\theta}}\right] = \theta - \delta \mathrm{P}\left[\chi^2_{(K+2,\lambda)}/\chi^2_{(T-K)} \leq cK/(T-K)\right]. \tag{4.5}$$

Consequently, if $\delta = 0$, the pre-test estimator is unbiased. This fortunate outcome aside, the size of the bias is affected by the probability of a random variable with a non-central F distribution being less than a constant, which is determined by the level of the test, the number of hypotheses, and the degree of hypothesis error, δ or λ. Since the probability is always equal to or less than one, the bias of the pre-test estimator is equal to or less than the bias of the restricted estimator (3.2).

4.1.2. *Sampling performance*

Since this estimator is used in much applied work, let us turn to its sampling performance under the squared error loss criterion. The risk function may be written, using (4.3) and following Judge and Bock (1978, p. 70), as

$$\begin{aligned}
\rho(\theta,\hat{\hat{\theta}}) &= E\left[(\hat{\hat{\theta}}-\theta)'(\hat{\hat{\theta}}-\theta)\right] \\
&= E\left[\left(\hat{\theta}-\theta-I_{(0,c)}(u)(\hat{\theta}-r)\right)'\left(\hat{\theta}-\theta-I_{(0,c)}(u)(\hat{\theta}-r)\right)\right] \\
&= E\left[(\hat{\theta}-\theta)'(\hat{\theta}-\theta)\right] - E\left[I_{(0,c)}(u)(\hat{\theta}-\theta)'(\hat{\theta}-\theta)\right] \\
&\quad + E\left[I_{(0,c)}(u)\right]\delta'\delta \\
&= \sigma^2 K + (2\delta'\delta - \sigma^2 K)\mathrm{P}\left[\frac{\chi^2_{(K+2,\lambda)}}{\chi^2_{(T-K)}} \leq \frac{cK}{T-K}\right] \\
&\quad - \delta'\delta \mathrm{P}\left[\frac{\chi^2_{(K+4,\lambda)}}{\chi^2_{(T-K)}} \leq \frac{cK}{T-K}\right],
\end{aligned} \tag{4.6}$$

or compactly as

$$\rho(\theta,\hat{\hat{\theta}}) = \sigma^2 K + (2\delta'\delta - \sigma^2 K)l(2) - \delta'\delta l(4),$$

where $1 > l(2) > l(4) > 0$. From the risk function (4.6) the following results are clear.

(1) If the restrictions are correct and $\underline{\delta} = \underline{0}$, the risk of the pre-test estimator is $\sigma^2 K[1 - l(2)]$, where $1 > (1 - l(2)) > 0$ for $0 < c < \infty$. Therefore, the pre-test estimator has a smaller risk than the least squares estimator at the origin and the

decrease in risk depends on the level of significance α and correspondingly the critical value of the test c.

(2) As the hypothesis error $\boldsymbol{\delta}$ or λ grows, the risk of the pre-test estimator increases, obtaines a maximum after exceeding the risk of the least squares estimator, and then monotonically decreases to approach $\sigma^2 K$, the risk of the least squares estimator.

(3) As the hypothesis error $\boldsymbol{\theta} - \boldsymbol{r} = \boldsymbol{\delta}$, and thus $\boldsymbol{\delta}'\boldsymbol{\delta}/2\sigma^2$ increases and approaches infinity, $l(\cdot)$ and $\boldsymbol{\delta}'\boldsymbol{\delta}l(\cdot)$ approach zero. Therefore, the risk of the pre-test estimator approaches $\sigma^2 K$, the risk of the unrestricted least squares estimator.

(4) The pre-test estimator risk function crosses the risk function of the least squares estimator in the $\boldsymbol{\delta}'\boldsymbol{\delta}/2\sigma^2$ parameter space within the bounds $K/4 \leq \boldsymbol{\delta}'\boldsymbol{\delta}/2\sigma^2 \leq K/2$.

The sampling characteristics of the preliminary test estimator are summarized in Figure 4.1 for various levels of α or c.

These results mean that the pre-test estimator does well relative to the least squares estimator if the hypotheses are correctly specified. However, in the $\lambda = \boldsymbol{\delta}'\boldsymbol{\delta}/2\sigma^2$ parameter space representing the range of hypothesis errors, the pre-test estimator is inferior to the least squares estimator over an infinite range

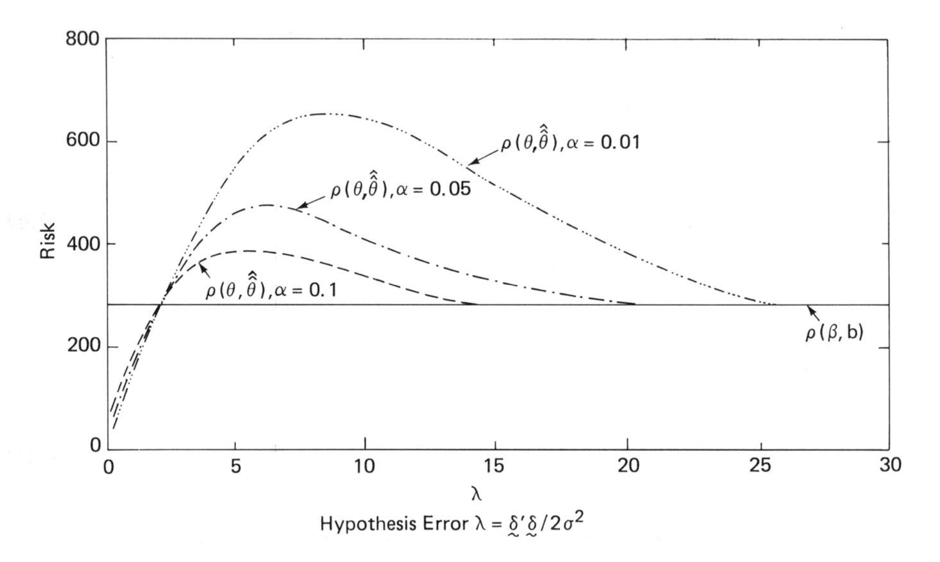

Figure 4.1. Risk functions for the least squares and restricted least squares estimators and typical risk functions for the pre-test estimator for various levels.

of the parameter space. Also, there is a range of the parameter space where the pre-test estimator has risk that is inferior to (greater than) both the unrestricted and restricted least squares estimators. No one estimator in the $\hat{\boldsymbol{\theta}}$, $\boldsymbol{\theta}^{+}$, and $\hat{\hat{\boldsymbol{\theta}}}$ family dominates the other competitors. In addition, in applied problems we seldom know the hypothesis errors nor the location of the correct λ in the $\boldsymbol{\delta}$ parameter space. Consequently, the choice of the estimator and the optimum α level are unresolved problems.

4.2. *Stochastic hypothesis pre-test estimator*

Since the stochastic prior and sample information provide two separate estimates of $R\boldsymbol{\theta}$, i.e. r and $R\hat{\boldsymbol{\theta}}$, Theil (1963) has proposed that we test the compatibility of the two estimates by the test statistic

$$u_1 = (\boldsymbol{r} - R\hat{\boldsymbol{\theta}})'[RS^{-1}R' + \Omega]^{-1}(\boldsymbol{r} - R\hat{\boldsymbol{\theta}})/\sigma^2, \tag{4.8}$$

which, if σ^2 is known and $\boldsymbol{\delta} = \boldsymbol{0}$, has a central chi-square distribution with J degrees of freedom. If $\boldsymbol{\delta} \neq \boldsymbol{0}$, then u_1 is distributed as a non-central chi-square with non-centrality parameter $\lambda_1 = \boldsymbol{\delta}'(RS^{-1}R' + \Omega)^{-1}\boldsymbol{\delta}/2\sigma^2$.

If we use the above test statistic to test the compatibility of the prior and sample information, the estimator chosen depends upon a preliminary test of significance and thereby produces the following pre-test estimator using (3.11):

$$\tilde{\boldsymbol{b}}^{**} = I_{(0,c)}(u_1)\boldsymbol{b}^{**} + I_{[c,\infty]}(u_1)\boldsymbol{b}, \tag{4.9}$$

where the $I_{(\cdot)}(u_1)$ are indicator functions and c is determined for a given level of α by $\int_c^\infty f(u_1)\mathrm{d}u_1 = \alpha$, where $f(u_1)$ is the density function of u_1, under the assumption that $\boldsymbol{\delta} = \boldsymbol{0}$.

Since the stochastic linear statistical model can be reformulated in a restricted linear model framework, many of the statistical implications of pre-testing developed in Section 4.1 carry over for the stochastic restricted pre-test estimator (4.8). The mean of $\tilde{\boldsymbol{b}}^{**}$ is

$$E[\tilde{\boldsymbol{b}}^{**}] = \boldsymbol{\beta} + l(2)S^{-1}R'(RS^{-1}R' + \Phi^{-1})^{-1}\delta, \tag{4.10}$$

where $l(2) < 1$. Consequently, the bias is $l(2)S^{-1}R'(RS^{-1}R' + \Phi^{-1})^{-1}\boldsymbol{\delta}$.

In terms of estimator comparisons, following Judge and Bock (1978) the stochastic restricted preliminary test estimator is better in a risk matrix or general mean square error sense than the unrestricted least squares estimator if the stochastic restriction error in the form of the non centrality parameter is

$$\lambda_1 = \boldsymbol{\delta}'[RS^{-1}R' + \Omega]^{-1}\boldsymbol{\delta}/2\sigma^2 \le 1/(2[2 - c/(c + (T-K)/J)]), \tag{4.11}$$

where $1/4$ might be considered the rough bound.

Alternatively, if the squared error loss criterion is used, then the equality of the risk of the stochastic restricted pre-test estimator and the least squares estimator occurs for a value of $\lambda_1 = \boldsymbol{\delta}'[RS^{-1}R' + \Omega]^{-1}\boldsymbol{\delta}/2\sigma^2$ within the following bounds:

$$\operatorname{tr} A/(2d_s[2 - l(4)/l(2)]) \leqslant \lambda_1 \leqslant \operatorname{tr} A/(2d_L[2 - l(4)/l(2)]), \tag{4.12}$$

where $A = (RS^{-1}R' + \Omega)^{-1}RS^{-1}R'$, $l(i) = \mathrm{P}(\chi^2_{(J+i,\lambda_1)} \leqslant c]$, with $0 < l(2) < l(4) < 1$ and d_s and d_L are the smallest and largest characteristic roots of A, respectively.

Since the results for both criteria depend on the critical value c or the level of the test α, the risk or risk matrix approaches that of the stochastic restricted estimator as $\alpha \to 0$ and $c \to \infty$. Conversely, as $\alpha \to 1$ and $c \to 0$ the risk or risk matrix of the pre-test estimator approaches that of the least squares estimator. Finally, for $\alpha < 1$ the risk or risk matrix approaches that of the least squares estimator as $\lambda_1 \to \infty$. As before the optimum level of the test is unresolved.

4.3. *Inequality hypothesis pre-test estimator*

In the context of (3.20) we continue to consider a single parameter and the following null and alternative hypotheses:

$$H_0: \theta > r; \qquad H_A: \theta < r. \tag{4.13}$$

As a basis for checking the compatibility of the sample information and a linear inequality hypothesis for θ, when σ^2 is known, consider the test statistic

$$(\hat{\theta} - r)/\sigma = u_2, \tag{4.14}$$

which is distributed as a normal random variable with mean δ/σ and variance 1. If it is assumed $\delta = \hat{\theta} - r = 0$, then u_2 is a standard normal random variable and the test structure may be formulated in terms of δ, with $H_0: \delta \geqslant 0$ and $H_A: \delta < 0$. Using test statistic (4.13), we use the following test mechanism or decision rule:

(i) Reject the hypothesis H_0 if $(\hat{\theta} - r)/\sigma = u_2 < c_2 \leqslant 0$ and use the maximum likelihood estimator $\hat{\theta}$, where c_2 is the critical value of the test from the standard normal table.

(ii) Accept the hypothesis H_0 if $u_2 = (\hat{\theta} - r)/\sigma > c_2$ and use the inequality restricted estimator

$$\theta^+ = I_{(-\infty, r)}(\hat{\theta})r + I_{[r,\infty)}(\hat{\theta})\hat{\theta}.$$

By accepting the hypothesis H_0, we take θ^+ as the estimate of θ and by rejecting H_0 the maximum likelihood estimate $\hat{\theta}$ is used. Consequently, when a preliminary test of the inequality hypothesis is made and a decision is taken based on the data at hand, the following pre-test estimator results:

$$\theta^{++} = I_{(-\infty, c_2)}(u_2)\hat{\theta} + I_{[c_2,\infty)}(u_2)\left[I_{(-\infty,r)}(\hat{\theta})r + I_{[r,\infty)}(\hat{\theta})\hat{\theta}\right]. \tag{4.15}$$

Recalling that $\delta = \theta - r$, and defining $w = (\hat{\theta} - \theta)/\sigma$ and $d = c_2 - \delta/\sigma$, the pre-test estimator (4.15) is

$$\begin{aligned}\theta^{++} = \hat{\theta} &+ \sigma\left[I_{(-\infty,d)}(w) - I_{(-\infty,-\delta/\sigma)}(w)\right]w \\ &- \left[I_{(-\infty,-\delta/\sigma)}(w) - I_{(-\infty,d)}(w)\right]\delta.\end{aligned} \tag{4.16}$$

4.3.1. *Mean of the inequality pre-test estimator*

When $-\infty < \delta < 0$ and θ is less than r and thus in agreement with the hypothesis H_A, and if $-\delta/\sigma > 0$ and $d > 0$, if we apply the Judge–Yancey (1977) corollaries 1 and 2 of Section 3.3 to eq. (4.16), the mean of the inequality restricted pre-test estimator is

$$\begin{aligned}E[\theta^{++}] = \theta &- (\sigma/\sqrt{2\pi})\left[P(\chi^2_{(2)} \geqslant \delta^2/\sigma^2) - P(\chi^2_{(2)} \geqslant d^2)\right] \\ &- (\delta/2)\left[P(\chi^2_{(1)} \geqslant \delta^2/\sigma^2) - P(\chi^2_{(1)} \geqslant d^2)\right].\end{aligned} \tag{4.17a}$$

For any given critical value of c_2 for $c_2 < 0$, if $r - \delta = \hat{\theta} = 0$, the $E[\theta^{++}] = \theta - (\sigma/\sqrt{2\pi})[1 - P(\chi^2_{(2)} \geqslant c_2^2)]$ and consequently has a negative bias. However, as $\delta \to -\infty$ the $E[\theta^{++}] = \theta$ and the pre-test estimator approaches θ, since in the limit the maximum likelihood estimator $\hat{\theta}$ will always be used. Furthermore if $c_2 = 0$ and $\delta = 0$, then $E[\theta^{++}] = \theta$; if $c \to -\infty$ and $\delta = 0$, then $E[\theta^{++}] \to \theta - \sigma/\sqrt{2\pi}$; if $c_2 \to -\infty$ and $\delta \to \infty$, then $E[\theta^{++}] \to \theta$.

When $0 < \delta/\sigma$ and $d < 0$, the mean of the pre-test estimator is

$$\begin{aligned}E[\theta^{++}] = \theta &- (\delta/2)\left[P(\chi^2_{(1)} \geqslant \delta^2/\sigma^2) - P(\chi^2_{(1)} \geqslant d^2)\right] \\ &- (\sigma/\sqrt{2\pi})\left[P(\chi^2_{(2)} \geqslant \delta^2/\sigma^2) - P(\chi^2_{(2)} \geqslant d^2)\right].\end{aligned} \tag{4.17b}$$

For a fixed $c_2 < 0$ if $\delta \to -\infty$, then $E[\theta^{++}] \to -\theta$. The unbiased outcome as $\delta \to \infty$ follows since in the limit the hypothesis (inequality restriction) is *rejected* all of the time and the maximum likelihood estimate is used for every sample.

These results imply that for any $c_2 < 0$ and any value of δ/σ the bias of the inequality pre-test estimator is equal to or less than that of the inequality restricted estimator. For a given $c_2 < 0$ the bias as a function of δ/σ is shown in Figure 3.1.

4.3.2. *Risk of the inequality pre-test estimator*

Given the test statistic and the test decision rule, the risk of the pre-test estimator under squared error loss may be written, using eq. (4.16), as

$$\rho(\theta^{++},\theta) = \sigma^2 + E\left[\sigma^2\left(I_{(-\infty,d)}(w)w^2 - I_{(-\infty,-\delta/\sigma)}(w)w^2\right)\right] + E\left[\delta^2\left(I_{(-\infty,-\delta/\sigma)}(w) - I_{(-\infty,d)}(w)\right)\right]. \tag{4.18a}$$

When $-\infty < c_2 < 0$ and $-\infty < \delta < 0$ or $c_2 + -\delta/\sigma > 0$ and use is made of the Judge and Yancey (1977) corollaries 2 and 3 used in Section 3.3, the risk of θ^{++} may be expressed as

$$\rho(\theta^{++},\theta) = \sigma^2 - (\sigma^2/2)\left[\mathrm{P}\left(\chi^2_{(3)} \geqslant d^2\right) - \mathrm{P}\left(\chi^2_{(3)} \geqslant \delta^2/\sigma^2\right)\right] - (\delta^2/2)\left[\mathrm{P}\left(\chi^2_{(1)} \geqslant \delta^2/\sigma^2\right) - \mathrm{P}\left(\chi^2_{(1)} \geqslant d^2\right)\right]. \tag{4.18b}$$

When $\delta = 0$, the $\rho(\theta^{++},\theta) = \sigma^2 - (\sigma^2/2)[\mathrm{P}(\chi^2_{(3)} \geqslant c_2^2) - 1]$, and the risk of the pre-test estimator is equal to or less than that of the maximum likelihood estimator. As the critical value $c_2 \to 0$, $\rho(\theta^{++},\theta) \to \sigma^2$, the maximum likelihood risk. As $c_2 \to -\infty$ and $\delta \to 0$, $\rho(\theta^{++},\theta) \to \sigma^2/2$, the risk of the inequality restricted estimator. Alternatively, with any fixed c_2, and as $\delta \to -\infty$, $\rho(\theta^{++},\theta) \to \sigma^2$, the maximum likelihood risk.

When $0 \leqslant \delta < \infty$ and $d = c_2 - \delta/\sigma \leqslant 0$, the risk of the pre-test estimator may be rewritten as

$$\rho(\theta^{++},\theta) = (\sigma^2/2)\left[\mathrm{P}\left(\chi^2_{(3)} \geqslant d^2\right) - \mathrm{P}\left(\chi^2_{(3)} \geqslant \delta^2/\sigma^2\right)\right] + \sigma^2 + (\delta^2/2)\left[\mathrm{P}\left(\chi^2_{(1)} \geqslant \delta^2/\sigma^2\right) - \mathrm{P}\left(\chi^2_{(1)} \geqslant d^2\right)\right]. \tag{4.18c}$$

When $\delta = 0$ and $-\infty < c_2 \leqslant 0$ the $\rho(\theta^{++},\theta) = (\sigma^2/2)[1 + \mathrm{P}(\chi^2_{(3)} \geqslant c_2^2)]$. As $c_2 \to 0$, $\rho(\theta^{++},\theta) \to \sigma^2$, the risk of the maximum likelihood estimator.

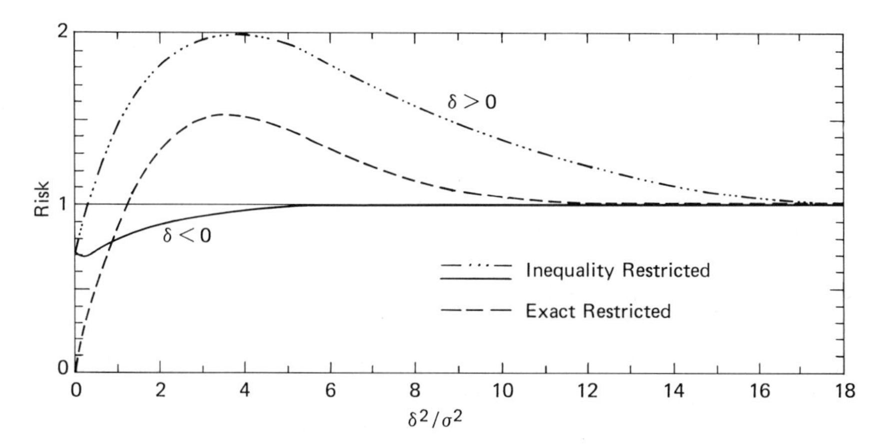

Figure 4.2. Risks for the maximum likelihood exact and inequality restricted pre-test estimators as a function of δ^2/σ^2, when σ^2 is known and equal to one, $\alpha = 0.05$, and $\rho(\hat{\theta}, \theta) \equiv 1$.

For any fixed δ as $c_2 \to -\infty$,

$$\rho(\theta^{++}, \theta) \to \delta^2\left[1/2\mathrm{P}\left(\chi^2_{(1)} \geqslant \delta^2/\sigma^2\right)\right] - (\sigma^2/2)\left[\mathrm{P}\left(\chi^2_{(3)} \geqslant \delta^2/\sigma^2\right)\right] + \sigma^2. \tag{4.18d}$$

The risk functions for the inequality pre-test estimator at the $\alpha = 0.05$ level of significance are shown in Figure 4.2 and are compared with those of the maximum likelihood, exact restricted, and inequality restricted least squares estimators. A comparison of the inequality restricted pre-test estimator, θ^{++}, and the traditional pre-test estimator [Bock, Yancey and Judge (1973)] is given in Figure 4.2 for each pre-test estimator. Note when $\delta \leqslant 0$, (i) the risk of the inequality pre-test estimator is less than the maximum likelihood estimator, and (ii) the risk of the traditional equality restricted pre-test estimator is less than that of the inequality pre-test estimator only over a small range of the parameter space. When σ^2 is unknown, the inequality test statistic counterpart, $(\hat{\theta} - r)/\hat{\sigma} = t$, is distributed as a central t random variable with $(T - K)$ degrees of freedom when $\delta = 0$. With this test statistic, the results are in general the same as those obtained for the standard normal test statistic.

4.4. *Bayesian pre-test estimators*

Bayesian pre-test estimators can be used in situations where there is uncertainty regarding the validity of different hypotheses. Within this context Zellner and

Vandaele (1975) demonstrate such estimators for a variety of statistical models where alternative hypotheses exist relative to the unknown coefficient vector. In particular, from a Bayesian point of view, when uncertainty exists relative to the model, we proceed by choosing a prior density and then develop a post-data description of the uncertainty. In traditional Bayesian significance testing, as discussed in Chapter 2 of this Handbook by Zellner, we usually compare the posterior probability of the null hypothesis with that of the alternative hypothesis, which may be of the nested variety. The ratio of the two alternative posterior probabilities (the posterior odds ratio) forms the basis for the decision and the null hypothesis is accepted if the ratio is larger than "some" limit. Zellner and Vandaele (1975) formulate a Bayesian pre-test estimator with posterior expected loss given by

$$E[L(\cdot)] = p_0(\tilde{\boldsymbol{\beta}} - \boldsymbol{r})'Q(\tilde{\boldsymbol{\beta}} - \boldsymbol{r}) + (1-p_0)[(\tilde{\boldsymbol{\beta}} - \bar{\boldsymbol{\beta}})'Q(\tilde{\boldsymbol{\beta}} - \bar{\boldsymbol{\beta}}) + E[(\boldsymbol{\beta} - \bar{\boldsymbol{\beta}})'Q(\boldsymbol{\beta} - \bar{\boldsymbol{\beta}})]], \tag{4.19}$$

and found a minimizing value of

$$\begin{aligned}\tilde{\boldsymbol{\beta}} &= \boldsymbol{r} + [1 - (K_0/(1+K_0))(\bar{\boldsymbol{\beta}} - \boldsymbol{r})] \\ &= (1-p_0)\bar{\boldsymbol{\beta}} + p_0\boldsymbol{r},\end{aligned} \tag{4.20}$$

where p_0 is the posterior probability of the hypothesis $H_0: \boldsymbol{\beta} = \boldsymbol{r}$, Q is a positive definite symmetric matrix, $\bar{\boldsymbol{\beta}}$ is the posterior mean of $\boldsymbol{\beta}$ under the hypothesis $H_1: \boldsymbol{\beta} \neq \boldsymbol{r}$, and $K_0 = p_0/(1-p_0)$ is the posterior odds ratio. In the usual Bayesian sense this pre-test estimator is consistent, admissible, and minimizes average risk. In contrast to the sampling theory pre-test estimator, $\tilde{\boldsymbol{\beta}}$ is a continuous function of the observations.

Leamer (1974) has posed the question: "Is there an acceptable prior that implies a post-data description of uncertainty that is essentially the same as that provided by a conventional search estimator?", and has answered by pointing out that a posterior mean is a weighted average of restricted and unrestricted estimates with the weight being a function of test statistics when positive probability is assigned to the null hypothesis. Leamer and Chamberlain (1976), building on the Zellner and Vandaele (1975) work, conclude that a conditional posterior mean under the standard assumptions for the linear model is a weighted average of the estimates resulting from all combinations of excluding and not excluding variables or, in other words, the restricted and unrestricted least squares estimates. Therefore a Bayesian by selecting a suitably defined prior may end, in many instances, with a posterior summary of model uncertainty that is essentially the same as a pre-testing sampling theorist.

4.5. *Variable selection estimators*

The problem of model specification and choice is discussed in Chapter 5 of this Handbook by Leamer. Therefore, in this chapter we limit our discussion to a few words about the omitted variable(s)–variable selection problem. In much applied work there is uncertainty concerning the appropriate dimension of the design matrix $X=[X_{\mathrm{I}}, X_{\mathrm{E}}]$, where X_{I} refers to the explanatory variables included in the model and X_{E} are the excluded variables or, in other words, the unselected variables that appear with a zero coefficient. The problem concerns the dimension of $X_{\mathrm{I}} \subset X$ or, in other words, the subset of variables to be included. As we know, if the investigator uses the complete set of variables X this collection may contain extraneous variables and the estimates are unbiased but may have large variance. Alternatively, estimates restricted to the nested model subset X_{I} have smaller variance but possibly large bias.

To aid in making a choice between the bias–precision extremes, a variety of variable selection criteria have been suggested under such acronyms as c_p [Mallows (1973)], *PC* [Amemiya (1976)], and AIC [Akaike (1974)]. These criteria and others have been summarized and evaluated by Hocking (1976), Amemiya (1976), and Judge et al. (1980). Although there is a certain intuitive appeal and logic to many of these informal, *ad hoc* model selection rules, we should keep in mind that they have a heuristic base, their sampling properties are unknown, and their practical use is demonstrated by numerical examples. In addition, most of the criteria involve unknown parameters and their operational counterparts vary depending on the sample estimate used to substitute for the unknown parameters. The criteria also involve, in one form or another, the residual error sum of squares corresponding to each model. Thus, the criterion can be identified with a test statistic and hence to various types of two or more stage pre-test estimators. These variable selection procedures lead to search rules that are inadmissible under a squared error measure and consequently have limited normative content since superior estimators have been demonstrated. Leamer (1978) has aptly characterized the conventional model search activity as "ad hoc inference with nonexperimental data".

When leaving the nested model case the statistical procedures for checking separate families of models that were originally proposed by Cox (1961, 1962) are relevant. Unfortunately, we know little about the small sample risk consequences for the tests proposed and the statistical results in general appear even more suspect than for the nested model case.

5. Conventional estimator inadmissibility and the Stein-rule alternatives

For estimation under squared error loss, Stein (1956) demonstrated that there are, in problems of more than two parameters, minimax estimators other than the maximum likelihood estimator. In 1961 James and Stein exhibited a minimax estimator which dominated the maximum likelihood estimator and thus demonstrated its inadmissibility. In this section we analyze the statistical implications of this family of biased–shrink estimators for a range of statistical models and loss functions.

5.1. *Estimation under squared error loss*

In this section we consider a range of Stein-like estimators for the orthonormal linear statistical model.

5.1.1. *The James and Stein estimator – orthonormal case*

Within the context of the orthonormal or K mean statistical model, assume σ^2 is known and therefore, without loss of generality, we may assume $\sigma^2 = 1$. Within this context the estimator developed by James and Stein, which is a function of the maximum likelihood estimator $\hat{\boldsymbol{\theta}}$, has the form

$$\tilde{\boldsymbol{\theta}} = \left[1 - a/\hat{\boldsymbol{\theta}}'\hat{\boldsymbol{\theta}}\right]\hat{\boldsymbol{\theta}}, \tag{5.1}$$

where $0 \leqslant a \leqslant 2(K-2)$. This rule makes the adjustment in the maximum likelihood estimator a smooth function of the data. The mean of the James and Stein estimator is

$$\begin{aligned} E[\tilde{\boldsymbol{\theta}}] &= E[\hat{\boldsymbol{\theta}}] - E\left[(a/\hat{\boldsymbol{\theta}}'\hat{\boldsymbol{\theta}})\hat{\boldsymbol{\theta}}\right] \\ &= \boldsymbol{\theta} - aE\left[1/\chi^2_{(K+2,\lambda)}\right]\boldsymbol{\theta}, \end{aligned} \tag{5.2}$$

where $\chi^2_{(K+2,\lambda)}$ is a non-central chi-square random variable with non-centrality parameter $\lambda = \boldsymbol{\theta}'\boldsymbol{\theta}/2$, and a theorem given by Judge and Bock (1978, p. 321) is used in the evaluation. Consequently, $\tilde{\boldsymbol{\theta}}$ is biased and has bias $aE[1/\chi^2_{(K+2,\lambda)}]\boldsymbol{\theta}$.

The risk of the James and Stein estimator under a squared error loss measure is

$$\begin{aligned} \rho(\boldsymbol{\theta},\tilde{\boldsymbol{\theta}}) &= E\left[(\tilde{\boldsymbol{\theta}} - \boldsymbol{\theta})'(\tilde{\boldsymbol{\theta}} - \boldsymbol{\theta})\right] \\ &= E\left[(\hat{\boldsymbol{\theta}} - \boldsymbol{\theta})'(\hat{\boldsymbol{\theta}} - \boldsymbol{\theta})\right] - 2a + 2a\boldsymbol{\theta}'E\left[\hat{\boldsymbol{\theta}}/\hat{\boldsymbol{\theta}}'\hat{\boldsymbol{\theta}}\right] + a^2E\left[1/\hat{\boldsymbol{\theta}}'\hat{\boldsymbol{\theta}}\right] \\ &= K - 2aE\left[\chi^2_{(K-2,\lambda)}/\chi^2_{(K-2,\lambda)}\right] + 2a\boldsymbol{\theta}'\boldsymbol{\theta}E\left[1/\chi^2_{(K+2,\lambda)}\right] \\ &\quad + a^2E\left[1/\chi^2_{(K,\lambda)}\right] \\ &= K - a[2(K-2) - a]E\left[1/\chi^2_{(K,\lambda)}\right], \end{aligned} \tag{5.3}$$

by theorems in Judge and Bock (1978, p. 322). Consequently, the risk of the James and Stein estimator is less than or equal to that of the maximum likelihood estimator if $0 \leqslant a \leqslant 2(K-2)$ and the strict inequality holds for all finite $\boldsymbol{\theta}'\boldsymbol{\theta}$, i.e. for $K \geqslant 3$ and $0 \leqslant a \leqslant 2(K-2)$, the $\rho(\boldsymbol{\theta},\tilde{\boldsymbol{\theta}}) \leqslant \rho(\boldsymbol{\theta},\hat{\boldsymbol{\theta}})$ for all $\boldsymbol{\theta}$. Given (5.3) the value of a which provides the minimal risk for $\tilde{\boldsymbol{\theta}}$ is found when $a = K-2$. Therefore, the optimal minimal risk James and Stein estimator is

$$\tilde{\boldsymbol{\theta}} = \left(1-(K-2)/\hat{\boldsymbol{\theta}}'\hat{\boldsymbol{\theta}}\right)\hat{\boldsymbol{\theta}}, \tag{5.1a}$$

and has risk

$$\rho(\boldsymbol{\theta},\tilde{\boldsymbol{\theta}}) = K-(K-2)^2 E\left[1/\chi^2_{(K,\lambda)}\right]. \tag{5.2a}$$

The risk for this estimator is smaller than that of the maximum likelihood estimator for all $\boldsymbol{\theta}$ such that $\lambda = (\boldsymbol{\theta}'\boldsymbol{\theta}/2) < \infty$. When $\boldsymbol{\theta} = \boldsymbol{0}$ the risk of the James and Stein estimator is 2, and increases to the risk of the maximum likelihood estimator $(\rho(\boldsymbol{\theta},\hat{\boldsymbol{\theta}}) = K)$ as $\boldsymbol{\theta}'\boldsymbol{\theta} \to \infty$. Thus, for values of $\boldsymbol{\theta}$ close to the origin the gain in risk is considerable. These risk results taken in conjunction with the bias results (5.2) imply that the trace of the covariance matrix $(\Sigma_{\tilde{\boldsymbol{\theta}}})$ for the James and Stein estimator is less than the trace of the covariance matrix $(\Sigma_{\hat{\boldsymbol{\theta}}})$ for the maximum likelihood alternative, i.e. $\operatorname{tr}\Sigma_{\tilde{\boldsymbol{\theta}}} \leqslant \operatorname{tr}\Sigma_{\hat{\boldsymbol{\theta}}} = K$. As Stein (1962) and others have noted, and as developed in Judge and Bock (1978, pp. 173–176), the James and Stein estimator (5.1) arises in quite natural fashion as an empirical Bayes estimator.

The James and Stein estimator (5.1) shrinks the maximum likelihood estimates toward a null mean vector. A more general formulation which introduces explicitly the arbitrary origin considers a mean vector $\boldsymbol{\theta}_0$ and an estimator of the form

$$\tilde{\boldsymbol{\theta}}_0 = \left[1-(K-2)/(\hat{\boldsymbol{\theta}}-\boldsymbol{\theta}_0)'(\hat{\boldsymbol{\theta}}-\boldsymbol{\theta}_0)\right](\hat{\boldsymbol{\theta}}-\boldsymbol{\theta}_0)+\boldsymbol{\theta}_0. \tag{5.1b}$$

This estimator has bias and risk characteristics consistent with the more conventional James and Stein estimator.

If σ^2 is unknown, the optimal James and Stein estimator may be written as

$$\tilde{\boldsymbol{\theta}}_s = \left[1-((K-2)/(T-K+2))(s/\hat{\boldsymbol{\theta}}'\hat{\boldsymbol{\theta}})\right]\hat{\boldsymbol{\theta}}, \tag{5.4}$$

and has risk

$$\rho(\boldsymbol{\theta},\sigma^2;\tilde{\boldsymbol{\theta}}_s) = K-\left((K-2)^2(T-K)\right)/(T-K+2)E\left[1/\chi^2_{(K,\lambda)}\right], \tag{5.5}$$

where $s/\sigma^2 = (T-K)\hat{\sigma}^2/\sigma^2$ has a $\chi^2_{(T-K)}$ distribution that is independent of $\hat{\boldsymbol{\theta}}$. Consequently, at the origin where $\boldsymbol{\theta} = \boldsymbol{0}$ the risk is $K-(T-K)(K-2)/(T-K+2)$ and the risk increases to K, the maximum likelihood risk, as $\boldsymbol{\theta}'\boldsymbol{\theta}/\sigma^2 \to \infty$. The evaluation of the mean and covariance of (5.4) may be developed as in (5.2) and its generalization to an arbitrary vector $\boldsymbol{\theta}_0$ proceeds as above. It is perhaps instructive to note that $\hat{\boldsymbol{\theta}}'\hat{\boldsymbol{\theta}}/\sigma^2$ is distributed as a $\chi^2_{(K,\lambda)}$ and thus the optimal James and Stein estimator (5.4) may be rewritten as

$$\begin{aligned}\tilde{\boldsymbol{\theta}}_s &= [1-((K-2)/(T-K+2))(s/\hat{\boldsymbol{\theta}}'\hat{\boldsymbol{\theta}})]\hat{\boldsymbol{\theta}} \\ &= [I-((T-K)(K-2))/((T-K+2)K)(1/u)]\hat{\boldsymbol{\theta}},\end{aligned} \tag{5.6}$$

where $u = \hat{\boldsymbol{\theta}}'\hat{\boldsymbol{\theta}}/(K\hat{\sigma}^2)$ is the likelihood ratio statistic which has an F distribution with K and $(T-K)$ degrees of freedom and non-centrality parameter $\lambda = \boldsymbol{\theta}'\boldsymbol{\theta}/2\sigma^2$. Thus, the pull back of the maximum likelihood estimator is determined by the data and the hypotheses vector $\boldsymbol{\theta}_0$ through the value of the test statistic. The larger the value of the test statistic the smaller the adjustment made in the maximum likelihood estimator.

When σ^2 is unknown the empirical Bayes counterpart is

$$\tilde{\boldsymbol{\theta}}_\beta = [1-((K-2)/(T-K))(s/\hat{\boldsymbol{\theta}}'\hat{\boldsymbol{\theta}})]\hat{\boldsymbol{\theta}} \tag{5.7}$$

and this estimator is dominated by the James and Stein estimator (5.4) since $(K-2)/(T-K) \neq (K-2)/(T-K+2)$ the optimal value of the constant a.

5.1.2. *Positive Stein rule*

Although the James and Stein rule is minimax, it is not admissible. Therefore, other alternative superior rules exist and one such rule is the Baranchik (1964) and Stein (1966) positive rule:

$$\begin{aligned}\tilde{\boldsymbol{\theta}}^{+} &= [1-\min(1, a/\hat{\boldsymbol{\theta}}'\hat{\boldsymbol{\theta}})]\hat{\boldsymbol{\theta}} \\ &= I_{[a,\infty)}(\hat{\boldsymbol{\theta}}'\hat{\boldsymbol{\theta}})[1-(a/\hat{\boldsymbol{\theta}}'\hat{\boldsymbol{\theta}})]\hat{\boldsymbol{\theta}},\end{aligned} \tag{5.8}$$

which uniformly improves on the James and Stein rule (5.1) and is minimax when $0 \leqslant a \leqslant 2(K-2)$. There is no one value of a for the positive rule estimator (5.8) that is optimal. However, Efron and Morris (1973, p. 124) have shown that rules with a in $[(K-2), 2(K-2)]$ dominate the rules with a in $[0,(K-2)]$. Bock (1975) has shown that a general estimator of the form $h(\hat{\boldsymbol{\theta}}'\hat{\boldsymbol{\theta}})\hat{\boldsymbol{\theta}}$, where h is a measurable real valued function, may be improved upon by its positive rule version $h(\hat{\boldsymbol{\theta}}'\hat{\boldsymbol{\theta}})I_{[0,\infty)}(h(\hat{\boldsymbol{\theta}}'\hat{\boldsymbol{\theta}}))\hat{\boldsymbol{\theta}}$. Judge and Bock (1978, pp. 186–187) show there is a family of positive rule estimators that dominate a spherically symmetric estimator.

5.1.3. A minimax–admissible rule

The James and Stein and the positive rule estimators are neither Bayes nor admissible. Although every Bayes estimator of $\boldsymbol{\theta}$ is admissible, not all Bayes estimators dominate the maximum likelihood rule. To fill this void Strawderman (1971) developed an admissible estimator that dominates the maximum likelihood estimator and thus is minimax. In the case where σ^2 is known the Bayes estimator is $K > 4$

$$\tilde{\boldsymbol{\theta}}^{s} = \Bigg[1-((K+2-2a_1)/\hat{\boldsymbol{\theta}}'\hat{\boldsymbol{\theta}}) \times\left\{1-2(K+2-2a_1)^{-1}\left[\int_0^1 \lambda^{(K-2a_1)/2}\exp[-(\lambda-1)\hat{\boldsymbol{\theta}}'\hat{\boldsymbol{\theta}}/2]\,\mathrm{d}\lambda\right]^{-1}\right\}\Bigg]\boldsymbol{\theta}. \tag{5.9}$$

This estimator belongs to the class of Baranchik estimators and is minimax and it is admissible since it is Bayes. Strawderman also showed there were no proper Bayes spherically symmetric minimax estimators for $K = 3$ or 4. In a Monte Carlo result for $K = 5$ and $a_1 = 1/2$, Judge and Bock (1978, pp. 220–221) found that the Strawderman estimator has risk equal to or less than the positive rule alternatives considered over a wide range of the parameter space and that the potential relative risk loss over the remaining part of the parameter space is quite small.

In the case that $a_1 = 2$, the prior distribution for θ that results in this Bayes estimator, has a measure of the form

$$\|\theta\|^{-(K-2)}.$$

It has been noted by Hinde (1978) that the estimator $\tilde{\boldsymbol{\theta}}^{s}_{(a_1=2)}$ dominates the optimal James–Stein estimator (5.1a).

5.1.4. Inadmissibility of the conventional pre-test estimator

In Section 4 we developed and evaluated the risk function for the biased conventional pre-test estimator $\hat{\hat{\boldsymbol{\theta}}}$ and noted that because of the discontinuous nature of the estimating rule, the estimator was inadmissible. In their article, Sclove, Morris and Radhakrishnan (1972) suggested another (pre-test) estimator that demonstrated the inadmissibility of the pre-test estimator $\hat{\hat{\boldsymbol{\theta}}}$. The reader may remember that in Section 4.1 we defined the conventional pre-test estimator as

$$\hat{\hat{\boldsymbol{\Theta}}} = I_{(0,c)}(u)\boldsymbol{\theta}_0 + I \qquad (u)\hat{\boldsymbol{\theta}}, \tag{5.10}$$

where $\boldsymbol{\theta}_0$ is the restricted or hypothesis vector $\boldsymbol{r}=\boldsymbol{\theta}_0$. Given this pre-test rule, Sclove et al. (1972) replaced the maximum likelihood estimator in (5.10) with the James and Stein, or more exactly with the Stein positive-rule, estimator. They then showed, under the squared error loss measure, that the estimator

$$\begin{aligned}\tilde{\tilde{\boldsymbol{\Theta}}}^{+} &= I_{(0,c)}(u)\boldsymbol{\theta}_0 + I_{[c,\infty)}(u)\tilde{\boldsymbol{\theta}}^{+} = I_{(0,c)}(u)\boldsymbol{\theta}_0 + I_{[c,\infty)}(u) \\ &\quad\times\left\{I_{(a^*,\infty)}(u)[1-a^*/u](\hat{\boldsymbol{\theta}}-\boldsymbol{\theta}_0)+\boldsymbol{\theta}_0\right\} \\ &= I_{[c,\infty)}(u)I_{[a^*,\infty)}(u)[1-a^*/u](\hat{\boldsymbol{\theta}}-\boldsymbol{\theta}_0)+\boldsymbol{\theta}_0 \end{aligned} \tag{5.11}$$

provides a minimax substitute for the conventional pre-test estimator, if the critical value c in (5.11) is less than or equal to a^*. When $c \geqslant a^*$, the estimator is no longer minimax but dominates the conventional estimator $\hat{\hat{\boldsymbol{\theta}}}$ for comparable values of c. The proofs of these propositions are given in detail in Judge and Bock (1978, pp. 191–194). The importance of this result is that the non-optimality of the conventional pre-test estimator is demonstrated. This means that the unfavorable properties for the pre-test estimator that were discussed in Section 4.1 are only part of the problem.

5.1.5. *Inadmissibility of the maximum likelihood inequality restricted and corresponding pre-test estimator*

Continue to consider the orthonormal statistical model with estimator performance evaluated under a squared error loss measure. For expository purposes we consider the case when the coefficient vector is of dimension $K=3$, the variance $\sigma^2=1$, and non-sample information is available in the form of a single inequality restriction or hypothesis, say, $\theta_1 \geqslant r_1 = 0$. Under this scenario and within the context of Section 4.1 the inequality restricted maximum likelihood estimator, when $K=3$, may be written as

$$\begin{aligned}\boldsymbol{\theta}^{+} &= I_{(-\infty,0)}(\hat{\theta}_1)\boldsymbol{\theta}^{*} + I_{(0,\infty)}(\hat{\theta}_1)\hat{\boldsymbol{\theta}} \\ &= I_{(-\infty,0)}(\hat{\theta}_1)\begin{bmatrix}0\\ \hat{\theta}_2\\ \hat{\theta}_3\end{bmatrix} + I_{[0,\infty)}(\hat{\theta}_1)\begin{bmatrix}\hat{\theta}_1\\ \hat{\theta}_2\\ \hat{\theta}_3\end{bmatrix}. \end{aligned} \tag{5.12}$$

The sampling properties of this estimator were discussed in Section 3.

As an alternative biased estimator, consider the following Stein-rule version of the inequality restricted estimator:

$$\tilde{\boldsymbol{\theta}}^{+}=I_{(-\infty,0)}(\hat{\theta}_1)\begin{bmatrix}0\\ \hat{\theta}_2\\ \hat{\theta}_3\end{bmatrix}+I_{[0,\infty)}(\hat{\theta}_1)\begin{bmatrix}\tilde{\theta}_1\\ \tilde{\theta}_2\\ \tilde{\theta}_3\end{bmatrix}, \tag{5.13}$$

where the maximum likelihood estimates of $\boldsymbol{\theta}$ are replaced by their corresponding James and Stein estimates $\tilde{\boldsymbol{\theta}}=[1-(1/\hat{\boldsymbol{\theta}}'\hat{\boldsymbol{\theta}})]\hat{\boldsymbol{\theta}}$ or their positive-rule counterpart. With a little algebra (5.13) may be written as

$$\tilde{\boldsymbol{\theta}}^{+}=\boldsymbol{\theta}^{+}-I_{[0,\infty)}(\hat{\theta}_1)(1/\hat{\boldsymbol{\theta}}'\hat{\boldsymbol{\theta}})\hat{\boldsymbol{\theta}}, \tag{5.13a}$$

where $\boldsymbol{\theta}^{+}$ is the conventional inequality restricted estimator (5.12).

Alternatively, if we replace all of the maximum likelihood estimates in (5.12) with their Stein-rule counterparts then the estimator becomes

$$\begin{aligned}\tilde{\tilde{\boldsymbol{\theta}}}^{+}&=I_{(-\infty,0)}(\hat{\theta}_1)\begin{bmatrix}0\\ \tilde{\theta}_2\\ \tilde{\theta}_3\end{bmatrix}+I_{[0,\infty)}(\hat{\theta}_1)\begin{bmatrix}\tilde{\theta}_1\\ \tilde{\theta}_2\\ \tilde{\theta}_3\end{bmatrix} \qquad (5.14)\\ &=\boldsymbol{\theta}^{+}-I_{[0,\infty)}(\hat{\theta}_1)(1/\hat{\boldsymbol{\theta}}'\hat{\boldsymbol{\theta}})\begin{bmatrix}\hat{\theta}_1\\ \hat{\theta}_2\\ \hat{\theta}_3\end{bmatrix}-I_{(-\infty,0)}(\hat{\theta}_1)(1/\hat{\boldsymbol{\theta}}'\hat{\boldsymbol{\theta}})\begin{bmatrix}0\\ \hat{\theta}_2\\ \hat{\theta}_3\end{bmatrix},\end{aligned}$$

and has $E[\tilde{\tilde{\boldsymbol{\theta}}}^{+}]\neq\boldsymbol{\theta}$. The risk of the Stein inequality estimator is

$$\begin{aligned}\rho(\tilde{\boldsymbol{\theta}}^{+},\boldsymbol{\theta})=\rho(\boldsymbol{\theta}^{+}\boldsymbol{\theta})&-2E\left[I_{[0,\infty)}(\hat{\theta}_1)(1/\hat{\boldsymbol{\theta}}'\hat{\boldsymbol{\theta}})\hat{\boldsymbol{\theta}}(\boldsymbol{\theta}^{+}-\boldsymbol{\theta})\right] \qquad (5.15)\\ &+E\left[I_{[0,\infty)}(\hat{\theta}_1)(1/\hat{\boldsymbol{\theta}}'\hat{\boldsymbol{\theta}})^2\hat{\boldsymbol{\theta}}'\hat{\boldsymbol{\theta}}\right].\end{aligned}$$

This means the risk of the Stein inequality estimator is less than or equal to the risk of the conventional inequality estimator if the last two terms of (5.15) are negative. Bock, Judge and Yancey (1980) show that this is indeed the case and thus demonstrate an estimator that dominates the conventional inequality estimator discussed in the literature. The risk characteristics of the alternative estimators are noted in Figure 5.1. In a similar way the estimator (5.14) can be shown to dominate the estimator (5.13) and the positive part version of (5.14) can be shown to dominate the James and Stein inequality estimator (5.14). It should be noted

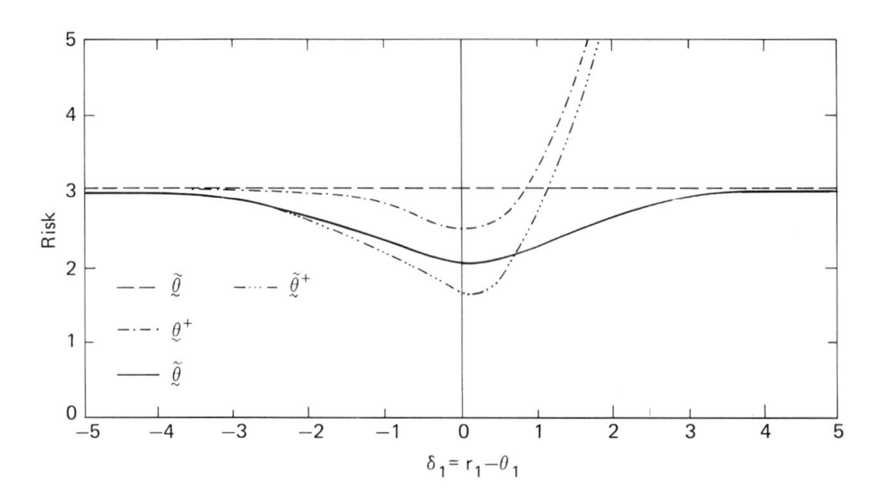

Figure 5.1. Risk functions for the maximum likelihood, James and Stein, and alternative inequality restricted estimators.

that if we are sure of the direction of the inequality restriction, the Stein inequality estimators have smaller risks than maximum likelihood and in fact dominate the conventional James and Stein estimator (5.1).

Another alternative inequality estimator results when Stein estimates replace the maximum likelihood estimates in the argument of the indicator function in estimator (5.14). Unfortunately, in this case the distribution of the James and Stein rule are unknown and therefore the risk determinations cannot be analytically evaluated. Monte Carlo results suggest that the risk characteristics of estimator (5.14) carry over to its completely Stein counterpart.

Bock, Judge and Yancey (1980) have also shown that the following version of the inequality restricted pre-test estimator,

$$\tilde{\boldsymbol{\theta}}^{++} = I_{(-\infty,c)}(\hat{\theta}_1)\begin{bmatrix}\tilde{\theta}_1\\ \tilde{\theta}_2\\ \tilde{\theta}_3\end{bmatrix} + I_{[0,\infty)}(\hat{\theta}_1)\begin{bmatrix}\tilde{\theta}_1^+\\ \tilde{\theta}_2^+\\ \tilde{\theta}_3^+\end{bmatrix}, \tag{5.16}$$

dominates under squared error loss the conventional inequality pre-test estimator,

$$\boldsymbol{\theta}^{++} = I_{(-\infty,c)}(\hat{\theta}_1)\begin{bmatrix}\hat{\theta}_1\\ \hat{\theta}_2\\ \hat{\theta}_3\end{bmatrix} + I_{[c,\infty)}(\hat{\theta}_1)\begin{bmatrix}\theta_1^+\\ \theta_2^+\\ \theta_3^+\end{bmatrix}, \tag{5.17}$$

which was discussed in Section 4 and demonstrates its inadmissibility.

5.2. Stein-like rules under weighted squared error loss

In economics many situations arise where interest centers on the estimation problem and thus the sampling characteristics of the estimator in the $\boldsymbol{\beta}$ space. Alternatively, even if we remain in the $\boldsymbol{\theta}$ space we may wish to weigh certain elements of the coefficient vector differently in the loss function. This means that for the general linear statistical model $\boldsymbol{y} = X\boldsymbol{\beta} + \boldsymbol{e}$ of Section 2 we are concerned with the estimator $\boldsymbol{\delta}(\boldsymbol{b})$ of the unknown vector $\boldsymbol{\beta}$, under the risk

$$\rho(\boldsymbol{\beta}, \sigma^2; \boldsymbol{\delta}) = E[(\boldsymbol{\delta}(\boldsymbol{b}) - \boldsymbol{\beta})'Q(\boldsymbol{\delta}(\boldsymbol{b}) - \boldsymbol{\beta})]/\sigma^2, \tag{5.18}$$

when Q is some $(K \times K)$ positive definite matrix. Under the reparameterized model of Section 2, where $\boldsymbol{\theta} = S^{-1/2}\boldsymbol{\beta}$, we may also transform the problem to that of evaluating the risk under

$$\rho(\boldsymbol{\theta}, \sigma^2; \boldsymbol{\delta}_0) = E[(\boldsymbol{\delta}_0(\hat{\boldsymbol{\theta}}) - \boldsymbol{\theta})'((S^{1/2})'QS^{1/2})(\boldsymbol{\delta}_0(\hat{\boldsymbol{\theta}}) - \boldsymbol{\theta})]/\sigma^2, \tag{5.19}$$

where $\boldsymbol{\delta}(\hat{\boldsymbol{\theta}}) = S^{-1/2}\boldsymbol{\delta}(S^{-1/2}\hat{\boldsymbol{\theta}})$.

For weighted squared error loss reflected by (5.18), where $\boldsymbol{b}$ is a normally distributed K dimensional random vector with mean $\boldsymbol{\beta}$ and covariance $\sigma^2(X'X)^{-1}$, Judge and Bock (1978, pp. 231–238) have shown that if $Q^{1/2}CQ^{1/2}$ and $Q^{1/2}BQ^{1/2}$ are positive definite matrices that commute with each other and with $Q^{1/2}(X'X)^{-1}Q^{1/2}$ and s/σ^2 has a chi-square distribution with $(T-K)$ degrees of freedom, then the estimator

$$\delta(\boldsymbol{b}, s) = \left[I_k - h\left(\frac{\boldsymbol{b}'B\boldsymbol{b}}{s}\right)C\right]\boldsymbol{b} \tag{5.20}$$

is minimax if certain conditions often fulfilled in practice are true. This means there is a family of minimax estimators that are superior to the maximum likelihood estimator under squared error loss. We now turn to some of the estimators in this minimax class.

5.2.1. A James and Stein-type minimax estimator

Out of the class of estimators defined by (5.16) if we let $Q = C = I$ and $B = X'X$ we have the estimator

$$\boldsymbol{\beta}_1^* = [1 - as/\boldsymbol{b}'X'X\boldsymbol{b}]\boldsymbol{b}, \tag{5.21}$$

or its generalized counterpart which shrinks toward a known K dimensional vector $\boldsymbol{\beta}_0$ instead of the conventional null vector. The mean of $\boldsymbol{\beta}_1^*$, by a theorem in Judge and Bock (1978, pp. 321–322), is

$$E[\boldsymbol{\beta}_1^*] = E\left[\left(1 - a\chi^2_{(T-K)}/\chi^2_{(K+2,\lambda)}\right)\right]\boldsymbol{\beta}, \tag{5.22}$$

where $2\lambda = \boldsymbol{\beta}'X'X\boldsymbol{\beta}/\sigma^2$. As shown in Judge and Bock (1978, p. 242), $\rho(\boldsymbol{\beta}, \sigma^2; \boldsymbol{\beta}_1^*) \leqslant \rho(\boldsymbol{\beta}, \sigma^2; \boldsymbol{b})$ for all $\boldsymbol{\beta}$ and σ^2 if

$$0 \leqslant a \leqslant 2\left(\operatorname{tr}(X'X)^{-1}\lambda_K - 2\right)/(T-K+2) \tag{5.23}$$

and

$$\operatorname{tr}(X'X)^{-1} > 2\lambda_K^{-1}, \tag{5.24}$$

where λ_K^{-1} is the largest characteristic root of $(X'X)^{-1}$. Note that if $X'X = I_K$, we have the James and Stein estimator discussed in Section 5.1. If $(\operatorname{tr}(X'X)^{-1}\lambda_K) \leqslant 2$, there is no value for $a > 0$ for which $\boldsymbol{\beta}_1^*$ dominates the least squares estimator. This means that an ill-conditioned $X'X$ matrix could affect whether or not $(\operatorname{tr}(X'X)^{-1}/\lambda_K) > 2$ and that the appearance of three or more regressors no longer assures that a minimax estimator of the form (5.21) exists. As Judge and Bock (1978, pp. 245–248) show, a positive rule variant of the estimator (5.21) exists and dominates it.

5.2.2. *An alternative estimator*

An alternative to (5.21) is an estimator of the form

$$\boldsymbol{\beta}_2^* = \left[I_K - a(X'X)s/\boldsymbol{b}'(X'X)\boldsymbol{b}\right]\boldsymbol{b}, \tag{5.25}$$

which has mean

$$E[\boldsymbol{\beta}_2^*] = \boldsymbol{\beta} - (X'X)\boldsymbol{\beta}aE\left[\chi^2_{(K+2,\lambda)}/\chi^2_{(T-K)}\right]. \tag{5.26}$$

Judge and Bock evaluate the risk for (5.25) and show that if

$$0 \leqslant a \leqslant 2(K-2)/(T-K+2)\lambda_K, \tag{5.27}$$

where λ_K is the smallest root of $(X'X)$, then β_2^* dominates $\boldsymbol{b}$. This means that compared to $\boldsymbol{\beta}_1^*$, this estimator does not have restrictive conditions on $\operatorname{tr}(X'X)^{-1}$ and λ_K and thus will always be minimax if the condition on a is maintained. As before, a positive rule variant of (5.25) exists and dominates it.

5.2.3. Berger's minimax estimator

Another estimator that is of the general form of the minimax estimator (5.20) has been suggested by Berger (1976a) and may be defined as follows:

$$\boldsymbol{\beta}_3^* = \left[I_K - a(X'X)s/\boldsymbol{b}'(X'X)^2\boldsymbol{b}\right]\boldsymbol{b}, \tag{5.28}$$

where $0 \leqslant a \leqslant 2(K-2)/(T-K+2)$ and s/σ^2 has a $\chi^2_{(T-K)}$ distribution. The mean of $\boldsymbol{\beta}_3^*$ is

$$\begin{aligned} E[\boldsymbol{\beta}_3^*] &= \boldsymbol{\beta} - aE\left[X'Xs/\boldsymbol{b}'(X'X)^2\boldsymbol{b}\right] \\ &= \left[I_K - a\sigma^2(T-K)E\left[\left(\boldsymbol{b}'(X'X)^2\boldsymbol{b}(X'X)^{-1} + \sigma^2\chi^2_{(2)}I_K\right)^{-1}\right]\right]\boldsymbol{\beta}, \end{aligned} \tag{5.29}$$

and $\boldsymbol{\beta}_3^*$ has risk

$$\begin{aligned} \rho(\boldsymbol{\beta}, \sigma^2; \boldsymbol{\beta}_3^*) = \operatorname{tr}(X'X)^{-1} + a\sigma^2(T-K)\{a(T-K+2) \\ -2(K-2)E\left[1/\boldsymbol{b}'(X'X)^2\boldsymbol{b}\right]\}, \end{aligned} \tag{5.30}$$

which is a minimum if $a = (K-2)/(T-K+2)$. A corresponding positive rule version of $\boldsymbol{\beta}_3^*$ exists and is given by Judge and Bock (1978, p. 252).

5.2.4. Admissible minimax estimator

When $\boldsymbol{b}$ has a normal distribution with mean vector $\boldsymbol{\beta}$ and covariance $\sigma^2(X'X)^{-1}$, with known $\sigma^2 = 1$, the Bock version of the Strawderman estimator which is minimax admissible, may be expressed as

$$\boldsymbol{\beta}_4^* = \left[1 - \frac{2c_a}{\boldsymbol{b}(X'X)\boldsymbol{b}}\left\{1 - \frac{\left(\frac{1}{2}\boldsymbol{b}'(X'X)\boldsymbol{b}\right)^{c_a}}{c_a!\left[\exp\left(\frac{1}{2}\boldsymbol{b}'(X'X)\boldsymbol{b}\right)\sum_{j=0}^{c_a-1}\left(\frac{1}{2}\boldsymbol{b}'X'X\boldsymbol{b}\right)^j/j!\right]}\right\}\right]b, \tag{5.31}$$

where $c_a = (K/2) + 1 - a_1$ is an integer and a_1 is chosen so that $3-(K/2) < a_1 < 2$ and $\operatorname{tr}(X'X)^{-1}\lambda_K \geqslant c_a + 2$. The conditions for (5.31) to be minimax proper Bayes are quite restrictive [Judge and Bock (1978)]. Fortunately, Berger (1976b) has suggested the following minimax admissible estimator which does not require restrictions on $\operatorname{tr}(X'X)^{-1}\lambda_K$:

$$\boldsymbol{\beta}_5^* = \left[I_K - \frac{(X'X)2c_b}{b'(X'X)^2\boldsymbol{b}} \right.$$

$$\left. \times \left\{ 1 - \frac{\left((\lambda_1^{-1}/2)\boldsymbol{b}'(X'X)^2\boldsymbol{b}\right)^{c_b}}{c_b!\left[\exp\left((\lambda_1^{-1}/2)\boldsymbol{b}'(X'X)^2\boldsymbol{b}\right) - \sum_{j=0}^{c_b-1}\left((\lambda_1^{-1}/2)\boldsymbol{b}'(X'X)^2\boldsymbol{b}\right)^j/j!\right]} \right\} \right] \boldsymbol{b}, \tag{5.32}$$

for $(K/2)-1 < c_b \leqslant K-2$, where λ_1^{-1} is the smallest root of $(X'X)^{-1}$ and c_b is an integer. To be a proper Bayes estimator $(K/2) < c_b \leqslant K-2$, which implies $K \geqslant 5$.

5.2.5. *Some risk comparisons*

In Section 5.2 we have discussed selected estimators from a family of estimators. We still do not know how these estimators compare in terms of risk over the parameter space. To get some idea of the relative sampling performances, one set of Monte Carlo sampling results is reported from Judge and Bock (1978, pp.

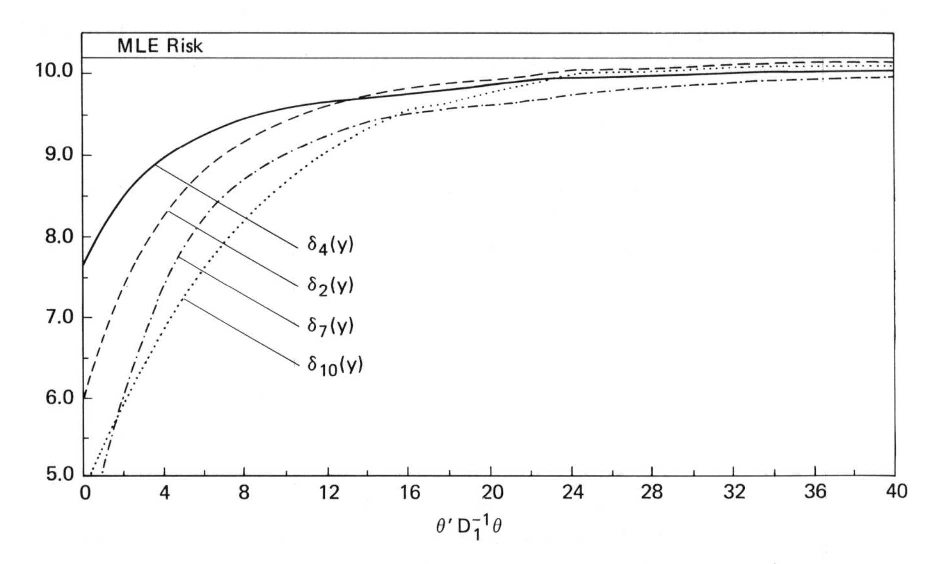

Figure 5.2. Empirical risk bounds for $\boldsymbol{\delta}_2(y)$, $\boldsymbol{\delta}_4(y)$, $\boldsymbol{\delta}_7(y)$, and $\boldsymbol{\delta}_{10}(y)$ (maximum risk bounds).

259–274). To obtain the empirical risks 300 five-parameter samples were drawn from a multinormal distribution with mean vector $\boldsymbol{\theta}$ and non-diagonal covariance matrix which had a characteristic root matrix:

$$D=\begin{bmatrix} 3.0 & & & & \\ & 2.5 & & & \\ & & 2.0 & & \\ & & & 1.5 & \\ & & & & 1.0 \end{bmatrix}. \tag{5.33}$$

The estimators compared were the positive rule counterparts of $\boldsymbol{\beta}_1^* = \delta_2(\boldsymbol{y})$, $\boldsymbol{\beta}_2^* = \delta_4(\boldsymbol{y})$, $\boldsymbol{\beta}_3^* = \delta_7(\boldsymbol{y})$, and $\boldsymbol{\beta}_5^* = \delta_{10}(\boldsymbol{y})$ and the empirical risk functions are given in Figure 5.2. These results indicate the Berger minimax estimator $\boldsymbol{\beta}_3^*$ has the smallest risk at the origin. However, the risk for this estimator quickly crosses the Berger–Strawderman estimator $\boldsymbol{\beta}_5^*$ and the risk for $\boldsymbol{\beta}_5^*$ is superior to that of all estimators compared over much of the parameter space. No one estimator is superior over the whole parameter space. At least these limited results indicate that the gains from having an admissible estimator are not large.

6. Some biased estimator alternatives for the stochastic regressor case

In many cases the economist must work with passively generated non-experimental data where the values of the design matrix X are not fixed in repeated samples. In this section we consider a sampling process where a sample of size T is drawn from a $K+1$ variate normal population with all parameters unknown. The tth random vector has the form

$$\begin{bmatrix} y_t \\ \boldsymbol{x}_t \end{bmatrix},$$

where $\boldsymbol{x}_t$ is a $(K\times 1)$ vector. Let the sample mean and covariance statistics based on the T independent random vectors be denoted by

$$\begin{bmatrix} \bar{y} \\ \bar{\boldsymbol{x}} \end{bmatrix} \quad \text{and} \quad \begin{bmatrix} S_y^2 & S_{\boldsymbol{x}y} \\ S_{\boldsymbol{x}y} & S_{\boldsymbol{xx}} \end{bmatrix},$$

where the population mean $\boldsymbol{\mu}$ and covariance Σ are unknown, and the sample correlation coefficient is $R^2 = S_{\boldsymbol{x}y}' S_{\boldsymbol{xx}}^{-1} S_{\boldsymbol{x}y}/S_y^2$. If we assume $y_{(T+1)}$ is unknown and we wish to predict $y_{(T+1)}$ from $\boldsymbol{x}_{(T+1)}$ by using a prediction function based on the original sample, the maximum likelihood prediction function is

$$\tilde{y} = \boldsymbol{x}_{(T+1)}' \tilde{\boldsymbol{\beta}} + \tilde{\beta}_0, \tag{6.1}$$

where $\tilde{\boldsymbol{\beta}} = S_{xx}^{-1}S_{xy}$ and $\tilde{\beta}_0 = \bar{y} - \bar{\boldsymbol{x}}'\tilde{\boldsymbol{\beta}}$. In gauging estimator performance we use the squared error loss measure

$$\phi(\mu, \Sigma; \hat{y}) = E\left[(\hat{y} - y_{(T+1)})^2\right]/\sigma^2. \tag{6.2}$$

Given this problem, Baranchik (1973) specified an alternative prediction estimator to $\tilde{y}$, when $K > 2$, that is found among the class of minimax estimators in the following theorem [Judge and Bock (1978, p. 278)], where $h(u) = c/u$, where c is a constant and $u = R^2/(1 - R^2)$.

A prediction estimator of the form

$$\hat{y} = \tilde{y} - h\left(R^2/(1 - R^2)\right)\left(\boldsymbol{x}_{(T+1)} - \bar{\boldsymbol{x}}\right)'\tilde{\boldsymbol{\beta}} \tag{6.3}$$

is minimax and dominates the maximum likelihood prediction function $\tilde{y}$ if

(i) $0 \leqslant uh(u) \leqslant 2(K-2)/(T-K+2)$ for $u \geqslant 0$;
(ii) the derivative of $uh(u)$ is non-negative for $u \geqslant 0$; and
(iii) $h(u) \leqslant (T-3)/(K-1)$ for $u \leqslant (K-1)/(T-K+2)$ when the derivative of $uh(u)$ is positive.

If $1 - h(u)$ is negative the positive rule version of the Baranchik estimator

$$\hat{y}_1 = \max\left[0, 1 - t(K-2)(1 - R^2)/\{R^2(T-K+2)\}\right]\left(\boldsymbol{x}_{(T+1)} - \bar{\boldsymbol{x}}\right)'\tilde{\boldsymbol{\beta}} + \bar{y}, \tag{6.4}$$

where $0 \leqslant t \leqslant 2$, should be used.

An alternative approach to determining a rule that is some function of the maximum likelihood prediction equation is to consider prediction functions of the form

$$\hat{y}_a = a\left(\boldsymbol{x}_{(T+1)} - \bar{x}\right)'\tilde{\boldsymbol{\beta}} + \bar{y}, \tag{6.5}$$

where a is any constant, and find the value of a that minimizes $E[(\hat{y}_a - y_{(T+1)})^2]$. King (1972, 1974) has investigated this problem and derived the following estimator:

$$\hat{y}_a = \left[(T-K+2)\rho^2/\{K + (T-2K-2)\rho^2\}\right]\left(\boldsymbol{x}_{(T+1)} - \bar{\boldsymbol{x}}\right)'\tilde{\boldsymbol{\beta}} + \bar{y}, \tag{6.6}$$

which unfortunately contains the unknown parameter ρ^2, the population multiple correlation coefficient. If, following King, the unknown parameter ρ^2 is replaced by a sample estimate R^2, (6.6), the estimator, is of the general form (6.3) and thus dominates the maximum likelihood estimator when the condition $4 \leqslant K \leqslant$

$(T-3)^2+1$ on T and K is satisfied. Analytically evaluating the risk for this estimator is a difficult task, but empirical risk functions developed by King (1974) indicate that the estimator compares quite favorably with the Baranchik estimator and its positive part for values of T and K normally found in practice.

In many econometric problems based on the stochastic regressor model, interest centers on the unknown parameter vector $\boldsymbol{\beta}$ and the performance of alternative estimators under the risk measure $E[(\hat{\boldsymbol{\beta}}-\boldsymbol{\beta})'(\hat{\boldsymbol{\beta}}-\boldsymbol{\beta})]$. Fortunately the theorems, lemmas, and risk derivations for the fixed regressor case given by Judge and Bock (1978, pp. 229–258) carry over directly for the stochastic regressor case.

7. Biased estimation with nearly collinear data

In the presence of a design matrix X that is nearly collinear, small changes in the values of $\boldsymbol{y}$, the vector of observations on the dependent variable, may result in dramatic changes in the values for $\boldsymbol{b}$, the unbiased least squares estimator of $\boldsymbol{\beta}$. Because of the negative inferential implications of this instability, estimators that are not subject to such extreme dependence on the value of $\boldsymbol{y}$ are considered in this section. In an attempt to mitigate the problem of imprecision we examine a class of biased estimators, known as ridge-type estimators, for various specifications of the weight matrix Q in the quadratic loss function

$$(\hat{\boldsymbol{\beta}}-\boldsymbol{\beta})'Q(\hat{\boldsymbol{\beta}}-\boldsymbol{\beta})/\sigma^2, \tag{7.1}$$

and compare them to the least squares estimator $\boldsymbol{b}$. For a more complete discussion of the identification and mitigation of the problem of multicollinearity the reader is referred to Judge, Griffiths, Hill and Lee (1980, ch. 12). A general survey of the literature devoted to ridge regression is given by Vinod (1978).

7.1. *A measure of "near" collinearity*

Certain biased estimators have arisen in attempts to solve problems of near collinearity in the design matrix X. This occurs when one or more columns of X are "nearly" equal to a linear combination of the other columns. One measure of near collinearity is the condition number of the matrix $X'X$ which is defined as the ratio of the largest over the smallest characteristic roots of $X'X$, i.e.

$$\lambda_1/\lambda_K. \tag{7.2}$$

The condition number does not change if all the independent variables are multiplied by the same scalar. Since

$$\lambda_1/\lambda_K \geqslant 1, \tag{7.3}$$

severe near collinearity is said to exist when the condition number, λ_1/λ_K, is very large. In such cases the data contains relatively little information about certain directions in the parameter space as compared to other directions [Thisted (1978a)]. As Thisted notes, if a direction is denoted by a K dimensional vector $\boldsymbol{c}$ whose Euclidean norm $\|c\|$ is one, then the admissible unbiased estimate of $\boldsymbol{c}'\boldsymbol{\beta}$ under squared error loss is $\boldsymbol{c}'\boldsymbol{b}$, the corresponding linear combination of the least squared coefficients. Thus, if $\boldsymbol{c}_1$ and $\boldsymbol{c}_2$ are directions and $\|\boldsymbol{c}_1'X\|$ is considerably smaller than $\|\boldsymbol{c}_2'X\|$, the variance of $\boldsymbol{c}_1'\boldsymbol{b}$ will be considerably larger than that of $\boldsymbol{c}_2'\boldsymbol{b}$. The condition number by itself may not be adequate to define multicollinearity. Instead, perhaps, as Silvey (1969) suggests, it may be preferable to look at all of the characteristic roots and the spectral decomposition to see if multicollinearity exists and if so its nature and extent. Building on Silvey's suggestion Belsley, Kuh and Welsch (1980, ch. 3) provide a set of condition indexes that identify one or more near dependences and adapts the Silvey regression–variance decomposition so that it can be used with the indexes to (i) isolate those variates that are involved and (ii) assess the degree to which the estimates are being distorted by near linear dependencies. Other measures of collinearity are discussed in Chapter 5 of this Handbook by Leamer.

Under near singularity the imprecision that exists in estimating some of the unknown parameters is reflected by the orthonormal statistical model (2.12). In this case the best linear unbiased estimator of $\boldsymbol{\alpha}$ is $\hat{\boldsymbol{\alpha}} = \Lambda^{-1}X'\boldsymbol{y}$, with covariance $\sigma^2\Lambda^{-1}$ and the variance of $\tilde{\alpha}_i$ is σ^2/λ_i, where λ_i is the ith characteristic root of $X'X$. Consequently, relatively precise estimation is possible for those parameters corresponding to the large characteristic roots. Alternatively, relatively imprecise estimation exists for those parameters corresponding to the small characteristic roots.

7.2. Ridge-type estimators

As we have seen in previous sections of this chapter, the transformation of an unbiased estimator often results in a biased estimator of the transformed parameter. In this context and in the face of nearly collinear design matrices, Hoerl and Kennard (1970) suggest biased estimators called "ridge regression estimators". They note that the average squared length of the least squares estimator $\boldsymbol{b}$ is too large, in the sense that $E[\boldsymbol{b}'\boldsymbol{b}] > \boldsymbol{\beta}'\boldsymbol{\beta}$, and Marquardt and Snee (1975) show that

$$E[\boldsymbol{b}'\boldsymbol{b}] = \boldsymbol{\beta}'\boldsymbol{\beta} + \sigma^2\operatorname{tr}(X'X)^{-1} > \boldsymbol{\beta}'\boldsymbol{\beta} + \sigma^2/\lambda_K, \tag{7.4}$$

where λ_K is the smallest characteristic root of $X'X$. Hoerl and Kennard use the results as a motivation for the use of biased estimators where the "shrinkage" factor is non-stochastic. They propose biased estimators of the form

$$\hat{\boldsymbol{\beta}}_c = [X'X + cI_K]^{-1}X'\boldsymbol{y}, \tag{7.5}$$

where c is a constant. In this family of estimators the replacement matrix $[X'X + cI_K]$, which replaces $X'X$ in the least squares estimator, has a lower condition number than $X'X$. These estimators have the property that their mean squared error is less than that of $\boldsymbol{b}$, the least squares estimator of β, for a properly chosen $c > 0$. Unfortunately, the appropriate value of c depended on the unknown parameters $\boldsymbol{\beta}$ and σ^2. For severe misspecification of c, $\hat{\boldsymbol{\beta}}_c$ would have a mean squared error larger than that of b and thus these estimators are not minimax for $Q = I_K$. The estimator $\hat{\boldsymbol{\beta}}_c$, though biased, does result in a more stable estimator for positive values of c.

The appropriate choice of c is no problem in the following Bayesian formulation. It is assumed that the prior distributions of $\boldsymbol{\beta}$ are normal with mean vector $\boldsymbol{0}$ and covariance matrix $\tau^2 I_K$, where τ^2 is a known positive constant. It is appropriate to choose $c = \sigma^2/\tau^2$ under squared error loss. In that case $\hat{\boldsymbol{\beta}}_c$ would be the Bayes estimator and preferable to $\boldsymbol{b}$ under the Bayes risk criterion. Lack of knowledge of the appropriate value of c leads to various specifications which depend on the value of $\boldsymbol{y}$ itself. In the remainder of this section we consider these sorts of estimators.

In passing we note that another way of characterizing the ridge estimator is that of an estimator which results from using the least squares criterion subject to the quadratic constraint $\boldsymbol{\beta}'\boldsymbol{\beta} = r$. Lacking analytical sampling theory results in this area, many of the Monte Carlo experiments for ridge estimators have made use of this type of formulation. The proper choice of r is of course a problem.

The original specification of ridge regression estimators was eventually to lead to the definition of "ridge-type" estimators: estimators of the form

$$\hat{\boldsymbol{\beta}}_A = A^{-1}X'\boldsymbol{y}, \tag{7.6}$$

where the condition number of the matrix A is less than that of $X'X$. Consequently, A may, as in the case of Stein-like estimators, be a stochastic matrix dependent on $\boldsymbol{y}$. Such estimators are more stable than the least squares estimator $\boldsymbol{b}$.

7.3. *Minimax ridge-type estimators*

If a ridge-type estimator dominates the least squares estimator, it is minimax. Such estimators exist only for $K \geqslant 3$ and we will assume that $K \geqslant 3$ for the remainder of this section.

If the loss function weight matrix Q is the identity matrix, there are no known minimax *ordinary* ridge regression estimators, of the type (7.5) with stochastic c, for extremely unstable design matrixes X. Thisted (1978b) notes that such rules have only been specified for design matrixes X in case that the minimaxity index,

$$\lambda_K^2 \sum_{j=1}^{K} \lambda_j^{-2}, \tag{7.7}$$

of $X'X$ is greater than 2. If the condition number of $X'X$, (λ_1/λ_K), is large, this inequality will not be satisfied. Work by Casella (1977) and Thisted (1977) indicates that it is unlikely that such types of minimax ridge rules exist when the minimaxity index of $X'X$ is less than or equal to two and $Q = I_K$. Thisted and Morris (1980) show that many of the ordinary ridge estimators that have been proposed in the literature are not minimax when the design matrix X is nearly collinear.

An example [Judge and Bock (1978)] of a minimax ridge-type estimator for $Q = I_K$ is

$$\delta_0(y) = \left[I_K - \frac{a_0 s^2 (X'X)^{-1}}{b'X'Xb} \right] \boldsymbol{b}, \tag{7.8}$$

where a_0 is a constant in the interval $[0, 2(T-K+2)^{-1}(\lambda_K^2 \sum_{i=1}^{K} \lambda_i^{-2} - 2)]$. Note that the estimator reduces to $\boldsymbol{b}$ if the interval is empty, that is, unless the minimaxity index of $X'X$ is greater than 2. However, even if $\delta_0(y)$ is distinct from $\boldsymbol{b}$, the amount of risk improvement of $\delta_0(y)$ over $\boldsymbol{b}$ diminishes as the condition number of $X'X$ grows large.

7.4. Generalized ridge estimators

A generalized ridge estimator may be defined as

$$\boldsymbol{b}^* = [X'X + C]^{-1} X'y, \tag{7.9}$$

where C is a positive definite matrix such that $(X'X)C = C(X'X)$. Note that C and $(X'X)$ are simultaneously diagonalizable. These are automatically known as "positive shrinkage estimators".

These generalized ridge estimators are not necessarily ridge-type estimators since under squared error loss, with $Q = I_K$, they do not necessarily improve on stability and may in fact make it worse. The Strawderman (1978) adaptive generalized ridge estimator is an example of an estimator out of this class that is minimax but the condition number for the replacement matrix is worse than that of $X'X$ in the least squares estimate $\boldsymbol{b}$.

For a loss function weight matrix $Q=(X'X)^m$, Hill (1979) has noted that an estimator of the form

$$\delta_m(y)=\left[I_K-a_m\frac{s^2(X'X)^{1-m}}{b'(X'X)^{2-m}b}\right]\boldsymbol{b} \tag{7.10}$$

has the ridge property and is minimax for all $X'X$ provided $m\geqslant 2$ and a_m is a constant in the interval $[0,(T-K+2)^{-1}2(K-2)]$. For $m=1$, we have $Q=X'X$ and the prediction loss function is implied. In this case δ_m is minimax but is not a ridge-type estimator.

7.5. A summary comment

In conclusion, we can say that under the usual specification of $Q=I_K$ in the quadratic loss, it does not appear to be possible to meet simultaneously the requirements of ridge-type and minimaxity when the X matrix is of the ill-conditioned form. In fact, Draper and Van Nostrand (1979) note that the amount of risk improvement for ridge estimators is strongly affected by the ill-conditioning of the $X'X$ matrix. No example of the simultaneous meeting of the requirements are known for $Q=X'X$. However, for $Q=(X'X)^m$ and $m\geqslant 2$ there are ridge-type estimators which dominate the least squares estimator for all specifications of $(T\times K)$ matrices X with rank K. Such loss functions heavily penalize estimation errors in those component parameters which can be estimated rather well. Finally, it should be noted that Smith and Campbell (1980) raise some questions as to the foundations of the ridge technique and discuss critically some current ridge practices. Discussions by Thisted and others of the Smith and Campbell article reflect in some sense the range of knowledge and practice in the ridge area.

8. Some final comments

Advances are made in theoretical econometrics by (i) changing the statistical model, (ii) changing the amount of information used, and (iii) changing the measure of performance. The specification and evaluation of the estimators discussed in this chapter have involved, in various degrees, departures from tradition in each of these areas.

Post data model evaluation procedures constitute, to a large degree, a rejection of the concept of a true statistical model for which statistical theory provides a basis for estimation and inference. In addition for the traditional regression model, although the maximum likelihood–least squares rule is the only unbiased estimator based on sufficient statistics it is plagued by the following problems: (i) two decades ago Stein proved that in estimation under squared error loss there is a better estimator, except in the case of one or two parameters, i.e. the estimator is inadmissible; (ii) the least squares rule does not take into account the often existing prior information or relationships among the coordinates; and (iii) when near collinearity is present the least squares rule is unstable and small changes in the observations result in very large changes in the estimates of the unknown coefficients.

Given these inferential pitfalls it seems natural to question the golden rule of unbiasedness and look at biased estimators as a possibility for improving estimator performance. We seek a rule which yields a "favorable" trade-off between bias and variance and thus accomplishes everywhere in the parameter space an overall reduction in mean square error. In the pursuit of this rule we have considered a range of alternatives concerning variants of traditional sampling theory estimators and the Stein and ridge families of estimators.

Within a sampling theory framework the biased estimators that combine sample information and non-sample information in the form of equality or stochastic restrictions *do not* retain, under a squared error loss measure, the minimax property. In addition, if we are uncertain of the prior information and perform a preliminary test of the non-sample information, based on the data at hand, the resulting estimators are inadmissible and are in fact inferior to the maximum likelihood estimator over a large range of the specification error parameter space. These pre-test estimators are commonly used in applied work, although little or no basis exists for choosing the optimum level of the test.

If non-sample information of an inequality form is used and the *direction* of the inequality information is correct, the resulting biased estimator has risk less than or equal to the conventional maximum likelihood estimator. Under the same requirement, the Stein versions of the inequality restricted estimator also dominate their conventional inequality, the James and Stein and the Stein positive-rule counterparts. If the direction of the inequality information is incorrect, the estimator is inferior to the maximum likelihood and Stein-rule risks over much of the parameter space.

The Stein-rule family of estimators which shrink the maximum likelihood estimates toward zero or some predetermined coordinate, enjoy good properties from both the sampling theory and Bayesian points of view. Since the operating characteristics of the Stein rules depend on means and variances of the observations and the unknown coefficients, the estimators are robust relative to the normality assumption. They also appear to be robust over a range of loss functions. Although there are several known minimax admissible rules, none of the rules analyzed herein dominate the positive Stein rule. The Stein family thus provide rules that are simple, efficient, and robust.

Ridge regression procedures which lead to biased estimators have been suggested as one means of "improving the conditioning of the design matrix" and coping with the multicollinearity problem. Strawderman (1978) and Judge and Bock (1978) have demonstrated a link between the adaptive generalized ridge estimator, the general minimax estimator, and the Stein estimators. To some this might suggest that the ridge estimators should be used even in the absence of the collinearity problems. However, as noted in Section 7, under conventional loss functions the general minimax and ridge-type estimators are not the solution to the multicollinearity problem and in general the mechanical application of ridge-type procedures that seek some improvement in estimator performance should be strongly questioned since ridge is not always better than its least squares competitor.

Finally, we should note that least squares and maximum likelihood estimation of non-linear statistical models lead in general to biased estimators. Some of these results are discussed elsewhere in this Handbook. Also, in general the Bayesian criterion leads to biased estimators. The Bayesian basis for estimation and inference is discussed in Sections 2.3, 3.4, and 4.4 of this chapter and in Chapter 2 of this Handbook by Zellner.

References

Akaike, H. (1974) "A New Look at the Statistical Identification Model", *IEEE*: *Transactions on Automatic Control*, 19, 716–723.

Amemiya, T. (1976) "Selection of Regressors", Technical Report no. 225, Stanford University.

Baranchik, A. J. (1964) "Multiple Regression and Estimation of the Mean of a Multivariate Normal Distribution", Technical Report No. 51, Department of Statistics, Stanford University, California.

Baranchik, A. J. (1973) "Inadmissibility of Maximum Likelihood Estimators in Some Multiple Regression Problems with Three or More Independent Variables", *Annals of Statistics*, 1, 312–321.

Belsley, D. A., E. Kuh, and R. E. Welsch (1980) *Regression Diagnostics*. New York: John Wiley & Sons.

Berger, J. (1976a) "Minimax Estimation of a Multivariate Normal Mean Unclear Arbitrary Quadratic Loss", *Journal of Multivariate Analysis*, 6, 256–264.

Berger, J. (1976b) "Admissible Minimax Estimation of a Multivariate Normal Mean with Arbitrary Quadratic Loss", *Annals of Statistics*, 4, 223–226.

Bock, M. E. (1975) "Minimax Estimators of the Mean of a Multivariate Normal Distribution", *Annals of Statistics*, 3, 209–218.

Bock, M. E., G. G. Judge and T. A. Yancey (1980) "Inadmissibility of the Inequality Estimator under Squared Error Loss", Working Paper, University of Illinois.

Bock, M. E., T. A. Yancey and G. G. Judge (1973) "The Statistical Consequences of Preliminary Test Estimators in Regression", *Journal of the American Statistical Association*, 68, 109–116.

Casella, G. (1977) "Minimax Ridge Estimation", Unpublished Ph.D. dissertation, Purdue University.

Cox, D. R. (1961) "Test of Separate Families of Hypotheses", in: *Proceedings of the Fourth Berkeley Symposium on Mathematical Statistics and Probability*, Vol. 1. Berkeley: University of California Press.

Cox, D. R. (1962) "Further Results on Tests of Separate Families of Hypothesis", *Journal of the Royal Statistical Society*, Levis, B, 24, 406–424.
Dempster, A. P. (1973) "Alternatives to Least Squares in Multiple Regression", in: Kabe and Gupta (eds.), *Multivariate Statistical Inference*. Amsterdam: North-Holland Publishing, pp. 25–40.
Dempster, A. P., M. Schatzoff and N. Wermuth (1977) "A Simulation Study of Alternatives to Ordinary Least Squares", *Journal of the American Statistical Association*, 72, 77–106.
Draper, N. R. and R. C. van Nostrand (1979) "Ridge Regression and James and Stein Estimation: Review and Comments", *Technometrics*, 21, 451–466.
Efron, B. and C. Morris (1973) "Stein's Estimation Rule and its Competitors—An Empirical Bayes Approach", *Journal of the American Statistical Association*, 68, 117–130.
Giles, D. E. A. and A. C. Rayner (1979) "The Mean Squared Errors of the Maximum Likelihood and Natural-Conjugate Bayes Regression Estimators", *Journal of Econometrics*, 11, 319–334.
Gunst, R. F. and R. L. Mason (1977) "Biased Estimation in Regression: An Evaluation Using Mean Square Error", *Technometrics*, 72, 616–628.
Hill, R. C. (1979) "The Sampling Characteristics of General Minimax and Ridge Type Estimators Under Multicollinearity", Research Paper, University of Georgia.
Hinde, R. (1978) "An Admissible Estimator Which Dominates the James-Stein Estimator", Research Paper 167, School of Economic and Financial Studies, Macquoue University.
Hocking, R. R. (1976) "The Analysis and Selection of Variables in Linear Regression", *Biometrics*, 32, 1–49.
Hoerl, A. E. and R. W. Kennard (1970) "Ridge Regression: Biased Estimation of Nonorthogonal Problems", *Technometrics*, 12, 55–67.
James, W. and C. Stein (1961) "Estimation with Quadratic Loss", in: *Proceedings of the Fourth Berkeley Symposium Mathematical Statistics and Probability*, vol. 1. Berkeley: University of California Press, pp. 361–379.
Judge, G. G. and M. E. Bock (1976) "A Comparison of Traditional and Stein Rule Estimators Under Weighted Squared Error Loss", *International Economic Review*, 17, 234–240.
Judge, G. G. and M. E. Bock (1978) *The Statistical Implications of Pre-Test and Stein-Rule Estimators in Econometrics*. Amsterdam: North-Holland Publishing Co.
Judge, G. G. and T. A. Yancey (1978), "Inequality Restricted Estimation Under Squared Error Loss", Working Paper Series, University of Georgia.
Judge, G. G., W. E. Griffiths, R. C. Hill and T. C. Lee (1980) *The Theory and Practice of Econometrics*. New York: John Wiley & Sons.
King, N. (1972) "An Alternative for the Linear Regression Equation When the Predictor Variable is Uncontrolled and the Sample is Small", *Journal of the American Statistical Association*, 67, 217–219.
King, N. (1974) "An Alternative for Multiple Regression when the Prediction Variables are Uncontrolled and the Sample Size is not Too Small", unpublished manuscript.
Leamer, E. E. (1974) "Fales Models and Post Data Model Evaluation", *Journal of the American Statistical Association*, 69, 122–131.
Leamer, E. E. (1978) *Specification Searches*. New York: John Wiley & Sons.
Leamer, E. E. and G. Chamberlain (1976) "A Bayesian Interpretation of Pre-Testing", *Journal of the Royal Statistical Society*, Ser. B, 38, 89–94.
Mallows, C. L. (1973) "Some Comments on C_p", *Technometrics*, 15, 661–676.
Marquardt, D. W. and R. D. Snee (1975) "Ridge Regression in Practice", *American Statistician*, 29, 3–19.
Mayer, L. S. and T. A. Willke (1973) "On Biased Estimations in Linear Models", *Technometrics*, 15, 497–508.
Rothenberg, T. J. (1973) *Efficient Estimation with A Priori Information*. New Haven: Yale University Press.
Sclove, S. L., C. Morris and R. Radhakrishnan (1972) "Non Optimality of Pre-Test Estimators for the Multinormal Mean", *Annals of Mathematical Statistics*, 43, 1481–1490.
Silvey, S. D. (1969) "Multicollinearity and Imprecise Estimation", *Journal of the Royal Statistical Society*, B, 31, 539–552.
Smith, G. and F. Campbell (1980) "A Critique of Some Ridge Regression Methods," *Journal of the American Statistical Association*, 75, 74–103.
Stein, C. (1955) "Inadmissibility of the Usual Estimator for the Mean of a Multivariate Normal

Distribution", in: *Proceedings of the Third Berkeley Symposium*, vol. 1. Berkeley: University of California Press, pp. 197–206.

Stein, C. (1962) "Confidence Sets for the Mean of the Multivariate Distribution", *Journal of the Royal Statistical Society*, Series B, 24, 265–296.

Stein, C. (1966) "An Approach to Recovery of Interblock Information in Incomplete Block Designs", in: F. N. David (ed.), *Research Papers in Statistics, Festschrift for J. Neyman*. New York: John Wiley & Sons, pp. 351–366.

Strawderman, W. (1971) "Proper Bayes Minimax Estimators of the Multivariate Normal Mean", *Annals of Mathematical Statistics*, 42, 385–388.

Strawderman, W. E. (1978) "Minimax Adaptive Generalized Ridge Regression Estimators", *Journal of the American Statistical Association*, 73, 623–627.

Swamy, P. A. V. B. and J. S. Mehta (1977) "Robustness of Theil's Mixed Regression Estimator", *The Canadian Journal of Statistics*, 5, 93–109.

Theil, H. (1963) "On the Use of Incomplete Prior Information in Regression Analysis", *Journal of the American Statistical Association*, 58, 401–414.

Theil, H. and A. Goldberger (1961) "Pure and Mixed Statistical Estimation in Economics", *International Economic Review*, 2, 65–78.

Thisted, R. (1977) "Ridge Regression, Minimax Estimation and Empirical Bayes Methods", Unpublished Ph.D. dissertation, Stanford University.

Thisted, R. (1978a) "Multicollinearity, Information, and Ridge Regression", Technical Report no. 66, Department of Statistics, University of Chicago.

Thisted, R. (1978b) "On Generalized Ridge Regressions", Technical Report no. 57, University of Chicago.

Thisted, R. and C. Morris (1980) "Theoretical Results for Adaptive Ordinary Ridge Regression Estimators", University of Chicago Technical Report no. 94.

Toro-Vizcorrondo, C. and T. D. Wallace (1968) "A Test of the Mean Square Error Criterion for Restrictions in Linear Regression", *Journal of the American Statistical Association*, 63, 558–572.

Vinod, H. D. (1978) "A Survey of Ridge Regression and Related Techniques for Improvements Over Ordinary Least Squares", *Review of Economics and Statistics*, 60, 121–131.

Zellner, A. (1980) "On Bayesian Regression Analysis with g-Prior Distributions", H. G. B. Alexander Research Foundation Paper, University of Chicago.

Zellner, A. and W. Vandaele (1975) "Bayes–Stein Estimators for *k*-Means", in: S. E. Fienberg and A. Zellner (eds.), *Studies in Bayesian Econometrics and Statistics*. Amsterdam: North-Holland Publishing Co., pp. 627–653.

Chapter 11

ESTIMATION FOR DIRTY DATA AND FLAWED MODELS

WILLIAM S. KRASKER*

Harvard University

EDWIN KUH and ROY E. WELSCH*

Massachusetts Institute of Technology

Contents

*This research was supported, in part, by the National Science Foundation, U.S. Department of Energy, and I.B.M.

Handbook of Econometrics, Volume I, Edited by Z. Griliches and M.D. Intriligator

1. Introduction

We are concerned with the econometric implications of the sensitivity to data of coefficient estimates, policy analyses, and forecasts in the context of a regression model. In contrast to the emphasis in standard treatments of the linear model paradigm described subsequently, we are interested in data, how they are generated, and particular data configurations in the context of a specified regression model. The focus of this chapter is on resistant estimation procedures and methods for evaluating the impact of particular data elements on regression estimates. While terminology is not yet firmly fixed in this rapidly evolving area, resistant estimation here is presumed to include classical robust estimation for location [Andrews et al. (1972)] or regression [Huber (1977)] and bounded-influence regression [Krasker and Welsch (1982a)]. "Classical robust" estimation reduces the effect of outliers in error space. Bounded-influence regression, in addition, limits the permissible impact of outliers in explanatory-variable space.

The time-honored point of departure in econometrics is the ordinary least squares (OLS) estimator $b=(X^{\mathrm{T}}X)^{-1}X^{\mathrm{T}}y$ for the linear regression model $y=X\beta+\varepsilon$, where y is the response variable data vector, X is the explanatory variable data matrix, β are coefficients to be estimated, and ε conditional on X is a random vector with $E(\varepsilon\varepsilon^{T})=\Sigma=\sigma^2 I$ and $E(\varepsilon)=0$. The widespread appeal of this model lies in its simplicity, its low computational cost, and the BLUE (Best Linear Unbiased Estimator) property shown by the Gauss–Markov theorem. When ε is normally distributed, there is the added theoretical imprimatur of maximum likelihood and attendant full efficiency. Also, for fixed X, exact small sample tests of significance are possible.

More elaborate estimators are needed when the simple assumptions that motivate OLS are considered invalid. Thus, generalized least squares (GLS) replaces OLS when $\Sigma\neq\sigma^2 I$ leading to the Aitkin estimator, $b=(X^{\mathrm{T}}\Sigma^{-1}X)^{-1}X^{\mathrm{T}}\Sigma^{-1}y$, when the errors are heteroscedastic or autocorrelated. GLS estimates are BLUE for known Σ and have desirable asymptotic properties when Σ has been consistently estimated.

When the explanatory variables cannot be viewed as fixed, the choice of estimator depends on the sources of random behavior and whatever further assumptions the econometrician considers tenable. Random behavior in the explanatory variables includes observational errors, endogenous variables that are part of a simultaneous-equation system, variance-component models, lagged endogenous variables, stochastic regressors with some joint distribution, and stochastic parameter variation. Failure to recognize these statistical attributes can lead to one or more of the following shortcomings: inefficiency, finite sample bias, inconsistency, and incorrect tests of significance. Generally speaking, correct estimation procedures differ from OLS/GLS when these circumstances prevail and estimators are tailored to whatever specific stochastic conditions are deemed the most important.

This perspective can be extended to encompass estimators that avoid undue reliance on small segments of the data when there are large but isolated departures from the maintained statistical hypotheses. Thus, reliable estimation sometimes calls for explicit consideration of the X matrix so as to limit the permissible influence of any one of its rows. At the same time, one would also like protection against occasional large ε. A class of resistant estimators that restricts unusually influential components of X and ε, called bounded-influence estimators, offers protection against several types of common specification problems and requires less restrictive assumptions about stochastic properties than those customarily required in the more complex regression structures enumerated above.

Robust regression has appeared in econometrics literature since the mid-1950s, mainly in the guise of Least Absolute Residuals, an estimator that minimizes the sum of the absolute values rather than the square of errors. According to a fine survey by Lester D. Taylor (1974): "LAR has the same illustrious progenitors as least squares (Gauss and Laplace)...but has historically never attracted much attention." Even though coefficient computations became practical through linear programming, as initially pointed out by Charnes, Cooper and Ferguson (1955), Karst (1958), Wagner (1959), and W. Fisher (1961), distribution theory has remained a problem, although a recent paper by Koenker and Basset (1978) provides asymptotic theory.

Two empirical studies suggest that in some cases LAR (or variants) may outperform OLS. Again, quoting Taylor: "What Meyer and Glauber (1964) did was first to estimate their investment models by LAR as well as least squares and then test the equations on post-sample data by using them to forecast the 9 (and sometimes 11) observations subsequent to the period of fit. They found that, with very few exceptions, the equations estimated by LAR outperformed the ones estimated by least squares even on criteria (such as the sum of squared forecast errors) with respect to which least squares is ordinarily thought to be optimal (p. 171)." Another study by Fair (1974) used approximations to LAR and adaptations of other robust estimators in a fifteen-equation macro model. His comparisons had an outcome similar to that of Meyer and Glauber: LAR outperformed OLS in post-sample forecasts.

While these isolated instances of empirical research are suggestive of potentially attractive results, resistant estimation (in its LAR garb or any other) has remained peripheral to mainstream econometric work because of computational costs as well as the absence of widely available code designed for this purpose, and the lack of convincing theoretical support. These deficiencies, along with more intense concerns about other econometric issues and widespread acceptance of OLS, help to explain the relative neglect of resistant regression.

Resistant estimators offer protection against certain fairly general model failures while preserving high efficiency in well-behaved situations. This approach differs from the more standard econometric approach where an alternative estimator is devised to cope with specific departures from a more standard specification.

There is an inevitable gap between a model and reality; it is one thing to write down a model and another to believe it. Three model/data problems are of immediate concern.[1] First, there may be "local errors", such as round-off errors or groupings of observations. Second, there may be "gross errors" in the data, e.g. incorrectly recorded numbers, keypunch errors, or observations made on the wrong quantity. Finally, the model itself is typically thought to be only an approximation. In regression, for example, the linearity of the model and the normality of the disturbance distribution are both good approximations, at best.

Local errors occur in virtually all data sets, if for no other reason than the fact that we work with only finitely many significant digits. However, local errors do not ordinarily cause serious problems for the classical regression procedures, so we will not be too concerned with them.

Gross errors occur more often in some types of data sets than in others. A time-series model using National Income Accounts data and a moderate number of observations is unlikely to contain data with gross errors (provided the numbers which are actually read into the computer are carefully checked). However, consider a large cross section for which the data were obtained by sending questionnaires to individuals. Some respondants will misinterpret certain questions, while others will deliberately give incorrect information. Further errors may result from the process of transcribing the information from the questionnaires to other forms; and then there are the inevitable keypunch errors. Even if the data collectors are careful, some fraction of the numbers which are ultimately fed into the computer will be erroneous.

The third category – the approximate nature of the model itself – is also a serious problem. Least squares can be very inefficient when the disturbance distribution is heavy tailed. Moreover, although the linear specification is often adequate over most of the range of the explanatory variables, it can readily fail for extreme values of the explanatory variables; unfortunately, the extreme values are typically the points which have the most influence on the least squares coefficient estimates.

Gross errors – even if they are a very small fraction of the data – can have an arbitrarily large effect on the distribution of least squares coefficient estimates. Similarly, a failure of the linear specification – even if it affects only the few observations which lie in extreme regions of the X space – can cause OLS to give a misleading picture of the pattern set by the bulk of the data.

[1] The discussion which follows goes back to Hampel (1968).

While general considerations about data appear in Chapter 27 by Griliches we need to examine in more detail those circumstances in which statistical properties of the data – in isolation or in relation to the model – counsel the use of resistant estimators. These will often be used as a check on the sensitivity of OLS or GLS estimates simply by noting if the estimates or predictions are sharply different. Sometimes they will be chosen as the preferred alternative to OLS or GLS.

The common practice in applied econometrics of putting a dummy variable into a regression equation to account for large residuals that are associated with unusual events, requires a closer look. The inclusion of a single dummy variable with zeros everywhere except in one period forces that period's residual to zero and is equivalent to deleting that particular row of data. The resulting distribution of residuals will then appear to be much better behaved. [See Belsley, Kuh and Welsch (1980, pp. 68–69).] Should dummy variables be used to downweight observations in this manner? Dummy variables are often an appealing way to increase estimation precision when there are strong *prior* reasons for their inclusion, such as strikes, natural disasters, or regular seasonal variation. Even then, dummy variables are often inadequate. When a strike occurs in a particular quarter, anticipations will influence earlier periods and unwinding the effects of the strike will influence subsequent periods. As an interesting alternative to OLS, one might wish to consider an algorithm that downweights observations smoothly according to reasonable resistant statistical criteria instead of introducing discrete dummy variables after the fact, which has the harsher effect of setting the row weight to zero.

Model-builders using macroeconomic time series are often plagued by occasional unusual events, leading them to decrease the weights to be attached to these data much in the spirit of resistant estimation. Even when there are good data and theory that correspond reasonably well to the process being modeled, there are episodic model failures. Since it is impractical to model reality in its full complexity, steps should be taken to prevent such model failures from contaminating the estimates obtainable from the "good" data. Some of these breakdowns are obvious, while others are not. At least some protection can be obtained through diagnostic tests. Where the aberrant behavior is random and transitory, estimators that restrict the influence of these episodes should be seriously considered. We do not view resistant estimation as a panacea: some types of model failure require different diagnostic tests and different estimators.

Other types of model difficulties are sometimes associated with cross sections, quite apart from the sample survey problems mentioned earlier. Cross-sectional data are often generated by different processes than those which generate time series. This hardly startling proposition is a belief widely shared by other econometricians, as evidenced by the proliferation of variance-components models which structure panel data error processes precisely with this distinction in mind. (See Chapter 22 by Chamberlain on panel data.)

To some extent these differences reflect the aggregation properties of the observational unit rather than different (i.e. intertemporal versus cross-sectional) behavior. Time series often are aggregates, while cross sections or panel data often are not. There is a tendency for aggregation to smooth out large random variations which are so apparent in disaggregated data. However, time series of speculative price changes for stock market shares, grains, and non-ferrous metals are often modeled as heavy-tailed Pareto–Levy distributions which are poorly behaved by our earlier definition. These constitute a significant exception, and there are doubtless other exceptions to what we believe, nevertheless, is a useful generalization.

Cross-sectional individual observations reflect numerous socio-demographic, spatial, and economic effects, some of which can reasonably be viewed as random additive errors and others as outlying observations among the explanatory variables; many of these are intertemporally constant, or nearly so. Such particularly large cross-sectional effects have four principal consequences in econometrics. One already mentioned is the burst of interest during the last twenty years in variance-component models. A second effect is the natural proclivity in empirical research to include a great many (relative to time series) subsidiary explanatory variables, i.e. socio-demographic and spatial variables of only minor economic interest. Their inclusion is designed to explain diverse behavior as much as possible, in the hope of improving estimation accuracy and precision. Third, the relative amount of explained variation measured by R^2 is characteristically lower in cross sections than in time series despite the many explanatory variables included. Fourth, anomalous observations are likely to appear in cross sections more often than in time series.

Thus, with individual or slightly aggregated observations, resistant estimation appears especially promising as an alternative estimator and diagnostic tool since ideosyncratic individual behavior – i.e. behavior explained poorly by the regression model or a normal error process – pervades cross-section data.

A strong trend exists for exploiting the information in large data sets based on sample surveys of individuals, firms, establishments, or small geographic units such as census tracts or countries. Often these are pooled time series and cross sections. A volume of more than 700 pages, containing 25 articles, was devoted to this subject alone [*Annales de l'Insee* (1978)].[2] Research based on social security records by Peter Diamond, Richard Anderson and Yves Balcer (1976) has 689,377 observations. This major evolution in the type of data used in econometrics is a consequence of several factors, not the least of which has been enormous reductions in computational costs.

[2]It includes empirical studies on investment by M. Atkinson and J. Mairesse with about 2300 observations and R. Eisner with 4800 observations; economic returns to schooling by G. Chamberlain with 2700 observations as well as an article on a similar topic by Z. Griliches, B. Hall and J. Hausman with 5000 observations.

Since survey data are notoriously prone to various kinds of mistakes, such as response or keypunch errors, it is essential to limit their effects on estimation. Some gross errors can be spotted by examining outliers in each particular data series, but it is often impossible to spot multivariate outliers. The isolated point in Figure 1.1 would not be observed by searches of this type. Thus, observational errors compound the effects of sporadic model failures in ways that are not overcome by large sample sizes (law of large numbers). Resistant estimation is a major innovation with the potential for reducing the impact of observational error on regression estimates.

To drive home the point that the likelihood of a slightly incorrect model and/or some bad data force us to change the way we look at those extremely large cross-section data sets, consider this example: via questionnaires, we obtain a sample from a certain population of individuals to estimate the mean value of some characteristic of that population, which is distributed with mean μ and standard deviation σ. However, there are "bad" observations occurring with probability ε in the sample due, for example, to keypunch errors, or forms sent to inappropriate people. The bad points are distributed with mean $\mu+\theta$ and standard deviation $k\sigma$. The mean squared error for the sample mean $\overline{X}_n$ is $\{(1-\varepsilon+\varepsilon k^2)+\theta^2\varepsilon(1-\varepsilon)\}\sigma^2/n$. Without loss of generality, suppose $\sigma=1$. Then if $\theta=1$, $k=2$, and $\varepsilon=0.05$ (which are not at all unreasonable), the mean squared error is $0.0025+1.20/n$. Obviously there is very little payoff to taking a sample larger than 1000 observations. Effort would be better spent improving the data.

Since bounded-influence estimators are designed to limit the influence that any small segment of the data can have on the estimated coefficients, it is not surprising that these estimators also contain diagnostic information (much as a first-order autoregressive coefficient is both part of the standard GLS transformation and also contains diagnostic/test information). Thus, the GLS compensation for heteroscedasticity, when computed by weighted least squares (WLS), has a parallel to an algorithm used in bounded-influence estimation (hereafter often

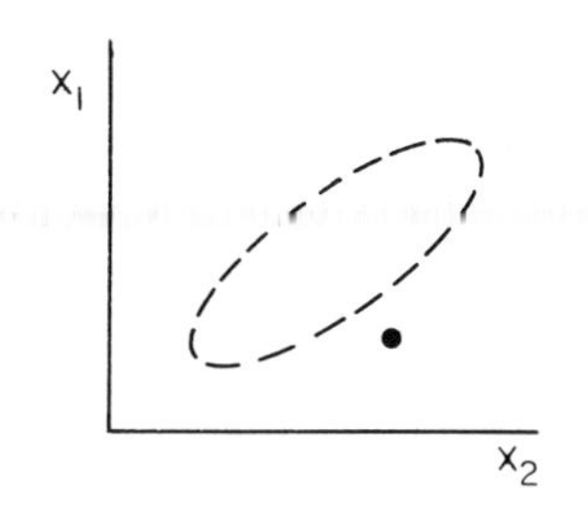

Figure 1.1

referred to as BIF) that gives weights to the rows of the data matrix: large error variances are downweighted in WLS while highly influential observations are downweighted in bounded-influence estimation. Hence, small weights in BIF point to influential data. Although computational complexity and costs are higher for BIF, they are decidely manageable.

Section 2 considers in more detail some model failures that can arise in practice. Section 3 describes recent developments in methods for the detection of influential data in regression. Section 4 is a sketch of the Krasker–Welsch BIF estimator. Section 5 raises several issues about inference in the resistant case. Section 6 considers some of the main theoretical foundations of robust and BIF estimation. Section 7 presents an example of BIF applied to the Harrison–Rubinfeld large cross-section hedonic price index. Section 8 gives some recent results on instrumental-variables bounded-influence estimation, and Section 9 discusses resistant estimation for time-series models.

2. Sources of model failure

In this section we discuss the ways in which the classical assumptions of the linear regression model are often violated. Our goal is to determine what types of data points must be downweighted in order to provide protection against model failures. Specifically, under what conditions should we downweight observations which have large residuals, "extreme" X rows, or both.

As we mentioned above, there are two categories of model failures that are potentially serious. The first consists of "gross errors", e.g. keypunch errors, incorrectly recorded numbers, or inherently low precision numbers. The second derives from the fact that the model itself is only an approximation. Typically an econometrician begins with a response (dependent) variable together with a list of explanatory (independent) variables with the full realization that there are in truth many more explanatory variables that might have been listed. Moreover, the true functional form is unknown, as is the true joint distribution of the disturbances.

A reasonable, conventional approach is to hypothesize a relatively simple model, which uses only a few of the enormous number of potential explanatory variables. The functional form is also chosen for simplicity; typically it is linear in the explanatory variables (or in simple functions of the explanatory variables). Finally, one assumes that the disturbances are i.i.d., or else that their joint distribution is described by some easily parameterized form of autocorrelation. All of these assumptions are subject to errors, sometimes very large ones.

We have described this procedure in detail in order to establish the proposition that *there is no such thing as a perfectly specified econometric model*. Proponents of robust estimation often recommend their robust estimators for "cases in which gross errors are possible", or "cases in which the model is not known exactly". With regard to gross errors the qualification is meaningful, since one can find data sets which are error-free. However, to point out that robust procedures are not needed when the model is known exactly is misleading because it suggests that an exactly known model is actually a possibility.

If the model is not really correct, what are we trying to estimate? It seems that this question has a sensible answer only if the model is a fairly good approximation, i.e. if the substantial majority of the observations are well described (in a stochastic sense) by the model. In this case, one can at least find coefficients such that the implied model describes the bulk of the data fairly well. The observations which do not fit that general pattern should then show up with large residuals. If the model does not provide a good description of the bulk of the data for any choice of coefficients, then it is not clear that the coefficient estimates can have any meaningful interpretation at all; and there is no reason to believe that bounded-influence estimators will be more useful than any other estimator, including ordinary least squares.

The hard questions always arise after one has found a fit for the bulk of the data, and located the outliers. To gain some insight, consider a data configuration which arises often enough in practice (Figure 2.1). Most of the data in Figure 2.1 lie in a rectangular area to the left; however, some of the observations lie in the circled region to the right. Line A represents the least-squares regression line, whereas line B would be obtained from a bounded-influence estimator, which restricts the influence of the circled points.

The fit given by line B at least allows us to see that the bulk of the data are well described by an upward-sloping regression line, although a small fraction of the observations, associated with large values of x, deviate substantially from this pattern. Line A, on the other hand, is totally misleading. The behavior of the bulk of the data is misrepresented and, worse yet, the circled outliers do not have large residuals and so might go unnoticed.

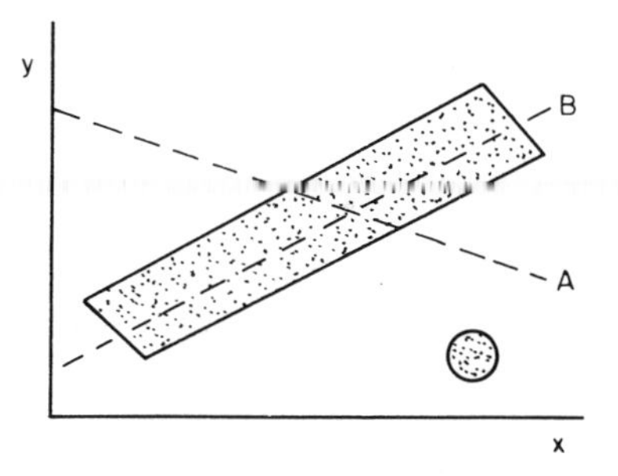

Figure 2.1

What happens as the number of observations in the circle grows large? Eventually, even the bounded-influence fit will pass near the circle. Indeed, an estimator which fits the "bulk" of the sample can hardly ignore the circled observations if they are a majority of the data. In this case there is no linear model which reasonably describes the vast majority of the observations, so that bounded-influence estimation would not help.

With the information provided by fit B, what should we do? There is no unique answer, for it depends on the purpose of the estimation. If our goal is merely to describe the bulk of the data, we might simply use the coefficients from the bounded-influence regression. If we were trying to forecast y conditional upon an x near the center of the rectangle, we would again probably want to use the bounded-influence fit.

If we want to forecast y conditional on an x near the circled points, the situation is entirely different. The circled points really provide all the data-based information we have in this case, and we would have to rely on them heavily. In practice, one would try to supplement these data with other sources of information.

Related to the last point is a well recognized circumstance among applied econometricians, namely that sometimes a small, influential subset of data contain most of the crucial information in a given sample. Thus, only since 1974 have relative energy prices shown large variability. If the post-1974 data have a different pattern from the pre-1974 data (most of the available observations) we might still prefer to rely on the post-1974 information. While this is a dramatic, identifiable (potential) change in regression regime where covariance analysis is appropriate, many less readily identifiable situations can arise in which a minority of the data contain the most useful information. Bounded-influence regression is one potentially effective way to identify these circumstances.

In short, one never simply throws away outliers. Often they are the most important observations in the sample. The reason for bounded-influence estimation is partly that we want to be sure of *detecting* outliers, to determine how they deviate from the general pattern. By trying to fit all the data well, under the assumption that the model is exactly correct, least squares frequently hides the true nature of the data.

3. Regression diagnostics

While realistic combinations of data, models, and estimators counsel that estimators restricting the permissible influence of any small segment of data be given serious consideration, it is also useful to describe a complementary approach designed to detect influential observations through regression diagnostics. While weights obtained from bounded-influence estimation have very important di-

Table 3.1
Notation

Population regression $y = X\beta + \varepsilon$	Estimated regression $y = Xb + e$ and $\hat{y} = Xb$
y: $n \times 1$ column vector for response variable	same
X: $n \times p$ matrix of explanatory variables	same
β: $p \times 1$ column vector of regression parameters	b: estimate of β
ε: $n \times 1$ column vector of errors	e: residual vector: $y - \hat{y}$
σ^2: error variance	s^2: estimated error variance
Additional notation	
x_i: ith *row* of X matrix	$b(i)$: estimate of β when ith row of X and y have been deleted
X_j: jth column of X matrix	$s^2(i)$: estimated error variance when ith row of X and y have been deleted
$X(i)$: X matrix with ith row deleted	

agnostic content, alternative diagnostics that are more closely related to traditional least-squares estimation provide valuable information and are easier to understand.

An influential observation is one that has an unusually large impact on regression outputs, such as the estimated coefficients, their standard errors, forecasts, etc. More generally, influential data are outside the pattern set by the majority of the data in the context of a model such as linear regression and an estimator (ordinary least squares, for instance). Influential points originate from various causes and appropriate remedies vary accordingly (including, but not restricted to, bounded-influence estimation). Diagnostics can assist in locating errors, allowing the user to report legitimate extreme data that greatly influence the estimated model, assessing model failures, and possibly direct research toward more reliable specifications.[3]

Two basic statistical measures, individually and in combination, characterize influential data: first, points in explanatory-variable (X)-space far removed from the majority of the X-data, and scaled residuals which are, of course, more familiar diagnostic fare. We now turn to influential X-data, or *leverage* points. As described above, an influential observation may originate from leverage, large regression residuals, or a combination of the two. A notational summary is given by Table 3.1. We note that $b = (X^{\mathrm{T}}X)^{-1}X^{\mathrm{T}}y$ for OLS and call $H = X(X^{\mathrm{T}}X)^{-1}X^{\mathrm{T}}$ the *hat matrix* with elements $h_{ik} = x_i(X^{\mathrm{T}}X)^{-1}x_k^{\mathrm{T}}$. Then $\hat{y} = Xb = Hy$ which is how the hat matrix gets its name. We can also describe the predicted values as $\hat{y}_i = \sum_{k=1}^{n} h_{ik} y_k$. Using the above results the last expression can be rewritten as

[3]This section is a condensed summary of material in chapter 2 of Belsley, Kuh and Welsch (1980).

$\hat{y}_i = \sum_{k \neq i} h_{ik} y_k + h_{ii} y_i$. Thus, the impact of y_i on $\hat{y}_i$ is controlled by the corresponding diagonal element $h_{ii} (\equiv h_i)$.

There is also a direct distance interpretation of h_i. Let $\tilde{X}$ be the X-matrix centered by column means $\bar{x}$, so that $\tilde{h}_i = (x_i - \bar{x})(\tilde{X}^{\mathrm{T}}\tilde{X})^{-1}(x_i - \bar{x})^{\mathrm{T}}$. It is apparent that large $\tilde{h}_i$ are relatively far removed from the center of the data measured in the $(\tilde{X}^{\mathrm{T}}\tilde{X})$ coordinate system. For simple bivariate regression

$$\tilde{h}_i = \frac{(x_i - \bar{x})^2}{\sum_{k=1}^{n} (x_k - \bar{x})^2},$$

so that a large hat matrix diagonal corresponds in the most transparent fashion to a point removed from the center of the data.

Since H is a projection matrix, it has the following properties:

(i) $0 \leqslant h_i \leqslant 1$
(ii) $\sum h_i = p$.

Thus a perfectly balanced X-matrix – one with equal leverage for all observations – is one for which $h_i = p/n$. As further elaborated in Belsley, Kuh and Welsch (1980), when h_i exceeds $2\,p/n$ (and certainly when it exceeds $3\,p/n$), we are inclined to consider that row of data as potentially influential.

Relatively large residuals have long been viewed as indicators of regression difficulties. Since for spherically distributed errors the least-squares error variance for the ith observation is $\sigma^2(1-h_i)$, we will scale residuals by $\hat{\sigma}^2(1-h_i)$. Instead of using s (the sample standard deviation) estimated from all the data to estimate σ, we prefer to use $s(i)$ (the sample standard deviation excluding the ith row) so that the denominator is stochastically independent of the numerator. We thus obtain the *studentized residual*:

$$e_i^* \equiv \frac{e_i}{s(i)\sqrt{1-h_i}}. \tag{3.1}$$

This has the t-distribution when ε is normally distributed and, interestingly, $1-h_i$ provides a link between the regular OLS residual, e_i, and the predicted residual:

$$y_i - x_i b(i) = \frac{e_i}{1-h_i}. \tag{3.2}$$

Furthermore, the standardized predicted residual is just the studentized residual (3.1). Since the least-squares estimator works to reduce large observed residuals, especially at leverage points, residuals (however scaled) need to be augmented by leverage information.

One can observe the influence of an individual data row on regression estimates by comparing OLS quantities based on the full data set with estimates obtained when one row of data at a time has been deleted. The two basic elements, hat matrix diagonals and studentized residuals, reappear in these regression quantities which more directly reflect influential data. We will restrict ourselves here to two such row deletion measures: the predicted response variable, or fitted values, and coefficients. Thus, for fitted values $\hat{y}_i = x_i b$, we have

$$x_i b - x_i b(i) = x_i(b - b(i)) = \frac{h_i e_i}{1-h_i}.$$

We measure this difference relative to the standard error of the fit here estimated by $s(i)\sqrt{h_i}$, giving a measure we have designated

$$DFFITS_i = \frac{h_i e_i/(1-h_i)}{s(i)\sqrt{h_i}} = \left[\frac{h_i}{1-h_i}\right]^{1/2} e_i^*. \tag{3.3}$$

It is evident that the ith data point:

(i) will have no influence even if $|e_i^*|$ is large provided h_i is small, reinforcing our belief that residuals alone are an inadequate diagnostic, and
(ii) that substantial leverage points can be a major source of influence on the fit even when $|e_i^*|$ is small.

A second direct measure of influence is the vector of estimated regression coefficients when the ith row has been deleted:

$$DFBETA_i \equiv b - b(i) = \frac{(X^{\mathrm{T}}X)^{-1}x_i^{\mathrm{T}}e_i}{1-h_i}. \tag{3.4}$$

This can be scaled by $s(i)\operatorname{diag}\sqrt{(X^{\mathrm{T}}X)^{-1}}$ yielding an expression called *DFBETAS*. The expression for *DFBETA* closely resembles Hampel's definition of the influence function as described subsequently in Section 4. It is clear from inspection that *DFBETA* and the corresponding influence (4.7) are unbounded for OLS. We observe once again that (conditional on X) the absence (presence) of a data row makes a more substantial difference when $|e_i^*|$ is large and/or h_i is large.

There is another way of viewing the influence of the ith data row that is based on fitted values and h_i. If we define $\hat{y}_i(i) = x_i b(i)$, then it can be shown that $\hat{y}_i$ for the full data set is the following weighted average of $\hat{y}_i(i)$ and y_i:

$$\hat{y}_i = (1-h_i)\hat{y}_i(i) + h_i y_i. \tag{3.5}$$

When leverage is substantial for the ith row, the predicted quantity depends heavily on the ith observation. In the example of Section 7, the largest hat matrix diagonal in a sample of 506 observations is 0.29, so that one-fifth of 1 percent of the data has a weight of nearly $1/3$ in determining that particular predicted value. Such imbalance is by no means uncommon in our experience.

When several data points in X-space form a relatively tight cluster that is distant from the bulk of the remaining data, the single row deletion methods described here might not work well, since influential subsets could have their effects masked by the presence of nearby points. Then various multiple subset deletion procedures (which can, however, become uncomfortably expensive for large data sets) described in Belsley, Kuh and Welsch (1980) may be used instead. We have also found that *partial regression leverage plots* (a scatter diagram of residuals from y regressed on all but the jth column of X plotted against the residuals of column X_j regressed on all but the jth column of X; its OLS slope is just b_j) contain much highly useful qualitative information about the "masking" problem alluded to here. However, when we turn to bounded-influence regression, we find that the weights provide an alternative and valid source of information about subset influences. This fact enhances the diagnostic appeal of BIF.

4. Bounded-influence regression

In this section we sketch the main ideas behind the Krasker–Welsch bounded-influence estimator. More details may be found in Krasker and Welsch (1982a).

The notation which we will find most useful for our treatment of bounded-influence estimation is

$$y_i = x_i\beta + u_i, \qquad i = 1,\dots,n. \tag{4.1}$$

For the "central model" we will suppose that the conditional distribution of u_i, given x_i, is $N(0,\sigma^2)$. For reasons which were discussed in detail in earlier sections, one expects small violations of this and all the other assumptions of the model. Our aim is to present an estimator which is not too sensitive to those violations.

To study asymptotic properties such as consistency and asymptotic normality of estimators for β, one usually assumes

$$\operatorname*{plim}_{n\to\infty} \frac{1}{n} X^{\mathrm{T}}X = Q \tag{4.2}$$

or, equivalently,

$$\operatorname*{plim}_{n\to\infty} \frac{1}{n} \sum_{i=1}^{n} x_i^{\mathrm{T}}x_i = Q, \tag{4.3}$$

where Q is non-singular. This will hold provided $E(x^{\mathrm{T}}x)$ exists and is non-singular, where the expectation is over the marginal distribution of x.

When these assumptions hold, the OLS estimator b has many desirable properties. In particular, b is asymptotically efficient, with

$$\sqrt{n}\,(b-\beta)\rightarrow N(0,\sigma^2 Q^{-1}) \tag{4.4}$$

in distribution.

As we have seen, least squares is quite sensitive to certain violations of the assumptions. Though formal, less sensitive alternatives have not been widely used in applied work, econometricians often do attempt to protect themselves against model or data failures. For example, a common practice in applied work is to:

(i) run an OLS regression,
(ii) examine the observations with large residuals to determine whether they should be treated separately from the bulk of the data, and
(iii) run another OLS regression with observations deleted, or dummy variables added, etc.

In Section 3 we learned that this practice is not fully satisfactory, since influential observations do not always have large least-squares residuals. Conversely, a large residual does not necessarily imply an influential observation. If we replace the word "residuals" in (ii) by "$|DFFITS|$", the three-step procedure is much improved; and one might ask whether there is any real payoff to using a more formal procedure. The answer is that the simple procedure of examining those observations with large $|DFFITS|$ is not too bad in small samples, but one can do considerably better in large samples. We can explain this as follows: for any reasonable estimator, the variability goes to zero as the sample size goes to infinity. On the other hand, a process which generates gross errors will often generate them as a certain *proportion* of the data, so that the bias caused by gross errors will *not* go to zero as the sample size increases. In these circumstances, *bias will often dominate variability in large samples*. If the concern is with mean squared error, one must therefore focus more on limiting bias as the sample size increases. In small samples it suffices to examine only highly influential observations, since gross errors which are not too influential will cause only a small bias relative to the variability. In large samples, where the variability is very small, we must be suspicious of even moderately influential observations, since even a small bias will be a large part of the mean squared error. If one used the informal three-step procedure outlined above, these considerations would lead us to delete a larger and larger fraction of the data as the sample size increased. As stated in the introduction, it is better to have a formal procedure which smoothly downweights observations according to how influential they are.

We will now introduce two concepts, the influence function Ω and the sensitivity γ, which are applicable to an arbitrary estimator $\hat{\beta}$. Essentially, the influence $\Omega(y_i, x_i)$ of an observation (y_i, x_i) approximates its effect (suitably normalized) on the estimator $\hat{\beta}$, and γ is the maximum possible influence of a single observation. Our formal definition of influence is based on what is called the "gross error model".

Consider a process which, with probability $1-\varepsilon$, generates a "good" data point (y_i, x_i) from the hypothesized joint distribution. However, with probability ε, the process breaks down and generates an observation identically equal to some fixed (y_0, x_0) [a $(p+1)$-vector which might have nothing to do with the hypothesized joint distribution]. That is, with probability ε, the process generates a "gross error" which is always equal to (y_0, x_0). Under these circumstances the estimator $\hat{\beta}$ will have an asymptotic bias, which we can denote by $C(\varepsilon, y_0, x_0)$. We are interested mainly in how $C(\varepsilon, y_0, x_0)$ varies as a function of (y_0, x_0) for small levels of contamination, ε. Therefore, we define

$$\Omega(y_0, x_0) = \lim_{\varepsilon \downarrow 0} \frac{C(\varepsilon, y_0, x_0)}{\varepsilon}. \tag{4.5}$$

Note that $C(\varepsilon, y_0, x_0)$ is approximately $\varepsilon\Omega(y_0, x_0)$ when ε is small, so that $\varepsilon\Omega(y_0, x_0)$ approximates the bias caused by ε-contamination at (y_0, x_0). Ω is called the *influence function* of the estimator $\hat{\beta}$. If Ω is a bounded function, $\hat{\beta}$ is called a *bounded-influence* estimator.

For the least-squares estimator b, one can show that the influence function for b is

$$\Omega(y, x) = (y - x\beta)Q^{-1}x^{\mathrm{T}}, \tag{4.6}$$

where Q was defined in (4.2). Note that b is *not* a bounded-influence estimator.

The next thing we will do is define the estimator's sensitivity, which we want to think of as the maximum possible influence (suitably normalized) of a single observation in a large sample. The most natural definition (and the one introduced by Hampel) is

$$\max_{y,x} \|\Omega(y, x)\|, \tag{4.7}$$

where $\|\cdot\|$ is the Euclidean norm. The problem with this definition is that it depends on the units of measurement of the explanatory variables. If we change the units in which the explanatory variables are measured, we trivially, but necessarily, redefine the parameters; and the new influence function will generally not have the same maximum as the original one.

Actually, we want more than invariance to the units of measurement. When we work with dummy variables, for example, there are always many equivalent formulations. We can obtain one from another by taking linear combinations of the dummy variables. The list of explanatory variables changes, but the p-dimensional subspace spanned by the explanatory variables stays the same. This suggests that the definition of an estimator's sensitivity should depend only on the p-dimensional subspace spanned by the explanatory variables and not on the particular choice of explanatory variables which appears in the regression.

We can gain some insight into a more reasonable definition of sensitivity by considering the change in the fitted values $\hat{y} = X\hat{\beta}$. The effect on $\hat{y}$ of a gross error (y, x) will be approximately $X\Omega(y, x)$. The norm of this quantity is

$$\|X\Omega(y,x)\| = \{\Omega(y,x)^{\mathrm{T}}X^{\mathrm{T}}X\Omega(y,x)\}^{1/2}. \tag{4.8}$$

When $\hat{\beta}$ is invariant (so that $\hat{y}$ depends only on the subspace spanned by the p explanatory variables), expression (4.8) will also be invariant.

While (4.8) provides invariance, it only considers the effects of the gross error (y, x) on the fitted value $x\hat{\beta}$. If we are interested in estimating what would happen for new observations on the explanatory variables x_* we would want to consider the effect of the gross error on the estimated value, $x_*\hat{\beta}$.

We will be concerned when the effect of the gross error, $x_*\Omega(y, x)$, is large relative to the standard error $(x_*Vx^{\mathrm{T}}{}_*)^{1/2}$ of $x_*\hat{\beta}$, where V denotes the asymptotic covariance matrix of $\hat{\beta}$ and $\Omega(y, x)$ is its influence function. These considerations lead us to consider

$$\max_{y,x} \frac{|x_*\Omega(y,x)|}{(x_*Vx^{\mathrm{T}}{}_*)^{1/2}} \tag{4.9}$$

as our measure of sensitivity for the particular explanatory variable observations, x_*. However, we often do not know in advance what x_* will be, so we consider the worst possible case and use

$$\max_{y,x}\max_{x_*} \frac{|x_*\Omega(y,x)|}{(x_*Vx^{\mathrm{T}}{}_*)^{1/2}} = \max_{y,x}\{\Omega^{\mathrm{T}}(y,x)V^{-1}\Omega(y,x)\}^{1/2} \equiv \gamma \tag{4.10}$$

as our definition of sensitivity. This definition of sensitivity is also invariant to the coordinate system.

An estimator $\hat{\beta}$ is called a bounded-influence estimator if its influence function, Ω, is a bounded function; or, equivalently, if its sensitivity, γ, is finite. The bounded-influence property is obviously desirable when gross errors or other departures from the assumptions of the model are possible. In this section we will study weighted least-squares (WLS) estimators with the bounded-influence property.

Though OLS is usually expressed in matrix notation:

$$b = (X^{\mathrm{T}}X)^{-1}X^{\mathrm{T}}y, \tag{4.11}$$

it is more convenient for our purposes to use an earlier notation, the "normal equations":

$$0 = \sum_{i=1}^{n} (y_i - x_i b)x_i^{\mathrm{T}}. \tag{4.12}$$

A WLS estimator $\hat{\beta}$ is an estimator of the form

$$0 = \sum_{i=1}^{n} w_i \cdot (y_i - x_i\hat{\beta})x_i^{\mathrm{T}}. \tag{4.13}$$

(This could be expressed in matrix form as $\hat{\beta} = (X^{\mathrm{T}}WX)^{-1}X^{\mathrm{T}}WY$, where W is a diagonal matrix.) The weight w_i will depend on y_i, x_i, and $\hat{\beta}$ and will also depend on the estimated scale $\hat{\sigma}$ (see Section 6). The $w_i = w(y_i, x_i, \hat{\beta})$ will usually be equal to one, although certain observations may have to be downweighted if the estimator is to have the bounded-influence property.

One can show that under general conditions the influence function of a weighted least squares estimator is

$$\Omega(y, x) = w(y, x, \beta)(y - x\beta)B^{-1}x^{\mathrm{T}} \tag{4.14}$$

for a certain $p \times p$ matrix B, and the estimator's asymptotic covariance matrix will be

$$\begin{aligned} V &= \mathrm{E}\Omega\Omega^{\mathrm{T}} \\ &= \sigma^2 B^{-1}\left[Ew(y, x, \beta)^2\left(\frac{y - x\beta}{\sigma}\right)^2 x^{\mathrm{T}}x\right](B^{-1})^{\mathrm{T}} \\ &= \sigma^2 B^{-1}A(B^{-1})^{\mathrm{T}}, \end{aligned} \tag{4.15}$$

where the expectation is over the joint distribution of (y, x) and A is defined as the $p \times p$ matrix in the square brackets. It follows that

$$\gamma = \max_{y,x} \left\{ \Omega(y,x)^{\mathrm{T}} V^{-1} \Omega(y,x) \right\}^{1/2}$$

$$= \max_{y,x} \left\{ w(y,x,\beta)^2 (y - x\beta)^2 x (B^{-1})^{\mathrm{T}} \frac{1}{\sigma^2} B^{\mathrm{T}} A^{-1} B B^{-1} x^{\mathrm{T}} \right\}^{1/2}$$

$$= \max_{y,x} w(y,x,\beta) \left| \frac{y - x\beta}{\sigma} \right| \{ x A^{-1} x^{\mathrm{T}} \}^{1/2}. \tag{4.16}$$

This is a good point at which to recapitulate. First of all, we adopted a definition of sensitivity which essentially reflects the maximum possible influence, on linear combinations of the estimator $\hat{\beta}$, of a single observation. Since ordinary least squares has infinite sensitivity, we considered the more general class of weighted least-squares (WLS) estimators. We then derived an expression for the sensitivity of an arbitrary WLS estimator, which has a nice interpretation. We see that, apart from the weights, the influence of (y, x) has two components. The first is the normalized residual $(y - x\beta)/\sigma$. The second is the quadratic expression $xA^{-1}x^{\mathrm{T}}$, which should be thought of as the square of a robust measure of the distance of x from the origin.

Suppose that we desire an estimator whose sensitivity γ is $\leqslant a$, where a is some positive number. One reasonable way to choose from among the various candidate estimators would be to find that estimator which is "as close as possible" to least squares, subject to the constraint $\gamma \leqslant a$. By this we mean that we will downweight an observation only if its influence would otherwise exceed the maximum allowable influence. An observation whose influence is below the maximum will be given a weight of one, as would *all* the observations under least squares. In this way we might hope to preserve much of the "central-model" efficiency of OLS, while at the same time protecting ourselves against gross errors. Formally, suppose we require $\gamma \leqslant a$ for $a > 0$. If, for a given observation (y_i, x_i), we have

$$\left| \frac{y_i - x_i \hat{\beta}}{\hat{\sigma}} \right| \{ x_i A^{-1} x_i^{\mathrm{T}} \}^{1/2} \leqslant a, \tag{4.17}$$

then we want $w(y_i, x_i, \hat{\beta}) = 1$. Otherwise, we will downweight this observation just enough so that its influence equals the maximum allowable influence, i.e. we set $w(y_i, x_i, \hat{\beta})$ so that

$$w(y_i, x_i, \hat{\beta}) \left| \frac{y_i - x_i \hat{\beta}}{\hat{\sigma}} \right| \{ x_i A^{-1} x_i^{\mathrm{T}} \}^{1/2} = a. \tag{4.18}$$

The weight function must therefore satisfy

$$w(y_i, x_i, \hat{\beta}) = \min\left\{1, \frac{a}{\left|\frac{y_i - x_i\hat{\beta}}{\hat{\sigma}}\right| \{x_i A^{-1} x_i^{\mathrm{T}}\}^{1/2}}\right\}. \tag{4.19}$$

Recall that under our "central model", the conditional distribution of $(y - x\beta)/\sigma$, given x, is $N(0,1)$. Let η denote a random variable whose distribution, given x, is $N(0,1)$. Plugging (4.19) into the expression for A, we find

$$\begin{aligned} A &= \mathrm{E}\min\left\{1, \frac{a}{\left|\frac{y - x\beta}{\sigma}\right| \{xA^{-1}x^{\mathrm{T}}\}^{1/2}}\right\}^2 \left(\frac{y - x\beta}{\sigma}\right)^2 x^{\mathrm{T}}x \\ &= \mathrm{E}_x\left[\mathrm{E}_{\eta|x}\min\left\{\eta^2, \frac{a^2}{xA^{-1}x^{\mathrm{T}}}\right\}\right] x^{\mathrm{T}}x \\ &= \mathrm{E}_x r\left(\frac{a}{(xA^{-1}x^{\mathrm{T}})^{1/2}}\right) x^{\mathrm{T}}x, \end{aligned} \tag{4.20}$$

where

$$r(t) = \mathrm{E}_{\eta|x}\min\{\eta^2, t^2\}. \tag{4.21}$$

One can show that the matrix A satisfying (4.20) will exist only if $a > \sqrt{p}$. This suggests the following estimator for β. Choose $a > \sqrt{p}$. Find A to satisfy

$$A = \frac{1}{n}\sum_{i=1}^{n} r\left(\frac{a}{(x_i A^{-1} x_i^{\mathrm{T}})^{1/2}}\right) x_i^{\mathrm{T}} x_i; \tag{4.22}$$

then find b^* to satisfy

$$0 = \sum_{i=1}^{n} \min\left\{1, \frac{a}{\left|\frac{y_i - x_i b^*}{\hat{\sigma}}\right| \{x_i A^{-1} x_i^{\mathrm{T}}\}^{1/2}}\right\} (y_i - x_i b^*) x_i^{\mathrm{T}}. \tag{4.23}$$

One can show that b^*, which is called the *Krasker–Welsch estimator*, has the following properties.

(1) b^* is consistent and asymptotically normal when the assumptions of the model hold.

(2) The sensitivity γ of b^* equals a.
(3) Among all weighted least-squares estimators for β with sensitivity $\leqslant a$, b^* satisfies a necessary condition for minimizing asymptotic variance (in the strong sense that its asymptotic covariance matrix differs from all others by a non-negative definite matrix).

To fully define this estimator we need to specify a. We know that $a > \sqrt{p}$, providing a lower bound. Clearly when $a = \infty$, the bounded-influence estimator reduces to least squares. In practice we want to choose the bound a so that the efficiency of BIF would not be too much lower than the least-squares efficiency if we had data ideal for the use of least squares. This usually means that X is taken as given and the error structure is normal. The relative efficiency then would be obtained by comparing the asymptotic variances $\sigma^2(X^TX)^{-1}$ and $\sigma^2 V(a)$ where $V(a) = n^{-1}\sigma^2 B^{-1}(a)A(a)B^{-1}(a)$.

There is no canonical way to compare two matrices. The trace, determinant, or largest eigenvalue could be used. For example, the relative efficiency could be defined as

$$e(a) = \left\{\frac{\det\left[\sigma^2(X^TX)^{-1}\right]}{\det\left[\sigma^2 V(a)\right]}\right\}^{1/p} \tag{4.24}$$

and then a found so that $e(a)$ equals, say, 0.95. This means we would be paying about a 5 percent insurance premium by using BIF in ideal situations for least-squares estimation. In return, we obtain protection in non-ideal situations.

The computations involved in obtaining a for a given relative efficiency are complex. Two approximations are available. The first assumes that the X data comes from a spherically symmetric distribution which implies that asymptotically both the OLS covariance matrix and $V(a)$ will be diagonal, say $\alpha(a)I$. Then we need only compare $\sigma^2 I$ to $\sigma^2\alpha(a)I$ which means the relative efficiency is just $e(a) = \alpha^{-1}(a)$. This is much easier to work with than (4.25) but makes unrealistic assumptions about the distribution of X.

The simplest approach is to examine the estimator in the location case. Then $V(a)$ and X^TX are scalars. It is then possible to compute the relative efficiencies because the BIF estimator reduces to a simple form. When the a value for location is found, say a_L, we then approximate the bound, a, for higher dimensions by using $a = a_L\sqrt{p}$. Further details may be found in Krasker and Welsch (1982a) and Peters, Samarov and Welsch (1982).

We would now like to show briefly how the concepts of bounded-influence relate to the regression diagnostics of Section 3. Full details may be found in Welsch (1982). Consider again the "gross error model" introduced above. Assume that our "good" data are (x_k, y_k), $k \neq i$, and the suspected bad observation is

(x_i, y_i). Then we can show that the potential influence [what would happen if we decided to use (x_i, y_i)] of the ith observation on $b(i)$ is

$$\Omega(x_i, y_i, b(i)) = (n-1)[X^{\mathrm{T}}(i)X(i)]^{-1}x_i^{\mathrm{T}}(y_i - x_i b(i)) \tag{4.25}$$

$$= (n-1)(X^{\mathrm{T}}X)^{-1}x_i^{\mathrm{T}}(y_i - x_i b)/(1-h_i)^2. \tag{4.26}$$

Note the presence of the predicted residual (3.2) in (4.25).

The analog to V in (4.16) turns out to be

$$V(i) = (n-1)s^2(i)[X^{\mathrm{T}}(i)X(i)]^{-1}. \tag{4.27}$$

Therefore, the norm for our measure of sensitivity (4.11) is

$$[\Omega^{\mathrm{T}}(y_i, x_i, b(i))V^{-1}(i)\Omega(y_i, x_i, b(i))]^{1/2}$$

which, after some matrix algebra, is just

$$\left((n-1)\frac{h_i}{1-h_i}\frac{(y_i - x_i b(i))^2}{s^2(i)}\right)^{1/2} \tag{4.28}$$

or

$$\left((n-1)\frac{h_i}{1-h_i}\frac{(y_i - x_i b)^2}{s^2(i)(1-h_i)^2}\right)^{1/2}. \tag{4.29}$$

Comparing this to (3.3), we obtain that (4.29) is equivalent to

$$(n-1)^{1/2}|DFFITS_i|/(1-h_i)^{1/2}. \tag{4.30}$$

To bound the influence, we require that

$$\max_i (n-1)^{1/2}|DFFITS_i|/(1-h_i)^{1/2} \leqslant a,$$

which clearly implies that

$$(n-1)^{1/2}|DFFITS_i| \leqslant a.$$

The simple choice of a for BIF discussed above was $a_L\sqrt{p}$. For location,

$$DFFITS_i = \frac{n^{1/2}}{n-1}\frac{e_i}{s(i)},$$

and we might consider $(n-1)^{1/2}|DFFITS_i|$ large if it exceeded 2. Hence, a_L around 2 is good for diagnostic purposes.

Clearly (4.30) could have been chosen as our basic diagnostic tool. However, *DFFITS* has a natural interpretation in the context of least squares and therefore we feel it is easier to understand and to use.

5. Aspects of robust inference

When we estimate the coefficient vector β in the linear model $y_i = x_i\beta + u_i$, it is usually because we want to draw inferences about some aspect of the conditional distribution of y given x. In forecasting, for example, we need a probability distribution for the response variable, conditional on a particular x. Alternatively, we might want to know how the conditional expectation of the response variable varies with x.

In this section we analyze the problems that are created for inference by the fact that the linear model will never be exactly correct. To be sure, failures of linearity that occur for *extreme* values of the x-vector will always show up in a bounded-influence regression. However, *gradual* curvature over the entire range of X is much more difficult to detect. Moreover, departures from linearity in extreme regions of the x-space are sometimes very difficult to distinguish from aberrant data. Unfortunately, there are applications in which the distinction can be important.

To illustrate this point, consider the data plotted in Figure 5.1, and suppose that we are trying to predict y, conditional upon $x = 4$. Obviously, the outlier is crucial. If these were known to be good data from a linear model, then the outlier would be allowed to have a large effect on the prediction. On the other hand, if the outliers were known to be erroneous or inapplicable for some reason, one would base inferences about $(y|x=4)$ on the remaining nine observations. The prediction for y would be substantially higher in the latter case; or, more precisely, the probability distribution would be centered at a larger value.

There is a third possibility: namely that the true regression line is slightly curved. With the data in Figure 5.1 even a small amount of curvature would make the outlier consistent with the rest of the sample. Were such curvature permitted, one would obtain a prediction for $(y|x=4)$ lying between the two just mentioned.

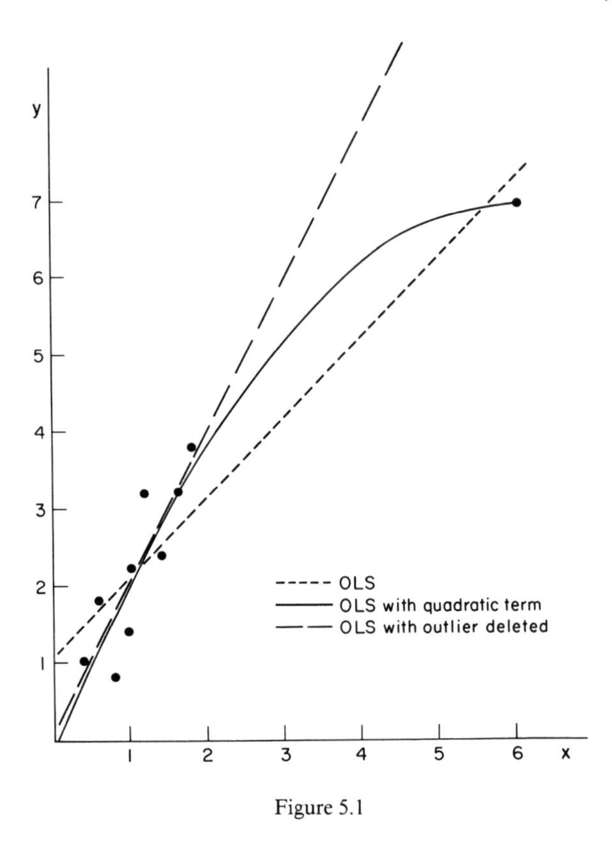

Figure 5.1

With data as simple as those in Figure 5.1, one could do a satisfactory job by computing three different fits and weighting them subjectively. The resulting probability distribution for y would be trimodal. However, this approach will not work with larger, multivariate regression problems. It seems that one has to use a model that builds-in the possibilities of bad data and curvature. As an illustration we will describe an approach proposed by Krasker (1981). Suppose that there are bad data occurring with probability ε, and that the good data are generated by a process satisfying

$$y_i = R(x_i, \theta) + u_i,$$

with $(u_i/\sigma | x_i, \theta, \sigma) \sim N(0,1)$.

For a bad (y_i, x_i) observation, suppose that $(y_i | x_i)$ has a density $h(y_i | x_i, \alpha)$ for some parameter α. In order to apply this approach one has to make specific choices for R and h. Krasker proposed a uniform distribution for h:

$$h(y|x,\alpha) \equiv \alpha \quad \text{for all } y.$$

Strictly speaking this is only a probability distribution when limited to an interval of length $1/\alpha$. The practical implication is that the parameter α ought to be small enough for all the y_i to lie in an interval of length $\leq 1/\alpha$. The uniform distribution reflects the notion that a gross error "could be anywhere".

For R, Krasker used

$$R(x,\beta,\gamma) = \beta_1 + \beta_2\left[\frac{\exp\{\gamma_2 x_2\}-1}{\gamma_2}\right] + \cdots + \beta_p\left[\frac{\exp\{\gamma_p x_p\}-1}{\gamma_p}\right].$$

This makes sense provided $\gamma_j \neq 0$ for all j. However.

$$\lim_{\gamma_j \to 0} \frac{\exp\{\gamma_j x_j\}-1}{\gamma_j} = x_j, \tag{5.1}$$

so that one can extend the definition of R even to $\gamma = 0$. Eq. (5.1) shows that when γ is small, $R(x,\beta,\gamma)$ is nearly linear in x.

Given x and the parameters, the density for y is the mixture

$$(1-\varepsilon)\frac{1}{\sigma}\phi\left(\frac{y-R(x,\beta,\gamma)}{\sigma}\right) + \varepsilon\alpha,$$

where ϕ is the density for $N(0,1)$. The likelihood function is

$$\prod_{i=1}^{n}\left\{(1-\varepsilon)\frac{1}{\sigma}\phi\left(\frac{y_i-R(x_i,\beta,\gamma)}{\sigma}\right) + \varepsilon\alpha\right\}.$$

The likelihood function will often be multimodal, for essentially the same reason that the subjective weighting of three fits for the data in figure 5.1 would lead to a trimodal probability distribution for y. Consequently, maximum likelihood is not adequate; one has to work with the entire likelihood function. Krasker's approach is Bayesian. Given a prior $p(\beta,\gamma,\sigma,\varepsilon,\alpha)$ for the parameters, one can find the posterior marginal distribution for any quantity of interest [such as some β_j or $R(x,\beta,\gamma)$] by numerical integration.

In most problems the priors on β and σ would be relatively diffuse. However, the prior information on the other parameters is sometimes crucial. This is particularly true for γ, which in a sense represents the amount of curvature in the regression surface. If, as is usually the case, the variables were transformed beforehand in order to make a linear model plausible, then one would choose a prior under which γ is near zero with high probability.

6. Historical background

In this section we provide a brief overview of the history and structure of bounded-influence estimation. For more background on classical robust estimation see Huber (1981) and Barnett and Lewis (1978, ch. 4). Koenker (1982) provides a good survey of robust procedures.

Huber (1973) proposed a form of robust regression based on a direct generalization of his work on the robust estimation of location parameters. His approach was to define

$$\rho_c(t)=\begin{cases} t^2/2 & |t|<c, \\ c|t|-c^2/2 & |t|\geq c, \end{cases} \tag{6.1}$$

and then minimize

$$d\sigma+\sum_{i=1}^{n}\sigma\rho_c[(y_i-x_i\beta)/\sigma] \tag{6.2}$$

with respect to β and σ. (The constant d is used to make the scale estimate consistent.) The influence function is

$$\Omega(y,x)=\psi_c[(y-x\beta)/\sigma]B^{-1}x^{\mathrm{T}}, \tag{6.3}$$

where $\psi_c(t)=\rho_c'(t)$ and B is a certain $p\times p$ matrix. Even though $\psi_c(\cdot)$ limits the effect of large residuals, the influence of (y,x) can be arbitrarily large because x^{T} multiplies $\psi_c(\cdot)$. This form of robust regression should be used with caution if there is a potential for outlying observations in the x data. Huber, and especially Hampel (1973), also stress this point.

Many other criterion functions like (6.1) can be considered [Holland and Welsch (1977)]. Those that have non-monotone $\psi(\cdot)$ functions are of special interest in regression because these functions are often zero for large residuals and hence remain zero when multiplied by x^{T}. However, they will not be bounded-influence estimators in all regions of the x-space and, because of the possibility of multiple solutions to (6.2), need a bounded-influence start to be effective.

Mallows (1973, 1975) proposed a way to construct bounded-influence estimators by, in essence, modifying (6.2) to read

$$d\sigma+\sum_{i=1}^{n}\sigma u(x_i)\rho_c[(y_i-x_i\beta)/\sigma] \tag{6.4}$$

for certain weights $u(x_i)$ which may depend on the entire X-matrix and not just

x_i. The influence function

$$\Omega(y,x)=u(x)\psi_c[(y-x\beta)/\sigma]B^{-1}x^{\mathrm{T}}, \tag{6.5}$$

where the B-matrix is not the same as in (6.3). If u is appropriately chosen, then Ω will be bounded. Some optimality results for this form are contained in Maronna, Bustos and Yohai (1979).

There is one problem that is immediately apparent. Outlying points in the X-space increase the efficiency of most estimation procedures. Any downweighting in X-space that does not include some consideration of how the y-values at these outlying observations fit the pattern set by the bulk of the data cannot be efficient.

In 1975, Schweppe [Handschin et al. (1975)] proposed essentially the form

$$d\sigma+\sum_{i=1}^{n}\sigma v^2(x_i)\rho_c[(y_i-x_i\beta_n)/\sigma v(x_i)], \tag{6.6}$$

with $v(x_i)=(1-h_i)^{1/2}$ and $h_i=x_i(X'X)^{-1}x_i^{\mathrm{T}}$. Again, (6.6) can provide bounded influence but with the additional property that if $(y-x\beta)/\sigma v(x)$ is small, the effect of $v(x)$ will be cancelled out. This has the potential to help overcome some of the efficiency problems outlined for the Mallows approach. Hill (1977) compared, via Monte Carlo, the Mallows and Schweppe forms along with several others and found that these two dominate, with the Schweppe form having an advantage.

Welsch (1977) tried a more direct approach to overcoming these efficiency problems. If the ith observation is an outlier in X-space (perhaps indicated by a large value of h_i), but (y_i, x_i) is consistent with the fit obtained from the rest of the data, $DFFITS_i$ [see (3.3)] would not be unduly large. Thus, Welsch proposed solving

$$\sum_{i=1}^{n}w_i(y_i-x_i\beta)x_i^{\mathrm{T}}=0,$$

where $w_i=w(DFFITS_i)$ and $w(\cdot)$ is a weight function such as

$$w_c(t)=\psi_c(t)/t. \tag{6.7}$$

This is just one iteratively-reweighted least-squares step of (6.6) with $v(x_i)=(1-h_i)/h_i^{1/2}$. Hinkley (1977), motivated by the jackknife, has proposed a similar class of estimators.

Without some additional criteria there is no way to choose among these approaches. A natural criterion, suggested by Hampel (1968), is to minimize the asymptotic variance subject to a bound on the influence function.

In the single parameter case this problem was solved by Hampel (1968). The multiparameter case was first considered in 1976 in several talks at Bell Laboratories by Hampel [see Hampel (1978) for further discussion] and at about the same time by Krasker during his doctoral research at M.I.T. [Krasker (1978, 1980)].

Krasker examined the very general class of estimators

$$0=\sum_{i=1}^{n}\phi(y_i,x_i,\beta_n) \tag{6.8}$$

for some function ϕ: $\mathbf{R}\times\mathbf{R}^p\times\mathbf{R}^p\to\mathbf{R}^p$ and showed that the estimator defined by

$$0=\sum_{i=1}^{n}\frac{\psi_c(|(y_i-x_i\beta_n)B^{-1}x_i^{\mathrm{T}}|)}{|(y_i-x_i\beta_n)B^{-1}x_i^{\mathrm{T}}|}(y_i-x_i\beta_n)x_i^{\mathrm{T}}, \tag{6.9}$$

with the side condition

$$B=\frac{-\partial}{\partial\beta}E\frac{\psi_c(|(y-x\beta)B^{-1}x^{\mathrm{T}}|)}{|(y-x\beta)B^{-1}x^{\mathrm{T}}|}(y-x\beta)x^{\mathrm{T}},$$

minimizes the trace of the asymptotic covariance matrix when a bound is placed on γ.

In Section 4 we restricted attention to estimators of the form

$$0=\sum_{i=1}^{n}w(y_i,x_i,\beta_n)(y_i-x_i\beta_n)x_i^{\mathrm{T}}, \tag{6.10}$$

where β_n is the estimate of β and the weight function w is non-negative, bounded, and continuous.

7. Bounded-influence estimates for a hedonic price index

Our example is drawn from a study by Harrison and Rubinfeld (1978), in which a hedonic price index for housing is estimated for use in a subsequent calculation of the marginal-willingness-to-pay for clean air.[4] Hedonic price indexes were introduced into recent econometrics literature by Griliches (1968). In essence, a hedonic price index is obtained from the fitted values in a regression where price is the response variable and the explanatory variables represent its qualitative

[4]Section 4.4 of Belsley, Kuh and Welsch (1980) provides a description of the Harrison–Rubinfeld problem as well as detailed regression diagnostics for it.

Table 7.1
Definition of model variables

Symbol	Definition
LMV	logarithm of the median value of owner-occupied homes
CRIM	per capita crime rate by town
ZN	proportion of a town's residential land zoned for lots greater than 25 000 square feet
INDUS	proportion of nonretail business acres per town
CHAS	Charles River dummy variable with value 1 if tract bounds on the Charles River
NOXSQ	nitrogen oxide concentration (parts per hundred million) squared
RM	average number of rooms squared
AGE	proportion of owner-occupied units built prior to 1940
DIS	logarithm of the weighted distances to five employment centers in the Boston region
RAD	logarithm of index of accessibility to radial highways
TAX	full-value property-tax rate (per $10 000)
PTRATIO	pupil–teacher ratio by town
B	$(Bk-0.63)^2$ where Bk is the proportion of blacks in the population
LSTAT	logarithm of the proportion of the population that is lower status

Table 7.2
OLS estimates: Housing-price equation

Variable	Coefficient estimate	Standard error	t-Statistic
INTERCEPT	9.758	0.150	65.23
CRIM	−0.0119	0.00124	−9.53
ZN	7.94×10^{-5}	5.06×10^{-4}	0.16
INDUS	2.36×10^{-4}	2.36×10^{-3}	0.10
CHAS	0.0914	0.0332	2.75
NOXSQ	−0.00639	0.00113	−5.64
RM	0.00633	0.00131	4.82
AGE	0.06×10^{-5}	5.26×10^{-4}	0.17
DIS	−0.191	0.0334	−5.73
RAD	0.0957	0.0191	5.00
TAX	-4.20×10^{-4}	1.23×10^{-4}	−3.42
PTRATIO	−0.0311	0.00501	−6.21
B	0.364	0.103	3.53
LSTAT	−0.371	0.0250	−14.83
$R^2=0.806$	$SER=0.182$		

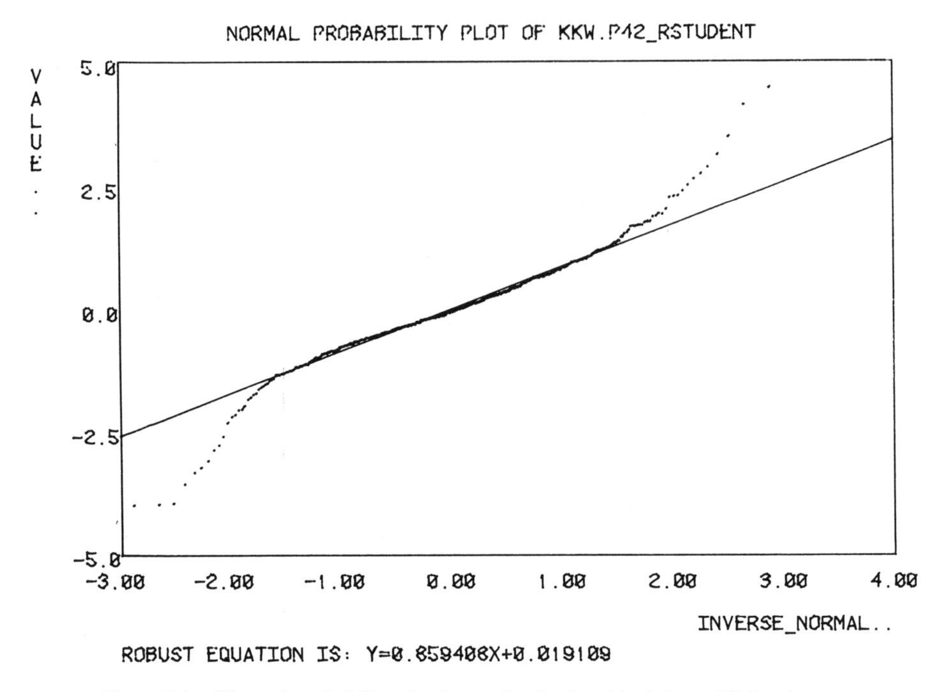

Figure 7.1. Normal probability plot for studentized residuals from OLS estimates; housing-price equation.

determinants. Harrison and Rubinfeld are principally interested in examining the impact of air pollution (as measured by the square of nitrogen oxide concentration, NOXSQ) on the price of owner-occupied homes. Thus, their hedonic price equation includes NOXSQ and thirteen other explanatory variables as indicators of qualities that affect the price of houses.

The basic data are a sample of 506 observations on census tracts in the Boston Standard Metropolitan Statistical Area (SMSA) in 1970[5] and the variables used are defined in Table 7.1. This study includes many socio-demographic variables at a relatively disaggregated level in common with many other current cross-section studies.

Table 7.2 reports least-squares estimates of eq. (7.1). The normal probability plot of studentized residuals in Figure 7.1 provides an instructive first diagnostic step. Substantial departures from normality are clearly evident, since large residuals are disproportionately present. Thus, at least for these observations, the potential exists for strongly influential observations.

[5]The original data together with a list of the census tracts appear in Belsley, Kuh and Welsch (1980).

7.1. *The model*

The hedonic housing-price model used by Harrison and Rubinfeld is

$$\begin{aligned} LMV = \beta_1 + \beta_2 CRIM + \beta_3 ZN + \beta_4 INDUS + \beta_5 CHAS + \beta_6 NOXSQ \\ + \beta_7 RM + \beta_8 AGE + \beta_9 DIS + \beta_{10} RAD + \beta_{11} TAX + \beta_{12} PTRATIO \\ + \beta_{13} B + \beta_{14} LSTAT + \varepsilon. \end{aligned} \tag{7.1}$$

A brief description of each variable is given in Table 7.1. Further details may be found in Harrison and Rubinfeld (1978).

7.2. *Partial plots*

Two partial-regression leverage plots (see the end of Section 3 for their description) reveal helpful information of both positive and negative import. Figure 7.2 for *NOXSQ*, a variable of major concern, reveals a scatter which is not obviously

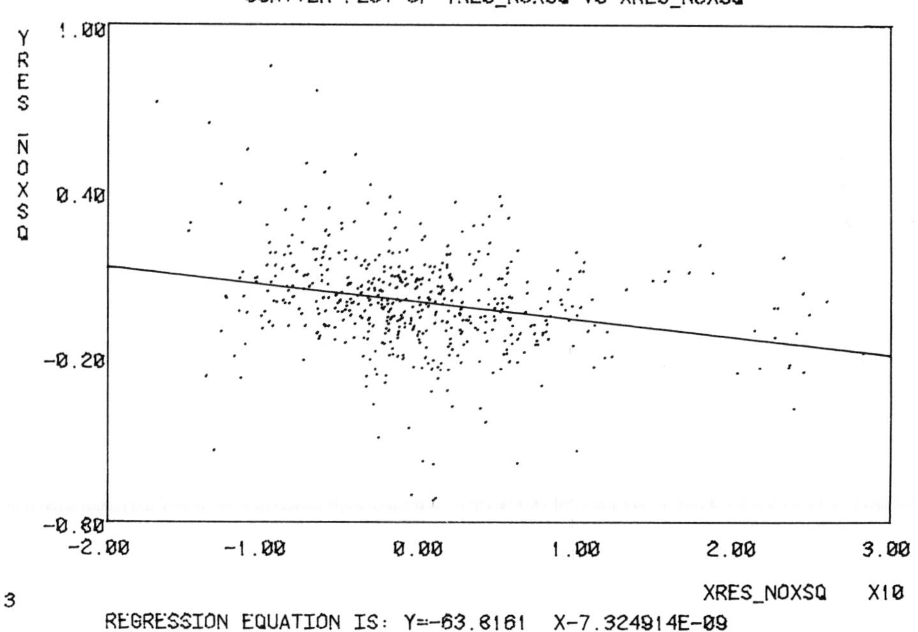

Figure 7.2. Partial-regression leverage plot for b_6 (*NOXSQ*), SE = 0.00113; housing-price equation.

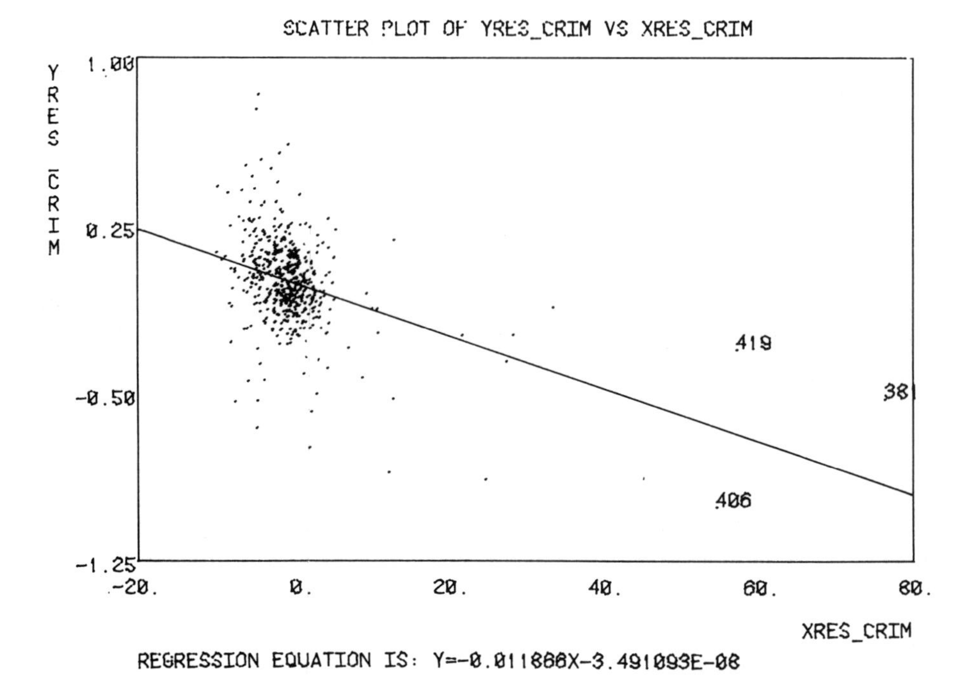

Figure 7.3. Partial-regression leverage plot for b_2 (*CRIM*), SE = 0.00124; housing-price equation.

dominated by extreme points. *CRIM*, shown in Figure 7.3 is another story. Three points near the center of Boston – 381 (Charlestown), 406 (South Boston), and 419 (Roxbury) – dominate this partial plot in terms of leverage.[6] Further investigation with deletion diagnostics, and then bounded-influence estimation, is clearly worthwhile. The willingness to pay for clean air function, W, mentioned in the first paragraph of this section, includes terms in both the predicted response variable and *NOXSQ*: $W_i = e^{\hat{y}_i}(-2b_6 NOX_i)$. We therefore ought to pay especially close attention to *DFFITS* and *DFBETAS*.

Table 7.3 contains row-deletion diagnostic information on 30 census tracts that exceeded stringent cutoffs for any of the following measures: hat-matrix diagonals, studentized residuals, *DFFITS*, or the *DFBETAS* for the two explanatory variables *NOXSQ* and *CRIM*.[7] *NOXSQ* is of special interest because its presence along with $\hat{y}$ (in this instance LMV) in the equation serves an important purpose

[6]The remaining plots, which were omitted to conserve space, do not portray strikingly influential observations comparable to that for *CRIM*.

[7]The cutoffs are more severe than those in Belsley, Kuh and Welsch (1980) since our objective here is to locate the most extreme influential points to compare them with related diagnostic information from bounded-influence estimation.

Table 7.3
Hat-matrix diagonals, studentized residuals, *DFFITS*, and *DFBETAS* for selected census tracts: housing-price equation[a]

Census tract	h_i	*RSTUDENT*	*DFFITS*	*DFBETAS* *NOXSQ*	*DFBETAS* *CRIM*
149	0.0485	1.773	0.4006	0.2609*	−0.0144
151	0.0456	0.980	0.2142	0.1612*	0.0026
153	0.0737	−1.312	−0.3701	−0.2013*	−0.0028
156	0.0840*	−0.427	−0.1292	−0.0650	0.0007
215	0.0579	2.910*	0.7214*	−0.0776	−0.0751
359	0.0520	1.250	0.2930	0.1364*	−0.0056
365	0.0891*	−2.747*	−0.8590*	−0.1780*	0.0649
366	0.0773	1.960	0.5671*	0.0549	−0.0875
368	0.0607	2.764*	0.7026*	−0.1952*	−0.0468
369	0.0982*	2.663*	0.8791*	−0.1256	−0.1105
372	0.0242	4.512*	0.7110*	−0.2709*	−0.1293
373	0.0532	4.160*	0.9856*	−0.1751*	−0.1264
381	0.2949*	2.559*	1.6551*	0.0975	1.5914*
386	0.0183	−2.564*	−0.3499	0.0113	−0.0567
398	0.0125	−3.212*	−0.3617	−0.0025	0.0928
399	0.0458	−3.301*	−0.7235*	−0.0168	−0.5774*
400	0.0213	−3.936*	−0.5808*	0.0132	0.1176
401	0.0225	−3.954*	−0.5995*	−0.0242	−0.3424*
402	0.0141	−3.988*	0.4766	−0.0204	−0.0636
406	0.1533*	2.141	−0.9112*	−0.0552	−0.8699*
410	0.0252	3.162*	0.5079*	−0.2685*	−0.0308
411	0.1116*	1.688	0.5983*	−0.1278	0.4130*
413	0.0477	3.520*	0.7878*	−0.3782*	−0.0060
414	0.0307	1.947	0.3470	−0.1554*	0.1783*
417	0.0387	−2.852*	−0.5724*	0.0473	0.1098
419	0.1843*	2.316	1.1009*	0.0352	1.0041*
427	0.0410	−1.956	−0.4012	0.1416*	0.0582
490	0.0514	−3.534*	−0.8225*	0.2957*	0.1797*
491	0.0527	−2.019	−0.4763	0.1760*	0.1107
506	0.0357	−3.070*	−0.5906*	−0.1193	−0.0547

[a]Starred values which exceed cutoff values: $h_i = 3(p/n) = 0.083$; *RSTUDENT* = 2.5; *DFFITS* = $3(\sqrt{p/n}) = 0.50$; *DFBETAS* = $3/\sqrt{n} = 0.133$.

of the authors, while *CRIM* is affected by some quite exceptional leverage points. The most influential point is clearly census tract 381 with a large hat-matrix diagonal and studentized residual. Its deletion alone causes $\hat{y}$ to shift by 1.66 standard deviations according to *DFFITS*.

Furthermore, the explanatory variables show severe imbalance in several instances. Since $p/n = 0.0275$ in the perfectly balanced case, the largest h_i of 0.2949 is more than ten times greater. Some other h_i are also severely out of balance. An excessive (relative to the normal distribution) number of large studentized residuals, first observed in the normal probability plot, are also apparent. This is a

Table 7.4
Distributions of diagnostic information from bounded-influence estimation of hedonic price index – centered and scaled robust distance based on *A*-matrix for *X* data

	99 percent EM	95 percent EM
13–14	0	1
12–13	0	0
11–12	1	0
10–11	0	1
9–10	0	1
8–9	0	0
7–8	2	0
6–7	0	1
5–6	1	0
4–5	0	1
3–4	4	6
2–3	18	16
1–2	129	128
0–1	351	351

situation where bounded-influence regression seems a natural alternative to least squares – initially as a diagnostic procedure and perhaps later to provide better regression estimates.

Two sets of bounded-influence estimates are presented next, for approximately 99 and 95 percent efficiency at the normal.[8] In the latter case, the limited influence allowed a given row of data would "cost" 5 percent efficiency if the normal error model prevails and leverage points correspond to completely valid elements of the model. The sensitivity bounds were $a_{0.99} = 12.0$ and $a_{0.95} = 8.0$, respectively. The 95 percent efficient model (hereafter 95 percent EM) will downweight a larger number of observations more heavily than the 99 percent EM, but if there are only a few very large influential points, even the less stringent 99 percent EM estimates will differ substantially from the OLS estimates.

We first examine the diagnostic content of the bounded-influence estimates. Table 7.4 describes the frequency distribution of robust distances. These are centered by the median distance and divided by 1.48 times the median absolute deviation from the median,[9] since the robust distances alone have no direct

[8]The TROLL program BIFMOD, written by Stephen Peters and Alexander Samarov, took a total of 19.5 CPU seconds on an IBM 370/168 for the more computer-intensive 95 percent efficiency model, at a cost well under $10. This total breaks down as follows: 1.9 CPU seconds for program loading and the OLS Initial start; 10.4 CPU seconds for 10 iterations to estimate the *A*-matrix; 7.2 CPU seconds for 54 iterations to estimate coefficients, scale, asymptotic covariance matrix and, of course, the weights. More details on the program and methods for computing efficiency may be found in Peters, Samarov and Welsch (1981).

[9]The 1.48 insures a consistent scale estimate for Gaussian data.

interpretation. Restricting comment to the 95 percent EM case we note that 479 observations, or 95 percent of the observations in all, are less than two robust standard errors from the median distance. Five observations are greatly distant from the median value of the robust distance; all of these correspond to large hat-matrix diagonals in the regression diagnostics with the exception of 415 which has a hat matrix diagonal of 0.067.

Table 7.5 presents individual diagnostic information for those observations with a final BIF weight less than one. Comment will be limited to the 95 percent EM and to a brief comparison with the regression diagnostics. First, the rank orderings of the largest hat-matrix diagonals and standardized robust distances are similar, but 415, 405, and 428 are now brought to our attention more forcefully. Second, while the two most influential bounded-influence observations (nos. 381 and 419) are also most influential according to *DFFITS* and the rank orderings are roughly alike, there are a few striking differences. The third most influential for BIF (no. 411) ranks 14th according to *DFFITS* and the third largest for *DFFITS* (no. 406) ranks 28th for BIF. Note that 406 has the third largest robust distance. From the structure of the bounded influence algorithm, the residuals associated with the BIF estimates must be extremely small for downweighting to be small in the presence of large leverage. It also appears that observations 370 and 415 were missed by the single row deletion diagnostics. This is especially true for 415 which does not appear in the first 100 rank ordered *DFFITS*.[10]

Regression coefficients for the bounded-influence estimates, their estimated standard errors, and t statistics (for the null hypothesis H_0: $\beta = 0$) appear in Table 7.6. The standard errors come from an asymptotic distribution, analogous to the way one finds standard errors for two-stage least squares or any non-linear estimator, or even for OLS when the disturbances are not exactly normal. Since the convergence to the asymptotic distribution can be slow when there are high-leverage X-rows, one must interpret these standard errors with some care. Of course, this comment applies also to OLS.

With a perfect model (including normal disturbances) one would expect the bounded-influence standard errors to exceed those of OLS, because bounded-influence will downweight some of the high-leverage observations which give OLS its central-model efficiency. However, bounded-influence can be *more* efficient than OLS if the disturbance distribution is heavy-tailed. We note that all coefficients change monotonically in the progression from least to most severe bounding of influence, i.e. from OLS to 99 percent EM to 95 percent EM. There is no certainty that monotonicity holds more generally. The coefficient of the key

[10]Multiple deletion methods, however, did turn up no. 415 as a potentially masked point [see Belsley, Kuh and Welsch (1980, p. 242)] thus confirming our expectation that BIF diagnostics offer effective alternatives to multiple deletion procedures.

Table 7.5
Bounded-influence diagnostic information about the most influential data tows: hedonic price index

99 percent EM				95 percent EM			
Standardized robust distance		Final Weights		Standardized robust distance		Final Weights	
Index	Value	Index	Value	Index	Value	Index	Value
381	11.14	381	0.231	381	13.68	381	0.086
419	7.98	419	0.301	419	10.07	419	0.103
406	7.10	373	0.489	406	9.02	411	0.165
411	5.01	411	0.511	411	6.28	369	0.208
369	3.68	369	0.517	415	4.42	373	0.228
365	3.30	365	0.558	369	3.73	365	0.252
415	3.26	413	0.579	405	3.57	368	0.254
156	3.12	490	0.591	428	3.36	413	0.264
343	2.99	368	0.618	365	3.33	490	0.298
366	2.90	399	0.670	156	3.30	366	0.307
163	2.84	372	0.673	399	3.13	399	0.317
153	2.72	215	0.736	163	2.97	372	0.334
371	2.70	401	0.771	366	2.97	401	0.360
284	2.49	366	0.777	153	2.94	370	0.397
405	2.47	400	0.836	343	2.89	414	0.399
428	2.47	406	0.853	371	2.84	415	0.403
164	2.46	506	0.856	143	2.65	400	0.433
162	2.45	417	0.865	164	2.57	215	0.437
143	2.44	410	0.940	284	2.56	417	0.440
157	2.42	370	0.978	370	2.56	506	0.464
370	2.39	491	0.989	157	2.55	410	0.481
				155	2.51	491	0.488
				162	2.45	402	0.520
				368	2.30	420	0.536
				161	2.06	371	0.548
				127	2.05	408	0.599
				124	2.00	467	0.635
				160	1.99	406	0.638
				373	1.94	375	0.661
				146	1.90	416	0.667
				258	1.89	8	0.673
				215	1.89	343	0.696
				147	1.88	398	0.725
				359	1.87	386	0.742
				148	1.82	149	0.748
				121	1.82	428	0.750
				123	1.81	367	0.786
				145	1.80	153	0.804
				125	1.80	427	0.829
				126	1.78	404	0.880
				319	1.77	182	0.896
				152	1.77	359	0.903
				122	1.76	412	0.923
				358	1.75	388	0.957

Table 7.6
Hedonic price index regression estimates

Estimated coefficients				Coefficient standard errors and t-statistics					
						Bounded-influence			
	Bounded-influence			OLS		99 percent EM		95 percent EM	
OLS	99 percent EM	95 percent EM	RHS variable	SE(b)	t-stat	SE(b)	t-stat	SE(b)	t-stat
-0.0119	-0.0143	-0.0158	*CRIM*	1.24×10^{-3}	-9.53	4.33×10^{-3}	-3.31	4.34×10^{-3}	-3.65
7.94×10^{-5}	7.52×10^{-5}	-2.39×10^{-5}	*ZN*	5.06×10^{-4}	0.16	3.63×10^{-4}	0.21	3.26×10^{-4}	-0.07
2.36×10^{-4}	3.98×10^{-4}	7.25×10^{-4}	*INDUS*	2.36×10^{-3}	0.10	1.68×10^{-3}	0.24	1.50×10^{-3}	0.48
0.0914	0.0863	0.0768	*CHAS*	0.0332	2.75	0.0301	2.87	0.0251	3.06
-6.39×10^{-3}	-5.86×10^{-3}	-4.84×10^{-3}	*NOXSQ*	1.13×10^{-3}	-5.64	1.18×10^{-3}	-4.97	1.04×10^{-3}	-4.67
6.33×10^{-3}	7.87×10^{-3}	0.0110	*RM*	1.31×10^{-3}	4.82	2.22×10^{-3}	3.55	1.67×10^{-3}	6.57
8.86×10^{-5}	-1.26×10^{-4}	-6.84×10^{-4}	*AGE*	5.26×10^{-4}	0.17	5.87×10^{-4}	-0.22	4.53×10^{-4}	-1.51
-0.191	-0.182	-0.165	*DIS*	0.0334	-5.73	0.0381	-4.78	0.0316	-5.21
0.0957	0.0922	0.0785	*RAD*	0.0191	5.00	0.0187	4.93	0.0152	5.15
-4.20×10^{-4}	-3.76×10^{-4}	-3.25×10^{-4}	*TAX*	1.23×10^{-4}	-3.42	1.14×10^{-4}	-3.30	9.56×10^{-5}	-3.40
-0.0311	-0.0305	-0.0290	*PTRATIO*	5.01×10^{-3}	-6.21	3.76×10^{-3}	-8.10	3.22×10^{-3}	-9.01
0.364	0.423	0.532	*B*	0.103	3.53	0.146	2.90	0.127	4.18
-0.371	-0.341	-0.284	*LSTAT*	0.0250	-14.83	0.0422	-8.09	0.0319	-8.91
9.76	9.71	9.64	*CONST*	0.150	65.23	0.156	62.35	0.132	73.18

variable *NOXSQ* follows the sequence −63.9; −58.6; −48.4 – differences that are large enough to cause big changes in the hedonic price calculations. However, since the fitted value also enters the willingness-to-pay function and other coefficients change, possibly in offsetting ways, a more detailed analysis would be needed to assess how much the bounded-influence calculations would ultimately affect the Harrison–Rubinfeld analysis.

Table 7.7

A. Difference between OLS and bounded-influence scaled by average of OLS and bounded-influence coefficient standard errors

RHS variable[a]	$\frac{b_{OLS}-b_{99}}{\frac{1}{2}(SE(b)_{OLS}+SE(b)_{99})}$	$\frac{b_{OLS-b95}}{\frac{1}{2}(SE(b)_{OLS}+SE(b)_{95})}$
**CRIM*	0.88	1.42
ZN	0.01	0.25
INDUS	0.08	0.25
CHAS	0.16	0.50
**NOXSQ*	0.46	1.43
**RM*	0.87	3.12
**AGE*	0.39	1.58
DIS	0.27	0.81
**RAD*	0.19	1.00
TAX	0.37	0.87
PTRATIO	0.15	0.52
**B*	0.48	1.46
**LSTAT*	0.89	3.05
CONST	0.32	0.86

B. Percentage difference between OLS and bounded-influence estimates

RHS variable[a]	$\frac{b_{OLS}-b_{99}}{b_{OLS}}$	$\frac{b_{OLS}-b_{95}}{b_{OLS}}$
**CRIM*	21%	33%
ZN	5	130
INDUS	69	207
CHAS	6	16
**NOXSQ*	8	24
**RM*	24	74
AGE	242	872
DIS	5	14
**RAD*	4	18
TAX	10	23
PTRATIO	2	7
**B*	16	46
**LSTAT*	8	23
CONST	0	1

[a]Values are starred when magnitude exceeds 1 for OLS − 95 percent EM differences, as explained in the text.

OLS and BIF estimates are compared in two ways in Table 7.7. First, Part A of Table 7.7 has their differences scaled by the average of the OLS and BIF coefficient standard errors. Percent differences, measured as the coefficient difference divided by the OLS coefficient, are shown in Part B of Table 7.7. Seven coefficients, including that for *NOXSQ*, change by one or more standard deviations using the 95 percent EM results; these changes have been starred. Two coefficients for the 95 percent EM model differ by more than three standard deviations. Next, the percent differences in Part B of Table 7.7 show that BIF estimation makes a large quantitative (as well as statistical) difference in the estimates, including those coefficients with the statistically most significant differences between them.

We noted above that $\hat{y}$ is an important part of the willingness-to-pay analysis based on the hedonic price index. More generally, the investigator is interested in comparing *ex-post* predictions among different procedures. As a first step we look at an index plot in Figure 7.4 of the point-by-point predicted value ratio, i.e. the $\hat{y}_i$ for OLS divided by $\hat{y}_i$ for the BIF 95 percent EM *after* conversion from logarithms into original units of dollar per dwelling unit. There are sizable numbers of large relative differences: the combined impact of coefficient changes

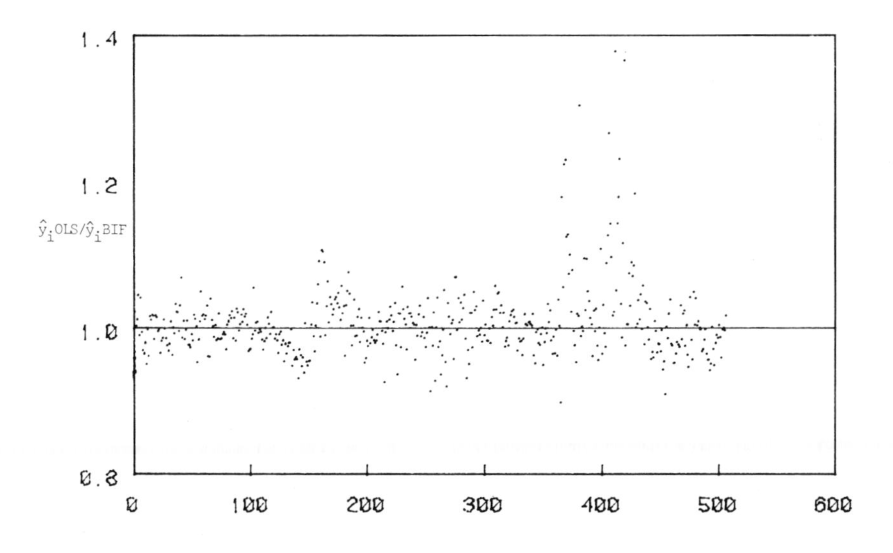

Figure 7.4. Ratio of OLS $\hat{y}_i$ to BIF $\hat{y}_i$: Census Tract Index for hedonic price equation.

sometimes *does* make a big difference to the predicted outcome. We also not that the largest differences occur when the BIF prediction is smaller than the OLS prediction which shows up in ratios well above unity; a number of predictions differ by more than 25 percent. Large differences are concentrated in the vicinity of Census Tracts 350–450. This points toward geographic concentration of large OLS–BIF differences, a fact on which we elaborate next in a related context.

A primary objective of bounding influence, we recall, is to limit bias, which tends to dominate variance in large samples. Upon noting that differences between OLS and BIF predictions are sometimes uncomfortably large, we ought to look for indications of systematic divergences between OLS and BIF estimates indicative of potential bias. This is readily achieved in the present instance by the scatter diagram in Figure 7.5 of BIF $\hat{y}_i$ against OLS $\hat{y}_i$ in logarithms that were used in the estimating equations. If the scatter is approximately uniformly distributed around a 45° line, bias can be treated as unimportant for practical purposes.

It is striking to observe in Figure 7.5 that the eight largest divergences all lie below the 45° line. It is furthermore clear that all but one of the most extreme points are below the average value of the response variable. Both effects are strongly suggestive of distortions or systematic differences indicative of potential bias in the OLS estimates. All the extreme points are in the center city, Boston.

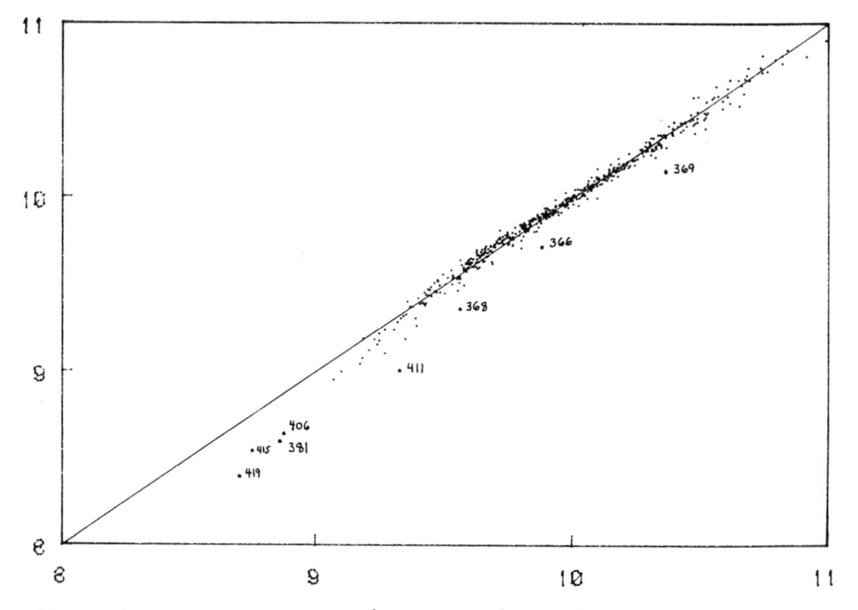

Figure 7.5. Scatter diagram of BIF $\hat{y}_i$ versus OLS $\hat{y}_i$: hedonic price equation.(*Note*: Both magnitudes are in natural logarithms.)

Observations 366, 368, and 369 are in Back Bay; Census Tract 381 is in Charlestown; 406 is in South Boston, 411 is in South Bay; 415 and 419 are in Roxbury. The close geographical proximity of these tracts is consistent with the hypothesis that different specifications hold for inner and outer city areas.[11]

We must emphasize that downweighted observations are not necessarily wrong, although some of them might be. Moreover, it is only under certain stringent assumptions that one could say that BIF is "better" than OLS. From the analysis done so far one can conclude only that certain observations seem to depart from the stochastic pattern of the bulk of the sample; moreover, those observations seem to be concentrated in the central city. Whether one should modify the model, or analyze the center-city observations separately, or simply assume that the anomalies are sampling error, is a question that cannot be decided by estimators alone. However, we feel it is highly advantageous to have an alternative fit that at least *reveals* the anomalies and the impact on estimated coefficients when their influence has been diminished.

How one chooses to treat influential observations, including the benign neglect of OLS, makes a sizable difference in this particular instance. Since many data sets that econometricians now prefer have related characteristics, this issue needs to be confronted.

8. Bounded-influence estimation with endogenous explanatory variables

So far we have dealt solely with cases in which, under the assumptions of the "central model", the conditional distribution of y_i given x_i is $N(x_i\beta, \sigma^2)$. In other words, we have assumed that there is a linear regression function whose parameters we wish to estimate. Quite commonly in economics, however, one can assume only that the conditional distribution of $y_i - x_i\beta$, given the values of the exogenous variables, is $N(0, \sigma^2)$. These assumptions are identical only if all the explanatory variables are exogenous, which is generally not true in simultaneous-equations models.

The most widely used estimator in simultaneous-equations models is two-stage least-squares (2SLS), which is a particular example of the instrumental variables (IV) approach which we will develop below. As we will see, 2SLS (or any IV estimator, for that matter) shares with ordinary least squares a very high sensitivity to failures of the underlying assumptions; and so we would like to have available a bounded-influence estimator for β. It turns out that the Krasker–Welsch estimator extends naturally, within the IV framework, to models with endogenous explanatory variables.

[11]The regression diagnostics in Belsley, Kuh and Welsch (1980, pp. 239 and 243) also pointed out systematic geographic concentrations of influential data.

We will suppose that we have an $n \times p$ matrix Z, whose columns are called instruments, and which (under the central model) satisfies the property mentioned above that the distribution of $y_i - x_i\beta$, given z_i, is $N(0, \sigma^2)$. The ordinary instrumental variables (IV) estimator for β is then defined by

$$b = (Z^{\mathrm{T}}X)^{-1}Z^{\mathrm{T}}y, \tag{8.1}$$

which can also be written as

$$0 = \sum_{i=1}^{n} (y_i - x_i b) z_i^{\mathrm{T}}. \tag{8.2}$$

2SLS is a special case of IV in which the instruments are formed by projecting the explanatory variables onto the space spanned by the model's exogenous variables. Let M be the $n \times p'$ matrix $(p' \geq p)$ of all the exogenous variables in the model, and define $Z = M(M^{\mathrm{T}}M)^{-1}M^{\mathrm{T}}X$. The 2SLS estimate is

$$b_{2\mathrm{SLS}} = (Z^{\mathrm{T}}Z)^{-1}Z^{\mathrm{T}}y, \tag{8.3}$$

which one can show is identical to the IV estimate in eq. (8.1) since $M(M^{\mathrm{T}}M)^{-1}M^{\mathrm{T}}$ is a projection matrix [see Theil (1971, p. 459)].

The influence function for the IV estimator turns out to be

$$\Omega(y, x, z) = (y - x\beta)Q^{-1}z^{\mathrm{T}}, \tag{8.4}$$

where

$$Q = Ez^{\mathrm{T}}x. \tag{8.5}$$

It is interesting to compare this influence function with the least-squares influence function in eq. (4.6). There, we noticed that the influence was a product of the disturbance $y - x\beta$ and a term $Q^{-1}x^{\mathrm{T}}$ which depended only on x and represented the "leverage" of that row of the X-matrix. For IV, the influence of a particular observation still depends on the disturbance. However, x now has no "independent" effect, but rather affects b *only* through the disturbance. The "leverage" term depends on z, not x (though some of the columns of X are also columns of Z). Even if the X-matrix were highly unbalanced, so that certain observations would have considerable least-squares leverage, it is possible for Z to be relatively balanced. This could happen, for example, if a "Wald Instrument" (consisting entirely of zeros and ones) were used as a column of Z.

A logical way to construct a bounded-influence estimator for β is to restrict attention to "weighted instrumental variables" (WIV) estimators of the form

$$0=\sum_{i=1}^{n} w_i\cdot(y_i-x_i\hat{\beta})z_i^{\mathrm{T}}. \tag{8.6}$$

The weights w_i will depend on y_i, x_i, z_i, $\hat{\beta}$, and $\hat{\sigma}$. One can show that the influence function is

$$\Omega(y,x,z)=w(y,x,z,\beta,\sigma)(y-x\beta)B^{-1}z^{\mathrm{T}} \tag{8.7}$$

for a certain non-singular $p\times p$ matrix B. The same argument which justified the Krasker–Welsch estimator can be used to suggest the following WIV estimator b^* for β. Choose $a>\sqrt{p}$. Find A to satisfy

$$A=\frac{1}{n}\sum_{i=1}^{n} r\left(\frac{a}{\left(z_iA^{-1}z_i^{\mathrm{T}}\right)^{1/2}}\right)z_i^{\mathrm{T}}z_i. \tag{8.8}$$

Finally, find b^* to satisfy

$$0=\sum_{i=1}^{n}\min\left\{1,\frac{a}{\left|\frac{y_i-x_ib^*}{\hat{\sigma}}\right|\left\{z_iA^{-1}z_i^{\mathrm{T}}\right\}^{1/2}}\right\}(y_i-x_ib^*)z_i^{\mathrm{T}}. \tag{8.9}$$

This estimator will have sensitivity equal to a, and analogous to the Krasker–Welsch estimator presented earlier, is in a sense as close to IV as possible subject to the constraint that the sensitivity be $\leq a$.

This bounded-influence weighted IV estimator provides the same powerful diagnostic information as the ordinary Krasker–Welsch estimator. However, many of its properties are not as well understood. We conjecture that it has maximum efficiency among all WIV estimators with the same sensitivity, though we have no proof as yet. Moreover, the process of approximating the distribution of b^* presents even more difficulties in the IV context than in ordinary regression. Further details may be found in Krasker and Welsch (1982b). An example of 2SLS regression diagnostics is contained in Kuh and Welsch (1980).

9. Resistant time-series estimation

In the previous sections we have said little about any special structure that might be present in the explanatory variable matrix, X. In particular, what are the consequences of having lagged endogenous and exogenous variables? Formally, the methods described above can be applied and useful diagnostic information obtained. However, potentially far more useful diagnostic information could be obtained if the time-series structure of the data were utilized.

Research on resistant estimation in the time-series context has lagged behind studies in resistant location and regression estimation; understandably so in view of the increased difficulties imposed by dependency between the data points. In the face of such complications, it seems imperative to study resistant time-series methods by first specifying simple outlier-generating models and focusing on single time series.

Martin (1979), generalizing the earlier work of Fox (1972), introduced two kinds of outliers for linear time-series models: innovations outliers (IO) and additive outliers (AO). Roughly speaking, IO correspond to problems with the error distribution, ε, and additive outliers to the gross errors discussed in previous sections.

The observational model considered by Martin (1980b) is of the following general form:

$$z_t = \mu + y_t + \delta_t, \qquad t = 1,2,\dots,n, \tag{9.1}$$

where μ is a location parameter, y_t is a zero-mean stationary ARMA (p, q) model:

$$y_t = \psi_1 y_{t-1} + \cdots + \psi_p y_{t-p} + \varepsilon_t + \theta_1 \varepsilon_{t-1} + \cdots + \theta_q \varepsilon_{t-q},$$

and ε_t has a symmetric distribution.

The IO model is obtained when $\delta_t \equiv 0$ and the ε_t are assumed to have a heavy-tailed non-Gaussian distribution. The AO model results when ε_t are Gaussian, δ_t independent of x_t, and $\text{prob}(\delta_t = 0) = 1 - \delta$ with δ positive and not too large. The problem of robust model selection for p-order autoregressive models was considered in Martin (1980a).

For autoregressive IO and AO models, Martin and Jong (1976), Denby and Martin (1979), and Martin (1980a) have shown that:

(a) The efficiency of the LS estimates in IO models decreases dramatically at "near-Gaussian" heavy-tailed distributions;
(b) Huber-type estimates (see Section 6) of the autoregressive parameters overcome the efficiency problem for finite variance IO models;
(c) both LS and Huber estimates lack resistance toward AO situations: and
(d) the time-series analog of the Mallows estimator (see Section 6) is resistant to both IO and AO situations.

Time-series analogs of the Krasker–Welsch estimator would also be resistant but no detailed comparisons have been made with the Mallows estimator. Thus, the bounded-influence approach is a good way to proceed for autoregressive models as well as regular regression models.

General ARMA models with additive outliers have been studied by Martin (1980b) and Martin, Samarov and Vandaele (1980). The later paper makes use of approximate conditional mean filters, a generalization of the Kalman filter. The method is, in fact, a special case of the following process [Huber (1980)].

(1) "Clean" the data by pulling outliers toward their fitted values (in the time series case by using a generalized Kalman filter).
(2) Apply LS to this adjusted data.
(3) Iterate (1) and (2) until convergence.

This method is most easily illustrated by using simple linear regression ($y_i = \alpha + \beta x_i + \varepsilon_i$) and the weights given in (4.19).

Assume that we have a fit ($\hat{y}$) to the data, i.e. a tentative regression line, an estimate of scale $\hat{\sigma}$, and an estimate of A [from (4.22)]. Then we "clean the data" by forming adjusted data,

$$\tilde{y}_i = \hat{y}_i + \left[\min\left(1, \frac{a}{\left|\frac{y_i - \hat{y}_i}{\hat{\sigma}}\right|\left(x_i A^{-1} x_i^{\mathrm{T}}\right)^{1/2}}\right)\right](y_i - \hat{y}_i), \tag{9.2}$$

and use LS on $\tilde{y}_i$(i.e. $\hat{\beta}_{\text{new}} = (X^{\mathrm{T}}X)^{-1}X^{\mathrm{T}}\tilde{y}$) to get the next tentative regression line.

Kleiner, Martin and Thompson (1979) treat the problem of robust estimation of power spectrum density. A robust method for dealing with seasonality in fitting ARMA models is presented in the robust seasonal adjustment procedure SABL (Seasonal Adjustment Bell Laboratories) due to Cleveland, Dunn and Terpening (1978a, 1978b).

10. Directions for further research

Economists often confront data and models (structural and stochastic) that are plagued by imperfections that can disproportionately influence estimates. Recent developments have led to model- and data-dependent weights for weighted least squares that bound the maximum permissible influence any row of data is allowed.

These iterative estimators have diagnostic content and, like other recently devised regression diagnostics, can be used to highlight data peculiarities and/or possible model failure. In addition, they can provide alternative estimates and predictions. The residuals from a bounded-influence fit are often more useful for assessing problems than the residuals from a least-squares fit. Bounded-influence procedures represent a significant advance over arbitrary use of dummy variables and/or judgmental elimination of data.

There are many interesting areas for further research. Bounded-influence estimators now exist for linear single equation models and linear instrumental variables models. Extensions to non-linear models, including logit and probit, remain to be fully worked out.

It is possible to consider bounding the influence of small departures from independent errors or small departures from a specified dependent error model. Work on resistant time-series estimation for full-scale econometric models is in its infancy. Extensions to more complex stochastic situations – variance component models for instance – would also be interesting.

The theoretical foundations for bounded influence are recent. We have focused on estimators that have a strong link to maximum likelihood (with a constraint to bound the influence) but other approaches – quantile estimation, pth power, and non-parametric regression – have their advocates. We feel that a new area of inquiry has been opened with basic assumptions that correspond more closely to the data and model properties encountered in the social sciences.

References

Andrews, D. F., P. J. Bickel, F. R. Hampel, P. J. Huber, W. H. Rogers and J. W. Tukey (1972) *Robust Estimates of Location*. Princeton University Press.

Annales de l'Insée, *The Econometrics of Panel Data* (1978) *Special Issue*, Issue 30–31, April–Sept. Paris.

Atkinson, Margaret and Jacques Mairesse (1978), "Length of life equipment in French manufacturing industries", *Annales de l'Insée*, 30–31, 23–48.

Barnett, V. and T. Lewis (1978) *Outliers in Statistical Data*. New York: John Wiley & Sons.

Belsley, David A., Edwin Kuh and Roy E. Welsch (1980) *Regression Diagnostics: Identifying Influential Data and Sources of Collinearity*. New York: John Wiley & Sons.

Chamberlain, Gary (1978) "Omitted Variable Bias in Panel Data: Estimating the Returns to Schooling", *Annales de l'Insée*, 30–31, 49–82.

Charnes, A., W. W. Cooper and R. O. Ferguson (1955) "Optimal Estimation of Executive Compensation by Linear Programming", *Management Science*, 1, 138–151.

Cleveland, W. S., D. M. Dunn and I. J. Terpenning (1978a) "A Resistant Seasonal Adjustment Procedure with Graphical Methods for Interpretation and Diagnosis", in: A. Zellner (ed.), *Seasonal Analysis of Economic Time Series*. U.S. Dept. of Commerce, Bureau of the Census.

Cleveland, W. S., D. M. Dunn and I. J. Terpenning (1978b) "The SABL Seasonal Adjustment Package-Statistical and Graphical Procedures", available from Computing Inf. Library, Bell Laboratories, Murray Hill, N. J.

Denby, L. and R. D. Martin (1979) "Robust Estimation of the First-Order Autoregressive Parameter", *Journal of the American Statistical Association*, 74, 140–146.

Diamond, Peter, Richard Anderson and Yves Balcer (1976) "A Model of Lifetime Earnings Patterns", in: *Report of the Consultant Panel on Social Security to the Congressional Research Service*, Appendix B, pp. 81–119.

Eisner, Robert (1978) "Cross Section and Time Series Estimates of Investment Functions", *Annales de l'Insée*, 30–31, 99–129.

Fair, Ray C. (1974) "On the Robust Estimation of Econometric Models", *Annals of Economic and Social Measurement*, 3.
Fisher, W. D. (1961) "A Note on Curve Fitting with Minimum Deviations by Linear Programming," *Journal of the American Statistical Association*. 56, 359–361.
Fox, A. J. (1972) "Outliers in Time Series", *Journal of the Royal Statistical Society*, B 34, 350–363.
Griliches, Z. (1968) "Hedonic Price Indexes for Automobiles: An Econometric Analysis of Quality Change", in: A. Zellner (ed.), *Readings in Economics and Statistics*. Little Brown.
Griliches, Zvi, Bronwyn H. Hall and Jerry A. Hausman (1978) "Missing Data and Self-Selection in Large Panels", *Annales de l'Insée*, 30–31, 137–176.
Hampel, F. R. (1968) "Contributions to the Theory of Robust Estimation", Ph.D. thesis, Berkeley.
Hampel, F. R. (1973) "Robust Estimation: A Condensed Partial Survey", *Zeitschrift für Wahrscheinlichkeitstheorie und Verw. Gebeite*, 27, 87–104.
Hampel, F. R. (1974), "The Influence Curve and Its Role in Robust Estimation", *Journal of the American Statistical Association*, 69, 383–393.
Hampel, F. R. (1978) "Optimally Bounding the Gross-Error-Sensitivity and the Influence of Position in Factor Space", in: 1978 *Proceedings of the Statistical Computing Section*. ASA, Washington, D.C. pp. 59–64.
Handschin, E., J. Kohlas, A. Fiechter and F. Schweppe (1975) "Bad Data Analysis for Power System State Estimation", *IEEE Transactions on Power Apparatus and Systems*, PAS-94, 2, 239–337.
Harrison, D. and D. L. Rubinfeld (1978) "Hedonic Prices and the Demand for Clean Air", *Journal of Environmental Economics and Management*, 5, 81–102.
Hill, R. W. (1977) "Robust Regression when there are Outliers in the Carriers", unpublished Ph.D. dissertation, Department of Statistics, Harvard University.
Hinkley, David V. (1977) "On Jackknifing in Unbalanced Situations", *Technometrics*, 19, 285–292.
Holland, P. W. and R. E. Welsch (1977) "Robust Regression Using Iteratively Reweighted Least-Squares", *Communications in Statistics*, A6, 813–827.
Huber, P. J. (1973) "Robust Regression: Asymptotics, Conjectures and Monte Carlo", *Annals of Statistics*, 1, 799–821.
Huber, Peter J. (1977) *Robust Statistical Procedures*. Philadelphia: SIAM.
Huber, P. J. (1981) *Robust Statistics*. New York: John Wiley & Sons.
Karst, O. J. (1958) "Linear Curve Fitting Using Least Deviations", *Journal of the American Statistical Association*, 53, 118–132.
Kleiner, R., R. D. Martin and D. J. Thomson (1979) "Robust Estimation of Power Spectra with Discussion", *Journal of the Royal Statistical Society*, B 41,
Koenker, Roger and Gilbert Basset, Jr. (1978) "Regression Quantiles", *Econometrica*, 46, 33–50.
Koenker, R. (1982) "Robust Methods in Econometrics", *Econometric Reviews* (to appear).
Krasker, William S. and Roy E. Welsch (1982a) "Efficient Bounded-Influence Regression Estimation", *Journal of the American Statistical Association*, 77, 595–604.
Krasker, William S. and Roy E. Welsch (1982b) "Resistant Estimation for Simultaneous-Equations Models Using Weighted Instrumental Variables", Unpublished manuscript, MIT Center for Computational Research in Economics and Management Science, Cambridge, MA.
Krasker, W. S. (1978) "Applications of Robust Estimation to Econometric Problems", unpublished Ph.D. thesis, Department of Economics, M.I.T.
Krasker, W. S. (1980) "Estimation in Linear Regression Models with Disparate Data Points", *Econometrica* 48, 1333–1346.
Krasker, W. (1981) "Robust Regression Inference", Unpublished manuscript, Graduate School of Business, Harvard University, Cambridge, MA.
Kuh, Edwin and Roy E. Welsch (1980), "Econometric Models and Their Assessment for Policy: Some New Diagnostics Applied to the Translog Energy Demand in Manufacturing", in: S. Gass (ed.), *Proceedings of the Workshop on Validation and Assessment Issues of Energy Models*. Washington, D.C.: National Bureau of Standards, pp. 445–475.
Mallows, C. L. (1973) "Influence Functions", talk at NBER Conference on Robust Regression, Cambridge, MA.
Mallows, C. L. (1975) "On Some Topics in Robustness", unpublished memorandum, Bell Telephone Laboratories, Murray Hill, New Jersey.
Marazzi (1980) "Robust Linear Regression Programs in ROBETH", Research Report No. 23,

Fachgruppe für Statistik, ETH, Zurich.
Maronna, R. A. (1976) "Robust M-estimators of Multivariate Location and Scatter", *Annals of Statistics* 4, 51–67.
Maronna, R. A. and V. J. Yohai (1980) "Asymptotic Behavior of General M-estimates for Regression and Scale with Random Carriers", *Zeitschrift für Wahrscheinlichkeitstheorie*, 58, 7–20.
Maronna, R. A., V. J. Yohai and O. Bustos (1979) "Bias- and Efficiency-Robustness of General M-estimators for Regression with Random Carriers", in: T. Gasser and M. Rosenblatt (eds.), *Smoothing Techniques for Curve Estimation*, Lecture Notes in Mathematics no. 757. New York: Springer Verlag, pp. 91–116.
Martin, R. D. (1979) "Robust Estimation for Time Series Autoregressions", in: R. L. Launer and G. Wilkinson (eds.), *Robustness in Statistics*. New York: Academic Press.
Martin, R. D. (1980a) "Robust Estimation of Autoregressive Models", in: Brillinger et al. (eds.), *Directions in Time Series*, Institute of Mathematical Statistics Publication.
Martin, R. D. (1980b) "Robust Methods for Time Series", in: *Proceedings of Nottingham International Time Series Meeting*. Amsterdam: North-Holland, Publishing Co.
Martin, R. D. and J. Jong (1976) "Asymptotic properties of robust generalized M-estimates for the first order autoregressive parameter", Bell Labs., Tech. Memo, Murray Hill, N.J.
Martin, R. D., A. Samarov and W. Vandaele (1981), "Robust Methods for ARIMA Models", M.I.T. Center for Computational Research in Economics and Management Science Technical Report no. 29, Cambridge, Mass.
McFadden, Daniel and Jeff Dubbin (1980) "An Econometric Analysis of Residential Electrical Appliance Holdings and Consumption", Department of Economics, Massachusetts Institute of Technology.
Meyer, J. R. and R. R. Glauber (1964) *Investment Decisions, Economic Forecasting and Public Policy*. Harvard Business School Press, Cambridge, Mass
Peters, Stephen C., Alexander Samarov and Roy E. Welsch (1982) "TROLL PROGRAM: BIF and BIFMOD", MIT Center for Computational Research in Economics and Management Science, Technical Report no. 30, Cambridge, MA.
Taylor, Lester D. (1974) "Estimation by Minimizing the Sum of Absolute Errors", in Paul Zarembka (ed.), *Frontiers in Econometrics*. New York: Academic Press, ch. 6.
Theil, H. (1971) *Principles of Econometrics*. New York: John Wiley & Sons, p. 459.
Wagner, H. M. (1959) "Linear Programming Techniques for Regression Analysis", *Journal of the American Statistical Association* 56, 206–212.
Welsch, R. E. (1977) "Regression Sensitivity Analysis and Bounded-Influence Estimation", Talk at NBER Conference on Criteria for Evaluation of Econometric Models, University of Michigan. Appears in J. Kmenta and J. Ramsey (1980) *Evaluation of Econometric Models*. New York: Academic Press, pp. 153–167.
Welsch, R. E. (1982) "Influence Functions and Regression Diagnostics", in: R. Launer and A. Siegel (eds.), *Modern Data Analysis*. New York: Academic Press, pp. 149–169.

Chapter 12

COMPUTATIONAL PROBLEMS AND METHODS

RICHARD E. QUANDT*

Princeton University

Contents

*I am indebted to David A. Belsley, Angus Deaton, Ray C. Fair, Stephen M. Goldfeld, Jerry A. Hausman, and Mark Plant for constructive comments.

Handbook of Econometrics, Volume I, Edited by Z. Griliches and M.D. Intriligator

1. Introduction

The very substantial growth in econometric and statistical theory in the last 30 years has been at least matched by the explosive growth of computer technology and computational methods and algorithms. For the average researcher 30 years ago it was a problem of some moment to need the inverse of a matrix of relatively small size, say 5×5. Many procedures that are routinely applied today were not even attempted, even if they had been thought of.

The impressive advances of hardware, software, and algorithmic technology since that time have significantly advanced the state of econometrics; they have, however, not been an unmixed blessing. On the one hand, new problems have emerged which can trap the unwary. On the other hand, there has occurred an increase in the capital/output ratio in research. It is difficult to escape the conclusion that, as a consequence, the average researcher today spends a higher fraction of his time in data management, computer-program writing and adaptation, in interpretation of masses of computed output and a lesser fraction of his time in reasoning about the underlying problem than did his predecessor.

The purpose of this chapter is to highlight some of the most important computational methods and problems of today. The emphasis is on algorithms and general procedures for solving problems and not on detailed implementation in concrete computer programs or systems. Hence, names familiar to many such as TSP, ESP, GREMLIN, TROLL, AUTOREG, SHAZAAM, etc. will not be discussed. For some classical approaches to numerical analysis the reader is referred to Hildebrand (1956). For detailed computer implementation see Carnahan, Luther and Wilkes (1969).

Section 2 is devoted to certain matrix methods involved in estimating the parameters of single and simultaneous equation models. Sections 3–7 cover various aspects of numerical optimization. These methods become relevant whenever the first-order conditions for a maximum are not linear in the parameters to be estimated. Section 3 gives a survey of the typical functions that are optimized. Section 4 discusses the basic theory of optimization. Section 5 covers special purpose algorithms and simplifications useful in econometrics; Section 6 considers some further aspects of algorithms. Section 7 deals with very particular difficulties encountered only in problems of certain types. Section 8 is devoted to numerical integration and Section 9 to random number generation.

The list is obviously incomplete and problems that are treated are covered only in broad outlines. An extensive bibliography refers the interested reader to many extensions.

2. Matrix methods

As is well known, many commonly used estimators of the coefficients of econometric equations are the solutions to equations of the form

$$A\hat{\beta} = c, \tag{2.1}$$

where $\hat{\beta}$ represents the k-element coefficient vector estimated, A is a $k \times k$ matrix (usually non-singular), c a k-element vector, and where A and c depend only on the data. Some examples are discussed below. For the pertinent econometric theory see Schmidt (1976), Theil (1971), and for computational aspects see Belsley (1974).

(1) Ordinary Least Squares. If the model is

$$Y = X\beta + u, \tag{2.2}$$

where Y and u are $n \times 1$ and X is $n \times k$ (and usually of rank k), then $A = X'X$ and $c = X'Y$.

If linear restrictions are imposed on β by

$$R\beta = r,$$

where R is $p \times k$ and of rank p, then $A = X'X$ as before and $c = X'Y + R'(R(X'X)^{-1}R')^{-1}(r - R(X'X)^{-1}X'Y)$. If the ridge estimator [Schmidt (1976)] is required instead, $c = X'Y$ as before but $A = X'X + sI$, where s is a constant.

(2) k-Class. Consider a full system of simultaneous equations

$$Y\Gamma + XB = U,$$

where Y and U are $n \times g$, Γ is $g \times g$ and non-singular, X is $n \times k$, and B is $k \times g$. To discuss single equations estimators consider the first equation of the system written as

$$y = Y_1\gamma + X_1\beta + u_{.1} = Z_1\delta + u_{.1},$$

where $Z_1 = [Y_1 \quad X_1]$, $\delta' = (\gamma' \quad \beta')$, and $u_{.1}$ is the first column of U. Then the following k-class estimators for δ are immediate from $A\hat{\delta} = c$. Let A be given by

$$A = \begin{bmatrix} Y_1'Y_1 - k_1V_1'V_1 & Y_1'X_1 \\ X_1'Y_1 & X_1'X_1 \end{bmatrix},$$

where $V_1 = Y_1 - X(X'X)^{-1}X'Y_1$ and c by

$$c = \begin{bmatrix} (Y_1' - k_2V_1')y \\ X_1'y \end{bmatrix}.$$

If $k_1 = k_2 = 1$, two-stage least squares results. If $k_1 = k_2 =$ the smallest eigenvalue λ of

$$|Y_1^{0\prime}Y_1^0 - Y_1^{0\prime}X_1(X_1'X_1)^{-1}X_1'Y_1^0 - \lambda(Y_1^{0\prime}Y_1^0 - Y_1^{0\prime}X(X'X)^{-1}X'Y_1^0)| = 0,$$

where $Y_1^0 = [y \quad Y_1]$, we obtain limited information maximum likelihood estimates. If $k_1 = k_2 = 1 + (k - k^* - g - 1)/n$, where k^* is the number of columns in X_1, we obtain Nagar's $O(n^{-1})$ unbiased estimator. Other estimators are obtained by choosing k_1 and k_2 to be unequal.

If W is a $(g_1 + k^*)\times n$ matrix of instruments uncorrelated with $u_{.1}$, instrumental variables estimators (as are the above) are given in general by setting $A = W'Z$ and $c = W'y$, which also includes the indirect least squares estimator.

(3) Three-stage least squares. Write the full system as

$$y_i = Z_i\delta_i + u_i, \qquad i = 1,\dots,g,$$

and define $y' = (y_1',\dots,y_g')$, $Z = \text{diag}(Z_i)$ a block-diagonal matrix with Z_i in the ith position, $\hat{\delta}_i$ as the two-stage least squares estimate of δ_i, and S the square matrix with (ij)th element $S_{ij} = (y_i - Z_i\hat{\delta}_i)'(y_j - z_j\hat{\delta}_j)/n$. Then if $A = Z'(S^{-1}\otimes X(X'X)^{-1}X')Z$ and $c = Z'(S^{-1}\otimes X(X'X)^{-1}X')y$, we have the three-stage least squares estimator.

2.1. *Methods for solving $A\hat{\beta} = c$*

The computation of each of the above estimators, as well as of many others, requires the inverse of A. Error in the inversion process accumulates as a result of rounding error in each computation. Rounding error, in turn, is due to the fact that the representation of numbers in a computer occupies a fixed number of places. In a binary computer floating point numbers are of the form $(\cdot a)(2^b)$, where a, the mantissa, and b, the characteristic, are binary integers stored in the computer and where the binary point "$\cdot$" and "2" are implied. The extent to which rounding error may affect the results is indicated by the condition number κ, which is the ratio of absolute value of the largest eigenvalue of A to the absolute value of the smallest [Golub (1969) and Jennings (1980)].[1] Various

[1]Since the matrix A is positive definite in all our examples, we may dispense with the absolute value in the definition of the condition number.

illustrative examples are provided by Golub (1969) and Wampler (1980). Consider as an example a case of OLS in which $A = X'X$ and assume

$$X = \begin{bmatrix} 1 & 1 & 1 & 1 \\ \varepsilon & 0 & 0 & 0 \\ 0 & \varepsilon & 0 & 0 \\ 0 & 0 & \varepsilon & 0 \\ 0 & 0 & 0 & \varepsilon \end{bmatrix}.$$

The eigenvalues of A are $4+\varepsilon^2$, ε^2, ε^2, and ε^2. If $\varepsilon < 2^{-t/2}$, where t is the number of binary digits in the mantissa of a floating point number, A will be a matrix with unity for each element and hence of rank 1 and not invertible. In general, the bound for the relative or proportionate error in the solution of an OLS problem is $\eta\kappa$, where η measures machine precision (e.g. 10^{-6}). Some principal matrix methods for controlling rounding error are discussed briefly below; for detailed application to econometric estimators see Belsley (1974), Golub (1969), and Wampler (1980). We illustrate the methods with reference to the ordinary regression model.

(1) Scaling. If the model is given by (2.2), it can also be written as $Y = Z\alpha + u$, where $Z = XB$ and B is a suitable diagonal matrix. The estimate for α is $\hat{\alpha} = (Z'Z)^{-1}Z'Y$ and $\hat{\beta} = B\hat{\alpha}$. Choosing b_{jj} as $[1/\sum_{i=1}^{n} x_{ij}^2]^{1/2}$ generally improves the conditioning of $Z'Z$.

(2) Cholesky factorization [Golub (1969), Klema (1973)]. If A is a positive definite matrix of order k, A may be factored so that

$$A = R'R, \tag{2.3}$$

where R is upper triangular. Error bounds for the factorization can be computed. Replacing A by $R'R$:

$$R'R\hat{\beta} = c,$$

and the solution proceeds in two steps: we first solve $R'\xi = c$ which is a triangular system and is solved easily; we next solve $R\hat{\beta} = \xi$ which is another triangular system. Cholesky factorizations for the $k \times k$ matrix A can be obtained in two ways [Golub (1969)]:

(a) Define

$$r_{11} = a_{11}^{1/2},$$
$$r_{1j} = a_{1j}/r_{11}, \qquad j = 2,\ldots,k,$$

and then let

$$r_{ii}=\left(a_{ii}-\sum_{p=1}^{i-1} r_{pi}^2\right)^{1/2}, \qquad i=2,\dots,k,$$

$$r_{ij}=\left(a_{ij}-\sum_{p=1}^{i-1} r_{pi}r_{pj}\right)\Big/ r_{ii}, \qquad \begin{array}{l} j=i+1,\dots,k, \\ i=2,\dots,k. \end{array}$$

(b) Define $a_{ij}^1=a_{ij}$ for all i, j. Then set

$$r_{pp}=\left(a_{pp}^p\right)^{1/2}, \qquad p=1,\dots,k,$$

$$r_{pj}=a_{pj}^p/r_{pp}, \qquad p=1,\dots,k, \quad j>k,$$

$$a_{ij}^{p+1}=a_{ij}^p-\frac{a_{pi}^p a_{pj}^p}{a_{pp}^p}, \qquad p=1,\dots,k, \quad i=p+1,\dots,k, \quad j\geqslant i.$$

The decompositions are themselves subject to rounding error and there is no guarantee that (b) can be completed even if A is positive definite.

(3) The QR decomposition [Belsley (1974), Golub (1969) and Jennings (1980)]. For all $n\times k$ matrices X there exists an $n\times n$ orthogonal matrix Q and a $k\times k$ upper triangular matrix R_1 such that

$$QX=\begin{bmatrix} R_1 \\ 0 \end{bmatrix}=R.$$

Partitioning $Q'=[Q_1 \quad Q_2]$ it follows that

$$X=Q_1R_1=Q'R.$$

The solution of $A\hat{\beta}=c$ in the ordinary least squares case is then particularly easy since $R_1'Q_1'Q_1R_1\hat{\beta}=R_1'Q_1'Y$ or $R_1\hat{\beta}=Q_1'Y$, which is triangular system. The relative error of its solution is small if the regression residuals are small and is given by $\eta_1\kappa^{1/2}+\eta_2\kappa(Y-X\hat{\beta})'(Y-X\hat{\beta})/\hat{\beta}'\hat{\beta}$, where η_1 and η_2 are functions of machine precision and κ is the condition number of $X'X$ [Jennings (1980)]. Moreover, $X'X=R'R$ and R is a Cholesky factorization of $X'X$. Two alternative methods are often employed to obtain the QR decomposition.

(a) The Householder transformation. Let $P=I-2vv'$, where v is a column vector and where $v'v=1$. Then P is a Householder transformation. Define $X^{(1)}=X$ and let $X^{(p+1)}=P^{(p)}X^{(p)}$, where $P^{(p)}=I-2v_pv_p'$, $v_p'v_p=1$, and v_p is chosen to make $X_{jp}^{(p+1)}=0$ for $j=p+1,\ldots,n$. Then $R=X^{(k+1)}$ and $Q=P^{(k)}P^{(k-1)}\ldots P^{(1)}$. For an application of Householder transformations to estimating regression coefficients subject to linear restrictions see Dent (1980).

(b) Gram–Schmidt orthogonalization. Two such procedures are in use: the classical and the modified methods. The former can be found in numerous algebra texts [Hoffman and Kunze (1961)]. The latter is preferred from the computational point of view, although in the absence of rounding errors they produce identical answers [Golub (1969)]. For the modified method replace $Q'R$ by PS, where S has unity on the diagonal and $P'P=$ diagonal. Now define

$$X^{(p)}=\left(p_1,\ldots,p_{p-1},x_p^{(p)},\ldots,x_k^{(p)}\right),$$

where p_i is the ith column of P and $x^{(p)}$ are columns defined below. Then at the pth step we let $p_p=x_p^{(p)}$ and set $d_p=p_p'p_p$, $s_{pr}=p_p'x_r^{(p)}/d_p$, and $x_r^{(p+1)}=x_r^{(p)}-s_{pr}p_p$ for $p+1\leqq r\leqq k$.

Some recent experimental results (Wampler (1980)) indicate that the QR method with either the Householder transformation or the modified Gram–Schmidt orthogonalization gives more accurate results than the Cholesky factorization. For application see Belsley (1974), Dent (1977), and Jennings (1980).

2.2. *Singular value decomposition* [Belsley (1974) and Chambers (1977)]

Any $n\times k$ matrix X can be decomposed as

$$X=U\Sigma V', \tag{2.4}$$

where the columns of U and V are orthonormal eigenvectors of XX' and of $X'X$, respectively, and where Σ is diagonal and contains the square roots (positive) of the eigenvalues of $X'X$ and XX'. If X has rank $r<k$, then (2.4) can be written with U as $n\times r$, Σ as $r\times r$, and V' as $r\times k$.

The singular value decomposition can be employed to compute the pseudoinverse of any matrix X, defined as X^+ satisfying (a) $XX^+X=X$, (b) $X^+XX^+=X^+$, (c) $(XX^+)'=XX^+$, and (d) $(X^+X)'=X^+X$. By substituting in (a) through (c) it can be shown that $X^+=V\Sigma^+U'$, where Σ^+ is the same as Σ except that its diagonal elements are the reciprocals of the non-zero diagonal elements of Σ.

Consider a regression model $Y=X\beta+u$ and the normal equations $X'X\hat{\beta}=X'Y$. Assume a case of exact multicollinearity so that the rank r of X satisfies $r<k$. Replacing X by its singular value decomposition leads to

$$VV'\hat{\beta}=X^+Y. \tag{2.5}$$

Substitution of $\hat{\beta} = X^{+}Y$ in the transformed normal equations (2.5) shows that they remain satisfied and that $X^{+}Y$ is a least squares estimate. It can be shown further that $\hat{\beta}$ has shortest length in the set of all least squares estimates. The singular value decomposition thus permits the computation of the shortest least squares coefficient vector in the presence of multicollinearity. It can also be employed for the computation, via the pseudoinverse, of least squares estimates subject to linear restrictions on the coefficients [Gallant and Gerig (1980)]. For the calculation of the singular value decomposition see Golub (1969) and Bussinger and Golub (1969).

2.3. Sparse matrix methods

In some applications, such as optimal control problems or in seemingly unrelated regression models, there may occur matrices in which the non-zero elements are a small fraction of the total number of elements. Computational efficiency can be gained by not storing and manipulating the matrices in their full size but only their non-zero elements and identification as to the location of these. The resulting techniques are called sparse matrix techniques [see Drud (1977/78) and Belsley (1980)]. Their use can result in dramatic reductions in computer time. Fair (1976) reports that the time required to evaluate the Jacobian in full-information maximum likelihood (see Section 3) was reduced by a factor of 28 when sparse methods were employed.

3. Common functions requiring optimization

The computation of econometric estimates characteristically requires the maximization or minimization of some function. Some of these possess first-order conditions that are linear in the parameters to be estimated, and the matrix techniques discussed in Section 2 have wide applicability in these cases. In many other instances, however, the first-order conditions for an optimum cannot be solved in closed form. In these cases one must either solve the equations representing the first-order conditions by numerical methods or apply numerical methods to the direct optimization of the function in question. The present section briefly outlines some of the principal types of objective functions.

3.1. Likelihood functions

Specific assumptions about the distribution of error terms characteristically permit the derivation of the likelihood function. Maximum likelihood estimates are desired because of their favorable asymptotic properties.

One of the most common models requiring numerical maximization for the attainment of maximum likelihood estimates is the linear simultaneous equations model

$$Y\Gamma + XB = U, \tag{3.1}$$

where Y is an $n\times g$ matrix of endogenous variables, X an $n\times k$ matrix of predetermined variables, U an $n\times g$ matrix of error terms, Γ a $g\times g$ non-singular matrix, and B a $k\times g$ matrix of coefficients. If it is assumed that the rows of U are distributed identically and independently as $N(0,\Sigma)$, where Σ is a $g\times g$ positive definite matrix, the likelihood function is

$$L=(2\pi)^{-gn/2}|\Sigma|^{-n/2}(\mathrm{abs}|\Gamma|)^{n}\exp\{-\tfrac{1}{2}\mathrm{tr}[\Sigma^{-1}(Y\Gamma+XB)'(Y\Gamma+XB)]\}, \tag{3.2}$$

where $|\cdot|$ denotes taking the determinant and where $|\Gamma|$ is the Jacobian of the transformation $U\to Y$ [Schmidt (1976)]. The logarithm of the condensed likelihood function is

$$\log L = \text{constant} - \frac{n}{2}\log|S| + \frac{n}{2}\log[\,|\Gamma|\,]^2, \tag{3.3}$$

where S has elements $s_{jk}=\sum_{i=1}^{n}\hat{u}_{ij}\hat{u}_{ik}$ and where $\hat{u}_{ij}$ is the ith residual in the jth equation. If the system is non-linear and is given by

$$f_j(y_i,x_i,\beta)=u_{ij},\qquad i=1,\dots,n;\quad j=1,\dots,g, \tag{3.4}$$

eq. (3.3) becomes

$$\log L = \text{constant} - \frac{n}{2}\log|S| + \tfrac{1}{2}\sum_{i=1}^{n}\log[\,|J_i|\,]^2, \tag{3.5}$$

where J_i is the Jacobian matrix corresponding to the ith observation with typical element $J_{ikl}=\partial u_{ik}/\partial y_{il}$. For a modification of (3.5) to perform robust estimation, see Fair (1974a). It should be noted that most linear simultaneous equations estimators that superficially might not be thought to be related to the maximization of (3.5) are in fact approximate solutions to the first-order conditions corresponding to (3.5) [Hendry (1976)].

Another very common example is provided by the ordinary regression model with error terms that obey a first-order Markov process $u_i=\rho u_{i-1}+\varepsilon_i$, $\varepsilon\sim N(0,\sigma^2 I)$. The log likelihood is

$$\log L = \text{constant} - \frac{n}{2}\log\sigma^2 + \tfrac{1}{2}\log(1-\rho^2) - \frac{1}{2\sigma^2}(Y-X\beta)'R'R(Y-X\beta), \tag{3.6}$$

where R is the matrix

$$R=\begin{bmatrix}(1-\rho^2)^{1/2} & 0 & 0 & \dots & 0 & 0\\ -\rho & 1 & 0 & \dots & 0 & 0\\ 0 & -\rho & 1 & \dots & 0 & 0\\ \dots & \dots & \dots & \dots & \dots & \dots\\ 0 & 0 & 0 & \dots & -\rho & 1\end{bmatrix}.$$

Additional specific likelihood functions are discussed in subsequent sections as necessary.

3.2. Generalized distance functions

A number of estimates are obtained by minimizing a suitable distance function. A simple example is the non-linear least squares estimator of the parameters of

$$y_i=f(x_i,\beta)+u_i, \tag{3.7}$$

obtained by minimizing

$$D=\sum_{i=1}^{n}(y_i-f(x_i,\beta))^2.$$

More complicated examples arise in simultaneous equation estimation.

If eqs. (3.4) are in reduced form,

$$y_j=g_j(x_j,\beta_j)+v_j,\qquad j=1,\dots,g, \tag{3.8}$$

where $y_j'=(y_{1j},\dots,y_{nj})$ and where x_j and β_j are the predetermined variables and coefficients in the jth equation, a non-linear two-stage estimator is given by minimizing

$$D=\left[Y_i-g_i(x_i,\beta_i)\right]'X(X'X)^{-1}X'\left[Y_i-g_i(x_i,\beta_i)\right].$$

Stacking the equations in (3.7) as

$$y=g(X,\beta)+V,$$

where $y'=(y_1',\dots,y_g')$, we obtain an analogue of three-stage least squares by minimizing.

$$D=[Y-g(X,\beta)]'\left[\hat{\Omega}^{-1}\otimes X(X'X)^{-1}X'\right][Y-g(X,\beta)],$$

where $\hat{\Omega}$ is a consistent estimate of Ω in $E(VV')=\Omega\otimes I$. [See Jorgenson and Laffont (1974), Berndt, Hall, Hall and Hausman (1974), and Parke (1979).]

3.3. *Functions in optimal control*

Consider a set of structural equations

$$f_j(y_i,x_i,z_i,\beta)=u_{ij},\qquad j=1,\ldots,g;\quad i=1,\ldots,n, \tag{3.9}$$

where the y_i are vectors of g endogenous variables to be controlled, x_i are vectors of exogenous variables, and z_i are vectors of control variables. Then the optimal control problem is to minimize some loss function $W(y_1,\ldots,y_n;\, x_1,\ldots,x_n;\, z_1,\ldots, z_n)$ subject to eqs. (3.9). A frequent assumption is that the loss function is quadratic as in

$$W=\sum_{i=1}^{n}(y_i-a_i)'K_i(y_i-a_i),$$

where the vectors a_i and matrices K_i are given [Fair (1974b), Chow (1975), and Chow and Megdal (1978)].

4. Algorithms for optimizing functions of many variables

4.1. *Introduction*

The present section deals with the fundamental ideas of optimization algorithms. Refinements and special problems encountered in individual cases are discussed in Sections 5, 6, and 7. For the sake of convenience we adopt the convention that functions are to be maximized; hence the problem is to

$$\text{maximize } F(x_1,\ldots,x_n) \tag{4.1}$$

with respect to the elements of the vector $x=(x_1,\ldots,x_n)$.[2] Under normal

[2]Obvious alterations of the algorithms to be discussed turn them into methods for minimizing functions. Equivalently, one may maximize $-F(x)$.

circumstances $F(x)$ is taken to be twice continuously differentiable; however, under some circumstances this assumption may be violated (see Section 7). Most often maximization is unconstrained and the present section is exclusively restricted to this case. Some techniques for dealing with constraints are discussed in Section 7. Since $\partial F/\partial x = 0$ is a necessary condition for maximizing $F(x)$, optimization methods can be adapted in a natural way to solving systems of equations.

Numerical methods of optimization characteristically assume that an initial value x^0 is given for vector of variables.[3] Algorithms are iterative procedures or sequences of steps with the kth step defined by

$$x^{k+1} = x^k + \lambda^k d^k, \tag{4.2}$$

where d^k is a direction vector and λ^k a suitable constant. Algorithms differ in the way in which they select λ^k and d^k.

The classification of algorithms could be based on numerous criteria. We adopt a simple classification according to whether the algorithm requires the evaluation of no derivatives, or of first partial derivatives, or of first as well as second partial derivatives.

Algorithms have many characteristics of interest and the choice of an algorithm represents a trade-off among these. Clearly, no "best" algorithm exists and the mix of characteristics possessed by an algorithm will vary from problem to problem to a greater or lesser extent. Two fundamental characteristics of algorithms are of interest here: (a) their robustness, i.e. the degree to which they are capable of providing an estimate $\hat{x}$ of the true maximum x^* such that $\|\hat{x} - x^*\| < \varepsilon$ for some prespecified positive ε, and (b) their cost. This latter measure is not uniquely given by the specification of the algorithm but is dependent on the actual charging scheme in effect for the various resources of a computer such as execution time, core, I/O requests, etc. Cost is frequently and heuristically taken to be proportional to the number of iterations (a concept not well defined when comparing different algorithms) or the number of function evaluations. In any event, the speed with which an algorithm can be expected to converge is a relevant consideration. An algorithm is said to be quadratically convergent if it attains the maximum of a quadratic function in a finite number of steps. Various criteria exist for defining the speed of convergence. One of these may be stated in terms of $c = \lim_k \sup|x^k - x^*|^{1/k}$. Convergence is sublinear, linear, or superlinear

[3]The choice of x^0 may itself be a non-trivial task. Clearly, even approximate information about the shape of the function is valuable in that convergence to the maximum is likely to be the faster the closer x^0 is to the location of the maximum. It is often asserted that in estimation problems x^0 must be a consistent estimate. This may well be essential for statistical reasons as in the computation of linearized maximum likelihood estimates [Rothenberg and Leenders (1964)], but is not necessary for computational reasons.

when x^k converges to x^* according to whether the asymptotic rate of convergence satisfies $c=1$, $0<c<1$, or $c=0$. Sublinear convergence to zero is provided by $1/k$, linear by 2^{-k}, and superlinear by k^{-k} [Brent (1973)]. The notion of quadratic convergence is important, for in the neighborhood of the maximum the function F is approximately quadratic in the following sense. Let the Hessian matrix of $F(x)$ be $G(x)=[\partial^2F(x)/\partial x_i\partial x_j]$ and let G satisfy the Lipschitz condition

$$|G(x^1)-G(x^2)| \leqq M\|x^1-x^2\|$$

for all x^1, x^2 in some domain R of F containing x^* in its interior, where $\|x^1-x^2\|$ is the Euclidean norm and M is a matrix of constants and where $|G(x^1)-G(x^2)|$ denotes a matrix the elements of which are the absolute values of $\partial^2F(x^1)/\partial x_i\partial x_j - \partial^2F(x^2)/\partial x_i\partial x_j$. Then

$$F(x)=F(x^*)+\tfrac{1}{2}(x-x^*)'G(x^*)(x-x^*)+Q(x) \tag{4.3}$$

for $x\in R$, where $|Q(x)|\leqq M\|x-x^*\|^3$. For x sufficiently near x^* the first two terms on the right-hand side of (4.3) provide a good approximation to $F(x)$.

4.2. *Methods employing no derivatives*

In principle, such methods are appealing because the computation of derivatives is almost always computationally costly. Nevertheless, relatively few algorithms of this type are in frequent use, particularly on problems of more than moderate size.

One class of derivative-free algorithms employs the notion of searching on a suitable grid of lattice points. A simple procedure is to start at some point x^0 and evaluate the function at x^0 and at the $2n$ lattice points $x^0\pm he_i$, where e_i $(i=1,\dots,n)$ is a vector with unity in the ith position and zeros elsewhere and where h is the preassigned lattice width. A step is taken from x^0 to x^1, where x^1 is the value of $x^0\pm he_i$ for which $F(x^1)=\sup F(x^0\pm he_i)$. The procedure is repeated starting from x^1 until no improvement is found for the given value of h. The value of h is then reduced and the search renewed. When h is finally reduced to the preassigned level of accuracy, the search is terminated and the last value of x taken as the location of the maximum. An algorithm in this class is that of Berman (1969).

Although the above algorithm is guaranteed to converge to a local maximum, in practice it is prohibitively expensive to employ. A different and more efficient version of search algorithms is that of Hooke and Jeeves (1961). The Hooke and Jeeves algorithm employs exploratory moves which are parallel to the coordinate axes and pattern moves which represent the average direction of several past moves together. If an exploratory move and a subsequent pattern move together result in function improvement, they are both accepted; otherwise only an exploratory move is made. Computation again begins with a prespecified value of h and ends when h has been reduced to the desired accuracy.

Search methods do have advantages over methods using (first and second) derivatives. These are the assurance of eventual convergence and their independence of the concavity or convexity of the function $F(x)$. Nevertheless, in practice they are not employed frequently. They tend to converge slowly even in the immediate vicinity of the location of a maximum and, as a rule, are computationally very expensive. An even more serious problem is that algorithms that change only one variable at a time may fail to converge altogether. Consider the simple algorithm that changes at each iteration one variable according to

$$F(x^{k+1}) = \max_{x} F\left(x_1^k,\dots,x_{k-1}^k, x, x_{k+1}^k,\dots,x_n^k\right).$$

Methods of this type are in common use; see for example the Cochrane–Orcutt iterations used to maximize (3.6). These methods frequently work well if precautions are taken to terminate iterations when function improvement becomes small. Nevertheless, the gradient may remain strictly positive over the path taken by an algorithm and Powell (1973) has given examples in which this algorithm could cycle indefinitely around the edges of a hypercube.

An alternative direct search method is the Simplex method of Nelder and Mead (1965).[4] The function is first evaluated at the $n+1$ vertices $x^0,\dots,x^n$ of an (irregular) simplex in the space R^n of variables. The corresponding function values, denoted by F_i $(i=0,\dots,n)$, are assumed to be ordered $F_n > F_{n-1} >,\dots,> F_0$. Among the points thus examined, x^n is currently the best, x^0 the worst. Compute the centroid c of the points not including the worst: $c=\sum_{j=1}^{n} x^j/n$. The steps of the algorithm are as follows:

(1) Reflect the simplex about the subsimplex given by $x^1,\dots,x^n$ by choosing a point $x^r = c+\alpha(c-x^0)$ where $\alpha>0$ is a coefficient chosen for the algorithm. If F_r, the function value corresponding to x^r, is such that $F_1 < F_r < F_n$, then x^r replaces x^0 and we return to Step 1.

(2) If $F_r > F_n$, then the simplex may profitably be stretched in the direction of x^r and an x^s is defined by $x^s = c+\beta(x^r-c)$, where $\beta>1$ is a coefficient chosen for the algorithm. If $F_s > F_n$, x^s replaces x^0. Otherwise x^r replaces x^0. In either event we return to Step 1.

[4]Not to be confused with the simplex method of linear programming. See also Swann (1972) and, for variants and computational experience, Parkinson and Hutchinson (1972).

(3) If $F_r < F_1$, then the simplex should be contracted. A positive $\gamma < 1$ is chosen and x^c set to $c + \gamma(x^0 - c)$ if $F_r < F_0$ and to $c + \gamma(x^r - c)$ if $F_r > F_0$. If $F_c > \max(F_r, F_0)$, x^c replaces x^0 and we return to Step 1. Otherwise the points other than the best point x^n are shrunk toward x^n by a preselected proportion and we return to Step 1.

The algorithm is useful because it does not require derivatives. Unfortunately, its performance depends on the values of the various (expansion, contraction, reflection) coefficients and it is not easy to develop sound intuition as to desirable values.

An even more useful algorithm is the conjugate gradient method of Powell (1964). The basic motivation of this method is its behavior in the case of quadratic functions and its application to more general functions rests on analogy, or at least the heuristic observation, that near a maximum well-behaved functions are approximately quadratic.[5]

Two direction vectors, p and q, are said to be conjugate relative to a symmetric matrix A if $p'Aq = 0$. The essence of the algorithm is a sequence of n linear searches of the function in n linearly independent, mutually conjugate directions. Assume that n such directions, $d_1^k, \dots, d_n^k$, are given at the beginning of the kth iteration and that the most recent estimate of the location of the maximum is x^k. The steps of an iteration are as follows.

(1) Calculate values ν_r $(r = 1, \dots, n)$ sequentially such that $F(x^k + \Sigma_{j=1}^r \nu_j d_j^k)$ is a maximum.

(2) Replace d_r^k by d_{r+1}^k $(r = 1, \dots, n-1)$.

(3) Replace d_n^k by $\Sigma_{j=1}^n \nu_j d_j^k$.

(4) Calculate ν such that $F(x^k + \Sigma_{j=1}^n \nu_j d_j^k + \nu(\Sigma_{j=1}^n \nu_j d_j^k))$ is a maximum and let x^{k+1} be given by $x^{k+1} = x^k + \Sigma_{j=1}^n \nu_j d_j^k + \nu(\Sigma_{j=1}^n \nu_j d_j^k)$.

The justification of the algorithm rests upon its convergence in the case of quadratic functions $F(x) = x'Ax + b'x + c$ and is established by the following theorems due to Powell (1964).

Theorem 4.1

Let $d_1, \dots, d_m$, $m \leqq n$, be mutually conjugate directions in a subspace of dimension m and let x^0 be the starting point in that subspace. Then the maximum of the quadratic function $F(x)$ in the subspace is found by searching along each direction only once.

[5]For details beyond those provided here see also Goldfeld and Quandt (1972), Brent (1973), Murray (1972), and Fletcher (1965).

Proof

The location of the maximum can be written $x^0+\sum_{i=1}^{m}\nu_i d_i$ and parameters ν_i are chosen so as to maximize $F(x^0+\sum_{i=1}^{m}\nu_i d_i)$. Substituting $x^0+\sum_{i=1}^{m}\nu_i d_i$ into the quadratic it can be seen that terms involving $d_i'Ad_j$ vanish by the assumption of conjugacy. Hence, the maximum with respect to ν_i does not depend on the value of any ν_j, $j\neq i$, proving the assertion.

Theorem 4.2

Let x^0 and x^1 be the locations of the maxima when the function is searched twice in the direction d from two starting points. Then the direction x^1-x^0 is conjugate to d.

Proof

Any point x is the location of the maximum in the direction d if

$$\frac{\partial}{\partial\nu}F(x+\nu d)=0\quad \text{at } \nu=0.$$

Performing the differentiation and alternately substituting x^1 and x^0 yields

$$\left(2(x^1)'A+b'\right)d=\left(2(x^0)'A+b'\right)d=0$$

or

$$(x^1-x^0)'Ad=0.$$

The convergence of the algorithm can then be proved by induction. Assume that on the kth iteration of the algorithm the last k directions searched were mutually conjugate. The x^k which is the starting point of the next iteration and the $x^k+\sum\nu_j d_j^k$ defined in it represent maxima involving the same search directions, hence their difference is also conjugate to the previously conjugate directions by the parallel property stated in Theorem 4.2. Thus, after two iterations two conjugate directions exist, and after n iterations n such directions will exist, each of which will have been searched once. Q.E.D.

The conjugate gradient method is usually initiated by taking the columns of an identity matrix as the search directions. In practice it is often a useful method, although it has been conjectured that for problems in excess of 10–15 variables it may not perform as well. The principal reason for this may be [see Zangwill (1967)] that at some iteration the optimal value of ν_i in the linear search may be zero. The resulting set of directions $d_1,\dots,d_n$ then become linearly dependent and henceforth the maximum can be found only over a proper subspace of the original n-space. Near linear dependence and slow convergence can occur if ν_i is approximately zero. There are at least three devices for coping with this, with no clear evidence as to which is preferable.

(1) If the search directions become nearly linearly dependent, we may reset them to the columns of the identity matrix.

(2) We may skip Step 3 of the algorithm and search again over the same n directions used previously.

(3) We may replace the matrix of direction vectors with a suitably chosen orthogonal matrix [Brent (1973)]. These vectors are computed on the assumption that $F(\cdot)$ is quadratic and negative definite as follows.

Let A be the matrix of the (approximating) quadratic function. A is generally unknown (although it could be obtained at significant cost by evaluating the Hessian of F). Let D be the matrix of direction vectors. Then, since the directions are mutually conjugate with respect to A,

$$D'AD = M, \tag{4.4}$$

where M is diagonal with negative diagonal elements. The linear search in each of the n directions may be accomplished by evaluating $F(x^k + \sum_{i=1}^{j} \nu_i d_i^k)$ at three points ν_j^1, ν_j^2, and ν_j^3 $(j = 1,\dots,n)$ and fitting a parabola to the function values (see Section 6). This involves computing the second differences of the function values which are easily shown to be

$$d_i'Ad_i = \mu_i, \tag{4.5}$$

where d_i is a column of D and μ_i is a diagonal element of M. Define $R = D(-M)^{1/2}$ and $H = A^{-1}$. Then $H = DM^{-1}D' = -RR'$. Since D and M are known from the iteration, H can be computed. It remains to compute Q such that

$$Q'HQ = M^{-1} \tag{4.6}$$

and the columns of Q are orthogonal eigenvectors of A. If the quadratic approximation is good, the resulting search directions are conjugate to a matrix that is approximately the true Hessian and, hence, convergence can be expected to be fast. In order to avoid bad rounding errors in the computation of eigenvectors for a badly conditioned matrix it may be desirable to find the singular value decomposition $Q'R'S$ of the matrix R', where Q is the matrix of directions sought.

4.3. Methods employing first and second derivatives

A reasonable starting point for very general methods is to approximate $F(x)$ by a second-order Taylor approximation about x^0:

$$F(x) \simeq F(x^0)+g(x^0)'(x-x^0)+\tfrac{1}{2}(x-x^0)'G(x^0)(x-x^0), \tag{4.7}$$

where $g(x^0)$ denotes the gradient of $F(x)$ evaluated at x^0. Maximizing F by setting its partial derivatives equal to zero yields

$$g(x^0)+G(x^0)(x-x^0)=0, \tag{4.8}$$

or, replacing x^0 by the current value of x at the kth iteration and replacing x by x^{k+1}, the new value sought is

$$x^{k+1}=x^k-\left[G(x^k)\right]^{-1}g(x^k), \tag{4.9}$$

which forms the basis of iterating according to Newton-type algorithms.[6] A very general class of algorithms is obtained by writing (4.9) as

$$x^{k+1}=x^k-\lambda^k H^k g(x^k), \tag{4.10}$$

where λ^k is a suitable constant and H^k is a matrix. Eq. (4.10) is of the same general form as (4.2) with $-H^k g(x^k)$ being the search direction. It can be shown that search direction d^k guarantees an improvement in the function if and only if it can be written as $-H^k g(x^k)$, with the matrix H^k being negative definite [Bard (1974)]. Numerous choices are available for λ^k as well as H^k; $\lambda^k=1$ and $H^k=[G(x^k)]^{-1}$ yields Newton's method. It is a method with the best asymptotic rate of convergence $c=0$.[7] It is, however, clearly expensive since it requires the evaluation of n first and $n(n+1)/2$ second derivatives. Moreover, (4.8) corresponds to a maximum only if the second-order conditions are satisfied, i.e. if $G(x^k)$ is a negative definite matrix. Obviously this may be expected to be the case if x^k is near the maximum; if not, and if $G(x^k)$ is not negative definite, iterating according to (4.9) will move the search in the "wrong" direction. A much simpler alternative is to set $H^k=-I$. The resulting method may be called the steepest ascent method. It locally always improves the value of the function but tends to

[6]Chow (1968, 1973) recommended this method for maximizing the likelihood for systems of simultaneous linear equations. Instead of directly maximizing the likelihood, he suggested the method for solving the first-order condition. It is also called the Newton–Raphson method. See also Hendry (1977) for various applications.

[7]See Parke (1979). Parke also discusses the asymptotic rates of convergence of the steepest ascent and univariate search methods. See also Dennis and Moré (1977).

behave badly near the optimum in that it tends to overshoot (indeed, for arbitrary fixed λ it is not guaranteed to converge) and near ridges in that it induces motion that is orthogonal to the contours of the function; these directions may well be nearly orthogonal to the desirable direction of search. Newton's method is useful precisely where the steepest ascent method is likely to fail. If $-G$ is positive definite, we have the decompositions

$$G=\sum_{i=1}^{n}\lambda_i P_i P_i', \qquad G^{-1}=\sum_{i=1}^{n}P_i P_i'/\lambda_i, \tag{4.11}$$

where the λ_i are the eigenvalues and the P_i the orthogonal eigenvectors of G. The eigenvectors point in the direction of the principal axes of the ellipsoid defined by $-y'Gy=$ constant and the quantities $(-1/\lambda_i)^{1/2}$ give their lengths. Since the eigenvectors are linearly independent, we can write $g=\sum_{i=1}^{n}\beta_i P_i$. Hence, the step defined by (4.9) can be expressed as

$$-\left[\sum_{i=1}^{n}P_i P_i'/\lambda_i\right]\sum_{j=1}^{n}\beta_j P_j=-\sum_{j=1}^{n}\beta_j P_j/\lambda_j.$$

If one of the λ's, say the kth, is very small, i.e. if the quadratic approximation defines ellipsoids that are highly elongated in the direction P_k, then the component P_k receives a weight proportional to $1/\lambda_k$ and the step will be nearly parallel to the ridge.

Several modifications exist for coping with the possibility that G might not be negative definite.

(1) Greenstadt (1967) replaces G by $-\sum_{i=1}^{n}|\lambda_i|P_i P_i'$.

(2) Marquardt (1963) suggests replacing G by $G-\alpha A$, where α is a small positive constant and A is a diagonal matrix with $a_{ii}=|G_{ii}|$ if $G_{ii}\neq 0$ and $a_{ii}=1$ otherwise.

(3) In maximum likelihood problems, in which $\log L$ is to be maximized, it may be possible to compute the value of $[E(\partial^2\log L/\partial\theta\partial\theta')]^{-1}$, where θ is the vector of variables with respect to which one wishes to maximize. Setting H^k equal to this matrix yields the method of scoring [Rao (1973), and Aitcheson and Silvey (1960)].

(4) In non-linear least squares problems [see eq. (3.7)] the objective function is $D=\sum_{i=1}^{n}(y_i-f(x_i,\beta))^2$. The second derivative matrix is

$$\frac{\partial^2 D}{\partial\beta\partial\beta'}=2\sum_{i=1}^{n}\left[\frac{\partial f(x_i,\beta)}{\partial\beta}\frac{\partial f(x_i,\beta)}{\partial\beta'}-(y_i-f(x_i,\beta))\frac{\partial^2 f(x_i,\beta)}{\partial\beta\partial\beta'}\right]. \tag{4.12}$$

If H^k is set equal to the first term of (4.12), it is guaranteed to be positive definite and the resulting method is known as the Gauss or Gauss–Newton method [Goldfeld and Quandt (1972), and Bard (1974)].

A quadratic hill-climbing algorithm due to Goldfeld, Quandt and Trotter (1966) attacks the non-negative-definiteness of G directly and replaces G by $G-\alpha I$, where α is chosen so that $G-\alpha I$ is negative definite. In practice $\alpha = 0$ when G is negative definite and $\alpha > \lambda_{\max}$, where $\lambda_{\max}$ is the largest eigenvalue of G, when G is not negative definite. The justification for the algorithm is based on the behavior of quadratic functions and is contained in the following theorems.[8] Let $Q(x)$ be an arbitrary quadratic function of x, let $Q'(x)$ denote the vector of first partial derivatives, and $Q''(x)$ the matrix of second partial derivatives. Define the iteration

$$x^{k+1} = x^k - \left(Q''(x^k) - \alpha I\right)^{-1} Q'(x^k) \tag{4.13}$$

and

$$r(\alpha) = \|x^{k+1} - x^k\|.$$

Then the following are true:

Theorem 4.3

For any α such that $Q''(x^k) - \alpha I$ is negative definite and any x such that $\|x - x^k\| = r(\alpha)$, $Q(x^{k+1}) \geqq Q(x^k)$.

Theorem 4.4

For all $Q'(x^k) \neq 0$, the radius $r(\alpha)$ is a strictly decreasing function of α for all $\alpha > \lambda_{\max}$.

Theorem 4.5

Let $R_\alpha = \{x \mid \|x - x^k\| \leqq r(\alpha)\}$, and assume $Q'(x^k) \neq 0$. Then the maximum of $Q(x)$ on R_α is at the boundary point x^{k+1} if $\alpha \geqq 0$ and at the interior point $x^k - [Q''(x^k)]^{-1} Q'(x^k)$ otherwise.

The algorithm thus basically works as follows: at each step $G(x^k)$ is examined for negative definiteness. If G is negative definite, a step equal to $-[G(x^k)]^{-1} g(x^k)$ is taken.[9] Otherwise the step taken is $-[G(x^k) - \alpha I]^{-1} g(x^k)$, where α is taken to be $\lambda_{\max} + \rho \|g(x^k)\|$. The quantity ρ is itself adjusted from

[8] For proof see Goldfeld, Quandt and Trotter (1966).

[9] In practice, the direction $-[G(x^k)]^{-1} g(x^k)$ is computed and a one-dimensional line search is performed since line searches are computationally efficient ways of improving the function value.

iteration to iteration since the radius $r(\alpha) \leqq \rho^{-1}$. At each step the actual improvement in the function is compared with the improvement in the quadratic Taylor series approximation to it; if the comparison is unfavorable, ρ is increased and the radius is shrunk. It should be noted that the resulting changes of α not only change the step size (which may be overridden anyway by a subsequent line search) but also change the direction of movement. In any event, the direction will tend to be intermediate between that of a Newton step ($\alpha = 0$) and that of a steepest ascent step ($\alpha \to \infty$). It also follows that if α is very large, convergence is certain, albeit slow since $x^{k+1} \simeq x^k + g(x^k)/\alpha$. The comparison of the present method with Greenstadt's suggests that the latter may make a non-optimal correction in the step if F has "wrong" curvature in some direction. Assume, for example, that $\lambda_1 = \sup \lambda_i > 0$. Using (4.11), the step according to the quadratic hill-climbing method is given by

$$x^{k+1} = x^k - \sum_i \left[\frac{P_i' g}{\lambda_i - \alpha}\right] P_i,$$

and according to Greenstadt's suggestion by

$$x^{k+1} = x^k - \sum_i \left[\frac{P_i' g}{|\lambda_i|}\right] P_i. \tag{4.14}$$

Since α is chosen so that $\lambda_1 - \alpha < 0$, we have $|1/(\lambda_1 - \alpha)| > |1/(\lambda_j - \alpha)|$ for all $j = 2, \dots, n$ and hence the step will contain the direction P_1 with relatively largest weight, a fact that need not hold for (4.14). Thus, the step will be relatively closer to the direction in which the function is convex [Powell (1971)].

A further refinement of the quadratic hill-climbing algorithm rests on the observation that recently successful directions of search may well be worth further searches. Thus, if the step from x^{k-1} to x^k is given by $x^k - x^{k-1} = \xi^k$, then the decomposition of any vector into its projection on ξ and its orthogonal complement permits the component parallel to ξ to be emphasized. To distinguish an actual x^k from arbitrary members of the coordinate system prevailing at the jth iteration, we use the notation $x(j)$. Thus, the coordinate system prevailing at the jth iteration may be transformed into a system prevailing at the $(j+1)$th by

$$x(j+1) = B_j x(j),$$

where $B_j = I + (1-\beta)M_j$ and where $0 < \beta < 1$ and $M_j = \xi_j(\xi_j'\xi_j)^{-1}\xi_j'$. A sequence of such transformations allows the original coordinate system and the one prevailing at the jth iteration to be related by $x(j) = Bx(0)$. Applying the hill-climbing algorithm thus alters the original procedure from maximizing at each step on a sphere to maximizing on a suitably oriented ellipsoid, since $x(j)'x(j)=x(0)'B'Bx(0)$. Writing the function in the jth coordinate system as $F(x(j))$ and differentiating, $\partial F(x(j))/\partial x(0)=B'\partial F(x(j))/\partial x(j)$. Hence, the gradient of $F(x(j))$ in terms of the original system is $g(x(j))=(B^{-1})'\partial F(x(j))/\partial x(0)$. By similar reasoning the Hessian is $(B^{-1})'G(x(j))B^{-1}$. It follows that the step taken can be expressed in the $x(j)$-coordinate system as $-[(B^{-1})'G(x(j))B^{-1}-\alpha I]^{-1}(B^{-1})'g(x(j))$. Premultiplying by B^{-1} yields the step in the $x(0)$-coordinate system and is $-(G(x(j))-\alpha B'B)^{-1}g(x(j))$ and is equivalent to replacing I in (4.13) by a positive definite matrix $B'B$.[10]

4.4. *Methods employing first derivatives*

A general theory of quadratically convergent algorithms has been given by Huang (1970).[11] The objective of Huang's theory is to derive a class of algorithms with the following properties: (a) searches at each iteration are one-dimensional; (b) the algorithms are quadratically convergent; (c) they calculate only function values and first derivatives; and (d) at the kth iteration they only employ information computed at the kth and $(k-1)$th iterations.

Requirement (a) states that at each iteration $k=1,2,\ldots$ a direction d^k be chosen and a scalar λ_k be determined such that

$$\partial F\left(x^k+\lambda_k d^k\right)/\partial\lambda_k=0. \tag{4.15}$$

This determines a displacement $\Delta x^k=\lambda_k d^k$ or $x^{k+1}=x^k+\lambda_k d^k$. Restricting attention [by Property (b)] to quadratic functions $F(x)=x'Ax+b'x+c$, it follows that

$$F(x^{k+1})=F(x^k)+g_k'\Delta x^k+\Delta x^{k\prime}A\Delta x^k, \tag{4.16}$$

where g_k denotes the gradient $g(x^k)$ at x^k. From the first order condition (4.15) it follows that $g_{k+1}'d^k=0$. Since $g_{k+1}=g_k+2A\Delta x^k$ and $\Delta x^k=\lambda_k d^k$, the optimal λ_k is given by

$$\lambda_k=-\frac{d^{k\prime}g_k}{2d^{k\prime}Ad^k}.$$

[10]A final modification pertains to the case when $g=0$ to the required accuracy without having achieved a negative definite $G(x^k)$. This is the case in which x^k is a saddlepoint. Although such cases may be only rarely encountered, the following proposition ensures that the algorithm does not terminate prematurely. If $g(x^k)=0$ and if $\lambda_{\max}>0$, the maximum of F within the sphere $\|x-x^k\|\leqq r$ occurs at $x^k\pm rP_{\max}$, where $P_{\max}$ is the eigenvector corresponding to the $\lambda_{\max}$. Thus, in such instances a search direction is provided by the appropriate eigenvector.

[11]See also detailed discussion in Powell (1971), Broyden (1972), and Dennis and Moré (1977).

Substituting for Δx^k and λ_k in (4.16) yields

$$F(x^{k+1}) - F(x^k) = -\frac{\left(d^{k\prime}g_k\right)^2}{2d^{k\prime}Ad^k},$$

which is positive if A is negative definite, thus ensuring that the function is monotone increasing over successive iterations. If it is further required that the successive search directions be conjugate with respect to A, quadratic convergence can be proved in straightforward fashion. Taking the search direction d^k to be a matrix multiple of the gradient

$$d^k = H^k g_k$$

produces an algorithm of the general form of (4.10) and for that reason these algorithms may be called quasi-Newton algorithms. The conjugacy condition places restrictions on the matrix H^k; however, the restrictions will be observed if H^k is updated according to

$$H^{k+1} = H^k + \theta_1 \Delta x^k \Delta x^{k\prime} + \theta_2 H^k \Delta g_k \Delta g_k' H^k + \theta_3\left[\Delta x^k \Delta g_k' H^k + H^k \Delta g_k \Delta x^{k\prime}\right], \tag{4.17}$$

where $\Delta g_k = g_{k+1} - g_k$. Different choices for θ_1, θ_2, and θ_3 yield different members of this class of algorithms. In any event, θ_2 and θ_3 must satisfy

$$1 + \theta_2 \Delta g_k' H^k \Delta g_k + \theta_3 \Delta x^{k\prime} \Delta g_k = 0. \tag{4.18}$$

At the start, H^1 is usually initialized to $-I$ (or I for minimization). Some of the alternatives are as follows.

(1) If $\theta_1 = 1/\Delta x^{k\prime}\Delta g_k$, $\theta_2 = -1/\Delta g_k' H^k \Delta g_k$, and $\theta_3 = 0$, the resulting algorithm is known as the Davidon–Fletcher–Powell (DFP) algorithm [Davidon (1959), and Fletcher and Powell (1963)]. In this case $F(x)$ is not required to be quadratic for convergence but very strict concavity conditions are required. If $F(x)$ is not concave, there is no assurance that convergence will take place. It should be noted that the quantity $g_k' H^k g_k$ increases monotonically over the iterations; since $g_k' H^k g_k$ is negative for concave functions, this implies that the search direction tends to become more nearly orthogonal to the gradient which can interfere with speedy convergence.

An important feature of DFP is contained in the following:

Theorem 4.6

If $F(x)$ is quadratic, then $H^n = G^{-1}$. The convergence of H to the inverse Hessian in the quadratic case is used in practice to obtain estimates of asymptotic variances and covariances in the presence of maximum likelihood estimation. However, care must exercised for if apparent convergence occurs in less than n iterations, H will not contain usable quantities. It is important, therefore, that computer implementations of DFP contain a restart facility by which computations can be (re)initiated not only with $H^1 = -I$, but $H^1 =$ a previously computed H^k.

(2) In order to make H^k approximate the inverse of the true Hessian $G(x^k)$, one may set

$$\theta_1 \Delta x^{k\prime} \Delta g_k + \theta_3 \Delta g_k' H^k \Delta g_k = 1 \tag{4.19}$$

and also require (4.18) to hold. Obviously, the DFP algorithm satisfies (4.19). In that case the required approximation holds because (4.18) and (4.19) imply $H^{k+1}\Delta g_k = \Delta x^k$, which is just the Taylor series approximation $g_{k+1} = g_k + G(x^k)\Delta x^k$. A special case is the rank-one correction formula, according to which

$$H^{k+1} = H^k + \frac{[\Delta x^k - H^k \Delta g_k][\Delta x^k - H^k \Delta g^k]'}{\Delta x^{k\prime}\Delta g_k - \Delta g_k' H^k \Delta g_k}.$$

Several other algorithms are defined by Huang and Powell and the reader is referred to Huang (1970) and Powell (1971) for details.[12]

Computational experience with many members of this class of algorithms is lacking. The best documented member of the class is almost certainly DFP. Overall its performance is good enough to make it a reasonable first choice for many problems, although it is both generally less robust than some variants of Newton's method (particularly near the maximum) and less efficient computationally than algorithms especially tailored to a problem [see Parke's (1979) Algorithm A] or possessing special *ad hoc* features such as MINOPT [Belsley (1980)]. The latter in particular makes a choice at each iteration whether to employ a steepest ascent step or a Newton step, allowing the former to guide computations at an initial point far from the optimum and the latter near it. It can thus perform more efficiently than DFP. In general, however, DFP and other members of the class must continue to be regarded as viable alternatives for many

[12]A recent new member of the class is due to Davidon (1975) and is designed to work without any line searches.

problems. Two related reasons why members of the quasi-Newton class may fail or perform poorly in practice are: (a) H^{k+1} may become (nearly) singular and (b) H^{k+1} may fail to provide a good approximation to G^{-1}. The latter affects the speed of convergence as a result of the following:

Theorem 4.7

If $F(x)$ is a negative definite quadratic function and x^* the location of the maximum, then $F(x^*)-F(x^{k+1}))\leqq[(K(R_k)-1)/(K(R_k)+1)](F(x^*)-F(x^k))$, where $K(R_k)$ is the condition number of the matrix $R_k=G^{1/2}H^kG^{1/2}$.

Hence, $K(R_k)$ should be small and decreasing which will be the case if H^k increasingly approximates G^{-1}. Oren and Luenberger (1974) designed a "self-scaling" algorithm in this class which guarantees that $K(R_{k+1})\leqq K(R_k)$, with updating formulae given by

$$v^k=\left(\Delta g_k' H^k \Delta g_k\right)^{1/2}\left[\frac{\Delta x^k}{\Delta g_k'\Delta x_k}-\frac{H^k\Delta g_k}{\Delta g_k' H^k\Delta g_k}\right],$$

$$H^{k+1}=\left[H^k-\frac{H^k\Delta g_k\Delta g_k' H^k}{\Delta g_k' H^k\Delta g_k}+\theta_{4k}v^kv^{k\prime}\right]\theta_{5k}+\frac{\Delta x^k\Delta x^{k\prime}}{\Delta x^{k\prime}\Delta g_k},$$

$$\theta_{5k}=\frac{\Delta x^{k\prime}\Delta g_k}{\Delta g_k' H^k\Delta g_k'}(1-\theta_{6k})+\frac{g_k'\Delta x^k}{g_k' H^k\Delta g_k}\theta_{6k},$$

where θ_{4k} and θ_{6k} are parameters to be chosen. In recent experiments various self-scaling and other quasi-Newton algorithms gave satisfactory results [Van der Hoek and Dikshoorn (1979)].

5. Special purpose algorithms and simplifications

There is no hard-and-fast dividing line between general and special purpose algorithms. In the present section we discuss some algorithms that are either especially suited for problems with a particular structure or contain more or less *ad hoc* procedures that appear to be useful in particular contexts.

5.1. Jacobi and Gauss–Seidel methods

Both of these procedures are designed to solve systems of (linear or non-linear) equations. In the context of maximizing a likelihood function, they are applied to solving the first-order conditions, the likelihood equations. Both Jacobi's method and the Gauss–Seidel method presuppose that the equation system can be solved in a particular manner. In the former case we require a solution

$$x = f(x), \tag{5.1}$$

where x is a vector of unknowns and $f(x)$ a vector-valued function. Jacobi's method iterates according to

$$x^{k+1} = f(x^k).$$

The Gauss–Seidel method is similar, except that the ith equation in (5.1), $i = 1,\dots,n$, is assumed to have the structure $x_i = f_i(x_1,\dots,x_{i-1}, x_{i+1},\dots,x_n)$. A further distinction may be obtained depending on whether in a given iteration of the algorithm all n x's are computed and then used only in the next iteration, or whether an x_i computed in a given iteration is used immediately in the computation of other x's in that same iteration. There is some reason to think that the latter procedure is more efficient [Fromm and Klein (1969)].

Jacobi's method was applied to Klein's Model I by Chow (1968). As shown in Section 3, the condensed log-likelihood function for a system of simultaneous linear equations $Y\Gamma + XB = U$ can be written as

$$L = \text{constant} - \frac{n}{2}\log|S| + \frac{n}{2}\log\left[|\Gamma|^2\right], \tag{5.2}$$

where Γ is the matrix of coefficients associated with the jointly dependent variables and S is the estimated covariance matrix of residuals with typical element

$$S_{ij} = \frac{1}{n}\sum_{k=1}^{n} u_{ik}u_{jk}.$$

S itself is a function of the parameters in Γ and B and setting derivatives of L with respect to the non-zero elements of Γ and B equal to zero yields equations of the form of (5.1). Jacobi's method or the Gauss–Seidel method are also routinely applied to solving non-linear systems of simultaneous equations as is required for the solution of stochastic control problems [Chow and Megdal (1978)] or for simulating non-linear econometric models after estimation [Duesenberry et al. (1969), and Fair (1976)].

The objectives of simulation may be to assess the sources of uncertainty and the quality of the predictions over several models or to estimate the effects and the uncertainty of various policy variables [Fair (1980a, 1980b)]. Simulations are stochastic if repeated trials are made in which either the error terms, or the coefficients employed in computing predictions, or exogenous variable values, or all of these are drawn from some appropriate distribution. Whatever simulation variant is chosen, the simulated endogenous variable values must be obtained by solving the system of econometric equations, which is typically non-linear.

A particularly interesting application is due to Fair (1979) in which models with rational expectations in bond and stock markets are simulated. In these models two layers of Gauss–Seidel alternate: for certain initial values of some variables, Gauss–Seidel is used to solve the system for the remaining ones. These solution values are used to obtain new values for the initial set of variables and the system is solved again for the remaining variables, etc.

Neither Jacobi's nor the Gauss–Seidel method can be expected to converge in general. A sufficient condition for convergence is that $f(x)$ be continuous and a contraction mapping; that is, given the distance function d over a compact region R, $f(x)$ is a contraction mapping if for $x \neq x^*$, $x, x^* \in R$, and $d(f(x), f(x^*)) < d(x, x^*)$. An example of such a contraction mapping is provided by Ito (1980) in connection with solving a two-market disequilibrium model with spillovers for values of the endogenous variables. The equations of such models are

$$\begin{aligned}
y^{\mathrm{d}} &= \alpha_{01}' x_1 + \alpha_1 (l - \tilde{l}^{\mathrm{s}}) + \varepsilon_1, \\
y^{\mathrm{s}} &= \alpha_{02}' x_2 + \alpha_2 (l - \tilde{l}^{\mathrm{d}}) + \varepsilon_2, \\
y &= \min(y^{\mathrm{d}}, y^{\mathrm{s}}), \\
l^{\mathrm{d}} &= \beta_{01}' z_1 + \beta_1 (y - \tilde{y}^{\mathrm{s}}) + \varepsilon_3, \\
l^{\mathrm{s}} &= \beta_{01}' z_2 + \beta_2 (y - \tilde{y}^{\mathrm{d}}) + \varepsilon_4, \\
l &= \min(l^{\mathrm{d}}, l^{\mathrm{s}}),
\end{aligned}$$

where $\tilde{y}^{\mathrm{d}} = \alpha_{01}' x_1 + \varepsilon_1$, $\tilde{y}^{\mathrm{s}} = \alpha_{02}' x_2 + \varepsilon_2$, $\tilde{l}^{\mathrm{d}} = \beta_{01}' z_1 + \varepsilon_3$, and $\tilde{l}^{\mathrm{s}} = \beta_{02}' z_2 + \varepsilon_4$. The x's and z's are exogenous variables and the ε's random errors. The y^{d}, y^{s}, l^{d}, and l^{s} are effective goods and labor demand and supply, respectively, the same symbols with a tilde ($\tilde{}$) represent notional demands and supplies, and y and l represent the actually transacted quantities. Ito shows that values of y and l may be calculated by Jacobi's method (for given values of x's, z's, and ε's) by starting with arbitrary y and l if $\alpha_1, \alpha_2, \beta_1, \beta_2 > 0$ and if $1 - \alpha_i \beta_j > 0$ for all $i = 1,2$ and $j = 1,2$.

Some algorithms that appear to be principally gradient methods exploit the idea of a Jacobi or a Gauss–Seidel iteration. Thus, for estimating the parameters of systems of simultaneous equations, Dagenais (1978) first computes a Jacobi iteration of the parameter vector and then further displaces the parameter vector in the direction of the difference between the original value and the Jacobi iterate.

This yields an iteration of the form $x^{k+1} = x^k + \lambda H^k g_k$, where H^k is a positive definite matrix and λ a scalar.

5.2. *Parke's Algorithm A*

An algorithm particularly suited for estimating the coefficients of linear or non-linear simultaneous equations by full-information maximum likelihood or by three-stage least squares is Parke's (1979) Algorithm A. Algorithms that are especially useful for simultaneous equation estimation have been used before. A case in point is the procedure implemented by Chapman and Fair (1972) for systems with autocorrelations of the residuals: their algorithm is a sequence of pairs of Newton steps in which the first operates only on the coefficients of the equations and the second on the autocorrelation coefficients.

Algorithm A performs sequences of searches at each iteration in order to exploit two empirical generalizations about the structure of simultaneous equations models: (a) that the coefficients in any one equation are more closely related than those in separate equations, and (b) that change in the values of the residuals of the equations usually has a substantial effect on the objective function. The algorithm uses searches, no derivatives, and performs numerous searches at each iteration; these facts make it superficially resemble the Powell (1964) class of algorithms.

The sequence of searches in an iteration may be briefly summarized as follows.

(a) For each equation in turn the coefficients of the equation are perturbed one by one (and in a particular order) with the constant term being continually readjusted so as to stabilize the residuals in the sense of holding the mean residual constant. Finally, the constant term itself is perturbed and then the change in the full set of coefficients for that equation is used as a search direction.

(b) After (a) is complete, the change in the coefficients for the system as a whole is used as a search direction.

(c) The last (equation-by-equation) search directions in (a) and the direction in (b) are searched again.

Searches in (a) are linear for linear equations but non-linear otherwise, since the constant term is not, in general, a linear function of the other coefficients when mean residuals are kept constant. The algorithm also provides for the case in which there are constraints on the coefficients.

General theorems about the convergence properties of Algorithm A are difficult to come by. On a small number of test problems the convergence rate of Algorithm A compares favorably with a simple steepest ascent or a simple univariate relaxation algorithm that searches parallel to the coordinate axes. No claim is made that Algorithm A's convergence rate can approximate that of Newton's method (although the latter is very much more expensive per iteration than the former), nor that Algorithm A will necessarily perform well on problems other than simultaneous equation estimation. Computational experience so far is fairly limited and appears to consist of estimates of two versions of the Fair (1976) model [see Fair and Parke (1980) and Parke (1979)]. In spite of the scant evidence the algorithm appears to be quite powerful in a rather sizeable model: in the model of Fair and Parke (1980), Algorithm A estimates 107 coefficients.

5.3. *The EM algorithm*

A particularly effective algorithm becomes possible in models involving incomplete data or latent or unobservable variables. The basic properties of the algorithm are given in Dempster, Laird and Rubin (1977); particular applications are treated in Hartley (1977a, 1977b) and Kiefer (1980).

The incomplete data problem may be stated as follows. Consider a random variable x with pdf $f(x|\theta)$ and assume the existence of a mapping from x to $y(x)$. It is assumed that x is not observed but is known to be in a set $X(y)$, where y represents the observed data. The y-data are incomplete in the sense that a y-observation does not unambiguously identify the corresponding x, but only $X(y)$. The y-data are generated by the density function

$$g(y|\theta)=\int_{X(y)}f(x|\theta)\,\mathrm{d}x. \tag{5.3}$$

A simple example is a multinomial model with k possible outcomes but with the restriction that for some pair of possible outcomes only their sum is observed. Another example is the switching regression model with the structure

$$\begin{aligned} y_i &= \beta_1'x_i + u_{1i} \quad \text{with probability } \lambda, \\ y_i &= \beta_2'x_i + u_{2i} \quad \text{with probability } 1-\lambda. \end{aligned} \tag{5.4}$$

In this model the x_i are exogenous variables, the β_j unknown parameters, the u_i the usual error terms, and the y_i the observed values of the dependent variables [see Hartley (1977a) and Kiefer (1980)]. The probability λ is unknown and we do not observe whether a particular y_i observation is generated by regime (5.4) or by (5.5). Other cases where the method is applicable are censored or truncated data, variance component estimation, estimation in disequilibrium models, etc.

The essential steps of the EM algorithm are the E-step and the M-step which are carried out at each iteration. At the kth iteration we have:

E-step: Given the current value θ^k of the parameter vector and the observed data y, calculate estimates for x^k as $\mathrm{E}(x|y,\theta^k)$.

M-step: Using the estimated values x^k, maximize the likelihood for the complete-data problem $\prod_{i=1}^{n} f(x_i^k|\theta)$ to determine θ^{k+1}.

The most important feature of the EM algorithm is that if it converges to a θ^*, then θ^* is a stationary point of the likelihood function $L(\theta)=\sum \log g(y_i|\theta)$. It has been suggested as a technique preferable to outright maximization of $L(\cdot)$ in instances (see Section 7) in which the likelihood function is unbounded in parameter space or, possibly, cases in which false apparent maxima exist. Whether these problems can typically be avoided by using the EM algorithm is not yet clear; nevertheless it is a powerful algorithm which may simplify as well as speed up convergence in the class of problems to which it is applicable. As an example we discuss the application to the switching regression model by Kiefer (1980).

Assume that n observations are generated by (5.4) with i.i.d. normal errors and the additional restriction that $\sigma_1^2=\sigma_2^2$. Let W_1 be a diagonal matrix of order n where the ith diagonal element w_i represents the expected weight of the ith observation in the first regime and let $W_2=I-W_1$. Then, maximizing the likelihood $\prod_{i=1}^{n} f(y_i|x_i,\theta)$, where $\theta=(\lambda,\beta_1,\beta_2,\sigma^2)$ yields

$$\hat{\beta}_j=(X'W_jX)^{-1}X'W_jY, \qquad j=1,2,$$

$$\hat{\sigma}^2=\sum_{j=1}^{2}(Y-X\beta_j)'W_j(Y-X\beta_j), \tag{5.5}$$

where X and Y are the matrices of observations on the x's and on y. Regarding regime choice as a Bernoulli experiment, λ is estimated as

$$\hat{\lambda}=\operatorname{tr}(W_1)/n.$$

Given these estimates for θ, representing the M-step, one can obtain new estimates for W_1 since for the ith observation

$$\begin{aligned}\mathrm{E}(w_i)&=(1)p(w_i=1|y_i)+(0)p(w_i=0|y_i)\\&=p(w_i=1|y_i)=\frac{\lambda g(y_i|w_i=1)}{\lambda g(y_i|w_i=1)+(1-\lambda)g(y_i|w_i=0)},\end{aligned} \tag{5.6}$$

by Bayes' Theorem. Since the right-hand side of (5.6) is easily computable, we can alternate between E and M steps as required.

5.4. *Simplified Jacobian computation*

If we are seeking FIML estimates for the coefficients of a system of simultaneous linear equations, the transformation from the pdf of the error terms to the pdf of the jointly dependent variables involves the Jacobian $|\Gamma|$ of the transformation as in eq. (5.2). In the event that the equation system is non-linear the term $(n/2)\log[|\Gamma|]^2$ in (5.2) is replaced by

$$\sum_{i=1}^{n} \log |J_i|, \tag{5.7}$$

where J_i is the Jacobian corresponding to the ith observation. Clearly, the evaluation of (5.7) is likely to be much more expensive than the corresponding term in a linear system. Parke (1979), Fair and Parke (1980), and Belsley (1979) report good success with approximations that do not compute all n terms in the summation of (5.7). Various alternatives can be employed, such as approximating (5.7) by $(n/2)(\log|J_1|+\log|J_n|)$ or by computing a somewhat larger number of distinct Jacobians and interpolating for the missing ones. Fair and Parke report an example in which computations start with the simpler approximation and switch to a somewhat more expensive one with six Jacobians being computed for 98 data points. Belsley employs three Jacobian terms. All authors report that the approximations work quite well. The two- and six-term Jacobian approximations lead to substantially similar coefficient estimates and the corresponding objective functions rank the coefficient vectors consistently. The three-term approximation produces essentially the same results as the full Jacobian. It is difficult to predict how this type of approximation will perform in general. The acceptability of the approximation will surely depend on the degree of non-linearity:[13] the greater the non-linearity the worse the approximation may be expected to be. The time saving in computation may, however, be appreciable enough to recommend the procedure in most if not all instances of non-linear models.

6. Further aspects of algorithms

The previous two sections dealt with general as well as with special purpose optimization algorithms in rather broad terms, i.e. in terms that emphasized the general strategy and the key ideas of the algorithms in question. Most of these algorithms share certain detailed aspects which have been neglected up to now. The present section considers some of the salient aspects in this category. We specifically discuss (a) the computation of derivatives, (b) the techniques of linear searches, (c) stopping criteria, and (d) the problem of multiple optima.

[13]For some measures of non-linearity see Beale (1960) and Guttman and Meeter (1965).

6.1. Computation of derivatives

As shown above, many algorithms require that at least the first partial derivatives of the function be calculated; Newton-type methods also require the computation of second partial derivatives. Derivatives may be calculated analytically, i.e. by writing computer programs that evaluate the formulae that result from formal differentiation of the function in question or numerically by finite differencing. The evidence is clear that, other things equal, the former is vastly preferable. Not only do the various convergence properties presume the use of analytic derivatives, but in terms of the required computer time analytic derivatives clearly dominate their numerical counterparts, particularly for Newton-type methods [Belsley (1980)]. Unfortunately, for all but the smallest problems the calculations of analytic derivatives is highly labor intensive and in practice numerical derivatives are often employed, although some computer programs for symbolic differentiation exist (e.g. FORMAC). For numerical evaluation at least two choices have to be made: (a) Should derivatives be evaluated symmetrically or unsymmetrically? (b) How should one choose the length of the interval over which function differences are computed for arriving at a derivative approximation? First partial derivatives at x^0 are given by

$$\frac{\partial F(x^0)}{\partial x_i}=\frac{F(x_1^0,\ldots,x_i^0+\varepsilon_i,\ldots,x_n^0)-F(x_1^0,\ldots,x_n^0)}{\varepsilon_i} \tag{6.1}$$

if evaluated unsymmetrically about x^0, and by

$$\frac{\partial F(x^0)}{\partial x_i}=\frac{F(x_1^0,\ldots,x_i^0+\varepsilon_i,\ldots,x_n^0)-F(x_1^0,\ldots,x_i^0-\varepsilon_i,\ldots,x_n^0)}{2\varepsilon_i} \tag{6.2}$$

if evaluated symmetrically. If the value of $F(x^0)$ is already available (i.e. having already been computed by the algorithm), (6.1) requires n and (6.2) $2n$ additional function evaluations. Second direct partial derivatives are

$$\frac{\partial^2 F(x^0)}{\partial x_i^2}=\frac{F(x_1^0,\ldots,x_i^0-\varepsilon_i,\ldots,x_n^0)-2F(x_1^0,\ldots,x_n^0)+F(x_1^0,\ldots,x_i^0+\varepsilon_i,\ldots,x_n^0)}{\varepsilon_i^2}. \tag{6.3}$$

Second cross partial derivatives are

$$\frac{\partial^2 F(x^0)}{\partial x_i \partial x_j} = \Big[F\big(x_1^0,\dots,x_i^0+\varepsilon_i,\dots,x_j^0+\varepsilon_j,\dots,x_n^0\big)$$
$$- F\big(x_1^0,\dots,x_j^0+\varepsilon_j,\dots,x_n^0\big)$$
$$- F\big(x_1^0,\dots,x_i^0+\varepsilon_i,\dots,x_n^0\big) + F\big(x_1^0,\dots,x_n^0\big)\Big]/\varepsilon_i\varepsilon_j \tag{6.4}$$

or

$$\frac{\partial^2 F(x^0)}{\partial x_i \partial x_j} = \Big[F\big(x_1^0,\dots,x_i^0+\varepsilon_i,\dots,x_j^0+\varepsilon_j,\dots,x_n^0\big)$$
$$- F\big(x_1^0,\dots,x_i^0-\varepsilon_i,\dots,x_j^0+\varepsilon_j,\dots,x_n^0\big)$$
$$- F\big(x_1^0,\dots,x_i^0+\varepsilon_i,\dots,x_j^0-\varepsilon_j,\dots,x_n^0\big)$$
$$+ F\big(x_1^0,\dots,x_i^0-\varepsilon_i,\dots,x_j^0-\varepsilon_j,\dots,x_n^0\big)\Big]/4\varepsilon_i\varepsilon_j. \tag{6.5}$$

The symmetric version (6.5) requires $(n-1)n/2$ more function evaluations than (6.4). The total number of function evaluations required for derivative calculations is:

	First derivatives	Second derivatives	Total
Unsymmetric	n	$(3n^2+n)/2$	$3n(n+1)/2$
Symmetric	$2n$	$2n^2$	$2n(n+1)$

Further compromises are clearly possible and often implemented, such as when first derivatives are calculated analytically and second derivatives numerically by differencing first derivatives.

An important question is how the values of ε_i and ε_j ought to be chosen. In practice they are chosen as small but arbitrary proportions of x_i^0 and x_j^0. For example, Chow and Megdal (1978) choose $\varepsilon_j = \max(\delta_1 x_j, \delta_2)$ where δ_1 and δ_2 are both 0.001. A procedure has been developed by Stewart (1967) for choosing the intervals optimally. Consider for simplicity a function $\phi(x)$ of a single variable and assume it to be the quadratic. Then, expanding $\phi(x+\varepsilon)$ about x and evaluating at $x=0$, the function can be written as

$$\phi(\varepsilon) = \alpha_0 + \alpha_1\varepsilon + \tfrac{1}{2}\alpha_2\varepsilon^2.$$

The first derivative can then be approximated by

$$\phi'(0) = \alpha_1 \simeq \frac{\phi(\varepsilon)-\alpha_0}{\varepsilon}. \tag{6.6}$$

The approximation (6.6) may be in error because it and all difference approximations to derivatives are Taylor series approximations and thus involve truncation error. The relative magnitude of this error is

$$\left|\frac{(\phi(\varepsilon)-\alpha_0)/\varepsilon-\alpha_1}{\alpha_1}\right|=\left|\frac{\alpha_2\varepsilon}{\alpha_1}\right|/2,$$

which clearly increases in the interval length ε. Another error is introduced if ε is small since in the numerator of (6.6) two numbers of comparable size are subtracted: in fixed word length computing serious rounding error may arise. We need to know an error bound $\bar{\eta}$ such that the computed value ϕ_c and true value ϕ are related by $\phi_c=\phi(1+\eta)$, where $|\eta|\leqq\bar{\eta}$. Then $\phi(\varepsilon)$ and α_0 can be computed as $\phi_c(\varepsilon)=\phi(\varepsilon)(1+\eta_1)$ and $\alpha_{0c}=\alpha_0(1+\eta_2)$, where $|\eta_1|$ and $|\eta_2|\leqq\bar{\eta}$. If ε is small so that $\alpha_0\simeq\phi(\varepsilon)$, $\phi_c(\varepsilon)-\alpha_{0c}\simeq\phi(\varepsilon)-\alpha_0+\eta_3\alpha_0$, where $|\eta_3|\leqq 2\bar{\eta}$. It follows that the relative cancellation error is

$$\left|\frac{(\phi_c(\varepsilon)-\alpha_{0c})-(\phi(\varepsilon)-\alpha_0)}{\phi(\varepsilon)-\alpha_0}\right|\leqq 2\bar{\eta}\left|\frac{\alpha_0}{\phi(\varepsilon)-\alpha_0}\right|,$$

which is decreasing in ε if $\phi(\varepsilon)-\alpha_0$ is increasing in ε. Stewart suggests choosing ε so that the errors from the two sources are equal, which can be determined as the solution of one of the cubic equations below, where $\varepsilon_1=|\varepsilon|$:

$$\begin{aligned}
\tfrac{1}{2}\alpha_2^2\varepsilon_1^3+|\alpha_2||\alpha_1|\varepsilon_1^2-4|\alpha_0||\alpha_1|\bar{\eta}&=0 \quad \text{if } \varepsilon>0,\\
-\tfrac{1}{2}\alpha_2^2\varepsilon_1^3+|\alpha_2||\alpha_1|\varepsilon_1^2-4|\alpha_0||\alpha_1|\bar{\eta}&=0 \quad \text{if } -2|\alpha_1|/|\alpha_2|\leqq\varepsilon\leqq 0,\\
\tfrac{1}{2}\alpha_2^2\varepsilon_1^3-|\alpha_2||\alpha_1|\varepsilon_1^2-4|\alpha_0||\alpha_1|\bar{\eta}&=0 \quad \text{if } \varepsilon\leqq -2|\alpha_1|/|\alpha_2|.
\end{aligned}\tag{6.7}$$

Simplified versions of (6.7) as well as optimal solutions based on symmetric approximations exist and are generally desirable.

The computational cost of second derivative methods is significant enough to make them less than fully practical for large problems. Fair (1973) has applied such methods to problems with up to 40 variables, but problems not much larger than these may well represent the practical upper bound for using Newton-type methods for most researchers. How to economize on second derivative evaluations has been a problem of high priority. A simple solution that works in practice is to evaluate the matrix of second partial derivatives not at every iteration but at, say, every second or third iteration. The degradation of convergence that may occur is often more than made up by the savings in function evaluations.[14] An

[14]If the same Hessian is used for a number of iterations and happens to be a good estimate of $E[\partial^2\log L/\partial\theta\partial\theta']$, the method is an approximation to the method of scoring.

important modification based on statistical theory is that of Berndt, Hall, Hall and Hausman (1974) and is applicable to the maximization of likelihood functions. The negative inverse of the matrix of second partial derivatives required for computation is a sample estimate of

$$-\left[\mathrm{E}\frac{\partial^2 \log L}{\partial\theta\partial\theta'}\right]^{-1} = \left[\mathrm{E}\frac{\partial \log L}{\partial\theta}\frac{\partial \log L}{\partial\theta'}\right]^{-1}.$$

The (negative) expected value of the Hessian is thus, at the optimum, the covariance matrix of the gradient. It may therefore be inexpensively approximated by using the first derivatives of the likelihood function. Given, for example, the system of simultaneous non-linear equations

$$f_i(y_i, x_i, \theta) = u_i, \qquad i = 1,\dots,n,$$

where f_i is a row vector with a component for each of g equations, y_i the vector of jointly dependent variables, and x_i the vector of predetermined variables, the log-likelihood function can be written analogously to (5.2) as

$$L = \text{const} - \tfrac{1}{2}\log|f'f| + \sum_{i=1}^{n} \log|J_i|,$$

where f is the $n \times g$ matrix containing as its rows f_i, and where $J_i = \partial f_i(y_i, x_i, \theta)/\partial y_i$. It is easy to show that the matrix of second partials can be approximated by

$$n\sum_{i=1}^{n}(p_i - q_i)(p_i - q_i)',$$

where $p_i = \partial \log|J_i|/\partial\theta$ and $q_i = (\partial f_i/\partial\theta)(\sum f_i' f_i)^{-1} f_i'$. The use of this approximation can be powerful by eliminating many of the function evaluations required for numerical derivatives. In addition, the approximation is positive definite and iterations will move uphill along the likelihood function. In practice it appears to work very well in many problems [Belsley (1980)], although it need not always provide a good approximation to the Hessian, particularly in small samples and at points not close to the optimum.

In spite of the several useful techniques discussed above, it may occur that a numerical approximation to the Hessian at the optimum is not negative definite. Although some algorithms may, in principle, converge to saddlepoints, this must generally be regarded as an unlikely event. The most plausible conclusion is that the numerical approximation to the Hessian has failed. Such an outcome is most frequent in cases where the function is extremely flat. It is clearly not an acceptable outcome in any event, but particularly not in the case of maximum likelihood estimation for then the negative inverse of the Hessian is used as an estimate of the asymptotic covariance matrix. A heuristic technique that may occasionally be employed with success in such cases is as follows. Choose alternative values of the intervals ε_i over some fairly wide range and evaluate the Hessian for each. For large ε_i the truncation error, and for small ε_i the cancellation error, may predominate. For extreme values of the ε_i the estimates of the Hessian are likely to be unstable in the sense that the values of ε_i that are near to each other do not yield Hessians that are comparable. There may exist a (problem dependent) range of ε_i over which the estimates of the Hessian appear to be stable. If such a range exists, it is likely to be associated with numerically more accurate estimates of the Hessian.

6.2. *Linear searches*

Many algorithms require at each iteration, say the kth, the computation of λ_k such that $F(x^k + \lambda_k d^k)$ is maximized. It is clearly too expensive to require that (4.15) be satisfied exactly. In fact, normally it appears not worthwhile to calculate λ_k very accurately because of the excessive number of function evaluations this tends to require. Three procedures are discussed briefly.

(1) Fibonacci search [Spang (1962)]. Assume that the location of a unique maximum after p cycles of the linear search is known to be between $x + \lambda_a^p d$ and $x + \lambda_b^p d$. Then two more test values, λ_1^p and λ_2^p, are selected $(\lambda_a^p < \lambda_1^p < {}_2^p < \lambda_b^p)$. If $F(x + \lambda_1^p d) > F(x + \lambda_2^p d)$, the maximum is between $x + \lambda_a^p d$ and $x + \lambda_2^p d$; the new lower and upper limits are $\lambda_a^{p+1} = \lambda_a^p$ and $\lambda_b^{p+1} = \lambda_2^p$, and the procedure is repeated with the new interval. (The corresponding actions are obvious if the inequality is reversed or if equality should be attained.) The values of λ_1 and λ_2 after the pth shrinking of the interval are obtained as

$$\lambda_1^p = \frac{U_{N-1-p}}{U_{N+1-p}}(\lambda_b^p - \lambda_a^p) + \lambda_a^p$$

and

$$\lambda_2^p = \frac{U_{N-p}}{U_{N+1-p}}(\lambda_b^p - \lambda_a^p) + \lambda_a^p,$$

where U_i denotes the ith Fibonacci number and N is a predetermined number of shrinkages. The rationale for Fibonacci search rests on the relationship between the number of function evaluations and the size of the region covered when the error in finding the maximum satisfies a certain bound. Assume specifically that $F(x+\lambda d)$ has a unique maximum in the interval $[0, \Lambda_N]$, where Λ_N allows the maximum to be found with an error bounded by unity in no more than N function evaluations. Define λ_N as $\inf(\Lambda_N)$. Then one can prove the following:

Theorem 6.1.

The sequence λ_N is the Fibonacci sequence. Thus, for a given error bound of unity, the area searched increases with N fastest for the Fibonacci sequence; conversely, for a bounded area the error declines fastest. However, in practice this requires a prior determination of N. If the initial interval is large, the temptation may be to use a large value of N which will result in a large number of function evaluations at each line search. The method has been successful in applications [Daganzo, Bouthelier and Sheffi (1977)].

(2) Successive quadratic approximation [Powell (1964)]. Evaluate the function at $x+\lambda_1 d$, $x+\lambda_2 d$, and $x+\lambda_3 d$ and determine the stationary point of the quadratic function of λ fitted exactly to those points. The stationary point occurs at

$$\lambda_4=\frac{1}{2}\frac{(\lambda_2^2-\lambda_3^2)F(x+\lambda_1 d)+(\lambda_3^2-\lambda_1^2)F(x+\lambda_2 d)+(\lambda_1^2-\lambda_2^2)F(x+\lambda_3 d)}{(\lambda_2-\lambda_3)F(x+\lambda_1 d)+(\lambda_3-\lambda_1)F(x+\lambda_2 d)+(\lambda_1-\lambda_2)F(x+\lambda_3 d)}.$$

If the stationary point is a maximum and if the proposed step is not greater than a preassigned tolerance, the λ corresponding to the smallest value of $F(x+\lambda d)$ is discarded and the computation repeated with the surviving three λ's. If λ_4 corresponds to a minimum or implies a step greater than the maximum allowed, λ_4 is chosen to correspond to the largest permitted step, the λ-value farthest from λ_4 is discarded and the computation repeated. If a computed λ_4 is within a preassigned ε-distance of λ_1, λ_2, and λ_3, it is accepted as the maximum and the line search is terminated. [A more elaborate cubic interpolation scheme is suggested by Davidon (1959). Dagenais (1978) first fits a quadratic from two function values and the gradient, adds a third point which is the maximum of the fitted quadratic, and then fits a cubic.]

(3) Powell (1971) and Berndt, Hall, Hall and Hausman (1974) recommend the following approximate procedure. First select an ε such that $0<\varepsilon<\frac{1}{2}$. Then find λ^k such that

$$\varepsilon\leq\frac{F(x^k+\lambda^k d^k)-F(x^k)}{\lambda^k g_k' d^k}\leq 1-\varepsilon. \tag{6.8}$$

Since the fraction in (6.8) approaches 1 as $\lambda^k \to 0$ and approaches zero or a negative number as $\lambda^k \to \infty$, a suitable value of λ^k exists and can be found by successive evaluations; the first λ^k that satisfies (6.8) is then used.

6.3. *Stopping criteria*

An important part of every algorithm is the criterion it employs for terminating computations. The ideal of reaching a point x^k such that $\partial F(x^k)/\partial x_i = 0$ $(i = 1,\dots,n)$ is not attainable in practice and the question is what compromise is most reasonable. In the neighborhood of the maximum any algorithm is likely to take small steps in the sense that $|x_i^{k+1} - x_i^k|$ is likely to be small for all values of i and in the sense that $|F(x^{k+1}) - F(x^k)|$ is likely to be small; accordingly both of these quantities have been employed as stopping criteria. In fact, it is theoretically possible for either difference to become small while the other is large; it is thus preferable to use both and continue iterating unless both criteria are satisfied. In addition, since at the exact maximum the gradient in zero, $g_k' g_k$ is also an obvious choice for judging closeness to the maximum. Perhaps the most common is to test the relative change in the variables and accordingly computations terminate if

$$\max \frac{|x_i^{k+1} - x_i^k|}{\max(\varepsilon_1, |x_i^k|)} \leqq \varepsilon_2,$$

where ε_1 and ε_2 are preassigned tolerances [see Powell (1964) and Berndt, Hall, Hall and Hausman (1974)]. A variant is to terminate if

$$2\left[\sum_{i=1}^{n} \left(x_i^{k+1} - x_i^k\right)^2\right]^{1/2} \leqq \varepsilon_1^{1/2}\left[\sum_{i=1}^{n} \left(x_i^{k+1}\right)^2\right]^{1/2} + \varepsilon_2,$$

where ε_1 is machine precision defined as $\beta^{1-\tau}$ (for truncated arithmetic) or $\beta^{1-\tau}/2$ (for rounded arithmetic), where β is the base of arithmetic and τ the number of floating point digits [Brent (1973)]. Some algorithms employ a combination of criteria and may terminate when any of the criteria are satisfied. Belsley (1980) criticizes the gradient criterion as not being scale independent and ignoring statistical considerations in the sense that a gradient component in the direction of an insignificant parameter has the same weight as one in the direction of a statistically significant one. He criticizes the relative variable-change criterion as treating all variables equally (although the criterion is scale invariant). The relative function-change criterion is not invariant with respect to the scaling of the function. Belsley suggests a weighted-gradient stopping criterion $-g_k' H^{-1} g_k < \varepsilon$, where H^{-1} is the inverse Hessian. A similar suggestion in a least squares context is contained in Dennis, Gay and Welsch (1979). This criterion is clearly scale invariant and in maximum likelihood estimation it can be interpreted to weight parameter estimates according to their (asymptotic) significance. Moreover, the criterion may be recognized as the Lagrange multiplier test statistic and accordingly iterations continue until the test statistic confirms that the gradient is small. Computational experience with the criterion appears to be quite satisfactory and it results in terminations that are relatively model independent.[15]

6.4. *Multiple optima*

All algorithms discussed so far (with the possible exception of appropriately designed grid searches) locate only local maxima. There is no guarantee that only one maximum exists or, if more than one exists, that the maximum found is the global maximum. There are, of course, numerous instances in which the likelihood function can be proved to be globally concave, in which case a unique maximum exists. Cases in point are the likelihood function associated with the classical normal regression model or that of the probit or logit models. For the latter two, in the simplest case, the dependent variable value at the ith observation, y_i, is dichotomous:

$$y_i = \begin{cases} 0 & \text{with probability } 1-\Phi(x_i'\beta), \\ 1 & \text{with probability } \Phi(x_i'\beta), \end{cases}$$

where x_i' is the (row) vector of independent variables, β a (column) vector of parameters, and Φ a distribution function, normal for the probit model and logistic for the logit model. The likelihood is

$$L = \prod_{i=1}^{n} [\Phi(x_i'\beta)]^{y_i}[1-\Phi(x_i'\beta)]^{1-y_i},$$

and it is not difficult to show that

$$\frac{\partial^2 \log L}{\partial\beta\partial\beta'} = -\sum_{i=1}^{n} \phi(x_i'\beta)x_i x_i',$$

where $\phi(z) = \mathrm{d}\Phi(z)/\mathrm{d}z$ is the probability density function, is negative definite [Dhrymes (1978)].

[15]A termination criterion that is qualitatively different from all of the ones discussed is the criterion employed in the Nelder and Mead (1965) algorithm. According to this criterion computations stop if the sample variance of the function values at the vertices of the current simplex falls below a preset tolerance.

In some cases the existence or the possibility of multiple maxima can be shown analytically. Some illustrative examples are provided by Goldfeld and Quandt (1972); a realistic example pertaining to pooling cross-section and time-series data is discussed by Maddala (1971). Numerous instances exist in which a global maximum has not necessarily been found [see Fair (1974a)]. When several maxima exist, it is important to attempt to find the global maximum since it is (customarily if not uniformly) implicit in the asymptotic justification for maximum likelihood estimation that the global maximum is attained.

Unfortunately, mostly *ad hoc* methods are employed for locating multiple optima. (a) The most common method is to select several (or many, if cost considerations permit) starting values for the vector of unknowns and to reoptimize repeatedly using the algorithms discussed previously. If all starting points lead to the same local optimum, the tendency is to declare it to be the unique maximum with substantial confidence. (b) Assume that a local maximum has been found at x^0. Goldfeld and Quandt (1972) propose to find a solution to $F(x) = F(x^0) + \varepsilon$ for small positive ε; if there exists a solution to the equation, x^0 cannot correspond to a global maximum. (c) A deflation method is suggested by Brown and Gearhart (1971) and explored by Salmon (1978) with a view towards solving large-scale econometric models. It appears feasible to apply the method to solving the first-order conditions of a maximum problem. The first-order conditions are

$$g(x) = 0 \tag{6.9}$$

and assume that (6.9) has been written as

$$x = \phi(x), \tag{6.10}$$

as would be required to obtain a solution by the Jacobi or Gauss–Seidel method. Let x^0 be a solution obtained by one of these methods. Define

$$\|x - x^0\|_p = \left[\sum_{i=1}^{n} (x_i - x_i^0)^p\right]^{1/p},$$

p normally > 1, and consider solving

$$g^*(x) = \frac{g(x)}{\|x - x^0\|_p} = x - \phi^*(x) = 0.$$

This suggests iterations according to $x^{k+1} = \phi^*(x^k)$, where

$$\phi^*(x) = x - \frac{x - \phi(x)}{\|x - x^0\|_p},$$

leading to iterations defined by

$$x^{k+1} = \left(1 - \frac{1}{\|x^k - x^0\|_p}\right) x^k + \frac{1}{\|x^k - x^0\|_p} \phi(x^k), \tag{6.11}$$

which shields the algorithm from the previously obtained solution x^0. Additional solutions give rise to similar deflators of $g(x)$ until no further solutions are found. Experience with the method appears to be limited to test examples and to solving the Australian Treasury N1F7 macro model with 128 equations. Although the method will share the difficulties of the Gauss–Seidel method and need not converge after deflation even if a second solution exists, it appears to be an interesting candidate for further study.

7. Particular problems in optimization

In addition to the standard features of optimization algorithms and problems, there occur in practice several particular problems that may be difficult to cope with. Most of these are associated with special models; accordingly, their treatment will be relatively brief.

7.1. Smoothing of non-differentiable functions

These are several related contexts in which non-differentiable likelihood functions may arise. One is the switching regression model

$$\begin{aligned} y_i &= \beta_i' x_i + u_{1i} \quad \text{if } \pi' z_i \leqq 0, \\ y_i &= \beta_2' x_i + u_{2i} \quad \text{if } \pi' z_i > 0, \end{aligned} \tag{7.1}$$

where β_1, β_2, and π are unobserved parameter vectors and x_i and z_i are exogenous variable vectors. Define $D = D(\pi' z_i) = 0$ if $\pi' z_i \leqq 0$ and $D_i = 1$ otherwise. Eqs. (7.1) may then be rewritten as

$$y_i = (1 - D_i)\beta_1' x_i + D_i \beta_2' x_i + u_{1i}(1 - D_i) + u_{2i} D_i. \tag{7.2}$$

If u_{1i} and u_{2i} are distributed as $N(0, \sigma_1^2)$ and $N(0, \sigma_2^2)$, the likelihood function is

$$L = (2\pi)^{-n/2} (\sigma_i^2)^{-1/2} \exp\left\{ - \sum_{i=1}^{n} (y_i - (1 - D_i)\beta_1' x_i - D_i \beta_2' x_i)^2 \Big/ 2\sigma_i^2 \right\}, \tag{7.3}$$

where σ_i^2 is defined as $\sigma_1^2(1-D_i)^2+\sigma_2^2D_i^2$. The unknowns in this problem are β_1, β_2, σ_1^2, σ_2^2, and the D_i which are discrete; hence, derivatives with respect to the D_i do not exist.

An alternative model, first suggested by Tishler and Zang (1977) and shown to be a special case of a general disequilibrium model by Goldfeld and Quandt (1978), is

$$\begin{aligned} y_{1i} &= \beta_1'x_{1i},\\ y_{2i} &= \beta_2'x_{2i},\\ y_i &= \min(y_{1i},y_{2i})+u_i, \end{aligned} \tag{7.4}$$

where y_{1i} and y_{2i} are not observed and y_i is observed. The likelihood function corresponding to (7.4) is

$$\begin{aligned} L = &\prod_{\beta_1'x_{1i}\leqq\beta_2'x_{2i}} \frac{1}{\sqrt{2\pi}\,\sigma}\exp\left\{-\frac{1}{2\sigma^2}(y_i-\beta_1'x_{1i})^2\right\}\\ &\times\prod_{\beta_1'x_{1i}>\beta_2'x_{2i}} \frac{1}{\sqrt{2\pi}\,\sigma}\exp\left\{-\frac{1}{2\sigma^2}(y_i-\beta_2'x_{2i})^2\right\}. \end{aligned} \tag{7.5}$$

Rewriting $\beta_1'x_{1i}-\beta_2'x_{2i}\leqq 0$ as $\pi'z_i\leqq 0$ shows that (7.5) is formally identical with (7.3) with the added restriction that $\sigma_1^2=\sigma_2^2$. The function (7.3) is not differentiable everywhere because of the discrete D_i and function (7.5) exhibits the same problem whenever the β's pass through values at which the sorting of observations between the two types of products changes. Various types of smoothing have been suggested to cope with this problem. The technique employed by Goldfeld and Quandt (1972) replaces D_i with the approximation

$$D_i \simeq \int_{-\infty}^{\pi'z_i}\frac{1}{\sqrt{2\pi}\,\sigma}\exp\left\{-\frac{1}{2}\frac{\xi^2}{\sigma^2}\right\}\mathrm{d}\xi, \tag{7.6}$$

where σ^2 is a new parameter that has been interpreted as a measure of the extent to which the approximation can reproduce the exact sorting of the data. Clearly, other distribution functions might be used in the place of (7.6). Tischler and Zang (1979) recommend several spline approximations of which the most promising appears to be the quintic given by

$$D_i = \begin{cases} 0 & \text{if } \pi'z_i\leqq-\alpha,\\ \dfrac{3}{16}\left(\dfrac{r}{\alpha}\right)^5-\dfrac{5}{8}\left(\dfrac{r}{\alpha}\right)^3+\dfrac{15}{16}\left(\dfrac{r}{\alpha}\right)+\dfrac{1}{2} & \text{if } -\alpha\leqq\pi'z_i\leqq\alpha,\\ 1 & \text{if } \alpha\leqq\pi'z_i, \end{cases} \tag{7.7}$$

where α is a new (positive) parameter to be determined. In practice both types of approximation appear to work well.

7.2. Unbounded likelihood functions and other false optima

Ordinarily the value of the likelihood function is bounded. In certain classes of models, usually dealing with unobserved or latent variables, the likelihood function may become unbounded. This is a serious difficulty in that any algorithm that happens to locate a neighborhood within which the function is unbounded is likely to break down; moreover, the point or points in parameter space at which unboundedness occur have no desirable statistical properties. Three examples are given.

The first is a case in which unboundedness may occur under special circumstances but is unlikely to do so in general. It is the well-known tobit model:

$$\begin{aligned} y_i &= \beta' x_i + u_i \quad && \text{if } \beta' x_i + u_i > 0, \\ y_i &= 0 && \text{if } \beta' x_i + u_i \leqq 0. \end{aligned} \tag{7.8}$$

The likelihood function is

$$L = \prod_{i \in I} \frac{1}{\sqrt{2\pi}\,\sigma} \exp\left\{ -\frac{1}{2}\left(\frac{y_i - \beta' x_i}{\sigma}\right)^2 \right\} \prod_{i \in \bar{I}} \left[1 - \Phi\left(\frac{\beta' x_i}{\sigma}\right) \right], \tag{7.9}$$

where I and $\bar{I}$ are the sets of indices for which the two inequalities in (7.8) hold, respectively, and where $\Phi(\cdot)$ is the standard cumulative normal distribution. Assume that β has k elements and that the number of elements in I is less than k. Then if the x_i $(i \in I)$ are linearly independent, there exists a value β^* such that $y_i - \beta^{*\prime} x_i \equiv 0$, $i \in I$. Assume further that $\beta^{*\prime} x_i \leqq 0$, $i \in \bar{I}$; i.e. that $\beta^{*\prime} x$ is a supporting hyperplane for the convex hull of the x_i, $i \in \bar{I}$. Then consider the sequence of points obtained by letting $\sigma \to 0$. The product of terms for $i \in I$ becomes unbounded, whereas the product of terms for $i \in \bar{I}$ remains bounded away from zero; hence $L \to \infty$.

The second example is the well-known switching regression model [Quandt (1972), and Quandt and Ramsey (1979)]:

$$y_i = \beta_i' x_{1i} + u_{1i} \quad \text{with probability } \lambda,$$

$$y_i = \beta_2' x_{1i} + u_{2i} \quad \text{with probability } 1 - \lambda.$$

The likelihood function with normal errors is

$$L=\prod_{i=1}^{n}\left[\frac{\lambda}{\sqrt{2\pi}\,\sigma_1}\exp\left\{-\frac{1}{2}\left(\frac{y_i-\beta_1'x_i}{\sigma_1}\right)^2\right\}\right.$$
$$\left.+\frac{1-\lambda}{\sqrt{2\pi}\,\sigma_2}\exp\left\{-\frac{1}{2}\left(\frac{y_i-\beta_2'x_i}{\sigma_2}\right)^2\right\}\right]. \tag{7.10}$$

Choose a β_1^* such that the kth observation $y_k-\beta_1^{*\prime}x_k\equiv 0$ and let $\sigma_1^2\to 0$. Then $\lambda\exp\{-(y_k-\beta^{*\prime}x_k)^2/2\sigma_1^2\}/\sqrt{2\pi}\,\sigma_1$ becomes unbounded, but $(1-\lambda)\exp\{-(y_i-\beta_2'x_i)^2/2\sigma_2^2\}/\sqrt{2\pi}\,\sigma_2$ remains bounded away from zero for all i; hence $L\to\infty$.

The third example is that of a simple disequilibrium model:

$$\begin{aligned} y_{1i}&=\beta_1'x_{1i}+u_{1i},\\ y_{2i}&=\beta_2'x_{2i}+u_{2i},\\ y_i&=\min(y_{1i},y_{2i}). \end{aligned} \tag{7.11}$$

The likelihood function with normal errors and u_{1i} and u_{2i} independent is

$$L=\prod_{i=1}^{n}\left[\frac{1}{\sqrt{2\pi}\,\sigma_1}\exp\left\{-\frac{1}{2}\left(\frac{y_i-\beta_1'x_{1i}}{\sigma_1}\right)^2\right\}\left(1-\Phi\left(\frac{y_i-\beta_2'x_{2i}}{\sigma_2}\right)\right)\right.$$
$$\left.+\frac{1}{\sqrt{2\pi}\,\sigma_2}\exp\left\{-\frac{1}{2}\left(\frac{y_i-\beta_2'x_{2i}}{\sigma_2}\right)^2\right\}\left(1-\Phi\left(\frac{y_i-\beta_1'x_{1i}}{\sigma_1}\right)\right)\right]. \tag{7.12}$$

An argument very similar to that employed in the case of the switching regression model can be used to show that the simple disequilibrium model has an unbounded likelihood function [Quandt (1978a)]. In most instances there is an infinity of points at which the likelihood function becomes unbounded: thus, for example, β_1^* in (7.10) can be chosen in infinitely many ways so as to make one residual exactly equal to zero. No completely satisfactory methods are known to avoid the computational problems. A device employed frequently is to constrain the variances by $\sigma_1^2=\alpha\sigma_2^2$, where α is selected *a priori*. This guarantees to solve the computational problem at the cost of introducing misspecification. Other methods [Hartley, (1977a, 1977b)] have suggested that adaptations of the EM algorithm may tend to avert the problem. Although the problem is not encountered very frequently in practice, it must be considered a difficult one.

Another problem with potentially equally bad computational consequences is the appearance of "false optima". A case in point is the likelihood function corresponding to (7.11) when $E(u_{1i}u_{2i})=\sigma_{12}\neq 0$. Define $r_{ij}=(y_i-\beta_j'x_i)/\sigma_j$ as a normalized residual. Then the following can be shown [Goldfeld and Quandt (1978)].

(a) If values of β_1, β_2, σ_1^2, and σ_2^2 have been selected such that $r_{i1} + r_{i2} < 0$ for all i, then the likelihood function increases as the correlation, ρ, between u_1 and u_2 approaches -1.

(b) If $r_{i2} - r_{i1} > 0$, when $\exp(-r_{i2}^2)/\sigma_2 > \exp(-r_{i1}^2)/\sigma_1$ and $r_{i2} - r_{i1} < 0$ otherwise, then the likelihood function increases as $\rho \to 1$.

The practical consequences of the result is that if values of parameters happen to be achieved that satisfy the above conditions during the iterations of an algorithm, then the algorithm may attempt to push the value of ρ arbitrarily close to ± 1. Since the likelihood function is not defined at $\rho = \pm 1$, computation normally fails at some point in such a neighborhood. Such a point is not a true local maximum since the likelihood function is defined only over the open interval $-1 < \rho < 1$, but it may computationally appear as one. There is as yet no obvious method to guard against the occurrence of this problem. One might wish to impose constraints so that the inequalities required by the difficulty do not hold. However, this may be unsatisfactory since the inequalites may be satisfied even if the true values of the β's and σ's are substituted.

7.3. *Constraints on the parameters*

Classical constrained optimization problems arise if there are equality or inequality constraints on the parameters arising from the intrinsic aspects of the problem. A case in point would be the requirement that the exponents of a Cobb–Douglas production function add to unity. Another example, discussed by MacKinnon (1979), is when economic considerations require that the Jacobian term in a likelihood function have a particular sign (even though it is the absolute value of the Jacobian that enters the likelihood function). Equality constraints, the more common case, may often be handled adequately by using the constraints to eliminate variables of optimization from the objective function.[16] In general, problems of this type need to be handled by the methods of non-linear programming. We briefly discuss only two classes of algorithms.

[16]This has the added advantage of reducing the number of variables in the optimization. It follows that whenever a subset of the first-order condition can be solved for a subset of variables, the (likelihood) function should be condensed. It is not hard to show that the negative inverse Hessian of the condensed log-likelihood function is the appropriate estimator of the asymptotic covariance matrix of the variables that have not been condensed out.

Consider the constrained optimization problem:

$$\begin{aligned} &\text{maximize} \quad F(x) \\ &\text{subject to} \quad \psi_i(x) \geqq 0, \qquad i=1,\dots,m. \end{aligned} \tag{7.13}$$

If none of the constraints in (7.13) is binding, then an iterative step may be taken as if one were dealing with an unconstrained problem. If, however, one or more constraints are binding, the question arises of how to choose a search direction. A particular approach to this problem is provided by the Rosen gradient projection method [Rosen (1960, 1961) and Walsh (1975)]. Assume that $m_1 \leqq m$ of the constraints are binding and consider the subspace spanned by the constraint gradients $\nabla\psi_i$, $i=1,\dots,m_1$. The key of Rosen's algorithm is to choose as the search direction the projection of the gradient of $F(x)$, g, on the orthogonal complement of the subspace spanned by the $\nabla\psi_i$. Denoting the matrix the columns of which are $\nabla\psi_i$ by Ψ, the search direction is $d=(I-\Psi(\Psi'\Psi)^{-1}\Psi')g$.

A different class of algorithms converts a constrained maximization problem into a sequence of unconstrained ones. This is usually accomplished by penalizing the objective function for (near) violations of the constraint by adding to it penalty or barrier functions [Fiacco and McCormick (1964), Osborne (1972), and Walsh (1975)]. Thus, one might define $\xi(\psi_1,\dots,\psi_m)=\sum_{i=1}^m \log\psi_i(x)$ and consider the unconstrained maximization of

$$W=F(x)+\gamma\xi(\psi(x))$$

for some positive number γ. Solve this unconstrained problem repeatedly for a sequence of $\gamma\to 0$. It can be shown under general conditions that the corresponding sequence of solutions, x, converges to the solution of (7.13). The obvious advantage of this approach is that it converts the original problem into one which is generally easier to solve. The disadvantage is that the single constrained optimization problem has been replaced by a sequence of unconstrained problems, which can result in high computational cost.

An interesting variant of this uses an augmented Lagrangean expression [Pierre and Lowe (1975)]. Consider the standard Lagrangean:

$$L(x,\alpha)=F(x)+\alpha'\psi(x), \tag{7.14}$$

where $\psi(x)$ is the vector with elements $\psi_i(x)$ and α a vector of constants. Form the augmented Lagrangean

$$L(x,\alpha,\beta_1,\beta_2)=L(x,\alpha)-\beta_1\sum_{i\in I_1}\psi_i(x)^2-\beta_2\sum_{i\in I_2}\psi_i(x)^2, \tag{7.15}$$

where β_1 and β_2 are preassigned weights and where $I_1 = \{i | \alpha_i > 0\}$, $I_2 = \{i | \alpha_i = 0$, and $\psi_i(x_i) \geqq 0\}$. The algorithm alternates between two steps: the maximization step in which an unconstrained maximum of (7.15) is found and an adjustment step in which the Lagrange multipliers are adjusted so as to equate the gradient of the augmented Lagrangean to that of the simple Lagrangean and in which β_1 and β_2 may be increased. The procedure obtains its justification from the following:

Theorem 7.1

If x^*, α^* solves the appropriate Kuhn–Tucker conditions, then x^* satisfies sufficient conditions for an unconstrained maximum for $L(x, \alpha^*, \beta)$ and sufficiently large β if and only if x^*, α^* solves the non-linear programming problem.

Constraints on the variables can also arise in a different manner. During the optimization the variables may stray into a region in which the function is not defined. This may occur in a number of different ways. In maximizing a function such as (7.10) it is not possible to condense out the variances and an algorithm may wish to take a trial step that would make a variance negative. Alternatively, an equation to be estimated may involve functional forms that are defined only for certain parameter values, say as in $y_i = \log(1 + \alpha x_i) + u_i$. Technically, another case is provided by a simultaneous linear equation model in which the number of endogenous and exogenous variables is greater than the number of observations, in which case the estimated covariance matrix of residuals will be singular and consequently the likelihood function undefined [Parke (1979)]; in this case, however, one would not actually attempt optimization.

At least three *ad hoc* remedies may be employed, although none of them is assured of success. First, in some cases it may be possible to reparameterize the variables. Thus, if in (7.10) σ_1^2 has a tendency to become negative, one may replace it by e^{ω}. This may exchange one problem for another: in the transformed space the likelihood function may become very flat. Alternatively, one may continually test whether the current values of the variables are attempting to enter a forbidden region and inhibit the algorithm from proceeding in such a direction, either by returning to the algorithm an extremely unfavorable (pseudo-) function value associated with the illegitimate point, or by shrinking the step size. The former technique gives seriously erroneous estimates of derivatives. The latter may slow down convergence considerably. In spite of these difficulties, these *ad hoc* techniques are often employed and often work reasonably well. What must be stressed, however, is that the latter two may be employed only to guard against attempts to evaluate the function at points at which it is not defined; they ought not be used to substitute for general non-linear programming in cases in which the constraints represent economic restrictions. There is no *a priori* reason to believe that just because an algorithm strays into a forbidden region in an economic sense, the location of the maximum will also be in a forbidden region. If several local constrained optima exist, a casual use of the constraint within an otherwise unconstrained maximization algorithm may well jeopardize locating the desired point.

8. Numerical integration

Normally, there are two econometric contexts in which numerical integration becomes necessary. The first is Bayesian analysis in which, say, the moments of the posterior density are required. An example is provided by Kloek and Van Dijk (1978) who deal with the linear simultaneous equation model:

$$Y\Gamma + ZB = U. \tag{8.1}$$

Denoting those elements of Γ and B that are not constant terms by θ and the prior density of θ by $p(\theta)$ and assuming (a) that the constant terms have uniform prior, (b) that the prior of the covariance matrix Σ is of the form $|\Sigma|^{-(G+1)/2}$, where G is the number of equations, and (c) that the constant terms and covariance matrix elements have been integrated out, the marginal posterior of θ can be shown to be proportional to

$$\kappa(\theta|Y,Z)p(\theta), \tag{8.2}$$

where $\kappa(\theta|Y,Z)$ depends on the structural parameters other than constant terms and on the observations. The moments are functions of θ, say $g(\theta)$, and can be written as

$$E[g(\theta)|Y,Z] = \frac{\int g(\theta)\kappa(\theta|Y,Z)p(\theta)\,\mathrm{d}\theta}{\int \kappa(\theta|Y,Z)p(\theta)\,\mathrm{d}\theta}. \tag{8.3}$$

Kloek and Van Dijk (1978) consider various alternatives for $p(\theta)$ in a small model such as the normal and beta distributions for which the integrals in (8.3) are not available in closed form.

The second context in which numerical integration is required is in finding maximum likelihood estimates in models with qualitative dependent variables, i.e. variables that are endogenous and involve essential elements of discreteness. Models of this type invariably contain some features that are not observable. Simple cases in point are the following models, where greek letters denote parameters, x's exogenous variables, i indexes observations, and $\Phi(\cdot)$ denotes the standard cumulative normal integral.

(a) The Probit Model:

$$y_i = 1 \quad \text{if } \beta' x_i + u_i > 0,$$
$$y_i = 0 \quad \text{otherwise},$$

where $u_i \sim N(0,1)$. The likelihood function is

$$L = \prod_{y_i = 1} (1 - \Phi(-\beta' x_i)) \prod_{y_i = 0} \Phi(-\beta' x_i).$$

(b) The Tobit Model which is stated in (7.8) with likelihood function (7.9).

(c) The Simple Disequilibrium Model which is stated in (7.11) with likelihood function (7.12).

In all of these likelihood functions $\Phi(\cdot)$ appears which is not available in closed form. Fortunately, simple and accurate approximations for the cumulative normal integral exist in the univariate case as given, for example, by the FORTRAN subroutines ERF and DERF. The problem becomes much more difficult in problems in which multiple integrals of multivariate densities are required [see Hausman and Wise (1978)]. An important example is provided by discrete choice models in which an individual must chose among m possibilities. Let the ith individual's utility from choosing alternative j be

$$U_{ij} = V(C_{ij}, \beta) + \varepsilon_{ij} = \overline{U}_{ij} + \varepsilon_{ij},$$

where $\overline{U}_{ij} = V(C_{ij}, \beta_i)$ represents the systematic part of utility and where C_{ij} are objective measures of the individual's and the alternative's characteristics, β are parameters, and ε_{ij} is a random variable. Then the probability that alternative k is chosen is

$$P_{ik} = \Pr\{\varepsilon_{ij} \leqq \overline{U}_{ik} - \overline{U}_{ij} + \varepsilon_{ik} \quad \text{for all } j \neq k\}.$$

If $h(\varepsilon_{i1}, \dots, \varepsilon_{im})$ is the joint density of the ε_{ij}, P_{ik} is

$$P_{ik} = \int_{-\infty}^{\infty} \int_{-\infty}^{\overline{U}_{ik} - \overline{U}_{i1} + \varepsilon_{ik}} \cdots \int_{-\infty}^{\overline{U}_{ik} - \overline{U}_{im} + \varepsilon_{ik}} h(\varepsilon_{i1}, \dots, \varepsilon_{im}) \, \mathrm{d}\varepsilon_{im} \dots \mathrm{d}\varepsilon_{i1} \, \mathrm{d}\varepsilon_{ik} \tag{8.4}$$

or

$$P_{ik} = \int_{-\infty}^{\overline{U}_{ik} - \overline{U}_{i1}} \cdots \int_{-\infty}^{\overline{U}_{ik} - \overline{U}_{im}} h^k(\eta_{i1k}, \dots, \eta_{imk}) \, \mathrm{d}\eta_{i1k} \dots \mathrm{d}\eta_{imk}, \tag{8.5}$$

where $\eta_{ijk} = \varepsilon_{ij} - \varepsilon_{ik}$ and where $h^k(\cdot)$ is the joint density of the $(m-1)$ η_{ijk}. Hence, with m alternatives, an $(m-1)$-fold integral must be evaluated. If there are n individuals, the likelihood of a sample of observations is

$$L = \prod_{i=1}^{n} P_{i1}^{y_{i1}} \dots P_{im}^{y_{im}}, \tag{8.6}$$

where $y_{ij} = 1$ if the ith individual chooses alternative j and zero otherwise and where the P_{ij} are replaced by the expressions in (8.4) or (8.5). Eq. (8.6) must then be maximized with respect to the parameter vector β; hence, every function evaluation requires the evaluation of multiple integrals. If the errors ε_{ij} are multivariate normally distributed with non-diagonal covariance matrix, as is assumed by Hausman and Wise (1978), the integrals must be obtained numerically. A similar situation arises in the multimarket disequilibrium model of the type given in Section 5: in general, the density function for the observable variables in an m-market disequilibrium model involves an m-fold integral of the m-variate normal distribution.

In general, one would expect to make greater demands for accuracy in the case of likelihood maximization for models with qualitative dependent variables than in the case of computing the moments of posterior densities. In the latter case it might be acceptable to have a 10 percent error from the point of view of providing economic interpretation for the results. In the case of likelihood maximization an average error of 10 percent in evaluating the likelihood function is likely to cause serious problems of convergence. Hence, methods that are suitable for one type of problem are not likely to be suitable for the other. In what follows we do not distinguish systematically between univariate and multivariate integration.

8.1. *Monte Carlo integration*

Assume that we require the integral

$$I = \int_a^b f(x)\,\mathrm{d}x. \tag{8.7}$$

Let $g(x)$ be a probability density defined over (a, b). Then

$$E(f(x)) = \int_a^b f(x) g(x)\,\mathrm{d}x,$$

and if $g(x)$ is uniform,

$$E(f(x)) = \frac{I}{b-a}.$$

If n points $x_1,\ldots,x_n$ are selected randomly, $\sum_{i=1}^{n} f(x_i)/n$ converges in probability to $E(f(x))$ and I can be approximated by I_a defined as

$$I_a = \frac{b-a}{n}\sum_{i=1}^{n} f(x_i). \tag{8.8}$$

Clearly, $E(I_a) = I$ and I_a is the sum of n i.i.d. random variables with mean I/n and variance $\omega^2 = (b-a)^2\sigma^2/n^2$, where $\sigma^2 = \operatorname{var} f(x_i)$ and may be estimated by the sample variance. By the Central Limit Theorem, with probability α,

$$|I_a - I| \leqq z_\alpha \frac{(b-a)\sigma}{\sqrt{n}},$$

where z_α satisfies $\Phi(z_\alpha) - \Phi(-z_\alpha) = \alpha$. The error decreases as $n^{-1/2}$ and is independent of the dimensionality of the integral [Hammersley and Handscomb (1964) and Shreider (1964)].

A variance reducing technique is importance sampling. Write (8.7) as

$$I = \int_a^b \frac{f(x)}{g(x)} g(x)\,\mathrm{d}x = E(f(x)/g(x)), \tag{8.9}$$

where $g(x)$ is a pdf over (a,b) as before. If points $x_1,\ldots,x_n$ are generated with pdf $g(x)$, I is now estimated by

$$I_a = \frac{1}{n}\sum \frac{f(x_i)}{g(x_i)}. \tag{8.10}$$

The variance of $f(x)/g(x)$ can be made to equal zero by setting $g(x) = |f(x)|/\int_a^b |f(x)|\,\mathrm{d}x$, which is not practical for it requires knowledge of the integral in question. As a practical matter, $g(x)$ is chosen so as to make the variation in $f(x)/g(x)$ small: the implication is that x will be sampled relatively more frequently in regions in which $f(x)$ is large or important. Examples of importance sampling are in Kloek and Van Dijk (1978) and Quandt (1978b).

8.2. *Polynomial approximations*

If a function $f(x)$ is approximated by a polynomial of degree n such that the approximating polynomial agrees with it at $n+1$ equally spaced points, it can be written as

$$f(x) \simeq \sum_{k=0}^{n} \lambda_k(x) f(x_k),$$

where the $\lambda_k(x)$ are the Lagrangean polynomial coefficients. A class of integration formulae employ the integral of the approximating polynomial and are called Newton–Cotes formulae. Simpson's Rule is a special case of Newton–Cotes integration. [For an application see Richard and Tompa (1980).]

Although simple to implement, Newton–Cotes formulae can be subject to serious error and there are cases where the approximation does not converge to the true integral as $n \to \infty$. More stable are Gaussian Quadrature formulae using n points of evaluation for which the approximation is exact if $f(x)$ is polynomial of degree $\leqq 2n-1$ [Hildebrand (1956) and Stroud and Secrest (1966)]. Gaussian formulae are obtained from approximating polynomials, the coefficients of which are derived by requiring $f(x)$ and the approximation to have the same values and derivatives at n points. In more than one dimension it is customary to take the Cartesian product of the points calculated by one-dimensional quadrature formulae. If n is the number of points used in one dimension, the integrand will have to be evaluated at n^k points, where k is the multiplicity of the integral; hence, multidimensional polynomial quadrature formulae tend to be too expensive in problems such as likelihood function maximization in which the integrals have to be computed many times. For a bivariate Simpson's Rule see Zellner (1971).

8.3. *Evaluation of multivariate normal integrals*

Possibly the most common multivariate integration problem is that of integrating the multivariate normal density. Let $N(x|0,\Sigma)$ be a multivariate normal density for the k-dimensional vector variable x with mean 0 and covariance matrix Σ. We require

$$I=\int_{l_1}^{\infty}\cdots\int_{l_k}^{\infty}N(x|0,\Sigma)\,\mathrm{d}x_1\ldots\mathrm{d}x_k. \tag{8.11}$$

Many fundamental relations concerning the bivariate, trivariate, and multivariate normal integrals are contained in Johnson and Kotz (1972). The bivariate case can be handled effectively by using Hausman and Wise's (1978) modification of a technique due to Owen (1956). For purposes of the method, one expresses the bivariate integral as

$$I(h,k;\rho)=\frac{1}{2\pi(1-\rho^2)^{1/2}}\times\int_{-\infty}^{h}\int_{-\infty}^{k}\exp\{-\tfrac{1}{2}(x^2-2\rho xy+y^2)/(1-\rho^2)\}\,\mathrm{d}x\,\mathrm{d}y.$$

If this is differentiated with respect to ρ, it can be integrated with respect to x and y and the result reintegrated with respect to ρ. The result is

$$\tfrac{1}{2}\Phi(h)+\tfrac{1}{2}\Phi(k)-T\left(h,\frac{k-\rho h}{h(1-\rho^2)^{1/2}}\right)-T\left(k,\frac{h-\rho k}{k(1-\rho^2)^{1/2}}\right)$$

$$\text{if } hk>0 \text{ or if } hk=0 \text{ and } h \text{ or } k \geqq 0;$$

$$I(h,k;\rho)=\tfrac{1}{2}\Phi(h)+\tfrac{1}{2}\Phi(k)-T\left(h,\frac{k-\rho h}{h(1-\rho^2)^{1/2}}\right)-T\left(k,\frac{h-\rho k}{k(1-\rho^2)^{1/2}}\right)-\frac{1}{2}$$

$$\text{if } hk<0 \text{ or if } hk=0 \text{ and } h \text{ or } k<0,$$

where

$$T(u,v)=-\frac{\tan^{-1}a}{2\pi}-\frac{1}{2\pi}\sum_{j=0}^{\infty}c_j v^{2j+1}$$

and

$$c_j=(-1)^j\frac{1}{2j+1}\left[1-\exp(-\tfrac{1}{2}u^2)\sum_{i=0}^{j}\frac{u^{2i}}{2^i i!}\right].$$

The Hausman–Wise modification works very fast and is suitable for use in optimizing likelihood functions.

Dutt (1976) represents I by Kendall's tetrachoric series which leads to

$$I=\left(\frac{1}{2}\right)^k-\left(\frac{1}{2}\right)^{k-1}\sum_{i=1}^{k}D^*_{1,i}+\left(\frac{1}{2}\right)^{k-2}\sum_{i=1}^{k}\sum_{k=2}^{k}D^*_{2,ij}+\cdots+D^*_{k,ij\cdots k},\quad i<j,$$

where the individual terms are defined in Dutt (1976) and involve Gaussian quadratures. The technique is recommended for up to $k=4$, beyond which it is likely to become expensive. Rules of thumb are suggested for selecting the degree N of the Hermite polynomials needed in the computation of the D^*'s. In practice it appears that poor choices of N can result either in costly computations or in poor approximations. In any event, for $k\geqq 3$ it is likely to be too expensive a method if integrals are required repetitively, as when a likelihood function is being maximized.

The multivariate normal integral (8.5) gives the probability that $\max_j[U_{ij}-U_{ik}]\leqq 0$. This observation allows Daganzo, Bouthelier and Sheffi (1977) to exploit an approximation due to Clark (1961). Consider random variables $x_1, x_2, \ldots, x_k$ distributed as $N(\mu, \Sigma)$. It is not difficult to obtain exact formulae for $E(\max(x_1, x_2))$, $\operatorname{var}(\max(x_1, x_2))$, and $\operatorname{cov}(\max(x_1, x_2), x_3)$. If one were to assume that $\max(x_1, x_2)$ is normally distributed, which it is not, then one can recursively calculate the moments of $\max(x_1, \ldots, x_k)$ and, by the (incorrect) distributional assumption, the univariate normal distribution which proxies that of $\max(x_1, \ldots, x_k)$, from which the required probability is easily obtained. The behavior of the Clark probabilities is tested by Manski and Lerman (1981) in choice problems involving three or five alternatives. They compare the computation of choice probabilities according to this method with a particular Monte Carlo procedure. The Clark probabilities agree remarkably well with the Monte Carlo results and are obtained with substantially smaller computational cost. It is estimated that for comparable accuracy, the Monte Carlo approach may be as much as 100 times more expensive. However, the Clark approximation tends to be unsatisfactory when the variances of the x's are relatively unequal [Danganzo (1979)]. Why the Clark probabilities are as accurate as they are in other cases, given that they are derived from a false assumption, is not known.

8.4. *Special cases of the multivariate normal integral*

The normal integrals occurring in discrete choice models may have simpler representations than (8.4) or (8.5) under certain circumstances. Two of these will be examined briefly [Hausman (1980)]. As before, the subscript i denotes the ith individual.

Case 1. If all ε_{ij}, $j = 1, \ldots, m$, in (8.4) are i.i.d. normal with mean zero and unit variance, (8.4) becomes

$$P_{ik} = \int_{-\infty}^{\infty} \frac{1}{\sqrt{2\pi}} \exp\left\{\frac{-\varepsilon_{ik}^2}{2}\right\} \prod_{j \neq k} \Phi\left(\bar{U}_{ik} - \bar{U}_{ij} + \varepsilon_{ik}\right) \mathrm{d}\varepsilon_{ik}, \tag{8.12}$$

which is substantially easier to evaluate than (8.4) without the simplifying assumption, since it requires only a single-dimensional numerical integral [with $\Phi(\cdot)$ being efficiently evaluated by a partial fraction expansion routinely available in program libraries].

Case 2. Consider the special case in which

$$\bar{U}_{ij} = z_{ij}'\beta, \tag{8.13}$$

where z'_{ij} is a vector of p observable variables and β a coefficient vector. Assume that β is a random vector distributed normally with mean β_μ and covariance matrix Σ_β. Let $\varepsilon_i = (\varepsilon_{i1},\dots,\varepsilon_{im})$ be normal with mean zero and covariance matrix Σ_ε and independent of β. The probability that the kth alternative is chosen is then

$$\begin{aligned} P_{ik} &= \Pr\{(z_{ik}-z_{ij})'\beta \geqq \eta_{ijk} \quad \forall j \neq k\} \\ &= \Pr\{(z_{ik}-z_{ij})'\beta_\mu \geqq \eta_{ijk} + (z_{ij}-z_{ik})'(\beta-\beta_\mu) \quad \forall j \neq k\}. \end{aligned} \tag{8.14}$$

Define the right-hand side in the last probability in (8.14) as ξ_{ijk}. The random vector $\xi'_{ik} = (\xi_{i1k},\dots,\xi_{imk})$ is normal with mean zero and covariance matrix $\Sigma_\xi = A\Sigma_\varepsilon A' + Z\Sigma_\beta Z'$, where

$$A = \begin{bmatrix} 1 & & & -1 & & & \\ & \cdot & & \cdot & & & \\ & & \cdot & \cdot & & & \\ & & \cdot & \cdot & & & \\ & & 1 & -1 & & & \\ & & & -1 & 1 & & \\ & & & -1 & \cdot & & \\ & & & \cdot & & \cdot & \\ & & & \cdot & & & \cdot \\ & & & -1 & & & 1 \end{bmatrix}$$

is $(m-1)\times m$ and has zeros except on the main diagonal and the kth column and where

$$z = \begin{bmatrix} (z_{i1}-z_{ik})' \\ \vdots \\ (z_{im}-z_{ik})' \end{bmatrix}$$

and is $(m-1)\times p$. If we assume that Σ_β and Σ_ε are both diagonal, it is easy to verify that Σ_ξ can be written as

$$\Sigma_\xi = QQ' + \Sigma_{\varepsilon(k)},$$

where $\Sigma_{\varepsilon(k)}$ is the matrix Σ_ε from which its kth row and column have been deleted and $Q = (z\Sigma_\beta^{1/2} \vdots \sigma_{\varepsilon_k} i)$, $i' = (1 \quad 1\dots1)$. Let $(v,w)' = (v_1,\dots,v_{k-1},v_{k+1},\dots,v_m,w_1,\dots,w_{p+1})$ be a vector of $m+p$ elements independently and normally distributed with mean zero and unit variance. Then

$$\xi_{ik} = \Sigma_{\varepsilon(k)}^{1/2} v - Qw,$$

and substituting on the right, (8.14) becomes

$$P_{ik} = \Pr\{(z_{ik} - z_{ij})'\beta_\mu \geqq \xi_{ijk} \quad \forall j \neq k\}$$
$$= \Pr\left\{ v_j \leqq \left[(z_{ik} - z_{ij})'^{-}\beta_\mu + \sum_{l=1}^{p+1} q_{jl}w_l \right] \Big/ \sigma_{\varepsilon_j} \quad \forall j \neq k \right\}. \tag{8.15}$$

If F and G are the distribution functions of scalars v and w, respectively, the following convolution formula holds [Marsaglia (1963)]:

$$\Pr\{v < c - w\} = \int \Pr\{v < c - w | w\} \mathrm{d}G = E[F(c - w)],$$

where c is a constant vector and the expectation is taken with respect to G. It follows that (8.15) is

$$P_{ik} = \frac{1}{(2\pi)^{(p+1)/2}} \int_{-\infty}^{\infty} \cdots \int_{-\infty}^{\infty} \prod_{j \neq k}$$
$$\times \Phi\left[\left((z_{ik} - z_{ij})'\beta_\mu + \sum_{l-1}^{p+1} q_{jl}w_l \right) \Big/ \sigma_{\varepsilon_j} \right] \times \exp\left\{ \frac{-t't}{2} \right\} \mathrm{d}t_1 \ldots \mathrm{d}t_{p+1}. \tag{8.16}$$

Hence, an $(m-1)$-fold numerical integral has been transformed into a $(p+1)$-fold numerical integral which is computationally advantageous if m is relatively large and p is small. [See Webster (1970) and particularly Hausman (1980) for details and extensions.]

It is an open question as to which method of integration of the multivariate normal density is best in several dimensions from the combined points of view of accuracy and cost. The issue is even less settled when multivariate integrals of other density functions are required.

9. The generation of random numbers

Monte Carlo or sampling experiments are the most common instances in econometrics in which it is necessary to generate (pseudo-) random numbers. An example is the following problem. Assume that an estimator $\hat{\theta}(x_1, \ldots, x_n)$ can be calculated from a sample $\{x_1, \ldots, x_n\}$, where the x_i are i.i.d. with pdf $f(x)$. If obtaining the sampling distribution of $\hat{\theta}$ analytically is an intractable problem, one may prefer to obtain experimental evidence about its behavior by repeated simulated drawings of samples of x's and by examination of the resulting $\hat{\theta}$'s. In order to enable one to perform such experiments, one must be able to sample from arbitrary distributions $f(x)$ [Cragg (1968)].

The present section is devoted to the principal features of generating random numbers and sampling from various distributions. It deals with the computational features of various techniques, but not with the principles of the design of sampling experiments [see, for example, Naylor (1971)]. Among the computational features of greatest interest are the accuracy with which $f(x)$ is sampled and the computational cost of the methods.

9.1. *The generation of uniformly distributed variables*

A fundamental step in generating a variable, x, with pdf $f(x)$, is first to generate u distributed uniformly on (0,1). A common method of generating x is based on the observation that the quantity $y = F(x) = \int_{-\infty}^{x} f(t)\mathrm{d}t$ is distributed as $U(0,1)$ for any continuous $f(x)$. Letting x, y denote random variables and $\bar{x}$, $\bar{y}$ particular values, the assertion follows from

$$\Pr\{y \leqq \bar{y}\} = \Pr\{F(x) \leqq \bar{y}\} = \Pr\{x \leqq F^{-1}(\bar{y})\} = F(F^{-1}(\bar{y})) = \bar{y}.$$

Given a sample of $u_1,\dots,u_n$ from $U(0,1)$, a corresponding sample $x_1,\dots,x_n$ for $f(x)$ is obtained by solving

$$u_i = F(x_i), \qquad i = 1,\dots,n. \tag{9.1}$$

Other uses of uniform deviates occur when some other function of uniform variables has the required distribution (see next subsection for generating normally distributed variables) or when Monte Carlo integration is to be performed (see Section 8).

The most commonly employed generators of $U(0,1)$ variables are based on the recurrence relation

$$R_{i+1} = \lambda R_i + \mu \pmod{M}, \tag{9.2}$$

where λ, μ, M and R_0 are integers chosen by the user. Uniform variables are obtained by calculating R_{i+1}/M, R_{i+2}/M, etc. Generators of the form of (9.2) are congruential generators; they are called mixed or linear if $\mu \neq 0$ and simple or multiplicative in the reverse case. All generators of type (9.2) have finite periods p such that $R_{i+p} = R_i$; the magnitude of p depends on λ, μ, and M. It is obviously desirable to choose these in such a manner as to make p large. The maximal p is easy to find if M is of the form 2^m. For $\mu \neq 0$, the maximal p is 2^m and is obtained if $\lambda = 1 \pmod 4$ and μ is odd; for $\mu = 0$ the maximal period is 2^{m-1} if $\lambda = 3$ or 5 (mod 8) and R_0 is odd [Newman and Odell (1971), Chambers (1977), and Atkinson (1980)]. In practice, for computers with a word-length of 32 bits, M is frequently chosen to be 2^{31} or $2^{31}-1$. The latter is particularly attractive since it is prime and has period $2^{31}-2$ which can be attained if λ is a primitive root of M [Hoaglin (1976)].

The extent to which numbers R_i/M can be thought to have been drawn from $U(0,1)$ can be tested by numerous statistical techniques, such as run tests, tests based on serial correlation properties including computation of the spectral density function, χ^2-tests, Kolmogorov–Smirnov tests, lattice tests, and many others. None of these may reveal adequately the following number-theoretic fact: that all n-tuples $(R_{i+1},\dots,R_{i+n})$ lie on at most a certain number of parallel hyperplanes in n-dimensional space. The extent to which the generated numbers approximate the uniform distribution will depend on the separation between the hyperplanes which can either be calculated explicitly or approximated from the n-dimensional spectrum of the R_i. Computations by Hoaglin (1976) indicate that for $M=2^{31}-1$ and $\mu=0$, suitable values of λ are 764261123, 1323257245, 1078318381, 1203248318, 397204094, 2027812808. Comparable information for $M=2^{31}$ does not yet appear to be available.

Several ways exist to improve the quality of random number generators. A simple device is to shuffle blocks of k successive random numbers $R=(R_{i+1},\dots,R_{i+k})'$ into a set $R^*=(R^*_{i+1},\dots,R^*_{i+k})'$ by $R^*=PR$, where P is a permutation matrix and may itself be altered from time to time. A somewhat more time-consuming but desirable technique is shuffling with replacement [Hill and Holland (1977)]. Two random number sequences $R_1,\dots,R_n$ and $S_1,\dots,S_n$ are generated from two different congruential generators. Select an integer k and set $m=2^k$. In the present application k will determine storage requirements and normally one would set $k\leq 8$. Set $T_1=R_1,\dots,T_m=R_m$. Now examine a predetermined set of k bits in S_1 which form an integer j' such that for $j=j'+1$, $1\leq j\leq m$. The first random number to be selected is then T_j. The content of T_j is replaced by R_{m+1} and the bits of S_2 are examined to select a new member of the array $T_1,\dots,T_m$. Thus, elements of the sequence S provide index values for selecting from among the elements of T which are then replenished by the next unused element of the sequence R.

9.2. *The generation of normally distributed variables*

Equation (9.1) is cumbersome to solve if $F(x_i)$ is the cumulative normal distribution and various alternative techniques are employed. A method for generating variables distributed approximately as $N(0,1)$ is by appealing to the Central Limit Theorem. The required variable might thus be computed as $\sum_{i=1}^{n}u_i/n-$

$0.5)/(1/\sqrt{12n})$. This method is currently regarded as not being sufficiently accurate.

Much more popular are transformations of pairs of uniform variables. One of the most frequently used of these is the Box–Müller transformation yielding two independent $N(0,1)$ variables x_1 and x_2 according to the transformation [Box and Müller (1958)]:

$$x_1 = (-2\log u_1)^{1/2}\sin(2\pi u_2),$$
$$x_2 = (-2\log u_1)^{1/2}\cos(2\pi u_2), \tag{9.3}$$

where u_1 and u_2 are independent $U(0,1)$. The joint pdf of u_1 and u_2 is $f(u_1, u_2) = 1$ and the pdf of x_1 and x_2 is $g(x_1, x_2) = |J^{-1}|f(u_1, u_2)$, where J is the Jacobian $\partial(x)/\partial(u)$. Its absolute value is easily shown to be $2\pi/u_1$, and it follows that $g(x_1, x_2) = \exp\{-(x_1^2 + x_2^2)/2\}/2\pi$.

Since (9.3) requires the evaluation of trigonometric functions in addition to logarithms, the Box–Müller method is fairly slow. What is more serious is that the exact distribution of variables generated from (9.3) with the aid of uniform variables obtained from a congruential generator such as (9.2) is not normal and contains 2λ discontinuities [Neave (1973)]. A more accurate and faster alternative to (9.3) is to generate x_1, x_2 according to the Marsaglia–Bray transformation:

$$x_1 = v_1\left[-2\log\left(v_1^2 + v_2^2\right)/\left(v_1^2 + v_2^2\right)\right]^{1/2},$$
$$x_2 = v_2\left[-2\log\left(v_1^2 + v_2^2\right)/\left(v_1^2 + v_2^2\right)\right]^{1/2}, \tag{9.4}$$

where v_1 and v_2 are $U(-1,1)$ and are conditioned by $v_1^2 + v_2^2 < 1$. An argument similar to that used above verifies that x_1 and x_2 are independent $N(0,1)$ variables [Marsaglia and Bray (1964)].

By far the most effective techniques are the decomposition and the acceptance–rejection methods, often used in combination. The basic rationale of these techniques are described by Newman and Odell (1971); various details and computational experience with different methods of this as well as of other types are given by Ahrens and Dieter (1972) and Kinderman and Ramage (1976).

Both techniques allow one to exploit the fact that it may be much easier to generate random variables from one distribution than from another. Assume we wish to sample from $f(x)$ over some interval (a, b) and that $\phi(x)$ is some other pdf over the same interval (from which it may be particularly easy to sample). Determine a coefficient α such that $f(x) \leq \alpha\phi(x)$ over (a, b). Then we may sample from $f(x)$ by the following procedure. (1) Draw an x from $\phi(x)$, say $\bar{x}$. (2) Draw a uniform $U(0,1)$ variate u. If $u \leq f(\bar{x})/\alpha\phi(\bar{x})$, then we accept $\bar{x}$. Otherwise we return to the first step. It is clear that the probabilities $\Pr\{x \leq x^0)$ will be proportional to $\int_a^{x^0} f(x)\mathrm{d}x$ for all x^0 with the factor of proportionality simply measuring the frequency with which step (2) leads to acceptance of x; hence, an easy-to-sample distribution is employed, instead of the possibly difficult-to-sample $f(x)$, to yield the same result.

An interesting variant is the ratio-of-uniforms method [Robertson and Walls (1980)]. To sample from pdf $f(x)$, define the region $R = \{(u, v) | 0 \leqq u \leqq f^{1/2}(v/u)\}$. Generate u and v uniformly over R. Defining $u = y$ and $v = xy$, it is easy to show that v/u is distributed with pdf $f(x)$. The required procedure then is (1) to generate u and v, (2) reject (u, v) if (u, v) does not fall in the region R, or (3) accept (u, v) and form $x = v/u$ otherwise. Efficiency requires that step (2) not be encountered too often. Robertson and Walls (1980) consider in detail sampling from the normal, Cauchy, t-, and gamma distributions.

A similar idea is exploited by decomposition methods in that they replace a difficult-to-sample distribution with a finite mixture of relatively easy-to-sample distributions. Consider $f(x)$ and $\phi_1(x)$, with $\phi_1(x)$ being easy-to-sample. Then choose $\alpha > 0$ such that $f(x) - \alpha\phi_1(x) \geqq 0$. Unless $f(x)$ and $\phi_1(x)$ coincide,

$$\int_{-\infty}^{\infty} (f(x) - \alpha\phi_1(x))\,\mathrm{d}x = 1 - \alpha > 0.$$

Hence, defining $\phi_2(x) = (f(x) - \alpha\phi_1(x))/(1-\alpha)$, it is clear that $f(x) = \alpha\phi_1(x) + (1-\alpha)\phi_2(x)$, an $(\alpha, 1-\alpha)$ weighted mixture of ϕ_1, ϕ_2. If $\phi_1(x)$ is easily sampled and if α can be chosen to be relatively large, we may sample from $f(x)$ efficiently by generating x from $\phi_1(x)$ with probability α and from $\phi_2(x)$ with probability $1-\alpha$. Clearly, the same type of decomposition may be applied to $\phi_2(x)$ and a decomposition scheme may set $f(x) = \sum_{j=1}^{m} \alpha_j \phi_j(x)$, $0 < \alpha_j < 1$, $j = 1, \dots, m$. A simple algorithm for sampling from $N(0,1)$ is given by Newman and Odell (1971). Let u_1, u_2, and u_3 denote independent $U(0,1)$ variates. The normal density is decomposed into four components with probabilities

$$\alpha_1 = 0.8638554642,$$
$$\alpha_2 = 0.110817965,$$
$$\alpha_3 = 0.002699796063,$$
$$\alpha_4 = 0.02262677245.$$

For each of the four possibilities we generate x as follows:

(1) $x = 2(u_1 + u_2 + u_3 - 1.5)$.

(2) $x = 1.5(u_1 + u_2 - 1)$.

(3) $x =$ the first normal variate from repeated Box–Müller transformations for which $|x| > 3$.

(4) Generate $x = 6u_1 - 3$ and $y = 0.3181471173u_2$ repeatedly if necessary until $y \leqq \psi(x)$, and then accept x, where

$$\psi(x) = \begin{cases} ae^{-x^2/2} - b(3 - x^2) - c(1.5 - |x|), & |x| < 1, \\ ae^{-x^2/2} - d(3 - |x|)^2 - c(1.5 - |x|)^2, & 1 \leqq |x| < 1.5, \\ ae^{-x^2/2} - d(3 - |x|)^2, & 1.5 \leqq |x| < 3, \\ 0, & \text{otherwise}, \end{cases}$$

where $a = 15.75192787$, $b = 4.263583239$, $c = 1.944694161$, and $d = 2.1317916185$. More complicated algorithms are described by Kinderman and Ramage (1976) and by Peterson and Kronmal (1982).

References

Ahrens, J. H. and U. Dieter (1972) "Computer Methods for Sampling from the Exponential and Normal Distributions", *Communications of the ACM*, 15, 873–882.

Aitcheson, J. and S. D. Silvey (1960) "Maximum Likelihood Estimation Procedures and Associated Tests of Significance", *Journal of the Royal Statistical Society*, Ser. B, 154–171.

Atkinson, A. C. (1980) "Tests of Pseudo-Random-Numbers", *Applied Statistics*, 29, 164–171.

Bard, Y. (1974) *Nonlinear Parameter Estimation*. New York: Academic Press.

Beale, E. M. L. (1960) "Confidence Regions in Non-linear Estimation", *Journal of the Royal Statistical Society*, Ser. B, 22, 41–76.

Belsley, D. A. (1974) "Estimation of Systems of Simultaneous Equations and Computational Specifications of GREMLIN," *Annals of Economic and Social Measurement*, 3, 551–614.

Belsley, D. A. (1979) "On the Computational Competitiveness of Full Information Maximum Likelihood and Three Stage Least Squares in the Estimation of Nonlinear, Simultaneous-Equations Models", *Journal of Econometrics*, 9, 315–342.

Belsley, D. A. (1980) "On the Efficient Computation of the Nonlinear Full-Information Maximum Likelihood Estimator", Technical Report no. 5, Center for Computational Research in Economics and Management Science, Vol. II, Cambridge, Mass.

Berman, G. (1979) "Lattice Approximations to the Minima of Functions of Several Variables", *Journal of the Association of Computing Machinery*, 16, 286–294.

Berndt, E. K., B. H. Hall, R. E. Hall and J. A. Hausman (1974) "Estimation and Inference in Nonlinear Structural Models", *Annals of Economic and Social Measurement*, 3, 653–666.

Box, G. E. P. and M. E. Müller (1958) "A Note on the Generation of Random Normal Deviates", *Annals of Mathematical Statistics*, 26, 610–611.

Brent, R. P. (1973) *Algorithms for Minimization without Derivatives*. Englewood Cliffs, N.J.: Prentice-Hall.

Brown, K. M. and W. B. Gearhart (1971) "Deflation Techniques for the Calculation of Further Solutions of a Nonlinear System", *Numerische Mathematik*, 16, 334–342.

Broyden, C. G. (1972) "Quasi-Newton Methods", in: W. Murray (ed.), *Numerical Methods for Unconstrained Optimization*. New York: Academic Press.

Bussinger, P. A. and G. H. Golub (1969) "Singular Value Decomposition of a Complex Matrix", *Communications of the ACM*, 12, 564–565.

Carnahan, B., H. A. Luther and J. D. Wilkes (1969) *Applied Numerical Methods*. New York: John Wiley & Sons.

Chambers, J. M. (1977) *Computational Methods for Data Analysis*. New York: John Wiley & Sons.
Chapman, D. R. and R. C. Fair (1972) "Full-Information Maximum Likelihood Program: User's Guide", Research Memo. no. 137, Econometric Research Program, Princeton University.
Chow, G. C. (1968) "Two Methods of Computing Full-Information Maximum Likelihood Estimates in Simultaneous Stochastic Equations", *International Economic Review*, 9, 100–112.
Chow, G. C. (1973) "On the Computation of Full Information Maximum Likelihood Estimates for Nonlinear Equation Systems", *The Review of Economics and Statistics*, LV, 104–109.
Chow, G. C. (1975) *Analysis and Control of Dynamic Economic Systems*. New York: John Wiley & Sons.
Chow, G. C. and S. B. Megdal (1978) "The Control of Large-Scale Nonlinear Econometric Systems", *IEEE Transactions on Automatic Control*, AC-23, 344–349.
Clark, C. E. (1961) "The Greatest of a Finite Set of Random Variables", *Operations Research*, 9, 145–162.
Cragg, J. C. (1968) "Some Effects of Incorrect Specifications on the Small Sample Properties of Several Simultaneous-Equation Estimators", *International Economic Review*, 9, 63–86.
Dagenais, M. G. (1978) "The Computation of FIML Estimates as Iterative Generalized Least Squares Estimates in Linear and Nonlinear Simultaneous Equations Models", *Econometrica*, 46, 1351–1362.
Daganzo, C. F. (1979) *Multinomial Probit*. New York: Academic Press.
Daganzo, C. F., F. Bouthelier and Y. Sheffi (1977) "Multinomial Probit and Qualitative Choice: A Computationally Efficient Algorithm", *Transportation Science*, II, 339–358.
Davidon, W. C. (1959) "Variable Metric Method for Minimization", AEC Research and Development Report ANL-5990 (Rev.).
Davidon, W. C. (1975) "Optimally Conditioned Optimization Algorithms Without Line Searches", *Mathematical Programming*, 9, 1–30.
Dempster, A. P., N. M. Laird and D. B. Rubin (1977) "Maximum Likelihood from Incomplete Data via the EM Algorithm", *Journal of the Royal Statistical Society*, Ser. B, 39, 1–38.
Dennis, J. E. Jr., D. M. Gay and R. E. Welsch (1979) "An Adaptive Nonlinear Least-Squares Algorithm", Technical Report TR-1, Alfred T. Sloan School of Management, Massachusetts Institute of Technology.
Dennis, J. E. and J. J. Moré (1977) "Quasi-Newton Methods, Motivation and Theory", *SIAM Review*, 9, 46–89.
Dent, W. T. (1976) "Information and Computation in Simultaneous Equation Systems", *Journal of Econometrics*, 4, 89–95.
Dent, W. T. (1977) "On Numerical Computation in Simultaneous Equation Systems", *Annals of Economic and Social Measurement*, 6, 123–125.
Dent, W. T. (1980) "On Restricted Estimation in Linear Models", *Journal of Econometrics*, 12, 49–58.
Dhrymes, P. J. (1978) *Introductory Econometrics*. New York: Springer-Verlag.
Drud, A. (1977/78) "An Optimization Code for Nonlinear Econometric Models Based on Sparse Matrix Techniques and Reduced Gradients", *Annals of Economic and Social Measurement*, 6, 563–580.
Duesenberry, J. S., G. Fromm, L. R. Klein and E. Kuh (eds.) (1969) *The Brookings Model: Some Further Results*. Chicago: Rand McNally.
Dutt, J. E. (1976) "Numerical Aspects of Multivariate Normal Probabilities in Econometric Models", *Annals of Economic and Social Measurement*, 5, 547–561.
Fair, R. C. (1974a) "On the Robust Estimation of Econometric Models", *Annals of Economic and Social Measurement*, 3, 667–678.
Fair, R. C. (1974b) "On the Solution of Optimal Control Problems as Maximization Problems", *Annals of Economic and Social Measurement*, 3, 135–154.
Fair, R. C. (1976) *A Model of Macroeconomic Activity, Volume II: The Empirical Model*. Cambridge: Ballinger.
Fair, R. C. (1979) "An Analysis of a Macro-economic Model with Rational Expectations in the Bond and Stock Markets", *American Economic Review*, 69, 539–552.
Fair, R. C. (1980a) "Estimating the Expected Predictive Accuracy of Econometric Models", *International Economic Review*, 21, 701–724.
Fair, R. C. (1980b) "Estimating the Uncertainty of Policy Effects in Nonlinear Models", *Econometrica*, 48, 1381–1391.

Fair, R. C. and W. R. Parke (1980) "Full Information Estimates of a Nonlinear Macroeconomic Model", *Journal of Econometrics*, 13, 269–292.

Fiacco, A. V. and G. P. McCormick (1964) "Sequential Unconstrained Minimization Technique for Nonlinear Programming, A Primal-Dual Method", *Management Science*, 10, 361–366.

Fletcher, R. (1965) "Function Minimization Without Evaluating Derivatives—A Review", *Computer Journal*, 8, 33–41.

Fletcher, R. and M. J. D. Powell (1963) "A Rapidly Convergent Descent Method for Minimization", *Computer Journal*, 6, 163–168.

Fromm, G. and L. R. Klein (1969) "Solutions of the Complete System," in: J. S. Duesenberry, G. Fromm, L. R. Klein and E. Kuh (eds.), *The Brookings Model: Some Further Results*. Chicago: Rand McNally.

Gallant, R. A. and T. M. Gerig (1980) "Computations for Constrained Linear Models", *Journal of Econometrics*, 12, 59–84.

Goldfeld, S. M. and R. E. Quandt (1972) *Nonlinear Methods in Econometrics*. Amsterdam: North-Holland Publishing Co.

Goldfeld, S. M. and R. E. Quandt (1978) "Some Properties of the Simple Disequilibrium Model with Covariance", *Economics Letters*, 1, 341–346.

Goldfeld, S. M., R. E. Quandt and H. F. Trotter (1966) "Maximization by Quadratic Hill-Climbing", *Econometrica*, 34, 541–551.

Goldfeld, S. M., R. E. Quandt and H. F. Trotter (1968) "Maximization by Improved Quadratic Hill-Climbing and Other Methods", Research Memo no. 95, Econometric Research Program, Princeton University.

Goldfeld, S. M. and R. E. Quandt (1979) "Recent Problems and Advances in Estimating Disequilibrium Models", Paper given at the Western Economic Association Meeting, Las Vegas.

Golub, G. H. (1969) "Matrix Decompositions and Statistical Calculations", in: *Statistical Computation*. New York: Academic Press, pp. 365–397.

Greenstadt, J. (1967) "On the Relative Efficiencies of Gradient Methods", *Mathematics of Computation*, 21, 360–367.

Guttman, I. and D. A. Meeter (1965) "On Beale's Measures of Non-Linearity", *Technometrics*, 7, 623–637.

Hammersley, J. M. and D. C. Handscomb (1964) *Monte Carlo Methods*. London: Methuen.

Hartley, M. J. (1977a) "On the Estimation of a General Switching Regression Model Via Maximum Likelihood Methods", Discussion Paper 415, Dept. of Economics, State University of New York at Buffalo.

Hartley, M. J. (1977b) "On the Calculation of the Maximum Likelihood Estimator for a Model of Markets in Disequilibrium", Discussion Paper 409, Dept. of Economics, State University of New York at Buffalo.

Hausman, J. A. (1980) "Les Modèles Probit de Choix Qualitatifs", *Cahiers du Seminaire d'Econométrie*, 21, 11–31.

Hausman, J. A. and D. A. Wise (1978) "A Conditional Probit Model For Qualitative Choice: Discrete Decisions Recognizing Interdependence and Heterogeneous Preferences", *Econometrica*, 46, 403–426.

Hendry, D. F. (1976) "The Structure of Simultaneous Equation Estimators", *Journal of Econometrics*, 4, 51–88.

Hendry, D. F. (1977) "Numerical Optimization Methods", London School of Economics, mimeo.

Hildebrand, F. B. (1956) *Introduction to Numerical Analysis*. New York: McGraw-Hill.

Hill, R. W. and P. W. Holland (1977) "Two Robust Alternatives to Least-Squares Regression", *Journal of the American Statistical Association*, 72, 828–833.

Hoaglin, D. C. (1976) "Theoretical Properties of Congruential Random Number Generators: An Empirical View", Department of Statistics, Memo. NS-340, Harvard University.

Hoffman, K. and R. Kunze (1961) *Linear Algebra*. Englewood Cliffs, N.J.: Prentice-Hall.

Huang, H. Y. (1970) "Unified approach to Quadratically Convergent Algorithms for Function Minimization", *Journal of Optimization Theory and Applications*, 5, 405–423.

Ito, T., (1980) "Methods of Estimation for Multimarket Disequilibrium Models", *Econometrica*, 48, 97–126.

Jennings, L. S. (1980) "Simultaneous Equations Estimation", *Journal of Econometrics*, 12, 23–39.

Johnson, N. L. and S. Kotz (1972) *Distributions in Statistics: Continuous Multivariate Distributions*. New York: John Wiley & Sons.
Jorgenson, D. W. and J. J. Laffont (1974) "Efficient Estimation of Nonlinear Simultaneous Equations with Additive Disturbances", *Annals of Economic and Social Measurement*, 3, 615–641.
Kiefer, N. M. (1980) "A Note on Switching Regressions and Logistic Discrimination", *Econometrica*, 48, 1065–1069.
Kinderman, A. J. and J. G. Ramage (1976) "Computer Generation of Normal Random Variables", *Journal of the American Statistical Association*, 71, 893–896.
Klema, V. (1973) "A Note on Matrix Factorization", *Annals of Economic and Social Measurement*, 2/3, 317–321.
Kloek, T. and H. K. van Dijk (1978) "Bayesian Estimates of Equation Systems Parameters: An Application of Integration by Monte Carlo", *Econometrica*, 46, 1–19.
Lootsma, F. A. (ed.) (1972) *Numerical Methods for Non-Linear Optimization*. New York: Academic Press.
MacKinnon, J. G. (1979) "Convenient Singularities and Maximum Likelihood Estimation", *Economics Letters*, 3, 41–44.
Maddala, G. S. (1971) "The Use of Variance Components Models in Pooling Cross Section and Time Series Data", *Econometrica*, 39, 341–358.
Manski, C. F. and S. R. Lerman (1981) "On the Use of Simulated Frequencies to Approximate Choice Probabilities", in: C. F. Manski and D. McFadden (eds.), *Structural Analysis of Discrete Data (with Econometric Applications)*. Cambridge, MA: MIT Press.
Marsaglia, G. (1963) "Expressing the Normal Distribution with Covariance Matrix $A + B$ in Terms of One with Covariance Matrix A", *Biometrika*, 50, 535–538.
Marsaglia, G. and T. A. Bray (1964) "A Convenient Method for Generating Normal Variables", *SIAM Review*, 6, 260–264.
Marquardt, D. W. (1963) "An Algorithm for Least Squares Estimation of Nonlinear Parameters", *SIAM Journal*, 11, 431–441.
McCarthy, M. D. and C. J. Palash (1977) "The Use of Almon- and Other Dummy-Variable Procedures to Increase the Efficiency of Maximization Algorithms in Economic Control", *Annals of Economic and Social Measurement*, 6, 225–230.
Murray, W. (1972) *Numerical Methods for Unconstrained Optimization*. New York: Academic Press.
Naylor, T. H. (1971) *Computer Simulation Experiments with Models of Economic Systems*. New York: John Wiley & Sons.
Neave, H. R. (1973) "On Using the Box–Müller Transformation with Multiplicative Congruential Pseudo-random Number Generators", *Applied Statistics*, 22, 92–97.
Nelder, J. A. and R. Mead (1965) "A Simplex Method for Function Minimization", *Computer Journal*, 7, 308–313.
Newman, T. G. and P. L. Odell (1971) *The Generation of Random Variates*. New York: Hafner.
Oren, S. S. and D. G. Luenberger (1974) "Self Scaling Variable Metric Algorithms, Part I", *Management Science*, 20, 845–862.
Osborne, M. R. (1972) "On Penalty and Barrier Function Methods in Mathematical Programming", in: S. Andersen, L. S. Jennings and D. M. Ryan (eds.), *Optimization*. St. Lucia: University of Queensland Press.
Owen, D. B. (1956) "Tables for Computing Bivariate Normal Probabilities", *Annals of Mathematical Statistics*, 27, 1075–1090.
Parke, W. R. (1979) "An Algorithm for Full Information Estimation", Ph.D. Dissertation, Yale University.
Parkinson, J. M. and D. Hutchinson (1972) "An Investigation into the Efficiency of Variants on the Simplex Method," in: F. A. Lootsman (ed.), *Numerical Methods for Non-Linear Optimization*. New York: Academic Press.
Peterson, A. V. and R. A. Kronmal (1982) "On Mixture Methods for the Computer Generation of Random Variables", *The American Statistician*, 36, 184–191.
Pierre, D. A. and M. J. Lowe (1975) *Mathematical Programming Via Augmented Lagrangians*. Reading, MA: Addison–Wesley.
Powell, M. J. D. (1964) "An Efficient Method for Finding the Minimum of a Function of Several Variables without Calculating Derivatives", *Computer Journal*, 7, 155–162.

Powell, M. J. D. (1971) "Recent Advances in Unconstrained Optimization", *Mathematical Programming*, 1, 26–57.
Powell, M. J. D. (1973) "On Search Directions for Minimization Algorithms", *Mathematical Programming*, 4, 193–201.
Powell, M. J. D. (1976) "Some Convergence Properties of the Conjugate Gradient Method", *Mathematical Programming*, 11, 42–49.
Quandt, R. E. (1972) "A New Approach to Estimating Switching Regressions", *Journal of the American Statistical Association*, 67, 306–310.
Quandt, R. E. (1978a) "Maximum Likelihood Estimation of Disequilibrium Models", in: *Pioneering Economics*. Padova: Cedam.
Quandt, R. E. (1978b) "Tests of the Equilibrium vs. Disequilibrium Hypotheses", *International Economics Review*, 19, 435–452.
Quandt, R. E. and J. B. Ramsey (1978) "Estimating Mixtures of Normal Distributions and Switching Regressions", *Journal of the American Statistical Association*, 73, 730–752.
Rao, C. R. (1973) *Linear Statistical Inference and Its Applications* (2nd edn.). New York: John Wiley & Sons.
Richard, J. F. and H. Tompa (1980) "On the Evaluation of Poly-*t* Density Functions", *Journal of Econometrics*, 12, 335–352.
Riddell, W. C. (1975) "Recursive Estimation Algorithms for Economic Research", *Annals of Economic and Social Measurement*, 4, 397–406.
Robertson, I. and L. A. Walls (1980) "Random Number Generators for the Normal and Gamma Distributions Using the Ratio of Uniforms Method", AERE-R 10032, Computer Science and Systems Division, AERE Harwell.
Rosen, J. B. (1960) "The Gradient Projection Method for Nonlinear Programming, Part I. Linear Constraints", *Journal of the Society of Industrial and Applied Mathematics*, 8, 181–217.
Rosen, J. B. (1961) "The Gradient Projection Method for Nonlinear Programming, Part II. Nonlinear Constraints", *Journal of the Society of Industrial and Applied Mathematics*, 9, 514–532.
Rothenberg, T. J. and C. T. Leenders (1964) "Efficient Estimation of Simultaneous Equation Systems", *Econometrica*, 32, 57–76.
Salmon, M. (1978) "The Detection of Successive Solutions to Nonlinear Econometric Models", CRES Working Paper R/WP 30, ISSN 0313 7414.
Schmidt, P. (1976) *Econometrics*. New York: Marcel Dekker.
Shreider, Y. A. (1964) *Method of Statistical Testing*. Amsterdam: Elsevier.
Spang, H. A., III (1962) "A Review of Minimization Techniques for Nonlinear Functions", *SIAM Review*, 4, 343–365.
Stewart, G. W. III (1967) "A Modification of Davidon's Minimization Method to Accept Difference Approximation of Derivatives" *Journal of the Association of Computing Machinery*, 14, 72–83.
Stroud, A. H. and D. Secrest (1966) *Gaussian Quadrature Formulas*. Englewood Cliffs: Prentice-Hall.
Swann, W. H. (1972) "Direct Search Methods", in: W. Murray (ed.), *Numerical Methods for Unconstrained Optimization*. New York: Academic Press, pp. 13–28.
Theil, H. (1971) *Principles of Econometrics*. New York: John Wiley & Sons.
Tischler, A. and I. Zang (1977) "Maximum Likelihood Method for Switching Regression Models Without A Priori Conditions", Tel Aviv University.
Tischler, A. and I. Zang (1979) "A Switching Regression Method Using Inequality Conditions", *Journal of Econometrics*, 11, 259–274.
Van der Hoek, G. and M. W. Dijkshoorn (1979) "A Numerical Comparison of Self Scaling Variable Metric Algorithms", Report 7910/0, Erasmus University, Rotterdam.
Walsh, G. R. (1975) *Methods of Optimization*. New York: John Wiley & Sons.
Wampler, R. H. (1980) "Test Procedures and Test Problems for Least Squares Algorithms", *Journal of Econometrics*, 12, 3–22.
Webster, J. T. (1970) "On the Application of the Method of Das in Evaluating a Multivariate Normal Integral", *Biometrika*, 57, 657–660.
Zangwill, W. I. (1967) "Minimizing a Function without Calculating Derivatives", *Computer Journal*, 10, 293–296.
Zellner, A. (1971) *An Introduction to Bayesian Inference in Econometrics*. New York: John Wiley & Sons.

LIST OF THEOREMS

INDEX